Problem Books in Mathematics

More information about this series at http://www.springer.com/series/714

Leszek Gasiński • Nikolaos S. Papageorgiou

Exercises in Analysis

Part 2: Nonlinear Analysis

Springer

Leszek Gasiński
Faculty of Mathematics
and Computer Science
Jagiellonian University
Kraków, Poland

Nikolaos S. Papageorgiou
Department of Mathematics
National Technical University
Athens, Greece

ISSN 0941-3502 ISSN 2197-8506 (electronic)
Problem Books in Mathematics
ISBN 978-3-319-80221-3 ISBN 978-3-319-27817-9 (eBook)
DOI 10.1007/978-3-319-27817-9

Mathematics Subject Classification (2010): 00A07; 46-XX; 30LXX, 28-XX, 60G46

Springer Cham Heidelberg New York Dordrecht London

Softcover reprint of the hardcover 1st edition 2016

Printed on acid-free paper

Springer International Publishing AG Switzerland is part of Springer Science+Business Media (www.springer.com)

Preface

This is the second part of the project initiated with "Exercises in Analysis. Part 1." In that volume we focused on some classical tools from Analysis. Specifically we considered Metric Spaces, Measure Theory, the interplay between Measure Theory and Topology, and Functional Analysis (Banach Spaces). In the present volume we turn our attention to the main topics of Nonlinear Analysis, which are useful in applications. So, we deal with the following subjects:

1. Function Spaces
2. Nonlinear and Multivalued Maps
3. Smooth and Nonsmooth Calculus
4. Degree Theory and Fixed Point Theory
5. Variational and Topological Methods

Each one of the above topics is a separate chapter. Each chapter starts with a comprehensive presentation of the theory and then has on the average 200 problems together with their solutions. The problems are marked with $\star$, $\star\star$, and $\star\star\star$ according to the level of difficulty of the problem ($\star$ for easy, $\star\star$ for reasonable, and $\star\star\star$ for difficult). We believe that people using tools of nonlinear analysis will find useful information either in the summary of the theory or in the problems. The topics studied in this volume cover a large part of what is known as Nonlinear Analysis.

Once again the authors express their gratitude to the editor Mrs. Elizabeth Loew for her cooperation and strong moral support in realizing this difficult project and for Krystyna Gasińska and Rafał Kawa for preparing the beautiful drawings of branches decorating the solutions of the problems in both volumes.

Kraków, Poland — Leszek Gasiński
Athens, Greece — Nikolaos S. Papageorgiou

Contents

Chapter 1

Function Spaces

1.1 Introduction

1.1.1 L^p-Spaces

Definition 1.1
Let (Ω, Σ, μ) be a measure space, $1 \leqslant p < +\infty$ and

$$\mathcal{L}^p(\Omega) = \Big\{ f\colon \Omega \longrightarrow \mathbb{R} :\ f \text{ is measurable and } \int_\Omega |f(\omega)|^p \, d\mu < +\infty \Big\}.$$

Consider the equivalence relation $\sim$ defined by

$$f \sim g \quad \text{if and only if} \quad f(\omega) = g(\omega) \quad \text{for } \mu\text{-a.a. } \omega \in \Omega$$

and then let $L^p(\Omega) = \mathcal{L}^p(\Omega)/_\sim$. On $L^p(\Omega)$ we consider the norm

$$\|f\|_p = \Big(\int_\Omega |f(\omega)|^p \, d\mu \Big)^{\frac{1}{p}}.$$

For $p = +\infty$, we define

$$\mathcal{L}^\infty(\Omega) = \Big\{ f\colon \Omega \longrightarrow \mathbb{R} :\ f \text{ is measurable and there exists } M > 0, \text{ such that } \mu\{\omega \in \Omega :\ |f(\omega)| > M\} = 0 \Big\}.$$

We set $L^\infty(\Omega) = \mathcal{L}^\infty(\Omega)/_\sim$ and on $L^\infty(\Omega)$ we consider the norm defined by

$$\|f\|_\infty = \inf \big\{ M > 0 :\ \mu\{\omega \in \Omega :\ |f(\omega)| > M\} = 0 \big\}.$$

L. Gasiński, N.S. Papageorgiou, *Exercises in Analysis: Part 2*, Problem Books in Mathematics, DOI 10.1007/978-3-319-27817-9_1

Remark 1.2
The normed spaces $(L^p(\Omega), \|\cdot\|_p)$ (with $1 \leqslant p < +\infty$) and $(L^\infty(\Omega), \|\cdot\|_\infty)$ are Banach spaces. If p, p' are positive real numbers satisfying $\frac{1}{p} + \frac{1}{p'} = 1$ (that is, $p+p' = pp'$ or $p' = \frac{p}{p-1}$), then we say that p and p' are a pair of ***conjugate exponents***. It is clear that $1 < p < +\infty$ and $1 < p' < +\infty$. A special case is when $p = p' = 2$. Note that as $p \searrow 1$, then $p' \nearrow +\infty$ and so 1 and $+\infty$ are also regarded as a pair of conjugate exponents.

Theorem 1.3 *(Hölder Inequality)*
If p and p' are conjugate exponents, $1 \leqslant p \leqslant +\infty$ and $f \in L^p(\Omega)$, $g \in L^{p'}(\Omega)$, then $fg \in L^1(\Omega)$ and $\|fg\|_1 \leqslant \|f\|_p \|g\|_{p'}$.
Moreover, equality holds if and only if $\alpha|f|^p = \beta|g|^{p'}$ μ-almost everywhere for some $\alpha, \beta \in \mathbb{R} \setminus \{0\}$.

Next we state some useful density results for the L^p-spaces.

Proposition 1.4
The space of simple functions in $L^p(\Omega)$ is dense in $L^p(\Omega)$ for all $1 \leqslant p \leqslant +\infty$.

Recall that for a Hausdorff topological space X, by $C_b(X)$ we denote the space of all bounded continuous functions $f\colon X \longrightarrow \mathbb{R}$. We know that $\big(C_b(X), \|\cdot\|_\infty\big)$ is a Banach space. The next results ensure the density of continuous functions in $L^p(\Omega)$.

Proposition 1.5
If (Ω, Σ, μ) is a measure space with a normal topological space Ω and a regular measure μ and $1 \leqslant p < +\infty$,
then $L^p(\Omega) \cap C_b(\Omega)$ is dense in $L^p(\Omega)$.

Proposition 1.6
If (Ω, Σ, μ) is a measure space with a locally compact topological space Ω and a Radon measure μ (see Definition I.4.9) and $1 \leqslant p < +\infty$,
then $C_c(\Omega)$ is dense in $L^p(\Omega)$.

Remark 1.7
In fact, if $\Omega \subseteq \mathbb{R}^N$ is an open set, then using mollification one can show that $C_c^\infty(\Omega)$ is dense in $L^p(\Omega)$ with $1 \leqslant p < +\infty$.

Definition 1.8
Let (Ω, Σ, μ) be a measure space with a topological space Ω, $\Sigma = \mathcal{B}(\Omega)$ (with $\mathcal{B}(\Omega)$ being the Borel σ-field on Ω) and $1 \leqslant p \leqslant +\infty$. A measurable function $f\colon \Omega \longrightarrow \mathbb{R} \cup \{\pm\infty\}$ is said to belong to $L^p_{\text{loc}}(\Omega)$ if and only if $f \in L^p(K)$ for every compact set $K \subseteq \Omega$.

Next we want to decide about the separability of the L^p-spaces. Recall that a measurable space (Ω, Σ) is said to be ***countably generated***, if there exists a sequence $\{A_n\}_{n\geqslant 1} \subseteq \Sigma$ such that $\Sigma = \sigma(\{A_n\}_{n\geqslant 1})$. The σ-algebra of all Lebesgue measurable sets in $\mathbb{R}^N$ is countably generated, since it can be generated by the countable family of cubes with centers in $\mathbb{Q}^N$ and rational side length. Similarly, if Ω is a separable metric space and $\Sigma = \mathcal{B}(\Omega)$, then Σ is countably generated.

Proposition 1.9
If (Ω, Σ, μ) *is a* σ*-finite measure space with* Σ *being countably generated and* $1 \leqslant p < +\infty$,
then the Banach space $L^p(\Omega)$ *is separable.*

Remark 1.10
Let $\{A_n\}_{n \geqslant 1} \subseteq \Sigma$ be the sequence generating Σ and let $\mathcal{A}$ be the smallest algebra containing $\{A_n\}_{n \geqslant 1}$. We consider all simple functions of the form

$$\sum_{k=1}^{n} c_k \chi_{E_k},$$

with $n \in \mathbb{N}$, $c_k \in \mathbb{Q}$, and $E_k \in \mathcal{A}$ with $\mu(E_k) < +\infty$ for all $k \in \{1, \ldots, n\}$. These simple functions form a countable set which is dense in $L^p(\Omega)$.

Next we consider some modes of convergence, closely related to the strong convergence in $L^p(\Omega)$ and examine their precise relations.

Definition 1.11
Let (Ω, Σ, μ) *be a measure space, let* $\{u_n : \Omega \longrightarrow \mathbb{R}\}_{n \geqslant 1}$ *be a sequence of measurable functions, and let* $u : \Omega \longrightarrow \mathbb{R}$ *be a measurable function.*
(a) *We say that the sequence* $\{u_n\}_{n \geqslant 1}$ ***converges to*** u ***in*** μ***-measure****, denoted by* $u_n \xrightarrow{\mu} u$ *if for every* $c > 0$

$$\lim_{n \to +\infty} \mu\big(\{\omega \in \Omega : |u_n(\omega) - u(\omega)| \geqslant c\}\big) = 0.$$

(b) *We say that the sequence* $\{u_n\}_{n \geqslant 1}$ ***converges to*** u ***almost uniformly****, denoted by* $u_n \xrightarrow{au} u$*, if for every* $\varepsilon > 0$*, there exists* $A \in \Sigma$ *with* $\mu(A) < \varepsilon$ *such that* $u_n \longrightarrow u$ *uniformly on* $\Omega \setminus A$.

Remark 1.12
If μ is a probability measure, then the convergence in μ-measure is called ***convergence in probability***. It is easy to verify that the convergence in measure satisfies the following properties:
(a) the limit is unique, i.e., if $u_n \xrightarrow{\mu} u$ and $u_n \xrightarrow{\mu} v$, then $u = v$;
(b) if $u_n \xrightarrow{\mu} u$, $v_n \xrightarrow{\mu} v$ and $\lambda, \vartheta \in \mathbb{R}$, then $\lambda u_n + \vartheta v_n \xrightarrow{\mu} \lambda u + \vartheta v$;
(c) if $u_n \xrightarrow{\mu} u$, then $u_n^+ \xrightarrow{\mu} u^+$, $u_n^- \xrightarrow{\mu} u^-$ and $|u_n| \xrightarrow{\mu} |u|$

Also recall that by the Egorov theorem, if (Ω, Σ, μ) is a finite measure space and $u_n \longrightarrow u$ μ-almost everywhere in Ω, then $u_n \xrightarrow{au} u$.

Proposition 1.13
(a) *If* $u_n \xrightarrow{\mu} u$,
then there exists a subsequence $\{u_{n_k}\}_{k\geqslant 1}$ *of* $\{u_n\}_{n\geqslant 1}$ *such that* $u_{n_k} \longrightarrow u$ *μ-almost everywhere.*
(b) *If* $\mu(\Omega) < +\infty$ *and* $u_n \longrightarrow u$ *μ-almost everywhere in* Ω,
then $u_n \xrightarrow{\mu} u$.

Remark 1.14
In part **(b)** of the above proposition, we cannot drop the hypothesis $\mu(\Omega) < +\infty$. Indeed, let $\Omega = \mathbb{R}$ be furnished with the Lebesgue measure λ. Let $u_n = \chi_{(n,n+1)}$ for $n \geqslant 1$. Then $u_n(\omega) \longrightarrow 0$ for each $\omega \in \mathbb{R}$. On the other hand, note that $\lambda(\{\omega \in \mathbb{R} : |u_n(\omega)| \geqslant 1\}) = 1$ and so we cannot have convergence in measure.

Proposition 1.15
If $u_n \xrightarrow{au} u$,
then $u_n \xrightarrow{\mu} u$ *and* $u_n \longrightarrow u$ *μ-almost everywhere.*

Proposition 1.16
If $u_n \longrightarrow u$ *in* $L^p(\Omega)$, *with* $1 \leqslant p \leqslant +\infty$ *(i.e.,* $\|u_n - u\|_p \longrightarrow 0$ *as* $n \to +\infty$*),*
then $u_n \xrightarrow{\mu} u$ *and we can find a subsequence* $\{u_{n_k}\}_{k\geqslant 1}$ *of* $\{u_n\}_{n\geqslant 1}$ *and a function* $h \in L^p(\Omega)$ *such that*

$$u_{n_k} \xrightarrow{au} u \quad \text{and} \quad |u_{n_k}(\omega)| \leqslant h(\omega) \quad \text{for } \mu\text{-a.a. } \omega \in \Omega.$$

Consider the space $L^0(\Omega)$ of all equivalence classes of measurable functions.

Proposition 1.17
If (Ω, Σ, μ) *is a finite measure space,*
then on $L^0(\Omega)$ *convergence in μ-measure is a metric convergence, that is,* $u_n \xrightarrow{\mu} u$ *if and only if* $dL^0(\Omega)(u_n, u) \longrightarrow 0$ *where* $dL^0(\Omega)$ *is the metric*

$$dL^0(\Omega)(u,v) = \int_\Omega \tfrac{|u-v|}{1+|u-v|}\, d\mu \qquad \forall\ u, v \in L^0(\Omega).$$

We recall the following notion which is central in the study of the L^p-spaces.

Definition 1.18
Let (Ω, Σ, μ) *be a measure space and let* $C \subseteq L^1(\Omega)$. *We say that the set* C *is* ***uniformly integrable***, *if the following holds:*
(a) *for every* $\varepsilon > 0$, *we can find* $A_\varepsilon \in \Sigma$ *such that*

$$\mu(A_\varepsilon) < +\infty \quad \text{and} \quad \sup_{u\in C} \int_{A_\varepsilon^c} |u|\, d\mu \leqslant \varepsilon;$$

(b) $$\lim_{\vartheta\to+\infty} \sup_{u\in C} \int_{\{|u|\geqslant\vartheta\}} |u|\, d\mu = 0.$$

Remark 1.19
Condition **(a)** in the above definition is automatically satisfied when $\mu(\Omega) < +\infty$. The Fatou lemma (see Theorem I.3.95) remains true, if the boundedness condition from below or above by an integrable function h is replaced by the requirement of uniform integrability. The result is often referred to as the ***generalized Fatou lemma***.

Theorem 1.20 *(Vitali Convergence Theorem)*
If $\{u_n\}_{n\geqslant 1} \subseteq L^p(\Omega)$ *is a sequence,* $u \in L^p(\Omega)$*, with* $1 \leqslant p < +\infty$
then $u_n \longrightarrow u$ *in* $L^p(\Omega)$ *if and only if* $u_n \xrightarrow{\mu} u$ *and the sequence* $\{|u_n|^p\}_{n\geqslant 1}$ *is uniformly integrable.*

Next we identify the dual of $L^p(\Omega)$, $1 \leqslant p < +\infty$.

Theorem 1.21 *(Riesz Representation Theorem)*
If $1 < p < +\infty$ *and* $\frac{1}{p} + \frac{1}{p'} = 1$ *(that is,* p *and* p' *are conjugate exponents),*
then for each $u^* \in L^p(\Omega)^*$ *there exists a unique* $y \in L^{p'}(\Omega)$ *such that*

$$\langle u^*, u\rangle = \int_\Omega uy\,dz \qquad \forall\, u \in L^p(\Omega)$$

(by $\langle\cdot,\cdot\rangle$ *we denote the duality brackets for the pair* $(L^p(\Omega)^*, L^p(\Omega)))$ *and* $\|u^*\|_* = \|y\|_{p'}$*. Therefore* $L^p(\Omega)^*$ *is isometrically isomorphic to* $L^{p'}(\Omega)$ *and for this reason we identify the two spaces, that is,*

$$L^p(\Omega)^* = L^{p'}(\Omega), \quad \textit{with}\ \frac{1}{p} + \frac{1}{p'} = 1.$$

Similarly, if (Ω, Σ, μ) *is* σ*-finite, then the result remains true if* $p = 1$*, that is,*

$$L^1(\Omega)^* = L^\infty(\Omega).$$

Remark 1.22
For the case $p = \infty$, the map $y \longmapsto u_y^* \in L^\infty(\Omega)^*$ defined by

$$u_y^*(u) = \int_\Omega uy\,d\mu$$

is always an isometric injection of $L^1(\Omega)$ into $L^\infty(\Omega)^*$, but it is not in general a surjection. In fact $L^\infty(\Omega)^*$ is the space of finitely additive measures.

Definition 1.23
Suppose that (Ω, Σ, μ) *is a* σ*-finite measure space.*
(a) $u^* \in L^\infty(\Omega)^*$ *is said to be* ***absolutely continuous with respect to*** μ*, if there exists* $y \in L^1(\Omega)$ *such that*

$$\langle u^*, u\rangle = \int_\Omega uy\,d\mu \qquad \forall\, u \in L^\infty(\Omega).$$

(b) $u^* \in L^\infty(\Omega)^*$ *is said to be* ***singular with respect to*** μ, *if there exists a sequence* $\{A_n\}_{n\geqslant 1} \subseteq \Sigma$ *such that* $A_{n+1} \subseteq A_n$ *for all* $n \geqslant 1$, $\mu(A_n) \searrow 0$ *and*

$$u^*(u) = u^*(\chi_{A_n} u) \qquad \forall\, u \in L^\infty(\Omega)$$

(i.e., the sets A_n *support* u^**).*

Remark 1.24
The absolutely continuous with respect to μ functionals are identified with the corresponding L^1-functions.

Theorem 1.25 *(Yosida–Hewitt Decomposition Theorem)*
If (Ω, Σ, μ) *is a* σ*-finite measure space and* $u^* \in L^\infty(\Omega)^*$,
then u^* *admits a unique decomposition* $u^* = u_a^* + u_s^*$, *where* u_a^* *is absolutely continuous with respect to* μ *and* u_s^* *is singular with respect to* μ.
Moreover, $\|u^*\|_* = \|u_a^*\|_1 + \|u_s^*\|_*$.

An immediate consequence of the Riesz representation theorem (see Theorem 1.21) is the following result.

Corollary 1.26
For every $1 < p < +\infty$, *the space* $L^p(\Omega)$ *is uniformly convex. So from the Milman–Pettis theorem (see Theorem I.5.89), it is also reflexive and has the* ***Kadec–Klee property****, namely "if* $\{u_n\}_{n\geqslant 1} \subseteq L^p(\Omega)$ *is a sequence such that* $u_n \xrightarrow{w} u$ *in* $L^p(\Omega)$ *and* $\|u_n\|_p \longrightarrow \|u\|_p$, *then* $u_n \longrightarrow u$ *in* $L^p(\Omega)$*" (cf. Definition I.5.88, Theorem I.5.89 and Proposition I.5.91).*

Definition 1.27
Let (Ω, Σ, μ) *be a measure space and let* p, p' *be conjugate exponents. If* $p = 1$ *or* $p = +\infty$, *then we assume in addition that the measure* μ *is* σ*-finite. A sequence* $\{u_n\}_{n\geqslant 1} \subseteq L^p(\Omega)$ ***converges weakly (weakly star*** *if* $p = +\infty$*) to some function* $u \in L^p(\Omega)$, *denoted by* $u_n \xrightarrow{w} u$ *(*$u_n \xrightarrow{w^*} u$ *if* $p = +\infty$*) if and only if*

$$\int_\Omega u_n y \, d\mu \longrightarrow \int_\Omega u y \, d\mu \qquad \forall\, y \in L^{p'}(\Omega)$$

(see Theorem 1.21).

Next we characterize weakly compact sets in $L^1(\Omega)$.

Theorem 1.28 *(Dunford–Pettis Theorem)*
If (Ω, Σ, μ) *is a measure space and* $C \subseteq L^1(\Omega)$,
then the set C *is relatively* w*-compact if and only if it is uniformly integrable (see Definition 1.18).*

For strong compactness in L^p, there are various types of criteria, some of which although quite general are difficult to apply to specific cases. For this reason, we restrict ourselves to $\mathbb{R}^N$, where we have more applicable conditions.

Theorem 1.29 *(Kolmogorov–Riesz Theorem)*
If $C \subseteq L^p(\mathbb{R}^N)$ *is a bounded set with* $1 \leqslant p < +\infty$ *and*

$$\lim_{|h|\to 0} \|\tau_h u - u\|_p = 0 \quad \text{uniformly in } u \in C,$$

where $\tau_h u(z) = u(z+h)$ *for all* $z, h \in \mathbb{R}^N$,
then for every $\Omega \subseteq \mathbb{R}^N$ *with finite Lebesgue measure, the set*

$$C|_\Omega = \{u|_\Omega : u \in C\}$$

is relatively compact in $L^p(\Omega)$.

The next inequality although straightforward, it is very useful in many circumstances.

Theorem 1.30 *(Chebyshev–Markov Inequality)*
If (Ω, Σ, μ) *is a measure space and* $u \in L^p(\Omega)$ *with* $1 \leqslant p < +\infty$,
then for every $c > 0$ *we have*

$$\mu(\{\omega \in \Omega : u(\omega) \geqslant c\}) \leqslant \tfrac{\|u\|_p^p}{c^p}.$$

Definition 1.31
Let $u \in L^1_{\text{loc}}(\mathbb{R}^N)$. *The* ***maximal function*** *(or* ***Hardy–Littlewood maximal function****) of* u *is defined by*

$$M(u)(x) = \sup_{r>0} \frac{1}{\lambda^N(B_r(x))} \int_{B_r(x)} |u(z)|\,dz \qquad \forall\, x \in \mathbb{R}^N,$$

where λ^N *denotes the Lebesgue measure on* $\mathbb{R}^N$ *and* $B_r(x) = \{z \in \mathbb{R}^N : |z - x| < r\}$.
For $R \in (0, +\infty]$ *we define*

$$M_R(u)(x) = \sup_{0<r<R} \frac{1}{\lambda^N(B_r(x))} \int_{B_r(x)} |u(z)|\,dz \qquad \forall\, x \in \mathbb{R}^N.$$

Evidently $M = M_\infty$.

Theorem 1.32
If $1 \leqslant p \leqslant +\infty$, *then*
(a) $u(x) \leqslant M(u)(x) < +\infty$ *for all* $u \in L^p(\mathbb{R}^N)$ *and for almost all* $x \in \mathbb{R}^N$;
(b) *if* $p = 1$, *then for any* $u \in L^1(\mathbb{R}^N)$ *and* $c > 0$,

$$\lambda^N(\{x \in \mathbb{R}^N : M(u)(x) \geqslant c\}) \leqslant \tfrac{3^N}{c}\|u\|_1;$$

(c) if $1 < p \leqslant +\infty$, then $M(u) \in L^p(\mathbb{R}^N)$ and there exists $c(N,p) > 0$ such that

$$\|M(u)\|_p \leqslant c(N,p)\|u\|_p \qquad \forall\, u \in L^p(\mathbb{R}^N).$$

Definition 1.33
Let $A \subseteq \mathbb{R}^N$ be a Lebesgue measurable set and let $u\colon A \longrightarrow \mathbb{R}$ be a measurable function. We say that u belongs to the space $L\log L(A)$ if and only if

$$\int_A |u(z)|\log(2+|u(z)|)\,dz < +\infty.$$

Remark 1.34
Note that $L\log L(\mathbb{R}^N) \subsetneq L^1(\mathbb{R}^N)$.

1.1.2 Lebesgue–Bochner Spaces

In this section, we deal with Banach space valued functions.

Definition 1.35
Let (Ω, Σ, μ) be a measure space and let X be a Banach space.
*(a) A **simple function**, is a function $s\colon \Omega \longrightarrow X$ of the form*

$$s(\omega) = \sum_{k=1}^{n} x_k \chi_{A_k}(\omega) \qquad \forall\, \omega \in \Omega,$$

with $n \in \mathbb{N}$, $\{x_k\}_{k=1}^n \subseteq X$ distinct vectors and $\{A_k\}_{k=1}^n \subseteq \Sigma$ are mutually disjoint.
*(b) A function $u\colon \Omega \longrightarrow X$ is said to be **strongly measurable**, if there exists a sequence $\{s_n\}_{n\geqslant 1}$ of simple functions such that*

$$\|u(\omega) - s_n(\omega)\|_X \longrightarrow 0 \quad \text{for } \mu\text{-almost all } \omega \in \Omega.$$

*(c) A function $u\colon \Omega \longrightarrow X$ is said to be **weakly measurable**, if for every $x^* \in X^*$ the function $\omega \longmapsto \langle x^*, u(\omega)\rangle$ is Σ-measurable (here by $\langle \cdot,\cdot\rangle$ we denote the duality brackets for the pair (X^*, X)).*
*(d) A function $u\colon \Omega \longrightarrow X^*$ is said to be w^***-measurable**, if for every $x \in X$ the function $\omega \longmapsto \langle u(\omega), x\rangle$ is Σ-measurable.*

The next theorem gives the relation between strong and weak measurability.

Theorem 1.36 *(Pettis Measurability Theorem)*
If (Ω, Σ, μ) is a finite measure space and X is a Banach space,
then a function $u\colon \Omega \longrightarrow X$ is strongly measurable if and only if
(a) u is weakly measurable; and
(b) there exists a set $A \in \Sigma$ with $\mu(A) = 0$ such that the set $u(\Omega \setminus A) \subseteq X$ is norm separable.

Remark 1.37
A function u satisfying statement **(b)** of the above theorem is said to be ***μ-essentially separably valued***. Also, a byproduct of the proof of this theorem is that u is strongly measurable if and only if u is the uniform limit of a sequence of countably valued μ-measurable functions.

Definition 1.38
Let (Ω, Σ, μ) be a measure space and let X be a Banach space.
*(a) A simple function $s\colon \Omega \longrightarrow X$ is **Bochner integrable**, if it has the form*

$$s(\omega) = \sum_{k=1}^{n} x_k \chi_{A_k}(\omega),$$

*with $n \in \mathbb{N}$, $\{x_k\}_{k=1}^n \subseteq X$ distinct vectors, $\{A_k\}_{k=1}^n \subseteq \Sigma$ mutually disjoint sets and $x_k = 0$ whenever $\mu(A_k) = +\infty$. Then for any $A \in \Sigma$ the **Bochner integral** of u over A is defined by*

$$\int_A s\, d\mu = \sum_{k=1}^{n} x_k \mu(A \cap A_k),$$

with $x_k\mu(A \cap A_k) = 0$, when $x_k = 0$ and $\mu(A \cap A_k) = +\infty$.
*(b) A strongly measurable function $u\colon \Omega \longrightarrow X$ is said to be **Bochner integrable**, if there exists a sequence $\{s_n\}_{n\geqslant 1}$ of integrable simple functions such that*

$$\|u(\omega) - s_n(\omega)\|_X \longrightarrow 0 \quad \text{for } \mu\text{-almost all } \omega \in \Omega$$

and

$$\int_\Omega \|u - s_n\|_X \, d\mu \longrightarrow 0 \quad \text{as } n \to +\infty.$$

*Then for any $A \in \Sigma$, the **Bochner integral** of u over A is defined by*

$$\int_A u\, d\mu = \lim_{n\to+\infty} \int_\Omega s_n\, d\mu.$$

Remark 1.39
It is routine to check that the last limit exists and is independent of the particular approximation sequence $\{s_n\}_{n\geqslant 1}$.

Proposition 1.40
<u>If</u> (Ω, Σ, μ) is a finite measure space and X is a Banach space,
<u>then</u> a strongly measurable function $u\colon \Omega \longrightarrow X$ is Bochner integrable if and only if $\|u(\cdot)\|_X \in L^1(\Omega)$.
Moreover, we have

$$\Big\| \int_\Omega u\, d\mu \Big\|_X \leqslant \int_\Omega \|u\|_X \, d\mu.$$

So, now we are in a position to introduce L^p-spaces for Banach space valued functions.

Definition 1.41
Let (Ω, Σ, μ) be a measure space and let X be a Banach space.
(a) *Let $1 \leqslant p < +\infty$. By $L^p(\Omega; X)$ we denote the space of all equivalence classes (for the relation $\sim$ of equality μ-almost everywhere), of the strongly measurable functions $u\colon \Omega \longrightarrow X$ such that $\|u(\cdot)\|_X \in L^p(\Omega)$. The norm is defined by*

$$\|u\|_p = \Big(\int_\Omega \|u\|_X^p \, d\mu\Big)^{\frac{1}{p}}.$$

(b) *By $L^\infty(\Omega; X)$ we define the space of all the equivalence classes (again for the relation $\sim$ of equality μ-almost everywhere), of the strongly measurable functions $u\colon \Omega \longrightarrow X$ such that $\|u(\cdot)\|_X \in L^\infty(\Omega)$. The norm is defined by*

$$\|u\|_p = \operatorname*{ess\,sup}_\Omega \|u(\cdot)\|_X.$$

Proposition 1.42
If (Ω, Σ, μ) is a measure space and X is a Banach space, then
(a) *$L^p(\Omega; X)$ is a Banach space for all $1 \leqslant p \leqslant +\infty$;*
(b) *the set of integrable simple functions is dense in $L^p(\Omega; X)$ for all $1 \leqslant p < +\infty$;*
(c) *if Ω is a separable metric space, μ is σ-finite Radon measure (see Definition I.4.9) and X is a separable Banach space, then $L^p(\Omega; X)$ is separable for all $1 \leqslant p < +\infty$.*

Definition 1.43
Let (Ω, Σ) be a measurable space and let X be a Banach space. A set function $m\colon \Sigma \longrightarrow X$ is a ***vector measure****, if for any sequence $\{A_n\}_{n \geqslant 1} \subseteq \Sigma$ of pairwise disjoint sets, we have*

$$m\Big(\bigcup_{n \geqslant 1} A_n\Big) = \sum_{n \geqslant 1} m(A_n)$$

in the norm topology of X.
The ***variation****, is the extended measure $|m|$ defined by*

$$|m|(A) = \sup_\pi \sum_{C \in \pi} \|m(C)\|_X$$

where the supremum is taken over the collection π of all finite Σ-partitions of A.
If $|m|(\Omega) < +\infty$, then we say that the vector measure m is of ***bounded variation.***

Remark 1.44
In the above definition, the convergence of the series $\sum_{n\geqslant 1} m(A_n)$ is unconditional in norm (since every subseries also converges). If $u\colon \Omega \longrightarrow X$ is Bochner integrable, then $\Sigma \ni A \longmapsto m(A) = \int_A u\, d\mu$ is a vector measure and

$$|m|(A) = \int_A \|u\|_X \, d\mu.$$

Definition 1.45
*A Banach space X has the **Radon–Nikodym Property** (RNP for short), if for every finite measure space (Ω, Σ, μ) and every vector measure $m\colon \Sigma \longrightarrow X$ of bounded variation such that $m \ll \mu$ (that is, m vanishes on μ-null sets), we can find $u \in L^1(\Omega; X)$ such that*

$$m(A) = \int_A u\, d\mu \qquad \forall\, A \in \Sigma.$$

Theorem 1.46
(a) Reflexive Banach spaces have the RNP.
(b) Separable dual Banach spaces have the RNP.

The notion of RNP leads to an extension of the Riesz representation theorem (see Theorem 1.21) to the Lebesgue–Bochner spaces $L^p(\Omega; X)$ $(1 \leqslant p < +\infty)$.

Theorem 1.47
If (Ω, Σ, μ) is a σ-finite measure space and X is a Banach space with X^* having the RNP,
then for all $1 \leqslant p < +\infty$, we have

$$L^p(\Omega; X)^* = L^{p'}(\Omega; X^*)$$

where $\frac{1}{p} + \frac{1}{p'} = 1$.

Next we examine $L^1(\Omega; X)^*$ when X^* fails to satisfy the RNP.

Definition 1.48
Let (Ω, Σ, μ) be a measure space, X a Banach space and $u, y\colon \Omega \longrightarrow X^*$ two w^*-measurable functions (see Definition 1.35(d)). We say that $u \sim y$, if

$$\langle u(\omega), x\rangle = \langle y(\omega), x\rangle \quad \text{for } \mu\text{-a.a. } \omega \in \Omega \text{ and all } x \in X.$$

Remark 1.49
In the above definition the exceptional μ-null set in general depends on $x \in X$. The relation $\sim$ is an equivalence relation.

Definition 1.50
By $L^\infty(\Omega; X^*_{w^*})$ we denote the linear space of the equivalence classes for the relation $\sim$ (see Definition 1.48) of functions $u\colon \Omega \longrightarrow X^*$ which are w^*-measurable and there exists $c > 0$ such that

$$\big|\langle u(\omega), x\rangle\big| \leqslant c\|x\|_X \quad \text{for } \mu\text{-almost all } \omega \in \Omega, \ \text{all } x \in X.$$

The exceptional μ-null set may depend on $x \in X$. The infimum of all such $c > 0$ is denoted by $\|u\|_{L^\infty(\Omega; X^*_{w^*})}$ and it is a norm on $L^\infty(\Omega; X^*_{w^*})$.

Remark 1.51
If X is separable and $u \in L^\infty(\Omega; X^*_{w^*})$, then the map $\omega \longmapsto \|u(\omega)\|_*$ belongs in $L^\infty(\Omega)$ and we have

$$\|u\|_{L^\infty(\Omega; X^*_{w^*})} = \operatorname*{ess\,sup}_{\Omega} \|u(\cdot)\|_*.$$

Theorem 1.52 *(Dinculeanu–Foias Theorem)*
<u>If</u> (Ω, Σ, μ) is a measure space and X is a Banach space,
<u>then</u> $L^1(\Omega; X)^* = L^\infty(\Omega; X^*_{w^*})$ and the duality brackets for this pair are defined by

$$\langle u, h\rangle = \int_\Omega \langle u(\omega), h(\omega)\rangle \, d\mu \qquad \forall\, u \in L^\infty(\Omega; X^*_{w^*}), \ h \in L^1(\Omega; X).$$

1.1.3 BV-Functions, Absolutely Continuous Functions, Spaces of Measures

We start with functions of one variable. As we will see, functions of bounded variation (BV-functions) are closely related to monotone functions (by monotone we understand both increasing or decreasing). Monotone functions need not be continuous. The next result reveals how discontinuous they can be.

Proposition 1.53
<u>If</u> I is an interval and $u\colon I \longrightarrow \mathbb{R}$ is monotone,
<u>then</u> u has at most countably many discontinuity points.
Conversely, given a countable set $D \subseteq \mathbb{R}$, there is a monotone function $u\colon \mathbb{R} \longrightarrow \mathbb{R}$ such that $\operatorname{disc} u = D$ (here $\operatorname{disc} u$ denotes the set of discontinuity points of u).

Remark 1.54
As a consequence of this proposition, we see that there is an increasing function $u\colon \mathbb{R} \longrightarrow \mathbb{R}$ continuous at all irrational points and discontinuous at all rational points.

The next proposition analyzes the inverse of an increasing function.

Proposition 1.55
<u>If</u> $I \subseteq \mathbb{R}$ is an interval bounded below, $u\colon I \longrightarrow \mathbb{R}$ is an increasing function, $J \subseteq \mathbb{R}$ is the smallest interval containing $u(I)$, and $w\colon J \longrightarrow \mathbb{R}$ is the function defined by

$$w(y) = \inf\big\{x \in I:\ u(x) \geqslant y\big\} \qquad \forall\, y \in J,$$

then
(a) *w is increasing and left continuous;*
(b) *w has a jump at some $y_0 \in J \setminus \{\sup_I u\}$ if and only if $u|_{(x_1,x_2)} \equiv y_0$ for some $x_1, x_2 \in I$, $x_1 < x_2$;*
(c) *$w(u(x)) \leqslant x$ for every $x \in I$ and the inequality is strict if and only if u is constant on some interval $[z, x) \subseteq I$, $z < x$;*
(d) *$w|_{(y_1,y_2)} \equiv x_0$ for some interval $(y_1, y_2) \subseteq I$, $y_1 < y_2$ and for some $x_0 \in \operatorname{int} I$ if and only if u has a jump at x_0 and $(y_1, y_2) \subseteq (u^-(x_0), u^+(x_0))$.*

In particular, if u is strictly increasing, then w is a left inverse of u and it is continuous.

Theorem 1.56 *(Lebesgue Theorem)*
If $I \subseteq \mathbb{R}$ is an interval and $u\colon I \longrightarrow \mathbb{R}$ is a monotone function,
then u is differentiable at all $x \in I \setminus D$, where D is a Lebesgue-null set.

In fact this theorem is sharp.

Proposition 1.57
Given any Lebesgue-null set $D \subseteq \mathbb{R}$, we can find a continuous monotone function which is not differentiable at every point of D.

Remark 1.58
In the above result, the set of points of nondifferentiability of u contains D and in general it may be strictly bigger. Recall that the Cantor function u is increasing and

$$u'(x) = 0 \quad \text{for almost all } x \in [0, 1].$$

In fact, there exists a strictly increasing function $u\colon [0, 1] \longrightarrow \mathbb{R}$ with zero derivative almost everywhere.

Given an interval $I \subseteq \mathbb{R}$, a ***partition*** of I is a finite set $P = \{x_k\}_{k=0}^n \subseteq I$ such that

$$x_0 < x_1 < \ldots < x_n.$$

Definition 1.59
Let $I \subseteq \mathbb{R}$ be an interval and let $u\colon I \longrightarrow \mathbb{R}$ be a function. The ***total variation*** *of u on I is defined by*

$$\operatorname{Var} u = \operatorname{Var}_I u = \sup \left\{ \sum_{k=1}^{n} \left| u(x_k) - u(x_{k-1}) \right| \right\},$$

where the supremum is taken over all partitions $P = \{x_k\}_{k=0}^n$ of I ($n \in \mathbb{N}$).

We say that u is of ***bounded variation*** *(**BV-function** for short), if $\operatorname{Var} u < +\infty$.*

We denote the space of functions of bounded variation, by $BV(I)$.

Remark 1.60
If one of the endpoints of I (say $b = \sup I$) belongs in I, then we can restrict ourselves to partitions $P = \{x_k\}_{k=0}^n$ with $x_n = b$. Also, we say that $u \in BV_{\text{loc}}(I)$, if $\operatorname{Var}_{[a,b]} u < +\infty$ for all intervals $[a,b] \subseteq I$. Evidently, if $I = [a,b]$, then $BV_{\text{loc}}(I) = BV(I)$. Finally, if $U \subseteq I$ is an open set, then $U = \bigcup_{n \geqslant 1} I_n$ with $\{I_n\}_{n \geqslant 1}$ disjoint open intervals and then for a function $u \colon U \longrightarrow \mathbb{R}$ we set

$$\operatorname{Var} u = \sum_{n \geqslant 1} \operatorname{Var}_{I_n} u.$$

Proposition 1.61
If $I \subseteq \mathbb{R}$ is an interval and $u \colon I \longrightarrow \mathbb{R}$, then
(a) *for every $c \in I$, $\sup_{x \in I} |u(x)| \leqslant |u(c)| + \operatorname{Var} u$, hence if $u \in BV(I)$, then u is bounded;*
(b) *for every $c \in I$, we have*

$$\operatorname{Var}_{I \cap (-\infty, c]} + \operatorname{Var}_{I \cap [c, +\infty)} = \operatorname{Var} u;$$

(c) *if I does not contain the right endpoint, then*

$$\lim_{x \nearrow \sup I} \operatorname{Var}_{I \cap (-\infty, x]} u = \operatorname{Var} u;$$

if I does not contain the left endpoint, then

$$\lim_{x \searrow \inf I} \operatorname{Var}_{I \cap [x, +\infty, x)} u = \operatorname{Var} u.$$

The next theorem establishes the relation between monotone functions and BV-functions.

Theorem 1.62
If $I \subseteq \mathbb{R}$ is an interval,
then $BV_{\text{loc}}(I)$ (respectively $BV(I)$) is the smallest vector space which contains all monotone (respectively bounded monotone) functions. Moreover, every element in $BV_{\text{loc}}(I)$ (respectively $BV(I)$) can be written as the difference of two increasing (respectively bounded increasing) functions. Hence every function $u \in BV_{\text{loc}}(I)$ is differentiable almost everywhere (see Theorem 1.56).

Note that $\operatorname{Var} u$ is not a norm on $BV(I)$, since $\operatorname{Var} u = 0$ implies that u is constant. In this direction we have the following result.

Proposition 1.63
If $I \subseteq \mathbb{R}$ is an interval and $c \in I$,
then the function $u \longmapsto \|u\|_{BV(I)} = |u(c)| + \operatorname{Var} u$ is a norm on the space $BV(I)$.

The question of completeness of $BV(I)$ with this norm (that is, whether $BV(I)$ is a Banach space) is answered positively with the help of the following compactness result.

Theorem 1.64 *(Helly Selection Theorem)*
If $I \subseteq \mathbb{R}$ *is an interval and* $\mathcal{D} \subseteq BV(I)$ *is an infinite subset satisfying*

$$|u(c)| + \operatorname{Var} u \leqslant M \qquad \forall\, u \in \mathcal{D},$$

for some $c \in I$ *and* $M > 0$,
then there exists a sequence $\{u_n\}_{n\geqslant 1} \subseteq \mathcal{D}$ *and* $u \in BV(I)$ *such that*

$$u_n(x) \longrightarrow u(x) \qquad \forall\, x \in I.$$

As a consequence of this result, we have the following corollary.

Corollary 1.65
If $I \subseteq \mathbb{R}$ *is an interval and* $c \in I$,
then $BV(I)$ *equipped with the norm* $\|u\|_{BV(I)} = |u(c)| + \operatorname{Var} u$ *is a Banach space.*

Definition 1.66
Let X, Y *be two nonempty sets,* $A \subseteq X$ *and* $u\colon X \longrightarrow Y$. *For every* $y \in Y$ *we define*

$$N_u(y; A) = \#\{x \in A : u(x) = y\}.$$

The map $N_u(\cdot; A)\colon Y \longrightarrow \mathbb{N}_0 \cup \{\infty\}$ *is called the* ***counting function*** *or the* ***Banach indicatrix*** *of* u *in the set* A.

Using this special function, we can characterize those continuous functions which are of bounded variation.

Theorem 1.67
If $I \subseteq \mathbb{R}$ *is an interval and* $u\colon I \longrightarrow \mathbb{R}$ *is continuous,*
then $N_u(\cdot; I)$ *is Borel measurable and*

$$\int_{\mathbb{R}} N_u(y; I)\, dy = \operatorname{Var} u.$$

Therefore $u \in BV(I)$ *if and only if* $N_u(\cdot; I) \in L^1(\mathbb{R})$.

Consider the Cantor function $g\colon [0,1] \longrightarrow [0,1]$. We know that it is continuous and increasing (hence $g \in BV([0,1])$ and $g'(x) = 0$ for all $x \in [0,1] \setminus C$ with C being the Cantor set, that is, $g' = 0$ almost everywhere). Note that g fails to satisfy the fundamental theorem of Lebesgue calculus since

$$\int_0^1 g'(s)\, ds = 0 < g(1) - g(0) = 1.$$

So, we want to identify the subspace of the space of functions of bounded variation for which the fundamental theorem of the Lebesgue calculus holds. This leads to absolutely continuous functions.

Definition 1.68
Let $I \subseteq \mathbb{R}$ be an interval and $u\colon I \longrightarrow \mathbb{R}$. We say that u is ***absolutely continuous*** *on I, if for every $\varepsilon > 0$, there exists $\delta > 0$ such that for every finite family $\{(a_k, b_k)\}_{k=1}^n$ of nonoverlapping subintervals of I with*

$$\sum_{k=1}^n (b_k - a_k) \leqslant \delta,$$

we have

$$\sum_{k=1}^n \left|u(b_k) - u(a_k)\right| \leqslant \varepsilon.$$

The space of all functions u which are absolutely continuous on I is denoted by $AC(I)$.

Remark 1.69
In the above definition $n \in \mathbb{N}$ is arbitrary. We can also define the space $AC_{\mathrm{loc}}(I)$, which contains of all functions $u\colon I \longrightarrow \mathbb{R}$ which are absolutely continuous on every $[a, b] \subseteq I$. If $I = [a, b]$, then $AC_{\mathrm{loc}}([a, b]) = AC([a, b])$.

Proposition 1.70
(a) *If $I \subseteq \mathbb{R}$ is an interval,*
then $AC_{\mathrm{loc}}(I) \subseteq BV_{\mathrm{loc}}(I)$ and every $u \in AC_{\mathrm{loc}}(I)$ is differentiable almost everywhere in I with $u' \in L^1_{\mathrm{loc}}(I)$.
(b) *If $I \subseteq \mathbb{R}$ is a bounded interval,*
then $AC(I) \subseteq BV(I)$ and every $u \in AC(I)$ is differentiable almost everywhere in I with $u' \in L^1(I)$.

There are bounded, continuous, monotone functions which fail to be absolutely continuous (recall the Cantor function). So, it is natural to ask what is missing from a continuous function which is not absolutely continuous. This is the so-called Lusin (N)-property.

Theorem 1.71
If $I \subseteq \mathbb{R}$ is an interval,
then $u \in AC_{\mathrm{loc}}(I)$ if and only if
(a) *u is continuous;*
(b) *u' exists almost everywhere and $u' \in L^1_{\mathrm{loc}}(I)$;*
(c) *u maps Lebesgue-null sets to Lebesgue-null sets (Lusin (N)-property).*

Absolutely continuous functions satisfy some of the basic results of differential calculus. The first result is the ***chain rule***.

Theorem 1.72 *(Chain Rule)*
If $I, J \subseteq \mathbb{R}$ are two intervals, $u\colon J \longrightarrow \mathbb{R}$ and $h\colon I \longrightarrow J$ are functions such that u, h and $u \circ h$ are differentiable almost everywhere and u maps Lebesgue-null sets to Lebesgue-null sets, then

$$(u \circ h)'(x) \;=\; u'(h(x))h'(x) \quad \textit{for almost all } x \in I,$$

where $u'(h(x))h'(x)$ is defined to be zero whenever $h'(x) = 0$ (even if u is not differentiable at $h(x)$).

As an easy consequence of this theorem, we have that absolutely continuous functions satisfy the ***change of variable formula***.

Theorem 1.73 *(Change of Variable Formula)*
If $u\colon [c,d] \longrightarrow \mathbb{R}$ is integrable and $h\colon [a,b] \longrightarrow [c,d]$ is differentiable almost everywhere in $[a,b]$,
then $u(h(\cdot))h'(\cdot)$ is integrable and the change of variation formula holds, namely

$$\int\limits_{h(s)}^{h(t)} u(y)\,dy \;=\; \int\limits_{s}^{t} u(h(x))h'(x)\,dx \qquad \forall\, s,t \in [a,b]$$

if and only if $u \circ h \in AC\,([a,b])$.

Now we turn our attention to multivariable functions.

Definition 1.74
Let $\Omega \subseteq \mathbb{R}^N$ be an open set. We say that $u \in L^1(\Omega)$ is of ***bounded variation (BV-function*** *for short), if for all $k \in \{1,\ldots,N\}$ there exists a finite signed measure $\mu_k\colon \mathcal{B}(\Omega) \longrightarrow \mathbb{R}$ ($\mathcal{B}(\Omega)$ being the Borel σ-algebra of Ω) such that*

$$\int\limits_{\Omega} u\frac{\partial \vartheta}{\partial z_k}\,dz \;=\; -\int\limits_{\Omega} \vartheta\,d\mu_k \qquad \forall\, \vartheta \in C_c^\infty(\Omega).$$

The space of functions of bounded variation is denoted by $BV(\Omega)$.

Remark 1.75
Using the language of distribution theory, we can say that μ_k is the ***distributional*** (or ***weak***) ***k-partial derivative*** of u and is denoted by $D_k u$. Also, we define

$$BV_{\mathrm{loc}}(\Omega) \;=\; \big\{u \in L^1_{\mathrm{loc}}(\Omega) : \; u \in BV(U),\; U \subset\subset \Omega\big\}.$$

Recall that $U \subset\subset \Omega$ means that $\overline{U}$ is compact and $\overline{U} \subseteq \Omega$. From the above definition we see that if $u \in BV(\Omega)$ and we set $Du = (D_1 u, \ldots, D_N u)$, then $Du \in M_b(\Omega;\mathbb{R}^N)$ ($M_b(\Omega;\mathbb{R}^N)$ being the space of Radon measures). For this measure the total variation is defined by

$$|Du|(A) \;=\; \sup_{\pi} \sum_{E\in\pi} |Du|(E),$$

where the supremum is taken over all partitions π of A into a finite number of pairwise disjoint Borel subsets of A. Since $M_b(\Omega;\mathbb{R}^N) = C_0\left(\Omega;\mathbb{R}^N\right)^*$ (see Sect. 1.1.3 below and Volume 1), we also have

$$\begin{aligned} |Du|(\Omega) &= \|Du\|_{M_b(\Omega;\mathbb{R}^N)} \\ &= \sup\left\{\sum_{k=1}^{N}\int_{\Omega}\vartheta_k\, d(D_k u) : \vartheta = (\vartheta_1,\ldots,\vartheta_N) \in C_0\left(\Omega;\mathbb{R}^N\right),\right. \\ &\qquad\qquad \left.\|\vartheta\|_{C_0(\Omega;\mathbb{R}^N)} \leqslant 1\right\} \\ &< +\infty. \end{aligned}$$

Recalling that $C_0^\infty\left(\Omega;\mathbb{R}^N\right)$ is dense in $C_0\left(\Omega;\mathbb{R}^N\right)$, we can write

$$|Du|(\Omega) = \sup\left\{\sum_{k=1}^{N}\int_{\Omega}\tfrac{\partial\vartheta_k}{\partial z_k}u\, dz : \vartheta \in C_0^\infty\left(\Omega;\mathbb{R}^N\right),\ \|\vartheta\|_{C_0^\infty(\Omega;\mathbb{R}^N)} \leqslant 1\right\}.$$

The right-hand side of the above equality determines the ***variation*** of u in Ω and is denoted by $\operatorname{Var} u$ or by $\operatorname{Var}(u;\Omega)$ if we want to emphasize the dependence on the domain Ω.

When $N = 1$, we see that we have two different definitions for the functions of bounded variation. One which is pointwise (see Definition 1.59) and the other global in nature given in Definition 1.74. It is natural to ask how are these two definitions related. The next theorem shows that they define the same function space.

Theorem 1.76
If $\Omega \subseteq \mathbb{R}$ is an open set and $u \in L^1(\Omega)$ is of bounded variation in the sense of Definition 1.59,
then u is of bounded variation in the sense of Definition 1.74 and

$$|Du|(\Omega) \leqslant \operatorname{Var} u.$$

Conversely, if u is of bounded variation in the sense of Definition 1.74,
then u admits a right continuous representative $\overline{u}$ which is of bounded variation in the sense of Definition 1.59 and we have

$$\operatorname{Var}\overline{u} = |Du|(\Omega).$$

Definition 1.77
*Let $\Omega \subseteq \mathbb{R}^N$ be an open set and let $A \subseteq \mathbb{R}^N$ be a Lebesgue measurable set. We say that A has **(locally) finite perimeter**, if $\chi_A \in BV(\Omega)$ (respectively $\chi_A \in BV_{\mathrm{loc}}(\Omega)$). The **perimeter** of A in Ω, denoted by $P(E;\Omega)$, is the variation of χ_A in Ω, i.e.,*

$$\begin{aligned} P(A;\Omega) &= \operatorname{Var}(\chi_A,\Omega) \\ &= \sup\left\{\sum_{k=1}^{N}\int_{A}\tfrac{\partial\vartheta_k}{\partial z_k}\, dz : \vartheta \in C_c^\infty(\Omega;\mathbb{R}^N),\ \|\vartheta\|_{C_0(\Omega;\mathbb{R}^N)} \leqslant 1\right\}. \end{aligned}$$

Proposition 1.78
If $\Omega \subseteq \mathbb{R}^N$ *is open and* $u \in BV_{\mathrm{loc}}(\Omega) \cap L^1(\Omega)$,
then $u \in BV(\Omega)$ *(i.e.,* $\operatorname{Var} u < +\infty$*) if and only if* $|Du|(\Omega) < +\infty$. *So, we can define*

$$\|u\|_{BV(\Omega)} = \|u\|_1 + |Du|(\Omega),$$

which is a norm on $BV(\Omega)$. *With this norm* $BV(\Omega)$ *is a Banach space.*

Proposition 1.79
If $\Omega \subseteq \mathbb{R}^N$ *is open,* $\{u_n\}_{n\geqslant 1} \subseteq L^1_{\mathrm{loc}}(\Omega)$ *is a sequence such that* $u_n \longrightarrow u$ *in* $L^1_{\mathrm{loc}}(\Omega)$,
then

$$|Du|(\Omega) \leqslant \liminf_{n\to+\infty} |Du_n|(\Omega).$$

Functions of bounded variation can be approximated by smooth functions.

Proposition 1.80
If $\Omega \subseteq \mathbb{R}^N$ *is open and* $u \in BV(\Omega)$,
then there exist functions $\{u_n\}_{n\geqslant 1} \subseteq BV(\Omega) \cap C^\infty(\Omega)$ *such that*

$$u_n \longrightarrow u \quad \text{in } L^1(\Omega) \quad \text{and} \quad |Du_n|(\Omega) \longrightarrow |Du|(\Omega).$$

Proposition 1.81
If $\Omega \subseteq \mathbb{R}^N$ *is bounded open with a Lipschitz boundary* $\partial\Omega$, $\{u_n\}_{n\geqslant 1} \subseteq BV(\Omega)$ *and*

$$\sup_{n\geqslant 1} \|u_n\|_{BV(\Omega)} < +\infty,$$

then there exist a subsequence $\{u_{n_k}\}_{k\geqslant 1}$ *of* $\{u_n\}_{n\geqslant 1}$ *and* $u \in BV(\Omega)$ *such that*

$$u_{n_k} \longrightarrow u \quad \text{in } L^1(\Omega).$$

The ***coarea formula*** relates the total variation measure $|Du|$ with the perimeter of the superlevel sets $U_t = \{z \in \Omega : \ u(z) > t\}$, $t \in \mathbb{R}$.

Theorem 1.82 *(Coarea Formula)*
If $\Omega \subseteq \mathbb{R}^N$ *is an open set and* $u \in BV(\Omega)$,
then the set U_t *has a finite perimeter for almost all* $t \in \mathbb{R}$ *and*

$$|Du|(\Omega) = \int_{-\infty}^{+\infty} P(U_t;\Omega)\,dt.$$

Conversely, if $u \in L^1(\Omega)$ *and* $\int_{-\infty}^{+\infty} P(U_t;\Omega)\,dt < +\infty$,
then $u \in BV(\Omega)$.

Definition 1.83
A ***rectangular cell*** *in* $\mathbb{R}^N$ *is a set of the form*

$$R = \prod_{k=1}^{N}(a_k, b_k).$$

For every $z \in \mathbb{R}$ *we write* $z = (\widehat{z}, z_k)$ *with* $\widehat{z} \in \widehat{R}^{N-1} = \prod_{\substack{i=1\\ i\neq k}}^{N} (a_i, b_i)$, $z_k \in (a_k, b_k)$ *and define*

$$u_k(\cdot) = u(\widehat{z}, \cdot) \qquad \forall\, k \in \{1, \ldots, N\}.$$

Theorem 1.84
If $u \in L^1_{\text{loc}}(\mathbb{R}^N)$,
then $u \in BV_{\text{loc}}(\mathbb{R}^N)$ *if and only if* $\int_{\widehat{R}} \operatorname{Var}_{(a_k,b_k)}(u_k)\, d\widehat{z} < +\infty$ *for each rectangular cell* $R \subseteq \mathbb{R}^N$ *and each* $k \in \{1, \ldots, N\}$.

Finally we show that composition with Lipschitz maps preserves bounded variation.

Proposition 1.85
If Ω *is an open set such that* $\lambda^N(\Omega) = +\infty$, $u \in BV(\Omega)$ *and* $g\colon \mathbb{R} \longrightarrow \mathbb{R}$ *is a Lipschitz continuous function with* $g(0) = 0$,
then $g \circ u \in BV(\Omega)$ *and*

$$\big|D(g \circ u)\big|(\Omega) \leqslant (\operatorname{Lip} g)|Du|(\Omega).$$

From the previous discussion of functions of bounded variation, we see that the spaces of measure play an important role. Spaces of measures are also important in the study of Young measures (see Sect. 3.1.4). So, it is worth having a look at such spaces.

Let X be a locally compact, σ-compact space (i.e., $X = \bigcup_{n\geqslant 1} K_n$ with K_n compact). We consider the following three spaces of continuous functions:

$$\begin{aligned}
C_c(X) &= \big\{u\colon X \longrightarrow \mathbb{R} :\ u \text{ is continuous and of compact support}\big\},\\
C_0(X) &= \big\{u\colon X \longrightarrow \mathbb{R} :\ u \text{ is continuous and vanishes at infinity}\big\},\\
C_b(X) &= \big\{u\colon X \longrightarrow \mathbb{R} :\ u \text{ is continuous and bounded}\big\}.
\end{aligned}$$

Recall that $u\colon X \longrightarrow \mathbb{R}$ is said to vanish at infinity if, for every $\varepsilon > 0$, there exists a compact set $K \subseteq X$ such that

$$|u(x)| \leqslant \varepsilon \qquad \forall\, x \in X \setminus K.$$

Evidently we have the following inclusions:

$$C_c(X) \subseteq C_0(X) \subseteq C_b(X).$$

If X is compact, then the three spaces coincide, while if X is not compact, then the above inclusions are strict. All three spaces are furnished with the supremum norm

$$\|u\|_\infty = \sup_{x\in X} |u(x)|.$$

Note that for $C_c(X)$ and $C_0(X)$, the supremum is actually a maximum (i.e., it is attained).

Proposition 1.86
The spaces $(C_0(X), \|\cdot\|_\infty)$ and $(C_b(X), \|\cdot\|_\infty)$ are Banach spaces (hence $C_0(X)$ is a closed subspace of $C_b(X)$), while $C_c(X)$ is dense in $C_0(X)$.

Definition 1.87
By $M_b(X)$ we denote the space of signed Radon measures μ on X such that $|\mu|$ is finite, The space $M_b(X)$ is furnished with the total variation norm defined by

$$\|\mu\|_{M_b(X)} = \int_X d|\mu| = |\mu|(X).$$

The normed space $(M_b(X), \|\cdot\|_{M_b(X)})$ is a Banach space.

Since $C_c(X)$ is dense in $C_0(X)$ (see Proposition 1.86), the two spaces have the same topological dual. Of course, we prefer to work with $C_0(X)$ which is a Banach space. The next theorem identifies $C_0(X)^*$.

Theorem 1.88
<u>If</u> X is a locally compact and σ-compact topological space, <u>then</u> $C_0(X)^* = M_b(X)$, i.e., for each $u^* \in C_0(X)^*$ there exists a unique $\mu \in M_b(X)$ such that

$$\langle u^*, u\rangle = \int_X u(x)\, d\mu(x) \qquad \forall\, u \in C_0(X)$$

and

$$\|u^*\|_* = \|\mu\|_{M_b(X)} = |\mu|(X).$$

We can have three different kinds of convergence of sequences in $M_b(X)$ depending on which of the three spaces $C_c(X)$, $C_0(X)$, or $C_b(X)$ we choose to put in duality with $M_b(X)$.

Definition 1.89
Let X be a locally compact and σ-compact topological space, let $\{\mu_n\}_{n\geqslant 1} \subseteq M_b(X)$ be a sequence and let $\mu \in M_b(X)$.
(a) *We say that the sequence $\{\mu_n\}_{n\geqslant 1}$* ***converges vaguely*** *to μ if and only if*

$$\int_X u\, d\mu_n \longrightarrow \int_X u\, d\mu \qquad \forall\, u \in C_c(X).$$

(b) We say that the sequence $\{\mu_n\}_{n\geqslant 1}$ ***converges weakly*** *to* μ *if and only if*

$$\int_X u\,d\mu_n \longrightarrow \int_X u\,d\mu \qquad \forall\, u \in C_0(X)$$

and we write $\mu_n \xrightarrow{w} \mu$.

(c) We say that the sequence $\{\mu_n\}_{n\geqslant 1}$ ***converges narrowly*** *to* μ *if and only if*

$$\int_X u\,d\mu_n \longrightarrow \int_X u\,d\mu \qquad \forall\, u \in C_b(X).$$

and we write $\mu_n \xrightarrow{n} \mu$.

Remark 1.90

Note that in the light of Theorem 1.88, the weak convergence in Definition 1.89(b) is in fact w^*-convergence in the Banach space $M_b(X)$. The name "weak convergence" is used by probabilists. So, including the norm convergence, we have four distinct modes of convergence for sequences in $M_b(X)$, which in general are related as follows:

$$\begin{gathered}\text{norm convergence}\\ \Downarrow\\ \text{narrow convergence}\\ \Downarrow\\ \text{weak convergence}\\ \Downarrow\\ \text{vague convergence.}\end{gathered}$$

Theorem 1.91

Let X *be a locally compact and* σ*-compact topological space and let* $\{\mu_n\}_{n\geqslant 1} \subseteq M_b(X)$ *be a sequence. Consider the following statements:*

(a) The sequence $\{\mu_n\}_{n\geqslant 1}$ *converges weakly.*

(b) The sequence $\{\mu_n\}_{n\geqslant 1}$ *converges narrowly.*

(c) The sequence $\{\mu_n\}_{n\geqslant 1}$ *is bounded in* $M_b(X)$ *and there is a dense subset* $D \subseteq C_0(X)$ *such that the sequence* $\left\{\int_X h\,d\mu_n\right\}_{n\geqslant 1}$ *converges for every* $h \in D$.

(d) For every $\varepsilon > 0$, *there exists a compact set* K_ε *such that*

$$|\mu_n|(K_\varepsilon^c) < \varepsilon \qquad \forall\, n \geqslant n_0.$$

(e) We have that $\mu_n \geqslant 0$, μ_n *converges weakly to* μ *and* $\mu_n(X) \longrightarrow \mu(X)$.

Then the following implications hold:

$$\begin{aligned} (a) &\iff (c)\\ \big[(a) \text{ and } (d)\big] &\iff (b)\\ (e) &\implies (b). \end{aligned}$$

Remark 1.92
The equivalence of **(a)** and **(c)** implies that on bounded subsets of $M_b(X)$ the weak topologies $w(M_b(X), C_0(X))$ and $w(M_b(X), C_c(X))$ coincide. In particular then, on bounded sets of $M_b(X)$, the notions of vague and weak convergence coincide. Property **(d)** is known as ***uniform tightness*** of $\{|\mu_n|\}_{n \geqslant n_0}$. If in addition, X is a complete metric space, we have

$$\textbf{(b)} \iff \textbf{(e)}.$$

Let $M_b^+(X)$ be the subset of $M_b(X)$ consisting of the (positive) Radon measures (see Definition I.4.9) and let $M_b^f(X)$ be the space of finite linear combinations of Dirac measures.

Proposition 1.93
If X is a locally compact and σ-compact topological space, $\{\mu_n\}_{n \geqslant 1} \subseteq M_b^+(X)$ is a sequence such that $\mu_n \xrightarrow{w} \mu$ with $\mu \in M_b(X)$,
then for any open set $U \subseteq X$, we have

$$\mu(U) \leqslant \liminf_{n \to +\infty} \mu_n(U);$$

for any compact set $K \subseteq X$, we have

$$\limsup_{n \to +\infty} \mu_n(K) \leqslant \mu(K);$$

for any relatively compact set $E \subseteq X$ with $\mu(\partial E) = 0$, we have

$$\lim_{n \to +\infty} \mu_n(E) = \mu(E).$$

Proposition 1.94
If X is a locally compact and σ-compact topological space and $\mu \in M_b(X)$,
then there exists a sequence $\{\mu_n\}_{n \geqslant 1} \subseteq M_b^f(X)$ converging narrowly to μ.

Finally, let X be a metrizable space and let us consider the weak topology $w_n = w(M_b(X), C_b(X))$ (n stand for narrow). Let $M_1^+(X)$ be the set of probability measures on X. We will examine $M_1^+(X)$ furnished with the relative w_n-topology. In what follows $\mathcal{U}_d(X)$ denotes the space of uniformly continuous functions for any compatible metric d on X.

Proposition 1.95
If X is a metrizable space, $\{\mu_\alpha\}_{\alpha \in I} \subseteq M_1^+(X)$ is a net and $\mu \in M_1^+(X)$,
then the following statements are equivalent:

(a) $\mu_\alpha \xrightarrow{w_n} \mu$.

(b) $\displaystyle\int_X u\, d\mu_\alpha \longrightarrow \int_X u\, d\mu$ *for all* $u \in C_b(X)$.

(c) $\displaystyle\int_X u\, d\mu_\alpha \longrightarrow \int_X u\, d\mu$ *for all* $u \in U_d(X)$.

(d) $\int_X u\, d\mu_\alpha \longrightarrow \int_X u\, d\mu$ *for all* $u \in \mathcal{D}$*, with* $\mathcal{D}$ *uniformly dense in* $U_d(X)$*.*

(e) $\limsup_\alpha \mu_\alpha(C) \leqslant \mu(C)$ *for all closed sets* $C \subseteq X$*.*

(f) $\mu(U) \leqslant \liminf_\alpha \mu_\alpha(U)$ *for all open sets* $U \subseteq X$*.*

(g) $\mu_\alpha(E) \longrightarrow \mu(E)$ *for all sets* $E \in \mathcal{B}(X)$ *with* $\mu(\partial E) = 0$*.*

Theorem 1.96

(a) *X is compact metrizable if and only if $\big(M_1^+(X), w_n\big)$ is compact metrizable.*

(b) *X is separable metrizable if and only if $\big(M_1^+(X), w_n\big)$ is separable metrizable.*

(c) *X is Polish if and only if $\big(M_1^+(X), w_n\big)$ is Polish.*

Theorem 1.97 *(Prohorov Theorem)*

If X is a Polish space,
then $\mathcal{A} \subseteq M_1^+(X)$ is relatively w_n-compact if and only if $\mathcal{A}$ is uniformly tight (i.e., for each $\varepsilon > 0$ there exists a compact set $K_\varepsilon \subseteq X$ such that $\mu(K_\varepsilon) > 1 - \varepsilon$ for all $\mu \in \mathcal{A}$).

1.1.4 Sobolev Spaces

First we deal with functions of one variable. So, let $I = (a, b)$ be an interval, possibly unbounded and $1 \leqslant p \leqslant +\infty$.

Definition 1.98

The Sobolev space $W^{1,p}(a, b)$ is defined by

$$W^{1,p}(a,b) = \Big\{u \in L^p(I) : \text{ there exists } h \in L^p(I) \text{ such that } \int_I u\vartheta'\, dt = -\int_I h\vartheta\, dt \text{ for all } \vartheta \in C_c^1(I)\Big\}.$$

*We write $u' = h$ (the **weak** or **distributional** derivative of u).*

Remark 1.99

$W^{1,p}(a, b)$ is furnished with the norm

$$\|u\|_{1,p} = \|u\|_p + \|u'\|_p$$

(or sometimes for $1 < p < +\infty$ with the equivalent norm $\big(\|u\|_p^p + \|u'\|_p^p\big)^{\frac{1}{p}}$) and it becomes a Banach space (or Hilbert if $p = 2$).

Theorem 1.100

If $u \in W^{1,p}(a, b)$ with $1 \leqslant p \leqslant +\infty$ and $I = (a, b)$ is an interval (possibly unbounded),
then there exists representation $\overline{u} \in C(\overline{I})$ of u such that

$$\overline{u}(x) - \overline{u}(v) = \int_v^x u'\, dt \qquad \forall\, x, v \in \overline{I}.$$

Remark 1.101
If $I = (a,b)$ is a bounded interval, then the representation $\overline{u}$ in the above theorem is absolutely continuous.

Theorem 1.102
For any $1 \leqslant p \leqslant +\infty$, there exists a bounded linear operator $P\colon W^{1,p}(a,b) \longrightarrow W^{1,p}(\mathbb{R})$ called ***extension operator*** *such that:*
(a) *$Pu|_{(a,b)} = u$ for all $u \in W^{1,p}(a,b)$;*
(b) *there exists $c > 0$ such that $\|Pu\|_{L^p(\mathbb{R})} \leqslant c\|u\|_{L^p(a,b)}$ for all $u \in W^{1,p}(a,b)$;*
(c) *there exists $c > 0$ such that $\|Pu\|_{W^{1,p}(\mathbb{R})} \leqslant c\|u\|_{W^{1,p}(a,b)}$ for all $u \in W^{1,p}(a,b)$.*

Theorem 1.103
If $u \in W^{1,p}(a,b)$ with $1 \leqslant p < +\infty$,
then we can find a sequence $\{u_n\}_{n\geqslant 1} \subseteq C_c^\infty(\mathbb{R})$ such that

$$u_n|_I \longrightarrow u \quad \text{in } W^{1,p}(a,b).$$

Remark 1.104
In general we cannot say that $C_c^\infty(I)$ is dense in $W^{1,p}(a,b)$. The closure of $C_c^\infty(I)$ in the Sobolev norm leads to a particular subspace of $W^{1,p}(a,b)$ (see Definition 1.105). This is in contrast to the L^p-spaces where $C_c^\infty(I)$ is dense in $L^p(I)$ for $1 \leqslant p < +\infty$ (see Remark 1.7).

Definition 1.105
Let $1 \leqslant p < +\infty$. We set $W_0^{1,p}(a,b)$ to be the closure of $C_c^\infty(I)$ in the Sobolev norm. We furnish $W_0^{1,p}(a,b)$ with the norm of $W^{1,p}(a,b)$ and we have a Banach subspace of $W^{1,p}(a,b)$.

Remark 1.106
Since $C_c^\infty(\mathbb{R})$ is dense in $W^{1,p}(\mathbb{R})$ (see Theorem 1.103), we have $W_0^{1,p}(\mathbb{R}) = W^{1,p}(\mathbb{R})$.

The next result characterizes $W_0^{1,p}(a,b)$ as a subspace of $W^{1,p}(a,b)$.

Proposition 1.107
If $u \in W^{1,p}(a,b)$,
then $u \in W_0^{1,p}(a,b)$ if and only if $u|_{\partial\Omega} = 0$.

Theorem 1.108 *(Poincaré Theorem)*
If $I = (a,b)$ is a bounded interval,
then there exists $c > 0$ such that

$$\|u\|_{1,p} \leqslant c\|u'\|_p \qquad \forall\, u \in W_0^{1,p}(a,b).$$

This means that the map $u \longmapsto \|u'\|_p$ is an equivalent norm in $W_0^{1,p}(a,b)$.

The next result explains the importance of Sobolev spaces in the study of boundary value problems.

Theorem 1.109 *(Sobolev Embedding Theorem)*
There exists a constant $c > 0$ (depending only on $\lambda^1(I) \leqslant \infty$) such that

$$\|u\|_\infty \leqslant c\|u\|_{1,p} \qquad \forall\, u \in W^{1,p}(a,b),\ 1 \leqslant p \leqslant +\infty.$$

So, $W^{1,p}(a,b)$ (with $1 \leqslant p \leqslant +\infty$) is embedded continuously into $L^\infty(I)$ (denoted by $W^{1,p}(a,b) \hookrightarrow L^\infty(I)$).
If $I = (a,b)$ is a bounded interval, then

(a) *for $1 < p \leqslant \infty$, $W^{1,p}(a,b)$ is embedded compactly in $C(\overline{I})$ (denoted by $W^{1,p}(a,b) \overset{c}{\hookrightarrow} C(\overline{I})$).*

(b) *$W^{1,1}(a,b)$ is embedded compactly in $L^q(I)$ for all $1 \leqslant q < +\infty$ (i.e., we have $W^{1,1}(a,b) \overset{c}{\hookrightarrow} L^q(I)$).*

These embeddings remain valid if $W^{1,p}(a,b)$ is replaced by $W_0^{1,p}(a,b)$.

Remark 1.110
The embedding of $W^{1,1}(a,b)$ into $C(\overline{I})$ is continuous, but never compact.

Theorem 1.109 leads to some useful equivalent norms for the Sobolev space $W^{1,p}(a,b)$ (with $1 \leqslant p \leqslant +\infty$).

Proposition 1.111
If $I = (a,b)$ is a bounded interval, $1 \leqslant p \leqslant +\infty$ and $1 \leqslant q \leqslant +\infty$,
then the mapping $u \longmapsto \|u\|_q + \|u'\|_p$ is an equivalent norm in $W^{1,p}(a,b)$.

The dual of the spaces $W_0^{1,p}(a,b)$ (with $1 \leqslant p < +\infty$) is denoted by $W^{-1,p'}(a,b)$ (where $\frac{1}{p} + \frac{1}{p'} = 1$) and admits the following representation.

Proposition 1.112
If $u^* \in W^{-1,p'}(a,b) = W_0^{1,p}(a,b)^*$,
then there exists two functions $h_0, h_1 \in L^{p'}(I)$ such that

$$\langle u^*, u\rangle = \int_I h_0 u\,dx + \int_I h_1 u'\,dx \qquad \forall\, u \in W_0^{1,p}(a,b)$$

and

$$\|u\|_* = \max\left\{\|h_0\|_{p'}, \|h_1\|_{p'}\right\}.$$

Moreover, if $I = (a,b)$ is a bounded interval, then we can take $h_0 = 0$.

Remark 1.113
According to this result u^* can be identified with the distribution $h_0 - h_1'$.

We can have Sobolev spaces of higher order.

Definition 1.114
For every integer $m \geqslant 2$ and for any $p \in [1, +\infty]$, by induction we can define the space

$$W^{m,p}(a,b) = \{u \in W^{m-1,p}(a,b) : \; u' \in W^{m-1,p}(a,b)\}.$$

We equip this space with the norm

$$\|u\|_{m,p} = \|u\|_p + \sum_{k=1}^{m} \|D^k u\|_p.$$

Then $\big(W^{m,p}(a,b), \|\cdot\|_{m,p}\big)$ is a Banach space.

Proposition 1.115
(a) *$W^{m,p}(a,b)$ (with $m \geqslant 1$, $1 \leqslant p < +\infty$) is separable.*
(b) *$W^{m,p}(a,b)$ (with $m \geqslant 1$, $1 < p < +\infty$) is reflexive and uniformly convex.*

Now we turn to functions of several variables.

Definition 1.116
Let $\Omega \subseteq \mathbb{R}^N$ be an open set and let $1 \leqslant p \leqslant +\infty$. The Sobolev space $W^{1,p}(\Omega)$ is defined by

$$W^{1,p}(\Omega) = \Big\{u \in L^p(\Omega) : \text{ there exist functions } h_1, \ldots, h_N \in L^p(\Omega) \text{ such that } \int_\Omega u \tfrac{\partial \vartheta}{\partial z_k}\, dz = -\int_\Omega h_k \vartheta \, dz \text{ for all } \vartheta \in C_c^\infty(\Omega) \text{ for all } k \in \{1, \ldots, N\}\Big\}.$$

Then we write

$$h_k = \tfrac{\partial u}{\partial z_k} \qquad \forall\, k \in \{1, \ldots, N\}$$

*(**weak partial derivative of** u) and*

$$Du = \left(\frac{\partial u}{\partial z_1}, \ldots, \frac{\partial u}{\partial z_N}\right).$$

The space $W^{1,p}(\Omega)$ is equipped with the norm

$$\|u\|_{1,p} = \|u\|_p + \|Du\|_{L^p(\Omega;\mathbb{R}^N)}$$

(or sometimes the equivalent norm $\big(\|u\|_p^p + \|Du\|^p_{L^p(\Omega;\mathbb{R}^N)}\big)^{\frac{1}{p}}$).
When $p = 2$, we write $H^1(\Omega) = W^{1,2}(\Omega)$.

The space $\big(W^{1,p}(\Omega), \|\cdot\|_{1,p}\big)$ is a Banach space (Hilbert if $p = 2$).

To obtain approximation results for Sobolev functions by smooth functions we use mollifiers.

Definition 1.117
A sequence of **mollifiers** $\{\vartheta_n\}_{n\geqslant 1}$ *is a sequence of functions* $\{\vartheta_n\}_{n\geqslant 1} \subseteq C_c^\infty(\mathbb{R}^N)$ *such that*

$$\vartheta_n \geqslant 0, \quad \operatorname{supp}\vartheta_n \subseteq \overline{B}_{\frac{1}{n}}, \quad \int\limits_{\mathbb{R}^N} \vartheta_n(z)\,dz = 1 \qquad \forall\, n \geqslant 1.$$

where $\overline{B}_{\frac{1}{n}} = \left\{z \in \mathbb{R}^N : |z| \leqslant \frac{1}{n}\right\}$*. Such a sequence can be obtained from a single function* ϑ_1 *as above, by putting* $\vartheta_n(z) = n^N\vartheta_1(nz)$ *for* $n \geqslant 2$*. Then the function* ϑ_1 *is called the* **convolution kernel**

Given $u \in L^1_{\mathrm{loc}}(\Omega)$*, the* **mollification** *(or* **regularization***) corresponding to the mollifiers* $\{\vartheta_n\}_{n\geqslant 1}$ *is given by*

$$u_n(z) \;=\; (\vartheta_n * u)(z) \;=\; \int\limits_{\mathbb{R}^N} u(z-y)\vartheta_n(y)\,dy \;=\; \int\limits_{\mathbb{R}^N} u(y)\vartheta_n(z-y)\,dy,$$

where we have extended u *to all* $\mathbb{R}^N$ *by setting* $u(z) = 0$ *for all* $z \in \mathbb{R}^N \setminus \Omega$*. We have*

$$\operatorname{supp} u_n \;\subseteq\; \operatorname{supp} u + \overline{B}_{\frac{1}{n}}, \quad u_n \in C_c^\infty(\mathbb{R}^N) \qquad \forall\, n \geqslant 1.$$

Theorem 1.118
Let $u \in L^1_{\mathrm{loc}}(\Omega)$ *and let* $\{u_n\}_{n\geqslant 1}$ *be a sequence of mollifications.*
(a) *If* $u \in C_c(\Omega)$*, then* $u_n \longrightarrow u$ *uniformly.*
(b) *If* $u \in L^p(\Omega)$ *(with* $1 \leqslant p < +\infty$*), then* $\|u_n\|_p \leqslant \|u\|_p$ *for all* $n \geqslant 1$ *and* $u_n \longrightarrow u$ *in* $L^p(\Omega)$*.*
(c) *If* $u \in W^{1,p}(\Omega)$ *(with* $1 \leqslant p < +\infty$*) and* $k \in \{1,\dots,N\}$*, then*

$$\frac{\partial u_n}{\partial z_k} \;=\; \frac{\partial \vartheta_n}{\partial z_k} * u \qquad \forall\, n \geqslant 1.$$

In particular $u_n \in C_c^\infty(\mathbb{R}^N)$*.*

Theorem 1.119 *(Meyers–Serrin Theorem)*
If $\Omega \subseteq \mathbb{R}^N$ *is any open set and* $1 \leqslant p < +\infty$
then $C^\infty(\Omega) \cap W^{1,p}(\Omega)$ *is dense in* $W^{1,p}(\Omega)$*.*

Remark 1.120
In general, if $u \in W^{1,p}(\Omega)$ (with $1 \leqslant p < +\infty$), there need not exist a sequence $\{u_n\}_{n\geqslant 1} \subseteq C_c^\infty(\mathbb{R}^N)$ such that $u_n|_\Omega \longrightarrow u$ in $W^{1,p}(\Omega)$.

However, we have the following result.

Proposition 1.121
If $u \in W^{1,p}(\Omega)$ *with* $1 \leqslant p < +\infty$*,*
then there exists a sequence $\{u_n\}_{n\geqslant 1} \subseteq C_c^\infty(\mathbb{R}^N)$ *such that*

$$u_n|_\Omega \longrightarrow u \quad \text{in } L^p(\Omega)$$

and

$$Du_n|_U \longrightarrow Du|_U \quad \text{in } L^p(U;\mathbb{R}^N), \text{ for all } U \subset\subset \Omega.$$

<u>*If*</u> $\Omega = \mathbb{R}^N$ *and* $u \in W^{1,p}(\mathbb{R}^N)$ *with* $1 \leqslant p < +\infty$,
<u>*then*</u> *there exists a sequence* $\{u_n\}_{n\geqslant 1} \subseteq C_c^\infty(\mathbb{R}^N)$ *such that*

$$u_n \longrightarrow u \quad \text{in } W^{1,p}(\mathbb{R}^N)$$

(i.e., $C_c^\infty(\mathbb{R}^N)$ *is dense in* $W^{1,p}(\mathbb{R}^N)$*).*

We can improve this result if we assume some regularity for the boundary $\partial\Omega$.

Proposition 1.122
<u>*If*</u> $\Omega \subseteq \mathbb{R}^N$ *has a* C^1*-boundary* $\partial\Omega$ *and* $u \in W^{1,p}(\Omega)$ *with* $1 \leqslant p < +\infty$,
<u>*then*</u> *there exists a sequence* $\{u_n\}_{n\geqslant 1} \subseteq C_c^\infty(\mathbb{R}^N)$ *such that*

$$u_n|_\Omega \longrightarrow u \quad \text{in } W^{1,p}(\Omega)$$

(i.e., $C_c^\infty(\mathbb{R}^N)|_\Omega$ *is dense in* $W^{1,p}(\Omega)$*).*

The next proposition provides a simple characterization of Sobolev functions.

Proposition 1.123
<u>*If*</u> $u \in L^p(\Omega)$ *with* $1 < p \leqslant +\infty$,
<u>*then*</u> *the following statements are equivalent:*
(a) $u \in W^{1,p}(\Omega)$.
(b) *There exists* $c > 0$ *such that* $\left|\int_\Omega u \frac{\partial \vartheta}{\partial z_k}\, dz\right| \leqslant c\|\vartheta\|_{p'}$ *for all* $\vartheta \in C_c^\infty(\Omega)$, *all* $k \in \{1, \ldots, N\}$.
(c) *There exists* $c > 0$ *such that for all* $U \subset\subset \Omega$ *and all* $h \in \mathbb{R}^N$ *with* $|h| < \operatorname{dist}(U, \partial\Omega)$, *we have*

$$\|\tau_h(u) - u\|_{L^p(\Omega)} \leqslant c|h|,$$

where $\tau_h(u)(z) = u(z+h)$.
Moreover, if $\Omega = \mathbb{R}^N$, *then*

$$\|\tau_h(u) - u\|_p \leqslant \|Du\|_p |h|.$$

Remark 1.124
In statements **(b)** and **(c)** we can take $c = \|Du\|_p$.
If $p = 1$, then

$$\textbf{(a)} \implies \textbf{(b)} \iff \textbf{(c)}.$$

In this case ($p = 1$), functions which satisfy **(b)** (or **(c)**) are functions of bounded variation (see Definition 1.74). So, we have

$$W^{1,1}(\Omega) \subseteq BV(\Omega).$$

Finally from Proposition 1.123 and in particular from the implication "**(a)$\Longrightarrow$(c)**," we infer that every $u \in W^{1,\infty}(\Omega)$ admits a continuous representative on Ω. In fact, if Ω is convex, then

$$\left|u(z) - u(z')\right| \leqslant \|Du\|_\infty |z - z'| \qquad \forall\, z, z' \in \Omega.$$

Theorem 1.125
If $\Omega \subseteq \mathbb{R}^N$ *is a bounded set with a Lipschitz boundary* $\partial\Omega$,
then there exists a linear operator $P\colon W^{1,p}(\Omega) \longrightarrow W^{1,p}(\mathbb{R}^N)$ *with* $1 \leqslant p \leqslant +\infty$ *such that:*
(a) $Pu|_\Omega = u$ *for all* $u \in W^{1,p}(\Omega)$;
(b) *there exists* $c > 0$ *such that* $\|Pu\|_{L^p(\mathbb{R}^N)} \leqslant c\|u\|_{L^p(\Omega)}$ *for all* $u \in W^{1,p}(\Omega)$;
(c) *there exists* $c > 0$ *such that* $\|Pu\|_{W^{1,p}(\mathbb{R}^N)} \leqslant c\|u\|_{W^{1,p}(\Omega)}$ *for all* $u \in W^{1,p}(\Omega)$.

From Proposition 1.121, we know that in general $C_c^\infty(\Omega)$ is not dense in $W^{1,p}(\Omega)$. This leads to the following definition.

Definition 1.126
Let $1 \leqslant p < +\infty$. *We define*

$$W_0^{1,p}(\Omega) = \overline{C_c^\infty(\Omega)}^{\|\cdot\|_{1,p}}.$$

When $p = 2$, *we write* $H_0^1(\Omega) = W_0^{1,2}(\Omega)$.

Evidently $W_0^{1,p}(\Omega) \subseteq W^{1,p}(\Omega)$ and it is a Banach space (Hilbert if $p = 2$) equipped with the $W^{1,p}(\Omega)$-norm.

Remark 1.127
From Proposition 1.121, we know that $W_0^{1,p}(\mathbb{R}^N) = W^{1,p}(\mathbb{R}^N)$. If $\Omega \subseteq \mathbb{R}^N$, then in general the inclusion $W_0^{1,p}(\Omega) \subseteq W^{1,p}(\Omega)$ is strict. However, if $\mathbb{R}^N \setminus \Omega$ is "thin" and $p < N$, then $W_0^{1,p}(\Omega) = W^{1,p}(\Omega)$ (for example when $\Omega = \mathbb{R}^N \setminus \{0\}$).

Proposition 1.128
If $\Omega \subseteq \mathbb{R}^N$ *is a set with a Lipschitz boundary* $\partial\Omega$ *and* $u \in W^{1,p}(\Omega) \cap C(\overline{\Omega})$ *with* $1 \leqslant p < +\infty$,
then the following statements are equivalent:
(a) $u \in W_0^{1,p}(\Omega)$.
(b) $u|_{\partial\Omega} = 0$.

To extend this result to all functions in $W_0^{1,p}(\Omega)$, we need to give meaning to $u|_{\partial\Omega}$ for any $W_0^{1,p}(\Omega)$. This is done with the notion of trace.

Definition 1.129
Let $\Omega \subseteq \mathbb{R}^N$ *be a bounded set with a Lipschitz boundary* $\partial\Omega$ *and let* $1 \leqslant p < +\infty$. *Then there exists a bounded linear operator* $\gamma\colon W^{1,p}(\Omega) \longrightarrow L^p(\partial\Omega; H^{N-1})$ *(where* H^{N-1} *is the* $(N-1)$*-dimensional Hausdorff measure on the topological submanifold* $\partial\Omega$ *of* $\mathbb{R}^N$*) such that:*

(a) $\gamma(u) = u|_{\partial\Omega}$ *for all* $u \in W^{1,p}(\Omega) \cap C(\overline{\Omega})$;

(b) $\int\limits_{\Omega} (\operatorname{div} \vartheta) u \, dz + \int\limits_{\Omega} (\vartheta, Du)_{\mathbb{R}^N} \, dz = \int\limits_{\partial\Omega} (\vartheta, n)_{\mathbb{R}^N} \gamma(u) \, dH^{N-1}$ *for all* $u \in W^{1,p}(\Omega)$, *where* n *denotes the outward unit normal on* $\partial\Omega$ *(since* $\partial\Omega$ *is Lipschitz, it exists* H^{N-1}*-almost everywhere);*

(c) $\ker \gamma = W_0^{1,p}(\Omega)$.

The operator γ *is called the* **trace operator** *and the function* $\gamma(u) \in L^p(\partial\Omega; H^{N-1})$ *is called the* **trace** *of* u *on* $\partial\Omega$ *and we understand it as representing the "boundary values" of* u.

Theorem 1.130 *(Green Identity)*
If $\Omega \subseteq \mathbb{R}^N$ *is a bounded set with a Lipschitz boundary* $\partial\Omega$ *and* $u \in H^2(\Omega)$, $y \in H^1(\Omega)$, *then*

$$\int\limits_{\Omega} (\Delta u) y \, dz + \int\limits_{\Omega} (Du, Dy)_{\mathbb{R}^N} \, dz = \int\limits_{\partial\Omega} \tfrac{\partial u}{\partial n} \gamma(y) \, dH^{N-1}.$$

Theorem 1.131 *(Poincaré Inequality)*
If $\Omega \subseteq \mathbb{R}^N$ *is a bounded set with a Lipschitz boundary* $\partial\Omega$ *and* $1 \leqslant p < +\infty$, *then there exists* $c = c(\Omega, p) > 0$ *such that*

$$\|u\|_p \leqslant c\|Du\|_p \qquad \forall\, u \in W_0^{1,p}(\Omega).$$

In particular $u \longmapsto \|Du\|_p$ *is an equivalent norm on the Sobolev space* $W_0^{1,p}(\Omega)$.

For the bigger space $W^{1,p}(\Omega)$ we have a related inequality.

Theorem 1.132 *(Poincaré–Wirtinger Inequality)*
If $\Omega \subseteq \mathbb{R}^N$ *is a bounded set with a Lipschitz boundary* $\partial\Omega$ *and* $1 \leqslant p < +\infty$, *then there exists* $c = c(\Omega, p) > 0$ *such that*

$$\|u - u_\Omega\|_p \leqslant c\|Du\|_p \qquad \forall\, u \in W^{1,p}(\Omega)$$

where $u_\Omega = \frac{1}{\lambda^N(\Omega)} \int\limits_{\Omega} u \, dz$.

Theorem 1.133 *(Gagliardo–Nirenberg–Sobolev Inequality)*
If $1 \leqslant p < N$ *and* $p^* = \frac{Np}{N-p}$, *then there exists* $c = c(\Omega, p) > 0$ *such that*

$$\|u\|_{p^*} \leqslant c\|Du\|_p \qquad \forall\, u \in W_0^{1,p}(\Omega).$$

Remark 1.134
Recall that $W_0^{1,p}(\mathbb{R}^N) = W^{1,p}(\mathbb{R}^N)$ (see Remark 1.127). So, the above result applies to the Sobolev space $W^{1,p}(\mathbb{R}^N)$ and it implies that $W^{1,p}(\mathbb{R}^N)$ is embedded continuously in $L^q(\mathbb{R}^N)$ for all $1 \leqslant q \leqslant p^*$.

When $\Omega \subseteq \mathbb{R}^N$ is bounded, then we have a more refined result.

Theorem 1.135 *(Rellich–Kondrachov Embedding Theorem)*
If $\Omega \subseteq \mathbb{R}^N$ *is a bounded set with a Lipschitz boundary* $\partial\Omega$, *then*

(a) *if* $1 \leqslant p < +\infty$, *then* $W^{1,p}(\Omega) \hookrightarrow L^q(\Omega)$ *for all* $q \in [1, p^*]$ *and* $W^{1,p}(\Omega) \overset{c}{\hookrightarrow} L^q(\Omega)$ *for all* $q \in [1, p^*)$.

(b) *if* $p = N$, *then* $W^{1,p}(\Omega) \overset{c}{\hookrightarrow} L^q(\Omega)$ *for all* $q \in [1, +\infty)$;

(c) *if* $p > N$, *then* $W^{1,p}(\Omega) \hookrightarrow C^{0,\alpha}(\overline{\Omega})$ *with* $\alpha = 1 - \frac{N}{p} \in (0,1)$ *(here* $C^{0,\alpha}(\overline{\Omega})$ *denotes the space of Hölder continuous functions with exponent* α, *that is, functions* $u \colon \overline{\Omega} \longrightarrow \mathbb{R}$ *satisfying*

$$|u(x) - u(y)| \leqslant c|x - y|^\alpha \qquad \forall\, x, y \in \overline{\Omega},$$

for some $c > 0$; *this is a Banach space furnished with the norm*

$$\|u\|_{C^{0,\alpha}(\overline{\Omega})} = \|u\|_\infty + \sup_{\substack{x,y \in \Omega \\ x \neq y}} \frac{|u(x) - u(y)|}{|x - y|^\alpha}).$$

If $W^{1,p}(\Omega)$ *is replaced by* $W_0^{1,p}(\Omega)$,
then the above embeddings hold for any bounded set $\Omega \subseteq \mathbb{R}^N$ *(with no additional conditions on the boundary).*

Remark 1.136
Because of this theorem, $p^* = \frac{Np}{N-p}$ (when $p < N$) is called the ***Sobolev critical exponent***. Often we extend this notion by setting $p^* = +\infty$ when $p \geqslant N$. If $\Omega = \mathbb{R}^N$, then we have:

- if $1 \leqslant p < N$, then $W^{1,p}(\mathbb{R}^N) \hookrightarrow L^q(\mathbb{R}^N)$ for all $q \in [1, p^*]$;
- if $p = N$, then $W^{1,p}(\mathbb{R}^N) \hookrightarrow L^q(\mathbb{R}^N)$ for all $q \in [1, +\infty)$;
- if $p > N$, then $W^{1,p}(\mathbb{R}^N) \hookrightarrow L^\infty(\mathbb{R}^N)$.

In this form the result is known as the ***Sobolev Embedding Theorem***. Note that none of the above embeddings is compact.

Theorem 1.135 leads to some equivalent norms for the Sobolev space $W^{1,p}(\Omega)$.

Proposition 1.137
If $\Omega \subseteq \mathbb{R}^N$ *is a bounded set with a Lipschitz boundary* $\partial\Omega$,
then the map $u \longmapsto \|u\|_q + \|Du\|_p$ *is an equivalent norm for the Sobolev space provided:*

- $1 \leqslant q \leqslant p^*$, *when* $1 \leqslant p < N$;
- $1 \leqslant q < +\infty$, *when* $p = N$;
- $1 \leqslant q \leqslant +\infty$, *when* $p > N$.

We can have Sobolev spaces of higher order.

Definition 1.138
Let $\Omega \subseteq \mathbb{R}^N$ *be an open set,* $m \in \mathbb{N}$ *with* $m \geqslant 2$ *and* $1 \leqslant p \leqslant +\infty$. *Inductively, we define the Sobolev space* $W^{m,p}(\Omega)$ *by*

$$W^{m,p}(\Omega) = \{u \in L^p(\Omega) : Du \in W^{m-1,p}(\Omega; \mathbb{R}^N)\}.$$

This is a Banach space for the norm

$$\|u\|_{W^{m,p}(\Omega)} = \sum_{|\alpha|\leqslant m} \|D^\alpha u\|_p,$$

where $\alpha = (\alpha_1, \ldots, \alpha_N)$ *is a multiindex,* $|\alpha| = \sum_{k=1}^{N} \alpha_k$ *and*

$$D^\alpha \vartheta = \frac{\partial^{|\alpha|}\vartheta}{\partial^{\alpha_1} z_1 \ldots \partial^{\alpha_N} z_N}.$$

Remark 1.139
An equivalent norm in $W^{m,p}(\Omega)$ is

$$u \longmapsto \|u\|_p + \sum_{|\alpha|=m} \|D^\alpha u\|_p.$$

Moreover, if $\Omega \subseteq \mathbb{R}^N$ is an open bounded set with a Lipschitz boundary $\partial\Omega$ and $mp > N$, then $W^{m,p}(\Omega) \overset{c}{\hookrightarrow} C^{k,\alpha}(\overline{\Omega})$ with k being the integer part of $m - \frac{N}{p}$ and $\alpha = m - \frac{N}{p} - k \in [0,1)$. Here

$$C^{k,\alpha}(\overline{\Omega}) = \{u \in C^k(\overline{\Omega}) : D^j u \in C^{0,\alpha}(\overline{\Omega}) \text{ for all multiindex } j \text{ with } |j| \leqslant k\}$$

(see Theorem 1.135 for the definition of Hölder continuous functions).

As in the one-dimensional case, the dual of $W_0^{1,p}(\Omega)$ is denoted by $W^{-1,p'}(\Omega)$ (where $\frac{1}{p} + \frac{1}{p'} = 1$). There is the following representation for this dual space.

Theorem 1.140
If $1 \leqslant p < +\infty$,
then $u^* \in W^{-1,p'}(\Omega) = W_0^{1,p}(\Omega)^*$ *if and only if there exist functions* $h_0, \ldots, h_N \in L^{p'}(\Omega)$ *such that*

$$\langle u^*, u\rangle = \int_\Omega h_0 u \, dz + \sum_{k=1}^{N} \int_\Omega h_k (D_k u) \, dz \qquad \forall\, u \in W_0^{1,p}(\Omega)$$

and

$$\|u^*\|_* = \max_{0\leqslant k\leqslant N} \|h_k\|_{p'}.$$

Moreover, if $\Omega \subseteq \mathbb{R}^N$ *is a bounded set, then we can take* $h_0 = 0$.

Remark 1.141
According to this theorem $u^* \in W^{-1,p'}(\Omega) = W_0^{1,p}(\Omega)^*$ can be identified with a distribution of the form $h_0 - \sum_{k=1}^{N} D_k h_k$, where $h_0, \ldots, h_N \in L^{p'}(\Omega)$.

Theorem 1.142 *(Chain Rule)*
If $1 \leqslant p < +\infty$, $\Omega \subseteq \mathbb{R}^N$ *is an open set of infinite Lebesgue measure,* $u \in W^{1,p}(\Omega)$ *and* $g\colon \mathbb{R} \longrightarrow \mathbb{R}$ *is a Lipschitz continuous function with* $g(0) = 0$ *when* Ω *has infinite Lebesgue measure,*
then $g \circ u \in W^{1,p}(\Omega)$ *and*

$$D(g \circ u)(z) = g^*(u(z))Du(z) \quad \text{for a.a. } z \in \Omega,$$

where $g^*\colon \mathbb{R} \longrightarrow \mathbb{R}$ *is any Lebesgue measurable function such that*

$$g^*(z) = g'(z) \quad \text{for a.a. } z \in \mathbb{R}.$$

The result remains true if we replace $W^{1,p}(\Omega)$ *by* $W_0^{1,p}(\Omega)$.

Proposition 1.143
(a) if $1 \leqslant p < +\infty$, *then* $W^{1,p}(\Omega)$ *is separable;*
(b) if $1 < p < +\infty$, *then* $W^{1,p}(\Omega)$ *is reflexive and uniformly convex.*
(c) if $\Omega \subseteq \mathbb{R}^N$ *is bounded and* $q \in [1, p^*]$ *when* $p < N, q \in [1, +\infty)$ *when* $p = N$ *and* $q \in [1, +\infty]$ *when* $p > N$, *then*

$$u \mapsto \|u\| = \|u\|_q + \|Du\|_p$$

is an equivalent norm on $W^{1,p}(\Omega)$.

1.1.5 Auxiliary Notions

Let us introduce some notions that will be needed in the forthcoming problems.

Definition 1.144
Let (Ω, Σ, μ) *be a measure space and let* $f\colon \Omega \times \mathbb{R} \longrightarrow \mathbb{R}$ *be a function. We say that* f *is a* ***Carathéodory function*** *if for all* $x \in \mathbb{R}$, *the function* $\Omega \ni \omega \longmapsto f(\omega, x) \in \mathbb{R}$ *is* Σ*-measurable and for* μ*-almost all* $\omega \in \Omega$, *the function* $\mathbb{R} \ni x \longmapsto f(\omega, x) \in \mathbb{R}$ *is continuous.*

Definition 1.145
Let (Ω, Σ, μ) *be a measure space and let* $f\colon \Omega \times \mathbb{R} \longrightarrow \mathbb{R}$ *be a Carathéodory function. The* ***Nemytskii*** *(or* ***superposition****) operator corresponding to* f *is the operator* N_f *which maps every* Σ*-measurable function* $u\colon \Omega \longrightarrow \mathbb{R}$ *to the measurable function* $N_f(u)(\cdot) = f(\cdot, u(\cdot))$.

1.2 Problems

Problem 1.1***
Let (Ω, Σ, μ) be a finite measure space and let $\{u_n\}_{n\geqslant 1} \subseteq L^1(\Omega)$ be a sequence such that $u_n \xrightarrow{w} u$ in $L^1(\Omega)$. Suppose that

$$u(\omega) \leqslant \liminf_{n\to+\infty} u_n(\omega) \quad \text{for } \mu\text{-a.a. } \omega \in \Omega.$$

Show that $u_n \longrightarrow u$ in $L^1(\Omega)$.

Problem 1.2**
Find a measure space (Ω, Σ, μ) and a sequence $\{u_n\}_{n\geqslant 1} \subseteq L^1(\Omega)$ which is uniformly integrable, but there is no $h \in L^1(\Omega)$ such that

$$|u_n(\omega)| \leqslant h(\omega) \quad \text{for } \mu\text{-a.a. } \omega \in \Omega \text{ all } n \geqslant 1.$$

Problem 1.3*
Let (Ω, Σ, μ) be a measure space and let $u \in L^p(\Omega)$ with $1 \leqslant p < +\infty$.
(a) Show that for every $c > 0$ we have

$$c^p \mu(\{\omega \in \Omega : u(\omega) \geqslant c\}) \leqslant \|u\|_p^p$$

(the Chebyshev–Markov inequality; see Theorem 1.30).
(b) Show that

$$\lim_{c\to+\infty} c^p \mu(\{\omega \in \Omega : u(\omega) \geqslant c\}) = 0.$$

Problem 1.4**
Let (Ω, Σ, μ) be a measure space and let $u_n \colon \Omega \longrightarrow \mathbb{R} \cup \{\pm\infty\}$ be Σ-measurable functions for $n \geqslant 1$ such that

$$|u_n(\omega)| \leqslant h(\omega) \quad \text{for } \mu\text{-a.a. } \omega \in \Omega, \text{ all } n \geqslant 1,$$

with $h \in L^p(\Omega)$, $1 \leqslant p < +\infty$. Show that the sequence $\{|u_n|^p\}_{n\geqslant 1} \subseteq L^1(\Omega)$ is uniformly integrable.

Problem 1.5**
Let (Ω, Σ, μ) be a measure space and let $C \subseteq L^1(\Omega)$ be a uniformly integrable set. Show that the following hold:
(a) C is $L^1(\Omega)$-bounded;
(b) for every $\varepsilon > 0$ we can find $\delta > 0$ such that

$$\text{if } A \in \Sigma \text{ with } \mu(A) \leqslant \delta, \text{ then } \sup_{u\in C} \int_A |u|\, d\mu \leqslant \varepsilon.$$

Problem 1.6**
Let (Ω, Σ, μ) be a finite measure space and let $C \subseteq L^1(\Omega)$ be a set. Show that, if:
(a) C is $L^1(\Omega)$-bounded; and
(b) for every $\varepsilon > 0$ we can find $\delta > 0$ such that

$$\text{if } A \in \Sigma \text{ with } \mu(A) \leqslant \delta, \text{ then } \sup_{u \in C} \int_A |u| \, d\mu \leqslant \varepsilon,$$

then the set C is uniformly integrable.
Moreover, show that if the measure space (Ω, Σ, μ) is not finite, then the above implication does not hold.

Problem 1.7**
Let (Ω, Σ, μ) be a measure space and let $C \subseteq L^1(\Omega)$ be a uniformly integrable set. Show that for every $\varepsilon > 0$ there exist $h_\varepsilon \in L^1(\Omega)_+$ and $\delta > 0$ such that

$$\text{if } A \in \Sigma \text{ is such that } \int_A h_\varepsilon \, d\mu \leqslant \delta, \text{ then } \sup_{u \in C} \int_A |u| \, d\mu \leqslant \varepsilon.$$

Problem 1.8**
Let (Ω, Σ, μ) be a measure space and let $1 \leqslant p < q < r \leqslant +\infty$. Show that $L^q(\Omega) \subseteq L^p(\Omega) + L^r(\Omega)$ (that is, every function in $L^q(\Omega)$ can be written as the sum of an $L^p(\Omega)$ and $L^r(\Omega)$ functions).

Problem 1.9**
Suppose that (Ω, Σ, μ) is a finite measure space and $u \in L^p(\Omega)$ for all $p \in [1, +\infty)$. Show that $u \in L^\infty(\Omega)$ if and only if $\lim_{p \to +\infty} \|u\|_p$ is finite (in fact $\lim_{p \to +\infty} \|u\|_p = \|u\|_\infty$).

Problem 1.10***
Let (Ω, Σ, μ) be a measure space and suppose that $1 \leqslant p < q < +\infty$. Show that $L^p(\Omega)$ is not contained in $L^q(\Omega)$ if and only if Σ has sets of arbitrarily small positive μ-measure.

Problem 1.11***
Let (Ω, Σ, μ) be a measure space and suppose that $1 \leqslant p < q < +\infty$. Show that $L^q(\Omega)$ is not contained in $L^p(\Omega)$ if and only if Σ has sets of arbitrarily big finite measure.

Problem 1.12**
Let (Ω, Σ, μ) be a measure space, $1 < p < +\infty$ and $C \subseteq L^p(\Omega)$ a nonempty, closed, and convex set. Consider $u \in L^p(\Omega) \setminus C$. Show that there exists $\widehat{h} \in C$ such that

$$\|u - \widehat{h}\|_p = \operatorname{dist}(u, C),$$

where $\operatorname{dist}(u, C) = \inf_{h \in C} \|u - h\|_p$.

Problem 1.13*
Find a measure space (Ω, Σ, μ) and a sequence of Σ-measurable functions $u_n\colon \Omega \longrightarrow \mathbb{R}$ which converges in μ-measure and μ-almost everywhere, but it does not converge weakly in $L^p(\Omega)$ for any $1 \leqslant p \leqslant +\infty$.

Problem 1.14**
Let (Ω, Σ, μ) be a measure space and let $\{u_n\}_{n\geqslant 1} \subseteq L^1(\Omega)$ be a sequence such that

$$u_n(\omega) \geqslant 0 \quad \text{for } \mu\text{-a.a. } \omega \in \Omega, \text{ all } n \geqslant 1,$$

$$u(\omega) \leqslant \liminf_{n\to+\infty} u_n(\omega) \quad \text{for } \mu\text{-a.a. } \omega \in \Omega,$$

with $u \in L^1(\Omega)$ and

$$\int_\Omega u_n \, d\mu = \int_\Omega u \, d\mu \qquad \forall\, n \geqslant 1.$$

Show that $u_n \longrightarrow u$ in $L^1(\Omega)$.

Problem 1.15**
Let (Ω, Σ, μ) be a measure space and let $C \subseteq L^1(\Omega)$ be a uniformly integrable set. Show that for every $\varepsilon > 0$, there exists $h_\varepsilon \in L^1(\Omega)_+$ such that

$$\sup_{u\in C} \int_{\{|u|\geqslant h_\varepsilon\}} |u| \, d\mu \leqslant \varepsilon.$$

Problem 1.16***
Let (Ω, Σ, μ) be a measure space and let $C \subseteq L^1(\Omega)$. Suppose that for every $\varepsilon > 0$, there exists $h_\varepsilon \in L^1(\Omega)_+$ such that

$$\sup_{u\in C} \int_{\{|u|\geqslant h_\varepsilon\}} |u| \, d\mu \leqslant \varepsilon.$$

Show that the set C is uniformly integrable.

Problem 1.17***
Suppose that (Ω, Σ, μ) is a measure space, $C \subseteq L^1(\Omega)$ is a bounded set and for every $\varepsilon > 0$, there exist $h_\varepsilon \in L^1(\Omega)_+$ and $\delta > 0$ such that

$$\text{if } A \in \Sigma \text{ is such that } \int_A h_\varepsilon \, d\mu \leqslant \delta, \text{ then } \sup_{u\in C} \int_A |u| \, d\mu \leqslant \varepsilon.$$

Show that C is uniformly integrable.

Remark. Problems 1.7, 1.15, 1.16, and 1.17 provide alternative definitions of uniform integrability and Problems 1.5 and 1.6 provide an alternative definition of uniform integrability within finite measure spaces.

Problem 1.18***
Let (Ω, Σ, μ) be a finite measure space and suppose that $\{u_n\}_{n\geqslant 1} \subseteq L^1(\Omega)$ is relatively sequentially w-compact. Show that $u_n \xrightarrow{w} u$ in $L^1(\Omega)$ if and only if

$$\int_\Omega |u+y|\,d\mu \leqslant \liminf_{n\to+\infty} \int_\Omega |u_n+y|\,d\mu \qquad \forall\, y \in L^1(\Omega).$$

Problem 1.19**
Let (Ω, Σ, μ) be a measure space and let $1 < p < +\infty$. Suppose that $\{u_n\}_{n\geqslant 1} \subseteq L^p(\Omega)$ is a bounded sequence such that $u_n \longrightarrow u$ in μ-measure or for almost all $\omega \in \Omega$. Show that $u \in L^p(\Omega)$ and $u_n \xrightarrow{w} u$ in $L^p(\Omega)$.

Problem 1.20**
Let (Ω, Σ, μ) be a σ-finite measure space, let $\{u_n\}_{n\geqslant 1} \subseteq L^1(\Omega)$ be a sequence and $u \in L^1(\Omega)$. Show that $u_n \xrightarrow{w} u$ in $L^1(\Omega)$ if and only if the following two conditions hold:

(a) $\sup\limits_{n\geqslant 1} \|u_n\|_1 < +\infty$; and

(b) $\int\limits_A u_n\,d\mu \longrightarrow \int\limits_A u\,d\mu$ for all $A \in \Sigma$.

Problem 1.21**
Let (Ω, Σ, μ) be a measure space and let $\{u_n\}_{n\geqslant 1} \subseteq L^1(\Omega)$ be a sequence such that $u_n \longrightarrow u$ in μ-measure or for μ-almost all $\omega \in \Omega$ for some $u \in L^1(\Omega)$. Show that

$$\lim_{n\to+\infty} \big|\|u_n\|_1 - \|u\|_1 - \|u_n - u\|_1\big| = 0$$

Problem 1.22***
Let (Ω, Σ, μ) be a measure space and let $\{u_n\}_{n\geqslant 1} \subseteq L^p(\Omega)$ (with $1 < p < +\infty$) be an L^p-bounded sequence such that $u_n(\omega) \longrightarrow u(\omega)$ for μ-almost all $\omega \in \Omega$. Show that

$$\lim_{n\to+\infty} \big[\|u_n\|_p^p - \|u_n - u\|_p^p\big] = \|u\|_p^p$$

Problem 1.23**
Let (Ω, Σ, μ) be a finite measure space and let $\{u_n\}_{n\geqslant 1} \subseteq L^1(\Omega)$ be a uniformly integrable sequence such that $u_n(\omega) \longrightarrow u(\omega)$ for μ-almost all $\omega \in \Omega$. Show that $u_n \longrightarrow u$ in $L^1(\Omega)$.

Problem 1.24**
Suppose that (Ω, Σ, μ) is a finite measure space, $u \in L^1(\Omega)$, $\{u_n\}_{n\geqslant 1} \subseteq L^1(\Omega)$ is a sequence such that $u_n \longrightarrow u$ in μ-measure and $u_n \xrightarrow{w} u$ in $L^1(\Omega)$. Show that $u_n \longrightarrow u$ in $L^1(\Omega)$,

Problem 1.25***
Let (Ω, Σ, μ) be a finite measure space, let $\{u_n\}_{n\geqslant 1} \subseteq L^1(\Omega)$ be a sequence such that $u_n \xrightarrow{w} u$ in $L^1(\Omega)$ for some $u \in L^1(\Omega)$, and let $\{y_n\}_{n\geqslant 1} \subseteq L^\infty(\Omega)$ be a bounded sequence such that $y_n(\omega) \longrightarrow y(\omega)$ for μ-almost all $\omega \in \Omega$. Show that $u_n y_n \xrightarrow{w} uy$ in $L^1(\Omega)$.

Problem 1.26**
Let (Ω, Σ, μ) be a finite measure space, $\varphi \colon \mathbb{R} \longrightarrow \overline{\mathbb{R}} = \mathbb{R} \cup \{+\infty\}$ a convex lower semicontinuous function which is not identically $+\infty$ and $u \in L^1(\Omega)$ with $u(\omega) \in \operatorname{dom}\varphi = \{x \in \mathbb{R} : \varphi(x) < +\infty\}$ for μ-almost all $\omega \in \Omega$. Show that

$$\varphi\Big(\frac{1}{\mu(\Omega)} \int_\Omega u \, d\mu\Big) \leqslant \frac{1}{\mu(\Omega)} \int_\Omega \varphi(u) \, d\mu.$$

(This is the so-called ***Jensen inequality***; cf. Theorem I.3.99.)

Problem 1.27**
Let (Ω, Σ, μ) be a measure space and let $1 \leqslant p \leqslant +\infty$. Use the Jensen inequality (see Problem 1.26) to prove the Hölder inequality, that is,

$$\int_\Omega |u||h| \, d\mu \leqslant \|u\|_p \|h\|_{p'} \qquad \forall\, u \in L^p(\Omega),\ h \in L^{p'}(\Omega)$$

(see Theorem 1.3).

Problem 1.28**
Let (Ω, Σ, μ) be a measure space, $p_k \in [1, +\infty]$ for all $k \in \{1, \ldots, n\}$, $\frac{1}{p_0} = \sum\limits_{k=1}^n \frac{1}{p_k}$ (as usual we set $\frac{1}{\infty} = 0$) and $u_n \in L^{p_k}(\Omega)$ for all $k \in \{1, \ldots, n\}$. Show that

$$\|u_1 \ldots u_n\|_{p_0} \leqslant \|u_1\|_{p_1} \cdots \|u_n\|_{p_n}.$$

Problem 1.29**
Let (Ω, Σ, μ) be a σ-finite measure space, $1 < p \leqslant +\infty$ and $u \colon \Omega \longrightarrow \mathbb{R}$ a Σ-measurable function such that $uh \in L^1(\Omega)$ for every $h \in L^{p'}(\Omega)$ (where $\frac{1}{p} + \frac{1}{p'} = 1$). Show that $u \in L^p(\Omega)$.

Problem 1.30**
Let (Ω, Σ, μ) be a measure space, $1 \leqslant p < q < r \leqslant +\infty$ and $u \in L^p(\Omega) \cap L^r(\Omega)$. Show that $u \in L^q(\Omega)$ and

$$\|u\|_q \leqslant \|u\|_p^\vartheta \|u\|_r^{1-\vartheta}$$

where $\vartheta \in (0, 1)$ is such that $\frac{1}{q} = \frac{\vartheta}{p} + \frac{1-\vartheta}{r}$.
(This is the so-called ***interpolation inequality***.)

Problem 1.31**
Let (Ω, Σ, μ) be a measure space, $1 \leqslant p < +\infty$, $\{u_n\}_{n\geqslant 1} \subseteq L^p(\Omega)$ and $\{h_n\}_{n\geqslant 1} \subseteq L^\infty(\Omega)$. Suppose that $u_n \longrightarrow$ in $L^p(\Omega)$, $\sup\limits_{n\geqslant 1} \|h_n\|_\infty < +\infty$ and

$$h_n(\omega) \longrightarrow h(\omega) \quad \text{for } \mu\text{-a.a. } \omega \in \Omega.$$

Show that $u_n h_n \longrightarrow uh$ in $L^p(\Omega)$.

Problem 1.32***
Let (Ω, Σ, μ) be a measure space, $1 \leqslant p < q < +\infty$ and $u \in L^p(\Omega) \cap L^q(\Omega)$. Show that the function $[p, q] \in r \longmapsto \|u\|_r \in \mathbb{R}$ is continuous.

Problem 1.33*
Let (Ω, Σ, μ) be a measure space and let $K \in \mathcal{L}\big(L^2(\Omega), L^2(\Omega)\big)$ be an isometry. Show that $K^*K = I_{L^2(\Omega)}$, where K^* is the adjoint of K.

Problem 1.34*
Let (Ω, Σ, μ) be a σ-finite measure space, $p \geqslant 1$ and $u \colon \Omega \longrightarrow \overline{\mathbb{R}}_+ = [0, +\infty]$ a Σ-measurable function. Show that

$$\int_\Omega u^p \, d\mu = p \int_0^{+\infty} \vartheta^{p-1} \mu(\{u \geqslant \vartheta\}) \, d\vartheta.$$

Problem 1.35**
Let $u \in L^1_{\text{loc}}(\mathbb{R}^N)$. Show that the Hardy–Littlewood maximal function $x \longmapsto M(u)(x)$ (see Definition 1.31) is lower semicontinuous.

Problem 1.36**
Suppose that $u \in L \log L(\mathbb{R}^N)$ (see Definition 1.33). Show that the Hardy–Littlewood maximal function $M(u)$ (see Definition 1.31) is integrable over Lebesgue measurable sets of finite Lebesgue measure.

Problem 1.37**
According to Corollary 1.26 for any measure space (Ω, Σ, μ) and any $p \in (1, +\infty)$ the space $L^p(\Omega)$ is uniformly convex, reflexive, and has the Kadec–Klee property. Have $L^1(\Omega)$ or $L^\infty(\Omega)$ (with w^*-convergence) the Kadec–Klee property? Justify your answer.

Problem 1.38**
Let (Ω, Σ, μ) be a measure space and let $1 \leqslant q \leqslant p \leqslant +\infty$. Let $\xi \colon \Omega \longrightarrow \mathbb{R}$ be a Σ-measurable function such that $\xi u \in L^q(\Omega)$ for all $u \in L^p(\Omega)$. Show that $\xi \in L^r(\Omega)$ with $r = \frac{pq}{p-q}$ if $p < +\infty$ and $r = q$ if $p = +\infty$.

Problem 1.39**
Show that weak convergence in $L^1(\Omega)$ does not imply convergence in μ-measure and convergence in μ-measure does not imply weak convergence in $L^1(\Omega)$.

Problem 1.40**
Is it true that a bounded sequence $\{u_n\}_{n\geqslant 1} \subseteq L^1(\mathbb{R})$ admits a weakly convergent subsequence? Justify your answer.

Problem 1.41***
Let (Ω, Σ) be a finite measurable space, let $f\colon \Omega \times \mathbb{R} \longrightarrow \mathbb{R}$ be a Carathéodory function, and let N_f be the Nemytskii operator corresponding to f (see Definitions 1.144 and 1.145). Show that:

(a) for every Σ-measurable function $u\colon \Omega \longrightarrow \mathbb{R}$, the function $\omega \longmapsto N_f(u)(\omega)$ is Σ-measurable too;

(b) if (Ω, Σ, μ) is a finite measure space and $u_n \xrightarrow{\mu} u$, then $N_f(u_n) \xrightarrow{\mu} N_f(u)$.

Problem 1.42**
Let (Ω, Σ, μ) be a σ-finite measure space, let $f\colon \Omega \times \mathbb{R} \longrightarrow \mathbb{R}$ be a Carathéodory function such that

$$|f(\omega, x)| \leqslant a(\omega) + c|x|^r \quad \text{for } \mu\text{-a.a. } \omega \in \Omega, \text{ all } x \in \mathbb{R},$$

with $a \in L^q(\Omega)$, $1 \leqslant q \leqslant +\infty$, $c > 0$ and $r > 0$, and let N_f be the Nemytskii operator corresponding to f (see Definitions 1.144 and 1.145). Show that:

(a) N_f maps $L^{qr}(\Omega)$ into $L^q(\Omega)$;

(b) $N_f\colon L^{qr}(\Omega) \longrightarrow L^q(\Omega)$ is continuous and bounded (maps bounded sets to bounded sets).

Problem 1.43***
Let (Ω, Σ, μ) be a finite measure space, let $f\colon \Omega \times \mathbb{R} \longrightarrow \mathbb{R}$ be a Carathéodory function and let N_f be the Nemytskii operator corresponding to f (see Definitions 1.144 and 1.145). Assume that N_f maps $L^p(\Omega)$ into $L^q(\Omega)$ with $p, q \in [1, +\infty)$. Show that N_f is continuous and bounded.

Problem 1.44**
Let (Ω, Σ, μ) be a finite measure space and let $1 < p < +\infty$. Assume that $\{u_n\}_{n\geqslant 1} \subseteq L^p(\Omega)$ is a sequence such that

$$\sup_{n\geqslant 1} \|u_n\|_p < +\infty$$

and

$$u_n(\omega) \longrightarrow u(\omega) \quad \text{for } \mu\text{-a.a. } \omega \in \Omega.$$

Show that $u_n \longrightarrow u$ in $L^q(\Omega)$ for all $q \in [1.p)$.

Problem 1.45**
Let (Ω, Σ, μ) be a measure space, $u \in L^1(\Omega)$ and let $\{u_n\}_{n\geqslant 1} \subseteq L^1(\Omega)$ be a sequence such that $u_n(\omega) \geqslant 0$ for μ-almost all $\omega \in \Omega$ and all $n \geqslant 1$ and $u_n(\omega) \longrightarrow u(\omega)$ for μ-almost all $\omega \in \Omega$. Assume that $\int_\Omega u_n\, d\mu \longrightarrow \int_\Omega u\, d\mu$. Show that for every $A \in \Sigma$, we have $\int_A u_n\, d\mu \longrightarrow \int_A u\, d\mu$.

Problem 1.46**

Let (Ω, Σ, μ) be a finite measure space and let $u, h\colon \Omega \longrightarrow \mathbb{R}_+ = [0, +\infty)$ be two Σ-measurable functions such that

$$\mu(\{h \geqslant \vartheta\}) \leqslant \frac{1}{\vartheta} \int_{\{h \geqslant \vartheta\}} u \, d\mu \qquad \forall\, \vartheta > 0.$$

Show that

$$\|h\|_p \leqslant \frac{p}{p-1} \|u\|_p \qquad \forall\, p \in (1, +\infty).$$

Problem 1.47**

Let (Ω, Σ, μ) be a σ-finite measure space, $h\colon \Omega \longrightarrow \mathbb{R}$ a Σ-measurable function, $1 \leqslant p < +\infty$ and consider the linear map $K(u) = uh$ for all $u \in L^p(\Omega)$. Suppose that

$$K(u) \in L^1(\Omega) \qquad \forall\, u \in L^p(\Omega).$$

Show that $h \in L^{p'}(\Omega)$ $(\frac{1}{p} + \frac{1}{p'} = 1)$.

Problem 1.48***

Show that $L^1(0,1)$ and $L^\infty(0,1)$ are not reflexive.

Problem 1.49***

Show that $L^1(0,1)$ is not the dual of any normed space V.

Problem 1.50**

Let $(\Omega_1, \Sigma_1, \mu_1)$ and $(\Omega_2, \Sigma_2, \mu_2)$ be two measure spaces and let $L \in \mathcal{L}\big(L^1(\Omega_1); L^1(\Omega_2)\big)$ (that is, a bounded linear operator from $L^1(\Omega_1)$ into $L^1(\Omega_2)$). Suppose that $C \subseteq L^1(\Omega_1)$ is a uniformly integrable set. Show that $L(C) \subseteq L^1(\Omega_2)$ is uniformly integrable.

Problem 1.51*

Let (Ω, Σ, μ) be a σ-finite measure space, $u \in L^1(\Omega)$ and suppose that for all $A \in \Sigma$ with $\mu(A) < +\infty$, we have $\int_A u \, d\mu \leqslant \eta$. Show that $\int_\Omega u \, d\mu \leqslant \eta$.

Problem 1.52**

(a) Let (Ω, Σ, μ) be a measure space and consider the set

$$C = \big\{u \in L^1(\Omega) : u(\omega) \geqslant 1 \text{ for a.a. } \omega \in \Omega\big\}.$$

Is the set $C \subseteq L^1(\Omega)$ weakly closed? Justify your answer.

(b) Let $u_n = \chi_{[n,n+1]}$. Is the sequence $\{u_n\}_{n \geqslant 1} \subseteq L^1(\mathbb{R})$ uniformly integrable? Justify your answer.

Problem 1.53*
Let (Ω, Σ, μ) be a finite measure space.

(a) Suppose that $C \subseteq L^p(\Omega)$ is a bounded set for some $p \in (1, +\infty]$. Show that the set C is uniformly integrable.

(b) Show that for $p = 1$ the above does not hold (i.e., L^1-boundedness does not imply uniform integrability).

(c) Show that conditions **(a)** and **(b)** in the definition of uniform integrability (see Definition 1.18) are independent (none of them implies the other).

Problem 1.54**
Let $u\colon [0,1] \longrightarrow \mathbb{R}$ be a nonnegative measurable function such that

$$\lambda(\{t \in [0,1] : u(t) \geqslant \vartheta\}) \leqslant \tfrac{1}{1+\vartheta^2} \qquad \forall\, \vartheta > 0$$

(where λ denotes the Lebesgue measure on $\mathbb{R}$). Determine those values of $p \in [1, +\infty)$ for which $u \in L^p(0,1)$ and find the minimum value of p for which $u \notin L^p(0,1)$.

Problem 1.55*
Show that $L^1(\mathbb{R}) \cap L^\infty(\mathbb{R})$ is dense in $L^p(\mathbb{R})$ for every $1 \leqslant p < +\infty$ but is not dense in $L^\infty(\mathbb{R})$.

Problem 1.56**
Let (Ω, Σ, μ) be a σ-finite measure space, $\xi \in L^\infty(\Omega)^*$, $A \in \Sigma$ and define $\xi_A \in L^\infty(\Omega)^*$ by setting $\xi_A(u) = \xi(\chi_A u)$. Show that, if $A_1, A_2 \in \Sigma$ and $A_1 \cap A_2 = \emptyset$, then $\|\xi_{A_1 \cup A_2}\|_* = \|\xi_{A_1}\|_* + \|\xi_{A_2}\|_*$.

Problem 1.57**
Let (Ω, Σ, μ) be a σ-finite measure space and let $\xi \in L^\infty(\Omega)^*$. Suppose that for every $\varepsilon > 0$, we can find $A \in \Sigma$ with $\mu(A) \leqslant \varepsilon$ and $\|\xi_A\|_* \geqslant \|\xi\|_* - \varepsilon$ (where $\xi_A(u) = \xi(\chi_A u)$ for all $u \in L^\infty(\Omega)$; see Problem 1.56). Show that ξ is singular (see Definition 1.23(b)).

Problem 1.58***
Let $\{u_n\colon [0,b] \longrightarrow \mathbb{R}\}_{n \geqslant 1}$ be a sequence of increasing (respectively decreasing) functions which converge pointwise to $u \in C([0,b])$. Show that u is also increasing (respectively decreasing) and $u_n \longrightarrow u$ uniformly on $[0,b]$.

Problem 1.59*
Let (Ω, Σ, μ) be a finite measure space and let $\{u_n\}_{n \geqslant 1} \subseteq L^2(\Omega)$ be a sequence such that

$$u_n \xrightarrow{w} u \ \text{ in } L^2(\Omega) \quad \text{and} \quad u_n^2 \xrightarrow{w} u^2 \ \text{ in } L^1(\Omega).$$

Show that

$$u_n \longrightarrow u \ \text{ in } L^2(\Omega).$$

Problem 1.60**
Show that in general weak measurability does not imply strong measurability.

Problem 1.61**
Let (Ω, Σ, μ) be a finite measure space, X a Banach space, $\{u_n : \Omega \longrightarrow X\}_{n\geqslant 1}$ a sequence of Bochner integrable functions such that

$$u_n(\omega) \longrightarrow u(\omega) \quad \text{for } \mu\text{-a.a. } \omega \in \Omega$$

and suppose that there exists $h \in L^1(\Omega)_+$ such that

$$\|u_n(\omega)\|_X \leqslant h(\omega) \quad \text{for } \mu\text{-a.a. } \omega \in \Omega, \text{ all } n \geqslant 1.$$

Show that u is Bochner integrable and $\int_\Omega u_n \, d\mu \longrightarrow \int_\Omega u \, d\mu$.
(This is the ***Lebesgue dominated convergence theorem for the Bochner integral***.)

Problem 1.62**
Let (Ω, Σ, μ) be a finite measure space, X a Banach space, $\{u_n : \Omega \longrightarrow X\}_{n\geqslant 1}$ a sequence of Bochner integrable functions and assume that the sequence $\{\int_\Omega \|u_n\|_X \, d\mu\}_{n\geqslant 1} \subseteq \mathbb{R}$ is bounded and

$$u_n(\omega) \longrightarrow u(\omega) \quad \text{for } \mu\text{-a.a. } \omega \in \Omega.$$

Show that u is Bochner integrable and

$$\int_\Omega \|u(\omega)\|_X \, d\mu \leqslant \liminf_{n\to+\infty} \int_\Omega \|u_n(\omega)\|_X \, d\mu$$

(a version of the Fatou lemma for the Bochner integral).

Problem 1.63***
Let (Ω, Σ, μ) be a finite measure space, X and Y two Banach spaces, $D \subseteq X$ a linear subspace, $L : D \longrightarrow Y$ a closed linear operator (recall that closed means that $\operatorname{Gr} L \subseteq X \times Y$ is closed) and $u : \Omega \longrightarrow X$ a Bochner integrable function such that $L \circ u : \Omega \longrightarrow Y$ is Bochner integrable too. Show that for all $A \in \Sigma$ we have

$$L\Big(\int_A u \, d\mu\Big) = \int_A L(u) \, d\mu.$$

Problem 1.64**
Let (Ω, Σ, μ) be a finite measure space, X a Banach space and $u, y : \Omega \longrightarrow X$ two Bochner integrable functionals such that

$$\int_A u(\omega) \, d\mu = \int_A y(\omega) \, d\mu \qquad \forall\, A \in \Sigma.$$

Show that

$$u(\omega) = y(\omega) \quad \text{for } \mu\text{-a.a. } \omega \in \Omega.$$

Problem 1.65**
Let (Ω, Σ, μ) be a finite measure space, X a Banach space and $u, y\colon \Omega \longrightarrow X$ two strongly measurable functions such that

$$\langle x^*, u(\omega)\rangle = \langle x^*, y(\omega)\rangle \quad \text{for } \mu\text{-a.a. } \omega \in \Omega, \text{ all } x^* \in X^*$$

(the exceptional μ-null set in the above equality may depend on $x^* \in X^*$). Show that $u(\omega) = y(\omega)$ for μ-almost all $\omega \in \Omega$.

Problem 1.66**
Let (Ω, Σ, μ) be a finite measure space, X a Banach space, $u\colon \Omega \longrightarrow X$ a Bochner integrable function, and $A \in \Sigma$. Show that

$$\tfrac{1}{\mu(A)} \int_A u(\omega)\, d\mu \in \overline{\operatorname{conv}}\, u(A).$$

Problem 1.67**
Let (Ω, Σ, μ) be a finite measure space, X a Banach space and $u\colon \Omega \longrightarrow X$ a Bochner integrable function. Show that $m\colon \Sigma \longrightarrow X$ defined by

$$m(A) = \int_A u\, d\mu \qquad \forall\, A \in \Sigma$$

is a vector measure (see Definition 1.43) and $m \ll \mu$ (that is, $\lim\limits_{\mu(A)\searrow 0} m(A) = 0$).

Problem 1.68**
Let (Ω, Σ, μ) be a finite measure space, X a Banach space, and $u \in L^1(\Omega; X)$. Show that $u = 0$ if and only if

$$\int_A u\, d\mu = 0 \qquad \forall\, A \in \Sigma.$$

Problem 1.69**
Let (Ω, Σ, μ) be a finite measure space, X a Banach space, $\Sigma_0 \subseteq \Sigma$ a sub-σ-algebra of Σ. Show that there exists a unique operator $E^{\Sigma_0} \in \mathcal{L}\big(L^1(\Omega, \Sigma; X); L^1(\Omega, \Sigma_0; X)\big)$ such that for each $A \in \Sigma_0$ and each $u \in L^1(\Omega, \Sigma; X)$, we have

$$\int_A u\, d\mu = \int_A E^{\Sigma_0} u\, d\mu$$

(E^{Σ_0} is called the ***conditional expectation*** with respect to the sub-σ-algebra Σ_0; cf. Definition I.3.161 and Proposition I.3.163 for $\mathbb{R}$-valued functions).

Problem 1.70**
Let (Ω, Σ, μ) be a measure space, X a Banach space and $u\colon \Omega \longrightarrow X^*$ a function such that

$$\langle u(\cdot), x\rangle \in L^1(\Omega) \qquad \forall\, x \in X.$$

Show that for every $A \in \Sigma$, there exists $x_A^* \in X^*$ such that

$$\langle x_A^*, x \rangle = \int_\Omega \langle u(\omega), x \rangle \, d\mu.$$

Problem 1.71***

(a) Let X, Y be two Banach spaces such that $X \hookrightarrow Y$ continuously and densely. Show that $Y^* \hookrightarrow X^*$ continuously and if X is reflexive, then $Y^* \hookrightarrow X^*$ continuously and densely.

(b) Let $T = [0, b]$ and let X, Y be two Banach spaces such that X is reflexive and $X \hookrightarrow Y$ continuously and densely. Suppose that $u \in L^\infty(T; X) \cap C(T; Y_w)$ (Y_w is the Banach space Y with the weak topology). Show that $u \in C(T, X_w)$.

Problem 1.72**

(a) Let X, Y, Z be three Banach spaces and assume that the embeddings $X \hookrightarrow Y \hookrightarrow Z$ are continuous, with the first being compact. Show that for every $\varepsilon > 0$ we can find $c_\varepsilon > 0$ such that

$$\|x\|_Y \leqslant \varepsilon \|x\|_X + c_\varepsilon \|x\|_Z \qquad \forall\, x \in X.$$

(b) Let $T = [0, b]$ and let X, Y, Z be three Banach spaces as in part **(a)**. Suppose that $K \subseteq L^p(T; X)$ ($1 \leqslant p \leqslant +\infty$) is a bounded set which is relatively compact in $L^p(T; Z)$. Show that K is relatively compact in $L^p(T; Y)$.

Problem 1.73**

Let (Ω, Σ, μ) be a finite measure space, X a reflexive Banach space, $1 < p < +\infty$, $\{u_n\}_{n \geqslant 1} \subseteq L^p(T; X)$ a bounded sequence such that

$$u_n(\omega) \longrightarrow u(\omega) \quad \text{for } \mu\text{-a.a. } \omega \in \Omega.$$

Show that $u \in L^p(\Omega; X)$ and

$$u_n \xrightarrow{w} u \quad \text{in } L^p(\Omega; X).$$

(This is the vectorial counterpart of Problem 1.19.)

Problem 1.74**

Let $T = [0, b]$ and let X be a Banach space. Suppose that $u \in L^p(T; X)$ with $1 \leqslant p < +\infty$. Show that

$$\lim_{h \to 0} \int_0^b \|u(t+h) - u(t)\|_X^p \, dt = 0,$$

where we set $u(s) = 0$ when $s \notin T$.

Problem 1.75***
Let $T = [0, b]$, X a Banach space, and $u, h \in L^1(T; X)$. Show that the following three statements are equivalent:
(a) $u(t) = x_0 + \int_0^t h(s)\,ds$ for almost all $t \in T$ and with $x_0 \in X$.
(b) for every $\vartheta \in C_c^\infty(0, b)$ we have

$$\int_0^b u(t)\vartheta'(t)\,dt = -\int_0^b h(t)\vartheta(t)\,dt;$$

(c) for every $x^* \in X^*$ we have

$$\tfrac{d}{dt}\langle x^*, u(t)\rangle = \langle x^*, h(t)\rangle \quad \text{for a.a. } t \in T.$$

Problem 1.76**
Let $I \subseteq \mathbb{R}$ be an interval and let $u\colon I \longrightarrow \mathbb{R}$ be a monotone function. Show that:
(a) for every subinterval $T \subseteq I$, we have

$$\mathrm{Var}_T u = \sup_T u - \inf_T u;$$

(b) $u \in BV(I)$ if and only if u is bounded.

Problem 1.77**
Suppose that $u \in C([0, 1])$ and that it is bounded. Is it true that $u \in BV([0, 1])$? Justify your answer.

Problem 1.78**
Let $u \in BV([a, b])$ and consider the function

$$f(a) = 0 \quad \text{and} \quad f(x) = \mathrm{Var}_{[a,x]} u \qquad \forall\, x \in (a, b].$$

Show that:
(a) $|u(x) - u(y)| \leqslant |f(x) - f(y)|$ for all $x, y \in [a, b]$;
(b) the points of continuity of u and f coincide.

Problem 1.79**
(a) Let $f\colon [a, b] \longrightarrow [0, +\infty)$ be an increasing function. Show that for every $h \in (0, b - a]$, we have

$$\tfrac{1}{h}\int_a^{b-h} \big(f(x + h) - f(x)\big)\,dx \leqslant f(b) - f(a).$$

(b) Let $I \subseteq \mathbb{R}$ be an interval and let $u\colon I \longrightarrow \mathbb{R}$ be a measurable function. Show that for every $h > 0$, we have

$$\tfrac{1}{h}\int_{I_h} \big|u(x + h) - u(x)\big|\,dx \leqslant \mathrm{Var}\, u,$$

where $I_h = \{x \in I : x + h \in I\}$.

Problem 1.80**

Let $u\colon [0,1] \longrightarrow \mathbb{R}$ be an increasing function. Show that u' is integrable and for all $0 < a < b < 1$ we have

$$\int_0^b u'(x)\,dx \leqslant u(b^+) - u(a^+).$$

Moreover, if u is in addition Lipschitz continuous, then equality holds.

Problem 1.81**

Let $I \subseteq \mathbb{R}$ be an interval and let $\{u_n\colon I \longrightarrow \mathbb{R}\}_{n\geqslant 1}$ be a sequence of functions such that

$$u_n(x) \longrightarrow u(x) \qquad \forall\, x \in \mathbb{R}.$$

Show that $\operatorname{Var} u \leqslant \liminf_{n\to+\infty} \operatorname{Var} u_n$.

Problem 1.82**

Let $u \in C^1(a,b)$. Show that $u \in BV([a,b])$ if and only if $u' \in L^1(a,b)$ and we have

$$\operatorname{Var}_{(a,b)} u = \|u'\|_1.$$

Problem 1.83*

Find a bounded function which belongs in $BV_{\mathrm{loc}}(\mathbb{R})$ but not in $BV(\mathbb{R})$.

Problem 1.84*

Let $I \subseteq \mathbb{R}$ be an interval and let $u \in BV(I)$ be continuous. Show that the set of values $y \in \mathbb{R}$ for which the set $u^{-1}(\{y\})$ is infinite in I has Lebesgue measure zero.

Problem 1.85***

Let $I \subseteq \mathbb{R}$ be an interval and let $u\colon I \longrightarrow \mathbb{R}$ be a function. Suppose that there exist a set $E \subseteq I$ (not necessarily measurable) and $M > 0$ such that the function u is differentiable at every $x \in E$ and

$$|u'(x)| \leqslant M \qquad \forall\, x \in E.$$

Show that

$$\lambda^*(u(E)) \leqslant M\lambda^*(E),$$

with λ^* denoting the one-dimensional Lebesgue outer measure.

Problem 1.86*

Let $I \subseteq \mathbb{R}$ be an interval and let $u\colon I \longrightarrow \mathbb{R}$ be a function. Suppose that $E \subseteq I$ is a set such that u is differentiable for all $x \in E$. Show that, if E is Lebesgue-null or if $u'(x) = 0$ for all $x \in E$, then $\lambda(u(E)) = 0$.

Problem 1.87**
Let $I \subseteq \mathbb{R}$ be an interval and let $u\colon I \longrightarrow \mathbb{R}$ be a Lebesgue measurable function. Suppose that $E \subseteq I$ is a Lebesgue measurable set such that $u'(x)$ exists at every $x \in E$. Show that $u(E)$ is Lebesgue measurable and

$$\lambda(u(E)) \leqslant \int_E |u'(x)|\, dx.$$

Problem 1.88*
Let $I \subseteq \mathbb{R}$ be an interval and suppose that $u\colon I \longrightarrow \mathbb{R}$ is a differentiable function such that $u' \in L^1(I)$. Show that for every $a, b \in I$, $a < b$ we have

$$\big|u(b) - u(a)\big| \leqslant \int_a^b |u'(x)|\, dx.$$

Problem 1.89*
Let $I = [a, b]$ and let $u \in AC(I)$. Assume that $u(I) \subseteq [\widehat{a}, \widehat{b}]$ and let $\varphi\colon [\widehat{a}, \widehat{b}] \longrightarrow \mathbb{R}$ be a Lipschitz continuous function. Show that $\varphi \circ u \in AC(I)$.

Problem 1.90**
Let $I = [a, b]$ and let $u \in AC(I)$ be a strictly increasing function. Suppose that $\varphi\colon [u(a), u(b)] \longrightarrow \mathbb{R}$ is an absolutely continuous function. Show that $\varphi \circ u$ is absolutely continuous.

Problem 1.91*
Let $I = (a, b)$ be a bounded open interval and let $u \in C^1(I)$. Show that the following statements are equivalent:
(a) $u \in AC(I)$.
(b) $\mathrm{Var}_I u < +\infty$.
(c) $u' \in L^1(I)$.
(d) If $m(A) = \int_A |u'|\, dx$ for every measurable set $A \subseteq I$, then $m \ll \lambda$.

Problem 1.92*
Let $h \in L^1(\mathbb{R})$ and define

$$u(x) = \int_{-\infty}^{x} h(s)\, ds \qquad \forall\, x \in \mathbb{R}.$$

Show that u is absolutely continuous and $u'(x) = h(x)$ for almost all $x \in \mathbb{R}$.

Problem 1.93**

Let $I \subseteq \mathbb{R}$ be an interval and let $u\colon I \longrightarrow \mathbb{R}$ be a function. Show that $u \in AC_{\text{loc}}(I)$ if and only if:

(a) u is continuous on I;

(b) u is differentiable for almost all $x \in I$ and $u' \in L^1_{\text{loc}}(I)$;

(c) $u(x) = u(c) + \int_c^x u'(s)\, ds$ for all $x, c \in I$ (fundamental theorem of calculus).

Problem 1.94**

Suppose that $I \subseteq \mathbb{R}$ is an interval, $u\colon I \longrightarrow \mathbb{R}$ is a function, and $E \subseteq I$ is a set such that u is differentiable on E and $u(E)$ is Lebesgue-null. Show that $u'(x) = 0$ for almost all $x \in E$.

Problem 1.95**

Let $I \subseteq \mathbb{R}$ be an interval and let $u \in AC(I)$. Show that u maps Lebesgue measurable sets to Lebesgue measurable sets.

Problem 1.96*

Let $I \subseteq \mathbb{R}$ be an interval, $c \in I$ and $u \in BV_{\text{loc}}(I)$. We consider the function

$$\widehat{h}(x) = \begin{cases} \operatorname{Var}_{[c,x]} u & \text{if } c \leqslant x, \\ -\operatorname{Var}_{[x,c]} u & \text{if } c > x. \end{cases}$$

Show that for all $x, y \in I$ with $x < y$, we have

$$\left|u(y) - u(x)\right| \leqslant \widehat{h}(y) - \widehat{h}(x) = \operatorname{Var}_{[x,y]} u.$$

Problem 1.97**

Let $I \subseteq \mathbb{R}$ be an interval and let $u \in BV(I)$. Show that:

(a) u is bounded and both limits $\lim\limits_{x \to (\sup I)^-} u(x)$ and $\lim\limits_{x \to (\inf I)^+} u(x)$ exist;

(b) $u' \in L^1(I)$ and

$$\int_I |u'|\, dx \leqslant \int_I |\widehat{h}'|\, dx \leqslant \sup_I \widehat{h} - \inf_I \widehat{h} = \operatorname{Var}_I u$$

(here $\widehat{h}$ is the function introduced in Problem 1.96).

Problem 1.98**

Let $u\colon [a,b] \longrightarrow \mathbb{R}$ be a monotone function (increasing or decreasing). Show that $u \in AC\left([a,b]\right)$ if and only if

$$\int_a^b |u'(s)|\, ds = \left|u(b) - u(a)\right|.$$

Problem 1.99**
Suppose that $I \subseteq \mathbb{R}$ is an interval, $u \in BV_{\mathrm{loc}}(I)$ and $[a,b] \subseteq I$. Show that $u \in AC([a,b])$ if and only if $\int\limits_a^b |u'(s)|\,ds = \mathrm{Var}_{[a,b]} u$.

Problem 1.100*
Let $I \subseteq \mathbb{R}$ be an interval and let $u, y \in AC_{\mathrm{loc}}(I)$. Show that for all $a, b \in I$, we have

$$\int\limits_a^b uy'\,dx + \int\limits_a^b u'y\,dx = u(b)y(b) - u(a)y(a)$$

(integration by parts formula).

Problem 1.101*
Let $I \subseteq \mathbb{R}$ be an interval and let $u\colon I \longrightarrow \mathbb{R}$ be a nonconstant function. We say that u is ***singular*** if and only if u is differentiable for almost all $x \in I$ and $u'(x) = 0$ for almost all $x \in I$. Show that u is singular if and only if there exists a Lebesgue measurable set $E \subseteq I$ such that $\lambda(I \setminus E) = 0$ and $\lambda(u(E)) = 0$ (λ being the Lebesgue measure on $\mathbb{R}$).

Problem 1.102**
Let $\Omega \subseteq \mathbb{R}^N$ be an open set and let $u \in L^1(\Omega)$. Suppose that there exists a sequence $\{u_n\}_{n\geqslant 1} \subseteq C^\infty(\Omega)$ such that $u_n \longrightarrow u$ in $L^1(\Omega)$ and

$$\eta = \lim_{n\to+\infty} \int\limits_\Omega |Du_n|\,dz < +\infty.$$

Show that $u \in BV(\Omega)$.

Problem 1.103***
Let $T = [0,b]$ and let X, X_1, X_2 be three Banach spaces such that X_1 and X_2 are reflexive and the embeddings $X_1 \hookrightarrow X \hookrightarrow X_2$ are continuous and the first of them is compact. For $1 < p, q < +\infty$ we define

$$W_{pq}(0,b) = \{u \in L^p(T;X_1) : u' \in L^q(T;X_2)\},$$

with u' being defined in the sense of vectorial distributions, that is,

$$\int\limits_0^b u(t)\vartheta'(t)\,dt = -\int\limits_0^b u'(t)\vartheta(t)\,dt \qquad \forall\, \vartheta \in C_c^\infty(0,b)$$

(see Definition 1.98). Show that:

(a) $W_{pq}(0,b)$ furnished with the norm

$$\|u\|_{pq} = \|u\|_{L^p(T;X_1)} + \|u'\|_{L^q(T;X_2)}$$

is a separable and reflexive Banach space;

(b) the embedding $W_{pq}(0,b) \hookrightarrow L^p(T;X)$ is compact.

Problem 1.104***

Let $T = [0, b]$ and let X be a separable reflexive Banach space and H a separable Hilbert space. Assume that the embedding $X \hookrightarrow H$ is continuous and dense. Identifying H with its dual H^* (pivot space), we have that the embeddings $X \hookrightarrow H \hookrightarrow X^*$ are continuous and dense (see Problem 1.71(a)). Such a triple (X, H, X^*) is known as an ***evolution triple***. Let $1 < p < +\infty$ and consider the space

$$W_p(0,b) \;=\; \{u \in L^p(T;X) :\; u' \in L^{p'}(T;X^*)\}$$

(where $\frac{1}{p}+\frac{1}{p'}=1$), with u' being defined in the sense of vectorial distributions, that is,

$$\int_0^b u(t)\vartheta'(t)\,dt \;=\; -\int_0^b u'(t)\vartheta(t)\,dt \qquad \forall\, \vartheta \in C_c^\infty(0,b)\,.$$

We equip $W_p(0,b)$ with the norm

$$\|u\|_{W_p} \;=\; \|u\|_{L^p(T;X)} + \|u'\|_{L^{p'}(T;X^*)}.$$

Show that:

(a) the embedding $W_p(0,b) \hookrightarrow C(T;H)$ is continuous and dense;

(b) if by $(\cdot,\cdot)$ we denote the inner product of H and by $\langle\cdot,\cdot\rangle$ the duality brackets for the pair (X^*, X), then

$$\tfrac{d}{dt}\,(u(t), y(t)) \;=\; \big\langle u'(t), y(t)\big\rangle + \big\langle y'(t), u(t)\big\rangle \quad \text{for a.a. } t \in T.$$

Problem 1.105**

Recall that $\varphi \in C_c^\infty(\mathbb{R}^N)$ is a ***convolution kernel*** if

$$\varphi \geqslant 0,\ \varphi \text{ is even},\ \operatorname{supp}\varphi \subseteq \overline{B}_1,\ \int_{\mathbb{R}^N} \varphi(z)\,dz = 1,$$

where $\overline{B}_1 = \{z \in \mathbb{R}^N :\ |z| \leqslant 1\}$. The family $\varphi_\varepsilon(z) = \frac{1}{\varepsilon^N}\varphi(\frac{z}{\varepsilon})$ for $\varepsilon > 0$ is called a family of ***mollifiers***. Also, if $\Omega \subseteq \mathbb{R}^N$ is an open set and $\mu \in M_b(\Omega)$, then

$$(\mu * \varphi_\varepsilon)(x) \;=\; \int_\Omega \varphi_\varepsilon(x-z)\,d\mu(z)$$

is called the ***convolution*** between φ_ε and μ. Show that the measures $\mu_\varepsilon = (\mu * \varphi_\varepsilon)\lambda^N$ converge vaguely in Ω to μ as $\varepsilon \searrow 0$ and for every Borel set $E \subseteq \Omega_{-\varepsilon} = \{z \in \Omega :\ \operatorname{dist}(z, \partial\Omega) > \varepsilon\}$, we have

$$\int_E |\mu * \varphi_\varepsilon|(z)\,dz \;\leqslant\; |\mu|(E_\varepsilon),$$

where $E_\varepsilon = \{z \in \Omega :\ \operatorname{dist}(z, E) < \varepsilon\}$.

Problem 1.106**
Suppose that $\Omega \subseteq \mathbb{R}^N$ is an open set, $K \subseteq \Omega$ is compact and $u \in BV(\Omega)$. Show that

$$\int_K |u * \varphi_\varepsilon - u|\, dx \leqslant \varepsilon |Du|(\Omega) \qquad \forall\, \varepsilon \in (0, \operatorname{dist}(K, \partial\Omega)).$$

Here $\{\varphi_\varepsilon\}_{\varepsilon>0}$ are mollifiers (see Definition 1.117).

Problem 1.107**
Suppose that $\Omega \subseteq \mathbb{R}^N$ is an open set, $\{u_n\}_{n\geqslant 1} \subseteq BV(\Omega)$ is a sequence of functions and $u \in BV(\Omega)$. We say that the sequence $\{u_n\}_{n\geqslant 1}$ ***w^*-converges to u in $BV(\Omega)$***, denoted $u_n \xrightarrow{w^*} u$ in $BV(\Omega)$, if $u_n \longrightarrow u$ in $L^1(\Omega)$ and $Du_n \xrightarrow{w} Du$ in the sense of Definition 1.89(b), i.e.,

$$\lim_{n\to+\infty} \int_\Omega \vartheta\, dDu_n = \int_\Omega \vartheta\, dDu \qquad \forall\, \vartheta \in C_0(\Omega).$$

Show that $u_n \xrightarrow{w^*} u$ in $BV(\Omega)$ if and only if the sequence $\{u_n\}_{n\geqslant 1} \subseteq BV(\Omega)$ is bounded and $u_n \longrightarrow u$ in $L^1(\Omega)$.

Problem 1.108*
Suppose that $\Omega \subseteq \mathbb{R}^N$ is an open set, $\{u_n\}_{n\geqslant 1} \subseteq BV(\Omega)$ is a sequence of functions and $u \in BV(\Omega)$. We say that the sequence $\{u_n\}_{n\geqslant 1}$ ***converges strictly*** to u in $BV(\Omega)$, if $u_n \longrightarrow u$ in $L^1(\Omega)$ and $|Du_n|(\Omega) \longrightarrow |Du|(\Omega)$. Show that strict convergence in $BV(\Omega)$ implies w^*-convergence in $BV(\Omega)$ (see Problem 1.107), but the converse is not in general true. Justify your answer.

Problem 1.109***
Suppose that X is a locally compact and σ-compact metric space, $\{\mu_n\}_{n\geqslant 1} \subseteq M_b(X)$, $\mu_n \geqslant 0$ for all $n \geqslant 1$ and $\mu_n \longrightarrow \mu$ vaguely (see Definition 1.89(a)). Show that for every lower semicontinuous function $\varphi\colon X \longrightarrow \overline{\mathbb{R}}_+ = \mathbb{R}_+ \cup \{+\infty\}$ and for every upper semicontinuous function $\psi\colon X \longrightarrow \mathbb{R}_+ = [0, +\infty)$ with compact support, we have

$$\liminf_{n\to+\infty} \int_X \varphi\, d\mu_n \geqslant \int_X \varphi\, d\mu \quad \text{and} \quad \limsup_{n\to+\infty} \int_X \psi\, d\mu_n \leqslant \int_X \psi\, d\mu.$$

Problem 1.110**
Suppose that X is a locally compact and σ-compact metric space, $\{u_n\}_{n\geqslant 1} \subseteq M_b(X)$ is a sequence such that $\mu_n \geqslant 0$ for all $n \geqslant 1$, $\mu \in M_b(X)$, $\mu \geqslant 0$ and

$$\mu_n(X) \longrightarrow \mu(X)$$

and

$$\mu(U) \leqslant \liminf_{n\to+\infty} \mu_n(U) \qquad \forall\, U \subseteq X \text{ open}.$$

Show that $\mu_n \longrightarrow \mu$ narrowly.

Problem 1.111***
Let X be a locally compact and σ-compact topological space and let $\{\mu_\alpha\}_{\alpha\in J}$ be a net in $M_b(X)$ which is bounded and $\mu_\alpha(B) \longrightarrow \mu(B)$ for all $B \in \mathcal{B}(X)$. Show that $\mu_\alpha \longrightarrow \mu$ narrowly (see Definition 1.89(c)).

Problem 1.112**
Suppose that X is a locally compact and σ-compact topological space, $\{\mu_n\}_{n\geq 1} \subseteq M_b(X)$ is a sequence and $\mu \in M_b(X)$. Show that $\mu_n \xrightarrow{w} \mu$ if and only if $\mu_n \longrightarrow \mu$ vaguely and $\sup\limits_{n\geqslant 1} |\mu_n|(X) < +\infty$

Problem 1.113**
Let X be a locally compact and σ-compact topological space. Show that the weak topology $w(M_b(X), C_0(X))$ (corresponding to the weak convergence of Definition 1.89(b)) is not equal to the weak topology $w(M_b(X), C_c(X))$ (corresponding to the vague convergence of Definition 1.89(a)).

Problem 1.114**
Suppose that X is a locally compact and σ-compact topological space, $\mu \in M_b(X)$ and $\{\mu_n\}_{n\geq 1} \subseteq M_b(X)$ is a sequence such that $\mu_n \longrightarrow \mu$ vaguely in $M_b(X)$ and $|\mu_n| \longrightarrow \vartheta$ vaguely in $M_b^+(X)$. Show that:
(a) $|\mu| \leqslant \vartheta$;
(b) for all relatively compact Borel sets $E \subseteq X$ with $\vartheta(\partial E) = 0$, we have $\mu_n(E) \longrightarrow \mu(E)$.

Problem 1.115**
Let $\Omega \subseteq \mathbb{R}^N$ be an open set, $\mu \in M_b(\Omega)$, $\varphi \in C_c^\infty(\mathbb{R}^N)$ and let $\{\varphi_\varepsilon\}_{\varepsilon>0}$ be the corresponding family of mollifiers (see Definition 1.117). Show that

$$\int_{\mathbb{R}^N} |\mu * \varphi_\varepsilon|(x)\, d\mu \longrightarrow \int_{\mathbb{R}^N} |\mu|(dx)$$

(see Problem 1.105).

Problem 1.116*
Suppose that X is a locally compact and σ-compact topological space, $\mu \in M_b(X)$ and $\{\mu_n\}_{n\geq 1} \subseteq M_b(X)$ is a sequence. Show that $\mu_n \longrightarrow \mu$ narrowly if and only if $\mu_n \xrightarrow{w} \mu$ and $\mu_n(X) \longrightarrow \mu(X)$.

Problem 1.117**
Show that the Banach space $BV([a,b])$ is not separable.

Problem 1.118*
Show that $BV([a,b]) \subseteq L^\infty(a,b)$ and

$$\|u\|_\infty \leqslant \tfrac{1}{b-a}\|u\|_1 + \|u'\|_{M_b(a,b)}$$

Problem 1.119**
Suppose that $\{u_n\}_{n\geqslant 1} \subseteq BV([a,b])$ is a bounded sequence. Show that there exists a subsequence which converges for almost all $x \in (a,b)$.

Problem 1.120**
Suppose that $1 \leqslant p < +\infty$. Show that the embedding $BV([a,b]) \hookrightarrow L^p(a,b)$ is compact.

Problem 1.121**
Show that for every $u \in BV([a,b])$, we have

$$\int_a^b \big|u(x) - \overline{u}\big|\, dx \;\leqslant\; \tfrac{b-a}{2}|u'|((a,b)),$$

where $\overline{u} = \frac{1}{b-a}\int_a^b u\, dx$ is the ***average*** of u on (a,b). Also, show that the above inequality can be strict.

Problem 1.122**
Let X be a Polish space and let $\{\mu_n\}_{n\geqslant 1} \subseteq M_1^+(X)$ be a sequence. Show that the sequence $\{\mu_n\}_{n\geqslant 1}$ is uniformly tight if and only if there exists a Borel function $\varphi\colon X \longrightarrow [0,+\infty]$ such that the set $\{\varphi \leqslant \eta\}$ is compact for every $\eta \in [0,+\infty)$, $\mu_n(\{\varphi = +\infty\}) = 0$ for all $n \geqslant 1$ and

$$\sup_{n\geqslant 1} \int_X \varphi(x)\, d\mu_n \;<\; +\infty.$$

Problem 1.123**
Suppose that X be a metrizable space, $C \subseteq X$ is a closed set, $U \subseteq X$ is an open set and $\eta \in \mathbb{R}$. Show that:
(a) the set $\big\{\mu \in M_1^+(X):\ \mu(C) \geqslant \eta\big\}$ is w_n-closed;
(b) the set $\big\{\mu \in M_1^+(X):\ \mu(U) > \eta\big\}$ is w_n-open;
where w_n denotes the narrow topology on $M_1^+(X)$.

Problem 1.124*
Let X be a metrizable space and consider the map $\xi\colon \big(C_b(X), \|\cdot\|_\infty\big) \times \big(M_1^+(X), w_n\big) \longrightarrow \mathbb{R}$ defined by

$$\xi(u,\mu) \;=\; \int_X u\, d\mu.$$

Show that ξ is continuous (recall that w_n denotes the narrow topology on $M_1^+(X)$).

Problem 1.125**
Let X be a metrizable space and let $D \subseteq X$ be a Borel subset. Let w_n denote the narrow topology on $M_1^+(X)$ and let $w_n(D)$ be the corresponding narrow topology on $M_1^+(D)$. Show that $w_n\big|_{M_1^+(D)} = w_n(D)$.

Problem 1.126*
Let X and Y be two metrizable spaces and let $h \in C(X;Y)$. We introduce the function $\widehat{h}\colon M_1^+(X) \longrightarrow M_1^+(Y)$ defined by

$$\widehat{h}(\mu)(A) \;=\; \mu(h^{-1}(A)) \qquad \forall\, A \in \mathcal{B}(Y)$$

(the ***push-forward*** of $M_1^+(X)$). Show that $\widehat{h} \in C\big(M_1^+(X); M_1^+(Y)\big)$ with both spaces $M_1^+(X)$ and $M_1^+(Y)$ furnished with their w_n-topologies (see Problem 1.123).

Problem 1.127**
Suppose that X is a Polish space, Y is a metrizable space, $h \in C(X;Y)$ and $\widehat{h}\colon M_1^+(X) \longrightarrow M_1^+(Y)$ is defined as in Problem 1.126. Show that if h is injective, then $\widehat{h}$ is injective too.

Problem 1.128**
Suppose that X and Y are two Polish spaces, $h \in C(X;Y)$ and $\widehat{h}\colon M_1^+(X) \longrightarrow M_1^+(Y)$ is as defined in Problem 1.126. Suppose that h is injective. Show that $\operatorname{im}\widehat{h} = M_1^+(h(X))$.

Problem 1.129*
Let X and Y be two Polish spaces and let $h\colon X \longrightarrow Y$ be a homeomorphism. Show that $\widehat{h}\colon M_1^+(X) \longrightarrow M_1^+(Y)$ (see Problem 1.126) is a homeomorphism too.

Problem 1.130***
Let μ be a Radon measure defined on $\mathcal{B}(\mathbb{R}^N)$ (with $\mathcal{B}(\mathbb{R}^N)$ being the Borel σ-field of $\mathbb{R}^N$). Let $L\colon C_0\left(\mathbb{R}^N\right) \longrightarrow \mathbb{R}$ be defined by

$$L(u) \;=\; \int\limits_{\mathbb{R}^N} u(x)\,d\mu(x) \qquad \forall\, u \in C_0\left(\mathbb{R}^N\right).$$

For every open set $U \subseteq \mathbb{R}^N$ and every compact set $K \subseteq \mathbb{R}^N$, we define the sets

$$\begin{aligned} \Gamma_U &= \left\{u \in C_0\left(\mathbb{R}^N\right) :\ 0 \leqslant u \leqslant 1,\ \operatorname{supp} u \subseteq U\right\}, \\ \Gamma_K &= \left\{u \in C_0\left(\mathbb{R}^N\right) :\ u \geqslant \chi_K\right\} \end{aligned}$$

(here χ_K is the characteristic function of K). Show that

$$\mu(U) \;=\; \sup_{u \in \Gamma_U} L(u) \quad \text{and} \quad \mu(K) \;=\; \inf_{u \in \Gamma_K} L(u).$$

Problem 1.131**
Suppose that X is a locally compact and σ-compact topological space, $\eta > 0$ and

$$C = \{\mu \in M_b(X) : \|\mu\|_{M_b(X)} = \eta\}.$$

Is the set C vaguely closed? Justify your answer.

Problem 1.132*
Let $I = (a,b)$ be a bounded open interval and let $u \in W^{1,1}(a,b)$. Show that

$$\|u\|_\infty \leqslant \frac{1}{\lambda(I)} \int_a^b |u|\,dx + \|u'\|_1,$$

where λ denotes the Lebesgue measure on $\mathbb{R}$.

Problem 1.133*
Show that if $1 < p < +\infty$ and $u \in W^{1,p}(a,b)$, then $u \in C^{0,\frac{1}{p'}}([a,b])$.

Problem 1.134**
Let I be an unbounded open interval in $\mathbb{R}$ and let $u \in W^{1,p}(I)$, with $1 \leqslant p < +\infty$. Show that

$$\lim_{\substack{|x| \to +\infty \\ x \in I}} u(x) = 0.$$

Problem 1.135**
Show that if $1 \leqslant p \leqslant +\infty$, I is an open interval in $\mathbb{R}$ and $u, v \in W^{1,p}(I)$, then $uv \in W^{1,p}(I)$ and we have

$$(uv)' = u'v + uv'.$$

Moreover, show that the following integration by parts formula hold

$$\int_y^x u'v\,dt + \int_y^x uv'\,dt = u(x)v(x) - u(y)v(y) \qquad \forall\, x, y \in \overline{I}$$

Problem 1.136*
Let $I = (a,b)$ be an open bounded interval, $1 < p < +\infty$ and let $\{u_n\}_{n\geqslant 1} \subseteq W_0^{1,p}(a,b)$ be a sequence such that $\{u_n'\}_{n\geqslant 1} \subseteq L^p(a,b)$ is bounded and

$$u_n(x) \longrightarrow u(x) \quad \text{for a.a. } x \in (a,b).$$

Show that $u_n \longrightarrow u$ uniformly on $\overline{I} = [a,b]$.

Problem 1.137**
Let $a, b \in \mathbb{R}$, $u \in W^{1,2}(a,b)$. Show that

$$\|u\|_2^2 \leqslant c\Big(\|u'\|_2^2 + \Big(\int_a^b u(x)\,dx\Big)^2\Big).$$

Problem 1.138**
Let $\Omega \subseteq \mathbb{R}^N$ be a bounded ***domain*** (i.e., bounded, connected and open set) with a Lipschitz boundary $\partial\Omega$ and let $V \subseteq W^{1,p}(\Omega)$ (with $1 < p < +\infty$) be a closed linear subspace such that the only constant function belonging to V is the zero function. Show that there exists $c > 0$ such that

$$\|u\|_p \leqslant c\|Du\|_p \qquad \forall\, u \in V$$

(this is a generalization of the Poincaré inequality; see Theorem 1.131).

Problem 1.139*
Let $\Omega \subseteq \mathbb{R}^N$ be a bounded domain with a Lipschitz boundary $\Gamma = \partial\Omega$ and let $\Gamma_0 \subseteq \Gamma$ be a subset of strictly positive surface (that is, Hausdorff) measure. We define

$$V = \{u \in W^{1,p}(\Omega) :\ \gamma(u)|_{\Gamma_0} = 0\}$$

(for $1 < p < +\infty$) with $\gamma\colon W^{1,p}(\Omega) \longrightarrow L^p(\Gamma)$ being the trace map. Show that there exists $c > 0$ such that

$$\|u\|_p \leqslant c\|Du\|_p \qquad \forall\, u \in V.$$

Problem 1.140**
Suppose that $\Omega \subseteq \mathbb{R}^N$ is a bounded open set, $1 < p < +\infty$, $\frac{1}{p} + \frac{1}{p'} = 1$, $u \in W^{1,p}(\Omega)$ and $y \in W^{1,p'}(\Omega)$. Show that $uy \in W^{1,1}(\Omega)$ and

$$D_k(uy) = u(D_k y) + y(D_k u) \qquad \forall\, k \in \{1, \ldots, N\}.$$

Problem 1.141**
Let $\Omega \subseteq \mathbb{R}^N$ be an open set and for all $1 \leqslant p < +\infty$ define

$$W^{1,p}_{\mathrm{loc}}(\Omega) = \{u \in L^1_{\mathrm{loc}}(\Omega) :\ u \in W^{1,p}(\Omega_0) \text{ for all open } \Omega_0 \subset\subset \Omega\}$$

(recall that $\Omega_0 \subset\subset \Omega$ means that $\overline{\Omega}_0 \subseteq \Omega$ and $\overline{\Omega}_0$ is compact). Let $u \in W^{1,p}_{\mathrm{loc}}(\Omega)$ and assume that $u \in L^p(\Omega)$, $Du \in L^p(\Omega; \mathbb{R}^N)$. Show that $u \in W^{1,p}(\Omega)$.

Problem 1.142**
Let $\Omega \subseteq \mathbb{R}^N$ be a domain (i.e., connected and open set) and let $u \in W^{1,p}(\Omega)$ (with $1 \leqslant p < +\infty$). Assume that $Du(z) = 0$ for almost all $z \in \Omega$. Show that u is constant.

Problem 1.143*
Let $\Omega \subseteq \mathbb{R}^N$ be a domain and let $u \in W^{1,p}_0(\Omega)$ (with $1 \leqslant p < +\infty$) with $Du \equiv 0$. Show that $u \equiv 0$.

Problem 1.144*
Find a continuous function on $I = (-1, 1)$ which does not belong to $W^{1,2}(I)$.

Problem 1.145***
Let $\Omega \subseteq \mathbb{R}^N$ (with $N \geqslant 2$) be an open set. Find a function $u \in H^1(\Omega)$ which does not have a continuous representative (so Theorem 1.100 does not extend to the multivariable case).

Problem 1.146**
Suppose that $\Omega \subseteq \mathbb{R}^N$ is a bounded open set, $1 < p < +\infty$, $K \subseteq W^{1,p}(\Omega)$ is a closed and convex set, $\{u_n\}_{n\geqslant 1} \subseteq K$ is a sequence, $u \in L^p(\Omega)$ and $y \in L^p(\Omega;\mathbb{R}^N)$. Suppose that

$$u_n \xrightarrow{w} u \quad \text{in } L^p(\Omega) \quad \text{and} \quad Du_n \xrightarrow{w} y \quad \text{in } L^p(\Omega;\mathbb{R}^N).$$

Show that $u \in K$ and $Du = y$.

Problem 1.147**
Let $\Omega, \Omega' \subseteq \mathbb{R}^N$ be two bounded open sets and let $\eta\colon \Omega' \longrightarrow \Omega$ be a C^1-diffeomorphism. We assume that $0 < c \leqslant |\det J_\eta(x)| \leqslant c_2$ for all $x \in \Omega'$, where $J_\eta(x)$ denotes the Jacobian of η at $x \in \Omega'$. Show that, if $u \in W^{1,p}(\Omega)$ (with $1 \leqslant p < +\infty$), then $\widehat{u} = u \circ \eta \in W^{1,p}(\Omega')$ and

$$D\widehat{u}(x) = J_\eta(x)^T Du(\eta(x)) \qquad \forall\, x \in \Omega'.$$

Problem 1.148**
Suppose that $\Omega \subseteq \mathbb{R}^N$ is an open set, $u\colon \Omega \longrightarrow \mathbb{R}$ is a locally Lipschitz function and $1 < p < +\infty$. Show that $u \in W^{1,p}_{\mathrm{loc}}(\Omega)$ (see Problem 1.141) and Du is the usual gradient of u.

Problem 1.149**
Suppose that $\Omega \subseteq \mathbb{R}^N$ is a bounded open set, $1 < p < +\infty$, $\{u_n\}_{n\geqslant 1} \subseteq W^{1,p}(\Omega)$ is a bounded sequence such that $u_n(z) \longrightarrow u(z)$ for almost all $z \in \Omega$. Show that $u \in W^{1,p}(\Omega)$ and $u_n \xrightarrow{w} u$ in $W^{1,p}(\Omega)$. Does the result remain true if $W^{1,p}(\Omega)$ is replaced by $W^{1,p}_0(\Omega)$?

Problem 1.150*
Let $\Omega \subseteq \mathbb{R}^N$ be an open set, $1 \leqslant p \leqslant +\infty$ and $u \in W^{1,p}(\Omega)$. Show that $u^+, u^-, |u| \in W^{1,p}(\Omega)$ and

$$Du^+ = \begin{cases} 0 & \text{for a.a. } z \in \{u \leqslant 0\}, \\ Du & \text{for a.a. } z \in \{0 < u\}, \end{cases}$$

$$Du^- = \begin{cases} -Du & \text{for a.a. } z \in \{u < 0\}, \\ 0 & \text{for a.a. } z \in \{0 \leqslant u\}, \end{cases}$$

$$D|u| = \begin{cases} -Du & \text{for a.a. } z \in \{u < 0\}, \\ 0 & \text{for a.a. } z \in \{u = 0\}, \\ Du & \text{for a.a. } z \in \{0 < u\}. \end{cases}$$

Also, show that the above conclusions remain valid if $W^{1,p}(\Omega)$ is replaced by $W^{1,p}_0(\Omega)$.

Problem 1.151*
Let $\Omega \subseteq \mathbb{R}^N$ be an open set, $1 \leqslant p \leqslant +\infty$ and $u, y \in W^{1,p}(\Omega)$ (respectively $W_0^{1,p}(\Omega)$). Show that $h_0 = \min\{u, y\} \in W^{1,p}(\Omega)$ (respectively $W_0^{1,p}(\Omega)$), $h_1 = \max\{u, y\} \in W^{1,p}(\Omega)$ (respectively $W_0^{1,p}(\Omega)$) and we have

$$Dh_0 = \begin{cases} Du & \text{for a.a. } z \in \{u \leqslant y\}, \\ Dy & \text{for a.a. } z \in \{y \leqslant u\}, \end{cases}$$
$$Dh_1 = \begin{cases} Dy & \text{for a.a. } z \in \{u \leqslant y\}, \\ Du & \text{for a.a. } z \in \{y \leqslant u\}, \end{cases}$$

Problem 1.152**
Suppose that $\Omega \subseteq \mathbb{R}^N$ is an open set, $1 < p < +\infty$ and $\{u_n\}_{n\geqslant 1} \subseteq W^{1,p}(\Omega)$ is a sequence such that

$$h = \sup_{n\geqslant 1} |Du_n| \in L^p(\Omega) \quad \text{and} \quad y = \sup_{n\geqslant 1} u_n \in L^p(\Omega).$$

Show that $y \in W^{1,p}(\Omega)$ and

$$|Dy(z)| \leqslant h(z) \quad \text{for a.a. } z \in \Omega.$$

Problem 1.153*
Show that in Problem 1.152, if $\Omega \subseteq \mathbb{R}$ is a bounded open set with a Lipschitz boundary and $\{u_n\}_{n\geqslant 1} \subseteq W_0^{1,p}(\Omega)$ is a sequence, then it suffices to assume that

$$h = \sup_{n\geqslant 1} |Du_n| \in L^p(\Omega),$$

to obtain that $y = \sup_{n\geqslant 1} u_n \in W^{1,p}(\Omega)$ and

$$|Dy(z)| \leqslant h(z) \quad \text{for a.a. } z \in \Omega.$$

Problem 1.154**
Suppose that $\Omega \subseteq \mathbb{R}^N$ is an open set, $1 \leqslant p \leqslant +\infty$ and let $\{u_n\}_{n\geqslant 1}, \{y_n\}_{n\geqslant 1} \subseteq W^{1,p}(\Omega)$ be two sequence and

$$u_n \longrightarrow u \quad \text{and} \quad y_n \longrightarrow y \quad \text{in } W^{1,p}(\Omega).$$

Show that

$$\min\{u_n, y_n\} \longrightarrow \min\{u, y\} \quad \text{and} \quad \max\{u_n, y_n\} \longrightarrow \max\{u, y\}$$
$$\text{in } W^{1,p}(\Omega).$$

Is the result valid if we replace $W^{1,p}(\Omega)$ by $W_0^{1,p}(\Omega)$?

Problem 1.155**
Suppose that $\Omega \subseteq \mathbb{R}^N$ is an open set, $1 \leqslant p < +\infty$, $u \in W^{1,p}(\Omega)$ and u vanishes outside a compact set $K \subseteq \Omega$. Show that $u \in W_0^{1,p}(\Omega)$.

Problem 1.156*
Suppose that $\Omega \subseteq \mathbb{R}^N$ is a bounded open set, $1 \leqslant p < +\infty$ and $u \in W^{1,p}(\Omega)$ is such that $\lim\limits_{z \to x} u(z) = 0$ for all $x \in \partial\Omega$. Show that $u \in W_0^{1,p}(\Omega)$.

Problem 1.157*
Suppose that $\Omega \subseteq \mathbb{R}^N$ is an open set, $1 \leqslant p \leqslant +\infty$ and $u \in W_0^{1,p}(\Omega)$, $u \geqslant 0$. Show that there exists a sequence of nonnegative functions $\{u_n\}_{n \geqslant 1} \subseteq C_c^\infty(\Omega)$ such that $u_n \longrightarrow u$ in $W^{1,p}(\Omega)$.

Problem 1.158**
Suppose that $\Omega \subseteq \mathbb{R}^N$ is an open set, $1 \leqslant p < +\infty$, $u \in W_0^{1,p}(\Omega)$, $y \in W^{1,p}(\Omega)$ and

$$0 \leqslant y(z) \leqslant u(z) \quad \text{for a.a. } z \in \Omega.$$

Show that $y \in W_0^{1,p}(\Omega)$.

Problem 1.159***
Let $u \in L^1_{\text{loc}}(a, b)$ and assume that $u' = 0$ (the distributional derivative). Show that there exists $\xi \in \mathbb{R}$ such that $u(x) = \xi$ for almost all $x \in (a, b)$.

Problem 1.160**
Let $\Omega \subseteq \mathbb{R}^N$ be a bounded open set, $1 \leqslant p \leqslant N$. Given $U \subseteq \Omega$ open, we define

$$\operatorname{cap}_p(U, \Omega) = \inf\left\{\|Du\|_p^p : u \in W_0^{1,p}(\Omega), \ u(z) \geqslant 1 \text{ for a.a. } z \in U\right\}$$

(the p-***capacity*** of U with respect to Ω). For any $A \subseteq \Omega$, we set

$$\operatorname{cap}_p(A, \Omega) = \inf\left\{\operatorname{cap}_p(U, \Omega) : U \text{ open}, \ A \subseteq U\right\}.$$

Show that for any $A \subseteq \Omega$, we have

$$\begin{aligned}\operatorname{cap}_p(A, \Omega) = \inf\{&\|Du\|_p^p : u \in W_0^{1,p}(\Omega) \text{ and} \\ &u(z) \geqslant 1 \text{ for a.a. } z \text{ in a neighborhood of } A\}.\end{aligned}$$

("$u(z) \geqslant 1$ for a.a. z in a neighborhood of A" means that we can find an open set U such that $A \subseteq U$ and $u(z) \geqslant 1$ for almost all $z \in U$).

Problem 1.161**
Suppose that $\Omega \subseteq \mathbb{R}^N$ is a bounded open set, $1 \leqslant p \leqslant N$ and $A \subseteq \Omega$. Let $\operatorname{cap}_p(A, \Omega)$ denote the p-capacity of A with respect to Ω (see Problem 1.160). Show that

$$\begin{aligned}\operatorname{cap}_p(A, \Omega) = \inf\{&\|Du\|_p^p : u \in W_0^{1,p}(\Omega), \ u \geqslant 0 \text{ and} \\ &u(z) = 1 \text{ for a.a. } z \text{ in a neighborhood of } A\}.\end{aligned}$$

Problem 1.162**
Suppose that $\Omega \subseteq \mathbb{R}^N$ is a bounded open set, $1 \leqslant p \leqslant N$ and let $K \subseteq \Omega$ is a compact set. Let $\operatorname{cap}_p(K, \Omega)$ denote the p-capacity of K with respect to Ω (see Problem 1.160). Show that

$$\operatorname{cap}_p(K, \Omega) = \inf\left\{\|Du\|_p^p : u \in C_c^\infty(\Omega),\ u(z) \geqslant 1 \text{ for all } z \in K\right\}.$$

Problem 1.163**
Suppose that $\Omega \subseteq \mathbb{R}^N$ is a bounded open set, $U \subseteq \Omega$ is an open set, $1 \leqslant p \leqslant N$ and $\operatorname{cap}_p(U, \Omega)$ denotes the p-capacity of U with respect to Ω (see Problem 1.160). Show that

$$\operatorname{cap}_p(U, \Omega) = \sup\left\{\operatorname{cap}_p(K, \Omega) : K \text{ is compact and } K \subseteq U\right\}.$$

Problem 1.164**
Let $\Omega \subseteq \mathbb{R}^N$ be a bounded open set with a Lipschitz boundary $\partial\Omega$, $1 \leqslant p < +\infty$ and let $c(p, \Omega) > 0$ be the smallest constant for which the Poincaré inequality holds (see Theorem 1.131). Set $c_* = \frac{1}{c(p,\Omega)}$. Also, let $h \in W^{1,p}(\Omega)$ and consider the affine subspace $W_h^{1,p}(\Omega)$ of $W^{1,p}(\Omega)$ defined by

$$W_h^{1,p}(\Omega) = \left\{u \in W^{1,p}(\Omega) : u - h \in W_0^{1,p}(\Omega)\right\}.$$

Suppose that $c \in (0, c_*)$. Show that there exist $\beta_1 > 0$ and $\beta_2 \geqslant 0$ such that

$$\|Du\|_p^p - c\|u\|_p^p \geqslant \beta_1\|u\|^p - \beta_2 \qquad \forall\, u \in W_h^{1,p}(\Omega).$$

Problem 1.165**
Suppose that $\Omega \subseteq \mathbb{R}^N$ is an open set, $1 < p < +\infty$ and $\varphi\colon L^p(\Omega) \longrightarrow \overline{\mathbb{R}} = \mathbb{R} \cup \{+\infty\}$ is a function such that:

(i) for every $\lambda \in \mathbb{R}$, we have $\left\{u \in L^p(\Omega) : \varphi(u) \leqslant \lambda\right\} \subseteq W^{1,p}(\Omega)$ and $\sup\left\{\|Du\|_p^p : u \in \{\varphi \leqslant \lambda\}\right\} < +\infty$;

(ii) $\varphi|_{W^{1,p}(\Omega)}$ is sequentially weakly lower semicontinuous.

Show that φ is lower semicontinuous in the strong topology of $L^p(\Omega)$.

Problem 1.166*
Suppose that $\Omega \subseteq \mathbb{R}^N$ is a bounded open set with a Lipschitz boundary $\partial\Omega$, $1 \leqslant p < +\infty$ and $\varphi\colon L^p(\Omega) \longrightarrow \overline{\mathbb{R}} = \mathbb{R} \cup \{+\infty\}$ is a lower semicontinuous functional such that for every $\lambda \in \mathbb{R}$, the sublevel set

$$L_\lambda = \left\{u \in L^p(\Omega) : \varphi(u) \leqslant \lambda\right\} \subseteq W^{1,p}(\Omega)$$

and it is bounded. Show that L_λ is compact in $L^p(\Omega)$ for every $\lambda \in \mathbb{R}$.

Problem 1.167**
Suppose that n is a unit vector in $\mathbb{R}^N$, $a > 0$ and $\Omega \subseteq \mathbb{R}^N$ is an open set such that

$$\Omega \subseteq \left\{z \in \mathbb{R}^N : |(z, n)_{\mathbb{R}^N}| \leqslant a\right\}$$

(so Ω is ***bounded in the direction of*** n). Show that

$$\|u\|_p^p \leqslant (2a)^p \left\|\frac{\partial u}{\partial n}\right\|_p^p \qquad \forall\, u \in W_0^{1,p}(\Omega)$$

(recall that $\frac{\partial u}{\partial n} = (Du, n)_{\mathbb{R}^N}$).

Problem 1.168**
Suppose that n be a unit vector in $\mathbb{R}^N$, $a > 0$ and $\Omega \subseteq \mathbb{R}^N$ is an open set such that

$$\Omega \subseteq \{z \in \mathbb{R}^N : |(z,n)_{\mathbb{R}^N}| \leqslant a\}.$$

Show that

$$\|u\|_{1,p} = \left(\|u\|_p^p + \|Du\|_p^p\right)^{\frac{1}{p}} \quad \text{and} \quad \|Du\|_p$$

are equivalent norms on $W_0^{1,p}(\Omega)$.

Remark. So, according to the above problem, the Poincaré inequality (see Theorem 1.131) is more generally true for open sets $\Omega \subseteq \mathbb{R}^N$ which are bounded in one direction.

Problem 1.169**
Show that for all $u \in W^{1,1}(0,1)$, we have

$$\|u - \overline{u}\|_\infty \leqslant \|u'\|_1,$$

with $\overline{u} = \int_0^1 u\,dx$ and in the above inequality the constant 1 is optimal, i.e.,

$$\sup_{\substack{u \in W^{1,1}(0,1) \\ \|u'\|_1 = 1}} \|u - \overline{u}\|_\infty = 1.$$

Also show that the above supremum is not achieved.

Problem 1.170*
Let $I = (a,b)$ be a bounded interval and let $1 \leqslant p, q \leqslant +\infty$. Show that there exists $c > 0$ such that

$$\|u\|_q \leqslant c\|u'\|_p \qquad \forall\, u \in W_0^{1,p}(a,b)$$

and the constant $c > 0$ is realized if $1 \leqslant q \leqslant +\infty$ and $1 < p \leqslant +\infty$.

Problem 1.171**
Let $\Omega \subseteq \mathbb{R}^N$ be an open set, $1 \leqslant p < +\infty$ and for all $k \in \mathbb{N}_0$ let $g_k \colon \mathbb{R} \longrightarrow \mathbb{R}$ be defined by

$$g_k(t) = \begin{cases} t & \text{if } |t| \leqslant k, \\ k\frac{t}{|t|} & \text{if } k < |t|. \end{cases}$$

Show that for all $u \in W^{1,p}(\Omega)$, we have

$$g_k \circ u \longrightarrow u \in W^{1,p}(\Omega).$$

Problem 1.172*
Let $\Omega \subseteq \mathbb{R}^N$ be a bounded open set and let $1 \leqslant p < +\infty$. Suppose that $h^* \in W^{-1,p'}(\Omega) \cap L^{p'}(\Omega)$. Show that

$$\langle h^*, u\rangle = \int_\Omega h^*(z)u(z)\,dz \qquad \forall\, u \in W_0^{1,p}(\Omega),$$

where $\langle\cdot,\cdot\rangle$ denotes the duality brackets for the pair $\big(W_0^{1,p}(\Omega), W^{-1,p'}(\Omega)\big)$ (see Theorem 1.140).

Problem 1.173**
Let $\Omega \subseteq \mathbb{R}^N$ be a bounded open set, $1 \leqslant p < +\infty$ and let $h^* \in W^{-1,p'}(\Omega)$ be such that $h^* \geqslant 0$ in the sense of distributions (i.e., $h^*(\varphi) \geqslant 0$ for all $\varphi \in C_c^\infty(\Omega)$, $\varphi \geqslant 0$). Show that

$$\big\langle h^*, u^+\big\rangle \geqslant 0 \qquad \forall\, u \in W_0^{1,p}(\Omega)$$

(see Problem 1.172)

Problem 1.174*
Show that the Meyers–Serrin theorem (see Theorem 1.119) is not true for the Sobolev space $W^{1,\infty}(\Omega)$.

Problem 1.175**
Show that the Poincaré inequality (see Theorem 1.131) is not true for unbounded domains.

Problem 1.176*
Suppose that $I = (a,b)$ is a bounded open interval, $1 \leqslant p \leqslant +\infty$ and $x_0 \in [a,b]$. We define

$$V = \big\{u \in W^{1,p}(a,b) : \ u(x_0) = 0\big\}.$$

Show that there exists $c > 0$ such that

$$\|u\|_p \leqslant c\|u'\|_p \qquad \forall\, u \in V.$$

Problem 1.177**
Let $u \in L^1_{\mathrm{loc}}\left(\mathbb{R}^N\right)$. Show that the following two statements are equivalent:
(a) u admits a Lipschitz continuous representative.
(b) The weak (distributional) gradient of u belongs to $L^\infty\left(\mathbb{R}^N;\mathbb{R}^N\right)$.

Problem 1.178*
Let $\Omega \in \mathbb{R}^N$ be an open set, $1 < p \leqslant +\infty$ and $u \in L^p(\Omega)$. Show that $u \in W^{1,p}(\Omega)$ if and only if

$$\Big|\int_\Omega u\tfrac{\partial\vartheta}{\partial z_k}\,dz\Big| \leqslant c\|\vartheta\|_{p'} \qquad \forall\, \vartheta \in C_c^\infty(\Omega),\ k \in \{1,\ldots,N\},$$

for some $c > 0$

Problem 1.179**
Let $1 \leqslant p < +\infty$, $R > 0$ and $\varphi \in C^\infty(0,\infty)$. We set

$$u(z) \; = \; \varphi(|z|) \qquad \forall \; z \in B_R,$$

where $B_R = \{z \in \mathbb{R}^N : \; |z| < R\}$. Suppose that

$$\lim_{r \searrow 0} r^{N-1}|\varphi(r)| = 0.$$

Show that $u \in W^{1,p}(B_R)$ if and only if $u \in L^p(B_R)$ and $\int\limits_0^R r^{N-1}|\varphi'(r)|^p \, dr < +\infty$.

Problem 1.180*
Let $\Omega = (0,1) \times (0,1) \subseteq \mathbb{R}^2$ and for $n \geqslant 1$ consider the functions

$$u_n(x,y) \; = \; \tfrac{1}{\sqrt{n}}(1-y)^n \sin(nx) \qquad \forall \; (x,y) \in \Omega.$$

Show that $u_n \xrightarrow{w} 0$ in $H^1(\Omega)$.

Problem 1.181*
Let (Ω, Σ, μ) be a measure space, $p, q \in [1,+\infty]$ and $\overline{B}_1^p = \{u \in L^p(\Omega) : \; \|u\|_p \leqslant 1\}$. Show that the set $\overline{B}_1^p \cap L^q(\Omega)$ is weakly closed in $L^q(\Omega)$.

Problem 1.182**
Suppose that (Ω, Σ, μ) is a finite measure space, $E \subseteq L^1(\Omega)$ is a nonempty, relatively weakly compact set, $h \in E$ and

$$C \; = \; \{u \in L^1(\Omega) : \; |u(z)| \leqslant |h(z)| \; \mu\text{-a.e. in } \Omega\}.$$

Show that the set $C \subseteq L^1(\Omega)$ is relatively weakly compact.

Problem 1.183**
Suppose that (Ω, Σ, μ) is a finite measure space, $1 < p < +\infty$, $u \in L^p(\Omega)$ and $\{u_n\}_{n \geqslant 1} \subseteq L^p(\Omega)$ is a sequence such that $u_n(z) \longrightarrow u(z)$ for μ-almost all $z \in \Omega$ and $\|u_n\|_p \longrightarrow \|u\|_p$. Show that $u_n \longrightarrow u$ in $L^p(\Omega)$.

Problem 1.184*
Assume that (Ω, Σ, μ) is a measure space, $1 < p < +\infty$, $\{u_n\}_{n \geqslant 1} \subseteq L^p(\Omega)$ is a sequence such that $u_n \xrightarrow{w} u$ in $L^p(\Omega)$ and

$$\limsup_{n \to +\infty} \|u_n\|_p \; \leqslant \; \|u\|_p.$$

Show that $u_n \longrightarrow u$ in $L^p(\Omega)$.

Problem 1.185**
Suppose that (Ω, Σ, μ) is a measure space, $u \in L^1(\Omega)$ and $\{u_n\}_{n\geqslant 1} \subseteq L^1(\Omega)$ is a sequence such that $u_n(z) \longrightarrow u(z)$ for μ-almost all $z \in \Omega$ and $\|u_n\|_1 \longrightarrow \|u\|_1$. Show that $u_n \longrightarrow u$ in $L^1(\Omega)$.

Problem 1.186*
Let $\{u_n\}_{n\geqslant 1} \supseteq L^1(0,1)$ be a sequence such that $u_n(t) \longrightarrow u(w)$ for almost all $t \in [0,1]$ with $u \in L^1(0,1)$. Is it true that $u_n \longrightarrow u$ in $L^1(0,1)$? Justify your answer.

Problem 1.187*
Let $u \in L^1(\mathbb{R})$ be such that $\int\limits_a^b u(t)\,dt = 0$ for all $a, b \in \mathbb{R}$. Show that $u(t) = 0$ for almost all $t \in \mathbb{R}$.

Problem 1.188*
Let $\{u_n\}_{n\geqslant 1} \subseteq W^{1,1}(0,1)$ be a sequence such that $u_n(t) \longrightarrow u(t)$ for almost all $t \in (0,1)$ and there exists $h \in L^1(0,1)$ for which we have $|u_n'(t)| \leqslant h(t)$ for almost all $t \in (0,1)$ and all $n \geqslant 1$. Show that $u_n(t) \longrightarrow u(t)$ uniformly on $[0,1]$.

Problem 1.189*
Let

$$\widehat{\lambda}_1 = \inf\left\{\frac{\|u'\|_p^p}{\|u\|_p^p} : u \in W_0^{1,p}(0,1),\ u \neq 0\right\}$$

Show that $\widehat{\lambda}_1 > 0$ and that there exists $\widehat{u}_1 \in W_0^{1,p}(0,1)$ such that $\widehat{\lambda}_1 = \frac{\|\widehat{u}'\|_p^p}{\|\widehat{u}\|_p^p}$. Moreover, we may assume that $\widehat{u}_1(t) \geqslant 0$ for all $t \in [0,1]$.

Problem 1.190*
Let $u\colon \mathbb{R}_+ \longrightarrow \mathbb{R}_+$ be a measurable function such that $u \in L^p(\mathbb{R}_+)$ (with $1 \leqslant p \leqslant +\infty$). Show that $\lim\limits_{t\to+\infty} \frac{1}{t}\int\limits_0^t u(s)\,ds = 0$ if and only if for every $\varepsilon > 0$, the set $C_\varepsilon = \{t \in \mathbb{R}_+ : u(t) \geqslant \varepsilon\}$ satisfies $\lim\limits_{\tau\to+\infty} \frac{\lambda(C_\varepsilon \cap [0,\tau])}{\tau} = 0$ (here λ denotes the Lebesgue measure on $\mathbb{R}$).

Problem 1.191**
Suppose that $I \subseteq \mathbb{R}$ is an open interval, $u \in L^2(I)$, $t_0 \in \overline{I}$ and $h(t) = \int\limits_{t_0}^t u(s)\,ds$ for all $t \in \overline{I}$. Show that $h \in C(\overline{I})$ and its weak derivative exists and equals u.

Problem 1.192**
Let $u \in H^1(\mathbb{R})$. Is it true that $u(t) \longrightarrow 0$ as $t \to \pm\infty$? Justify your answer.

Problem 1.193**
Let $N \geqslant 2$ and $u \in H^1(\mathbb{R}^N)$. Is it true that $u(z) \longrightarrow 0$ as $|z| \longrightarrow +\infty$? Justify your answer.

Problem 1.194*
Let $\Omega \subseteq \mathbb{R}^N$ be a bounded domain with a Lipschitz boundary $\partial\Omega$. Show that there exists a constant $c > 0$ depending only on Ω such that

$$\|u\|_p \leqslant c\big(\|Du\|_p + \big|\int_\Omega u\,dz\big|\big) \qquad \forall\, u \in W^{1,p}(\Omega)$$

(where $1 < p < +\infty$).

Problem 1.195*
Let $u \in W^{1,2}(a,b)$. Show that $\|u\|_\infty \leqslant c\|u\|$, with $c = \max\big\{\frac{1}{\sqrt{b-a}}, \sqrt{b-a}\big\}$.

Problem 1.196*
Let $\xi\colon W_0^{1,2}(0,1)$ be the linear functional defined by

$$\xi(u) = \int_0^1 u(t)\,dt \qquad \forall\, u \in W_0^{1,2}(0,1).$$

Show that $\xi \in W^{-1,2}(0,1) = W_0^{1,2}(0,1)^*$ and determine the element of $W_0^{1,2}(0,1)$ which corresponds to ξ by the Riesz–Fréchet representation theorem.

Problem 1.197*
Let $1 \leqslant p \leqslant +\infty$ and let $\overline{B}_1(p) = \big\{u \in W^{1,p}(0,1) : \|u\| \leqslant 1\big\}$. Show that for $p \in (1,+\infty]$, the set $\overline{B}_1(p)$ is compact in $L^p(0,1)$, while $\overline{B}_1(1)$ is not closed in $L^1(0,1)$.

Problem 1.198**
Let $u \in W^{1,p}(0,1)$ with $1 \leqslant p < +\infty$. Show that

$$u'(t) = 0 \quad \text{for a.a. } t \in Z,$$

where $Z = \{t \in (0,1) : u(t) = 0\}$.

Problem 1.199**
Suppose that (Ω, Σ, μ) is a σ-finite measure space, $1 < p < +\infty$, $\{u_n\}_{n\geqslant 1} \subseteq L^p(\Omega)$ is a bounded sequence such that $u_n \longrightarrow u$ in $L^1(\Omega)$. Show that $u_n \xrightarrow{w} u$ in $L^p(\Omega)$.

Problem 1.200*
Suppose that (Ω, Σ, μ) is a σ-finite measure space, $1 \leqslant p < +\infty$ and $1 \leqslant q \leqslant +\infty$. Show that the set

$$C = \{u \in L^p(\Omega) \cap L^q(\Omega) : \|u\|_q \leqslant 1\}$$

is closed in $L^p(\Omega)$.

Problem 1.201*
Let $u \in L^p(\mathbb{R})$ and let $u_n(t) = u(t+n)$ for all $t \in \mathbb{R}$, all $n \geqslant 1$. Show that $u_n \xrightarrow{w} 0$ in $L^p(\mathbb{R})$.

Problem 1.202**
Suppose that (Ω, Σ, μ) is a σ-finite measure space, $h\colon \Omega \longrightarrow \mathbb{R}$ is a Σ-measurable function and consider the set

$$C = \big\{u \in L^\infty(\Omega) : u(z) \geqslant h(z) \text{ for } \mu\text{-a.a. } z \in \Omega\big\}.$$

Show that the set C is w^*-closed in $L^\infty(\Omega)$.

Problem 1.203**
Suppose that (Ω, Σ, μ) is a σ-finite measure space, $h_1, h_2 \in L^\infty(\Omega)$ and consider the set

$$C_0 = \big\{u \in L^\infty(\Omega) : h_1(z) \leqslant u(z) \leqslant h_2(z) \text{ for } \mu\text{-a.a. } z \in \Omega\big\}.$$

Show that the set C_0 is w^*-compact in $L^\infty(\Omega)$.

Problem 1.204**
Let $u \in L^\infty(\mathbb{R}^N)$ and let $\{\vartheta_n\}_{n\geqslant 1} \subseteq C_c^\infty(\mathbb{R}^N)$ be a sequence of mollifiers (see Definition 1.117). Consider a sequence $\{\xi_n\}_{n\geqslant 1} \subseteq L^\infty(\mathbb{R}^N)$ such that

$$\|\xi_n\|_\infty \leqslant 1 \qquad \forall\, n \geqslant 1 \quad \text{and} \quad \xi_n(z) \longrightarrow \xi(z) \quad \text{for a.a. } z \in \mathbb{R}^N.$$

Set $h_n = \vartheta_n \star (\xi_n u)$ for $n \geqslant 1$ and $h = \xi u$. Show that:
(a) $h_n \xrightarrow{w^*} h$ in $L^\infty(\mathbb{R}^N)$;
(b) for every ball $B \subseteq \mathbb{R}^N$, we have $\|h_n - h\|_{L^1(B)} \longrightarrow 0$.

Problem 1.205***
Let $\Omega \subseteq \mathbb{R}^N$ be a nonempty open set. Show that $C_c^\infty(\Omega)$ is w^*-dense in $L^\infty(\Omega)$.

Problem 1.206**
Let $\Omega \subseteq \mathbb{R}^N$ be a nonempty open set and let $u \in L^1(\Omega)$. Show that

$$\|u\|_1 = \sup\Big\{\int_\Omega u\vartheta\, dz : \vartheta \in C_c^\infty(\Omega),\ \|\vartheta\|_\infty \leqslant 1\Big\}.$$

Problem 1.207**
Suppose that (Ω, Σ, μ) is a measure space, $u \in L^1(\Omega)$ and $\{u_n\}_{n\geqslant 1} \subseteq L^1(\Omega)$ is a sequence such that $u_n \geqslant 0$ for $n \geqslant 1$,

$$\liminf_{n\to+\infty} u_n(z) \geqslant u(z) \quad \text{for a.a. } z \in \Omega$$

and

$$\limsup_{n\to+\infty} \int_\Omega u_n(z)\, d\mu \leqslant \int_\Omega u\, d\mu < +\infty.$$

Show that $u_n \longrightarrow u$ in $L^1(\Omega)$.

Problem 1.208**
Let (Ω, Σ, μ) be a probability measure space and let $u \in L^1(\Omega)$ be such that $u \geqslant 0$ and $\|u\|_1 = 1$. Show that $\int_\Omega u \ln u \, d\mu \geqslant 0$.

Problem 1.209*
Suppose that (Ω, Σ, μ) is a σ-finite measure space, X, Y are two Banach spaces and $u \colon \Omega \times X \longrightarrow Y$ is a Carathéodory map which satisfies the following condition:

UC: "For every $\varepsilon > 0$ and every $c > 0$, there are $\delta = \delta(\varepsilon, c) > 0$ and a μ-null set $D = D(\varepsilon, c) \in \Sigma$ such that

$$\|u(z, x) - u(z, v)\|_Y \leqslant \varepsilon$$

for all $z \in \Omega \setminus D$ and all $x, v \in X$ with $\|x\|_X, \|v\|_X \leqslant C$ and $\|x - v\|_X \leqslant \delta$."

Let N_u be the Nemytskii map corresponding to u (see Definition 1.145). Suppose that N_u maps $L^\infty(\Omega; X)$ into $L^\infty(\Omega; Y)$. Show that $N_u \colon L^\infty(\Omega; X) \longrightarrow L^\infty(\Omega; Y)$ is continuous.

Problem 1.210**
Suppose that (Ω, Σ, μ) is a σ-finite measure space, X, Y are two Banach spaces with X being separable, $u \colon \Omega \times X \longrightarrow Y$ is a Carathéodory map, N_u is the Nemytskii map corresponding to u and it is continuous from $L^\infty(\Omega; X)$ into $L^\infty(\Omega; Y)$. Show that condition UC from Problem 1.209 is satisfied.

1.3 Solutions

Solution of Problem 1.1

Without any loss of generality, we may assume that $u \equiv 0$ and $\mu(\Omega) = 1$. From the Dunford–Pettis theorem (see Theorem 1.28), we know that the sequence $\{u_n\}_{n\geqslant 1}$ is uniformly integrable. So, given $\varepsilon > 0$, we can find $\delta > 0$ such that

$$\text{if } \mu(A) \leqslant \delta \text{ then } \int_A |u_n|\, d\mu \leqslant \varepsilon \qquad \forall\, n \geqslant 1 \tag{1.1}$$

For every integer $k \geqslant 1$, let

$$A_k = \big\{\omega \in \Omega : \inf_{n\geqslant k} u_n(\omega) \geqslant -\varepsilon\big\}.$$

Since by hypothesis we have

$$0 \leqslant \liminf_{n\to+\infty} u_n(\omega) \quad \text{for } \mu\text{-a.a. } \omega \in \Omega,$$

we can find $k_0 \geqslant 1$ big such that $\mu(A_{k_0}^c) \leqslant \delta$. By hypothesis $u_n \xrightarrow{w} u = 0$ in $L^1(\Omega)$. So, we can find $k \geqslant k_0$ such that

$$\Big|\int_{A_{k_0}} u_n\, d\mu\Big| \leqslant \varepsilon \qquad \forall\, n \geqslant k.$$

Then for $n \geqslant k$ we have

$$\begin{aligned} \int_\Omega |u_n|\, d\mu &= \int_{A_{k_0}} |u_n|\, d\mu + \int_{A_{k_0}^c} |u_n|\, d\mu \\ &\leqslant \int_{A_{k_0}} |u_n+\varepsilon|\, d\mu + \int_{A_{k_0}} \varepsilon\, d\mu + \int_{A_{k_0}^c} |u_n|\, d\mu. \end{aligned} \tag{1.2}$$

Note that

$$\int_{A_{k_0}^c} |u_n|\, d\mu \leqslant \varepsilon \quad \text{and} \quad \int_{A_{k_0}} \varepsilon\, d\mu \leqslant \varepsilon \tag{1.3}$$

(from (1.1) and recall that we have assumed that $\mu(\Omega) = 1$). In addition, recalling the definition of the set A_{k_0}, we have

$$\int_{A_{k_0}} |u_n+\varepsilon|\, d\mu = \int_{A_{k_0}} (u_n+\varepsilon)\, d\mu \leqslant \Big|\int_{A_{k_0}} u_n\, d\mu\Big| + \varepsilon \leqslant 2\varepsilon \tag{1.4}$$

(since $n \geqslant k$). Using (1.3) and (1.4) in (1.2), we obtain

$$\int_\Omega |u_n|\, d\mu \leqslant 4\varepsilon \qquad \forall\, n \geqslant k,$$

so $u_n \longrightarrow u = 0$ in $L^1(\Omega)$.

Solution of Problem 1.2

Let $\Omega = [0,1]$, Σ the Lebesgue σ-field on $[0,1]$ and $\mu = \lambda$ the Lebesgue measure on $[0,1]$. We define

$$u_n = \tfrac{2^n}{n}\chi_{[\frac{1}{2^n},\frac{1}{2^{n-1}}]} \qquad \forall\, n \geqslant 1.$$

We have

$$\int_0^1 |u_n|\, d\lambda = \tfrac{2^n}{n}\tfrac{1}{2^n} = \tfrac{1}{n} \longrightarrow 0,$$

so

$$u_n \longrightarrow 0 \quad \text{in } L^1(\Omega).$$

Hence the sequence $\{u_n\}_{n\geqslant 1} \subseteq L^1(\Omega)$ is uniformly integrable (see Theorem 1.28).

If we could find $h \in L^1(\Omega)$ such that

$$u_n(\omega) \leqslant h(\omega) \quad \text{for } \lambda\text{-a.a. } \omega \in [0,1], \text{ all } n \geqslant 1,$$

then

$$\sum_{n\geqslant 1} u_n(\omega) \leqslant h(\omega) \quad \text{for } \mu\text{-a.a. } \omega \in [0,1]. \tag{1.5}$$

On the other hand, we have

$$\int_0^1 \Big(\sum_{n\geqslant 1} u_n\Big)\, d\lambda = \sum_{n\geqslant 1} \tfrac{1}{n} = +\infty. \tag{1.6}$$

Comparing (1.5) and (1.6) we reach a contradiction.

Solution of Problem 1.3

(a) Let $A_c = \{\omega \in \Omega :\ u(\omega) \geqslant c\}$ for $c > 0$. Then

$$\|u\|_p^p = \int_\Omega |u|^p\, d\mu \geqslant \int_{A_c} u^p\, d\mu \geqslant \int_{A_c} c^p\, d\mu = c^p\mu(A_c).$$

(b) Let $A_c = \{\omega \in \Omega :\ u(\omega) \geqslant c\}$ for $c > 0$. Then $\{A_c\}_{c>0}$ is a family of sets decreasing in c (i.e., if $c_1 < c_2$, then $A_{c_1} \supseteq A_{c_2}$) and $\mu(A_c) \searrow 0$ as $c \nearrow +\infty$ (as $u \in L^p(\Omega)$). Thus

$$\lim_{c\to+\infty} c^p\mu(A_c) = \lim_{c\to+\infty} \int_{A_c} c^p\, d\mu \leqslant \lim_{c\to+\infty} \int_{A_c} |u|^p\, d\mu = 0.$$

Solution of Problem 1.4
From the assumptions, it is obvious that $u_n \in L^p(\Omega)$ and so $|u_n|^p \in L^1(\Omega)$ for all $n \geqslant 1$. We will show that the sequence $\{|u_n|^p\}_{n\geqslant 1} \subseteq L^1(\Omega)$ is uniformly integrable. Let

$$m(A) = \int_A h^p \, d\mu \qquad \forall \, A \in \Sigma.$$

Then m is a finite measure on Σ and $m \ll \mu$. So, given $\varepsilon > 0$, we can find $\delta > 0$ such that

$$\text{if } A \in \Sigma \text{ with } \mu(A) \leqslant \delta, \text{ then } \int_A h^p \, d\mu \leqslant \varepsilon. \tag{1.7}$$

Also, we have

$$\lim_{c \to +\infty} \int_{\{\frac{1}{c}<h<c\}} h^p \, d\mu = \int_{\{0<h<+\infty\}} h^p \, d\mu = \int_\Omega h^p \, d\mu \tag{1.8}$$

(recall that since $h \in L^p(\Omega)$, we have $\mu(\{h = +\infty\}) = 0$). Let $\varepsilon > 0$ and let $\delta > 0$ be as postulated by (1.7). Then we have

$$\text{if } A \in \Sigma \text{ with } \mu(A) \leqslant \delta, \text{ then } \int_A |u_n|^p \, d\mu \leqslant \varepsilon \qquad \forall \, n \geqslant 1. \tag{1.9}$$

From the Chebyshev–Markov inequality (see Theorem 1.30), for every $c > 0$, we have

$$\mu\big(\{|u_n|^p \geqslant c\}\big) \leqslant \tfrac{1}{c} \int_\Omega |u_n|^p \, d\mu \leqslant \tfrac{1}{c} \int_\Omega h^p \, d\mu.$$

It follows that we can find $M > 0$ such that

$$\mu\big(\{|u_n|^p \geqslant c\}\big) \leqslant \delta \qquad \forall \, c \geqslant M, \ n \geqslant 1,$$

so

$$\int_{\{|u_n|^p \geqslant c\}} |u_n|^p \, d\mu \leqslant \varepsilon \qquad \forall \, c \geqslant M, \ n \geqslant 1$$

(see (1.9)). So, we have satisfied requirement **(b)** in Definition 1.18. Next, from (1.8) and for $c > 0$ large, we have

$$\int_\Omega h^p \, d\mu - \int_{\{\frac{1}{c}<h<c\}} h^p \, d\mu = \int_{\Omega \setminus \{\frac{1}{c}<h<c\}} h^p \, d\mu \leqslant \varepsilon. \tag{1.10}$$

Let

$$A_\varepsilon = \big\{\omega \in \Omega : \tfrac{1}{c} < h(\omega) < c\big\}.$$

Then $A_\varepsilon \in \Sigma$ and by the Chebyshev–Markov inequality, we see that $\mu(A_\varepsilon) < +\infty$. We have

$$\int_{A_\varepsilon^c} |u_n|^p \, d\mu \leqslant \int_{A_\varepsilon^c} h^p \, d\mu \leqslant \varepsilon \qquad \forall\, n \geqslant 1$$

(see (1.10)). So, we have satisfied requirement **(a)** in Definition 1.18. We conclude that the sequence $\{|u_n|^p\}_{n\geqslant 1} \subseteq L^1(\Omega)$ is uniformly integrable.

Solution of Problem 1.5

(a) Let $\varepsilon > 0$. Since the set C is uniformly integrable, we can find a set $A_\varepsilon \in \Sigma$ such that

$$\mu(A_\varepsilon) < +\infty \quad \text{and} \quad \sup_{u\in C} \int_{A_\varepsilon^c} |u| \, d\mu \leqslant \varepsilon \tag{1.11}$$

(see Definition 1.18(a)). Also, we can find $c > 0$ big such that

$$\sup_{u\in C} \int_{\{|u|\geqslant c\}} |u| \, d\mu \leqslant \varepsilon \tag{1.12}$$

(see Definition 1.18(b)). Then we have

$$\begin{aligned} \sup_{u\subset C} \int_\Omega |u| \, d\mu &= \sup_{u\subset C} \Big(\int_{A_\varepsilon\cap\{|u|\geqslant c\}} |u| \, d\mu + \int_{A_\varepsilon\cap\{|u|< c\}} |u| \, d\mu + \int_{A_\varepsilon^c} |u| \, d\mu \Big) \\ &\leqslant 2\varepsilon + c\mu(A_\varepsilon), \end{aligned}$$

so the set C is L^1-bounded.

(b) Let $\varepsilon > 0$. Let $c > 0$ be as postulated in (1.12) and let $\delta = \frac{\varepsilon}{c}$. Then for every $A \in \Sigma$ with $\mu(A) \leqslant \delta$ and $u \in C$, we have

$$\begin{aligned} \int_A |u| \, d\mu &= \int_{A\cap\{|u|\geqslant c\}} |u| \, d\mu + \int_{A\cap\{|u|< c\}} |u| \, d\mu \\ &\leqslant \int_{\{|u|\geqslant c\}} |u| \, d\mu + c\mu(A) \leqslant \varepsilon + \varepsilon = 2\varepsilon \end{aligned}$$

(see (1.12) and recall the choice of $A \in \Sigma$) and so statement **(b)** holds.

Solution of Problem 1.6

First note that in a finite measure space, part **(a)** of the definition of uniform integrability is satisfied for any set $C \subseteq L^1(\Omega)$ (take $A_\varepsilon = \Omega$ in Definition 1.18(a)). We need to show part **(b)** of this definition, namely that

$$\lim_{\vartheta \to +\infty} \sup_{u \in C} \int\limits_{\{|u| \geqslant \vartheta\}} |u| \, d\mu = 0.$$

Let

$$a_n = \sup_{u \in C} \int\limits_{\{|u| \geqslant n\}} |u| \, d\mu \qquad \forall\, n \geqslant 1.$$

First we note that the sequence $\{a_n\}_{n \geqslant 1}$ is well defined (as from hypothesis **(a)**, the set C is $L^1(\Omega)$-bounded), decreasing, nonnegative, so also convergent, let us say $\lim\limits_{n \to +\infty} a_n = c \geqslant 0$.

To proceed by contradiction, suppose that $c > 0$. So there exists $N \geqslant 1$ such that for all $n \geqslant N$ there exists $u_n \in C$ such that

$$\tfrac{c}{2} \leqslant \int\limits_{\{|u_n| \geqslant n\}} |u_n| \, d\mu \leqslant 2c. \tag{1.13}$$

Let $\varepsilon = \frac{c}{3} > 0$. From assumption **(b)**, we know that there exists $\delta > 0$ such that for all sets $A \in \Sigma$ with $\mu(A) \leqslant \delta$, we have

$$\sup_{u \in C} \int\limits_A |u| \, d\mu \leqslant \varepsilon. \tag{1.14}$$

Let $n_0 \geqslant N$ be such that $\frac{2c}{n_0} < \delta$ and let $A_{n_0} = \{|u_{n_0}| \geqslant n_0\}$. Then, from (1.13), we have

$$\mu(A_{n_0}) = \tfrac{1}{n_0} \int\limits_{A_{n_0}} n_0 \, d\mu \leqslant \tfrac{1}{n_0} \int\limits_{A_{n_0}} |u_{n_0}| \, d\mu \leqslant \frac{2c}{n_0} < \delta.$$

Thus, using (1.13) and (1.14), we have

$$\tfrac{c}{2} \leqslant \int\limits_{A_{n_0}} |u_{n_0}| \, d\mu \leqslant \sup_{u \in C} \int\limits_{A_{n_0}} |u| \, d\mu \leqslant \varepsilon = \tfrac{c}{3},$$

a contradiction. Thus $c = 0$ and the proof of uniform integrability of the set C is complete.

Next we will show that for an arbitrary measure space, the implication does not hold. Let $\Omega = \mathbb{R}$, Σ the Lebesgue σ-field of $\mathbb{R}$ and $\mu = \lambda$ the Lebesgue measure. Consider the sequence $\{\chi_{[n,n+1)}\}_{n \geqslant 1} \subseteq L^1(\Omega)$ of characteristic functions of the intervals $[n, n+1)$. Then

$$\int_{\mathbb{R}} \chi_{[n,n+1)}\, d\lambda = 1 \qquad \forall\, n \geqslant 1, \tag{1.15}$$

so the sequence $\{\chi_{[n,n+1)}\}_{n\geqslant 1}$ is $L^1(\Omega)$-bounded, thus condition **(a)** of the problem holds.

Next for any $\varepsilon > 0$, let $\delta = \varepsilon$. For any $A \in \Sigma$ with $\lambda(A) \leqslant \delta$, we have

$$\int_A |\chi_{[n,n+1)}|\, d\lambda \leqslant \lambda(A) \leqslant \varepsilon,$$

so also condition **(b)** of the problem holds.

We will show that the sequence $\{\chi_{[n,n+1)}\}_{n\geqslant 1} \subseteq L^1(\Omega)$ is not uniformly integrable. As part **(b)** of the Definition 1.18 is obviously satisfied, we will show that part **(a)** fails. Let $A_\varepsilon \subseteq \mathbb{R}$ be any set such that $\lambda(A_\varepsilon) < +\infty$. We will show that in fact

$$\sup_{n\geqslant 1} \int_{A_\varepsilon^c} |\chi_{[n,n+1)}|\, d\lambda = 1. \tag{1.16}$$

For this purpose, let $\eta \in (0,1)$ be arbitrary. Note that there exists $n_0 \geqslant 1$ such that

$$\lambda(A_\varepsilon^c \cap [n_0, n_0+1)) \geqslant 1 - \eta,$$

as otherwise we would have

$$\lambda(A_\varepsilon \cap [1,+\infty)) = \sum_{n\geqslant 1} \lambda(A_\varepsilon \cap [n,n+1)) \geqslant \sum_{n\geqslant 1} \eta = +\infty,$$

a contradiction). Then

$$\begin{aligned} \sup_{n\geqslant 1} \int_{A_\varepsilon^c} |\chi_{[n,n+1)}|\, d\lambda &\geqslant \int_{A_\varepsilon^c} |\chi_{[n_0,n_0+1)}|\, d\lambda \\ &= \lambda(A_\varepsilon^c \cap [n_0, n_0+1)) \geqslant 1 - \eta. \end{aligned}$$

Because $\eta \in (0,1)$ was arbitrary, we obtain

$$\sup_{n\geqslant 1} \int_{A_\varepsilon^c} |\chi_{[n,n+1)}|\, d\lambda \geqslant 1.$$

Because of (1.15) we obtain (1.16). So the sequence $\{\chi_{[n,n+1)}\}_{n\geqslant 1}$ is not uniformly integrable in $L^1(\mathbb{R})$.

Solution of Problem 1.7

From Problem 1.5 we know that the set C is L^1-bounded and given $\varepsilon > 0$, we can find $\delta > 0$ such that for every $A \in \Sigma$ with $\mu(A) \leqslant \delta$ we have

$$\sup_{u\in C} \int_A |u|\, d\mu \;\leqslant\; \tfrac{\varepsilon}{2}.$$

The uniform integrability of C implies that we can find a set $D_\varepsilon \in \Sigma$ with $\mu(D_\varepsilon) < +\infty$ such that

$$\sup_{u\in C} \int_{D_\varepsilon^c} |u|\, d\mu \;\leqslant\; \tfrac{\varepsilon}{2}$$

(see Definition 1.18). Let $h_\varepsilon = \chi_{D_\varepsilon}$. If $A \in \Sigma$ satisfies

$$\mu(A \cap D_\varepsilon) \;=\; \int_A h_\varepsilon\, d\mu \;\leqslant\; \delta,$$

then for all $u \in C$, we have

$$\int_A |u|\, d\mu \;=\; \int_{A\cap D_\varepsilon} |u|\, d\mu + \int_{A\cap D_\varepsilon^c} |u|\, d\mu \;\leqslant\; \tfrac{\varepsilon}{2} + \tfrac{\varepsilon}{2} \;=\; \varepsilon.$$

Solution of Problem 1.8

Let $u \in L^q(\Omega)$ and let

$$A \;=\; \big\{\omega \in \Omega : |u(\omega)| > 1\big\}.$$

Let us set

$$y \;=\; u\chi_A \quad \text{and} \quad h = u\chi_{A^c}.$$

We have

$$|y|^p \;=\; |u|^p\chi_A \;\leqslant\; |u|^q\chi_A$$

(since $p < q$), so $y \in L^p(\Omega)$. Similarly, we have

$$|h|^r \;=\; |u|^r\chi_{A^c} \;\leqslant\; |u|^q\chi_{A^c}$$

(since $q < r$), so $h \in L^r(\Omega)$. Note that $u = y + h$.

Solution of Problem 1.9

Let $p \in [1, +\infty)$. Then from the Chebyshev–Markov inequality (see Theorem 1.30), we have

$$\mu\big(\{|u| \geqslant c\}\big) \;\leqslant\; \tfrac{1}{c^p}\|u\|_p^p, \tag{1.17}$$

so

$$c\mu\big(\{|u| \geqslant c\}\big)^{\frac{1}{p}} \leqslant \|u\|_p$$

and thus

$$c \leqslant \liminf_{p\to+\infty} \|u\|_p, \text{ provided } \mu\big(\{|u| \geqslant c\}\big) > 0. \tag{1.18}$$

Therefore, if $\lim\limits_{p\to+\infty} \|u\|_p$ is finite, from (1.17) it is clear that $u \in L^\infty(\Omega)$ and from (1.18) we have

$$\|u\|_\infty \leqslant \lim_{p\to+\infty} \|u\|_p.$$

Then note that

$$\|u\|_p^p = \int_\Omega |u|^p \, d\mu \leqslant \|u\|_\infty^p \mu(\Omega),$$

so

$$\lim_{p\to+\infty} \|u\|_p \leqslant \|u\|_\infty$$

and thus $\|u\|_p \longrightarrow \|u\|_\infty$ as $p \to +\infty$.

From the above argument we see that if $\lim\limits_{p\to+\infty} \|u\|_p$ is not finite, then $u \notin L^\infty(\Omega)$ and so we have proved the equivalence claimed by the problem.

Solution of Problem 1.10

"$\Longrightarrow$": Suppose that $L^p(\Omega)$ is not contained in $L^q(\Omega)$. So, we can find $u \in L^p(\Omega)$ such that

$$\int_\Omega |u|^q \, d\mu = +\infty. \tag{1.19}$$

For each integer $n \geqslant 1$, we define

$$A_n = \big\{\omega \in \Omega : \, |u(\omega)| > n\big\} \in \Sigma.$$

From the Chebyshev–Markov inequality (see Theorem 1.30), we have

$$\mu(A_n) \leqslant \tfrac{1}{n^p} \int_\Omega |u|^p \, d\mu \longrightarrow 0.$$

So, it suffices to show that $\mu(A_n) > 0$ for all $n \geqslant 1$. Suppose that we can find $n \in \mathbb{N}$ such that $\mu(A_n) = 0$. Then

$$\int_\Omega |u|^q \, d\mu = \int_{\{|u|\leqslant n\}} |u|^q \, d\mu \leqslant n^{q-p} \int_{\{|u|\leqslant n\}} |u|^p \, d\mu < \infty,$$

which contradicts (1.19). Therefore Σ has sets of arbitrarily small positive μ-measure.
"$\Longleftarrow$": Suppose that Σ has sets of arbitrarily small positive μ-measure. So, we can find a sequence $\{A_n\}_{n\geqslant 1} \subseteq \Sigma$ of pairwise disjoint sets such that

$$\mu(A_n) > 0 \qquad \forall\, n \geqslant 1$$

and

$$\mu(A_n) \searrow 0.$$

Let $\xi_n \nearrow +\infty$ be such that

$$\sum_{n\geqslant 1} \xi_n^q \mu(A_n) = +\infty \quad \text{and} \quad \sum_{n\geqslant 1} \xi_n^p \mu(A_n) < +\infty.$$

Let

$$u = \sum_{n\geqslant 1} \xi_n \chi_{A_n}.$$

Then evidently $u \in L^p(\Omega) \setminus L^q(\Omega)$.

Solution of Problem 1.11

"$\Longrightarrow$": Suppose that $L^q(\Omega)$ is not contained in $L^p(\Omega)$. So, we can find $u \in L^q(\Omega)$ such that

$$\int_\Omega |u|^p \, d\mu = +\infty. \tag{1.20}$$

For every integer $n \geqslant 1$, we define

$$C_n = \left\{\omega \in \Omega : \tfrac{1}{n+1} < |u(\omega)| \leqslant \tfrac{1}{n}\right\} \in \Sigma$$

and

$$\widehat{C} = \bigcup_{n\geqslant 1} C_n = \left\{\omega \in \Omega : 0 < |u(\omega)| \leqslant 1\right\}.$$

Suppose that $\mu(\widehat{C}) < +\infty$. Then

$$\begin{aligned} \int_\Omega |u|^p \, d\mu &= \int_{\{|u|\leqslant 1\}} |u|^p \, d\mu + \int_{\{|u|>1\}} |u|^p \, d\mu \\ &\leqslant \mu(\widehat{C}) + \int_{\{|u|>1\}} |u|^q \, d\mu < +\infty \end{aligned}$$

(since $p < q$ and $u \in L^q(\Omega)$). This contradicts (1.20). Therefore $\mu(\widehat{C}) = +\infty$. Note that

$$+\infty > \int_\Omega |u|^q \, d\mu \geqslant \int_{\{\frac{1}{n+1} < |u| \leqslant \frac{1}{n}\}} |u|^q \, d\mu \geqslant \tfrac{1}{(n+1)^q} \mu(C_n) \qquad \forall\, n \geqslant 1. \tag{1.21}$$

So, if we set

$$E_n = \bigcup_{k=1}^{n} C_k \qquad \forall\, n \geqslant 1,$$

then

$$\mu(E_n) < +\infty \qquad \forall\, n \geqslant 1$$

(see (1.21)) and

$$\mu(E_n) \nearrow +\infty = \mu(\widehat{C}).$$

This implies that Σ has sets of arbitrarily big finite measure.
"$\Longleftarrow$": Suppose that Σ has sets of arbitrarily big finite measure. We can find a sequence $\{A_n\}_{n\geqslant 1} \subseteq \Sigma$ of pairwise disjoint sets such that $\mu(A_n) \nearrow +\infty$. Let $\xi_n \searrow 0$ be such that

$$\sum_{n\geqslant 1} \xi_n^q \mu(A_n) < +\infty \quad \text{and} \quad \sum_{n\geqslant 1} \xi_n^p \mu(A_n) = +\infty.$$

Let

$$u = \sum_{n\geqslant 1} \xi_n \chi_{A_n}.$$

Then evidently $u \in L^q(\Omega) \setminus L^p(\Omega)$.

Solution of Problem 1.12
Let $\{h_n\}_{n\geqslant 1} \subseteq C$ be a sequence such that $\|u - h_n\|_p \searrow \operatorname{dist}(u, C)$. Then clearly the sequence $\{h_n\}_{n\geqslant 1} \subseteq L^p(\Omega)$ is bounded and so, by virtue of the reflexivity of $L^p(\Omega)$ (see Corollary 1.26), by passing to a suitable subsequence if necessary, we may assume that

$$h_n \xrightarrow{w} \widehat{h} \quad \text{in } L^p(\Omega)$$

for some $\widehat{h} \in L^p(\Omega)$ (see Theorem I.5.76(c)). Since the set C is closed and convex, by the Mazur theorem (see Theorem I.5.58), it is also weakly closed and so $\widehat{h} \in C$. Also, the weak lower semicontinuity of the norm functional (see Proposition I.5.56(c)), implies that

$$\|u - \widehat{h}\|_p \leqslant \liminf_{n\to+\infty} \|u - h_n\|_p = \operatorname{dist}(u, C),$$

so

$$\|u - \widehat{h}\|_p = \operatorname{dist}(u, C).$$

Solution of Problem 1.13

Let $\Omega = [0,1]$, Σ the Lebesgue σ-field of $[0,1]$ and $\mu = \lambda$ the Lebesgue measure. Let

$$u_n = n\chi_{(0,\frac{1}{n})}.$$

Evidently $u_n \longrightarrow 0$ in λ-measure (see Definition 1.11) and λ-almost everywhere. However, if we choose $h \equiv 1 \in L^{p'}(0,1)$ for every $p \in [1, +\infty]$ (recall $\frac{1}{p} + \frac{1}{p'} = 1$), then

$$\int_0^1 u_n h \, d\lambda = 1 \qquad \forall\, n \geqslant 1,$$

so the sequence $\{u_n\}_{n \geqslant 1}$ does not converge weakly in $L^p(\Omega)$ for any $p \in [1, +\infty]$.

Solution of Problem 1.14

We know that

$$u - u_n = (u - u_n)^+ - (u_n - u)^- \qquad \forall\, n \geqslant 1.$$

Since by hypothesis $\int_\Omega u_n \, d\mu = \int_\Omega u \, d\mu$ for all $n \geqslant 1$, it follows that

$$\int_\Omega (u - u_n)^+ \, d\mu = \int_\Omega (u - u_n)^- \, d\mu \qquad \forall\, n \geqslant 1. \tag{1.22}$$

Recall that

$$|u - u_n| = (u - u_n)^+ + (u_n - u)^- \qquad \forall\, n \geqslant 1.$$

So, from (1.22) we infer that

$$\int_\Omega |u - u_n| \, d\mu = 2 \int_\Omega (u - u_n)^+ \, d\mu \qquad \forall\, n \geqslant 1.$$

The hypothesis that $u(\omega) \leqslant \liminf_{n \to +\infty} u_n(\omega)$ for μ-almost all $\omega \in \Omega$, implies that

$$(u - u_n)^+(\omega) \longrightarrow 0 \quad \text{for } \mu\text{-a.a. } \omega \in \Omega.$$

In addition, we have

$$\begin{aligned}(u-u_n)^+(\omega) &= \max\{(u-u_n)(\omega),\,0\}\\ &\leqslant \max\{u(\omega),\,0\}\\ &= u^+(\omega)\end{aligned}$$

(since $u_n \geqslant 0$ for all $n \geqslant 1$). Since $u^+ \in L^1(\Omega)$, we can apply the Lebesgue dominated convergence theorem (see Theorem I.3.94) to the sequence $\{|u-u_n|\}_{n\geqslant 1}$ and conclude that

$$u_n \longrightarrow u \quad \text{in } L^1(\Omega).$$

Solution of Problem 1.15

Let $\varepsilon > 0$. From the definition of uniform integrability (see Definition 1.18), we know that there exists $A_\varepsilon \in \Sigma$ such that

$$\mu(A_\varepsilon) < +\infty \quad \text{and} \quad \sup_{u\in C} \int_{A_\varepsilon^c} |u|\,d\mu \leqslant \tfrac{\varepsilon}{2}$$

and

$$\lim_{\vartheta\to+\infty} \sup_{u\in C} \int_{\{|u|\geqslant\vartheta\}} |u|\,d\mu = 0,$$

so in particular for some $\widehat{\vartheta} > 0$, we have

$$\sup_{u\in C} \int_{\{|u|\geqslant\widehat{\vartheta}\}} |u|\,d\mu \leqslant \tfrac{\varepsilon}{2}.$$

Let $h_\varepsilon = \widehat{\vartheta}\chi_{A_\varepsilon}$. Then, for every $u \in C$, we have

$$\begin{aligned}\int_{\{|u|\geqslant h_\varepsilon\}} |u|\,d\mu &= \int_{A_\varepsilon\cap\{|u|\geqslant h_\varepsilon\}} |u|\,d\mu + \int_{A_\varepsilon^c\cap\{|u|\geqslant h_\varepsilon\}} |u|\,d\mu\\ &\leqslant \int_{\{|u|\geqslant\widehat{\vartheta}\}} |u|\,d\mu + \int_{A_\varepsilon^c} |u|\,d\mu \leqslant \tfrac{\varepsilon}{2} + \tfrac{\varepsilon}{2} = \varepsilon.\end{aligned}$$

Solution of Problem 1.16

Let $\varepsilon > 0$. By assumption, we can choose $h_\varepsilon \in L^1(\Omega)_+$ such that

$$\sup_{u \in C} \int_{\{|u| \geqslant h_\varepsilon\}} |u| \, d\mu \leqslant \frac{\varepsilon}{2}. \tag{1.23}$$

By the Lebesgue dominated convergence theorem (see Theorem I.3.94), we have

$$\lim_{\vartheta \to +\infty} \int_\Omega h_\varepsilon \chi_{\{|h_\varepsilon| \geqslant \vartheta\}} \, d\mu \longrightarrow 0$$

and so there exists $\widehat{\vartheta} > 0$ such that

$$\int_\Omega h_\varepsilon \chi_{\{|h_\varepsilon| \geqslant \vartheta\}} \, d\mu \leqslant \frac{\varepsilon}{2} \qquad \forall \, \vartheta > \widehat{\vartheta}. \tag{1.24}$$

For any $u \in C$ and $\vartheta > 0$ we have

$$\{|u| < h_\varepsilon\} \cap \{|u| \geqslant \vartheta\} \subseteq \{h_\varepsilon > \vartheta\},$$

so, using also (1.23) and (1.24), we have

$$\begin{aligned} \int_{\{|u| \geqslant \vartheta\}} |u| \, d\mu &= \int_{\{|u| \geqslant h_\varepsilon\} \cap \{|u| \geqslant \vartheta\}} |u| \, d\mu + \int_{\{|u| < h_\varepsilon\} \cap \{|u| \geqslant \vartheta\}} |u| \, d\mu \\ &\leqslant \int_{\{|u| \geqslant h_\varepsilon\}} |u| \, d\mu + \int_{\{|h_\varepsilon| \geqslant \vartheta\}} h_\varepsilon \, d\mu \\ &= \int_{\{|u| \geqslant h_\varepsilon\}} |u| \, d\mu + \int_\Omega h_\varepsilon \chi_{\{|h_\varepsilon| \geqslant \vartheta\}} \, d\mu = \frac{\varepsilon}{2} + \frac{\varepsilon}{2} = \varepsilon \qquad \forall \, \vartheta > \widehat{\vartheta}. \end{aligned} \tag{1.25}$$

So, from (1.25) and since $\varepsilon > 0$ was arbitrary, we infer that

$$\lim_{\vartheta \to +\infty} \sup_{u \in C} \int_{\{|u| \geqslant \vartheta\}} |u| \, d\mu = 0,$$

which proves the statement of part **(b)** in the definition of uniform integrability (see Definition 1.18). Let $\varepsilon > 0$. Choose $h_\varepsilon \in L^1(\Omega)_+$ such that

$$\sup_{u \in C} \int_{\{|u| \geqslant h_\varepsilon\}} |u| \, d\mu \leqslant \frac{\varepsilon}{2}. \tag{1.26}$$

Next note that by the Lebesgue dominated convergence theorem (see Theorem I.3.94), we have

$$\lim_{c \searrow 0} \int_\Omega \min\{h_\varepsilon, c\} \, d\mu = 0.$$

So, for some $c = c(\varepsilon) > 0$ sufficiently small, we will have

$$\int_{\{h_\varepsilon < c\}} h_\varepsilon \, d\mu \leqslant \int_\Omega \min\{h_\varepsilon, c\} \, d\mu \leqslant \frac{\varepsilon}{2}. \tag{1.27}$$

Using the Chebyshev–Markov inequality (see Theorem 1.30), we have

$$\mu(\{h_\varepsilon \geqslant c\}) \leqslant \frac{1}{c} \int_\Omega h_\varepsilon \, d\mu < +\infty.$$

So, let $A_\varepsilon = \{h_\varepsilon \geqslant c\}$. Then

$$\sup_{u \in C} \int_{A_\varepsilon^c} |u| \, d\mu = \sup_{u \in C} \Big(\int_{\{h_\varepsilon < c\} \cap \{|u| \geqslant h_\varepsilon\}} |u| \, d\mu + \int_{\{h_\varepsilon < c\} \cap \{|u| < h_\varepsilon\}} |u| \, d\mu \Big)$$

$$\leqslant \sup_{u \in C} \int_{\{|u| \geqslant h_\varepsilon\}} |u| \, d\mu + \sup_{u \in C} \int_{\{h_\varepsilon < c\}} h_\varepsilon \, d\mu \leqslant \frac{\varepsilon}{2} + \frac{\varepsilon}{2} = \varepsilon$$

(see (1.26) and (1.27)). This is the statement of part **(a)** in the definition of uniform integrability (see Definition 1.18). Therefore we have proved that the set $C \subseteq L^1(\Omega)$ is uniformly integrable.

Solution of Problem 1.17

Given $\varepsilon > 0$ let $h_\varepsilon \in L^1(\Omega)_+$ and $\delta > 0$ be as postulated by the hypothesis. If $M \geqslant \frac{1}{\delta} \sup_{u \in C} \|u\|_1$, then

$$\|u\|_1 \geqslant \int_{\{|u| \geqslant M h_\varepsilon\}} |u| \, d\mu \geqslant M \int_{\{|u| \geqslant M h_\varepsilon\}} h_\varepsilon \, d\mu,$$

so

$$\int_{\{|u| \geqslant M h_\varepsilon\}} h_\varepsilon \, d\mu \leqslant \frac{1}{M} \sup_{u \in C} \|u\|_1 \leqslant \delta,$$

thus by hypothesis

$$\sup_{u \in C} \int_{\{|u| > M h_\varepsilon\}} |u| \, d\mu \leqslant \varepsilon.$$

Invoking Problem 1.16, we conclude that the set $C \subseteq L^1(\Omega)$ is uniformly integrable.

Solution of Problem 1.18

"$\Longrightarrow$": Assume that $u_n \xrightarrow{w} u$ in $L^1(\Omega)$. Then for every $y \in L^1(\Omega)$, we have

$$u_n + y \xrightarrow{w} u + y \quad \text{in } L^1(\Omega).$$

From the weak lower semicontinuity of the norm functional (see Proposition I.5.56(c)), we have

$$\|u + y\|_1 \leqslant \liminf_{n \to +\infty} \|u_n + y\|_1,$$

so

$$\int_\Omega |u + y| \, d\mu \leqslant \liminf_{n \to +\infty} \int_\Omega |u_n + y| \, d\mu.$$

"$\Longleftarrow$": Assume that

$$\int_\Omega |u + y| \, d\mu \leqslant \liminf_{n \to +\infty} \int_\Omega |u_n + y| \, d\mu \qquad \forall\, y \in L^1(\Omega).$$

By hypothesis there is a subsequence of $\{u_n\}_{n \geqslant 1}$ which converges weakly in $L^1(\Omega)$ to some $\widehat{u} \in L^1(\Omega)$. Let

$$\Omega_+ = \{\omega \in \Omega : \widehat{u}(\omega) > u(\omega)\} \quad \text{and} \quad \Omega_- = \{\omega \in \Omega : \widehat{u}(\omega) \leqslant u(\omega)\}$$

From the Dunford–Pettis theorem (see Theorem 1.28), we know that the sequence $\{u_n - u\}_{n \geqslant 1} \subseteq L^1(\Omega)$ is uniformly integrable. So, given $\varepsilon > 0$, we can find $\vartheta > 0$ big such that

$$\int_{\{|u_n - u| \geqslant \vartheta\}} |u_n - u| \, d\mu \leqslant \varepsilon \qquad \forall\, n \geqslant 1. \tag{1.28}$$

Let us set

$$y = -u - \vartheta \chi_{\Omega_+} + \vartheta \chi_{\Omega_-}.$$

Since μ is finite, we see that $y \in L^1(\Omega)$ and

$$|u_n + y| \leqslant \vartheta - (u_n - u)\chi_{\Omega_+} + (u_n - u)\chi_{\Omega_-} + 2|u_n - u|\chi_{\{|u_n - u| \geqslant \vartheta\}},$$

so

$$\begin{aligned}
\vartheta\mu(\Omega) &= \int_\Omega |u + y| \, d\mu \leqslant \liminf_{n \to +\infty} \int_\Omega |u_n + y| \, d\mu \\
&\leqslant \vartheta\mu(\Omega) - \int_{\Omega_+} (\widehat{u} - u) \, d\mu + \int_{\Omega_-} (\widehat{u} - u) \, d\mu + 2\varepsilon \\
&= \vartheta\mu(\Omega) - \int_\Omega |\widehat{u} - u| \, d\mu + 2\varepsilon
\end{aligned}$$

and thus

$$\int_\Omega |\widehat{u} - u| \, d\mu \leqslant 2\varepsilon.$$

As $\varepsilon > 0$ was arbitrary, we see that $\widehat{u} = u$. This implies that for the whole sequence, we have $u_n \xrightarrow{w} u$ in $L^1(\Omega)$.

Solution of Problem 1.19

Since $1 < p < +\infty$, we know that the space $L^p(\Omega)$ is reflexive (see Corollary 1.26). Since by hypothesis the sequence $\{u_n\}_{n\geqslant 1} \subseteq L^p(\Omega)$ is bounded, it is relatively w-compact and so by the Eberlein–Smulian theorem (see Theorem I.5.78), we can find a subsequence $\{u_{n_k}\}_{k\geqslant 1}$ of $\{u_n\}_{n\geqslant 1}$ and a function $\widehat{u} \in L^p(\Omega)$ such that $u_{n_k} \xrightarrow{w} \widehat{u}$ in $L^p(\Omega)$. By the Mazur theorem (see Corollary I.5.59), we know that there exists a sequence $\{h_k\}_{k\geqslant 1} \subseteq L^p(\Omega)$ consisting of convex combinations of the $\{u_{n_k}\}_{k\geqslant 1}$ such that $h_k \longrightarrow \widehat{u}$ in $L^p(\Omega)$. Then by passing to a suitable subsequence if necessary, we conclude that $h_k \longrightarrow \widehat{u}$ in μ-measure or for μ-almost all $\omega \in \Omega$. On the other hand, since $u_n \longrightarrow u$ in μ-measure or for μ-almost all $\omega \in \Omega$, we also have that $h_k \longrightarrow u$ in μ-measure or for μ-almost all $\omega \in \Omega$. Therefore $\widehat{u} = u$. This uniqueness of the weak limit implies that for the original sequence, we have $u_n \xrightarrow{w} u$ in $L^p(\Omega)$.

Solution of Problem 1.20

"$\Longrightarrow$": Assume that $u_n \xrightarrow{w} u$ in $L^1(\Omega)$. Note that for every $A \in \Sigma$, we have $\chi_A \in L^\infty(\Omega) = L^1(\Omega)^*$ (Riesz representation theorem; see Theorem 1.21). So, we have

$$\int_A u_n \, d\mu = \int_\Omega u_n \chi_A \, d\mu \longrightarrow \int_\Omega u \chi_A \, d\mu = \int_A u \, d\mu.$$

Recall that every weakly convergent sequence is norm bounded (see Proposition I.5.56). So, we have established both **(a)** and **(b)**.
"$\Longleftarrow$": Assume that **(a)** and **(b)** hold. By virtue of **(b)**, for every simple function $s \in L^\infty(\Omega)$ we have

$$\int_\Omega u_n s \, d\mu \longrightarrow \int_\Omega u s \, d\mu. \tag{1.29}$$

Given any $h \in L^\infty(\Omega)$ and $\varepsilon > 0$, we can find a simple function $s \in L^\infty(\Omega)$ such that

$$\|h - s\|_\infty < \tfrac{\varepsilon}{2M}, \tag{1.30}$$

where $M = \max\big\{\|u\|_1, \sup\limits_{n\geqslant 1}\|u_n\|_1\big\}$ (see Proposition 1.4). Then we have

$$\begin{aligned}\Big|\int\limits_\Omega (u_n-u)h\,d\mu\Big| &\leqslant \Big|\int\limits_\Omega (u_n-u)(h-s)\,d\mu\Big| + \Big|\int\limits_\Omega (u_n-u)s\,d\mu\Big| \\ &\leqslant \|u_n-u\|_1\|h-s\|_\infty + \Big|\int\limits_\Omega (u_n-u)s\,d\mu\Big| \leqslant \varepsilon + \Big|\int\limits_\Omega (u_n-u)s\,d\mu\Big|\end{aligned}$$

(see (1.30)), so

$$\limsup_{n\to+\infty}\Big|\int\limits_\Omega (u_n-u)h\,d\mu\Big| \leqslant \varepsilon$$

(see (1.29)). Since $\varepsilon>0$ is arbitrary, we conclude that

$$\int\limits_\Omega u_n h\,d\mu \longrightarrow \int\limits_\Omega uh\,d\mu \qquad \forall\, h\in L^\infty(\Omega),$$

so $u_n \xrightarrow{w} u$ in $L^p(\Omega)$.

Solution of Problem 1.21
We have

$$\big|\|u_n\|_1 - \|u\|_1 - \|u_n-u\|_1\big| \leqslant \int\limits_\Omega \big||u_n|-|u|-|u_n-u|\big|\,d\mu \qquad \forall\, n\geqslant 1.$$

Note that

$$\big||u_n|-|u|-|u_n-u|\big| \longrightarrow 0 \quad \text{in } \mu\text{-measure or for } \mu\text{-almost all } \omega\in\Omega. \tag{1.31}$$

Also, we have

$$\big||u_n|-|u|-|u_n-u|\big| \leqslant \big||u_n|-|u_n-u|\big| + |u| \leqslant 2|u|. \tag{1.32}$$

Then (1.31) and (1.32) permit the use of the Lebesgue dominated convergence theorem (see Theorem I.3.94 and Proposition 1.13(a)). So, we obtain

$$\big|\|u_n\|_1 - \|u\|_1 - \|u_n-u\|_1\big| \longrightarrow 0$$

Solution of Problem 1.22
By the Fatou lemma (see Theorem I.3.95), we have

$$\|u\|_p \leqslant \liminf_{n\to+\infty} \|u_n\|_p < +\infty.$$

Let $\varepsilon > 0$ be given. We show that there exists $c_\varepsilon > 0$ such that

$$\big||a+b|^p - |a|^p - |b|^p\big| \leqslant \varepsilon|a|^p + c_\varepsilon|b|^p \qquad \forall\, a,b \in \mathbb{R}. \tag{1.33}$$

If $b = 0$, then (1.33) clearly holds. So, let $b \neq 0$. Then we can divide both sides by $|b|^p$ and reduce the problem into showing the following inequality

$$\big||a+1|^p - |a|^p - 1\big| \leqslant \varepsilon|a|^p + c_\varepsilon \qquad \forall\, a \in \mathbb{R}. \tag{1.34}$$

Observe that

$$\lim_{a\to+\infty} \tfrac{||a+1|^p - |a|^p - 1|}{|a|^p} = 0.$$

Then from this limit we have at once (1.34) and therefore (1.33) too.

We use (1.33) with $a = u_n - u$ and $b = u$ and we have

$$\big||u_n|^p - |u_n-u|^p - |u|^p\big| \leqslant \varepsilon|u_n-u|^p + c_\varepsilon|u|^p \qquad \forall\, n \geqslant 1. \tag{1.35}$$

Let $h_n = |u_n|^p - |u_n-u|^p - |u|^p \in L^1(\Omega)$. Then from (1.35), we have

$$|h_n| - \varepsilon|u_n - u|^p \leqslant c_\varepsilon|u|^p.$$

Hence by the Fatou lemma (see Theorem I.3.95), we have

$$\limsup_{n\to+\infty} \int_\Omega \big[|h_n| - \varepsilon|u_n-u|^p\big]\, d\mu \leqslant 0. \tag{1.36}$$

We have

$$|h_n| \leqslant \big(|h_n| - \varepsilon|u_n-u|^p\big) + \varepsilon|u_n-u|^p \qquad \forall\, n \geqslant 1,$$

so

$$\int_\Omega |h_n|\, d\mu \leqslant \int_\Omega \big(|h_n| - \varepsilon|u_n-u|^p\big)\, d\mu + \varepsilon M \qquad \forall\, n \geqslant 1,$$

with $M = \sup\limits_{n\geqslant 1} \|u_n - u\|_p^p$ and thus

$$\limsup_{n\to+\infty} \int_\Omega |h_n|\, d\mu \leqslant \varepsilon M$$

(see (1.36)). Since $\varepsilon > 0$ is arbitrary, we conclude that

$$h_n \longrightarrow 0 \quad \text{in } L^1(\Omega),$$

so

$$\left[\|u_n\|_p^p - \|u_n - u\|_p^p\right] \longrightarrow \|u\|_p^p.$$

Solution of Problem 1.23

By the Dunford–Pettis theorem (see Theorem 1.28), we can find a subsequence $\{u_{n_k}\}_{k\geqslant 1}$ of $\{u_n\}_{n\geqslant 1}$ such that

$$u_{n_k} \xrightarrow{w} \widehat{u} \quad \text{in } L^1(\Omega),$$

with $\widehat{u} \in L^1(\Omega)$. Then we have

$$\int_A u_{n_k}\, d\mu \longrightarrow \int_A \widehat{u}\, d\mu \qquad \forall\, A \in \Sigma \tag{1.37}$$

(see also Problem 1.20). On the other hand, since $u_n \longrightarrow u$ for μ-almost all $\omega \in \Omega$ and $\{u_n\}_{n\geqslant 1} \subseteq L^1(\Omega)$ is uniformly integrable, by the generalized Fatou lemma (see Remark 1.19), we have

$$\int_A u_n\, d\mu \longrightarrow \int_A u\, d\mu \qquad \forall\, A \in \Sigma,$$

so

$$\int_A u\, d\mu = \int_A \widehat{u}\, d\mu \qquad \forall\, A \in \Sigma$$

(see (1.37)) and thus $u = \widehat{u}$.

This "uniqueness" of the limit implies that for the original sequence we have $u_n \xrightarrow{w} u$ in $L^1(\Omega)$. Then invoking Problem 1.1, we conclude that $u_n \longrightarrow u$ in $L^1(\Omega)$.

Solution of Problem 1.24

From Proposition 1.13 we know that at least for a subsequence $\{u_{n_k}\}_{k\geqslant 1}$ of $\{u_n\}_{n\geqslant 1}$, we have

$$u_{n_k}(\omega) \longrightarrow u(\omega) \quad \text{for } \mu\text{-a.a. } \omega \in \Omega.$$

Evidently $u_{n_k} \xrightarrow{w} u$ in $L^1(\Omega)$. From the Dunford–Pettis theorem (see Theorem 1.28), we have that the sequence $\{u_{n_k}\}_{k\geqslant 1}$ is uniformly integrable. Then Problem 1.23 implies that $u_{n_k} \longrightarrow u$ in $L^1(\Omega)$. The uniqueness of the limit u implies that for the original sequence, we have $u_n \longrightarrow u$ in $L^1(\Omega)$.

Solution of Problem 1.25

Note that the sequence $\{u_n\}_{n\geqslant 1} \subseteq L^1(\Omega)$ is bounded. Then by the Hölder inequality (see Theorem 1.3), the sequence $\{u_n y_n\}_{n\geqslant 1} \subseteq L^1(\Omega)$ is bounded. So, according to Problem 1.20 in order to prove that $u_n y_n \xrightarrow{w} uy$ in $L^1(\Omega)$, it suffices to show that for every $A \in \Sigma$, we have

$$\int_A u_n y_n \, d\mu \longrightarrow \int_A uy \, d\mu. \tag{1.38}$$

Let $\varepsilon > 0$ and let $A \in \Sigma$. By the Dunford–Pettis theorem (see Theorem 1.28), we know that the sequence $\{u_n\}_{n\geqslant 1} \subseteq L^1(\Omega)$ is uniformly integrable. So, by virtue of Problem 1.5, we can find $\delta > 0$ such that

$$\text{if } B \in \Sigma \text{ and } \mu(B) \leqslant \delta, \text{ then } \sup_{n\geqslant 1} \int_B |u_n| \, d\mu \leqslant \tfrac{\varepsilon}{4M_1}, \tag{1.39}$$

with $M_1 = \sup\limits_{n\geqslant 1} \|y_n\|_\infty$. Since μ is finite and $y_n(\omega) \longrightarrow y(\omega)$ for μ-almost all $\omega \in \Omega$, by the Egorov theorem (see Theorem I.3.76), we can find $C_\varepsilon \in \Sigma$ such that $\mu(C_\varepsilon) \leqslant \delta$ and $y_n \longrightarrow y$ uniformly on $\Omega \setminus C_\varepsilon$. So, we can find $n_0 \in \mathbb{N}$ such that

$$\|y_n - y\|_{L^\infty(\Omega\setminus C_\varepsilon)} \leqslant \tfrac{\varepsilon}{4M_2} \qquad \forall\, n \geqslant n_0, \tag{1.40}$$

with $M_2 = \sup\limits_{n\geqslant 1} \|u_n\|_1$ and

$$\Big| \int_{A\setminus C_\varepsilon} (u_n - u) y \, d\mu \Big| \leqslant \tfrac{\varepsilon}{4} \qquad \forall\, n \geqslant n_0 \tag{1.41}$$

(recall that $u_n \xrightarrow{w} u$ in $L^1(\Omega)$). Then we have

$$\begin{aligned}
&\Big| \int_A u_n y_n \, d\mu - \int_A uy \, d\mu \Big| \\
&= \Big| \int_{A\setminus C_\varepsilon} (u_n - u) y \, d\mu + \int_{A\setminus C_\varepsilon} u_n (y_n - y) \, d\mu + \int_{A\cap C_\varepsilon} u_n y_n \, d\mu - \int_{A\cap C_\varepsilon} uy \, d\mu \Big| \\
&\leqslant \Big| \int_{A\setminus C_\varepsilon} (u_n - u) y \, d\mu \Big| + \|y_n - y\|_{L^\infty(\Omega\setminus C_\varepsilon)} M_2 \\
&\quad + M_1 \int_{C_\varepsilon} |u_n| \, d\mu + \|y\|_\infty \int_{C_\varepsilon} |u| \, d\mu \\
&\leqslant \tfrac{\varepsilon}{4} + \tfrac{\varepsilon}{4} + \tfrac{\varepsilon}{4} + \tfrac{\varepsilon}{4} = \tfrac{\varepsilon}{4} \qquad \forall\, n \geqslant n_0.
\end{aligned}$$

Therefore (1.38) holds and we conclude that $u_n y_n \xrightarrow{w} uy$ in $L^1(\Omega)$.

Solution of Problem 1.26

We know that there exists a sequence of affine functions $\vartheta_n(x) = a_n x + c_n$ such that $\varphi = \sup\limits_{n\geqslant 1} \vartheta_n$. Evidently $\vartheta_n \circ u \in L^1(\Omega)$ and

$$\vartheta_n\Big(\tfrac{1}{\mu(\Omega)} \int\limits_\Omega u\,d\mu\Big) \;=\; \tfrac{1}{\mu(\Omega)} \int\limits_\Omega \vartheta_n(u)\,d\mu \;\leqslant\; \tfrac{1}{\mu(\Omega)} \int\limits_\Omega \varphi(u)\,d\mu \qquad \forall\, n \geqslant 1,$$

so

$$\varphi\Big(\tfrac{1}{\mu(\Omega)} \int\limits_\Omega u\,d\mu\Big) \;=\; \sup_{n\geqslant 1} \vartheta_n\Big(\tfrac{1}{\mu(\Omega)} \int\limits_\Omega u\,d\mu\Big) \;\leqslant\; \tfrac{1}{\mu(\Omega)} \int\limits_\Omega \varphi(u)\,d\mu.$$

Solution of Problem 1.27

Evidently, without any loss of generality, we may assume that $u \geqslant 0$, $h \geqslant 0$. Also, note that the case $p = 1$, $p' = +\infty$ (or $p = +\infty$, $p' = 1$) is trivial. So, we may as well assume that $1 < p, p' < +\infty$.

Let

$$A \;=\; \big\{\omega \in \Omega :\; h(\omega) > 0\big\} \quad \text{and} \quad C \;=\; \Omega \setminus A \;=\; \big\{\omega \in \Omega :\; h(\omega) = 0\big\}.$$

We have $A, C \in \Sigma$ and

$$\int\limits_\Omega u^p\,d\mu \;=\; \int\limits_A u^p\,d\mu + \int\limits_C u^p\,d\mu,$$

$$\int\limits_\Omega h^p\,d\mu \;=\; \int\limits_A h^p\,d\mu \quad \text{and} \quad \int\limits_\Omega uh\,d\mu \;=\; \int\limits_A uh\,d\mu.$$

So, in order to prove the Hölder inequality stated in the problem, we may assume that $\Omega = A$. On (Ω, Σ) we consider the new measure $m = h^{p'}\mu$. Also let

$$y(\omega) \;=\; \frac{u(\omega)}{h(\omega)^{\frac{p'}{p}}} \qquad \forall\, \omega \in \Omega$$

(this is well define as $h > 0$). We have

$$\tfrac{1}{m(\Omega)} \int\limits_\Omega y\,dm \;=\; \Big(\int\limits_\Omega h^{p'}\,d\mu\Big)^{-1} \int\limits_\Omega u h^{-\frac{p'}{p}} h^{p'}\,d\mu \;=\; \frac{\int_\Omega uh\,d\mu}{\int_\Omega h^{p'}\,d\mu}. \tag{1.42}$$

Consider the continuous convex function $\varphi\colon \mathbb{R} \longrightarrow \mathbb{R}$ defined by $\varphi(t) = |t|^p$ for all $t \in \mathbb{R}$. We have

$$\int\limits_{\Omega} \varphi(y)\, dm = \int\limits_{\Omega} u^p\, d\mu. \tag{1.43}$$

From the Jensen inequality (see Problem 1.26), we have

$$\varphi\Big(\tfrac{1}{m(\Omega)} \int\limits_{\Omega} y\, dm\Big) \leqslant \tfrac{1}{m(\Omega)} \int\limits_{\Omega} \varphi(y)\, dm,$$

so

$$\frac{\left(\int_\Omega uh\, d\mu\right)^p}{\left(\int_\Omega h^{p'}\, d\mu\right)^{p-1}} \leqslant \int\limits_{\Omega} u^p\, d\mu$$

(see (1.42) and (1.43)) and thus

$$\int\limits_{\Omega} uh\, d\mu \leqslant \|u\|_p \|h\|_{p'}.$$

Solution of Problem 1.28

Let $I_n = \{1, \ldots, n\}$ and $F = \{k \in I_n : \ p_k < +\infty\}$. Then we have

$$\|u_1 \ldots u_n\|_{p_0} \leqslant \Big\| \prod_{k \in F} u_k \Big\|_{p_0} \prod_{k \notin F} \|u_k\|_\infty$$

and

$$\tfrac{1}{p_0} = \sum_{k \in F} \tfrac{1}{p_k}.$$

Therefore, without any loss of generality, we may assume that $F = I_n$ (that is, $p_k < +\infty$ for all $k \in \{1, \ldots, n\}$).

First suppose that $n = 2$. Then we have $\frac{p_0}{p_1} + \frac{p_0}{p_2} = 1$. So, applying the Hölder inequality (see Theorem 1.3 and Problem 1.27) with conjugate exponents $p = \frac{p_1}{p_0}$, $p' = \frac{p_2}{p_0}$ and functions $\widehat{u}_1 = |u_1|^{p_0}$, $\widehat{u}_2 = |u_2|^{p_0}$, we have

$$\int\limits_{\Omega} \widehat{u}_1 \widehat{u}_2\, d\mu = \int\limits_{\Omega} |u_1|^{p_0} |u_2|^{p_0}\, d\mu \leqslant \Big(\int\limits_{\Omega} |u_1|^{p_1}\, d\mu\Big)^{\frac{p_0}{p_1}} \Big(\int\limits_{\Omega} |u_2|^{p_2}\, d\mu\Big)^{\frac{p_0}{p_2}},$$

which proves the problem when $n = 2$.

The general case $n > 2$, is proved by induction. To this end suppose that the result holds for $n - 1$. We set

$$\tfrac{1}{\tau} = \sum_{k=2}^{n} \tfrac{1}{p_k}.$$

Then $\frac{1}{p_0} = \frac{1}{p_1} + \frac{1}{\tau}$ and we have

$$\|u_1 \dots u_n\|_{p_0} \leqslant \|u_1\|_{p_1} \|u_2 \dots u_n\|_\tau$$

(since the result holds for $n = 2$). Thus by the induction hypothesis, we get $\|u_1 \dots u_n\|_{p_0} \leqslant \|u_1\|_{p_1} \dots \|u_n\|_{p_n}$.

Solution of Problem 1.29

Since μ is σ-finite, we can find an increasing sequence $\{\Omega_n\}_{n\geqslant 1} \subseteq \Sigma$ such that

$$\Omega = \bigcup_{n\geqslant 1} \Omega_n \quad \text{and} \quad \mu(\Omega_n) < +\infty \qquad \forall\, n \geqslant 1.$$

Let

$$A_n = \{\omega \in \Omega_n : |u(\omega)| \leqslant n\} \qquad \forall\, n \geqslant 1$$

and set

$$u_n = u\chi_{A_n} \qquad \forall\, n \geqslant 1.$$

We define the linear functionals $\widehat{u}$ and $\widehat{u}_n$ (for $n \geqslant 1$) on $L^{p'}(\Omega)$ by setting

$$\widehat{u}(h) = \int_\Omega uh\, d\mu \quad \text{and} \quad \widehat{u}_n(h) = \int_\Omega u_n h\, d\mu \qquad \forall\, h \in L^{p'}(\Omega),\ n \geqslant 1.$$

Evidently $\widehat{u}_n \in L^{p'}(\Omega)^*$ for all $n \geqslant 1$. Moreover, we have

$$u_n h \longrightarrow uh \quad \text{for } \mu\text{-a.a. } \omega \in \Omega$$

and

$$|u_n h| \leqslant |uh| \in L^1(\Omega) \qquad \forall\, n \geqslant 1.$$

So, we can apply the Lebesgue dominated convergence theorem (see Theorem I.3.94) and have

$$\int_\Omega u_n h\, d\mu \longrightarrow \int_\Omega uh\, d\mu$$

thus

$$\widehat{u}_n(h) \longrightarrow \widehat{u}(h).$$

Hence by the uniform boundedness principle (see Theorem I.5.39), we have $\widehat{u} \in L^{p'}(\Omega)^*$ and so we can find $M > 0$ such that

$$\Big|\int_\Omega uh\, d\mu\Big| \leqslant M \qquad \forall\, h \in L^{p'}(\Omega),\ \|h\|_{p'} \leqslant 1,$$

thus

$$\|u\|_p \leqslant M,$$

hence $u \in L^p(\Omega)$.

Solution of Problem 1.30

We apply the Hölder inequality (see Theorem 1.3 and Problem 1.27) with exponents $\frac{p}{\vartheta q}$, $(\frac{p}{\vartheta q})'$ and functions $|u|^{\vartheta q}$ and $|u|^{(1-\vartheta)q}$. Note that $|u|^{\vartheta q} \in L^{\frac{p}{\vartheta q}}(\Omega)$ and since $r = \frac{(1-\vartheta)qp}{p-\vartheta q}$ we have that $|u|^{(1-\vartheta)q} \in L^{\frac{p}{p-\vartheta q}}(\Omega)$ and $\frac{p}{p-\vartheta q} = (\frac{p}{\vartheta q})'$. We have

$$\begin{aligned} \int_\Omega |u|^q \, d\mu &= \int_\Omega |u|^{\vartheta q} |u|^{(1-\vartheta)q} \, d\mu \\ &\leqslant \Big(\int_\Omega |u|^{\vartheta q (\frac{p}{\vartheta q})} \, d\mu \Big)^{\frac{\vartheta q}{p}} \Big(\int_\Omega |u|^{(1-\vartheta)q \frac{p}{p-\vartheta q}} \, d\mu \Big)^{\frac{p-\vartheta q}{p}} \\ &= \Big(\int_\Omega |u|^p \, d\mu \Big)^{\frac{\vartheta q}{p}} \Big(\int_\Omega |u|^r \, d\mu \Big)^{\frac{p-\vartheta q}{p}}, \end{aligned}$$

so

$$\|u\|_q \leqslant \|u\|_p^\vartheta \|u\|_r^{1-\vartheta}.$$

Solution of Problem 1.31

We have

$$\begin{aligned} \int_\Omega |u_n h_n - uh|^p \, d\mu &= \int_\Omega \big|(u_n - u)h_n + u(h_n - h)\big|^p \, d\mu \\ &\leqslant 2^{p-1} \int_\Omega |u_n - u|^p |h_n|^p \, d\mu + 2^{p-1} \int_\Omega |u(h_n - h)|^p \, d\mu \\ &\leqslant 2^{p-1} M \|u_n - u\|_p^p + 2^{p-1} \int_\Omega |u(h_n - h)|^p \, d\mu, \end{aligned} \tag{1.44}$$

where $M = \sup\limits_{n \geqslant 1} \|h_n\|_\infty^p < +\infty$. Note that

$$(u_n h)(\omega) \longrightarrow (uh)(\omega) \quad \text{for } \mu\text{-a.a. } \omega \in \Omega$$

and

$$|u(h_n - h)|^p \leqslant 2M|u|^p \in L^1(\Omega). \tag{1.45}$$

So, we can apply the Lebesgue dominated convergence theorem (see Theorem I.3.94) and have

$$\int_{\Omega} |u(h_n - h)|^p \, d\mu \longrightarrow 0.$$

Therefore, if in (1.44) we pass to the limit as $n \to +\infty$ and use (1.45) and the fact that $u_n \longrightarrow u$ in $L^p(\Omega)$, we obtain

$$u_n h_n \longrightarrow uh \quad \text{in } L^p(\Omega).$$

Solution of Problem 1.32

For every $r \in [p, q]$, we have

$$|u(\omega)|^r \leqslant |u(\omega)|^p + |u(\omega)|^q = h(\omega) \quad \text{for } \mu\text{-a.a. } \omega \in \Omega.$$

Indeed, if $|u(\omega)| \leqslant 1$, then $|u(\omega)|^r \leqslant |u(\omega)|^p$ since $p \leqslant r$, while if $|u(\omega)| > 1$, then $|u(\omega)|^r \leqslant |u(\omega)|^q$ since $r \leqslant q$. Suppose that $\{r_n\}_{n\geqslant 1} \subseteq [p, q]$ is a sequence such that $r_n \longrightarrow r_0 \in [p, q]$. Then

$$|u(\omega)|^{r_n} \longrightarrow |u(\omega)|^{r_0} \quad \text{for } \mu\text{-a.a. } \omega \in \Omega \quad \text{as } n \to +\infty$$

and

$$|u(\omega)|^{r_n} \leqslant h(\omega) \quad \text{for } \mu\text{-a.a. } \omega \in \Omega \text{ and all } n \geqslant 1.$$

Since $h \in L^1(\Omega)$ (recall that $u \in L^p(\Omega) \cap L^q(\Omega)$), we can apply the Lebesgue dominated convergence theorem (see Theorem I.3.94) and conclude that

$$\|u\|_{r_n} \longrightarrow \|u\|_{r_0} \quad \text{as } n \to +\infty.$$

Solution of Problem 1.33

Let $(u, y) \in L^2(\Omega) \times L^2(\Omega)$. We have

$$\|K(u + y)\|_2 = \|u + y\|_2$$

(since K is an isometry), so

$$\|K(u)\|_2^2 + 2\int_{\Omega} K(u)K(y)\, d\mu + \|K(y)\|_2^2$$

$$= \|u\|_2^2 + 2\int_{\Omega} uy\, d\mu + \|y\|_2^2$$

and thus

$$\int_\Omega K(u)K(y)\,d\mu = \int_\Omega uy\,d\mu.$$

If by $(\cdot,\cdot)_{L^2}$ we denote the inner product in $L^2(\Omega)$, we have

$$\int_\Omega K(u)K(y)\,d\mu = (K(u),K(y))_{L^2} = (K^*K(u),y)_{L^2} = \int_\Omega (K^*K)(u)y\,d\mu,$$

so

$$\int_\Omega (K^*K)(u)y\,d\mu = \int_\Omega uy\,d\mu \qquad \forall\, y\in L^2(\Omega),$$

thus

$$(K^*K)(u) = u \qquad \forall\, u\in L^2(\Omega),$$

and hence $K^*K = I_{L^2(\Omega)}$.

Solution of Problem 1.34
From the Fubini Theorem, we have

$$p\int_0^{+\infty} \vartheta^{p-1}\mu(\{u\geqslant\vartheta\})\,d\vartheta = p\int_0^{+\infty}\vartheta^{p-1}\Big(\int_\Omega \chi_{\{u\geqslant\vartheta\}}\,d\mu\Big)\,d\vartheta$$

$$= p\int_\Omega\Big(\int_0^{+\infty}\vartheta^{p-1}\chi_{\{u\geqslant\vartheta\}}\,d\vartheta\Big)\,d\mu = p\int_\Omega\Big(\int_0^{u(\omega)}\vartheta^{p-1}\,d\vartheta\Big)\,d\mu = \int_\Omega u^p\,d\mu.$$

Solution of Problem 1.35
For every $r>0$, let

$$A_r(u)(x) = \frac{1}{\lambda^N(B_r(x))}\int_{B_r(x)} |u(z)|\,dz.$$

Also, let $\eta\geqslant 0$ and consider the set

$$U_\eta = \big\{x\in\mathbb{R}^N:\ M(u)(x)>\eta\big\}.$$

If we show that U_η is open, then $M(u)(\cdot)$ is lower semicontinuous. So, let $x \in U_\eta$. Then by virtue of Definition 1.31, there exists $r > 0$ such that $A_r(u)(x) > \eta$. If $\varepsilon > 0$ and $|y - x| < \varepsilon$, then $B_{r+\varepsilon}(y) \supseteq B_r(x)$ and so

$$\int_{B_r(x)} |u|\,dz \leqslant \int_{B_{r+\varepsilon}(y)} |u|\,dz,$$

thus

$$\eta < \left(\tfrac{r}{r+\varepsilon}\right)^N M(u)(x) \leqslant A_{r+\varepsilon}(u)(y) \leqslant M(u)(y),$$

hence

$$B_\varepsilon(x) \subseteq U_\eta$$

and finally we get that U_η is open. Therefore, $M(u)(\cdot)$ is lower semicontinuous.

Solution of Problem 1.36

Let $A \subseteq \mathbb{R}^N$ be a Lebesgue measurable set with finite Lebesgue measure. From Problem 1.34, we have

$$\int_A M(u)(z)\,dz = \int_0^{+\infty} \lambda^N(\{M(u) \geqslant \vartheta\})\,d\vartheta$$

$$\leqslant \lambda^N(A) + \int_1^{+\infty} \lambda^N(\{M(u) \geqslant \vartheta\})\,d\vartheta. \tag{1.46}$$

Let

$$u_\vartheta(z) = \begin{cases} 0 & \text{if } |u(z)| < \frac{\vartheta}{2}, \\ u(z) & \text{if } |u(z)| \geqslant \frac{\vartheta}{2}. \end{cases}$$

We have $u \in L^1(\mathbb{R}^N)$. Note that

$$|u| \leqslant |u_\vartheta| + \tfrac{\vartheta}{2},$$

so

$$M(u) \leqslant M(u_\vartheta) + \tfrac{\vartheta}{2}$$

(see Definition 1.31), thus

$$\{M(u) \geqslant \vartheta\} \supseteq \{M(u_\vartheta) \geqslant \tfrac{\vartheta}{2}\}.$$

Using Theorem 1.32(b) on the function $u_\vartheta \in L^1(\mathbb{R}^N)$, we have

$$\lambda^N(\{M(u) \geqslant \vartheta\}) \leqslant \tfrac{3^N 2}{\vartheta}\|u_\vartheta\|_1 = \tfrac{3^N 2}{\vartheta} \int_{\{|u| \geqslant \frac{\vartheta}{2}\}} |u|\,dz. \tag{1.47}$$

Returning to (1.46), using (1.47) and the Fubini theorem, we obtain

$$\begin{aligned}\int\limits_A M(u)(z)\,dz &\leqslant \lambda^N(A) + 3^N 2 \int\limits_1^{+\infty} \tfrac{1}{\vartheta}\Big(\int\limits_{\{|u|\geqslant\frac{\vartheta}{2}\}} |u(z)|\,dz\Big)\,d\vartheta \\ &\leqslant \lambda^N(A) + 3^N 2 \int\limits_{\mathbb{R}^N} |u(z)|\Big(\int\limits_1^{2|u(z)|} \tfrac{1}{\vartheta}\,d\vartheta\Big)\,dz \\ &= \lambda^N(A) + 3^N 2 \int\limits_{\mathbb{R}^N} |u(z)| \ln(2|u(z)|)\,dz \;<\; +\infty.\end{aligned}$$

Solution of Problem 1.37

The answer is negative in both cases.

For $p = 1$, let $\Omega = (0, 2\pi)$ with Σ being the Lebesgue σ-field and $\mu = \lambda$ being the Lebesgue measure. According to Problem 1.1, in order to produce a weakly convergent but not strongly convergent sequence of L^1-functions, we need to consider a sequence which oscillates violently around its weak limit. For this reason we consider the functions

$$u_n(z) = 4 + \sin nz \qquad \forall\, z \in (0, 2\pi),\ n \geqslant 1.$$

Evidently the sequence $\{u_n\}_{n\geqslant 1} \subseteq L^1(0, 2\pi)$ is bounded and for all measurable sets $A \subseteq (0, 2\pi)$, we have

$$\int\limits_A (4 + \sin nz)\,dz \longrightarrow 4\mu(A),$$

so

$$u_n \xrightarrow{w} 4 \quad \text{in } L^1(0, 2\pi)$$

(see Problem 1.20). Also $\|u\|_1 \longrightarrow 8\pi = \|4\|_1$. Moreover, note that

$$\|u_n - 4\|_1 = \int\limits_0^{2\pi} |\sin nz|\,dz = 2n \int\limits_0^{\frac{\pi}{n}} \sin nz\,dz = 4 \qquad \forall\, n \geqslant 1,$$

so $u_n \not\longrightarrow 4$ in $L^1(0, 2\pi)$.

Let $p = +\infty$. Consider $\Omega = (0, 1)$ with Σ being the Lebesgue σ-field and $\mu = \lambda$ being the Lebesgue measure. Let $\{u_n\}_{n\geqslant 1}$ be the sequence of functions defined by

$$u_n(z) = \begin{cases} 0 & \text{if } 0 \leqslant z \leqslant \frac{1}{n}, \\ 1 & \text{if } \frac{1}{n} < z \leqslant 1, \end{cases}$$

for all $n \geqslant 1$. Clearly $u_n \xrightarrow{w^*} 1$ in $L^\infty(0,1)$ and $\|u_n\|_\infty \longrightarrow 1$. However, we have

$$\|u_n - 1\|_\infty = 1 \qquad \forall\, n \geqslant 1,$$

so $u_n \not\longrightarrow 1$ in $L^\infty(0,1)$.

Solution of Problem 1.38

Consider the linear operator $L\colon L^p(\Omega) \longrightarrow L^q(\Omega)$ defined by

$$L(u) = \xi u \qquad \forall\, u \in L^p(\Omega).$$

Let $\{u_n\}_{n\geqslant 1} \subseteq L^p(\Omega)$ be such that $u_n \longrightarrow u$ in $L^p(\Omega)$ and assume that

$$L(u_n) = \xi u_n \longrightarrow h \quad \text{in } L^q(\Omega).$$

By passing to a suitable subsequence if necessary, we may assume that

$$u_n(\omega) \longrightarrow u(\omega) \quad \text{and} \quad (\xi u_n)(\omega) \longrightarrow h(\omega) \quad \text{for } \mu\text{-a.a. } \omega \in \Omega.$$

So, we have $h = \xi u$ and then the closed graph theorem (see Theorem I.5.50) implies that $L \in \mathcal{L}(L^p(\Omega); L^q(\Omega))$. Hence we can find $c > 0$ such that

$$\|\xi u\|_q \leqslant c\|u\|_p \qquad \forall\, u \in L^p(\Omega). \tag{1.48}$$

If $1 \leqslant p < +\infty$, then from (1.48) we have

$$\int\limits_\Omega |\xi|^q |y|\, d\mu \leqslant c^q \|y\|_{\frac{p}{q}} \qquad \forall\, y \in L^{\frac{p}{q}}(\Omega),$$

so $y \longmapsto \int\limits_\Omega |\xi|^q y\, dz$ is a linear continuous functional on $L^{\frac{p}{q}}(\Omega)$ and thus $|\xi|^q \in L^{(\frac{p}{q})'}(\Omega)$ (by the Riesz representation theorem; see Theorem 1.21).

Note that $(\frac{p}{q})' = \frac{p}{p-q}$. Therefore $\xi \in L^r(\Omega)$ with $r = \frac{pq}{p-q}$.

If $p = +\infty$, then in (1.48) we choose $u \equiv 1$ and we have $\|\xi\|_q \leqslant c < +\infty$, which means that $\xi \in L^q(\Omega)$.

Solution of Problem 1.39

First we show that weak convergence in $L^1(\Omega)$ does not imply convergence in μ-measure. Consider $\Omega = (0, 2\pi)$ with the Lebesgue σ-field and the Lebesgue measure λ. Let $\{u_n\}_{n\geqslant 1} \subseteq L^1(0,2\pi)$ be the sequence defined by

$$u_n(z) = \cos(nz) \qquad \forall\, z \in (0, 2\pi),\ n \geqslant 1.$$

We have $u_n \xrightarrow{w} 0$ in $L^1(0,2\pi)$ (see Problem I.4.132 and its solution). On the other hand, we have

$$\lambda(\{z \in (0,2\pi) : |\cos(nz)| \geqslant \tfrac{1}{2}\}) = \tfrac{4}{3}\pi,$$

so $u_n \not\longrightarrow 0$ in measure (see Definition 1.11(a)).

Now we show that convergence in μ-measure does not imply weak convergence in $L^1(\Omega)$. Consider $\Omega = (0,1)$ with the Lebesgue σ-field and the Lebesgue measure. Let $\{u_n\}_{n\geqslant 1} \subseteq L^1(0,1)$ be the sequence defined by

$$u_n(z) = \begin{cases} n & \text{if } z \in [0, \frac{1}{n}], \\ 0 & \text{if } z \in (\frac{1}{n}, 1], \end{cases}$$

for all $n \geqslant 1$. Evidently $u_n \longrightarrow 0$ in measure (see Definition 1.11(a)). On the other hand, note that

$$\int_0^1 u_n\, dz = 1 \qquad \forall\, n \geqslant 1$$

and so the sequence $\{u_n\}_{n\geqslant 1}$ does not converge weakly to zero in $L^1(0,1)$. Indeed, let $h \equiv 1$ on $[0,1]$. Then $h \in L^\infty(0,1)$ and

$$\langle u_n, h\rangle = \int_0^1 u_n h\, dz = \int_0^1 u_n\, dz = 1 \not\longrightarrow 0.$$

Solution of Problem 1.40

No. Consider $\Omega = \mathbb{R}$ with the Lebesgue σ-field and the Lebesgue measure λ. Let $u(z) = \chi_{(0,1)}(z)$ and consider the sequence

$$u_n(z) = u(z+n) \qquad \forall\, z \in \mathbb{R},\ n \geqslant 1.$$

Evidently

$$\int_{\mathbb{R}} |u_n|\, dz = 1 \qquad \forall\, n \geqslant 1,$$

so the sequence $\{u_n\}_{n\geqslant 1} \subseteq L^1(\mathbb{R})$ is bounded. Arguing by contradiction, suppose that we could find a subsequence $\{u_{n_k}\}_{k\geqslant 1}$ of $\{u_n\}_{n\geqslant 1}$ which converges weakly in $L^1(\mathbb{R})$. Let $h \in L^\infty(\mathbb{R})$ be defined by

$$h(z) = \sum_{k\geqslant 1} (-1)^k \chi_{(-n_k, -n_k+1)}(z) \qquad \forall\, z \in \mathbb{R}.$$

Then we have

$$\langle u_{n_k}, h\rangle = \int_{\mathbb{R}} u_{n_k} h\,dz = (-1)^k \qquad \forall\, k \geqslant 1$$

and this does not converge.

Solution of Problem 1.41

(a) Let $\{s_n\}_{n\geqslant 1}$ be a sequence of simple functions such that $s_n(\omega) \longrightarrow u(\omega)$ for all $\omega \in \Omega$. Evidently, for each $n \geqslant 1$, the function $\omega \longmapsto f(\omega, s_n(\omega))$ is Σ-measurable. Since by the continuity of $f(\omega, \cdot)$ we have

$$f(\omega, s_n(\omega)) \longrightarrow f(\omega, s(\omega)) \qquad \forall\, \omega \in \Omega,$$

we conclude that the function $\omega \longmapsto f(\omega, u(\omega)) = N_f(u)(\omega)$ is Σ-measurable.

(b) Let $u_n \xrightarrow{\mu} u$ (see Definition 1.11) Without any loss of generality, we may assume that $f(\omega, 0) = 0$ for μ-almost all $\omega \in \Omega$ and that $u \equiv 0$. Indeed, otherwise we replace $f(\omega, x)$ by

$$\widehat{f}(\omega, x) = f(\omega, x + u(\omega)) - f(\omega, u(\omega)).$$

So, our aim is to show that given $\varepsilon > 0$, we can find $n_0 = n_0(\varepsilon) \geqslant 1$ such that

$$\mu(\{\omega \in \Omega : |f(\omega, u_n(\omega))|\} \geqslant \varepsilon) \leqslant \varepsilon \qquad \forall\, n \geqslant n_0 \tag{1.49}$$

(see Definition 1.11). Let

$$\Omega_k = \{\omega \in \Omega : |f(\omega, x)| < \varepsilon \text{ for all } |x| \leqslant \tfrac{1}{k}\}.$$

Clearly the sequence $\{\Omega_k\}_{k\geqslant 1} \subseteq \Sigma$ is increasing and $\bigcup\limits_{k=1}^{\infty} \Omega_k = \Omega$, so $\mu(\Omega_k) \to \mu(\Omega)$. Thus, we can find $k_0 \geqslant 1$ such that

$$\mu(\Omega \setminus \Omega_{k_0}) \leqslant \tfrac{\varepsilon}{2}.$$

Let

$$C_n = \{\omega \in \Omega : |u_n(\omega)| \leqslant \tfrac{1}{k_0}\}.$$

Since $u_n \xrightarrow{\mu} 0$, we can find $n_0 = n_0(\varepsilon) \geqslant 1$ such that

$$\mu(\Omega \setminus C_n) \leqslant \tfrac{\varepsilon}{2} \qquad \forall\, n \geqslant n_0.$$

Let

$$E_n = \{\omega \in \Omega : |f(\omega, u_n(\omega))| \leqslant \varepsilon\}.$$

Then $C_n \cap \Omega_{k_0} \subseteq E_n$. So, we have

$$\mu(\Omega \setminus E_n) \leqslant \mu(\Omega \setminus C_n) + \mu(\Omega \setminus \Omega_{k_0}) \leqslant \tfrac{\varepsilon}{2} + \tfrac{\varepsilon}{2} = \varepsilon \qquad \forall\, n \geqslant n_0,$$

thus $N_f(u_n) \xrightarrow{\mu} N_f(u)$.

Solution of Problem 1.42

(a) From the assumption

$$|f(\omega,x)| \leqslant a(\omega) + c|x|^r \quad \text{for } \mu\text{-a.a. } \omega \in \Omega,\ \text{all } x \in \mathbb{R} \tag{1.50}$$

and using the Minkowski inequality (see Theorem I.3.101), we have

$$\|N_f(u)\|_q \leqslant \|a\|_q + c\||u|^r\|_q = \|a\|_q + c\|u\|_{qr}. \tag{1.51}$$

So, N_f maps $L^{qr}(\Omega)$ into $L^q(\Omega)$.

(b) From (1.51) it is clear that N_f is bounded (i.e., maps bounded sets in $L^{qr}(\Omega)$ to bounded sets in $L^q(\Omega)$).
Next, suppose that $u_n \longrightarrow u$ in $L^{qr}(\Omega)$. By passing to a suitable subsequence if necessary, we may assume that

$$u_n(\omega) \longrightarrow u(\omega) \quad \text{for } \mu\text{-a.a. } \omega \in \Omega$$

and

$$|u_n(\omega)| \leqslant h(\omega) \quad \text{for } \mu\text{-a.a. } \omega \in \Omega,\ \text{all } n \geqslant 1,$$

with $h \in L^{qr}(\Omega)$ (see Propositions 1.16 and 1.15). From (1.50), we have

$$|f(\omega,u_n(\omega))| \leqslant a(\omega) + c|h(\omega)|^r = g(\omega) \quad \text{for } \mu\text{-a.a. } \omega \in \Omega,$$

with $g \in L^q(\Omega)$. Since $f(\omega,u_n(\omega)) \longrightarrow f(\omega,u(\omega))$ for μ-almost all $\omega \in \Omega$, we can apply the Lebesgue dominated convergence theorem (see Theorem I.3.94) and have (for the original sequence) that

$$N_f(u_n) \longrightarrow N_f(u) \quad \text{in } L^q(\Omega).$$

So the function $u \longmapsto N_f(u)$ is continuous.

Solution of Problem 1.43

Fiest we show the continuity of N_f.

As in the solution of Problem 1.41(b), without any loss of generality, we may assume that

$$f(\omega,0) = 0 \quad \text{for } \mu\text{-a.a. } \omega\in\Omega$$

and reduce the problem to proving the continuity of N_f at the origin. Arguing by contradiction, suppose that N_f is not continuous at $u=0$. So, we can find a sequence $\{u_n\}_{n\geqslant 1}\subseteq L^p(\Omega)$ such that $u_n \longrightarrow 0$ in $L^p(\Omega)$ and $N_f(u_n)\not\longrightarrow 0$ in $L^q(\Omega)$. By passing to a suitable subsequence if necessary, we may assume that

$$\sum_{n\geqslant 1}\|u_n\|_p^p < +\infty \quad \text{and} \quad \int_\Omega |f(\omega,u_n(\omega))|^q\,d\mu \geqslant \beta > 0 \qquad \forall\, n\geqslant 1. \tag{1.52}$$

Let

$$A_n = \left\{\omega\in\Omega:\ \big|f(\omega,u_n(\omega))\big| \geqslant \left(\tfrac{\beta}{3\mu(\Omega)}\right)^{\frac{1}{q}}\right\} \qquad \forall\, n\geqslant 1. \tag{1.53}$$

From Problem 1.41 and its solution, we know that $A_n\in\Sigma$ and $\mu(A_n)\to 0$.

We construct a decreasing sequence $\{\varepsilon_k\}_{k\geqslant 1}$ of positive numbers and a subsequence $\{u_{n_k}\}_{k\geqslant 1}$ of $\{u_n\}_{n\geqslant 1}$ as follows.

Let $\varepsilon_1=\mu(\Omega)$ and $u_{n_1}=u_1$.

Next choose $\varepsilon_2\in(0,\frac{\varepsilon_1}{2})$ such that

$$\int_C |f(\omega,u_{n_1}(\omega))|^q\,d\mu \leqslant \tfrac{\beta}{3} \qquad \forall\, C\in\Sigma, \text{ with } \mu(C)\leqslant 2\varepsilon_2.$$

Then choose $n_2\geqslant 1$ such that $\mu(A_{n_2})\leqslant\varepsilon_2$.

Now that we have ε_2 and u_{n_2}, we choose $\varepsilon_3\in(0,\frac{\varepsilon_2}{2})$ such that

$$\int_C |f(\omega,u_{n_2}(\omega))|^q\,d\mu \leqslant \tfrac{\beta}{3} \qquad \forall\, C\in\Sigma, \text{ with } \mu(C)\leqslant 2\varepsilon_3.$$

Then choose $n_3\geqslant 1$ such that $\mu(A_{n_3})\leqslant\varepsilon_3$.

We continue this way. Let

$$E_{n_k} = A_{n_k}\setminus\bigcup_{i=k+1}^{\infty}A_{n_i}.$$

The sets $\{E_{n_k}\}_{k\geqslant 1}$ are pairwise disjoint. We set

$$u(\omega) = \begin{cases} u_{n_k}(\omega) & \text{if } \omega\in E_{n_k},\ k=1,2,\ldots, \\ 0 & \text{otherwise.}\end{cases}$$

From (1.52) we see that $u\in L^p(\Omega)$. Then by hypothesis $N_f(u)\in L^q(\Omega)$.

On the other hand, let

$$\begin{aligned}\eta_k &= \int_{E_{n_k}} |f(\omega,u(\omega))|^q\,d\mu = \int_{E_{n_k}} |f(\omega,u_{n_k}(\omega))|^q \\ &= \int_{A_{n_k}} |f(\omega,u_{n_k}(\omega))|^q\,d\mu - \int_{A_{n_k}\setminus E_{n_k}} |f(\omega,u_{n_k}(\omega))|^q\,d\mu \end{aligned} \tag{1.54}$$

Note that

$$\int_{A_{n_k}} |f(\omega, u_{n_k}(\omega))|^q d\mu = \int_{\Omega} |f(\omega, u_{n_k}(\omega))|^q d\mu - \int_{\Omega \setminus B_{n_k}} |f(\omega, u_{n_k}(\omega))|^q d\mu$$

$$\geqslant \beta - \tfrac{\beta}{3} = \tfrac{2\beta}{3}$$

(see (1.52) and (1.53)) and

$$\int_{A_{n_k} \setminus E_{n_k}} |f(\omega, u_{n_k}(\omega))|^q \, d\mu < \tfrac{\beta}{3},$$

since $A_{n_k} \setminus E_{n_k} \subseteq \bigcup_{i \geqslant k+1} A_{n_i}$ and so

$$\mu(A_{n_k} \setminus E_{n_k}) \leqslant \sum_{i \geqslant k+1} \mu(A_{n_i}) \leqslant 2\varepsilon_{k+i}.$$

Returning to (1.54), we have $\eta_k \geqslant \frac{\beta}{3}$ and so

$$\int_{\Omega} |f(\omega, u_{n_k}(\omega))|^q \, d\mu = \sum_{k \geqslant 1} \eta_k = +\infty,$$

a contradiction.

Remark. In fact one can show that 1.51 holds when $r = \frac{p}{q}$. Moreover, if N_f maps $L^p(\Omega)$ into $L^\infty(\Omega)$, then there exists a function $a \in L^\infty(\Omega)$ such that

$$|f(\omega, x)| \leqslant a(\omega) \quad \text{for } \mu\text{-a.a. } \omega \in \Omega, \text{ all } x \in \mathbb{R}$$

(see Krasnoselskii [13, p. 27]). Moreover, the result is also true for σ-finite measure spaces (see Krasnoselskii [13, p. 22]).

Solution of Problem 1.44

From Problem 1.19 we know that $u \in L^p(\Omega)$. By the Egorov theorem (see Theorem I.3.76), given $\varepsilon > 0$, we can find $A_\varepsilon \in \Sigma$ with $\mu(A_\varepsilon) \leqslant \varepsilon$ such that $u_n \longrightarrow u$ uniformly on $\Omega \setminus A_\varepsilon$. So, we can find $n_0 = n_0(\varepsilon) \geqslant 1$ such that

$$|u_n(\omega) - u(\omega)| \leqslant \frac{\varepsilon}{\mu(\Omega)^{\frac{1}{q}}} \qquad \forall\, \omega \in \Omega \setminus A_\varepsilon,\ n \geqslant n_0.$$

For $n \geqslant n_0$, we have

$$\begin{aligned} \|u_n - u\|_q^q &= \int_\Omega |u_n - u|^q \, d\mu = \int_{\Omega \setminus A_\varepsilon} |u_n - u|^q \, d\mu + \int_{A_\varepsilon} |u_n - u|^q \, d\mu \\ &\leqslant \varepsilon^q + \|u_n - u\|_p^q \mu(A_\varepsilon)^{1-\frac{q}{p}} \leqslant \varepsilon^q + M^q \varepsilon^{1-\frac{q}{p}} \end{aligned}$$

(by the Hölder inequality; see Theorem 1.3 and Problem 1.27), where $M = \max\left\{\sup_{n\geqslant 1} \|u_n\|_p, \|u\|_p\right\} < +\infty$, so

$$\limsup_{n\to+\infty} \|u_n - u\|_q^q \leqslant \varepsilon^q + M^q \varepsilon^{1-\frac{q}{p}}.$$

Since $\varepsilon > 0$ is arbitrary, we let $\varepsilon \searrow 0$ to conclude that

$$u_n \longrightarrow u \quad \text{in } L^q(\Omega),$$

for all $q \in [1, p)$.

Solution of Problem 1.45

For any $A \in \Sigma$ and $n \geqslant 1$ we have

$$\int_A u_n \, d\mu = \int_\Omega u_n \, d\mu - \int_{A^c} u_n \, d\mu,$$

so

$$\begin{aligned} \limsup_{n\to+\infty} \int_A u_n \, d\mu &= \int_\Omega u \, d\mu - \liminf_{n\to+\infty} \int_{A^c} u_n \, d\mu \\ &\leqslant \int_\Omega u \, d\mu - \int_{A^c} u \, d\mu = \int_A u \, d\mu \end{aligned} \tag{1.55}$$

(by the Fatou lemma; see Theorem I.3.95). On the other hand, by the Fatou lemma, we have

$$\int_A u \, d\mu \leqslant \liminf_{n\to+\infty} \int_A u_n \, d\mu. \tag{1.56}$$

So, finally, we conclude that $\int_A u_n \, d\mu \longrightarrow \int_A u \, d\mu$.

Solution of Problem 1.46

Evidently we may assume that $u \in L^p(\Omega)$ (otherwise the inequality holds trivially). Let $h_n = \min\{h, n\}$ for $n \geqslant 1$. We have $\{h_n \geqslant \vartheta\} \subseteq \{h \geqslant \vartheta\}$ and if $\vartheta > n$, then $\{h_n \geqslant \vartheta\} = \emptyset$. Using Problem 1.34, the hypotheses, the Fubini theorem and the Hölder inequality (see Theorem 1.3 and Problem 1.27), we have

$$\begin{aligned}
\|h_n\|_p^p &= \int\limits_\Omega h_n^p \, d\mu \;=\; p \int\limits_0^n \vartheta^{p-1} \mu(\{h_n \geqslant \vartheta\}) \, d\vartheta \\
&\leqslant p \int\limits_0^n \vartheta^{p-2} \Big(\int\limits_{\{h_n \geqslant \vartheta\}} u \, d\mu \Big) \, d\vartheta \\
&= p \int\limits_\Omega \Big(\int\limits_0^{h_n(\omega)} \vartheta^{p-2} \, d\vartheta \Big) u \, d\mu \;=\; \tfrac{p}{p-1} \int\limits_\Omega u h_n^{p-1} \, d\mu \\
&\leqslant \tfrac{p}{p-1} \|u\|_p \Big(\int\limits_\Omega h_n^{(p-1)p'} \, d\mu \Big)^{\frac{1}{p'}} \;=\; \tfrac{p}{p-1} \|u\|_p \|h_n\|_p^{p-1}
\end{aligned}$$

(recall that $p' = \frac{p}{p-1}$), so

$$\|h_n\|_p \leqslant \tfrac{p}{p-1} \|u\|_p \qquad \forall\, n \geqslant 1.$$

We let $n \to +\infty$ and use the Lebesgue monotone convergence theorem (see Theorem I.3.92), to conclude that

$$\|h\|_p \leqslant \tfrac{p}{p-1} \|u\|_p.$$

Solution of Problem 1.47

Let $\{u_n\}_{n \geqslant 1} \subseteq L^p(\Omega)$ be a sequence such that $u_n \longrightarrow u$ in $L^p(\Omega)$. By passing to a subsequence if necessary, we may assume that

$$u_n(\omega) \longrightarrow u(\omega) \quad \text{for } \mu \text{ a.a. } \omega \subset \Omega$$

and

$$|u_n(\omega)| \leqslant \xi(\omega) \quad \text{for } \mu\text{-a.a. } \omega \in \Omega, \text{ all } n \geqslant 1,$$

with $\xi \in L^p(\Omega)$. Note that

$$(u_n h)(\omega) \longrightarrow (uh)(\omega) \quad \text{for } \mu\text{-a.a. } \omega \in \Omega$$

and

$$|(u_n h)(\omega)| \leqslant |(\xi h)(\omega)| \quad \text{for } \mu\text{-a.a. } \omega \in \Omega, \text{ all } n \geqslant 1$$

and by hypothesis $\xi h \in L^1(\Omega)$. Therefore, by virtue of the Lebesgue dominated convergence theorem (see Theorem I.3.94), for the original sequence, we have

$$u_n h \longrightarrow uh \quad \text{in } L^1(\Omega),$$

so

$$K(u_n) \longrightarrow K(u) \quad \text{in } L^1(\Omega).$$

Thus, if $L\colon L^p(\Omega) \longrightarrow \mathbb{R}$ is defined by $L(u) = \int\limits_\Omega uh\,d\mu$, then $L \in L^p(\Omega)^*$. Therefore by the Riesz representation theorem, there exists a unique $\widehat{h} \in L^{p'}(\Omega)$ such that

$$L(u) = \int\limits_\Omega u\widehat{h}\,d\mu \qquad \forall\, u \in L^p(\Omega),$$

so

$$\int\limits_\Omega u(\widehat{h} - h)\,d\mu = 0 \qquad \forall\, u \in L^p(\Omega)$$

and thus $\widehat{h} = h \in L^{p'}(\Omega)$.

Solution of Problem 1.48

Consider the map $K\colon C([0,1]) \longrightarrow \mathbb{R}$ defined by

$$K(u) = u(0) \qquad \forall\, u \in C([0,1]).$$

Clearly K is a linear continuous functional on $C([0,1])$ (that is, $K \in C([0,1])^*$). Since $C([0,1]) \subseteq L^\infty(0,1)$, by the Hahn–Banach theorem (see Theorem I.5.24), we can find $\widehat{K} \in L^\infty(0,1)^*$ such that $\widehat{K}|_{C([0,1])} = K$. If $L^1(0,1)$ is reflexive, then $L^\infty(0,1)^* = L^1(0,1)$ (recall that $L^1(0,1)^* = L^\infty(0,1)$; see Theorem 1.21). So, we can find $h \in L^1(0,1)$ such that

$$\widehat{K}(u) = \int_0^1 uh\,d\mu \qquad \forall\, u \in L^\infty(0,1),$$

so

$$\widehat{K}(u) = \int_0^1 uh\,d\mu = u(0) \qquad \forall\, u \in C([0,1]).$$

Let

$$u_n(t) = \max\{1 - nt, 0\} \qquad \forall\, t \in [0,1].$$

Then $u_n \in C([0,1])$ and

$$\widehat{K}(u_n) = u_n(0) = 1 \qquad \forall\, n \geqslant 1. \tag{1.57}$$

Note that $u_n(t) \longrightarrow 0$ for all $t > 0$. So, by the Lebesgue dominated convergence theorem (see Theorem I.3.94), we have

$$\widehat{K}(u_n) = \int_0^1 u_n h \, dt \longrightarrow 0,$$

which contradicts (1.57). This proves the nonreflexivity of $L^1(0,1)$.

Finally, if $L^\infty(0,1)$ is reflexive, then so is $L^1(0,1)$ (see Theorem I.5.76), a contradiction.

Solution of Problem 1.49

We argue by contradiction. So, suppose that there exists a normed space V such that $V^* = L^1(0,1)$. Then from the Alaoglu theorem (see Theorem I.5.66), we have that $\overline{B}_1^{L^1}$ is w^*-compact (where $\overline{B}_1^{L^1} = \{u \in L^1(0,1) : \|u\|_1 \leqslant 1\}$). We claim that $\operatorname{ext} \overline{B}_1^{L^1} = \emptyset$ (see Definition I.5.127). To see this, let $u \in \overline{B}_1^{L^1}$ and let $b \in (0,1)$ be such that

$$\int_0^b |u| \, ds = \int_b^1 |u| \, ds = \tfrac{\|u\|_1}{2}.$$

Since $t \longmapsto \int_0^t |u| \, ds$ is continuous such that b exists. We have

$$u = \tfrac{1}{2}(2u\chi_{[0,b]}) + \tfrac{1}{2}(2u\chi_{[b,1]}).$$

Note that $2u\chi_{[0,b]}, 2u\chi_{[b,1]} \in \overline{B}_1^{L^1}$. So, we infer that u is not extremal and since $u \in \overline{B}_1^{L^1}$ was arbitrary, we conclude that $\operatorname{ext} \overline{B}_1^{L^1} = \emptyset$. On the other hand, by the Krein–Milman theorem (see Theorem I.5.128), we must have $\operatorname{ext} \overline{B}_1^{L^1} \neq \emptyset$, a contradiction.

Solution of Problem 1.50

We know that $L \in \mathcal{L}\big(L^1(\Omega_1)_w, L^1(\Omega_2)_w\big)$, where $L^1(\Omega_k)_w$ denotes the Lebesgue space $L^1(\Omega_k)$ furnished with the weak topology, for $k = 1,2$ (see Proposition I.5.61). So, we have

$$L(\overline{C}^w) \subseteq \overline{L(C)}^w.$$

From the Dunford–Pettis theorem (see Theorem 1.28), we have that $\overline{C}^w$ is w-compact in $L^1(\Omega_1)$. Hence $L(\overline{C}^w)$ is w-compact in $L^1(\Omega_2)$ and so

$$L(\overline{C}^w) = \overline{L(C)}^w,$$

thus the set $L(C)$ is relatively w-compact in $L^1(\Omega_2)$ and hence the set $L(C) \subseteq L^1(\Omega_2)$ is uniformly integrable (again from the Dunford–Pettis theorem).

Solution of Problem 1.51

Without any loss of generality we may assume that $u \geqslant 0$. Since the measure space is σ-finite, we can find a sequence $\{A_n\}_{n\geqslant 1} \subseteq \Sigma$ such that $\mu(A_n) < +\infty$, $A_n \subseteq A_{n+1}$ for all $n \geqslant 1$ and $\Omega = \bigcup\limits_{n\geqslant 1} A_n$. We set

$$u_n = \chi_{A_n} u \in L^1(\Omega) \qquad \forall\, n \geqslant 1.$$

Evidently $u_n \nearrow u$ and so by the Lebesgue monotone convergence theorem (see Theorem I.3.92), we have

$$\int\limits_\Omega u_n \, d\mu \nearrow \int\limits_\Omega u \, d\mu.$$

By hypothesis $\int\limits_\Omega u_n \, d\mu = \int\limits_{A_n} u \, d\mu \leqslant \eta$ for all $n \geqslant 1$, hence $\int\limits_\Omega u \, d\mu \leqslant \eta$.

Solution of Problem 1.52

(a) Note that the set C is convex. So, by the Mazur theorem (see Theorem I.5.58) to show that C is weakly closed, it suffices to show that it is strongly closed. So, let $\{u_n\}_{n\geqslant 1} \subseteq C$ be a sequence such that $u_n \longrightarrow u$ in $L^1(\Omega)$. Then by passing to a subsequence if necessary, we may assume that $u_n(\omega) \longrightarrow u(\omega)$ for μ-almost all $\omega \in \Omega$. Since $u_n(\omega) \geqslant 1$ for μ-almost all $\omega \in \Omega$, we have that $u(\omega) \geqslant 1$ for μ-almost all $\omega \in \Omega$ and so $u \in C$. Therefore C is weakly closed.

(b) No. Part **(b)** of Definition 1.18 is clearly satisfied, but we will show that part **(a)** of this definition does not hold. Proceeding by contradiction, assume that the sequence $\{u_n = \chi_{[n,n+1]}\}_{n\geqslant 1} \subseteq L^1(\mathbb{R})$ is uniformly integrable. Let $\varepsilon = \frac{1}{2}$. Then there exists a set $A \in \Sigma$ such that $\mu(A) < +\infty$ and

$$\sup_{n\geqslant 1} \int\limits_{A^c} \chi_{[n,n+1]}(x) \, dx \leqslant \tfrac{1}{2}.$$

In particular, we have

$$\tfrac{1}{2} \geqslant \int_{A^c} \chi_{[n,n+1]}(x)\,dx = \lambda(A^c \cap [n,n+1]) \qquad \forall\, n \geqslant 1$$

(λ being the Lebesgue measure in $\mathbb{R}$). Then

$$\lambda(A \cap [n,n+1]) = 1 - \lambda(A^c \cap [n,n+1]) \geqslant 1 - \tfrac{1}{2} = \tfrac{1}{2}$$

and so

$$\lambda(A) \geqslant \sum_{n \geqslant 1} \lambda(A \cap [n,n+1]) = +\infty,$$

a contradiction. Thus the sequence $\{\chi_{[n,n+1]}\}_{n\geqslant 1} \subseteq L^1(\mathbb{R})$ is not uniformly integrable.

Solution of Problem 1.53

(a) Let $p \in (1,+\infty)$. Since $\mu(\Omega) < +\infty$, we have $L^p(\Omega) \subseteq L^1(\Omega)$ for all $p \in (1,+\infty)$. So, the set C is L^1-bounded. Also, for every $A \in \Sigma$, by the Hölder inequality (see Theorem 1.3 and Problem 1.27), we have

$$\int_A |u|\,d\mu \leqslant \|u\|_p \|\chi_A\|_{p'} \leqslant M\mu(A)^{\frac{1}{p'}}, \tag{1.58}$$

with $M = \sup_{u \in C} \|u\|_p < +\infty$. So, given $\varepsilon > 0$, if we choose $\delta = (\frac{\varepsilon}{M})^{p'} > 0$, we have

$$\int_A |u|\,d\mu \leqslant \varepsilon \qquad \forall\, u \in C,\ A \in \Sigma \text{ with } \mu(A) \leqslant \delta$$

(see (1.58)), so the set C is uniformly integrable (see Problem 1.6).

If $p = +\infty$, then reasoning is even simpler (we do not need use the Hölder inequality in the above proof).

(b) Let $\{u_n\}_{n\geqslant 1} \subseteq L^1(0,1)$ be the sequence defined by

$$u_n(x) = \begin{cases} n & \text{if } x \in (0, \frac{1}{n}), \\ 0 & \text{if } x \in [\frac{1}{n}, 1), \end{cases}$$

for all $n \geqslant 1$. Then $\|u_n\|_1 = 1$ for all $n \geqslant 1$, so the sequence $\{u_n\}_{n\geqslant 1} \subseteq L^1(0,1)$ is bounded. But choosing $\varepsilon = \frac{1}{2}$, for any $\delta > 0$ we can find $n_0 \geqslant 1$ such that $\frac{1}{n_0} < \delta$ and then for $A = (0, \frac{1}{n_0})$ and for u_{n_0}, we have

$$\lambda(A) < \delta \quad \text{and} \quad \int_A |u_{n_0}|\,d\lambda = 1 > \varepsilon$$

(λ being the Lebesgue measure in $(0,1)$). Thus the sequence $\{u_n\}_{n\geqslant 1} \subseteq L^1(0,1)$ is not uniformly integrable.

(c) First note that in a finite measure space (Ω, Σ, μ), the condition of part **(a)** of the definition of uniform integrability (see Definition 1.18) is satisfied for any set $C \subseteq L^1(\Omega)$, so the condition of part **(b)** of this definition is the only one which is needed for the set C to be uniformly integrable.
We will show that in an infinite measure space parts **(a)** and **(b)** of the definition are independent. Let $\Omega = \mathbb{R}$, Σ the Lebesgue σ-field on $[0,1]$ and $\mu = \lambda$ the Lebesgue measure on $[0,1]$. First consider the sequence $\{u_n\}_{n\geqslant 1} \subseteq L^1(\mathbb{R})$ defined by

$$u_n(x) = \begin{cases} n & \text{if } x \in (0, \frac{1}{n}), \\ 0 & \text{if } x \in \mathbb{R} \setminus (0, \frac{1}{n}), \end{cases}$$

for all $n \geqslant 1$. Similarly, as in the solution of part **(b)** of this problem, we easily show that the sequence $\{u_n\}_{n\geqslant 1}$ is $L^1(\mathbb{R})$-bounded (thus satisfies part **(a)** of the definition of uniform integrability) and does not satisfy part **(b)** of this definition. On the other hand, take the sequence $\{u_n = \chi_{[n,n+1)}\}_{n\geqslant 1} \subseteq L^1(\mathbb{R})$ of characteristic functions of the intervals $[n, n+1)$. As in the last part of the solution of Problem 1.6, we can show that it satisfies part **(b)** but not part **(a)** of Definition 1.18.

Solution of Problem 1.54

Let $p \in [1,2)$. We have

$$\begin{aligned} \sum_{n\geqslant 1} \lambda(\{u^p \geqslant n\}) &= \sum_{n\geqslant 1} \lambda(\{u \geqslant n^{\frac{1}{p}}\}) \\ &\leqslant \sum_{n\geqslant 1} \frac{1}{1+n^{\frac{2}{p}}} < \sum_{n\geqslant 1} \frac{1}{n^{\frac{2}{p}}} < +\infty, \end{aligned}$$

so $u \in L^p(0,1)$.

On the other hand, let $u(t) = \frac{1}{\sqrt{t}} - 1$ for all $t \in [0,1]$. Then $u \notin L^2(0,1)$, but

$$\begin{aligned} \lambda(\{u \geqslant \vartheta\}) &= \lambda(\{t \in (0,1) : \tfrac{1}{\sqrt{t}} - 1 \geqslant \vartheta\}) \\ &= \lambda(\{t \in (0,1) : t \leqslant \tfrac{1}{(1+\vartheta)^2}\}) \\ &= \tfrac{1}{(1+\vartheta)^2} < \tfrac{1}{1+\vartheta^2} \qquad \forall\, \vartheta > 0. \end{aligned}$$

Therefore $u \notin L^p(0,1)$ for all $p \geqslant 2$. So, the minimum value of p is 2.

Solution of Problem 1.55

Note that $L^1(\mathbb{R}) \cap L^\infty(\mathbb{R}) \subseteq L^p(\mathbb{R})$ for every $1 \leqslant p < +\infty$. Indeed, if $u \in L^1(\mathbb{R}) \cap L^\infty(\mathbb{R})$, then

$$\|u\|_p^p \leqslant \|u\|_\infty^{p-1}\|u\|_1,$$

so $u \in L^p(\mathbb{R})$.

For any $u \in L^p(\mathbb{R})$, we set

$$u_n = \chi_{(-n,n)}u \in L^1(\mathbb{R}) \cap L^\infty(\mathbb{R})$$

and we have $\|u - u_n\|_p \longrightarrow 0$.

Next we show that $L^1(\mathbb{R}) \cap L^\infty(\mathbb{R})$ is not dense in $L^\infty(\mathbb{R})$. Let $u \equiv 1$. Then $u \in L^\infty(\mathbb{R})$ and proceeding by contradiction suppose that there exists $v \in L^1(\mathbb{R}) \cap L^\infty(\mathbb{R})$ such that $\|v - u\|_\infty \leqslant \frac{1}{2}$. Thus

$$\begin{aligned} \operatorname*{ess\,inf}_{x\in\mathbb{R}} v(x) &= 1 - \big(1 - \operatorname*{ess\,inf}_{x\in\mathbb{R}} v(x)\big) \geqslant 1 - \operatorname*{ess\,sup}_{x\in\mathbb{R}}(1 - v(x)) \\ &\geqslant 1 - \operatorname*{ess\,sup}_{x\in\mathbb{R}} |1 - v(x)| = 1 - \|u - v\|_\infty \geqslant \tfrac{1}{2}, \end{aligned}$$

so obviously $v \notin L^1(\mathbb{R})$, a contradiction.

Solution of Problem 1.56

Recall that for every $\xi \in L^\infty(\Omega)^*$, we have

$$\|\xi\|_* = \sup\big\{|\xi(u)| : \|u\|_\infty \leqslant 1\big\}.$$

So, given $\varepsilon > 0$, we can find $u_1, u_2 \in L^\infty(\Omega)$ with $\|u_1\|_\infty, \|u_2\|_\infty \leqslant 1$ such that

$$\|\xi_{A_1}\|_* - \varepsilon \leqslant \xi_{A_1}(u_1) \quad \text{and} \quad \|\xi_{A_2}\|_* - \varepsilon \leqslant \xi_{A_2}(u_2). \tag{1.59}$$

We define

$$u(\omega) = \begin{cases} u_1(\omega) & \text{if } \omega \in A_1, \\ u_2(\omega) & \text{if } \omega \in \Omega \setminus A_1. \end{cases}$$

We have

$$\begin{aligned} \xi_{A_1\cup A_2}(u) &= \xi(\chi_{A_1\cup A_2}u) = \xi(\chi_{A_1}u + \chi_{A_2}u) \\ &= \xi(\chi_{A_1}u) + \xi(\chi_{A_2}u) \geqslant \|\xi_{A_1}\|_* + \|\xi_{A_2}\|_* - 2\varepsilon \end{aligned} \tag{1.60}$$

(recall that $A_1 \cap A_2 = \emptyset$ and see (1.59)). But note that $\|u\|_\infty \leqslant 1$, so from (1.60) we have

$$\|\xi_{A_1\cup A_2}\|_* \geqslant \|\xi_{A_1}\|_* + \|\xi_{A_2}\|_* - 2\varepsilon.$$

Let $\varepsilon \searrow 0$, to conclude that

$$\|\xi_{A_1 \cup A_2}\|_* \geqslant \|\xi_{A_1}\|_* + \|\xi_{A_2}\|_*. \tag{1.61}$$

Next, for every $u \in L^\infty(\Omega)$ with $\|u\|_\infty \leqslant 1$ we have

$$\begin{aligned} |\xi_{A_1 \cup A_2}(u)| &= |\xi(\chi_{A_1 \cup A_2} u)| = |\xi(\chi_{A_1} u) + \xi(\chi_{A_2} u)| \\ &\leqslant |\xi_{A_1}(u)| + |\xi_{A_2}(u)| \end{aligned}$$

(since $A_1 \cap A_2 = \emptyset$), so

$$\|\xi_{A_1 \cap A_2}\|_* \leqslant \|\xi_{A_1}\|_* + \|\xi_{A_2}\|_*. \tag{1.62}$$

From (1.61) and (1.62), we conclude that $\|\xi_{A_1 \cap A_2}\|_* = \|\xi_{A_1}\|_* + \|\xi_{A_2}\|_*$.

Solution of Problem 1.57

For every $n \geqslant 1$, let $A_n \in \Sigma$ be such that

$$\mu(A_n) \leqslant \tfrac{1}{2^n} \quad \text{and} \quad \|\xi_{A_n}\|_* \geqslant \|\xi\|_* - \tfrac{1}{2^n}.$$

By virtue of Problem 1.56, we have

$$\|\xi_{A_n^c}\|_* \leqslant \tfrac{1}{2^n} \qquad \forall\, n \geqslant 1.$$

Let $C_n = \bigcup\limits_{k \geqslant n+1} A_k \in \Sigma$. Then

$$\mu(C_n) \leqslant \tfrac{1}{2^n} \qquad \forall\, n \geqslant 1.$$

Let $u \in L^\infty(\Omega)$ be such that $u|_{C_n} = 0$. Then for $k \geqslant n+1$, we have $u = \chi_{A_k^c} u$ and so

$$|\xi(u)| = |\xi_{A_k^c}(u)| \leqslant \|\xi_{A_k^c}\|_* \|u\|_\infty \leqslant \tfrac{1}{2^n} \|u\|_\infty,$$

so ξ is supported by $\{C_n\}_{n \geqslant 1}$ and $C_{n+1} \subseteq C_n$ for all $n \geqslant 1$ and $\mu(C_n) \searrow 0$. Thus ξ is singular.

Solution of Problem 1.58

To fix things, we assume that each u_n is increasing. First we are going to show that u is increasing too. Arguing by contradiction, suppose that u is not increasing on $[0,b]$. Then we can find $x_1, x_2 \in [0,b]$, $x_1 < x_2$ such that $u(x_1) > u(x_2)$. Let $h = u(x_1) - u(x_2) > 0$. Since $u_n \longrightarrow u$ pointwise, we can find $n_0 \geqslant 1$ such that

$$|u_n(x_1) - u(x_1)| < \tfrac{h}{2} \quad \text{and} \quad |u_n(x_2) - u(x_2)| < \tfrac{h}{2} \qquad \forall\, n \geqslant n_0.$$

Then

$$\begin{aligned} u_n(x_2) - u_n(x_1) &< u(x_2) + \tfrac{h}{2} + \tfrac{h}{2} - u(x_1) \\ &= h - \big(u(x_1) - u(x_2)\big) \;=\; h - h \;=\; 0 \qquad \forall\, n \geqslant n_0. \end{aligned} \tag{1.63}$$

On the other hand, we have

$$0 \leqslant u_n(x_2) - u_n(x_1) \qquad \forall\, n \geqslant 1. \tag{1.64}$$

Comparing (1.63) and (1.64), we reach a contradiction. This proves that u is increasing.

Next we show that the convergence on $[0,b]$ is in fact uniform. We already know that u is increasing. Moreover, since $u \in C([0,b])$, it is uniformly continuous. So, given $\varepsilon > 0$, we can find $x_1, \ldots, x_m$ such that $0 = x_1 < x_2 < \ldots < x_m = b$ and

$$0 \leqslant u(x_k) - u(x_{k-1}) \leqslant \varepsilon \qquad \forall\, k \in \{2,3,\ldots,m\}.$$

Thus using also the monotonicity of u, we have

$$0 \leqslant u(x_k) - u(x) \leqslant \varepsilon \qquad \forall\, x \in [x_{k-1}, x_k],\ k \in \{2,3,\ldots,m\}. \tag{1.65}$$

Since $u_n \longrightarrow u$ pointwise on $[0,b]$, for all $n \geqslant n_0 = n_0(\varepsilon) \geqslant 1$, we can have

$$|u_n(x_k) - u(x_k)| \leqslant \varepsilon \qquad \forall\, k \in \{1,2,\ldots,m\}. \tag{1.66}$$

Let $x \in [0,b]$. Then $x \in [x_{k-1}, x_k]$ for some $k \in \{2,3,\ldots,m\}$. So, we have

$$u_n(x) \leqslant u_n(x_k) \leqslant u(x_k) + \varepsilon \leqslant u(x) + 2\varepsilon \qquad \forall\, n \geqslant n_0. \tag{1.67}$$

(see (1.65) and (1.66)). Similarly, we show

$$u(x) - 2\varepsilon \leqslant u_n(x) \qquad \forall\, n \geqslant n_0. \tag{1.68}$$

From (1.67) and (1.68) and since $n_0 \geqslant 1$ is independent of $x \in [0,b]$, we conclude that $u_n \longrightarrow u$ uniformly on $[0,b]$.

Solution of Problem 1.59

We have

$$\begin{aligned} \|u_n - u\|_2^2 &= \int_\Omega (u_n - u)^2 \, d\mu \\ &= \int_\Omega u_n^2 \, d\mu - 2 \int_\Omega u_n u \, d\mu + \int_\Omega u^2 \, d\mu \qquad \forall\, n \geqslant 1. \end{aligned} \tag{1.69}$$

Since by hypothesis $u_n^2 \xrightarrow{w} u^2$ in $L^1(\Omega)$, by choosing $h \equiv 1 \in L^\infty(\Omega)$, we have

$$\int_\Omega u_n^2 \, d\mu = \int_\Omega u_n^2 h \, d\mu \longrightarrow \int_\Omega u^2 h \, d\mu = \int_\Omega u^2 \, d\mu. \tag{1.70}$$

Also since $u_n \xrightarrow{w} u$ in $L^2(\Omega)$, then $u \in L^2(\Omega)$ and so

$$\int_\Omega u_n u \, d\mu \longrightarrow \int_\Omega u^2 \, d\mu. \tag{1.71}$$

So, if in (1.69) we pass to the limit as $n \to +\infty$ and we use (1.70) and (1.71), then we have $u_n \longrightarrow u$ in $L^2(\Omega)$.

Solution of Problem 1.60

The Pettis measurability theorem (see Theorem 1.36) implies that we need to consider functions with values in a nonseparable Banach space. So, let $X = l^2([0,1])$. This is a nonseparable Hilbert space. Let $\{e_t\}_{t \in [0,1]}$ be an orthonormal basis for X. We consider the function $u\colon [0,1] \longrightarrow l^2([0,1])$ defined by

$$u(t) = e_t \qquad \forall\, t \in [0,1].$$

This function is weakly measurable. Indeed, for every $u^* \in l^2([0,1])^* = l^2([0,1])$ we have

$$(u^*, u(t))_X = (u^*, e_t)_X = 0 \quad \text{for a.a. } t \in [0,1].$$

On the other hand, if $A \subseteq [0,1]$, then the set $u([0,1] \setminus A)$ is separable if and only if $[0,1] \setminus A$ is countable, which means that A cannot be Lebesgue-null. So, by virtue of the Pettis measurability theorem (see Theorem 1.36), u is not strongly measurable.

Solution of Problem 1.61

Note that

$$\|u(\omega)\|_X \leqslant h(\omega) \quad \text{for } \mu\text{-a.a. } \omega \in \Omega.$$

Also, for every $n \geqslant 1$ the function $\omega \longmapsto \|u(\omega) - u_n(\omega)\|_X$ is Σ-measurable and we have

$$\|u(\omega) - u_n(\omega)\|_X \leqslant 2h(\omega) \quad \text{for } \mu\text{-a.a. } \omega \in \Omega, \text{ all } n \geqslant 1.$$

Then the Lebesgue dominated convergence theorem (see Theorem I.3.94) implies that

$$\int_\Omega \|u(\omega) - u_n(\omega)\|_X \, d\mu \longrightarrow 0. \tag{1.72}$$

Since u_n is Bochner integrable, by Definition 1.38(b), we can find a simple function $s_n \colon \Omega \longrightarrow X$ such that

$$\int\limits_{\Omega} \|u_n(\omega) - s_n(\omega)\|_X \, d\mu \leqslant \tfrac{1}{n} \qquad \forall\, n \geqslant 1. \tag{1.73}$$

Then we have

$$\begin{aligned} &\int\limits_{\Omega} \|u(\omega) - s_n(\omega)\|_X \, d\mu \\ &\leqslant \int\limits_{\Omega} \|u(\omega) - u_n(\omega)\|_X \, d\mu + \int\limits_{\Omega} \|u_n(\omega) - s_n(\omega)\|_X \, d\mu \longrightarrow 0 \end{aligned} \tag{1.74}$$

(see (1.72) and (1.73)), so u is Bochner integrable (see Definition 1.38(b)).

Also, we have

$$\begin{aligned} &\Big\| \int\limits_{\Omega} u(\omega)\, d\mu - \int\limits_{\Omega} u_n(\omega)\, d\mu \Big\|_X \\ &\leqslant \Big\| \int\limits_{\Omega} u(\omega)\, d\mu - \int\limits_{\Omega} s_n(\omega)\, d\mu \Big\|_X + \Big\| \int\limits_{\Omega} s_n(\omega)\, d\mu - \int\limits_{\Omega} u_n(\omega)\, d\mu \Big\|_X \\ &\leqslant \int\limits_{\Omega} \|u(\omega) - s_n(\omega)\|_X \, d\mu + \int\limits_{\Omega} \|s_n(\omega) - u(\omega)\|_X \, d\mu \longrightarrow 0 \end{aligned}$$

(see Proposition 1.40, (1.73) and (1.74)). Therefore we conclude that

$$\int\limits_{\Omega} u_n \, d\mu \longrightarrow \int\limits_{\Omega} u \, d\mu \quad \text{in } X.$$

Solution of Problem 1.62

Clearly $u \colon \Omega \longrightarrow X$ is weakly measurable. Also from the Pettis measurability theorem (see Theorem 1.36), since u_n is strongly measurable for all $n \geqslant 1$, we can find $A_n \in \Sigma$ with $\mu(A_n) = 0$ such that $u_n(\Omega \setminus A_n)$ is separable in X for all $n \geqslant 1$. Also, let $D \in \Sigma$ be the μ-null set such that

$$u_n(\omega) \longrightarrow u(\omega) \qquad \forall\, \omega \in \Omega \setminus D.$$

We set $C = D \cup \bigcup\limits_{n \geqslant 1} A_n$. Then $C \in \Sigma$ and $\mu(C) = 0$. Also, we define

$$Y = \overline{\operatorname{span}} \bigcup_{n \geqslant 1} u_n(\Omega \setminus A_n),$$

which is a separable Banach subspace of X and $u(\Omega \setminus C) \subseteq Y$. Then once again the Pettis measurability theorem (see Theorem 1.36) implies that u is strongly measurable. From Proposition 1.40, we know that for every $n \geqslant 1$ we have $\|u_n(\cdot)\|_X \in L^1(\Omega)$ and so by the Fatou lemma (see Theorem I.3.95), we have

$$\int_\Omega \|u(\omega)\|_X \, d\mu \leqslant \liminf_{n \to +\infty} \int_\Omega \|u_n(\omega)\|_X \, d\mu,$$

thus u is Bochner integrable (see Proposition 1.40).

Solution of Problem 1.63

Since u is Bochner integrable, the function $\omega \longmapsto \|u(\omega)\|_X$ is Σ-measurable. So

$$A_0 = \{\omega \in \Omega : \|u(\omega)\|_X > 0\} \in \Sigma.$$

From Remark 1.37, we know that given $\varepsilon > 0$, we can find countable valued measurable functions $h_\varepsilon \colon A_0 \longrightarrow X$ and $g_\varepsilon \colon A_0 \longrightarrow Y$ and a μ-null set $D \subseteq \Sigma$ such that

$$\sup_{\omega \in \Omega \setminus D} \|u(\omega) - h_\varepsilon(\omega)\|_X < \tfrac{\varepsilon}{2}$$

and

$$\sup_{\omega \in \Omega \setminus D} \|L(u(\omega)) - g_\varepsilon(\omega)\|_Y < \tfrac{\varepsilon}{2}.$$

Let $\{C_n\}_{n \geqslant 1} \subseteq \Sigma \cap A_0$ be a common refinement of the subdivisions of A_0 defining h_ε and g_ε and let $\omega_n \in C_n$ for all $n \geqslant 1$. We introduce the function $u_\varepsilon \colon \Omega \longrightarrow X$ defined by

$$u_\varepsilon(\omega) = \begin{cases} u(\omega_n) & \text{if } \omega \in C_n,\ n \geqslant 1, \\ 0 & \text{if } \omega \in \Omega \setminus A_0. \end{cases}$$

Then we have

$$\int_\Omega \|u(\omega) - u_\varepsilon(\omega)\|_X \, d\mu < \varepsilon\mu(\Omega) \tag{1.75}$$

and

$$\int_\Omega \|L(u(\omega)) - L(u_\varepsilon(\omega))\|_Y \, d\mu < \varepsilon\mu(\Omega). \tag{1.76}$$

Also, for every $A \in \Sigma$, we have

$$\int_A u_\varepsilon(\omega) \, d\mu = \sum_{n \geqslant 1} u(\omega_n)\mu(C_n \cap A) = \lim_{N \to +\infty} \sum_{n=1}^{N} u(\omega_n)\mu(C_n \cap A) \tag{1.77}$$

and

$$\int_A L(u_\varepsilon(\omega))\,d\mu = \sum_{n\geqslant 1} L(u(\omega_n))\mu(C_n\cap A)$$
$$= \lim_{N\to+\infty}\sum_{n=1}^{N} L(u(\omega_n))\mu(C_n\cap A). \tag{1.78}$$

But by hypothesis L is closed. Therefore from (1.77) and (1.78) it follows that

$$\Big(\int_A u_\varepsilon\,d\mu, \int_A L(u_\varepsilon)\,d\mu\Big) \in \operatorname{Gr} L. \tag{1.79}$$

Now let $\varepsilon_n \searrow 0$ and let $u_n = u_{\varepsilon_n}$. Then from (1.75) and (1.76) (they are also valid if Ω is replaced by A), we have

$$\int_A u_n\,d\mu \longrightarrow \int_A u\,d\mu \quad\text{and}\quad \int_A L(u_n)\,d\mu \longrightarrow \int_A L(u)\,d\mu, \tag{1.80}$$

and from (1.79), we have

$$\int_A L(u_n)\,d\mu = L\Big(\int_A u_n\,d\mu\Big) \qquad \forall\, n\geqslant 1.$$

Then (1.80) and the closedness of L imply that

$$\int_A L(u)\,d\mu = L\Big(\int_A u\,d\mu\Big).$$

Solution of Problem 1.64

By replacing u by $u-y$, we see that we may assume that $y=0$. Moreover, by virtue of the Pettis measurability theorem (see Theorem 1.36) without any loss of generality we may assume that X is a separable Banach space. Then the dual unit ball $\overline{B}_1^* = \big\{x^*\in X^* : \|x^*\|_* \leqslant 1\big\}$ furnished with the relative w^*-topology, is compact and metrizable. Therefore, we can find a sequence $\{x_n^*\}_{n\geqslant 1}\subseteq \overline{B}_1^*$ which is w^*-dense. Then using Problem 1.63, we have

$$\langle x_n^*, \int_A u\,d\mu\rangle = \int_A \langle x_n^*, u\rangle\,d\mu \qquad \forall\, n\geqslant 1,\ A\in\Sigma,$$

so

$$\int_A \langle x_n^*, u\rangle\,d\mu = 0$$

(recall we have assumed that $y = 0$) and thus

$$\langle x_n^*, u(\omega) \rangle = 0 \qquad \forall\, \omega \in \Omega \setminus D_n, \text{ with } \mu(D_n) = 0,\ n \geqslant 1.$$

Let $D = \bigcup\limits_{n \geqslant 1} D_n$. Then $D \in \Sigma$ and $\mu(D) = 0$. Also, we have

$$\langle x_n^*, u(\omega) \rangle = 0 \qquad \forall\, \omega \in \Omega \setminus D,\ n \geqslant 1,$$

so

$$\|u(\omega)\|_X = 0 \qquad \forall\, \omega \in \Omega \setminus D$$

(recall that the sequence $\{x_n^*\}_{n \geqslant 1}$ is w^*-dense in $\overline{B}_1^*$), thus finally

$$u(\omega) = y(\omega) \quad \text{for } \mu\text{-a.a. } \omega \in \Omega.$$

Solution of Problem 1.65

As in the solution of Problem 1.64, we may assume that $y = 0$ and that X is separable. So, $\overline{B}_1^* = \{x^* \in X^* : \|x^*\|_* \leqslant 1\}$ equipped with the relative w^*-topology is compact metrizable. So, we can find a sequence $\{x_n^*\}_{n \geqslant 1} \subseteq \overline{B}_1^*$ which is w^*-dense. By hypothesis, we have

$$\langle x_n^*, u(\omega) \rangle = 0 \qquad \forall\, \omega \in \Omega \setminus D_n,\ \mu(D_n) = 0,\ n \geqslant 1$$

(recall that $y = 0$). Let $D = \bigcup\limits_{n \geqslant 1} D_n$. Then $D \in \Sigma$, $\mu(D) = 0$ and

$$\langle x_n^*, u(\omega) \rangle = 0 \qquad \forall\, \omega \in \Omega \setminus D,\ n \geqslant 1.$$

Exploiting the w^*-density of the sequence $\{x_n^*\}_{n \geqslant 1} \subseteq \overline{B}_1^*$, we conclude that

$$\langle x^*, u(\omega) \rangle = 0 \qquad \forall\, \omega \in \Omega \setminus D,\ x^* \in \overline{B}_1^*,$$

so

$$\|u(\omega)\|_X = 0 \qquad \forall\, \omega \in \Omega \setminus D,$$

thus

$$u(\omega) = 0 \quad \text{for } \mu\text{-a.a. } \omega \in \Omega.$$

Solution of Problem 1.66

We argue by contradiction. So, suppose that $\frac{1}{\mu(A)} \int\limits_A u(\omega)\, d\mu \notin \overline{\operatorname{conv}}\, u(A)$. Then by the strong separation theorem (see Theorem I.5.29), we can find $x^* \in X^* \setminus \{0\}$ and $\eta \in \mathbb{R}$ such that

$$\langle x^*, \frac{1}{\mu(A)} \int\limits_A u\, d\mu \rangle \;<\; \eta \;<\; \langle x^*, u(\omega) \rangle \qquad \forall\, \omega \in A,$$

so

$$\frac{1}{\mu(A)} \int\limits_A \langle x^*, u \rangle\; d\mu \;<\; \eta \;<\; \langle x^*, u(\omega) \rangle \qquad \forall\, \omega \in A$$

(see Problem 1.63) and thus

$$\int\limits_A \langle x^*, u \rangle\; d\mu \;<\; \eta\mu(A) \;<\; \int\limits_A \langle x^*, u \rangle\; d\mu,$$

a contradiction.

Solution of Problem 1.67

Let $\{A_n\}_{n \geqslant 1} \subseteq \Sigma$ be a sequence of pairwise disjoint sets. From Proposition 1.40, we know that

$$\| \int\limits_{A_n} u\, d\mu \|_X \;\leqslant\; \int\limits_{A_n} \|u\|_X\, d\mu \qquad \forall\, n \geqslant 1.$$

Therefore in the series

$$\sum_{n \geqslant 1} m(A_n) \;=\; \sum_{n \geqslant 1} \int\limits_{A_n} u\, d\mu$$

each term is dominated by $\int\limits_{A_n} \|u\|_X\, d\mu$ and from the properties of the Lebesgue integral $A \longmapsto \int\limits_A \|u\|_X\, d\mu$ is a finite measure on Σ. Therefore the series $\sum\limits_{n \geqslant 1} m(A_n)$ is absolutely convergent. Also, we have

$$\begin{aligned} &\|m(\bigcup_{n \geqslant 1} A_n) - \sum_{n=1}^{k} \int\limits_{A_n} u\, d\mu\|_X \\ &= \|\int\limits_{\bigcup\limits_{n \geqslant k+1} A_n} u\, d\mu\|_X \;\leqslant\; \int\limits_{\bigcup\limits_{n \geqslant k+1} A_n} \|u\|_X\, d\mu \;\longrightarrow\; 0 \quad \text{as } k \to +\infty, \end{aligned}$$

so

$$m\Big(\bigcup_{n\geqslant 1} A_n\Big) \;=\; \sum_{n\geqslant 1} m(A_n)$$

and thus m is a vector measure.

Since $\|u(\cdot)\|_X \in L^1(\Omega)$ (see Proposition 1.40), from the absolute continuity of the Lebesgue integral, we have

$$\lim_{\mu(A)\searrow 0} \int_A \|u\|_X \, d\mu \;=\; 0.$$

But

$$\|m(A)\|_X \;\leqslant\; \int_A \|u\|_X \, d\mu \qquad \forall\, A \in \Sigma.$$

Therefore

$$\lim_{\mu(A)\searrow 0} m(A) \;=\; 0 \quad \text{in } X.$$

Solution of Problem 1.68

"$\Longrightarrow$": This is obvious.

"$\Longleftarrow$": By the Pettis measurability theorem (see Theorem 1.36), we may assume that X is separable. Then $X^*_{w^*}$ is Souslin, in particular then w^*-separable. So, we can find a sequence $\{x_n^*\}_{n\geqslant 1} \subseteq X^*$ which is w^*-dense in X^*. Using Problem 1.63, we have

$$0 \;=\; \Big\langle x_n^*, \int_A u \, d\mu \Big\rangle \;=\; \int_A \langle x_n^*, u\rangle \, d\mu \qquad \forall\, A \in \Sigma,\ n \geqslant 1,$$

so

$$\langle x_n^*, u(\omega)\rangle \;=\; 0 \qquad \forall\, \omega \in \Omega \setminus D_n,\ \mu(D_n) = 0,\ n \geqslant 1.$$

Let $D = \bigcup_{n\geqslant 1} D_n$. Then $D \in \Sigma$, $\mu(D) = 0$ and

$$\langle x_n^*, u(\omega)\rangle \;=\; 0 \qquad \forall\, \omega \in \Omega \setminus D,\ n \geqslant 1,$$

thus $u \equiv 0$ (since the sequence $\{x_n^*\}_{n\geqslant 1}$ is w^*-dense in X^*).

Solution of Problem 1.69

Fix $A \in \Sigma$ and define $\xi \colon \Sigma_0 \longrightarrow [0, \mu(\Omega)]$ by

$$\xi(C) \;=\; \mu(C \cap A) \qquad \forall\, C \in \Sigma_0.$$

Clearly $\xi \ll \mu$ and so by the Radon–Nikodym theorem (see Theorem I.3.152), we can find $\vartheta_A \in L^1(\Omega, \Sigma_0)$ such that

$$\xi(C) = \mu(C \cap A) = \int_C \vartheta_A \, d\mu \qquad \forall\, C \in \Sigma_0.$$

Let $s\colon \Omega \longrightarrow X$ be a simple function, that is,

$$s(\omega) = \sum_{k=1}^{n} x_k \chi_{A_k}(\omega) \qquad \forall\, \omega \in \Omega,$$

where $x_k \in X$, $A_k \in \Sigma$ for $k \in \{1, \ldots, n\}$. We define

$$E^{\Sigma_0}(s) = \sum_{k=1}^{n} x_k \vartheta_{A_k}.$$

So, the conditional expectation operator is well defined and clearly linear on the simple functions which are dense in $L^1(\Omega, \Sigma; X)$. Moreover, we have

$$\begin{aligned} \left\|E^{\Sigma_0}(s)\right\|_1 &\leqslant \int_\Omega \Big(\sum_{k=1}^{n} \|x_k\|_X \vartheta_{A_k}\Big) \, d\mu \\ &\leqslant \Big\|\sum_{k=1}^{n} \|x_k\|_X \chi_{A_k}\Big\|_1 = \|s\|_1. \end{aligned}$$

Therefore, E^{Σ_0} has a contractive linear extension to all of $L^1(\Omega, \Sigma; X)$ which we still denote by E^{Σ_0}. Evidently

$$\int_A u \, d\mu = \int_A E^{\Sigma_0} u \, d\mu \qquad \forall\, A \in \Sigma_0.$$

Solution of Problem 1.70

Consider the linear operator $K\colon X \longrightarrow L^1(\Omega)$ defined by

$$K(x)(\cdot) = \langle u(\cdot), x\rangle \qquad \forall\, x \in X.$$

We claim that K has closed graph. To this end, suppose that $\{x_n\}_{n\geqslant 1} \subseteq X$ is a sequence such that $x_n \longrightarrow x$ in X and $K(x_n) \longrightarrow h$ in $L^1(\Omega)$. By passing to a suitable subsequence if necessary, we may assume that

$$K(x_n)(\omega) \longrightarrow h(\omega) \quad \text{for } \mu\text{-a.a. } \omega \in \Omega.$$

On the other hand,

$$\langle u(\omega), x_n\rangle \longrightarrow \langle u(\omega), x\rangle \quad \text{for } \mu\text{-a.a. } \omega \in \Omega.$$

Therefore $h(\omega) = \langle u(\omega), x\rangle$ and so $\operatorname{Gr} K \subseteq X \times L^1(\Omega)$ is closed. By the closed graph theorem (see Theorem I.5.50), we obtain that $K \in \mathcal{L}(X; L^1(\Omega))$. Also, the map $T_A \colon L^1(\Omega) \longrightarrow \mathbb{R}$ defined by

$$T_A(h) = \int_A h \, d\mu \qquad \forall\, h \in L^1(\Omega)$$

is linear continuous. Therefore $T_A \circ K \in X^*$ and so there exists $x_A^* \in X^*$ such that

$$\langle x_A^*, x\rangle = \int_A \langle u(\omega), x\rangle \, d\mu \qquad \forall\, x \in X.$$

Solution of Problem 1.71

(a) Since $X \hookrightarrow Y$ continuously, we can find $c_1 > 0$ such that

$$\|x\|_Y \leqslant c_1 \|x\|_X \qquad \forall\, x \in X. \tag{1.81}$$

Let $y^* \in Y^*$. Then

$$\big|\langle y^*, x\rangle_Y\big| \leqslant \|y^*\|_{Y^*}\|x\|_Y \leqslant c_1\|y^*\|_{Y^*}\|x\|_X \qquad \forall\, x \in X, \tag{1.82}$$

(see (1.81)). Let $\widehat{y}^* = y^*|_X$. Then from (1.82) we have

$$\|\widehat{y}^*\|_{X^*} \leqslant c_1\|y^*\|_{Y^*}. \tag{1.83}$$

We show that $\widehat{y}^* = 0$ implies that $y^* = 0$. To this end, for $x \in X$ we have

$$0 = \langle \widehat{y}^*, x\rangle_X = \langle y^*, x\rangle_Y$$

(recall that $\widehat{y}^* = y^*|_X$).
The density of the embedding $X \hookrightarrow Y$ implies that $y^* = 0$. Therefore, the map $i^* \colon Y^* \longrightarrow X^*$ defined by $i^*(y^*) = \widehat{y}^*$ is continuous injective and so we conclude that $Y^* \hookrightarrow X^*$ continuously (see (1.83)).

Now suppose that X is reflexive and X is dense in Y. We will show that Y^* is dense in X^*. Arguing by contradiction, suppose that $\overline{Y^*}^{\|\cdot\|_*} \neq X^*$. Then we can find $x_0 \in X^{**} = X$ (recall that X is reflexive), $x_0 \neq 0$ such that

$$\langle y^*, x_0\rangle = 0 \qquad \forall\, y^* \in Y^*. \tag{1.84}$$

Since $x_0 \in X \subseteq Y$, from (1.84) we infer that $x_0 = 0$, a contradiction.

(b) From part **(a)**, we know that the embedding $Y^* \hookrightarrow X^*$ is continuous and dense. According to Remark 1.37, we can find a sequence $\{s_n \colon \Omega \longrightarrow X\}_{n\geqslant 1}$ of Σ-measurable and countably valued functions such that

$$s_n(t) \longrightarrow u(t) \quad \text{uniformly in } t \in T. \tag{1.85}$$

We have

$$\|s_n(t)\|_X \leqslant c_2\|s_n\|_{L^\infty(T;X)} \qquad \forall\, t \in T,\ n \geqslant 1, \tag{1.86}$$

for some $c_2 > 0$ and

$$\langle y^*, s_n(t)\rangle_X \longrightarrow \langle y^*, u(t)\rangle_X \qquad \forall\, y^* \in Y^* \subseteq X^*,\ t \in T. \tag{1.87}$$

Then we can write

$$\big|\langle y^*, s_n(t)\rangle_X\big| \leqslant \|y^*\|_{X^*}\|s_n(t)\|_X \leqslant c_2\|y^*\|_{X^*}\|s_n\|_{L^\infty(T;X)}$$

(see (1.86)), so

$$\big|\langle y^*, u(t)\rangle_X\big| \leqslant c_2\|y^*\|_{X^*}\|u\|_{L^\infty(T;X)} \qquad \forall\, t \in T \tag{1.88}$$

(see (1.85) and (1.87)) and thus

$$u(t) \in X \quad \text{and} \quad \|u(t)\|_X \leqslant c_2\|u\|_{L^\infty(T;X)} \tag{1.89}$$

(since the embedding $Y^* \hookrightarrow X^*$ is continuous and dense).
Let $x^* \in X^*$. We can find a sequence $\{y_n^*\}_{k\geqslant 1} \subseteq Y^*$ such that $y_k^* \longrightarrow x^*$ in X^*. Let $t_n \longrightarrow t$. Since by hypothesis $u \in C(T; Y_w)$, we have

$$\langle y_k^*, u(t_n)\rangle_X = \langle y_k^*, u(t_n)\rangle_Y \longrightarrow \langle y_k^*, u(t)\rangle_Y \quad \text{as } n \to +\infty$$
$$\forall\, k \geqslant 1,\ t \in T.$$

Also, we have

$$\langle y_k^*, u(t)\rangle_Y = \langle y_k^*, u(t)\rangle_X \longrightarrow \langle x^*, u(t)\rangle_X \quad \text{as } k \to +\infty. \tag{1.90}$$

From (1.88) and (1.90) it follows that we can find a sequence $\{k(n)\}_{n\geqslant 1}$ increasing (not necessarily strictly) to $+\infty$ such that

$$\big\langle y_{k(n)}^*, u(t_n)\big\rangle_X \longrightarrow \langle x^*, u(t)\rangle_X \quad \text{as } n \to +\infty. \tag{1.91}$$

Finally we have

$$\begin{aligned}
&\big|\langle x^*, u(t_n)\rangle_X - \langle x^*, u(t)\rangle_X\big| \\
&\leqslant \big|\langle x^*, u(t_n)\rangle_X - \big\langle y_{k(n)}^*, u(t_n)\big\rangle_X\big| + \big|\big\langle y_{k(n)}^*, u(t_n)\big\rangle_X - \langle x^*, u(t)\rangle_X\big| \\
&\leqslant \|x^* - y_{k(n)}^*\|_{X^*}\|u(t_n)\|_X + \big|\big\langle y_{k(n)}^*, u(t_n)\big\rangle_X - \langle x^*, u(t)\rangle_X\big| \\
&\leqslant c_2\|x^* - y_{k(n)}^*\|_{X^*}\|u\|_{L^\infty(T;X)} + \big|\big\langle y_{k(n)}^*, u(t_n)\big\rangle_X - \langle x^*, u(t)\rangle_X\big|
\end{aligned}$$

as $n \to +\infty$ (see (1.89) and (1.91)), so $u \in C(T; X_w)$.

Solution of Problem 1.72

(a) Suppose that the claim of the problem is not true. Then we can find $\varepsilon > 0$ and a sequence $\{x_n\}_{n\geqslant 1} \subseteq X$ such that

$$\|x_n\|_Y > \varepsilon\|x_n\|_X + n\|x_n\|_Z \qquad \forall\, n \geqslant 1.$$

Let $y_n = \frac{x_n}{\|x_n\|_X}$ for all $n \geqslant 1$. We have

$$\|y_n\|_Y > \varepsilon + n\|y_n\|_Z \qquad \forall\, n \geqslant 1. \tag{1.92}$$

Note that $\|y_n\|_X = 1$ for all $n \geqslant 1$ and since the embedding $X \hookrightarrow Y$ is compact, we have

$$\|y_n\|_Z \longrightarrow 0 \tag{1.93}$$

(see (1.92)) and the sequence $\{y_n\}_{n\geqslant 1} \subseteq Y$ is relatively compact. Therefore we can find a subsequence $\{y_{n_k}\}_{k\geqslant 1}$ of $\{y_n\}_{n\geqslant 1}$ such that

$$y_{n_k} \longrightarrow y \quad \text{in } Y,$$

so

$$y_{n_k} \longrightarrow y \quad \text{in } Z$$

since the embedding $Y \hookrightarrow Z$ is continuous. Thus $y = 0$ (see (1.93)).
On the other hand, from (1.92) and by passing to the limit as $n \to +\infty$, we obtain

$$\|y\|_Y \geqslant \varepsilon > 0,$$

a contradiction.

(b) Since the set K is relatively compact in $L^p(T; Z)$, it is totally bounded (see Definition I.1.70 and Theorem I.1.71). So, given $\varepsilon > 0$ we can find $u_1, \ldots, u_n \in K$ such that for every $u \in K$, there exists $k \in \{1, \ldots, n\}$ such that $\|u - u_k\|_{L^p(T;Z)} < \varepsilon$. By virtue of part **(a)**, given $\xi > 0$, we can find $c_\xi > 0$ such that

$$\begin{aligned}\|u - u_k\|_{L^p(T;Y)} &\leqslant \xi\|u - u_k\|_{L^p(T;X)} + c_\xi\|u - u_k\|_{L^p(T;Z)}\\ &\leqslant \xi \mathrm{diam}_{L^p(T;X)}(K) + c_\xi\varepsilon.\end{aligned} \tag{1.94}$$

Given $\varepsilon' > 0$, let $\xi = \frac{\varepsilon'}{2\mathrm{diam}_{L^p(T;X)}(K)}$ and $\varepsilon = \frac{\varepsilon'}{2c_\xi}$. Then from (1.94) we have

$$\|u - u_k\|_{L^p(T;Y)} \leqslant \varepsilon',$$

which implies that the set K is relatively compact in $L^p(T;Y)$.

Solution of Problem 1.73

From Theorems 1.46 and 1.47, we know that the space $L^p(\Omega;X)$ is reflexive. So, by passing to a subsequence if necessary, we may assume that

$$u_n \xrightarrow{w} \widehat{u} \quad \text{in } L^p(\Omega;X), \tag{1.95}$$

with $\widehat{u} \in L^p(\Omega;X)$. Also, by the Mazur theorem (see Corollary I.5.59), there is a sequence $\{y_n\}_{n\geqslant 1} \subseteq L^p(\Omega;X)$ consisting of convex combinations of the u_n's such that

$$y_n \longrightarrow \widehat{u} \quad \text{in } L^p(\Omega;X).$$

Hence by passing to a subsequence if necessary, we have

$$y_n(\omega) \longrightarrow \widehat{u}(\omega) \quad \text{for } \mu\text{-a.a. } \omega \in \Omega,$$

so $\widehat{u} = u$.

The uniqueness of the limit implies that for the original sequence, we have

$$u_n \xrightarrow{w} u \quad \text{in } L^p(\Omega;X)$$

(see (1.95)).

Solution of Problem 1.74

Fot all $h \in \mathbb{R}$ consider the translation operator τ_h defined by

$$\tau_h(u)(t) = u(t+h) \qquad \forall\, u \in L^p(T;X),\ t \in T.$$

We have

$$\|\tau_h(u)\|_{L^p(T;X)} = \Big(\int_0^b \|u(t+h)\|_X^p\,dt\Big)^{\frac{1}{p}}$$

$$\leqslant \Big(\int_0^b \|u(t)\|_X^p\,dt\Big)^{\frac{1}{p}} = \|u\|_{L^p(T;X)} \qquad \forall\, h \in \mathbb{R}. \tag{1.96}$$

Let $A \subseteq T$ be a Lebesgue measurable set and let $x \in X$. Set

$$u_A(t) = \chi_A(t)x \qquad \forall\, t \in T.$$

We have

$$\|\tau_h(u_A) - u_A\|_{L^p(T;X)} = \Big(\int_0^b \|u_A(t+h) - u_A(t)\|_X^p \, dt\Big)^{\frac{1}{p}}$$
$$\leqslant 2|h|\|x\|_X = 2|h|b^{-\frac{1}{p}}\|u_A\|_{L^p(T;X)},$$

so

$$\lim_{h\to 0} \Big(\int_0^b \|u_A(t+h) - u_A(t)\|^p \, dt\Big)^{\frac{1}{p}} = 0.$$

Since τ_h is linear, for every simple function s, we have

$$\|\tau_h(s) - s\|_{L^p(T;X)} \leqslant 2|h|b^{-\frac{1}{p}}\|s\|_{L^p(T;X)}.$$

But from Proposition 1.42, we know that simple functions are dense in $L^p(T;X)$ (where $1 \leqslant p < +\infty$). So, we conclude that

$$\lim_{h\to 0} \int_0^b \|u(t+h) - u(t)\|_X^p \, dt = 0 \qquad \forall\, u \in L^p(T;X).$$

Solution of Problem 1.75

"(a) $\Longrightarrow$ (b)": This implication follows from integration by parts.
"(a) $\Longrightarrow$ (c)": This is obvious.
"(c) $\Longrightarrow$ (b)": Let $\vartheta \in C_c^\infty(0,b)$. From integration by parts, we have

$$\int_0^b \langle x^*, u(t)\rangle\, \vartheta'(t)\, dt = -\int_0^b \langle x^*, h(t)\rangle\, \vartheta(t)\, dt,$$

so

$$\langle x^*, \int_0^b u(t)\vartheta'(t)\, dt + \int_0^b h(t)\vartheta(t)\, dt\rangle = 0 \qquad \forall\, x^* \in X^*$$

(see Problem 1.63) and thus

$$\int_0^b u(t)\vartheta'(t)\, dt = -\int_0^b h(t)\vartheta(t)\, dt.$$

"**(b)** $\Longrightarrow$ **(a)**": Let $\widehat{u}(t) = \int_0^t h(s)\,ds$ for all $t \in T$. Clearly $\widehat{u}\colon T \longrightarrow X$ is absolutely continuous (see Definition 1.68 with the absolute value of $\mathbb{R}$, replaced by the norm of X) and $\widehat{u}' = h$. Let $y = u - \widehat{u}$. By hypothesis we have

$$\int_0^b (y(t) + \widehat{u}(t))\vartheta'(t)\,dt = -\int_0^b h(t)\vartheta(t)\,dt \qquad \forall\, \vartheta \in C_c^\infty(0,b),$$

so

$$\int_0^b y(t)\vartheta'(t)\,dt = 0 \qquad \forall\, \vartheta \in C_c^\infty(0,b). \tag{1.97}$$

We will be done if we show that y is a constant function. Let $\eta \in C_c^\infty(0,b)$ be such that $\int_0^b \eta\,dt = 1$. Then any $\vartheta \in C_c^\infty(0,b)$ can be written as

$$\vartheta = \lambda\eta + \psi', \quad \text{with } \lambda = \int_0^b \vartheta(t)\,dt \quad \text{and} \quad \psi \in C_c^\infty(0,b).$$

To see this, note that

$$\int_0^b (\vartheta - \lambda\eta)\,dt = 0$$

and so it follows that the primitive of $\vartheta - \lambda\eta$ vanishing at $t = 0$ belongs in $C_c^\infty(0,b)$. Since $\psi' = \vartheta - \lambda\eta$, it follows that ψ is this primitive. Then we have

$$\begin{aligned}
\lambda\int_0^b y(t)\eta(t)\,dt &= \int_0^b y(t)(\vartheta(t) - \psi'(t))\,dt \\
&= \int_0^b y(t)\vartheta(t)\,dt - \int_0^b y(t)\psi'(t)\,dt \\
&= \int_0^b y(t)\vartheta(t)\,dt
\end{aligned}$$

(see (1.97)), so

$$\int_0^b (y(t) - \xi)\vartheta(t)\,dt = 0 \qquad \forall\, \vartheta \in C_c^\infty(0,b),$$

where $\xi = \int\limits_0^b y(t)\eta(t)\,dt$ (recall that $\lambda = \int\limits_0^b \vartheta(t)\,dt$) and thus

$$\int\limits_0^b \langle x^*, y(t){-}\xi\rangle\, \vartheta(t) = \Big\langle x^*, \int\limits_0^b (y(t){-}\xi)\vartheta(t)\,dt\Big\rangle = 0$$

$$\forall\, \vartheta \in C_c^\infty(0,b)\,,\, x^* \in X^*$$

(see Problem 1.63). Since $\langle x^*, y(\cdot) - \xi\rangle \in L^1(T)$ and $C_c^\infty(0,b)$ is dense in $L^1(T)$, we obtain

$$\langle x^*, y(t) - \xi\rangle \;=\; 0 \qquad \forall\, t \in T \setminus N(x^*),$$

for some Lebesgue-null subset $N(X^*)$ of T. As before, using the Pettis measurability theorem (see Theorem 1.36), we may assume that X is separable. This implies that $X^*_{w^*}$ (the dual space X^* furnished with w^*-topology) is Souslin, hence separable. So, we can find a sequence $\{x_k^*\}_{k\geqslant 1} \subseteq X^*$ which is w^*-dense. Let $N = \bigcup\limits_{k\geqslant 1} N(x_k^*)$. Then $N \subseteq T$ is Lebesgue-null and

$$\langle x_k^*, y(t) - \xi\rangle \;=\; 0 \qquad \forall\, k \geqslant 1,\; t \in T \setminus N,$$

so

$$y(t) \;=\; \xi \quad \text{for a.a. } t \in T.$$

Solution of Problem 1.76

(a) To fix things, we assume that u is increasing. For every partition $P = \{x_k\}_{k=0}^n$ of T, we have

$$\begin{aligned}\sum_{k=1}^n \big|u(x_k) - u(x_{k-1})\big| \;&=\; \sum_{k=1}^n \big(u(x_k) - u(x_{k-1})\big)\\ &=\; u(x_n) - u(x_0) \;\leqslant\; \sup_T u - \inf_T u\end{aligned}$$

(since u is increasing). Taking the supremum over all partitions P of T, we obtain

$$\operatorname{Var}_T u \;\leqslant\; \sup_T u - \inf_T u \tag{1.98}$$

(see Definition 1.59).

Next we prove that the opposite inequality holds. So, consider the partition $P = \{a, b\}$ of T with $\inf T \leqslant a < b \leqslant \sup T$ (we assume that T is not a singleton or otherwise there is nothing to prove). Then

$$u(b) - u(a) \;=\; \big|u(b) - u(a)\big| \;\leqslant\; \operatorname{Var}_T u.$$

If $\sup T \in T$, then we choose $b = \sup T$ and we have $u(b) = \sup\limits_T u$ (recall that u is increasing). If $\sup T \notin T$, then we let $b \nearrow \sup T$ and so $u(b) \nearrow \sup\limits_T u$. Similarly, for the point a. So, we infer that

$$\sup_T u - \inf_T u \leqslant \mathrm{Var}_T u$$

and so

$$\mathrm{Var}_T u = \sup_T u - \inf_T u.$$

(see (1.98)).

(b) It follows immediately from **(a)** (see Definition 1.59).

Solution of Problem 1.77

No. Consider the function

$$u(x) = \begin{cases} x \sin \frac{1}{x} & \text{if } x \in (0,1], \\ 0 & \text{if } x = 0. \end{cases}$$

Then $u \in C([0,1])$ and u is bounded. However, we claim that u is not of bounded variation. Let $x_k = \frac{1}{(k+\frac{1}{2})\pi}$ for $k \geqslant 1$. Then

$$u(x_k) = \frac{(-1)^k}{(k+\frac{1}{2})\pi} \qquad \forall\, k \geqslant 1$$

and we have

$$\mathrm{Var}_{[0,1]} u > \sum_{k=1}^{n} \frac{2}{k\pi} \longrightarrow +\infty,$$

so $u \notin BV([0,1])$.

Solution of Problem 1.78

(a) Let $P = \{a, b\}$. Then

$$\big|u(b) - u(a)\big| \leqslant \mathrm{Var}_{[a,b]} u.$$

Hence, if $x < y$, then

$$\big|u(x) - u(y)\big| \leqslant \mathrm{Var}_{[x,y]} u = f(y) - f(x)$$

and if $y < x$, then

$$\big|u(y) - u(x)\big| \leqslant \mathrm{Var}_{[y,x]} u = f(x) - f(y).$$

Thus **(a)** holds.

(b) From **(a)**, it follows that if

$$\begin{aligned} \operatorname{Cont}(u) &= \{z \in [a,b] : z \text{ is a point of continuity of } u\} \\ \operatorname{Cont}(f) &= \{z \in [a,b] : z \text{ is a point of continuity of } f\}, \end{aligned}$$

then

$$\operatorname{Cont}(f) \subseteq \operatorname{Cont}(u).$$

Next, let $\widehat{x} \in \operatorname{Cont}(u) \cap [a,b)$. Then, given $\varepsilon > 0$, we can find $\delta = \delta(\varepsilon) > 0$ such that

$$\text{if } |x - \widehat{x}| < \delta, \text{ then } \big|u(x) - u(\widehat{x})\big| < \tfrac{\varepsilon}{2}. \tag{1.99}$$

Let $P = \{\widehat{x} = x_0, x_1, \ldots, x_n = b\}$ be a partition of $[\widehat{x}, b]$ such that

$$\operatorname{Var}_{[\widehat{x},b]} u - \tfrac{\varepsilon}{2} < \sum_{k=1}^{n} \big|u(x_k) - u(x_{k-1})\big|.$$

Let $x_{\frac{1}{2}} \in (\widehat{x}, x_1)$ be such that

$$x_{\frac{1}{2}} - \widehat{x} < \delta.$$

Then, we have

$$\begin{aligned} \operatorname{Var}_{[\widehat{x},b]} u - \tfrac{\varepsilon}{2} &< \big|u(x_1) - u(\widehat{x})\big| + \sum_{k=2}^{n} \big|u(x_k) - u(x_{k-1})\big| \\ &\leqslant \big|u(x_1) - u(x_{\frac{1}{2}})\big| + \big|u(x_{\frac{1}{2}}) - u(\widehat{x})\big| + \sum_{k=2}^{n} \big|u(x_k) - u(x_{k-1})\big| \\ &< \big|u(x_1) - u(x_{\frac{1}{2}})\big| + \tfrac{\varepsilon}{2} + \sum_{k=2}^{n} \big|u(x_k) - u(x_{k-1})\big| \end{aligned} \tag{1.100}$$

(see (1.99)). Note that $\{x_{\frac{1}{2}}, x_1, \ldots, x_n = b\}$ is a partition of $[x_{\frac{1}{2}}, b]$. So, from (1.100) we have

$$\operatorname{Var}_{[\widehat{x},b]} u - \varepsilon < \operatorname{Var}_{[x_{\frac{1}{2}},b]} u,$$

thus

$$\operatorname{Var}_{[\widehat{x},b]} u - \operatorname{Var}_{[x_{\frac{1}{2}},b]} u < \varepsilon$$

and hence

$$\operatorname{Var}_{[\widehat{x},x_{\frac{1}{2}}]} u < \varepsilon.$$

Hence we have

$$f(x_{\frac{1}{2}}) - f(\widehat{x}) = \operatorname{Var}_{[a,x_{\frac{1}{2}}]} u - \operatorname{Var}_{[a,\widehat{x}]} u = \operatorname{Var}_{[\widehat{x},x_{\frac{1}{2}}]} u < \varepsilon,$$

so f is right continuous at $\widehat{x}$.
Similarly, if $\widehat{x} \in \operatorname{Cont}(u) \cap (a,b]$, then as above we can show that f is left continuous at $\widehat{x}$.

Therefore, we conclude that $\widehat{x} \in \operatorname{Cont}(f)$. So

$$\operatorname{Cont}(u) \subseteq \operatorname{Cont}(f)$$

and finally

$$\operatorname{Cont}(u) = \operatorname{Cont}(f).$$

Solution of Problem 1.79

(a) For $0 < h \leqslant \frac{b-a}{2}$, we have

$$\begin{aligned}
\frac{1}{h}\int_a^{b-h} (f(x+h) - f(x))\,dx &= \frac{1}{h}\Big(\int_a^{b-h} f(x+h)\,dx - \int_a^{b-h} f(x)\,dx\Big) \\
&= \frac{1}{h}\Big(\int_{a+h}^{b} f(y)\,dy - \int_a^{b-h} f(x)\,dx\Big) = \frac{1}{h}\Big(\int_{b-h}^{b} f(y)\,dy - \int_a^{a+h} f(x)\,dx\Big) \\
&\leqslant \frac{1}{h}\,(f(b)h - f(a)h) = f(b) - f(a),
\end{aligned}$$

while for $\frac{b-a}{2} < h \leqslant b-a$, we have

$$\begin{aligned}
\frac{1}{h}\int_a^{b-h} (f(x+h) - f(x))\,dx &= \frac{1}{h}\Big(\int_a^{b-h} f(x+h)\,dx - \int_a^{b-h} f(x)\,dx\Big) \\
&\frac{1}{h}\Big(\int_{a+h}^{b} f(y)\,dy - \int_a^{b-h} f(x)\,dx\Big) \leqslant \frac{b-a-h}{h}\,(f(b) - f(a)) \leqslant f(b) - f(a).
\end{aligned}$$

(b) We assume that $u \in BV(I)$ or otherwise the inequality is obvious. Let $[a,b] \subseteq I$ be such that $0 < h \leqslant b-a$. If f is as in Problem 1.78, we have

$$\begin{aligned}
\frac{1}{h}\int_a^{b-h} \big|u(x+h) - u(x)\big|\,dx &\leqslant \frac{1}{h}\int_a^{b-h} \big(f(x+h) - f(x)\big)\,dx \\
&\leqslant f(b) = \operatorname{Var}_{[a,b]} u
\end{aligned} \tag{1.101}$$

(see Problem 1.78(a) and **(a)** above). Consider a sequence of intervals $\{[a_n, b_n]\}_{n \geqslant 1}$ in I such that

$$a_n \searrow \inf I \quad \text{and} \quad b_n \nearrow \sup I.$$

If $\operatorname{diam} I_h > 0$, then we can find $n_0 \geqslant 1$ such that $0 < h < b_n - a_n$ for all $n \geqslant n_0$ and so

$$\frac{1}{h}\int_{a_n}^{b_n-h} \big|u(x+h)-u(x)\big|\,dx \;\leqslant\; \operatorname{Var}_{[a_n,b_n]} u \;\leqslant\; \operatorname{Var} u \qquad \forall\, n \geqslant n_0$$

(see (1.101)). Let $n \to +\infty$, to conclude that

$$\frac{1}{h}\int_{I_h} \big|u(x+h)-u(x)\big|\,dx \;\leqslant\; \operatorname{Var} u.$$

If $\operatorname{diam} I_h = 0$, then the left-hand side is zero and so the inequality is trivially true.

Solution of Problem 1.80

From the Lebesgue theorem (see Theorem 1.56), we know that $u'(x)$ exists for almost all $x \in [0,1]$. So, there is a Lebesgue-null set $D \subseteq [0,1]$ such that

$$u'(x) \;=\; \lim_{n\to+\infty} \frac{u(x+\frac{1}{n})-u(x)}{\frac{1}{n}} \qquad \forall\, x \in [0,1]\setminus D,$$

so u' is Lebesgue measurable.

Moreover, from the Fatou lemma (see Theorem I.3.95), for any $0 < a < b < 1$, we have

$$\int_a^b u'(x)\,dx \;\leqslant\; \liminf_{n\to+\infty} \int_a^b \frac{u(x+\frac{1}{n})-u(x)}{\frac{1}{n}}\,dx. \tag{1.102}$$

Then we can find $n_0 \geqslant 1$ such that

$$\int_a^b u(x+\tfrac{1}{n})\,dx \;=\; \int_{a+\frac{1}{n}}^{b+\frac{1}{n}} u(x)\,dx \qquad \forall\, n \geqslant n_0. \tag{1.103}$$

Hence, we have

$$\begin{aligned}\int_a^b \frac{u(x+\frac{1}{n})-u(x)}{\frac{1}{n}}\,dx &= n\int_{a+\frac{1}{n}}^{b+\frac{1}{n}} u(x)\,dx - n\int_a^b u(x)\,dx\\ &= n\int_b^{b+\frac{1}{n}} u(x)\,dx - n\int_a^{a+\frac{1}{n}} a(x)\,dx \qquad \forall\, n \geqslant n_0\end{aligned} \tag{1.104}$$

(see (1.103)). Note that

$$n \int\limits_b^{b+\frac{1}{n}} u(x)\,dx \longrightarrow u(b^+) \quad \text{and} \quad \int\limits_a^{a+\frac{1}{n}} a(x)\,dx \longrightarrow u(a^+).$$

So, if in (1.104) we pass to the limit as $n \to +\infty$, we obtain

$$\int\limits_a^b u'(x)\,dx \leqslant u(b^+) - u(a^+) \tag{1.105}$$

(see (1.102)). We have that

$$u'(x) \geqslant 0 \quad \text{for a.a. } x \in [0,1].$$

Therefore, if in (1.105) we let $a \searrow 0$ and $b \nearrow 1$, we obtain

$$\int\limits_0^1 u'(x)\,dx \leqslant u(1) - u(0) < +\infty,$$

so $u' \in L^1(0,1)$.

Finally, if u is Lipschitz continuous, then

$$\frac{\left|u(x+\frac{1}{n}) - u(x)\right|}{\frac{1}{n}} \leqslant k,$$

for some $k > 0$ (the Lipschitz constant). So, applying the Lebesgue dominated convergence theorem (see Theorem I.3.94), we conclude that

$$\int\limits_0^1 u'(x)\,dx = u(1) - u(0).$$

Solution of Problem 1.81

Evidently, without any loss of generality, we may assume that $\operatorname{Var} u > 0$ (otherwise, the inequality is trivial). Let $\eta \in (0, \operatorname{Var} u)$ and let $P = \{x_0, \ldots, x_m\}$ be a partition of I such that

$$\eta < \sum_{i=1}^m \left|u(x_k) - u(x_{k-1})\right|. \tag{1.106}$$

By virtue of the pointwise convergence of $\{u_n\}_{n\geqslant 1}$ to u, given $\varepsilon > 0$, we can find $n_0 = n_0(\varepsilon) \geqslant 1$ such that

$$\left|u_n(x_k) - u(x_k)\right| \leqslant \tfrac{\varepsilon}{2m} \qquad \forall\, n \geqslant n_0,\ k \in \{0, \ldots, m\}. \tag{1.107}$$

Then from (1.106) and (1.107), we have

$$\begin{aligned}\eta &< \sum_{k=1}^{m} \big|u(x_k) - u(x_{k-1})\big| \leqslant \sum_{k=1}^{m} \big(\big|u_n(x_k) - u_n(x_{k-1})\big| + \tfrac{\varepsilon}{m}\big) \\ &= \sum_{k=1}^{m} \big|u_n(x_k) - u_n(x_{k-1})\big| + \varepsilon \leqslant \operatorname{Var} u_n + \varepsilon \qquad \forall\, n \geqslant n_0,\end{aligned}$$

so

$$\eta \leqslant \liminf_{n\to+\infty} \operatorname{Var} u_n + \varepsilon.$$

Let $\eta \longrightarrow (\operatorname{Var} u)^-$ and $\varepsilon \searrow 0$, to conclude that

$$\operatorname{Var} u \leqslant \liminf_{n\to+\infty} \operatorname{Var} u_n.$$

Solution of Problem 1.82

Let $\{x_0, \ldots, x_m\}$ be a partition of (a, b), i.e., $a < x_0 < x_1 < \ldots < x_m < b$. For every $k \in \{1, \ldots, m\}$, we have

$$\big|u(x_k) - u(x_{k-1})\big| = \Big|\int_{x_{k-1}}^{x_k} u'(x)\, dx\Big| \leqslant \int_{x_{k-1}}^{x_k} |u'(x)|\, dx,$$

so

$$\sum_{k=1}^{m} \big|u(x_k) - u(x_{k-1})\big| \leqslant \int_{x_0}^{x_m} |u'(x)|\, dx \leqslant \int_a^b |u'(x)|\, dx = \|u'\|_1$$

and thus

$$\operatorname{Var}_{(a,b)} u \leqslant \|u'\|_1. \tag{1.108}$$

Next we show that the opposite inequality holds. Let $a < c < d < b$. Since $u \in C^1(a, b)$, we have that $u'|_{[c,d]}$ is uniformly continuous. So, given $\varepsilon > 0$, we can find $\delta > 0$ such that

$$\text{if } x, y \in [c, d] \text{ and } |x - y| < \delta, \text{ then } \big|u'(x) - u'(y)\big| < \varepsilon. \tag{1.109}$$

Let $P = \{x_0 = c, x_1, \ldots, x_{m-1}, x_m = d\}$ be a partition of $[c, d]$ such that $x_k - x_{k-1} < \delta$ for all $k \in \{1, \ldots, m\}$. Then for every $x \in [x_{k-1}, x_k]$, we have

$$\begin{aligned}u(x_k) - u(x_{k-1}) &= \int_{x_{k-1}}^{x_k} u'(y)\, dy \\ &= \int_{x_{k-1}}^{x_k} \big(u'(y) - u'(x)\big)\, dy + (x_k - x_{k-1})u'(x),\end{aligned}$$

so

$$|u'(x)| \leqslant \frac{|u(x_k)-u(x_{k-1})|}{x_k-x_{k-1}} + \frac{1}{x_k-x_{k-1}} \int\limits_{x_{k-1}}^{x_k} |u'(y)-u'(x)|\, dy$$
$$< \frac{|u(x_k)-u(x_{k-1})|}{x_k-x_{k-1}} + \varepsilon.$$

Integrating over $[x_{k-1}, x_k]$ and then summing for $k = 1, \ldots, m$, we obtain

$$\int\limits_c^d |u'(x)|\, dx \leqslant \sum_{k=1}^m \big|u(x_k) - u(x_{k-1})\big| + \varepsilon(d-c)$$
$$\leqslant \operatorname{Var}_{[c,d]} u + \varepsilon(d-c)$$
$$\leqslant \operatorname{Var}_{(a,b)} u + \varepsilon(b-c).$$

Let $d \nearrow b$ and $c \searrow a$, to obtain

$$\|u'\|_1 \leqslant \operatorname{Var}_{(a,b)} u + \varepsilon(b-a).$$

Finally, we let $\varepsilon \searrow 0$ to conclude that $\|u'\|_1 \leqslant \operatorname{Var}_{(a,b)} u$, so $\operatorname{Var}_{(a,b)} u = \|u'\|_1$ (see (1.108)).

Solution of Problem 1.83

Consider the function $u(x) = \sin x$ for $x \in \mathbb{R}$. Then, by the mean value theorem, we have

$$\big|u(x) - u(y)\big| \leqslant |x-y| \qquad \forall\, x, y \in \mathbb{R},$$

so

$$u \in BV_{\mathrm{loc}}(\mathbb{R}).$$

On the other hand, if $x = (k+1)\frac{\pi}{2}$ for $k = 0, \ldots, 2m$, then

$$\big|u(x_k) - u(x_{k-1})\big| = 1 \qquad \forall\, k \in \{1, \ldots, 2m\},$$

so $u \notin BV(\mathbb{R})$.

Solution of Problem 1.84

By virtue of Theorem 1.67, the Banach indicatrix of u is integrable. Hence

$$N_u(y; I) < +\infty \quad \text{for a.a. } y \in \mathbb{R},$$

so

$$u^{-1}(\{y\}) \text{ is finite for a.a. } y \in \mathbb{R}.$$

Solution of Problem 1.85

Without any loss of generality, we may assume that $E \subseteq \operatorname{int} I$. Given $\varepsilon > 0$ for each $n \geqslant 1$, let

$$\begin{aligned} E_n &= \{x \in E : \lambda^*(u(T)) \leqslant (M+\varepsilon)\lambda^*(T) \text{ for all intervals } T \subseteq I \\ &\qquad \text{with } x \in T \text{ and } 0 < \lambda(T) < \tfrac{1}{n}\}. \end{aligned}$$

Evidently the sequence $\{E_n\}_{n\geqslant 1}$ is increasing, i.e., $E_n \subseteq E_{n+1}$ for all $n \geqslant 1$.

We will show that

$$E = \bigcup_{n\geqslant 1} E_n. \tag{1.110}$$

Clearly we have $E \supseteq \bigcup_{n\geqslant 1} E_n$. Now, let $x \in E$. We have $|u'(x)| \leqslant M$ and so we can find $\delta > 0$ such that

$$\big|u(y) - u(x)\big| \leqslant (M+\varepsilon)|y-x| \qquad \forall\, y \in I,\ |y-x| < \delta. \tag{1.111}$$

Then for $y, y' \in I$ with $|y-y'| < \delta$ and $y < x < y'$, we have

$$\begin{aligned} \big|u(y) - u(y')\big| &\leqslant \big|u(y) - u(x)\big| + \big|u(x) - u(y')\big| \\ &\leqslant (M+\varepsilon)(x-y) + (M+\varepsilon)(y'-x) = (M+\varepsilon)(y'-y) \end{aligned}$$

(see (1.111)), so

$$x \in E_n \qquad \forall\, n > \tfrac{1}{\delta}.$$

This proves (1.110).

Next fix $n \geqslant 1$. We will show that

$$\lambda^*(u(E_n)) \leqslant (M+\varepsilon)(\lambda^*(E_n) + \varepsilon). \tag{1.112}$$

From the definition of the outer measure λ^*, we can find an open set $U_n \supseteq E_n$ such that

$$\lambda(U_n) \leqslant \lambda^*(E_n) + \varepsilon, \tag{1.113}$$

with λ being the Lebesgue measure on $\mathbb{R}$. Replacing U_n with $U_n \cap (\operatorname{int} I)$, we see that we may assume that $U_n \subseteq \operatorname{int} I$. We have

$$U_n = \bigcup_{k\geqslant 1} T_k^{(n)},$$

where $\{T_k^{(n)}\}_{k\geqslant 1}$ are pairwise disjoint intervals with $\lambda(T_k^{(n)}) < \frac{1}{n}$. Let

$$\mathcal{Y} = \{k \in \mathbb{N} : T_k^{(n)} \cap E_n \neq \emptyset\}.$$

From the definition of E_n, we have

$$\lambda^*(u(T_k^{(n)})) \leqslant (M+\varepsilon)\lambda^*(T_k^{(n)}) \qquad \forall\, k \in \mathcal{Y},$$

so

$$\begin{aligned}
\lambda^*(u(E_n)) &\leqslant \lambda^*\Big(\bigcup_{k\in\mathcal{Y}} u(T_k^{(n)})\Big) \leqslant \sum_{k\in\mathcal{Y}} \lambda^*(u(T_k^{(n)})) \\
&\leqslant (M+\varepsilon)\sum_{k\in\mathcal{Y}} \lambda(T_k^{(n)}) = (M+\varepsilon)\lambda\Big(\bigcup_{k\in\mathcal{Y}} T_k^{(n)}\Big) \\
&\leqslant (M+\varepsilon)\lambda(U_n) \leqslant (M+\varepsilon)\big(\lambda^*(E_n)+\varepsilon\big)
\end{aligned}$$

(recall that the sets $\{T_k^{(n)}\}_{k\geq 1}$ are pairwise disjoint and use (1.113)). This proves (1.112).

Recalling that the sequence $\{E_n\}_{n\geqslant 1}$ is increasing (hence so is $\{u(E_n)\}_{n\geqslant 1}$), if in (1.112) we pass to the limit as $n \to +\infty$, we obtain

$$\lambda^*(u(E)) \leqslant (M+\varepsilon)(\lambda^*(E)+\varepsilon).$$

Finally let $\varepsilon \searrow 0$, to conclude that $\lambda^*(u(E)) \leqslant M\lambda^*(E)$.

Solution of Problem 1.86

First suppose that E is a Lebesgue-null set. Let

$$E_n = \big\{x \in E : |u'(x)| \leqslant n\big\} \qquad \forall\, n \geqslant 1.$$

From Problem 1.85, we have

$$\lambda^*(u(E_n)) \leqslant n\lambda^*(E_n) \leqslant n\lambda^*(E) = 0 \qquad \forall\, n \geqslant 1.$$

Exploiting the countable subadditivity of λ^*, we have

$$\begin{aligned}
\lambda^*(u(E)) &= \lambda^*\Big(u\Big(\bigcup_{n\geqslant 1} E_n\Big)\Big) = \lambda^*\Big(\bigcup_{n\geqslant 1} u(E_n)\Big) \\
&\leqslant \sum_{n\geqslant 1} \lambda^*(u(E_n)) = 0.
\end{aligned}$$

Next suppose that $u'(x) = 0$ for all $x \in E$. Then the result follows from Problem 1.85 with $M = 0$.

Solution of Problem 1.87

Exploiting the regularity of the Lebesgue measure λ in $\mathbb{R}$ (see Theorem I.4.12 and recall that λ is σ-finite), we can write that

$$E = D \cup \bigcup_{n\geqslant 1} K_n,$$

where D is a Lebesgue-null set and each K_n is compact. From Problem 1.86, we have

$$\lambda(u(D)) \;=\; 0.$$

Also, the differentiability of u implies that for each $n \geqslant 1$, the function $u|_{K_n} : K_n \longrightarrow \mathbb{R}$ is continuous, hence $u(K_n) \subseteq \mathbb{R}$ is compact (see Proposition I.1.74), thus in particular a Lebesgue measurable set. Therefore the set

$$u(E) \;=\; u\big(D \cup \bigcup_{n \geqslant 1} K_n\big) \;=\; u(D) \cup u\big(\bigcup_{n \geqslant 1} K_n\big) \;=\; u(D) \cup \bigcup_{n \geqslant 1} u(K_n)$$

is Lebesgue measurable.

Suppose that $\lambda(E) < +\infty$. We fix $n \geqslant 1$ and for every $k \geqslant 1$, we define

$$E_n^k \;=\; \big\{x \in E : \tfrac{k-1}{2^n} \leqslant |u'(x)| < \tfrac{k}{2^n}\big\}.$$

we have $E = \bigcup\limits_{k \geqslant 1} E_n^k$ and

$$\begin{aligned}
\lambda(u(E)) \;&=\; \lambda\big(u\big(\bigcup_{k \geqslant 1} E_n^k\big)\big) \;=\; \lambda\big(\bigcup_{k \geqslant 1} u(E_n^k)\big) \;\leqslant\; \sum_{k \geqslant 1} \lambda(u(E_n^k)) \\
&\leqslant\; \sum_{k \geqslant 1} \tfrac{k}{2^n} \lambda(E_n^k) \;=\; \sum_{k \geqslant 1} \tfrac{k-1}{2^n} \lambda(E_n^k) + \tfrac{1}{2^n} \sum_{k \geqslant 1} \lambda(E_n^k) \\
&\leqslant\; \sum_{k \geqslant 1} \int_{E_n^k} |u'(x)|\, dx + \tfrac{1}{2^n} \lambda(E) \;=\; \int_E |u'(x)|\, dx + \tfrac{1}{2^n} \lambda(E)
\end{aligned}$$

(see Problem 1.85). Let $n \to +\infty$ to conclude that

$$\lambda(u(E)) \;\leqslant\; \int_E |u'(x)|\, dx.$$

Now assume that $\lambda(E) = +\infty$. In this case for every $k \in \mathbb{Z}$, let

$$E_k \;=\; E \cap [k, k+1].$$

Then $\lambda(E_k) < +\infty$ and so from the previous part of the solution we have

$$\begin{aligned}
\lambda(u(E)) \;&=\; \lambda\big(u\big(\bigcup_{k \in \mathbb{Z}} E_k\big)\big) \;=\; \lambda\big(\bigcup_{k \in \mathbb{Z}} u(E_k)\big) \\
&\leqslant\; \sum_{k \in \mathbb{Z}} \lambda(u(E_k)) \;\leqslant\; \sum_{k \in \mathbb{Z}} \int_{E_k} |u'(x)|\, dx \;=\; \int_E |u'(x)|\, dx.
\end{aligned}$$

Solution of Problem 1.88

Since $u\colon I \longrightarrow \mathbb{R}$ is differentiable, it is continuous and so $u([a,b])$ is an interval. We have

$$\big|u(b) - u(a)\big| \leqslant \lambda(u([a,b])).$$

On the other hand, from Problem 1.87, we have

$$\lambda\big(u([a,b])\big) \leqslant \int\limits_a^b |u'(x)|\,dx,$$

so

$$\big|u(b) - u(a)\big| \leqslant \int\limits_a^b |u'(x)|\,dx.$$

Solution of Problem 1.89

Let $\xi > 0$ be the Lipschitz constant of φ. Let $\varepsilon > 0$. Since $u \in AC(I)$, we can find $\delta > 0$ such that for any finite family $\big\{[a_k, b_k]\big\}_{k=1}^m$ of nonoverlapping subintervals of $I = [a,b]$, we have

$$\text{if } \sum_{k=1}^m (b_k - a_k) \leqslant \delta, \text{ then } \sum_{k=1}^m \big|u(b_k) - u(a_k)\big| \leqslant \tfrac{\varepsilon}{\xi}$$

(see Definition 1.68). Then

$$\sum_{k=1}^m \big|(\varphi \circ u)(b_k) - (\varphi \circ u)(a_k)\big| \leqslant \xi \sum_{k=1}^m \big|u(b_k) - u(a_k)\big| \leqslant \xi \tfrac{\varepsilon}{\xi} = \varepsilon,$$

so $\varphi \circ u \in AC(I)$.

Solution of Problem 1.90

Since $\varphi\colon [u(a), u(b)] \longrightarrow \mathbb{R}$ is absolutely continuous, given $\varepsilon > 0$, we can find $\delta > 0$ such that any family $\big\{[c_k, d_k)\big\}_{k=1}^m$ of nonoverlapping subintervals of $[u(a), u(b)]$, we have

$$\text{if } \sum_{k=1}^m (d_k - c_k) \leqslant \delta, \text{ then } \sum_{k=1}^m \big|\varphi(d_k) - \varphi(c_k)\big| \leqslant \varepsilon. \tag{1.114}$$

Since $u \in AC(I)$, for this $\delta > 0$, there exists $\eta > 0$ such that for any family $\{[a_k, b_k)\}_{k=1}^m$ of nonoverlapping subintervals of I, we have

$$\text{if } \sum_{k=1}^m (b_k - a_k) \leqslant \eta, \text{ then } \sum_{k=1}^m \left|u(b_k) - u(a_k)\right| \leqslant \delta. \tag{1.115}$$

Note that since u is strictly increasing $\{[c_k = u(a_k), d_k = u(b_k))\}_{k=1}^m$ are nonoverlapping subintervals of $[u(a), u(b)]$. So, from (1.114) and (1.115), we infer that

$$\text{if } \sum_{k=1}^m (b_k - a_k) \leqslant \eta, \text{ then } \sum_{k=1}^m \left|(\varphi \circ u)(b_k) - (\varphi \circ u)(a_k)\right| \leqslant \varepsilon,$$

so $\varphi \circ u \in AC(I)$.

Solution of Problem 1.91
"(a) $\Longrightarrow$ (b)": Follows from Proposition 1.70(b).
"(b) $\Longrightarrow$ (c)": Follows from Problem 1.82.
"(c) $\Longrightarrow$ (d)": This is a consequence of the absolute continuity of the Lebesgue integral.
"(d) $\Longrightarrow$ (a)": Since $m \ll \lambda$, we have that given $\varepsilon > 0$, we can find $\delta > 0$ such that

$$\text{if } A \subseteq I \text{ is measurable and } \lambda(A) \leqslant \delta, \text{ then } m(A) \leqslant \varepsilon.$$

Now let $\{(a_k, b_k)\}_{k=1}^m$ be nonoverlapping subintervals of I such that

$$\sum_{k=1}^m (b_k - a_k) \leqslant \delta.$$

We set $A = \bigcup_{k=1}^m [a_k, b_k)$. We have

$$\sum_{k=1}^m \left|u(b_k) - u(a_k)\right| \leqslant \sum_{k=1}^m \int_{a_k}^{b_k} |u'(x)|\, dx = \int_A |u'(x)|\, dx = m(A) \leqslant \varepsilon,$$

so $u \in AC(I)$.

Solution of Problem 1.92

Let $m(A) == \int_A h(s)\,ds$ for every Borel set A. From the absolute continuity of the Lebesgue integral we have that $m \ll \lambda$. We set $u(x) = m\big((-\infty, x)\big)$. Note that m is bounded (since $h \in L^1(\mathbb{R})$). So, $m \ll \lambda$ is equivalent to saying that for every $\varepsilon > 0$, there exists $\delta > 0$ such that for every Borel set $A \subseteq \mathbb{R}$ with $\lambda(A) \leqslant \delta$ we have $|m(A)| \leqslant \varepsilon$. Hence u is absolutely continuous. Then from Theorem 1.71 we have

$$\int_y^x h(s)\,ds = \int_y^x u'(s)\,ds \qquad \forall\, x, y \in \mathbb{R},$$

so

$$h(x) = u'(x) \quad \text{for a.a. } x \in \mathbb{R}.$$

Solution of Problem 1.93

"$\Longrightarrow$": **(a)** and **(b)** follow from Theorem 1.71. Let $[a, b] \subseteq I$ with $c \in [a, b]$. We define

$$y(x) = u(x) - \Big(u(c) + \int_c^x u'(s)\,ds\Big) \qquad \forall\, x \in [a, b].$$

From Problem 1.92, we know that $y \in AC\,([a, b])$ and $y'(x) = 0$ for all $x \in [a, b] \setminus D$, with $\lambda(D) = 0$. Also Theorem 1.71 implies that $\lambda(y(D)) = 0$, while Problem 1.86 implies that $\lambda\big(y([a, b] \setminus D)\big) = 0$. Therefore

$$\lambda\big(y([a, b])\big) = 0.$$

The continuity of y implies that $y([a, b])$ is an interval and hence y is a constant function. Since $y(c) = 0$, we infer that $y \equiv 0$ and so

$$u(x) = u(c) + \int_c^x u'(s)\,ds \qquad \forall\, x \in [a, b].$$

Since $a, b, c \in I$ are arbitrary, we conclude that **(c)** holds.
"$\Longleftarrow$": Follows from Problem 1.92.

Solution of Problem 1.94

Let $A = \big\{x \in E : \ |u'(x)| > 0\big\}$ and for every $n \geqslant 1$, we define

$$A_n = \big\{x \in A : \ |u(y) - u(x)| > \tfrac{|y-x|}{n} \text{ for all } y \in I \text{ with } |y - x| < \tfrac{1}{n}\big\}.$$

Then $A = \bigcup_{n\geqslant 1} A_n$. Fix $n \geqslant 1$ and let $T \subseteq \mathbb{R}$ be any interval of length less than $\frac{1}{n}$. We set $B = A_n \cap T$. Note that $\lambda(u(B)) = 0$. So, given $\varepsilon > 0$ we can find a sequence of intervals $\{I_k\}_{k\geqslant 1}$ such that

$$u(B) \subseteq \bigcup_{k\geqslant 1} I_k \quad \text{and} \quad \sum_{k\geqslant 1} \lambda(I_k) < \varepsilon.$$

Set $B_k = u^{-1}(I_k) \cap B$ for $k \geqslant 1$. Then $B \subseteq \bigcup_{k\geqslant 1} B_k$ and so we have

$$\begin{aligned}
\lambda^*(B) &\leqslant \sum_{k\geqslant 1} \lambda^*(B_k) \leqslant \sum_{k\geqslant 1} \sup_{x,y\in B_k} |x-y| \\
&\leqslant n \sum_{k\geqslant 1} \sup_{x,y\in B_k} \big|u(x)-u(y)\big| \\
&\leqslant n \sum_{k\geqslant 1} I_k \leqslant n\varepsilon
\end{aligned}$$

(since $B_k \subseteq A_n \cap I$ and $u(B_k) \subseteq I_k$). Let $\varepsilon \searrow 0$ (recall that $n \geqslant 1$ is fixed), to conclude that the set B is Lebesgue-null. Hence $\lambda(A_n) = 0$. Since $n \geqslant 1$ is arbitrary, we conclude that $\lambda(A) = 0$ and so $u'(x) = 0$ for almost all $x \in E$.

Solution of Problem 1.95

Let $E \subseteq I$ be a Lebesgue measurable set. Since $u \in AC(I)$, from Theorem 1.71, we know that

$$E = E_0 \cup D,$$

with $u'(x)$ existing at every $x \in E_0$ and D being a Lebesgue-null set. Problem 1.87 implies that $u(E_0)$ is Lebesgue measurable. Also, by Theorem 1.71, the set $u(D)$ is Lebesgue-null. Therefore the set

$$u(E) = u(E_0 \cup D) = u(E_0) \cup u(D)$$

is Lebesgue measurable.

Solution of Problem 1.96

Let $x, y \in I$ with $x < y$. We have

$$\operatorname{Var}_{[x,y]} u = \begin{cases} \operatorname{Var}_{[c,y]} u - \operatorname{Var}_{[c,x]} u = -\widehat{h}(x) + \widehat{h}(y) & \text{if } c \leqslant x < y, \\ \operatorname{Var}_{[x,c]} u - \operatorname{Var}_{[y,c]} u = -\widehat{h}(x) + \widehat{h}(y) & \text{if } x < y \leqslant c, \\ \operatorname{Var}_{[x,c]} u + \operatorname{Var}_{[c,y]} u = -\widehat{h}(x) + \widehat{h}(y) & \text{if } x \leqslant c \leqslant y. \end{cases}$$

From Definition 1.59 we have

$$\left|u(x) - u(y)\right| \leqslant \operatorname{Var}_{[x,y]} u,$$

so

$$\left|u(x) - u(y)\right| \leqslant \widehat{h}(y) - \widehat{h}(x).$$

Solution of Problem 1.97

(a) From Theorem 1.62 we know that u can be written as the difference of two bounded increasing functions. Therefore u is bounded and both limits $\lim\limits_{x\to(\sup I)^-} u(x)$ and $\lim\limits_{x\to(\inf I)^+} u(x)$ exist.

(b) Let $\widehat{h}$ be the function introduced in Problem 1.96. From that problem we know that $\widehat{h}$ is increasing, therefore differentiable for almost all $x \in I$ (by the Lebesgue theorem; see Theorem 1.56). Also, since $u \in BV(I)$, $u'(x)$ exists for almost all $x \in I$ (see Theorem 1.62). Now, fix a point $x \in \operatorname{int} I$ where both u and $\widehat{h}$ are differentiable. Then for $y > x$ we have

$$\tfrac{|u(y)-u(x)|}{y-x} \leqslant \tfrac{\widehat{h}(y)-\widehat{h}(x)}{y-x}$$

(see Problem 1.96). Letting $y \searrow x$, we obtain

$$|u'(x)| \leqslant \widehat{h}'(x),$$

so

$$\int\limits_I |u'(x)|\,dx \leqslant \int\limits_I \widehat{h}(x)\,dx \leqslant \sup_I \widehat{h} - \inf_I \widehat{h} = \operatorname{Var}_I u$$

(see Problem 1.80 and Proposition 1.61).

Solution of Problem 1.98

"$\Longrightarrow$": To fix things, assume that u is increasing. From Problem 1.93 we have

$$\int\limits_a^b u'(s)\,ds = u(b) - u(a),$$

so

$$\int\limits_a^b |u'(s)|\,ds = \left|u(b) - u(a)\right|$$

(since u is increasing).

"$\Longleftarrow$": Let $y(x) = \int\limits_a^x u'(s)\,ds$ for all $x \in [a,b]$. Then y is increasing and from Problem 1.80 we have

$$y(x) - y(v) = \int\limits_v^x u'(s)\,ds \leqslant u(x) - u(v) \qquad \forall\, a \leqslant v < x \leqslant b,$$

so $u - y$ is increasing on $[a,b]$. Using the hypothesis we have

$$u(b) - u(a) = \int\limits_a^b u'(s)\,ds = y(b) - y(a)$$

(note that $y(a) = 0$), so

$$(u-y)(b) = (u-y)(a)$$

and thus $u - y$ is constant on $[a,b]$. But $y \in AC([a,b])$ (absolute continuity of the Lebesgue integral). Therefore $u \in AC([a,b])$.

Solution of Problem 1.99

"$\Longrightarrow$": Let $P = \{x_0 = a, x_1, \ldots, x_m = b\}$ be a partition of $[a,b]$. We have

$$\begin{aligned}\sum_{k=1}^m \left|u(x_k) - u(x_{k-1})\right| &= \sum_{k=1}^m \Big|\int\limits_{x_{k-1}}^{x_k} u'(s)\,ds\Big| \\ &\leqslant \sum_{k=1}^m \int\limits_{x_{k-1}}^{x_k} |u'(s)|\,ds = \int\limits_a^b |u'(s)|\,ds.\end{aligned}$$

Taking supremum over all partitions, we have

$$\mathrm{Var}_{[a,b]} u \leqslant \int\limits_a^b |u'(s)|\,ds.$$

This together with Problem 1.97(b) implies that

$$\mathrm{Var}_{[a,b]} u = \int\limits_a^b |u'(s)|\,ds.$$

"$\Longleftarrow$": Let $\widehat{h}$ be the increasing function introduced in Problem 1.96. From Problem 1.97 and our hypothesis, we have

$$\int\limits_a^b |\widehat{h}'|\,dx = \widehat{h}(b) - \widehat{h}(a),$$

so $\widehat{h} \in AC([a,b])$ (see Problem 1.98 and recall that $\widehat{h}$ is increasing). Using Problem 1.96, for any partition $\{x_0, \ldots, x_m\}$ of $[a,b]$, we have

$$\sum_{k=1}^{m} \left|u(x_k) - u(x_{k-1})\right| \leqslant \sum_{k=1}^{m} \left(\widehat{h}(x_k) - \widehat{h}(x_{k-1})\right),$$

so $u \in AC([a,b])$ (since $\widehat{h} \in AC([a,b])$).

Solution of Problem 1.100

We claim that $uy \in AC_{\mathrm{loc}}(I)$. Indeed, for every $x, v \in I$, we have

$$\begin{aligned}\left|(uy)(x) - (uy)(v)\right| &\leqslant \left|u(x)y(x) - u(x)y(v)\right| + \left|u(x)y(v) - u(v)y(v)\right| \\ &\leqslant \|u\|_\infty \left|y(x) - y(v)\right| + \|y\|_\infty \left|u(x) - u(v)\right|.\end{aligned}$$

From this follows readily that $uv \in AC_{\mathrm{loc}}(I)$. Then according to Problem 1.93, we have

$$(uy)(b) - (uy)(a) = \int_a^b (uy)'\,dx,$$

so

$$u(b)y(b) - u(a)y(a) = \int_a^b u'y\,dx + \int_a^b uy'\,dx.$$

Solution of Problem 1.101

"$\Longrightarrow$": Let $E = \{x \in I : u'(x) = 0\}$. Evidently the set $E \subseteq I$ is Lebesgue measurable and since u is by hypothesis singular, we have that $\lambda(I \setminus E) = 0$. So, invoking Problem 1.86, we have that $\lambda(u(E)) = 0$.

"$\Longleftarrow$": According to Problem 1.94, we have that $u'(x) = 0$ for almost all $x \in E$. Since $\lambda(I \setminus E) = 0$, we have $u'(x) = 0$ for almost all $x \in I$.

Solution of Problem 1.102

By hypothesis $\sup\limits_{n \geqslant 1} |Du_n|(\Omega) < +\infty$. Since $C_0\left(\Omega; \mathbb{R}^N\right)^* = M_b(\Omega; \mathbb{R}^N)$ (see Theorem 1.88) and $C_0\left(\Omega; \mathbb{R}^N\right)$ is a separable Banach space, bounded sets in $M_b(\Omega; \mathbb{R}^N)$

with the w^*-topology are relatively compact (weakly compact in the terminology of Definition 1.89) and metrizable. Therefore, by passing to a suitable subsequence if necessary, we may assume that

$$(Du_n)\lambda^N \xrightarrow{w} \mu \tag{1.116}$$

with $\mu \in M_b(\Omega; \mathbb{R}^N)$ (see Definition 1.89(b) and Remark 1.90). We have $|\mu|(\Omega) \leqslant \eta$. From the classical integration by parts formula, we have

$$\int_\Omega u_n \frac{\partial \vartheta}{\partial z_k}\, dz = -\int_\Omega \vartheta \frac{\partial u_n}{\partial z_k}\, dz \qquad \forall\, \vartheta \in C_c^\infty(\Omega),\ k \in \{1, \ldots, N\},$$

so

$$\int_\Omega u \frac{\partial \vartheta}{\partial z_k}\, dz = -\int_\Omega \vartheta\, d\mu_k \qquad \forall\, \vartheta \in C_c^\infty(\Omega),\ k \in \{1, \ldots, N\}$$

(see (1.116)). Thus $u \in BV(\Omega)$ (see Definition 1.74).

Solution of Problem 1.103

(a) Let $\{u_n\}_{n\geqslant 1} \subseteq W_{pq}(0,b)$ be a Cauchy sequence. Then $\{u_n\}_{n\geqslant 1} \subseteq L^p(T; X_1)$ and $\{u_n'\}_{n\geqslant 1} \subseteq L^q(T; X_2)$ are both Cauchy sequences. So, we can find $h \in L^p(T; X_1)$ and $g \in L^q(T; X_2)$ such that

$$u_n \longrightarrow h \quad \text{in } L^p(T; X_1) \quad \text{and} \quad u_n' \longrightarrow g \quad \text{in } L^q(T; X_2).$$

Note that

$$\int_0^b u_n \vartheta'\, dt = -\int_0^b u_n' \vartheta\, dt \qquad \forall\, \vartheta \in C_c^\infty(0,b),$$

so

$$\int_0^b h \vartheta'\, dt = -\int_0^b g \vartheta\, dt \qquad \forall\, \vartheta \in C_c^\infty(0,b),$$

thus $h' = g$. This proves that $W_{pq}(0,b)$ is a Banach space. Note that $W_{pq}(0,b)$ can be viewed as a closed subspace of $L^p(T; X_1) \times L^q(T; X_2)$. But from Proposition 1.42(c) and Theorems 1.46 and 1.47, we have that $L^p(T; X_1)$ and $L^p(T; X_2)$ are both separable and reflexive. Therefore $L^p(T; X_1) \times L^q(T; X_2)$ is separable and reflexive and so we can conclude that $W_{pq}(0,b)$ is separable and reflexive.

(b) Let $\{u_n\}_{n\geqslant 1} \subseteq W_{pq}(0,b)$ be a bounded sequence. We need to show that the sequence $\{u_n\}_{n\geqslant 1}$ admits a subsequence which converges strongly in $L^p(T; X)$. The reflexivity of $W_{pq}(0,b)$ (see part **(a)**) implies that by passing to a subsequence if necessary, we may assume that

$$u_n \xrightarrow{w} u \quad \text{in } W_{pq}(0,b),$$

so

$$u_n \xrightarrow{w} u \quad \text{in } L^p(T; X_1) \quad \text{and} \quad u_n' \xrightarrow{w} u' \quad \text{on } L^q(T; X_2).$$

By considering $u_n - u$ instead of u_n (for $n \geqslant 1$) without any loss of generality, we may assume that $u = 0$. From Problem 1.75 we know that the embedding $W_{pq}(0, b) \hookrightarrow C(T; X_2)$ is continuous. So, we can find $c_1 > 0$ such that

$$\|u_n(t)\|_{X_2} \leqslant c_1 \qquad \forall\, t \in T,\ n \geqslant 1. \tag{1.117}$$

We show that $u_n(t) \longrightarrow 0$ in X_2 for all $t \in T$. We do this for $t = 0$, the proof for other $t \in T$ being similar. We have

$$u_n(0) = u_n(t) - \int_0^t u_n'(s)\, ds$$

(see Problem 1.75). Integrating this equation on $[0, \tau]$ (for $\tau \in T$) and using the Fubini theorem, we obtain

$$\begin{aligned} u_n(0) &= \tfrac{1}{\tau}\Big(\int_0^\tau u_n(t)\, dt - \int_0^\tau \int_0^t u_n'(s)\, ds\, dt\Big) \\ &= \tfrac{1}{\tau}\int_0^\tau u_n(t)\, dt - \tfrac{1}{\tau}\int_0^\tau (\tau - s) u_n'(s)\, ds \\ &= \eta_n + \xi_n, \end{aligned}$$

where

$$\eta_n = \tfrac{1}{\tau}\int_0^\tau u_n(t)\, dt \quad \text{and} \quad \xi_n = -\tfrac{1}{\tau}\int_0^\tau (\tau - s) u_n'(s)\, ds \qquad \forall\, n \geqslant 1.$$

Given $\varepsilon > 0$, we can find $\tau \in T$ small such that

$$\|\xi_n\|_{X_2} \leqslant \int_0^\tau \|u_n'(s)\|_{X_2}\, ds \leqslant \tfrac{\varepsilon}{2} \qquad \forall\, n \geqslant 1$$

(since the sequence $\{\|u_n'(\cdot)\|_{X_2}\}_{n \geqslant 1}$ is uniformly integrable as a bounded sequence in $L^q(T; X_2)$; see Problem 1.53). Since $u_n \xrightarrow{w} 0$ in $L^p(T; X_1)$, we see that $\eta_n \xrightarrow{w} 0$ in X_1, hence $\eta_n \longrightarrow 0$ in X_2 (recall that the embedding $X_1 \hookrightarrow X$ is compact and the embedding $X \hookrightarrow X_2$ is continuous). Therefore, we can find $n_0 = n_0(\varepsilon) \in \mathbb{N}$ such that

$$\|\eta_n\|_{X_2} \leqslant \tfrac{\varepsilon}{2} \qquad \forall\, n \geqslant n_0.$$

So, finally we have

$$\|u_n(0)\|_{X_2} \leqslant \|\eta_n\|_{X_2} + \|\xi_n\|_{X_2} \leqslant \varepsilon \qquad \forall\, n \geqslant n_0,$$

so

$$u_n(0) \longrightarrow 0 \quad \text{in } X_2.$$

Therefore, we have proved that $u_n(t) \longrightarrow 0$ in X_2 for all $t \in T$, hence from (1.117) and the Lebesgue dominated convergence theorem (see Problem 1.61), we have

$$u_n \longrightarrow 0 \quad \text{in } L^p(T;X_2). \tag{1.118}$$

From Problem 1.72(a), we know that given $\delta > 0$, we can find $c_\delta > 0$ such that

$$\begin{aligned}\|u_n\|_{L^p(T;X)} &\leqslant \delta\|u_n\|_{L^p(T;X_1)} + c_\delta\|u_n\|_{L^p(T;X_2)} \\ &\leqslant \delta c_2 + c_\delta\|u_n\|_{L^p(T;X_2)} \qquad \forall\, n \geqslant 1,\end{aligned}$$

for some $c_2 > 0$ (recall that the sequence $\{u_n\}_{n\geqslant 1} \subseteq W_{pq}(0,b)$ is bounded), so

$$\limsup_{n\to+\infty} \|u_n\|_{L^p(T;X)} \leqslant \delta c_2$$

(see (1.118)) and thus

$$u_n \longrightarrow 0 \quad \text{in } L^p(T;X)$$

(since δ is arbitrary). So, we conclude that the embedding $W_{pq}(0,b) \hookrightarrow L^p(T;X)$ is compact.

Solution of Problem 1.104

(a) Similarly as in the solution of Problem 1.103(a), we show that $W_p(0,b)$ is a separable reflexive Banach space. From the definition of the evolution triple (X,H,X^*) and Proposition 1.42(b), we have that the embedding $C^1(T;X) \hookrightarrow W_p(0,b)$ is continuous and dense.

For $u,y \in C^1(T;H)$ we have

$$\tfrac{d}{dt}\,(u(t),y(t)) = \big(u'(t),y(t)\big) + \big(u(t),y'(t)\big) \qquad \forall\, t \in T,$$

so for $0 \leqslant s \leqslant t \leqslant \tau$, we have

$$(u(t),y(t)) - (u(s),y(s)) = \int_s^t \Big(\big(u'(\tau),y(\tau)\big) + \big(u(\tau),y'(\tau)\big)\Big)\,d\tau. \tag{1.119}$$

Note that

$$(\cdot,\cdot)\,\Big|_{H\times X} = \langle\cdot,\cdot\rangle\,.$$

Therefore, from (1.119), for all $u, y \in C^1(T; X)$ and all $0 \leqslant s \leqslant t \leqslant b$, we have

$$(u(t), y(t)) - (u(s), y(s)) = \int_s^t \left(\langle u'(\tau), y(\tau)\rangle + \langle u(\tau), y'(\tau)\rangle\right) d\tau. \tag{1.120}$$

Choose a function $\vartheta \in C^1(\mathbb{R})$ with $\vartheta(t) = 1$, $\vartheta(s) = 0$ and $|\vartheta| + |\vartheta'| \leqslant c$, with $c > 0$. Let $y = \vartheta u$. Then from (1.120) and the Hölder inequality (see Theorem 1.3 and Problem 1.27), we have

$$\begin{aligned}\|u(t)\|_H^2 &\leqslant \int_s^t \left(2\left\langle u'(t), \vartheta(t)u(t)\right\rangle + \vartheta'(t)\left\langle u(t), u(t)\right\rangle\right) dt \\ &\leqslant c_1 \|u'\|_{p'} \|u\|_p \leqslant c_2 \|u\|_{W_p}^2 \qquad \forall\, t \in T, u \in C^1(T; X),\end{aligned}$$

for some $c_1, c_2 > 0$, so

$$\|u\|_{C(T;H)} \leqslant c_3 \|u\|_{W_p} \qquad \forall\, u \in C^1(T; X), \tag{1.121}$$

with $c_3 = \sqrt{c_2}$. Since the embedding $C^1(T; X) \hookrightarrow W_p(0, b)$ is dense, from (1.121) we infer that the embedding $W_p(0, b) \hookrightarrow C(T; H)$ is continuous.
Finally the embedding is dense, because the X-valued polynomials are dense in both $W_p(0, b)$ and $C(T; H)$.

(b) We have seen in part **(a)** (see (1.120)) that the identity holds for all $u, y \in C^1(T; X)$. Exploiting the density of $C^1(T; X)$ into $W_p(0, b)$ we conclude that the identity is also true for all $u, y \in W_p(0, b)$.

Solution of Problem 1.105

Using the Fubini theorem and symmetry of the convolution kernel, we have

$$\int_\Omega (\mu * \varphi_\varepsilon) h \, dz = \int_\Omega (h * \varphi_\varepsilon) \, d\mu \tag{1.122}$$

when $h \in L^1(\Omega)$ and either $\operatorname{supp} \mu \subseteq \Omega_{-\varepsilon}$ or $h = 0$ outside $\Omega_{-\varepsilon}$. So, if $h \in C_c(\Omega)$, then from (1.122) and since $h * \varphi_\varepsilon \longrightarrow h$ in $L^1(\Omega)$ as $\varepsilon \searrow 0$, we have

$$(\mu * \varphi_\varepsilon)\lambda^N \longrightarrow \mu \quad \text{vaguely in } \Omega \text{ as } \varepsilon \searrow 0$$

(see Definition 1.89(a)). Also we have

$$\begin{aligned}\int_E |\mu * \varphi_\varepsilon|(x)\,dx &= \int_E \Big|\int_\Omega \varphi_\varepsilon(x-z)\,d\mu(z)\Big|\,dx \\ &\leqslant \int_E\int_\Omega \varphi_\varepsilon(x-z)\,d|\mu|(z)\,dx = \int_\Omega\int_E \varphi_\varepsilon(x-z)\,dx\,d|\mu|(z) \\ &= \int_{E_\varepsilon}\int_E \varphi_\varepsilon(x-z)\,dx\,d|\mu|(z) \leqslant |\mu|(E_\varepsilon).\end{aligned}$$

Solution of Problem 1.106

By virtue of Proposition 1.80, without any loss of generality, we may assume that $u \in C^\infty(\Omega)$. We have

$$u(x-\varepsilon y) - u(x) = -\varepsilon \int_0^1 (Du(x-\varepsilon t y), y)_{\mathbb{R}^N}\,dt \qquad \forall\, x \in K,\ y \in \overline{B}_1,$$

where $\overline{B}_1 = \{z \in \mathbb{R}^N : |z| \leqslant 1\}$, so

$$\begin{aligned}\int_K \big|u(x-\varepsilon y) - u(x)\big|\,dx &\leqslant \varepsilon \int_K\int_0^1 \big|Du(x-\varepsilon t y)\big|\,dt\,dx \\ &= \varepsilon \int_0^1\int_K \big|Du(x-\varepsilon t y)\big|\,dx\,dt \leqslant \varepsilon |Du|(\Omega),\end{aligned}$$

for $\varepsilon \in \big(0, \operatorname{dist}(K, \partial\Omega)\big)$. We multiply both sides with $\varphi(y)$ and integrate over $\mathbb{R}^N$. Using the Fubini theorem once more, we obtain

$$\int_K \Big(\int_{\mathbb{R}^N} \big|u(x-\varepsilon y) - u(x)\big|\varphi(y)\,dy\Big)\,dx \leqslant \varepsilon |Du|(\Omega) \tag{1.123}$$

(recall that $\int_{\mathbb{R}^N} \varphi(y)\,dy = 1$). Note that

$$(u * \varphi_\varepsilon)(x) - u(x) = \int_{\mathbb{R}^N} \big(u(x-\varepsilon y) - u(x)\big)\varphi(y)\,dy.$$

Therefore from (1.123) we have

$$\int_K |u * \varphi_\varepsilon - u|\,dx \leqslant \varepsilon |Du|(\Omega) \qquad \forall\, \varepsilon \in (0, \operatorname{dist}(K, \partial\Omega)).$$

Solution of Problem 1.107

"$\Longrightarrow$": This implication follows from the uniform boundedness principle (see Theorem I.5.39 and recall that $C_0\left(\Omega;\mathbb{R}^N\right)^* = M_b(\Omega;\mathbb{R}^N)$; see Theorem 1.88).

"$\Longleftarrow$": We need only to show that $Du_n \xrightarrow{w} Du$ in $M_b(\Omega;\mathbb{R}^N)$ (see Definition 1.89(b)).

Note that the space $C_0\left(\Omega;\mathbb{R}^N\right)$ is separable. Hence bounded sets in $M_b(\Omega;\mathbb{R}^N)$ with the relative w^*-topology are relatively compact metrizable (see Theorem I.5.85(a)). So, we may assume that at least for a subsequence, we have

$$Du_n \xrightarrow{w} \mu \quad \text{in } M_b(\Omega;\mathbb{R}^N)$$

(see Definition 1.89(b)). We have

$$\int\limits_\Omega u_n \frac{\partial \vartheta}{\partial z_k}\, dz = -\int\limits_\Omega \vartheta\, dD_k u_n \qquad \forall\, \vartheta \in C_c^1(\Omega),\ k \in \{1,\ldots,N\},$$

so

$$\int\limits_\Omega u \frac{\partial \vartheta}{\partial z_k}\, dz = -\int\limits_\Omega \vartheta\, d\mu_k \qquad \forall\, \vartheta \in C_c^1(\Omega),\ k \in \{1,\ldots,N\}$$

and thus $\mu = Du$. Hence $Du_n \xrightarrow{w} Du$ in $M_b(\Omega;\mathbb{R}^N)$.

Solution of Problem 1.108

That strict convergence implies w^*-convergence, follows from Problem 1.107.

To show that the converse is not in general true, consider

$$u_n(x) = \tfrac{1}{n}\sin(nx) \qquad \forall\, x \in (0,2\pi),\ n \geqslant 1.$$

Then we have

$$u_n \xrightarrow{w^*} 0 \quad \text{in } BV(\Omega)$$

(see Problem 1.107), but $u_n'\big((0,2\pi)\big) = 4$ for all $n \geqslant 1$. So, we cannot have strict convergence.

Solution of Problem 1.109

Let

$$\varphi_\lambda(x) = \inf_{y\in X}\left(\varphi(y) + \lambda d_X(x,y)\right) \qquad \forall\, x \in X$$

(here d_X denotes the metric of the space X). Let $x, v \in X$. We have

$$\begin{aligned}\varphi_\lambda(x) &\leqslant \varphi(y) + \lambda d_X(x,y)\\ &\leqslant \varphi(y) + \lambda d_X(x,v) + \lambda d_X(v,y) \qquad \forall\, y \in X\end{aligned}$$

(by the triangle inequality), so

$$\varphi_\lambda(x) - \big(\varphi(y) + \lambda d_X(v,y)\big) \leqslant \lambda d_X(x,v) \qquad \forall\, y \in X.$$

Taking supremum over $y \in X$, we obtain

$$\varphi_\lambda(x) - \varphi_\lambda(v) \leqslant \lambda d_X(x,v).$$

Interchanging the roles of $x, v \in X$, we finally have

$$\big|\varphi_\lambda(x) - \varphi_\lambda(v)\big| \leqslant \lambda d_X(x,v),$$

so φ_λ is λ-Lipschitz continuous for each λ. Note that

$$\varphi_\lambda(x) \leqslant \varphi(x) \qquad \forall\, x \in X,\ \lambda > 0.$$

So, if $\varphi_\lambda(x) \nearrow +\infty$, we have $\varphi(x) = +\infty$. Also, clearly the sequence $\{\varphi_\lambda(x)\}_{\lambda>0}$ is increasing. So, if $\lim\limits_{\lambda\to+\infty} \varphi_\lambda(x)$ is finite, then let $x_k \in X$ be such that

$$\varphi(x_k) + k d_X(x_k, x) - \tfrac{1}{k} \leqslant \varphi_\lambda(x) \qquad \forall\, x \in X,$$

so the sequence $\{k d_X(x_k,x)\}_{k\geqslant 1} \subseteq \mathbb{R}$ is bounded and $d_X(x_k,x) \longrightarrow 0$ as $k \to +\infty$. Thus, we have

$$\varphi(x) \leqslant \liminf_{k\to+\infty} \varphi(x_k) \leqslant \lim_{\lambda\to+\infty} \varphi_\lambda(x)$$

and so

$$\varphi_\lambda(x) \nearrow \varphi(x) \quad \text{as } \lambda \nearrow +\infty \qquad \forall\, x \in X.$$

Now, let $\vartheta \in C_c(X)$ be such that $0 \leqslant \vartheta \leqslant 1$. Then $\vartheta\varphi_\lambda \in C_c(X)$ and so

$$\int\limits_X \vartheta\varphi_\lambda\, d\mu = \lim_{n\to+\infty} \int\limits_X \vartheta\varphi_\lambda\, d\mu_n \leqslant \liminf_{n\to+\infty} \int\limits_X \varphi_\lambda\, d\mu_n$$

(see Definition 1.89(a)). Taking the supremum over all such $\vartheta \in C_c(X)$ and letting $\lambda \to +\infty$, we obtain

$$\int\limits_X \varphi\, d\mu \leqslant \liminf_{n\to+\infty} \int\limits_X \varphi\, d\mu_n.$$

Now, let $\psi\colon X \longrightarrow \mathbb{R}_+$ be upper semicontinuous and of compact support. In this case we define

$$\psi_\lambda(x) = \sup\big\{\psi(y) - \lambda d_X(x,y) : y \in X\big\}.$$

We have that ψ_λ is λ-Lipschitz continuous and $\psi_\lambda \searrow \psi$ as $\lambda \to +\infty$. Since by hypothesis $\operatorname{supp}\psi$ is compact and ψ is bounded, we can find a relatively compact neighborhood U of $\operatorname{supp}\psi$ which contains the supports of ψ_λ for $\lambda > 0$ large. Then

$$\int\limits_X \psi\, d\mu_n \leqslant \int\limits_X \psi_\lambda\, d\mu_n \longrightarrow \int\limits_X \psi_\lambda\, d\mu \quad \text{as } n \to +\infty, \text{ for } \lambda > 0 \text{ large},$$

so

$$\limsup_{n\to+\infty}\int_X \psi\,d\mu_n \leqslant \int_X \psi_\lambda\,d\mu \qquad \forall\,\lambda>0 \text{ large}$$

and thus

$$\limsup_{n\to+\infty}\int_X \psi\,d\mu_n \leqslant \int_X \psi\,d\mu$$

(by the Lebesgue monotone convergence theorem; see Theorem I.3.92).

Solution of Problem 1.110

For any $v\in C(X)$, $v\geqslant 0$ we have

$$\int_X v\,d\mu_n = \int_0^{+\infty}\mu_n(\{v>\lambda\})\,d\lambda \qquad \forall\,n\geqslant 1$$

(see Problem 1.34), so

$$\begin{aligned}\liminf_{n\to+\infty}\int_X v\,d\mu_n &= \liminf_{n\to+\infty}\int_0^{+\infty}\mu_n(\{v>\lambda\})\,d\lambda\\ &\geqslant \int_0^{+\infty}\liminf_{n\to+\infty}\mu_n(\{v>\lambda\})\,d\lambda \geqslant \int_0^{+\infty}\mu(\{v>\lambda\})\,d\lambda = \int_X v\,d\mu\end{aligned} \tag{1.124}$$

(by the Fatou lemma and by hypothesis). Now let $u\in C_b(X)$. By replacing u by $au+\zeta$ for suitable $a,\zeta\geqslant 0$ if necessary, without any loss of generality, we may assume that $0\leqslant u\leqslant 1$. We have

$$\liminf_{n\to+\infty}\int_X u\,d\mu_n \geqslant \int_X u\,d\mu \tag{1.125}$$

and

$$\liminf_{n\to+\infty}\int_X (1-u)\,d\mu_n \geqslant \int_X (1-u)\,d\mu \tag{1.126}$$

(see (1.124)). Also we have

$$\int_X u\,d\mu_n + \int_X (1-u)\,d\mu_n = \mu_n(X) \longrightarrow \mu(X) = \int_X u\,d\mu + \int_X (1-u)\,d\mu. \tag{1.127}$$

From (1.125), (1.126) and (1.127) it follows that

$$\int_X u\, d\mu_n \longrightarrow \int_X u\, d\mu \quad \text{and} \quad \int_X (1-u)\, d\mu_n \longrightarrow \int_X (1-u)\, d\mu,$$

so $\mu_n \longrightarrow \mu$ narrowly (see Definition 1.89(c)).

Solution of Problem 1.111

Suppose that

$$\|\mu_\alpha\|_{M_b(X)} \leqslant M \qquad \forall\, \alpha \in J$$

(see Definition 1.87). Then $\|\mu\|_{M_b(X)} \leqslant M$. Let $u \in C_b(X)$ and $\varepsilon > 0$. We may assume that $|u| \leqslant 1$. We can find $n \geqslant 1$ and $c_1, \ldots, c_n \in [-1, 1]$, such that

$$c_1 = -1,\ c_n = 1,\ 0 < c_{k+1} - c_k < \varepsilon \quad \text{and} \quad |\mu|(\{|u| = c_k\}) = 0,$$

for all $k \in \{1, \ldots, n\}$. We define

$$h(x) \;=\; c_k \quad \text{when } u(x) \in [c_k, c_{k+1}],\ \text{for } k \in \{1, \ldots, n\}.$$

Then we have

$$\big|u(x) - h(x)\big| \;<\; \varepsilon \qquad \forall\, x \in X. \tag{1.128}$$

Since by hypothesis

$$\lim_{\alpha \in J} \mu_\alpha\big(\{c_k \leqslant u < c_{k+1}\big) \;=\; \mu\big(\{c_k \leqslant u < c_{k+1}\}\big)$$

and

$$\{c_k \leqslant u < c_{k+1}\} \;=\; \{u < c_{k+1}\} \setminus \{u < c_k\} \qquad \forall\, k \in \{1, \ldots, \},$$

we obtain

$$\Big|\int_X h\, d\mu_\alpha - \int_X h\, d\mu\Big| \;<\; \varepsilon \qquad \forall\, \alpha \geqslant \alpha_0, \tag{1.129}$$

for some $\alpha_0 \in J$. From (1.128), (1.129), we see that

$$\Big|\int_X u\, d\mu_\alpha - \int_X u\, d\mu\Big| \;\leqslant\; (2M + 1)\varepsilon \qquad \forall\, \alpha \geqslant \alpha_0,$$

so $\mu_\alpha \longrightarrow \mu$ narrowly.

Solution of Problem 1.112
"$\Longrightarrow$": Since $C_c(X) \subseteq C_0(X)$, from Definition 1.89, we see that

$$\text{if } \mu_n \xrightarrow{w} \mu, \text{ then } \mu_n \longrightarrow \mu \text{ vaguely.}$$

Moreover, from Theorem 1.88, we know that $M_b(X) = C_0(X)^*$. Since $C_0(X)$ with the supremum norm is a Banach space, from the uniform boundedness principle (see Corollary I.5.40), we have

$$\sup_{n\geqslant 1} \|\mu_n\|_{M_b(X)} = \sup_{n\geqslant 1} |\mu_n|(X) < +\infty$$

(see Definition 1.87).
"$\Longleftarrow$": From Proposition 1.86 we know that $C_c(X)$ is dense in the Banach space $C_0(X)$. So, given $u \in C_0(X)$, we can find $\widehat{u} \in C_c(X)$ such that

$$\|\widehat{u} - u\|_\infty \leqslant \varepsilon. \tag{1.130}$$

Let $M = \sup_{n\geqslant 1} |\mu_n|(X) < +\infty$. We have

$$\begin{aligned}
&\Big|\int_X u\,d\mu_n - \int_X u\,d\mu\Big| \\
&\leqslant \Big|\int_X u\,d\mu_n - \int_X \widehat{u}\,d\mu_n\Big| + \Big|\int_X \widehat{u}\,d\mu_n - \int_X \widehat{u}\,d\mu\Big| + \Big|\int_X \widehat{u}\,d\mu - \int_X u\,d\mu\Big| \\
&\leqslant \|\widehat{u} - u\|_\infty M + \Big|\int_X \widehat{u}\,d\mu_n - \int_X \widehat{u}\,d\mu\Big| + \|\widehat{u} - u\|_\infty |\mu|(X) \\
&\leqslant (M + |\mu|(X))\varepsilon + \Big|\int_X \widehat{u}\,d\mu_n - \int_X \widehat{u}\,d\mu\Big|. \qquad (1.131)
\end{aligned}$$

Since $\mu_n \longrightarrow \mu$ vaguely (see Definition 1.89(a)) and $\widehat{u} \in C_c(X)$ we have

$$\int_X \widehat{u}\,d\mu_n \longrightarrow \int_X \widehat{u}\,d\mu.$$

So, if in (1.131) we pass to the limit as $n \to +\infty$, then

$$\limsup_{n\to+\infty} \Big|\int_X u\,d\mu_n - \int_X u\,d\mu\Big| \leqslant (M + |\mu|(X))\varepsilon.$$

But $\varepsilon > 0$ is arbitrary, so we let $\varepsilon \searrow 0$ to conclude that

$$\int_X u\,d\mu_n \longrightarrow \int_X u\,d\mu,$$

thus $\mu_n \xrightarrow{w} \mu$ (see Definition 1.89(b)).

Solution of Problem 1.113

The argument in the solution of Problem 1.112 reveals that the two topologies coincide on bounded subsets of $M_b(X)$. So, we need to consider unbounded sets in order to illustrate the difference of the two topologies. Let $X = (0, +\infty)$. Of course X is locally compact and σ-compact metric space. Let $\mu_n = n\delta_n$. Then

$$\mu_n \longrightarrow 0 \quad \text{vaguely.}$$

On the other hand, let $u \in C_0(X)$ be such that $u \sim \frac{1}{x}$ in a neighborhood of $+\infty$. Then

$$\langle \mu_n, u \rangle = \int_X u \, d\mu_n \longrightarrow 1$$

and so $\{ \mu_n \}_{n \geqslant 1}$ is not weakly convergent to 0 (see Definitions 1.89(a) and (b)).

Solution of Problem 1.114

(a) Let $U \subseteq X$ be an open set and let $u \in C_c(U)$ be a function such that $\|u\|_\infty \leqslant 1$. We have

$$\Big| \int_U u \, d\mu_n \Big| \leqslant \int_U |u| \, d|\mu_n|,$$

so

$$\Big| \int_U u \, d\mu \Big| \leqslant \int_U |u| \, d\vartheta$$

(since $\mu_n \longrightarrow \mu$, $|\mu_n| \longrightarrow \vartheta$ both vaguely; see Definition 1.89(a)). Taking the supremum over $u \in C_c(X)$ with $\|u\|_\infty \leqslant 1$, we obtain

$$\|\mu\|_{M_b(U)} = |\mu|(U) \leqslant \vartheta(U) = \|\vartheta\|_{M_b(X)}. \tag{1.132}$$

Finally recall that for every $E \in \mathcal{B}(X)$ we have

$$|\mu|(E) = \inf \big\{ |\mu|(U) : \ E \subseteq U, \ U \text{ is open} \big\} \tag{1.133}$$

and

$$\vartheta(E) = \inf \big\{ \vartheta(U) : \ E \subseteq U, \ U \text{ is open} \big\}. \tag{1.134}$$

From (1.132), (1.133) and (1.134) we conclude that $|\mu| \leqslant \vartheta$.

(b) Evidently sequences $\{\mu_n^+\}_{n\geqslant 1}$ and $\{\mu_n^-\}_{n\geqslant 1}$ are bounded in $M_b^+(X)$. So, by passing to a suitable subsequence if necessary, we may assume that

$$\mu_n^+ \xrightarrow{w} \tau_+ \quad \text{and} \quad \mu_n^- \xrightarrow{w} \tau_- \quad \text{in } M_b^+(X).$$

Since $\mu_n \longrightarrow \mu$ vaguely, it follows that

$$\mu = \tau_+ - \tau_-. \tag{1.135}$$

Because $\mu_n^\pm \leqslant |\mu_n|$ for all $n \geqslant 1$, we infer that $\tau_\pm \leqslant \vartheta$. Invoking Proposition 1.93, we have

$$\mu_n^\pm(E) \longrightarrow \tau_\pm(E),$$

so

$$\mu_n(E) \longrightarrow \mu(E)$$

(see (1.135)).

Solution of Problem 1.115

Using the Fubini theorem, we have (see Problem 1.105)

$$\begin{aligned}\int\limits_{\mathbb{R}^N} |\mu * \varphi_\varepsilon|(x)\,dx &= \tfrac{1}{\varepsilon^N}\int\limits_{\mathbb{R}^N}\Big|\int\limits_{\mathbb{R}^N}\varphi\big(\tfrac{x-z}{\varepsilon}\big)\,\mu(dz)\Big|\,dx \\ \leqslant \tfrac{1}{\varepsilon^N}\int\limits_{\mathbb{R}^N}\int\limits_{\mathbb{R}^N}\varphi\big(\tfrac{x-z}{\varepsilon}\big)|\mu|(dz)\,dx &= \int\limits_{\mathbb{R}^N}\Big(\int\limits_{\mathbb{R}^N}\tfrac{1}{\varepsilon^N}\varphi\big(\tfrac{x-z}{\varepsilon}\big)\,dx\Big)|\mu|(dz) \\ &= \int\limits_{\mathbb{R}^N}|\mu|(dx) \qquad \forall\, \varepsilon > 0.\end{aligned} \tag{1.136}$$

From Problem 1.105 we know that

$$(\mu * \varphi_\varepsilon)\lambda^N \longrightarrow \mu \quad \text{vaguely as } \varepsilon \searrow 0 \tag{1.137}$$

(see Definition 1.89(a)). The map $\mu \longmapsto \int\limits_{\mathbb{R}^N}|\mu|$ is vaguely lower semicontinuous (weak lower semicontinuity of the norm functional in a Banach space; see Proposition I.5.56(c)). So, from (1.137) we have

$$\liminf_{\varepsilon\to 0}\int\limits_{\mathbb{R}^N}|\mu * \varphi_\varepsilon|(x)\,dx \geqslant \int\limits_{\mathbb{R}^N}|\mu|(dx),$$

so

$$\int\limits_{\mathbb{R}^N}|\mu * \varphi_\varepsilon|(x)\,dx \longrightarrow \int\limits_{\mathbb{R}^N}|\mu|(dx) \quad \text{as } \varepsilon \searrow 0.$$

Solution of Problem 1.116

"$\Longrightarrow$": This implication follows immediately from Definition 1.89.

"$\Longleftarrow$": Let $\mu_n \xrightarrow{w} \mu$ and $\mu_n(X) \longrightarrow \mu(X)$. Let $u \in C_b(X)$, $\varepsilon > 0$ and let $K \subseteq X$ be a compact set such that $\mu(X \setminus K) \leqslant \varepsilon$. Consider $h \in C_c(X)$ such that $0 \leqslant h \leqslant 1$, $h|_K = 1$. We have

$$\begin{aligned}
&\Big|\int_X u\,d\mu_n - \int_X u\,d\mu\Big| \\
&\leqslant \Big|\int_X u\,d\mu_n - \int_X uh\,d\mu_n\Big| + \Big|\int_X uh\,d\mu_n - \int_X uh\,d\mu\Big| + \Big|\int_X uh\,d\mu - \int_X u\,d\mu\Big| \\
&\leqslant \|u\|_\infty \int_X (1-h)\,d\mu_n + \Big|\int_X uh\,d\mu_n - \int_X uh\,d\mu\Big| + \|u\|_\infty \int_X (1-h)\,d\mu,
\end{aligned}$$

so

$$\limsup_{n\to+\infty} \Big|\int_X u\,d\mu_n - \int_X u\,d\mu\Big| \leqslant 2\varepsilon\|u\|_\infty.$$

Since $\varepsilon > 0$ is arbitrary, we let $\varepsilon \searrow 0$ to conclude that

$$\int_X u\,d\mu_n \longrightarrow \int_X u\,d\mu$$

(recall that $\mu_n(X) \longrightarrow \mu(X)$), so $\mu_n \longrightarrow \mu$ narrowly.

Solution of Problem 1.117

For every $x \in (a, b)$ we define the function

$$h_x(t) = \begin{cases} 0 & \text{if } t \leqslant x, \\ 1 & \text{if } x < t. \end{cases}$$

Evidently $h_x \in BV([a, b])$ and the weak (distributional) derivative h_x' coincides with the Dirac measure δ_x. Suppose that $BV([a, b])$ is separable. Then the set

$$\mathcal{Y} = \{h_x : x \in (a, b)\}$$

is separable too. But for $x \neq u$, we have

$$\begin{aligned}
\|h_x - h_u\|_{BV} &\geqslant \|h_x' - h_u'\|_{M_b(a,b)} = \|\delta_x - \delta_u\|_{M_b(a,b)} \\
&= \|\delta_x\|_{M_b(a,b)} + \|\delta_u\|_{M_b(a,b)} = 2,
\end{aligned}$$

which means that $\mathcal{Y}$ is not separable, a contradiction.

Solution of Problem 1.118

We know that the weak (distributional) derivative u' of u belongs in $M_b(a,b)$ and we may assume that u is right continuous (see Definition 1.74 and Theorem 1.76). Then for every $x, y \in (a,b)$ we have

$$u(y) - u(x) = \int\limits_{(x,y]} du',$$

so

$$|u(y)| \leqslant |u(x)| + |u'|((x,y]) \leqslant |u'(x)| + \|u\|_{M_b(a,b)}.$$

Integrating with respect to x, we obtain

$$\|u\|_\infty \leqslant \tfrac{1}{b-a}\|u\|_1 + \|u'\|_{M_b(a,b)}.$$

Solution of Problem 1.119

Theorem 1.76 implies that

$$\sup_{n\geqslant 1} \|u_n'\|_{M_b(a,b)} = M < +\infty.$$

Also, Problem 1.118 implies that the sequence $\{u_n\}_{n\geqslant 1}$ is bounded in $L^\infty(a,b)$. So, by passing to a suitable subsequence if necessary, we have that

$$u_n' \longrightarrow \mu \quad \text{in } M_b(a,b), \quad \text{with } \mu \in M_b(a,b) \tag{1.138}$$

and for every $x_0 \in (a,b)$ such that $u_n'(\{x_0\}) = \mu(\{x_0\}) = 0$, we have

$$u_n(x_0) \longrightarrow \eta \in \mathbb{R}. \tag{1.139}$$

Let

$$u(x) = \begin{cases} \eta - \mu\big((x,x_0)\big) & \text{if } x < x_0, \\ \eta + \mu\big((x_0,x)\big) & \text{if } x_0 \leqslant x. \end{cases}$$

Every u_n for almost all $x \in (a,b)$ can be written as

$$u_n(x) = \begin{cases} u_n(x_0) + u_n'\big((x,x_0)\big) & \text{if } x < x_0, \\ u_n(x_0) + u_n'\big((x_0,x)\big) & \text{if } x_0 \leqslant x. \end{cases}$$

From (1.138) and (1.139) and Proposition 1.93, we have

$$u_n(x) \longrightarrow u(x) \quad \text{for a.a. } x \in (a,b).$$

Solution of Problem 1.120

Evidently every $u \in BV([a,b])$ belongs in $L^p(a,b)$ (see Proposition 1.61(a) or Problem 1.118). Suppose that $\{u_n\}_{n\geqslant 1} \subseteq BV([a,b])$ is a bounded sequence. Then by virtue of Problem 1.119 and by passing to a subsequence if necessary, we may assume that

$$u_n(x) \longrightarrow u(x) \quad \text{for a.a. } x \in (a,b), \tag{1.140}$$

with $u \in BV([a,b])$ (see the solution of Problem 1.119). We have

$$\big|u_n(x) - u(x)\big|^p \leqslant \|u_n - u\|_\infty^p \leqslant M \qquad \forall\, n \geqslant 1, \tag{1.141}$$

for some $M > 0$ (see Problem 1.118). From (1.140) and (1.141), we see that we can apply the Lebesgue dominated convergence theorem (see Theorem I.3.94) and conclude that

$$\|u_n - u\|_p^p = \int_a^b |u_n - u|^p\,dx \longrightarrow 0,$$

so the embedding $BV([a,b]) \hookrightarrow L^p(a,b)$ is compact.

Solution of Problem 1.121

First assume that $u \in C^1(a,b)$. Then for every $x, y \in (a,b)$ we have

$$\big|u(x) - u(y)\big| = \Big|\int_x^y u'\,dt\Big| \leqslant \begin{cases} \int_x^y |u'|\,dt & \text{if } x \leqslant y, \\ \int_y^x |u'|\,dt & \text{if } y < x. \end{cases}$$

Then

$$\begin{aligned} \int_a^b |u(x) - \overline{u}|\,dx &\leqslant \tfrac{1}{b-a} \int_a^b\int_a^b \big|u(x) - u(y)\big|\,dx\,dy \\ &\leqslant \tfrac{2}{b-a} \iint_{\{x\leqslant y\}} \Big(\int_x^y |u'(t)|\,dt\Big)\,dx\,dy. \end{aligned} \tag{1.142}$$

Let

$$\varphi(x) = \int_a^x |u'(t)|\, dt.$$

Then

$$\begin{aligned}
\tfrac{2}{b-a} \iint_{\{x \leqslant y\}} \Big(\int_x^y |u'|\, dt \Big)\, dx\, dy &= \tfrac{2}{b-a} \int_a^b \Big(\int_a^y (\varphi(y) - \varphi(x))\, dx \Big)\, dy \\
= \tfrac{2}{b-a} \int_a^b \Big(\int_a^y (x-a)|u'|(x)\, dx \Big)\, dy &= \tfrac{2}{b-a} \int_a^b (y-a)(b-y)|u'|(y)\, dy,
\end{aligned} \tag{1.143}$$

where to obtain the last two equalities we have used integration by parts. Note that

$$(y-a)(b-y) \leqslant \tfrac{(b-a)^2}{4}.$$

So, from (1.142) and (1.143), we obtain

$$\int_a^b |u(x) - \overline{u}|\, dx \leqslant \tfrac{b-a}{2} \int_a^b |u'|\, dx.$$

For the general case of $u \in BV([a,b])$, use Proposition 1.80 which says that $BV([a,b]) \cap C^1(a,b)$ is dense in $BV([a,b])$.

To see that the inequality can be strict, let $(a,b) = (0,1)$ and consider

$$u(x) = \begin{cases} -1 & \text{if } 0 \leqslant x \leqslant \frac{1}{3}, \\ 1 & \text{if } \frac{1}{3} < x \leqslant 1. \end{cases}$$

Then $\overline{u} = \frac{1}{3}$, $\int_0^1 |u(x) - \overline{u}|\, dx = \frac{8}{9}$ and $|u'|(0,1) = 2$.

Solution of Problem 1.122

"$\Longrightarrow$": Suppose that the sequence $\{\mu_n\}_{n \geqslant 1}$ is uniformly tight (see Remark 1.92). Let $\{K_m\}_{m \geqslant 1}$ be an increasing sequence of compact sets such that

$$1 - \tfrac{1}{2^m} < \mu_n(K_m) \qquad \forall\, n \geqslant 1,\ m \geqslant 1.$$

We introduce the function $\varphi \colon X \longrightarrow [0, +\infty]$ defined by

$$\varphi|_{K_1} = 1, \quad \varphi|_{K_{m+1} \setminus K_m} = m \qquad \forall\, m \geqslant 1 \quad \text{and} \quad \varphi|_{X \setminus \bigcup\limits_{m \geqslant 1} K_m} = +\infty.$$

Then for all $n \geqslant 1$ we have

$$\int_X \varphi \, d\mu_n \;=\; \mu_n(K_1) + \sum_{m \geqslant 1} m\mu_n(K_{m+1} \setminus K_m) \;\leqslant\; 1 + \sum_{m \geqslant 1} \tfrac{m}{2^m},$$

so

$$\sup_{n \geqslant 1} \int_X \varphi \, d\mu_n \;<\; +\infty.$$

"$\Longleftarrow$": Suppose that there exists a Borel function $\varphi \colon X \longrightarrow [0, +\infty]$ such that the set $\{ \varphi \leqslant \eta \}$ is compact for every $\eta \in [0, +\infty)$, $\mu_n(\{ \varphi = +\infty \}) = 0$ for all $n \geqslant 1$ and

$$\sup_{n \geqslant 1} \int_X \varphi(x) \, d\mu_n \;<\; +\infty.$$

In this case the uniform tightness of the sequence $\{ \mu_n \}_{n \geqslant 1}$ follows at once from Chebyshev–Markov inequality (see Theorem 1.30), which says that

$$\mu_n(\{ \varphi \geqslant \eta \}) \;\leqslant\; \tfrac{1}{\eta} \int_X \varphi \, d\mu_n \qquad \forall \, n \geqslant 1.$$

The properties of φ imply the uniform tightness of the sequence $\{ \mu_n \}_{n \geqslant 1}$.

Solution of Problem 1.123

(a) The characteristic function $\chi_C \colon X \longrightarrow [0, +\infty)$ defined by

$$\chi_C(x) \;=\; \begin{cases} 1 & \text{if } x \in C, \\ 0 & \text{if } x \notin C \end{cases}$$

is upper semicontinuous. From the solution of Problem 1.109, we know that we can find a sequence $\{u_k \colon X \longrightarrow [0, +\infty)\}_{k \geqslant 1}$ of Lipschitz continuous functions such that

$$u_k(x) \searrow \chi_C(x) \qquad \forall \, x \in X.$$

Let $\{\mu_\alpha\}_{\alpha \in J} \subseteq M_1^+(X)$ be a net such that

$$\mu_\alpha(C) \;\geqslant\; \eta \qquad \forall \, \alpha \in J \quad \text{and} \quad \mu_\alpha \;\overset{w_n}{\longrightarrow}\; \mu \;\; \text{in } M_1^+(X) \tag{1.144}$$

(see Definition 1.89(c)). We have

$$\eta \;\leqslant\; \mu_\alpha(C) \;=\; \int_X \chi_C \, d\mu_\alpha \;\leqslant\; \int_X u_k \, d\mu_\alpha \qquad \forall \, k \geqslant 1, \; \alpha \in J. \tag{1.145}$$

Replacing u_k by $\min\{u_k, k\}$ if necessary, we may assume that $u_k \in C_b(X)$. Then for every $k \geqslant 1$, we have

$$\int_X u_k \, d\mu_\alpha \underset{\alpha \in J}{\longrightarrow} \int_X u_k \, d\mu$$

(see (1.144) and Proposition 1.95), so

$$\eta \leqslant \int_X u_k \, d\mu \qquad \forall \, k \geqslant 1$$

(see (1.145)) and thus

$$\eta \leqslant \int_X \chi_C \, d\mu = \mu(C)$$

(by the Lebesgue monotone convergence theorem; see Theorem I.3.92). This proves that the set $\{\mu \in M_1^+(X) : \mu(C) \geqslant \eta\}$ is w_n-closed.

(b) When $U \subseteq X$ is an open set, the characteristic function $\chi_U : X \longrightarrow [0, +\infty)$ is lower semicontinuous. So, according to the solution of Problem 1.109, we can find a sequence $\{u_k : X \longrightarrow [0, +\infty)\}_{k \geqslant 1}$ of Lipschitz continuous functions such that

$$u_k(x) \nearrow \chi_C \qquad \forall \, x \in X.$$

Let $\{\mu_\alpha\}_{\alpha \in J} \subset M_1^+(X)$ be a net such that

$$\eta \geqslant \mu_\alpha(U) \qquad \forall \, \alpha \in J \quad \text{and} \quad \mu_\alpha \xrightarrow{w_n} \mu \quad \text{in } M_1^+(X). \tag{1.146}$$

We have

$$\eta \geqslant \mu_\alpha(U) = \int_X \chi_U \, d\mu_\alpha \geqslant \int_X u_k \, d\mu_\alpha \qquad \forall \, k \geqslant 1. \tag{1.147}$$

For every $k \geqslant 1$, we have

$$\int_X u_k \, d\mu_\alpha \underset{\alpha \in J}{\longrightarrow} \int_X u_k \, d\mu$$

(see (1.146)), so

$$\eta \geqslant \int_X u_k \, d\mu \qquad \forall \, k \geqslant 1$$

(see (1.147)) and thus

$$\eta \geqslant \mu(U) = \int_X \chi_U \, d\mu$$

(by the Lebesgue monotone convergence theorem; see Theorem I.3.92). So, we have proved that the set $\{\mu \in M_1^+(X) : \mu(U) \leqslant \eta\}$ is w_n-closed and so its complement $\{\mu \in M_1^+(X) : \mu(U) > \eta\}$ is w_n-open.

Solution of Problem 1.124

Let $\{(u_\alpha, \mu_\alpha)\}_{\alpha \in J} \subseteq C_b(X) \times M_1^+(X)$ be a net such that

$$u_\alpha \xrightarrow{\|\cdot\|_\infty} u \quad \text{in } C_b(X) \quad \text{and} \quad \mu_\alpha \xrightarrow{w_n} \mu \quad \text{in } M_1^+(X) \tag{1.148}$$

(see Definition 1.89(c)). We have

$$\begin{aligned} &\Big|\int_X u_\alpha \, d\mu_\alpha - \int_X u \, d\mu\Big| \\ &\leqslant \Big|\int_X u_\alpha \, d\mu_\alpha - \int_X u \, d\mu_\alpha\Big| + \Big|\int_X u \, d\mu_\alpha - \int_X u \, d\mu\Big| \\ &\leqslant \|u_\alpha - u\|_\infty + \Big|\int_X u \, d\mu_\alpha - \int_X u \, d\mu\Big| \end{aligned}$$

(recall that $\mu_\alpha(X) = 1$ for all $\alpha \in J$), so

$$\lim_{\alpha \in J} \Big|\int_X u_\alpha \, d\mu_\alpha - \int_X u \, d\mu\Big| = 0$$

(see (1.148)), hence ξ is continuous.

Solution of Problem 1.125

Let $\{\mu_\alpha\}_{\alpha \in J} \subseteq M_1^+(D)$ be a net and let $\mu \in M_1^+(D)$. Suppose that

$$\mu_\alpha \xrightarrow{w_n(D)} \mu \quad \text{in } M_1^+(D) \tag{1.149}$$

(see Definition 1.89(c)). Let $u \in C_b(X)$. Evidently $u|_D \in C_b(D)$. So, we have

$$\int_X u \, d\mu_\alpha = \int_D u|_D \, d\mu_\alpha \longrightarrow \int_D u|_D \, d\mu = \int_X u \, d\mu,$$

so

$$\mu_\alpha \xrightarrow{w_n} \mu \quad \text{in } M_1^+(X).$$

On the other hand, suppose that

$$\mu_\alpha \xrightarrow{w_n} \mu \quad \text{in } M_1^+(X). \tag{1.150}$$

Let V be an open subset of D with the relative X-topology. So, we can find an open subset U of X such that $V = D \cap U$. We have

$$\begin{aligned} \liminf_{\alpha \in J} \mu_\alpha(V) &= \liminf_{\alpha \in J} \mu_\alpha(D \cap U) = \liminf_{\alpha \in J} \mu_\alpha|_D(U) \\ &\geqslant \mu|_D(U) = \mu(D \cap U) = \mu(V) \end{aligned}$$

(see (1.150) and Proposition 1.95(f)). Once again Proposition 1.95 implies that $\mu_\alpha \xrightarrow{w_n(D)} \mu$ and this proves that $w_n|_{M_1^+(D)} = w_n(D)$.

Solution of Problem 1.126

Let $\{\mu_\alpha\}_{\alpha \in J} \subseteq M_1^+(Y)$ be a net such that $\mu_\alpha \xrightarrow{w_n} \mu$ in $M_1^+(Y)$ (see Definition 1.89(c)). Let $v \in C_b(Y)$ and note that $u = v \circ h \in C_b(X)$. Using a change of variables, we have

$$\int_Y v \, d\widehat{h}(\mu_\alpha) = \int_X u \, d\mu_\alpha \longrightarrow \int_X u \, d\mu = \int_Y v \, d\widehat{h}(\mu)$$

(since $u \in C_b(X)$ and $\mu_\alpha \xrightarrow{w_n} \mu$ in $M_1^+(Y)$), so

$$\widehat{h}(\mu_\alpha) \xrightarrow{w_n} \widehat{h}(\mu) \quad \text{in } M_1^+(Y)$$

and thus $\widehat{h}$ is continuous.

Solution of Problem 1.127

Let $\mu, \lambda \in M_1^+(X)$, $\mu \neq \lambda$. Since X is a Polish space, both probability measures μ and λ are Radon (see Definition I.4.9). Therefore,we can find a compact set $K \subseteq X$ such that $\mu(K) \neq \lambda(K)$. The set $h(K) \subseteq Y$ is compact (see Proposition I.1.74), hence Borel. Because h is injective, we have $K = h^{-1}(h(K))$. Therefore

$$\widehat{h}(\mu)(h(K)) = \mu(K) \neq \lambda(K) = \widehat{h}(h(K)),$$

so

$$\widehat{h}(\mu) \neq \widehat{h}(\lambda)$$

and thus $\widehat{h}$ is injective.

Solution of Problem 1.128

By the Kuratowski theorem (see Theorem I.4.44), for any Borel subset $E \subseteq X$, we have that $h(E) \subseteq Y$ is Borel too. Therefore, the set $E \subseteq X$ is Borel if and only if the set $h(E) \subseteq h(X)$ is Borel too.

Let $\mu \in M_1^+(X)$. Then $\widehat{h}(\mu) \in M_1^+(h(X))$. On the other hand, if $\vartheta \in M_1^+(h(X))$, then $\lambda(E) = \vartheta(h(E))$ for all $E \in \mathcal{B}(X)$ (with $\mathcal{B}(X)$ being the Borel σ-field of X) defines a Borel measure on X and $\widehat{h}(\lambda) = \vartheta$. From Problem 1.125, we know that

$$w_n\big|_{M_1^+(h(X))} = w_n(h(X))$$

(with w_n being the narrow topology on $M_1^+(Y)$). So, we conclude that $\widehat{h}\colon M_1^+(X) \longrightarrow M_1^+(Y)$ is bijective and continuous (see Problems 1.127 and 1.126) and so $\operatorname{im}\widehat{h} = M_1^+(h(X))$.

Solution of Problem 1.129

From Problems 1.126, 1.127, 1.128, we know that $\widehat{h}^{-1}\colon M_1^+(Y) \longrightarrow M_1^+(X)$ is bijective and continuous. Moreover, if $\vartheta \in M_1^+(Y)$, then for every $E \in \mathcal{B}(X)$, we have

$$\widehat{h^{-1}}(\vartheta)(E) = \vartheta((h^{-1})^{-1}(E)) = \vartheta(h(E)) = \widehat{h}^{-1}(\vartheta)(E),$$

so

$$\widehat{h}^{-1} = \widehat{h^{-1}}$$

and thus $\widehat{h}$ is a bicontinuous, bijective, hence a homeomorphism.

Solution of Problem 1.130

We can find a sequence $\{K_n\}_{n\geqslant 1}$ of compact sets in U such that

$$K_n \subseteq K_{n+1} \quad \text{and} \quad d_{\mathbb{R}^N}(K_{n+1}^c, K_n) = \delta_n > 0 \qquad \forall\, n \geqslant 1$$

and

$$U = \bigcup_{n\geqslant 1} K_n.$$

Let $u_n \in C_0\left(\mathbb{R}^N\right)$ be such that

$$u_n\big|_{K_n} = 1, \quad \operatorname{supp} u_n \subseteq K_{n+1} \quad \text{and} \quad 0 \leqslant u_n \leqslant 1.$$

Evidently $\{u_n\}_{n\geqslant 1} \subseteq \Gamma_U$. We have

$$\sup_{u\in\Gamma_U} L(u) \leqslant \mu(U) = \lim_{n\to+\infty} \int_{K_n} d\mu \leqslant \lim_{n\to+\infty} \int_{K_{n+1}} u_n\, d\mu \leqslant \sup_{u\in\Gamma_U} L(u)$$

(by the Lebesgue monotone convergence theorem; see Theorem I.3.92), so

$$\mu(U) = \sup_{u \in \Gamma_U} L(u).$$

Next, let $\{C_n\}_{n \geqslant 1}$ be a sequence of compact supersets of K such that

$$C_{n+1} \subseteq C_n \quad \text{and} \quad d_{\mathbb{R}^N}(C_{n+1}, C_n^c) = \delta_n > 0 \qquad \forall\, n \geqslant 1$$

and

$$K = \bigcap_{n \geqslant 1} C_n.$$

Let $u_n \in C_0(\mathbb{R}^N)$ be such that

$$u_n|_{C_{n+1}} = 1, \quad \operatorname{supp} u_n \subseteq C_n \quad \text{and} \quad 0 \leqslant u_n \leqslant 1.$$

We have that $\{u_n\}_{n \geqslant 1} \subseteq \Gamma_K$ and

$$\mu(K) = \inf_{n \geqslant 1} \mu(C_n) = \lim_{n \to +\infty} \int_{C_n} d\mu$$

(by the Lebesgue monotone convergence theorem; see Theorem I.3.92). So, given $\varepsilon > 0$, we can find $n_0 = n_0(\varepsilon) \in \mathbb{N}_0$, such that

$$\mu(C_n) \leqslant \mu(K) + \varepsilon \qquad \forall\, n \geqslant n_0$$

and thus

$$\inf_{u \in \Gamma_K} L(u) \leqslant \int_{C_n} u_n \, d\mu \leqslant \mu(C_n) \leqslant \mu(K) + \varepsilon \leqslant \inf_{u \in \Gamma_K} L(u) + \varepsilon$$

$$\forall\, n \geqslant n_0.$$

Let $\varepsilon \searrow 0$, to conclude that

$$\mu(K) = \inf_{u \in \Gamma_K} L(u).$$

Solution of Problem 1.131

No. Let $X = [-1, 1]$, $\eta = 2$ and let $\mu_n = \delta_{\frac{1}{n}} - \delta_{-\frac{1}{n}}$ for $n \geqslant 1$. Then

$$\mu_n \longrightarrow 0 \quad \text{vaguely and } \|\mu_n\|_{M_b([-1,1])} = 2$$

(see Definition 1.89(a)). So the set $C = \{\mu \in M_b([-1,1]) : \|\mu\|_{M_b([-1,1])} = 2\}$ is not vaguely closed.

Solution of Problem 1.132

From Theorem 1.100, we know that u has a representative $u \in C(\overline{I})$ and

$$u(x) - u(y) = \int_y^x u'(t)\,dt \qquad \forall\, x, y \in \overline{I},$$

so

$$\big|u(x) - u(y)\big| \leqslant \Big|\int_y^x |u'(t)|\,dt\Big|$$

and thus

$$|u(x)| \leqslant |u(y)| + \int_a^b |u'(t)|\,dt.$$

Integrating this inequality with respect to y on $I = (a,b)$ we obtain

$$\lambda(I)|u(x)| \leqslant \int_a^b |u(y)|\,dy + \lambda(I)\|u'\|_1,$$

so

$$|u(x)| \leqslant \tfrac{1}{\lambda(I)} \int_a^b |u(y)|\,dy + \|u'\|_1.$$

Solution of Problem 1.133

By virtue of Theorem 1.100, we have

$$\big|u(x) - u(y)\big| \leqslant \int_y^x |u'(t)|\,dt \qquad \forall\, x, y \in [a,b],$$

so

$$\begin{aligned} \big|u(x) - u(y)\big| &\leqslant \Big(\int_y^x |u'(t)|^p\,dt\Big)^{\frac{1}{p}} |x-y|^{\frac{1}{p'}} \\ &\leqslant \|u'\|_p |x-y|^{\frac{1}{p'}} \qquad \forall\, x, y \in [a,b] \end{aligned}$$

(by the Hölder inequality; see Theorem 1.3 and Problem 1.27), thus $u \in C^{0,\frac{1}{p'}}([a,b])$ (recall that $\frac{1}{p} + \frac{1}{p'} = 1$; see Theorem 1.135 for the definition of Hölder continuous functions).

Solution of Problem 1.134

By virtue of Theorem 1.103, we can find a sequence $\{u_n\}_{n\geqslant 1} \subseteq C_c^\infty(\mathbb{R})$ such that $u_n|_I \longrightarrow u$ in $W^{1,p}(I)$. From the continuity of the embedding $W^{1,p}(a,b) \hookrightarrow L^\infty(I)$ (see Theorem 1.109), we have

$$\|u_n|_I - u\|_{L^\infty(I)} \longrightarrow 0.$$

So, given $\varepsilon > 0$, we can find $n_0 \in \mathbb{N}$ such that

$$\|u_n|_I - u\|_{L^\infty(I)} \leqslant \varepsilon \qquad \forall\, n \geqslant n_0.$$

Since $u_{n_0} \in C_c^\infty(\mathbb{R})$, for $|x|$ large we have $u_{n_0}(x) = 0$. Hence

$$|u(x)| \leqslant \varepsilon \qquad \forall\, x \in I, \text{ with } |x| \text{ large}$$

and so $\lim\limits_{\substack{|x| \to +\infty \\ x \in I}} u(x) = 0.$

Solution of Problem 1.135

From Theorem 1.109, we know that $u, v \in L^\infty(I)$ and so $uv \in L^p(I)$. Next, we show that the weak (distributional) derivative $(uv)' \in L^p(I)$. To this end, first we assume that $1 \leqslant p < +\infty$. Invoking Theorem 1.103, we can find two sequences $\{u_n\}_{n\geqslant 1}, \{v_n\}_{n\geqslant 1} \subseteq C_c^\infty(\mathbb{R})$ such that

$$u_n|_I \longrightarrow u \quad \text{and} \quad v_n|_I \longrightarrow v \quad \text{in } W^{1,p}(I),$$

so

$$u_n|_I \longrightarrow u \quad \text{and} \quad v_n|_I \longrightarrow v \quad \text{in } L^\infty(I)$$

(see Theorem 1.109) and thus

$$(u_n v_n)|_I \longrightarrow uv \quad \text{in } L^\infty(I) \text{ and in } L^p(I).$$

We have

$$(u_n v_n)' = u_n' v_n + u_n v_n' \longrightarrow u'v + uv' \quad \text{in } L^p(I),$$

so

$$u_n v_n \longrightarrow uv \quad \text{in } W^{1,p}(I)$$

and thus

$$(uv)' = u'v + uv'.$$

Integrating this last equality, we obtain

$$(uv)(x) - (uv)(y) = \int_y^x u'v\,dt + \int_y^x uv'\,dt \qquad \forall\, x, y \in \overline{I}.$$

Next we consider the case $p = +\infty$. So, let $u, v \in W^{1,\infty}(I)$. Hence $uv \in L^\infty(I)$ and $u'v + uv' \in L^\infty(I)$. We need to show that

$$(uv)' = u'v + uv'$$

(in the weak (distributional) sense), that is,

$$\int_I (uv)\vartheta'\,dt = -\int_I (u'v + uv')\vartheta\,dt \qquad \forall\, \vartheta \in C_c^1(I). \tag{1.151}$$

To this end, let $T \subseteq I$ be a bounded interval such that $\operatorname{supp}\vartheta \subseteq T$. Then $u, v \in W^{1,p}(T)$ for all $p \in [1, +\infty)$ and so from the first part of the solution, we have that

$$\int_T uv\vartheta'\,dt = -\int_T (u'v + uv')\vartheta\,dt,$$

so

$$\int_I uv\vartheta'\,dt = -\int_I (u'v + uv')\vartheta\,dt$$

(since $\operatorname{supp}\vartheta \subseteq T$). It follows that (1.151) holds.

Solution of Problem 1.136

Since the sequence $\{u_n'\}_{n\geqslant 1} \subseteq L^p(a,b)$ is bounded, by the Poincaré inequality (see Theorem 1.108), we have that the sequence $\{u_n\}_{n\geqslant 1} \subseteq W_0^{1,p}(a,b)$ is bounded. Proposition 1.115 implies that the space $W_0^{1,p}(a,b)$ is reflexive, so, passing to a subsequence if necessary (see the Eberlein–Smulian theorem; Theorem I.5.78), we have

$$u_n \xrightarrow{w} u \quad \text{in } W_0^{1,p}(a,b).$$

Then, for the original sequence, we have

$$u_n \longrightarrow u \quad \text{in } C(\overline{I})$$

(see Theorem 1.109).

Solution of Problem 1.137
Let $x_1, x_2 \in (a,b)$ with $x_1 \leqslant x_2$. We have

$$\begin{aligned} u(x_2)^2 + u(x_1)^2 - 2u(x_1)u(x_2) &= \big(u(x_2) - u(x_1)\big)^2 \\ = \Big(\int_{x_1}^{x_2} u'(t)\,dt\Big)^2 &\leqslant (b-a)\int_{x_1}^{x_2} u'(t)^2\,dt \end{aligned}$$

(see Theorem 1.100 and use the Jensen inequality (see Problem 1.26)). Integrating with respect to x_1 over I, we obtain

$$(b-a)u(x_2)^2 + \int_a^b u(x_1)^2\,dx_1 - 2u(x_2)\int_a^b u(x_1)\,dx_1 \leqslant (b-a)^2\int_a^b u'(t)^2\,dt.$$

A new integration with respect to x_2 over I, leads to

$$2(b-a)\int_a^b u(x)^2\,dx - 2\Big(\int_a^b u(x)\,dx\Big)^2 \leqslant (b-a)^3\int_a^b u'(t)^2\,dt,$$

so

$$\begin{aligned} \|u\|_2^2 &\leqslant \tfrac{(b-a)^2}{2}\|u'\|_2^2 + \tfrac{1}{b-a}\Big(\int_a^b u(x)\,dx\Big)^2 \\ &\leqslant c\Big(\|u'\|_2^2 + \Big(\int_a^b u(x)\,dx\Big)^2\Big), \end{aligned}$$

with $c = \max\left\{\frac{(b-a)^2}{2}, \frac{1}{b-a}\right\}$.

Solution of Problem 1.138

Arguing by contradiction, suppose that we can find a sequence $\{u_n\}_{n\geqslant 1} \subseteq V$ such that

$$n\|Du_n\|_p < \|u_n\|_p \qquad \forall\, n \geqslant 1. \tag{1.152}$$

Let $y_n = \frac{u_n}{\|u_n\|_p}$ for $n \geqslant 1$. Then from (1.152), we have

$$\|Dy_n\|_p < \tfrac{1}{n} \qquad \forall\, n \geqslant 1,$$

so

$$Dy_n \longrightarrow 0 \quad \text{in } L^p(\Omega;\mathbb{R}^N). \tag{1.153}$$

Also, note that $\|y_n\|_p = 1$ for $n \geqslant 1$. Therefore the sequence $\{y_n\}_{n\geqslant 1} \subseteq W^{1,p}(\Omega)$ is bounded and so by passing to a suitable subsequence if necessary, we can have

$$y_n \xrightarrow{w} y \quad \text{in } W^{1,p}(\Omega) \quad \text{and} \quad y_n \longrightarrow y \in L^p(\Omega)$$

(see Proposition 1.115 and Theorem 1.135). We have $\|y\|_p = 1$ and

$$\|Dy\|_p \leqslant \lim_{n\to+\infty} \|Dy_n\|_p = 0$$

(see (1.153)), so

$$y \equiv \xi \in \mathbb{R}$$

(recall that Ω is connected).

Also $y \in V$ (since it is w-closed by the Mazur theorem; see Theorem I.5.58). Therefore by hypothesis $y \equiv 0$, which contradicts the fact that $\|y\|_p = 1$. So, there exists $c > 0$ such that

$$\|u\|_p \leqslant \|Du\|_p \qquad \forall\, u \in V.$$

Solution of Problem 1.139

Since γ is a continuous linear operator (see Definition 1.129), $V \subseteq W^{1,p}(\Omega)$ is a closed linear subspace. Let $u \in V$ and suppose that $u \equiv \xi \in \mathbb{R}$. We have $\gamma(u)|_{\Gamma_0} = \xi$ (since Γ_0 has strictly positive surface measure) and so $\xi = 0$. Therefore the only constant function in V is the zero function. So, we can use Problem 1.138 and conclude that there exists $c > 0$ such that $\|u\|_p \leqslant c\|Du\|_p$ for all $u \in V$.

Solution of Problem 1.140

First we assume that $u \in C^1(\Omega)$. Since $y \in W^{1,p'}(\Omega)$, by the Meyers–Serrin theorem (see Theorem 1.119), we can find a sequence $\{y_n\}_{n\geqslant 1} \subseteq C^1(\Omega)$ such that $y_n \longrightarrow y$ in $W^{1,p'}(\Omega)$. For every $\vartheta \in C_c^\infty(\Omega)$ and every $k \in \{1,\ldots,N\}$, we have

$$\int_\Omega u y_n (D_k \vartheta)\, dz \;=\; -\int_\Omega \big(u(D_k y_n) + y_n(D_k u)\big)\vartheta\, dz \qquad \forall\, n \geqslant 1.$$

Recall that $y_n \longrightarrow y$ in $W^{1,p'}(\Omega)$. So, we have

$$y_n \longrightarrow y \quad\text{and}\quad D_k y_n \longrightarrow D_k y \quad \text{in } L^{p'}(\Omega) \qquad \forall\, k \in \{1,\ldots,N\}.$$

Hence, it follows that

$$\int_\Omega \big(u(D_k y_n) + y_n(D_k u)\big)\vartheta\, dz \;\longrightarrow\; \int_\Omega \big(u(D_k y) + y(D_k u)\big)\vartheta\, dz$$

and

$$\int_\Omega u y_n (D_k \vartheta)\, dz \;\longrightarrow\; \int_\Omega u y (D_k \vartheta)\, dz \qquad \forall\, k \in \{1,\ldots,N\}.$$

Therefore, in the limit as $n \to +\infty$, we obtain

$$\int_\Omega u y (D_k \vartheta)\, dz \;=\; -\int_\Omega \big(u(D_k y) + y(D_k u)\big)\vartheta\, dz$$
$$k \in \{1,\ldots,N\},\ \vartheta \in C_c^\infty(\Omega),$$

so

$$D_k(uy) \;=\; u(D_k y) + y(D_k u) \qquad \forall\, k \in \{1,\ldots,N\}$$

and so $uv \in W^{1,1}(\Omega)$.

Now, we remove the restriction that $u \in C^1(\Omega)$. Theorem 1.119 implies that we can find a sequence $\{u_n\}_{n\geqslant 1} \subseteq C^1(\Omega)$ such that

$$u_n \longrightarrow u \quad \text{in } W^{1,p}(\Omega).$$

Note that, for all $k \in \{1,\ldots,N\}$, we have

$$u_n y \longrightarrow uy \quad\text{and}\quad D_k(u_n y) \longrightarrow D_k(uy) \quad \text{in } L^1(\Omega).$$

So, $\{u_n y\}_{n\geqslant 1} \subseteq W^{1,1}(\Omega)$ is a Cauchy sequence, hence

$$u_n y \longrightarrow uy \quad \text{in } W^{1,1}(\Omega).$$

But from the first part of the solution, we have

$$D_k(u_n y) \;=\; y(D_k u_n) + u_n(D_k y) \qquad \forall\, n \geqslant 1,\ k \in \{1,\ldots,N\}.$$

So, in the limit as $n \to +\infty$, we obtain

$$D_k(uy) \;=\; y(D_k u) + u(D_k y) \qquad \forall\, k \in \{1, \ldots, N\}.$$

Solution of Problem 1.141
We choose an increasing sequence of open sets $\{U_n\}_{n \geqslant 1}$ such that

$$U_n \;\subset\subset\; U_{n+1} \;\subset\subset\; \Omega \quad \text{and} \quad \Omega \;=\; \bigcup_{n \geqslant 1} U_n.$$

Let $\mathcal{P} \;=\; \{\psi_n\}_{n \geqslant 1}$ be a smooth partition of unity subordinate to the covering $\{U_{n+1} \setminus \overline{U}_{n-1}\}_{n \geqslant 1}$ (where $U_0 = \emptyset$; see Definition I.1.111). For each $n \geqslant 1$, let φ_n be the finite sum of $\psi \in \mathcal{P}$ such that

$$\operatorname{supp} \psi \;\subseteq\; U_{n+1} \setminus \overline{U}_{n-1} \quad \text{and} \quad \operatorname{supp} \psi \not\subseteq \overline{U}_n^c.$$

Then $\varphi_n \in C_c^\infty(U_{n+1} \setminus \overline{U}_{n-1})$ for $n \geqslant 1$ and $\sum\limits_{n \geqslant 1} \varphi_n \equiv 1$ in Ω. Let $\varepsilon > 0$ and for each $n \geqslant 1$ choose $\vartheta_n \in C_c^\infty(U_{n+1} \setminus \overline{U}_{n-1})$ such that

$$\|\vartheta_n - \varphi_n u\|_{1,p} \;\leqslant\; \tfrac{\varepsilon}{2^n}. \tag{1.154}$$

Let $\vartheta = \sum\limits_{n \geqslant 1} \vartheta_n \in C^\infty(\Omega)$. We have

$$\|\vartheta - u\|_{1,p} \;=\; \Big\|\sum_{n \geqslant 1} \vartheta_n - \sum_{n \geqslant 1} \varphi_n u\Big\|_X \;\leqslant\; \sum_{n \geqslant 1} \|\vartheta_n - \varphi_n u\|_{1,p} \;\leqslant\; \varepsilon$$

(see (1.154)), so $u \in W^{1,p}(\Omega)$ (see Theorem 1.119).

Solution of Problem 1.142
From Theorem 1.119, we know that we can find a sequence $\{u_n\}_{n \geqslant 1} \subseteq C^\infty(\Omega)$ such that

$$u_n \;\longrightarrow\; u \quad \text{in } W^{1,p}(\Omega).$$

Let $B \subset\subset \Omega$ be a ball. Since $Du_n \longrightarrow 0$ in $L^p(B; \mathbb{R}^N)$, from the Poincaré–Wirtinger inequality (see Theorem 1.132), we have

$$u_n - (u_n)_B \;\longrightarrow\; 0 \quad \text{in } L^p(B)$$

(recall that $(u_n)_B = \frac{1}{\lambda^N(B)} \int\limits_B u_n\, dz$). By passing to a suitable subsequence if necessary, we have

$$u_n(z) \;\longrightarrow\; c \;=\; \lim_{n \to +\infty} (u_n)_B \quad \text{for a.a. } z \in B.$$

Therefore, $u|_B$ is constant. Since Ω is connected, we conclude that u is constant in Ω.

Solution of Problem 1.143

We can view u as a function in $W_0^{1,p}(\mathbb{R}^N)$ whose gradient vanishes. Then according to Problem 1.142, we have that $u \equiv \xi \in \mathbb{R}^N$. We must have that $\xi = 0$, since $\xi\lambda^N(\mathbb{R}^N) = \|u\|_p^p < +\infty$.

Solution of Problem 1.144

Let $\lambda \in (0,1]$ and let $u_\lambda(x) = |x|^\lambda$. Then $u_\lambda'(x) = \lambda|x|^{\lambda-2}x$. So,

$$\int_{-1}^{1} u_\lambda'(x)^2\,dx = \lambda^2 \int_{-1}^{1} x^{2(\lambda-1)}\,dx$$

and the latter integral is finite if and only if $2-2\lambda < 1$, hence $\lambda > \frac{1}{2}$. Therefore $|x|^\lambda$ belongs in $W^{1,2}(I)$ if and only if $\lambda > \frac{1}{2}$. In particular the continuous function $u(x) = \sqrt{|x|}$ does not belong to $W^{1,2}(I)$.

Solution of Problem 1.145

Let $\varrho \in (0,1)$ and let

$$\Omega = B_\varrho = \left\{z = (z_1, z_2) \in \mathbb{R}^2 : |z| = \sqrt{z_1^2 + z_2^2} < \varrho\right\}.$$

For $\lambda \in \mathbb{R}$, we consider the function

$$u_\lambda(z) = \left|\ln|z|\right|^\lambda \qquad \forall\, z \in \mathbb{R}^2 \setminus \{(0,0)\}.$$

Let $r = |z|$ and $\widehat{u}_\lambda(r) = u_\lambda(z)$ (that is, $\widehat{u}_\lambda(r) = |\ln r|^\lambda$). Note that

$$|Du_\lambda|^2 = (u_\lambda'(r))^2,$$

thus

$$\int_{B_\varrho} u_\lambda(z)^2\,dz = 2\pi \int_0^\varrho (\ln r)^{2\lambda} r\,dr$$

and

$$\int\limits_{B_\varrho} |Du_\lambda(z)|^2\, dz \;=\; 2\pi \int\limits_0^\varrho \tfrac{\lambda^2}{r^2}(\ln r)^{2(\lambda-1)} r\, dr.$$

Adding the last two equalities and making the change of variables $t = -\ln r$, we obtain

$$\int\limits_{B_\varrho} \left(u_\lambda^2 + |Du_\lambda|^2\right) dz \;=\; 2\pi \int\limits_{-\ln\varrho}^{+\infty} t^{2\lambda} e^{-2t}\, dt + 2\pi\lambda^2 \int\limits_{-\ln\varrho}^{+\infty} \frac{dt}{t^{2-2\lambda}}\, dt.$$

So, we see that $u \in H^1(B_\varrho)$ if and only if $\lambda < \frac{1}{2}$. So, for $\lambda \in (0, \frac{1}{2})$ the function $u_\lambda \in H^1(\Omega)$ and blows up to $+\infty$ at zero. Therefore u_λ cannot have a continuous representative.

Solution of Problem 1.146

By hypothesis, we have

$$(u_n, Du_n) \xrightarrow{w} (u, y) \quad \text{in } L^p(\Omega) \times L^p(\Omega;\mathbb{R}^N).$$

By the Mazur theorem (see Corollary I.5.59), we can find $\{\lambda_{n,k}\}_{k=1}^{M_n} \subseteq [0,1]$ such that

$$\sum_{k=1}^{M_n} \lambda_{n,k} \;=\; 1$$

and

$$h_n \;=\; \sum_{k=1}^{M_n} \lambda_{n,k}(u_n.Du_n) \;\longrightarrow\; (u, y) \quad \text{in } L^p(\Omega) \times L^p(\Omega;\mathbb{R}^N).$$

Note that $w_n = \sum\limits_{k=1}^{M_n} \lambda_{n,k} u_k \in K$ and it is a Cauchy sequence in $W^{1,p}(\Omega)$. So, we can find $\widehat{w} \in K$, such that

$$w_n \;\longrightarrow\; \widehat{w} \quad \text{in } W^{1,p}(\Omega).$$

Evidently, $\widehat{w} = u$ and $D\widehat{w} = y = Du$.

Solution of Problem 1.147

From Theorem 1.119, we know that there exists a sequence $\{u_n\}_{n \geqslant 1} \subseteq C^1(\Omega)$ such that

$$u_n \;\longrightarrow\; u \quad \text{in } W^{1,p}(\Omega). \tag{1.155}$$

We set $\widehat{u}_n = u_n \circ \eta$ for all $n \geqslant 1$. Evidently $\widehat{u}_n \in C^1(\Omega')$. Also, let $\widehat{\vartheta} \in C_c^\infty(\Omega')$ and set $\vartheta = \widehat{\vartheta} \circ \eta^{-1}$. Then $\vartheta \in C_c^\infty(\Omega)$. Using the change of variable formula, for every $k \in \{1, \ldots, N\}$, we have

$$\int_{\Omega'} \widehat{u}_n (D_k \widehat{\vartheta})\, dx \;=\; \int_{\Omega} u_n \big(D\vartheta,\ D\eta(\eta^{-1}(x)) e_k\big)_{\mathbb{R}^N} |\det J_{\eta^{-1}}(z)|\, dz,$$

with $J_{\eta^{-1}}(z)$ denoting the Jacobian matrix of the C^1-function η^{-1} at $z \in \Omega$. Passing to the limit as $n \to +\infty$ and using (1.155), for all $k \in \{1, \ldots, N\}$, we obtain

$$\begin{aligned}
\int_{\Omega'} \widehat{u} (D_k \widehat{\vartheta})\, dx &= \int_{\Omega} u \big(D\vartheta,\ D\eta(\eta^{-1}(z)) e_k\big)_{\mathbb{R}^N} |\det J_{\eta^{-1}}(z)|\, dz \\
&= -\int_{\Omega} \vartheta \big(Du,\ D\eta(\eta^{-1}(z)) e_k\big)_{\mathbb{R}^N} |\det J_{\eta^{-1}}(z)|\, dz \\
&= -\int_{\Omega'} \widehat{\vartheta} \big(D(u \circ \eta)(x),\ (D\eta) e_k\big)_{\mathbb{R}^N}\, dx \\
&= -\int_{\Omega'} \widehat{\vartheta} \big(J_\eta(x)^T Du(\eta(x)),\ e_k\big)_{\mathbb{R}^N}\, dx
\end{aligned}$$

(by integration by parts). Since $\vartheta \in C_c^\infty(\Omega')$ is arbitrary, we obtain

$$D\widehat{u}(x) \;=\; J_\eta(x)^T Du(\eta(x)) \qquad \forall\, x \in \Omega'.$$

Solution of Problem 1.148

Let $\Omega_1 \subset\subset \Omega$. Multiplying with a cut-off function $\vartheta \in C_c^\infty(\Omega)$, $0 \leqslant \vartheta \leqslant 1$ and $\vartheta|_{\Omega_1} = 1$, without any loss of generality, we may assume that u is Lipschitz and bounded on $\mathbb{R}^N$. Let φ be a convolution kernel (see Definition 1.117) and let $\{\varphi_\varepsilon\}_{\varepsilon>0}$ be the corresponding family of mollifiers (i.e., $\varphi_\varepsilon(z) = \frac{1}{\varepsilon^N}\varphi(\frac{z}{\varepsilon})$ for all $z \in \Omega$; see Definition 1.117). We set $u_n = \varphi_{\frac{1}{n}} * u$. It is well known that for all $n \geqslant 1$, we have

$$u_n \in C^\infty(\mathbb{R}^N), \quad \|u_n\|_\infty \leqslant \|u\|_\infty \quad \text{and} \quad u_n \longrightarrow u \text{ uniformly on } \Omega_1.$$

Because

$$D_k u_n \;=\; \varphi_{\frac{1}{n}} * D_k u \;\longrightarrow\; D_k u \quad \text{for a.a. } z \in \Omega, \text{ and all } k \in \{1, \ldots, N\}$$

and

$$\|D_k u_n\|_\infty \leqslant \|D_k u\|_\infty \qquad \forall\, k \in \{1, \ldots, N\},$$

using the Lebesgue dominated convergence theorem (see Theorem I.3.94), we have

$$\int_{\Omega'} |u_n - u|^p \, dz \longrightarrow 0$$

and

$$\int_{\Omega'} \big|Du_n - (D_1 u, \ldots, D_N u)\big|^p \, dz \longrightarrow 0.$$

This means that $u \in W^{1,p}(\Omega_1)$ (see Meyers–Serrin theorem; Theorem 1.119), hence $u \in W^{1,p}_{\mathrm{loc}}(\Omega)$. Moreover, Du is the usual gradient of u.

Solution of Problem 1.149

Since by hypothesis the sequence $\{u_n\}_{n\geqslant 1} \subseteq W^{1,p}(\Omega)$ is bounded, it follows that it is also bounded in $L^p(\Omega)$. This fact and because $u_n(z) \longrightarrow u(z)$ for almost all $z \in \Omega$, implies that $u_n \xrightarrow{w} u$ in $L^p(\Omega)$ (see Problem 1.19). Similarly, the sequence $\{Du_n\}_{n\geqslant 1} \subseteq L^p(\Omega;\mathbb{R}^N)$ is bounded and exploiting the reflexivity of $L^p(\Omega;\mathbb{R}^N)$, we can find a subsequence $\{Du_{n_k}\}_{k\geqslant 1}$ of $\{Du_n\}_{n\geqslant 1}$ such that $Du_{n_k} \longrightarrow h$ in $L^p(\Omega;\mathbb{R}^N)$. Problem 1.146 implies that $h = Du$ and the uniqueness of the weak limit implies that for the original sequence, we have $Du_n \longrightarrow Du$ in $L^p(\Omega;\mathbb{R}^N)$. Evidently, the above reasoning remains valid if $W^{1,p}(\Omega)$ is replaced by $W^{1,p}_0(\Omega)$.

Solution of Problem 1.150

Consider the Lipschitz continuous function $g\colon \mathbb{R} \longrightarrow \mathbb{R}$ defined by

$$g(x) = x^+ = \max\{x, 0\} \qquad \forall\, x \in \mathbb{R}.$$

Then $u^+ = g \circ u$ and from the chain rule (see Theorem 1.142), we have

$$Du^+(z) = g'(u(z))Du(z), \tag{1.156}$$

so $Du^+ \in L^p(\Omega;\mathbb{R}^N)$ and thus $u^+ \in W^{1,p}(\Omega)$. Similarly, we show that $u^- \in W^{1,p}(\Omega)$. Recall that $|u| = u^+ + u^-$ and conclude $|u| \in W^{1,p}(\Omega)$.

From (1.156) we see that

$$Du^+ = \begin{cases} 0 & \text{for a.a. } z \in \{u \leqslant 0\}, \\ Du & \text{for a.a. } z \in \{0 < u\}. \end{cases}$$

Similarly for Du^- and $D|u|$.

Since the chain rule is also true on $W_0^{1,p}(\Omega)$ (see Theorem 1.142), all the above arguments remain true if $W^{1,p}(\Omega)$ is replaced by $W_0^{1,p}(\Omega)$.

Solution of Problem 1.151

Note that

$$h_0 = u - (u-y)^+ \quad \text{and} \quad h_1 = (u-y)^+ + y.$$

The result follows now from Problem 1.150.

Solution of Problem 1.152

Let $y_k = \max\limits_{1 \leqslant n \leqslant k} u_n$. From Problem 1.151, we know that $y_k \in W^{1,p}(\Omega)$ for every $k \geqslant 1$ and

$$|Dy_k(z)| \leqslant \max_{1 \leqslant n \leqslant k} |Du_n(z)| \leqslant h(z) \quad \text{for a.a. } z \in \Omega, \text{ and all } k \geqslant 1. \tag{1.157}$$

The sequence $\{y_k\}_{k \geqslant 1}$ is increasing and

$$y_k(z) \longrightarrow y(z) \quad \text{for a.a. } z \in \Omega.$$

So, by the Lebesgue monotone convergence theorem (see Theorem I.3.92), we know that

$$y_k \longrightarrow y \quad \text{in } L^p(\Omega).$$

Also from the reflexivity of $L^p(\Omega;\mathbb{R}^N)$ (recall that $1 < p < +\infty$) and (1.157) (recall that $h \in L^p(\Omega)$), we see that by passing to a subsequence if necessary, we may assume that

$$Dy_k \xrightarrow{w} g \quad \text{in } L^p(\Omega;\mathbb{R}^N).$$

From Problem 1.146, we infer that $g = Dy$ and $y \in W^{1,p}(\Omega)$. Moreover, from (1.157) and the Mazur theorem; see Corollary I.5.59) we conclude that

$$|Dy(z)| \leqslant h(z) \quad \text{for a.a. } z \in \Omega.$$

Solution of Problem 1.153

From (1.157) and the Poincaré inequality (see Theorem 1.131), we have that the sequence $\{y_k\}_{k \geqslant 1} \subseteq W_0^{1,p}(\Omega)$ (see the solution of Problem 1.151) is bounded. Because

$$y_k(z) \longrightarrow y(z) \quad \text{for a.a. } z \in \Omega,$$

we see that for the whole sequence, we have

$$y_k \xrightarrow{w} y \quad \text{in } W_0^{1,p}(\Omega),$$

so

$$y \in W_0^{1,p}(\Omega) \quad \text{and} \quad |Dy(z)| \leqslant h(z) \quad \text{for a.a. } z \in \Omega$$

(by the Mazur theorem; see Corollary I.5.59).

Solution of Problem 1.154

Because

$$\min\{u,y\} = u - (u-y)^+ \quad \text{and} \quad \max\{u,y\} = (u-y)^+ + y,$$

we see that it suffices to show that

$$\text{if } u_n \longrightarrow u \text{ in } W^{1,p}(\Omega), \text{ then } u_n^+ \longrightarrow u \text{ in } W^{1,p}(\Omega). \tag{1.158}$$

First note that

$$|u_n^+ - u^+| \leqslant |u_n - u| \qquad \forall\, n \geqslant 1,$$

so

$$u_n^+ \longrightarrow u^+ \quad \text{in } L^p(\Omega). \tag{1.159}$$

Let $g = \chi_{(0,+\infty)}$ and assume that $1 \leqslant p < +\infty$. From Problem 1.150, we have

$$\|Du_n^+ - Du^+\|_p^p = \int_\Omega |g(u_n)Du_n - g(u)Du|^p\,dz$$

$$\leqslant \|Du_n - Du\|_p^p + \int_\Omega |Du|^p |g(u_n) - g(u)|^p\,dz \longrightarrow 0,$$

so

$$Du^+ \longrightarrow Du^+ \quad \text{in } L^p(\Omega;\mathbb{R}^N)$$

and thus

$$u_n^+ \longrightarrow u^+ \quad \text{in } W^{1,p}(\Omega)$$

(see (1.159)). Therefore (1.158) holds and from this follows the solution of the problem.

If $p = +\infty$, then note that

$$|u_n^+ - u^+| \leqslant |u_n - u| \quad \text{and} \quad |Du_n^+ - Du^+| \leqslant |Du_n - Du|$$

(see Problem 1.150). So

$$\|u_n^+ - u^+\|_\infty \leqslant \|u_n - u\|_\infty \quad \text{and} \quad \|Du_n^+ - Du^+\|_\infty \leqslant \|Du_n - Du\|_\infty$$

and the result follows.

The same argument works in $W_0^{1,p}(\Omega)$.

Solution of Problem 1.155
Let $\Omega_1 \subseteq \mathbb{R}^N$ be a bounded open set such that

$$K \subseteq \Omega_1 \subset\subset \Omega.$$

Let $\varphi \in C_c^\infty(\mathbb{R}^N)$ be such that $\varphi|_K \equiv 1$ and $\varphi|_{\Omega_1^c} \equiv 0$ (a cut-off function). We have

$$\varphi u = u.$$

By Proposition 1.121, we can find a sequence $\{u_n\}_{n\geqslant 1} \subseteq C_c^\infty(\mathbb{R}^N)$ such that

$$u_n \longrightarrow u \quad \text{in } L^p(\Omega) \quad \text{and} \quad Du_n \longrightarrow Du \quad \text{in } L^p(\Omega;\mathbb{R}^N).$$

We have

$$\varphi u_n \in C_c^\infty(\Omega) \qquad \forall\, n \geqslant 1$$

and

$$\varphi u_n \longrightarrow \varphi u = u \quad \text{in } W^{1,p}(\Omega),$$

so $u \in W_0^{1,p}(\Omega)$ (see Definition 1.126).

Solution of Problem 1.156
Since $u = u^+ - u^-$, without any loss of generality, we may assume that $u \geqslant 0$. Let $u_n = (u - \frac{1}{n})^+ \in W^{1,p}(\Omega)$ for $n \geqslant 1$ (see Problem 1.150). By hypothesis $\lim\limits_{z\to x} u(z) = 0$ for all $x \in \partial\Omega$, which implies that u_n vanishes outside a compact subset of Ω. Invoking Problem 1.155, we conclude that $u_n \in W_0^{1,p}(\Omega)$. As $u_n - \frac{1}{n} \longrightarrow u$ in $W^{1,p}(\Omega)$, from Problem 1.154, we have that

$$u_n = (u_n - \tfrac{1}{n})^+ \longrightarrow u^+ = u \quad \text{in } W^{1,p}(\Omega).$$

As $W_0^{1,p}(\Omega)$ is a closed subspace of $W^{1,p}(\Omega)$, we conclude that $u \in W_0^{1,p}(\Omega)$.

Solution of Problem 1.157

Since $u \in W_0^{1,p}(\Omega)$, from Definition 1.126 we know that there exists a sequence $\{\widehat{u}_n\}_{n\geqslant 1} \subseteq C_c^\infty(\Omega)$ such that

$$\widehat{u}_n \longrightarrow u \quad \text{in } W^{1,p}(\Omega).$$

From Problem 1.154, we have that

$$u_n = \widehat{u}_n^+ \longrightarrow u^+ = u \quad \text{in } W^{1,p}(\Omega).$$

Solution of Problem 1.158

Since $u \in W_0^{1,p}(\Omega)$, according to Definition 1.126, we can find a sequence $\{u_n\}_{n\geqslant 1} \subseteq C_c^\infty(\Omega)$ such that

$$u_n \longrightarrow u \quad \text{in } W^{1,p}(\Omega).$$

Let $y_n = \min\{y, u_n\} \in W^{1,p}(\Omega)$ for every $n \geqslant 1$ (see Problem 1.151). Since $y \geqslant 0$, we see that for every $n \geqslant 1$, y_n has compact support. Then Problem 1.155 implies that $y_n \in W_0^{1,p}(\Omega)$ for all $n \geqslant 1$. Moreover, Problem 1.154 implies that

$$y_n \longrightarrow \min\{y, u\} = y \quad \text{in } W^{1,p}(\Omega),$$

so $y \in W_0^{1,p}(\Omega)$.

Solution of Problem 1.159

Since the distributional derivative $u' = 0$, we have

$$\int_a^b u\vartheta'\,dx = 0 \qquad \forall\, \vartheta \in C_c^\infty(a,b). \tag{1.160}$$

Let $W = \{\vartheta' : \vartheta \in C_c^\infty(a,b)\}$. Note that for every $\vartheta \in C_c^\infty(a,b)$, we have $\vartheta' \in C_c^\infty(a,b)$ and

$$\int_a^b \vartheta'(s)\,ds = \vartheta(b) - \vartheta(a) = 0.$$

Conversely, let $\eta \in C_c^\infty(a,b)$ be such that $\int_a^b \eta(s)\,ds = 0$. Let

$$\vartheta(x) = \int_a^x \eta(s)\,ds \qquad \forall\, x \in (a,b).$$

Then $\vartheta \in C^\infty(a,b)$ and $\vartheta' = \eta$. Since $\eta \in C_c^\infty(a,b)$, we may assume that $\eta = 0$ outside $[c,d]$, with $a < c < d < b$. Hence $\vartheta(x) = 0$ for all $x \in (a,c]$, while for all $x \in (d,b)$, we have

$$\vartheta(x) = \int_a^x \eta(s)\,ds = \int_a^b \eta(s)\,ds = 0,$$

so $\vartheta \in C_c^\infty(a,b)$. Therefore, we have proved that

$$W = \{\vartheta' : \vartheta \in C_c^\infty(a,b)\} = \{\eta \in C_c^\infty(a,b) : \int_a^b \eta(s)\,ds = 0\}.$$

Now, let $\vartheta \in C_c^\infty(a,b)$ be such that $\int_a^b \vartheta\,dx = 1$. Then, for any $\varphi \in C_c^\infty(a,b)$, the function

$$\psi(x) = \varphi(x) - (\int_a^b \varphi\,dx)\vartheta(x) \qquad \forall\, x \in (a,b), \tag{1.161}$$

satisfies

$$\psi \in C^\infty(a,b) \quad \text{and} \quad \int_a^b \psi\,dx = 0.$$

Then $\psi \in W$ and we can use it in (1.160). We obtain

$$\int_a^b u\psi\,dx = 0,$$

so

$$\int_a^b u\varphi\,dx = \int_a^b \varphi\,dx \int_a^b u\vartheta\,dx \tag{1.162}$$

(see (1.161)). Let $\xi = \int_a^b u\vartheta\,dx$. Then from (1.162), we have

$$\int_a^b (u-\xi)\varphi\,dx = 0.$$

Since $\varphi \in C_c^\infty(a,b)$ is arbitrary, we conclude that $u(x) = \xi$ for almost all $x \in (a,b)$.

Solution of Problem 1.160
Let

$$D(p) = \{u \in W_0^{1,p}(\Omega),\ u(z) \geqslant 1 \text{ for a.a. } z \text{ in a neighborhood of } A\},$$

$$\widehat{C}_p = \inf_{u \in D(p)} \|Du\|_p^p \tag{1.163}$$

and let $u \in \widehat{C}_p$. Then, there exists an open set U such that $A \subseteq U$ and $u(z) \geqslant 1$ for almost all $z \in U$. Thus

$$\operatorname{cap}_p(A, \Omega) \leqslant \operatorname{cap}_p(U, \Omega) \leqslant \|Du\|_p^p,$$

and

$$\operatorname{cap}_p(A, \Omega) \leqslant \widehat{C}_p \tag{1.164}$$

(see (1.163)). To show the opposite inequality, we may assume that $\operatorname{cap}_p(A, \Omega) < +\infty$ (otherwise, there is nothing to prove). Then, given $\varepsilon > 0$ we can find an open set $U_\varepsilon > 0$ such that $A \subseteq U_\varepsilon$ and

$$\operatorname{cap}_p(U_\varepsilon, \Omega) \leqslant \operatorname{cap}_p(A, \Omega) + \varepsilon. \tag{1.165}$$

Then, we can find $u_\varepsilon \in W_0^{1,p}(\Omega)$ with $u_\varepsilon(z) \geqslant 1$ for almost all $z \in U_\varepsilon$ such that

$$\|Du_\varepsilon\|_p^p \leqslant \operatorname{cap}_p(U_\varepsilon, \Omega) + \varepsilon,$$

so

$$\|Du_\varepsilon\|_p^p \leqslant \operatorname{cap}_p(A, \Omega) + 2\varepsilon$$

(see (1.165)) and thus

$$\widehat{C}_p \leqslant \operatorname{cap}_p(A, \Omega) + 2\varepsilon.$$

Let $\varepsilon \searrow 0$, to conclude that $\widehat{C}_p \leqslant \operatorname{cap}_p(A, \Omega)$. So finally we have

$$\widehat{C}_p = \operatorname{cap}_p(A, \Omega).$$

Solution of Problem 1.161
Let $\tau \colon \mathbb{R} \longrightarrow \mathbb{R}$ be the function defined by

$$\tau(x) = \min\{1, x^+\} \qquad \forall\, x \in \mathbb{R}$$

(recall that $x^+ = \max\{x, 0\}$). Then τ is nonexpansive. Also, if $u \in W_0^{1,p}(\Omega)$ satisfies $u \geqslant 1$ for almost all z in a neighborhood of A, then from Theorem 1.142, we know that $\tau(u) \in W_0^{1,p}(\Omega)$ and clearly $\tau(u) = 1$ for almost all z in a neighborhood of A. Also, we have

$$\|D\tau(u)\|_p^p \leqslant \|Du\|_p^p \tag{1.166}$$

(see Theorem 1.142). Therefore, if

$$\begin{aligned}\widetilde{C}_p &= \inf\{\|Du\|_p^p : u \in W_0^{1,p}(\Omega),\ u \geqslant 0 \text{ and} \\ &\qquad u(z) = 1 \text{ for a.a. } z \text{ in a neighborhood of } A\},\end{aligned}$$

then

$$\widetilde{C}_p \leqslant \|D\tau(u)\|_p^p \leqslant \|Du\|_p^p$$

(see (1.166)). Taking infima with respect to u, we have

$$\widetilde{C}_p \leqslant \widehat{C}_p$$

(where $\widehat{C}_p$ is as in the solution of Problem 1.160 (see (1.163))).

The opposite inequality is easy to see, since the infimum in the definition of $\widehat{C}_p$ can be taken only over nonnegative functions.

Solution of Problem 1.162

Let $u \in W_0^{1,p}(\Omega)$ be such that $u \geqslant 1$ for almost all z in a neighborhood of K. Then, we can find an open set U such that $K \subseteq U$ and $u(z) \geqslant 1$ for almost all $z \in U$. The compactness of K implies that we can find $\varepsilon > 0$ such that if

$$K_\varepsilon = \{z \in \Omega : \operatorname{dist}(z, K) < \varepsilon\},$$

then $K_\varepsilon \subseteq U$. Hence

$$u(z) \geqslant 1 \quad \text{for a.a. } z \in K_\varepsilon.$$

Let φ be a convolution kernel and let $\{\varphi_\varepsilon\}_{\varepsilon>0}$ be the corresponding family of mollifiers (see Definition 1.117). Then we can find $u_n = u * \varphi_{\frac{1}{n}} \in C_c^\infty(\Omega)$, for $n \geqslant 1$, such that

$$u_n \geqslant 1 \quad \text{for a.a. } z \text{ in a neighborhood of } K$$

and

$$\|Du_n\|_p \longrightarrow \|Du\|_p.$$

It follows that

$$\begin{aligned}\operatorname{cap}_p(K,\Omega) &= \{\|Du\|_p^p : u \in C_c^\infty(\Omega),\ u(z) \geqslant 1 \\ &\qquad \text{for all } z \text{ in a neighborhood of } K\} \\ &\geqslant \{\|Du\|_p^p : u \in C_c^\infty(\Omega),\ u(z) \geqslant 1 \text{ for all } z \in K\}.\end{aligned}$$

Then from the definition of $\mathrm{cap}_p(K,\Omega)$ (see Problem 1.163), we conclude that equality must hold.

Solution of Problem 1.163

Evidently $\mathrm{cap}_p(\cdot,\Omega)$ is monotone (see Problem 1.160). So, we have

$$\mathrm{cap}_p(U,\Omega) \geqslant \sup\big\{\mathrm{cap}_p(K,\Omega) : K \text{ is compact and } K\subseteq U\big\}. \tag{1.167}$$

We need to show the opposite inequality. We may assume that

$$\sup\big\{\mathrm{cap}_p(K,\Omega) : K \text{ is compact and } K\subseteq U\big\} < +\infty.$$

Let $\{K_n\}_{n\geqslant 1}$ be an increasing sequence of compact subsets of U such that $\bigcup\limits_{n\geqslant 1} K_n = U$. Using Problem 1.162, for every $n\geqslant 1$, we can find $u_n \in C_c^\infty(\Omega)$ such that

$$u_n \geqslant 1 \quad \text{on } K_n \quad \text{and} \quad \|Du_n\|_p^p \leqslant \mathrm{cap}_p(K_n,\Omega)+\tfrac{1}{n} \qquad \forall\, n\geqslant 1. \tag{1.168}$$

Evidently the sequence $\{u_n\}_{n\geqslant 1}\subseteq W_0^{1,p}(\Omega)$ is bounded (see Theorem 1.131). So, by passing to a subsequence if necessary, we may assume that

$$u_n \xrightarrow{w} u \quad \text{in } W_0^{1,p}(\Omega). \tag{1.169}$$

Then $u(z)\geqslant 1$ for almost all $z\in U$ and we have

$$\begin{aligned} \mathrm{cap}_p(U,\Omega) &\leqslant \|Du\|_p^p \\ &\leqslant \liminf_{n\to+\infty}\|Du_n\|_p^p \leqslant \liminf_{n\to+\infty}\mathrm{cap}_p(K_n,\Omega) \\ &\leqslant \sup\big\{\mathrm{cap}_p(K,\Omega) : K \text{ is compact and } K\subseteq U\big\}. \end{aligned} \tag{1.170}$$

(see Problem 1.160 and (1.169)). Using also (1.170), we finally conclude that

$$\mathrm{cap}_p(U,\Omega) = \sup\big\{\mathrm{cap}_p(K,\Omega) : K \text{ is compact and } K\subseteq U\big\}.$$

Solution of Problem 1.164

First, note that for every $a,b>0$ and $\lambda\in(0,1)$, we have

$$(a+b)^p \leqslant (1-\lambda)^{1-p}a^p+\lambda^{1-p}b^p. \tag{1.171}$$

To show (1.171), we use the convexity of the function $(0,+\infty)\longmapsto x^p$ and obtain

$$\big((1-\lambda)\lambda a+\lambda(1-\lambda)b\big)^p \leqslant (1-\lambda)\lambda^p a^p+\lambda(1-\lambda)^p b^p.$$

Now, dividing both sides by $(1-\lambda)^p\lambda^p$, we obtain (1.171).

We fix $u \in W_h^{1,p}(\Omega)$. From the Poincaré inequality (see Theorem 1.131), we have

$$c_*\|u-h\|_p^p \leqslant \|Du-Dh\|_p^p \leqslant \int_\Omega (|Du|+|Dh|)^p\,dz. \tag{1.172}$$

Using (1.171), for every $\lambda \in (0,1)$, we have

$$\begin{aligned} c_*\|u\|_p^p &\leqslant c_*(1-\lambda)^{1-p}\|u-h\|_p^p + c_*\lambda^{1-p}\|h\|_p^p \\ &\leqslant (1-\lambda)^{1-p}\int_\Omega (|Du|+|Dh|)^p\,dz + c_*\lambda^{1-p}\|h\|_p^p \\ &\leqslant (1-\lambda)^{2-2p}\|Du\|_p^p + (1-\lambda)^{1-p}\lambda^{1-p}\|Dh\|_p^p + c_*\lambda^{1-p}\|h\|_p^p. \end{aligned} \tag{1.173}$$

Therefore, for every $\lambda, \delta \in (0,1)$, we have

$$\begin{aligned} \|Du\|_p^p - c\|u\|_p^p &= \delta\|Du\|_p^p + (1-\delta)\|Du\|_p^p - c\|u\|_p^p \\ &\geqslant \delta\|Du\|_p^p + \big((1-\delta)c_*(1-\lambda)^{2p-2} - c\big)\|u\|_p^p \\ &\quad -(1-\lambda)^{p-1}\lambda^{1-p}\|Dh\|_p^p - c_*(1-\lambda)^{2p-2}\lambda^{1-p}\|h\|_p^p. \end{aligned} \tag{1.174}$$

(see (1.173)). Since $c < c_*$, we can choose $\lambda, \delta \in (0,1)$ such that

$$(1-\delta)c_*(1-\lambda)^{2p-2} > c.$$

Then from (1.174), we conclude that

$$\|Du\|_p^p - c\|u\|_p^p \geqslant \beta_1\|u\|^p - \beta_2,$$

for some $\beta_1 > 0$, $\beta_2 \geqslant 0$.

Solution of Problem 1.165

We need to show that for every $\lambda \in \mathbb{R}$, the sublevel set

$$L_\lambda = \{u \in L^p(\Omega) : \varphi(u) \leqslant \lambda\}$$

is strongly closed. So, let $\{u_n\}_{n\geqslant 1} \subseteq L_\lambda$ be a sequence such that $u_n \longrightarrow u$ in $L^p(\Omega)$. By virtue of hypothesis **(i)**, $u_n \in W^{1,p}(\Omega)$ for $n \geqslant 1$ and the sequence $\{Du_n\}_{n\geqslant 1} \subseteq L^p(\Omega;\mathbb{R}^N)$ is bounded. Therefore the sequence $\{u_n\}_{n\geqslant 1} \subseteq W^{1,p}(\Omega)$ is bounded and due to the reflexivity of $W^{1,p}(\Omega)$ by passing to a suitable subsequence if necessary, we may assume that $u_n \xrightarrow{w} \widehat{u}$ in $W^{1,p}(\Omega)$. Clearly $\widehat{u} = u$. Also, hypothesis (ii) implies that

$$\varphi(u) \leqslant \liminf_{n\to+\infty} \varphi(u_n) \leqslant \lambda,$$

so $u \in L_\lambda$ and so L_λ is closed in $L^p(\Omega)$. This proves that φ is lower semicontinuous for the strong topology of $L^p(\Omega)$.

Solution of Problem 1.166

Consider a sequence $\{u_n\}_{n\geqslant 1} \subseteq L_\lambda$. By hypothesis $\{u_n\}_{n\geqslant 1} \subseteq W^{1,p}(\Omega)$ and it is bounded. Then the Rellich–Kondrachov embedding theorem (see Theorem 1.135) implies that at least for a subsequence, we can have

$$u_n \longrightarrow u \quad \text{in } L^p(\Omega),$$

so

$$\varphi(u) \leqslant \liminf_{n\to+\infty} \varphi(u_n) \leqslant \lambda$$

and thus $u \in L_\lambda$. This proves that L_λ is compact in $L^p(\Omega)$.

Solution of Problem 1.167

From Definition 1.126, we see that it is enough to prove the inequality for all $u \in C_c^\infty(\Omega)$. So, let $u \in C_c^\infty(\Omega)$ and extend it to all of $\mathbb{R}^N$ be setting u equal to 0 outside Ω. Without any loss of generality, we can choose a coordinate system such that $n = e_1$. Then for every $z \in \Omega$, we have

$$u(z) = u(z) - u(-a, z_2, \dots, z_N) = \int_{-a}^{z_1} D_{z_1} u(s, z_2, \dots, z_N)\, ds,$$

so

$$\begin{aligned} |u(z)|^p &\leqslant \Big(\int_{-a}^{z_1} \big|D_{z_1} u(s, z_2, \dots, z_N)\big|\, ds\Big)^p \\ &\leqslant |z_1 + a|^{p-1} \int_{-a}^{a} \big|D_{z_1} u(s, z_2, \dots, z_N)\big|^p\, ds \end{aligned}$$

(by the Jensen inequality; see Problem 1.26). Integrating this inequality in z_1, we obtain

$$\int_{-a}^{a} \big|u(z_1, z_2, \dots, z_N)\big|^p\, dz_1 \leqslant (2a)^p \int_{-a}^{a} \big|D_{z_1} u(z_1, z_2, \dots, z_N)\big|^p\, dz_1.$$

Integrating in the other directions also, we conclude that

$$\|u\|_p^p \leqslant (2a)^p \left\|\frac{\partial u}{\partial n}\right\|_p^p.$$

Solution of Problem 1.168

From Problem 1.167, we know that for all $u \in C_c^\infty(\Omega)$ we have

$$\|u\|_p^p \leqslant (2a)^p \left\|\frac{\partial u}{\partial n}\right\|_p^p \leqslant (2a)^p \|Du\|_p^p,$$

so

$$\|u\|_p^p + \|Du\|_p^p \leqslant \left((2a)^p + 1\right) \|Du\|_p^p.$$

On the other hand, directly from the definition of the Sobolev norm, we have

$$\|Du\|_p^p \leqslant \|u\|_{1,p}^p.$$

Finally from the density of the embedding $C_c^\infty(\Omega) \hookrightarrow W_0^{1,p}(\Omega)$ (see Definition 1.126) we conclude that the two norms are equivalent in $W_0^{1,p}(\Omega)$.

Solution of Problem 1.169

Recall that $u \in C([0,1])$ (see Theorem 1.100) and so from the mean value theorem for integrals, we know that there exists $\xi \in [0,1]$ such that

$$u(\xi) = \int_0^1 u(x)\, dx = \overline{u}.$$

Then we have

$$\left|u(x) - \overline{u}\right| = \left|u(x) - u(\xi)\right| \leqslant \left|\int_\xi^x |u'(s)|\, ds\right| \qquad \forall\, x \in (0,1)$$

(see Theorem 1.100), so

$$\|u - \overline{u}\|_\infty \leqslant \int_0^1 |u'(s)|\, ds = \|u'\|_1.$$

Next, let $\{u_n\}_{n\geqslant 1} \subseteq C^1([0,1])$ be a sequence such that

$$u_n'(x) \geqslant 0 \qquad \forall\, x \in [0,1],\ n \geqslant 1$$

and

$$u_n(1) = 1, \quad u_n\big|_{[0,1-\frac{1}{n}]} = 0 \qquad \forall\, n \geqslant 1.$$

We have

$$\overline{u}_n = \int_{1-\frac{1}{n}}^{1} u_n(x)\,dx,$$

so

$$|\overline{u}_n| \leqslant \tfrac{1}{n} \qquad \forall\, n \geqslant 1.$$

It follows that

$$\|u_n - \overline{u}_n\|_\infty \geqslant u_n(1) - \overline{u}_n \geqslant 1 - \tfrac{1}{n} \qquad \forall\, n \geqslant 1.$$

In addition, we have

$$\|u_n'\|_1 = \int_0^1 u_n'(x)\,dx = u_n(1) - u_n(0) = 1.$$

Therefore, we infer that

$$\sup_{\substack{u \in W^{1,1}(0,1) \\ \|u'\|_1 = 1}} \|u - \overline{u}\|_\infty = 1.$$

Finally arguing by contradiction suppose that the above supremum is realized by some function $u_0 \in W^{1,1}(0,1)$, i.e.,

$$\|u_0 - \overline{u}_0\|_\infty = 1 \quad \text{and} \quad \|u_0'\|_1 = 1.$$

Without any loss of generality, we may assume that there exists $x_0 \in [0,1]$ such that

$$u_0(x_0) - \overline{u}_0 = 1. \tag{1.175}$$

Also, note that

$$\overline{u}_0 = \int_0^1 \big(u_0(x) - \min_{[0,1]} u_0\big)\,dz + \min_{[0,1]} u_0 \geqslant \min_{[0,1]} u_0 = u_0(y_0), \tag{1.176}$$

for some $y_0 \in [0,1]$. From (1.175) and (1.176), it follows that

$$u_0(x_0) - u_0(y_0) \geqslant 1. \tag{1.177}$$

On the other hand, from Theorem 1.100, we have

$$u_0(x_0) - u_0(y_0) \leqslant \int_0^1 |u_0'(x)|\,dx = 1$$

(recall that $\|u_0'\|_1 = 1$), so

$$u_0(x_0) - u_0(y_0) = 1$$

(see (1.177)), thus $\overline{u} = u_0(y_0)$ and hence

$$u_0 \equiv \min_{[0,1]} u_0,$$

which contradicts (1.175). Therefore, the supremum is not realized.

Solution of Problem 1.170

The inequality

$$\|u\|_q \leqslant c\|u'\|_p \qquad \forall\, u \in W_0^{1,p}(a,b)$$

follows from the Poincaré inequality (see Theorem 1.108) and Proposition 1.111.

Let

$$\eta = \inf\left\{\|u'\|_p : u \in W_0^{1,p}(a,b),\ \|u\|_q = 1\right\}$$

and let $\{u_n\}_{n\geqslant 1} \subseteq W_0^{1,p}(a,b)$ be a minimizing sequence. Then from Theorem 1.108, we see that the sequence $\{u_n\}_{n\geqslant 1} \subseteq W_0^{1,p}(a,b)$ is bounded. So, passing to a subsequence if necessary, we may assume that

$$u_n \xrightarrow{w} u \quad \text{in } W_0^{1,p}(a,b), \quad \text{if } 1 < p < +\infty,$$

or

$$u_n' \xrightarrow{w^*} u' \quad \text{in } L^\infty(a,b), \quad \text{if } p = +\infty.$$

Invoking Theorem 1.109(a), we have that

$$u_n \longrightarrow u \quad \text{in } C([a,b])$$

(recall that $p > 1$). Then we have

$$\|u'\|_p \leqslant \eta \quad \text{and} \quad \|u\|_p = 1,$$

which implies that η is realized.

Solution of Problem 1.171

From Theorem 1.142, we know that $g_k \circ u \in W^{1,p}(\Omega)$ for all $k \geqslant 1$. We have

$$\|g_k \circ u - u\|_p^p = \int_{\{u>k\}} |k-u|^p\,dz + \int_{\{u<-k\}} |k+u|^p\,dz$$

$$\leqslant \int_{\{u>k\}} |u|^p\,dz + \int_{\{u<-k\}} |u|^p\,dz = \int_{\{|u|>k\}} |u|^p\,dz \longrightarrow 0$$

(since $|u|^p \in L^1(\Omega)$). For the gradients, we have

$$\begin{aligned} \|D(g_k \circ u) - Du\|_p^p &\leqslant \int_\Omega |g_k'(u) - 1|^p |Du|^p \, dz \\ &= \int_{\{|u|>k\}} |Du|^p \, dz \longrightarrow 0 \end{aligned}$$

(see Theorem 1.142 and since $|Du|^p \in L^1(\Omega)$). So, we conclude that $g_k \circ u \longrightarrow u$ in $W^{1,p}(\Omega)$.

Solution of Problem 1.172

Let $u \in W_0^{1,p}(\Omega)$. According to Definition 1.126, we can find a sequence $\{u_n\}_{n\geqslant 1} \subseteq C_c^\infty(\Omega)$ such that

$$u_n \longrightarrow u \quad \text{in } W^{1,p}(\Omega).$$

For every $n \geqslant 1$, we have

$$\langle h^*, u_n \rangle = \int_\Omega h^*(z) u_n(z) \, dz.$$

Passing to the limit as $n \to +\infty$ and using the Lebesgue dominated convergence theorem (see Theorem I.3.94), we obtain

$$\langle h^*, u \rangle = \int_\Omega h^*(z) u(z) \, dz.$$

Solution of Problem 1.173

Let $u \in W_0^{1,p}(\Omega)$ and let $\{u_n\}_{n\geqslant 1} \subseteq C_c^\infty(\Omega)$ be such that

$$u_n \longrightarrow u \quad \text{in } W_0^{1,p}(\Omega)$$

(see Definition 1.126) and so

$$u_n^+ \longrightarrow u^+ \quad \text{in } W_0^{1,p}(\Omega) \tag{1.178}$$

(see Problem 1.154 and its solution). Evidently, for each $n \geqslant 1$, the function u_n^+ has compact support. So, using regularization with mollifiers $\{\varphi_\varepsilon\}_{\varepsilon>0}$ (see Definition 1.117), we can find a sequence $\{y_{nk}\}_{k\geqslant 1} \subseteq C_c^\infty(\Omega)$, $y_{nk} \geqslant 0$ such that

$$y_{nk} \longrightarrow u_n^+ \quad \text{in } W_0^{1,p}(\Omega) \quad \text{as } k \to +\infty \qquad \forall\, n \geqslant 1 \tag{1.179}$$

(see Problem 1.157). Then (1.178) and (1.179) imply that there exists a nondecreasing sequence $\{k(n)\}_{n \geqslant 1}$ such that $k(n) \to +\infty$ and

$$y_{nk(n)} \longrightarrow u^+ \quad \text{in } W_0^{1,p}(\Omega) \quad \text{as } n \to +\infty.$$

By hypothesis, we have

$$h^*(y_{nk(n)}) = \langle h^*, y_{nk(n)} \rangle \geqslant 0 \qquad \forall\, n \geqslant 1,$$

so

$$\langle h^*, u^+ \rangle \geqslant 0 \qquad \forall\, u \in W_0^{1,p}(\Omega).$$

Solution of Problem 1.174

Let $\Omega = (-1,1)$ and consider the function

$$u(x) = \begin{cases} 0 & \text{if } x \leqslant 0, \\ x & \text{if } x > 0. \end{cases}$$

Evidently $u \in AC(-1,1)$ and the distributional derivative of u is given by

$$u'(x) = \begin{cases} 0 & \text{if } x < 0, \\ 1 & \text{if } x > 0. \end{cases}$$

Let $\vartheta \in C^\infty(-1,1)$ be such that

$$\|u' - \vartheta'\|_\infty < \varepsilon.$$

Hence, if $x < 0$, we have $|\vartheta'(x)| < \varepsilon$, while if $x > 0$, then $|\vartheta'(x) - 1| < \varepsilon$ and so $\vartheta'(x) > 1 - \varepsilon$. By continuity, we obtain

$$\vartheta'(0) \leqslant \varepsilon \quad \text{and} \quad \vartheta'(0) \geqslant 1 - \varepsilon.$$

If $\varepsilon \in (0, \frac{1}{2})$, we have a contradiction. This shows that $u \in W^{1,\infty}(-1,1)$ cannot be approximated in $W^{1,\infty}(\Omega)$ by smooth functions.

Solution of Problem 1.175

According to Problem 1.168, we need to consider a truly unbounded set (that is, Ω cannot be bounded in any direction). So, let $\Omega = \mathbb{R}^N$ and consider $\vartheta \in C_c^\infty(\mathbb{R}^N)$ such that $\vartheta(z) = 1$ if $|z| \leqslant 1$, $\vartheta(z) = 0$ if $|z| \geqslant 2$ and $0 \leqslant \vartheta \leqslant 1$. Let $\vartheta_k(z) = \vartheta(\frac{z}{k})$ for all $k \geqslant 1$. Then assuming $N < p$, we have

$$\|D\vartheta_k\|_p \longrightarrow 0$$

and

$$\|\vartheta_k\|_p \geqslant \lambda^N(B_k) \longrightarrow +\infty.$$

So, the Poincaré inequality fails.

Solution of Problem 1.176
Let $u \in V$. For every $x \in I$ we have

$$u(x) = u(x_0) + \int_{x_0}^{x} u'(s)\,ds = \int_{x_0}^{x} u'(s)\,ds$$

(see Theorem 1.100). If $p = 1$ or $p = +\infty$, then we immediately get

$$\|u\|_p \leqslant c\|u'\|_p.$$

If $1 < p < +\infty$, then note that V is a closed linear subspace of $W^{1,p}(a,b)$ and the only constant function in V, is the zero function. So, from Problem 1.138, we get

$$\|u\|_p \leqslant c\|u'\|_p.$$

Solution of Problem 1.177
"(a) $\Longrightarrow$ (b)": Let $\vartheta \in C_c^\infty(\mathbb{R}^N)$ and let $n \in \mathbb{R}^N$ with $|n| = 1$. We have

$$\Big|\int_{\mathbb{R}^N} u(z)\tfrac{\partial\vartheta}{\partial n}(z)\,dz\Big| = \lim_{t\searrow 0}\Big|\int_{\mathbb{R}^N} u(z)\tfrac{\vartheta(z+tn)-\vartheta(z)}{t}\,dz\Big|$$

$$= \lim_{t\searrow 0}\Big|\int_{\mathbb{R}^N} \tfrac{u(z-tn)-u(z)}{t}\vartheta\,dz\Big| \leqslant \operatorname{Lip}(u)\|\vartheta\|_1$$

($\operatorname{Lip}(u)$ being the Lipschitz constant of u). Then from Theorem 1.21 (see also Proposition 1.6), we have that there exists a unique function $\eta_n \in L^\infty(\mathbb{R}^N)$ such that

$$\int_{\mathbb{R}^N} u\tfrac{\partial\vartheta}{\partial n}\,dz = -\int_{\mathbb{R}^N} \eta_n\vartheta\,dz \qquad \forall\, \vartheta \in C_c^\infty(\mathbb{R}^N).$$

Note that the left-hand side is linear in n. So, we can find $\widehat{\xi} \in L^\infty\left(\mathbb{R}^N;\mathbb{R}^N\right)$ such that

$$\int_{\mathbb{R}^N} u\,(D\vartheta, n)_{\mathbb{R}^N}\,dz = -\int_{\mathbb{R}^N} (\widehat{\xi}, n)_{\mathbb{R}^N}\,dz,$$

so

$$\widehat{\xi} = Du \in L^\infty\left(\mathbb{R}^N;\mathbb{R}^N\right).$$

"(b) $\Longrightarrow$ (a)": Let $\{\varphi_\varepsilon\}_{\varepsilon>0}$ be a family of mollifiers (see Definition 1.117). We define $u_\varepsilon = u * \varphi_\varepsilon$. Then $u_\varepsilon \in C^\infty(\mathbb{R}^N)$ and

$$\|Du_\varepsilon\|_\infty = \|(Du) * \varphi_\varepsilon\|_\infty \leqslant \|Du\|_\infty \qquad \forall\, \varepsilon > 0.$$

We have that

$$u_\varepsilon \longrightarrow u \quad \text{in } L^1_{\text{loc}}\left(\mathbb{R}^N\right) \quad \text{as } \varepsilon \searrow 0.$$

So, we can find a sequence $\{\varepsilon_n\}_{n\geqslant 1} \in (0,+\infty)$ such that $\varepsilon_n \searrow 0$ and

$$u_{\varepsilon_n}(z) \longrightarrow u(z) \quad \text{for a.a. } z \in \Omega.$$

Let $D \subseteq \mathbb{R}^N$ be the Lebesgue null set outside which we have pointwise convergence. Let $E = \mathbb{R}^N \setminus D$. For $z, x \in E$, we have

$$\begin{aligned} \left|u(z) - u(x)\right| &= \lim_{n\to+\infty} \left|u_{\varepsilon_n}(z) - u_{\varepsilon_n}(x)\right| \\ &\leqslant \sup_{n\geqslant 1} \|Du_{\varepsilon_n}\|_\infty |z - x| \leqslant \|Du\|_\infty |z - x|, \end{aligned}$$

so $u|_E$ is Lipschitz continuous.

Note that E is dense in $\mathbb{R}^N$ (recall that $\lambda^N(D) = \lambda^N(\mathbb{R}^N \setminus E) = 0$). So, we conclude that u admits a Lipschitz continuous representative.

Solution of Problem 1.178

"$\Longrightarrow$": From Definition 1.116 and the Hölder inequality (see Theorem 1.3 and Problem 1.27), we have

$$\Big|\int_\Omega u\frac{\partial\vartheta}{\partial z_k}\,dz\Big| = \Big|\int_\Omega \frac{\partial u}{\partial z_k}\vartheta\,dz\Big| \leqslant \Big\|\frac{\partial u}{\partial z_k}\Big\|_p \|\vartheta\|_{p'} \leqslant c\|\vartheta\|_{p'} \quad \forall\, k \in \{1,\ldots,N\},$$

with $c = \|Du\|_p$.

"$\Longleftarrow$": We consider functions $\xi_k\colon C_c^\infty(\Omega) \longrightarrow \mathbb{R}$ for $k \in \{1,\ldots,N\}$ defined by

$$\xi_k(\vartheta) = \int_\Omega u\frac{\partial\vartheta}{\partial z_k}\,dz.$$

Using the Hölder inequality, we see that this is a continuous linear functional on $C_c^\infty(\Omega)$ with the $L^{p'}$-norm. The density of the embedding $C_c^\infty(\Omega) \hookrightarrow L^{p'}(\Omega)$, implies that $\xi_k \in L^{p'}(\Omega)^*$. So, by Theorem 1.21, there exists unique $h_k \in L^p(\Omega)$ such that

$$\xi_k(\vartheta) = \int_\Omega h_k\vartheta\,dz \qquad \forall\, \vartheta \in L^{p'}(\Omega),$$

so

$$\int_{\Omega} u\frac{\partial \vartheta}{\partial z_k}\,dz \;=\; \int_{\Omega} h_k\vartheta\,dz \qquad \forall\,\vartheta\in C_c^\infty(\Omega),\ k\in\{1,\ldots,N\},$$

thus

$$\frac{\partial u}{\partial z_k} \;=\; -h_k \;\in\; L^p(\Omega) \qquad \forall\,k\in\{1,\ldots,N\}$$

and hence $u\in W^{1,p}(\Omega)$.

Solution of Problem 1.179

We show that for all $k\in\{1,\ldots,N\}$ and $z\neq 0$ we have

$$\frac{\partial u}{\partial z_k}(z) \;=\; \varphi'(|z|)\frac{z_k}{|z|}. \tag{1.180}$$

Let $\vartheta\in C_c^\infty(B_R)$ and let $\varepsilon>0$ be small. For all $k\in\{1,\ldots,N\}$, we have

$$\begin{aligned}
\int_{B_R} u\frac{\partial \vartheta}{\partial z_k}\,dz \;&=\; \int_{B_R\setminus B_\varepsilon} u\frac{\partial \vartheta}{\partial z_k}\,dz + \int_{B_\varepsilon} u\frac{\partial \vartheta}{\partial z_k}\,dz\\
&=\; -\int_{B_R\setminus B_\varepsilon} \frac{\partial u}{\partial z_k}\vartheta\,dz - \int_{\partial B_\varepsilon} u\vartheta\frac{z_k}{|z|}\,d\sigma + \int_{B_\varepsilon} u\frac{\partial \vartheta}{\partial z_k}\,dz\\
&=\; -\int_{B_R} \frac{\partial u}{\partial z_k}\vartheta\,dz + \int_{B_\varepsilon} \frac{\partial u}{\partial z_k}\vartheta\,dz + \int_{B_\varepsilon} u\frac{\partial \vartheta}{\partial z_k}\,dz - \int_{\partial B_\varepsilon} u\vartheta\frac{z_k}{|z|}\,d\sigma
\end{aligned} \tag{1.181}$$

(by the Green theorem and σ being the surface measure). Note that $\frac{\partial u}{\partial z_k}\vartheta,\, u\frac{\partial \vartheta}{\partial z_k}\in L^1(B_R)$. So, we have

$$\lim_{\varepsilon\searrow 0}\int_{B_\varepsilon} \frac{\partial u}{\partial z_k}\vartheta\,dz \;=\; \lim_{\varepsilon\searrow 0}\int_{B_\varepsilon} u\frac{\partial \vartheta}{\partial z_k}\,dz \;=\; 0.$$

Also, if $\omega_{N-1}=\sigma(\partial B_R)$ ($\sigma=H^{N-1}$ is the $(N-1)$-dimensional Hausdorff (surface) measure), then

$$\Big|\int_{\partial B_\varepsilon} u\vartheta\frac{z_k}{|z|}\,d\sigma\Big| \;\leqslant\; \omega_{N-1}\|\vartheta\|_\infty\varepsilon^{N-1}\varphi(\varepsilon) \;\longrightarrow\; 0 \quad \text{as } \varepsilon\searrow 0.$$

So, if in (1.181) we pass to the limit as $\varepsilon\searrow 0$, then we have (1.180) from which it follows that $|Du(z)|=|\varphi'(|z|)|$. From this inequality, we deduce at once the claim of the problem.

Solution of Problem 1.180
Evidently $u_n \longrightarrow 0$ in $L^\infty(\Omega)$, hence in $L^2(\Omega)$ too. Also, we have

$$\tfrac{\partial u_n}{\partial x} = \sqrt{n}(1-y)^n \cos(nx), \quad \tfrac{\partial u_n}{\partial y} = -\sqrt{n}(1-y)^{n-1}\sin(nx),$$

so

$$\sup_{n\geqslant 1} \iint_\Omega |Du_n(x,y)|^2\,dx\,dy < +\infty.$$

Since $H^1(\Omega)$ is a Hilbert space, we infer that for the original sequence $\{u_n\}_{n\geqslant 1}$, we have

$$u_n \xrightarrow{w} 0 \quad \text{in } H^1(\Omega).$$

Solution of Problem 1.181
Evidently the set $\overline{B}_1^p \cap L^q(\Omega)$ is convex. So, by the Mazur theorem (see Theorem I.5.58), to establish the claim of the problem, it suffices to show that the set $\overline{B}_1^p \cap L^q(\Omega)$ is strongly closed in $L^q(\Omega)$. So, let $\{u_n\}_{n\geqslant 1} \subseteq \overline{B}_1^p \cap L^q(\Omega)$ be a sequence such that $u_n \longrightarrow u$ in $L^q(\Omega)$. Passing to a subsequence if necessary, we may assume that $u_n(z) \longrightarrow u(z)$ for μ-almost all $z \in \Omega$. If $p = +\infty$, then $|u(z)| \leqslant 1$ for μ-almost all $z \in \Omega$ and so $u \in \overline{B}_1^\infty \cap L^q(\Omega)$. If $1 \leqslant p < +\infty$, then by the Fatou lemma (see Theorem I.3.95), we have

$$\|u\|_p^p = \int_\Omega \liminf_{n\to+\infty} |u_n|^p\,d\mu \leqslant \liminf_{n\to+\infty} \int_\Omega |u_n|^p\,d\mu \leqslant 1,$$

so $u \in \overline{B}_1^p \cap L^q(\Omega)$. This proves that the set $\overline{B}_1^p \cap L^q(\Omega)$ is closed in $L^q(\Omega)$.

Solution of Problem 1.182
Evidently the set $C \subseteq L^1(\Omega)$ is bounded since

$$\sup_{u\in C} \|u\|_1 \leqslant \sup_{h\in E} \|h\|_1 < +\infty.$$

Also, since by hypothesis the set $E \subseteq L^1(\Omega)$ is uniformly integrable (see the Dunford–Pettis theorem; Theorem 1.28), given $\varepsilon > 0$, we can find $\delta > 0$ such that

$$\text{if } A \in \Sigma \text{ with } \mu(A) \leqslant \delta, \text{ then } \int_A |h|\,d\mu \leqslant \varepsilon \text{ for all } h \in E.$$

From the definition of C, it follows that

$$\text{if } A \in \Sigma \text{ with } \mu(A) \leqslant \delta, \text{ then } \int_A |u|\,d\mu \leqslant \varepsilon \text{ for all } u \in C.$$

Then from Problem 1.6, we conclude that the set $C \subseteq L^1(\Omega)$ is uniformly integrable.

Solution of Problem 1.183

Let $h \in L^{p'}(\Omega)$ (where $\frac{1}{p} + \frac{1}{p'} = 1$). Then using Problem 1.53(a) and the Vitali convergence theorem (see Theorem 1.20), we have

$$\int_{\Omega} u_n h \, d\mu \longrightarrow \int_{\Omega} uh \, d\mu.$$

Since $h \in L^{p'}(\Omega)$ is arbitrary, it follows that

$$u_n \xrightarrow{w} u \quad \text{in } L^p(\Omega).$$

Since by hypothesis we have $\|u_n\|_p \longrightarrow \|u\|_p$, we can use the Kadec–Klee property (see Corollary 1.26) and conclude that

$$u_n \longrightarrow u \quad \text{in } L^p(\Omega).$$

Solution of Problem 1.184

From the weak lower semicontinuity of the norm functional in a Banach space (see Proposition I.5.56(c)), we have

$$\|u\|_p \leqslant \liminf_{n \to +\infty} \|u_n\|_p,$$

so $\|u_n\|_p \longrightarrow \|u\|_p$ (use the hypothesis), thus $u_n \longrightarrow u$ in $L^p(\Omega)$ (see Corollary 1.26).

Solution of Problem 1.185

From Problem 1.21, we have

$$\big| \|u_n\|_1 - \|u\|_1 - \|u_n - u\|_1 \big| \longrightarrow 0.$$

Since by hypothesis $\|u_n\|_1 \longrightarrow \|u\|_1$, it follows that $\|u_n - u_1\|_1 \longrightarrow 0$.

Solution of Problem 1.186

No. To see this, let $u_n(t) = n\chi_{[0,\frac{1}{n}]}(t)$, for all $t \in [0,1]$. Then $\{u_n\}_{n\geqslant 1} \subseteq L^1(0,1)$ and $u_n(t) \longrightarrow 0$ for almost all $t \in [0,1]$, However, $\|u_n\|_1 = 1$ for all $n \geqslant 1$.

Solution of Problem 1.187

We know that, if the set $U \subseteq \mathbb{R}$ is open, then U can be written as a countable union of open intervals. Hence $\int\limits_U u(t)\,dt = 0$. If $A \subseteq \mathbb{R}$ is a Lebesgue measurable set, exploiting the regularity of the Lebesgue measure (see Definition I.4.9), we can find a sequence of open sets $\{U_n\}_{n\geq 1} \subseteq 2^{\mathbb{R}}$ such that $A \subseteq U_n$ and $\lambda(U_n) \longrightarrow \lambda(A)$ (where λ is the Lebesgue measure on $\mathbb{R}$). Then from the Lebesgue dominated convergence theorem, we have

$$\int\limits_{U_n} u\,dt \longrightarrow \int\limits_A u\,dt,$$

so

$$\int\limits_A u\,dt = 0 \qquad \forall\ A \subseteq \mathbb{R} \text{ measurable.}$$

Let $A_+ = \{t \in \mathbb{R} :\ u^+(t) > 0\}$ and $A_- = \{t \in \mathbb{R} :\ u^-(t) > 0\}$. Both are Lebesgue measurable and we have

$$\int\limits_{A_+ \cup A_-} |u|\,dt = 0,$$

so $\lambda(A_+ \cup A_-) = 0$, hence $u(t) = 0$ for almost all $t \in \mathbb{R}$.

Solution of Problem 1.188

By Theorem 1.100 and by identifying each u_n with its absolutely continuous representation, we have

$$u_n(t) - u_n(\tau) = \int\limits_\tau^t u_n'(s)\,ds \qquad \forall\ n \geqslant 1,\ t, \tau \in [0,1],\ \tau \leqslant t,$$

so

$$|u_n(t) - u_n(\tau)| \leqslant \int\limits_\tau^t h(s)\,ds,$$

hence the sequence $\{u_n\}_{n\geqslant 1} \subseteq C([0,1])$ is equicontinuous. Since $u_n(t) \longrightarrow u(t)$ for all $t \in (0,1)$, we can apply the Arzela–Ascoli theorem (see Theorem I.2.181) and have that

$$u_n \longrightarrow u \quad \text{in } C([0,1]),$$

so $u_n(t) \longrightarrow u(t)$ uniformly on $[0,1]$.

Solution of Problem 1.189

Let $\{u_n\}_{n\geqslant 1} \subseteq W_0^{1,p}(0,1)$ be a minimizing sequence for $\widehat{\lambda}_1$. Exploiting the 0-homogeneity of the quotient $\frac{\|u'\|_p^p}{\|u\|_p^p}$, we may assume that

$$\|u_n\|_p = 1 \qquad \forall\, n \geqslant 1 \quad \text{and} \quad \|u_n'\|_p^p \searrow \widehat{\lambda}_1.$$

Then it is clear that the sequence $\{u_n\}_{n\geqslant 1} \subseteq W_0^{1,p}(0,1)$ is bounded. So, by the Sobolev embedding theorem (see Theorem 1.109), we may assume that

$$u_n \xrightarrow{w} \widehat{u}_1 \quad \text{in } W_0^{1,p}(0,1) \quad \text{and} \quad u_n \longrightarrow \widehat{u}_1 \quad \text{in } C([0,1]),$$

so $\|\widehat{u}_1\|_p = 1$, hence $\widehat{u}_1 \neq 0$. Since $u_n' \xrightarrow{w} \widehat{u}_1$ in $L^p(0,1)$, we have

$$\|\widehat{u}_1\|_p^p \leqslant \liminf_{n\to+\infty} \|u_n'\|_p^p = \widehat{\lambda}_1,$$

so $\|\widehat{u}_1'\|_p^p = \widehat{\lambda}_1$, $\|\widehat{u}_1\|_p = 1$. Evidently, if we replace $\widehat{u}_1$ by $|\widehat{u}_1|$, we still have

$$|\widehat{u}_1| \in W_0^{1,p}(0,1), \quad \|\widehat{u}_1\|_p = 1, \quad \|\widehat{u}_1\|_p^p = \widehat{\lambda}_1.$$

Therefore, we may assume that $\widehat{u}_1(t) \geqslant 0$ for all $t \in [0,1]$.

Solution of Problem 1.190

"$\Longrightarrow$": We have

$$\frac{\lambda(C_\varepsilon \cap [0,\tau])}{\tau} \leqslant \frac{1}{\tau}\int_0^\tau \frac{u(t)}{\varepsilon}\,dt = \frac{1}{\varepsilon\tau}\int_0^\tau u(t)\,dt \longrightarrow 0 \quad \text{as } \tau \to +\infty.$$

"$\Longleftarrow$": Fix any $\varepsilon > 0$. For $\tau > 0$, we have

$$\begin{aligned}\frac{1}{\tau}\int_0^\tau u(t)\,dt &= \frac{1}{\tau}\int_{C_\varepsilon\cap[0,\tau]} u(t)\,dt + \frac{1}{\tau}\int_{[0,\tau]\setminus C_\varepsilon} u(t)\,dt \\ &\leqslant \frac{1}{\tau}\int_{C_\varepsilon\cap[0,\tau]} u(t)\,dt + \varepsilon. \qquad (1.182)\end{aligned}$$

If $1 < p < +\infty$, then

$$\frac{1}{\tau} \int\limits_{C_\varepsilon \cap [0,\tau]} u(t)\, dt \leqslant \frac{1}{\tau^{1-\frac{1}{p'}}} \frac{\lambda(C_\varepsilon \cap [0,\tau])}{\tau} \Big(\int\limits_{\mathbb{R}+} u(t)^p \, dt \Big)^{\frac{1}{p}},$$

where $\frac{1}{p} + \frac{1}{p'} = 1$ (using the Hölder inequality (see Theorem 1.3 and Problem 1.27)), so

$$\lim_{\tau \to +\infty} \frac{1}{\tau} \int\limits_0^\tau u(t)\, dt \leqslant \varepsilon \qquad \forall\, \varepsilon > 0$$

(see (1.182)), thus

$$\lim_{\tau \to +\infty} \frac{1}{\tau} \int\limits_0^\tau u(t)\, dt = 0.$$

If $p = 1$, then

$$\frac{1}{\tau} \int\limits_{C_\varepsilon \cap [0,\tau]} u(t)\, dt \leqslant \frac{1}{\tau} \int\limits_{\mathbb{R}_+} u(t)\, dt = \frac{\|u\|_1}{\tau},$$

thus

$$\lim_{\tau \to +\infty} \frac{1}{\tau} \int\limits_0^\tau u(t)\, dt = 0.$$

Finally, if $p = +\infty$, then

$$\frac{1}{\tau} \int\limits_{C_\varepsilon \cap [0,\tau]} u(t)\, dt \leqslant \frac{\lambda(C_\varepsilon \cap [0,\tau])}{\tau} \|u\|_\infty,$$

so

$$\lim_{\tau \to +\infty} \frac{1}{\tau} \int\limits_0^\tau u(t)\, dt = 0.$$

Solution of Problem 1.191

Continuity follows from the fundamental theorem of the Lebesgue calculus. Let $\vartheta \in C_c^\infty(I)$ and fix $a, b \in \overline{I}$ such that $\operatorname{supp} \vartheta \subseteq [a, b]$. We have

$$\begin{aligned}\int\limits_I h\vartheta'\,dt &= -\int\limits_a^{t_0}\Big(\int\limits_t^{t_0} u(s)\vartheta'(t)\,ds\Big)\,dt + \int\limits_{t_0}^{b}\Big(\int\limits_{t_0}^{t} u(s)\vartheta'(t)\,ds\Big)\,dt \\ &= -\int\limits_a^{t_0}\Big(\int\limits_a^{s} u(s)\vartheta'(t)\,dt\Big)\,ds + \int\limits_{t_0}^{b}\Big(\int\limits_s^{b} u(s)\vartheta'(t)\,dt\Big)\,ds \\ &= -\int\limits_a^{t_0} u(s)\Big(\int\limits_a^{s} \vartheta'(t)\,dt\Big)\,ds + \int\limits_{t_0}^{b} u(s)\Big(\int\limits_s^{b} \vartheta'(t)\,dt\Big)\,ds \\ &= -\int\limits_a^{b} u(t)\vartheta(t)\,dt,\end{aligned}$$

so $h' = u$ as a weak derivative.

Solution of Problem 1.192

Yes. For all $t, \tau \in \mathbb{R}$, $\tau \leqslant t$, we have

$$u(t)^2 - u(\tau)^2 = 2\int\limits_\tau^t u(s)u'(s)\,ds,$$

so

$$|u(t)^2 - u(\tau)^2| \leqslant 2\Big(\int\limits_\tau^t u(s)^2\,ds\Big)^{\frac{1}{2}}\Big(\int\limits_\tau^t u'(s)^2\,ds\Big)^{\frac{1}{2}} \tag{1.183}$$

(by the Cauchy–Schwarz inequality). Since $u \in H^1(\mathbb{R})$, we have

$$\int\limits_\tau^t u^2\,ds \longrightarrow 0 \quad \text{and} \quad \int\limits_\tau^t (u')^2\,ds \longrightarrow 0 \quad \text{as } \tau \to +\infty,$$

so $\lim\limits_{t\to+\infty} u(t)^2 = \eta \geqslant 0$ (see (1.183)). If $\eta > 0$, then we pick $\varepsilon \in (0, \eta)$ and have

$$u(t)^2 > \eta - \varepsilon \qquad \forall\, t > M = M(\varepsilon) > 0.$$

Therefore, we have

$$\int\limits_{\mathbb{R}} u(s)^2\,ds \geqslant \int\limits_M^{+\infty} u(s)^2\,ds > \int\limits_M^{+\infty} (\eta - \varepsilon)\,ds = +\infty,$$

a contradiction to the fact that $u \in L^2(\mathbb{R})$.

Solution of Problem 1.193

No. Let $N = 3$ and consider the function

$$u_r(z) \;=\; \begin{cases} 1 & \text{if } \; |z - z_0| < \frac{r}{2} \\ \frac{r}{|z-z_0|} - 1 & \text{if } \; \frac{r}{2} \leqslant |z - z_0| \leqslant r \end{cases} \qquad \forall\, z \in \overline{B}_r(z_0). \tag{1.184}$$

Note that $u_r|_{\partial B_r(z_0)} \equiv 0$ and $u_r|_{\partial B_{\frac{r}{2}}(z_0)} \equiv 1$. Hence $u_r \in C_0(\overline{B}_r(z_0))$ and $0 \leqslant u_r \leqslant 1$. Moreover, we have

$$|\nabla u_r(z)| \;=\; \frac{r}{|z-z_0|^2} \;\leqslant\; \frac{2}{r} \qquad \forall\, z \in \overline{B}_r(z_0) \setminus B_{\frac{r}{2}}(z_0). \tag{1.185}$$

For $k \geqslant 1$, let $z_k = \begin{pmatrix} k \\ 0 \\ 0 \end{pmatrix}$ and $r_k = \frac{1}{k^2}$. We set

$$u(z) \;=\; \begin{cases} u_{r_k}(z) & \text{if } \; z \in \overline{B}_{r_k}(z_k),\; k \geqslant 1, \\ 0 & \text{if } \; z \in \mathbb{R}^3 \setminus \bigcup\limits_{k \geqslant 1} \overline{B}_{r_k}(z_k), \end{cases}$$

with u_{r_k} as in (1.184) with $r = r_k$. We have

$$\int\limits_{\mathbb{R}^3} u^2\, dz \;=\; \sum_{k \geqslant 1} \int\limits_{\overline{B}_{r_k}(z_k)} u^2_{z_k}\, dz \;<\; \frac{4\pi}{3} \sum_{k \geqslant 1} r_k^3 \;=\; \frac{4\pi}{3} \sum_{k \geqslant 1} \frac{1}{k^6} \;<\; +\infty$$

(recall that $0 \leqslant u_{r_k} \leqslant 1$). Also, we have

$$\int\limits_{\mathbb{R}^3} |Du|^2\, dz \;=\; \sum_{k \geqslant 1} \int\limits_{\overline{B}_{r_k}(z_k)} |\nabla u_{r_k}|^2\, dz \;<\; \frac{8\pi}{3} \sum_{k \geqslant 1} r_k \;=\; \frac{8\pi}{3} \sum_{k \geqslant 1} \frac{1}{k^2} \;<\; +\infty$$

(see (1.185)). Therefore, we infer that $u \in H^1(\mathbb{R})$ and it is bounded. However

$$|z_k| \;=\; k \;\longrightarrow\; +\infty \quad \text{and} \quad u(z_k) \;=\; 1 \qquad \forall\, k \geqslant 1.$$

Solution of Problem 1.194

Let $\overline{u} = \frac{1}{\lambda^N(\Omega)} \int\limits_{\Omega} u\, dz$ with λ^N being the Lebesgue measure on $\mathbb{R}^N$. Let $\widehat{u} = u - \overline{u}$. From the Poincaré–Wirtinger inequality (see Theorem 1.132), we have

$$\|\widehat{u}\|_p \;\leqslant\; c_1 \|Du\|_p \qquad \forall\, u \in W^{1,p}(\Omega), \tag{1.186}$$

for some $c_1 = c_1(\Omega) > 0$. Then using the Minkowski inequality (see Theorem I.3.101), we have

$$\begin{aligned} \|u\|_p &\leqslant \|\widehat{u}\|_p + \|\overline{u}\|_p \leqslant c_1\|Du\|_p + \Big|\int_\Omega u\,dz\Big| \\ &\leqslant c\big(\|Du\|_p + \Big|\int_\Omega u\,dz\Big|\big), \end{aligned}$$

with $c = \max\{c_1, 1\} > 0$.

Solution of Problem 1.195

The inequality follows from the Sobolev embedding theorem (see Theorem 1.109). We need to establish the claimed value of $c > 0$. Since $u \in C([a,b])$, from the mean value theorem for integrals, we can find $\eta \in [a,b]$ such that

$$u(\eta)^2 = \tfrac{1}{b-a}\int_a^b u(s)^2\,ds = \tfrac{1}{b-a}\|u\|_2^2. \tag{1.187}$$

We have

$$\begin{aligned} u(t) &= u(\eta) + \int_\eta^t u'(s)\,ds \leqslant \tfrac{1}{\sqrt{b-a}}\|u\|_2 + \int_a^b |u'(s)|\,ds \\ &\leqslant \tfrac{1}{\sqrt{b-a}}\|u\|_2 + \sqrt{b-a}\|u'\|_2 \leqslant c\big(\|u\|_2 + \|u'\|_2\big) \qquad \forall\, t \in [a,b], \end{aligned}$$

with $c = \max\big\{\frac{1}{\sqrt{b-a}}, \sqrt{b-a}\big\}$ (see (1.187) and use the Cauchy–Schwarz inequality). Since $t \in [a,b]$ is arbitrary, it follows that $\|u\|_\infty \leqslant c\|u\|$, with $c = \max\big\{\frac{1}{\sqrt{b-a}}, \sqrt{b-a}\big\}$.

Solution of Problem 1.196

Note that

$$|\xi(u)| = \Big|\int_0^1 u(t)\,dt\Big| \leqslant \sqrt{2}\|u\|_2 \leqslant \sqrt{2}c\|u'\|_2,$$

for some $c > 0$ (by the Cauchy–Schwartz inequality and the Poincaré inequality (see Theorem 1.108)), so $\xi \in W^{-1,2}(0,1) = W_0^{1,2}(0,1)^*$. By the Riesz–Fréchet representation theorem (see Theorem I.5.100) there exists a unique $h \in W_0^{1,2}(0,1)$ such that

$$\int_0^1 h'u'\,dt = \int_0^1 u\,dt \qquad \forall\, u \in W_0^{1,2}(0,1).$$

Assume momentarily that h is twice continuously differentiable. Then integrating by parts, we have

$$h'(1)u(1) - h'(0)u(0) - \int_0^1 h''u\,dt = \int_0^1 u\,dt \qquad \forall\, u \in W_0^{1,2}(0,1),$$

so

$$\int_0^1 (h''(t) - 1)u(t)\,dt = 0 \qquad \forall\, u \in W_0^{1,2}(0,1),$$

thus

$$h''(t) = 1 \quad \text{for a.a. } t \in (0,1), \quad h(0) = h(1) = 0,$$

hence

$$h(t) = \tfrac{1}{2}(t^2 - t) \qquad \forall\, t \in (0,1).$$

The uniqueness of h implies that this is the desired element of $W_0^{1,2}(0,1)$.

Solution of Problem 1.197

Let $p \in (1, +\infty]$. From the Sobolev embedding theorem (see Theorem 1.109), we know that $\overline{B}_1(p)$ is relatively compact in $L^p(0,1)$. So, we need to show that it is closed in $L^p(0,1)$. To this end, let $\{u_n\}_{n\geqslant 1} \subseteq \overline{B}_1(p)$ be a sequence such that $u_n \longrightarrow u$ in $L^p(0,1)$. Then the sequence $\{u_n'\}_{n\geqslant 1} \subseteq L^p(0,1)$ is bounded and so the sequence $\{u_n\}_{n\geqslant 1} \subseteq W^{1,p}(\Omega)$ is bounded and thus we have

$$u_n \xrightarrow{w} u \quad \text{in } W^{1,p}(0,1),$$

so

$$\|u\| \leqslant \liminf_{n\to+\infty} \|u_n\| \leqslant 1$$

and hence $u \in B_1(p)$ which proves that $B_1(p) \subseteq L^p(0,1)$ is closed.

Now, let $p = 1$. We consider a sequence of functions $\{u_n\}_{n\geqslant 1} \subseteq W^{1,1}(0,1)$ defined by

$$u_n(t) = \begin{cases} 0 & \text{if } t \in [0, \frac{1}{2}], \\ n(t - \frac{1}{2}) & \text{if } t \in (\frac{1}{2}, \frac{1}{2} + \frac{1}{n}), \\ 1 & \text{if } t \in [\frac{1}{2} + \frac{1}{n}, 1] \end{cases} \qquad \forall\, n \geqslant 1.$$

We have $\|u_n\|_1 \leqslant \frac{1}{2}$ and $\|u_n'\|_1 \leqslant 1$ for all $n \geqslant 2$. Therefore $\{\frac{2}{3}u_n\}_{n\geqslant 2} \subseteq \overline{B}_1(1)$. Note that $\frac{2}{3}u_n \longrightarrow \frac{2}{3}u$ in $L^1(0,1)$ with

$$u(t) = \begin{cases} 0 & \text{if } t \in (0, \frac{1}{2}), \\ 1 & \text{if } t \in (\frac{1}{2}, 1). \end{cases}$$

However, $u \notin W^{1,1}(0,1)$ and so $\overline{B}_1(1)$ is not closed in $L^1(0,1)$.

Solution of Problem 1.198

Let $g \in C^1(\mathbb{R})$ be a function such that

$$|g(t)| \leqslant 1, \quad |g'(t)| \leqslant c \qquad \forall\, t \in \mathbb{R},$$

for some $c > 0$ and such that

$$g(t) = \begin{cases} 1 & \text{if } t \geqslant 1, \\ t & \text{if } |t| \leqslant \frac{1}{2}, \\ -1 & \text{if } t \geqslant -1. \end{cases}$$

We set

$$h_n(t) = \tfrac{1}{n} g(nu(t)) \qquad \forall\, t \in (0,1).$$

The chain rule (see Theorem 1.142) implies that

$$h_n \in W^{1,p}(0,1) \quad \text{and} \quad h_n'(t) = g'(nu(t))u'(t) \quad \text{for a.a. } t \in (0,1).$$

Then we have

$$h_n'(t) \longrightarrow \xi(t) = \begin{cases} 0 & \text{if } u(t) \neq 0 \\ u'(t) & \text{if } u(t) = 0 \end{cases} \quad \text{for a.a. } t \in (0,1). \tag{1.188}$$

Note that $u_n \in L^\infty(0,1)$ for all $n \geqslant 1$ (see the Sobolev embedding theorem; Theorem 1.109) and $\|h_n\|_\infty \longrightarrow 0$. For every $\vartheta \in C_c^\infty(0,1)$ we have

$$\int_0^1 h_n \vartheta' \, dt = -\int_0^1 h_n' \vartheta \, dt \qquad \forall\, n \geqslant 1,$$

so

$$\int_0^1 \xi\vartheta \, dt = 0 \qquad \forall\, \vartheta \in C_c^\infty(0,1),$$

thus

$$\xi(t) = 0 \quad \text{for a.a. } t \in (0,1)$$

and hence

$$u'(t) = 0 \quad \text{for a.a. } t \in \{u = 0\}.$$

Solution of Problem 1.199

The hypothesis that $u_n \longrightarrow u$ in $L^1(\Omega)$ implies that at least for a subsequence, we have

$$u_n(z) \longrightarrow u(z) \quad \text{for } \mu\text{-a.a. } z \in \Omega.$$

Then we can use Problem 1.19 and conclude that for the original sequence, we have $u_n \xrightarrow{w} u$ in $L^p(\Omega)$.

Solution of Problem 1.200

Let $\{u_n\}_{n\geqslant 1} \subseteq C$ be a sequence such that $u_n \longrightarrow u$ in $L^p(\Omega)$. By passing to a suitable subsequence if necessary, we may assume that $u_n(z) \longrightarrow u(z)$ for almost all $z \in \Omega$. Then by the Fatou lemma (see Theorem I.3.95), we have that $u \in L^q(\Omega)$ and $\|u\|_q \leqslant 1$. Therefore $u \in C$ and so we have proved that the set C is closed in $L^p(\Omega)$.

Solution of Problem 1.201

Let $I \subseteq \mathbb{R}$ be a bounded interval and consider the sequence of functions $\{\chi_I u_n\}_{n\geqslant 1}$. Then

$$(\chi_I u_n)(t) \longrightarrow 0 \quad \text{for a.a. } t \in \mathbb{R}.$$

So, using the Vitali convergence theorem (see Theorem 1.20), we have

$$\int_{\mathbb{R}} \chi_I u_n \, dt \longrightarrow 0,$$

so for every step function $s \in L^{p'}(\mathbb{R})$, we have

$$\int_{\mathbb{R}} s u_n \, dt \longrightarrow 0.$$

But recall that step functions are dense in $L^{p'}(\mathbb{R})$. So, we get

$$\int_{\mathbb{R}} h u_n \, dt \longrightarrow 0 \qquad \forall\, h \in L^{p'}(\mathbb{R}),$$

thus $u_n \xrightarrow{w} 0$ in $L^p(\mathbb{R})$.

Solution of Problem 1.202
Let

$$\widehat{C} = \Big\{u \in L^\infty(\Omega) : \int_\Omega u\vartheta \, dz \geqslant \int_\Omega h\vartheta \, dz \quad \text{for all } \vartheta \in L^1(\Omega), \text{ with } h\vartheta \in L^1(\Omega), \ \vartheta \geqslant 0\Big\}.$$

We claim that $C = \widehat{C}$. Evidently, we have $C \subseteq \widehat{C}$. So, let $u \in \widehat{C}$ and let $\Omega = \bigcup\limits_{n \geqslant 1} \Omega_n$ with $\Omega_n \in \Sigma$, $\mu(\Omega_n) < +\infty$ (recall that the measure space (Ω, Σ, μ) is σ-finite). Let

$$\widehat{\Omega}_n = \Omega_n \cap \{|h| \leqslant n\} \in \Sigma \qquad \forall\, n \geqslant 1.$$

Let $A = \{u < h\} \in \Sigma$ and choose $\vartheta = \chi_{\Omega_n \cap A}$. Then $\vartheta \in L^1(\Omega)$, $\vartheta \geqslant 0$ and $h\vartheta \in L^1(\Omega)$. Since $u \in \widehat{C}$ we have

$$\int_{\Omega_n \cap A} u \, d\mu \geqslant \int_{\Omega_n \cap A} h \, d\mu,$$

so

$$\int_{\Omega_n \cap A} (h - u) \, d\mu \leqslant 0,$$

thus

$$\mu(\Omega_n \cap A) = 0 \qquad \forall\, n \geqslant 1,$$

hence $\mu(A) = 0$. From this, we get that $u \geqslant h$, so $C = \widehat{C}$.

Now for every fixed $\vartheta \in L^1(\Omega)$, with $h\vartheta \in L^1(\Omega)$, $\vartheta \geqslant 0$, the set

$$C(\vartheta) = \Big\{u \in L^\infty(\Omega) : \int_\Omega u\vartheta \, d\mu \geqslant \int_\Omega h\vartheta \, d\mu\Big\}$$

is clearly w^*-closed in $L^\infty(\Omega)$. Hence the set $C = \widehat{C} = \bigcap_{\vartheta} C(\vartheta)$ is w^*-closed in $L^\infty(\Omega)$.

Solution of Problem 1.203

Evidently $C_0 \subseteq L^\infty(\Omega)$ is bounded. Also, note that

$$C_0 = \{u \in L^\infty(\Omega) : u \geqslant h_1\} \cap \{u \in L^\infty(\Omega) : -u \geqslant -h_2\}.$$

Using Problem 1.202, we see that the set C_0 is w^*-closed. Hence from the Alaoglu theorem (see Theorem I.5.66), we conclude that the set C_0 is w^*-compact.

Solution of Problem 1.204

(a) Let $v \in L^1(\mathbb{R}^N)$. We have

$$\int_{\mathbb{R}^N} h_n v\, dz = \int_{\mathbb{R}^N} \xi_n u(\vartheta_n \star v)\, dz = \int_{\mathbb{R}^N} \xi_n u(\vartheta_n \star v - v))\, dz + \int_{\mathbb{R}^N} \xi_n uv\, dz,$$

so

$$\begin{aligned} \Big|\int_{\mathbb{R}^N} h_n v\, dz - \int_{\mathbb{R}^N} hv\, dz\Big| &= \Big|\int_{\mathbb{R}^N} \xi_n u(\vartheta_n \star v - v)\, dz + \int_{\mathbb{R}^N} (\xi_n - \xi) uv\, dz\Big| \\ &\leqslant \|u\|_\infty \|\vartheta_n \star v - v\|_1 + \|u\|_\infty \|(\xi_n - \xi)v\|_1. \end{aligned} \tag{1.189}$$

From Theorem 1.118(b), we know that

$$\|\vartheta_n \star v - v\|_1 \longrightarrow 0. \tag{1.190}$$

Also, using the Lebesgue dominated convergence theorem, we have

$$\|(\xi_n - \xi)v\|_1 \longrightarrow 0. \tag{1.191}$$

Returning to (1.189) and using (1.190) and (1.191), we obtain

$$\int_{\mathbb{R}^N} h_n v\, dz \longrightarrow \int_{\mathbb{R}^N} hv\, dz \qquad \forall\, v \in L^1(\mathbb{R}^N),$$

so $h_n \xrightarrow{w^*} h$ in $L^\infty(\mathbb{R}^N)$.

(b) Let $B = \overline{B}_r(z_0)$ with $z_0 \in \mathbb{R}^N$, $r > 0$ and let $\chi = \chi_{\overline{B}_{r+1}(z_0)}$. We set

$$\widehat{h}_n = \vartheta_n \star (\xi_n \chi u) \qquad \forall\, n \geqslant 1.$$

From Definition 1.117, we know that

$$\operatorname{supp}(\widehat{h}_n - h_n) \subseteq \overline{B_{\frac{1}{n}}(0) + B_{r+1}(z_0)^c},$$

so $\widehat{h}_n = h$ on $\overline{B}_r(z_0)$. Then, we have

$$\int\limits_B |\widehat{h}_n - h|\, dz = \int\limits_B |\widehat{h}_n - \chi h|\, dz \leqslant \int\limits_{\mathbb{R}^N} |\widehat{h}_n - \chi h|\, dz$$
$$\leqslant \int\limits_{\mathbb{R}^N} |\vartheta_n \star (\xi_n - \xi)\chi u|\, dz + \int\limits_{\mathbb{R}^N} |(\vartheta_n \star \chi u) - \chi u|\, dz$$
$$\leqslant \int\limits_{\mathbb{R}^N} |(\xi_n - \xi)\chi u|\, dz + \int\limits_{\mathbb{R}^N} |(\vartheta_n \star \chi u) - \chi u|\, dz \longrightarrow 0$$

(see Theorem 1.118).

Solution of Problem 1.205

Let $u \in L^\infty(\Omega)$ and extend it to all of $\mathbb{R}^N$ be setting it to be equal to 0 on $\mathbb{R}^N \setminus \Omega$. Denote this extension by $\overline{u}$. Evidently $\overline{u} \in L^\infty(\mathbb{R}^N)$. Let

$$\Omega_n = \left\{ z \in \Omega : \operatorname{dist}(z, \partial\Omega) > \tfrac{2}{n},\ |z| < n \right\}.$$

Let $\xi_n = \chi_{\Omega_n}$ for $n \geqslant 1$ and $\xi = \chi_\Omega$. Evidently $\xi_n(z) \longrightarrow \xi(z)$ for almost all $z \in \mathbb{R}^N$. Let $\{\vartheta_n\}_{n\geqslant 1} \subseteq C_c^\infty(\mathbb{R}^N)$ be a sequence of mollifiers and define

$$h_n = \vartheta_n \star (\xi_n \overline{u}) \qquad \forall\, n \geqslant 1.$$

We have that $h_n \in C_c^\infty(\mathbb{R}^N)$ and from Problem 1.204, we have

$$\int\limits_B |h_n - \overline{u}|\, dz \longrightarrow 0 \qquad \forall\, B \subseteq \mathbb{R}^N \text{ ball.}$$

So, we can find a subsequence (depending in general on the ball B), such that

$$h_n(z) \longrightarrow \overline{u}(z) \quad \text{for a.a. } z \in B.$$

Since $\mathbb{R}^N = \bigcup\limits_{n\geqslant 1} \overline{B}_n(0)$, by a diagonal process, we obtain a subsequence $\{h_{n_k}\}_{k\geqslant 1}$ such that

$$h_{n_k}(z) \longrightarrow \overline{u}(z) \quad \text{for a.a. } z \in \mathbb{R}^N.$$

Then for every $v \in L^1(\Omega)$ and using the Vitali convergence theorem (see Theorem 1.20), we have

$$\int_\Omega h_{n_k} v\,dz \longrightarrow \int_\Omega \overline{u} v\,dz,$$

thus

$$h_{n_k} \xrightarrow{w^*} u \quad \text{in } L^\infty(\Omega).$$

Since $h_{n_k} \in C_c^\infty(\Omega)$, we conclude that $C_c^\infty(\Omega)$ is w^*-dense in $L^\infty(\Omega)$.

Solution of Problem 1.206

We know that

$$\|u\|_1 = \sup\Big\{\int_\Omega u\vartheta\,dz : \vartheta \in L^\infty(\Omega),\ \|\vartheta\|_\infty \leqslant 1\Big\}. \tag{1.192}$$

From Problem 1.205 and its solution we know that for every $\vartheta \in L^\infty(\Omega)$, we can find a sequence $\{h_n\}_{n\geqslant 1} \subseteq C_c^\infty(\Omega)$ such that

$$\|h_n\|_\infty \leqslant \|\vartheta\|_\infty \qquad \forall\, n \geqslant 1 \quad \text{and} \quad h_n \xrightarrow{w^*} \vartheta \quad \text{in } L^\infty(\Omega).$$

Then we have

$$\int_\Omega u h_n\,dz \longrightarrow \int_\Omega u\vartheta\,dz.$$

So, from (1.192), it follows that

$$\|u\|_1 = \sup\Big\{\int_\Omega u\vartheta\,dz : \vartheta \in C_c^\infty(\Omega),\ \|\vartheta\|_\infty \leqslant 1\Big\}.$$

Solution of Problem 1.207

From the Fatou lemma (see Theorem I.3.95), we have

$$\int_\Omega u\,d\mu \leqslant \int_\Omega \liminf_{n\to+\infty} u_n\,d\mu \leqslant \liminf_{n\to+\infty} \int_\Omega u_n\,d\mu,$$

so

$$d_n = \int_\Omega u_n\,d\mu \longrightarrow \int_\Omega u\,d\mu = d. \tag{1.193}$$

Using Problem 1.14, we have

$$\tfrac{1}{d_n} u_n \longrightarrow \tfrac{1}{d} u \quad \text{in } L^1(\Omega). \tag{1.194}$$

From (1.193) and (1.194), it follows that $u_n \longrightarrow u$ in $L^1(\Omega)$.

Solution of Problem 1.208

Consider the function $\varphi(x) = x \ln x$ for all $x \geqslant 0$. This is a convex continuous function. Then from the Jensen inequality (see Problem 1.26), we have

$$\varphi\Big(\int\limits_\Omega u\,d\mu\Big) = \Big(\int\limits_\Omega u\,d\mu\Big) \ln \Big(\int\limits_\Omega u\,d\mu\Big) = 0 \leqslant \int\limits_\Omega \varphi(u)\,d\mu = \int\limits_\Omega u \ln u\,d\mu.$$

Solution of Problem 1.209

Let $v \in L^\infty(\Omega; X)$ and let $c = 2\|v\|_\infty$. If $\delta_0 < \frac{c}{2}$ and $\|w-v\|_\infty < \delta$ with $w \in L^\infty(\Omega; X)$, then $\|w\|_\infty \leqslant c$. Let $\varepsilon > 0$ be given and let $\delta = \delta(\varepsilon, c) > 0$ be as postulated by condition UC. We may assume that $\delta \leqslant \delta_0$. If $\|w - v\|_\infty \leqslant \delta$, then

$$\|w(z) - v(z)\|_X \leqslant \delta \qquad \forall\, z \in \Omega \setminus \widetilde{D}, \tag{1.195}$$

with $\widetilde{D} \in \Sigma$ such that $\mu(\widetilde{D}) = 0$ (see condition UC). Also, we have

$$\max\big\{\|w(z)\|_X, \|v(z)\|_X\big\} \leqslant c \qquad \forall\, z \in \Omega \setminus \widetilde{D}', \tag{1.196}$$

with $\widetilde{D}' \in \Sigma$ such that $\mu(\widetilde{D}') = 0$. So, if we consider $\widehat{D} = \widetilde{D} \cup \widetilde{D}'$, then $\mu(\widehat{D}) = 0$ and both (1.195) and (1.196) hold. Using condition UC, we have

$$\|u(z, w(z)) - u(z, v(z))\|_Y \leqslant \varepsilon \qquad \forall\, z \in \Omega \setminus (\widehat{D} \cup D),$$

with $\mu(D) = 0$, so

$$\|N_u(w) - N_u(v)\|_\infty \leqslant \varepsilon,$$

thus the map $v \longmapsto N_u(v)$ is continuous from $L^\infty(\Omega; X)$ into $L^\infty(\Omega; Y)$.

Solution of Problem 1.210

Let $c > 0$ and let $E_c = \{x \in X : \|x\|_X \leqslant c\}$. The separability of X implies that we can find a countable dense subset $E'_c \subseteq E_c$. We define

$$\xi_{c,\delta}(z) = \sup_{\substack{x, v \in E_c \\ \|x-v\|_X \leqslant \delta}} \|u(z,x) - u(z,v)\|_Y = \sup_{\substack{x, v \in E'_c \\ \|x-v\|_X \leqslant \delta}} \|u(z,x) - u(z,v)\|_Y,$$

so $\xi_{c,\delta}$ is Σ-measurable.

Arguing by contradiction, suppose that condition UC (see Problem 1.209) is not true. Then we can find $\varepsilon > 0$ and $c > 0$ such that for all $n \geqslant 1$ and all μ-null sets $D \in \Sigma$ we have

$$\sup_{z \in \Omega \setminus D} \xi_{c,\frac{1}{n}}(z) > \varepsilon.$$

Let $S_n = \big\{z \in \Omega : \xi_{c,\frac{1}{n}}(z) > \varepsilon\big\}$ for all $n \geqslant 1$. Then $S_n \in \Sigma$ and $\mu(S_n) > 0$ for all $n \geqslant 1$. So, we can find $G_n \in \Sigma$ with $\mu(G_n) > 0$ and $x_n, w_n \in E_c$ such that

$$u(z,x_n) - u(z,w_n) \geqslant \varepsilon \qquad \forall\, z \in G_n, \quad \|x_n - w_n\|_X \leqslant \tfrac{1}{n}.$$

Without any loss of generality, we assume that

$$\mu(G_{n+1}) < \tfrac{1}{2}\mu(G_n) \qquad \forall\, n \geqslant 1.$$

Let $G'_n = G_n \setminus \bigcup\limits_{k \geqslant n+1} G_k \in \Sigma$. These sets are pairwise disjoint and $\mu(G'_n) > 0$. We define

$$x(z) = \begin{cases} x_n & \text{if } z \in G'_n\ n \geqslant 1, \\ 0 & \text{if } z \in \Omega \setminus \bigcup\limits_{n \geqslant 1} G'_n \end{cases}$$

$$w_m(z) = \begin{cases} x_n & \text{if } z \in G'_n\ n \in \{1, \ldots, m-1\} \\ w_n & \text{if } z \in G'_n\ n \geqslant m \\ 0 & \text{if } z \in \Omega \setminus \bigcup\limits_{n \geqslant 1} G'_n \end{cases} \qquad \forall\, m \geqslant 1.$$

Then $\|x(z) - w_m(z)\|_X \leqslant \frac{1}{m}$ on Ω, but $\|u(z,x(z)) - u(z,w_m(z))\|_Y \geqslant \varepsilon$ on $\bigcup\limits_{n \geqslant 1} G'_n$, which contradicts the continuity of N_u.

Bibliography

[1] Adams, R.A., Fournier, J.F.: Sobolev Spaces. Academic, Amsterdam (2003)

[2] Ambrosio, L., Da Prato, G., Mennucci, A.: Introduction to Measure Theory and Integration. Scuola Normale Superiore Pisa (2011)

[3] Ambrosio, L., Fusco, N., Pallara, D.: Functions of Bounded Variation and Free Discontinuity Problems. Oxford University Press, New York (2000)

[4] Brézis, H.: Functional Analysis, Sobolev Spaces and Partial Differential Equations. Springer, New York (2011)

[5] Choquet, G.: Lectures on Analysis. Vol. II: Representation Theory. W.A. Benjamin, New York (1969)

[6] Denkowski, Z., Migórski, S., Papageorgiou, N.S.: An Introduction to Nonlinear Analysis: Theory. Kluwer Academic, Boston (2003)

[7] DiBenedetto, E.: Real Analysis. Birkhäuser, Boston (2002)

[8] Diestel, J., Uhl, J.: Vector Measures. American Mathematical Society, Providence (1977)

[9] Dunford, N., Schwartz, J.T.: Linear Operators. Part I. General Theory. Wiley-Interscience, New York (1988)

[10] Evans, L.C., Gariepy, R.F.: Measure Theory and Fine Properties of Functions. CRC, Boca Raton (1992)

[11] Folland, G.B.: Real Analysis. Modern Techniques and Their Applications. Wiley-Interscience, New York (1999)

[12] Gasiński L., Papageorgiou, N.S.: Nonlinear Analysis. Chapman & Hall/CRC, Boca Raton (2006)

[13] Krasnosel'skii, M.A.: Topological Methods in the Theory of Nonlinear Integral Equations. Pergamon, New York (1964)

[14] Leoni, G.: A First Course in Sobolev Spaces. American Mathematical Society, Providence (2009)

[15] Malliavin, P.: Integration and Probability. Springer, New York (1995)

[16] Maz'ja, V.G.: Sobolev Spaces. Springer, Berlin (1985)

[17] Natanson, I.P.: Theory of Functions of a Real Variable, vol. I. Fredrick Ungar Publishing Co., New York (1964)

[18] Ziemer, W.P.: Weakly Differentiable Functions. Sobolev Spaces and Functions of Bounded Variation. Springer, New York (1989)

Chapter 2

Nonlinear and Multivalued Maps

2.1 Introduction

2.1.1 Compact, Completely Continuous, and Proper Maps

Definition 2.1
Let X and Y be two Banach spaces, let $D \subseteq X$ be a set, and let $f: D \longrightarrow Y$ be a map.

(a) *We say that f is **compact**, if it is continuous and maps bounded subsets of D to relatively compact sets (i.e., if $B \subseteq D$ is bounded, then $\overline{f(B)} \subseteq Y$ is compact). We denote the set of all compact maps $f: D \longrightarrow Y$ by $K(D;Y)$. If $D = X$, then we set $\mathcal{L}_c(X;Y) = \mathcal{L}(X;Y) \cap K(X;Y)$.*

(b) *We say that f is **completely continuous**, if for every sequence $\{u_n\}_{n\geqslant 1} \subseteq D$ such that $u_n \xrightarrow{w} u \in D$ in X, we have $f(u_n) \longrightarrow f(u)$ in Y (i.e., f maps weakly convergent sequences in D to strongly convergent sequences in Y).*

(c) *We say that f is a **finite rank map**, if $f(D)$ lies in a finite dimensional subspace of Y. We denote the set of all finite rank maps $f: D \longrightarrow Y$ by $K_f(D;Y)$. If $D = X$, then we set $\mathcal{L}_f(X;Y) = \mathcal{L}(X;Y) \cap K_f(X;Y)$.*

(d) *We say that f is a **weakly compact map**, if it maps bounded sets in D to relatively weakly compact subsets of Y (that is, if $B \subseteq D$ is bounded, then $\overline{f(B)}^w \subseteq Y$ is weakly compact). We denote the set of all weakly compact maps $f: D \longrightarrow Y$ by $K_{wc}(D;Y)$. If $D = X$, then we set $\mathcal{L}_{wc}(X;Y) = \mathcal{L}(X;Y) \cap K_{wc}(X;Y)$.*

Remark 2.2
Evidently we have $K_f(X;Y) \subseteq K(X;Y)$. A completely continuous linear operator $A: X \longrightarrow Y$ is sometimes called ***Dunford–Pettis operator***. Clearly the completely continuous maps are continuous.

L. Gasiński, N.S. Papageorgiou, *Exercises in Analysis: Part 2*, Problem Books in Mathematics, DOI 10.1007/978-3-319-27817-9_2

Proposition 2.3
If X *and* Y *are two Banach spaces and* $K \in \mathcal{L}_c(X;Y)$,
then K *is completely continuous.*

The converse of this result is not in general true.

Proposition 2.4
If X *is a reflexive Banach space,* $D \subseteq X$ *is nonempty closed and* $f: D \longrightarrow Y$ *is completely continuous,*
then $f \in K(D;Y)$.

Combining Propositions 2.3 and 2.4, we infer the following.

Corollary 2.5
If X *is a reflexive Banach space,* Y *is a Banach space and* $K \in \mathcal{L}(X;Y)$,
then $K \in \mathcal{L}_c(X;Y)$ *if and only if* K *is completely continuous.*

The next theorem presents a useful approximation property that compact maps have and explains why those maps are the right class to extend the finite dimensional theory.

Theorem 2.6
If X *and* Y *are two Banach spaces,* $D \subseteq X$ *is a bounded closed set and* $f: D \longrightarrow Y$ *is a map,*
then $f \in K(D;Y)$ *if and only if* f *is the uniform limit of finite rank maps*

Remark 2.7
In fact given $\varepsilon > 0$, we can find a finite rank map $f_\varepsilon: D \longrightarrow Y$ such that

$$\|f(u) - f_\varepsilon(u)\|_Y < \varepsilon \qquad \forall\, u \in D$$

and

$$\overline{f_\varepsilon(D)} \subseteq \overline{\operatorname{conv}}\, f(D).$$

Compact maps exhibit a nice extension property, which is a consequence of the following theorem, which generalizes the well-known Tietze extension theorem (see Theorem I.2.138).

Theorem 2.8 *(Dugundji Extension Theorem)*
If X *is a metric space,* Y *is a locally convex space,* $D \subseteq X$ *is a nonempty closed set, and* $f: D \longrightarrow Y$ *is a continuous map,*
then there exists a continuous map $\widehat{f}: X \longrightarrow Y$ *such that*

$$\widehat{f}|_D = f \quad \textit{and} \quad \widehat{f}(X) \subseteq \operatorname{conv} f(D).$$

Using the above theorem together with the Mazur theorem on the convex hull of compact sets (according to which the convex hull of a compact set in X is relatively compact; see Theorem I.5.86), we have the following result.

Theorem 2.9
If X and Y are two Banach spaces, $D \subseteq X$ is a nonempty closed and bounded set and $f \in K(D;Y)$,
then there exists $\widehat{f} \in K(X;Y)$ such that

$$\widehat{f}|_D = f \quad \text{and} \quad \widehat{f}(X) \subseteq \operatorname{conv} f(D).$$

The space $\mathcal{L}_c(X;Y)$ is closed and so

$$\overline{\mathcal{L}_f(X;Y)}^{\|\cdot\|_{\mathcal{L}}} \subseteq \mathcal{L}_c(X;Y).$$

In general this inclusion can be strict.

Definition 2.10
*A Banach space Y is said to have the **approximation property** (the **AP** for short), if for every Banach space X we have*

$$\overline{\mathcal{L}_f(X;Y)}^{\|\cdot\|_{\mathcal{L}}} = \mathcal{L}_c(X;Y).$$

Remark 2.11
The Hilbert spaces and the Banach spaces $c_0 = \big\{\widehat{u} = \{u_k\}_{k\geqslant 1} : u_k \longrightarrow 0\big\}$ and l^p (with $p \in [1, +\infty)$) have the AP.

Theorem 2.12 *(Schauder Theorem)*
If X and Y are two Banach spaces and $K \in \mathcal{L}(X;Y)$, then
(a) $K \in \mathcal{L}_c(X;Y)$ if and only if $K^* \subset \mathcal{L}_c(Y^*;X^*)$;
(b) $K \in \mathcal{L}_f(X;Y)$ if and only if $K^* \in \mathcal{L}_f(Y^*;X^*)$.

The next notion is important because it restricts the size of the solution set of an operator equation of the form $f(u) = y$.

Definition 2.13
*Let X and Y be two Hausdorff topological spaces and let $f \in C(X;Y)$. We say that f is **proper**, if for every compact set $C \subseteq Y$, the set $f^{-1}(C) \subseteq X$ is compact.*

This notion is related to the coercivity of f which is important in variational analysis.

Proposition 2.14
If X and Y are two Banach spaces and $f \in C(X;Y)$,
then the following conditions are equivalent:
(a) f is proper.
(b) f is a closed map (i.e., maps closed sets in X to closed sets in Y). Moreover, if X and Y are finite dimensional, then the above conditions are also equivalent to the following one.
(c) f is coercive (i.e., $\|f(u)\|_Y \longrightarrow +\infty$ as $\|u\|_X \longrightarrow +\infty$).

In some special cases for maps between infinite dimensional Banach spaces, coercivity implies properness.

Proposition 2.15

If X and Y are two Banach spaces, $f \in C(X;Y)$ is coercive (i.e., $\|f(u)\|_Y \longrightarrow +\infty$ as $\|u\|_X \longrightarrow +\infty$) and at least one of the following conditions holds:
(i) $f = g + h$ where g is proper and h is compact; or
(ii) X is reflexive and f is completely continuous (see Definition 2.1(b)),
then f is proper (see Definition 2.13).

The notions of compactness and properness are related as follows.

Proposition 2.16

If X is a Banach space, $D \subseteq X$ is a bounded closed set and $f \in K(D;X)$, then $I_X - f$ is proper (I_X being the identity on X).

Now we turn our attention to the space $\mathcal{L}_c(X;Y)$ and in particular to the spectral properties of such operators.

Proposition 2.17

If X and Y are two Banach spaces, then $\mathcal{L}_c(X;Y)$ furnished with the operator norm is a Banach space

Definition 2.18

*Let X be a complex Banach space and let $A \in \mathcal{L}(X)$. The **spectrum** $\sigma(A)$ of A is defined by*

$$\sigma(A) = \{\lambda \in \mathbb{C} : \lambda I_X - A \text{ is not invertible}\}.$$

*The set $\mathbb{C} \setminus \sigma(A)$ is called the **resolvent set** of A and it is denoted by $\varrho(A)$. Also, the operator $R(\lambda) = (\lambda I_X - A)^{-1}$ is called the **resolvent** of A.*

Remark 2.19

If we insist on real Banach spaces, then some important results of the spectral theory fail. For this reason we consider complex Banach spaces. For example, the next result is no longer true, if X is a real Banach space.

Proposition 2.20

If X is a complex Banach space and $A \in \mathcal{L}(X)$, then $\sigma(A) \neq \emptyset$ and in particular $\sigma(A)$ is a nonempty compact subset of $\{\lambda \in \mathbb{C} : |\lambda| \leqslant \|A\|_{\mathcal{L}}\}$.

Definition 2.21

Let X be a complex Banach space and let $A \in \mathcal{L}(X)$.

*(a) We say that $\lambda \in \mathbb{C}$ is an **eigenvalue** of A, if there is $u \in X$, $u \neq 0$ such that $(\lambda I_X - A)(u) = 0$. Then u is an **eigenvector** corresponding to the eigenvalue λ. The subspace $(\lambda I_X - A)^{-1}(0)$ is the **eigenspace** corresponding to the eigenvalue λ. The set of all eigenvalues of A is called the **point spectrum** of A and is denoted by $\sigma_p(A)$.*

(b) *If* $\lambda \in \sigma(A) \setminus \sigma_p(A)$ *and* $R(\lambda I_X - A)$ *is not dense in* X, *then we say that* λ *is in the* ***residual spectrum*** *of* A.

Remark 2.22
The reason that we single out the residual spectrum is that it does not occur for a large class of operators, for example, for self-adjoint operators on a Hilbert space.

Theorem 2.23
If X *is an infinite dimensional complex Banach space and* $A \in \mathcal{L}_c(X)$, *then*
(a) $\sigma(A)$ *is a countably compact set (see Definition I.2.88) with* 0 *as the only possible limit point;*
(b) $\sigma(A) = \sigma_p(A) \cup \{0\}$;
(c) *if* $\lambda \in \sigma_p(A) \setminus \{0\}$, *then the eigenspace corresponding to* λ *is finite dimensional and*

$$\dim \ker (\lambda I_X - A) \;=\; \operatorname{codim} R(\lambda I_X - A)$$

(recall that $\operatorname{codim} R(\lambda I_X - A) = \dim \left(X/_{R(\lambda I_X - A)}\right)$*).*

We can expand **(c)** in the above theorem.

Proposition 2.24
If X *is a Banach space,* $A \in \mathcal{L}_c(X)$ *and* λ *is a nonzero scalar, then*
(a) $\ker (\lambda I_X - A)$ *is finite dimensional;*
(b) $R(\lambda I_X - A)$ *is closed and* $R(\lambda I_X - A) = \ker (\lambda I_X^* - A^*)^\perp$;
(c) $\ker (\lambda I_X - A) = \{0\}$ *if and only if* $R(\lambda I_X - A) = X$;
(d) $\dim \ker (\lambda I_X - A) = \dim \ker (\lambda I_X^* - A^*)$.

Remark 2.25
Statement **(c)** says that $\lambda I_X - A$ is injective if and only if it is surjective. This is a well-known fact for operators between finite dimensional spaces. So, according to this statement, the equation

$$(\lambda I_X - A)(u) \;=\; y$$

has a solution for every $y \in X$ if and only if the equation

$$(\lambda I_X - A)(u) \;=\; 0$$

admits only the trivial solution. In this form, the result is known as the ***Fredholm alternative theorem***.

For compact self-adjoint operators defined on a separable Hilbert space, we have the following result.

Theorem 2.26 *(Spectral Theorem)*
If H *is an infinite dimensional separable Hilbert space and* $A \in \mathcal{L}_c(H)$,

then there exists an orthonormal basis $\{e_n\}_{n\geqslant 1}$ of H consisting of eigenvectors of A such that for every $u \in H$ we have

$$A(u) = \sum_{k\geqslant 1} \lambda_k (u, e_k)_H e_k,$$

with λ_k being the eigenvalue corresponding to e_k and $(\cdot,\cdot)_H$ denoting the inner product of H.

Definition 2.27
Let H be a Hilbert space and let $A \in \mathcal{L}(H)$.
*(a) We say that A is **normal**, if $AA^* = A^*A$;*
*(b) We say that A is **unitary**, if $A^* = A^{-1}$ (i.e., $AA^* = A^*A = I_H$);*
*(c) We say that A is **projection**, if $A^2 = A$ and it is an **orthogonal projection**, if it is a projection and $R(A) = (\ker A)^{\perp}$.*

Proposition 2.28
If H is a complex separable Hilbert space and $A \in \mathcal{L}_c(H)$ is normal, then there exists an orthonormal basis $\{e_n\}_{n\geqslant 1}$ of H consisting of eigenvectors of A such that for every $u \in H$, we have

$$A(u) = \sum_{k\geqslant 1} \lambda_k (u, e_k)_H e_k,$$

with λ_k being the eigenvalue corresponding to e_k and $(\cdot,\cdot)_H$ denoting the inner product of H.

Definition 2.29
*Let X and Y be two Banach spaces and let $L \in \mathcal{L}(X;Y)$. We say that L is a **Fredholm operator**, if $\ker L$ is finite dimensional and $R(L) = L(X)$ is finite codimensional (that is, $\dim(Y/_{R(L)}) < +\infty$). The number $i(L) = \dim\ker L - \dim(Y/_{R(L)})$ is called the **index** of L. The set of Fredholm operators $L\colon X \longrightarrow Y$ is denoted by $\mathrm{Fred}\,(X;Y)$.*

Remark 2.30
If $L \in \mathrm{Fred}\,(X;Y)$, then $R(L) \subseteq Y$ is closed. Moreover, we have

$$X = \ker L \oplus V$$

and $L|_V$ is an isomorphism of V onto $L(X)$. $\mathrm{Fred}\,(X;Y)$ is an open subset of $\mathcal{L}(X;Y)$ and the map $L \longmapsto i(L)$ is continuous (hence, it is constant on each connected component of $\mathrm{Fred}\,(X;Y)$). Finally every $L \in \mathrm{Fred}\,(X;Y)$ is invertible modulo finite rank operators, that is, there exists $T \in \mathcal{L}(Y;X)$ such that both $L \circ T - I_Y$ and $T \circ L - I_X$ are finite rank operators.

Finally we introduce a notion that allows us to go beyond compact maps.

Definition 2.31

Let X be a Banach space and let $\mathcal{B}$ be the family of bounded subsets of X.

(a) *The* **Kuratowski measure of noncompactness** $\alpha\colon \mathcal{B} \longrightarrow \mathbb{R}_+$ *is defined by*

$$\alpha(C) = \inf\big\{d>0: \ C \text{ can be covered by a finite number of sets of diameter} \leqslant d\big\}.$$

(b) *The* **Hausdorff measure of noncompactness** $\beta\colon \mathcal{B} \longrightarrow \mathbb{R}_+$ *is defined by*

$$\beta(C) = \inf\big\{r>0: \ C \text{ can be covered by a finite number of balls of radius } r\big\}.$$

Let $D \subseteq X$, let Y be another Banach space and let $f\colon D \longrightarrow Y$ be a continuous and bounded (i.e., maps bounded sets to bounded sets) map. In what follows by γ_X (respectively γ_Y) we denote either α or β on X (respectively on Y).

(c) *We say that f is a* **k-set-Lipschitz map**, *if for every bounded set $C \subseteq D$ we have*

$$\gamma_Y(f(C)) \leqslant k\gamma_X(C).$$

(d) *We say that f is a* **k-set contraction**, *if it is k-set Lipschitz with $k<1$. We denote by $SC_k(D;Y)$ the family of all such maps.*

(e) *We say that f is a* **condensing**, *if for every bounded set $C \subseteq D$ with $\gamma_X(C)>0$, we have $\gamma_Y(f(C)) < \gamma_X(C)$. We denote by $S(D;Y)$ the family of all such maps.*

Proposition 2.32

If X is Banach space and $\gamma\colon \mathcal{B} \longrightarrow \mathbb{R}_+$ is either α or β, then

(a) *$\gamma(C)=0$ if and only if $\overline{C}$ is compact (regularity);*

(b) *$\gamma(\lambda C)=|\lambda|\gamma(C)$ for all $\lambda \in \mathbb{R}$ and $\gamma(C_1+C_2) \leqslant \gamma(C_1)+\gamma(C_2)$ (seminorm);*

(c) *if $C_1 \subseteq C_2$, then $\gamma(C_1) \leqslant \gamma(C_2)$ (monotonicity);*

(d) *$\gamma(C_1 \cup C_2) = \max\{\gamma(C_1), \gamma(C_2)\}$ (semiadditivity);*

(e) *$\gamma(C) = \gamma(\operatorname{conv} C)$ and $\gamma(C)=\gamma(\overline{C})$.*

Another measure of noncompactness that we will use is related to the weak topology on X.

Definition 2.33

Let X be a Banach space and let $\mathcal{B}$ be the family of bounded subsets of X. The weak measure of noncompactness $\xi\colon \mathcal{B} \longrightarrow [0,+\infty)$ is defined by

$$\xi(C) = \inf\big\{\varepsilon>0: \ \text{there exists a } w\text{-compact set } D \subseteq X \text{ such that } C \subseteq D+\varepsilon\overline{B}_1\big\},$$

where $\overline{B}_1 = \{u \in X: \ \|u\|_X \leqslant 1\}$.

The properties of ξ are similar to those of α and β.

Proposition 2.34
If X is a Banach space and $\xi\colon \mathcal{B} \longrightarrow \mathbb{R}_+$ is the weak measure of noncompactness, then
(a) $\xi(C)=0$ *if and only if* $\overline{C}^w$ *is w-compact;*
(b) $\xi(\lambda C)=|\lambda|\xi(C)$ *for all* $\lambda\in\mathbb{R}$ *and* $\xi(C_1+C_2)\leqslant\xi(C_1)+\xi(C_2)$*;*
(c) if $C_1\subseteq C_2$*, then* $\xi(C_1)\leqslant\xi(C_2)$*;*
(d) $\xi(C_1\cup C_2)=\max\{\xi(C_1),\xi(C_2)\}$*;*
(e) $\xi(C)=\xi(\operatorname{conv} C)$ *and* $\xi(C)=\xi(\overline{C}^w)$*.*

2.1.2 Multifunctions

Throughout this section, we will use the following notation for certain hyperspaces. So, let X be a Hausdorff topological space. We define

$$\begin{aligned} P_f(X) &= \{A\subseteq X:\ A \text{ is nonempty and closed}\}\\ P_k(X) &= \{A\subseteq X:\ A \text{ is nonempty and compact}\}\\ \widehat{P}_f(X) &= P_f(X)\cup\{\emptyset\}. \end{aligned}$$

Moreover, if X is a normed space, then

$$\begin{aligned} P_{fc}(X) &= \{A\in P_f(X):\ A \text{ is convex}\}\\ P_{kc}(X) &= \{A\in P_k(X):\ A \text{ is convex}\}\\ P_{wkc}(X) &= \{A\subseteq X:\ A \text{ is nonempty, } w\text{-compact and convex}\}\\ P_{bf(c)}(X) &= \{A\subseteq X:\ A \text{ is nonempty, bounded, closed (and convex)}\}. \end{aligned}$$

Definition 2.35
Let X and Y be two sets, let $F\colon X\longrightarrow 2^Y\setminus\{\emptyset\}$ be a multifunction and let $D\subseteq Y$. We define
(a) The ***weak inverse image*** *of D under F is the set*

$$F^-(D) = \{u\in X:\ F(u)\cap D\neq\emptyset\}.$$

(b) The ***strong inverse image*** *of D under F is the set*

$$F^+(D) = \{u\in X:\ F(u)\subseteq D\}.$$

Using these notions, we can define some continuity concepts for multifunctions between two Hausdorff topological spaces.

Definition 2.36
Let X and Y be two Hausdorff topological spaces and let $F\colon X\longrightarrow 2^Y$ be a multifunction.
(a) We say that F is ***upper semicontinuous*** *at $u_0\in X$ (**usc** at x_0 for short), if for every open set $V\subseteq Y$ such that $F(u_0)\subseteq V$, we can find $U\in\mathcal{N}(u_0)$ (the filter of neighborhoods of u_0), such that $F(u)\subseteq V$ for all $u\in U$. We say that F is* ***upper semicontinuous*** *(**usc** for short), if it is upper semicontinuous at any $u\in X$.*

(b) *We say that F is* **lower semicontinuous** *at $u_0 \in X$ (***lsc** *at x_0 for short), if for every open set $V \subseteq Y$ such that $F(u_0) \cap V \neq \emptyset$, we can find $U \in \mathcal{N}(u_0)$, such that $F(u) \cap V \neq \emptyset$ for all $u \in U$. We say that F is* **lower semicontinuous (lsc** *for short), if it is lower semicontinuous at any $u \in X$.*

(c) *We say that F is* **continuous** *(or* **Vietoris continuous***) at $u_0 \in X$, if it is both upper and lower semicontinuous at u_0. We say that F is* **continuous** *(or* **Vietoris continuous***) if it is continuous at any $u \in X$.*

These definitions lead to the following alternative descriptions of the above continuity notions for multifunctions.

Proposition 2.37
If X and Y be two Hausdorff topological spaces and $F\colon X \longrightarrow 2^Y$ is a multifunction, then the following statements are equivalent:
(a) *F is upper semicontinuous.*
(b) *For every closed set $C \subseteq Y$, the set $F^-(C)$ is closed in X.*
(c) *If $x \in X$, $\{x_\alpha\}_{\alpha \in J}$ is a net in X, $x_\alpha \longrightarrow x$, $V \subseteq Y$ is an open set such that $F(x) \subseteq V$, then there exists $\alpha_0 \in J$, such that for all $\alpha \in J$, $\alpha \geqslant \alpha_0$ we have $F(x_\alpha) \subseteq V$.*
(d) *For every open set $U \subseteq Y$, the set $F^+(U)$ is open in X.*

Proposition 2.38
If X and Y be two Hausdorff topological spaces and $F\colon X \longrightarrow 2^Y$ is a multifunction, then the following statements are equivalent:
(a) *F is lower semicontinuous.*
(b) *For every closed set $C \subseteq Y$, the set $F^+(C)$ is closed in X.*
(c) *If $x \in X$, $\{x_\alpha\}_{\alpha \in J}$ is a net in X, $x_\alpha \longrightarrow x$, $V \subseteq Y$ is an open set such that $F(x) \cap V \neq \emptyset$, then there exists $\alpha_0 \in J$, such that for all $\alpha \in J$, $\alpha \geqslant \alpha_0$ we have $F(x_\alpha) \cap V \neq \emptyset$.*
(d) *If $x \in X$, $\{x_\alpha\}_{\alpha \in J}$ is a net in X, $x_\alpha \longrightarrow x$ and $y \in F(x)$, then for every $\alpha \in J$ we can find $y_\alpha \in F(x_\alpha)$, such that $y_\alpha \longrightarrow y$ in Y.*
(e) *For every open set $U \subseteq Y$, the set $F^-(U)$ is open in X.*

Proposition 2.39
If X and Y are two Hausdorff topological spaces and $F\colon X \longrightarrow 2^Y$ is a multifunction, then the following statements are equivalent:
(a) *F is continuous.*
(b) *For every closed set $C \subseteq Y$, the sets $F^+(C)$ and $F^-(C)$ are both closed in X.*
(c) *If $x \in X$, $\{x_\alpha\}_{\alpha \in J}$ is a net in X, $x_\alpha \longrightarrow x$, $V, W \subseteq Y$ are open sets such that $F(x) \subseteq V$ and $F(x) \cap W \neq \emptyset$, then there exists $\alpha_0 \in J$, such that for all $\alpha \in J$, $\alpha \geqslant \alpha_0$ we have $F(x_\alpha) \subseteq V$ and $F(x_\alpha) \cap W \neq \emptyset$.*

Definition 2.40
Let X and Y be two Hausdorff topological spaces and let $F: X \longrightarrow 2^Y$ be a multifunction. The ***graph*** *of F is the set*

$$\operatorname{Gr} F = \{(x, y) \in X \times Y : \ y \in F(x)\}.$$

We say that F is ***closed****, at $u_0 \in X$, if for every net $\{(x_\alpha, y_\alpha)\}_{\alpha \in J} \subseteq \operatorname{Gr} F$ such that $(x_\alpha, y_\alpha) \longrightarrow (x_0, y_0)$ in $X \times Y$, we have $y_0 \in F(x_0)$. We say that F is* ***closed****, if F is closed at every $x \in X$ (that is, $\operatorname{Gr} F \subseteq X \times Y$ is closed).*

Proposition 2.41

(a) If X and Y are two Hausdorff topological spaces with Y being regular and $F: X \longrightarrow P_f(Y)$ is an upper semicontinuous multifunction,
then F is closed.

(b) If X is a complete metric space, Y is a Hausdorff topological space with topology τ, $F: X \longrightarrow P_k(Y)$ is an upper semicontinuous multifunction and there is a metrizable topology τ_1 of Y such that $\tau_1 \subseteq \tau$,
then F is continuous on a residual set (that is, on the complement of a set of first Baire category).

For multifunctions with compact values, we can drop the regularity condition on Y. This is a consequence of the following proposition.

Proposition 2.42
If X and Y are two Hausdorff topological spaces and $F: X \longrightarrow 2^Y \setminus \{\emptyset\}$ is a multifunction,
then F is $P_k(Y)$-valued and upper semicontinuous if and only if for every net $\{(x_\alpha, y_\alpha)\}_{\alpha \in J} \subseteq \operatorname{Gr} F$ such that $x_\alpha \longrightarrow x$ in X, $\{y_\alpha\}_{\alpha \in J}$ has a limit point in $F(x)$.

With this result, we can have a version of Proposition 2.41(a) in which the regularity of the space Y is no longer required.

Proposition 2.43
If X and Y are two Hausdorff topological spaces and $F: X \longrightarrow P_k(Y)$ is an upper semicontinuous multifunction,
then F is closed.

Now we introduce a metric structure on the range space Y and for every $y \in Y$, we examine the function $x \longmapsto \operatorname{dist}_Y(y, F(x))$.

Proposition 2.44
If X is a Hausdorff topological space, (Y, d_Y) is a metric space and $F: X \longrightarrow 2^Y \setminus \{\emptyset\}$ is a multifunction,
then F is lower semicontinuous if and only if for every $y \in Y$, the function $x \longmapsto \operatorname{dist}_Y(y, F(x))$ is upper semicontinuous (as an $\mathbb{R}_+$-valued function).

There is no corresponding result for upper semicontinuous multifunctions. We only have the following result.

Proposition 2.45
If X is a Hausdorff topological space, (Y, d_Y) is a metric space and $F\colon X \longrightarrow 2^Y \setminus \{\emptyset\}$ is a locally compact multifunction (i.e., for every $x \in X$, we can find $U \in \mathcal{N}(x)$ such that $\overline{F(U)} \in P_k(Y)$) and for every $y \in Y$, the function $x \longmapsto \mathrm{dist}_Y(y, F(x))$ is lower semicontinuous,
then F is upper semicontinuous.

If on Y we introduce linear structure, then there is another important functional that we can associate with a multifunction $F\colon X \longrightarrow 2^Y \setminus \{\emptyset\}$, namely the support function.

Definition 2.46
Let Y be a normed space, Y^* its topological dual and $D \subseteq Y$ a nonempty set. The ***support function*** *of D, $\sigma_D\colon Y^* \longrightarrow \overline{\mathbb{R}} = \mathbb{R} \cup \{+\infty\}$ is defined by*

$$\sigma_D(y^*) = \sup_{y \in D} \langle y^*, y \rangle ,$$

where by $\langle \cdot, \cdot \rangle$ we denote the duality brackets for the pair (Y^*, Y).

Proposition 2.47
If X is a Hausdorff topological space, Y is a normed space furnished with its weak topology and $F\colon X \longrightarrow 2^Y \setminus \{\emptyset\}$ is an upper semicontinuous multifunction,
then for every $y^* \in Y^*$, the function $x \longmapsto \sigma_{F(x)}(y^*)$ is upper semicontinuous.

To have the converse of the above proposition, we need to strengthen the conditions on the multifunction F.

Proposition 2.48
If X is a Hausdorff topological space, Y is a normed space and $F\colon X \longrightarrow P_{wkc}(Y)$ is a multifunction such that for all $y^* \in Y^*$, the function $x \longmapsto \sigma_{F(x)}(y^*)$ is upper semicontinuous,
then F is upper semicontinuous from X into Y equipped with the weak topology.

In the presence of a metric structure on Y, we can have a pseudometric structure on the hyperspace 2^Y.

Definition 2.49
Let (Y, d_Y) be a metric space and let $C, D \in 2^Y$ be two sets. We define:
(a) $h^*(C, D) = \sup_{c \in C} \mathrm{dist}_Y(c, D)$ *(the* ***excess*** *of C over D).*
(b) $h(C, D) = \max \{h^*(C, D), h^*(D, C)\}$ *(the* ***Hausdorff distance*** *of C from D).*

Remark 2.50
Evidently h is a generalized pseudometric on 2^Y (that is, h is a pseudometric that can also take the value $+\infty$). Moreover, $h(C,D)=0$ if and only if $\overline{C}=\overline{D}$ and so $P_f(Y)\cup\{\emptyset\}$ is a generalized metric space, with $\emptyset$ an isolated point. Then we call h the ***Hausdorff metric***. The Hausdorff metric h is not a topological construction, that is, the Hausdorff metric topology on $P_f(Y)$ is not determined by the topology of Y. It is easy to check that

$$h(\lambda C,\lambda D) = |\lambda| h(C,D) \qquad \forall\,\lambda\in\mathbb{R}$$

and if Y is a normed space, then

$$h(C_1+C_2,D_1+D_2) \leqslant h(C_1,D_1)+h(C_2,D_2)$$

and

$$h(\lambda C_1+(1-\lambda)C_2,\lambda D_1+(1-\lambda)D_2) \leqslant \lambda h(C_1,D_1)+(1-\lambda)h(C_2,D_2),$$

for all $\lambda\in[0,1]$. Therefore, we have

$$h(\overline{\operatorname{conv}}\,C,\overline{\operatorname{conv}}\,D) \leqslant h(C,D).$$

The same inequalities are also valid for h^*. Also, note that

$$h^*(C,D) = \sup_{y\in Y}\big(\operatorname{dist}(y,C)-\operatorname{dist}(y,D)\big)$$

and is Y is a normed space and $C,D\in P_{bfc}(Y)$, then

$$h^*(C,D) = \sup\{\sigma_C(y^*)-\sigma_D(y^*):\ \|y^*\|_*\leqslant 1\}$$

An immediate consequence of Definition 2.49 are the following two useful expressions for the Hausdorff metric h.

Proposition 2.51
(a) *If* (Y,d_Y) *is a metric space and* $C,D\in 2^Y\setminus\{\emptyset\}$,
then $h(C,D)=\sup\limits_{y\in Y}\big|\operatorname{dist}_Y(y,C)-\operatorname{dist}_Y(y,D)\big|$;
(b) *If* Y *is a normed space and* $C,D\in P_{bfc}(Y)$,
then $h(C,D)=\sup\limits_{\|y^*\|\leqslant 1}\big|\sigma_C(y^*)-\sigma_D(y^*)\big|$ *(the* ***Hörmander formula****).*
(c) *If* Y *is a normed space and* $C\subseteq Y$ *is a nonempty, convex set*
then for every $v\in Y$, *we have*

$$\operatorname{dist}_Y(v,C) = \sup\big\{\langle v^*,v\rangle-\sigma_C(v^*):\ v^*\in Y^*,\ \|v^*\|_*\leqslant 1\big\}$$

(see Problem 3.88).

(d) *If* Y *is a Banach space and* $C \subseteq Y$,
then $C \in P_{wkc}(Y)$ *if and only if the map* $v^* \longmapsto \sigma_C(v^*)$ *is* $m(Y^*, Y)$*-continuous with* $m(Y^*, Y)$ *being the Mackey topology.*

Assuming extra structure on Y, we can say more about the metric space $(P_f(Y), h)$.

Proposition 2.52

(a) *If* (Y, d_Y) *is a complete metric space,*
then $(P_f(Y), h)$ *is a complete metric space too.*

(b) $P_{bf}(Y)$ *is a closed subset of* $(P_f(Y), h)$ *and if* (Y, d_Y) *is a separable metric space, then so is* $(P_f(Y), h)$.

(c) *If* Y *is a normed space,*
then $P_{kc}(Y) \subseteq P_{bfc}(Y)$ *and* $P_k(Y) \subseteq P_{bf}(Y)$ *are all closed subsets of the metric space* $(P_f(Y), h)$.

Using these metric notions, we can introduce new continuity concepts for a multifunction F, in general distinct from those of Definition 2.36.

Definition 2.53

Let X *be a Hausdorff topological space and let* (Y, d_Y) *be a metric space.*

(a) *A multifunction* $F\colon X \longrightarrow 2^Y \setminus \{\emptyset\}$ *is said to be* h-***upper semicontinuous*** *at* $x_0 \in X$ *(*h-**usc** *at* x_0 *for short), if the function* $x \longmapsto h^*(F(x), F(x_0))$ *is continuous at* x_0. *We say that* F *is* h-***upper semicontinuous*** *(*h-**usc** *for short), if it is* h*-upper semicontinuous at every* $x \in X$.

(b) *A multifunction* $F\colon X \longrightarrow 2^Y \setminus \{\emptyset\}$ *is said to be* h-***lower semicontinuous*** *at* $x_0 \in X$ *(*h-**lsc** *at* x_0 *for short), if the function* $x \longmapsto h^*(F(x_0), F(x))$ *is continuous at* x_0. *We say that* F *is* h-***lower semicontinuous*** *(*h-**lsc** *for short), if it is* h*-lower semicontinuous at every* $x \in X$.

(c) *A multifunction* $F\colon X \longrightarrow 2^Y \setminus \{\emptyset\}$ *is said to be* h-***continuous*** *at* $x_0 \in X$, *if it is both* h*-lower and* h*-upper semicontinuous at* $x_0 \in X$ *(that is, the multifunction function* $F\colon X \longrightarrow (2^Y \setminus \{\emptyset\}, h)$ *is continuous at* x_0*). We say that* F *is* h-***continuous***, *if it is* h*-continuous at every* $x \in X$.

How are these continuity notions related to those introduced in Definition 2.36?

Proposition 2.54

If X *is a Hausdorff topological space,* (Y, d_Y) *is a metric space, and* $F\colon X \longrightarrow 2^Y \setminus \{\emptyset\}$ *is upper semicontinuous,*
then F *is* h*-upper semicontinuous.*

Proposition 2.55

If X *is a Hausdorff topological space,* (Y, d_Y) *is a metric space, and* $F\colon X \longrightarrow 2^Y \setminus \{\emptyset\}$ *is* h*-lower semicontinuous,*
then F *is lower semicontinuous.*

In general the converse implications are not true (see Problem 2.67). For compact valued multifunctions, the situation is better.

Proposition 2.56
If X is a Hausdorff topological space, (Y, d_Y) is a metric space, and $F\colon X \longrightarrow P_k(Y)$ is a multifunction, then
(a) F is upper semicontinuous if and only if F is h-upper semicontinuous.
(b) F is lower semicontinuous if and only if F is h-lower semicontinuous.
(c) F is continuous if and only if F is h-continuous.

Now we turn our attention to the fundamental question of the existence of a continuous selection.

Definition 2.57
Let X and Y be two Hausdorff topological spaces and let $F\colon X \longrightarrow 2^Y \setminus \{\emptyset\}$ be a multifunction. A **continuous selection** *(or* **selector***) of F is a continuous function $f\colon X \longrightarrow Y$ such that $f(x) \in F(x)$ for all $x \in X$.*

We have the following existence result.

Theorem 2.58 *(Michael Selection Theorem)*
If X is a paracompact space, Y is a Banach space and $F\colon X \longrightarrow P_{fc}(Y)$ is a lower semicontinuous multifunction,
then F admits a continuous selection.

When X is a metric space (thus also a paracompact space) and Y is a separable Banach space, we can produce a whole sequence of continuous selectors of F which is dense in $F(x)$ for all $x \in X$.

Proposition 2.59
If X is a metric space, Y is a separable Banach space and $F\colon X \longrightarrow P_{fc}(Y)$ is a lower semicontinuous multifunction,
then there exists a sequence $\{f_n\}_{n\geqslant 1}$ of continuous selectors of F such that

$$F(x) = \overline{\{f_n(x)\}_{n\geqslant 1}} \qquad \forall\, x \in X.$$

In general Theorem 2.58 is optimal in the sense that the hypotheses of the theorem cannot be relaxed. However, if Y is a separable Banach space, then we can enlarge the range space of F.

Proposition 2.60
If X is a paracompact, perfectly normal topological space (see Definitions I.2.142 and I.2.137), Y is a separable Banach space, and $F\colon X \longrightarrow 2^Y \setminus \{\emptyset\}$ is a lower semicontinuous multifunction with convex values such that for all $x \in X$ either $\operatorname{int} F(x) \neq \emptyset$ or $F(x)$ is finite dimensional,
then F admits a continuous selection.

So far, we have focused on the topological properties of multifunctions. Next we turn our attention to the measure theoretic ones.

Definition 2.61
Let (Ω, Σ) be a measurable space, X a metric space and $F\colon \Omega \longrightarrow 2^X$ a multifunction.
(a) *We say that F is* **measurable**, *if for every open set $U \subseteq X$, we have $F^-(U) \in \Sigma$ (see Definition 2.35(a)).*
(b) *We say that F is* **graph measurable**, *if $\operatorname{Gr} F \in \Sigma \times \mathcal{B}(X)$ ($\mathcal{B}(X)$ being the Borel σ-field of X).*

Remark 2.62
The ***domain*** of F is defined to be the set
$$\operatorname{dom} F \;=\; \big\{\omega \in \Omega:\; F(\omega) \neq \emptyset\big\}.$$
Evidently, if F is measurable, then $\operatorname{dom} F \in \Sigma$. So, when dealing with measurable multifunctions F, we can always assume that $\operatorname{dom} F = \Omega$.

The next theorem summarizes what is known about the measurability of multifunctions with closed values.

Theorem 2.63
Let (Ω, Σ) be a measurable space, (X, d_X) a separable metric space and $F\colon \Omega \longrightarrow P_f(X)$ is a multifunction. We consider the following properties of F:
(1) $F^-(D) \in \Sigma$ for every $D \in \mathcal{B}(X)$ ($\mathcal{B}(X)$ being the Borel σ-field of X).
(2) $F^-(C) \in \Sigma$ for every closed set $C \subseteq X$.
(3) F is measurable (see Definition 2.61(a)).
(4) For every $x \in X$, the function $\omega \longrightarrow \operatorname{dist}(x, F(x))$ is Σ-measurable.
(5) F is graph measurable.
We have the following relations among the above properties:
(a) (1) $\Longrightarrow$ (2) $\Longrightarrow$ (3) $\Longleftrightarrow$ (4) $\Longrightarrow$ (5).
(b) *If X is σ-compact, then* (2) $\Longleftrightarrow$ (3).
(c) *If Σ is complete (that is, $\Sigma = \widehat{\Sigma}$ ($\widehat{\Sigma}$ being the universal σ-field; see Definition I.4.45)) and X is complete, then* (1) $\Longleftrightarrow$ (2) $\Longleftrightarrow$ (3).

Now we pass to the problem of existence of selections. In this case we look for a measurable selections.

Theorem 2.64 *(Kuratowski–Ryll Nardzewski Selection Theorem)*
<u>If</u> (Ω, Σ) is a measurable space, X is a Polish space and $F\colon \Omega \longrightarrow P_f(X)$ is a measurable multifunction,
<u>then</u> F admits a Σ-measurable selection.

Remark 2.65
In fact we can be slightly more general and instead assume that X is a separable metrizable space (that is, we drop the completeness hypothesis on X) and that F has complete values.

We can improve the above selection theorem (compare with Proposition 2.59).

Theorem 2.66
If (Ω, Σ) *is a measurable space,* X *is a Polish space and* $F\colon \Omega \longrightarrow P_f(X)$ *is a multifunction,*
then the following statements are equivalent:
(a) F *is measurable.*
(b) *There exists a sequence of* Σ*-measurable selectors* $\{f_n\colon \Omega \longrightarrow X\}_{n\geqslant 1}$ *of the multifunction* F *such that*

$$F(\omega) = \overline{\{f_n(\omega)\}}_{n\geqslant 1} \qquad \forall\, \omega \in \Omega.$$

This result leads to the following improved version of Theorem 2.63.

Theorem 2.67
Let (Ω, Σ) *be a measurable space,* (X, d_X) *a separable metric space and* $F\colon \Omega \longrightarrow P_f(X)$ *a multifunction. We consider the following statements:*
(1) $F^-(D) \in \Sigma$ *for every* $D \in \mathcal{B}(X)$ *(with* $\mathcal{B}(X)$ *being the Borel* σ*-field of* X*).*
(2) $F^-(C) \in \Sigma$ *for every closed set* $C \subseteq X$*.*
(3) F *is measurable (see Definition 2.61(a)).*
(4) For every $x \in X$*, the function* $\omega \longmapsto \operatorname{dist}(x, F(\omega))$ *is* Σ*-measurable.*
(5) There exists a sequence $\{f_n\colon \Omega \longrightarrow X\}_{n\geqslant 1}$ *of* Σ*-selectors of* F *such that*

$$F(\omega) = \overline{\{f_n(\omega)\}}_{n\geqslant 1} \qquad \forall\, \omega \in \Omega.$$

(6) F *is graph measurable.*
We have the following relations among the above properties:
(a) $(1) \implies (2) \implies (3) \implies (4) \implies (6)$.
(b) *If* X *is complete (that is,* X *is a Polish space), then* $(3) \iff (4) \iff (5)$.
(c) *If* X *is* σ*-compact, then* $(2) \iff (3)$.
(d) *If* Σ *is complete (that is,* $\Sigma = \widehat{\Sigma}$ *(*$\widehat{\Sigma}$ *being the universal* σ*-field)) and* X *is a Polish space, then all properties (1) to (6) are equivalent.*

In Theorem 2.64 (see also Theorem 2.66), we require that F has closed values. We have a second measurable selection theorem for multifunctions which need not to be closed valued.

Theorem 2.68 *(Yankov–von Neumann–Aumann Selection Theorem)*
If (Ω, Σ) *is a complete measurable space (that is,* $\Sigma = \widehat{\Sigma}$ *(*$\widehat{\Sigma}$ *being the universal* σ*-field)),* X *is a Souslin space and* $F\colon \Omega \longrightarrow 2^X \setminus \{\emptyset\}$ *is a graph measurable multifunction (see Definition 2.61),*
then F *admits a* Σ*-measurable selection.*

Remark 2.69
Recall that a Hausdorff topological space X is said to be ***Souslin***, if there is a Polish space Y and a continuous surjection from Y onto X. For locally compact spaces, the notions of Souslin space and Polish space coincide. A separable Banach space with the weak topology is a Souslin nonmetrizable space. Similarly, if X^* is the topological

dual of a separable Banach space and we furnish X^* with the w^*-topology (denoted by $X^*_{w^*}$), then $X^*_{w^*}$ is a Souslin nonmetrizable space. In fact the above theorem can be strengthened by saying that there is a whole sequence $\{f_n\}_{n\geqslant 1}$ of Σ-measurable selections of F such that $F(\omega) \subseteq \overline{\{f_n(\omega)\}}_{n\geqslant 1}$ for all $\omega \in \Omega$.

The notion of decomposability which we will introduce next formally resembles that of convexity and is important in the theory of measurable multifunctions.

Definition 2.70
Let (Ω, Σ, μ) be a σ-finite measure space, X a separable Banach space, $L^0(\Omega; X) = \{u\colon \Omega \longrightarrow X : u \text{ is } \Sigma\text{-measurable}\}$ and $D \subseteq L^0(\Omega; X)$. We say that the set D is ***decomposable****, if for every triple $(A, x_1, x_2) \in \Sigma \times D \times D$, we have*

$$\chi_A x_1 + \chi_{A^c} x_2 \ \in \ D.$$

Remark 2.71
Since $\chi_{A^c} = 1 - \chi_A$, we see that the notion of decomposability reminds us that of convexity. Only now the coefficients instead of scalars are functions.

Decomposable sets are closely related to measurable multifunctions.

Theorem 2.72
<u>If</u> (Ω, Σ, μ) is a σ-finite measure space, X is a separable Banach space and the set $D \subseteq L^p(\Omega; X)$ (with $1 \leqslant p < +\infty$) is closed,
<u>then</u> the set D is decomposable if and only if there exists a unique measurable multifunction $F\colon \Omega \longrightarrow P_f(X)$ such that

$$D \ = \ S^p_F \ = \ \{u \in L^p(\Omega; X) : \ u(\omega) \in F(\omega) \ \mu\text{-a.e.}\}.$$

Theorem 2.73
<u>If</u> (Ω, Σ, μ) is a finite nonatomic measure space (see Definition I.3.40), X is a Banach space and $D \subseteq L^p(\Omega; X)$ (with $1 \leqslant p < +\infty$) is a nonempty, decomposable and w-closed set,
<u>then</u> D is convex.

The next theorem is another remarkable consequence of decomposability and is a basic tool in variational analysis.

Theorem 2.74
<u>If</u> (Ω, Σ, μ) is a σ-finite measure space, X is a separable Banach space, $h\colon \Omega \times X \longrightarrow \overline{\mathbb{R}} = \mathbb{R} \cup \{+\infty\}$ is measurable, $F\colon \Omega \longrightarrow 2^X \setminus \{\emptyset\}$ is graph measurable, $\xi(u) = \int\limits_\Omega h(\omega, u(\omega))\, d\mu$ for all $u \in L^p(\Omega; X)$ (with $1 \leqslant p \leqslant +\infty$) and it is defined (maybe $+\infty$) for all $u \in S^p_F$ and there exists $u_0 \in S^p_F$ such that $\xi(u_0) > -\infty$,
<u>then</u>

$$\sup_{u \in S^p_F} \xi(u) \ = \ \int\limits_\Omega \sup_{x \in F(\omega)} h(\omega, x)\, d\mu.$$

Definition 2.75
Let (Ω, Σ, μ) be a σ-finite measure space, X a separable Banach space, and $F\colon \Omega \longrightarrow 2^X \setminus \{\emptyset\}$ a multifunction. We say that F is ***L^p-integrably bounded*** *if $1 < p \leqslant +\infty$ or simply* ***integrably bounded*** *if $p = 1$, when there exists $h \in L^p(\Omega)$ such that*

$$|F(\omega)| = \sup_{x \in F(\omega)} \|x\|_X \leqslant h(\omega) \quad \text{for } \mu\text{-a.a. } \omega \in \Omega.$$

Theorem 2.76
If (Ω, Σ, μ) is a σ-finite measure space, X is a separable Banach space and $F\colon \Omega \longrightarrow P_f(X)$ is a graph measurable and integrably bounded multifunction,
then the set $S_F^1 = \{u \in L^1(\Omega; X) : u(\omega) \in F(\omega) \text{ for } \mu\text{-a.a. } \omega \in \Omega\}$ is nonempty, w-compact, and convex if and only if

$$F(\omega) \in P_{wkc}(X) \quad \text{for } \mu\text{-a.a. } \omega \in \Omega.$$

Decomposability leads to a continuous selection theorem for nonconvex valued multifunctions (compare with Theorem 2.58).

Theorem 2.77
If (Ω, Σ, μ) is a σ-finite measure space, V is a separable metric space, X is a separable Banach space, and $F\colon V \longrightarrow P_f\left(L^1(\Omega; X)\right)$ is lower semicontinuous and has decomposable values,
then F admits a continuous selection.

Let $T = [0, b]$ (a bounded closed interval in $\mathbb{R}$) and let X be a separable Banach space. We introduce the following weaker norm on the Lebesgue–Bochner space $L^1(T; X)$.

Definition 2.78
The weak norm $\|\cdot\|_w$ on $L^1(T; X)$ is defined by

$$\|u\|_w = \sup_{0 \leqslant t \leqslant t' \leqslant b} \Big\| \int_t^{t'} u(s)\, ds \Big\|_X \qquad \forall\, u \in L^1(T; X).$$

Remark 2.79
An equivalent way to define the weak norm is to set

$$\|u\|_w = \sup_{0 \leqslant t \leqslant b} \Big\| \int_0^{t} u(s)\, ds \Big\|_X \qquad \forall\, u \in L^1(T; X)$$

We want to compare the weak topology on $L^1(T; X)$ with the weak norm topology. To do this, we introduce the following notion.

Definition 2.80
We say that the set $E \subseteq L^1(T;X)$ has ***property U****, if*
(a) E is uniformly integrable;
(b) for every $\varepsilon > 0$, we can find $K_\varepsilon \in P_k(X)$ such that for every $u \in E$ there exists a measurable set $T_{u,\varepsilon} \subseteq T$ with $\lambda(T \setminus T_{u,\varepsilon}) < \varepsilon$ (λ being the Lebesgue measure on $\mathbb{R}$) such that $u(t) \in K_\varepsilon$ for all $t \in T_{u,\varepsilon}$.

Remark 2.81
If $E \subseteq L^1(T;X)$ has property U, then E is relatively w-compact in $L^1(T;X)$.

Proposition 2.82
If $E \subseteq L^1(T;X)$ has property U,
then the w-topology and $\|\cdot\|_w$-topology coincide on E.

Definition 2.83
Let Y be a Polish space and let $F\colon T \times Y \longrightarrow P_{wkc}(X)$ be a multifunction. We say that F is an ***h-Carathéodory multifunction****, if*
(a) for every $y \in Y$, the multifunction $t \longmapsto F(t,y)$ is measurable;
(b) for almost all $t \in T$, the multifunction $y \longmapsto F(t,y)$ is h-continuous;
(c) for every $C \in P_k(Y)$, we can find $\varphi_C \in L^1(T)$ such that

$$|F(t,y)| = \sup_{x \in F(t,y)} \|x\|_X \leqslant \varphi_C(t) \quad \text{for a.a. } t \in T, \text{ all } y \in C.$$

Given a nonempty compact set $K \subseteq C(T;Y)$, let $\Gamma\colon K \longrightarrow P_{wkc}\big(L^1(T;X)\big)$ be defined by

$$\Gamma(y) = S^1_{F(\cdot,y(\cdot))} \qquad \forall\, y \in K.$$

Let CS^w_Γ (respectively $CS^w_{\operatorname{ext}\Gamma}$) denote the set of selectors of Γ (respectively of $\operatorname{ext}\Gamma$) which are continuous from K into $(L^1(T;X), \|\ \|_w)$ (usually denoted by $L^1_w(T;X)$).

Theorem 2.84
If $F\colon T \times Y \longrightarrow P_{wkc}(X)$ is an h-continuous Carathéodory multifunction and $K \subseteq C(T;Y)$ is a compact set, then

$$CS^w_\Gamma = \overline{CS^w_{\operatorname{ext}\Gamma}}^{\|\cdot\|_w}.$$

Definition 2.85
Let (Ω, Σ, μ) be a σ-finite measure space, X a separable Banach space and $F\colon \Omega \longrightarrow 2^X \setminus \{\emptyset\}$ a multifunction with $S^1_F \neq \emptyset$. The ***set-valued integral*** *of F (or* ***Aumann integral*** *of F) is defined by*

$$\int_\Omega F(\omega)\, d\mu = \Big\{ \int_\Omega u\, d\mu : \ u \in S^1_F \Big\}.$$

Next we introduce some convergence notions for sets and multifunctions, which arise often in variational analysis.

Definition 2.86
Let (X, d_X) be a metric space, $A \in P_f(X)$ and $\{A_n\}_{n\geqslant 1} \subseteq P_f(X)$ a sequence of sets. We say that the sequence $\{A_n\}_{n\geqslant 1}$ converges to A in the ***Hausdorff sense****, denoted by $A_n \xrightarrow{h} A$ or $h\text{-}\lim\limits_{n\to+\infty} A_n = A$ if $h(A_n, A) \longrightarrow 0$, with h being the Hausdorff metric on $P_f(X)$ (see Definition 2.49(c)).*

Remark 2.87
This mode of set-convergence is actually convergence in the metric space $(P_f(X), h)$.

Definition 2.88
Let X be a Hausdorff topological space with τ denoting the topology of X. Let $\{A_n\}_{n\geqslant 1} \subseteq 2^X \setminus \{\emptyset\}$. We define

$$\begin{aligned}
\tau\text{-}\liminf_{n\to+\infty} A_n &= \big\{x \in X:\ x = \tau\text{-}\lim_{n\to+\infty} x_n,\ x_n \in A_n,\ n \geqslant 1\big\},\\
\tau\text{-}\limsup_{n\to+\infty} A_n &= \big\{x \in X:\ x = \tau\text{-}\lim_{k\to+\infty} x_{n_k},\ x_{n_k} \in A_{n_k},\ n_1 < n_2 < \ldots\big\}.
\end{aligned}$$

*We call $\tau\text{-}\liminf\limits_{n\to+\infty} A_n$ the τ-**Kuratowski limit inferior** of the sequence $\{A_n\}_{n\geqslant 1}$ and $\tau\text{-}\limsup\limits_{n\to+\infty} A_n$ the τ-**Kuratowski limit superior** of the sequence $\{A_n\}_{n\geqslant 1}$. If $A = \tau\text{-}\liminf\limits_{n\to+\infty} A_n = \tau\text{-}\limsup\limits_{n\to+\infty} A_n$, then we say that A is the τ-**Kuratowski limit** of the sequence $\{A_n\}_{n\geqslant 1}$ and we write $A = \tau\text{-}\lim\limits_{n\to+\infty} A_n$ or $A_n \xrightarrow{K_\tau} A$.*

Remark 2.89
Evidently $\tau\text{-}\liminf\limits_{n\to+\infty} A_n \subseteq \tau\text{-}\limsup\limits_{n\to+\infty} A_n$ and the inclusion can be strict. If X is a metric space with metric d_X, then we drop the letter τ and have

$$\begin{aligned}
\liminf_{n\to+\infty} A_n &= \big\{x \in X:\ \lim_{n\to+\infty} \operatorname{dist}(x, A_n) = 0\big\},\\
\limsup_{n\to+\infty} A_n &= \big\{x \in X:\ \liminf_{n\to+\infty} \operatorname{dist}(x, A_n) = 0\big\}.
\end{aligned}$$

Moreover, if X is a first countable space (see Definition I.2.24), then

$$\tau\text{-}\limsup_{n\to+\infty} A_n = \bigcap_{k\geqslant 1} \overline{\bigcup_{n\geqslant k} A_n}.$$

Note that in general if the topology τ is clear from the context, for notational simplicity we drop the letter τ. This mode of convergence is useful in the context of locally compact spaces, but exhibits serious problems if X is an infinite dimensional Banach space. For this reason in the next definition, we mix the topologies.

Definition 2.90
Let X be a Banach space. By w (respectively s) we denote the weak (respectively strong) topology on X. Let $\{A_n\}_{n\geqslant 1} \subseteq 2^X \setminus \{\emptyset\}$ be a sequence. We say that the

sequence $\{A_n\}_{n\geqslant 1}$ *converges to* A *in the* ***Mosco sense****, denoted by* $A_n \xrightarrow{M} A$*, if we have*

$$A = s\text{-}\liminf_{n\to+\infty} A_n = w\text{-}\limsup_{n\to+\infty} A_n.$$

Remark 2.91
Evidently we have that $A_n \xrightarrow{M} A$ if and only if $A_n \xrightarrow{K_w} A$ and $A_n \xrightarrow{K_s} A$.

This mode of set convergence is more effective in the context of reflexive Banach spaces.

Definition 2.92
Let (X, d_X) *be a metric space,* $A \in P_f(X)$ *and let* $\{A_n\}_{n\geqslant 1} \subseteq P_f(X)$ *be a sequence. We say that the sequence* $\{A_n\}_{n\geqslant 1}$ *converges to* A *in the* ***Wijsman sense****, denoted by* $A_n \xrightarrow{W} A$*, if we have*

$$\operatorname{dist}(x, A_n) \longrightarrow \operatorname{dist}(x, A) \qquad \forall\, x \in X.$$

Remark 2.93
From Proposition 2.51(a), it is clear that, if $A_n \xrightarrow{h} A$, then $A_n \xrightarrow{W} A$.

Definition 2.94
Let X *be a Banach space,* $A \in P_{fc}(X)$ *and let* $\{A_n\}_{n\geqslant 1} \subseteq P_{fc}(X)$ *be a sequence. We say that the sequence* $\{A_n\}_{n\geqslant 1}$ ***converges weakly*** *(or* ***scalarly****) to* A*, denoted by* $A_n \xrightarrow{w} A$*, if for all* $x^* \in X^*$*, we have*

$$\sigma_{A_n}(x^*) \longrightarrow \sigma_A(x^*)$$

(see Definition 2.46).

Remark 2.95
From the Hörmander formula (see Proposition 2.51(b)), we see that for a sequence $\{A_n\}_{n\geqslant 1} \subseteq P_{fc}(X)$, if $A_n \xrightarrow{h} A$, then $A_n \xrightarrow{w} A$.

Recall that a sequence in $L^1(\Omega)$, which converges weakly but not strongly, oscillates wildly around its weak limit (see Problem 1.1). For vector-valued functions, we have an extremality condition which prevents such a behavior.

Proposition 2.96
If (Ω, Σ, μ) *is a finite measure space and* $\{u_n\}_{n\geqslant 1} \subseteq L^1(\Omega; \mathbb{R}^N)$ *is a sequence such that* $u_n \xrightarrow{w} u$ *in* $L^1(\Omega; \mathbb{R}^N)$ *and* $u(\omega) \in \operatorname{ext}\left\{\overline{\operatorname{conv}} \limsup_{n\to+\infty} \{u_n(\omega)\}\right\}$ *for almost all* $\omega \in \Omega$,
then $u_n \longrightarrow u$ *in* $L^1(\Omega; \mathbb{R}^N)$.

The next proposition elaborates further the above result.

Proposition 2.97
If (Ω, Σ, μ) *is a finite measure space,* $u \in L^1(\Omega; \mathbb{R}^N)$, $\{u_n\}_{n \geqslant 1} \subseteq L^1(\Omega; \mathbb{R}^N)$ *is a sequence such that* $u_n \xrightarrow{w} u$ *in* $L^1(\Omega; \mathbb{R}^N)$ *and there exists a measurable multifunction* $F: \Omega \longrightarrow P_{fc}(\mathbb{R}^N)$ *such that:*
(i) $\operatorname{dist}(u_n(\omega), F(\omega)) \longrightarrow 0$ *for* μ*-almost all* $\omega \in \Omega$;
(ii) $u(\omega) \in \operatorname{ext} F(\omega)$ *for* μ*-almost all* $\omega \in \Omega$,
then $u_n \longrightarrow u$ *in* $L^1(\Omega; \mathbb{R}^N)$.

2.1.3 Maximal Monotone Maps and Generalizations

A function $u: \mathbb{R} \longrightarrow \mathbb{R}$ is said to be nondecreasing if

$$x \leqslant y \implies u(x) \leqslant u(y).$$

Unfortunately this definition involves the order of $\mathbb{R}$ and so it cannot be extended to maps between Banach spaces. However, we can define alternatively the monotonicity of u, by saying that

$$(x-y) \cdot (u(x) - u(x)) \geqslant 0 \qquad \forall\, x, y \in \mathbb{R}.$$

This definition avoids the order structure of $\mathbb{R}$ and can be easily extended to maps from a Banach space X into its dual X^* by replacing the product with the duality brackets for the pair (X^*, X).

So, let X be a Banach space, X^* its topological dual and let $\langle \cdot, \cdot \rangle$ denote the duality brackets for the pair (X^*, X). Consider the map $A: X \supseteq D(A) \longrightarrow 2^{X^*}$. Here

$$D(A) = \big\{x \in X : A(x) \neq \emptyset\big\} \quad \text{(the \textbf{\textit{domain}} of } A).$$

Also,

$$\operatorname{Gr} A = \big\{(x, x^*) \in X \times X^* : x^* \in A(x)\big\} \quad \text{(the \textbf{\textit{graph}} of multifunction } A).$$

Based on the initial remark, we make the following definitions.

Definition 2.98
Let X *be a Banach space and let* $A: X \supseteq D(A) \longrightarrow 2^{X^*}$ *be a multivalued map.*
(a) We say that A *is* ***monotone****, if*

$$\langle u^* - v^*, u - v\rangle \geqslant 0 \qquad \forall\, (u, u^*), (v, v^*) \in \operatorname{Gr} A.$$

(b) We say that A *is* ***strictly monotone****, if*

$$\langle u^* - v^*, u - v\rangle > 0 \qquad \forall\, (u, u^*), (v, v^*) \in \operatorname{Gr} A,\ u \neq v.$$

(c) We say that A *is* ***strongly monotone****, if*

$$\langle u^* - v^*, u - v\rangle \geqslant c\|u - v\|_X^2 \qquad \forall\, (u, u^*), (v, v^*) \in \operatorname{Gr} A,$$

with some $c > 0$.

*(d) We say that A is **uniformly monotone**, if there exists a continuous function $\vartheta\colon \mathbb{R}_+ \longrightarrow \mathbb{R}_+$ which is strictly increasing, $\vartheta(0) = 0$, $\vartheta(t) \longrightarrow +\infty$ as $t \to +\infty$ and*

$$\langle u^* - v^*, u - v\rangle \geqslant \vartheta(\|u-v\|_X)\|u-v\|_X \quad \forall\,(u,u^*),(v,v^*) \in \operatorname{Gr} A.$$

*(e) We say that A is **coercive**, if $D(A)$ is bounded or $D(A)$ is unbounded and*

$$\inf_{u^* \in A(u)} \|u^*\|_* \longrightarrow +\infty \quad \text{as } \|u\|_X \to +\infty,\ u \in D(A).$$

*(f) We say that A is **strongly coercive**, if $D(A)$ is bounded or $D(A)$ is unbounded and*

$$\frac{\inf\limits_{(u,u^*)\in \operatorname{Gr} A} \langle u^*, u\rangle}{\|u\|_X} \longrightarrow +\infty \quad \text{as } \|u\|_X \to +\infty,\ u \in D(A).$$

Remark 2.99

From the above definition it is clear that we have

strong monotonicity

$\Downarrow$

uniform monotonicity

$\Downarrow$

strict monotonicity

$\Downarrow$

monotonicity

and also

uniform monotonicity

$\Downarrow$

strong coercivity

$\Downarrow$

coercivity.

The next definition identifies a subclass of monotone maps which is central in the theory and in applications.

Definition 2.100

*Let X be a Banach space and let $A\colon X \supseteq D(A) \longrightarrow 2^{X^*}$ be a monotone map. We say that A is **maximal monotone**, if*

$$\big[\langle u^* - v^*, u - v\rangle \geqslant 0 \qquad \forall\,(u,u^*) \in \operatorname{Gr} A\big] \implies (v,v^*) \in \operatorname{Gr} A.$$

Remark 2.101

The above definition says that A is maximal monotone if and only if $\operatorname{Gr} A$ is maximal with respect to inclusion among the graphs of monotone maps.

Definition 2.102
Let X be a Banach space and let $A\colon X \supseteq D(A) \longrightarrow 2^{X^*}$ be a map. We say that the map A is **locally bounded** *at $u \in D(A)$, if there exist $\varrho > 0$ and $M > 0$ such that*

$$\|v^*\|_* \leqslant M \qquad \forall\, v^* \in A(v),\ v \in D(A) \cap \overline{B}_\varrho(u).$$

We say that A is **bounded,** *if it maps bounded sets in $X \cap D(A)$ to bounded sets in X^*. Also, if $C \subseteq X$ and $u \in C$, then we say that u is an* **absorbing point** *of C, if*

$$X = \bigcup_{\lambda>0} \lambda(C - u).$$

Remark 2.103
If $\operatorname{int} C \neq \emptyset$, then every interior point is absorbing. However, the set C can have absorbing points even if $\operatorname{int} C = \emptyset$.

Proposition 2.104
If X is a Banach space, $A\colon X \supseteq D(A) \longrightarrow 2^{X^*}$ is a monotone map, and $u \in D(A)$ is an absorbing point of $D(A)$,
then A is locally bounded at u.

Proposition 2.105
If X is a Banach space, $A\colon X \supseteq D(A) \longrightarrow 2^{X^*}$ is a maximal monotone map and $\operatorname{int} D(A) \neq \emptyset$,
then $A|_{\operatorname{int} D(A)}$ is upper semicontinuous from X with the norm topology into X^* with the w^*-topology.

Proposition 2.106
If X is a Banach space and $A\colon X \supseteq D(A) \longrightarrow 2^{X^*}$ is a maximal monotone map,
then for every $u \in D(A)$, the set $A(u) \subseteq X^*$ is nonempty, convex, and w^*-closed.

Combining Propositions 2.104 and 2.106, we obtain the following result.

Corollary 2.107
If X is a reflexive Banach space, $A\colon X \supseteq D(A) \longrightarrow 2^{X^*}$ is a maximal monotone map, and $u \in D(A)$ is absorbing,
then $A(u)$ is nonempty, convex, and w-compact.
In particular if $D(A) = X$, then A has values in $P_{wkc}(X^*)$.

The next result is a partial converse of Proposition 2.105.

Proposition 2.108
If X is a Banach space, $A\colon X \supseteq D(A) \longrightarrow 2^{X^*}$ is a monotone map with $D(A) = X$, has w^*-closed and convex values and for every $u, h \in X$, the multifunction $t \longmapsto A(u + th)$ is upper semicontinuous from $[0, 1]$ into X^* with the w^*-topology,
then A is maximal monotone.

Remark 2.109
If A is upper semicontinuous from X with the norm topology into X^* with the w^*-topology, then the previous proposition holds.

Definition 2.110
Let $A\colon X \longrightarrow X^*$ be a single valued map with $D(A) = X$.
(a) We say that A is ***demicontinuous,*** *if for any sequence $\{x_n\}_{n\geqslant 1} \subseteq X$ with $x_n \longrightarrow x$, we have $A(x_n) \xrightarrow{w^*} A(x)$.*
(b) We say that A is ***hemicontinuous,*** *if for any $x, h \in X$, the map $t \longmapsto A(x + th)$ is continuous from $[0, 1]$ into X^* with the w^*-topology.*

As a consequence of Proposition 2.108, we have

Corollary 2.111
If X is a Banach space and $A\colon X \longrightarrow X^*$ is a monotone and hemicontinuous map, then A is maximal monotone.

Definition 2.112
Let X be a Banach space. The map $\mathcal{F}\colon X \longrightarrow 2^{X^*}$ defined by

$$\mathcal{F}(x) \;=\; \{x^* \in X^* : \langle x^*, x\rangle = \|x\|_X^2 = \|x^*\|_*^2\}$$

is called the ***(normalized) duality map*** *on X.*

Remark 2.113
As a consequence of the Hahn–Banach theorem (see Corollary I.5.26), we have that $D(\mathcal{F}) = X$. Moreover, $\mathcal{F}$ has closed and convex values.

The duality map is essentially dependent on the norm of X. In fact, two equivalent norms on X need not generate the same duality map. The next proposition shows that the geometry of X and X^* is crucial concerning the properties of $\mathcal{F}$.

Proposition 2.114
(a) If X is a reflexive Banach space and X^* is strictly convex (see Definition I.5.168 and Remark I.5.169),
then $\mathcal{F}$ is single valued, bounded, odd, demicontinuous, and maximal monotone.
(b) If X is a reflexive Banach space and both X and X^* are strictly convex,
then $\mathcal{F}$ is strictly monotone.
(c) If X is a reflexive Banach space and X^* is locally uniformly convex,
then $\mathcal{F}$ is continuous.
(d) If X^* is uniformly convex,
then $\mathcal{F}$ is uniformly continuous on bounded sets.

Remark 2.115
The Troyanski renorming theorem is a basic tool in the theory of maximal monotone maps (see Theorem I.5.192). We recall that this theorem says that every reflexive

Banach space X be can equivalently renormed so that both X and X^* are locally uniformly convex and have Fréchet differentiable norms. Note that equivalently renorming X and X^* does not affect important properties of the map A, such as maximal monotonicity and coercivity (simple or strong).

Using the duality map, we can have a criterion for the maximality of a map $A\colon X \supseteq D(A) \longrightarrow 2^{X^*}$.

Theorem 2.116
If X is a reflexive Banach space with X and X^* both strictly convex (see Definition I.5.168 and Remark I.5.169) and $A\colon X \supseteq D(A) \longrightarrow 2^{X^*}$ is monotone, then A is maximal monotone if and only if $R(A + \mathcal{F}) = X^*$.

Maximal monotone operators are very useful because they exhibit remarkable surjectivity properties.

Definition 2.117
Let X be a Banach space and let $A\colon X \supseteq D(A) \longrightarrow 2^{X^*}$ be a multivalued map. We define $A^{-1}\colon X^* \supseteq D^* \longrightarrow 2^X$ by setting

$$\operatorname{Gr} A^{-1} = \big\{(x^*, x) \in X^* \times X : \ (x, x^*) \in \operatorname{Gr} A\big\}.$$

Using this notion, we have the following criterion for the surjectivity of a maximal monotone map.

Theorem 2.118
If X is a reflexive map and $A\colon X \supseteq D(A) \longrightarrow 2^{X^*}$ is maximal monotone, then A is surjective if and only if A^{-1} is locally bounded.

Evidently, if A is coercive (see Definition 2.98(e)), then A^{-1} is locally bounded. So, from Theorem 2.118 we deduce the following fundamental surjectivity result for maximal monotone maps.

Theorem 2.119
If X is a reflexive Banach space and $A\colon X \supseteq D(A) \longrightarrow 2^{X^*}$ is maximal monotone and coercive, then A is surjective.

Corollary 2.120
If X is a reflexive Banach space and $A\colon X \supseteq D(A) \longrightarrow X^*$ is monotone, hemicontinuous and coercive, then A is surjective.

In finite dimensional spaces, we can drop the monotonicity requirement.

Proposition 2.121
If X is a finite dimensional Banach space and $A\colon X \supseteq D(A) \longrightarrow P_{kc}(X^*)$ is upper semicontinuous and strongly coercive (see Definition 2.98(f)), then A is surjective.

The surjectivity results lead to some useful single-valued approximations of a maximal monotone map. These approximations are more useful in the context of pivot Hilbert spaces (a Hilbert space is ***pivot***, if it is identified with its dual).

Definition 2.122
Let H be a pivot Hilbert space (i.e., $H^* = H$) and let $A\colon H \supseteq D(A) \longrightarrow 2^H$ be a maximal monotone map. For every $\lambda > 0$, we define

$$\begin{aligned} J_\lambda^A &= (I_H + \lambda A)^{-1} \quad \textit{(the } \textbf{resolvent} \textit{ of } A\textit{)} \\ A_\lambda &= \tfrac{1}{\lambda}(I_H - J_\lambda^A)^{-1} \quad \textit{(the } \textbf{Yosida approximation} \textit{ of } A\textit{).} \end{aligned}$$

Proposition 2.123
If H is a pivot Hilbert space, $A\colon H \supseteq D(A) \longrightarrow 2^H$ is a maximal monotone map and $\lambda > 0$, then
(a) $D(J_\lambda^A) = D(A_\lambda) = H$.
(b) J_λ^A is nonexpansive (i.e., Lipschitz continuous with Lipschitz constant 1).
(c) $A_\lambda(x) \in A(J_\lambda^A(x))$ for all $x \in H$.
(d) A_λ is monotone and Lipschitz continuous with Lipschitz constant $\frac{1}{\lambda}$.
(e) $\|A_\lambda(x)\|_H \leqslant \|A^0(x)\|_H$ for all $x \in D(A)$, where $A^0(x) \in A(x)$ is the unique element of $A(x)$ with smallest norm.
(f) $\lim\limits_{\lambda \searrow 0} A_\lambda(x) = A^0(x)$ for all $x \in D(A)$.
(g) $\overline{D(A)}$ is convex and $\lim\limits_{\lambda \searrow 0} J_\lambda^A(x) = \operatorname{proj}_{\overline{D(A)}}(x)$ for all $x \in H$ (here $\operatorname{proj}_{\overline{D(A)}}$ denotes the metric projection on the closed convex set $\overline{D(A)}$).

Remark 2.124
If $x \notin D(A)$, then $\|A_\lambda(x)\|_H \longrightarrow +\infty$ and $\lambda \searrow 0$.

Perturbation results for maximal monotone operators are important in applications. We start with two results in Hilbert spaces.

Theorem 2.125
If H is a pivot Hilbert space (i.e., $H^* = H$) and $A\colon H \supseteq D(A) \longrightarrow 2^H$, $B\colon H \supseteq D(B) \longrightarrow 2^H$ are two maximal monotone maps such that $D(A) \cap D(B) \neq \emptyset$ and

$$0 \leqslant (y, B_\lambda(x))_H \qquad \forall\, (x, y) \in \operatorname{Gr} A,\ \lambda > 0,$$

then $A + B$ is maximal monotone.

Theorem 2.126
If H is a pivot Hilbert space (i.e., $H^* = H$) and $A\colon H \supseteq D(A) \longrightarrow 2^H$, $B\colon H \supseteq D(B) \longrightarrow 2^H$ are two maximal monotone maps such that

$$0 \in \operatorname{int}\big(D(A) - D(B)\big),$$

then $A + B$ is maximal monotone.

For maps defined on a reflexive Banach space with values in its dual, we have the following result.

Theorem 2.127
If X is a reflexive Banach space and $A\colon X \supseteq D(A) \longrightarrow 2^{X^*}$, $B\colon X \supseteq D(B) \longrightarrow 2^{X^*}$ are two maximal monotone maps such that

$$\operatorname{int} D(A) \cap D(B) \neq \emptyset,$$

then $A + B$ is maximal monotone.

Remark 2.128
Since $\operatorname{int} D(A) - D(B) \subseteq \operatorname{int}\big(D(A) - D(B)\big)$, we see that the hypothesis of Theorem 2.126 is weaker than that of Theorem 2.127. Note that $0 \in \operatorname{int}(D(A) - D(B))$ can happen even if $\operatorname{int} D(A) = \operatorname{int} D(B) = \emptyset$.

Next we pass to generalizations of the notion of maximal monotonicity.

Definition 2.129
Let X be a reflexive Banach space and let $A\colon X \longrightarrow 2^{X^*}$ be a multifunction. We say that A is ***pseudomonotone****, if*

(a) *$D(A) = X$ and for every $x \in X$, $A(x) \in P_{wkc}(X)$;*
(b) *A is upper semicontinuous from every finite dimensional subspace of X into X^* with the weak topology;*
(c) *if $x_n \xrightarrow{w} x$ in X, $x_n^* \in A(x_n)$ for all $n \geqslant 1$ and $\limsup\limits_{n\to+\infty} \langle x_n^*, x_n - x\rangle \leqslant 0$, then for each $v \in X$, we can find $y^*(v) \in A(x)$ such that*

$$\langle y^*(v), x - v\rangle \leqslant \liminf_{n\to+\infty} \langle x_n^*, x_n - v\rangle\,.$$

Remark 2.130
A completely continuous map $A\colon X \longrightarrow X^*$ is pseudomonotone. Also, if X is finite dimensional, then any continuous operator $A\colon X \longrightarrow X^*$ is pseudomonotone.

Definition 2.131
Let X be a reflexive Banach space and let $A\colon X \supseteq D(A) \longrightarrow 2^{X^*}$ be a multifunction. We say that A is ***generalized pseudomonotone****, if for every sequences $\{x_n\}_{n\geqslant 1} \subseteq X$, $\{x_n^*\}_{n\geqslant 1} \subseteq X^*$ such that $x_n \xrightarrow{w} x$ in X and $x_n^* \xrightarrow{w} x^*$ in X^* with $x_n^* \in A(x_n)$ for all $n \geqslant 1$ and $\limsup\limits_{n\to+\infty} \langle x_n^*, x_n - x\rangle \leqslant 0$, we have*

$$x^* \in A(x) \quad \text{and} \quad \langle x_n^*, x_n\rangle \longrightarrow \langle x^*, x\rangle\,.$$

Proposition 2.132
If X is a reflexive Banach space and $A\colon X \longrightarrow 2^{X^*}$ is a pseudomonotone map, then A is generalized pseudomonotone.

There is a version of the converse to this proposition.

Proposition 2.133
If X is a reflexive Banach space and $A\colon X \longrightarrow P_{fc}(X^*)$ is a bounded and generalized pseudomonotone map, then A is pseudomonotone.

Proposition 2.134
If X is a reflexive Banach space and $A\colon X \supseteq D(A) \longrightarrow P_{fc}(X^*)$ is a maximal monotone map with $D(A) = X$, then A is pseudomonotone.

The property of pseudomonotonicity is preserved by addition.

Proposition 2.135
If X is a reflexive Banach space and $A_1, A_2\colon X \longrightarrow 2^{X^*}$ are two pseudomonotone maps, then $A_1 + A_2$ is pseudomonotone.

Pseudomonotone maps exhibit remarkable surjectivity properties and for this reason are important in applications.

Theorem 2.136
If X is a reflexive Banach space and $A\colon X \longrightarrow 2^{X^*}$ is a pseudomonotone and strongly coercive map (see Definition 2.98(f)), then A is surjective.

A notion closely related to pseudomonotonicity and to generalized pseudomonotonicity is that of an $(S)_+$-operator, which is particularly useful in the application of variational methods. In particular it is the right property to check the Palais–Smale condition for a functional $\varphi \in C^1(X)$ (see Definition 5.45(a) in Chap. 5).

Definition 2.137
*Let X be a reflexive Banach space, $C \subseteq X$ a nonempty subset and let $A\colon C \longrightarrow X^*$ be a map. We say that A is an $(S)_+$-**map**, if for every sequence $\{u_n\}_{n\geqslant 1} \subseteq C$ such that $x_n \xrightarrow{w} x$ in X and $\limsup\limits_{n\to+\infty} \langle A(x_n), x_n - x\rangle \leqslant 0$, we have $x_n \longrightarrow x$ in X.*

The next proposition summarizes some basic properties of $(S)_+$-maps.

Proposition 2.138
Let X be a reflexive Banach space and let $C \subseteq X$ be a nonempty subset.

(a) If $A\colon C \longrightarrow X^*$ is an $(S)_+$-map and $\lambda > 0$, then λA is an $(S)_+$-map too.

(b) *If* $A, B: C \longrightarrow X^*$ *are two* $(S)_+$*-maps and one of them is also demicontinuous,* *then* $A + B$ *is an* $(S)_+$*-map too.*

(c) *If* $A: C \longrightarrow X^*$ *is an* $(S)_+$*-map and* $B: X \longrightarrow X^*$ *is a monotone map,* *then* $A + B: C \longrightarrow X^*$ *is an* $(S)_+$*-map.*

(d) *If* $A: C \longrightarrow X^*$ *is an* $(S)_+$*-map and* $B: X \longrightarrow X^*$ *is a completely continuous map,*

then $A + B: C \longrightarrow X^*$ *is an* $(S)_+$*-map.*

2.1.4 Accretive Maps

So far we have considered maps from a Banach space X into its dual X^*. Now, we focus on maps from X into itself and introduce the notion of accretive maps. Such operators are closely related to the generation theory of semigroups of operators (linear and nonlinear alike).

Definition 2.139
Let X *be a Banach space and let* $A: X \supseteq D(A) \longrightarrow 2^X$ *be a multivalued map.*

(a) *We say that* A *is* ***accretive****, if for every* $(x, u), (y, v) \in \operatorname{Gr} A$ *we can find* $x^* \in \mathcal{F}(x - y)$ *such that* $\langle x^*, u - v\rangle \geqslant 0$ *(where* $\mathcal{F}: X \longrightarrow 2^{X^*}$ *is the normalized duality map; see Definition 2.112).*

(b) *We say that* A *is* ***maximal accretive****, if its graph is maximal with respect to inclusion among the graphs of all accretive maps.*

(c) *We say that* A *is* m***-accretive****, if* $R(I_X + A) = X$*.*

(d) *We say that* A *is* ***dissipative*** *(respectively* m***-dissipative****), if* $-A$ *is accretive (respectively* m*-accretive).*

Remark 2.140
If H is a pivot Hilbert space (i.e., $H^* = H$), then $\mathcal{F} = I_H$ and A is accretive if and only if A is monotone. Moreover, the notions of maximal accretivity and m-accretivity coincide and correspond to maximal monotonicity (see Definition 2.100 and Theorem 2.116).

Proposition 2.141
If X *is a Banach space and* $A: X \supseteq D(A) \longrightarrow 2^X$ *is a multivalued map,* *then* A *is an accretive map if and only if*

$$\|x - y + \lambda(u - v)\|_X \geqslant \|x - y\|_X \qquad \forall\, \lambda > 0,\ (x, u), (y, v) \in \operatorname{Gr} A.$$

In analogy to Definition 2.112, we introduce the following single-valued approximation of the identity and of the map itself.

Definition 2.142
Let X be a Banach space and let $A\colon X \supseteq D(A) \longrightarrow 2^X$ be an accretive map. For $\lambda > 0$ we define

$$\begin{aligned} J_\lambda(x) &= (I_X + \lambda A)^{-1}(x), \\ A_\lambda(x) &= \tfrac{1}{\lambda}(I_X - J_\lambda)(x), \\ D(A_\lambda) &= D_\lambda \;=\; R(I_X + \lambda A), \\ |A(x)| &= \inf_{u \in A(x)} \|u\|_X. \end{aligned}$$

As an easy consequence of Proposition 2.141, we have

Proposition 2.143
If X is a Banach space and $A\colon X \supseteq D(A) \longrightarrow 2^X$ is a multivalued map,
then A is accretive if and only if for every $\lambda > 0$, J_λ is single-valued and nonexpansive, i.e.,

$$\|J_\lambda(x) - J_\lambda(y)\|_X \leqslant \|x - y\|_X \qquad \forall\, x, y \in D_\lambda.$$

Moreover, for any accretive map $A\colon X \supseteq D(A) \longrightarrow 2^X$, we have
(a) *$A_\lambda\colon R(I_X + \lambda A) \longrightarrow X$ is single-valued, accretive and Lipschitz continuous;*
(b) *$A_\lambda(u) \in A(J_\lambda(u))$ for all $u \in R(I_X + \lambda A)$;*
(c) *$\|A_\lambda(u)\|_X \leqslant \inf\limits_{v \in A(y)} \|v\|_X$ for all $v \in D(A) \cap R(I_X + \lambda A)$;*
(d) *$\lim\limits_{\lambda \searrow 0} J_\lambda(u) = u$ for all $u \in D(A) \cap \big(\bigcap\limits_{\lambda>0} R(I_X + \lambda A)\big)$.*

Proposition 2.144
If X is a Banach space and $A\colon X \supseteq D(A) \longrightarrow 2^X$ is an m-accretive map,
then A is maximal accretive and $R(I_X + \lambda A) = X$ for all $\lambda > 0$.

Definition 2.145
Let X be a Banach space and let $A\colon X \supseteq D(A) \longrightarrow 2^X$ be a multivalued map. The ***minimal section*** *of A is the map $A^0\colon X \supseteq D(A^0) \longrightarrow 2^X$ defined by*

$$A^0(x) = \big\{u \in A(x) : \|u\|_X = |A(x)|\big\}.$$

Proposition 2.146
If X is a reflexive and strictly convex Banach space, the dual X^* is strictly convex too (see Definition I.5.168 and Remark I.5.169) and $A\colon X \supseteq D(A) \longrightarrow 2^X$ is maximal accretive,
then $D(A^0) = D(A)$ and A^0 is single-valued.

The importance of A^0 is illustrated by the next theorem.

Theorem 2.147
If X is a Banach space with uniformly convex dual X^* and $A, B\colon X \supseteq D \longrightarrow 2^X$ are two m-accretive maps such that

$$A^0(x) \cap B^0(x) \neq \emptyset \qquad \forall\, x \in D,$$

then $A = B$. Hence, if $A^0 = B^0$, then $A = B$.

The next theorem is analogous to Theorem 2.125.

Theorem 2.148
If X is a Banach space with uniformly convex dual X^* and $A\colon X \supseteq D(A) \longrightarrow 2^X$, $B\colon X \supseteq D(B) \longrightarrow 2^X$ are two m-accretive maps such that: ⊎
(i) $D(A) \cap D(B) \neq \emptyset$,
(ii) $\langle \mathcal{F}(B_\lambda(x)), y\rangle \geqslant 0$ *for all* $(x, y) \in \operatorname{Gr} A$ *and* $\lambda > 0$,
then $A + B$ *is* m*-accretive.*

As we already mentioned m-accretive maps are related to the generation of contraction semigroups of operators.

Definition 2.149
*Let X be a Banach space and let $\{S(t)\}_{t\geqslant 0} \subseteq \mathcal{L}(X)$. We say that the family $\{S(t)\}_{t\geqslant 0}$ is a C_0-**semigroup**, if the following is true:*
(a) $S(0) = I_X$,
(b) $S(t + \tau) = S(t) \circ S(\tau)$ *for all* $t, \tau \geqslant 0$ *(semigroup property),*
(c) for every $x \in X$, $S(t)x \longrightarrow x$ *as* $t \searrow 0$.
If $\|S(t)\|_{\mathcal{L}} \leqslant 1$ *for all* $t \geqslant 0$, *then* $\{S(t)\}_{t\geqslant 0}$ *is called a* ***contraction semigroup***.

Definition 2.150
Let X be a Banach space and let $\{S(t)\}_{t\geqslant 0}$ be a C_0-semigroup. The ***infinitesimal generator*** *of the semigroup is a linear operator A defined by*

$$A(x) = \lim_{t\searrow 0} \frac{S(t)x - x}{t} \qquad \forall\, x \in D(A),$$

with $D(A) = \Big\{x \in X : \lim\limits_{t\searrow 0} \frac{S(t)x-x}{t} \text{ exists}\Big\}$.

Proposition 2.151
If X is a Banach space and $\{S(t)\}_{t\geqslant 0}$ is a C_0-semigroup with infinitesimal operator A,
then A is closed and densely defined.

Remark 2.152
If $A \in \mathcal{L}(X)$, then A is the infinitesimal generator of

$$S(t) = e^{At} = \sum_{k\geqslant 0} \frac{(At)^k}{k!}.$$

For an unbounded linear operator A to be the infinitesimal generator of a C_0-semigroup, the precise conditions are provided by the following theorem.

Theorem 2.153 *(Hille–Yosida Theorem)*
Necessary and sufficient conditions for a linear unbounded operator A on a Banach space X to be the infinitesimal generator of a C_0-semigroup satisfying that

$$\|S(t)\|_{\mathcal{L}} \leqslant Me^{\omega t} \qquad \forall\, t \geqslant 0,$$

for some $M \geqslant 1$ and $\omega \geqslant 0$, are
(a) A is closed and densely defined;
(b) $(\lambda I_X - A)^{-1}$ exists for $\lambda > \omega$;
(c) $\|(\lambda I_X - A)^{-m}\|_{\mathcal{L}} \leqslant \frac{M}{(\lambda-\omega)^m}$ for $\lambda > \omega$, $m \geqslant 1$.

Remark 2.154
The family of the operators $R(\lambda) = (\lambda I_X - A)^{-1}$ for $\lambda > \omega$ is called the ***resolvent*** of A.

Proposition 2.155
If X is a Banach space and $\{S(t)\}_{t\geqslant 0} \subseteq \mathcal{L}(X)$ is a contraction semigroup with infinitesimal generator A,
then for any $x \in X$,

$$S(t)x = \lim_{n\to+\infty} \left(I_X - \tfrac{t}{n}A\right)^{-n} x = \lim_{n\to+\infty} \left(\tfrac{n}{t}R(\tfrac{n}{t})\right)^n x.$$

We can have also nonlinear semigroups.

Definition 2.156
*Let X be a Banach space and let $C \subseteq X$ be a nonempty subset. A family of nonlinear maps $\{S(t)\colon C \longrightarrow C\}_{t\geqslant 0}$ is said to be a **semigroup of nonexpansive maps** on C, if*
(a) $\|S(t)x - S(t)y\|_X \leqslant \|x-y\|_X$ for all $t \geqslant 0$, $x, y \in C$;
(b) $S(t+\tau) = S(t) \circ S(\tau)$ for all $t, \tau \geqslant 0$ (semigroup property),
(c) $S(0) = I_C$,
(d) $\lim_{t\searrow 0} S(t)x = x$ for every $x \in C$.

Remark 2.157
The map $(t, x) \longmapsto S(t)x$ is jointly continuous from $\mathbb{R}_+ \times C$ into C.

The main generation theorem for nonlinear semigroups reads as follows.

Theorem 2.158
If $A\colon X \supseteq D(A) \longrightarrow 2^X$ is an m-accretive operator,
then for every $x \in \overline{D(A)}$,

$$S(t)x = \lim_{n\to+\infty} \left(I_X + \tfrac{t}{n}A\right)^{-n} x$$

exists uniformly in t on compact sets in $\mathbb{R}_+$, $\{S(t)\}_{t\geqslant 0}$ is a semigroup of nonexpansive maps and for each $x \in D(A)$ and $t > 0$, we have

$$\|S(t)x - x\|_X \leqslant t|A(x)|$$

(see Definition 2.142). Moreover, if X and X^* are both uniformly convex and $u_0 \in D(A)$, then

(a) *the map* $t \longmapsto S(t)u_0$ *is right differentiable on* $\mathbb{R}_+ = [0,+\infty)$ *and we have*

$$\frac{d^+}{dt^+}S(t)u_0 \;=\; AS(t)u_0 \qquad \forall\, t \geqslant 0;$$

(b) *the map* $t \longmapsto \|AS(t)u_0\|_X$ *is nonincreasing on* $\mathbb{R}_+$ *and the map* $t \longmapsto AS(t)u_0$ *is right continuous at every* $t \geqslant 0$;

(c) $\frac{d}{dt}S(t)u_0 = AS(t)u_0$ *exists and is continuous on* $\mathbb{R}_+$ *except for a countable set of point.*

Definition 2.159
Let X *be a Banach space.*

(a) *A nonlinear semigroup* $\{S(t)\colon C \longrightarrow C\}_{t\geqslant 0}$ *of nonexpansive maps on* C *is said to be* ***compact****, if for every* $t > 0$, $S(t)$ *is compact.*

(b) *A nonlinear semigroup* $\{S(t)\colon C \longrightarrow C\}_{t\geqslant 0}$ *of nonexpansive maps on* C *is said to be* ***equicontinuous****, if for every bounded set* $B \subseteq X$*, the family* $\{S(\cdot)x :\ x \in B\}$ *is equicontinuous at every* $t > 0$.

Remark 2.160
Recall that $S(0) = I_C$. So, if in Definition 2.159(a), $S(t)$ is compact for all $t \geqslant 0$, then the set C imbeds compactly in X. In particular, if $C = X$, then X is finite dimensional.

Proposition 2.161
If X *is a Banach space and* $A\colon X \supseteq D(A) \longrightarrow 2^X$ *is accretive,*
then for all $x \in X$ *and all* $\lambda, \mu > 0$*, we have*

$$J_\lambda(x) \;=\; J_\mu\big(\tfrac{\mu}{\lambda}x + \tfrac{\lambda-\mu}{\lambda}J_\lambda(x)\big)$$

(resolvent identity).

Remark 2.162
Using the resolvent identity we see that J_λ is compact for all $\lambda > 0$ if and only if it is compact for some $\lambda > 0$.

2.1.5 Miscellaneous Results

Theorem 2.163
If X *is a Banach space and* $C \subseteq X$*,*
then C *is relatively compact if and only if there is a sequence* $\{u_n\}_{n\geqslant 1} \subseteq X$ *converging to the origin such that* $C \subseteq \overline{\operatorname{conv}}\,\{u_n\}_{n\geqslant 1}$.

Theorem 2.164
A set $K \subseteq c_0$ *is relatively compact if and only if it is bounded and the limit* $\lim\limits_{n\to+\infty} u_n$ *exists uniformly for* $\{u_n\}_{n\geqslant 1} \in K$.

Theorem 2.165 *(Lyapunov Convexity Theorem)*

(a) *If* (Ω, Σ) *is a measurable space,* X *is a finite dimensional Banach space and* $m\colon \Sigma \longrightarrow X$ *is a vector measure which is nonatomic,*
then the set $m(\Sigma) \subseteq X$ *is compact and convex.*

(b) *If* (Ω, Σ) *is a measurable space,* X *is a Banach space with the RNP and* $m\colon \Sigma \longrightarrow X$ *is a vector measure which is nonatomic and of bounded variation,*
then the set $\overline{m(\Sigma)} \subseteq X$ *is compact and convex.*

The Lusin theorem (see Theorem I.3.77) says that an $\mathbb{R}$-valued Borel function is actually continuous outside a "small" set. Below we provide an abstract version of this result for maps with values in a separable metric space.

Theorem 2.166 *(Lusin Theorem II)*
If X *is a Polish space,* Y *is a separable metric space,* $f\colon X \longrightarrow Y$ *is a Borel measurable function and* μ *is a finite Borel measure on* X,
then for every $\varepsilon > 0$*, we can find a compact set* $K_\varepsilon \subseteq X$ *such that* $\mu(X \setminus K_\varepsilon) < \varepsilon$ *and* $f|_{K_\varepsilon}$ *is continuous.*

Remark 2.167
Recall that every finite Borel measure on a Polish space is Radon (see Theorem I.4.12).

Proposition 2.168
If (Ω, Σ, μ) *is a finite measure space and* $\mathcal{D}$ *is a family of* Σ*-measurable* $\mathbb{R}_+$*-valued functions,*
then there exists a unique (modulo μ*-a.e. equality)* Σ*-measurable function* $h\colon \Omega \longrightarrow \mathbb{R}_+$ *such that:*

(a) $f(\omega) \leqslant h(\omega)$ *for* μ*-almost all* $\omega \in \Omega$ *and all* $f \in \mathcal{D}$;

(b) *if* $g\colon \Omega \longrightarrow \mathbb{R}_+$ *is* Σ*-measurable and* $f(\omega) \leqslant g(\omega)$ *for* μ*-almost all* $\omega \in \Omega$ *and* $f \in \mathcal{D}$*, then* $h(\omega) \leqslant g(\omega)$ *for* μ*-almost all* $\omega \in \Omega$.

Moreover, there exists a sequence $\{f_n\}_{n\geqslant 1} \subseteq \mathcal{D}$ *such that*

$$h(\omega) = \sup_{n\geqslant 1} f_n(\omega) \quad \text{for } \mu\text{-a.a. } \omega \in \Omega.$$

Finally, if $\mathcal{D}$ *is upward directed (that is, if* $f_1, f_2 \in \mathcal{D}$*, then there exists* $\widehat{f} \in \mathcal{D}$ *such that* $f_1 \leqslant \widehat{f}$ *and* $f_2 \leqslant \widehat{f}$*), then the sequence* $\{f_n\}_{n\geqslant 1}$ *can be chosen to be increasing.*

2.2 Problems

Problem 2.1*
Let X and Y be two Banach spaces and let $A\colon X \longrightarrow Y$ be a linear map such that for each bounded sequence $\{u_n\}_{n\geqslant 1} \subseteq X$, the sequence $\{A(u_n)\}_{n\geqslant 1} \subseteq Y$ admits a strongly convergent subsequence. Show that $A \in \mathcal{L}_c(X;Y)$.

Problem 2.2*
Suppose that X and Y are two Banach spaces, $D \subseteq X$ is a nonempty subset and $\{f_\alpha\colon D \longrightarrow Y\}_{\alpha\in J}$ is a net of compact maps (i.e., $\{f_\alpha\}_{\alpha\in J} \subseteq K(D;Y)$) such that $f_\alpha \longrightarrow f$ uniformly on bounded subsets of D. Show that $f \in K(D;Y)$.

Problem 2.3*
Suppose that X is a Banach space, $D \subseteq X$ is a nonempty bounded and closed subset, and $f\colon D \longrightarrow X$ is a compact map. Suppose that given $\varepsilon > 0$, we can find $u_\varepsilon \in D$ such that $\|u_\varepsilon - f(u_\varepsilon)\|_X < \varepsilon$. Show that f has a fixed point in D, i.e., there exists $v \in D$ such that $v = f(v)$.

Problem 2.4**
Let X and Y be two Banach spaces and let $A \in \mathcal{L}_c(X;Y)$. Show that:
(a) the subspace $A(X) \subseteq Y$ is separable;
(b) if $A(X)$ is of second category in itself, then $A \in \mathcal{L}_f(X;Y)$.

Problem 2.5**
Let X and Y be two metric spaces and let $f \in C(X;Y)$. Show that the following two conditions are equivalent:
(a) Every sequence $\{u_n\}_{n\geqslant 1} \subseteq X$ such that $f(u_n) \longrightarrow y \in Y$ admits a subsequence $\{u_{n_k}\}_{k\geqslant 1}$ such that $u_{n_k} \longrightarrow u \in X$.
(b) f is a closed map and for every $y \in Y$, the set $f^{-1}(y) \subseteq Y$ is compact.

Problem 2.6**
Let X and Y be two metric spaces and suppose that $f \in C(X;Y)$ satisfies one of the equivalent properties **(a)** or **(b)** of Problem 2.5. Show that f is proper.

Problem 2.7**
Let X and Y be two metric spaces with X being compact. Show that every function $f \in C(X;Y)$ is closed and proper.

Problem 2.8**
Let $\{\vartheta_n\}_{n\geqslant 1} \subseteq \mathbb{R}$ be a sequence such that $\vartheta_n \longrightarrow 0$ and let $K\colon l^2 \longrightarrow l^2$ be the linear operator defined by

$$K(u) = \{\vartheta_n x_n\}_{n\geqslant 1} \qquad \forall\, u = \{x_n\}_{n\geqslant 1} \in l^2.$$

Show that $K \in \mathcal{L}_c(l^2)$.

Problem 2.9*
Suppose that X is a Banach space, $A \in \mathcal{L}_c(X)$ and $A^2 = A$ (that is, A is a projection; see Definition 2.27(c)). Show that $A \in \mathcal{L}_f(X)$.

Problem 2.10*
Suppose that X is a Banach space, $D \subseteq X$ is a nonempty closed and bounded subset, and $f\colon D \longrightarrow X$ is a condensing map (that is, $f \in S(D)$). Show that $I_X - f$ is proper.

Problem 2.11***
Let X be an infinite dimensional Banach space and let $B_1 = \{u \in X : \|u\|_X = 1\}$. Show that $\alpha(B_1) = 2$ and $\beta(B_1) = 1$.

Problem 2.12**
Suppose that X is a Banach space, $\mathcal{B}$ is the family of bounded subsets of X, $\{C_n\}_{n\geqslant 1} \subseteq \mathcal{B}$ is a decreasing sequence of sets (i.e., $C_{n+1} \subseteq C_n$ for all $n \geqslant 1$), and $\lim\limits_{n\to+\infty} \gamma(C_n) = 0$, where $\gamma = \alpha$ or $\gamma = \beta$. Show that the set $\bigcap\limits_{n\geqslant 1} \overline{C}_n$ is nonempty and compact.

Problem 2.13**
Suppose that X is a reflexive Banach space, Y is a Banach space, $A \in \mathcal{L}_c(X;Y)$, and $C \subseteq X$ is a nonempty, bounded, closed, and convex set. Show that the set $A(C) \subseteq Y$ is compact.

Problem 2.14*
Suppose that X and Y are two Banach spaces, $A \in \mathcal{L}_c(X;Y)$ and $L \in \mathcal{L}(Y;X)$. Show that $AL \in \mathcal{L}_c(Y)$ and $LA \in \mathcal{L}_c(X)$.

Problem 2.15**
Let X and Y be two Banach spaces and let $A \in \mathcal{L}(X;Y)$. Suppose that there exists $c > 0$ such that

$$\|A(u)\|_Y \geqslant c\|u\|_X \qquad \forall\, u \in X.$$

Show that A is compact if and only if X is finite dimensional.

Problem 2.16**
Let X be a Banach space and let $A \in \mathcal{L}_c(X) \setminus \mathcal{L}_f(X)$. Show that $0 \in \overline{A(\partial B_1)}$ (recall that $\partial B_1 = \{x \in X : \|x\|_X = 1\}$).

Problem 2.17*
Let $S\colon l^2 \longrightarrow l^2$ be the forward shift operator, i.e.,

$$S(u_1, u_2, \ldots) = (0, u_1, u_2, \ldots) \qquad \forall\, \{u_n\}_{n\geqslant 1} \in l^2.$$

Evidently $S \in \mathcal{L}(l^2)$. Is S compact? Justify your answer.

Problem 2.18**

Suppose that X and Y are two Banach spaces, $A \in \mathcal{L}(X;Y)$ and A is also continuous from X with the weak topology into Y with the strong topology. Show that $A \in \mathcal{L}_f(X;Y)$.

Problem 2.19*

Let X be a reflexive Banach space and let $A \in \mathcal{L}_c(X)$. Show that there exists $u_0 \in X$ with $\|u_0\|_X \leqslant 1$ such that $\|A\|_{\mathcal{L}} = \|A(u_0)\|_X$.

Problem 2.20**

Find a Banach space X and $A \in \mathcal{L}(X)$ such that A is not compact but A^2 is.

Problem 2.21***

Let H be a Hilbert space and let $A \in \mathcal{L}(H)$. Show that the following properties are equivalent:

(a) $A \in \mathcal{L}_c(H)$.

(b) For every orthonormal basis $\{e_\alpha\}_{\alpha\in J}$ of H and $\varepsilon > 0$, the set

$$\{\alpha \in J : \ |(A(e_\alpha), e_\alpha)_H| \geqslant \varepsilon\}$$

is finite.

(c) There is a sequence $\{A_n\}_{n\geqslant 1} \subseteq \mathcal{L}_f(H)$ such that $\|A - A_n\|_{\mathcal{L}} \longrightarrow 0$.

Problem 2.22**

Let (S, Σ_1, μ) and (T, Σ_2, ξ) be two finite separable measure spaces and let $K \in L^2(T \times S)$. Consider the integral operator $L\colon L^2(S) \longrightarrow L^2(T)$ defined by

$$L(u)(t) = \int_S K(t,s)u(s)\,d\mu \qquad \forall\, t \in T.$$

Show that L is compact.

Problem 2.23**

Let $h \in C([0,1])$, $h \neq 0$ and let $A\colon C([0,1]) \longrightarrow C([0,1])$ be defined by

$$A(u)(t) = h(t)u(t) \qquad \forall\, u \in C([0,1]),\ t \in [0,1].$$

Show that A is well defined, linear, continuous but not compact.

Problem 2.24**

Show that l^1 is embedded continuously in l^2 and examine if the embedding is compact.

Problem 2.25**

Let X be a reflexive Banach space. Show that every $A \in \mathcal{L}(X; l^1)$ is compact.

Problem 2.26***

Show that the space $\mathcal{L}_c(l^2)$ is not reflexive.

Problem 2.27**
Let X and Y be two Banach spaces with X of infinite dimension and let $A \in \mathcal{L}_c(X;Y)$. Show that there exists a sequence $\{u_n\}_{n\geqslant 1} \subseteq X$ such that $\|u_n\|_X = 1$ for all $n \geqslant 1$ and $\|A(u_n)\|_Y \longrightarrow 0$.

Problem 2.28**
Let X be a reflexive Banach space and let $A \in \mathcal{L}(c_0;X)$. Show that A is compact (recall that

$$c_0 = \{\{u_n\}_{n\geqslant 1} : \{u_n\}_{n\geqslant 1} \text{ is a real sequence such that } u_n \longrightarrow 0\}$$

with the norm $\|\{u_n\}_{n\geqslant 1}\|_{c_0} = \sup\limits_{n\geqslant 1} |u_n|$).

Problem 2.29**
Suppose that X and Y are two Banach spaces, $A \in \mathcal{L}_c(X;Y)$ and $R(A) \subseteq Y$ is closed. Show that $A \in \mathcal{L}_f(X;Y)$. In addition, if $\dim \ker A < +\infty$, then X is finite dimensional.

Problem 2.30**
Let X and Y be two Banach spaces and let $A \in \mathcal{L}_{wc}(X;Y)$ (see Definition 2.1(d)). Show that $R(A)$ is closed if and only if $R(A)$ is reflexive.

Problem 2.31***
Let X and Y be two Banach spaces and let $A \in \mathcal{L}(X;Y)$. Show that the following four properties are equivalent:

(a) A is weakly compact (see Definition 2.1(d)).

(b) $A^{**}(X^{**}) \subseteq Y$.

(c) $A^*\colon Y^*_{w^*} \longrightarrow X^*_w$ is continuous (here by $Y^*_{w^*}$ we denote the space Y^* furnished with the w^*-topology and by X^*_w, the space X^* furnished with its w-topology).

(d) $A^*\colon Y^* \longrightarrow X^*$ is weakly compact.

Problem 2.32***
Let X and Y be two Banach spaces and let $A \in \mathcal{L}(X;Y)$. Show that $A \in \mathcal{L}_c(X;Y)$ if and only if there exists a sequence $\{u_n^*\}_{n\geqslant 1} \subseteq X^*$ such that $\|u_n^*\|_{X^*} \longrightarrow 0$ and

$$\|A(u)\|_Y \leqslant \sup_{n\geqslant 1} |\langle x_n, u\rangle| \qquad \forall\, u \in X$$

Problem 2.33**
Let X be a reflexive Banach space and let $f\colon X \longrightarrow X^*$ be a coercive map satisfying the condition: "for any sequence $\{u_n\}_{n\geqslant 1} \subseteq X$ such that $u_n \xrightarrow{w} u$ in X and $\lim\limits_{n\to+\infty} \langle f(u_n) - f(u), u_n - u\rangle = 0$, we have that $u_n \longrightarrow u$ in X." Show that f is proper.

Problem 2.34**
Let X and Y be two Banach spaces and let $A \in \mathcal{L}(X;Y)$. Show that $A \in \mathcal{L}_c(X;Y)$ if and only if A factors with compact factors through a closed subspace of c_0.

Problem 2.35**
Let X be a Banach space and let $A \in \mathcal{L}(c_0, X)$. Show that $A \in \mathcal{L}_{wc}(c_0, X)$ if and only if $A \in \mathcal{L}_c(c_0; X)$.

Problem 2.36**
Suppose that X is a reflexive Banach space, Y is a Banach space, $A \in \mathcal{L}_c(X; Y)$, $C \subseteq X$ is a nonempty, bounded, closed, and convex set and $y \in Y$. Show that there exists $u_0 \in C$ such that

$$\|A(u_0) - y\|_Y = \inf_{u \in C} \|A(u) - y\|_Y.$$

Problem 2.37**
Let X and Y be two Banach spaces and let $A \in \mathcal{L}(X; Y)$. Show that $A \in \mathcal{L}_f(X; Y)$ if and only if $A^* \in \mathcal{L}_f(Y^*; X^*)$. Moreover $\dim R(A) = \dim R(A^*)$.

Problem 2.38**
Let Y be a Banach space and let $A \in \mathcal{L}(l^1; X)$. Show that A is weakly compact if and only if the sequence $\{A(e_n)\}_{n \geqslant 1} \subseteq Y$ is relatively w-compact (here $\{e_n\}_{n \geqslant 1}$ denotes the standard Schauder basis of l^1).

Problem 2.39**
Suppose that X is a Banach space, $\{u_n\}_{n \geqslant 1} \subseteq X$ is a relatively w-compact sequence and $A \colon l^1 \longrightarrow X$ is defined by

$$A(x) = \sum_{n \geqslant 1} x_n u_n \qquad \forall\, x = \{x_n\}_{n \geqslant 1} \in l^1.$$

Show that A is weakly compact.

Problem 2.40**
Let X be an infinite dimensional Banach space and let $A \in \mathcal{L}_c(X)$. Show that $0 \in \sigma(A)$.

Problem 2.41**
Let $A \in \mathcal{L}(L^2(0,1))$ be defined by

$$A(u)(t) = tu(t) \qquad \forall\, t \in [0,1].$$

Show that $\sigma_p = \emptyset$ (that is, A has no eigenvalues) and $[0,1] \subseteq \sigma(A)$.

Problem 2.42**
Let H be a Hilbert space and let $A \in \mathcal{L}(H)$ be self-adjoint. Show that:

(a) $\|A\|_{\mathcal{L}} = \sup\limits_{\|u\|_H \leqslant 1} (A(u), u)_H$;

(b) all eigenvalues of A are real and eigenvectors corresponding to distinct eigenvalues are orthogonal.

Problem 2.43**
Suppose that H is a Hilbert space, $A \in \mathcal{L}(H)$ is a self-adjoint operator and $\lambda \in \mathbb{C}$. Show that $\lambda \in \sigma(A)$ if and only if

$$\inf_{\|u\|_H=1} \|(\lambda I_H - A)(u)\|_H = 0.$$

Problem 2.44**
Suppose that H is a Hilbert space, $A \in \mathcal{L}(H)$ is a self-adjoint operator and

$$m = \inf_{\|u\|_H=1} (A(u), u)_H \quad \text{and} \quad M = \sup_{\|u\|_H=1} (A(u), u)_H .$$

Show that $\sigma(A) \subseteq [m, M]$ and $m, M \in \sigma(A)$.

Problem 2.45*
Suppose that X is a Banach space and $A \in \mathcal{L}(X)$ is such that $A \neq 0$, $A \neq I_X$ and A is a projection (see Definition 2.27(c)). Show that $\sigma_p(A) = \sigma(A) = \{0, 1\}$.

Problem 2.46**
Suppose that X is a Banach space, $A \in \mathcal{L}(X)$ and $\xi \in \mathbb{C} \setminus \{0\}$. Assume that the series $\sum\limits_{n\geqslant 0} \xi^{-(n+1)} A^n(u)$ converges for every $u \in X$. Show that $\xi \in \varrho(A)$.

Problem 2.47**
Let $X = L^p(0,1)$ with $1 \leqslant p < +\infty$ and let $A\colon X \longrightarrow X$ be defined by

$$A(u)(t) = \int_0^t u(s)\,ds \qquad \forall\, t \in (0,1).$$

Show that $A \in \mathcal{L}_c(X)$ and find $\sigma(A)$ and $\sigma_p(A)$.

Problem 2.48**
Let H be a Hilbert space and let $A \in \mathcal{L}(H)$ be a self-adjoint isomorphism of H which is positive (i.e., $(A(u), u)_H \geqslant 0$ for all $u \in H$). Show that

$$\langle u, h \rangle = (A(u), h)_H \qquad \forall\, u, h \in H$$

defines a new inner product on H and $|u| = \langle u, u \rangle^{\frac{1}{2}}$ is an equivalent norm on H.

Problem 2.49*
Let X and Y be two Banach spaces and let $L \in \mathcal{L}(X;Y)$. Suppose that $T, S \in \mathcal{L}(Y;X)$ are such that $L \circ T = I_Y + K$ and $S \circ L = I_X + \widehat{K}$ with K and $\widehat{K}$ compact operators. Show that $L \in \operatorname{Fred}(X;Y)$.

Problem 2.50*
Let X and Y be two Banach spaces, $L \in \operatorname{Fred}(X;Y)$ and $K \in \mathcal{L}_c(X;Y)$. Show that $L + K \in \operatorname{Fred}(X;Y)$ and $i(K) = i(L + K)$.

Problem 2.51**
Let X be a Hausdorff topological space and let $\varphi\colon X \longrightarrow \mathbb{R}$ be a function. Show that:
(a) φ is a lower semicontinuous function if and only if the multifunction $u \longmapsto E_\varphi(u)$ is upper semicontinuous (where $E_\varphi(u) = \{\lambda \in \mathbb{R} : \varphi(u) \leqslant \lambda\}$);
(b) φ is an upper semicontinuous function if and only if the multifunction $u \longmapsto E_\varphi(u)$ is lower semicontinuous.

Problem 2.52**
Let X and Y be two Hausdorff topological spaces and let $F\colon X \longrightarrow 2^Y \setminus \{\emptyset\}$ be a multifunction with connected values which is upper semicontinuous or lower semicontinuous. Show that F maps connected sets to connected sets.

Problem 2.53*
Let X and Y be two Hausdorff topological spaces and let $F\colon X \longrightarrow 2^Y \setminus \{\emptyset\}$ be a multifunction. We introduce the multifunction $\overline{F}\colon X \longrightarrow P_f(X)$ defined by

$$\overline{F}(u) = \overline{F(u)} \qquad \forall\, u \in X.$$

Show that F is lower semicontinuous if and only if $\overline{F}$ is lower semicontinuous. Is the result true for upper semicontinuous functions? Justify your answer.

Problem 2.54**
Suppose that X is a Hausdorff topological space, Y is a normed space and $F\colon X \longrightarrow 2^Y \setminus \{\emptyset\}$ is a lower semicontinuous multifunction. Show that the multifunctions $u \longmapsto \operatorname{conv} F(u)$ and $u \longmapsto \overline{\operatorname{conv}}\, F(u)$ are both lower semicontinuous.

Problem 2.55*
Suppose that X is a Hausdorff topological space, Y is a Banach space and $F\colon X \longrightarrow P_k(Y)$ is an upper semicontinuous multifunction. Show that the multifunction $u \longmapsto \overline{\operatorname{conv}}\, F(u)$ is upper semicontinuous too.

Problem 2.56**
Let X and Y be two Hausdorff topological spaces and let $F\colon X \longrightarrow 2^Y \setminus \{\emptyset\}$ be a multifunction. Show that:
(a) if F is $P_k(Y)$-valued and upper semicontinuous and $K \in P_k(X)$, then $F(K) \in P_k(Y)$;
(b) if $F\colon X \longrightarrow 2^Y \setminus \{\emptyset\}$ is closed, $G\colon X \longrightarrow P_k(Y)$ is upper semicontinuous and $F(u) \cap G(u) \neq \emptyset$ for all $u \in X$, then $X \ni u \longmapsto H(u) = F(u) \cap G(u) \in 2^Y \setminus \{\emptyset\}$ is upper semicontinuous.
(c) Is the intersection of two lower semicontinuous multifunctions necessary a lower semicontinuous multifunction? Justify your answer.

Problem 2.57**
Let (X, d_X) be a metric space and suppose that $\varphi \in C(X)$ is in the closure of $\{\operatorname{dist}(\cdot, E) : E \in P_f(X)\}$ for the topology of pointwise convergence. Assume that $D = \{x \in X : \varphi(x) = 0\} \neq \emptyset$ and for each $u \in X$ we have $\operatorname{dist}(u, D) \leqslant \varphi(u)$. Show that $\varphi(\cdot) = \operatorname{dist}(\cdot, D)$

Problem 2.58**
Let (X, d_X) be a metric space and let $\{C_n\}_{n\geqslant 1} \subseteq P_f(X)$ be a sequence such that $C_n \xrightarrow{h} C \in P_k(X)$. Show that

$$C = \bigcap_{m\geqslant 1} \overline{\bigcup_{n\geqslant m} C_n} = \bigcap_{\varepsilon>0} \bigcup_{m\geqslant 1} \bigcap_{n\geqslant m} (C_n)_\varepsilon,$$

where $(C_n)_\varepsilon = \{u \in X : \operatorname{dist}(u, C_n) < \varepsilon\}$.

Problem 2.59**
Let (X, d_X) be a metric space and let $K \subseteq (P_f(X), h)$ be compact. Show that the set $\bigcup_{C\in K} C \subseteq X$ is closed.

Problem 2.60**
Let (X, d_X) be a metric space and let $\mathcal{F}$ be the family of finite subsets of X. Show that $\mathcal{F}$ is dense in $(P_f(X), h)$ if and only if X is totally bounded.

Problem 2.61*
Suppose that X, Y, and V are three Hausdorff topological spaces, $\varphi \colon X \times Y \longrightarrow V$ is a continuous map, and $M \colon X \longrightarrow 2^Y \setminus \{\emptyset\}$ is a lower semicontinuous multifunction. We set

$$F(u) = \varphi(u, M(u)) \qquad \forall\, u \in X.$$

Show that $u \longmapsto F(u)$ is lower semicontinuous form X into $2^V \setminus \{\emptyset\}$.

Problem 2.62*
Suppose that X, Y, and V are three metric spaces, $\varphi \colon X \times Y \longrightarrow V$ is a continuous map, and $M \colon X \longrightarrow P_k(Y)$ is an upper semicontinuous multifunction. We set

$$F(u) = \varphi(u, M(u)) \qquad \forall\, u \in X.$$

Show that $u \longmapsto F(u)$ is upper semicontinuous form X into $P_k(V)$.

Problem 2.63*
Let (X, d_X) be a metric space and suppose that $(P_f(X), h)$ is compact. Show that X is compact.

Problem 2.64**
Suppose that X and Y are two Hausdorff topological spaces, $\varphi \colon X \times Y \longrightarrow \mathbb{R}^* = \mathbb{R} \cup \{\pm\infty\}$ is a lower semicontinuous function and $F \colon X \longrightarrow 2^Y \setminus \{\emptyset\}$ is a lower semicontinuous multifunction. Let

$$v(u) = \sup_{y\in F(u)} \varphi(u, y)$$

(the ***value function***). Show that $v \colon X \longrightarrow \mathbb{R}^*$ is lower semicontinuous.

Problem 2.65**
Let X and Y be two Hausdorff topological spaces and let $F\colon X \longrightarrow P_f(Y)$ be a multifunction which is closed (see Definition 2.40) and locally compact (see Proposition 2.45). Show that F is upper semicontinuous.

Problem 2.66**
Suppose that X and Y are two Hausdorff topological spaces, $\varphi\colon X \times Y \longrightarrow \mathbb{R}$ is a continuous function, and $F\colon X \longrightarrow P_k(X)$ is a continuous multifunction. We define

$$\begin{aligned} v(u) &= \sup_{y \in F(u)} \varphi(u,y) \qquad \forall\, u \in X, \\ S(u) &= \{y \in F(x) :\, v(u) = \varphi(u,y)\} \qquad \forall\, u \in X. \end{aligned}$$

Show that $v\colon X \longrightarrow \mathbb{R}$ is continuous and $S\colon X \longrightarrow P_k(Y)$ is upper semicontinuous.

Problem 2.67*
Show that in general a lower semicontinuous multifunction need not be h-lower semicontinuous and an h-upper semicontinuous multifunction need not be upper semicontinuous (cf. Propositions 2.54 and 2.55).

Problem 2.68*
Suppose that X is a Hausdorff topological space, (Y, d_Y) is a metric space, and $F\colon X \longrightarrow P_f(Y)$ is an h-upper semicontinuous multifunction. Show that F is closed.

Problem 2.69*
Suppose that X is a Hausdorff topological space, (Y, d_Y) is a metric space and $F\colon X \longrightarrow 2^Y \setminus \{\emptyset\}$ is an h-upper semicontinuous multifunction. Show that for every $y \in X$ the function $u \longmapsto \varphi_y(u) = \operatorname{dist}_Y(y, F(u))$ is lower semicontinuous.

Problem 2.70**
Let (X, d_X) be a complete metric space and let $F\colon X \longrightarrow P_k(X)$ be a multifunction such that

$$h(F(u), F(v)) \leqslant k d_X(u,v) \qquad \forall\, u, v \in X,$$

with $k > 0$. For any compact set $K \subseteq X$, we set $R_F(K) = \bigcup_{u \in K} F(u)$. Show that:

(a) $R_F(K) \in P_k(X)$;
(b) $R_F\colon P_k(X) \longrightarrow P_k(X)$ is Lipschitz continuous with Lipschitz constant k (on $P_k(X)$ we consider the Hausdorff metric corresponding to d_X).

Problem 2.71**
Suppose that X and Y are two Hausdorff topological spaces, $F\colon X \longrightarrow 2^Y \setminus \{\emptyset\}$ is a lower semicontinuous multifunction, and $G\colon X \longrightarrow 2^Y \setminus \{\emptyset\}$ is a multifunction with open graph. Suppose that

$$F(u) \cap G(u) \neq \emptyset \qquad \forall\, u \in X.$$

Show that the multifunction $u \longmapsto F(u) \cap G(u)$ is lower semicontinuous.

Problem 2.72**

Let X be a paracompact, perfectly normal topological space, let Y be a Banach space, and let $F\colon X \longrightarrow 2^Y \setminus \{\emptyset\}$ be a lower semicontinuous multifunction with convex values. Show that for every $r > 0$, there exists a continuous map $h\colon X \longrightarrow Y$ such that

$$\mathrm{dist}_Y(h(u), F(u)) < r \qquad \forall\, u \in X.$$

Problem 2.73**

Suppose that X is a compact Hausdorff topological space, Y is a locally convex vector space, and $F\colon X \longrightarrow 2^Y \setminus \{\emptyset\}$ is a multifunction with convex values. Assume that for every $y \in Y$, the set $F^-(\{y\}) = \{u \in X : \; y \in F(u)\}$ is open. Show that there exists a continuous map $f\colon X \longrightarrow Y$ such that $f(u) \in F(u)$ for all $u \in X$.

Problem 2.74**

Let X and Y be two Banach spaces and let $A \in \mathcal{L}(X;Y)$ be surjective. Show that there exists a continuous map $f\colon Y \longrightarrow X$ such that $A(f(y)) = y$ for all $y \in Y$.

Problem 2.75*

Suppose that X is a paracompact space, Y is a Banach space, $C \in P_f(X)$ and $F\colon X \longrightarrow P_{fc}(Y)$ is a lower semicontinuous multifunction. Show that any continuous selection of $F|_C$ can be extended to a continuous selection of F.

Problem 2.76*

Suppose that X is metric space, $\varphi\colon X \longrightarrow \mathbb{R}$ is an upper semicontinuous function and $\psi\colon X \longrightarrow \mathbb{R}$ is a lower semicontinuous function such that $\varphi(u) < \psi(u)$ for all $u \in X$. Show that there exists a continuous function $f\colon X \longrightarrow \mathbb{R}$ such that

$$\varphi(u) < f(u) < \psi(u) \qquad \forall\, u \in X.$$

Problem 2.77**

Suppose that X is a Hausdorff topological space, Y is a topological vector space, $F\colon X \longrightarrow 2^Y \setminus \{\emptyset\}$ is a lower semicontinuous multifunction, $\varphi \in C(X;Y)$ and U is an open neighborhood of the origin in Y. Assume that

$$H(u) = F(u) \cap \big(\varphi(u) + U\big) \neq \emptyset \qquad \forall\, u \in X.$$

Show that the multifunction $u \longmapsto H(u)$ is lower semicontinuous from X into Y.

Problem 2.78**

Suppose that X is a Hausdorff topological space, Y is a topological vector space, and $F, G\colon X \longrightarrow 2^Y \setminus \{\emptyset\}$ are two lower semicontinuous multifunctions. Show that the multifunction $X \ni u \longmapsto H(u) = F(u) + G(u) \in 2^Y \setminus \{\emptyset\}$ is lower semicontinuous.

Problem 2.79**

Let X be a paracompact space and let Y be a normed space. We say that a multifunction $F\colon X \longrightarrow 2^Y \setminus \{\emptyset\}$ is ***almost lower semicontinuous*** (*a-lsc* for short), if for every $\varepsilon > 0$ and every $u \in X$, there exists $U \in \mathcal{N}(u)$ (where $\mathcal{N}(u)$ is the filter of neighborhoods of u) such that

$$\bigcap_{u' \in U} F(u')_\varepsilon \neq \emptyset$$

(recall that $F(u')_\varepsilon = \{y \in Y : \operatorname{dist}_Y(y, F(u')) < \varepsilon)\}$). Show that, if F is an almost lower semicontinuous multifunction and has convex values, then for every $\varepsilon > 0$ there exists $f_\varepsilon \in C(X;Y)$ such that $\operatorname{dist}_Y(f_\varepsilon(u), F(u)) < \varepsilon$ for all $u \in X$.

Problem 2.80*
Suppose that X is a paracompact space, Y is a Banach space, $F\colon X \longrightarrow P_{fc}(Y)$ is a lower semicontinuous multifunction and $(\widehat{u}, \widehat{y}) \in \operatorname{Gr} F$. Show that there exists a continuous selector $\widehat{f}$ of F such that $\widehat{f}(\widehat{u}) = \widehat{y}$.

Problem 2.81*
Suppose that X and Y are two Hausdorff topological spaces, $D \subseteq X$ is a set and $F\colon X \longrightarrow 2^Y \setminus \{\emptyset\}$ is a multifunction which is lower semicontinuous on $\overline{D} \setminus D$. Show that $F(\overline{D}) \subseteq \overline{F(D)}$.

Problem 2.82*
Show that the Cartesian product of two lower semicontinuous multifunctions is lower semicontinuous. Is this also true for upper semicontinuous multifunctions? Justify your answer.

Problem 2.83***
Suppose that (X, d_X) is a metric space, Y is a Banach space and $F\colon X \longrightarrow 2^Y \setminus \{\emptyset\}$ is an upper semicontinuous multifunction with convex values. Show that for every $\varepsilon > 0$, we can find a locally Lipschitz function $f_\varepsilon\colon X \longrightarrow Y$ such that

$$f_\varepsilon(X) \subseteq \operatorname{conv} F(X) \quad \text{and} \quad h^*(\operatorname{Gr} f_\varepsilon, \operatorname{Gr} F) < \varepsilon.$$

Problem 2.84***
Suppose that X is a metric space, Y is a Banach space, $W \subseteq X$ is an open set, $K \subseteq W$ is a compact set and $F\colon \overline{W} \longrightarrow 2^Y \setminus \{\emptyset\}$ is an upper semicontinuous multifunction with convex values. Show that for every $\varepsilon > 0$, there exists an open neighborhood V_ε of K and a locally Lipschitz function $f_\varepsilon\colon V_\varepsilon \longrightarrow \operatorname{conv} F(K)$ with finite dimensional range such that

$$f_\varepsilon(u) \in F(K \cap B_\varepsilon(u)) + \varepsilon B_1 \qquad \forall\, u \in V_\varepsilon.$$

Problem 2.85**
Suppose that X is a compact topological space and $F\colon X \longrightarrow P_f(X)$ is an upper semicontinuous multifunction. Show that there exists a nonempty closed set $C \subseteq X$ such that $F(C) = C$.

Problem 2.86**
Find a compact valued multifunction which is closed but not upper semicontinuous.

Problem 2.87**
Let X be a Banach space and let $K \subseteq X$ be a nonempty set which is boundedly compact (respectively boundedly w-compact), i.e., for every $r > 0$, the set $K \cap \overline{B}_r$ is compact (respectively w-compact). Let

$$\mathrm{proj}_K(u) = \{h \in K : \|u - h\|_X = \mathrm{dist}(u, K)\}.$$

Show that the multifunction $u \longmapsto \mathrm{proj}_K(u)$ (known as the ***metric projection*** on K) is upper semicontinuous from X into X (respectively from X into X_w, where X_w denotes the Banach space X furnished with the weak topology).

Problem 2.88**
Suppose that X is a Banach space, X^* is its topological dual, $\mathcal{F}\colon X \longrightarrow 2^{X^*} \setminus \{\emptyset\}$ is the duality map. Show that $\mathcal{F}$ is upper semicontinuous from X with norm topology into X^*_{w*} (X^*_{w*} denoting X^* with the w^*-topology).

Problem 2.89**
Let X be a metric space and let $F\colon X \longrightarrow P_k(X)$ be a lower semicontinuous and closed multifunction with connected values. Show that F is continuous.

Problem 2.90*
Produce an upper semicontinuous multifunction with closed and convex values which does not have a continuous selector.

Problem 2.91**
Let T and X be two separable metric spaces which are Borel sets in their respective completions (that is, T and X are Borel spaces) and let $F\colon T \longrightarrow P_{fc}(X)$ be a measurable multifunction with σ-compact values. Show that for every closed set $C \subseteq X$, we have $F^-(C) \in \mathcal{B}(X)$.

Problem 2.92*
Suppose that (Ω, Σ) is a measurable space, X is a Polish space and $\varphi\colon X \longrightarrow \Omega$ is a map such that:
(i) for every $\omega \in \Omega$, $\varphi^{-1}(\{\omega\}) \in P_f(X)$; and
(ii) for every open set $V \subseteq X$, $\varphi(V) \in \Sigma$.
Show that there exists a Σ-measurable map $f\colon \Omega \longrightarrow X$ such that $\varphi(f(\omega)) = \omega$ for all $\omega \in \Omega$.

Problem 2.93**
Suppose that (Ω, Σ) is a measurable space, X is a separable metric space and $F\colon \Omega \longrightarrow P_k(X)$ is a multifunction. Show that F is measurable if and only if $F^-(C) \in \Sigma$ for every closed set $C \subseteq X$.

Problem 2.94**
Suppose that (Ω, Σ) is a measurable space, X is a separable metric space, and $F\colon \Omega \longrightarrow P_f(X)$ is a measurable multifunction. Show that for every compact set $K \subseteq X$, we have $F^-(K) \in \Sigma$.

Problem 2.95**

Suppose that (Ω, Σ) is a measurable space, X is a separable metric space, Y is a metric space, $\varphi\colon \Omega \times X \longrightarrow Y$ is a Carathéodory map (i.e., for all $u \in X$, the function $\omega \longmapsto \varphi(\omega, u)$ is Σ-measurable, while for every $\omega \in \Omega$, the function $u \longmapsto \varphi(\omega, u)$ is continuous) and $U \subseteq Y$ is a nonempty open set. Show that the multifunction $F\colon \Omega \longrightarrow 2^X$ defined by

$$F(\omega) = \{u \in X : \varphi(\omega, u) \in U\}.$$

is measurable.

Problem 2.96**

Let X and Y be two σ-compact metric spaces and let $F\colon X \longrightarrow 2^Y \setminus \{\emptyset\}$ be a closed multifunction. Show that F is measurable.

Problem 2.97***

Suppose that (Ω, Σ) is a measurable space, X is a separable metric space and $\{F_n\colon \Omega \longrightarrow P_f(X)\}_{n\geqslant 1}$ is a sequence of measurable multifunctions. Suppose that at least one of them is compact valued. Show that the map $\omega \longmapsto H(\omega) = \bigcap_{n\geqslant 1} F_n(\omega)$ is measurable.

Problem 2.98**

Suppose that (Ω, Σ) is a measurable space, X is a Polish space, Y is a metric space, $\varphi\colon \Omega \times X \longrightarrow Y$ is a Carathéodory map, $U\colon \Omega \longrightarrow P_f(X)$ is a measurable multifunction, and

$$G(\omega) = \varphi(\omega, U(\omega)) \qquad \forall\, \omega \in \Omega.$$

Show that $G\colon \Omega \longrightarrow Y$ is measurable.

Problem 2.99**

Suppose that (Ω, Σ) is a measurable space, X is a Polish space, $F\colon \Omega \longrightarrow P_k(X)$ is a measurable multifunction, and $g\colon \Omega \longrightarrow X$ is a Σ-measurable function. Show that there exists a Σ-measurable selector f of F such that

$$\mathrm{dist}(g(\omega), F(\omega)) = d_X(g(\omega), f(\omega)) \qquad \forall\, \omega \in \Omega.$$

Problem 2.100*

Suppose that (Ω, Σ) is a measurable space, X is a σ-compact metric space, and $F\colon \Omega \longrightarrow P_f(X)$ is a multifunction such that $F^-(K) \in \Sigma$ for every $K \in P_k(X)$. Show that F is measurable.

Problem 2.101***

Suppose that (Ω, Σ) is a complete measurable space, X is a Polish space, Y is a σ-compact Polish space, and $F\colon \Omega \longrightarrow P_f(X \times Y)$ is a multifunction. We define the multifunction $G\colon \Omega \times X \longrightarrow P_f(Y)$ by setting

$$G(\omega, u) = \{y \in Y : (u, y) \in F(\omega)\} \qquad \forall\, (\omega, u) \in \Omega \times X.$$

Show that F is measurable if and only if G is measurable.

Problem 2.102*

Suppose that (Ω, Σ) is a complete measurable space, X is a separable Banach space and $F\colon \Omega \longrightarrow 2^X \setminus \{\emptyset\}$ is a graph measurable multifunction (see Definition 2.61(b)). Show that for every $u^* \in X^*$, the map

$$\omega \longmapsto \sigma_{F(\omega)}(u^*) \;=\; \sup_{u \in F(\omega)} \langle u^*, u \rangle$$

is Σ-measurable.

Problem 2.103***

Suppose that (Ω, Σ) is a measurable space, X is a separable Banach space, and $F\colon \Omega \longrightarrow P_{wkc}(X)$ is a multifunction. Show that F is measurable if and only if for every $u^* \in X^*$, the map $\omega \longmapsto \sigma_{F(\omega)}(u^*)$ in Σ-measurable.

Problem 2.104**

Suppose that (Ω, Σ) is a complete measurable space, X is a Souslin space $\varphi\colon \Omega \times X \longrightarrow \mathbb{R}^*$ is a jointly measurable function and $F\colon \Omega \longrightarrow 2^X \setminus \{\emptyset\}$ is a graph measurable multifunction. We define

$$m(\omega) \;=\; \inf_{u \in F(\omega)} \varphi(\omega, u) \qquad \forall\, \omega \in \Omega.$$

Show that $m\colon \Omega \longrightarrow \mathbb{R}^*$ is Σ-measurable.

Problem 2.105**

(a) Suppose that (Ω, Σ) is a measurable space, X is a separable Banach space and $F\colon \Omega \longrightarrow P_f(X)$ is a measurable multifunction. Show that multifunctions $\omega \longmapsto \operatorname{conv} F(\omega)$ and $\omega \longmapsto \overline{\operatorname{conv}}\, F(\omega)$ are both measurable.

(b) Suppose that (Ω, Σ) is a complete measurable space, X is a separable Banach space and $F\colon \Omega \longrightarrow 2^X \setminus \{\emptyset\}$ is a graph measurable multifunction. Show that the multifunction $\omega \longmapsto \overline{\operatorname{conv}}\, F(\omega)$ is measurable.

Problem 2.106*

Suppose that (Ω, Σ) is a measurable space, X is a separable Banach space, and $u\colon \Omega \longrightarrow X$ and $\varrho\colon \Omega \longrightarrow (0, +\infty)$ are Σ-measurable functions. Show that the multifunction

$$\Omega \ni \omega \longmapsto B_{\varrho(\omega)}(u(\omega)) = \{y \in X : \; \|y - u(\omega)\|_X < \varrho(\omega)\}$$

is measurable.

Problem 2.107**

Suppose that (Ω, Σ) is a measurable space, X is a compact metric space, and $\varphi\colon \Omega \times X \longrightarrow \mathbb{R}$ is a Carathéodory function (see Problem 2.95). Show that the multifunction $F\colon \Omega \longrightarrow 2^X$ defined by

$$F(\omega) \;=\; \{u \in X : \; \varphi(\omega, u) = 0\}$$

is measurable.

Problem 2.108**

We suppose that (Ω, Σ) and $(T, \mathcal{T})$ are two measurable spaces with Σ being complete (i.e., $\Sigma = \widehat{\Sigma}$; see Definition I.4.45), X is a Souslin space, $F\colon \Omega \longrightarrow 2^X \setminus \{\emptyset\}$, and $G\colon \Omega \longrightarrow 2^T \setminus \{\emptyset\}$ are two graph measurable multifunctions, $h\colon \Omega \times X \longrightarrow T$ is a $(\Sigma \times \mathcal{B}, \mathcal{T})$-measurable map and

$$h(\omega, F(\omega)) \cap G(\omega) \neq \emptyset \qquad \forall\, \omega \in \Omega.$$

Show that there exists a Σ-measurable selector $f\colon \Omega \longrightarrow X$ of F such that

$$h(\omega, f(\omega)) \in G(\omega) \qquad \forall\, \omega \in \Omega.$$

Problem 2.109*

Suppose that (Ω, Σ) is a complete measurable space, X is a Polish space and $\{F_n\colon \Omega \longrightarrow P_f(X)\}_{n\geqslant 1}$ is a sequence of measurable multifunctions. Show that the multifunctions

$$\omega \longmapsto G(\omega) = \overline{\bigcup_{n\geqslant 1} F_n(\omega)} \quad \text{and} \quad \omega \longmapsto H(\omega) = \bigcap_{n\geqslant 1} F_n(\omega)$$

are both measurable.

Problem 2.110**

Suppose that (Ω, Σ, μ) is a σ-finite measure space, X is a separable Banach space and $F\colon \Omega \longrightarrow 2^X \setminus \{\emptyset\}$ is a multifunction with $\operatorname{Gr} F \in \Sigma_\mu \times \mathcal{B}(X)$ (Σ_μ being the μ-completion of Σ; see Definition I.4.45). Show that for $1 \leqslant p \leqslant +\infty$, we have $S_F^p \neq \emptyset$ (see Theorem 2.72) if and only if

$$\inf_{u \in F(\omega)} \|u\|_X \leqslant \vartheta(\omega) \quad \text{for } \mu\text{-a.a. } \omega \in \Omega,$$

with $\vartheta \in L^p(\Omega)_+$.

Problem 2.111**

Suppose that (Ω, Σ, μ) is a complete σ-finite measure space, X is a Polish space and $\{F_n\colon \Omega \longrightarrow P_f(X)\}_{n\geqslant 1}$ is a sequence of measurable multifunctions. Show that the multifunctions

$$\omega \longmapsto \liminf_{n\to+\infty} F_n(\omega) \quad \text{and} \quad \omega \longmapsto \limsup_{n\to+\infty} F_n(\omega)$$

(see Definition 2.88) are both measurable.

Problem 2.112**

Suppose that (Ω, Σ, μ) is a σ-finite measure space, X is a separable Banach space and $F\colon \Omega \longrightarrow 2^X \setminus \{\emptyset\}$ a graph measurable multifunction such that $S_F^p \neq \emptyset$ with $1 \leqslant p < \infty$. Show that we can find a sequence $\{f_n\}_{n\geqslant 1} \subseteq S_F^p$ such that

$$F(\omega) \subseteq \overline{\{f_n(\omega)\}_{n\geqslant 1}} \quad \text{for } \mu\text{-a.a. } \omega \in \Omega.$$

Problem 2.113**
Suppose that (Ω, Σ, μ) is a σ-finite measure space, X is a separable Banach space, $F\colon \Omega \longrightarrow P_f(X)$ a graph measurable multifunction such that $S_F^p \neq \emptyset$ with $1 \leqslant p < +\infty$, $\{f_n\}_{n\geqslant 1} \subseteq S_F^p$ is a sequence such that

$$F(\omega) = \overline{\{f_n(\omega)\}_{n\geqslant 1}} \quad \text{for } \mu\text{-a.a. } \omega \in \Omega$$

(see Problem 2.112), $f \in S_F^p$ and $\varepsilon > 0$. Show that there exists a finite Σ-partition $\{A_1, \ldots, A_l\}$ of Ω such that

$$\Big\| f - \sum_{k=1}^{l} \chi_{A_k} f_k \Big\|_X < \varepsilon.$$

Problem 2.114**
Suppose that (Ω, Σ, μ) is a σ-finite measure space, X is a separable Banach space and $F, G\colon \Omega \longrightarrow P_f(X)$ are two measurable multifunctions such that $S_F^p, S_G^p \neq \emptyset$ with some $1 \leqslant p \leqslant +\infty$. Set $H(\omega) = \overline{F(\omega) + G(\omega)}$. Show that $S_H^p = \overline{S_F^p + S_G^p}$.

Problem 2.115**
Suppose that (Ω, Σ, μ) is a σ-finite measure space, X is a separable Banach space, $F\colon \Omega \longrightarrow 2^X \setminus \{\emptyset\}$ a graph measurable multifunction such that $S_F^p \neq \emptyset$ with $1 \leqslant p < +\infty$. Show that

$$\overline{\operatorname{conv}}\, S_F^p = S_{\overline{\operatorname{conv}}\, F}^p.$$

Problem 2.116**
Suppose that (Ω, Σ, μ) is a σ-finite measure space, X is a separable Banach space, $F\colon \Omega \longrightarrow 2^X \setminus \{\emptyset\}$ a graph measurable multifunction such that $S_F^p \neq \emptyset$ with $1 \leqslant p < +\infty$. Show that $\overline{S_F^p} = S_{\overline{F}}^p$.

Problem 2.117***
Suppose that (Ω, Σ, μ) is a σ-finite nonatomic measure space, X, Y are two Banach spaces with Y having the RNP (see Definition 1.45), $D \subseteq L^p(\Omega; X)$ (with $1 \leqslant p < +\infty$) is a decomposable set (see Definition 2.70) and $T \in \mathcal{L}(L^p(\Omega; X); Y)$. Show that the set $\overline{T(D)} \subseteq Y$ is convex.

Problem 2.118***
Suppose that (Ω, Σ, μ) is a σ-finite nonatomic measure space, X is a Banach space, $D \subseteq L^p(\Omega; X)$ (with $1 \leqslant p < +\infty$) is a decomposable set which is w-closed. Show that the set D is convex.

Problem 2.119**
Suppose that (Ω, Σ, μ) is a σ-finite nonatomic measure space, X is a separable Banach space, and $F\colon \Omega \longrightarrow 2^X \setminus \{\emptyset\}$ is a graph measurable multifunction, such that $S_F^p \neq \emptyset$ (with $1 \leqslant p < +\infty$). Show that $\overline{S_F^p}^w = S_{\overline{\operatorname{conv}}\, F}^p$ (here by w we denote the weak topology on the Lebesgue–Bochner space $L^p(\Omega; X)$).

Problem 2.120**
Suppose that (Ω, Σ, μ) is a σ-finite measure space, X is a separable Banach space and $F, G\colon \Omega \longrightarrow 2^X \setminus \{\emptyset\}$ are two graph measurable multifunction, such that $S_F^p = S_G^p$ for some $1 \leqslant p \leqslant +\infty$. Show that $F(\omega) = G(\omega)$ for μ-almost all $\omega \in \Omega$.

Problem 2.121**
Suppose that (Ω, Σ, μ) is a σ-finite measure space, X is a separable Banach space, and $F\colon \Omega \longrightarrow 2^X \setminus \{\emptyset\}$ is a graph measurable multifunction.

(a) Suppose that S_F^p (with $1 \leqslant p < +\infty$) is nonempty and closed. Show that $F(\omega) \in P_f(X)$ for μ-almost all $\omega \in \Omega$.

(b) Suppose that μ is nonatomic and S_F^p (with $1 \leqslant p < +\infty$) is nonempty, closed, and convex. Show that $F(\omega) \in P_{fc}(X)$ for μ-almost all $\omega \in \Omega$.

Problem 2.122**
Suppose that (Ω, Σ, μ) is a nonatomic, σ-finite measure space, X is a separable Banach space and $F\colon \Omega \longrightarrow 2^X \setminus \{\emptyset\}$ is a graph measurable multifunction with $S_F^1 \neq \emptyset$. Show that the set $\operatorname{cl} \int\limits_\Omega F\, d\mu \subseteq X$ is convex (see Definition 2.85).

Problem 2.123*
Suppose that (Ω, Σ, μ) is a σ-finite measure space, X is a separable Banach space and $F\colon \Omega \longrightarrow P_{wkc}(X)$ is a graph measurable and integrably bounded multifunction. Show that $\int\limits_\Omega F(\omega)\, d\mu \in P_{wkc}(X)$.

Problem 2.124*
Suppose that (Ω, Σ, μ) is a σ-finite measure space, X is a separable Banach space and $F\colon \Omega \longrightarrow 2^X \setminus \{\emptyset\}$ is a graph measurable multifunction with $S_F^1 \neq \emptyset$. Show that

$$\sigma_{S_F^1}(h^*) \;=\; \int\limits_\Omega \sigma_{F(\omega)}(h^*(\omega))\, d\mu \qquad \forall\, h^* \in L^\infty(\Omega; X^*_{w^*}).$$

Problem 2.125**
Suppose that V is a Polish space, X is a separable metric space, $\mu\colon \mathcal{B}(V) \longrightarrow \mathbb{R}$ is a finite Borel measure on V and $F\colon V \longrightarrow P_f(X)$ is a measurable multifunction. Show that for every $\varepsilon > 0$, there exists a compact set $K_\varepsilon \subseteq V$ with $\mu(V \setminus K_\varepsilon) < \varepsilon$ such that $F|_{K_\varepsilon}$ is closed.

Problem 2.126**
Suppose that V is a Polish space, X is a separable metric space, $\mu\colon \mathcal{B}(V) \longrightarrow \mathbb{R}$ is a finite Borel measure on V and $F\colon V \longrightarrow P_f(X)$ is a measurable multifunction. Show that for every $\varepsilon > 0$, there exists a compact set $K_\varepsilon \subseteq V$ with $\mu(V \setminus K_\varepsilon) < \varepsilon$ such that $F|_{K_\varepsilon}$ is lower semicontinuous.

Problem 2.127*
Suppose that (Ω, Σ) and $(T, \mathcal{T})$ are two measurable spaces, X is a separable metric space and $F\colon \Omega \times T \longrightarrow P_f(X)$ is a $\Sigma \times \mathcal{T}$-measurable multifunction. Fix $t_0 \in T$ and

let u_0 be a Σ-measurable selector of $F(\cdot, t_0)$. Show that given $\varepsilon > 0$, we can find a $\Sigma \times \mathcal{T}$-measurable selector $\widehat{f}$ of F such that

$$d_X(\widehat{f}(\omega, t_0), u_0(\omega)) \leqslant \varepsilon \qquad \forall\, \omega \in \Omega.$$

Problem 2.128*
Suppose that (Ω, Σ) is a complete measurable space, X is a Souslin space and $F\colon \Omega \longrightarrow 2^X$ is a graph measurable multifunction. Show that for every $D \in \mathcal{B}(X)$, we have $F^-(D) \in \Sigma$.

Problem 2.129**
Suppose that (Ω, Σ) is a complete measurable space, X is a Polish space and $F\colon \Omega \longrightarrow 2^X \setminus \{\emptyset\}$ is a graph measurable multifunction. Show that the multifunction $\omega \longmapsto \partial F(\omega)$ is measurable and $\omega \longmapsto \operatorname{int} F(\omega)$ is graph measurable (where $\partial F(\omega)$ denotes the boundary of $F(\omega)$).

Problem 2.130**
Suppose that (Ω, Σ, μ) is a σ-finite measure space, X is a separable Banach space and $F\colon \Omega \longrightarrow 2^X \setminus \{\emptyset\}$ is a graph measurable multifunction with open values such that $S_F^1 \neq \emptyset$. Show that for every set $A \in \Sigma$, the set $\int_A F\, d\mu \subseteq X$ is open.

Problem 2.131**
Suppose that (Ω, Σ, μ) is a finite measure space, X is a separable Banach space and $F\colon \Omega \longrightarrow 2^X \setminus \{\emptyset\}$ is a graph measurable multifunction such that $S_F^1 \neq \emptyset$, $F(\omega)$ is convex for all $\omega \in \Omega$ and

$$\operatorname{int} F(\omega) \neq \emptyset \quad \text{for } \mu\text{-a.a. } \omega \in \Omega.$$

Show that

$$\operatorname{int} \int_A F\, d\mu = \int_A \operatorname{int} F\, d\mu \qquad \forall\, A \in \Sigma.$$

Problem 2.132**
Suppose that (Ω, Σ, μ) is a finite measure space, X is a Banach space and $D \subseteq L^1(\Omega; X)$ is a bounded and decomposable set. Show that D is uniformly integrable.

Problem 2.133**
Let $T = [0, b]$ and let $F\colon T \longrightarrow P_{kc}(\mathbb{R}^N)$ be a graph measurable and integrably bounded multifunction. For every $t \in T$ we set

$$V(t) = \int_0^t F(s)\, ds.$$

Show that $V\colon T \longrightarrow P_{kc}(\mathbb{R}^N)$ is h-continuous and continuous and the set CS_V of continuous selectors of V is nonempty and compact in $C(T; \mathbb{R}^N)$.

Problem 2.134*
Let X be a Banach space and let $\{A_n\}_{n\geqslant 1} \subseteq 2^X \setminus \{\emptyset\}$ be a sequence of sets. Show that for every $u \in X$, we have

$$\limsup_{n\to+\infty} \operatorname{dist}(u, A_n) \leqslant \operatorname{dist}(u, s\text{-}\liminf_{n\to+\infty} A_n).$$

Problem 2.135*
Suppose that X is a Banach space, $C \subseteq X$ is a nonempty, w-closed set such that for every $r > 0$, we have $C \cap \overline{B}_r \in P_{wk}(X)$ (where $\overline{B}_r = \{u \in X : \|u\|_X \leqslant r\}$) and $\{A_n\}_{n\geqslant 1} \subseteq 2^X \setminus \{\emptyset\}$ is a sequence of sets such that $A_n \subseteq C$ for all $n \geqslant 1$. Show that for every $u \in X$, we have

$$\operatorname{dist}(u, w\text{-}\limsup_{n\to+\infty} A_n) \leqslant \liminf_{n\to+\infty} \operatorname{dist}(u, A_n).$$

Problem 2.136*
Suppose that X is a reflexive Banach space, $A \in P_{fc}(X)$, $\{A_n\}_{n\geqslant 1} \subseteq P_{fc}(X)$ is a sequence such that $A_n \xrightarrow{M} A$ and $\{u_n\}_{n\geqslant 1} \subseteq X$ is a sequence such that $u_n \longrightarrow u$ in X. Show that $\operatorname{dist}(u_n, A_n) \longrightarrow \operatorname{dist}(u, A)$.

Problem 2.137**
Suppose that (Ω, Σ) is a measurable space, X is a σ-compact metric space and $\{F_n : \Omega \longrightarrow P_f(X)\}_{n\geqslant 1}$ is a sequence of measurable multifunctions with

$$\limsup_{n\to+\infty} F_n(\omega) \neq \emptyset \qquad \forall\, \omega \in \Omega.$$

Show that the multifunction $\omega \longmapsto H(\omega) = \limsup\limits_{n\to+\infty} F_n(\omega)$ is measurable.

Problem 2.138*
Suppose that X is a Banach space, $A \in 2^X \setminus \{\emptyset\}$, $\{A_n\}_{n\geqslant 1} \subseteq 2^X \setminus \{\emptyset\}$ is a sequence such that

$$\limsup_{n\to+\infty} \operatorname{dist}(u, A_n) \leqslant \operatorname{dist}(u, A) \qquad \forall\, u \in X.$$

Show that $A \subseteq s\text{-}\liminf\limits_{n\to+\infty} A_n$.

Problem 2.139**
Assume that X is a Banach space, $C \in P_{wk}(X)$ and $\{A_n\}_{n\geqslant 1} \subseteq 2^X \setminus \{\emptyset\}$ is a sequence such that $A_n \subseteq C$ for all $n \geqslant 1$. Show that $w\text{-}\limsup\limits_{n\to+\infty} A_n \neq \emptyset$ and

$$\sigma_{w\text{-}\limsup\limits_{n\to+\infty} A_n}(u^*) = \limsup_{n\to+\infty} \sigma_{A_n}(u^*) \qquad \forall\, u^* \in X^*.$$

Problem 2.140*
Suppose that X is a Banach space, $A \in 2^X \setminus \{\emptyset\}$ and $\{A_n\}_{n\geqslant 1} \subseteq 2^X \setminus \{\emptyset\}$ is a sequence such that

$$\limsup_{n\to+\infty} \sigma_{A_n}(u^*) \leqslant \sigma_A(u^*) \qquad \forall\, u^* \in X^*.$$

Show that $w\text{-}\limsup\limits_{n\to+\infty} A_n \subseteq \overline{\operatorname{conv}}\, A$.

Problem 2.141***
Suppose that (Ω, Σ, μ) is a σ-finite measure space, X is a Banach space, $h \in L^p(\Omega; X)$, $\{h_n\}_{n\geqslant 1} \subseteq L^p(\Omega; X)$ (with $1 \leqslant p < +\infty$) is a sequence such that $h_n \xrightarrow{w} h$ in $L^p(\Omega; X)$ and

$$h_n(\omega) \in G(\omega) \in P_{wk}(X) \quad \text{for } \mu\text{-a.a. } \omega \in \Omega.$$

Show that

$$h(\omega) \in \overline{\operatorname{conv}}\ w\text{–}\limsup_{n\to+\infty} \{h_n(\omega)\} \quad \text{for } \mu\text{-a.a. } \omega \in \Omega.$$

Problem 2.142***
Suppose that (Ω, Σ, μ) is a σ-finite measure space, X and Y are two separable Banach spaces and $F\colon \Omega \times X \longrightarrow P_{fc}(Y)$ is a multifunction such that:

(i) for all $u \in X$, the multifunction $\omega \longmapsto F(\omega, u)$ is graph measurable;
(ii) for μ-almost all $\omega \in \Omega$, the multifunction $u \longmapsto F(\omega, u)$ has a graph which is closed in $X \times Y_w$ (where Y_w denotes the Banach space Y with the w-topology);
(iii) $F(\omega, u) \subseteq G(\omega)$ for μ-almost all $\omega \in \Omega$, all $u \in X$ with a graph measurable and integrably bounded multifunction $G\colon \Omega \longrightarrow P_{wkc}(Y)$.

Show that $S^1_{F(\cdot, v(\cdot))} \neq \emptyset$ for any $v \in L^1(\Omega; X)$.

Problem 2.143*
Suppose that (Ω, Σ, μ) is a σ-finite measure space, X is a separable Banach spaces, and $\{h_n\}_{n\geqslant 1} \subseteq L^1(\Omega; X)$ is a sequence which is uniformly integrable and such that

$$\overline{\{h_n(\omega)\}_{n\geqslant 1}}^{w} \in P_{wk}(X) \quad \text{for } \mu\text{-a.a. } \omega \in \Omega.$$

Show that the sequence $\{h_n\}_{n\geqslant 1} \subseteq L^1(\Omega; X)$ is relatively w-compact.

Problem 2.144*
Suppose that (Ω, Σ, μ) is a finite measure space, $h \in L^1(\Omega)$, $\{h_n\}_{n\geqslant 1} \subseteq L^1(\Omega)$ is a sequence such that $h_n \xrightarrow{w} h$ in $L^1(\Omega)$ and assume that at least one of the following conditions holds:

(i) $h(\omega) \leqslant \liminf_{n\to+\infty} h_n(\omega)$ for μ-almost all $\omega \in \Omega$; or
(ii) $\limsup_{n\to+\infty} h_n(\omega) \leqslant h(\omega)$ for μ-almost all $\omega \in \Omega$.

Show that $h_n \longrightarrow h$ in $L^1(\Omega)$.

Problem 2.145**
Show that in general for a sequence $\{A_n\}_{n\geqslant 1} \subseteq P_f(X)$ in a Banach space X, the set $w\text{–}\limsup_{n\to+\infty} A_n$ need not be closed or weakly closed.

Problem 2.146*
Let X be a Banach space and let $\{A_n\}_{n\geqslant 1} \subseteq P_{fc}(X)$ be a sequence such that

$$A_{n+1} \subseteq A_n \quad (\text{respectively } A_{n+1} \supseteq A_n) \qquad \forall\, n \geqslant 1.$$

Show that $A_n \xrightarrow{M} \bigcap_{n\geqslant 1} A_n$ (respectively $A_n \xrightarrow{M} \overline{\bigcup_{n\geqslant 1} A_n}$).

Problem 2.147**
Suppose that (Ω, Σ, μ) is a σ-finite measure space, X and Y are two separable Banach spaces and $F\colon \Omega \times X \longrightarrow 2^Y \setminus \{\emptyset\}$ is a multifunction such that:

(i) F is measurable on $\Omega \times X$;
(ii) for μ-almost all $\omega \in \Omega$, the multifunction $u \longmapsto F(\omega, u)$ is lower semicontinuous;
(iii) there exists $\xi \in L^1(\Omega)$ such that

$$|F(\omega, u)| \;=\; \sup_{y\in F(\omega,u)} \|y\|_Y \;\leqslant\; \xi(\omega) \quad \text{for } \mu\text{-a.a. } \omega \in \Omega, \text{ all } u \in X.$$

Show that the multifunction $h \longmapsto S^1_{F(\cdot,h(\cdot))}$ is lower semicontinuous from $L^1(\Omega; X)$ into $L^1(\Omega; Y)$.

Problem 2.148*
Let X be a Banach space and let $A\colon X \supseteq D(A) \longrightarrow 2^{X^*}$ be a monotone map with $D(A) \neq \emptyset$. Show that there exists a maximal monotone extension $\widehat{A}\colon X \supseteq D(\widehat{A}) \longrightarrow 2^{X^*}$ of A (that is, a maximal monotone map $\widehat{A}\colon X \supseteq D(\widehat{A}) \longrightarrow 2^{X^*}$ such that $\operatorname{Gr} A \subseteq \operatorname{Gr} \widehat{A}$).

Problem 2.149**
Let X be a Banach space and let $A\colon X \supseteq D(A) \longrightarrow 2^{X^*}$ be a maximal monotone map. Show that $\operatorname{Gr} A$ is closed in $X_w \times X^*$ and in $X \times X^*_{w^*}$ (here X_w (respectively $X^*_{w^*}$) denotes the Banach space X (respectively X^*) furnished with the weak (respectively weak*) topology). Can we say that $\operatorname{Gr} A$ is closed in $X_w \times X^*_{w^*}$? Justify your answer.

Problem 2.150*
Suppose that $T = (0, b)$, H is a Hilbert space, and $u_0 \in H$. We define $A\colon L^2(T; H) \supseteq D(A) \longrightarrow L^2(T; H)$ by

$$A(u) \;=\; \tfrac{d}{dt}u \qquad \forall\, u \in D(A) = \big\{u \in W^{1,2}(T; H):\; u(0) = u_0\big\}$$

(see Problem 1.104). Show that A is maximal monotone.

Problem 2.151**
Let X be a finite dimensional Banach space and let $A\colon X \longrightarrow X^*$ be a monotone map with $D(A) = X$ which is also hemicontinuous (see Definition 2.110(b)). Show that A is continuous.

Problem 2.152**
Let X be a Banach space and let $A\colon X \longrightarrow X^*$ be a monotone map with $D(A) = X$. Show that A is demicontinuous if and only if it is hemicontinuous.

Problem 2.153*
Let $\Omega \subseteq \mathbb{R}^N$ be a bounded open set with a Lipschitz boundary $\partial\Omega$, $1 < p < +\infty$ and let $a\colon \mathbb{R}^N \longrightarrow \mathbb{R}^N$ be a continuous and monotone map satisfying the following growth condition

$$|a(y)| \;\leqslant\; c(1 + |y|^{p-1}) \qquad \forall\, y \in \mathbb{R}^N,$$

for some $c > 0$. For every $u \in W_0^{1,p}(\Omega)$, let $A(u) = -\operatorname{div} a(Du)$. Show that $A\colon W_0^{1,p}(\Omega) \longrightarrow W^{-1,p'}(\Omega) = W_0^{1,p}(\Omega)^*$ (where $\frac{1}{p} + \frac{1}{p'} = 1$) is bounded (that is, maps bounded sets to bounded ones), continuous, and maximal monotone.

Problem 2.154**
Let $\Omega \subseteq \mathbb{R}^N$ be a bounded open set with a Lipschitz boundary $\partial\Omega$, $1 < p < +\infty$ and let $a\colon \mathbb{R}^N \longrightarrow \mathbb{R}^N$ be a continuous and monotone map satisfying the following conditions

$$|a(y)| \leqslant c(1 + |y|^{p-1}) \qquad \forall\, y \in \mathbb{R}^N,$$

for some $c > 0$ and

$$(a(y), y)_{\mathbb{R}^N} \geqslant \widehat{c}|y|^p \qquad \forall\, y \in \mathbb{R}^N,$$

for some $\widehat{c} > 0$. Let $g \in W^{-1,p'}(\Omega) = W_0^{1,p}(\Omega)^*$. Show that there exists $u \in W_0^{1,p}(\Omega)$ such that

$$-\operatorname{div} a(Du) = g \quad \text{in } W^{-1,p'}(\Omega).$$

Problem 2.155*
Let X be a reflexive Banach space and let $A\colon X \supseteq D(A) \longrightarrow X^*$ be a linear monotone map. Show that the map A is maximal monotone if and only if $\operatorname{Gr} A$ is maximal among all linear monotone subsets (graphs) of $X \times X^*$.

Problem 2.156**
Let X be a reflexive Banach space and let $A\colon X \longrightarrow X^*$ be a linear demicontinuous monotone map. Show that $A \in \mathcal{L}(X; X^*)$.

Problem 2.157**
Let X be a Banach space and let $A\colon X \supseteq D(A) \longrightarrow 2^{X^*}$ be a monotone map which is lower semicontinuous at $\widehat{u} \in D(A) \subseteq X$ into $X^*_{w^*}$ (where $X^*_{w^*}$ denotes the space X^* furnished with the w^*-topology). Show that $A(\widehat{u})$ is a singleton.

Problem 2.158***
Let X be a separable Banach space and let $A\colon X \supseteq D(A) \longrightarrow 2^{X^*}$ be a maximal monotone map with $\operatorname{int} D(A) \neq \emptyset$. Show that the set

$$\Gamma = \{u \in D(A) : \ A(u) \text{ is not a singleton}\}$$

is of first category.

Problem 2.159**
Show that the duality map of a Banach space X is linear if and only if X is a Hilbert space.

Problem 2.160***
Suppose that X is a reflexive Banach space, $A\colon X \supseteq D(A) \longrightarrow 2^{X^*}$ is a monotone map, and C is a closed convex set such that $D(A) \subseteq C$. Show that there exists a maximal monotone map $\widehat{A}\colon X \supseteq D(\widehat{A}) \longrightarrow 2^{X^*}$ such that $\operatorname{Gr} A \subseteq \operatorname{Gr} \widehat{A}$ and $D(\widehat{A}) \subseteq C$.

Problem 2.161**
Let X be a reflexive Banach space with X and X^* both locally uniformly convex (which is always possible thanks to the Troyanski renorming theorem; see Theorem I.5.192 or Remark 2.115) and let $A\colon X \supseteq D(A) \longrightarrow 2^{X^*}$ be a maximal monotone map. Show that for every $u \in X$ and for every $\lambda > 0$, there exists unique $u_\lambda \in D(A)$ such that $0 \in \lambda A(u_\lambda) + \mathcal{F}(u_\lambda - u)$.

Remark. According to this problem, for every $\lambda > 0$ we can define the map $J_\lambda\colon X \longrightarrow D(A)$ by setting

$$J_\lambda(u) \;=\; u_\lambda \;\in\; D(A)$$

and then the map $A_\lambda\colon X \longrightarrow X^*$ by setting

$$A_\lambda(u) \;=\; -\tfrac{1}{\lambda}\mathcal{F}(u_\lambda - u) \;=\; \tfrac{1}{\lambda}\mathcal{F}(u - u_\lambda).$$

The map J_λ is called the ***resolvent of*** A and A_λ is called the ***Yosida approximation of*** A. They are extensions to Banach spaces of the items introduced in Definition 2.122.

Problem 2.162**
Suppose that X is a reflexive Banach space, $A\colon X \supseteq D(A) \longrightarrow 2^{X^*}$ is a maximal monotone map, $\{(u_n, u_n^*)\}_{n\geqslant 1} \subseteq \operatorname{Gr} A$ is a sequence such that $u_n \xrightarrow{w} u$ in X and $u_n^* \xrightarrow{w} u^*$ in X^* and assume that

$$\limsup_{n,k\to+\infty} \langle u_n^* - u_k^*, u_n - u_k\rangle \;\leqslant\; 0 \quad \text{or} \quad \limsup_{n\to+\infty} \langle u_n^* - u^*, u_n - u\rangle \;\leqslant\; 0.$$

Show that $(u, u^*) \in \operatorname{Gr} A$ and $\langle u_n^*, u_n\rangle \longrightarrow \langle u^*, u\rangle$.

Problem 2.163***
Let X be a reflexive Banach space with X and X^* both locally uniformly convex. Show that for every $\lambda > 0$, the Yosida approximation $A_\lambda\colon X \longrightarrow X^*$ (see the Remark after Problem 2.161) is everywhere defined, single valued, monotone, bounded, demicontinuous and satisfies

$$A_\lambda(u) \;\in\; A(J_\lambda(u)) \qquad \forall\, u \in X$$

and $J_\lambda\colon X \longrightarrow D(A)$ is continuous.

Problem 2.164**
Let X be a reflexive Banach space with X and X^* both locally uniformly convex and let $A\colon X \supseteq D(A) \longrightarrow 2^{X^*}$ be a maximal monotone map. Show that

$$A_\lambda(u) \;\longrightarrow\; A^0(u) \quad \text{in } X^* \quad \text{as } \lambda \searrow 0 \qquad \forall\, u \in D(A),$$

with $A^0(u) \in A(u)$ such that

$$\|A^0(u)\|_{X^*} \;=\; \inf_{u^*\in A(u)} \|u^*\|_{X^*}$$

(the ***minimal section of*** A) and

$$J_\lambda(u) \longrightarrow u \quad \text{in } X \quad \text{as } \lambda \searrow 0 \qquad \forall\, u \in \overline{\operatorname{conv}}\, D(A).$$

Problem 2.165**
Let X be a reflexive Banach space and let $A\colon X \supseteq D(A) \longrightarrow 2^{X^*}$ be a maximal monotone map. Show that $\overline{D(A)}$ and $\overline{R(A)}$ are both convex.

Problem 2.166**
Let $A\colon \mathbb{R}^N \longrightarrow \mathbb{R}^N$ be a surjective monotone map. Show that $\lim\limits_{|u|\to+\infty} |A(u)| = +\infty$.

Problem 2.167*
Let $A\colon \mathbb{R}^N \longrightarrow \mathbb{R}^N$ be a monotone map on $B_{r+\varepsilon} = \{u \in \mathbb{R}^N : |u| < r+\varepsilon\}$ (where $r, \varepsilon > 0$). Show that the set $A(\overline{B}_r) \subseteq \mathbb{R}^N$ is bounded.

Problem 2.168**
Suppose that $T = (a,b)$, H is a Hilbert space, $u_0 \in H$ and $A\colon L^2(T;H) \supseteq D(A) \longrightarrow L^2(T;H)$ is defined by

$$A(u) = \tfrac{d}{dt}u \qquad \forall\, u \in D(A),$$

with $D(A) = \{u \in W^{1,2}(T;H) : u(0) = u_0\}$ (see Problem 1.104). Find the resolvent J_1^A (see Definition 2.122).

Problem 2.169**
Let X be a reflexive Banach space and let $A\colon X \longrightarrow X^*$ be an everywhere defined uniformly monotone and hemicontinuous map. Show that A is surjective.

Problem 2.170**
Suppose that X is a reflexive Banach space and $A\colon X \supseteq D(A) \longrightarrow 2^{X^*}$ is a bounded map (that is, A maps bounded sets to bounded sets) which satisfies the following condition: if $\{(u_n, u_n^*)\}_{n\geqslant 1} \subseteq \operatorname{Gr} A$ is a sequence such that $u_n \xrightarrow{w} u$ in X, $u_n^* \xrightarrow{w^*} u^*$ in X^* and $\limsup\limits_{n\to+\infty} \langle u_n^*, u_n - u\rangle \leqslant 0$, then $(u,u^*) \in \operatorname{Gr} A$. Show that A is upper semicontinuous from every finite dimensional subspace of X into X_w^* (where X_w^* denotes the space X^* furnished with the w-topology).

Problem 2.171**
Suppose that X is a reflexive Banach space and $A\colon X \supseteq D(A) \longrightarrow 2^{X^*}$ and $C\colon X \supseteq D(C) \longrightarrow X^*$ are two monotone maps with $D(C) = X$. Assume that the map $u \longmapsto (A+C)(u)$ is maximal monotone. Show that A is maximal monotone.

Problem 2.172**
Suppose that (Ω, Σ, μ) is a σ-finite measure space, H is a Hilbert space, $A\colon H \supseteq D(A) \longrightarrow 2^H$ is a maximal monotone map with $(0,0) \in \operatorname{Gr} A$ and

$\widehat{A}\colon L^2(\Omega;H) \supseteq D(\widehat{A}) \longrightarrow L^2(\Omega;H)$ is the realization of A on the Hilbert space $L^2(\Omega;H)$, that is,

$$\widehat{A}(u) = \big\{h \in L^2(\Omega;H):\ h(\omega) \in A(u(\omega)) \text{ for } \mu\text{-a.a. } \omega \in \Omega\big\}$$

for all $u \in D(\widehat{A}) = \big\{v \in L^2(\Omega;H):\ S^2_{A(v(\cdot))} \neq \emptyset\big\}$. Show that $\widehat{A}$ is maximal monotone and find its Yosida approximation $\widehat{A}_\lambda$, for $\lambda > 0$. If $\mu(\Omega) < +\infty$, then we can drop the requirement that $(0,0) \in \operatorname{Gr} A$.

Problem 2.173*

Given two maximal monotone maps A and C which are defined everywhere, is it true that $R(A+C) = R(A) + R(C)$? Justify your answer.

Problem 2.174**

Suppose that (Ω,Σ) is a complete measurable space, X is a separable reflexive Banach space and for every $\omega \in \Omega$, the map $A(\omega)\colon X \supseteq D(A(\omega)) \longrightarrow 2^{X^*}$ is maximal monotone. Show that the following two properties are equivalent:

(a) The multifunction $\omega \longmapsto \operatorname{Gr} A(\omega)$ is measurable from Ω into $P_f(X \times X^*)$.

(b) For every $u^* \in X^*$, the map $\omega \longmapsto (A(\omega)+\mathcal{F})^{-1}(u^*)$ is Σ-measurable from Ω into X.

Problem 2.175**

Suppose that (Ω,Σ) is a complete measurable space, X is a separable reflexive Banach space and

$$\{A(\omega)\colon X \supseteq D(A(\omega)) = D(\omega) \longrightarrow 2^{X^*}\}_{\omega\in\Omega}$$

is a family of maximal monotone maps satisfying property **(a)** (or equivalently property **(b)**) in Problem 2.174. Show that:

(1) for every $E \in \mathcal{B}(X)$, we have $D^-(E) = \{\omega \in \Omega : D(\omega) \cap E \neq \emptyset\} \in \Sigma$;

(2) for every $u \in X$, the multifunction $\omega \longmapsto A(\omega)(u)$ is measurable from Ω into X^*;

(3) for every $u \in X$, we have $\Omega(u) = \{\omega \in \Omega:\ u \in D(\omega)\} \in \Sigma$ and the minimal section map $\omega \longmapsto A^0(\omega)(u)$ (see Problem 2.164) is Σ-measurable from $\Omega(u)$ into X^*.

Problem 2.176**

Suppose that X is a reflexive Banach space, $A\colon X \supseteq D(A) \longrightarrow 2^{X^*}$ is a maximal monotone surjective map with $D(A) = X$ and $C \subseteq X$ is a nonempty, closed, and convex set. Show that the set $A(C) \subseteq X^*$ is closed.

Problem 2.177**

Suppose that X is a strictly convex (see Definition I.5.168 and Remark I.5.169), reflexive Banach space with a strictly convex dual X^*, $A\colon X \supseteq D(A) \longrightarrow 2^{X^*}$ and $C\colon X \supseteq D(C) \longrightarrow 2^{X^*}$ are two maximal monotone maps such that $D(A) \cap D(C) \neq \emptyset$. Show that for every $h^* \in X^*$, the set of solutions $\{u_\lambda\}_{\lambda>0}$ of the equations $h^* \in (A + C_\lambda + \mathcal{F})(u_\lambda)$ remains bounded in X as $\lambda \searrow 0$.

Problem 2.178**
Let X be a Banach space and let $u, y \in X$. Show that the following two properties are equivalent:
(a) $\|u\|_X \leqslant \|u + \lambda y\|_X$ for all $\lambda > 0$.
(b) There exists $u^* \in \mathcal{F}(u)$ such that $\langle u^*, y \rangle \geqslant 0$.

Problem 2.179**
Suppose that X is a reflexive Banach space with a strictly convex dual X^* (see Definition I.5.168 and Remark I.5.169), $\mathcal{F}\colon X \longrightarrow X^*$ is the duality map, $C \subseteq X$ is a nonempty and convex set and $u_0 \in C$. Show that $\|u_0\|_X = \inf\limits_{u \in C} \|u\|_X$ if and only if

$$\langle \mathcal{F}(u_0), u_0 \rangle \leqslant \langle \mathcal{F}(u_0), u \rangle \qquad \forall\, u \in C.$$

Problem 2.180*
Suppose that X is a reflexive Banach space with both X and X^* being strictly convex, $A\colon X \supseteq D(A) \longrightarrow 2^{X^*}$ is a maximal monotone multifunction, $\{\lambda_n\}_{n\geqslant 1} \subseteq \mathbb{R}$ and $\{u_n\}_{n\geqslant 1} \subseteq X$ are two sequences such that

$$\lambda_n \longrightarrow 0, \quad u_n \xrightarrow{w} u \quad \text{in } X \quad \text{and} \quad A_{\lambda_n}(u_n) \xrightarrow{w} y^* \quad \text{in } X^*$$

and

$$\limsup_{n,m \to +\infty} \langle A_{\lambda_n}(u_n) - A_{\lambda_m}(u_m), u_n - u_m \rangle \leqslant 0.$$

Show that $(u, y^*) \in \operatorname{Gr} A$ and

$$\limsup_{n,m \to +\infty} \langle A_{\lambda_n}(u_n) - A_{\lambda_m}(u_m), u_n - u_m \rangle = 0.$$

Problem 2.181***
Suppose that X is a strictly convex reflexive Banach space with a strictly convex dual X^* (see Definition I.5.168 and Remark I.5.169), $A\colon X \supseteq D(A) \longrightarrow 2^{X^*}$ and $C\colon X \supseteq D(C) \longrightarrow 2^{X^*}$ are two maximal monotone maps such that $D(A) \cap D(C) \neq \emptyset$. Show that if for every $\lambda > 0$, $u_\lambda \in X$ is the unique solution of the operator inclusion

$$h^* \in (A + C_\lambda + \mathcal{F})(u_\lambda)$$

and $\{C_\lambda(u_\lambda)\}_{\lambda>0} \subseteq X^*$ is bounded as $\lambda \searrow 0$, then $h^* \in R(A + C + \mathcal{F})$.

Problem 2.182*
Let X be a reflexive Banach space and let $A\colon X \supseteq D(A) \longrightarrow 2^{X^*}$ be a maximal monotone map. We set

$$m(u) = \inf_{u^* \in A(u)} \|u^*\|_*$$

(with $\inf \emptyset = +\infty$). Show that the function $m\colon X \longrightarrow \overline{\mathbb{R}} = \mathbb{R} \cup \{+\infty\}$ is lower semicontinuous.

Problem 2.183**
Suppose that X is a reflexive Banach space, $A\colon X \supseteq D(A) \longrightarrow 2^{X^*}$ and $C\colon X \supseteq D(C) \longrightarrow X^*$ are two monotone maps such that $D(C) = X$, $R(A+C) = X^*$ and $(A+C)^{-1}$ is continuous. Show that A is maximal monotone.

Problem 2.184**
Let X be a Banach space and let $\mathcal{F}$ be its duality map. Show that X is reflexive if and only if $R(\mathcal{F}) = X^*$.

Problem 2.185*
Let X be a locally uniformly convex Banach space with a strictly convex dual X^* (see Definition I.5.168 and Remark I.5.169). Show that the duality map $\mathcal{F}\colon X \longrightarrow X^*$ is of type $(S)_+$ (see Definition 2.137).

Problem 2.186**
Suppose that X is a reflexive Banach space, $A\colon X \longrightarrow X^*$ is a demicontinuous map of type $(S)_+$, and $K\colon X \longrightarrow X^*$ is a compact map. Show that the map $T\colon X \longrightarrow X^*$ defined by

$$T(u) \;=\; (A+K)(u) \qquad \forall\, u \in X$$

is generalized pseudomonotone (see Definition 2.131).

Problem 2.187**
Suppose that X is a reflexive Banach space, $A\colon X \longrightarrow X^*$ is a generalized pseudomonotone map with $D(A) = X$ and $K\colon X \longrightarrow X^*$ is a map which is sequentially continuous from X_w (where X_w denotes the Banach space X furnished with the weak topology) into X^*_w (where X^*_w denotes the Banach space X^* furnished with the weak topology) and such that the function $u \longmapsto \varphi(u) = \langle K(u), u\rangle$ is sequentially weakly lower semicontinuous. Let $\{u_n\}_{n\geqslant 1} \subseteq X$ be a sequence such that $u_n \xrightarrow{w} u$ in X, $(A+K)(u_n) \xrightarrow{w} u^*$ in X^* and $\limsup\limits_{n\to+\infty} \langle (A+K)(u_n), u_n - u\rangle \leqslant 0$. Show that $u^* \in (A+K)(u)$.

Problem 2.188*
Suppose that X is a reflexive Banach space, $A\colon X \supseteq D(A) \longrightarrow 2^{X^*}$ is a generalized pseudomonotone map and $C \subseteq X$ is a nonempty, bounded, and weakly closed set. Show that the set

$$A(C) \;=\; \{u^* \in X^* : \; u^* \in A(u) \text{ for some } u \in C\}$$

is strongly closed in X^*.

Problem 2.189**
Let H be a Hilbert space and let $K\colon H \longrightarrow H$ be a compact map. Show that the map $u \longmapsto u + K(u)$ is generalized pseudomonotone.

Problem 2.190**
Let X be a reflexive Banach space and let $K\colon X \longrightarrow X^*$ be a compact map. Is K necessarily generalized pseudomonotone? Justify your answer.

Problem 2.191*
Let X be a reflexive Banach space and let $A\colon X \longrightarrow X^*$ be a monotone map which is sequentially weakly continuous. Show that the function $u \longmapsto \varphi(u) = \langle A(u), u\rangle$ is sequentially weakly lower semicontinuous.

Problem 2.192***
Let $\Omega \subseteq \mathbb{R}^N$ be a bounded open set with a Lipschitz boundary and let $G\colon \Omega \times \mathbb{R}^N \longrightarrow \mathbb{R}$ satisfies:

(i) for all $y \in \mathbb{R}^N$, the function $z \longmapsto G(z, y)$ is measurable;

(ii) for almost all $z \in \Omega$, the function $y \longmapsto G(z, y)$ is C^1, strictly convex and $G(z, 0) = 0$;

(iii) there exist $\widehat{a} \in L^\infty(\Omega)$ and $1 < p < +\infty$ such that

$$|\nabla_y G(z, y)| \leqslant \widehat{a}(z)(1 + |y|^{p-1})$$

for almost all $z \in \Omega$ and all $y \in \mathbb{R}^N$;

(iv) we have

$$(\nabla_y G(z, y), y)_{\mathbb{R}^N} \leqslant pG(z, y)$$

for almost all $z \in \Omega$ and all $y \in \mathbb{R}^N$;

(v) there exists $c_0 > 0$ such that

$$c_0|y|^p \leqslant pG(z, y)$$

for almost all $z \in \Omega$ and all $y \in \mathbb{R}^N$.

Let $a(z, y) = \nabla_y G(z, y)$ and let $A\colon W^{1,p}(\Omega) \longrightarrow W^{1,p}(\Omega)^*$ be the nonlinear map defined by

$$\langle A(u), h\rangle = \int_\Omega (a(z, Du(z)), Dh(z))_{\mathbb{R}^N}\, dz \qquad \forall\, u, h \in W^{1,p}(\Omega).$$

Show that A is maximal monotone and of type $(S)_+$.

Remark. If $G(z, y) = G(y) = \frac{1}{p}|y|^p$ for all $y \in \mathbb{R}^N$ with $1 < p < +\infty$, then $a(y) = |y|^{p-2}y$ for all $y \in \mathbb{R}^N$ and so A corresponds to the p-Laplacian differential operator with Neumann boundary condition.

Problem 2.193**
Suppose that H is a Hilbert space, $A\colon H \supseteq D(A) \longrightarrow 2^H$ is a maximal monotone map, and $C\colon H \supseteq D(C) \longrightarrow H$ is a monotone map with $D(C) \subseteq H$ closed and satisfies

$$\|C(u) - C(y)\|_H \leqslant k\|u - y\|_H \qquad \forall\, u, y \in D(C),$$

for some $k \in (0, 1)$. Show that the map $u \longmapsto (A + C)(u)$ is maximal monotone.

Problem 2.194*
Let X be a reflexive Banach space and let $A\colon X \longrightarrow X^*$ be a demicontinuous map of type $(S)_+$. Show that A is pseudomonotone.

Problem 2.195**

Let X be a reflexive Banach space and let $A\colon X \longrightarrow X^*$ be a demicontinuous, strongly coercive (see Definition 2.98(f)) and bounded map which is of type $(S)_+$. Show that A is surjective.

Problem 2.196**

Let $\Omega \subseteq \mathbb{R}^N$ (with $N \leqslant 3$) be a bounded open set and let $A\colon H_0^1(\Omega) \longrightarrow H^{-1}(\Omega) = H_0^1(\Omega)^*$ be defined by

$$\langle A(u), h\rangle = \int_\Omega \sum_{k=1}^N (\sin u(z))(D_k u(z))h(z)\,dz \qquad \forall\, u, h \in H_0^1(\Omega).$$

Show that A is completely continuous.

Problem 2.197***

Let H be a Hilbert space and let $A\colon H \supseteq D(A) \longrightarrow H$ be a linear operator (not necessarily bounded) which is monotone (positive) and symmetric. Show that A is self-adjoint if and only if A is maximal monotone.

Problem 2.198**

Suppose that H is a Hilbert space, $A\colon H \supseteq D(A) \longrightarrow H$ and $T\colon H \supseteq D(T) \longrightarrow H$ are two linear monotone operators such that A is self-adjoint and $A \subseteq T$ (i.e., $\operatorname{Gr} A \subseteq \operatorname{Gr} T$). Show that $A = T$.

Problem 2.199*

Let X be a Banach space with a strictly convex dual X^* and let $A\colon X \supseteq D(A) \longrightarrow 2^X$ be a maximal accretive map. Show that for every $u \in D(A)$, we have $A(u) \in P_{fc}(X)$.

Problem 2.200**

(a) Let X be a Banach space and let $A\colon X \supseteq D(A) \longrightarrow 2^X$ be a maximal accretive map. Show that $\operatorname{Gr} A \subseteq X \times X$ is closed.

(b) Let X be a Banach space with locally uniformly convex dual X^* and let $A\colon X \supseteq D(A) \longrightarrow 2^X$ be a maximal accretive map. Show that $\operatorname{Gr} A$ is closed in $X \times X_w$ (where X_w denotes the Banach space X furnished with the weak topology).

Problem 2.201**

Let X be a uniformly convex Banach space with a uniformly convex dual X^* and let $A\colon X \supseteq D(A) \longrightarrow 2^X$ be an m-accretive operator. Show that for all $u \in D(A)$ we have

$$\lim_{\lambda \searrow 0} A_\lambda(u) = A^0(u) \quad \text{and} \quad \lim_{\lambda \searrow 0} A^0(J_\lambda(u)) = A^0(u).$$

Problem 2.202**
Let X be a reflexive Banach space with uniformly convex dual X^* and let $A: X \supseteq D(A) \longrightarrow 2^X$ be a maximal accretive map, which is locally bounded. Show that A is upper semicontinuous from X into X_w (where X_w denotes the Banach space X furnished with the weak topology).

Problem 2.203**
Suppose that X is a Banach space, $u: \mathbb{R} \longrightarrow X$ is a map which is almost everywhere weakly differentiable and the map $t \longmapsto \|u(t)\|_X$ is almost everywhere differentiable. Show that for almost all $t \in \mathbb{R}$, we have

$$\|u(t)\|_X \tfrac{d}{dt}\|u(t)\|_X = \langle u^*, u'(t)\rangle \qquad \forall\, u^* \in \mathcal{F}(u(t)).$$

Problem 2.204*
Let X be a Banach space and let $A: X \longrightarrow X$ be a continuous accretive map such that $D(A) = X$. Show that A is m-accretive.

Problem 2.205**
Let X be a reflexive Banach space with a uniformly convex dual X^* and let $A: X \longrightarrow P_{fc}(X)$ be an accretive map which is upper semicontinuous from X into X_w (where X_w denotes the Banach space X furnished with the weak topology). Show that A is maximal accretive.

Problem 2.206*
Suppose that X is a uniformly convex Banach space with uniformly convex dual X^*, $C \subseteq X$ is a nonempty and closed set and $\{S(t): C \longrightarrow C\}_{t\geqslant 0}$ is a semigroup of nonexpansive maps (see Definition 2.156) with infinitesimal generator A and which has the following property: "if $\{u_n\}_{n\geqslant 1} \subseteq D(A)$ is a sequence such that $\|u_n\|_X \longrightarrow +\infty$, then $\|A(u_n)\|_X \longrightarrow +\infty$." Show that for every $u \in D(A)$, we have $\sup\limits_{t\geqslant 0} \|S(t)u_0\|_X < +\infty$.

Problem 2.207*
Suppose that X is a Banach space, $A: X \supseteq D(A) \longrightarrow 2^X$ is an m-accretive map and $\{S(t): \overline{D(A)} \longrightarrow \overline{D(A)}\}_{t\geqslant 0}$ is the semigroup of nonexpansive maps generated by A (see Theorem 2.158). Show that for every $u \in \overline{D(A)}$ and $t > 0$, we have

$$\|S(t)u - u\|_X \leqslant \tfrac{2}{t} \int\limits_0^t \|S(\tau)u - u\|_X \, d\tau.$$

Problem 2.208*
Suppose that X is a Banach space, $A: X \supseteq D(A) \longrightarrow 2^X$ is an m-accretive map and $\{S(t): \overline{D(A)} \longrightarrow \overline{D(A)}\}_{t\geqslant 0}$ is the nonlinear semigroup of nonexpansive maps generated by A (see Theorem 2.158). Show that for every $u \in \overline{D(A)}$, every $t > 0$ and every $\lambda > 0$, we have

$$\|S(t)u - u\|_X \leqslant \left(2 + \tfrac{t}{\lambda}\right) \|u - J_\lambda(u)\|_X$$

and

$$\|u - J_\lambda(u)\|_X \leqslant \tfrac{2}{t}\left(1 + \tfrac{\lambda}{t}\right) \int_0^t \|S(\tau)u - u\|_X \, d\tau.$$

Problem 2.209***
Suppose that X is a Banach space, $A\colon X \supseteq D(A) \longrightarrow 2^X$ is an m-accretive map and $\{S(t)\colon \overline{D(A)} \longrightarrow \overline{D(A)}\}_{t\geqslant 0}$ is the nonlinear semigroup of nonexpansive maps generated by A (see Theorem 2.158). Show that the following two properties are equivalent:

(a) The nonlinear semigroup $\{S(t)\}_{t\geqslant 0}$ is compact (see Definition 2.159(a)).

(b) For every $\lambda > 0$, the map J_λ is compact and $\{S(t)\}_{t\geqslant 0}$ is equicontinuous (see Definition 2.159(b)).

Problem 2.210**
Let X be a Banach space and let $A\colon X \supseteq D(A) \longrightarrow 2^X$ be an m-accretive map. Suppose that $\{u_n\}_{n\geqslant 1} \subseteq D(A)$ is a sequence such that $u_n \longrightarrow u$ in X and $A_{\lambda_n}(u_n) \longrightarrow h$ with $\lambda_n \searrow 0$. Show that $(u,h) \in \operatorname{Gr} A$. Also show that if X is reflexive with locally uniformly convex dual, then we may assume that $A_{\lambda_n}(u_n) \xrightarrow{w} h$ (instead of $A_{\lambda_n}(u_n) \longrightarrow h$).

Problem 2.211**
Let X be a Banach space and let $A\colon X \supseteq D(A) \longrightarrow 2^X$ be an m-accretive map. Show that the following two properties are equivalent:

(a) For every $\lambda > 0$, J_λ is compact.

(b) For every $\eta > 0$, the sublevel set $L_\eta = \{u \in D(A) : \|u\|_X + |A(u)| \leqslant \eta\}$ is relatively compact in X.

Problem 2.212*
Let $L^2_{2\pi}(\mathbb{R})$ be the space of $L^2(\mathbb{R})$-functions which are 2π-period. We equip $L^2_{2\pi}(\mathbb{R})$ with the norm $\|u\| = \|u|_{[0,2\pi]}\|_2$. It becomes a Hilbert space. Let $A\colon L^2_{2\pi}(\mathbb{R}) \supseteq D(A) \longrightarrow L^2_{2\pi}(\mathbb{R})$ be defined by $A(u) = u'$ for every $u \in D(A)$, where $D(A) = \big\{u \in L^2_{2\pi}(\mathbb{R}) : \ u' \in L^2_{2\pi}(\mathbb{R})\big\}$. Show that A is maximal monotone and it generates a C_0-semigroup $\{S(t)\}_{t\geqslant 1}$ which is not compact.

Problem 2.213**
Suppose that X is a Banach space, (E, d_E) is a compact metric space, and $F\colon E \longrightarrow P_{w^*kc}(X^*)$ is an upper semicontinuous multifunction from E into $X^*_{w^*}$ (with $X^*_{w^*}$ being the Banach space X^* furnished with the w^*-topology) and

$$\gamma(u) = \inf_{v^* \in F(u)} \|v^*\|_* \qquad \forall\, u \in E \quad \text{and} \quad \eta = \inf_{u\in E} \gamma(u).$$

Show that for every $\varepsilon > 0$, we can find a continuous map $\xi\colon E \longrightarrow X$ such that for every $u \in E$ and every $v^* \in F(u)$ we have

$$\|\xi(u)\|_X \leqslant 1 \quad \text{and} \quad \langle v^*, \xi(u)\rangle \geqslant \eta - \varepsilon.$$

Problem 2.214**

Suppose that X is a reflexive Banach space, Y, Z are two Banach spaces with X being continuously embedded in Z and $L \in \mathcal{L}_c(X;Y)$. Show that for every $\varepsilon > 0$, there exists $c_\varepsilon > 0$ such that

$$\|L(u)\|_Y \leqslant \varepsilon\|u\|_X + c_\varepsilon\|u\|_Z \qquad \forall\, u \in X.$$

Problem 2.215**

Let $\Omega \subseteq \mathbb{R}^N$ be a bounded domain with a Lipschitz boundary $\partial\Omega$. Suppose that $p \in (1, N)$ and $q \in (1, \frac{Np-p}{N-p})$. Show that for every $\varepsilon > 0$, there exists $c_\varepsilon > 0$ such that

$$\|u\|_{L^p(\partial\Omega)} \leqslant \varepsilon\|Du\|_p + c_\varepsilon\|u\|_{L^p(\Omega)} \qquad \forall\, u \in W^{1,p}(\Omega).$$

Problem 2.216**

Suppose that X is a Banach space, X^* is its topological dual, $U \subseteq X$ is a nonempty open set and $A\colon U \longrightarrow 2^{X^*}\setminus\{\emptyset\}$ is a monotone map with convex and w^*-closed values which is upper semicontinuous from X into $X^*_{w^*}$ (by $X^*_{w^*}$ we denote the Banach space X^* furnished with the w^*-topology). Show that A is maximal monotone.

Problem 2.217**

Suppose that X is a Banach space, X^* is its topological dual, $A\colon X \longrightarrow 2^{X^*}\setminus\{\emptyset\}$ is a monotone, locally bounded map with convex values and $\operatorname{Gr} A$ is closed in $X \times X^*_{w^*}$ (by $X^*_{w^*}$ we denote the Banach space X^* furnished with the w^*-topology). Show that A is maximal monotone.

Problem 2.218*

Let X be a Banach space and let $A\colon X \supseteq D(A) \longrightarrow 2^{X^*}$ be a strongly coercive and surjective map. Show that A^{-1} is locally bounded.

2.3 Solutions

Solution of Problem 2.1

Evidently for each bounded set $B \subseteq X$, the set $\overline{A(B)} \subseteq Y$ is compact. We need to show that A is also continuous (see Definition 2.1(a)). Arguing by contradiction, suppose that A is not continuous. Then for each $n \geqslant 1$, we can find $u_n \in X$ such that

$$\|A(u_n)\|_Y > n\|u_n\|_X.$$

Then the sequence $\{\frac{u_n}{\|u_n\|_X}\}_{n\geqslant 1}$ is bounded (it is on the unit sphere of X). But the sequence $\{A(\frac{u_n}{\|u_n\|_X})\}_{n\geqslant 1}$ is unbounded, a contradiction to the hypothesis.

Solution of Problem 2.2

Clearly f is continuous. Let $B \subseteq D$ be a nonempty bounded set. Given $\varepsilon > 0$, we can find $\alpha_0 = \alpha_0(\varepsilon, B) \in J$ such that

$$\|f_\alpha(u) - f(u)\|_Y < \tfrac{\varepsilon}{2} \qquad \forall\, \alpha \geqslant \alpha_0,\ u \in B. \tag{2.1}$$

We fix $\alpha \in J$, $\alpha \geqslant \alpha_0$. We have that $f_\alpha(B) \subseteq Y$ is relatively compact, in particular then totally bounded (see Definition I.1.70 and Theorem I.1.71). So, we can find $N = N(\varepsilon, B) \geqslant 1$ and $y_1, \ldots, y_N \in Y$ such that

$$f_\alpha(B) \subseteq \bigcup_{n=1}^{N} B_{\frac{\varepsilon}{2}}(y_n), \tag{2.2}$$

where $B_{\frac{\varepsilon}{2}}(y_n) = \{y \in Y : \|y - y_n\|_Y < \frac{\varepsilon}{2}\}$. Let $u \in B$. We can find $n_0 \in \{1, \ldots, N\}$ such that

$$\|f_\alpha(u) - y_{n_0}\|_Y < \tfrac{\varepsilon}{2}$$

(see (2.2)), so

$$\|f(u) - y_{n_0}\|_Y \leqslant \|f(u) - f_\alpha(u)\|_Y + \|f_\alpha(u) - y_{n_0}\|_Y < \tfrac{\varepsilon}{2} + \tfrac{\varepsilon}{2} = \varepsilon$$

(see (2.1) and (2.2)), thus

$$f(B) \subseteq \bigcup_{n=1}^{N} B_\varepsilon(y_n)$$

(since $u \in B$ is arbitrary). This means that $f(B)$ is totally bounded, hence relatively compact. We conclude that $f \in K(D;Y)$ (see Definition 2.1(a)).

Solution of Problem 2.3

By hypothesis, for every $n \geqslant 1$, we can find $u_n \in D$ such that

$$\|u_n - f(u_n)\|_X < \tfrac{1}{n}. \tag{2.3}$$

Since f is compact (see Definition 2.1(a)), passing to a subsequence if necessary, we may assume that $f(u_n) \longrightarrow v \in \overline{F(D)}$. It follows that $u_n \longrightarrow v$ (see (2.3)) and since D is closed, we have $v \in D$. The continuity of f implies that $f(u_n) \longrightarrow f(v)$. Hence $v = f(v)$ with $v \in D$.

Solution of Problem 2.4

(a) Let $\overline{B}_1$ be the closed unit ball in X. We have

$$A(X) = \bigcup_{n\geqslant 1} nA(\overline{B}_1). \tag{2.4}$$

But since $A \in \mathcal{L}_c(X;Y)$ (see Definition 2.1(a)), we have that the set $\overline{A(\overline{B}_1)}$ is compact, hence separable. It follows that $\bigcup\limits_{n\geqslant 1} n\overline{A(\overline{B}_1)}$ is separable and so we conclude that $A(X)$ is separable (see (2.4)).

(b) Since $A(X)$ is of second category in itself (see Definition I.1.25), from the open mapping theorem (see Theorem I.5.47), we have that $\overline{A(\overline{B}_1)}$ is a neighborhood of the origin in $(A(X), \|\cdot\|_Y)$. The compactness of A (see Definition 2.1(a)) implies that $\overline{A(\overline{B}_1)}$ is compact. So, $(A(X), \|\cdot\|_Y)$ is locally compact, hence finite dimensional (see Proposition I.5.9(a)). Therefore $A \in \mathcal{L}_f(X;Y)$ (see Definition 2.1(c)).

Solution of Problem 2.5

"**(a)** $\Longrightarrow$ **(b)**": Let $\{u_n\}_{n\geqslant 1} \subseteq f^{-1}(y)$ be a sequence. Then $f(u_n) = y$ for all $n \geqslant 1$ and so from property **(a)**, we can find a subsequence $\{u_{n_k}\}_{k\geqslant 1}$ such that $u_{n_k} \longrightarrow u \in X$. The continuity of f implies that $f(u_{n_k}) \longrightarrow f(u)$ in Y. Hence $y = f(u)$. Therefore $u \in f^{-1}(y)$ and this shows that $f^{-1}(y)$ is compact. Let $C \subseteq X$ be closed and let $y \in \overline{f(C)}$. We can find a sequence $\{u_n\}_{n\geqslant 1} \subseteq C$ such that $f(u_n) \longrightarrow y$ in Y. According to property **(a)**, we can find a subsequence $\{u_{n_k}\}_{k\geqslant 1}$ such that $u_{n_k} \longrightarrow u \in C$ (recall that the set C is closed). The continuity of f implies that $f(u_{n_k}) \longrightarrow f(u)$ in Y, hence $y = f(u) \in f(C)$. This proves that $f(C)$ is closed and so f is a closed map (see Proposition 2.14).

"**(b)** $\Longrightarrow$ **(a)**": Let $\{u_n\}_{n\geqslant 1} \subseteq X$ be a sequence such that $f(u_n) \longrightarrow y \in Y$. Let $C_n = \{u_k : k \geqslant n\}$ for $n \geqslant 1$. Since f is closed, we have $f(\overline{C}_n) = \overline{f(C_n)}$. Since $f(u_n) \longrightarrow y$, we have

$$\{y\} = \bigcap_{n\geqslant 1} \overline{f(C_n)} = \bigcap_{n\geqslant 1} f(\overline{C}_n).$$

So, if $x \in f^{-1}(y)$, then

$$f(x) \in \bigcap_{n \geqslant 1} f(\overline{C}_n).$$

Then the closed sets $D_n = f^{-1}(y) \cap \overline{C}_n$ for $n \geqslant 1$, are nonempty and have the finite intersection property (see Definition I.2.80). The compactness of $f^{-1}(y)$ implies that

$$f^{-1}(y) \cap \Big(\bigcap_{n \geqslant 1} \overline{C}_n \Big) \neq \emptyset$$

(see and Theorem I.2.81). So, there is a limit point of the sequence $\{u_n\}_{n \geqslant 1}$ which belongs in $f^{-1}(y)$.

Solution of Problem 2.6

Let $C \subseteq Y$ be a compact set and let $\{u_n\}_{n \geqslant 1} \subseteq f^{-1}(C)$ be a sequence. Since $\{f(u_n)\}_{n \geqslant 1} \subseteq C$, we can find a subsequence $\{u_{n_k}\}_{k \geqslant 1}$ such that $f(u_{n_k}) \longrightarrow y \in C$. From property **(a)** of Problem 2.5, we know that we can find a subsequence $\{u_{n_{k_m}}\}_{m \geqslant 1}$ of $\{u_{n_k}\}_{k \geqslant 1}$ such that $u_{n_{k_m}} \longrightarrow v \in X$. The continuity of f implies that $f(u_{n_{k_m}}) \longrightarrow f(v) \in C$. Then $y = f(v) \in C$ and so $v \in f^{-1}(C)$ which establishes the compactness of $f^{-1}(C)$. Therefore f is proper (see Definition 2.13).

Solution of Problem 2.7

Let $C \subseteq X$ be a closed subset. The compactness of X implies that C is compact too (see Proposition I.1.69(b)). Then from the continuity of f, we have that the set $f(C) \subseteq Y$ is compact (see Proposition I.1.74), hence closed too (see Proposition 2.14(b)). Thus we have proved that f is a closed map (see Proposition 2.14). Also, for every $y \in Y$, the set $f^{-1}(y) \subseteq X$ is closed (due to the continuity of f), hence compact in X (since X is compact; see Proposition I.1.69(b)). So, we have verified property **(b)** of Problem 2.5. Invoking Problem 2.6, we conclude that f is proper (see Definition 2.13).

Solution of Problem 2.8

Evidently $K \in \mathcal{L}(l^2)$. For every $n \geqslant 1$, we consider $K_n \in \mathcal{L}_f(l^2)$ defined by

$$K_n(u) = (\vartheta_1 x_1, \vartheta_2 x_2, \ldots, \vartheta_n x_n, 0, \ldots) \qquad \forall\, u = \{x_n\}_{n \geqslant 1} \in l^2.$$

Since by hypothesis $\vartheta_n \longrightarrow 0$, given $\varepsilon > 0$, we can find $n_0 \geqslant 1$ such that

$$|\vartheta_n| \leqslant \varepsilon \qquad \forall\, n \geqslant n_0.$$

We have

$$\|(K - K_n)(u)\|_{l^2} = \Big(\sum_{k \geqslant n+1} \vartheta_k^2 x_k^2 \Big)^{\frac{1}{2}} \leqslant \varepsilon \|u\|_{l^2} \qquad \forall\, n \geqslant n_0,\ u \in l^2,$$

so

$$\|K - K_n\|_{\mathcal{L}} \leqslant \varepsilon \qquad \forall\, n \geqslant n_0,$$

thus $K_n \longrightarrow K$ and hence $K \in \mathcal{L}_c(l^2)$ (see Problem 2.2).

Solution of Problem 2.9

Let $V = A(X) \subseteq X$ and consider $A|_V : V \longrightarrow V$. If $v \in V$, then we can find $x \in X$ such that $v = A(x)$. Hence $A(v) = A^2(x) = A(x) = v$. If follows that $A|_V = I_V$, which implies that V is finite dimensional (see Theorem I.5.22). Therefore $A \in \mathcal{L}_f(X)$ (see Definition 2.1(c)).

Solution of Problem 2.10

Let $C \subseteq D$ be a compact subset and let $E = (I_X - f)^{-1}(C)$. The continuity of $I_X - f$ implies that E is closed in X. Also, note that $E \subseteq C + f(E)$ and so from Proposition 2.32, we have

$$\gamma(E) \leqslant \gamma(C + f(E)) \leqslant \gamma(C) + \gamma(f(E)) = \gamma(f(E)). \tag{2.5}$$

If E is not compact, then $\gamma(E) > 0$ (see Proposition 2.32) and since γ is condensing (see Definition 2.31(e)), we have

$$\gamma(f(E)) < \gamma(E). \tag{2.6}$$

Comparing (2.5) and (2.6) we reach a contradiction and so we conclude that E is compact. This means that $I_X - f$ is proper (see Definition 2.13).

Solution of Problem 2.11

First let us show that $\alpha(B_1) = 2$ (see Definition 2.31). Note that by virtue of Proposition 2.32(e), we may replace B_1 by $\partial B_1 = \{x \in X : \ \|x\|_X = 1\}$. From Definition 2.31, it is clear that $\alpha(\partial B_1) \leqslant 2$. Suppose that $\alpha(\partial B_1) < 2$. This means that

$$\partial B_1 = \bigcup_{n=1}^{N} C_n,$$

with $\operatorname{diam} C_n < 2$ for all $n \in \{1, \ldots, N\}$. Since $\overline{\bigcup_{n=1}^{N} C_n} = \bigcup_{n=1}^{N} \overline{C}_n$, we may assume that the sets C_n are closed. Let V be an N-dimensional subspace of X. Then

$$\partial B_1 \cap V = \bigcup_{n=1}^{N} (C_n \cap V).$$

Therefore $\{C_1 \cap V, C_2 \cap V, \ldots, C_N \cap V\}$ is a closed cover of $\partial B_1 \cap V$ (the unit sphere of V) and so at least for one $n_0 \in \{1, \ldots, N\}$, the set $C_{n_0} \cap V$ must contain a pair of antipodal points. Thus we have

$$2 = \operatorname{diam} (C_{n_0} \cap V) \leqslant \operatorname{diam} C_{n_0},$$

a contradiction. This proves that $\alpha(B_1) = 2$.

Next let us show that $\beta(B_1) = 1$. Again from Definition 2.31, we have $\beta(B_1) \leqslant 1$. Suppose that $\beta(B_1) = \lambda < 1$. Let $\varepsilon > 0$ be such that $\lambda + \varepsilon < 1$. From Definition 2.31, we know that we can find $u_1, \ldots, u_N \in X$ such that

$$B_1 \subseteq \bigcup_{n=1}^{N} B_{\lambda+\varepsilon}(u_n) = \bigcup_{n=1}^{N} (u_n + (\lambda+\varepsilon)B_1).$$

Then using Proposition 2.32, we have

$$\lambda = \gamma(B_1) \leqslant (\lambda+\varepsilon)\gamma(B_1) = (\lambda+\varepsilon)\lambda < \lambda,$$

a contradiction. This proves that $\beta(B_1) = 1$.

Solution of Problem 2.12

We do the solution for $\gamma = \alpha$ and since $\beta \leqslant \alpha$ (see Definition 2.31), the result is also true for $\gamma = \beta$. For each $n \geqslant 1$ choose $u_n \in C_n$ and let $C = \{u_n\}_{n \geqslant 1}$. Since by hypothesis $\alpha(C_n) \longrightarrow 0$, given $\varepsilon > 0$, we can find $n_0 \geqslant 1$ such that

$$\alpha(\overline{C}_n) \leqslant \varepsilon \qquad \forall\, n \geqslant n_0. \tag{2.7}$$

We have

$$\begin{aligned} \alpha(C) &= \alpha\big(\{u_1, \ldots, u_{n_0}\} \cup \{u_n\}_{n \geqslant n_0+1}\big) \\ &= \alpha\big(\{u_n\}_{n \geqslant n_0+1}\big) \leqslant \alpha(C_{n_0}) \leqslant \varepsilon \end{aligned}$$

(see Proposition 2.32, (2.7) and recall that the sequence $\{C_n\}_{n \geqslant 1}$ is decreasing). Since $\varepsilon > 0$ is arbitrary, we let $\varepsilon \searrow 0$ and obtain $\alpha(C) = 0$, which means that $\overline{C}$ is compact. Therefore, by passing to a subsequence if necessary, we may assume that $u_n \longrightarrow u$ in

X. Evidently $u \in \overline{C}_n$ for all $n \geqslant 1$ and so $u \in \bigcap_{n\geqslant 1} \overline{C}_n$. Therefore $\bigcap_{n\geqslant 1} \overline{C}_n$ is nonempty. Moreover $\bigcap_{n\geqslant 1} \overline{C}_n$ is closed and

$$\alpha\Big(\bigcap_{n\geqslant 1} \overline{C}_n\Big) \leqslant \alpha(\overline{C}_n) = \alpha(C_n) \qquad \forall\, n \geqslant 1,$$

so

$$\alpha\Big(\bigcap_{n\geqslant 1} \overline{C}_n\Big) = 0,$$

thus $\bigcap_{n\geqslant 1} \overline{C}_n$ is compact.

Solution of Problem 2.13

We know that $\overline{A(C)}$ is compact. So, it suffices to show that $A(C) \subseteq Y$ is closed. Let $\{y_n\}_{n\geqslant 1} \subseteq A(C)$ be a sequence such that $y_n \longrightarrow y$ in Y. We have $y_n = A(u_n)$ with $u_n \in C$ for all $n \geqslant 1$. Since X is reflexive and $C \subseteq X$ is bounded, by the Eberlein–Smulian theorem (see Theorem I.5.78) and passing to a suitable subsequence if necessary, we may assume that $u_n \xrightarrow{w} u$ in X. So, $u \in \overline{C}^w$. But the convexity of C and the Mazur theorem (see Theorem I.5.58) imply that $u \in \overline{C} = C$. Also, since $A \in \mathcal{L}(X;Y)$, we have that $A \in \mathcal{L}(X_w;Y_w)$ (here X_w and Y_w denote the Banach spaces X and Y with their weak topologies). Therefore $A(u_n) \xrightarrow{w} A(u)$ in Y. Recall that $A(u_n) = y_n \longrightarrow y$ in Y. It follows that $y = A(u) \in A(C)$ and so $A(C) \subseteq Y$ is closed, hence compact.

Solution of Problem 2.14

Let $\{y_n\}_{n\geqslant 1} \subseteq Y$ be a bounded sequence. The continuity of L implies that the sequence $\{L(y_n)\}_{n\geqslant 1} \subseteq X$ is bounded. Since A is compact, it follows that the sequence $\{A(L(y_n))\}_{n\geqslant 1}$ admits a strongly convergent subsequence and this shows that $AL \in \mathcal{L}_c(Y)$ (see Problem 2.1). Let $\{x_n\}_{n\geqslant 1} \subseteq X$ be a bounded sequence. Since $A \in \mathcal{L}_c(X;Y)$, the sequence $\{A(x_n)\}_{n\geqslant 1}$ admits a strongly convergent subsequence $\{A(x_{n_k})\}_{k\geqslant 1}$. But then $\{L(A(x_{n_k}))\}_{k\geqslant 1}$ is strongly convergent too and so by Problem 2.1, we conclude that $LA \in \mathcal{L}_c(X)$.

Solution of Problem 2.15

Hypothesis on A implies that A is injective. Let $V = A(X) \subseteq Y$. Then $A\colon X \longrightarrow V$ is bijective and so we can define $A^{-1}\colon V \longrightarrow X$ which is linear. Moreover, from the hypothesis on A, we have

$$\|A^{-1}(y)\|_X \leqslant \tfrac{1}{c}\|y\|_Y \qquad \forall\, y \in V,$$

so

$$A^{-1} \in \mathcal{L}(V;X).$$

If A is compact (see Definition 2.1(a)), then $I_X = A^{-1}A \in \mathcal{L}_c(X)$ (see Problem 2.14), hence X is finite dimensional (being locally compact; see Proposition I.5.9(a)).

On the other hand, if X is finite dimensional, then every $A \in \mathcal{L}(X;Y)$ is compact.

Solution of Problem 2.16

We argue indirectly. So, suppose that the claim of the problem is not true. Then we have

$$\inf_{u\in\partial B_1} \|A(u)\|_X > 0.$$

So, we can find $c > 0$ such that $\|A(u)\|_X \geqslant c\|u\|_X$ for all $u \in X$. From Problem 2.15, it follows that X is finite dimensional, a contradiction as $A \notin \mathcal{L}_f(X)$. This proves that $0 \in \overline{A(\partial B_1)}$.

Solution of Problem 2.17

No. Let e_n be the standard n-th basis element of l^2. Then $\|e_n\|_{l^2} = 1$ for all $n \geqslant 1$ and

$$\|S(e_n) - S(e_m)\|_{l^2} = \sqrt{2} \qquad \forall\, n,m \geqslant 1,\ n \neq m.$$

Therefore the sequence $\{S(e_n)\}_{n\geqslant 1}$ has no convergence subsequence and consequently $S \notin \mathcal{L}_c(X)$ (see Definition 2.1).

Solution of Problem 2.18

The hypothesis on A implies that we can find $u_1^*, \ldots, u_n^* \in X^*$ such that if

$$V = \big\{u \in X :\ |\langle u_k^*, u\rangle| < 1 \text{ for all } k = 1, \ldots, n\big\},$$

then $A(V) \subseteq B_1^Y = \{y \in Y :\ \|y\|_Y < 1\}$. Then we have $A|_{\bigcap\limits_{k=1}^{n} (u_k^*)^{-1}(0)} = 0$. Note that $\bigcap\limits_{k=1}^{n} (u_k^*)^{-1}(0)$ is finite codimensional. So, we can write

$$X = \Big(\bigcap_{k=1}^{n}(u_k^*)^{-1}(0)\Big) \oplus Z,$$

with $\dim Z < +\infty$. Observe that $A(X) = A(Z)$ and the latter is finite dimensional. Therefore $A \in \mathcal{L}_f(X;Y)$ (see Definition 2.1).

Solution of Problem 2.19

We know that

$$\|A\|_{\mathcal{L}} = \sup_{\|u\|_X \leqslant 1} \|A(u)\|_X.$$

From Problem 2.13, we know that the set $A(\overline{B}_1) \subseteq X$ is compact. Since the norm functional is continuous, from the Weierstrass theorem (see Theorem I.1.75), we know that we can find $u_0 \in X$ with $\|u_0\|_X \leqslant 1$ such that

$$\|A(u_0)\|_X = \sup_{\|u\|_X \leqslant 1} \|A(u)\|_X = \|A\|_{\mathcal{L}}.$$

Solution of Problem 2.20

Let $X = l^p$ with $1 \leqslant p < \infty$ and let $\{e_n\}_{n\geqslant 1}$ be the standard Schauder basis of l^p. Let $A \in \mathcal{L}(l^p)$ be defined by

$$A(u_1, u_2, \ldots) = (0, u_1, 0, u_3, 0, u_5, \ldots) \qquad \forall\ \{u_n\}_{n\geqslant 1} \in l^2.$$

Then $A(e_{2n-1}) = e_{2n}$ for all $n \geqslant 1$. Hence

$$\|A(e_n) - A(e_m)\|_{l^2} = 2^{\frac{1}{p}} \qquad \forall\ n, m \text{ odd},\ n \neq m,$$

so the sequence $\{A(e_n)\}_{n\geqslant 1}$ has no strongly convergent subsequence and thus A is not compact (see Definition 2.1(a)).

On the other hand,

$$A^2(u_1, u_2, \ldots) = A(0, u_1, 0, u_3, 0, u_5, \ldots) = (0, 0, \ldots) \qquad \forall\ \{u_n\}_{n\geqslant 1} \in l^2,$$

that is, $A^2 = 0$ and so it is trivially compact.

Solution of Problem 2.21

"(a) $\Longrightarrow$ (b)": Suppose that the implication is not true. So, we can find an orthonormal basis $\{e_\alpha\}_{\alpha\in J}$ and $\varepsilon>0$ such that the set $\{\alpha\in J:\ |(A(e_\alpha),e_\alpha)_H|\geqslant\varepsilon\}$ is infinite. Thus, we can find an orthonormal sequence $\{e_n\}_{n\geqslant 1}\subseteq\{e_\alpha\}_{\alpha\in J}$ such that

$$|(A(e_n),e_n)_H|\ \geqslant\ \varepsilon\qquad\forall\, n\geqslant 1.$$

The compactness of A (see Definition 2.1(a)) implies that we can find a subsequence $\{e_{n_k}\}_{k\geqslant 1}$ of $\{e_n\}_{n\geqslant 1}$ such that $A(e_{n_k})\longrightarrow u$ in H. By throwing away a finite number of elements of this sequence, we may assume that

$$\|A(e_{n_k})-u\|_H\ <\ \tfrac{\varepsilon}{2}\qquad\forall\, k\geqslant 1,$$

so

$$\begin{aligned}|(A(e_{n_k}),e_{n_k})_H-(u,e_{n_k})_H| &\leqslant\ \|A(e_{n_k})-u\|\,\|e_{n_k}\|_H\\ &=\ \|A(e_{n_k})-u\|_H\ <\ \tfrac{\varepsilon}{2}\qquad\forall\, k\geqslant 1\end{aligned}$$

and so

$$|(u,e_{n_k})_H|\ >\ \tfrac{\varepsilon}{2}\qquad\forall\, k\geqslant 1.$$

But this contradicts the Bessel inequality (see Theorem I.5.105(a)).

"(b) $\Longrightarrow$ (c)": Let $n\geqslant 1$ and let $\mathcal{Y}$ be the family of all orthonormal sets $\{e_\alpha\}_{\alpha\in J}$ in H such that

$$|(A(e_\alpha),e_\alpha)|\ \geqslant\ \tfrac{1}{n}\qquad\forall\,\alpha\in J.$$

By property **(b)**, J is finite. The family $\mathcal{Y}$ is partially ordered by inclusion. Let $\mathcal{D}\subseteq\mathcal{Y}$ be a chain (that is, a linear ordered subset of $\mathcal{Y}$). The union of the orthonormal set in $\mathcal{D}$ is still an element of $\mathcal{Y}$ and of course is an upper bound for $\mathcal{D}$. Thus invoking the Kuratowski–Zorn lemma (see Theorem 4.120), we infer that $\mathcal{Y}$ has a maximal element $\{e_\alpha\}_{\alpha\in L}$. Evidently L is finite. Let $Y=\operatorname{span}\{e_\alpha\}_{\alpha\in L}$. Then Y is finite dimensional and

$$|(A(u),u)_H|\ <\ \tfrac{1}{n}\qquad\forall\, u\in Y^\perp,\ \|u\|_H=1$$

or otherwise $\{e_\alpha\}_{\alpha\in L}\cup\{u\}\in\mathcal{Y}$, which contradicts the maximality of $\{e_\alpha\}_{\alpha\in L}$. Let $P_n\in\mathcal{L}(H)$ be the orthogonal projection on Y (see Definition 2.27(c)) and let $u=(I_H-P_n)x$ with $x\in H$. We have

$$|(A(I_H-P_n)x,(I_H-P_n)x)_H|\ <\ \tfrac{1}{n},$$

so

$$|((I_H-P_n)A(I_H-P_n)x,x)_H|\ <\ \tfrac{1}{n}\qquad\forall\, x\in H,\ \|x\|_H\leqslant 1$$

(see Proposition I.5.161) and thus

$$\|(I_H-P_n)A(I_H-P_n)\|_{\mathcal{L}}\ <\ \tfrac{2}{n}.$$

The linear operator $A_n = AP_n + P_nA - P_nAP_n \in \mathcal{L}(H)$ has finite rank (i.e., $A_n \in \mathcal{L}_f(H)$; see Definition 2.1(c)) and from last inequality we see that $\|A - A_n\|_{\mathcal{L}} < \frac{2}{n}$.

"(c) $\Longrightarrow$ (a)": This implication follows from Problem 2.2 and Remark 2.2.

Solution of Problem 2.22

Evidently $L \in \mathcal{L}(L^2(S); L^2(T))$ and $\|L\|_{\mathcal{L}} \leqslant \|K\|_{L^2(T\times S)}$. Let $\{v_n\}_{n\geqslant 1}$ be an orthonormal basis for $L^2(T)$. For fixed $s \in S$, we expand $K(\cdot, s)$ with respect to this basis. So, we have

$$K(t,s) = \sum_{n\geqslant 1} v_n(t)h_n(s).$$

From the property of the Fourier coefficients, we have $h_n \in L^2(S)$. Using the Parseval relation (see Theorem I.5.105(b)), we have

$$\int_T |K(t,s)|^2\, ds = \sum_{n\geqslant 1} |h_n(s)|^2.$$

Integrating over S, we obtain

$$\int_S\int_T |K(t,s)|^2\, d\xi\, d\mu = \sum_{n\geqslant 1}\int_S |h_n(s)|^2\, d\mu.$$

We define

$$K_m(t,s) = \sum_{n=1}^{m} v_n(t)h_n(s)$$

and

$$L_m(u)(t) = \int_S K_m(t,s)u(s)\, ds \qquad \forall\, u \in L^2(S).$$

We see that $L_m \in \mathcal{L}_f(L^2(S); L^2(T))$ and we have

$$\|L - L_m\|_{\mathcal{L}}^2 \leqslant \int_T\int_S |K(t,s) - K_m(t,s)|^2\, d\mu\, d\xi \qquad \forall\, m \geqslant 1,$$

so $L_m \longrightarrow L$ in $\mathcal{L}(L^2(S); L^2(T))$. Thus operator L is compact (see Definition 2.1(a) and Problem 2.2).

Solution of Problem 2.23

Note that $hu \in C([0,1])$ for any $u \in C([0,1])$. So, A is well defined and of course linear. We have

$$\begin{aligned}\|A(u)\|_{C([0,1])} &= \sup_{t\in[0,1]} |A(u)(t)| = \sup_{t\in[0,1]} |h(t)u(t)| \\ &\leqslant \|h\|_{C([0,1])}\|u\|_{C([0,1])},\end{aligned}$$

so $A \in \mathcal{L}(C([0,1]), C([0,1]))$.

Suppose that A is compact (see Definition 2.1(a)). Then $A(\overline{B}_1) \subseteq C([0,1])$ is relatively compact (recall that $\overline{B}_1 = \{u \in C([0,1]) : \|u\|_{C([0,1])} \leqslant 1\}$). Let $\eta \in (0,1)$ and let $n \geqslant 1$ be such that $\frac{1}{n} < \eta$. We define

$$u_n(t) = \begin{cases} 0 & \text{if } 0 \leqslant t < \eta - \frac{1}{n}, \\ n(t-\eta)+1 & \text{if } \eta - \frac{1}{n} \leqslant t \leqslant \eta, \\ 1 & \text{if } \eta < t \leqslant 1, \end{cases}$$

for all $t \in [0,1]$. Then $hu_n \in A(\overline{B}_1)$ for all $n > \frac{1}{\eta}$. Since $A(\overline{B}_1) \subseteq C([0,1])$ is relatively compact, by the Arzela–Ascoli theorem (see Theorem I.2.181) it is equicontinuous (uniformly since $[0,1]$ is compact). So, for every $\varepsilon > 0$, there exists $\delta = \delta(\varepsilon) > 0$, such that for all $t, s \in [0,1]$ with $|t-s| \leqslant \delta$, we have

$$\big|h(t)u_n(t) - h(s)u_n(s)\big| < \varepsilon \qquad \forall\, n \geqslant 1,\ \tfrac{1}{n} < \eta.$$

We can find $n_\varepsilon \geqslant 1$ such that $\frac{1}{n} < \min\{\eta, \delta\}$ for all $n \geqslant n_\varepsilon$. For $t = \eta - \frac{1}{n}$, $s = \eta$, we have that $|t-s| = \frac{1}{n} < \delta$ and so

$$\big|h(\eta - \tfrac{1}{n})u_n(\eta - \tfrac{1}{n}) - h(\eta)u_n(\eta)\big| < \varepsilon \qquad \forall\, n \geqslant n_\varepsilon.$$

From the definition of $\{u_n\}_{n\geqslant 1}$ it follows that $|h(\eta)| < \varepsilon$ for all $\varepsilon > 0$. Let $\varepsilon \searrow 0$ to conclude that $h(\eta) = 0$ for $\eta \in (0,1)$. From the continuity of h, we have $h(0) = h(1) = 0$, that is, $h \equiv 0$, a contradiction to our hypothesis. Therefore $A(\overline{B}_1) \subseteq C([0,1])$ is not relatively compact which means that A is not a compact operator.

Solution of Problem 2.24

Let $u = \{u_n\}_{n\geqslant 1} \in l^1$. We have

$$\|u\|_{l^2} = \Big(\sum_{n\geqslant 1} u_n^2\Big)^{\frac{1}{2}} \leqslant \sum_{n\geqslant 1} |u_n| = \|u\|_{l^1}. \tag{2.8}$$

This shows that l^1 is embedded continuously into l^2. Let i denote the embedding operator. From (2.8) we see that $\|i\|_{\mathcal{L}} \leqslant 1$. On the other hand, if $e_1 = (1,0,0,\ldots)$, then

$$1 = \|i(e_1)\|_{l^2} \leqslant \|i\|_{\mathcal{L}}\|e_1\|_{l^1} = \|i\|_{\mathcal{L}}$$

and so we conclude that $\|i\|_{\mathcal{L}} = 1$.

Let $\{e_n\}_{n\geqslant 1}$ be the complete Schauder basis for l^1. We have

$$\{e_n\}_{n\geqslant 1} = \{i(e_n)\}_{n\geqslant 1} \subseteq i(\overline{B}_1)$$

(where $\overline{B}_1 = \{u \in l^1 : \|u\|_{l^1} \leqslant 1\}$), so i cannot be compact (see Definition 2.1(a) and Problem 2.1).

Solution of Problem 2.25

From the Schur property of l^1 (see Remark I.5.57), we know that a subset of l^1 is relatively w-compact if and only if it is norm totally bounded (see Definition I.1.70). Since X is reflexive, the closed unit ball $\overline{B}_1 = \{u \in X : \|u\| \leqslant 1\}$ is w-compact. We know that A is weak-to-weak continuous. Hence $A(\overline{B}_1) \subseteq l^1$ is w-compact (see Proposition I.2.82), thus by the Schur property mentioned above, the set $A(\overline{B}_1)$ is relatively norm compact. This implies that A is compact.

Solution of Problem 2.26

In the space

$$c_0 = \{\{u_n\}_{n\geqslant 1} : \{u_n\}_{n\geqslant 1} \text{ is a real sequence such that } u_n \longrightarrow 0\}$$

we consider the norm $\|\{u_n\}_{n\geqslant 1}\|_{c_0} = \sup\limits_{n\geqslant 1} |u_n|$. Given $\vartheta = \{\vartheta_n\}_{n\geqslant 1} \in c_0$, from Problem 2.8, we know that $A_\vartheta(u) = \{\vartheta_n u_n\}_{n\geqslant 1}$ for every $u = \{u_n\}_{n\geqslant 1} \in l^2$ is a compact linear operator. Consider the map $\xi\colon c_0 \longrightarrow \mathcal{L}_c(l^2)$ defined by

$$\xi(\vartheta) = A_\vartheta \qquad \forall\, \vartheta \in c_0.$$

Clearly ξ is linear and for $\vartheta', \vartheta \in c_0$, we have

$$\begin{aligned}
\|A_{\vartheta'} - A_\vartheta\|_{\mathcal{L}} &= \sup_{\|u\|_{l^2}\leqslant 1} \|A_{\vartheta'}(u) - A_\vartheta(u)\|_{l^2} \\
&= \sup_{\|u\|_{l^2}\leqslant 1} \Big(\sum_{n\geqslant 1} (\vartheta'_n - \vartheta_n)^2 u_n^2\Big)^{\frac{1}{2}} \\
&\leqslant \|\vartheta' - \vartheta\|_{c_0} \Big(\sup_{\|u\|_{l^2}\leqslant 1} \sum_{n\geqslant 1} u_n^2\Big)^{\frac{1}{2}} = \|\vartheta' - \vartheta\|_{c_0},
\end{aligned}$$

so $\xi \in \mathcal{L}(c_0; \mathcal{L}_c(l^2))$.

We claim that it is an isometry. To see this, it is enough to check that if $\|\vartheta\|_{c_0} = 1$, then $\|A_\vartheta\|_{\mathcal{L}} = 1$. To this end note that, if $\|u\|_{l^2} = 1$, then

$$\|A_\vartheta(u)\|_{l^2} = \Big(\sum_{n\geqslant 1} \vartheta_n^2 u_n^2\Big)^{\frac{1}{2}} \leqslant \|\vartheta\|_{c_0} \Big(\sum_{n\geqslant 1} u_n^2\Big)^{\frac{1}{2}} = \|\vartheta\|_{c_0} \|u\|_{l^2} = 1,$$

so $\|A_\vartheta\|_{\mathcal{L}} \leqslant 1$. On the other hand, given $\varepsilon > 0$, choose n_0 such that $|\vartheta_{n_0}| > 1 - \varepsilon$ (recall that $\|\vartheta\|_{c_0} = 1$). Let $e_{n_0} \in l^2$ be the standard basic element. Then $A_\vartheta(e_{n_0}) = \vartheta_{n_0} e_{n_0}$ and so

$$\|A_\vartheta(e_{n_0})\|_{l^2} = |\vartheta_{n_0}| > 1 - \varepsilon.$$

Let $\varepsilon \searrow 0$ to conclude that $\|A_\vartheta\|_{\mathcal{L}} \geqslant 1$.

Thus we have shown that $\|A_\vartheta\|_{\mathcal{L}} = 1$. This proves that ξ is an isometry. So, c_0 is isometrically isomorphic to a subspace of $\mathcal{L}_c(l^2)$. Since c_0 is not reflexive, we conclude that $\mathcal{L}_c(l^2)$ cannot be reflexive.

Solution of Problem 2.27

We argue indirectly. So, suppose that the conclusion of the problem is not true. Then we can find $c > 0$ such that

$$\|A(u)\|_Y \geqslant c\|u\|_X \qquad \forall\, u \in X.$$

Since A is compact, Problem 2.15 implies that X is finite dimensional, a contradiction to our hypothesis.

Solution of Problem 2.28

We know that $A^* \in \mathcal{L}(X^*; c_0^*)$. Since X^* is reflexive and $c_0^* = l^1$, from Problem 2.25, we infer that A^* is compact. Then Theorem 2.12 implies that A is compact.

Solution of Problem 2.29

Let $V = R(A)$. Then V is a Banach space and $A\colon X \longrightarrow V$ is surjective. So, by the open mapping theorem (see Theorem I.5.47), we can find $c > 0$ such that $cB_1^V \subseteq A(B_1^X)$ (where $B_1^X = \big(u \in X : \|u\|_X < 1\big)$ and $B_1^V = \big(u \in V : \|u\|_V < 1\big)$). Since A is compact, we have that B_1^V is relatively compact and so V is finite dimensional (being locally compact). This means that $A \in \mathcal{L}_f(X;Y)$. Let Z be a topological complement of $\ker A$ (i.e., Z is closed and $X = \ker A \oplus Z$). Then $\widehat{A} = A|_Z$ is bijective from Z onto $R(A)$. So, $\dim Z = \dim R(A) < +\infty$ (as proved earlier). Therefore we conclude that the Banach space X is finite dimensional.

Solution of Problem 2.30

"$\Longrightarrow$": Suppose that $R(A) \subseteq Y$ is closed. Then by the open mapping theorem (see Theorem I.5.47), A is an open map from X onto $V = R(A)$. Therefore, $A(\overline{B}_1^X)$ is a relatively w-compact set which contains an open neighborhood of the origin (namely the set $A(B_1^X)$). Therefore the set $\overline{B}_1^V = \{v \in V : \|v\|_V \leqslant 1\}$ is w-compact. So, by Theorem I.5.73, the space V is reflexive.

"$\Longleftarrow$": Recall that every reflexive subspace of Y is closed.

Solution of Problem 2.31

"**(a)** $\Longrightarrow$ **(b)**": Let

$$\overline{B}_1^X = \{u \in X : \|u\|_X \leqslant 1\} \quad \text{and} \quad \overline{B}_1^{X^{**}} = \{h \in X^{**} : \|h\|_{X^{**}} \leqslant 1\}.$$

From the Goldstine theorem (see Theorem I.5.70), we know that $\overline{\overline{B}_1^X}^{w^*} = \overline{B}_1^{X^{**}}$, with w^* being the w^*-topology on the space X^{**}. Recall that $A^{**} \in \mathcal{L}(X^{**}_{w^*}; Y^{**}_{w^*})$, where $X^{**}_{w^*}$ (respectively $Y^{**}_{w^*}$) denotes the space X^{**} (respectively Y^{**}) furnished with the w^*-topology. Also, the w^*-topology of Y^{**} restricted to Y, is the w-topology of Y. Therefore, since A is weakly compact (see Definition 2.1(d)), we have $\overline{A(\overline{B}_1^X)}^{w^*} \subseteq Y$, hence $A(\overline{\overline{B}_1^X}^{w^*}) = A(\overline{B}_1^{X^{**}}) \subseteq Y$ from which we conclude that $A(X^{**}) \subseteq Y$.

"**(b)** $\Longrightarrow$ **(c)**": Let $y^*_\alpha \xrightarrow{w^*} 0$ in Y^* and let $h \in X^{**}$. By hypothesis $A^{**}(h) \in Y$, so we have

$$\langle A^*(y^*_\alpha), h\rangle = \langle y^*_\alpha, A^{**}(h)\rangle \longrightarrow 0$$

and thus $A^*(y^*_\alpha) \xrightarrow{w} 0$ in X^*. This prove the continuity of $A^* \colon Y^*_{w^*} \longrightarrow X^*_w$.
"**(c)** $\Longrightarrow$ **(d)**": From the Alaoglu theorem (see Theorem I.5.66), we know that $\overline{B}_1^{Y^*} = \{y^* \in Y^* : \|y^*\|_{Y^*} \leqslant 1\}$ is w^*-compact. So, by hypothesis **(c)**, we have that $A^*(\overline{B}_1^{Y^*}) \subseteq X^*$ is w-compact, which means that A^* is weakly compact.

"**(d)** $\Longrightarrow$ **(a)**": From the implications established thus far, we have that $A^{**} \colon X^{**}_{w^*} \longrightarrow Y^{**}_w$ is continuous. Therefore the set $A^{**}(\overline{B}_1^{X^{**}}) \subseteq Y^{**}$ is w-compact. The space Y is strongly closed in Y^{**} and so by the Mazur theorem (see Theorem I.5.58), it is also weakly closed in Y^{**}. It follows that the set $A^{**}(\overline{B}_1^{X^{**}}) \cap Y \subseteq Y^{**}$ is w-compact and since $w^* \subseteq w$ on Y^{**}, we infer that the set $A^{**}(\overline{B}_1^{X^{**}}) \cap Y \subseteq Y^{**}$ is w^*-compact. We know that

$$A(\overline{B}_1^X) = A^{**}(\overline{B}_1^X) \subseteq A^{**}(\overline{B}_1^{X^{**}}) \cap Y$$

and the w^*-topology of Y^{**} restricted on Y is the w-topology of Y.

So, we have that the set $A(\overline{B}_1^X) \subseteq Y$ is relatively w-compact, hence A is a weakly compact operator.

Solution of Problem 2.32

"$\Longrightarrow$": Assume that $A \in \mathcal{L}_c(X;Y)$. From Theorem 2.12 we have that $A^* \in \mathcal{L}_c(Y^*;X^*)$. Therefore $A^*(\overline{B}_1^{Y^*}) \subseteq X^*$ is relatively compact (recall that $\overline{B}_1^{Y^*} = \{y^* \in Y^* : \|y^*\|_{Y^*} \leqslant 1\}$). So, we can find a sequence $\{u_n^*\}_{n\geqslant 1} \subseteq X^*$ such that $\|u_n^*\|_{X^*} \longrightarrow 0$ and $A^*(\overline{B}_1^{Y^*}) \subseteq \overline{\text{conv}}\,\{u_n^*\}_{n\geqslant 1}$ (see Theorem 2.163). For each $u \in X$, we have

$$\begin{aligned}\|A(u)\|_Y &= \sup_{\|y^*\|_{Y^*}\leqslant 1} \big|\langle y^*, A(u)\rangle\big| \\ &= \sup_{\|y^*\|_{Y^*}\leqslant 1} \big|\langle A^*(y^*), u\rangle\big| \leqslant \sup_{n\geqslant 1}\big|\langle u_n^*, u\rangle\big|.\end{aligned}$$

"$\Longleftarrow$": Consider $T \in \mathcal{L}(X;c_0)$ (see Problem 2.28), defined by

$$T(u) = \{\langle u_n^*, u\rangle\}_{n\geqslant 1}.$$

Let $h = \{\|u_n^*\|_{X^*}\}_{n\geqslant 1} \in c_0$. We have that $T(\overline{B}_1^X) \subseteq [-h,h]$ (here by $[-h,h]$ we denote the order interval in c_0 determined by $\pm h$). The order interval $[-h,h]$ is compact in c_0 (see Theorem 2.164). Therefore $T(\overline{B}_1^X) \subseteq c_0$ is relatively compact. Let $V = T(X)$ and define the operator $L\colon V \longrightarrow Y$ by

$$L(T(u)) = A(u).$$

This operator is well defined. Indeed, if $T(u) = T(x)$, then

$$\langle u_n^*, u - x\rangle = 0 \qquad \forall\, n \geqslant 1,$$

so

$$\|A(u) - A(x)\|_Y \leqslant \sup_{n\geqslant 1}\big|\langle u_n^*, u - x\rangle\big| = 0,$$

thus $A(u) = A(x)$. Moreover, note that

$$\|L(T(u))\|_Y = \|A(u)\|_Y \leqslant \sup_{n\geqslant 1}|\langle u_n^*, u\rangle| = \|T(u)\|_{c_0},$$

so $L \in \mathcal{L}(X;Y)$. Since $A(\overline{B}_1^X) = L(T(\overline{B}_1^X))$ and the latter is relatively compact in Y, we conclude that $A \in \mathcal{L}_c(X;Y)$.

Solution of Problem 2.33

Let $C \subseteq X^*$ be a compact set. We need to show that the set $f^{-1}(C) \subseteq X$ is compact. To this end let $\{u_n\}_{n\geqslant 1} \subseteq F^{-1}(C)$ be a sequence. For every $n \geqslant 1$, let $h_n = f(u_n) \in C$. Since $C \subseteq X^*$ is compact, passing to a subsequence if necessary, we may assume that $h_n \longrightarrow h$ in X^*. The coercivity of f implies that the sequence $\{u_n\}_{n\geqslant 1} \subseteq X$

is bounded. Because X is reflexive, by the Eberlein–Smulian theorem (see Theorem I.5.78), passing to another subsequence if necessary, we may assume that $u_n \xrightarrow{w} u$ in X. We have

$$\langle f(u_n) - f(u), u_n - u \rangle \; = \; \langle h_n - h, u_n - u \rangle \longrightarrow 0,$$

so, from the property of f, we have $u_n \longrightarrow u$ in X. Thus the set $f^{-1}(C)$ is compact, hence f is proper (see Definition 2.13).

Solution of Problem 2.34

"$\Longrightarrow$": By Problem 2.32, we can find a sequence $\{u_n^*\}_{n\geqslant 1} \subseteq X^*$ such that $u_n^* \longrightarrow 0$ in X^* and

$$\|A(u)\|_Y \; \leqslant \; \sup_{n\geqslant 1} |\langle u_n^*, u\rangle| \qquad \forall\, u \in X. \tag{2.9}$$

We can assume that $u_n^* \neq 0$ for every $n \geqslant 1$ and define

$$y_n^* \; = \; \frac{u_n^*}{\|u_n^*\|^{\frac{1}{2}}} \qquad \forall\, n \geqslant 1.$$

Then $\|y_n^*\|_{X^*} = \|u_n^*\|_*^{\frac{1}{2}} \longrightarrow 0$. So, we can define the linear operator $T\colon X \longrightarrow c_0$ by setting

$$T(u) \; = \; \{\langle y_n^*, u\rangle\}_{n\geqslant 1} \qquad \forall\, u \in X.$$

Note that

$$\|T(u)\|_{c_0} \; = \; \sup_{n\geqslant 1} |\langle y_n^*, u\rangle| \; \leqslant \; c\|u\|_X \qquad \forall\, u \in X,$$

for some $c > 0$, thus $T \in \mathcal{L}(X; c_0)$.

Moreover, Problem 2.32 implies that $T \in \mathcal{L}_c(X; c_0)$. We set $V = \overline{T(X)}$ and clearly we have $T \in \mathcal{L}_c(X; V)$. We consider the linear operator $L\colon T(X) \longrightarrow Y$ defined by $L(T(u)) = A(u)$. Note that, if $T(u) = T(v)$, then $\langle u_n^*, u - v\rangle = 0$ for all $n \geqslant 1$ and so from (2.9), we have that $A(u) = A(v)$, hence L is well defined. Moreover, we have

$$\|L(T(u))\|_Y \; = \; \|A(u)\|_Y \; \leqslant \; \sup_{n\geqslant 1} |\langle u_n^*, u\rangle| \; \leqslant \; \widehat{c}\, |\langle y_n^*, u\rangle| \; \leqslant \; \widehat{c}\|T(u)\|_{c_0}$$

with $\widehat{c} = \sup_{n\geqslant 1} \|u_n^*\|_*^{\frac{1}{2}}$ (see (2.9) and recall the definition of T), so $L \in \mathcal{L}(T(X); Y)$. Therefore L has a continuous extension $\widehat{L}$ on $V = \overline{T(X)} \subseteq c_0$. Clearly $A = L \circ T$.

"$\Longleftarrow$": This is an immediate consequence of Problem 2.14.

Solution of Problem 2.35

From Problem 2.31 we know that $A\colon c_0 \longrightarrow X$ is weakly compact (see Definition 2.1(d)) if and only if $A^*\colon X^* \longrightarrow c_0^* = l^1$ is weakly compact. From the Schur property (see Remark I.5.57), it follows that $A^*\colon X^* \longrightarrow l^1$ is weakly compact if and only if it is compact. Invoking the Schauder theorem (see Theorem 2.12) we conclude that $A^*\colon X^* \longrightarrow l^1$ is compact if and only if $A\colon c_0 \longrightarrow X$ is compact.

Solution of Problem 2.36

Let $\{u_n\}_{n\geqslant 1} \subseteq C$ be a minimizing sequence, i.e.,

$$\|A(u_n) - y\|_Y \longrightarrow \inf_{u\in C} \|A(u) - y\|_Y.$$

From Problem 2.13 we know that the set $A(C) \subseteq Y$ is compact. So, we can find a subsequence $\{u_{n_k}\}_{k\geqslant 1}$ of $\{u_n\}_{n\geqslant 1}$ and $u_0 \in C$ such that $A(u_{n_k}) \longrightarrow A(u_0)$ in Y. Then

$$\|A(u_{n_k}) - y\|_Y \longrightarrow \|A(u_0) - y\|_Y,$$

so

$$\|A(u_0) - y\|_Y = \inf_{u\in C} \|A(u) - y\|_Y.$$

Solution of Problem 2.37

Suppose that $A \in \mathcal{L}_f(X;Y)$ and let $n = \dim R(A)$. Let $\{v_1, \ldots v_n\}$ be a basis of $R(A)$. For every $u \in X$, we can write

$$A(u) = \sum_{k=1}^{n} \xi_k(u) v_k. \tag{2.10}$$

The coefficients $\xi_k(u)$ are uniquely determined and clearly ξ_k are linear functions of u, which are bounded since

$$|\xi_k(u)| \leqslant c\|A(u)\|_Y \leqslant c\|A\|_{\mathcal{L}}\|u\|_X$$

for some $c > 0$. Therefore, we can find $v_k^* \in X^*$ for $k = 1, \ldots, n$ such that $\xi_k(u) = \langle v_k^*, u\rangle$. Hence (2.10) becomes

$$A(u) = \sum_{k=1}^{n} \langle v_k^*, u\rangle\, v_k.$$

For any $y^* \in Y^*$ we have

$$\begin{aligned}\langle A^*(y^*), u\rangle &= \langle y^*, A(u)\rangle = \sum_{k=1}^{n} \langle v_k^*, u\rangle \langle y^*, v_k\rangle \\ &= \Big\langle \sum_{k=1}^{n} \langle y^*, v_k\rangle v_k^*, u\Big\rangle.\end{aligned}$$

This is true for every $u \in X$. Therefore

$$A^*(y^*) = \sum_{k=1}^{n} \langle y^*, v_k\rangle v_k^*,$$

so $R(A^*)$ is spanned by $\{v_1^*, \ldots, v_n^*\}$. This shows that $A^* \in \mathcal{L}_f(Y^*; X^*)$ and $\dim R(A^*) \leqslant \dim R(A)$.

In the above argument we replace A by A^* and obtain

$$A^{**} \in \mathcal{L}_f(X; Y) \quad \text{and} \quad \dim R(A^{**}) \leqslant \dim R(A^*).$$

Hence

$$\dim R(A^{**}) \leqslant \dim R(A).$$

But recalling that A is a restriction of A^{**}, we conclude that $\dim R(A) = \dim R(A^*)$.

Solution of Problem 2.38

Let $C = \{A(e_n) : n \geqslant 1\}$. Then $C \subseteq A(\overline{B}_1^{l^1})$ (where $\overline{B}_1^{l^1} = \{u \in l^1 : \|u\|_{l^1} \leqslant 1\}$). So, if A is weakly compact (see Definition 2.1(d)), the set $A(\overline{B}_1^{l^1})$ is relatively weakly compact, hence the same is true for C.

Conversely, suppose that $C \subseteq Y$ is a relatively w-compact set. Let $u = \{u_n\}_{n\geqslant 1} \subseteq l^1$ be such that $\|u\|_{l^1} \leqslant 1$. Let

$$h_n = \sum_{k=1}^{n} u_k A(e_k) \qquad \forall\, n \geqslant 1.$$

Then $\{h_n\}_{n\geqslant 1} \subseteq \overline{\operatorname{conv}}\, C$ and the latter is w-compact since the set $C \subseteq Y$ is relatively w-compact (see Theorem I.5.86). We have

$$h_n = A\Big(\sum_{k=1}^{n} u_k e_k\Big) \longrightarrow A(u) \quad \text{in } Y,$$

so $A(u) \in \overline{\operatorname{conv}}\, C$ and thus $A(\overline{B}_1^{l^1}) \subseteq \overline{\operatorname{conv}}\, C$, which implies that the set $A(\overline{B}_1^{l^1})$ is relatively w-compact. Therefore we conclude that A is weakly compact.

Solution of Problem 2.39

Note that there exists $M > 0$ such that $\|u_n\|_X \leqslant M$ for all $n \geqslant 1$. So, the series $\sum\limits_{n\geqslant 1} x_n u_n$ is norm convergent in X for each $x = \{x_n\}_{n\geqslant 1} \in l^1$. Therefore $A \in \mathcal{L}(l^1; X)$. We see that

$$A(e_n) = u_n \qquad \forall\, n \geqslant 1.$$

Therefore, we can apply Problem 2.38 to conclude that A is weakly compact (see Definition 2.1(d)).

Solution of Problem 2.40

Arguing by contradiction, suppose that $0 \notin \sigma(A)$ (see Definition 2.18). Then A is invertible and so by setting

$$|u| = \|A(u)\|_X \qquad \forall\, u \in X$$

we have defined an equivalent norm on X, whose open unit ball is $A(B_1)$. But this is relatively compact. Therefore $(X, |\cdot|)$ is locally compact, hence X is finite dimensional (see Proposition I.5.9(a)), a contradiction.

Solution of Problem 2.41

Let $\lambda \in \mathbb{C}$ and assume that for some $u \in L^2(0,1)$ we have

$$(\lambda - t)u(t) = 0 \quad \text{for a.a. } t \in (0,1).$$

Thus $u(t) = 0$ for almost all $t \in (0,1)$ and so $\lambda \notin \sigma_p(A)$ (see Definition 2.21).

Next, let $\lambda \in [0,1]$ and let $\varepsilon > 0$ be such that

$$[\lambda, \lambda+\varepsilon] \subseteq [0,1] \quad \text{or} \quad [\lambda-\varepsilon, \lambda] \subseteq [0,1].$$

To fix things, we assume that $[\lambda, \lambda+\varepsilon] \subseteq [0,1]$ (the analysis being analogous when the second inclusion holds). We define

$$u_\varepsilon(t) = \begin{cases} \frac{1}{\sqrt{\varepsilon}} & \text{if } t \in [\lambda, \lambda+\varepsilon], \\ 0 & \text{if } t \in [0,1] \setminus [\lambda, \lambda+\varepsilon]. \end{cases}$$

We have

$$\int_0^1 u_\varepsilon(t)^2\, dt = \int_\lambda^{\lambda+\varepsilon} \tfrac{1}{\varepsilon}\, dt = 1$$

and so $u_\varepsilon \in \partial B_1^{L^2} = \{u \in L^2(0,1) : \|u\|_2 = 1\}$. Then

$$\|(\lambda I_{L^2(0,1)} - A)(u_\varepsilon)\|_2^2 = \tfrac{1}{\varepsilon} \int_\lambda^{\lambda+\varepsilon} (\lambda - t)^2 \, dt = \varepsilon,$$

so

$$(\lambda I_{L^2(0,1)} - A)(u_\varepsilon) \longrightarrow 0 \quad \text{in } L^2(0,1) \quad \text{as } \varepsilon \searrow 0.$$

This implies that $\lambda I_{L^2(0,1)} - A$ is not invertible. Indeed, otherwise we would have

$$\begin{aligned} \|u_\varepsilon\|_2 &= \|(\lambda I_{L^2(0,1)} - A)^{-1}(\lambda I_{L^2(0,1)} - A)(u_\varepsilon)\|_2 \\ &\leqslant \|(\lambda I_{L^2(0,1)} - A)^{-1}\|_{\mathcal{L}} \|(\lambda I_{L^2(0,1)} - A)(u_\varepsilon)\|_2 \longrightarrow 0 \quad \text{as } \varepsilon \searrow 0, \end{aligned}$$

a contradiction to the definition of u_ε. So $\lambda \in \sigma(A)$ (see Definition 2.18) and thus $[0,1] \subseteq \sigma(A)$

Solution of Problem 2.42

(a) Let $u \in H$ be such that $\|u\|_H \leqslant 1$. We have

$$|(A(u), u)_H| \leqslant \|A(u)\|_H \|u\|_H \leqslant \|A\|_{\mathcal{L}} \|u\|_H^2,$$

so

$$\sup_{\|u\|_H \leqslant 1} |(A(u), u)_H| \leqslant \|A\|_{\mathcal{L}}. \tag{2.11}$$

Now, let $\xi = \sup\limits_{\|u\|_H \leqslant 1} |(A(u), u)_H|$. We have

$$|(A(u), u)_H| \leqslant \xi \|u\|_H^2 \qquad \forall\, u \in H.$$

For $u \in H \setminus \{0\}$, we introduce $\lambda = \left(\frac{\|A(u)\|_H}{\|u\|_H}\right)^{\frac{1}{2}}$ and set $y = \frac{1}{\lambda}A(u)$. We have

$$\begin{aligned} &\|A(u)\|_H^2 \\ &= (A(u), A(u))_H = \left(A(\lambda u), \tfrac{1}{\lambda}A(u)\right)_H = (A(\lambda u), y)_H \\ &= \tfrac{1}{4}\left((A(\lambda u + y), \lambda u + y)_H - (A(\lambda u - y), \lambda u - y)_H \right) \\ &\leqslant \tfrac{1}{4}\xi\left(\|\lambda u + y\|_H^2 + \|\lambda u - y\|_H^2\right) \\ &= \tfrac{1}{2}\xi\left(\|\lambda u\|_H^2 + \|y\|_H^2\right) = \tfrac{1}{2}\xi\left(\lambda^2\|u\|_H^2 + \tfrac{1}{\lambda^2}\|A(u)\|_H^2\right) \end{aligned} \tag{2.12}$$

(from the polarization identity since A is self-adjoint and from the parallelogram law; see Remark I.5.94). From the definition of $\lambda > 0$, we see that $\lambda\|u\|_H = \frac{1}{\lambda}\|A(u)\|_H$. Therefore

$$0 = \left(\lambda\|u\|_H - \tfrac{1}{\lambda}\|A(u)\|_H\right)^2,$$

so

$$\lambda^2\|u\|_H^2 + \tfrac{1}{\lambda^2}\|A(u)\|_H^2 \;=\; 2\|u\|_H\|A(u)\|_H.$$

Using this equality in (2.12), we obtain

$$\|A(u)\|_H^2 \;\leqslant\; \xi\|u\|_H\|A(u)\|_H,$$

so

$$\|A(u)\|_H \;\leqslant\; \xi\|u\|_H$$

and thus $\|A\|_{\mathcal{L}} \leqslant \xi$. From this and (2.11), we conclude that

$$\|A\|_{\mathcal{L}} \;=\; \sup_{\|u\|_H\leqslant 1} (A(u),u)_H\,.$$

(b) Since A is self-adjoint, for every $u \in H$, we have that $(A(u),u)_H \in \mathbb{R}$. Let λ be an eigenvalue of A with eigenfunction u (see Definition 2.21). We have

$$(A(u),u)_H \;=\; (\lambda u,u)_H \;=\; \lambda\|u\|_H^2,$$

so

$$\lambda \;=\; \tfrac{(A(u),u)_H}{\|u\|_H^2} \;\in\; \mathbb{R}.$$

Thus, every eigenvalue of A is real.
Next, let λ,μ be two distinct eigenvalues with corresponding eigenvectors $u,v \in H\setminus\{0\}$, respectively. Then

$$A(u) \;=\; \lambda u \quad\text{and}\quad A(v) \;=\; \mu v.$$

We have

$$(A(u),v)_H \;=\; \lambda\,(u,v)_H$$

and

$$(A(u),v)_H \;=\; (u,A(v))_H \;=\; \mu\,(u,v)_H$$

(since A is self-adjoint), so

$$(\lambda-\mu)\,(u,v)_H \;=\; 0,$$

thus $(u,v)_H = 0$. This shows that eigenvectors are orthogonal.

Solution of Problem 2.43
"$\Longrightarrow$": Let $\lambda \in \sigma(A)$ (see Definition 2.18).
It suffices to show that if $\inf\limits_{\|u\|_H=1} \|(\lambda I_H - A)(u)\|_H > 0$ then the operator $\lambda I_H - A$ is invertible. So, suppose that

$$\inf_{\|u\|_H=1} \|(\lambda I_H - A)(u)\|_H \;=\; c \;>\; 0.$$

Then

$$\|(\lambda I_H - A)(u)\|_H \geqslant c\|u\|_H \qquad \forall\, u \in H. \tag{2.13}$$

From (2.13) it is clear that $\lambda I_H - A$ is injective. So, if we show that $\lambda I_H - A$ is also surjective, then by the Banach theorem (see Theorem I.5.48), we would have that $\lambda I_H - A$ is invertible. We show surjectivity of $\lambda I_H - A$ in two steps. First we show that $(\lambda I_H - A)(H)$ is dense in H and then we show that $(\lambda I_H - A)(H)$ is closed.
To show that $(\lambda I_H - A)(H)$ is dense in H, we argue indirectly. So, suppose that $(\lambda I_H - A)(H)$ is not dense in H. Then we can find $\widehat{h} \in H \setminus \{\emptyset\}$ such that

$$\big((\lambda I_H - A)(u), \widehat{h}\big)_H = 0 \qquad \forall\, u \in H.$$

Exploiting the fact that A is self-adjoint (see Definition I.5.108(b)), we have

$$\big((\lambda I_H - A)(u), \widehat{h}\big)_H = \big(u, (\overline{\lambda} I_H - A)(\widehat{h})\big)_H \qquad \forall\, u \in H,$$

so

$$(\overline{\lambda} I_H - A)(\widehat{h}) = 0, \quad \text{with } \widehat{h} \neq 0,$$

thus $\overline{\lambda}$ is an eigenvalue of A (see Definition 2.21. But from Problem 2.42(b), we know that all eigenvalues of A are real. Hence $\overline{\lambda} \in \mathbb{R}$ and so

$$(\lambda I_H - A)(\widehat{h}) = 0, \quad \text{with } \widehat{h} \neq 0,$$

which contradicts (2.13). This proves the density of $(\lambda I_H - A)(H)$ in H. Next we show that $(\lambda I_H - A)(H) \subseteq H$ is closed. So, let $\{u_n\}_{n \geqslant 1} \subseteq H$ be a sequence such that

$$(\lambda I_H - A)(u_n) \longrightarrow h.$$

From (2.13), for all $n, m \geqslant 1$, we have

$$c\|u_n - u_m\|_H \leqslant \|(\lambda I_H - A)(u_n - u_m)\|_H \longrightarrow 0 \quad \text{as } n, m \to +\infty,$$

so $\{u_n\}_{n \geqslant 1} \subseteq H$ is a Cauchy sequence. Thus, we have $u_n \longrightarrow u$ in H for some $u \in H$. From the continuity of $\lambda I_H - A$, it follows that

$$(\lambda I_H - A)(u_n) \longrightarrow (\lambda I_H - A)(u) \quad \text{in } H,$$

so $h = (\lambda I_H - A)(u)$ and thus $h \in R(\lambda I_H - A)$. This proves that $R(\lambda I_H - A)$ is closed. Therefore $R(\lambda I_H - A) = H$.

"$\Longleftarrow$": If $\lambda \in \varrho(A)$, then $(\lambda I_H - A)^{-1} \in \mathcal{L}(H)$ and for $u \in \partial B_1$, we have

$$\begin{aligned} 1 &= \|u\|_H = \|(\lambda I_H - A)^{-1}(\lambda I_H - A)(u)\|_H \\ &\leqslant \|(\lambda I_H - A)^{-1}\|_{\mathcal{L}}\|(\lambda I_H - A)(u)\|_H \end{aligned}$$

(where $\partial B_1 = \{h \in H : \|h\|_H = 1\}$), so

$$\|(\lambda I_H - A)^{-1}\|_{\mathcal{L}}^{-1} \leqslant \|(\lambda I_H - A)(u)\|_H \qquad \forall\, u \in \partial B_1.$$

Solution of Problem 2.44

From Problem 2.42(b) we know that $\sigma(A) \in \mathbb{R}$. Let $\vartheta > 0$. We will show that $M + \vartheta \notin \sigma(A)$. According to Problem 2.43, it suffices to show that

$$\inf_{\|u\|_H=1} \|((M+\vartheta)I_H - A)(u)\|_H > 0.$$

For $u \in H$, with $\|u\|_H = 1$, we have

$$((M+\vartheta)u - A(u), u)_H = (M+\vartheta) - (A(u), u)_H \tag{2.14}$$

$$\geqslant (M+\vartheta) - M = \vartheta > 0 \tag{2.15}$$

(recall that $\|u\|_H = 1$ and see the definition of M). Also we have

$$((M+\vartheta)u - A(u), u)_H \leqslant \|((M+\vartheta)I_H - A)(u)\|_H. \tag{2.16}$$

From (2.14) and (2.16) it follows that

$$0 < \vartheta \leqslant \inf_{\|u\|_H=1} \|((M+\vartheta)I_H - A)(u)\|_H,$$

so $M + \vartheta \notin \sigma(A)$ (see Problem 2.43).

In a similar fashion, we show that $m - \vartheta \notin \sigma(A)$. Therefore $\sigma(A) \subseteq [m, M]$.

Next we show that $M \in \sigma(A)$. Note that $\sigma(A + \vartheta I_H) = \sigma(A) + \vartheta$ and so by replacing A by $A + \vartheta I_H$, we may assume without only loss of generality that $0 \leqslant m \leqslant M$. Then by virtue of Problem 2.42(a), we have

$$\|A\|_{\mathcal{L}} = M. \tag{2.17}$$

According to Problem 2.43, it suffices to show that $\inf\limits_{\|u\|_H=1} \|(MI_H - A)(u)\|_H = 0$. Let $\{u_n\}_{n\geqslant 1} \subseteq H$ be a sequence such that $\|u_n\|_H = 1$ for all $n \geqslant 1$ and

$$(A(u_n), u_n)_H \longrightarrow M = \|A\|_{\mathcal{L}} \tag{2.18}$$

(see (2.17)). We have

$$\begin{aligned} 0 &\leqslant \|(MI_H - A)(u_n)\|_H^2 = (Mu_n - A(u_n), Mu_n - A(u_n))_H \\ &= M^2\|u_n\|_H^2 - 2M\,(A(u_n), u_n)_H + \|A(u_n)\|_H^2 \\ &\leqslant 2M^2 - 2M\,(A(u_n), u_n)_H \longrightarrow 0 \end{aligned}$$

(since A is self-adjoint and using (2.18)), so

$$\inf_{\|u\|_H=1} \|(MI_H - A)(u)\|_H = 0.$$

and thus $M \in \sigma(A)$.

Similarly we show that $m \in \sigma(A)$.

Solution of Problem 2.45

Let $Y = A(X)$ and $V = (I_X - A)(X)$. Then Y and V are nontrivial closed vector subspaces of X and $X = Y \oplus V$. If $y \in Y$ and $v \in V$ are both nonzero, then $A(y) = y$ and $A(v) = 0$, so 0 and 1 are eigenvalues of A (i.e., $\{0,1\} \subseteq \sigma_p(A)$; see Definition 2.21).

Next let $\lambda \in \mathbb{C} \setminus \{0,1\}$. If $u \in X$, then $u = y + v$ with $y \in Y$, $v \in V$ (in a unique way). Then

$$(\lambda I_X - A)(u) = \lambda u - A(u) = (\lambda - 1)y + \lambda v.$$

So, if $(\lambda I_X - A)(u) = 0$, then $(\lambda - 1)y = 0$ and $\lambda v = 0$. Since $\lambda \notin \{0,1\}$, we see that $y = v = 0$ and so $u = 0$. This proves that $\lambda I_X - A$ is injective.

Also, if $u = y + v$ with $y \in Y$, $v \in V$, then for $x = \frac{y}{\lambda-1} + \frac{v}{\lambda}$ we have

$$(\lambda I_X - A)(x) = u,$$

so $\lambda I_X - A$ is surjective and thus $\lambda I_X - A$ is invertible (from the Banach theorem; see Theorem I.5.48). Thus $\sigma(A) \subseteq \{0,1\}$ and so $\sigma_p(A) = \sigma(A) = \{0,1\}$ (see Definitions 2.18 and 2.21).

Solution of Problem 2.46

Let

$$L(u) = \sum_{n \geqslant 0} \xi^{-(n+1)} A^n(u)$$

and

$$L_m(u) = \sum_{n=0}^{m} \xi^{-(n+1)} A^n(u) \qquad \forall\, u \in X,\ m \geqslant 1.$$

Clearly $L_m \in \mathcal{L}(X)$ and $L_m(u) \longrightarrow L(u)$ in X. Then by the uniform boundedness principle (see Corollary I.5.40), we infer that $L \in \mathcal{L}(X)$. For each $u \in X$, we have

$$\begin{aligned}(\xi I_X - A)L(u) &= \lim_{m\to+\infty} \Big((\xi I_X - A)\sum_{n=0}^{m} \xi^{-(n+1)} A^n(u)\Big) \\ &= \lim_{m\to+\infty} \Big(\sum_{n=0}^{m} \xi^{-n} A^n(u) - \sum_{n=0}^{m} \xi^{-(n+1)} A^{n+1}(u)\Big) \\ &= \lim_{m\to+\infty} \big(u - \xi^{-(m+1)} A^{m+1}(u)\big).\end{aligned}$$

By hypothesis $\lim\limits_{m\to+\infty} \xi^{-(m+1)} A^{m+1}(u) = 0$. Therefore it follows that

$$(\xi I_X - A)L(u) = u.$$

In a similar fashion we show that

$$L(\xi I_X - A)(u) = u.$$

From the two last relations we conclude that $(\xi I_X - A)^{-1} = L \in \mathcal{L}(X)$, hence $\xi \in \varrho(A)$ (see Definition 2.18).

Solution of Problem 2.47

Evidently A is well defined and linear. Also, by the Hölder inequality (see Theorem 1.3 and Problem 1.27), we have

$$\big|A(u)(t) - A(u)(\tau)\big| \leqslant |t-\tau|^{\frac{1}{p'}} \|u\|_p \qquad \forall\, t, \tau \in (0,1)$$

(with $\frac{1}{p} + \frac{1}{p'} = 1$). If $1 < p < +\infty$, then $p' < +\infty$ and so we can apply the Arzela–Ascoli theorem (see Theorem I.2.181) and conclude that $A \in \mathcal{L}_c(X)$.

If $p = 1$, then $p' = +\infty$ and then we apply the Kolomogorov–Riesz theorem (see Theorem 1.29) to conclude that $A \in \mathcal{L}_c(X)$.

Next we determine $\sigma(A)$ and $\sigma_p(X)$ (see Definitions 2.18 and 2.21). First we show that $0 \notin \sigma_p(A)$. To this end, assume that $A(u) = 0$. Then

$$\int_0^1 u(s)\chi_{[a,b]}(s)\, ds = 0 \qquad \forall\, a, b \in [0,1].$$

Exploiting the density of simple functions in $L^p(0,1)$ (with $1 \leqslant p \leqslant +\infty$), we conclude that $u \equiv 0$.

Also, let $\lambda \neq 0$ and $u \in C([0,1])$. We set $h = (\lambda I_X - A)^{-1}(u)$. Then $y(t) = \int\limits_0^t h(s)\,ds$ satisfies

$$y \in C^1([0,1]) \quad \text{and} \quad y' = (\lambda I_X - A)^{-1}(u),$$

so

$$\lambda y' - A(y') = \lambda y' - y = u, \quad \text{with } y(0) = 0$$

and thus

$$h(t) = \tfrac{1}{\lambda}u(t) - \tfrac{1}{\lambda^2}\int\limits_0^t e^{(t-s)\lambda}u(s)\,ds.$$

Exploiting the density of $C([0,1])$ in $L^p(0,1)$ (with $1 \leqslant p < +\infty$), we conclude that the last relation is in fact valid for all $u \in L^p(0,1)$. Therefore $\lambda \notin \sigma(A)$ and so

$$\sigma(A) = \{0\} \quad \text{and} \quad \sigma_p(A) = \emptyset.$$

Solution of Problem 2.48

Since A is an isomorphism, we see that $0 \notin \sigma(A)$ (see Definition 2.18). Then from Problem 2.43 we have

$$c\|u\|_H^2 \leqslant (A(u),u)_H \qquad \forall\, u \in H, \tag{2.19}$$

for some $c > 0$. Therefore $\langle A(u),u\rangle > 0$ for all $u \neq 0$ and it follows that $\langle\cdot,\cdot\rangle$ is a new inner product on H.

Let $|u| = \langle u,u\rangle^{\frac{1}{2}}$. Then

$$|u|^2 \leqslant \|A\|_{\mathcal{L}}\|u\|_H^2 \qquad \forall\, u \in H. \tag{2.20}$$

From (2.19) and (2.20) it follows that

$$c\|u\|_H^2 \leqslant |u|^2 \leqslant \|u\|_{\mathcal{L}}\|u\|_H^2 \qquad \forall\, u \in H,$$

so $\|\cdot\|_H$ and $|\cdot|$ are equivalent norms on H.

Solution of Problem 2.49

Note that $\ker L \subseteq \ker(S \circ L) = \ker(I_X + \widehat{K})$ and $\ker(I_X + \widehat{K})$ is finite dimensional (see Proposition 2.24). Also, we have

$$R(L) \supseteq R(L \circ T) = R(I_Y + K)$$

and $R(I_Y + K)$ is finite codimensional, hence so is $R(L)$ (see Proposition 2.24). This shows that $L \in \operatorname{Fred}(X;Y)$ (see Definition 2.29).

Solution of Problem 2.50

We can find $T \in \mathcal{L}(X;Y)$ such that

$$L \circ T = I_Y + K_1 \quad \text{and} \quad T \circ L = I_X + K_2,$$

with $K_1 \in \mathcal{L}_c(Y;Y)$ and $K_2 \in \mathcal{L}_c(X;X)$ (see Remark 2.30). Then we have

$$\begin{aligned} (L+K) \circ T &= I_Y + K_1 + K \circ T, \\ T \circ (L+K) &= I_X + K_2 + T \circ K. \end{aligned}$$

Since $K_1 + K \circ T \in \mathcal{L}_c(Y;Y)$ and $K_2 + T \circ K \in \mathcal{L}_c(X;X)$, from Problem 2.49, we infer that $L+K \in \operatorname{Fred}(X;Y)$. Then consider the map $[0,1] \ni t \longmapsto L(t) = L + tK$. From the first part of the solution, we have that $L(t) \in \operatorname{Fred}(X;Y)$ for all $t \in [0,1]$. Moreover, from Remark 2.30, it follows that $i(L) = i(L+K)$.

Solution of Problem 2.51

(a) "$\Longrightarrow$": Let φ be a lower semicontinuous function and let $V \subseteq \mathbb{R}$ be an open set. According to Proposition 2.37(d) we need to show that the set $E_\varphi^+(V)$ is open (where $E_\varphi^+(V) = \{u \in X : E_\varphi(u) \subseteq V\}$). Let $u \in E_\varphi^+(V)$. Then the inclusion $E_\varphi(u) \subseteq V$ means that $V \supseteq (\lambda, +\infty)$ for some $\lambda < \varphi(u)$. The lower semicontinuity of φ implies that we can find a neighborhood U of u such that $\lambda < \varphi(h)$ for all $h \in U$. Then $E_\varphi(h) \subseteq V$ for all $h \in U$ and so we conclude that $u \longmapsto E_\varphi(u)$ is upper semicontinuous.

"$\Longleftarrow$": Since E_φ is upper semicontinuous, for every $\lambda \in \mathbb{R}$, the set

$$\begin{aligned} E_\varphi^+((\lambda, +\infty)) &= \{u \in X : E_\varphi(u) \subseteq (\lambda, +\infty)\} \\ &= \{u \in X : \varphi(u) > \lambda\} \end{aligned}$$

is open. This means that φ is a lower semicontinuous function.

(b) "$\Longrightarrow$": Let φ be an upper semicontinuous function and let $V \subseteq \mathbb{R}$ be an open set. In this case, we need to show that $E_\varphi^-(V) = \{u \in X : E_\varphi(u) \cap V \neq \emptyset\}$ is open (see Proposition 2.38(e)). Let $u \in E_\varphi^-(V)$. Then the relation $E_\varphi(u) \cap V \neq \emptyset$ means that we can find $\lambda \in V$ such that $\varphi(u) \leqslant \lambda$. In fact the openness of V implies that we can always choose $\lambda \in V$ such that $\varphi(u) < \lambda$. Then the upper

semicontinuity of the function φ implies that we can find a neighborhood U of u such that $\varphi(h) < \lambda$ for all $h \in U$. Hence $E_\varphi(h) \cap V \neq \emptyset$ for all $h \in U$. This implies that the set $E_\varphi^-(V)$ is open, therefore E_φ is lower semicontinuous.
"$\Longleftarrow$": Since E_φ is lower semicontinuous, for every $\lambda \in \mathbb{R}$, the set

$$\begin{aligned} E_\varphi^-((-\infty,\lambda)) &= \{u \in X : E_\varphi(u) \cap (-\infty,\lambda) \neq \emptyset\} \\ &= \{u \in X : \varphi(u) < \lambda\} \end{aligned}$$

is open. This means that φ is an upper semicontinuous function.

Solution of Problem 2.52

First we consider the case of F being lower semicontinuous (see Definition 2.36(b)). So, let $C \subseteq X$ be a connected set and let $V_1, V_2 \subseteq X$ be two open sets such that

$$F(C) \subseteq V_1 \cup V_2, \quad F(C) \cap V_1 \neq \emptyset \quad \text{and} \quad F(C) \cap V_2 \neq \emptyset. \tag{2.21}$$

We need to show that $F(C) \cap V_1 \cap V_2 \neq \emptyset$. To this end, suppose that

$$C \cap F^-(V_1) \cap F^-(V_2) = \emptyset. \tag{2.22}$$

Since F is lower semicontinuous, the sets $F^-(V_1)$ and $F^-(V_2)$ are both open and from (2.21) we infer that

$$C \subseteq F^-(V_1) \cup F^-(V_2), \quad C \cap F^-(V_1) \neq \emptyset \quad \text{and} \quad C \cap F^-(V_2) \neq \emptyset. \tag{2.23}$$

Combining (2.22) and (2.23), we see that we have a contradiction to the connectedness of C. Therefore (2.22) cannot happen and so we have

$$C \cap F^-(V_1) \cap F^-(V_2) \neq \emptyset.$$

Let $u \in C \cap F^-(V_1) \cap F^-(V_2)$. We have

$$F(u) \subseteq V_1 \cup V_2, \quad F(u) \cap V_1 \neq \emptyset \quad \text{and} \quad F(u) \cap V_2 \neq \emptyset.$$

So, from the connectedness of $F(u)$ we infer that

$$F(u) \cap V_1 \cap V_2 \neq \emptyset,$$

so $F(C) \cap V_1 \cap V_2 \neq \emptyset$, which proves the connectedness of $F(C)$.

When F is upper semicontinuous, then the proof if similar, working this time with a "disconnection" consisting of closed sets.

Solution of Problem 2.53

Note that for every nonempty open set $V \subseteq Y$ we have

$$F(u) \cap V \neq \emptyset \quad \text{if and only if} \quad \overline{F(u)} \cap V \neq \emptyset.$$

Therefore F is lower semicontinuous if and only if $\overline{F}$ is lower semicontinuous (see Definition 2.36).

For upper semicontinuous multifunctions the result is not true. To see this, let $X = Y = \mathbb{R}$ and consider the multifunction

$$F(u) = (u-1, u+1) \qquad \forall\, u \in \mathbb{R}.$$

Then

$$F^+((-1,1)) = \{0\}$$

(see Definition 2.35(b)) and so F is not upper semicontinuous. On the other hand,

$$\overline{F}(u) = [u-1, u+1] \qquad \forall\, u \in \mathbb{R}$$

and for every $y \in \mathbb{R}$

$$\varphi_y(u) = \operatorname{dist}_Y(y, \overline{F}(u)) = \begin{cases} u-1-y & \text{if } y \leqslant u-1, \\ 0 & \text{if } u-1 < y < u+1, \\ y-(u+1) & \text{if } u+1 \leqslant y. \end{cases}$$

Clearly φ_y is lower semicontinuous. So, Proposition 2.45 implies that $\overline{F}$ is upper semicontinuous.

Solution of Problem 2.54

Let $V \subseteq Y$ be an open set and let $y \in \operatorname{conv} F(u) \cap V$. Then we have

$$y = \sum_{k=1}^{n} \lambda_k y_k \in V, \quad \text{with } y_1, \ldots, y_n \in F(u),\ \lambda_1, \ldots, \lambda_n \subseteq [0,1],$$

$$\sum_{k=1}^{n} \lambda_k = 1.$$

Let $\varepsilon > 0$ be small such that

$$B_\varepsilon(y) = \{y' \in Y : \|y' - y\|_Y < \varepsilon\} \subseteq V.$$

Note that $F(u) \cap B_\varepsilon(y_k) \neq \emptyset$ for all $k \in \{1, \ldots, n\}$. Then since F is lower semicontinuous, we can find $U_k \in \mathcal{N}(u)$ ($\mathcal{N}(u)$ being the filter of neighborhoods of u) such that for every $u' \in U_k$ we have $F(u') \cap B_\varepsilon(y_k) \neq \emptyset$, with $k \in \{1, \ldots, n\}$. We set $U = \bigcap_{k=1}^{n} U_k$. For every $u' \in U$, let $y'_k \in F(u') \cap B_\varepsilon(y_k)$ and set $y' = \sum_{k=1}^{n} \lambda_k y'_k$. Then

$$\|y' - y\|_Y = \Big\|\sum_{k=1}^{n} \lambda_k (y'_k - y)\Big\|_Y \leqslant \sum_{k=1}^{n} \lambda_k \|y'_k - y_k\|_Y < \varepsilon,$$

so $y' \in \operatorname{conv} F(u') \cap V$ and thus the set $\operatorname{conv} F^-(V)$ is open. Hence $u \longmapsto \operatorname{conv} F(u)$ is lower semicontinuous.

Then using Problem 2.53, we conclude that $u \longmapsto \overline{\operatorname{conv}}\, F(u)$ is lower semicontinuous too.

Solution of Problem 2.55

Since F has compact values, for every $u \in X$ we have $\overline{\operatorname{conv}}\, F(u) \in P_{kc}(Y)$ (by the Mazur theorem; see Theorem I.5.86). Then by virtue of Proposition 2.56(a), it suffices to show that F is h-upper semicontinuous (see Definition 2.53(a)). So, suppose that $\{x_\alpha\}_{\alpha \in J} \subseteq X$ is a net such that $x_\alpha \longrightarrow x$ in X. Then

$$h^*\left(\overline{\operatorname{conv}}\, F(u_\alpha), \overline{\operatorname{conv}}\, F(u)\right) \leqslant h^*\left(F(u_\alpha), F(u)\right)$$

(see Remark 2.50), so using the h-upper semicontinuity of F (see Proposition 2.56(a)), we have

$$h^*\left(\overline{\operatorname{conv}}\, F(u_\alpha), \overline{\operatorname{conv}}\, F(u)\right) \xrightarrow{\alpha \in J} 0.$$

Thus the multifunction $u \longmapsto \overline{\operatorname{conv}}\, F(u)$ is h-upper semicontinuous, hence upper semicontinuous too.

Solution of Problem 2.56

(a) Let $\{y_\alpha\}_{\alpha \in J} \subseteq F(K)$ be a net. Then $y_\alpha \in F(u_\alpha)$ for all $\alpha \in J$ with some net $\{u_\alpha\}_{\alpha \in J} \subseteq K$. The compactness of K implies that we can find a subnet $\{u_\beta\}_{\beta \in I}$ of $\{u_\alpha\}_{\alpha \in J}$ such that $u_\beta \longrightarrow u \in K$. Note that $\{(u_\beta, y_\beta)\}_{\beta \in I} \subseteq \operatorname{Gr} F$. Invoking Proposition 2.42, we infer that $\{y_\beta\}_{\beta \in I}$ admits a cluster point in $F(u) \subseteq F(K)$. This proves the compactness of the set $F(K)$.

(b) According to Proposition 2.37, it suffices to show that for every closed set $C \subseteq Y$, we have that the set $H^-(C) = \{u \in X : H(u) \cap C \neq \emptyset\}$ is closed in X. To this end, let $\{u_\alpha\}_{\alpha \in J} \subseteq H^-(C)$ be a net such that $u_\alpha \longrightarrow u$ in X. Let

$$K = \{u\} \cup \{u_\alpha : \alpha \in J\} \in P_k(X).$$

From **(a)** we have that $G(K) \in P_k(Y)$. Let $y_\alpha \in H(u_\alpha) \cap C$. Then $\{y_\alpha\}_{\alpha \in J} \subseteq G(K)$ and so we can find a subnet $\{y_\beta\}_{\beta \in I}$ of $\{y_\alpha\}_{\alpha \in J}$ such that $y_\beta \longrightarrow y$ in Y. Since the set $C \subseteq Y$ is closed, we have that $y \in C$. Also, using Proposition 2.43, we see that the multifunction $u \longmapsto F(u) \cap G(u)$ is closed, hence $y \in F(u) \cap G(u)$. Therefore $y \in H(u) \cap C$ and so $u \in H^-(C)$. This shows that $H^-(C) \subseteq X$ is closed, hence the multifunction H is upper semicontinuous (see Proposition 2.37).

(c) No. To see this, let $X = Y = [0,1]$ and consider the multifunctions

$$\begin{aligned} F(u) &= \begin{cases} \{\frac{1}{n} : n \geqslant 1\} & \text{if } u \neq 0, \\ \{0\} & \text{if } u = 0, \end{cases} \\ G(u) &= \begin{cases} (0,1] \setminus \{\frac{1}{n} : n \geqslant 2\} & \text{if } u \neq 0, \\ [0,1] & \text{if } u = 0. \end{cases} \end{aligned}$$

Both multifunctions are lower semicontinuous and

$$(F \cap G)(u) = \begin{cases} \{1\} & \text{if } u \neq 0, \\ \{0\} & \text{if } u = 0, \end{cases}$$

which is not lower semicontinuous (see Proposition 2.38) because for the closed set $C = \{1\} \subseteq X$, the set $F^+(C) = (0,1] \subseteq Y$ is not closed.

Solution of Problem 2.57

Evidently, it suffices to show that $\varphi(u) \leqslant \operatorname{dist}(u, D)$ for all $u \in X$. Arguing by contradiction, suppose that we can find $u \in X$ such that $\operatorname{dist}(u, D) < \varphi(u)$. Let $\varepsilon = \varphi(u) - \operatorname{dist}(u, D) > 0$ and choose $h \in D$ such that

$$d_X(u,h) < \operatorname{dist}(u, D) + \tfrac{\varepsilon}{3}.$$

From the hypothesis concerning φ, we can find $E \in P_f(X)$ such that

$$|\varphi(h) - \operatorname{dist}(h, E)| < \tfrac{\varepsilon}{3} \quad \text{and} \quad |\varphi(u) - \operatorname{dist}(u, E)| < \tfrac{\varepsilon}{3}.$$

We have

$$\begin{aligned} \varphi(u) &< \operatorname{dist}(u, E) + \tfrac{\varepsilon}{3} \leqslant d_X(u,h) + \operatorname{dist}(h, E) + \tfrac{\varepsilon}{3} \\ &\leqslant d_X(u,h) + \varphi(h) + \tfrac{2\varepsilon}{3} \\ &= d_X(u,h) + \tfrac{2\varepsilon}{3} < \operatorname{dist}(u, D) + \varepsilon \end{aligned}$$

(since $h \in D$), which contradicts the definition of ε. We conclude that

$$\varphi(u) = \operatorname{dist}(u, D) \qquad \forall\, u \in X.$$

Solution of Problem 2.58

The hypothesis that $C_n \xrightarrow{h} C$ (see Definition 2.86) implies that given $\varepsilon > 0$, we can find $n_0 = n_0(\varepsilon) \geqslant 1$ such that

$$C \subseteq (C_n)_\varepsilon \quad \text{and} \quad C_n \subseteq C_\varepsilon \qquad \forall\, n \geqslant n_0. \tag{2.24}$$

From the second inclusion in (2.24), we have that

$$\bigcup_{n \geqslant n_0} C_n \subseteq C_\varepsilon,$$

so

$$\overline{\bigcup_{n \geqslant n_0} C_n} \subseteq C_{2\varepsilon},$$

thus

$$\bigcap_{m \geqslant 1} \overline{\bigcup_{n \geqslant m} C_n} \subseteq C_{2\varepsilon} \qquad \forall\, \varepsilon > 0$$

and hence

$$\bigcap_{m \geqslant 1} \overline{\bigcup_{n \geqslant m} C_n} \subseteq \overline{C} = C \tag{2.25}$$

(since $C \in P_f(X)$). From the first inclusion in (2.24), we obtain

$$C \subseteq \bigcap_{\varepsilon > 0} \bigcup_{m \geqslant 1} \bigcap_{n \geqslant m} (C_n)_\varepsilon. \tag{2.26}$$

Finally let $u \in \bigcap\limits_{\varepsilon > 0} \bigcup\limits_{m \geqslant 1} \bigcap\limits_{n \geqslant m} (C_n)_\varepsilon$. This means that for every $\varepsilon > 0$, we can find $m_0 = m_0(\varepsilon) \geqslant 1$ such that for all $n \geqslant m_0$ we have $u \in (C_n)_\varepsilon \subseteq \big(\bigcup\limits_{n \geqslant m_0} C_n\big)_\varepsilon$. Since $\varepsilon > 0$ is arbitrary, we infer that

$$u \in \overline{\bigcup_{n \geqslant m_0} C_n},$$

so

$$\bigcap_{\varepsilon > 0} \bigcup_{m \geqslant 1} \bigcap_{n \geqslant m} (C_n)_\varepsilon \subseteq \bigcap_{m \geqslant 1} \overline{\bigcup_{n \geqslant m} C_n}. \tag{2.27}$$

Combining (2.25), (2.26) and (2.27), we reach the equality claimed by the problem.

Solution of Problem 2.59

Let $y \in X \setminus (\bigcup_{C \in K} C)$. Then for every $C \in K$, $y \in X \setminus C$ and so $\mathrm{dist}(y, C) = m_C > 0$. This means that $h(\{y\}, C) \geqslant m_C > 0$. Consider the function $\xi \colon (P_f(X), h) \longrightarrow \mathbb{R}_+$ defined by

$$\xi(D) = h(\{h\}, D) \qquad \forall\, D \in P_f(X).$$

This function is continuous and so $\inf_K \xi$ is attained. Therefore

$$h(\{y\}, C) \geqslant \varepsilon > 0 \qquad \forall\, C \in K,$$

for some $\varepsilon > 0$. This means that $y \notin (\bigcup_{C \in K} C)_\varepsilon$, so

$$\mathrm{dist}(y, \bigcup_{C \in K} C) > 0.$$

Since $y \in X \setminus (\bigcup_{C \in K} C)$ is arbitrary, we conclude that the set $\bigcup_{C \in K} C$ is closed.

Solution of Problem 2.60

"$\Longrightarrow$": Let $\mathcal{F}$ is dense in $(P_f(X), h)$. Note that if $C \in P_f(X)$ and $F \subseteq X$ is finite, then F is an ε-net for the set C if and only if $C \subseteq F_\varepsilon$. Therefore, if $h(F, C) < \varepsilon$ (see Definition 2.49), then F is an ε-net for C. Let

$$B_h(C, \varepsilon) = \{D \in P_f(X) : h(D, C) < \varepsilon\}.$$

If $B_h(C, \varepsilon) \cap \mathcal{F} \neq \emptyset$ for all $\varepsilon > 0$, then C is totally bounded (see Definition I.1.70). In particular this is also true if $C = X$.

"$\Longleftarrow$": Since X is totally bounded, every $C \in P_f(X)$ is also totally bounded. Therefore, for every $C \in P_f(X)$ and every $\varepsilon > 0$, we can find $F \in \mathcal{F}$ such that $C \subseteq F_\varepsilon$. We can also have $F \subseteq C_\varepsilon$, because if this inclusion is not true, then we can find $\widehat{u} \in F$ such that $\mathrm{dist}(\widehat{u}, C) \geqslant \varepsilon$. This implies that $C \cap B_\varepsilon(\widehat{u}) = \emptyset$ (recall that $B_\varepsilon(\widehat{u}) = \{u \in X : d_X(u, \widehat{u}) < \varepsilon\}$) and so $F' = F \setminus \{\widehat{u}\}$ is still an ε-net for C and so, we can replace F by F'. Then $h(F, C) < \varepsilon$ and so $\mathcal{F}$ is dense in $(P_f(X), h)$.

Solution of Problem 2.61

Let $\{u_\alpha\}_{\alpha \in J} \subseteq X$ be a net such that $u_\alpha \longrightarrow u$ in X and let $v = F(u)$. According to Proposition 2.38(d), in order to establish the lower semicontinuity of the multifunction F, we need to find $v_\alpha \in F(u_\alpha)$ for all $\alpha \in J$ such that $v_\alpha \longrightarrow v$ in V. From the definition of F we have that

$$v = \varphi(u, y), \quad \text{with } y \in M(u).$$

Since by hypothesis M is lower semicontinuous, we can find $y_\alpha \in M(u_\alpha)$ for $\alpha \in J$ such that $y_\alpha \longrightarrow y$ in Y (see Proposition 2.38). The continuity of φ implies that

$$v_\alpha = \varphi(u_\alpha, y_\alpha) \longrightarrow \varphi(u, y) = v.$$

Since $v_\alpha = F(u_\alpha)$, we have proved the lower semicontinuity of F.

Solution of Problem 2.62

According to Proposition 2.37, it suffices to show that for every closed set $C \subseteq Y$, the set $F^-(C) = \{u \in X : F(u) \cap C \neq \emptyset\}$ is closed. To this end, let $\{u_n\}_{n\geqslant 1} \subseteq F^-(C)$ be a sequence such that $u_n \longrightarrow u$ in X. From the definition of F, we can find $y_n \in M(u_n)$ such that $\varphi(u_n, y_n) \in C$ for all $n \geqslant 1$. Note that $y_n \in M(\{u\} \cup \{u_n : n \geqslant 1\})$ and from Problem 2.56(a), we know that $M(\{u\} \cup \{u_n : n \geqslant 1\}) \in P_k(Y)$ and so we can find a subsequence $\{y_{n_k}\}_{k\geqslant 1}$ of $\{y_n\}_{n\geqslant 1}$ such that $y_{n_k} \longrightarrow y$ in Y. Invoking Proposition 2.43, we have that $y \in M(u)$. Also, the continuity of φ, implies that

$$\varphi(u_{n_k}, y_{n_k}) \longrightarrow \varphi(u, y) \in C \quad \text{in } V.$$

Since $y \in M(u)$, it follows that $u \in F^-(C)$ and so $F^-(C)$ is closed. This implies the upper semicontinuity of the multifunction F.

Solution of Problem 2.63

Let $\{u_n\}_{n\geqslant 1} \subseteq X$. For each $n \geqslant 1$, let $C_n = \{u_n\} \in P_f(X)$ for $n \geqslant 1$ and since by hypothesis $(P_f(X), h)$ is compact, we can find a subsequence $\{C_{n_k}\}_{k\geqslant 1}$ of $\{C_n\}_{n\geqslant 1}$ and a set $C \in P_f(X)$ such that

$$h(C_{n_k}, C) \longrightarrow 0$$

(see Definition 2.49). Given $\varepsilon > 0$, we can find $k_0 = k_0(\varepsilon) \geqslant 1$ such that

$$C \subseteq (C_{n_k})_\varepsilon \quad \text{and} \quad C_{n_k} \subseteq C_\varepsilon \qquad \forall\, k \geqslant k_0. \tag{2.28}$$

Note that $(C_{n_k})_\varepsilon = B_\varepsilon(u_{n_k})$. Hence $\operatorname{diam} C \leqslant 2\varepsilon$, so $C = \{u\}$ (i.e., C is singleton), thus

$$u_{n_k} \in B_\varepsilon(u) \qquad \forall\, k \geqslant k_0$$

(see (2.28)) and so $u_{n_k} \longrightarrow u$ in X. This proves the compactness of X.

Solution of Problem 2.64

We need to show that for every $\lambda \in \mathbb{R}$, the set $L_\lambda = \{u \in X : v(y) \leqslant \lambda\}$ is closed (see Definition I.2.46). So, let $\{u_\alpha\}_{\alpha \in J} \subseteq L_\lambda$ be a net such that $u_\alpha \longrightarrow u$ in X. Given any $\varepsilon > 0$, we can find $y \in F(u)$ such that

$$v(u) - \varepsilon \leqslant \varphi(u, y). \tag{2.29}$$

The lower semicontinuity of F implies that we can find $y_\alpha \in F(u_\alpha)$ for $\alpha \in J$ such that $y_\alpha \longrightarrow y$ in Y (see Proposition 2.38). We have

$$\varphi(u_\alpha, y_\alpha) \leqslant v(u_\alpha) \leqslant \lambda \qquad \forall\, \alpha \in J,$$

so

$$\varphi(u, y) \leqslant \liminf_{\alpha \in J} \varphi(u_\alpha, y_\alpha) \leqslant \lambda$$

(since φ is lower semicontinuous), thus $v(u) - \varepsilon \leqslant \lambda$ (see (2.29)).

Since $\varepsilon > 0$ was arbitrary, we let $\varepsilon \searrow 0$, to conclude that $v(u) \leqslant \lambda$, so $u \in L_\lambda$ and thus L_λ is closed. We conclude that v is a lower semicontinuous function.

Solution of Problem 2.65

Let $\{(u_\alpha, y_\alpha)\}_{\alpha \in J} \subseteq \operatorname{Gr} F$ be a net and assume that $u_\alpha \longrightarrow u$ in X. Since F is locally compact, we can find $U \in \mathcal{N}(u)$ (where $\mathcal{N}(u)$ is the filter of neighborhoods of u) such that $\overline{F(U)} \in P_k(Y)$. We can find $\alpha_0 \in J$ such that $u_\alpha \in U$ for all $\alpha \geqslant \alpha_0$. Then $\{y_\alpha\}_{\alpha \geqslant \alpha_0} \subseteq Y$ is relatively compact, hence $\{y_\alpha\}_{\alpha \in J}$ has a cluster point in $\overline{F(U)}$. We can find a subnet $\{y_\beta\}_{\beta \in I}$ of $\{y_\alpha\}_{\alpha \in J}$ and $y \in \overline{F(U)}$ such that $y_\beta \longrightarrow y$ in Y. Note that $\{(u_\beta, y_\beta)\}_{\beta \in J} \subseteq \operatorname{Gr} F$ and since $u_\beta \longrightarrow u$ in X, $y_\beta \longrightarrow y$ in Y and F is closed, we have $(u, y) \in \operatorname{Gr} F$, that is, $y \in F(u)$. So, invoking Proposition 2.42, we conclude that F is upper semicontinuous.

Solution of Problem 2.66

From Problem 2.64 we already know that v is lower semicontinuous. So, it suffices to show that v is upper semicontinuous. To this end, for every $\lambda \in \mathbb{R}$, let

$$U_\lambda = \{u \in X : v(u) \geqslant \lambda\}.$$

We need to show that U_λ is closed (see Definition I.2.46). So, let $\{u_\alpha\}_{\alpha \in J} \subseteq U_\lambda$ be a net and assume that $u_\alpha \longrightarrow u$ in X. Since φ is continuous and F is $P_k(Y)$-valued, for every $\alpha \in J$, we can find $y_\alpha \in F(u_\alpha)$ such that

$$v(u_\alpha) = \varphi(u_\alpha, y_\alpha) \geqslant \lambda. \tag{2.30}$$

Invoking Proposition 2.42, we can find a subnet $\{y_\beta\}_{\beta\in I}$ of $\{y_\alpha\}_{\alpha\in J}$ and $y \in F(u)$ such that $y_\beta \longrightarrow y$ in Y. Then the continuity of φ implies that

$$\varphi(u_\beta, y_\beta) \longrightarrow \varphi(u, y) \leqslant v(u),$$

so

$$\lambda \leqslant v(u)$$

(see (2.30)), thus $u \in U_\lambda$ and hence U_λ is closed. This proves that v is also upper semicontinuous, therefore it is continuous.

Clearly S is $P_k(Y)$-valued. For every $u \in X$, let

$$G(u) = \{y \in Y : v(u) = \varphi(u, y)\}.$$

The continuity of v and φ implies that G is closed (see Definition 2.40). We have

$$S(u) = G(u) \cap F(u).$$

Invoking Problem 2.56(b), we conclude that S is upper semicontinuous.

Solution of Problem 2.67

Let $X = [0,1]$ and $Y = \mathbb{R}^2_+$. Consider the multifunction $F\colon X \longrightarrow 2^Y \setminus \{\emptyset\}$ defined by

$$F(u) = \{(t, ut) : t \geqslant 0\}.$$

Clearly F is lower semicontinuous (see Proposition 2.38(d)), but F is not h-lower semicontinuous (see Definitions 2.53(b) and 2.49(b)).

Next, let $X = [0,1]$ and $Y = \mathbb{R}$. Consider the multifunction $F\colon X \longrightarrow 2^Y \setminus \{\emptyset\}$ defined by

$$F(u) = \begin{cases} [0,1] & \text{if } u \in [0,1), \\ [0,1) & \text{if } u = 1. \end{cases}$$

Clearly F is h-upper semicontinuous (see Definitions 2.53(a) and 2.49(b)). However, note that $F^+((-1,1)) = \{1\}$ (see Definition 2.35(b)) and so F is not upper semicontinuous at 1 (see Definition 2.36(a) and Proposition 2.37).

Solution of Problem 2.68

Let $\{(u_\alpha, y_\alpha)\}_{\alpha\in J} \subseteq \operatorname{Gr} F$ be a net and assume that $(u_\alpha, y_\alpha) \longrightarrow (u, y)$ in $X \times Y$. We have

$$\operatorname{dist}_Y(y_\alpha, F(u)) \leqslant h^*(F(u_\alpha), F(u)) \longrightarrow 0$$

(see Definitions 2.53(a) and 2.49), so $\operatorname{dist}_Y(y, F(u)) = 0$, thus $y \in F(u)$ and hence F is closed (see Definition 2.40).

Solution of Problem 2.69

We need to show that for every $\lambda \geqslant 0$, the set

$$L_\lambda(y) = \{u \in X : \varphi_y(u) \leqslant \lambda\}$$

is closed. To this end, let $\{u_\alpha\}_{\alpha\in J} \subseteq L_\lambda(y)$ be a net such that $u_\alpha \longrightarrow u$ in X. For every $v_\alpha \in F(u_\alpha)$ we have

$$\begin{aligned}\operatorname{dist}_Y(y, F(u)) &\leqslant d_Y(y, v_\alpha) + \operatorname{dist}_Y(v_\alpha, F(u)) \\ &\leqslant d_Y(y, v_\alpha) + h^*(F(u_\alpha), F(u))\end{aligned}$$

(see Definition 2.49), so

$$\varphi_y(u) \leqslant \varphi_y(u_\alpha) + h^*(F(u_\alpha), F(u)),$$

thus

$$\varphi_y(u) \leqslant \lambda + h^*(F(u_\alpha), F(u))$$

and from the h-upper semicontinuity of F, we get

$$\varphi_y(u) \leqslant \lambda.$$

This means that $u \in L_\lambda(u)$, which proves that the set $L_\lambda(u)$ is closed. Hence the function $u \longmapsto \varphi_y(u)$ is lower semicontinuous.

Solution of Problem 2.70

(a) This is a consequence of Proposition 2.56(a) and Problem 2.56(a).

(b) From the definition of the Hausdorff metric (see Definition 2.49), we have

$$\begin{aligned} h^*(R_F(K), R_F(K')) &= \sup_{h \in R_F(K)} \inf_{y \in R_F(K')} d_X(h,y) \\ = \sup_{h \in R_F(K)} \inf_{v \in K'} h^*(h, F(v)) &= \sup_{u \in K} \inf_{v \in K'} h^*(F(u), F(v)) \\ \leqslant \sup_{u \in K} \inf_{v \in K'} h(F(u), F(v)) \leqslant k \sup_{u \in K} \inf_{v \in K'} d_X(u,v) &= kh^*(K, K'). \end{aligned}$$

In a similar way, we show that

$$h^*(R_F(K'), R_F(K)) \leqslant kh^*(K', K).$$

Hence

$$h(R_F(K), R_F(K')) \leqslant kh(K, K')$$

(see Definition 2.49).

Solution of Problem 2.71

Let $V \subseteq Y$ be an open set. We need to show that $(F \cap G)^-(V) \subseteq X$ is open (see Proposition 2.38). So, let $u \in (F \cap G)^-(V)$ and $y \in F(u) \cap G(u) \cap V$. Then $(u, y) \in \operatorname{Gr} G \cap (X \times V)$. Since by hypothesis $\operatorname{Gr} G \subseteq X \times Y$ is open, we can find $U_1(u) \in \mathcal{N}(u)$ and $V_1(y) \in \mathcal{N}(y)$ (where $\mathcal{N}(u)$ and $\mathcal{N}(y)$ are the filters of neighborhoods of u in X and y in Y respectively) such that

$$U_1(x) \times V_1(y) \subseteq \operatorname{Gr} G \cap (X \times V). \tag{2.31}$$

Note that $F(u) \cap V_1(y) \neq \emptyset$ and recall that F is lower semicontinuous. So, we can find $U_2(u) \in \mathcal{N}(u)$ such that

$$F(u') \cap V_1(y) \neq \emptyset \qquad \forall\, u' \in U_2(u). \tag{2.32}$$

We set $U(u) = U_1(u) \cap U_2(u) \in \mathcal{N}(u)$. Then for all $u' \in U(u)$ we have

$$F(u') \cap V_1(y) \neq \emptyset \quad \text{and} \quad U(u) \times V_1(y) \subseteq \operatorname{Gr} G \cap (X \times Y)$$

(see (2.31) and (2.32)), so

$$F(u') \cap G(u') \cap V \neq \emptyset \qquad \forall\, u' \in U(u),$$

thus the set $(F\cap G)^-(V)\subseteq X$ is open and hence $u\longmapsto (F\cap G)(u)$ is lower semicontinuous.

Solution of Problem 2.72

Consider the multifunction $H\colon X\longrightarrow 2^Y\setminus\{\emptyset\}$ defined by

$$H(u) = F(u)_r = \{y\in Y:\ \mathrm{dist}_Y(y,F(u))<r\} \qquad \forall\, u\in X.$$

Evidently H has convex and open values. We claim that H is lower semicontinuous (see Definition 2.36(b)). To this end, let $\{u_\alpha\}_{\alpha\in J}\subseteq X$ be a net such that $u_\alpha\longrightarrow u$ in X and let $y\in H(u)$. If $y\in F(u)$, then exploiting the lower semicontinuity of F (see Proposition 2.38) we can find $y_\alpha\in F(u_\alpha)$ for $\alpha\in J$ such that $y_\alpha\longrightarrow y$. Clearly $y_\alpha\in H(u_\alpha)$. If $y\notin F(u)$, then we can find $v\in F(u)$ such that $\|v-y\|_Y<r$. Let $v_\alpha\in F(u_\alpha)$ be such that $v_\alpha\longrightarrow v$ (again this is possible due to the lower semicontinuity of F). Then we can find $\alpha_0\in J$ such that

$$\|v_\alpha-y\|_Y < r \qquad \forall\,\alpha\geqslant\alpha_0,$$

so

$$\mathrm{dist}_Y(y,F(u_\alpha)) < r \qquad \forall\,\alpha\geqslant\alpha_0,$$

thus $y\in H(u_\alpha)$ for all $\alpha\geqslant\alpha_0$ and hence H is lower semicontinuous (see Proposition 2.37). Invoking Proposition 2.60 (note that $\operatorname{int} H(u)\neq\emptyset$ for all $u\in X$), we can find a continuous function $h\colon X\longrightarrow Y$ such that

$$h(u) \in H(u) \qquad \forall\, u\in X,$$

so

$$\mathrm{dist}_Y(h(u),F(u)) < r \qquad \forall\, u\in X.$$

Solution of Problem 2.73

Recall that a compact Hausdorff space is paracompact (see Definition I.2.142 and Theorem I.2.144). Also, the collection $\{F^-(\{y\})\}_{y\in Y}$ is an open cover of X. The compactness of X implies that we can find a finite set $\{y_1,\ldots,y_n\}\subseteq Y$ such that $\{F^-(\{y_1\}),\ldots,F^-(\{y_1\})\}$ is an open cover of X. The paracompactness of X implies that there is a continuous partition of unity $\{s_1,\ldots,s_n\}$ subordinate to this finite open cover (see Definition I.2.146 and Theorem I.2.147). Let

$$f(u) = \sum_{k=1}^n s_k(u)y_k.$$

Then f is continuous. If $s_k(u) \neq 0$, then $u \in F^-(\{y_k\})$ and so $y_k \in F(u)$. The convexity of the values of F implies that $f(u) \in F(u)$ for all $u \in X$.

Solution of Problem 2.74

Let $G\colon Y \longrightarrow 2^X \setminus \{\emptyset\}$ be the multifunction defined by

$$G(y) \;=\; A^{-1}(y) \;=\; \{u \in X : \; A(u) = y\}. \tag{2.33}$$

Since $A \in \mathcal{L}(X;Y)$ is surjective, by the open mapping theorem (see Theorem I.5.47), A is open. For every open set $V \subseteq X$, we have that the set $G^-(V) = A(V) \subseteq Y$ is open. So, by Proposition 2.38, G is lower semicontinuous. Also, from (2.33) it is clear that G has values in $P_{fc}(X)$. We can apply the Michael selection theorem (see Theorem 2.58) and find a continuous map $f\colon Y \longrightarrow X$ such that

$$f(y) \;\in\; G(y) \qquad \forall\, y \in Y,$$

so

$$A(f(y)) \;=\; y \qquad \forall\, y \in Y.$$

Solution of Problem 2.75

Suppose that $f\colon C \longrightarrow Y$ is a continuous selection of $F|_C$. Let $\widehat{F}\colon X \longrightarrow P_{fc}(Y)$ be the multifunction defined by

$$\widehat{F}(u) \;=\; \begin{cases} f(u) & \text{if } \; u \in C, \\ F(u) & \text{if } \; u \in X \setminus C. \end{cases}$$

Clearly $\widehat{F}$ is lower semicontinuous (see Proposition 2.38(d)). Also, it has values in $P_{fc}(Y)$. So, we can apply the Michael selection theorem (see Theorem 2.58) and conclude that there exists a continuous selection $\widehat{f}\colon X \longrightarrow Y$ of $\widehat{F}$. Evidently $\widehat{f}$ extends f on all of X.

Solution of Problem 2.76

For every $u \in X$, let $F(u) = (\varphi(u), \psi(u)) \subseteq \mathbb{R}$. We claim that F is lower semicontinuous. Indeed, let $u_n \longrightarrow u$ in X and let $y \in F(u)$. Exploiting the upper semicontinuity of φ and the lower semicontinuity of ψ, we see that we can find $n_0 \geqslant 1$ such that

$$y \;\in\; (\varphi(u_n), \psi(u_n)) \qquad \forall\, n \geqslant n_0,$$

so F is lower semicontinuous. Applying Proposition 2.60, we obtain a continuous selection $f\colon X \longrightarrow \mathbb{R}$ of F. Then $\varphi(u) < f(u) < \psi(u)$ for all $u \in X$.

Solution of Problem 2.77

According to Proposition 2.38, we need to show that for every open set $V \subseteq Y$, the set $H^-(V) = \{u \in X : H(u) \cap V \neq \emptyset\} \subseteq X$ is open. So, let $y_0 \in H(u_0) \cap V$. Then $y_0 \in (\varphi(u_0) + U) \cap V$. Thus there exists an open symmetric neighborhood W of the origin in V such that

$$y_0 + W + W \subseteq \big(\varphi(u_0) + U\big) \cap V,$$

so $y_0 + W \subseteq V$.

Let $D = \varphi^{-1}(\varphi(u_0) + W)$. Since φ is continuous, we see that $D \subseteq X$ is open and $u_0 \in D$.

Next, we will show that

$$y_0 + W \subseteq \varphi(u) + U \qquad \forall\, u \in D. \tag{2.34}$$

To prove this, let $w \in W$ and $u \in D$. Then

$$h = \varphi(u_0) - \varphi(u) \in W.$$

Since $y_0 + W + W \subseteq \varphi(u_0) + U$, we can find $\widehat{u} \in U$ such that $y_0 + w + h = \varphi(u_0) + \widehat{u}$ and so

$$y_0 + w = \varphi(u) + \widehat{u} \in \varphi(u) + U,$$

which proves (2.34). Since $y_0 \in F(u_0) \cap (y_0 + W)$, we have $u_0 \in F^-(y_0 + W)$ and since F is lower semicontinuous, we have that the set $F^-(y_0 + W)$ is open. Let

$$E = D \cap F^-(y_0 + W).$$

Then E is an open neighborhood of u_0. If $u \in E$, then

$$\begin{aligned} H(u) \cap V &= F(u) \cap \big(\varphi(u) + U\big) \cap V \\ &\supseteq F(u) \cap \big(\varphi(u) + U\big) \cap (y_0 + W) \\ &= F(u) \cap (y_0 + W) \neq \emptyset \end{aligned}$$

(see (2.34) and recall that $y_0 + W \subseteq V$), so H is lower semicontinuous.

Solution of Problem 2.78

We consider the multifunction $\Gamma\colon X \longrightarrow 2^{Y \times Y} \setminus \{\emptyset\}$ defined by

$$\Gamma(u) = F(u) \times G(u) \qquad \forall\, u \in X.$$

Let $V_1 \times V_2 \subseteq Y \times Y$ be a basic open set. Then

$$\Gamma^-(V_1 \times V_2) = F^-(V_1) \cap G^-(V_2),$$

so $\Gamma^-(V_1 \times V_2) \subseteq X$ is open (since both F and G are lower semicontinuous) and thus Γ is lower semicontinuous. Let $\varphi\colon Y \times Y \longrightarrow Y$ be defined by

$$\varphi(y_1, y_2) = y_1 + y_2 \qquad \forall\, y_1, y_2 \in Y.$$

We know that φ is continuous. Then we have $H = \varphi \circ \Gamma$. It is easily seen that the composition of two lower semicontinuous multifunctions is lower semicontinuous too. Therefore H is lower semicontinuous.

Solution of Problem 2.79

Given $\varepsilon > 0$, for every $u \in X$, we can find $V_u \in \mathcal{N}(u)$ such that

$$\bigcap_{u' \in V_u} F(u')_\varepsilon \neq \emptyset.$$

The collection $\{V_u\}_{u \in X}$ is an open cover of X. Since X is paracompact (see Definition I.2.142), there is a locally finite refinement $\{W_j\}_{j \in J}$ of $\{V_u\}_{u \in X}$ (see Definition I.2.146 and Theorem I.2.147). For each $j \in J$, let $u_j \in X$ be such that $W_j \subseteq V_{u_j}$. Also, let $\{\xi_j\}_{j \in J}$ be a corresponding continuous partition of unity subordinate to $\{W_j\}_{j \in J}$ (see Definition I.1.111). For every $j \in J$, we choose $y_j \in \bigcap\limits_{u' \in W_j} F(u')_\varepsilon$ and then define $f_\varepsilon\colon X \longrightarrow Y$ by setting

$$f_\varepsilon(u) = \sum_{j \in J} \xi_j(u) y_j.$$

Clearly $f_\varepsilon \in C(X;Y)$ and since F is convex valued, we have

$$f_\varepsilon(u) \in \operatorname{conv} F(u)_\varepsilon = F(u)_\varepsilon \qquad \forall\, u \in X.$$

Solution of Problem 2.80

Let $\widehat{F}\colon X \longrightarrow P_{fc}(Y)$ be the multifunction defined by

$$\widehat{F}(u) = \begin{cases} \{\widehat{y}\} & \text{if } u = \widehat{u}, \\ F(u) & \text{if } u \neq \widehat{u}. \end{cases}$$

It is easy to see that $\widehat{F}$ is lower semicontinuous (see Proposition 2.38) and has values in $P_{fc}(Y)$. So, we can apply the Michael selection theorem (see Theorem 2.58) and find a continuous function $\widehat{f}\colon X \longrightarrow Y$ such that $\widehat{f}(u) \in \widehat{F}(u)$ for all $u \in X$. Evidently $\widehat{f}$ is a continuous selector of $\widehat{F}$ such that $\widehat{f}(\widehat{u}) = \widehat{y}$.

Solution of Problem 2.81
For $u_0 \in D$, we have $F(u_0) \subseteq F(D) \subseteq \overline{F(D)}$. Suppose that $u_0 \in \overline{D} \setminus D$ and let $y \in F(u_0)$. For each open set $V \subseteq Y$ containing y, we have that $F^-(V) \subseteq X$ is an open set containing u_0 (see Proposition 2.38). Since $u_0 \in \overline{D}$, we can find $u \in D \cap F^-(V)$. Then $F(u) \cap V \neq \emptyset$ and so $F(D) \cap V \neq \emptyset$. This shows that $y \in \overline{F(D)}$ and so we conclude that $F(\overline{D}) \subseteq \overline{F(D)}$.

Solution of Problem 2.82
See the solution of Problem 2.78 for the Cartesian product of two lower semicontinuous multifunctions. The statement is not true for upper semicontinuous multifunctions. To see this let $X = Y = \mathbb{R}$ and consider the multifunction $F\colon \mathbb{R} \longrightarrow 2^{\mathbb{R}} \setminus \{\emptyset\}$ defined by

$$F(u) = (-1,1) \qquad \forall\, u \in \mathbb{R}.$$

Let $H\colon \mathbb{R} \longrightarrow 2^{\mathbb{R}\times\mathbb{R}} \setminus \{\emptyset\}$ be the multifunction defined by $H = I_{\mathbb{R}} \times F$. Clearly $I_{\mathbb{R}}$ and F are upper semicontinuous multifunctions. On the other hand, if $B_1 = \{h \in \mathbb{R}^2 : |h| < 1\}$, then $(I_{\mathbb{R}} \times F)(0) \subseteq B_1$. However, we cannot find a neighborhood U of 0 in $\mathbb{R}$ such that $(I_{\mathbb{R}} \times F)(U) \subseteq B_1$. Therefore the map $u \longmapsto H(u) = (u, F(u))$ is not upper semicontinuous.

Solution of Problem 2.83
We fix $\varepsilon > 0$. Because F is upper semicontinuous, for every $u \in X$, we can find $\delta = \delta(\varepsilon, u) \in (0, \frac{\varepsilon}{2})$ such that, if $u' \in B_\delta(u) = \{u' \in X : d_X(u', u) < \delta\}$, then $F(u') \subseteq F(u) + \frac{\varepsilon}{2} B_1$, where $B_1 = \{y \in Y : \|y\|_Y < \varepsilon\}$ (see Proposition 2.54). The family $\{B_{\frac{\delta}{4}}(u)\}_{u\in X}$ is an open cover of X. Recall that a metric space is paracompact (see Definition I.2.142 and Theorem I.2.144). So, we can find a locally finite refinement $\{U_\alpha\}_{\alpha\in J}$ of $\{B_{\frac{\delta}{4}}(u)\}_{u\in X}$ (see Definition I.2.146 and Theorem I.2.147). There exists a locally Lipschitz partition of unity $\{\xi_\alpha\}_{\alpha\in J}$ subordinate to this cover $\{U_\alpha\}_{\alpha\in J}$ (see Definition I.1.111). For each $\alpha \in J$, we choose $(u_\alpha, y_\alpha) \in \operatorname{Gr} F \cap (U_\alpha \times Y)$ and define

$$f_\varepsilon(u) = \sum_{\alpha\in J} \xi_\alpha(u) y_\alpha \qquad \forall\, u \in X.$$

Since the cover $\{U_\alpha\}_{\alpha\in J}$ is locally finite, f_ε is well defined and locally Lipschitz. Moreover, it is clear that

$$f_\varepsilon(u) \in \operatorname{conv} F(X) \qquad \forall\, u \in X.$$

Next, we fix $u \in X$. Recalling that $\{U_\alpha\}_{\alpha\in J}$ is locally finite, we have that $0 < \xi_\alpha(u)$ for all $\alpha \in J(u) \subseteq J$ with $J(u)$ finite. For every $\alpha \in J(u)$, let $u_\alpha \in X$

be such that $U_\alpha \subseteq B_{\frac{\delta_\alpha}{4}}(u_\alpha)$, with $\delta_\alpha = \delta(\varepsilon, u_\alpha) > 0$. Let $\beta \in J(u_\alpha)$ and set $\delta_\beta = \max\{\delta_\alpha : \alpha \in J(u)\}$. Then $u_\alpha \in B_{\frac{\delta_\beta}{4}}(u_\beta)$ and so $U_\alpha \subseteq B_{\delta_\beta}(u_\beta)$. Hence for every $\alpha \in J(u)$, we have

$$y_\alpha \in F(U_\alpha) \subseteq F(U_\beta) + \tfrac{\varepsilon}{2}B_1.$$

Since the latter set is convex, we have

$$f_\varepsilon(u) \in F(U_\beta) + \tfrac{\varepsilon}{2}B_1.$$

Thus we can find $y_\beta \in F(u_\beta)$ such that $\|f_\varepsilon(u) - y_\beta\|_Y < \frac{\varepsilon}{2}$ and so

$$d_{X\times Y}\big((u, f_\varepsilon(u)), (u_\beta, y_\beta)\big) = d_X(u, u_\beta) + \|f_\varepsilon(u) - y_\beta\|_Y < \varepsilon,$$

so $(u, f_\varepsilon(u)) \in \operatorname{Gr} F + \varepsilon B_1$. Since $u \in X$ is arbitrary, we conclude that $h^*(\operatorname{Gr} f_\varepsilon, \operatorname{Gr} F) < \varepsilon$ (see Definition 2.49).

Solution of Problem 2.84

We keep the notation from the solution of Problem 2.83. The family $\{B_\delta(u)\}_{u\in K}$ is an open cover of the compact set K. So, we can find a finite subcover $\{B_{\delta_1}(u_1), \ldots, B_{\delta_m}(u_m)\}$ with $\delta_i = \delta_i(\varepsilon, u_i) > 0$ for $i \in \{1, \ldots, m\}$. We can find a locally Lipschitz partition of unity $\{\xi_1, \ldots, \xi_m\}$ subordinate to this finite subcover (see Definition I.1.111). Also, let $y_i \in F(u_i)$ for $i \in \{1, \ldots, m\}$ and define

$$f_\varepsilon(u) = \sum_{i=1}^m \xi_i(u) y_i \qquad \forall\, u \in V_\varepsilon = \bigcup_{i=1}^m B_{\delta_i}(u_i).$$

Then $f_\varepsilon \colon V_\varepsilon \longrightarrow Y$ is locally Lipschitz and $f_\varepsilon(u) \in \operatorname{conv} F(K) \cap Y_\varepsilon$ with $Y_\varepsilon = \operatorname{span}\{y_1, \ldots, y_m\}$, hence f_ε has a finite dimensional range. As in the solution of Problem 2.83, we see that

$$f_\varepsilon(u) \in F(K \cap B_\varepsilon(u)) + \varepsilon B_1 \qquad \forall\, u \in V_\varepsilon.$$

Solution of Problem 2.85

Let $X_1 = F(X)$ and $X_{n+1} = F(X_n)$ for all $n \geqslant 1$. From Problem 2.56(a), we know that for each $n \geqslant 1$, the set X_n is compact. Also, since $X_1 \subseteq X$, we see that $X_{n+1} \subseteq X_n$ for all $n \geqslant 1$. Let $C = \bigcap_{n\geqslant 1} X_n$. Since $\{X_n\}_{n\geqslant 1}$ has the finite intersection property (being decreasing; see Definition I.2.80 and Theorem I.2.81), it follows that $C \neq \emptyset$. We have

$$F(C) \subseteq \bigcap_{n \geqslant 1} F(X_n) \subseteq \bigcap_{n \geqslant 1} X_{n-1} = C$$

(setting $X_0 = X$). Suppose that $u_0 \in C \setminus F(C)$. Note that C is closed, hence $F(C) \subseteq X$ is closed too (see Problem 2.56(a)). Recall that a compact space is regular. So, we can find open sets $U_1, U_2 \subseteq X$ such that

$$u_0 \in U_1, \quad F(C) \subseteq U_2 \quad \text{and} \quad U_1 \cap U_2 = \emptyset. \tag{2.35}$$

Since F is upper semicontinuous, we have that the set $F^+(U_2) \subseteq X$ is open (see Proposition 2.37) and $C = \bigcap\limits_{n \geqslant 1} X_n \subseteq F^+(U_2)$. We have

$$X \setminus F^+(U_2) \subseteq \bigcup_{n \geqslant 1} (X \setminus X_n). \tag{2.36}$$

The set $X \setminus F^+(U_2) \subseteq X$ is closed, hence compact. From (2.36), it follows that we find a finite subcover of the open cover $\{(X \setminus X_n)\}_{n \geqslant 1}$. So, we can say that

$$X \setminus F^+(U_2) \subseteq \bigcup_{n=1}^{m} (X \setminus X_n),$$

so

$$\bigcap_{n=1}^{m} X_n \subseteq F^+(U_2),$$

thus

$$X_n \subseteq F^+(U_2) \qquad \forall\, n \geqslant m$$

hence

$$X_{n+1} = F(X_n) \subseteq U_2 \qquad \forall\, n \geqslant m$$

and so $C \subseteq U_2$. But $u_0 \in C$ and $u_0 \notin U_2$ (see (2.35)), we reach a contradiction. Therefore, we conclude that $F(C) = C$.

Solution of Problem 2.86

In the light of Problem 2.65, we need to consider a noncompact domain. So, let $X = Y = \mathbb{R}_+$ and consider the multifunction defined by

$$F(u) = \begin{cases} \{0\} & \text{if } u = 0, \\ [0, u] \cup \{\frac{1}{u}\} & \text{if } u > 0. \end{cases}$$

Evidently, for every $u \geqslant 0$, we have $F(u) \in P_k(Y)$. Also $\operatorname{Gr} F$ is closed. But F is not upper semicontinuous at 0. Indeed, note that $F(0) \subseteq [0, 1)$, but $F^+([0, 1)) = \{0\}$ which is not open in $Y = \mathbb{R}_+$. Therefore F is not upper semicontinuous (see Proposition 2.37).

Solution of Problem 2.87

We will consider the boundedly w-compact case. The other case can be established similarly. First note that for every $u \in X$, we have $\operatorname{proj}_K(u) \neq \emptyset$. Indeed, let $\{h_n\}_{n\geqslant 1} \subseteq K$ be a minimizing sequence, that is,

$$\|u - h_n\|_X \searrow \operatorname{dist}(u, K). \tag{2.37}$$

From (2.37) we see that the sequence $\{h_n\}_{n\geqslant 1} \subseteq K$ is bounded. So, we can find $r > 0$ such that $\{h_n\}_{n\geqslant 1} \subseteq K \cap \overline{B}_r$. Since K is by hypothesis boundedly w-compact, we have that the sequence $\{h_n\}_{n\geqslant 1} \subseteq X$ is relatively w-compact. So, by the Eberlein–Smulian theorem (see Theorem I.5.78), passing to a subsequence if necessary, we may assume that $h_n \xrightarrow{w} h \in K \cap \overline{B}_r$. Exploiting the weak lower semicontinuity of the norm functional (see Proposition I.5.56(c)), we have

$$\|u - h\|_X \leqslant \liminf_{n\to+\infty} \|u - h_n\|_X,$$

so

$$\|u - h\|_X = \operatorname{dist}(u, K)$$

(see (2.37) and recall that $h \in K$), thus $h \in \operatorname{proj}_K(u)$ and hence $\operatorname{proj}_K(u) \neq \emptyset$.

Now let $C \subseteq X$ be a w-closed set. According to Proposition 2.37, we need to show that the set $\operatorname{proj}_K^-(C) = \{u \in X : \operatorname{proj}_K(u) \cap C \neq \emptyset\}$ is closed. So, let $\{u_n\}_{n\geqslant 1} \subseteq \operatorname{proj}_K^-(C)$ be a sequence such that $u_n \longrightarrow u$ in X. We have that $\operatorname{proj}_K(u_n) \cap C \neq \emptyset$ for all $n \geqslant 1$. We can find $h_n \in C$ such that

$$\operatorname{dist}(u_n, K) = \|u_n - h_n\|_X \qquad \forall\, n \geqslant 1. \tag{2.38}$$

We know that

$$\operatorname{dist}(u_n, K) \longrightarrow \operatorname{dist}(u, K). \tag{2.39}$$

From (2.38) and (2.39), it follows that the sequence $\{h_n\}_{n\geqslant 1} \subseteq K \cap C$ is bounded. So, there exists $r > 0$ such that $\{h_n\}_{n\geqslant 1} \subseteq K \cap \overline{B}_r$. Because K is boundedly w-compact, invoking the Eberlein Smulian theorem and by passing to a suitable subsequence if necessary, we can have

$$h_n \xrightarrow{w} h \in K \cap C \cap \overline{B}_r$$

(since C is w-closed). Also, the weak lower semicontinuity of the norm functional (see Proposition I.5.56(c)) implies that

$$\|u - h\|_X \leqslant \liminf_{n\to+\infty} \|u_n - h_n\|_X,$$

so

$$\|u - h\|_X = \operatorname{dist}(u, K)$$

(see (2.38), (2.39) and recall that $h \in K \cap C$), thus $u \in \operatorname{proj}_K^-(C)$ and hence proj_K is upper semicontinuous from X into X_w.

Solution of Problem 2.88

Note that $\mathcal{F}$ is closed from X into $X^*_{w^*}$. Indeed, let $\{(u_\alpha, u^*_\alpha)\}_{\alpha\in J} \subseteq \operatorname{Gr}\mathcal{F}$ be a net such that $u_\alpha \longrightarrow u$ in X and $u^*_\alpha \xrightarrow{w^*} u^*$ in X^*. Then $\|u_\alpha\|_X \longrightarrow \|u\|_X$ and

$$\|u_\alpha\|_X^2 \;=\; \|u^*_\alpha\|_*^2 \;=\; \langle u^*_\alpha, u_\alpha\rangle \;\longrightarrow\; \langle u^*, u\rangle\,,$$

so

$$\|u_\alpha\|_X^2 \;=\; \langle u^*, u\rangle \;\leqslant\; \|u^*\|_{X^*}\|u\|_X \tag{2.40}$$

On the other hand, the weak* lower semicontinuity of the norm functional of X^* (see Proposition I.5.65(c)) implies that

$$\|u^*\|_{X^*} \;\leqslant\; \liminf_{\alpha\in J} \|u^*_\alpha\|_{X^*} \;=\; \liminf_{\alpha\in J} \|u_\alpha\|_X \;=\; \|u\|_X. \tag{2.41}$$

From (2.40) and (2.41) it follows that

$$\|u^*\|_{X^*} \;=\; \|u\|_X \;=\; \langle u^*, u\rangle\,,$$

so $u^* \in \mathcal{F}(u)$ and hence $\mathcal{F}$ is closed.

From the definition of $\mathcal{F}$ (see Definition 2.112) and the Alaoglu theorem (see Theorem I.5.66), we see that $\mathcal{F}$ is locally compact from X into $X^*_{w^*}$. So, invoking Problem 2.65, we conclude that $\mathcal{F}$ is upper semicontinuous from X into $X^*_{w^*}$.

Solution of Problem 2.89

Since F is compact valued, according to Proposition 2.56(a), it suffices to show that F is h-upper semicontinuous (see Definition 2.53(a)). Arguing indirectly, suppose that F is not h-upper semicontinuous. Then we can find a sequence $\{u_n\}_{n\geqslant 1} \subseteq X$ with $u_n \longrightarrow u$ in X and $\varepsilon > 0$ such that

$$F(u_n) \;\not\subseteq\; F(u) + \varepsilon B_1 \qquad \forall\, n \geqslant 1. \tag{2.42}$$

From (2.42) it follows that we can find $h_n \in F(u_n)$ such that

$$\operatorname{dist}(h_n, F(u)) \;\geqslant\; \varepsilon \qquad \forall\, n \geqslant 1.$$

Since F is lower semicontinuous and compact valued, it is also h-lower semicontinuous (see Proposition 2.56(b)). So, we can find $n_0 \in \mathbb{N}$ such that

$$F(u) \;\subseteq\; F(u_n) + \varepsilon B_1 \qquad \forall\, n \geqslant n_0. \tag{2.43}$$

Let $h \in F(u)$. From (2.43) we see that we can find $\widehat{h}_n \in F(u_n)$ for $n \geqslant n_0$ such that

$$\|\widehat{h}_n - h\|_X \;<\; \varepsilon \qquad \forall\, n \geqslant n_0,$$

so

$$\operatorname{dist}(\widehat{h}_n, F(u)) < \varepsilon \qquad \forall\, n \geqslant n_0. \tag{2.44}$$

Recall that F has connected values. Then the continuity of the map $y \longmapsto \operatorname{dist}(y, F(u))$ on $F(u_n)$ for $n \geqslant n_0$, together with (2.43) and (2.44) imply that there exists $h_n' \in F(u_n)$ such that

$$\operatorname{dist}(h_n', F(u)) = \varepsilon \qquad \forall\, n \geqslant n_0 \tag{2.45}$$

(the Bolzano theorem; see Theorem I.1.90). So, we have that

$$\{h_n'\}_{n \geqslant n_0} \subseteq F(u) + \varepsilon \overline{B}_1 \in P_k(\mathbb{R}^N).$$

By passing to a subsequence if necessary, we may assume that

$$h_n' \longrightarrow h' \quad \text{in } \mathbb{R}^N.$$

But $(u_n, h_n') \in \operatorname{Gr} F$ and the latter by hypothesis is closed. So $h' \in F(u)$. On the other hand, from (2.45), we have

$$\operatorname{dist}(h', F(u)) = \varepsilon > 0,$$

so $h' \notin F(u)$, a contradiction. Therefore F is h-upper semicontinuous, hence continuous.

Solution of Problem 2.90

Let $X = Y = \mathbb{R}$ and consider the multifunction $F\colon \mathbb{R} \longrightarrow P_{fc}(\mathbb{R})$ defined by

$$F(u) = \begin{cases} -1 & \text{if } u < 0, \\ [-1,1] & \text{if } u = 0, \\ 1 & \text{if } u > 0 \end{cases}$$

(the sign multifunction). Clearly F is upper semicontinuous (see Definition 2.36(a)) but does not have a continuous selector.

Solution of Problem 2.91

From Theorem 2.63, we know that $\operatorname{Gr} F \in \mathcal{B}(T \times X)$. Let $\widehat{T}$ and $\widehat{X}$ be the metric completions of T and X respectively. Since by hypothesis T and X are Borel spaces, we have $\operatorname{Gr} F \in \mathcal{B}(\widehat{T} \times \widehat{X})$. Let $C \in P_f(X)$ and let $\widehat{C}$ be the $\widehat{X}$-closure of C. Then $C = \widehat{C} \cap X$. We have

$$\begin{aligned} F^-(C) &= \{t \in T : F(t) \cap C \neq \emptyset\} \\ &= \left\{t \in T : F(t) \cap \widehat{C} \neq \emptyset\right\} \\ &= \operatorname{proj}_{\widehat{T}}\big(\operatorname{Gr} F \cap (\widehat{T} \times \widehat{C})\big) \in \mathcal{B}(X) \end{aligned}$$

(by the Yankov–von Neumann–Aumann projection theorem; see Theorem I.4.65).

Solution of Problem 2.92

Consider the multifunction $F\colon \Omega \longrightarrow 2^X$ defined by

$$F(\omega) \;=\; \varphi^{-1}(\{\omega\}) \;=\; \{u \in X:\ \varphi(u)=\omega\}. \tag{2.46}$$

Therefore by hypothesis **(i)**, F has values in $P_f(X)$. Also, if $V \subseteq X$ is open, then

$$F^-(V) \;=\; \{\omega\in\Omega:\ F(\omega)\cap V\neq\emptyset\} \;=\; \varphi(V) \;\in\; \Sigma,$$

so F is measurable (see Definition 2.61). Invoking the Kuratowski–Ryll Nardzewski selection theorem (see Theorem 2.64), we can find a Σ-measurable map $f\colon \Omega\longrightarrow X$ such that

$$f(\omega) \;\in\; F(\omega) \qquad \forall\, \omega\in\Omega.$$

From (2.46) we see that $\varphi(f(\omega))=\omega$ for all $\omega\in\Omega$.

Solution of Problem 2.93

"$\Longrightarrow$": Let $C\subseteq X$ be a closed set and let $U=X\setminus C$. We know that U is an F_σ-set. In particular, we have

$$U \;=\; \bigcup_{n\geqslant 1}\widehat{C}_n \quad\text{with } \widehat{C}_n=\big\{x\in X:\ \operatorname{dist}(x,C)\geqslant \tfrac{1}{n}\big\} \qquad \forall\, n\geqslant 1.$$

We have

$$F^-(C) \;=\; \Omega\setminus F^+(U) \;=\; \Omega\setminus F^+\big(\bigcup_{n\geqslant 1}\widehat{C}_n\big). \tag{2.47}$$

It is easy to see that we always have

$$\bigcup_{n\geqslant 1}F^+(\widehat{C}_n) \;\subseteq\; F^+\big(\bigcup_{n\geqslant 1}\widehat{C}_n\big). \tag{2.48}$$

We will show that the opposite inclusion also holds. Arguing indirectly, suppose that we can find $\omega\in F^+\big(\bigcup\limits_{n\geqslant 1}\widehat{C}_n\big)$ such that $F(\omega)\cap\widehat{C}_n^c\neq\emptyset$ for all $n\geqslant 1$. Let $u_n\in F(\omega)\cap\widehat{C}_n^c$ for all $n\geqslant 1$. Since $F(\omega)\in P_k(X)$, we may assume that $u_n\longrightarrow u\in F(\omega)\subseteq U$. We know that $u_n\in\widehat{C}_n^c$ for all $n\geqslant 1$. Hence we have

$$\operatorname{dist}(u_n,C) \;<\; \tfrac{1}{n} \qquad \forall\, n\geqslant 1$$

and so $\operatorname{dist}(u,C)=0$, therefore $u\in C$, a contradiction. So, the inclusion in (2.48) is in fact an equality of the two sets and we have

$$F^+\big(\bigcup_{n\geqslant 1}\widehat{C}_n\big) \;=\; \bigcup_{n\geqslant 1}F^+(\widehat{C}_n) \;=\; \bigcup_{n\geqslant 1}\big(\Omega\setminus F^-(\widehat{C}_n^c)\big) \;\in\; \Sigma$$

(since F is measurable), thus $F^-(C)\in\Sigma$ (see (2.47)).

"$\Longleftarrow$": This follows from Theorem 2.63(a).

Solution of Problem 2.94

Since X is a separable metric space, it can be viewed as a dense subset of a compact metric space Y (in fact, X is homeomorphic to a subset of the Hilbert cube $[0,1]^{\mathbb{N}}$ which is compact by the Tichonov theorem; see Theorem I.2.91). Let $G\colon \Omega \longrightarrow P_k(Y)$ be defined by

$$G(\omega) = \overline{F(\omega)}^{Y} \qquad \forall\, \omega \in \Omega.$$

Then for every compact set $K \subseteq X$, we have

$$\begin{aligned} F^-(K) &= \{\omega \in \Omega:\ F(\omega) \cap K \neq \emptyset\} \\ &= \{\omega \in \Omega:\ G(\omega) \cap X \cap K \neq \emptyset\} \ = \ G^-(K). \end{aligned}$$

Since G is compact valued in Y and measurable, from Problem 2.93, we have that $G^-(K) \in \Sigma$ and so $F^-(K) \in \Sigma$.

Solution of Problem 2.95

We know that φ is jointly measurable. Let $C \subseteq X$ be a closed set and let $\{u_n\}_{n\geqslant 1} \subseteq C$ be a sequence dense in C. We have

$$\begin{aligned} F^-(C) &= \{\omega \in \Omega:\ F(\omega) \cap C \neq \emptyset\} \\ &= \{\omega \in \Omega:\ \varphi(\omega, u) \in U \text{ for some } u \in C\} \\ &= \{\omega \in \Omega:\ \varphi(\omega, u_{n_0}) \in U \text{ for some } n_0 \geqslant 1\} \\ &= \bigcup_{n\geqslant 1} \{\omega \in \Omega:\ \varphi(\omega, u_n) \in U\} \ \in \ \Sigma \end{aligned}$$

(since U is open and $\varphi(\omega,\cdot)$ is continuous), so F is measurable (see Theorem 2.63(a)).

Solution of Problem 2.96

Since X and Y are two σ-compact metric spaces, both are separable and

$$\begin{aligned} X &= \bigcup_{n\geqslant 1} C_n \quad \text{with } C_n \subseteq X \text{ compact for all } n \geqslant 1, \\ Y &= \bigcup_{m\geqslant 1} K_m \quad \text{with } K_m \subseteq Y \text{ compact for all } m \geqslant 1. \end{aligned}$$

Let $D \subseteq Y$ be a closed set. We have

$$F^-(D) = \bigcup_{n,m\geqslant 1} \operatorname{proj}_X\big((C_n \times (D \cap K_m)) \cap \operatorname{Gr} F\big).$$

Since F is a closed multifunction (see Definition 2.40), we have that the set

$$(C_n \times (D \cap K_m)) \cap \operatorname{Gr} F \subseteq X \times Y$$

is compact for all $n, m \geqslant 1$. Recall that proj_X is continuous. Hence the set $\operatorname{proj}_X\big((C_n \times (D \cap K_m)) \cap \operatorname{Gr} F\big)$ is compact for each $n, m \geqslant 1$, thus a Borel set. Therefore $F^-(D) \subseteq X$ is a Borel set and so F is measurable (see Theorem 2.63(a)).

Solution of Problem 2.97

Without any loss of generality, we may assume that $H(\omega) \neq \emptyset$ for all $\omega \in \Omega$. First we assume that all the multifunctions F_n are compact valued. We start by showing that the map $\omega \longmapsto H_2(\omega) = F_1(\omega) \cap F_2(\omega)$ is measurable. To this end, let $P(\omega) = F_1(\omega) \times F_2(\omega)$. Note that for every set $D \subseteq X \times Y$, we have

$$P^-(D) = F_1^-(\operatorname{proj}_1 D) \cap F_2^-(\operatorname{proj}_2 D),$$

with proj_1 (respectively proj_2) being the projection on the first (respectively second) factor of the product space. So, if $D \subseteq X \times Y$ is an open set, then $P^-(D) \in \Sigma$ as

$$P^-(D) = \{\omega \in \Omega : F_1(\omega) \cap \operatorname{proj}_1 D \neq \emptyset\} \cap \{\omega \in \Omega : F_2(\omega) \cap \operatorname{proj}_2 D \neq \emptyset\},$$

since the projections are open maps and F_1, F_2 are by hypothesis measurable multifunctions (see Definition 2.61(a)). So, P is measurable and compact valued.

Let $\Delta_2 \subseteq X \times X$ be the diagonal set, that is

$$\Delta_2 = \{(u, u) : u \in X\}.$$

For every closed set $\widehat{C} \subseteq X \times X$, we have

$$H_2^-(\widehat{C}) = \Big\{\omega \in \Omega : P(\omega) \cap \Delta_2 \cap \widehat{C} \neq \emptyset\Big\} \in \Sigma$$

(see Problem 2.93), so H_2 is measurable (see Theorem 2.63).

By induction, we see that the multifunction

$$\omega \longmapsto H_n(\omega) = \bigcap_{k=1}^{n} F_k(\omega)$$

is measurable for every $n \geqslant 1$. For every $\omega \in \Omega$, we have

$$h^*(H_n(\omega), H(\omega)) = \sup_{u \in H_n(\omega)} \operatorname{dist}(u, H(\omega))$$

(see Definition 2.49). Since $H_n(\omega) \in P_k(X)$, we can find $u_n \in H_n(\omega)$ such that

$$h^*(H_n(\omega), H(\omega)) = \operatorname{dist}(u_n, H(\omega)) \qquad \forall\, n \geqslant 1.$$

Note that $\{u_n\}_{n\geqslant 1} \subseteq F_1(\omega) \in P_k(X)$. Passing to a subsequence if necessary, we may assume that $u_n \longrightarrow u$ in X. Evidently $u \in H(\omega)$ and so

$$h^*(H_n(\omega), H(\omega)) \longrightarrow 0,$$

so

$$\operatorname{dist}(u, H_n(\omega)) \longrightarrow \operatorname{dist}(u, H(\omega)) \qquad \forall\, \omega \in \Omega,\ u \in X.$$

Since H_n is measurable, the map $\omega \longmapsto \operatorname{dist}(u, F_n(\omega))$ is Σ-measurable for every $n \geqslant 1$, hence the map $\omega \longmapsto \operatorname{dist}(u, H(\omega))$ is Σ-measurable for every $u \in X$. This implies the measurability of H, when all multifunctions F_n are compact valued.

Now, assume that at least on F_n is compact valued. Let $\widehat{X}$ be the metrizable compactification of X (recall that X being separable, is homeomorphic to a subset of the Hilbert cube $[0,1]^{\mathbb{N}}$). Let $\widehat{F}_n \colon \Omega \longrightarrow P_k(\widehat{X})$ be defined by

$$\widehat{F}_n(\omega) = \overline{F_n(\omega)} \qquad \forall\, \omega \in \Omega$$

(the closure in $\widehat{X}$). From the first part of the solution we have that, if

$$\widehat{H}(\omega) = \bigcap_{n\geqslant 1} \widehat{F}_n(\omega) \qquad \forall\, \omega \in \Omega,$$

then $\widehat{H}$ is measurable. Since $F_{n_0} = \widehat{F}_{n_0}$ for some $n_0 \geqslant 1$, it follows that $H = \widehat{H}$ and so H is measurable.

Solution of Problem 2.98

From Theorem 2.66, we know that there is a sequence $\{h_n \colon \Omega \longrightarrow X\}_{n\geqslant 1}$ of Σ-measurable selectors of U such that

$$U(\omega) = \overline{\{h_n(\omega)\}_{n\geqslant 1}} \qquad \forall\, u \in \Omega.$$

Let $V \subseteq Y$ be an open set. Then

$$\begin{aligned} G^-(V) &= \{\omega \in \Omega : \ \varphi(\omega, U(\omega)) \cap V \neq \emptyset\} \\ &= \bigcup_{n\geqslant 1} \{\omega : \ \varphi(\omega, h_n(\omega)) \in V\} \end{aligned} \tag{2.49}$$

(due to the continuity of $\varphi(\omega, \cdot)$). But φ being Carathéodory, is jointly measurable. Hence, for every $n \geqslant 1$, the map $\omega \longmapsto \varphi(\omega, h_n(\omega))$ is Σ-measurable. Therefore

$$\{\omega \in \Omega : \ \varphi(\omega, h_n(\omega)) \in V\} \in \Sigma \qquad \forall\, n \geqslant 1,$$

so $G^-(V) \in \Sigma$ (see (2.49)) and hence G is measurable (see Definition 2.61(a)).

Solution of Problem 2.99

Let

$$G(\omega) = \{u \in F(\omega) : d_X(g(\omega), u) = \operatorname{dist}(g(\omega), F(\omega))\}.$$

Since F is $P_k(X)$-valued we see that $G(\omega) \in P_k(X)$ for all $\omega \in \Omega$. Also, let

$$\varphi(\omega, u) = d_X(g(\omega), u) - \operatorname{dist}(g(\omega), F(\omega)).$$

The measurability of F and the Σ-measurability of g imply that $\varphi(\omega, u)$ is a Carathéodory function. Let

$$L_n(\omega) = \{u \in X : \varphi(\omega, u) < \tfrac{1}{n}\}.$$

From Problem 2.95 we know that L_n is measurable, hence the multifunction $\omega \longmapsto \overline{L}_n(\omega) = \overline{L_n(\omega)}$ is measurable. Problem 2.97 implies that the multifunction $\omega \longmapsto \overline{L}_n(\omega) \cap F(\omega)$ is measurable and $P_k(X)$-valued for every $n \geqslant 1$. Once again Problem 2.97 implies that the multifunction $\omega \longmapsto G(\omega) = \bigcap\limits_{n\geqslant 1} (\overline{L}_n(\omega) \cap F(\omega))$ is measurable and $P_k(X)$-valued. Apply the Kuratowski–Ryll Nardzewski selection theorem (see Theorem 2.64), to find a Σ-measurable function $f : \Omega \longrightarrow X$ such that

$$f(\omega) \in G(\omega) \qquad \forall\, \omega \in \Omega,$$

so

$$f(\omega) \in F(\omega) \quad \text{and} \quad \operatorname{dist}(g(\omega), F(\omega)) = d_X(g(\omega), f(\omega)) \qquad \forall\, \omega \in \Omega.$$

Solution of Problem 2.100

Since X is σ-compact, we have

$$X = \bigcup_{n\geqslant 1} K_n \quad \text{with } K_n \in P_k(X) \qquad \forall\, n \geqslant 1.$$

Let $C \subseteq X$ be a closed set. Then

$$F^-(C) = F^-\Big(\bigcup_{n\geqslant 1}(C \cap K_n)\Big) = \bigcup_{n\geqslant 1} F^-(C \cap K_n) \in \Sigma$$

(since $C \cap K_n \in P_k(X)$ for all $n \geqslant 1$), so F is measurable (see Theorem 2.63(a)).

Solution of Problem 2.101

"$\Longrightarrow$": Let F be measurable. According to Problem 2.100, it suffices to show that for every $K \in P_k(Y)$, we have $G^-(K) \in \Sigma \times \mathcal{B}(X)$. Note that

$$\begin{aligned} G^-(K) &= \{(\omega, u) \in \Omega \times X : \ G(\omega, u) \cap K \neq \emptyset\} \\ &= \big\{(\omega, u) \in \Omega \times X : \text{ there exists } y \in Y \\ &\qquad \text{such that } (u, y) \in F(\omega) \cap (X \times K)\big\}. \end{aligned}$$

Let $\Omega_K = \{\omega \in \Omega : \ F(\omega) \cap (X \times K) \neq \emptyset\}$. Since F is measurable, we have $\Omega_K \in \Sigma$ (see Theorem 2.63). Let $H : \Omega_K \longrightarrow P_f(X \times Y)$ be defined by

$$H(\omega) = F(\omega) \cap (X \times K) \qquad \forall\, \omega \in \Omega_K. \tag{2.50}$$

Then we have

$$\operatorname{Gr} H = \operatorname{Gr} F \cap (\Omega \times X \times K) \in \Sigma \times \mathcal{B}(X) \times \mathcal{B}(Y)$$

(see Theorem 2.63), so H is measurable (again by Theorem 2.63). Thus Theorem 2.66 implies that we can find $(\Sigma \cap \Omega_K)$-measurable maps

$$f_n : \Omega_K \longrightarrow X \quad \text{and} \quad g_n : \Omega_K \longrightarrow Y \qquad \forall\, n \geqslant 1,$$

such that

$$H(\omega) = \overline{\{f_n(\omega), g_n(\omega)\}}_{n \geqslant 1} \qquad \forall\, \omega \in \Omega.$$

Let $D_K = \big\{(\omega, u) \in \Omega_K \times X : \ u \in \overline{\{f_n(\omega)\}}_{n \geqslant 1}\big\}$.

We will show that $G^-(K) = D_K$. From the definitions of the two sets, we see that

$$G^-(K) \subseteq D_K.$$

Next let $(\omega, u) \in D_K$. Then we can find a subsequence $\{f_{n_k}(\omega)\}_{k \geqslant 1}$ of $\{f_n(\omega)\}_{n \geqslant 1}$ such that $f_{n_k}(\omega) \longrightarrow u$ in X. We consider the corresponding subsequence $\{g_{n_k}(\omega)\}_{k \geqslant 1}$ of $\{g_n(\omega)\}_{n \geqslant 1}$. From (2.50) we see that $\{g_{n_k}(\omega)\}_{k \geqslant 1} \subseteq K$ and so, passing to a further subsequence if necessary, we may assume that $g_{n_k}(\omega) \longrightarrow y \in K$ in Y. Therefore $(u, y) \in H(\omega)$, hence $(\omega, u) \in G^-(K)$. This shows that $D_K \subseteq G^-(K)$ so we get $G^-(K) = D_K$.

Note that

$$D_K = \big\{(\omega, u) \in \Omega \times X : \ \inf_{n \geqslant 1} d_X(u, f_n(\omega)) = 0\big\} \in \Sigma \times \mathcal{B}(X),$$

so $G^-(K) \in \Sigma \times \mathcal{B}(X)$ (as $G^-(K) = D_K$) and thus G is measurable (see Problem 2.100).

"$\Longleftarrow$": Let G be measurable. Note that

$$\operatorname{Gr} F = \operatorname{Gr} G \in \Sigma \times \mathcal{B}(X) \times \mathcal{B}(Y) = \Sigma \times \mathcal{B}(X \times Y).$$

So, by Theorem 2.63(a), F is measurable.

Solution of Problem 2.102

From the Yankov–von Neumann–Aumann selection theorem (see Theorem 2.68 and Remark 2.69), we can find a sequence $\{f_n : \Omega \longrightarrow X\}_{n\geqslant 1}$ of Σ-measurable selectors of F such that

$$F(\omega) \subseteq \overline{\{f_n(\omega)\}_{n\geqslant 1}} \qquad \forall\, \omega \in \Omega.$$

Then $\sigma_{F(\omega)}(u^*) = \sup\limits_{n\geqslant 1} \langle u^*, f_n(\omega)\rangle$ for all $\omega \in \Omega$ and so the function $\omega \longmapsto \sigma_{F(\omega)}(u^*)$ is Σ-measurable for every $u^* \in X^*$ (see Corollary I.3.69).

Solution of Problem 2.103

"$\Longrightarrow$": By the Mazur theorem (see Theorem I.5.58), the values of F are in $P_{fc}(X)$. So, according to Theorem 2.66, we can find a sequence $\{f_n : \Omega \longrightarrow X\}_{n\geqslant 1}$ of Σ-measurable selectors of F such that

$$F(\omega) = \overline{\{f_n(\omega)\}_{n\geqslant 1}} \qquad \forall\, \omega \in \Omega.$$

Then for every $u^* \in X^*$, we have

$$\sigma_{F(\omega)}(u^*) = \sup_{n\geqslant 1} \langle u^*, f_n(\omega)\rangle,$$

so the map $\omega \longmapsto \sigma_{F(\omega)}(u^*)$ in Σ-measurable (see Corollary I.3.69).

"$\Longleftarrow$": Let $\overline{B}_1^{X^*} = \{u^* \in X^* : \|u^*\|_{X^*} \leqslant 1\}$. Since X is separable, the set $\overline{B}_1^{X^*}$ furnished with the relative w^*-topology is compact (by the Alaoglu theorem; see Theorem I.5.66) and metrizable (see Theorem I.5.85). Note that

$$X^* = \bigcup_{n\geqslant 1} n\overline{B}_1^{X^*},$$

hence $X^*_{w^*}$ (the space X^* furnished with the w^*-topology) is separable. Then X^* is separable in all topologies τ such that $(X^*_\tau)^* = X$. In particular then this is the case for $\tau = m(X^*, X)$ (the Mackey topology on X^* for the pair (X^*, X), i.e., the strongest, locally convex topology τ on X^* for which we have $(X^*_\tau)^* = X$). Since F is $P_{wkc}(X)$-valued, the function $u^* \longmapsto \sigma_{F(\omega)}(u^*)$ is m-continuous (see Proposition 2.51(d)). From Proposition 2.51(c), we know that for all $u \in X$, we have

$$\operatorname{dist}(u, F(\omega)) = \sup\big\{ \langle u^*, u\rangle - \sigma_{F(\omega)}(u^*) : \|u^*\|_{X^*} \leqslant 1\big\}.$$

So, if $\{u_n^*\}_{n\geqslant 1} \subseteq \overline{B}_1^{X^*}$ is an m-dense sequence, then

$$\operatorname{dist}(u, F(\omega)) = \sup_{n\geqslant 1} \big(\langle u_n^*, u\rangle - \sigma_{F(\omega)}(u_n^*)\big),$$

thus the map $\omega \longmapsto \text{dist}(u, F(\omega))$ is Σ-measurable and hence F is measurable (see Theorem 2.63(a)).

Solution of Problem 2.104

Let $\lambda \in \mathbb{R}$. Then, $m(\omega) < \lambda$ if and only if there exists $u \in F(\omega)$ such that $\varphi(\omega, u) < \lambda$. Therefore, we have

$$\{\omega \in \Omega : \, m(\omega) < \lambda\} \;=\; \text{proj}_\Omega\big(\{(\omega, u) : \, \varphi(\omega, u) < \lambda\} \cap \text{Gr}\, F\big).$$

The joint measurability of φ, the graph measurability of F and the Yankov–von Neumann–Aumann projection theorem (see Theorem I.4.65) imply that

$$\text{proj}_\Omega\big(\{(\omega, u) : \, \varphi(\omega, u) < \lambda\} \cap \text{Gr}\, F\big) \;\in\; \widehat{\Sigma} \;=\; \Sigma$$

(since Σ is complete), so

$$\{\omega \in \Omega : \, m(\omega) < \lambda\} \;\in\; \Sigma \qquad \forall \, \lambda \in \mathbb{R}$$

and thus the map $m\colon \Omega \longrightarrow \mathbb{R}^*$ is Σ-measurable (cf. Proposition I.3.63).

Solution of Problem 2.105

(a) From Theorem 2.66, we know that there exists a sequence $\{f_n : \Omega \longrightarrow X\}_{n \geqslant 1}$ of Σ-measurable selectors of F such that

$$F(\omega) \;=\; \overline{\{f_n(\omega)\}_{n \geqslant 1}} \qquad \forall \, \omega \in \Omega.$$

Also, let

$$\Delta \;=\; \Big\{\{r_k\}_{k \geqslant 1} : \, r_k \in \mathbb{Q}, \; r_k \geqslant 0 \text{ and all but a finite number equal zero and } \sum_{k \geqslant 1} r_k = 1\Big\}.$$

Let

$$\Gamma \;=\; \Big\{h\colon \Omega \longrightarrow X : \; h(\omega) = \sum_{n \geqslant 1} r_n f_n(\omega) \text{ for all } \omega \in \Omega \text{ with } \{r_n\}_{n \geqslant 1} \in \Delta\Big\}.$$

Note that the set Δ is countable, hence so is Γ and each $h \in \Gamma$ is a Σ-measurable selector of $\omega \longmapsto \text{conv}\, F(\omega)$. Moreover, for every $\omega \in \Omega$, we have

$$\overline{\Gamma(\omega)} \;=\; \overline{\text{conv}}\, F(\omega),$$

so the multifunction $\omega \longmapsto \overline{\Gamma(\omega)}$ is measurable (see Theorem 2.66) and thus the multifunction $\omega \longmapsto \overline{\text{conv}}\, F(\omega)$ is measurable.
Since for every nonempty open set $U \subseteq X$, we have

$$\text{conv}\, F(\omega) \cap U \neq \emptyset \quad \text{if and only if} \quad \overline{\text{conv}}\, F(\omega) \cap U \neq \emptyset,$$

so it follows that the multifunction $\omega \longmapsto \text{conv}\, F(\omega)$ is measurable too.

(b) This is proved as the measurability of $\omega \longmapsto \overline{\text{conv}}\, F(\omega)$ in part **(a)**, using this time the Yankov–von Neumann–Aumann selection theorem (see Theorem 2.68).

Solution of Problem 2.106

Let $\{\widehat{u}_n\}_{n \geqslant 1}$ be a sequence which is dense in the open unit ball of X. Let

$$h_n(\omega) = u(\omega) + \varrho(\omega)\widehat{u}_n \qquad \forall\, \omega \in \Omega,\ n \geqslant 1.$$

Then for each $n \geqslant 1$, the map $\omega \longmapsto h_n(\omega)$ is Σ-measurable and

$$\overline{B_{\varrho(\omega)}(u(\omega))} = \overline{\{h_n(\omega)\}}_{n \geqslant 1} \qquad \forall\, \omega \in \Omega,$$

so the multifunction $\omega \longmapsto \overline{B_{\varrho(\omega)}(u(\omega))}$ is measurable (see Theorem 2.66) and thus the multifunction $\omega \longmapsto B_{\varrho(\omega)}(u(\omega))$ is measurable.

Solution of Problem 2.107

For every $n \geqslant 1$, let

$$F_n(\omega) = \left\{u \in X : |\varphi(\omega, u)| < \tfrac{1}{n}\right\}.$$

Problem 2.95 implies that for each $n \geqslant 1$, the multifunction $\omega \longmapsto F_n(\omega)$ is measurable, hence so is the multifunction $\omega \longmapsto \overline{F_n(\omega)} = \left\{u \in X : |\varphi(\omega, u)| \leqslant \tfrac{1}{n}\right\}$. Note that

$$F(\omega) = \bigcap_{n \geqslant 1} F_n(\omega),$$

so, invoking Problem 2.97, we conclude that F is measurable.

Solution of Problem 2.108

Let $\Gamma(\omega) = \{u \in F(\omega) : h(\omega, u) \in G(\omega)\}$. By hypothesis $\Gamma(\omega) \neq \emptyset$ for all $\omega \in \Omega$. Also, let $h_0 \colon \Omega \times X \longrightarrow \Omega \times T$ be defined by

$$h_0(\omega, u) = (\omega, h(\omega, u)) \qquad \forall\, (\omega, u) \in \Omega \times X.$$

Clearly h_0 is $(\Sigma \times \mathcal{B}(X), \Sigma \times \mathcal{T})$-measurable (see Definition I.3.53). We have

$$\operatorname{Gr}\Gamma = \operatorname{Gr}F \cap h_0^{-1}(\operatorname{Gr}G).$$

Since h_0 is measurable and F, G are both graph measurable (see Definition 2.61(b)), we have

$$\operatorname{Gr}\Gamma \in \Sigma \times \mathcal{B}(X).$$

So, we can apply the Yankov–von Neumann–Aumann selection theorem (see Theorem 2.68) and find a Σ-measurable function $f\colon \Omega \longrightarrow X$ such that

$$f(\omega) \in \Gamma(\omega) \qquad \forall\, \omega \in \Omega,$$

so

$$f(\omega) \in F(\omega) \quad \text{and} \quad h(\omega, f(\omega)) \in G(\omega) \qquad \forall\, \omega \in \Omega.$$

Solution of Problem 2.109
Let $U \subseteq X$ be an open set. Then

$$G^-(U) = \{\omega \in \Omega : \bigcup_{n\geqslant 1}(F_n(\omega) \cap U) \neq \emptyset\} = \bigcup_{n\geqslant 1} F_n^-(U) \in \Sigma,$$

so the multifunction $\omega \longmapsto G(\omega)$ is measurable.

The multifunction H has closed values and

$$\operatorname{Gr}H = \bigcap_{n\geqslant 1} \operatorname{Gr}F_n \in \Sigma \times \mathcal{B}(X),$$

so the multifunction H is measurable (see Theorem 2.67).

Solution of Problem 2.110
"$\Longrightarrow$": Obvious.

"$\Longleftarrow$": From the Yankov–von Neumann–Aumann selection theorem (see Theorem 2.68 and Remark 2.69), we can find a sequence $\{f_n\colon \Omega \longrightarrow X\}_{n\geqslant 1}$ of Σ_μ-measurable selectors of F such that

$$F(\omega) \subseteq \overline{\{f_n(\omega)\}}_{n\geqslant 1} \qquad \forall\, \omega \in \Omega.$$

Then

$$m(\omega) = \inf_{u\in F(\omega)} \|u\|_X = \inf_{n\geqslant 1} \|f_n(\omega)\|_X \qquad \forall\, \omega \in \Omega,$$

so the map m is Σ_μ-measurable and hence $m \in S_F^p$ (as $m \leqslant \vartheta$). Let $\varepsilon \in L^p(\Omega)_+$ with $\varepsilon(\omega) > 0$ for all $\omega \in \Omega$ and consider the multifunction

$$H_\varepsilon(\omega) \;=\; \{u \in F(\omega):\ \|u\|_X \leqslant m(\omega) + \varepsilon(\omega)\}\,.$$

Then $\operatorname{Gr} H_\varepsilon \in \Sigma_\mu \times \mathcal{B}(X)$.

So, we can find a Σ-measurable map $f\colon \Omega \longrightarrow X$ such that $f(\omega) \in H_\varepsilon(\omega)$ for μ-almost all $\omega \in \Omega$ (see the Yankov–von Neumann–Aumann selection theorem; Theorem 2.68). Since

$$\|f(\omega)\|_X \;\leqslant\; m(\omega) + \varepsilon(\omega) \quad \text{for } \mu\text{-a.a. } \omega \in \Omega,$$

we conclude that $f \in S_F^p$ (recall that $m + \varepsilon \in S_F^p$) and so $S_F^p \neq \emptyset$.

Solution of Problem 2.111

From Remark 2.89 we know that

$$H_l(\omega) \;=\; \liminf_{n\to+\infty} F_n(\omega) \;=\; \big\{u \in X:\ \lim_{n\to+\infty} \operatorname{dist}(u, F_n(\omega)) = 0\big\}, \tag{2.51}$$

$$H_u(\omega) \;=\; \limsup_{n\to+\infty} F_n(\omega) \;=\; \big\{u \in X:\ \liminf_{n\to+\infty} \operatorname{dist}(u, F_n(\omega)) = 0\big\}. \tag{2.52}$$

Both multifunctions are closed valued (see Remark 2.89). For every $n \geqslant 1$, the function

$$\varphi_n(\omega, u) \;=\; \operatorname{dist}(u, F_n(\omega)) \qquad \forall\ (\omega, u) \in \Omega \times X$$

is a Carathéodory function, hence jointly measurable (i.e., $\Sigma \times \mathcal{B}(X)$-measurable). Therefore

$$h_l(\omega, u) \;=\; \liminf_{n\to+\infty} \varphi_n(\omega, u) \quad \text{and} \quad h_u(\omega, u) \;=\; \limsup_{n\to+\infty} \varphi_n(\omega, u)$$

are both $\Sigma \times \mathcal{B}(X)$-measurable. From (2.51)-(2.51), we have

$$\begin{aligned}
\operatorname{Gr} H_l &= \{(\omega, u) \in \Omega \times X:\ h_l(\omega, u) = h_u(\omega, u) = 0\} \;\in\; \Sigma \times \mathcal{B}(X),\\
\operatorname{Gr} H_u &= \{(\omega, u) \in \Omega \times X:\ h_l(\omega, u) = 0\} \;\in\; \Sigma \times \mathcal{B}(X),
\end{aligned}$$

so H_l and H_u are both measurable multifunctions (see Theorem 2.67).

Solution of Problem 2.112

By the Yankov–von Neumann–Aumann selection theorem (see Theorem 2.68 and Remark 2.69), we can find a sequence $\{g_m\}_{m\geqslant 1}$ of Σ-measurable selectors of F such that

$$F(\omega) \subseteq \overline{\{g_m(\omega)\}_{m\geqslant 1}} \quad \text{for } \mu\text{-a.a. } \omega \in \Omega.$$

Since the measure space is σ-finite, we can find a sequence $\{\Omega_n\}_{n\geqslant 1} \subseteq \Sigma$ of pairwise disjoint sets such that $\mu(\Omega_n) < +\infty$ for all $n \geqslant 1$ and $\Omega = \bigcup\limits_{n\geqslant 1} \Omega_n$. Let $h \in S_F^p$ and define

$$C_{kmn} = \{\omega \in \Omega : \ k-1 \leqslant g_m(\omega) < k\} \cap \Omega_n$$

and

$$f_{kmn} = \chi_{C_{kmn}} g_m + \chi_{C_{kmn}^c} h \in S_F^p.$$

Then, we have

$$F(\omega) \subseteq \overline{\{f_{kmn}(\omega)\}_{k,m,n\geqslant 1}} \qquad \forall\, \omega \in \Omega.$$

Solution of Problem 2.113

Let $\delta \in L^1(\Omega)$ be such that $\delta(\omega) > 0$ for all $\omega \in \Omega$ and

$$\int_\Omega \delta(\omega)\, d\mu < \tfrac{2\varepsilon^p}{3}.$$

Let $\{C_n\}_{n\geqslant 1} \subseteq \Sigma$ be a sequence of pairwise disjoint sets such that $\Omega = N \cup \big(\bigcup\limits_{n\geqslant 1} C_n\big)$ with N being μ-null and

$$\|f(\omega) - f_n(\omega)\|_X^p < \delta(\omega) \qquad \forall\, \omega \in C_n.$$

As

$$\begin{aligned}
&\sum_{k\geqslant 1} \int_{C_k} \|f_k(\omega)\|_X^p\, d\mu \\
&\leqslant 2^{p-1} \sum_{k\geqslant 1} \int_{C_k} \|f_k(\omega) - f(\omega)\|_X^p\, d\mu + 2^{p-1} \sum_{k\geqslant 1} \int_{C_k} \|f(\omega)\|_X^p\, d\mu \\
&\leqslant 2^{p-1} \int_\Omega \delta(\omega)\, d\mu + 2^{p-1} \int_\Omega \|f(\omega)\|_X^p\, d\mu,
\end{aligned}$$

we see that series $\sum\limits_{k\geqslant 1}\int\limits_{C_k} \|f_k(\omega)\|_X^p \, d\mu$ is convergent. Let $N \in \mathbb{N}$ be such that

$$\sum_{k\geqslant N+1}\int_{C_k} \|f_k(\omega)\|_X^p \, d\mu \; < \; \tfrac{\varepsilon^p}{2^p \cdot 3}$$

and

$$\sum_{k\geqslant N+1}\int_{C_k} \|f(\omega)\|_X^p \, d\mu \; < \; \tfrac{\varepsilon^p}{2^p \cdot 3}.$$

We define the finite Σ-partition of Ω by setting

$$A_1 \; = \; C_1 \cup \Big(\bigcup_{k\geqslant N+1} C_k \Big) \quad \text{and} \quad A_m \; = \; C_m \qquad \forall\, 2 \leqslant m \leqslant N.$$

We have

$$\Big\|f - \sum_{k=1}^{N} \chi_{A_k} f_k\Big\|_p^p \; \leqslant \; \int_{\Omega} \delta \, d\mu + \sum_{k\geqslant N+1} 2^{p-1}\big(\|\chi_{C_k} f\|_p^p + \|\chi_{C_k} f_k\|_p^p\big) \; < \; \varepsilon^p.$$

Solution of Problem 2.114

According to Theorem 2.66, we can find two sequences $\{f_n \colon \Omega \longrightarrow X\}_{n\geqslant 1}$ and $\{g_m \colon \Omega \longrightarrow X\}_{m\geqslant 1}$ of Σ-measurable selectors of F and G respectively such that

$$F(\omega) \; = \; \overline{\{f_n(\omega)\}_{n\geqslant 1}} \quad \text{and} \quad G(\omega) \; = \; \overline{\{g_m(\omega)\}_{m\geqslant 1}} \qquad \forall\, \omega \in \Omega.$$

We have

$$H(\omega) \; = \; \overline{\{f_n(\omega) + g_m(\omega)\}_{n,m\geqslant 1}},$$

so the multifunction H is measurable and $P_f(X)$-valued (see Theorem 2.66).

Note that since H is closed valued, then S_H^p is closed. So, we have

$$\overline{S_F^p + S_G^p} \; \subseteq \; S_H^p. \tag{2.53}$$

We need to show that the opposite inclusion also holds. To this end, let $h \in S_H^p$ and $\varepsilon > 0$. According to Problem 2.113, we can find a finite Σ-partition $\{A_1, \ldots, A_l\}$ of Ω and positive integers $n_1, \ldots, n_l$ and $m_1, \ldots, m_l$ such that

$$\Big\|h - \sum_{k=1}^{l} \chi_{A_k}(f_{n_k} + g_{n_k})\Big\|_p \; < \; \varepsilon,$$

so $h \in \overline{S_F^p + S_G^p}$. Thus, we have

$$S_H^p \; \subseteq \; \overline{S_F^p + S_G^p}. \tag{2.54}$$

From (2.53) and (2.54) we conclude that $S_H^p = \overline{S_F^p + S_G^p}$.

Solution of Problem 2.115

Clearly $S^p_{\overline{\text{conv}}\, F}$ is closed, convex and so we have

$$\overline{\text{conv}}\, S^p_F \subseteq S^p_{\overline{\text{conv}}\, F}. \tag{2.55}$$

Suppose that the inclusion in (2.55) is strict. So, we can find $h \in S^p_{\overline{\text{conv}}\, F} \setminus \overline{\text{conv}}\, S^p_F$. The strong separation theorem for convex sets (see Theorem I.5.29) implies that we can find $h^* \in L^p(\Omega; X)^* = L^{p'}(\Omega; X^*_{w^*})$ (with $\frac{1}{p} + \frac{1}{p'} = 1$; see Theorem 1.47) such that

$$\sigma_{\overline{\text{conv}}\, S^p_F}(h^*) < \langle h^*, h\rangle \tag{2.56}$$

(see Definition 2.46), where by $\langle\cdot,\cdot\rangle$ we denote the duality brackets for the pair $(L^{p'}(\Omega; X^*_{w^*}), L^p(\Omega; X))$, i.e.,

$$\langle h^*, h\rangle = \int_\Omega \langle h^*(\omega), h(\omega)\rangle_X \, d\mu,$$

where by $\langle\cdot,\cdot\rangle_X$ we denote the duality brackets for the pair (X^*, X). We have

$$\begin{aligned}
\sigma_{\overline{\text{conv}}\, S^p_F}(h^*) &= \sigma_{S^p_F}(h^*) = \sup_{f\in S^p_F} \langle h^*, f\rangle \\
&= \sup_{f\in S^p_F} \int_\Omega \langle h^*(\omega), f(\omega)\rangle_X \, d\mu = \int_\Omega \sup_{u\in F(\omega)} \langle h^*(\omega), u\rangle_X \, d\mu \\
&= \int_\Omega \sigma_{F(\omega)}(h^*(\omega))\, d\mu = \int_\Omega \sigma_{\overline{\text{conv}}\, F(\omega)}(h^*(\omega))\, d\mu \\
&< \int_\Omega \langle h^*(\omega), h(\omega)\rangle_X \, d\mu
\end{aligned}$$

(see Theorem 2.74 and (2.56)), which contradicts the fact that $h \in S^p_{\overline{\text{conv}}\, F}$. This proves that in (2.55) we have equality.

Solution of Problem 2.116

Since $S^p_F \subseteq L^p(\Omega; X)$ is closed, we have

$$\overline{S^p_F} \subseteq S^p_F. \tag{2.57}$$

Let $f \in S^p_{\overline{F}}$ and for every $n \geqslant 1$ we define

$$V_n(\omega) = \left\{u \in F(\omega) : \|u - f(\omega)\|_X < \tfrac{1}{n}\right\}.$$

By modifying $V_n(\omega)$ on a μ-null set, we may assume that $V_n(\omega) \neq \emptyset$ for all $\omega \in \Omega$. Note that, if $\varphi(\omega, u) = \|u - f(\omega)\|_X$, then φ is a Carathéodory function, hence jointly measurable and we have

$$\operatorname{Gr} V_n = \{(\omega, u) \in \Omega \times X : \varphi(\omega, u) < \tfrac{1}{n}\} \cap \operatorname{Gr} F,$$

so $\operatorname{Gr} V_n \in \Sigma \times \mathcal{B}(X)$. Invoking the Yankov–von Neumann–Aumann selection theorem (see Theorem 2.68 and Remark 2.69), we can find a sequence $\{f_n : \Omega \longrightarrow X\}_{n \geqslant 1}$ of Σ-measurable functions such that

$$f_n(\omega) \in V_n(\omega) \quad \text{for } \mu\text{-a.a. } \omega \in \Omega, \text{ all } n \geqslant 1,$$

so

$$\|f_n(\omega) - f(\omega)\|_X < \tfrac{1}{n} \quad \text{for } \mu\text{-a.a. } \omega \in \Omega, \text{ all } n \geqslant 1$$

and $f_n \in S_F^p$ for all $n \geqslant 1$, thus

$$f_n(\omega) \longrightarrow f(\omega) \quad \text{for } \mu\text{-a.a. } \omega \in \Omega. \tag{2.58}$$

We have

$$\|f_n(\omega)\|_X \leqslant \|f(\omega)\|_X + 1 \quad \text{for } \mu\text{-a.a. } \omega \in \Omega, \text{ all } n \geqslant 1, \tag{2.59}$$

with $f \in L^p(\Omega; X)$. Because of (2.58) and (2.59) and the Lebesgue dominated convergence theorem, we have

$$f_n \longrightarrow f \quad \text{in } L^p(\Omega; X),$$

so $f \in \overline{S_F^p}$, hence we have equality in (2.57).

Solution of Problem 2.117

By working on each component of Ω of finite μ-measure, we see that we may assume that μ is finite. Let $f, g \in D$ and define $m : \Sigma \longrightarrow Y$ by setting

$$m(A) = T(\chi_A(f - g)) \qquad \forall A \in \Sigma.$$

We show that m is a vector measure (see Definition 1.43). To this end, let $\{A_n\}_{n \geqslant 1} \subseteq \Sigma$ be a sequence of pairwise disjoint sets and let $A = \bigcup_{n \geqslant 1} A_n$. We set $C_n = \bigcup_{k \geqslant n+1} A_k \in \Sigma$. From the linearity of T, for every $n \geqslant 1$, we have

$$\begin{aligned} m(A) &= T(\chi_A(f - g)) = T\Big(\sum_{k=1}^{n} \chi_{A_k}(f - g) + \chi_{C_n}(f - g)\Big) \\ &= \sum_{k=1}^{n} T(\chi_{A_k}(f - g)) + T(\chi_{C_n}(f - g)), \end{aligned}$$

so

$$\Big\|m(A) - \sum_{k=1}^{n} T(\chi_{A_k}(f-g))\Big\|_Y = \Big\|m(A) - \sum_{k=1}^{n} m(A_k)\Big\|_Y$$
$$= \|T(\chi_{C_n}(f-g))\|_Y. \tag{2.60}$$

But $C_n \searrow \emptyset$ and so

$$\chi_{C_n}(f-g) \longrightarrow 0 \quad \text{in } L^p(\Omega; X).$$

Then the continuity of T implies that

$$\|T(\chi_{C_n}(f-g))\|_Y \longrightarrow 0,$$

so m is a vector measure (see (2.60) and Definition 2.70).

If $\mu(A) = 0$, then $\chi_A(f-g) = 0$ and so $m(A) = 0$. This implies that $m \ll \mu$. Moreover, the nonatomicity of μ implies that m is nonatomic (see Definition I.3.40). Note that

$$\|m(A)\|_Y = \|T(\chi_A(f-g))\|_Y \leqslant \|T\|_{\mathcal{L}} \|\chi_A(f-g)\|_{L^p(\Omega;X)}. \tag{2.61}$$

If

$$\vartheta(A) = \int_A \|f(\omega) - g(\omega)\|_X^p \, d\mu \qquad \forall\, A \in \Sigma,$$

then ϑ is a measure on (Ω, Σ) of bounded variation and from (2.61) we have

$$|m|(\Omega) \leqslant \|T\|_{\mathcal{L}} \vartheta(\Omega)^{\frac{1}{p}}$$

(see Definition 1.43), so m is a vector measure of bounded variation.

Therefore, so far we have that $m \colon \Sigma \longrightarrow Y$ is a nonatomic vector measure of bounded variation. Because Y has the RNP, we can apply the Lyapunov convexity theorem (see Theorem 2.165) and infer that $\overline{m(\Sigma)} \subseteq Y$ is convex and compact. Then the set

$$\overline{\bigcup_{A \in \Sigma} T(\chi_A(f-g)) + T(g)} = \overline{\bigcup_{A \in \Sigma} T(\chi_A(f-g)) + T(g)}$$

is convex. Using the linearity of T, we have

$$T(\chi_A(f-g)) + T(g) = T(\chi_A f + \chi_{A^c} g).$$

The decomposability of D (see Definition 2.70) implies that $\chi_A f + \chi_{A^c} g \in D$ for all $A \in \Sigma$, so

$$\overline{\bigcup_{A \in \Sigma} T(\chi_A f + \chi_{A^c} g)} \subseteq \overline{T(D)}.$$

It follows that for all $\lambda \in [0,1]$, we have

$$T(\lambda f + (1-\lambda) g) \in \overline{T(D)},$$

so the set $\overline{T(D)}$ is convex (recall that $f, g \in D$).

Solution of Problem 2.118

By working on each component of Ω of finite μ-measure, we see that we may assume that μ is finite. Let $h \in \operatorname{conv} D$. Then

$$h = \sum_{k=1}^{n} \lambda_k f_k, \quad \text{with } \lambda_k \in [0,1],\ \sum_{k=1}^{n} \lambda_k = 1,\ f_k \in D \text{ for } k \in \{1,\ldots,n\}.$$

Recall that $L^p(\Omega;X)^* = L^{p'}(\Omega;X^*_{w^*})$ (with $\frac{1}{p}+\frac{1}{p'}=1$; see Theorem 1.47) and let $\langle\cdot,\cdot\rangle$ denote the duality brackets for this dual pair. Let $U_w(h)$ be a basic weak neighborhood of h defined by

$$U_w(h) = \{g \in L^p(\Omega;X) : |\langle u_k^*, g-h\rangle| < \varepsilon \text{ for all } k \in \{1,\ldots,m\}\},$$

with $m \geqslant 1$, $u_k^* \in L^{p'}(\Omega;X^*_{w^*})$ for all $k \in \{1,\ldots,m\}$ and $\varepsilon > 0$. We consider the continuous linear map $T\colon L^p(\Omega;X) \longrightarrow \mathbb{R}^m$ defined by

$$T(h) = (\langle u_1^*, h\rangle, \ldots, \langle u_m^*, h\rangle).$$

From Problem 2.117 we have that $T(D)$ is convex (no closure is needed since in finite dimensional Banach spaces the Lyapunov theorem is exact; see Theorem 2.165). Therefore, we can find $f \in D$ such that $T(f) = T(h)$ and so $D \cap U_w(h) \neq \emptyset$. Hence $h \in \overline{D}^w = D$ (since by definition D is w-closed). Then we conclude that $D = \operatorname{conv} D$.

Solution of Problem 2.119

Note that the set $\overline{S_F^p}^w$ remains decomposable (see Definition 2.70). Also, it is w-closed. Hence according to Problem 2.118 it is convex. Also, from Problem 2.115, we have $\overline{\operatorname{conv}}\, S_F^p = S^p_{\overline{\operatorname{conv}}\, F}$. Therefore

$$\overline{\operatorname{conv}}\, S_F^p \subseteq \overline{S_F^p}^w. \tag{2.62}$$

But by the Mazur theorem (see Theorem I.5.58), a convex set is closed if and only if it is w-closed. So, from (2.62) it follows that $\overline{S_F^p}^w = \overline{\operatorname{conv}}\, S_F^p$, thus

$$\overline{S_F^p}^w = S^p_{\overline{\operatorname{conv}}\, F}.$$

Solution of Problem 2.120

We argue indirectly. So, suppose that $H(\omega) = F(\omega) \setminus G(\omega)$ is not empty on a set of positive μ-measure. Then

$$A = \operatorname{dom} H = \{\omega \in \Omega : H(\omega) \neq \emptyset\} \in \Sigma_\mu.$$

Since $H\colon A \longrightarrow 2^X \setminus \{\emptyset\}$ is graph measurable (see Definition 2.61(b)), form the Yankov–von Neumann–Aumann selection theorem (see Theorem 2.68), we can find a Σ_A-measurable map $h\colon A \longrightarrow X$ such that

$$h(\omega) \in H(\omega) \quad \text{for } \mu\text{-a.a. } \omega \in A$$

(here $\Sigma_A = \Sigma \cap A$). Since μ is σ-finite, we can find a sequence $\{\Omega_n\}_{n\geqslant 1} \subseteq \Sigma$ with $\mu(\Omega_n) < +\infty$ for all $n \geqslant 1$ such that $\Omega = \bigcup\limits_{n\geqslant 1} \Omega_n$. Also for every $m \geqslant 1$, let

$$A_m = \{\omega \in A:\ m-1 \leqslant h(\omega) < m\} \in (\Sigma_\mu)_A.$$

We set $C_{mn} = A_m \cap \Omega_n$. Then $A = \bigcup\limits_{n,m\geqslant 1} C_{mn}$. Since $\mu(A) > 0$, we can find $m, n \geqslant 1$ such that $\mu(C_{mn}) > 0$. Let $f \in S_F^p$ and let $\vartheta = \chi_{C_{mn}} h + \chi_{C_{mn}^c} f \in S_F^p$ (by decomposability; see Definition 2.70) but $\vartheta \notin S_G^p$, a contradiction to the hypothesis that $S_F^p = S_G^p$.

Solution of Problem 2.121

(a) From Problem 2.116 we have $S_F^p = \overline{S_F^p} = S_{\overline{F}}^p$. Invoking Problem 2.120, we infer that

$$F(\omega) = \overline{F}(\omega) \quad \text{for } \mu\text{-a.a. } \omega \in \Omega,$$

hence $F(\omega) \in P_f(X)$ for μ-almost all $\omega \in \Omega$.

(b) From the Mazur theorem (see Theorem I.5.58) and Problem 2.119, we have

$$S_F^p = \overline{S_F^p}^w = S_{\overline{\operatorname{conv}}\, F}^p,$$

so

$$F(\omega) = \overline{\operatorname{conv}}\, F(\omega) \quad \text{for } \mu\text{-a.a. } \omega \in \Omega$$

(see Problem 2.120) and thus $F(\omega) \in P_{fc}(X)$ for μ-almost all $\omega \in \Omega$.

Solution of Problem 2.122

Let $T \in \mathcal{L}(L^1(\Omega; X); X)$ be defined by

$$T(h) = \int_\Omega h(\omega)\, d\mu \qquad \forall\, h \in L^1(\Omega; X).$$

From Definition 2.85 we know that

$$T(S^1_F) = \int_\Omega F(\omega)\, d\mu.$$

Problem 2.117 implies that $\overline{T(S^1_F)}$ is convex (note that in present case we do not need X to have the RNP, since the vector measure is from the definition of T in integral form). Therefore, we conclude that the set $\mathrm{cl} \int_\Omega F\, d\mu \subseteq X$ is convex.

Solution of Problem 2.123

Recall that $\int_\Omega F(\omega)\, d\mu = T(S^1_F)$, where $T \in \mathcal{L}(L^1(\Omega; X); X)$ is defined by

$$T(h) = \int_\Omega h(\omega)\, d\mu \qquad \forall\, h \in S^1_F$$

(see Definition 2.85). From Theorem 2.76, we have that $S^1_F \subseteq L^1(\Omega; X)$ is w-compact and convex. Hence

$$T(S^1_F) = \int_\Omega F(\omega)\, d\mu \in P_{wkc}(X).$$

Solution of Problem 2.124

Let $\langle \cdot, \cdot \rangle$ denote the duality brackets for the pair

$$(L^\infty(\Omega; X^*_{w^*}) = L^1(\Omega; X)^*, L^1(\Omega; X))$$

(see Theorem 1.52). Also, by $\langle \cdot, \cdot \rangle_X$ we denote the duality brackets for the dual pair (X^*, X). According to Definition 2.46, we have

$$\begin{aligned} \sigma_{S^1_F}(h^*) &= \sup_{f \in S^1_F} \langle h^*, f \rangle = \sup_{f \in S^1_F} \int_\Omega \langle h^*(\omega), f(\omega) \rangle_X \, d\mu \\ &= \int_\Omega \sup_{u \in F(\omega)} \langle h^*(\omega), u \rangle_X = \int_\Omega \sigma_{F(\omega)}(h^*(\omega))\, d\mu. \end{aligned}$$

(see Theorem 2.74)

Solution of Problem 2.125

We fix $\varepsilon > 0$ and let $\{\widehat{u}_n\}_{n\geqslant 1} \subseteq X$ be a sequence which is dense in X. Let

$$\varphi_n(v) \;=\; \operatorname{dist}(\widehat{u}_n, F(v)) \qquad \forall\, v \in V,\; n \geqslant 1.$$

By the Lusin theorem (see Theorem 2.166), we can find compact sets $K_n \subseteq V$ with $\mu(V \setminus K_n) < \frac{\varepsilon}{2^n}$ for all $n \geqslant 1$, such that $\varphi_n|_{K_n}$ is continuous. Let $K_\varepsilon = \bigcap\limits_{n\geqslant 1} K_n$. Evidently $K_\varepsilon \subseteq V$ is compact and $\mu(V \setminus K_\varepsilon) < \varepsilon$. Also, for every $n \geqslant 1$, $\varphi_n|_{K_\varepsilon}$ is continuous. Let $\{(v_m, u_m)\}_{m\geqslant 1} \subseteq \operatorname{Gr} F \cap (K_\varepsilon \times X)$ be a sequence such that $(v_m, u_m) \longrightarrow (\widetilde{v}, \widetilde{u})$. We fix $\delta > 0$ and choose $n \geqslant 1$ such that $d_X(\widehat{u}_n, \widetilde{u}) < \delta$. Then we can find $m_0 = m_0(n) \geqslant 1$ such that

$$\varphi_n(v_m) \;=\; \operatorname{dist}(\widehat{u}_n, F(v_m)) \;\leqslant\; d_X(\widehat{u}_n, u_m) \;<\; \delta \qquad \forall\, m \geqslant m_0.$$

The continuity of $\varphi_n|_{K_\varepsilon}$ implies that $\operatorname{dist}(\widehat{u}_n, F(\widetilde{v})) \leqslant \delta$ and so $\operatorname{dist}(\widetilde{u}, F(\widetilde{v})) < 2\delta$. Since $\delta > 0$ is arbitrary, we let $\delta \searrow 0$ and obtain $\operatorname{dist}(\widetilde{u}, F(\widetilde{v})) = 0$, thus $(\widetilde{v}, \widetilde{u}) \in \operatorname{Gr} F$ (recall that F has closed values) and this proves that $F|_{K_\varepsilon}$ is closed (see Definition 2.40).

Solution of Problem 2.126

Theorem 2.66 implies that there exists a sequence $\{f_n \colon V \longrightarrow X\}_{n\geqslant 1}$ of Borel measurable functions such that

$$F(v) \;=\; \overline{\{f_n(v)\}_{n\geqslant 1}} \qquad \forall\, v \in V. \tag{2.63}$$

By the Lusin theorem (see Theorem 2.166), for every $\varepsilon > 0$, we can find a compact set $K_\varepsilon \subseteq V$ with $\mu(V \setminus K_\varepsilon) < \varepsilon$ such that $f_n|_{K_\varepsilon}$ is continuous for every $n \geqslant 1$ (see the solution of Problem 2.125). Then because of (2.63) for every closed set $C \subseteq X$, we have that the set

$$F|_{K_\varepsilon}^{+}(C) \;=\; \bigcap_{n\geqslant 1} f_n|_{K_\varepsilon}^{-1}(C)$$

is closed. Therefore $F|_{K_\varepsilon}$ is lower semicontinuous (see Proposition 2.38).

Solution of Problem 2.127

By Theorem 2.66 there exists a sequence $\{f_n\}_{n\geqslant 1}$ of $\Sigma \times \mathcal{T}$-measurable selectors of F such that

$$F(\omega, t) \;=\; \overline{\{f_n(\omega, t)\}_{n\geqslant 1}} \qquad \forall\, (\omega, t) \subset \Omega \times \mathcal{T}. \tag{2.64}$$

For every $n \geqslant 1$, let

$$\Omega_n \;=\; \{\omega \in \Omega :\; d_X(u_0(\omega), f_n(\omega, t_0)) \leqslant \varepsilon\}. \tag{2.65}$$

From (2.64) we have $\Omega = \bigcup_{n\geqslant 1} \Omega_n$. Let

$$D_1 = \Omega_1 \quad \text{and} \quad D_n = \Omega_n \setminus \bigcup_{k=1}^{n-1} D_k \qquad \forall\, n \geqslant 2.$$

We have $\{D_n\}_{n\geqslant 1} \subseteq \Sigma$. We set

$$\widehat{f}(\omega, t) = \sum_{n\geqslant 1} \chi_{D_n}(\omega) f_n(\omega, t).$$

Then $\widehat{f}$ is a $\Sigma \times \mathcal{T}$-measurable selector of F and

$$d_X(\widehat{f}(\omega, t_0), u_0(\omega)) \leqslant \varepsilon \qquad \forall\, \omega \in \Omega.$$

(see (2.65)).

Solution of Problem 2.128

By definition, we have

$$\begin{aligned} F^-(D) &= \{\omega \in \Omega :\ F(\omega) \cap D \neq \emptyset\} \\ &= \operatorname{proj}_\Omega \big(\operatorname{Gr} F \cap (\Omega \times D)\big) \ \in\ \Sigma \end{aligned}$$

(by the Yankov–von Neumann–Aumann projection theorem; see Theorem I.4.65).

Solution of Problem 2.129

Note that $\omega \longmapsto F(\omega)$ and $\omega \longmapsto F^c(\omega) = X \setminus F(\omega)$ are both graph measurable (see Definition 2.61(b)). Then from Problem 2.128 we see that they are measurable and then so are $\omega \longmapsto \overline{F(\omega)}$ and $\omega \longmapsto \overline{F^c(\omega)}$. Therefore the map $\omega \longmapsto \partial F(\omega) = \overline{F(\omega)} \cap \overline{F^c(\omega)}$ is graph measurable and then since ∂F is closed valued, from Theorem 2.67(d), we conclude that $\omega \longmapsto \partial F(\omega)$ is measurable.

Next note that $\operatorname{int} F(\omega) = F(\omega) \setminus \partial F(\omega)$, hence

$$\operatorname{Gr}(\operatorname{int} F) = \operatorname{Gr} F \cap (\operatorname{Gr} \partial F)^c \ \in\ \Sigma \times \mathcal{B}(X)$$

and thus $\omega \longmapsto \operatorname{int} F(\omega)$ is graph measurable.

Solution of Problem 2.130

We may assume that $A = \Omega$. Also, let $f \in S^1_F$. Then by replacing $F(\omega)$ by $F(\omega) - f(\omega)$, which is still graph measurable and open valued, we may assume that $0 \in F(\omega)$ for all $\omega \in \Omega$. Let $\xi_F(\omega) = \operatorname{dist}(0, F(\omega)^c)$. Since F is graph measurable (see Definition 2.61(b)), so is $\omega \longrightarrow F(\omega)^c$ which is closed valued, hence ξ_F is Σ_μ-measurable (see Theorem 2.67(a)). Note that $\xi_F(\omega) > 0$ for all $\omega \in \Omega$. So, we can find $\varepsilon > 0$ and $A \in \Sigma$ with $\mu(A) > 0$ such that $\xi_F(\omega) \geqslant \varepsilon$ for all $\omega \in A$. This means that, if $B_\varepsilon = \{u \in X : \|u\|_X < \varepsilon\}$ then

$$\mu(A)B_\varepsilon \subseteq \int_A F\,d\mu \subseteq \int_\Omega F\,d\mu$$

(since $0 \in F(\omega)$ for all $\omega \in \Omega$), so the set $\int_A F\,d\mu \subseteq X$ is open.

Solution of Problem 2.131

Without any loss of generality, we may assume that $\operatorname{int} F(\omega) \neq \emptyset$ for all $\omega \in \Omega$. We fix $A \in \Sigma$ with $\mu(A) > 0$ and we show that the set $\int_A \operatorname{int} F\,d\mu$ is dense in $\int_A F\,d\mu$. To this end let $f \in S^1_F$ and $\varepsilon > 0$. We introduce the multifunction $H\colon \Omega \longrightarrow 2^X \setminus \{\emptyset\}$ defined by

$$H(\omega) = \Big\{u \in \operatorname{int} F(\omega) : \|f(\omega) - u\|_X < \tfrac{\varepsilon}{\mu(A)}\Big\}.$$

If $\varphi(\omega, u) = \|f(\omega) - u\|_X$, then φ is a Carathéodory function, hence jointly measurable and so

$$\operatorname{Gr} H = \operatorname{Gr} \operatorname{int} F \cap \Big\{(\omega, u) \in \Omega \times X : \varphi(\omega, u) < \tfrac{\varepsilon}{\mu(A)}\Big\} \in \Sigma_\mu \times \mathcal{B}(X)$$

(see Problem 2.129 and recall that Σ_μ stands for the μ-completion of Σ). Invoking the Yankov–von Neumann–Aumann selection theorem (see Theorem 2.68) we can find a Σ-measurable map $h\colon \Omega \longrightarrow X$ such that

$$h(\omega) \in H(\omega) \quad \text{for } \mu\text{-a.a. } \omega \in \Omega.$$

We have

$$h(\omega) \in \operatorname{int} F(\omega) \quad \text{and} \quad \|f(\omega) - h(\omega)\|_X < \tfrac{\varepsilon}{\mu(A)} \quad \text{for } \mu\text{-a.a. } \omega \in \Omega,$$

so $h \in S^1_{\operatorname{int} F}$ (recall that $f \in S^1_F$) and

$$\Big\| \int_A (f(\omega) - h(\omega))\,d\mu \Big\|_X < \varepsilon,$$

so $\int_A \operatorname{int} F\,d\mu$ is dense in $\int_A F\,d\mu$.

From Problem 2.130, we know that $\int_A \operatorname{int} F\,d\mu$ is open and convex in X. So

$$\int_A \operatorname{int} F\,d\mu = \operatorname{int} \overline{\int_A \inf F\,d\mu} = \operatorname{int} \overline{\int_A F\,d\mu} \supseteq \operatorname{int} \int_A F\,d\mu. \tag{2.66}$$

On the other hand, the openness of $\int_A \operatorname{int} F\,d\mu$ implies

$$\int_A \operatorname{int} F\,d\mu \;\subseteq\; \operatorname{int} \int_A F\,d\mu. \tag{2.67}$$

From (2.66) and (2.67) we conclude that $\operatorname{int} \int_A F\,d\mu = \int_A \operatorname{int} F\,d\mu$ for all $A \in \Sigma$.

Solution of Problem 2.132

Let $|D| = \{\|f(\cdot)\|_X :\ f \in D\} \subseteq L^1(\Omega)$ and let $h = \operatorname{ess\,sup} |D|$. We know that we can find a sequence $\{f_n\}_{n\geqslant 1} \subseteq D$ such that

$$h(\omega) \;=\; \sup_{n\geqslant 1} \|f_n(\omega)\|_X \quad \text{for } \mu\text{-a.a. } \omega \in \Omega$$

(see Proposition 2.168). The decomposability of D (see Definition 2.70) implies that we can have $\{\|f_n(\cdot)\|_X\}_{n\geqslant 1}$ to be increasing. Also, the boundedness of D and the Lebesgue monotone convergence theorem, imply that $h \in L^1(\Omega)_+$. Since

$$\|f(\omega)\|_X \;\leqslant\; h(\omega) \quad \text{for } \mu\text{-a.a. } \omega \in \Omega \text{ and all } f \in D,$$

we conclude that D is uniformly integrable (see Definition 1.18).

Solution of Problem 2.133

From Theorem 2.76 we know that $S^1_F \in P_{wkc}\big(L^1(\Omega;\mathbb{R}^N)\big)$. Therefore

$$V(t) \;=\; \int_0^t F(s)\,ds \;\in\; P_{kc}(\mathbb{R}^N) \qquad \forall\, t \in T$$

(see Definition 2.85). For $t, \tau \in [0,b]$ with $\tau < t$, we have

$$h(V(t),V(\tau)) \;=\; \sup_{|y^*|\leqslant 1} |\sigma_{V(t)}(y^*) - \sigma_{V(\tau)}(y^*)|$$

$$\leqslant\; \sup_{|y^*|\leqslant 1} \int_\tau^t \sigma_{F(s)}(y^*)\,ds \;\leqslant\; \int_\tau^t h(s)\,ds$$

(see the Hörmander formula (Proposition 2.51(b)) and recall that F is integrably bounded; see Definition 2.75) with $h \in L^1(T)$, so V is h-continuous, hence continuous too (since is $P_{kc}(\mathbb{R}^N)$-valued; see Proposition 2.56(c)).

The Michael selection theorem (see Theorem 2.58) implies that $CS_V \neq \emptyset$. Let $\{v_n\}_{n\geqslant 1} \subseteq CS_V$ be a sequence. Then we can find a sequence $\{f_n\}_{n\geqslant 1} \subseteq S_F^1$ such that

$$v_n(t) = \int_0^t f_n(s)\,ds \qquad \forall\, t \in T,\ n \geqslant 1.$$

From Theorem 2.76 we know that by passing to a suitable subsequence, we may assume that

$$f_n \xrightarrow{w} f \quad \text{in } L^1(T;\mathbb{R}^N), \tag{2.68}$$

with $f \in S_F^1$. Since F is integrably bounded, we have

$$|F(t)| \leqslant h(t) \quad \text{for a.a. } t \in T,$$

with $h \in L^1(T)$. So, S_F^1 is uniformly integrable (see Definition 1.18). Moreover, using the Lusin theorem on h (see Theorem 2.166), we see that condition **(b)** of Definition 2.80 is satisfied. So, S_F^1 has property U (see Definition 2.80) and then by Proposition 2.82, the weak topology and the $\|\cdot\|_w$-topology (see Definition 2.78) coincide. Let

$$v(t) = \int_0^t f(s)\,ds \qquad \forall\, t \in T.$$

Then $v \in CS_V$ and we have

$$\begin{aligned}
\|v_n - v\|_\infty &= \max_{t\in T} |v_n(t) - v(t)| = \max_{t\in T} \Big|\int_0^t (f_n(s) - f(s))\,ds\Big| \\
&\leqslant c\|f_n - f\|_w \qquad \forall\, n \geqslant 1,
\end{aligned}$$

for some $c > 0$ (see Definition 2.78 and Remark 2.79). Because of (2.68) we have $\|f_n - f\|_w \longrightarrow 0$ and so $\|v_n - v\|_\infty \longrightarrow 0$, which proves the compactness of $CS_V \subseteq C(T;\mathbb{R}^N)$.

Remark. An alternative solution can be based on the Arzela–Ascoli theorem (see Theorem I.2.181).

Solution of Problem 2.134

If $s\text{-}\liminf\limits_{n\to+\infty} A_n = \emptyset$, then $\operatorname{dist}(u, s\text{-}\liminf\limits_{n\to+\infty} A_n) = +\infty$ and the desired inequality holds. So, we assume that $s\text{-}\liminf\limits_{n\to+\infty} A_n \neq \emptyset$. Let $v \in s\text{-}\liminf\limits_{n\to+\infty} A_n$. Then we can find a sequence $\{v_n\}_{n\geqslant 1}$ with $v_n \in A_n$ for all $n \geqslant 1$ such that $v_n \longrightarrow v$ in X (see Definition 2.88). We have

$$\operatorname{dist}(u, A_n) \leqslant \|u - v_n\|_X,$$

so

$$\limsup_{n\to+\infty} \operatorname{dist}(u, A_n)) \leqslant \|u-v\|_X. \tag{2.69}$$

Since $v \in s\text{-}\liminf\limits_{n\to+\infty} A_n$ is arbitrary, from (2.69) we conclude that

$$\limsup_{n\to+\infty} \operatorname{dist}(u, A_n) \leqslant \operatorname{dist}(u, s\text{-}\liminf_{n\to+\infty} A_n).$$

Solution of Problem 2.135

Let $u \in X$ and let $r = \liminf\limits_{n\to+\infty} \operatorname{dist}(u, A_n)$. If $r = +\infty$, then the desired inequality holds. So, let us assume that $r < +\infty$. Let $\{v_n\}_{n\geqslant 1}$ be a sequence such that $v_n \in A_n$ for all $n \geqslant 1$ and

$$\|u-v_n\|_X \leqslant \operatorname{dist}(u, A_n) + \tfrac{1}{n} \qquad \forall\, n \geqslant 1. \tag{2.70}$$

We have $\{v_n\}_{n\geqslant 1} \subseteq C \cap \overline{B}_{(r+\|u\|_X+1)} \in P_{wk}(X)$. So, by virtue of the Eberlein–Smulian theorem (see Theorem I.5.78), passing to a subsequence if necessary, we have

$$v_n \xrightarrow{w} v \quad \text{in } X \quad \text{and} \quad v \in w\text{-}\limsup_{n\to+\infty} A_n. \tag{2.71}$$

Exploiting the weak lower semicontinuity of the norm functional (see Proposition I.5.56(c)), we have

$$\|u-v\|_X \leqslant \liminf_{n\to+\infty} \|u-v_n\|_X \leqslant \liminf_{n\to+\infty} \operatorname{dist}(u, A_n)$$

(see (2.70)), so

$$\operatorname{dist}(u, w\text{-}\limsup_{n\to+\infty} A_n) \leqslant \liminf_{n\to+\infty} \operatorname{dist}(u, A_n)$$

(see (2.71)).

Solution of Problem 2.136

Because X is reflexive and $A \in P_{fc}(X)$, we can find $v \in A$ such that

$$\operatorname{dist}(u, A) = \|u-v\|_X. \tag{2.72}$$

Since by hypothesis $A_n \xrightarrow{M} A$, we can find $v_n \in A_n$ for all $n \geqslant 1$ such that $v_n \longrightarrow v$ in X (see Definitions 2.90 and 2.88). We have

$$\operatorname{dist}(u_n, A_n) \leqslant \|u_n - v_n\|_X \qquad \forall\, n \geqslant 1,$$

so

$$\limsup_{n\to+\infty} \operatorname{dist}(u_n, A_n) \leqslant \|u-v\|_X = \operatorname{dist}(u, A) \tag{2.73}$$

(see (2.72)). On the other hand, from triangle inequality, we have

$$\operatorname{dist}(u_n, A_n) \leqslant \|u-u_n\|_X + \operatorname{dist}(u, A_n) \qquad \forall\, n \geqslant 1. \tag{2.74}$$

The hypothesis that $A_n \xrightarrow{M} A$, the reflexivity of X and Problems 2.134 and 2.135 imply that

$$\operatorname{dist}(u, A_n) \longrightarrow \operatorname{dist}(u, A).$$

So, if in (2.74) we pass to the limit as $n \to +\infty$, then

$$\operatorname{dist}(u, A) \leqslant \liminf_{n\to+\infty} \operatorname{dist}(u_n, A_n). \tag{2.75}$$

From (2.73) and (2.75), we conclude that $\operatorname{dist}(u_n, A_n) \longrightarrow \operatorname{dist}(u, A)$.

Solution of Problem 2.137

From Remark 2.89, we know that the multifunction $\omega \longmapsto H(\omega) = \limsup_{n\to+\infty} F_n(\omega)$ is $P_f(X)$-valued. So, according to Problem 2.100, it suffices to show that for all $K \in P_k(X)$, we have $H^-(K) \in \Sigma$. Note that

$$H^-(K) = \bigcap_{m\geqslant 1} \bigcap_{k\geqslant 1} \bigcup_{n\geqslant k} F_n^-(K_{\frac{1}{m}})$$

(see Remark 2.87), where $K_{\frac{1}{m}} = \{u \in X : \operatorname{dist}(u, K) < \frac{1}{m}\}$. But for each $n, m \geqslant 1$, we have $F_n^-(K_{\frac{1}{m}}) \in \Sigma$ (since F_n is measurable; see Definition 2.61(a)). Hence $H^-(K) \in \Sigma$ and so H is measurable.

Solution of Problem 2.138

Let $u \in A$. Then from the hypothesis we have $\operatorname{dist}(u, A_n) \longrightarrow 0$ and this by virtue of Definition 2.88 implies that $u \in s\text{-}\liminf_{n\to+\infty} A_n$. So, we conclude that $A \subseteq s\text{-}\liminf_{n\to+\infty} A_n$.

Solution of Problem 2.139

Since $A_n \subseteq C \in P_{wk}(X)$ for all $n \geqslant 1$, using the Eberlein–Smulian theorem (see Theorem I.5.78) and Definition 2.88 we conclude that $w\text{–}\limsup\limits_{n\to+\infty} A_n \neq \emptyset$.

Let $h \in w\text{–}\limsup\limits_{n\to+\infty} A_n$. Then we can find a subsequence $\{A_{n_k}\}_{k\geqslant 1}$ of $\{A_n\}_{n\geqslant 1}$ and $h_{n_k} \in A_{n_k}$ for all $k \geqslant 1$ such that $h_{n_k} \xrightarrow{w} h$ in X. For every $u^* \in X^*$, we have

$$\langle u^*, h_{n_k}\rangle \leqslant \sigma_{A_{n_k}}(u^*) \qquad \forall\, k \geqslant 1$$

(see Definition 2.46), so

$$\langle u^*, h\rangle \leqslant \limsup_{k\to+\infty} \sigma_{A_{n_k}}(u^*) \leqslant \limsup_{n\to+\infty} \sigma_{A_n}(u^*),$$

thus

$$\sigma_{\substack{w\text{–}\limsup A_n\\ n\to+\infty}}\big(u^*\big) \leqslant \limsup_{n\to+\infty} \sigma_{A_n}(u^*). \tag{2.76}$$

On the other hand, given $u^* \in X^*$, we can find $h_n \in A_n$ such that

$$\sigma_{A_n}(u^*) - \tfrac{1}{n} \leqslant \langle u^*, h_n\rangle \qquad \forall\, n \geqslant 1 \tag{2.77}$$

(see Definition 2.46). We have $\{h_n\}_{n\geqslant 1} \subseteq C \in P_{wk}(X)$. The Eberlein–Smulian theorem implies that we can find a subsequence $\{h_{n_k}\}_{k\geqslant 1}$ of $\{h_n\}_{n\geqslant 1}$, such that

$$h_{n_k} \xrightarrow{w} h \quad \text{in } X.$$

It follows that $h \in w\text{–}\limsup\limits_{n\to+\infty} A_n$. Then from (2.77) we have

$$\limsup_{n\to+\infty} \sigma_{A_n}(u^*) \leqslant \sigma_{\substack{w\text{–}\limsup A_n\\ n\to+\infty}}\big(u^*\big). \tag{2.78}$$

From (2.76) and (2.78), we conclude that

$$\sigma_{\substack{w\text{–}\limsup A_n\\ n\to+\infty}}\big(u^*\big) = \limsup_{n\to+\infty} \sigma_{A_n}(u^*) \qquad \forall\, u^* \in X^*.$$

Solution of Problem 2.140

Let $h \in w\text{–}\limsup\limits_{n\to+\infty} A_n$. Then we can find a subsequence $\{A_{n_k}\}_{k\geqslant 1}$ of $\{A_n\}_{n\geqslant 1}$ and $h_{n_k} \in A_{n_k}$ for all $k \geqslant 1$ such that $h_{n_k} \xrightarrow{w} h$ in X. For every $u^* \in X^*$ we have

$$\langle u^*, h_{n_k}\rangle \longrightarrow \langle u^*, h\rangle,$$

so

$$\langle u^*, h\rangle \leqslant \limsup_{n\to+\infty} \sigma_{A_{n_k}}(u^*) \leqslant \sigma_A(u^*)$$

(see Definition 2.46), thus $h \in \overline{\operatorname{conv}}\, A$ and hence $w\text{–}\limsup\limits_{n\to+\infty} A_n \subseteq \overline{\operatorname{conv}}\, A$.

Solution of Problem 2.141

From the Mazur theorem (see Theorem I.5.58), we know that

$$h(\omega) \in \overline{\operatorname{conv}} \bigcup_{n \geqslant k} h_n(\omega) \qquad \forall\, k \geqslant 1.$$

So, for every $k \geqslant 1$, $u^* \in X^*$ and $\omega \in \Omega \setminus N$ with $\mu(N) = 0$, we have

$$\langle u^*, h(\omega) \rangle \leqslant \sigma_{\bigcup\limits_{n \geqslant k} h_n(\omega)}(u^*) = \sup_{n \geqslant k} \langle u^*, h_n(\omega) \rangle ,$$

so

$$\langle u^*, h(\omega) \rangle \leqslant \limsup_{n \to +\infty} \langle u^*, h_n(\omega) \rangle = \sigma_{\substack{w\text{–}\limsup \{h_n(\Omega)\} \\ n \to +\infty}}(u^*)$$

(see Problem 2.139), thus

$$h(\omega) \in \overline{\operatorname{conv}}\, w\text{–}\limsup_{n \to +\infty} \{h_n(\omega)\} \quad \text{for } \mu\text{-a.a. } \omega \in \Omega.$$

Solution of Problem 2.142

Note that hypotheses **(i)** and **(ii)** do not imply the joint measurability of F (see Hu–Papageorgiou [8, p. 227]). So, the conclusion cannot be deduced from the Yankov–von Neumann–Aumann selection theorem (see Theorem 2.68). Let $v \in L^1(\Omega; X)$. We can find a sequence of simple functions $\{s_n\}_{n \geqslant 1} \subseteq L^1(\Omega; X)$ such that

$$\|v(\omega) - s_n(\omega)\|_X \longrightarrow 0 \quad \text{for } \mu\text{-a.a. } \omega \in \Omega$$

(see Definition 1.35(b)). Then because of hypothesis **(i)** for every $n \geqslant 1$, the multifunction $\omega \longmapsto F(\omega, s_n(\omega))$ is graph measurable (see Definition 2.61(b)). So, we can apply the Yankov–von Neumann–Aumann selection theorem (see Theorem 2.68) and obtain $h_n \in S^1_{F(\cdot, s_n(\cdot))}$. But from hypothesis **(iii)** (see Definitions 2.61(c) and 2.75) we know that $S^1_{F(\cdot, s_n(\cdot))} \subseteq S^1_G$ and the latter is w-compact in $L^1(\Omega; Y)$ (see Theorem 2.76). So, by passing to a suitable subsequence if necessary, we may assume that $h_n \xrightarrow{w} h$ in $L^1(\Omega; Y)$. From Problem 2.140, we have

$$\begin{aligned} h(\omega) &\in \overline{\operatorname{conv}}\, w\text{–}\limsup_{n \to +\infty} \{h_n(\omega)\} \\ &\subseteq \overline{\operatorname{conv}}\, w\text{–}\limsup_{n \to +\infty} F(\omega, s_n(\omega)) \\ &\subseteq \overline{\operatorname{conv}}\, F(\omega, v(\omega)) = F(\omega, v(\omega)) \quad \text{for } \mu\text{-a.a. } \omega \in \Omega \end{aligned}$$

(see hypothesis **(ii)**), so $h \in S^1_{F(\cdot, v(\cdot))}$ and thus $S^1_{F(\cdot, v(\cdot))} \neq \emptyset$.

Solution of Problem 2.143

Let m denote the Mackey topology $m(X^*, X)$ on X^* (i.e., the strongest, locally convex topology τ on X^* for which we have $(X^*_\tau)^* = X$). The space X^*_m (i.e., the space X^* furnished with the Mackey topology) is separable. Let $\{u_n^*\}_{n\geqslant 1} \subseteq X^*$ be an m-dense sequence. For every $u, v \in X$, we set

$$d_X(u,v) = \sum_{n\geqslant 1} \frac{|\langle u_n^*, u-v\rangle|}{1+|\langle u_n^*, u-v\rangle|}.$$

This is a metric on X and it can be easily verified that the d_X-topology is weaker than the weak topology and the two coincide on w-compact subsets of X (see Theorem I.2.84).

Let $H(\omega) = \overline{\{h_n(\omega)\}_{n\geqslant 1}}^w \in P_{wk}(X)$. Then $H(\omega) = \overline{h_n(\omega)}^d$ for μ-almost all $\omega \in \Omega$ and since X_d is separable, from Theorem 2.67 we infer that H is measurable into X_d. Then $\operatorname{Gr} H \in \Sigma \times \mathcal{B}(X_d) = \Sigma \times \mathcal{B}(X_w) = \Sigma \times \mathcal{B}(X)$ and so we can apply Theorem 2.76 and have that $S^1_H \subseteq L^1(\Omega; X)$ is w-compact. Therefore the sequence $\{h_n\}_{n\geq 1} \subseteq L^1(\Omega; X)$ is relatively w-compact.

Solution of Problem 2.144

We do the solution when condition **(i)** holds. The solution is similar if condition **(ii)** holds. Let $F(\omega) = [h(\omega), +\infty)$. Then F is a measurable multifunction from Ω into $P_{fc}(\mathbb{R})$. We have

$$\operatorname{dist}_{\mathbb{R}^N}(h_n(\omega), F(\omega)) = (h - h_n)^+(\omega) \longrightarrow 0 \quad \text{for } \mu\text{-a.a. } \omega \in \Omega$$

(due to condition **(i)**). Also we have

$$h(\omega) \in \operatorname{ext} F(\omega) \quad \text{for } \mu\text{-a.a. } \omega \in \Omega.$$

Therefore we can apply Proposition 2.97 and conclude that $h_n \longrightarrow h$ in $L^1(\Omega)$.

Solution of Problem 2.145

First we show that the set $w\text{-}\limsup\limits_{n\to+\infty} A_n$ need not be closed in X. Let X be a Banach space with a separable dual X^* (therefore, X is separable too; see Theorem I.5.82). Let $\overline{B}_1 = \{u \in X : \|u\|_X \leqslant 1\}$ and $S_1 = \partial B_1 = \{u \in X : \|u\|_X = 1\}$. We know that $0 \in \overline{S_1}^w$ and since the weak topology on $\overline{B}_1$ is metrizable, we can find a sequence $\{u_n\}_{n\geq 1} \subseteq S_1$ such that $u_n \xrightarrow{w} 0$ in X. Let $A_n = \{u_1, \ldots, u_n\}$ for all $n \geqslant 1$. Then

$$w\text{-}\limsup_{n\to+\infty} A_n = \{u_n\}_{n\geq 1} \cup \{0\}$$

and this set is not strongly closed.

Next we show that $w\text{–}\limsup\limits_{n\to+\infty} A_n$ need not be w-closed. To this end let $X = l^1$ and let $A_n = S_1$ for all $n \geqslant 1$. Then $w\text{–}\limsup\limits_{n\to+\infty} A_n$ is not w-closed. Indeed, recall that $0 \in \overline{S_1}^w$, but by the Schur property (see Remark I.5.57) there is no sequence in S_1 w-converging to 0 and this implies that $w\text{–}\limsup\limits_{n\to+\infty} A_n$ is not w-closed.

Solution of Problem 2.146

First we treat the "decreasing" case. Let $A = \bigcap\limits_{n\geqslant 1} A_n$. From Remark 2.89 we have that $A = s\text{–}\liminf\limits_{n\to+\infty} A_n$. If $w\text{–}\limsup\limits_{n\to+\infty} A_n = \emptyset$, then $A = \emptyset$ and we are done. So, assume that $w\text{–}\limsup\limits_{n\to+\infty} A_n \neq \emptyset$ and let $u \in w\text{–}\limsup\limits_{n\to+\infty} A_n$. We can find a subsequence $\{A_{n_k}\}_{k\geqslant 1}$ of $\{A_n\}_{n\geqslant 1}$ and $u_{n_k} \in A_{n_k}$ such that $u_{n_k} \xrightarrow{w} u$ in X. Since $\{A_n\}_{n\geqslant 1}$ is decreasing, we have $u \in \bigcap\limits_{k\geqslant 1} A_{n_k} = \bigcap\limits_{n\geqslant 1} A_n$, hence $u \in A$ and so $w\text{–}\limsup\limits_{n\to+\infty} A_n \subseteq A$. We conclude that

$$A_n \xrightarrow{M} \bigcap_{n\geqslant 1} A_n$$

(see Definition 2.90). For the "increasing" case, we have

$$s\text{–}\liminf_{n\to+\infty} A_n = \overline{\bigcup_{n\geqslant 1} A_n} \tag{2.79}$$

(see Remark 2.89). The set $\overline{\bigcup\limits_{n\geqslant 1} A_n}$ is closed, convex, hence it is w-closed, which implies that

$$w\text{–}\limsup_{n\to+\infty} A_n \subseteq \overline{\bigcup_{n\geqslant 1} A_n},$$

so

$$A_n \xrightarrow{M} \overline{\bigcup_{n\geqslant 1} A_n}$$

(see (2.79)).

Solution of Problem 2.147

Let $h \in L^1(\Omega; X)$ and $v \in L^1(\Omega; Y)$. Then

$$\begin{aligned}\operatorname{dist}_{L^1(\Omega;Y)}(v, S^1_{F(\cdot,h(\cdot))}) &= \inf_{y \in S^1_{F(\cdot,h(\cdot))}} \|v - y\|_{L^1(\Omega;Y)} \\ = \inf_{y \in S^1_{F(\cdot,h(\cdot))}} \int_\Omega \|v(\omega) - y(\omega)\|_Y \, d\mu &= \int_\Omega \inf_{z \in F(\omega,h(\omega))} \|v(\omega) - z\|_Y \, d\mu \\ = \int_\Omega \operatorname{dist}_Y(v(\omega), F(\omega, h(\omega))) \, d\mu & \end{aligned} \tag{2.80}$$

(see hypothesis **(i)** and Theorem 2.74). According to Proposition 2.44 it suffices to show that for every $v \in L^1(\Omega; Y)$, the function $h \longmapsto \operatorname{dist}_{L^1(\Omega;Y)}(v, S^1_{F(\cdot,h(\cdot))})$ is upper semicontinuous. So, for every $v \in L^1(\Omega; Y)$ and $\lambda \geqslant 0$ we need to show that the set

$$U_\lambda = \left\{ h \in L^1(\Omega; X) : \operatorname{dist}_{L^1(\Omega;Y)}(v, S^1_{F(\cdot,h(\cdot))}) \geqslant \lambda \right\}$$

is closed (see Proposition I.2.54). To this end, let $\{h_n\}_{n \geqslant 1} \subseteq U_\lambda$ be a sequence such that $h_n \longrightarrow h$ in $L^1(\Omega; X)$. By passing to a suitable subsequence, we may assume that

$$h_n(\omega) \longrightarrow h(\omega) \quad \text{for } \mu\text{-a.a. } \omega \in \Omega. \tag{2.81}$$

Then we have

$$\begin{aligned}\lambda &\leqslant \limsup_{n \to +\infty} \operatorname{dist}_{L^1(\Omega;Y)}(v, S^1_{F(\cdot,h_n(\cdot))}) \\ &= \limsup_{n \to +\infty} \int_\Omega \operatorname{dist}_Y(v(\omega), S^1_{F(\omega,h_n(\omega))}) \, d\mu \\ &\leqslant \int_\Omega \limsup_{n \to +\infty} \operatorname{dist}_Y(v(\omega), F(\omega, h_n(\omega))) \, d\mu \\ &\leqslant \int_\Omega \operatorname{dist}_Y(v(\omega), F(\omega, h(\omega))) \, d\mu \\ &= \operatorname{dist}_{L^1(\Omega;Y)}(v, S^1_{F(\cdot,h(\cdot))})\end{aligned}$$

(by the Fatou lemma, hypotheses **(iii) and (ii)**, Proposition 2.44, (2.81), and (2.80)), so $h \in U_\lambda$ and thus the function $h \longmapsto \operatorname{dist}_{L^1(\Omega;Y)}(v, S^1_{F(\cdot,h(\cdot))})$ is upper semicontinuous on $L^1(\Omega; X)$. This proves that the multifunction $h \longmapsto S^1_{F(\cdot,h(\cdot))}$ is lower semicontinuous from $L^1(\Omega; X)$ into $L^1(\Omega; Y)$.

Solution of Problem 2.148

Let $\mathcal{Y} = \{E\colon X \supseteq D(E) \longrightarrow 2^{X^*} : E$ is monotone and $\operatorname{Gr} A \subseteq \operatorname{Gr} E\}$. Evidently $\mathcal{Y} \neq \emptyset$. We introduce a partial order $\preceq$ on $\mathcal{Y}$ as follows:

$$E_1 \preceq E_2 \quad \overset{\text{def}}{\Longleftrightarrow} \quad \operatorname{Gr} E_1 \subseteq \operatorname{Gr} E_2,$$

for all $E_1, E_2 \in \mathcal{Y}$. Let $\mathcal{C}$ be a chain in $\mathcal{Y}$. The map with graph equal to $\bigcup\limits_{E \in \mathcal{C}} \operatorname{Gr} E$ is an upper bound for $\mathcal{C}$. Invoking the Kuratowski–Zorn lemma (see Theorem 4.120), we can find a maximal element $\widehat{A} \in \mathcal{Y}$. Clearly then $\widehat{A}\colon X \supseteq D(\widehat{A}) \longrightarrow 2^{X^*}$ is maximal monotone (see Definition 2.100) and $\operatorname{Gr} A \subseteq \operatorname{Gr} \widehat{A}$.

Solution of Problem 2.149

Let $\{(u_\alpha, u_\alpha^*)\}_{\alpha \in J} \subseteq \operatorname{Gr} A$ be a net and assume that

$$u_\alpha \xrightarrow{w} u \quad \text{in } X \quad \text{and} \quad u_\alpha^* \longrightarrow u^* \quad \text{in } X^*. \tag{2.82}$$

The monotonicity of A implies that

$$\langle u_\alpha^* - v^*, u_\alpha - v\rangle \geqslant 0 \qquad \forall\, (v, v^*) \in \operatorname{Gr} A$$

(see Definition 2.98(a)), so

$$\langle u^* - v^*, u - v\rangle \geqslant 0 \qquad \forall\, (v, v^*) \in \operatorname{Gr} A$$

(see (2.82)) and thus $(u, u^*) \in \operatorname{Gr} A$ (from the maximality of A; see Definition 2.100). This proves that $\operatorname{Gr} A$ is closed in $X_w \times X^*$.

Similarly, we show the closedness of $\operatorname{Gr} A$ in $X \times X^*_{w^*}$.

For the second part of the problem, the answer is No. To see this, let $X = l^2$ (a Hilbert space), let $C = \overline{B}_1$ and let p_C be the metric projection (that is, for any $u \in X$, $p_C(u) = \{h \in C : \|u - h\|_{l^2} = \inf\limits_{h' \in C} \|u - h'\|_{l^2}\}$). We know that p_C is single valued and nonexpansive (i.e., $\|p_C(u) - p_C(v)\|_{l^2} \leqslant \|u - v\|_{l^2}$ for all $u, v \in X$) (see Proposition I.5.99). We have that the map $u \longmapsto (I_X - p_C)(u)$ is continuous and monotone (see Definition 2.98(a)), since

$$\begin{aligned}
&\langle (I_X - p_C(u) - (I_X - p_C))(v), u - v\rangle \\
\geqslant\ & \|u - v\|_{l^2}^2 - \|p_C(u) - p_C(v)\|_{l^2} \|u - v\|_{l^2} \\
\geqslant\ & \|u - v\|_{l^2}^2 - \|u - v\|_{l^2}^2 \ = \ 0.
\end{aligned}$$

Invoking Corollary 2.111, we infer that $u \longmapsto (I_X - p_C)(u)$ is maximal monotone (see Definition 2.100). Let $\{e_n\}_{n \geqslant 1}$ be the standard orthonormal basis of $X = l^2$ and consider the sequence $\{u_n\}_{n \geqslant 1}$ defined by $u_n = e_1 + e_{2n}$ for all $n \geqslant 1$. Then

$$\big\{(u_n, (1 - \tfrac{1}{\sqrt{2}}) u_n)\big\}_{n \geqslant 1} \subseteq \operatorname{Gr} (I_X - p_C)$$

and we have

$$(u_n, (1 - \tfrac{1}{\sqrt{2}})u_n) \xrightarrow{w} (e_1, (1 - \tfrac{1}{\sqrt{2}})e_1)$$

(recall that $e_n \xrightarrow{w} 0$). Note that $(e_1, (1 - \frac{1}{\sqrt{2}})e_1) \notin \operatorname{Gr} A$.

Solution of Problem 2.150

Let $u, v \in D(A)$. We have

$$\begin{aligned}\langle A(u) - A(v), u - v\rangle &= \int_0^b \big(u'(t) - v'(t), u(t) - v(t)\big)_H \, dt \\ &= \tfrac{1}{2}\int_0^b \tfrac{d}{dt}\|u(t) - v(t)\|_H^2 \, dt = \tfrac{1}{2}\big(\|u(b) - v(b)\|_H^2 - \|u(0) - v(0)\|_H^2\big) \\ &= \tfrac{1}{2}\|u(b) - v(b)\|_H^2 \geqslant 0,\end{aligned}$$

so A is monotone (see Definition 2.98(a)).

According to Theorem 2.116, in order to show the maximality of A, it suffices to show that $R(A + I_{L^2(T;H)}) = L^2(T;H)$. So, we fix $h \in L^2(T;H)$ and consider the abstract Cauchy problem

$$\begin{cases} u'(t) + u(t) = h(t) & \text{for a.a. } t \in T, \\ u(0) = u_0. \end{cases} \tag{2.83}$$

Let $g(t) = e^t h(t)$ for $t \in T$. Evidently $g \in L^2(T;H)$. We define

$$y(t) = u_0 + \int_0^t g(s)\, ds.$$

We have $y \in W^{1,2}(T;H) \subseteq C(T;H)$ and $y(0) = y_0$. Setting $u(t) = e^{-t}y(t)$ for $t \in T$, we have $u'(t) = -e^{-t}y(t) + e^{-t}y'(t) = -u(t) + h(t)$. Therefore $u \in W^{1,2}(T;H)$, $u(0) = u_0$ and it solves (2.83). This then proves the maximality of A.

Solution of Problem 2.151

Let $\{u_n\}_{n\geqslant 1} \subseteq X$ be a sequence such that $u_n \longrightarrow u$ in X. The map A is locally bounded (see Definition 2.102 and Proposition 2.104). Therefore the sequence

$\{A(u_n)\}_{n\geqslant 1} \subseteq X^*$ is bounded and so it has a cluster point $h^* \in X^*$. The monotonicity of A implies that

$$\langle h^* - A(v), u - v \rangle \geqslant 0 \qquad \forall\, v \in X.$$

Let $v = ty + (1-t)u$ with $t \in [0,1]$ and $y \in X$. The hemicontinuity of A implies

$$\langle h^* - A(y), u - y \rangle \geqslant 0 \qquad \forall\, y \in X. \tag{2.84}$$

From Corollary 2.111, we know that A is maximal monotone (see Definition 2.100). So, (2.84) implies that $h^* = A(u)$. This means that for the original sequence, we have $A(u_n) \longrightarrow A(u)$ and so we conclude that A is continuous.

Solution of Problem 2.152

"$\Longrightarrow$": Assume that A is demicontinuous (see Definition 2.98(a)) and monotone. It is clear from Definition 2.110, that A is hemicontinuous too.

"$\Longleftarrow$": Assume that A is monotone and hemicontinuous. Let $\{u_n\}_{n\geqslant 1} \subseteq X$ be a sequence such that $u_n \longrightarrow u$ in X. Let $t_n = \|u_n - u\|_X^{\frac{1}{2}}$ and for $h \in X$, let $h_n = u + t_n h$. We have

$$\langle A(u_n), h \rangle = \tfrac{1}{t_n} \langle A(u_n), h_n - u_n \rangle + \tfrac{1}{t_n} \langle A(u_n), u_n - u \rangle . \tag{2.85}$$

Note that the sequence $\{A(u_n)\}_{n\geqslant 1} \subseteq X^*$ is bounded (see Proposition 2.104) and so we have

$$\tfrac{1}{t_n} \langle A(u_n), u_n - u \rangle = \Big\langle A(u_n), \tfrac{u_n - u}{\|u_n - u\|_X^{\frac{1}{2}}} \Big\rangle \longrightarrow 0. \tag{2.86}$$

Also, the hemicontinuity of A implies that

$$A(h_n) \xrightarrow{w^*} A(u) \quad \text{in } X^*,$$

so

$$\tfrac{1}{t_n} \langle A(h_n), h_n - u_n \rangle \longrightarrow \langle A(u), h \rangle . \tag{2.87}$$

From the monotonicity of A, we have

$$\langle A(u_n), h_n - u_n \rangle \leqslant \langle A(h_n), h_n - u_n \rangle \qquad \forall\, n \geqslant 1,$$

so

$$\limsup_{n\to+\infty} \tfrac{1}{t_n} \langle A(u_n), h_n - u_n \rangle \leqslant \langle A(u), h \rangle \tag{2.88}$$

(see (2.87)). Returning to (2.85), passing to the limit as $n \to +\infty$ and using (2.86) and (2.88), we obtain

$$\limsup_{n\to+\infty} \langle A(u_n), h \rangle \leqslant \langle A(u), h \rangle \qquad \forall\, h \in X,$$

so

$$\langle A(u_n), h \rangle \longrightarrow \langle A(u), h \rangle \qquad \forall\, h \in X$$

(see Proposition 2.104 and Problem 2.149), thus

$$A(u_n) \xrightarrow{w^*} A(u) \quad \text{in } X^*$$

and hence A is demicontinuous (see Definition 2.110).

Solution of Problem 2.153
For every $u \in W_0^{1,p}(\Omega)$, we have that $Du \in L^p(\Omega;\mathbb{R}^N)$ and so from the growth condition on a, we have

$$a(Du(\cdot)) \in L^{p'}(\Omega;\mathbb{R}^N).$$

From Theorem 1.140 it follows that $-\operatorname{div} a(Du(\cdot)) \in W^{-1,p'}(\Omega)$. We denote this operator by $A\colon W_0^{1,p}(\Omega) \longrightarrow W^{-1,p'}(\Omega)$ and we have

$$\langle A(u), h \rangle = \langle -\operatorname{div} a(Du), h \rangle = \int_\Omega (a(Du), Dh)_{\mathbb{R}^N}\, dz \qquad \forall\, h \in W_0^{1,p}(\Omega)$$

(since $h|_{\partial\Omega} = 0$; see Definition 1.129). If $E \subseteq W_0^{1,p}(\Omega)$ is bounded, then

$$\text{the set } \{Du :\ u \in E\} \subseteq L^p(\Omega;\mathbb{R}^N) \text{ is bounded}$$

and so from the growth condition on a it follows that

$$\text{the set } \{a(Du) :\ u \in E\} \subseteq L^{p'}(\Omega;\mathbb{R}^N) \text{ is bounded.} \tag{2.89}$$

So, we have

$$\begin{aligned} |\langle A(u), h \rangle| &= \Big| \int_\Omega (a(Du), Dh)_{\mathbb{R}^N} \Big| \leqslant \int_\Omega |a(Du)|\,|Dh|\, dz \\ &\leqslant c_1 \|Dh\|_{L^p(\Omega;\mathbb{R}^N)} \qquad \forall\, h \in W_0^{1,p}(\Omega), \end{aligned}$$

for some $c_1 > 0$ (see (2.89)), so the set $A(E) \subseteq W^{-1,p'}(\Omega)$ is bounded.

Let $\{u_n\}_{n\geqslant 1} \subseteq W_0^{1,p}(\Omega)$ be a sequence such that $u_n \longrightarrow u$ in $W_0^{1,p}(\Omega)$. Then for all $h \in W_0^{1,p}(\Omega)$ we have

$$\begin{aligned} |\langle A(u_n) - A(u), h \rangle| &\leqslant \int_\Omega |a(Du_n) - a(Du)|\,|Dh|\, dz \\ &\leqslant \|a(Du_n) - a(Du)\|_{L^{p'}(\Omega;\mathbb{R}^N)} \|Dh\|_{L^p(\Omega;\mathbb{R}^N)} \\ &= \|a(Du_n) - a(Du)\|_{L^{p'}(\Omega;\mathbb{R}^N)} \|h\|_{W_0^{1,p}(\Omega)} \end{aligned}$$

(see Theorem 1.131), so

$$\|A(u_n) - A(u)\|_{W^{-1,p'}(\Omega)} \leqslant \|a(Du_n) - a(Du)\|_{L^{p'}(\Omega;\mathbb{R}^N)}.$$

But by virtue of the Krasnoselskii theorem (see Problem 1.42), we have

$$\|a(Du_n) - a(Du)\|_{L^{p'}(\Omega;\mathbb{R}^N)} \longrightarrow 0,$$

so A is continuous. Finally for all $u, v \in W_0^{1,p}(\Omega)$, we have

$$\langle A(u) - A(v), u - v\rangle = \int_\Omega (a(Du) - a(Dv), Du - Dv)_{\mathbb{R}^N}\, dz \geqslant 0$$

(since a is monotone; see Definition 2.98(a)). Therefore A is continuous and monotone, thus according to Corollary 2.111, also maximal monotone (see Definition 2.100).

Solution of Problem 2.154

Let $A\colon W_0^{1,p}(\Omega) \longrightarrow W^{-1,p'}(\Omega)$ be the nonlinear map defined by

$$\langle A(u), h\rangle = \langle -\mathrm{div}\,(Du), h\rangle = \int_\Omega (a(Du), Dh)_{\mathbb{R}^N}\, dz \qquad \forall\, h \in W_0^{1,p}(\Omega).$$

From Problem 2.153 we know that A is maximal monotone (see Definition 2.100). From the Poincaré inequality (see Theorem 1.131), we know that we can have

$$\|u\|_{1,p} = \|Du\|_{L^p(\Omega;\mathbb{R}^N)} \qquad \forall\, u \in W_0^{1,p}(\Omega).$$

From the assumption on a, we have

$$\frac{\langle A(u),u\rangle}{\|u\|_{1,p}} \geqslant \widehat{c}\frac{\|u\|_{1,p}^p}{\|u\|_{1,p}} = \widehat{c}\|u\|_{1,p}^{p-1},$$

so A is strongly coercive (see Definition 2.98(f)). Invoking Theorem 2.119, we conclude that there exists $u \in W_0^{1,p}(\Omega)$ such that $A(u) = g$.

Solution of Problem 2.155

"$\Longrightarrow$": This follows immediately from the definition of maximal monotonicity (see Remark 2.101).

"$\Longleftarrow$": Suppose that $\mathrm{Gr}\, A$ is maximal among all linear monotone graphs and suppose that it is not maximal monotone. Then according to Definition 2.100 there exists $(u, u^*) \in (X \times X^*) \setminus \mathrm{Gr}\, A$ such that

$$\langle u^* - h^*, u - h\rangle \geqslant 0 \qquad \forall\, (h, h^*) \in \mathrm{Gr}\, A.$$

Then the linear space span $(\mathrm{Gr}\, A \cup (u, u^*))$ is a linear monotone graph extending $\mathrm{Gr}\, A$, a contradiction to our hypothesis. So, A is a maximal monotone map.

Solution of Problem 2.156

Arguing indirectly suppose that $A \notin \mathcal{L}(X; X^*)$. Then we can find a sequence $\{u_n\}_{n\geqslant 1} \subseteq X$ and $\varepsilon > 0$ such that

$$u_n \longrightarrow 0 \quad \text{in } X \quad \text{and} \quad \|A(u_n)\|_{X^*} \geqslant \varepsilon \qquad \forall\, n \geqslant 1. \tag{2.90}$$

Let $t_n = \frac{1}{\|u_n\|_X^{\frac{1}{2}}}$ for all $n \geqslant 1$ and set $h_n = t_n u_n$ for all $n \geqslant 1$. Clearly

$$\|h_n\|_X = \|u_n\|_X^{\frac{1}{2}} \longrightarrow 0.$$

Hence by the demicontinuity of A, we have that $A(h_n) \xrightarrow{w} 0$. Thus there exists $M > 0$ such that

$$\|A(h_n)\|_{X^*} \leqslant M \qquad \forall\, n \geqslant 1. \tag{2.91}$$

On the other hand, we have

$$\|A(h_n)\|_{X^*} = t_n \|A(u_n)\|_{X^*} \geqslant t_n \varepsilon \qquad \forall\, n \geqslant 1$$

(see (2.90)), so

$$\|A(h_n)\|_{X^*} \longrightarrow +\infty,$$

which contradicts (2.91).

Solution of Problem 2.157

We argue indirectly. So, suppose that we can find $u^*, v^* \in A(\widehat{u})$, $u^* \neq v^*$. Then there exists $u \in X$ such that $\varepsilon = \langle u^* - v^*, u \rangle > 0$. Let

$$\widehat{u}_n = \widehat{u} + \tfrac{1}{n} u \qquad \forall\, n \geqslant 1.$$

Then $\widehat{u}_n \longrightarrow \widehat{u}$ in X and because A is by hypothesis lower semicontinuous at $\widehat{u}$ into $X^*_{w^*}$, for some big $n_0 \geqslant 1$, we have that

$$S_n = A(\widehat{u}_n) \cap \left\{h^* \in X^* : |\langle u^* - h^*, u\rangle| < \tfrac{\varepsilon}{2}\right\} \neq \emptyset \qquad \forall\, n \geqslant n_0$$

(see Definition 2.36). Let $h_n \in S_n$ for $n \geqslant n_0$. Then since A is monotone (see Definition 2.98(a)), we have

$$
\begin{aligned}
0 &\leqslant \langle h_n^* - v^*, \widehat{u}_n - \widehat{u} \rangle = \tfrac{1}{n} \langle h_n^* - v^*, u \rangle \\
&= \tfrac{1}{n} \langle h_n^* - u^*, u \rangle + \tfrac{1}{n} \langle u^* - v^*, u \rangle \\
&< \tfrac{\varepsilon}{2n} - \tfrac{\varepsilon}{n} < 0
\end{aligned}
$$

for big $n \geqslant 1$, a contradiction.

Solution of Problem 2.158

From Proposition 2.105 we know that A is upper semicontinuous from $\operatorname{int} D(A)$ into $X^*_{w^*}$ (where $X^*_{w^*}$ denotes the space X^* furnished with the w^*-topology). Moreover, from Propositions 2.104 and 2.106 and the Alaoglu theorem (see Theorem I.5.66), we see that $A|_{\operatorname{int} D(A)}$ has nonempty, convex, and w^*-compact values. Since X is separable, we can find a sequence $\{u_n\}_{n \geqslant 1} \subseteq X$ which is dense in X. Then for $u^*, x^* \in X^*$, we set

$$
d_{X^*}(u^*, v^*) = \sum_{n \geqslant 1} \tfrac{1}{2^n} \tfrac{|\langle u^* - v^*, u_n \rangle|}{1 + |\langle u^* - v^*, u_n \rangle|}.
$$

This is a metric on X^* and the topology it generates is weaker than the w^*-topology. So, we can apply Proposition 2.41(b) and infer that the set

$$
\{u \in \operatorname{int} D(A) : \ A \text{ is not lower semicontinuous at } u\}
$$

is of first category. Then Problem 2.157 implies that the set Γ is of first category.

Solution of Problem 2.159

"$\Longrightarrow$": We assume that the duality map $\mathcal{F}$ is linear (see Definition 2.112). Then for all $u, v \in X$, we have

$$
\|u \pm v\|_X^2 = \langle \mathcal{F}(u \pm v), u \pm v \rangle = \|u\|_X^2 \pm \langle \mathcal{F}(u), v \rangle \pm \langle \mathcal{F}(v), u \rangle + \|v\|_X^2
$$

(from the linearity of $\mathcal{F}$), so

$$
\|u + v\|_X^2 + \|u - v\|_X^2 = 2 \left(\|u\|_X^2 + \|v\|_X^2 \right).
$$

So, the space X satisfies the parallelogram law (see Remark I.5.94), hence X must be a Hilbert space.

"$\Longleftarrow$": A Hilbert space has a differentiable norm and we can easily check that for all $u, h \in X$, we have

$$
\begin{aligned}
\langle \mathcal{F}(u), h \rangle &= \tfrac{1}{2} \tfrac{d}{dt} \|u + th\|_X^2 |_{t=0} = \tfrac{1}{2} \tfrac{d}{dt} (u + th, u + th)_X |_{t=0} \\
&= \tfrac{1}{2} \tfrac{d}{dt} \left(\|u\|_X^2 + 2t (u, h)_X + t^2 \|h\|_X^2 \right) |_{t=1} = (u, h)_X,
\end{aligned}
$$

so $\mathcal{F}(u) = u$ and so $\mathcal{F} = I_X$.

Solution of Problem 2.160

Invoking the Troyanski renorming theorem (see Theorem I.5.192 or Remark 2.115), without any loss of generality we may assume that both X and X^* are locally uniformly convex (see Definition I.5.168(b) and Remark I.5.169). As in the situation of Problem 2.148, using the Kuratowski–Zorn lemma (see Theorem 4.120), we can show that A admits a monotone extension $\widehat{A}$ which is maximal in the class of all monotone maps with domain in C. We will show that $\widehat{A}$ is in fact maximal monotone (see Definition 2.100). According to Theorem 2.116, it suffices to show that $R(A+\mathcal{F}) = X^*$. Let $h^* \in X^*$. From Theorem 3.117, we know that there exists $u \in C$ such that

$$\langle \mathcal{F}(u) + v^* - h^*, v - u \rangle \geqslant 0 \qquad \forall\, (v, v^*) \in \operatorname{Gr} \widehat{A}.$$

Since $\widehat{A}$ has no monotone extension on $C \times X^*$, we infer that

$$\text{if } u \in D(\widehat{A}) \text{ and } h^* - \mathcal{F}(u) \in \widehat{A}(u), \text{ then } h^* \in \widehat{A}(u) + \mathcal{F}(u).$$

Since $h^* \in X^*$ is arbitrary, we conclude that $R(\widehat{A} + \mathcal{F}) = X^*$ and so $\widehat{A}$ is maximal monotone.

Solution of Problem 2.161

From Theorem 2.119 we know that there exists $u_\lambda \in D(A)$ such that

$$0 \in \lambda A(u_\lambda) + \mathcal{F}(u_\lambda - u).$$

We show that this solution $u_\lambda \in D(A)$ is in fact unique. So, suppose that $u_\lambda, v_\lambda \in D(A)$ satisfy

$$\lambda u_\lambda^* + \mathcal{F}(u_\lambda - u) = 0, \quad \text{with } u_\lambda^* \in A(u_\lambda), \tag{2.92}$$

and

$$\lambda v_\lambda^* + \mathcal{F}(v_\lambda - u) = 0, \quad \text{with } v_\lambda^* \in A(v_\lambda). \tag{2.93}$$

From the monotonicity of A and $\mathcal{F}$ (see Proposition 2.114) and (2.92), (2.93), we have

$$0 \leqslant \lambda \langle u_\lambda^* - v_\lambda^*, u_\lambda - v_\lambda \rangle = \langle \mathcal{F}(v_\lambda - u) - \mathcal{F}(u_\lambda - u), u_\lambda - v_\lambda \rangle = 0,$$

so

$$\langle \mathcal{F}(u_\lambda - u) - \mathcal{F}(v_\lambda - u), u_\lambda - v_\lambda \rangle = 0. \tag{2.94}$$

But $\mathcal{F}$ is strictly monotone (see Definition 2.98(b) and Proposition 2.114(b)). Then from (2.94) we infer that $u_\lambda = v_\lambda$, which proves the uniqueness of the solution.

Solution of Problem 2.162

We do the solution under the assumption that

$$\limsup_{n,k\to+\infty} \langle u_n^* - u_k^*, u_n - u_k\rangle \leqslant 0. \tag{2.95}$$

The other case can be treated similarly.

The monotonicity of A and (2.95) imply that

$$\limsup_{n,k\to+\infty} \langle u_n^* - u_k^*, u_n - u_k\rangle = 0. \tag{2.96}$$

Let $\{(u_m, u_m^*)\}_{m\geqslant 1}$ be a subsequence of $\{(u_k, u_k^*)\}_{k\geqslant 1} \subseteq \operatorname{Gr} A$ such that

$$\lim_{m\to+\infty} \langle u_m^*, u_m\rangle = \limsup_{k\to+\infty} \langle u_k^*, u_k\rangle = \xi. \tag{2.97}$$

Let $\eta_{nm} = \langle u_n^* - u_m^*, u_n - u_m\rangle$ and $\eta_n = \langle u_n^*, u_n\rangle - \langle u_n^*, u\rangle - \langle u^*, u_n\rangle + \xi$. We have

$$\eta_{nm} \longrightarrow \eta_n \quad \text{as } m \to +\infty$$

(see (2.96)) and

$$\eta_n \longrightarrow 2\xi - 2\langle u^*, u\rangle$$

(see (2.97)). So, we can find an increasing (not necessarily strictly) sequence $\{n(m)\}_{m\geqslant 1}$ such that

$$\eta_{n(m)m} \longrightarrow 2\xi - 2\langle u^*, u\rangle \quad \text{as } m \to +\infty \tag{2.98}$$

(see Problem I.1.175). But from (2.96) we have

$$\eta_{n(m)m} \longrightarrow 0 \quad \text{as } m \to +\infty,$$

so $\xi = \langle u^*, u\rangle$ (see (2.98)), thus

$$\langle u_n^*, u_n\rangle \longrightarrow \langle u^*, u\rangle .$$

The monotonicity of A implies that

$$\langle u^* - h^*, u - h\rangle \geqslant 0 \qquad \forall\, (h, h^*) \in \operatorname{Gr} A,$$

so $(u, u^*) \in \operatorname{Gr} A$ (since A is maximal monotone; see Definition 2.100)).

Solution of Problem 2.163

By definition

$$A_\lambda(u) = \tfrac{1}{\lambda}\mathcal{F}(u - J_\lambda(u)) \qquad \forall\, u \in X.$$

So clearly A_λ is everywhere defined and single valued (see Proposition 2.114(a)). Moreover, from Problem 2.161, we see that

$$A_\lambda(u) \in A(J_\lambda(u)) \qquad \forall\, u \in X.$$

Now, let $u, v \in X$. We have

$$\begin{aligned}&\langle A_\lambda(u)-A_\lambda(v), u-v\rangle = \tfrac{1}{\lambda}\langle \mathcal{F}(u-J_\lambda(u))-\mathcal{F}(v-J_\lambda(v)), u-v\rangle\\ =\ &\tfrac{1}{\lambda}\langle \mathcal{F}(u-J_\lambda(u))-\mathcal{F}(v-J_\lambda(v)), u-J_\lambda(u)-(v-J_\lambda(v))\rangle\\ &+\langle A_\lambda(u)-A_\lambda(v), J_\lambda(u)-J_\lambda(v)\rangle,\end{aligned}$$

with $A_\lambda(u) \in A(J_\lambda(u))$ and $A_\lambda(v) \in A(J_\lambda(v))$. Therefore exploiting the monotonicity of A and $\mathcal{F}$ (see Proposition 2.114), we have

$$\langle A_\lambda(u)-A_\lambda(v), u-v\rangle \geqslant 0,$$

so A_λ is monotone (see Definition 2.98(a)).

For every $(h, h^*) \in \operatorname{Gr} A$, from the monotonicity of A, we have

$$\begin{aligned}&\langle h^*, J_\lambda(u)-h\rangle \leqslant \langle A_\lambda(u), J_\lambda(u)-h\rangle\\ &= -\tfrac{1}{\lambda}\langle \mathcal{F}(J_\lambda(u)-u), J_\lambda(u)-u\rangle - \tfrac{1}{\lambda}\langle \mathcal{F}(J_\lambda(u)-u), u-h\rangle\end{aligned} \tag{2.99}$$

(recall that $A_\lambda(u) \in A(J_\lambda(u))$ and see Problem 2.161), so

$$\|J_\lambda(u)-u\|_X^2 \leqslant -\lambda\langle h^*, J_\lambda(u)-h\rangle - \langle \mathcal{F}(J_\lambda(u)-u), u-h\rangle \tag{2.100}$$

(see Definition 2.112), thus

$$\|J_\lambda(u)-u\|_X \leqslant \lambda\|h^*\|_{X^*} + \|u-h\|_X,$$

hence

$$\|J_\lambda(u)\|_X \leqslant \lambda\|h^*\|_{X^*} + 2\|u\|_X + \|h\|_X,$$

which proves that J_λ is bounded (that is, maps bounded sets to bounded sets).

Since $A_\lambda(u) = \frac{1}{\lambda}\mathcal{F}(u-J_\lambda(u))$, it follows that A_λ is bounded too (see Definition 2.112).

Next let $\{u_n\}_{n\geqslant 1} \subseteq X$ be a sequence such that $u_n \longrightarrow u$ in X and set $y_n = J_\lambda(u_n)$ and $y_n^* = A_\lambda(u_n)$ for all $n \geqslant 1$. Then

$$\lambda y_n^* + \mathcal{F}(y_n-u_n) = 0 \qquad \forall\, n \geqslant 1. \tag{2.101}$$

So, we have

$$\begin{aligned}&\langle \mathcal{F}(y_n-u_n)-\mathcal{F}(y_k-u_k), u_k-u_n\rangle\\ =\ &\langle \mathcal{F}(y_n-u_n)-\mathcal{F}(y_k-u_k), y_n-u_n-(y_k-u_k)\rangle\\ &+\lambda\langle y_n^*-y_k^*, y_n-y_k\rangle \qquad \forall\, n \geqslant 1.\end{aligned} \tag{2.102}$$

Note that

$$\langle \mathcal{F}(y_n-u_n)-\mathcal{F}(y_k-u_k), u_k-u_n\rangle \longrightarrow 0 \quad \text{as } n, k \to +\infty$$

(see (2.101)). Moreover, both summands in the right-hand side of (2.102) are nonnegative (due to the monotonicity of $\mathcal{F}$ and A respectively). So, we obtain

$$\lim_{n,k\to+\infty} \langle \mathcal{F}(y_n-u_n)-\mathcal{F}(y_k-u_k), y_n-u_n-(y_k-u_k)\rangle = 0,$$

and

$$\lim_{n,k\to+\infty} \langle y_n^* - y_k^*, y_n - y_k \rangle = 0.$$

The boundedness of the maps J_λ, A_λ and $\mathcal{F}$ implies that by passing to a suitable subsequence if necessary, we may assume that

$$y_n \xrightarrow{w} y \text{ in } X, \quad y_n^* \xrightarrow{w} y^* \text{ in } X^* \quad \text{and} \quad \mathcal{F}(y_n - u_n) \xrightarrow{w} v^* \in X^*.$$

Then Problem 2.162 implies that

$$(y, y^*) \in \operatorname{Gr} A, \quad \mathcal{F}(y-u) = v^* \quad \text{and} \quad \lambda y^* + \mathcal{F}(y-u) = 0,$$

so

$$y = J_\lambda(u) \quad \text{and} \quad y^* = A_\lambda(u).$$

Therefore

$$J_\lambda(u_n) \xrightarrow{w} J_\lambda(u) \text{ in } X \quad \text{and} \quad A_\lambda(u_n) \xrightarrow{w} A_\lambda(u) \text{ in } X^*.$$

The second convergence proves the demicontinuity of A_λ. From the first convergence, Proposition 2.114 and the Kadec–Klee property, we have the continuity of J_λ.

Solution of Problem 2.164

From the solution of Problem 2.163 (see (2.99)), with $u = h$, we have

$$\|J_\lambda(u) - u\|_X \leqslant \lambda\|u^*\|_{X^*} \qquad \forall\, (u, u^*) \in \operatorname{Gr} A,$$

so

$$\|A_\lambda(u)\|_{X^*} = \tfrac{1}{\lambda}\|J_\lambda(u) - u\|_X \leqslant \|u^*\|_{X^*}$$

(see Remark after Problem 2.161 and Definition 2.112), thus

$$\|A_\lambda(u)\|_{X^*} \leqslant \|A^0(u)\|_{X^*} \qquad \forall\, \lambda > 0.$$

The reflexivity of X, implies that we can find a sequence $\lambda_n \searrow 0$ such that

$$A_{\lambda_n}(u) \xrightarrow{w} y^* \quad \text{in } X^*, \tag{2.103}$$

so

$$\|y^*\|_{X^*} \leqslant \|A^0(u)\|_{X^*}. \tag{2.104}$$

Also since

$$\|J_{\lambda_n}(u) - u\|_X \leqslant \sup_{n\geqslant 1} \lambda_n \|A^0(u)\|_{X^*} < +\infty, \tag{2.105}$$

we may assume that

$$\mathcal{F}(J_{\lambda_n}(u) - u) \xrightarrow{w} h^* \quad \text{in } X^*. \tag{2.106}$$

Then for all $v \in D(A)$ we have

$$\|J_{\lambda_n}(u) - u\|_X^2 \leqslant -\lambda_n \left\langle A^0(v), J_{\lambda_n}(u) - u\right\rangle - \langle \mathcal{F}(J_{\lambda_n}(u) - u), u - v\rangle$$

(cf. (2.100) in the solution of Problem 2.163), so

$$\limsup_{n\to+\infty} \|J_{\lambda_n}(u) - u\|_X^2 \leqslant \langle h^*, u - v\rangle$$

(see (2.105) and (2.106)). Evidently this is true for every $u, v \in \overline{\operatorname{conv}}\, D(A)$. Set $u = v$, to have

$$J_{\lambda_n}(u) \longrightarrow u \quad \text{in } X. \tag{2.107}$$

From the monotonicity of A, we have

$$0 \leqslant \langle w^* - A_{\lambda_n}(u), w - J_{\lambda_n}(u)\rangle \qquad \forall\, (w, w^*) \in \operatorname{Gr} A$$

(as $A_{\lambda_n}(u) \in A(J_{\lambda_n}(u))$; see Problem 2.163) so

$$0 \leqslant \langle w^* - y^*, w - u\rangle \qquad \forall\, (w, w^*) \in \operatorname{Gr} A$$

(see (2.103) and (2.107)), thus $y^* \in A(u)$ (since A is maximal monotone; see Definition 2.100)), hence

$$\|A^0(u)\|_{X^*} \leqslant \|y^*\|_{X^*} \leqslant \|A^0(u)\|_{X^*}$$

(see (2.104)), which implies that

$$\|A^0(u)\|_{X^*} = \|y^*\|_{X^*} \leqslant \|A^0(u)\|_{X^*}$$

and so $y^* = A^0(u)$. We conclude that $A_\lambda(u) \longrightarrow A^0(u)$ as $\lambda \searrow 0$ for all $u \in D(A)$.

Solution of Problem 2.165

Since the notion of maximal monotonicity is independent of the particular norms on X and X^*, invoking the Troyanski renorming theorem (see Theorem I.5.192 or Remark 2.115) without any loss of generality, we may assume that both X and X^* are locally uniformly convex (see Definition I.5.168(b) and Remark I.5.169). From Problem 2.164 we know that for every $u \in \overline{\operatorname{conv}}\, D(A)$, we have

$$J_\lambda(u) \longrightarrow u \quad \text{in } X \quad \text{as } \lambda \searrow 0.$$

Recall that $J_\lambda(u) \in D(A)$. Therefore $u \in \overline{D(A)}$. Hence $\overline{D(A)} = \overline{\operatorname{conv}}\, D(A)$ and thus the set $\overline{D(A)}$ is convex.

Note that $R(A) = D(A^{-1})$ and A^{-1} remains maximal monotone (recall that $A^{-1}(u^*) = \{u \in X : (u, u^*) \in \operatorname{Gr} A\}$). Therefore $\overline{R(A)} = \overline{D(A^{-1})} \subseteq X^*$ is convex too.

Solution of Problem 2.166

We argue indirectly. So, suppose that the conclusion does not hold. Then we can find a sequence $\{u_n\}_{n\geqslant 1} \subseteq X$ and $M > 0$ such that

$$|u_n| \longrightarrow +\infty \quad \text{and} \quad |A(u_n)| \leqslant M \qquad \forall\, n \geqslant 1.$$

We may assume that $A(u_n) \longrightarrow h^*$ in $\mathbb{R}^N$. Set $y_n = \frac{u_n}{|u_n|}$ for all $n \geqslant 1$. Then $|y_n| = 1$ for all $n \geqslant 1$ and so we may assume that $y_n \longrightarrow y$ in $\mathbb{R}^N$, $|y| = 1$. From the monotonicity of A, we have

$$0 \leqslant \big(A(u_n) - A(h), \tfrac{u_n}{|u_n|} - \tfrac{h}{|u_n|}\big)_{\mathbb{R}^N} \qquad \forall\, h \in \mathbb{R}^N,\ n \geqslant 1,$$

so

$$0 \leqslant (h^* - A(h), y)_{\mathbb{R}^N} \qquad \forall\, h \in \mathbb{R}^N. \tag{2.108}$$

Since by hypothesis A is surjective, for every $n \geqslant 1$, we can find $v_n \in \mathbb{R}^N$ such that $A(v_n) = ny$. Then from (2.108), we have

$$0 \leqslant (h^* - A(v_n), y)_{\mathbb{R}^N} \qquad \forall\, n \geqslant 1,$$

so

$$0 \leqslant (h^* - ny, y)_{\mathbb{R}^N} \qquad \forall\, n \geqslant 1,$$

thus

$$n|y|^2 \leqslant (h^*, y)_{\mathbb{R}^N} \leqslant |h^*||y|$$

and hence

$$n|y| \leqslant |h^*| \qquad \forall\, n \geqslant 1,$$

a contradiction since $|y| = 1$. Therefore $\lim\limits_{|u|\to+\infty} |A(u)| = +\infty$.

Solution of Problem 2.167

Arguing by contradiction, let $\{u_n\}_{n\geqslant 1} \subseteq \overline{B}_r$ be a sequence such that $u_n \longrightarrow u$ in $\mathbb{R}^N$ and $|A(u_n)| \longrightarrow +\infty$. Let $y_n = \frac{A(u_n)}{|A(u_n)|}$ for $n \geqslant 1$. Then $|y_n| = 1$ for all $n \geqslant 1$ and so we may assume that $y_n \longrightarrow y$ in $\mathbb{R}^N$, with $|y| = 1$. The monotonicity of A implies that

$$0 \leqslant \big(\tfrac{A(u_n)}{|A(u_n)|} - \tfrac{A(v)}{|A(u_n)|}, u_n - v\big)_{\mathbb{R}^N} = \big(y_n - \tfrac{A(v)}{|A(u_n)|}, u_n - v\big)_{\mathbb{R}^N}$$

for all $v \in B_{r+\varepsilon}$ and all $n \geqslant 1$, so

$$0 \leqslant (y, u-v)_{\mathbb{R}^N} \qquad \forall\, v \in B_{r+\varepsilon}$$

and thus

$$(y, v)_{\mathbb{R}^N} \leqslant (y, u)_{\mathbb{R}^N} \qquad \forall\, v \in B_{r+\varepsilon}.$$

Let $v = u + \frac{\varepsilon}{2}y$. Then we reach a contradiction. So, the set $A(\overline{B}_r) \subseteq \mathbb{R}^N$ is bounded.

Solution of Problem 2.168

From Problem 2.150 we know that A is maximal monotone (see Definition 2.100). So, J_1^A can be defined (see Definition 2.122) and $\operatorname{dom} J_1^A = L^2(T;H)$ (see Proposition 2.123(a)). Moreover, from the solution of Problem 2.150, we know that for every $u \in L^2(T;H)$, if

$$v(t) = e^{-t}u_0 + \int_0^t e^{t-s}u(s)\,ds \qquad \forall\, t \in [0,b],$$

then $v \in W^{1,2}(T;H)$, $v(0) = u_0$ and $u(t) = v(t) + v'(t)$ for almost all $t \in T$. Hence $u = (I_{L^2(T;H)} + A)v$, so $J_1^A(u) = v$ (see Definition 2.122 and Proposition 2.123).

Solution of Problem 2.169

Since A is everywhere defined and uniformly monotone (see Definition 2.98(d)), we have

$$\langle A(u) - A(v), u - v\rangle \geqslant \vartheta(|u-v|)|u-v| \qquad \forall\, u, v \in X, \tag{2.109}$$

with a strictly increasing function $\vartheta\colon \mathbb{R}_+ \longrightarrow \mathbb{R}_+$ such that $\vartheta(0) = 0$ and $\vartheta(r) \longrightarrow +\infty$ as $r \to +\infty$. Then $\vartheta^{-1}\colon \mathbb{R}_+ \longrightarrow \mathbb{R}_+$ exists and it is strictly increasing too, with $\vartheta^{-1}(0) = 0$. Let

$$u^* = A(u) \quad \text{and} \quad v^* = A(v) \quad \text{with } u, v \in X.$$

Then from (2.109), we have

$$\|A^{-1}(u^*) - A^{-1}(v^*)\|_X \leqslant \vartheta^{-1}(\|u^* - v^*\|_*),$$

so A^{-1} is continuous on $R(A)$. Let $B_r^* = \{u^* \in X^* : \|u^*\|_* < r\}$. Then for all $u^* \in B_r^*$ we have

$$\|A^{-1}(u^*)\|_X \leqslant \vartheta^{-1}(r) + \|A^{-1}(0)\|_X,$$

so A^{-1} is bounded from X^* into X. Since A is maximal monotone (see Corollary 2.111) we can apply Theorem 2.118 to conclude that A is surjective.

Solution of Problem 2.170

Let V be a finite dimensional subspace of X. Let $u \in D(A) \cap V$ and let U be a weakly open set in X^* containing $A(u)$. We need to show that $A|_V^+(U) = \{y \in V : A(y) \subseteq U\}$ is open in V (see Definition 2.36). Arguing by contradiction, suppose that $A|_V^+(U)$ is not open in V. Then we can find a sequence $\{u_n\}_{n\geqslant 1} \subseteq V$ such that $u_n \longrightarrow u$ and $A(u_n) \cap U^c \neq \emptyset$ for all $n \geqslant 1$. Let $u_n^* \in A(u_n) \cap U^c$. The boundedness of A implies that the sequence $\{u^*\}_{n\geqslant 1} \subseteq X^*$ is bounded. Then the reflexivity of X (hence of X^* too) and the Eberlein–Smulian theorem (see Theorem I.5.78) imply that at least for a subsequence, we have $u_n^* \xrightarrow{w} u^*$ in X^*. Then

$$\langle u_n^*, u_n - u\rangle \longrightarrow 0,$$

so $(u, u^*) \in \operatorname{Gr} A$ (see the condition which is assumed on A). Also U^c is w-closed in X^*, hence $u^* \in U^c$. Therefore $u^* \in A(u) \cap U^c$, a contradiction to the fact that $A(u) \subseteq U$. This proves the desired upper semicontinuity property of the map A.

Solution of Problem 2.171

Let $u \in X$ and $u^* \in X^*$ be such that

$$0 \leqslant \langle u^* - h^*, u - h\rangle \qquad \forall\, (h, h^*) \in \operatorname{Gr} A. \tag{2.110}$$

Given $v^* \in (A + C)(v)$, we have

$$v^* = h^* + C(v), \tag{2.111}$$

for some $h^* \in A(v)$. Then we have

$$\begin{aligned} \langle u^* + C(u) - v^*, u - v\rangle &= \langle u^* + C(u) - h^* - C(v), u - v\rangle \\ = \langle u^* - h^*, u - v\rangle + \langle C(u) - C(v), u - v\rangle &\geqslant 0 \end{aligned} \tag{2.112}$$

(see (2.111) and (2.110)). Because by hypothesis $A+C$ is a maximal monotone map (see Definition 2.100), from (2.112) we infer that

$$u \in D(A + C) \quad \text{and} \quad u^* + C(u) \in (A + C)(u).$$

This implies that $u^* \in A(u)$ and so we have the maximal monotonicity of A (see (2.110)).

Solution of Problem 2.172

The monotonicity of $\widehat{A}$ is a direct consequence of the monotonicity of A. Then according to Theorem 2.116, in order to show maximality of $\widehat{A}$, it suffices to show that $R(\widehat{A}+I_{L^2(\Omega;H)}) = L^2(\Omega;H)$. So, let $h \in L^2(\Omega;H)$ and exploiting the maximal monotonicity of A, set

$$v(\omega) = (I_H + A)^{-1}(h(\omega)) \qquad \forall\, \omega \in \Omega.$$

Evidently v is Σ-measurable. Also, by hypothesis $(0,0) \in \operatorname{Gr} A$, we have $J_1^A(0) = 0$ and so from Proposition 2.123(b), it follows that

$$\|v(\omega)\|_H \leqslant \|h(\omega)\|_H \quad \text{for } \mu\text{-a.a. } \omega \in \Omega,$$

so $v \in L^2(\Omega;H)$. Therefore $R(A+I_H) = L^2(\Omega;H)$ and we have proved the maximality of $\widehat{A}$. From the above argument it is clear that for every $\lambda > 0$, we have

$$J_\lambda^{\widehat{A}}(u)(\cdot) = J_\lambda^A(u(\cdot)) \qquad \forall\, u \in L^2(\Omega;H).$$

Hence we infer that

$$\widehat{A}_\lambda(u)(\cdot) = A_\lambda(u(\cdot)) \qquad \forall\, u \in L^2(\Omega;H)$$

(see Definition 2.122). Finally, if $\mu(\Omega) < +\infty$, then constant functions belong to $L^2(\Omega;H)$ and so from the fact that J_1^A is nonexpansive (see Proposition 2.123(b)), we have that $v \in L^2(\Omega;H)$. So, in that case we can drop the requirement that $(0,0) \in \operatorname{Gr} A$.

Solution of Problem 2.173

No. We always have $R(A+C) \subseteq R(A) + R(C)$ and the inclusion can be strict. For example, let $H = \mathbb{R}^2$ and let $A, C\colon \mathbb{R}^2 \longrightarrow \mathbb{R}^2$ be the operators that perform rotation by $\frac{\pi}{2}$ and by $-\frac{\pi}{2}$ respectively. Then $R(A+C) = \{0\}$, while $R(A)+R(C) = \mathbb{R}^2$.

Solution of Problem 2.174

Without any loss of generality we may assume that both X and X^* are locally uniformly convex (by the Troyanski renorming theorem; see Theorem I.5.192 or Remark 2.115) and so the duality map $\mathcal{F}\colon X \longrightarrow X^*$ is a homeomorphism (see Proposition 2.114). Let $\xi\colon X \times X^* \longrightarrow X^* \times X$ be the map defined by $\xi(u,u^*) = (u^* + \mathcal{F}(u), u)$. Then ξ is a homeomorphism. For every $\omega \in \Omega$, we have

$$\xi(\operatorname{Gr} A(\omega)) = \operatorname{Gr}\,(A(\omega)+\mathcal{F})^{-1}.$$

Then property **(a)** is equivalent to saying that

$$\begin{aligned}&\{(\omega,u^*,u)\in\Omega\times X^*\times X:\, u=(A(\omega)+\mathcal{F})^{-1}(u^*)\}\\&\qquad\in\Sigma\times\mathcal{B}(X^*)\times\mathcal{B}(X) \;=\; \Sigma\times\mathcal{B}(X^*\times X),\end{aligned}$$

which is precisely property **(b)**.

Solution of Problem 2.175

(1) Let $E\in\mathcal{B}(X)$. Then $E\times X^*\in\mathcal{B}(X)\times\mathcal{B}(X^*)=\mathcal{B}(X\times X^*)$. We have

$$\begin{aligned}D^-(E) &= \{\omega\in\Omega:\; D(\omega)\cap E\neq\emptyset\}\\ &=\{\omega\in\Omega:\; \operatorname{Gr}A(\omega)\cap(E\times X^*)\in\Sigma\}\end{aligned}$$

(see property **(a)** in Problem 2.174).

(2) Let $E^*\in\mathcal{B}(X^*)$. Then $\{u\}\times E^*\in\mathcal{B}(X)\times\mathcal{B}(X^*)=\mathcal{B}(X\times X^*)$ and so

$$\begin{aligned}&\{\omega\in\Omega:\; A(\omega)(u)\cap E^*\neq\emptyset\}\\&=\{\omega\in\Omega:\; \operatorname{Gr}A(\omega)\cap(\{u\}\times E^*)\neq\emptyset\}\;\in\;\Sigma,\end{aligned}$$

so the multifunction $\omega\longmapsto A(\omega)(u)$ is measurable from Ω into X^*.

(3) Fix $u\in X$. We have

$$\Omega(u) \;=\; \{\omega\in\Omega:\; A(\omega)(u)\cap X^*\neq\emptyset\},$$

so $\Omega(u)\in\Sigma$ (see part **(2)**). Then the multifunction $\omega\longmapsto A(\omega)(u)$ is measurable form $\Omega(u)$ into $P_{fc}(X^*)$. So, Theorem 2.66 implies the existence of a sequence $\{v_n^*\colon\Omega(u)\longrightarrow X^*\}_{n\geqslant 1}$ of $\Sigma\cap\Omega(u)$-measurable functions such that

$$A(\omega)(u) \;=\; \overline{\{v_n^*(\omega)\}_{n\geqslant 1}} \qquad \forall\,\omega\in\Omega.$$

Then we have

$$\|A^0(\omega)(u)\|_* \;=\; \inf_{n\geqslant 1}\|v_n^*(\omega)\|_* \qquad \forall\,\omega\in\Omega,$$

so the map $\omega\longmapsto\|A^0(\omega)(u)\|_*$ is $\Sigma\cap\Omega(u)$-measurable. Note that

$$\begin{aligned}\operatorname{Gr}A^0(\cdot)(u) \;&=\; \{(\omega,u^*)\in\Omega(u)\times X^*:\; u^*\in A(\omega)(u),\\&\qquad \|u^*\|_*=\|A^0(\omega)(u)\|_*\}\\&\in\; (\Sigma\cap\Omega(u))\times\mathcal{B}(X^*)\end{aligned}$$

(see part **(2)**), so $\omega\longmapsto A^0(\omega)(u)$ is $\Sigma\cap\Omega(u)$-measurable (since Σ is complete).

Solution of Problem 2.176

Let $\{u_n^*\}_{n\geqslant 1} \subseteq A(C)$ be a sequence such that $u_n^* \longrightarrow u^*$ in X^*. We have that $u_n^* \in A(u_n)$ with $u_n \in C$. Recall that A^{-1} is maximal monotone (see Definition 2.100) from $R(A) = X^*$ (recall that A is surjective) onto X. Proposition 2.104 implies that the sequence $\{A^{-1}(u_n^*)\}_{n\geqslant 1}$ is bounded. Hence the sequence $\{u_n\}_{n\geqslant 1} \subseteq C$ is bounded and because X is reflexive, by passing to a suitable subsequence if necessary, we may assume that $u_n \xrightarrow{w} u$ in X. Since the set $C \subseteq X$ is closed and convex, it is also weakly closed (by the Mazur theorem; see Theorem I.5.58) and so $u \in C$. Also, by virtue of Problem 2.149, we have that $(u, u^*) \in \operatorname{Gr} A$ and so finally $u^* \in A(C)$. This proves that the set $A(C) \subseteq X^*$ is closed.

Solution of Problem 2.177

For any $\widehat{u} \in D(A) \cap D(C)$, the elements $h_\lambda^* \in (A + C_\lambda + \mathcal{F})(\widehat{u})$ for $\lambda > 0$, are bounded in $\lambda \in (0, 1]$ since

$$\|C_\lambda(\widehat{u})\|_* \leqslant \|C^0(\widehat{u})\|_*$$

(see the solution of Problem 2.164). Exploiting the monotonicity of the maps A and C_λ (see Problem 2.163), we have

$$\left(\|u_\lambda\|_X - \|\widehat{u}\|_X\right)^2 \leqslant \langle h^* - h_\lambda^*, u_\lambda - \widehat{u}\rangle ,$$

so

$$\|u_\lambda\|_X^2 \leqslant \left(\|h^* - h_\lambda^*\|_* + 2\|\widehat{u}\|_X\right)\|u_\lambda\|_X + \left(\|h^* - h_\lambda^*\|_* - \|\widehat{u}\|_X\right)\|\widehat{u}\|_X,$$

hence the family $\{u_\lambda\}_{\lambda>0}$ remains bounded in X as $\lambda \searrow 0$.

Solution of Problem 2.178

"**(a)** $\Longrightarrow$ **(b)**": Let $\lambda > 0$ and let $u_\lambda^* \in \mathcal{F}(u + \lambda y)$. We set $h_\lambda^* = \frac{u_\lambda^*}{\|u_\lambda^*\|_*}$. We have

$$\begin{aligned} \|u\|_X \leqslant \|u + \lambda y\|_X &= \frac{\langle u_\lambda^*, u + \lambda y\rangle}{\|u_\lambda^*\|_*} = \langle h_\lambda^*, u + \lambda y\rangle \\ &= \langle h_\lambda^*, u\rangle + \lambda\langle h_\lambda^*, y\rangle \leqslant \|u\|_X + \lambda\langle h_\lambda^*, y\rangle . \end{aligned} \tag{2.113}$$

Since $\|h_\lambda^*\|_* = 1$, the net $\{h_\lambda^*\}_{\lambda>0}$ admits a w^*-cluster point h^*, with $\|h^*\|_* \leqslant 1$. Also, we have

$$\|u\|_X \leqslant \langle h^*, u\rangle \quad \text{and} \quad \langle h^*, y\rangle \geqslant 0 \tag{2.114}$$

(see (2.113)). Therefore

$$\|u\|_X \leqslant \langle h^*, u\rangle \leqslant \|u\|_X$$

(see (2.114) and recall that $\|h^*\|_* \leqslant 1$), so $\|u\|_X = \langle h^*, u\rangle$ and thus $\|h^*\|_* = 1$. Then

$$u^* = \|u\|h^* \in \mathcal{F}(u) \quad \text{and} \quad \langle u^*, y\rangle \geqslant 0.$$

"**(b)** $\Longrightarrow$ **(a)**": For every $\lambda > 0$, we have

$$\begin{aligned} \|u\|_X^2 &= \langle u^*, u\rangle \leqslant \langle u^*, u\rangle + \lambda\langle u^*, y\rangle = \langle u^*, u + \lambda y\rangle \\ &\leqslant \|u^*\|_*\|u + \lambda y\|_X = \|u\|_X\|u + \lambda y\|_X, \end{aligned}$$

so $\|u\|_X \leqslant \|u + \lambda y\|_X$ for all $\lambda > 0$.

Solution of Problem 2.179

"$\Longrightarrow$": For every $u \in C$ and every $\lambda \in [0, 1]$, we have

$$\|u_0\|_X \leqslant \|u_0 + \lambda(u - u_0)\|_X$$

(recall that C is convex). Invoking Problem 2.178 and since $\mathcal{F}$ is single valued (see Proposition 2.114(a)), we have

$$\langle \mathcal{F}(u_0), u - u_0\rangle \geqslant 0 \qquad \forall\, u \in C,$$

so $\langle \mathcal{F}(u_0), u_0\rangle \leqslant \langle \mathcal{F}(u_0), u\rangle$ for all $u \in C$.

"$\Longleftarrow$": We have

$$\|u_0\|_X^2 = \langle \mathcal{F}(u_0), u_0\rangle \leqslant \langle \mathcal{F}(u_0), u\rangle \leqslant \|u_0\|_X\|u\|_X \qquad \forall\, u \in C,$$

so

$$\|u_0\|_X \leqslant \|u\|_X \qquad \forall\, u \in C.$$

Solution of Problem 2.180

From Problem 2.161, we know that

$$(J_{\lambda_n}(u_n), A_{\lambda_n}(u_n)) \in \operatorname{Gr} A \qquad \forall\, n \geqslant 1.$$

Note that the boundedness of the sequence $\{A_{\lambda_n}(u_n)\}_{n\geqslant 1} \subseteq X^*$ implies that

$$\|J_{\lambda_n}(u_n) - u_n\|_X \longrightarrow 0$$

(see the solution of Problem 2.164). Hence Problem 2.162 implies that

$$(u, y) \in \operatorname{Gr} A \quad \text{and} \quad \limsup_{n,m\to+\infty} \langle A_{\lambda_n}(u_n) - A_{\lambda_m}(u_m), u_n - u_m\rangle = 0.$$

Solution of Problem 2.181

From Problem 2.177 and the hypothesis, we know that we can find $M > 0$ and $\lambda^* > 0$ such that

$$\|C_\lambda(u_\lambda)\|_* \leqslant M \quad \text{and} \quad \|u_\lambda\|_* \leqslant M \qquad \forall\, \lambda \in (0, \lambda^*). \tag{2.115}$$

So, if $\lambda_n \searrow 0$, then because of (2.115) and the reflexivity of X, we may assume that

$$C_n(u_n) \;=\; C_{\lambda_n}(u_{\lambda_n}) \xrightarrow{w} v^* \;\text{ in } X^*, \quad \mathcal{F}(u_n) \xrightarrow{w} g^* \;\text{ in } X^* \tag{2.116}$$

and

$$u_n = u_{\lambda_n} \;\xrightarrow{w}\; u \quad \text{in } X. \tag{2.117}$$

We have

$$h^* \;=\; u_n^* + C_n(u_n) + \mathcal{F}(u_n) \;\text{ with } u_n^* \in A(u_n) \qquad \forall\, n \geqslant 1. \tag{2.118}$$

For $n, m \geqslant 1$, we have

$$\begin{aligned} \langle u_n^* - u_m^*, u_n - u_m\rangle + \langle C_n(u_n) - C_m(u_m), u_n - u_m\rangle \\ + \langle \mathcal{F}(u_n) - \mathcal{F}(u_m), u_n - u_m\rangle \;=\; 0. \end{aligned} \tag{2.119}$$

Because of the monotonicity of the duality map (see Proposition 2.114) and the monotonicity of A (by hypothesis), we have

$$\limsup_{n,m\to+\infty} \langle C_n(u_n) - C_m(u_m), u_n - u_m\rangle \;\leqslant\; 0. \tag{2.120}$$

Then from (2.120) and Problem 2.180, we have

$$(u, v^*) \;\in\; \operatorname{Gr} C \quad \text{and} \quad \lim_{n,m\to+\infty} \langle C_n(u_n) - C_m(u_m), u_n - u_m\rangle \;=\; 0. \tag{2.121}$$

From (2.119) and (2.121), it follows that

$$\lim_{n,m\to+\infty} \langle u_n^* + \mathcal{F}(u_n) - (u_m^* + \mathcal{F}(u_m)), u_n - u_m\rangle \;=\; 0,$$

so

$$\limsup_{n,m\to+\infty} \langle \mathcal{F}(u_n) - \mathcal{F}(u_m), u_n - u_m\rangle \;\leqslant\; 0 \tag{2.122}$$

(since A is monotone). From (2.122) and Problem 2.162, we infer that

$$g^* \;=\; \mathcal{F}(u) \quad \text{and} \quad \lim_{n,m\to+\infty} \langle u_n^* - u_m^*, u_n - u_m\rangle \;=\; 0. \tag{2.123}$$

From (2.115) and (2.118), we see that the sequence $\{u_n^*\}_{n\geqslant 1} \subseteq X^*$ is bounded and so, passing to a subsequence if necessary, we may assume that

$$u_n^* \xrightarrow{w} u^* \quad \text{in } X^*. \tag{2.124}$$

Using (2.123), (2.124) and Problem 2.162, we have

$$(u, u^*) \in \operatorname{Gr} A. \tag{2.125}$$

Passing to the limit as $n \to +\infty$ in (2.118) and using (2.121), (2.123) and (2.125), we obtain

$$h^* = u^* + v^* + \mathcal{F}(u),$$

with $u^* \in A(u)$, $v^* \in C(u)$. Therefore $h^* \in R(A + C + \mathcal{F})$.

Solution of Problem 2.182

We need to show that for every $\lambda \geqslant 0$, the sublevel set

$$L_\lambda = \{u \in X : m(u) \leqslant \lambda\}$$

is closed. To this end, let $\{u_n\}_{n\geqslant 1} \subseteq L_\lambda$ be a sequence such that $u_n \longrightarrow u$ in X. Since A is maximal monotone (see Definition 2.100) for every $n \geqslant 1$, we have $A(u_n) \in P_{fc}(X^*)$ (see Corollary 2.107). Since X^* is reflexive (recall that X is reflexive), we can find $u_n^* \in A(u_n)$ such that $m(u_n) = \|u_n^*\|_*$ for all $n \geqslant 1$. We have $\|u_n^*\|_* \leqslant \lambda$ for all $n \geqslant 1$ and so by passing to a suitable subsequence if necessary, we may assume that $u_n^* \xrightarrow{w} u^*$ in X^*. From Problem 2.149 we have $(u, u^*) \in \operatorname{Gr} A$, from the weak lower semicontinuity of the norm functional (see Proposition I.5.56(c)), we have

$$\|u^*\|_* \leqslant \liminf_{n\to+\infty} \|u_n^*\|_* = \liminf_{n\to+\infty} m(u_n) \leqslant \lambda,$$

so $m(u) \leqslant \lambda$ (since $u^* \in A(u)$), thus $u \in L_\lambda$ and hence L_λ is closed. This proves lower semicontinuity of m.

Solution of Problem 2.183

Let $u \in X$ and $u^* \in X^*$ be such that

$$0 \leqslant \langle u^* - h^*, u - h\rangle \qquad \forall\, (h, h^*) \in \operatorname{Gr} A. \tag{2.126}$$

From (2.126) and the monotonicity of C, we have

$$0 \leqslant \langle u^* + C(u) - (h^* + C(h)), u - h\rangle \qquad \forall\, (h, h^*) \in \operatorname{Gr} A. \tag{2.127}$$

Let $v^* \in X^*$, $t > 0$, $u_t^* = u^* + C(u) + tv^*$, $u_t = (A+C)^{-1}(u_t^*)$. Since $R(A + C) = X^*$, we can find $y_t^* \in A(u_t)$ such that $y_t^* + C(u_t) = u_t^*$. In (2.127) we choose $h = u_t$ and $h^* = y_t^*$. Then

$$0 \leqslant \langle v^*, u_t - u\rangle .$$

But by hypothesis $(A+C)^{-1}$ is continuous. So, if $t \searrow 0$, then

$$0 \leqslant \langle v^*, u_0 - u\rangle \quad \text{with } u_0 = (A+C)^{-1}(u^* + C(u)). \tag{2.128}$$

Since $v^* \in X^*$ is arbitrary, from (2.128) we infer that $u_0 = u$. Therefore $u^* \in A(u)$, which by virtue of (2.126) implies the maximal monotonicity of A.

Solution of Problem 2.184

"$\Longrightarrow$": If X is reflexive, then $\overline{B}_1 = \{u \in X : \|u\|_X \leqslant 1\}$ is w-compact. So, given $u^* \in X^*$, we can find $u_0 \in \overline{B}_1$ such that

$$\langle u^*, u_0\rangle = \sup\{\langle u^*, u\rangle : \|u\|_X \leqslant 1\} = \|u^*\|_*,$$

so $u^* \in \mathcal{F}(\|u^*\|_* u_0)$ (see Definition 2.112), hence $R(\mathcal{F}) = X^*$.

"$\Longleftarrow$": Suppose that X is not reflexive Then $\overline{B}_1$ is not w-compact and so by the James theorem (see Theorem I.5.74) we can find $u_0^* \in X^*$ such that $\sup\big\{\langle u_0^*, u\rangle : \|u\|_X \leqslant 1\big\}$ is not attained. Then $u_0^* \notin R(\mathcal{F})$ and so $R(\mathcal{F}) \neq X^*$.

Solution of Problem 2.185

From Proposition 2.114, we know that $\mathcal{F}$ is single valued. Let $u_n \xrightarrow{w} u$ in X and assume that

$$\limsup_{n\to+\infty} \langle \mathcal{F}(u_n), u_n - u\rangle \leqslant 0. \tag{2.129}$$

We have

$$\left(\|u_n\|_X - \|u\|_X\right)^2 \leqslant \langle \mathcal{F}(u_n) - \mathcal{F}(u), u_n - u\rangle,$$

so

$$\|u_n\|_X \longrightarrow \|u\|_X$$

(see (2.129) and recall that $u_n \xrightarrow{w} u$ in X), thus $u_n \longrightarrow u$ (by the Kadec–Klee property; see Corollary 1.26) and hence $\mathcal{F}$ is of type $(S)_+$ (see Definition 2.137).

Solution of Problem 2.186

According to Definition 2.131 we consider a sequence $\{u_n\}_{n\geqslant 1} \subseteq X$ such that $u_n \xrightarrow{w} u$ in X, $T(u_n) \xrightarrow{w} u^*$ in X^* and $\limsup\limits_{n\to+\infty} \langle T(u_n), u_n - u\rangle \leqslant 0$. Since K is compact, the sequence $\{K(u_n)\}_{n\geqslant 1} \subseteq X^*$ is relatively compact and so by passing to a suitable subsequence if necessary, we may assume that

$$K(u_n) \longrightarrow v^* \quad \text{in } X^*. \tag{2.130}$$

Then we have

$$A(u_n) \xrightarrow{w} u^* - v^* \quad \text{in } X^*.$$

Also

$$\langle K(u_n), u_n - u\rangle \longrightarrow 0$$

(see (2.130)) and

$$\limsup_{n\to+\infty} \langle A(u_n), u_n - u\rangle \leqslant 0 \tag{2.131}$$

(from the choice of the sequence $\{u_n\}_{n\geqslant 1}$), so

$$u_n \longrightarrow u \quad \text{in } X$$

(since A is of type $(S)_+$; see Definition 2.137 and (2.131)). Then we have $K(u_n) \longrightarrow K(u)$ (since K is compact, hence continuous too) and $A(u_n) \xrightarrow{w} A(u)$ in X^* (since A is demicontinuous). Therefore $v^* = A(u) + K(u) = T(u)$ and we have that

$$\langle T(u_n), u_n\rangle \longrightarrow \langle T(u), u\rangle\,,$$

that is, T is generalized pseudomonotone (see Definition 2.131).

Solution of Problem 2.187

Let $T = A + K$. By hypothesis we have that $u_n \xrightarrow{w} u$ in X, $T(u_n) \xrightarrow{w} u^*$ in X^* and

$$\limsup_{n\to+\infty} \langle T(u_n), u_n - u\rangle \leqslant 0. \tag{2.132}$$

From the hypothesis on K, we have that $K(u_n) \xrightarrow{w} K(u)$ in X^*. Then

$$A(u_n) \xrightarrow{w} u^* - K(u) \quad \text{in } X^*. \tag{2.133}$$

Also we have

$$\begin{aligned}\limsup_{n\to+\infty} \langle A(u_n), u_n\rangle &= \limsup_{n\to+\infty} \left(\langle T(u_n), u_n\rangle - \langle K(u_n), u_n\rangle\right)\\ &\leqslant \limsup_{n\to+\infty} \langle T(u_n), u_n\rangle - \liminf_{n\to+\infty} \langle K(u_n), u_n\rangle\\ &\leqslant \langle u^* - K(u), u\rangle\end{aligned}$$

(see (2.132) and use the semicontinuity of φ), so

$$\begin{aligned}\limsup_{n\to+\infty} \langle A(u_n), u_n - u\rangle &\leqslant \limsup_{n\to+\infty} \langle A(u_n), u_n\rangle - \liminf_{n\to+\infty} \langle A(u_n), u\rangle\\ &\leqslant \langle u^* - K(u), u\rangle - \langle u^* - K(u), u\rangle = 0\end{aligned}$$

(see (2.133)). But recall that A is generalized pseudomonotone. Hence from the above inequality and Definition of 2.131, it follows that $A(u) = u^* - K(u)$, so $u^* = (A + K)(u)$.

Solution of Problem 2.188

Let $\{u_n^*\}_{n \geqslant 1} \subseteq A(C)$ be a sequence such that $u_n^* \longrightarrow u^*$ in X^*. Then we have

$$u_n^* \in A(u_n) \quad \text{with } u_n \in C \qquad \forall\, n \geqslant 1.$$

The set C is w-compact in X since the space X is reflexive. So, by the Eberlein–Smulian theorem (see Theorem I.5.78) we may assume that $u_n \xrightarrow{w} u$ in X. We have $u \in C$ and

$$\lim_{n \to +\infty} \langle u_n^*, u_n - u \rangle = 0.$$

Since A is generalized pseudomonotone (see Definition 2.131), we have $(u, u^*) \in \operatorname{Gr} A$, hence $u^* \in A(u)$ with $u \in C$, which proves that the set $A(C) \subseteq X$ is closed.

Solution of Problem 2.189

According to Problem 2.186, it suffices to show that the identity map I_H is of type $(S)_+$ (see Definition 2.137). So, let $\{u_n\}_{n \geqslant 1} \subseteq H$ be a sequence such that $u_n \xrightarrow{w} u$ in H and

$$\limsup_{n \to +\infty} (u_n, u_n - u)_H \leqslant 0.$$

So, using also the weak lower semicontinuity of the norm functional (see Proposition I.5.56(c)), we have

$$\|u\|_H^2 \leqslant \liminf_{n \to +\infty} \|u_n\|_H^2 \leqslant \limsup_{n \to +\infty} \|u_n\|_H^2 \leqslant \|u\|_H^2,$$

thus $\|u_n\|_H \longrightarrow \|u\|_H$ and from the Kadec–Klee property in Hilbert spaces (see Corollary 1.26), we get $u_n \longrightarrow u$. Therefore I_H is of type $(S)_+$ and so we can use Problem 2.186 to conclude that the map $u \longmapsto u + K(u)$ is generalized pseudomonotone (see Definition 2.131).

Solution of Problem 2.190

No. Let $X = H = l^2$. This is a Hilbert space and $X^* = H = l^2$. We consider the map $K\colon l^2 \longrightarrow l^2$ defined by

$$K(\widehat{u}) = (2 - \|\widehat{u}\|_{l^2}, 0, 0, \ldots) \qquad \forall\, \widehat{u} = \{u_k\}_{k\geqslant 1} \in l^2.$$

Evidently K is of finite rank (see Definition 2.1(c)) and so it is compact (see Remark 2.2). Let $\widehat{u}_n = \{u_k^n\}_{k\geqslant 1} = \{\delta_{nk}\}_{k\geqslant 1}$, with δ_{nk} being the Kronecker symbol, i.e.,

$$\delta_{nk} = \begin{cases} 1 & \text{if } n = k, \\ 0 & \text{if } n \neq k, \end{cases} \qquad n, k \geqslant 1.$$

Then $\widehat{u}_n \in l^2$ for all $n \geqslant 1$ and $\widehat{u}_n \xrightarrow{w} \widehat{u} = 0$ in l^2. Also

$$K(\widehat{u}_n) \longrightarrow \widehat{h} = (1, 0, 0, \ldots) \quad \text{in } l^2$$

and

$$(K(\widehat{u}_n), \widehat{u}_n)_H \longrightarrow 0.$$

However, $\widehat{h} \neq K(\widehat{u}) = (2, 0, 0, \ldots)$ and so K is not generalized pseudomonotone (see Definition 2.131).

Solution of Problem 2.191

Let $u_n \xrightarrow{w} u$ in X. The monotonicity of A implies that

$$\langle A(u), u_n - u\rangle \leqslant \langle A(u_n), u_n - u\rangle \qquad \forall\, n \geqslant 1,$$

so

$$0 \leqslant \liminf_{n\to+\infty} \langle A(u_n), u_n - u\rangle\,. \tag{2.134}$$

From the sequential weak continuity of A, we have

$$\liminf_{n\to+\infty} \langle A(u_n), u_n - u\rangle = \liminf_{n\to+\infty} \langle A(u_n), u_n\rangle - \langle A(u), u\rangle\,,$$

so

$$\varphi(u) \leqslant \liminf_{n\to+\infty} \varphi(u_n)$$

(see (2.134)), thus φ is sequentially weakly lower semicontinuous.

Solution of Problem 2.192

The maximal monotonicity of A can be established similarly as in the solution of Problem 2.153. So, we need to show that A is of type $(S)_+$ (see Definition 2.137). To this end, let $\{u_n\}_{n\geqslant 1} \subseteq W^{1,p}(\Omega)$ be such that

$$u_n \xrightarrow{w} u \quad \text{in } W^{1,p}(\Omega) \quad \text{and} \quad \limsup_{n\to+\infty} \langle A(u_n), u_n - u\rangle \leqslant 0. \tag{2.135}$$

From (2.135) we have

$$\limsup_{n\to+\infty} \langle A(u_n) - A(u), u_n - u\rangle \leqslant 0. \tag{2.136}$$

The monotonicity of A implies that

$$\liminf_{n\to+\infty} \langle A(u_n) - A(u), u_n - u\rangle \geqslant 0. \tag{2.137}$$

From (2.136) and (2.137), we have

$$\lim_{n\to+\infty} \langle A(u_n) - A(u), u_n - u\rangle = 0. \tag{2.138}$$

Let

$$\xi_n(z) = (a(z, Du_n(z)) - a(z, Du(z)), Du_n(z) - Du(z))_{\mathbb{R}^N}.$$

From the convexity of $G(z,\cdot)$, we have that $a(z,\cdot)$ is monotone (see Definition 2.98(a)) and so

$$\xi_n(z) \geqslant 0 \quad \text{for a.a. } z \in \Omega, \text{ all } n \geqslant 1.$$

Also, from (2.138), we have

$$\int_\Omega \xi_n(z)\,dz \longrightarrow 0,$$

so

$$\xi_n \longrightarrow 0 \quad \text{in } L^1(\Omega). \tag{2.139}$$

From (2.139) and by passing to a suitable subsequence if necessary, we may assume that

$$\xi_n(z) \longrightarrow 0 \quad \text{for a.a. } z \in \Omega \tag{2.140}$$

and

$$\xi_n(z) \leqslant k(z) \quad \text{for a.a. } z \in \Omega, \tag{2.141}$$

with $k \in L^1(\Omega)$. From the convexity of $G(z,\cdot)$, we have

$$(a(z,y), y)_{\mathbb{R}^N} \geqslant G(z,y) \geqslant \tfrac{c_0}{p}|y|^p \quad \text{for a.a. } z \in \Omega, \text{ all } y \in \mathbb{R}^N \tag{2.142}$$

(see hypothesis **(v)**). Using (2.140)–(2.142) and hypothesis **(iii)**, for almost all $z \in \Omega$ and all $n \geqslant 1$, we have

$$\begin{aligned} k(z) \geqslant \xi_n(z) &= (a(z, Du_n(z)) - a(z, Du(z)), Du_n(z) - Du(z))_{\mathbb{R}^N} \\ &\geqslant \tfrac{c_0}{p}\left(|Du_n(z)|^p + |Du(z)|^p\right) \\ &\quad - \left(\widehat{a}(z)(1 + |Du(z)|^{p-1})\right)|Du_n(z)| \\ &\quad - \left(\widehat{a}(z)(1 + |Du_n(z)|^{p-1})\right)|Du(z)|. \end{aligned} \tag{2.143}$$

From (2.143) it follows that for all $z \in \Omega \setminus D$ with $|D|_N = 0$ (by $|\cdot|_N$ we denote the Lebesgue measure on $\mathbb{R}^N$), the sequence $\{Du_n(z)\}_{n\geqslant 1} \subseteq \mathbb{R}^N$ is bounded. Hence, by passing to a suitable subsequence if necessary (in general the subsequence depends on $z \in \Omega \setminus D$), we will have

$$Du_n(z) \longrightarrow \eta(z) \quad \text{in } \mathbb{R}^N \qquad \forall\, z \in \Omega \setminus D,$$

so

$$a(z, Du_n(z)) \longrightarrow a(z, \eta(z)) \qquad \forall\, z \in \Omega \setminus D_1,$$

with $D_1 \supseteq D$, $|D_1|_N = 0$. Passing to the limit as $n \to +\infty$ and using (2.140)–(2.141), we obtain

$$(a(z, \eta(z)) - a(z, Du(z)), \eta(z) - Du(z))_{\mathbb{R}^N} = 0 \qquad \forall\, z \in \Omega \setminus D_1. \tag{2.144}$$

Since $G(z,\cdot)$ is strictly convex, we have that $a(z,\cdot)$ is strictly monotone. So, from (2.144) it follows that $\eta(z) = Du(z)$ for almost all $z \in \Omega$. So, by the Urysohn criterion for the convergence of sequences (see Problem I.1.3), for the original sequence, we have

$$Du_n(z) \longrightarrow Du(z) \quad \text{for a.a. } z \in \Omega. \tag{2.145}$$

Let $E \subseteq \Omega$ be a Lebesgue measurable set. Then from (2.143), using the Hölder inequality (see Theorem 1.3 and Problem 1.27) and the boundedness of the sequence $\{Du_n\}_{n\geqslant 1} \subseteq L^p(\Omega; \mathbb{R}^N)$ (see (2.135)), we have

$$\frac{c_0}{p} \int_E |Du_n|^p \, dz \leqslant c_2 \Big(\int_E |Du| \, dz + |C|_N^{\frac{1}{p'}} + \Big(\int_E |Du|^p \, dz \Big)^{\frac{1}{p}} + \Big(\int_E |Du|^p \, dz \Big)^{\frac{1}{p'}} \Big)$$

for some $c_2 >$ (here $\frac{1}{p} + \frac{1}{p'} = 1$). So the sequence $\{|Du_n|^p\}_{n\geqslant 1} \subseteq L^1(\Omega)$ is uniformly integrable (see Definition 1.18 and Problem 1.6).

This fact, (2.145) and the Vitali theorem (the extended dominated convergence theorem; see Theorem I.3.128 and Proposition I.3.130) imply that

$$\|Du_n\|_p \longrightarrow \|Du\|_p. \tag{2.146}$$

Recall that

$$Du_n \xrightarrow{w} Du \quad \text{in } L^p(\Omega; \mathbb{R}^N) \tag{2.147}$$

(see (2.135)). Since $L^p(\Omega; \mathbb{R}^N)$ is uniformly convex, it has the Kadec–Klee property (see Corollary 1.26). So, from (2.146) and (2.147) it follows that

$$Du_n \longrightarrow Du \quad \text{in } L^p(\Omega; \mathbb{R}^N). \tag{2.148}$$

On the other hand, from (2.135) and the Rellich–Kondrachov embedding theorem (see Theorem 1.135), we have

$$u_n \longrightarrow u \quad \text{in } L^p(\Omega). \tag{2.149}$$

From (2.148) and (2.149) we conclude that $u_n \longrightarrow u$ in $W^{1,p}(\Omega)$ and so we infer that A is of type $(S)_+$.

Solution of Problem 2.193

According to Theorem 2.116, it suffices to show that $R(A+C+I_H)=H$. Let $h \in H$ and

$$E(u) = (A+I_H)^{-1}(h-C(u)) \qquad \forall\, u \in D(C).$$

We know that $(A+I_H)^{-1}$ is nonexpansive (see Proposition 2.123(b)). So, we have

$$\|E(u)-E(y)\|_H \leqslant k\|u-y\|_H \qquad \forall\, u,y \in D(C).$$

Since by hypothesis $D(C) \subseteq H$ is closed, we can apply the Banach fixed point theorem (see Theorem I.1.49) and find $\widehat{v} \in D(C)$ such that $E(\widehat{v}) = \widehat{v}$. Then $h \in (A+C+I_H)(\widehat{v})$. Because $h \in H$ is arbitrary, we infer that $R(A+C+I_H)=H$ and so we conclude that $A+C$ is maximal monotone (see Definition 2.100).

Solution of Problem 2.194

Let $\{u_n\}_{n\geqslant 1} \subseteq X$ be a sequence such that $u_n \xrightarrow{w} u$ in X and

$$\limsup_{n\to+\infty} \langle A(u_n), u_n-u\rangle \leqslant 0.$$

Since A is of type $(S)_+$ (see Definition 2.137), we have that $u_n \longrightarrow u$ in X (see Definition 2.137). Because of the demicontinuity of A we have $A(u_n) \xrightarrow{w} A(u)$ in X^*. Hence for every $y \in X$, we have

$$\langle A(u), u-y\rangle \leqslant \liminf_{n\to+\infty} \langle A(u_n), u_n-y\rangle,$$

so A is pseudomonotone (see Definition 2.129).

Solution of Problem 2.195
From Problem 2.194, we know that A is pseudomonotone. Then strong coercivity of A and Theorem 2.136 imply that A is surjective.

Solution of Problem 2.196
Let $\{u_n\}_{n\geqslant 1} \subseteq H_0^1(\Omega)$ be a sequence such that $u_n \xrightarrow{w} u$ in $H_0^1(\Omega)$. Since $N \leqslant 3$, we have that $H_0^1(\Omega)$ is embedded compactly in $L^4(\Omega)$ (see the Rellich–Kondrachov embedding theorem; Theorem 1.135). So, we have $u_n \longrightarrow u$ in $L^4(\Omega)$. We need to show that $A(u_n) \longrightarrow A(u)$ in $H^{-1}(\Omega)$. Arguing by contradiction, suppose that the last convergence in $H^{-1}(\Omega)$ is not true. Then we can find $\varepsilon > 0$ and a sequence $\{v_n\}_{n\geqslant 1} \subseteq H_0^1(\Omega)$ such that

$$\|v_n\|_{H_0^1(\Omega)} \leqslant 1 \quad \text{and} \quad \langle A(u_n) - A(u), v_n\rangle \geqslant \varepsilon \qquad \forall\, n \geqslant 1. \tag{2.150}$$

From (2.150) we see that by passing to a subsequence if necessary, we may assume that $v_n \xrightarrow{w} v$ in $H_0^1(\Omega)$, so

$$v_n \longrightarrow v \quad \text{in } L^4(\Omega).$$

For all $n \geqslant 1$ and $k \in \{1, \ldots, N\}$, we have

$$\begin{aligned} & (\sin u_n)(D_k u_n)v_n - (\sin u)(D_k u)v_n \\ = \;& (\sin u_n - \sin u)(D_k u_n)v_n + (\sin u)(D_k u_n)(v_n - v) \\ & +(\sin u)(D_k u_n - D_k u)v + (\sin u)(D_k u)(v - v_n). \end{aligned} \tag{2.151}$$

Using the mean value theorem, we have

$$|\sin u_n - \sin u| \leqslant |u_n - u|.$$

Then invoking the generalized Hölder inequality (see Theorem I.3.105), we have

$$\begin{aligned} \Big|\int_\Omega (\sin u_n - \sin u)(D_k u_n)v_n\, dz\Big| &\leqslant \|u_n - u\|_4 \|D_k u_n\|_2 \|v_n\|_4 \\ &\leqslant c\|u_n - u\|_4 \|u_n\|_{H_0^1(\Omega)} \|v_n\|_{H_0^1(\Omega)}, \end{aligned}$$

with $c > 0$, so

$$\int_\Omega (\sin u_n - \sin u)(D_k u_n)v_n\, dz \longrightarrow 0.$$

Also, since $v_n \xrightarrow{w} v$ in $H_0^1(\Omega)$ and $D_k u_n \xrightarrow{w} D_k u$ in $L^2(\Omega)$, then

$$\begin{aligned}
\int\limits_\Omega (\sin u)(D_k u_n)(v_n - v)\,dz &\longrightarrow 0,\\
\int\limits_\Omega (\sin u)(D_k u_n - D_k u)v\,dz &\longrightarrow 0,\\
\int\limits_\Omega (\sin u)(D_k u)(v - v_n)\,dz &\longrightarrow 0.
\end{aligned}$$

We return to (2.151), integrate over Ω (for every $k \in \{1, \ldots, N\}$) and obtain

$$\langle A(u_n) - A(u), v_n\rangle \longrightarrow 0,$$

which contradicts (2.150). This proves the complete continuity of A (see Definition 2.1(b)).

Solution of Problem 2.197

"$\Longrightarrow$": Suppose that A is self-adjoint (see Definition I.5.108(b)). According to Theorem 2.116, it suffices to show that $R(A + I_H) = H$. First we show that $R(A + I_H)$ is dense in H. So, let $h \in R(A + I_H)^\perp$. Then

$$(h, u)_H + (h, A(u))_H = 0 \qquad \forall\, u \in D(A),$$

so

$$(h, A(u))_H = -(h, u)_H \qquad \forall\, u \in D(A)$$

and thus $h \in D(A^*)$ and $A^*(h) = -h$. Since by hypothesis A is self-adjoint (i.e., $A = A^*$), we have $h \in D(A)$ and $A(h) = -h$. The monotonicity of A implies that

$$0 \leqslant (A(h), h)_H = -\|h\|_H^2,$$

so $h = 0$ and so $R(A + I_H)$ is dense in H.

Next we show that $R(A + I_H)$ is closed. To this end, let $\{h_n\}_{n\geqslant 1} \subseteq R(A + I_H)$ and suppose that $h_n \longrightarrow h$ in H. Let $\{u_n\}_{n\geqslant 1} \subseteq D(A)$ be such that

$$h_n = A(u_n) + u_n \qquad \forall\, n \geqslant 1, \tag{2.152}$$

so

$$\|u_n - u_m\|_H \leqslant \|h_n - h_m\|_H \qquad \forall\, n, m \geqslant 1$$

(see Proposition 2.123(b)), so $\{u_n\}_{n\geqslant 1} \subseteq H$ is a Cauchy sequence and thus $u_n \longrightarrow u$ in H. From (2.152), we have

$$A(u_n) \longrightarrow h - u \quad \text{in } H.$$

But A being self-adjoint, is closed and so $A(u) = h - u$, so $h = A(u) + u$, thus $h \in R(A + I_H)$ and hence the set $R(A + I_H) \subseteq H$ is closed. We conclude that $R(A + I_H) = H$ and so A is maximal monotone (see Definition 2.100).

"$\Longleftarrow$": Suppose that A is maximal monotone. First we show that A is densely defined (i.e., $\overline{D(A)} = H$). So, let $h \in D(A)^\perp$. Then $(h, u)_H = 0$ for all $u \in D(A)$. Since A is maximal monotone, we have $R(A + I_H) = H$ (see Theorem 2.116). Hence we can find $u \in D(A)$ such that $h = A(u) + u$. We have

$$\|u\|_H^2 \leqslant (A(u), u) + \|u\|_H^2 = (h, u)_H = 0$$

(recall that A is monotone), so $u = 0$ and thus $h = 0$. This shows that the operator A is densely defined.

Since by hypothesis A is symmetric, we have $A \subseteq A^*$ (i.e., $\operatorname{Gr} A \subseteq \operatorname{Gr} A^*$). Therefore, in order to have that A^* is self-adjoint, it is enough to show that $D(A^*) \subseteq D(A)$. We fix $v \in D(A^*)$. Since $R(A + I_H) = H$ (from the maximal monotonicity of A), we can find $u, z \in D(A)$ such that

$$A(u) + u = A^*(v) + v \quad \text{and} \quad A(z) + z = v - u. \tag{2.153}$$

Since A is symmetric, then so is $A + I_H$ and we have

$$\begin{aligned} (A^*(v) + v, z)_H &= (A(u) + u, z)_H \\ = (A(z) + z, u)_H &= (v - u, u)_H \end{aligned} \tag{2.154}$$

(see (2.153) and use the symmetry of $A + I_H$). From the definition of the adjoint operator, we have

$$(A^*(v), z)_H = (A(z), v)_H .$$

It follows that

$$\begin{aligned} (v - u, v)_H &= (A(z) + z, v)_H = (z, A^*(v) + v)_H \\ &= (z, A(u) + u)_H = (v - u, u)_H \end{aligned}$$

(see (2.153), (2.154) and use the symmetry of $A + I_H$), so $\|v - u\|_H^2 = 0$, thus $v = u \in D(A)$ and hence $D(A^*) \subset D(A)$. This proves that $D(A) = D(A^*)$, which shows that A is self-adjoint.

Solution of Problem 2.198

From Problem 2.197 we have that A is maximal monotone (see Definition 2.100). Since T is monotone and $\operatorname{Gr} A \subseteq \operatorname{Gr} T$, we must have $\operatorname{Gr} A = \operatorname{Gr} T$ (see Remark 2.101) and so $A = T$.

Solution of Problem 2.199

From Proposition 2.114(a) we know that the duality map $\mathcal{F}\colon X\longrightarrow X^*$ is single valued. The maximality of A (see Definition 2.139(b)) implies that

$$A(u) = \bigcap_{(x,y)\in \mathrm{Gr}\, A} \{v \in X : \langle \mathcal{F}(u-x), v-y\rangle \geqslant 0\}$$

(recall that A is accretive; see Definition 2.139(a)). In this intersection, every set is closed and convex. Therefore $A(u) \in P_{fc}(X)$.

Solution of Problem 2.200

(a) Let $\{(u_n, v_n)\}_{n\geqslant 1} \subseteq \mathrm{Gr}\, A$ be a sequence such that $u_n \longrightarrow u$ and $v_n \longrightarrow v$ in X. According to Definition 2.139(a), for every $(x, y) \in \mathrm{Gr}\, A$, we can find $h_n^* \in \mathcal{F}(u_n - x)$ such that

$$\langle h_n^*, v_n - y\rangle \geqslant 0. \tag{2.155}$$

From Problem 2.88 we know that the duality map is upper semicontinuous from X into $X^*_{w^*}$ (where $X^*_{w^*}$ denotes the Banach space X^* furnished with the w^*-topology). Then since $K = \{u_n - x : n \geqslant 1\} \cup \{u - x\} \in P_k(X)$, from Problem 2.56(a), we have that the set $\mathcal{F}(K) \subseteq X^*$ is w^*-compact. Because $\{h_n^*\}_{n\geqslant 1} \subseteq \mathcal{F}(K)$, we can find a subnet $\{h_\alpha^*\}_{\alpha\in J}$ of $\{h_n^*\}_{n\geqslant 1}$ such that $h_n^* \xrightarrow{w^*} h^*$ in X^* and $h^* \in \mathcal{F}(u - x)$. From (2.155) we obtain

$$\langle h^*, v - x\rangle \geqslant 0 \qquad \forall\, (x, y) \in \mathrm{Gr}\, A.$$

The maximality of A implies that $(u, v) \in \mathrm{Gr}\, A$ and so $\mathrm{Gr}\, A$ is closed in $X \times X$.

(b) Since the space X^* is locally uniformly convex (see Definition I.5.168), it is also strictly convex (see Remark I.5.169). By Proposition 2.114, the duality map $\mathcal{F}\colon X \longrightarrow X^*$ is single valued and continuous. Let $\{(u_\alpha, v_\alpha)\}_{\alpha\in J} \subseteq \mathrm{Gr}\, A$ be a net such that

$$u_\alpha \longrightarrow u \;\text{ in } X \quad\text{and}\quad v_\alpha \xrightarrow{w} v \;\text{ in } X. \tag{2.156}$$

For every $(x, y) \in \mathrm{Gr}\, A$, we have

$$\langle \mathcal{F}(u_\alpha - x), v_\alpha - y\rangle \geqslant 0,$$

so

$$\langle \mathcal{F}(u - x), v - y\rangle \geqslant 0$$

(see (2.156)) and thus $(u, v) \in \mathrm{Gr}\, A$ (due to maximality of A).

Solution of Problem 2.201

First note that X being uniformly convex, it is reflexive (see the Milman–Pettis theorem; Theorem I.5.89). We will show that for every sequence $\{\lambda_n\}_{n\geqslant 1}$ such that $\lambda_n \searrow 0$, the sequence $\{A_{\lambda_n}(u)\}_{n\geqslant 1}$ admits a subsequence strongly convergent to $A^0(u)$. By the Urysohn criterion for the convergence of subsequences (see Problem I.1.3), this implies that $A_\lambda(u) \longrightarrow A^0(u)$ as $\lambda \searrow 0$.

From Proposition 2.143, we see that the sequence $\{A_{\lambda_n}(u)\}_{n\geqslant 1} \subseteq X$ is bounded. The reflexivity of X implies that the sequence $\{A_{\lambda_n}(u)\}_{n\geqslant 1}$ is relatively w-compact in X. Then the Eberlein–Smulian theorem (see Theorem I.5.78) implies that we can find a subsequence $\{A_{\lambda_{n_k}}(u)\}_{k\geqslant 1}$ of $\{A_{\lambda_n}(u)\}_{n\geqslant 1}$ such that

$$A_{\lambda_{n_k}}(u) \xrightarrow{w} v \quad \text{in } X. \tag{2.157}$$

From Proposition 2.143(b), we know that

$$A_{\lambda_{n_k}}(u) \in A(J_{\lambda_{n_k}}(u)) \qquad \forall\, k \geqslant 1. \tag{2.158}$$

From Proposition 2.143(d), we have

$$J_{\lambda_{n_k}}(u) \longrightarrow u \quad \text{in } X. \tag{2.159}$$

Because A is m-accretive (see Definition 2.139(c)) it is also maximal accretive (see Proposition 2.144). So, from (2.157), (2.158), (2.159) and Problem 2.200(b), we have $(u,v) \in \operatorname{Gr} A$. Then

$$\|A^0(u)\|_X \leqslant \|v\|_X. \tag{2.160}$$

Also, we have

$$\|v\|_X \leqslant \liminf_{k\to+\infty} \|A_{\lambda_{n_k}}(u)\|_X \leqslant \limsup_{k\to+\infty} \|A_{\lambda_{n_k}}(u)\|_X \leqslant \|A^0(u)\|_X \tag{2.161}$$

(see Proposition 2.143(c)). From (2.160) and (2.161), we have $\|v\|_X = \|A^0(u)\|_X$. From Problem 2.199 we know that $A(u) \in P_{fc}(X)$. Therefore $v = A^0(u)$ and so

$$\|A_{\lambda_{n_k}}(u)\|_X \longrightarrow \|A^0(u)\|_X \quad \text{and} \quad A_{\lambda_{n_k}}(u) \xrightarrow{w} A^0(u) \quad \text{in } X,$$

so

$$A_{\lambda_{n_k}}(u) \longrightarrow A^0(u) \quad \text{in } X$$

(by the Kadec–Klee property; see Corollary 1.26)). Using once more Proposition 2.143, we have

$$\|A^0(J_\lambda(u))\|_X \leqslant \|A_\lambda(u)\|_X \leqslant \|A^0(u)\|_X \qquad \forall\, u \in D(A),\ \lambda > 0.$$

Then the same argument as above, via the Kadec–Klee property, gives

$$A^0(J_\lambda(u)) \longrightarrow A^0(u) \quad \text{as } \lambda \searrow 0 \qquad \forall\, u \in D(A).$$

Solution of Problem 2.202

We argue indirectly. So, suppose that A is not upper semicontinuous as claimed by the problem. Then we can find a weakly open set $U \subseteq X$ such that

$$A^+(U) \;=\; \{u \in D(A):\; A(u) \subseteq U\}$$

is not open in $D(A)$ (see Proposition 2.37). This means that we can find $u \in D(A) \cap A^+(U)$ and a sequence $\{u_n\}_{n\geqslant 1} \subseteq D(A)$ such that $u_n \longrightarrow u$ in X and $v_n \in A(u_n)$ such that $v_n \notin U$ for all $n \geqslant 1$. By hypothesis A is locally bounded (see Definition 2.102). So, the sequence $\{v_n\}_{n\geqslant 1} \subseteq X$ is bounded and due to the reflexivity of X, by passing to a subsequence if necessary, we may assume that $v_n \xrightarrow{w} v \in X \setminus U$. Since $(u_n, v_n) \in \operatorname{Gr} A$ for every $n \geqslant 1$, Problem 2.200(b) implies that $v \in A(u) \subseteq U$, a contradiction.

Solution of Problem 2.203

For every $t, s \in \mathbb{R}$ and every $u^* \in \mathcal{F}(u(t))$, we have

$$\langle u^*, u(t)\rangle \;=\; \|u(t)\|_X^2 \quad \text{and} \quad \langle u^*, u(s)\rangle \;\leqslant\; \|u(t)\|_X \|u(s)\|_X.$$

It follows that

$$\langle u^*, u(s) - u(t)\rangle \;\leqslant\; \|u(t)\|_X \left(\|u(s)\|_X - \|u(t)\|_X\right).$$

Let $t \in \mathbb{R}$ be a point of differentiability of $\|u(\cdot)\|_X$ and of weak differentiability of u. Dividing by $s - t$ and letting $s \to t$, we obtain

$$\big\langle u^*, u'(t)\big\rangle \;\leqslant\; \|u(t)\|_X \tfrac{d}{dt}\|u(t)\|_X \quad \text{as } s \to t^+$$

and

$$\big\langle u^*, u'(t)\big\rangle \;\geqslant\; \|u(t)\|_X \tfrac{d}{dt}\|u(t)\|_X \quad \text{as } s \to t^-.$$

So, finally we have

$$\big\langle u^*, u'(t)\big\rangle \;=\; \|u(t)\|_X \tfrac{d}{dt}\|u(t)\|_X \quad \text{for a.a. } t \in T, \text{ all } u^* \in \mathcal{F}(u(t)).$$

Solution of Problem 2.204
According to Definition 2.139(c), we need to show that for every $h \in X$, the operator equation $u + A(u) = h$ admits a solution $u \in X$.

Let

$$\widehat{A}(u) \;=\; (I_X + A)(u) + h \qquad \forall\, u \in X.$$

Evidently $\widehat{A}$ is continuous accretive (see Definition 2.139) with $D(\widehat{A}) = X$. For fixed $u_0 \in X$, we consider the following abstract Cauchy problem

$$\begin{cases} \frac{du}{dt}(t) + \widehat{A}(u(t)) \;=\; 0, \\ u(0) = u_0. \end{cases} \tag{2.162}$$

From Theorem 2.158, we know that problem (2.162) has a unique solution $u \in C^1((0,+\infty);X)$. We have

$$\tfrac{d}{dt}\left(u(t+h) - u(t)\right) = -A\left(u(t+h)\right) + A\left(u(t)\right) - u(t+h) + u(t) \qquad \forall\, t > 0,$$

so

$$\|u(t+h) - u(t)\|_X \tfrac{d}{dt}\|u(t+h) - u(t)\|_X \;\leqslant\; -\|u(t+h) - u(t)\|_X^2 \qquad \forall\, t \geqslant 0$$

(see Problem 2.203 and recall that A is accretive), thus

$$\|u(t+h) - u(t)\|_X \;\leqslant\; e^{-t}\|u(h) - u_0\|_X \qquad \forall\, t \geqslant 0. \tag{2.163}$$

Similarly, from the equation

$$\tfrac{d}{dt}\left(u(t) - u_0\right) \;=\; -A(u(t)) - u(t) + h,$$

we obtain

$$\begin{aligned} &\|u(t) - u_0\|_X \tfrac{d}{dt}\|u(t) - u_0\|_X \\ \leqslant\;& -\|u(t) - u_0\|_X^2 + \|u(t) - u_0\|_X \left(\|A(u_0)\|_X + \|u_0\|_X + \|h\|_X\right), \end{aligned}$$

so

$$\|u(t) - u_0\|_X \;\leqslant\; (1 + e^{-t})\left(\|A(u_0)\|_X + \|u_0\|_X + \|h\|_X\right) \qquad \forall\, t \geqslant 0. \tag{2.164}$$

From (2.163) and (2.164), we infer that $\lim\limits_{t\to+\infty} u(t) = \widehat{u} \in X$ exists and $\lim\limits_{t\to+\infty} \frac{du}{dt}(t) = 0$. Therefore, if in (2.162) we pass to the limit as $t \to +\infty$, we obtain $(I_X + A)\widehat{u} = h$, so $R(I_X + A) = X$ and thus A is m-accretive (see Definition 2.139).

Solution of Problem 2.205

We argue by contradiction. So, suppose that A is not maximal accretive (see Definition 2.139(b)). Then we can find $(\widehat{u}, \widehat{v}) \notin \operatorname{Gr} A$ such that

$$0 \leqslant \langle \mathcal{F}(x - \widehat{u}), y - \widehat{v} \rangle \qquad \forall\, (x, y) \in \operatorname{Gr} A. \tag{2.165}$$

The set $A(\widehat{u})$ is nonempty, closed, and convex. Since $\widehat{v} \notin A(\widehat{u})$, we can find $h^* \in X^*$, $h^* \neq 0$ such that

$$\sup_{v \in A(\widehat{u})} \langle h^*, v \rangle < \langle h^*, \widehat{v} \rangle .$$

The duality map $\mathcal{F}$ is monotone, continuous and coercive (see Proposition 2.114). So, Theorem 2.119 implies that $R(\mathcal{F}) = X^*$. Therefore we can find $h \in X$ such that $\mathcal{F}(h) = h^*$. So, we have

$$\langle \mathcal{F}(h), v \rangle < \langle \mathcal{F}(h), \widehat{v} \rangle \qquad \forall\, v \in A(\widehat{u}). \tag{2.166}$$

Set $U = \{y \in X : \langle \mathcal{F}(h), y - \widehat{v} \rangle < 0\}$. Evidently $U \subseteq X$ is weakly open and $A(\widehat{u}) \subseteq U$ (see (2.166)). Let $t \in (0,1)$ and let $\widehat{u}_t = \widehat{u} + th$, $\widehat{v}_t \in A(\widehat{u}_t)$. By hypothesis A is upper semicontinuous from X into X_w. So, for $t \in (0,1)$ small, $\widehat{v}_t \in U$. Then

$$\langle \mathcal{F}(h), \widehat{v}_t - \widehat{v} \rangle < 0. \tag{2.167}$$

On the other hand, from (2.165), we have

$$0 \leqslant \langle \mathcal{F}(\widehat{u}_t - \widehat{u}), \widehat{v}_t - \widehat{v} \rangle = t \langle \mathcal{F}(h), \widehat{v}_t - \widehat{v} \rangle ,$$

so

$$\langle \mathcal{F}(h), \widehat{v}_t - \widehat{v} \rangle \geqslant 0,$$

a contradiction to (2.167). This shows that A is maximal accretive.

Solution of Problem 2.206

Arguing indirectly, suppose that we can find $u_0 \in D(A)$ and a sequence $\{t_n\}_{n \geqslant 1} \subseteq (0, +\infty)$ such that

$$\|S(t_n)u_0\|_X \geqslant n \qquad \forall\, n \geqslant 1.$$

Since $t \longmapsto S(t)u_0$ is continuous, we must have $t_n \longrightarrow +\infty$. Then from the hypothesis of the problem, we have

$$\|AS(t_n)u_0\|_X \longrightarrow +\infty.$$

On the other hand, from Theorem 2.158, we know that the map $t \longmapsto \|AS(t)u_0\|_X$ is nonincreasing, a contradiction. So, we conclude that $\sup\limits_{t\geqslant 0} \|S(t)u_0\|_X < +\infty$.

Solution of Problem 2.207

We have

$$\begin{aligned}
\Big\|S(t)u - \tfrac{1}{t}\int_0^t S(\tau)u\,d\tau\Big\|_X &= \tfrac{1}{t}\Big\|\int_0^t (S(t)u - S(\tau)u)\,d\tau\Big\|_X \\
\leqslant \tfrac{1}{t}\int_0^t \big\|S(\tau)(S(t-\tau)u) - S(\tau)u\big\|_X\,d\tau &\leqslant \tfrac{1}{t}\int_0^t \|S(t-\tau)u - u)\|_X\,d\tau \\
= \tfrac{1}{t}\int_0^t \|S(r)u - u\|_X\,dr && (2.168)
\end{aligned}$$

(see Definitions 2.156(b) and (a)). Using the triangle inequality, we have

$$\begin{aligned}
\|S(t)u - u\|_X &\leqslant \Big\|S(t)u - \tfrac{1}{t}\int_0^t S(\tau)u\,d\tau\Big\|_X + \tfrac{1}{t}\Big\|\int_0^t (S(\tau)u - u)\,d\tau\Big\|_X \\
&\leqslant \tfrac{2}{t}\int_0^t \|S(\tau)u - u\|_X\,d\tau
\end{aligned}$$

(see (2.168)).

Solution of Problem 2.208

From the triangle inequality, for every $u \in \overline{D(A)}$, $x \in D(A)$, $y \in A(x)$ and $t > 0$, we have

$$\begin{aligned}
\|S(t)u - u\|_X &\leqslant \|S(t)u - S(t)x\|_X + \|S(t)x - x\|_X + \|y - x\|_X \\
&\leqslant 2\|u - x\|_X + t\|y\|_X && (2.169)
\end{aligned}$$

(see Definition 2.156 and Theorem 2.158). In (2.169), we let $x = J_\lambda(u)$ and $y = A_\lambda(u)$ (recall that $A_\lambda(u) \in A(J_\lambda(u))$). So, we obtain

$$\|S(t)u - u\|_X \leqslant \left(2 + \tfrac{t}{\lambda}\right)\|J_\lambda(u) - u\|_X$$

(see Definition 2.142). For all $x \in D(A)$, all $y \in A(x)$ and all $t > 0$, we have

$$\|S(t)u - x\|_X \leqslant \|u - x\|_X + \tfrac{1}{\lambda}\int_0^t \big(\|S(\tau)u - x - \lambda y\|_X - \|S(\tau)u - x\|_X\big)\, d\tau \quad \forall\, \lambda > 0.$$

Choosing $x = J_\lambda(u)$ and $y = A_\lambda(u)$, we obtain

$$\|S(t)u - J_\lambda(u)\|_X \leqslant \|u - J_\lambda(u)\|_X + \tfrac{1}{\lambda}\int_0^t \big(\|S(\tau)u - u\|_X - \|S(\tau)u - J_\lambda(u)\|_X\big)\, d\tau,$$

so

$$\|u - J_\lambda(u)\|_X \leqslant \tfrac{\lambda}{t}\|S(t)u - u\|_X + \tfrac{2}{t}\int_0^t \|S(\tau)u - u\|_X\, d\tau$$

and thus

$$\|u - J_\lambda(u)\|_X \leqslant \tfrac{2}{t}\left(1 + \tfrac{\lambda}{t}\right)\int_0^t \|S(\tau)u - u\|_X\, d\tau$$

(see Problem 2.207).

Solution of Problem 2.209

"(a) $\Longrightarrow$ (b)": Suppose that the nonlinear semigroup $\{S(t)\}_{t \geqslant 0}$ is compact (see Definition 2.159(a)). From Remark 2.160, we know that $\{S(t)\}_{t \geqslant 0}$ is equicontinuous. Recall that $J_\lambda \colon X \longrightarrow D(A)$ (see Definition 2.142). So, from Theorem 2.158, for all $u \in X$, we have

$$\|S(t)J_\lambda(u) - J_\lambda(u)\|_X \leqslant t\,|A(J_\lambda(u))| \leqslant t\|A_\lambda(u)\|_X = \tfrac{t}{\lambda}\|u - J_\lambda(u)\|_X$$

(recall that $A_\lambda(u) \in A(J_\lambda(u))$ and see Definition 2.142), so

$$\lim_{t \searrow 0} S(t)J_\lambda = J_\lambda \text{ uniformly on bounded subsets of } X$$

(see Proposition 2.143). But $S(t)J_\lambda$ is compact. Hence so is the limit J_λ.

"(b) $\Longrightarrow$ (a)": Suppose that for every $\lambda > 0$, the map J_λ is compact and $\{S(t)\}_{t \geqslant 0}$ is equicontinuous (see Definition 2.159(b)). From Problem 2.208, for all $u \in \overline{D(A)}$, all $t > 0$ and all $\lambda > 0$, we have

$$\|J_\lambda(S(t)u) - S(t)u\|_X \leqslant \tfrac{4}{\lambda}\int_0^\lambda \|S(t+\tau)u - S(t)u\|_X\, d\tau.$$

The equicontinuity hypothesis on $\{S(t)\}_{t\geqslant 0}$ implies that for every bounded set $B \subseteq \overline{D(A)}$, we can find a function $\xi_B \colon \mathbb{R}_+ \longrightarrow \mathbb{R}_+$ such that $\xi_B(r) \longrightarrow 0$ as $r \searrow 0$ and for all $\tau \geqslant 0$, all $u \in B$, we have

$$\|S(t+\tau)u - S(t)u\|_X \leqslant \xi_B(\tau)$$

(see Definition 2.159(b)). So, for all $u \in B$, we have

$$\|J_\lambda(S(t)u) - S(t)u\|_X \leqslant 4 \sup_{r\in[0,\lambda]} \xi_B(r),$$

thus

$$J_\lambda(S(t)u) \longrightarrow S(t)u \quad \text{as } \lambda \searrow 0 \text{ uniformly on bounded sets of } \overline{D(A)}.$$

But $J_\lambda \circ S$ is compact for every $\lambda > 0$. So, Problem 2.2 implies that $S(t)$ is compact for all $t > 0$.

Solution of Problem 2.210

For every $n \geqslant 1$, we have

$$\begin{aligned} \|u - J_{\lambda_n}(u_n)\|_X &\leqslant \|u - u_n\|_X + \|u_n + J_{\lambda_n}(u_n)\|_X \\ &\leqslant \|u - u_n\|_X + \lambda_n\|A_{\lambda_n}(u_n)\|_X \leqslant \|u - u_n\|_X + \lambda c, \end{aligned}$$

for some $c > 0$ independent of $n \geqslant 1$ (see Definition 2.142 and recall that the sequence $\{A_{\lambda_n}(u_n)\}_{n\geqslant 1} \subseteq X$ is bounded), so

$$\|u - J_{\lambda_n}(u_n)\|_X \longrightarrow 0. \tag{2.170}$$

We know that $A_{\lambda_n}(u_n) \in A(J_{\lambda_n}(u_n))$ for all $n \geqslant 1$ (see Proposition 2.143). From (2.170) and since $A_{\lambda_n}(u_n) \longrightarrow h$ in X, invoking Problem 2.200(a), we conclude that $(u,h) \in \operatorname{Gr} A$. The last part of the problem is a consequence of Problem 2.200(b).

Solution of Problem 2.211

"**(a) $\Longrightarrow$ (b)**": From Proposition 2.143(c), we know that

$$\|A_\lambda(u)\|_X = \tfrac{1}{\lambda}\|u - J_\lambda(u)\|_X \leqslant \eta \qquad \forall\, u \in L_\eta,\ \lambda > 0,$$

so

$$\|u - J_\lambda(u)\|_X \leqslant \lambda\eta,$$

thus

$$\lim_{\lambda \searrow 0} J_\lambda = I_X \quad \text{uniformly on } L_\eta.$$

Since by hypothesis J_λ is compact, it follows that $I_X|_{L_\eta}$ is compact (see Problem 2.2) and so L_η is relatively compact in X.

"**(b) $\Longrightarrow$ (a)**": Let $B \subseteq X$ be a bounded set. Then $J_\lambda(B)$ is bounded for every $\lambda > 0$ (recall that J_λ in nonexpansive). For any $u \in X$ and $\lambda > 0$, we have

$$\begin{aligned} \|J_\lambda(u)\|_X + |A(J_\lambda(u))| &\leqslant \|J_\lambda(u)\|_X + \|A_\lambda(u)\|_X \\ &= \|J_\lambda(u)\|_X + \tfrac{1}{\lambda}\|u - J_\lambda(u)\|_X. \end{aligned}$$

So, there exists $\eta > 0$ such that $J_\lambda(B) \subseteq L_\eta$, hence $\overline{J_\lambda(B)}$ is compact in X. Since $B \subseteq X$ is an arbitrary bounded set, we conclude that J_λ is compact.

Solution of Problem 2.212

To show the maximal monotonicity of A, it suffices to show that

$$R(A + I_{L^2_{2\pi}(\mathbb{R})}) = L^2_{2\pi}(\mathbb{R})$$

(see Theorem 2.116). So, for every $h \in L^2_{2\pi}(\mathbb{R})$, we consider the periodic problem

$$\begin{cases} u'(t) + u(t) = h(t) & \text{for a.a. } t \in T, \\ u(0) = u(2\pi). \end{cases} \tag{2.171}$$

It is well known that (2.171) has a unique solution $u \in W^{1,2}(0, 2\pi)$ (variation of constant formula). Therefore A is maximal monotone (see Definition 2.100). The Hille–Yosida theorem (see Theorem 2.153) implies that A is the infinitesimal generator of the C_0-semigroup defined by

$$S(t)u(\tau) = u(t - \tau) \qquad \forall\, u \in L^2_{2\pi}(\mathbb{R}),\ t \geqslant 0.$$

so, for each $t \geqslant 0$, $S(t)$ is an isometry and so it cannot be compact for $t > 0$.

Solution of Problem 2.213

Without any loss of generality, we may assume that $\eta > 0$. Let $\varepsilon \in (0, \eta)$. If $\overline{B}^*_{\eta - \frac{\varepsilon}{2}} = \{v^* \in X^* : \|v^*\|_* \leqslant \eta - \frac{\varepsilon}{2}\}$, then

$$\overline{B}^*_{\eta - \frac{\varepsilon}{2}} \cap F(u) = \emptyset \qquad \forall\, u \in E.$$

From the separation theorem for convex sets (see Theorem I.5.29), we see that for all $u \in E$, we can find $\xi_u \in X$ such that $\|\xi_u\|_X = 1$ and

$$\langle h^*, \xi_u \rangle \leqslant \langle v^*, \xi_u \rangle \qquad \forall\, h^* \in \overline{B}^*_{\eta - \frac{\varepsilon}{2}},\ v^* \in F(u).$$

Then

$$\eta - \varepsilon \leqslant \langle v^*, \xi_u \rangle \,.$$

The upper semicontinuity of F implies that we can find an open neighborhood $U(u)$ of u such that

$$\langle v^*, \xi_u \rangle > \eta - \varepsilon \qquad \forall\, u' \in U(u),\ v^* \in F(u').$$

Since the set E is compact and $\{U(u)\}_{u\in E}$ is an open cover of E, we can find a finite subcover $\{U(u_1), \ldots, U(u_m)\}$. Let $\{v_1, \ldots, v_m\} \subseteq X$ be such that

$$\|v_k\|_X = 1 \quad \text{and} \quad \langle v^*, v_k \rangle > \eta - \varepsilon$$

for all $k \in \{1, \ldots, m\}$, $u \in U(u_k)$ and $v^* \in F(u)$. If $\lambda_k(u) = \operatorname{dist}_E(u, \partial U(u_k))$, we define

$$\mu_k(u) = \frac{\lambda_k(u)}{\sum\limits_{k=1}^{m} \lambda_k} \quad \text{and} \quad \xi(u) = \sum_{k=1}^{m} \mu_k(u) v_k.$$

This is the desired continuous map $\xi \colon E \longrightarrow X$.

Solution of Problem 2.214

We argue by contradiction. So, suppose that we can find $\varepsilon_0 > 0$ and a sequence $\{u_n\}_{n\geqslant 1} \subseteq X$ such that

$$\|L(u_n)\|_Y > \varepsilon_0 \|u_n\|_X + n\|u_n\|_Z. \tag{2.172}$$

Replacing u_n with $\frac{u_n}{\|u_n\|_X}$ if necessary, we may assume that $\|u_n\|_X = 1$ for all $n \geqslant 1$. Since X is reflexive, passing to a subsequence if necessary, we may assume that

$$u_n \xrightarrow{w} u \quad \text{in } X,$$

so

$$u_n \xrightarrow{w} u \quad \text{in } Z \tag{2.173}$$

(since the embedding $X \hookrightarrow Z$ is continuous). As $L \in \mathcal{L}_c(X;Y)$, from Proposition 2.3, we have

$$L(u_n) \longrightarrow L(u) \quad \text{in } Y. \tag{2.174}$$

From (2.172), we have

$$\|u_n\|_Z < \tfrac{1}{n}\|L(u_n)\|_Y,$$

so $\|u\|_Z \leqslant 0$ (see (2.173) and (2.174)), thus $u = 0$. So, we have $L(u_n) \longrightarrow 0$ in Y, while from (2.172), we have $\varepsilon_0 \leqslant \liminf \|L(u_n)\|_Y$, a contradiction. This completes the solution of the problem.

Solution of Problem 2.215

Let $X = W^{1,p}(\Omega)$, $Y = L^q(\partial\Omega)$, $Z = L^p(\Omega)$ and let $L = \gamma \in \mathcal{L}_c(W^{1,p}(\Omega), L^q(\partial\Omega))$ (the trace map; see Definition 1.129). So, we can use Problem 2.214 and then given $\varepsilon > 0$, we can find $\widehat{c}_\varepsilon > 0$ such that for all $u \in W^{1,p}(\Omega)$, we have

$$\|u\|_{L^q(\partial\Omega)} \leqslant \varepsilon\big(\|u\|_p + \|Du\|_p\big) + \widehat{c}\|u\|_p = \varepsilon\|Du\|_p + c_\varepsilon\|u\|_p,$$

with $c_\varepsilon = \widehat{c}_\varepsilon + \varepsilon > 0$.

Solution of Problem 2.216

Let $(v, v^*) \in U \times X^*$ and suppose that

$$\langle v^* - u^*, v - u\rangle \geqslant 0 \qquad \forall\, (u, u^*) \in \operatorname{Gr} A. \tag{2.175}$$

Suppose that $v^* \notin A(v)$. Then by the strong separation theorem (see Theorem I.5.29 and recall that $A(v)$ is nonempty, convex, and w^*-closed), we can find $h \in X \setminus \{0\}$ such that

$$F(v) \subseteq \{y^* \in X^* : \langle y^*, h\rangle < \langle v^*, h\rangle\} = W. \tag{2.176}$$

The set W is w^*-open. Since F is upper semicontinuous from X into $X^*_{w^*}$, we can find a norm open neighborhood D of v such that $A(D) \subseteq W$ (see Definition 2.36(a)). For $t > 0$ small we have $v + th \in D$ and so $A(v + th) \subseteq W$. Returning to (2.175) and choosing $(u, u^*) = (v + th, u^*)$ with $u^* \in A(v + th)$ we have

$$0 \leqslant \langle v^* - u^*, -th\rangle = -t\,\langle v^* - u^*, h\rangle\,,$$

so

$$\langle v^*, h\rangle \leqslant \langle u^*, h\rangle$$

and thus $u^* \notin W$, a contradiction. So, $v^* \in A(v)$ and this proves the maximal monotonicity of A (see Definition 2.100).

Solution of Problem 2.217

First note that A has w^*-closed values. Also, since A is locally bounded, by the Alaoglu theorem (see Theorem I.5.66), it is locally compact (see Proposition 2.45). So, we can use Problem 2.65 and infer that A is upper semicontinuous from X into $X^*_{w^*}$. Then we can invoke Problem 2.216 and conclude that A is maximal monotone.

Solution of Problem 2.218

Let $u^* \in X^*$ and let $\{u_n^*\}_{n\geqslant 1} \subseteq X^*$ be a sequence such that $u_n^* \longrightarrow u^*$ in X^*. Let $u_n \in D(A)$ be such that $u_n^* \in A(u_n)$ for all $n \geqslant 1$. Since A is strongly coercive (see Definition 2.98(f)), we can find a function $\eta\colon \mathbb{R}_+ \longrightarrow \mathbb{R}$ with $\eta(r) \longrightarrow +\infty$ as $r \to +\infty$ such that

$$\langle u_n^*, u_n\rangle \geqslant \eta(\|u_n\|)\|u_n\|_X \qquad \forall\, n \geqslant 1,$$

so

$$\eta(\|u_n\|_X) \leqslant \|u_n^*\|_* \leqslant c_1 \qquad \forall\, n \geqslant 1,$$

for some $c_1 > 0$, thus

$$\|u_n\|_X \leqslant c_2 \qquad \forall\, n \geqslant 1,$$

for some $c_2 > 0$ and hence A^{-1} is locally bounded at $u^* \in X^*$ (see Definition 2.102). Since $u^* \in X^*$ was arbitrary, we conclude that A^{-1} is locally bounded.

Bibliography

[1] Aliprantis, C.D., Border, K.C.: Infinite Dimensional Analysis. A Hitchhiker's Guide. Springer, Berlin (2006)

[2] Aubin, J.-P., Frankowska, H.: Set-Valued Analysis. Birkhäuser, Boston (2009)

[3] Barbu, V.: Nonlinear Semigroups and Differential Equations in Banach Spaces. Noordhoff International Publishing, Leiden (1976)

[4] Brézis, H.: Opérateurs Maximaux Monotones et Semi-Groupes de Contractions dans les Espaces de Hilbert. American Elsevier Publishing Co., Inc., New York (1973)

[5] Browder, F.E.: Nonlinear Operators and Nonlinear Equations of Evolution in Banach Spaces. In: Proc. Sympos. Pure Math., vol. XVIII, Part 2, Chicago, 1968. American Mathematical Society, Providence (1976)

[6] Denkowski, Z., Migórski, S., Papageorgiou, N.S.: An Introduction to Nonlinear Analysis: Applications. Kluwer Academic, Boston (2003)

[7] Gasiński L., Papageorgiou, N.S.: Nonlinear Analysis. Chapman & Hall/CRC, Boca Raton (2006)

[8] Hu, S., Papageorgiou, N.S.: Handbook of Multivalued Analysis. Vol. I. Theory. Kluwer Academic, Dordrecht (1997)

[9] Hu, S., Papageorgiou, N.S.: Handbook of Multivalued Analysis. Vol. II. Applications. Kluwer Academic, Dordrecht (2000)

[10] Kato, T.: Perturbation Theory for Linear Operators. Springer, Berlin (1995)

[11] Klein, E., Thompson, A.C.: Theory of Correspondences. Wiley, New York (1984)

[12] Papageorgiou, N.S., Kyritsi-Yiallourou, S.Th.: Handbook of Applied Analysis. Springer, New York (2009)

[13] Pascali, D., Sburlan, S.: Nonlinear Mappings of Monotone Type. Sijthoff & Noordhoff International Publishers, Alphen aan den Rijn (1978)

[14] Reed, M., Simon, B.: Methods of Modern Mathematical Physics. Vol. I. Functional Analysis. Academic, New York (1980)

[15] Showalter, R.E.: Monotone Operators in Banach Space and Nonlinear Partial Differential Equations. American Mathematical Society, Providence, RI (1997)

[16] Vainberg, M.M.: Variational Method and Method of Monotone Operators in the Theory of Nonlinear Equations. Wiley, New York (1973)

[17] Zeidler, E.: Nonlinear Functional Analysis and Its Applications, vols. II/A and II/B. Springer, New York (1990)

Chapter 3

Smooth and Nonsmooth Calculus

3.1 Introduction

In this chapter, X and Y are Banach spaces.

3.1.1 Gâteaux and Fréchet Derivatives

Definition 3.1

*(a) A map $\varphi\colon X \longrightarrow Y$ is said to be **Gâteaux differentiable at** $u \in X$ if and only if there exists $A \in \mathcal{L}(X;Y)$ such that*

$$\lim_{t \to 0} \tfrac{\varphi(u+th)-\varphi(u)}{t} = A(h) \qquad \forall\, h \in X.$$

*Then A is called the **Gâteaux derivative** of φ at $u \in X$ and it is denoted by $\varphi'_G(u)$.*

*(b) A map $\varphi\colon X \longrightarrow Y$ is said to be **Fréchet differentiable at** $u \in X$ if and only if there exists $A \in \mathcal{L}(X;Y)$ such that*

$$\varphi(u+h) - \varphi(u) = A(h) + e(u,h) \qquad \forall\, h \in X,$$

*with $\lim\limits_{h \to 0} \tfrac{\|e(u,h)\|_Y}{\|h\|_X} = 0$. Then A is called the **Fréchet derivative** of φ at $u \in X$ and it is denoted by $\varphi'(u)$.*

Remark 3.2

Evidently, if φ is Fréchet differentiable at $u \in X$, then φ is Gâteaux differentiable at $u \in X$, but the converse is not in general true. For this reason often Fréchet differentiable maps are called simply ***differentiable***. Note that a function φ is Fréchet

L. Gasiński, N.S. Papageorgiou, *Exercises in Analysis: Part 2*, Problem Books in Mathematics, DOI 10.1007/978-3-319-27817-9_3

differentiable at $u \in X$ if and only if

$$\lim_{t \to 0} \frac{\varphi(u+th)-\varphi(u)}{t} = \langle \varphi'(u), h \rangle$$

and the convergence is uniform for all $h \in X$ in a bounded subset of X.

The relation between these two concepts of differentiability is given in the next proposition.

Proposition 3.3
If $\varphi\colon X \longrightarrow Y$ *is Gâteaux differentiable at all points in some neighborhood of* $u \in X$ *and* $u \longmapsto \varphi'_G(u)$ *is continuous as a map from* X *into* $\mathcal{L}(X;Y)$,
then φ *is Fréchet differentiable at* $u \in X$ *and* $\varphi'(u) = \varphi'_G(u)$.

Example 3.4
Let H be a Hilbert space and let $\varphi\colon H \longrightarrow \mathbb{R}$ be a function which is Gâteaux differentiable at $u \in X$. By the Riesz representation theorem, we know that there exists unique $\nabla\varphi(u) \in H$ such that

$$\varphi'_G(u)(h) = (\nabla\varphi(u), h)_H \qquad \forall\, h \in H.$$

The element $\nabla\varphi(u)$ is called the ***gradient*** of φ at $u \in H$. Note that $\nabla\varphi\colon H \longrightarrow H$.

Proposition 3.5
If $\varphi\colon X \longrightarrow Y$ *is Fréchet differentiable at* $u \in X$,
then φ *is continuous at* $u \in X$.

Remark 3.6
The result is no longer true if the Fréchet differentiability at $u \in X$ is replaced by the Gâteaux differentiability. We can find a map $\varphi\colon X \longrightarrow Y$ which is Gâteaux differentiable at $u \in X$ but is not continuous there.

Equivalent renorming of X and Y does not affect Fréchet differentiability.

Proposition 3.7
If the spaces X *and* Y *are equivalently renormed,*
then the differentiability of $\varphi\colon X \longrightarrow Y$ *at* $u \in X$ *is unaffected.*
In particular, if X *and* Y *are both finite dimensional, then we may choose any pair of norms.*

Definition 3.8
Let $U \subseteq X$ *be an open set, let* $\varphi\colon X \longrightarrow Y$ *be a map which is Gâteaux differentiable at every* $u \in U$ *and let the map* $U \ni u \longmapsto \varphi'_G(u) \in \mathcal{L}(X;Y)$ *be continuous. Then we say that* $\varphi \in C^1(U)$.

Proposition 3.9
<u>If</u> $U \subseteq X$ is an open set and $\varphi = (\varphi_1, \varphi_2)\colon U \longrightarrow Y_1 \times Y_2$ is a map, where Y_1, Y_2 are two Banach spaces,
<u>then</u> φ is Fréchet differentiable at $u \in X$ if and only if the coordinate maps $\varphi_1\colon U \longrightarrow Y_1$ and $\varphi_2\colon U \longrightarrow Y_2$ are both Fréchet differentiable at $u \in X$.

Definition 3.10
*Let X_1, X_2 be two Banach spaces and let $\varphi\colon X_1 \times X_2 \longrightarrow Y$. For any $\widehat{u}_2 \in X_2$ consider the map $\varphi_1\colon X_1 \ni u_1 \longmapsto \varphi(u_1, \widehat{u}_2) \in Y$. Suppose that φ_1 is Gâteaux (respectively Fréchet) differentiable at $\widehat{u}_1 \in X$. Then $(\varphi_1)'_G(\widehat{u}_1)$ (respectively $\varphi_1'(\widehat{u}_1)$) is called the **partial Gâteaux** (respectively **Fréchet**) **derivative** of φ at $(\widehat{u}_1, \widehat{u}_1)$ with respect to the first variable and it is denoted by*

$$(\varphi'_{u_1})_G(\widehat{u}_1, \widehat{u}_2) \quad (\text{respectively } \varphi'_{u_1}(\widehat{u}_1, \widehat{u}_2)).$$

Similarly we define the partial derivative of φ with respect to the second variable.

Remark 3.11
If $\varphi'_G(\widehat{u}_1, \widehat{u}_2)$ exists, then the partial Gâteaux derivatives $(\varphi'_{u_1})_G(\widehat{u}_1, \widehat{u}_2)$ and $(\varphi'_{u_2})_G(\widehat{u}_1, \widehat{u}_2)$ exist and we have

$$\varphi'_G(\widehat{u}_1, \widehat{u}_2)(h_1, h_2) \;=\; (\varphi'_{u_1})_G(\widehat{u}_1, \widehat{u}_2)h_1 + (\varphi'_{u_2})_G(\widehat{u}_1, \widehat{u}_2)h_2. \tag{3.1}$$

Similarly for Fréchet differentiable maps. So, in particular if $\varphi \in C^1(X_1 \times X_2)$ (see Definition 3.8), then the partial derivatives exist and are continuous.

Proposition 3.12
<u>If</u> X_1, X_2 are two Banach spaces, $\varphi\colon X_1 \times X_2 \longrightarrow Y$ is a map, $(\varphi'_{u_1})_G$ exists in a neighborhood of $(\widehat{u}_1, \widehat{u}_2) \in X_1 \times X_2$ and the map $(\varphi'_{u_2})\colon X_1 \times X_2 \longrightarrow \mathcal{L}(X_2; Y)$ is continuous at $(\widehat{u}_1, \widehat{u}_2)$,
<u>then</u> $\varphi'_G(\widehat{u}_1, \widehat{u}_2)$ exists and (3.1) holds.
Moreover, if in addition $\varphi'_{u_1}(\widehat{u}_1, \widehat{u}_2)$ exists, then $\varphi'(\widehat{u}_1, \widehat{u}_2)$ exists too and

$$\varphi'(\widehat{u}_1, \widehat{u}_2)(h_1, h_2) \;=\; \varphi'_{u_1}(\widehat{u}_1, \widehat{u}_2)h_1 + \varphi'_{u_2}(\widehat{u}_1, \widehat{u}_2)h_2.$$

Corollary 3.13
<u>If</u> X, Y_1, Y_2 are Banach spaces, $U \subseteq X$ is an open set and $\varphi = (\varphi_1, \varphi_2)\colon U \longrightarrow Y$ is a map (where $Y = Y_1 \times Y_2$),
<u>then</u> $\varphi \in C^1(U; Y)$ if and only if $\varphi_1 \in C^1(U; Y_1)$ and $\varphi_2 \in C^1(U; Y_2)$.

Remark 3.14
Suppose that X and Y are finite dimensional with $m = \dim X$ and $N = \dim Y$. Let $U \subseteq X$ be an open set and let $\varphi\colon U \longrightarrow Y$ be a map. We fix a basis $\{e_1, \ldots, e_m\}$ of X and a basis $\{v_1 \ldots, v_N\}$ of Y. Partial derivatives provide a convenient way to represent the derivative of φ in terms of $\mathbb{R}$-valued functions. We have

$$\varphi \;=\; \sum_{k=1}^{N} \varphi_k v_k$$

and then

$$\frac{\partial\varphi}{\partial u_i}(u) = \varphi'(u)(e_i) = \sum_{k=1}^{N} \frac{\partial\varphi_k}{\partial u_i}(u)v_k \qquad \forall\, u \in \mathbb{R}^N,\ i \in \{1,\ldots,m\}.$$

Here there are mN $\mathbb{R}$-valued partial derivatives $\frac{\partial\varphi_k}{\partial u_i}(u)$ for $k \in \{1,\ldots,N\}$ and $i \in \{1,\ldots,m\}$. We arrange these mN real numbers as an $N \times m$-matrix by setting

$$J\varphi(u) = \begin{bmatrix} \frac{\partial\varphi_1}{\partial u_1}(u) & \cdots & \frac{\partial\varphi_1}{\partial u_m}(u) \\ \vdots & \ddots & \vdots \\ \frac{\partial\varphi_N}{\partial u_1}(u) & \cdots & \frac{\partial\varphi_N}{\partial u_m}(u) \end{bmatrix} = \begin{bmatrix} \varphi_1'(u)(e_1) & \cdots & \varphi_1'(u)(e_m) \\ \vdots & \ddots & \vdots \\ \varphi_N'(u)(e_1) & \cdots & \varphi_N'(u)(e_m) \end{bmatrix}.$$

This matrix is known as the ***Jacobian matrix of*** φ. Clearly $J\varphi$ depends on the choice of the bases in X and Y. The Jacobian matrix for φ at $u \in U$ can exist even if φ is not Fréchet differentiable at $u \in U$. However, if φ is Fréchet differentiable at $u \in U$, then $\varphi'(u) \in \mathcal{L}(X;Y)$ is represented by $J\varphi(u)$ with respect to the given bases in X and Y. Therefore, if $y = \varphi'(u)(w)$, then

$$y = \begin{bmatrix} y_1 \\ \vdots \\ y_N \end{bmatrix} = \begin{bmatrix} \varphi_1'(u)(e_1) & \cdots & \varphi_1'(u)(e_m) \\ \vdots & \ddots & \vdots \\ \varphi_N'(u)(e_1) & \cdots & \varphi_N'(u)(e_m) \end{bmatrix} \begin{bmatrix} w_1 \\ \vdots \\ w_m \end{bmatrix} = J\varphi(u)w.$$

The mean value theorem is an important tool in Real Analysis. In the present more general context of vector valued functions, it takes the following form.

Theorem 3.15
If $\varphi\colon [a,b] \longrightarrow Y$ *is continuous and differentiable in* (a,b) *and we have*

$$\|\varphi'(c)\|_Y \leqslant M \qquad \forall\, c \in (a,b),$$

then $\|\varphi(b) - \varphi(a)\|_Y \leqslant M(b-a)$.

Another result very close to the classical mean value theorem is the following.

Theorem 3.16
If $U \subseteq X$ *is an open set,* $u, v \in U$ *are such that* $[u,v] \subseteq U$ *(where* $[u,v] = \{tu + (1-t)v :\ t \in [0,1]\}$*) and* $\varphi\colon U \longrightarrow \mathbb{R}$ *is a Fréchet differentiable function, then there exist* $\widehat{u} \in (u,v)$ *(where* $(u,v) = \{tu + (1-t)v :\ t \in (0,1)\}$*) such that*

$$\varphi(v) - \varphi(u) = \left\langle \varphi'(\widehat{u}), v - u \right\rangle,$$

with $\langle\cdot,\cdot\rangle$ *denoting the duality brackets for the pair* (X^*, X).

Remark 3.17
If $X = \mathbb{R}^m$, then this result can be written as

$$\varphi(v) - \varphi(u) = \sum_{k=1}^{m} \frac{\partial\varphi}{\partial x_k}(\widehat{u})(v_k - u_k).$$

Proposition 3.18
If $U \subseteq X$ *is an open connected set,* $\varphi\colon U \longrightarrow \mathbb{R}$ *is a Fréchet differentiable function and* $\varphi'(u) = 0$ *for all* $u \in U$,
then φ *is constant.*

Theorem 3.19 *(Chain Rule)*
If $U \subseteq X$ *and* $V \subseteq Y$ *are open sets,* Z *is a third Banach space,* $\varphi\colon U \longrightarrow X$ *is Gâteaux differentiable at* $u \in U$, $\varphi(U) \subseteq V$ *and* $\psi\colon V \longrightarrow Z$ *is Fréchet differentiable at* $\varphi(u)$,
then $g = \psi \circ \varphi\colon U \longrightarrow Z$ *is Gâteaux differentiable at* $u \in U$ *and*

$$g'_G(u) = \psi'(\varphi(u)) \circ \varphi'_G(u).$$

Moreover, if φ *is Fréchet differentiable at* u, *then so is* g *and*

$$g'(u) = \psi'(\varphi(u)) \circ \varphi'(u).$$

Remark 3.20
If in the above proposition the spaces X and Y are finite dimensional, then

$$J_{\psi\circ\varphi}(u) = J_{\psi}(\varphi(u))J_{\varphi}(u).$$

Moreover, the composition of two C^1-maps (see Definition 3.8) is a C^1-map.

Definition 3.21
Let $\varphi\colon X \longrightarrow \mathbb{R}$ *be a map. We say that* $u \in X$ *is a* ***local minimum*** *(respectively* ***maximum****) of* φ *if there exists a neighborhood* U *of* u *such that*

$$\varphi(u) \leqslant \varphi(v) \quad (\text{respectively } \varphi(u) \geqslant \varphi(v)) \qquad \forall\, v \in U.$$

We say that $u \in X$ *is a* ***local extremum*** *of* φ *if it is a local minimum or local maximum of* φ.

Remark 3.22
Clearly a global minimum (respectively maximum) is also local minimum (respectively maximum) but the converse is not true.

Proposition 3.23
If $U \subseteq X$ *is an open set,* $C \subseteq U$ *is a convex set and* $\varphi\colon U \longrightarrow \mathbb{R}$ *is a map such that* $\varphi|_C$ *has a local minimum at* $\widehat{u} \in C$ *and it is Gâteaux differentiable at* $\widehat{u}$,
then $\varphi'_G(\widehat{u})(y - \widehat{u}) \geqslant 0$ *for all* $y \in C$.

Remark 3.24
Evidently, if $\widehat{u}$ is a local maximum of $\varphi|_C$, then the inequality of the conclusion holds in the opposite direction.

Proposition 3.25
If $U \subseteq X$ *is an open set and* $\varphi\colon U \longrightarrow \mathbb{R}$ *has a local extremum at* $u \in U$ *and it is Gâteaux differentiable there,*
then $\varphi'_G(u) = 0$.

The next theorem examines what can be said about a map φ, if $\varphi'(u)$ is invertible.

Theorem 3.26 *(Inverse Mapping Theorem)*
If $U \subseteq X$ *is an open set,* $\varphi \in C^1(U;Y)$, $u_0 \in U$, $\varphi'(u_0)$ *is invertible and* $\varphi'(u_0)^{-1} \in \mathcal{L}(Y;X)$,
then φ *is a local homeomorphism, i.e., there exists an open set* $V \subseteq U$ *containing* u_0 *such that* $\varphi|_V$ *is a homeomorphism onto* $\varphi(V)$*, an open neighborhood of* $y_0 = \varphi(u_0)$*. Moreover, there is possibly a smaller open set* $W \subseteq V$ *containing* u_0 *such that* $\varphi|_U^{-1} \in C^1(\varphi(W);X)$ *and*

$$(\varphi|_U^{-1})'(\varphi(u)) = \varphi'(u)^{-1} \qquad \forall\, u \in W.$$

In fact $\varphi|_U^{-1}$ *is as smooth as* φ*, i.e., if* $\varphi \in C^m(U)$*, then* $\varphi|_U^{-1} \in C^m(\varphi(W))$.

The above result has some interesting consequences in the finite dimensional case.

Proposition 3.27
If $U \subseteq \mathbb{R}^m$ *is an open set,* $u_0 \in U$, $\varphi \in C^1(U;\mathbb{R}^N)$ *and* $y_0 = \varphi(u_0)$,
then ***(a)*** *if* $m \leqslant N$ *and* $\operatorname{rank}\varphi'(u_0) = m$*, we can find neighborhoods* U_0 *of* u_0 *and* V_0 *of* y_0 *and a Fréchet differentiable map* $h\colon V_0 \longrightarrow \mathbb{R}^N$ *such that*

$$(h \circ \varphi)(u) = i(u) \qquad \forall\, u \in U_0,$$

where $i\colon \mathbb{R}^m \longrightarrow \mathbb{R}^N$ *is the canonical injection, i.e.,*

$$i(u_1,\ldots,u_m) = (u_1,\ldots,u_m,0,\ldots,0) \qquad \forall\, u = (u_1,\ldots,u_m) \in \mathbb{R}^m.$$

(b) *if* $m \geqslant N$ *and* $\operatorname{rank}\varphi'(u_0) = N$*, we can find a neighborhood* U_0 *of* u_0 *and a Fréchet differentiable map* $f\colon U_0 \longrightarrow U$ *such that*

$$f(u_0) = u_0 \quad \text{and} \quad (\varphi \circ f)(u) = p(u) \qquad \forall\, u \in U_0,$$

where $p\colon \mathbb{R}^m \longrightarrow \mathbb{R}^N$ *is the canonical projection*

$$p(u_1,\ldots,u_m) = (u_1,\ldots,u_N) \qquad \forall\, u = (u_1,\ldots,u_m) \in \mathbb{R}^m.$$

Next we examine the following problem. Suppose that X, Y, Z are Banach spaces and $\varphi\colon X \times Y \longrightarrow Z$ is such that $\varphi(u_0, y_0) = 0$. Is it possible to find, at least locally, a map $u \longmapsto y = h(u)$ such that $\varphi(u, h(u)) = 0$? Here h should be continuous and even differentiable provided φ itself is a regular one. Moreover, we want, at least locally, h to be the only map satisfying this. This is the content of the next theorem.

Theorem 3.28 *(Implicit Function Theorem)*
If X, Y, Z *are three Banach spaces,* $W \subseteq X \times Y$ *is an open set,* $(u_0, y_0) \in W$, $\varphi \in C^1(W; Z)$ *with* $\varphi(u_0, y_0) = 0$ *and* $\varphi'_y(u_0, y_0) \in \mathcal{L}(X; Y)$ *is an isomorphism, then there exist open sets* $U \subseteq X$, $V \subseteq Y$ *and a Fréchet differentiable map* $h: U \longrightarrow V$ *such that* $u_0 \in U$, $y_0 \in V$, $U \times V \subseteq W$,

$$\varphi(u, h(u)) = 0 \quad \forall\, u \in U \tag{3.2}$$

and

$$h'(u) = -\varphi'_y(u, h(u))^{-1}\varphi'_u(u, h(u)) \quad \forall\, u \in U.$$

Moreover, for every $u \in U$, $h(u)$ *is the only solution of (3.2) in* V.

Next we consider extrema of functions under constraints (side conditions). In this case one tries to reduce the problem to an unconstrained one by using the idea of Lagrange, that is, we incorporate in the cost functional the constraints and minimize without any constraints the functional

$$\psi(u, \lambda_0, y^*) = \lambda_0\varphi(u) + \langle y^*, h(u)\rangle,$$

where φ is the original cost functional, h is the constraint map and $\{\lambda_0, y^*\}$ are the so-called ***Lagrange multipliers***.

Theorem 3.29
If X, Y *are two Banach spaces,* $u_0 \in X$, U *is an open neighborhood of* u_0 *and* $\varphi \in C^1(U)$, $h \in C^1(U; Y)$ *with* $h(u_0) = 0$ *and* $R(\varphi'(u_0))$ *is closed and*

$$\varphi(u_0) = \inf\{\varphi(u) : u \in U,\ h(u) = 0\},$$

then there exist $\lambda_0 \in \mathbb{R}$ *and* $y^* \in Y^*$ *not all zero such that*

$$\lambda_0\varphi'(u_0) + h'(u_0)^*(y^*) = 0 \quad \text{in } X^*.$$

Moreover, if $R(h'(u_0)) = Y$, *then* $\lambda_0 \neq 0$.

Remark 3.30
If $Y = \mathbb{R}$, then the surjectivity condition in the last part of the above theorem, becomes a nondegeneracy condition, that is, $h'(u_0) \neq 0$. In both cases we may assume that $\lambda_0 = 1$.

Recall that if $\varphi: X \longrightarrow Y$, then $\varphi'(u) \in \mathcal{L}(X; Y)$. So, if we want to introduce a second order derivative, we need to consider the space $\mathcal{L}(X; \mathcal{L}(X; Y))$. Moreover, if we pass to even higher order derivatives, then we are faced with even more complicated spaces. We overcome this difficulty by identifying these spaces with spaces of multilinear maps. For this reason, next we briefly discuss multilinear maps.

Definition 3.31
Let $X_1, \ldots, X_n$ and Y be Banach spaces. A map $A\colon X_1 \times \ldots \times X_n \longrightarrow Y$ is said to be ***multilinear*** *or* n-***linear****, if for every $k \in \{1, \ldots, n\}$ and every family of vectors $u_i^* \in X_i$ for $i \neq k$, the map*

$$X_k \ni u_k \longmapsto A(u_i^*, \ldots, u_{k-1}^*, u_k, u_{k+1}^*, \ldots, u_n^*) \in Y$$

is linear.

On the product space $X = X_1 \times \ldots \times X_n$, usually we consider the norm

$$\|\widehat{u}\|_X = \max_{1 \leqslant k \leqslant n} \|u_k\|_{X_k} \qquad \forall\, \widehat{u} = (u_1, \ldots, u_n) \in X.$$

We can characterize continuous multilinear maps, in a way analogous to the one used for linear operators.

Proposition 3.32
If $X_1, \ldots, X_n$ and Y are Banach spaces and $A\colon X_1 \times \ldots \times X_n \longrightarrow Y$ is a multilinear map,
then the following properties are equivalent:
(a) *A is continuous on $X = X_1 \times \ldots \times X_n$.*
(b) *A is continuous at the origin of X.*
(c) *There exists $c > 0$ such that*

$$\|A(\widehat{u})\|_X \leqslant c\|u_1\|_{X_1} \cdots \|u_n\|_{X_n} \qquad \forall\, \widehat{u} = (u_1, \ldots, u_n) \in X.$$

Remark 3.33
In what follows by $\mathcal{L}(X_1, \ldots, X_n; Y)$ we denote the family of continuous multilinear maps from $X = X_1 \times \ldots \times X_n$ into Y. This is a normed space with norm

$$\|A\|_{\mathcal{L}(X;Y)} = \sup\big\{\|A(\widehat{u})\|_Y : \widehat{u} = (u_1, \ldots, x_n) \in X,\ \|u_k\|_{X_k} \leqslant 1 \text{ for all } k \in \{1, \ldots, n\}\big\}.$$

Then $\|A\|_{\mathcal{L}(X;Y)}$ is the smallest constant $c > 0$ for which Proposition 3.32(c) holds. As for the linear operators, we can easily check that $\big(\mathcal{L}(X;Y), \|\cdot\|_{\mathcal{L}(X;Y)}\big)$ is a Banach space.

Now, let $\mathcal{L}_1(X;Y) = \mathcal{L}(X;Y)$ and inductively we define

$$\mathcal{L}_k(X;Y) = \mathcal{L}(X; \mathcal{L}_{k-1}(X;Y)) \qquad \forall\, k \geqslant 2.$$

The next identification theorem makes possible the introduction of higher order derivatives.

Theorem 3.34
The spaces $\mathcal{L}_k(X;Y)$ and $\mathcal{L}(X^k;Y)$ are isometrically isomorphic.

Definition 3.35
We say that the n-linear map $A\colon X^n \longrightarrow X$ is symmetric, if for every permutation $\sigma \in S_n$ we have

$$A(u_1, \ldots, u_n) \;=\; A(u_{\sigma(1)}, \ldots, u_{\sigma(n)}) \qquad \forall\, u_1, \ldots, u_n \in X.$$

We denote the family of symmetric n-linear map by $\mathcal{L}_s(X^n;Y)$. Evidently this is a linear subspace of $\mathcal{L}(X^n;Y)$.

Proposition 3.36
$\mathcal{L}_s(X^n;Y)$ is a closed subspace of $\mathcal{L}(X^n;Y)$ (hence a Banach space with the norm $\|\cdot\|_{\mathcal{L}(X^n;Y)}$).

Proposition 3.37
If $U \subseteq X$ is an open set and $\varphi\colon U \longrightarrow Y$ is a map,
then φ is $(n+1)$-differentiable at $u_0 \in U$ $(n \geqslant 1)$ if and only if $\varphi^{(n)}$ (the n-th derivative of φ) is differentiable at $u_0 \in U$. In this case we have

$$\begin{aligned}(\varphi^{(n)})'(u_0)h(h_1, \ldots, h) \;&=\; \varphi^{(n+1)}(u_0)(h, h_1, \ldots, h_n)\\ &\qquad \forall\, h, h_1, \ldots, h_n \in X.\end{aligned}$$

Proposition 3.38
If $U \subseteq X$ is an open set and $\varphi\colon U \longrightarrow Y$ is n-times differentiable at $u_0 \in U$,
then $\varphi^{(n)}(u_0) \in \mathcal{L}_s(X^n;Y)$.

Remark 3.39
If $X = \mathbb{R}^N$, $Y = \mathbb{R}$ and $\varphi\colon \mathbb{R}^N \longrightarrow \mathbb{R}$ is twice differentiable. then

$$\varphi^{(2)}(u_0)(h_1, h_2) \;=\; (H\varphi(u_0)h_1, h_2)_{\mathbb{R}^N} \qquad \forall\, h_1, h_2 \in \mathbb{R}^N,$$

where $H\varphi(u_0) = \big(\frac{\partial^2 \varphi}{\partial u_i \partial u_j}(u_0)\big)_{q \leqslant i,j \leqslant n}$. This symmetric matrix is known as the ***Hessian matrix*** of φ at u_0.

Using the second derivative, we can have a necessary condition for a local minimum. Of course an analogous criterion holds for local maximum simply by replacing φ by $-\varphi$.

Proposition 3.40
If $U \subseteq X$ is an open set, $u_0 \in U$, $\varphi\colon U \longrightarrow Y$ is twice differentiable at u_0 and u_0 is a local minimizer of φ,
then $\varphi^{(2)}(u_0)(h,h) \geqslant 0$ for all $h \in X$.

Remark 3.41
This is only a necessary condition for local minimum. It is not sufficient as the simple example of $\varphi\colon \mathbb{R} \longrightarrow \mathbb{R}$, $\varphi(u) = u^3$ at $u_0 = 0$ illustrates.

Definition 3.42

(a) *If $U \subseteq X$ and $V \subseteq Y$ are open sets and $\varphi\colon U \longrightarrow V$ is a map, then we say that φ is a* ***C^k-diffeomorphism*** *($k \geqslant 1$), if φ is bijective, $\varphi \in C^k(U;Y)$ and $\varphi^{-1} \in C^k(V;X)$.*

(b) *If $U \subseteq X$ is an open set, $\varphi\colon U \longrightarrow \mathbb{R}$ is differentiable at $u_0 \in U$, then we say that $u_0 \in U$ is a* ***critical point*** *of φ if $\varphi'(u_0) = 0$.*

(c) *If $U \subseteq X$ is an open set, $\varphi\colon U \longrightarrow \mathbb{R}$ is twice differentiable at $u_0 \in U$ and u_0 is a critical point of φ, then we say that u_0 is a* ***nondegenerate critical point*** *of φ, if $\varphi^{(2)}(u_0) \in \mathcal{L}_2(X;\mathbb{R}) = \mathcal{L}(X;X^*)$ is a topological isomorphism.*

Next we recall a few basic facts about vector fields.

Definition 3.43

Let $U \subseteq X$ be an open set. A continuous map $F\colon U \longrightarrow X$ is called a ***vector field****. If $F \in C^k(U;X)$, $k \geqslant 1$, then we say that F is a* ***C^k-vector field****. If I is an open interval and $\sigma\colon I \longrightarrow U$ is a differentiable map such that*

$$\sigma'(t) = F(\sigma(t)) \qquad \forall\, t \in I, \tag{3.3}$$

then we say that σ is an ***integral curve*** *of the vector field F. If σ cannot be extended to an integral curve defined on open interval strictly containing I, then we say that σ is a* ***maximal integral curve*** *of F.*

Remark 3.44

From (3.3) it is clear that $\sigma \in C^1(I;X)$ (note that $F \circ \sigma$ is continuous). Note that an integral curve σ is simply a situation of the first order differential equation $u' = F(u)$. So, if F is a C^k-vector field, then $\sigma \in C^{k+1}(T;X)$. As we will see in the sequel, of special interest are locally Lipschitz vector fields.

Since the vector field is time-independent, integral curves are translation invariant. In what follows $T_{t_0}\colon \mathbb{R} \longrightarrow \mathbb{R}$ is the translation function defined by

$$T_{t_0}(t) = t - t_0.$$

Proposition 3.45

<u>If</u> $F\colon U \longrightarrow X$ is a vector field, $\sigma\colon I \longrightarrow U$ is an integral curve of F and $t_0 \in \mathbb{R}$, <u>then</u> $\sigma \circ T_{t_0}$ is an integral curve of F defined on $I + t_0$ and $\sigma \circ T_{t_0}$ is maximal if and only if σ is maximal.

The next theorem confirms the importance of locally Lipschitz vector fields. For such vector fields there is always a unique integral curve passing from a given initial condition.

Theorem 3.46

<u>If</u> $F\colon U \longrightarrow X$ is a locally Lipschitz vector field, $u_0 \in U$ and $t_0 \in \mathbb{R}$, <u>then</u> there exists a unique maximal integral curve of F satisfying $\sigma(t_0) = u_0$.

Remark 3.47
If σ is the unique maximal integral curve defined on I and satisfying $\sigma(t_0) = u_0$, then the unique maximal integral curve $\widehat{\sigma}$ satisfying $\widehat{\sigma}(t_1) = u_0$ is defined on $I + (t_1 - t_0)$ and $\widehat{\sigma} = \sigma \circ T_{t_1 - t_0}$.

An easy case of interest is that of linear vector fields. So, let $A \in \mathcal{L}(X)$ and let

$$F(u) \;=\; A(u) \qquad \forall\, u \in X.$$

Evidently F is Lipschitz continuous.

Proposition 3.48
<u>If</u> $\sigma(t) = e^{tA}$ for all $t \in \mathbb{R}$,
<u>then</u> $\sigma'(t) = e^{tA} \circ A = A \circ e^{tA}$ for all $t \in \mathbb{R}$ (i.e., σ is global integral curve of the linear vector field $F = A$).
Moreover, if $u \in X$, then $\mathbb{R} \ni t \longmapsto e^{tA} u_0$ is the maximal integral curve of A passing from u_0 at time $t_0 = 0$.

Definition 3.49
*Let $F\colon U \longrightarrow X$ be a vector field. We say that $u_0 \in U$ is an **equilibrium point** of F, if $F(u_0) = 0$. All other points of U are said to be **regular**.*

Remark 3.50
If $t_0 \in \mathbb{R}$ and $u_0 \in U$ is an equilibrium point for F, then the constant map $\sigma\colon \mathbb{R} \longrightarrow U$ defined by $\sigma(t) - u_0$ for all $t \in \mathbb{R}$ is the maximal integral curve of F passing from u_0 at time $t = t_0$.

In what follows by $\widetilde{\sigma}(t, t_0, u_0)$ we denote the maximal integral curve of F such that $\widetilde{\sigma}(t_0) - u_0$. By I_{t_0,u_0} we denote the maximal time interval on which $\widetilde{\sigma}$ is defined. Often we call $\widetilde{\sigma}(\cdot, t_0, u_0)$ the flow of F through (t_0, u_0). So, if $F = A \in \mathcal{L}(X)$ (the linear vector field), then $\widetilde{\sigma}$ is defined on $\mathbb{R} \times \mathbb{R} \times X$ and it is given by

$$\widetilde{\sigma}(t, t_0, u_0) \;=\; e^{(t-t_0)A} u_0.$$

Next we examine the dependence of the integral curve on the initial condition.

Theorem 3.51
<u>If</u> $F\colon U \longrightarrow X$ is an M-Lipschitz vector field and σ_1, σ_2 are integral curves defined on $[t_0, t_1]$ such that $\sigma_1(t_0) = u_0^1$ and $\sigma_2(t_0) = u_0^2$, <u>then</u>

$$\|\sigma_1(t) - \sigma_2(t)\|_X \;\leqslant\; \|u_0^1 - u_0^2\|_X e^{M(t-t_0)}.$$

Theorem 3.52
<u>If</u> $\widetilde{\sigma}(\cdot, t_0, u_0)$ is defined on $[a, b]$ $(a < b)$,
<u>then</u> there exists a neighborhood V of u_0 such that for all $y_0 \in V$, $\widetilde{\sigma}(\cdot, t_0, y_0)$ is also defined on $[a, b]$.

Proposition 3.53
The flow $(t, t_0, u_0) \longmapsto \widetilde{\sigma}(t, t_0, u_0)$ *is defined on an open set in* $\mathbb{R}^2 \times X$ *and it is continuous there.*

Definition 3.54

(a) *A vector field* F *for which all maximal integral curves are defined on* $\mathbb{R}$ *is said to be* ***complete****.*

(b) *If* $F: U \longrightarrow X$ *is a vector field, then a* C^1*-function* $\varphi: U \longrightarrow \mathbb{R}$ *such that* $\varphi \circ \sigma =$ *constant for any integral curve* σ *of* F *is said to be a* ***first integral*** *of* F*.*

3.1.2 Convex Functionals and Variational Inequalities

Through this section X denotes a Banach space and $\overline{\mathbb{R}} = \mathbb{R} \cup \{+\infty\}$.

Definition 3.55
Let $\varphi: X \longrightarrow \overline{\mathbb{R}}$ *be a map. The* ***effective domain*** *of* φ *is the set*

$$\operatorname{dom} \varphi = \{u \in X : \varphi(u) < +\infty\}.$$

If $\operatorname{dom} \varphi \neq \emptyset$*, then we say that* φ *is* ***proper****. The* ***epigraph*** *of* φ *is the set*

$$\operatorname{epi} \varphi = \{(u, \lambda) \in X \times \mathbb{R} : \varphi(u) \leqslant \lambda\}.$$

Given $\lambda \in \mathbb{R}$*, we introduce*

$$\varphi^\lambda = \{u \in X : \varphi(u) \leqslant \lambda\}, \quad \varphi_\lambda = \{u \in X : \varphi(u) \geqslant \lambda\}$$

and the ***indicator function*** *of* C *defined by*

$$i_C(u) = \begin{cases} 0 & \text{if } u \in C, \\ +\infty & \text{if } u \notin C. \end{cases}$$

We say that φ *is* ***convex****, if for all* $u_1, u_2 \in \operatorname{dom} \varphi$ *and all* $\lambda \in [0, 1]$*, we have*

$$\varphi(\lambda u_1 + (1 - \lambda) u_2) \leqslant \lambda \varphi(u_1) + (1 - \lambda) \varphi(u_2).$$

We say that φ *is* ***lower semicontinuous*** *(or* ***closed****), if the set* $\varphi^\lambda \subseteq X$ *is closed for every* $\lambda \in \mathbb{R}$*.*

Remark 3.56
We can have a local definition of lower semicontinuity and say that φ is ***lower semicontinuous*** at $u_0 \in X$, if

$$\varphi(u_0) \leqslant \liminf_{u \to u_0} \varphi(u) = \sup_{\varepsilon > 0} \inf_{\|u - u_0\|_X < \varepsilon} \varphi(u).$$

Of course φ is lower semicontinuous if and only if it is lower semicontinuous at every $u_0 \in X$. We say that φ is ***upper semicontinuous*** (at $u_0 \in X$), if $-\varphi$ is lower semicontinuous (at $u_0 \in X$). Clearly, $\varphi: X \longrightarrow \overline{\mathbb{R}}$ is convex if and only if $\operatorname{epi} \varphi \subseteq X \times \mathbb{R}$

is convex. If φ is convex, then for every $\lambda \in \mathbb{R}$, the set φ^λ is convex. The converse is not true as the simple example of the function $\varphi(u) = \sqrt{|u|}$ for $u \in \mathbb{R}$ illustrates. If $C \subseteq X$ is closed (respectively convex), then i_C is lower semicontinuous (respectively convex). By $\Gamma_0(X)$ we denote the cone of proper, lower semicontinuous, and convex functions.

Proposition 3.57
If $\varphi\colon X \longrightarrow \overline{\mathbb{R}}$, then the following statements are equivalent:
(a) *φ is lower semicontinuous.*
(b) $\varphi(u_0) \leqslant \liminf\limits_{u\to u_0}$ *for all $u_0 \in X$.*
(c) $\operatorname{epi}\varphi \subseteq X \times \mathbb{R}$ *is closed.*

Given a function $\varphi\colon X \longrightarrow \overline{\mathbb{R}}$, we can easily see that the closure of the epigraph is also epigraph of a function. This leads to the following definition.

Definition 3.58
Given a function $\varphi\colon X \longrightarrow \overline{\mathbb{R}}$, the ***closure*** *of φ (respectively the* ***convex closure*** *of φ) is the function $\operatorname{cl}\varphi$ (respectively $\overline{\operatorname{conv}}\,\varphi$) defined by $\operatorname{epi}(\operatorname{cl}\varphi) = \overline{\operatorname{epi}\varphi}$ (respectively $\operatorname{epi}(\overline{\operatorname{conv}}\,\varphi) = \overline{\operatorname{conv}}\,(\operatorname{epi}\varphi)$).*

Proposition 3.59
If $\varphi\colon X \longrightarrow \overline{\mathbb{R}}$ is a map,
then $\operatorname{cl}\varphi$ is the biggest lower semicontinuous function minorizing φ and

$$\operatorname{cl}\varphi(u_0) = \sup_{\varepsilon>0} \inf_{\|u-u_0\|_X<\varepsilon} \varphi(u) = \lim_{\varepsilon\searrow 0} \inf_{\|u-u_0\|_X<\varepsilon} \varphi(u).$$

The next theorem summarizes the remarkable continuity properties that convex functions have.

Theorem 3.60
If $\varphi\colon X \longrightarrow \overline{\mathbb{R}}$ is a proper convex function,
then the following properties are equivalent:
(a) *φ is bounded above in a neighborhood of $u_0 \in X$.*
(b) *φ is continuous at $u_0 \in X$.*
(c) *φ is Lipschitz on some neighborhood of $u_0 \in X$.*
(d) $\operatorname{int}\operatorname{epi}\varphi \neq \emptyset$.
(e) $\operatorname{int}\operatorname{dom}\varphi \neq \emptyset$ *and $\varphi|_{\operatorname{int}\operatorname{dom}\varphi}$ is locally Lipschitz.*
Moreover, if one (so all) of the above properties hold, then

$$\begin{aligned} \operatorname{int}\operatorname{epi}\varphi &= \{(u,\lambda) \in X \times \mathbb{R} : u \in \operatorname{int}\operatorname{dom}\varphi, \varphi(u) < \lambda\}, \\ \operatorname{int}\operatorname{dom}\varphi &= \{u \in X : \textit{there exists } \lambda \in \mathbb{R} \textit{ such that } (u,\lambda) \in \operatorname{int}\operatorname{epi}\varphi\}. \end{aligned}$$

Definition 3.61
Given a nonempty set $C \subseteq X$, we define the **core** *of C to be the set*

$$\begin{aligned} \operatorname{core} C \;=\; \big\{ u \in X : \ & \text{for every } h \in X \text{ with } \|h\|_X = 1, \\ & \text{there exists } t_0 = t_0(h) > 0, \\ & \text{such that } u + th \in C \text{ for all } t \in [0, t_0] \big\}. \end{aligned}$$

Remark 3.62
The core of C is in fact the set of absorbing points of C (see Definition 2.102). We have that $\operatorname{int} C \subseteq \operatorname{core} C$ and the inclusion can be strict. In fact we can have $\operatorname{int} C = \emptyset$, $\operatorname{core} C = C$. The core of C is also called the ***algebraic interior*** of C. If $C \subseteq X$ is nonempty, closed, and convex, then $0 \in \operatorname{core} C$ if and only if $0 \in \operatorname{int} C$ and if and only if C is absorbing (i.e., $\bigcup\limits_{n \geqslant 1} nC = X$). If X is finite dimensional and $C \subseteq X$ is nonempty and convex, then $\operatorname{core} C = \operatorname{int} C$.

Proposition 3.63
If $\varphi \colon X \longrightarrow \overline{\mathbb{R}}$ is proper, lower semicontinuous, and convex,
then the following properties are equivalent:
(a) φ is continuous at $u_0 \in X$.
(b) $u_0 \in \operatorname{int} \operatorname{dom} \varphi$.
(c) $u_0 \in \operatorname{core} \operatorname{dom} \varphi$.
In particular φ is continuous (in fact locally Lipschitz) on $\operatorname{core} \operatorname{dom} \varphi$.

In finite dimensions the situation is simpler.

Proposition 3.64
If X is finite dimensional and $\varphi \colon X \longrightarrow \overline{\mathbb{R}}$ is convex,
then $\varphi|_{\operatorname{core} \operatorname{dom} \varphi}$ is continuous (in fact locally Lipschitz).
Moreover, if $U \subseteq \mathbb{R}^N$ is an open convex set and $\{ \varphi_n \colon U \longrightarrow \mathbb{R} \}_{n \geqslant 1}$ is a sequence of convex functions converging pointwise to φ, then φ is also convex, the convergence is uniform on every closed bounded subset of U and the sequence $\{ \varphi_n \}_{n \geqslant 1}$ is equi-Lipschitz there.

Definition 3.65
Let $\varphi \colon X \longrightarrow \overline{\mathbb{R}}$, $u \in \operatorname{dom} \varphi$ and $h \in X$. The ***directional derivative*** *of φ at u in the direction h is defined by*

$$\varphi'(u; h) \;=\; \lim_{\lambda \searrow 0} \tfrac{\varphi(u + \lambda h) - \varphi(u)}{\lambda}$$

when this limit exists.

If φ is convex and $u \in \operatorname{core} \operatorname{dom} \varphi$, then the map $\lambda \longmapsto \frac{\varphi(u+\lambda h) - \varphi(u)}{\lambda}$ is nondecreasing on $(0, +\infty)$. So, we have

Proposition 3.66
If $\varphi \colon X \longrightarrow \overline{\mathbb{R}}$ *is proper, convex and* $u \in \operatorname{core} \operatorname{dom} \varphi$,
then the directional derivative $\varphi'(u;\cdot)$ *is everywhere finite and sublinear (i.e., is positively homogeneous and subadditive). Moreover, if* u *is a point of continuity for* φ, *then* $\varphi'(u;\cdot)$ *is Lipschitz continuous.*

Remark 3.67
Evidently, if $\varphi'(u;\cdot) \in X^*$, then $\varphi'(u;\cdot) = \varphi'_G(u)(\cdot)$ (the Gâteaux derivative of φ at u).

Since a point $u \in X$ of Gâteaux differentiability of φ belongs in the $\operatorname{core} \operatorname{dom} \varphi$, invoking Proposition 3.63, we obtain the following result.

Proposition 3.68
If $\varphi \colon X \longrightarrow \overline{\mathbb{R}}$ *is a proper, lower semicontinuous, and convex map which is Gâteaux differentiable at* $u_0 \in X$,
then φ *is continuous at* $u_0 \in X$.

Remark 3.69
We recall that in general Gâteaux differentiability at a point, does not imply continuity at this point.

Definition 3.70
Let $\varphi \colon X \longrightarrow \overline{\mathbb{R}}$ *be a proper function and* $u_0 \in \operatorname{dom} \varphi$. *The* ***subdifferential*** *of* φ *at* $u_0 \in X$ *is the set* $\partial\varphi(u_0) \subseteq X^*$ *(possibly empty) defined by*

$$\partial\varphi(u_0) = \{u^* \in X^* : \langle u^*, u - u_0\rangle \leqslant \varphi(u) - \varphi(u_0) \text{ for all } u \in X\}.$$

If $u_0 \notin \operatorname{dom} \varphi$, *then we set* $\partial\varphi(u_0) = \emptyset$. *The elements of the set* $\partial\varphi(u_0)$ *are called* ***subgradients*** *of* φ *at* u_0.

The notion of subdifferential is particulary useful in the context of convex functions.

Theorem 3.71
If $\varphi \colon X \longrightarrow \overline{\mathbb{R}}$ *is convex and* u *is a point of continuity of* φ,
then $\partial\varphi(u) \subseteq X^*$ *is nonempty,* w^**-compact, convex and*

$$\varphi'(u;h) = \max_{u^* \in \partial\varphi(u)} \langle u^*, h\rangle = \sigma_{\partial\varphi(u)}(h)$$

(see Definition 2.46).

Remark 3.72
When φ is also lower semicontinuous (that is, $\varphi \in \Gamma_0(X)$), the condition that u is a continuity point of φ, is equivalent to saying that $u \in \operatorname{core} \operatorname{dom} \varphi$ (see Proposition 3.63).

It is natural to ask what is the precise relation between the subdifferential of φ and the Gâteaux differential of φ when the latter exists. Thanks to Proposition 3.68 we have the following result.

Theorem 3.73
If $\varphi\colon X \longrightarrow \overline{\mathbb{R}}$ is a proper and convex map and $u \in X$ is a point of continuity of φ, then φ is Gâteaux differentiable at u if and only if $\partial\varphi(u)$ is a singleton.

When φ is convex and continuous at $u_0 \in X$, then we can describe the differentiability of φ at u_0 without mentioning the derivative.

Proposition 3.74
If $\varphi\colon X \longrightarrow \overline{\mathbb{R}}$ is a proper and convex map and $u_0 \in X$ is a point of continuity of φ, then the following properties are equivalent:
(a) *φ is Fréchet differentiable at u_0.*
(b) *For every $\varepsilon > 0$, there exists $\delta > 0$ such that*

$$\varphi(u_0 + h) + \varphi(u_0 - h) - 2\varphi(u_0) \leqslant \varepsilon\|h\|_X \qquad \forall\ \|h\|_X \leqslant \delta.$$

(c) *For every $\varepsilon > 0$, there exists $\delta > 0$ such that*

$$\varphi(u_0 + h) + \varphi(u_0 - h) - 2\varphi(u_0) \leqslant \varepsilon\delta \qquad \forall\ \|h\|_X \leqslant \delta.$$

There is an analogous result for the Gâteaux derivative.

Proposition 3.75
If $\varphi\colon X \longrightarrow \overline{\mathbb{R}}$ is a proper and convex map and $u_0 \in X$ is a point of continuity of φ, then φ is Gâteaux differentiable at u_0 if and only if for every $\varepsilon > 0$ and every $h \in X$ with $\|h\|_X = 1$, there exists $\delta = \delta(\varepsilon, h) > 0$ such that

$$\varphi(u_0 + th) + \varphi(u_0 - th) - 2\varphi(u_0) \leqslant \varepsilon t \qquad \forall\ t \in [0, \delta].$$

These two propositions have interesting consequences concerning the differentiability properties of convex functions. The first theorem is about the Gâteaux differentiability of φ.

Theorem 3.76 *(Mazur Theorem)*
If X is a separable Banach space, $U \subseteq X$ is an open and convex set and $\varphi\colon U \longrightarrow \mathbb{R}$ is a continuous and convex map,
then φ is Gâteaux differentiable on a dense G_δ-subset of U.

For the Fréchet differentiability of φ, there is the following theorem.

Theorem 3.77
If X^* is separable, $U \subseteq X$ is an open and convex set and $\varphi\colon U \longrightarrow \mathbb{R}$ is a continuous and convex map,
then φ is Fréchet differentiable on a dense G_δ-subset of U.

Remark 3.78
In general, for any Banach space X, for an open, convex set $U \subseteq X$ and for a continuous, convex map $\varphi \colon U \longrightarrow \mathbb{R}$, the set of points of Fréchet differentiability of φ is a (possibly empty) G_δ-subset of U.

Theorems 3.76 and 3.77 lead to the following definition.

Definition 3.79
A Banach space X is said to be an ***Asplund space*** *(respectively* ***weak Asplund****), if every continuous convex function defined on a nonempty, open, and convex set $U \subseteq X$ is Fréchet (Gâteaux) differentiable at every point of a dense G_δ-subset of U.*

Remark 3.80
Separable Banach spaces are weak Asplund spaces (see Theorem 3.76). Banach spaces with separable dual are Asplund spaces (see Theorem 3.77).

The existence of all partial derivatives does not necessarily imply differentiability of the function. This changes for convex functions.

Proposition 3.81
If $U \subseteq \mathbb{R}^N$ is an open set, $\varphi \colon U \longrightarrow \mathbb{R}$ is a convex map and all partial derivatives of φ at $u_0 \in U$ exist,
then φ is Fréchet differentiable at u_0.

The next proposition is a nonsmooth generalization of the classical Fermat rule for minimization and illustrates the importance of the notion of convex subdifferential in optimization.

Proposition 3.82
If $\varphi \colon X \longrightarrow \overline{\mathbb{R}}$ is a proper and convex map,
then $u_0 \in X$ is a global minimizer of φ if and only if $0 \in \partial\varphi(u_0)$.

Duality is a central theme in the theory of convex functions and in that direction the next notion is crucial.

Definition 3.83
Given a proper function $\varphi \colon X \longrightarrow \overline{\mathbb{R}}$, the ***conjugate*** *of φ is the function $\varphi^* \colon X^* \longrightarrow \overline{\mathbb{R}}$ defined by*

$$\varphi^*(u^*) = \sup_{u \in X} \left(\langle u^*, u \rangle - \varphi(u) \right).$$

We see that φ^* is proper, lower semicontinuous, and convex (that is, $\varphi^* \in \Gamma_0(X^*)$) being the supremum of affine continuous functions. We can define the conjugate of φ^*. This is the ***second conjugate*** *of φ, it is denoted by φ^{**} and it is defined on X^{**}.*

Proposition 3.84
If $\varphi\colon X \longrightarrow \overline{\mathbb{R}}$ *is a proper function, then*
(a) $\langle u^*, u\rangle \leqslant \varphi(u) + \varphi^*(u^*)$ *for all* $u \in X$, $u^* \in X^*$ ***(Young–Fenchel inequality)***;
(b) $\langle u^*, u\rangle = \varphi(u) + \varphi^*(u^*)$ *if and only if* $u^* \in \partial\varphi(u)$;
(c) *if* $\varphi \leqslant \psi$, *then* $\psi^* \leqslant \varphi^*$;
(d) $\varphi^{**}|_X \leqslant \varphi$;
(e) $\varphi^* = (\operatorname{cl}\varphi)^* = (\overline{\operatorname{conv}}\,\varphi)^*$;
(f) $\varphi^{**}|_X = \overline{\operatorname{conv}}\,\varphi$.

These observations lead to the following fundamental result.

Theorem 3.85
If $\varphi\colon X \longrightarrow \overline{\mathbb{R}}$ *is a proper function,*
then $\varphi^{**}|_X = \varphi$ *if and only if* $\varphi \in \Gamma_0(X)$.

Remark 3.86
The above theorem says that for a proper function $\varphi\colon X \longrightarrow \overline{\mathbb{R}}$ we have that $\varphi \in \Gamma_0(X)$ if and only if it is the supremum of a family of affine continuous functions.

Definition 3.87
Let $\varphi, \psi\colon X \longrightarrow \overline{\mathbb{R}}$ *be two proper functions. The* ***infimal convolution*** *of* φ *and* ψ *is defined by*

$$(\varphi \boxdot \psi)(u) \;=\; \inf_{v\in X}\left\{\varphi(v) + \psi(u-v)\right\}.$$

Remark 3.88
Geometrically, the infimal convolution of φ and ψ is the biggest extended real valued function whose epigraph contains $\operatorname{epi}\varphi + \operatorname{epi}\psi$. If φ and ψ are convex, then so is $\varphi \boxdot \psi$, but if φ and ψ are lower semicontinuous, then $\varphi \boxdot \psi$ need not be lower semicontinuous. Evidently, if $C \subseteq X$ is closed and convex, then for the distance function $\operatorname{dist}(\cdot, C)$ we have $\operatorname{dist}(\cdot, C) = \|\cdot\|_X \boxdot i_C$.

Let us mention some useful conjugation rules:

- $\varphi(\lambda u) \longrightarrow \varphi^*(\frac{u^*}{\lambda})$ for $\lambda \neq 0$;
- $\varphi(u+h) \longrightarrow \varphi^*(u^*) - \langle u^*, h\rangle$;
- $\lambda\varphi(u) \longrightarrow \lambda\varphi^*(\frac{u^*}{\lambda})$ for $\lambda > 0$;
- $\varphi^*(0) = -\inf\limits_{u\in X} \varphi(u)$;
- $\big(\inf\limits_{i\in I}\varphi_i\big)^* = \sup\limits_{i\in I}\varphi_i^*$ and $\big(\sup\limits_{i\in I}\varphi_i\big)^* \leqslant \inf\limits_{i\in I}\varphi_i^*$;
- $(\varphi + c)^* = \varphi^* - c$ for all $c \in \mathbb{R}$.

Proposition 3.89
(a) *If* $\varphi, \psi \colon X \longrightarrow \overline{\mathbb{R}}$ *are two proper functions, then* $(\varphi \boxdot \psi)^* = \varphi^* + \psi^*$.
(b) *If* $\varphi, \psi \colon X \longrightarrow \overline{\mathbb{R}}$ *are two proper and convex functions and*

$$\operatorname{dom} \varphi \cap \operatorname{int} \operatorname{dom} \psi \neq \emptyset,$$

then $(\varphi + \psi)^* = \varphi^* \boxdot \psi^*$.

The next result is important in convex optimization.

Theorem 3.90 *(Fenchel Duality Theorem)*
If $\varphi \colon X \longrightarrow \overline{\mathbb{R}}$ *and* $\psi \colon Y \longrightarrow \overline{\mathbb{R}}$ *are two proper convex functions,* $A \in \mathcal{L}(X;Y)$ *and*

$$p = \inf_{u \in X} (\varphi(u) + \psi(A(u))) \tag{3.4}$$

(primal value) and

$$d = \sup_{y^* \in Y^*} (-\varphi^*(A^*(y^*)) - \psi^*(-y^*)) \tag{3.5}$$

(dual value),
then $d \leqslant p$ *(weak duality) and if*

$$A(\operatorname{dom} \varphi) \cap \operatorname{int} \operatorname{dom} \psi \neq \emptyset$$

or alternatively, if $\varphi \in \Gamma_0(X)$, $\psi \in \Gamma_0(Y)$ *and*

$$0 \in \operatorname{core} (\operatorname{dom} \psi - A(\operatorname{dom} \varphi)),$$

then $p = d$ *(strong duality) and the supremum in (3.5) is realized if finite.*

Remark 3.91
In the above setting, (3.4) is called ***primal problem*** and (3.5) is the ***dual problem***.

Next we state some fundamental properties of the subdifferential map and derive the main rules of the subdifferential calculus.

Theorem 3.92
If $\varphi \in \Gamma_0(X)$, *then* $\partial\varphi \colon X \supseteq D(\partial\varphi) \longrightarrow 2^{X^*}$ *is maximal monotone.*

However, not every maximal monotone operator from X into X^* arises as the subdifferential of a convex function. So, we want to isolate that property which characterizes subdifferential maps among all maximal monotone operators. This leads to the following definition.

Definition 3.93
A map $A\colon X \supseteq D(A) \longrightarrow 2^{X^*}$ is said to be ***cyclically monotone****, if for all $n \geqslant 1$ we have*

$$0 \leqslant \sum_{k=1}^{n} \langle u_k^*, u_k - u_{k+1} \rangle \qquad \forall\ (u_1, u_1^*), \ldots, (u_n, u_n^*) \in \operatorname{Gr} A,\ u_{n+1} = u_1.$$

We say that A is ***maximal cyclically monotone****, if it is cyclically monotone and there is no cyclically monotone map $A'\colon X \supseteq D(A') \longrightarrow 2^{X^*}$ such that $\operatorname{Gr} A \subsetneq \operatorname{Gr} A'$*

Theorem 3.94
<u>If</u> $A\colon X \supseteq D(A) \longrightarrow 2^{X^*}$ is a map,
<u>then</u> $A = \partial\varphi$ for some $\varphi \in \Gamma_0(X)$ if and only if A is maximal cyclically monotone. Moreover, A determines φ up to an additive constant.

Definition 3.95
A proper and convex function $\varphi\colon X \longrightarrow \overline{\mathbb{R}}$ is said to be ***strictly convex****, if for all $u, v \in \operatorname{dom}\varphi$ with $u \neq v$ and all $\lambda \in (0,1)$, the following inequality holds*

$$\varphi(\lambda u + (1-\lambda)v) \ < \ \lambda\varphi(u) + (1-\lambda)\varphi(v).$$

Remark 3.96
For a proper, strictly convex function $\varphi\colon X \longrightarrow \overline{\mathbb{R}}$, the subdifferential map $\partial\varphi$ is strictly monotone.

Proposition 3.97
<u>If</u> $\varphi\colon X \longrightarrow \overline{\mathbb{R}}$ and $\psi\colon Y \longrightarrow \overline{\mathbb{R}}$ are proper and convex maps and $A \in \mathcal{L}(X;Y)$,
<u>then</u> for every $u \in X$, we have

$$\partial\varphi + A^*(\partial\psi(A(u))) \ \subseteq \ \partial(\varphi + \psi \circ A)(u)$$

and equality holds if

$$A(\operatorname{dom}\varphi) \cap \operatorname{int}\operatorname{dom}\psi \ \neq \ \emptyset$$

or alternatively if $\varphi \in \Gamma_0(X)$, $\psi \in \Gamma_0(Y)$ and

$$0 \ \in \ \operatorname{core}\left(\operatorname{dom}\psi - A(\operatorname{dom}\varphi)\right).$$

Proposition 3.98
<u>If</u> $A \in \mathcal{L}(X;Y)$ and $\varphi\colon Y \longrightarrow \overline{\mathbb{R}}$ is a proper and convex function,
<u>then</u> $A^*\partial\varphi(A(u)) \subseteq \partial(\varphi \circ A)(u)$ for all $u \in X$ and equality holds if

$$R(A) \cap \operatorname{int}\operatorname{dom}\varphi \ \neq \ \emptyset.$$

Definition 3.99
Let $\varphi\colon X \longrightarrow \overline{\mathbb{R}}$ be a proper function, $u_0 \in \operatorname{dom}\varphi$ and $\varepsilon > 0$. The ε-subdifferential of φ at u_0 is the set $\partial_\varepsilon\varphi(u_0) \subseteq X^*$ defined by

$$\partial_\varepsilon\varphi(u_0) \ = \{u^* \in X^* :\ \langle u^*, u - u_0 \rangle \leqslant \varphi(u) - \varphi(u_0) + \varepsilon \text{ for all } u \in X\}.$$

Remark 3.100
This innocent looking perturbation of the exact subdifferential (see Definition 3.70) leads to some remarkable properties, that distinguish $\partial_\varepsilon\varphi$ from the exact subdifferential $\partial\varphi$. For $\varepsilon > 0$, $\partial_\varepsilon\varphi(u)$ is a global notion (in contrast to $\partial\varphi(u)$ which is local), that is, $\partial_\varepsilon\varphi(u)$ is sensitive to variations of φ far from $u \in X$.

The next proposition establishes the main difference between $\partial_\varepsilon\varphi(u)$ and $\partial\varphi(u)$. The approximate subdifferential is always nonempty.

Proposition 3.101
If $\varphi \in \Gamma_0(X)$, $u_0 \in \operatorname{dom}\varphi$ *and* $\varepsilon > 0$,
then $\partial_\varepsilon\varphi(u_0) \subseteq X^*$ *is nonempty,* w^**-closed, convex, and*

$$\varphi'(u_0;h) = \lim_{\varepsilon\searrow 0} \sup_{u^*\in\partial_\varepsilon\varphi(u_0)} \langle u^*, h\rangle \qquad \forall\, h \in X.$$

Theorem 3.102
If $\varphi \in \Gamma_0(X)$, $u_0 \in \operatorname{dom}\varphi$, $\varepsilon, \lambda > 0$ *and* $u_0^* \in \partial_\varepsilon\varphi(u_0)$,
then there exist $u_\varepsilon \in \operatorname{dom}\varphi$ *and* $u_\varepsilon^* \in X^*$ *such that*

$$u_\varepsilon^* \in \partial\varphi(u_\varepsilon), \quad \|u_\varepsilon - u_0\|_X \leqslant \tfrac{\varepsilon}{\lambda} \quad \text{and} \quad \|u_\varepsilon^* - u_0^*\|_{X^*} \leqslant \sqrt{\varepsilon}.$$

The next theorem is a useful test for differentiability.

Theorem 3.103 *(Smulian Theorem)*
If $\varphi\colon X \longrightarrow \overline{\mathbb{R}}$ *is a proper and convex function and* $u_0 \in X$ *is a continuity point for* φ,
then the following properties are equivalent:

(a) φ *is Fréchet differentiable at* u_0.

(b) *If* $\{u_n\}_{n\geqslant 1} \subseteq X$ *is a sequence such* $u_n \longrightarrow u_0$ *in* X *and* $u_0^* \in \partial\varphi(u_0)$, *then there exists* $n_0 \geqslant 1$ *and a sequence* $\{u_n^*\}_{n\geqslant 0} \subseteq X^*$ *such that* $u_n^* \in \partial\varphi(u_n)$ *and* $u_n^* \longrightarrow u_0^*$ *in* X^*.

(c) *If* $\{u_n\}_{n\geqslant 1} \subseteq X$ *and* $\{u_n^*\}_{n\geqslant 1} \subseteq X^*$ *are sequences such that* $u_n \longrightarrow u_0$ *in* X, $u_n^* \in \partial\varphi(u_n)$ *for* $n \geqslant 1$ *and* $u_0^* \in \partial\varphi(u_0)$, *then* $u_n^* \longrightarrow u_0^*$ *in* X^*.

(d) *If* $\{u_n\}_{n\geqslant 1}, \{v_n\}_{n\geqslant 1} \subseteq X$, $\{u_n^*\}_{n\geqslant 1}, \{v_n^*\}_{n\geqslant 1} \subseteq X^*$ *and* $\{\varepsilon_n\}_{n\geqslant 1} \subseteq \mathbb{R}$ *are sequences such that* $u_n \longrightarrow u_0$ *and* $v_n \longrightarrow v_0$ *in* X, $\varepsilon_n \searrow 0$, $u_n^* \in \partial_{\varepsilon_n}(u_n)$ *and* $v_n^* \in \partial_{\varepsilon_n}(v_n)$ *for* $n \geqslant 1$, *then* $u_n^* - v_n^* \longrightarrow 0$ *in* X^*.

(e) *If* $\{\varepsilon_n\}_{n\geqslant 1} \subseteq \mathbb{R}$ *and* $\{u_n^*\}_{n\geqslant 1} \subseteq X^*$ *are sequences such that* $\varepsilon_n \searrow 0$, $u_n^* \in \partial_{\varepsilon_n}\varphi(u_0)$ *for* $n \geqslant 1$ *and* $u_0^* \in \partial\varphi(u_0)$, *then* $u_n^* \longrightarrow u_0^*$ *in* X^*.

Remark 3.104
The same result is true for Gâteaux differentiability at u_0, provided all the strong convergences in X^* are replaced by w^*-convergence.

In analogy to Definition 2.98(e) and (f), we provide the following definitions for proper functions.

Definition 3.105
Let $\varphi\colon X \longrightarrow \overline{\mathbb{R}}$ be a proper function.
(a) We say that φ is **coercive**, *if* $\lim\limits_{\|u\|_X \to +\infty} \varphi(u) = +\infty.$
(b) We say that φ is **strongly coercive**, *if* $\lim\limits_{\|u\|_X \to +\infty} \frac{\varphi(u)}{\|u\|_X} = +\infty.$

Proposition 3.106
If $U \subseteq \mathbb{R}^N$ is a nonempty open set and $\varphi\colon U \longrightarrow \mathbb{R}$ is a convex function, then φ is differentiable almost everywhere on U with respect to the Lebesgue measure on $\mathbb{R}^N$.

In infinite dimensional Banach spaces, the situation is more involved since there is no analog of the Lebesgue measure on the infinite dimensional Banach spaces. Nevertheless, classes of null sets can be defined as follows.

Definition 3.107
Let X be a Banach space and let $D \in \mathcal{B}(X)$ (with $\mathcal{B}(X)$ being the Borel σ-field of X). We say that D is **Haar null set**, *if there exists a not necessarily unique Radon probability measure m on X such that*

$$m(u + D) = 0 \qquad \forall\, u \in X.$$

More generally, an arbitrary set $D \subseteq X$ is **Haar null set**, *if it is contained in a Haar null Borel set. The measure m is a* **test measure** *for D.*

Theorem 3.108
If X is a Banach space with separable dual and $\varphi\colon X \longrightarrow \mathbb{R}$ is a continuous and convex map,
then φ is Fréchet differentiable outside a Haar null set.

Now we turn our attention to integral functionals. So, in what follows let (Ω, Σ, μ) be a σ-finite measure space and let X be a separable Banach space. We consider integral functionals of the form

$$I_\varphi(u) = \int_\Omega \varphi(\omega, u(\omega))\, d\mu,$$

where $\varphi\colon \Omega \times X \longrightarrow \overline{\mathbb{R}}$ and $u\colon \Omega \longrightarrow X$ belongs in some vector space of functions.

Definition 3.109
We say that $\varphi\colon \Omega \times X \longrightarrow \overline{\mathbb{R}}$ is a **normal integrand**, *if*
(a) φ is $\Sigma \times \mathcal{B}(X)$-measurable; and
(b) for μ-almost all $\omega \in \Omega$, the map $x \longmapsto \varphi(\omega, x)$ is proper and lower semicontinuous.
We say that φ is a **convex normal integrand**, *if it is normal integrand and for μ-almost all $\omega \in \Omega$, the map $x \longmapsto \varphi(\omega, x)$ is convex.*

Theorem 3.110

(a) *If φ is a normal integrand and $I_\varphi\colon L^1(\Omega;X) \longrightarrow [-\infty,+\infty]$ is finite at least at one point $u_0 \in L^1(\Omega;X)$,*
then $(I_\varphi)^* = I_{\varphi^*}$.

(b) *If φ is a convex normal integrand, I_φ is finite at $u_0 \in L^1(\Omega;X)$ and I_{φ^*} is finite at $u_0^* \in L^\infty(\Omega;X^*_{w^*})$,*
then both I_φ and I_{φ^*} are proper and lower semicontinuous functionals which are conjugate to each other.

Theorem 3.111

If φ is a convex normal integrand, $1 \leqslant p < +\infty$, $\frac{1}{p}+\frac{1}{p'}=1$, $I_\varphi\colon L^p(\Omega;X) \longrightarrow \overline{\mathbb{R}}$ is finite at least at one point $u_0 \in L^p(\Omega;X)$,
then for every $u \in L^p(\Omega;X)$, we have

$$u^* \in \partial I_\varphi(u) \subseteq L^{p'}(\Omega;X^*_{w^*}) \iff [u^*(\omega) \in \partial\varphi(\omega,u(\omega)) \text{ for } \mu\text{-a.a. } \omega \in \Omega].$$

Definition 3.112

*Let X be a Banach space and let $C \subseteq X$ be a nonempty and convex set with $u \in \overline{C}$. The **tangent cone** to C at u is the set*

$$T_C(u) = \overline{\bigcup_{\lambda>0} \tfrac{1}{\lambda}(C-u)}.$$

*The **normal cone** to C at u is the set*

$$\begin{aligned} N_C(u) &= \{u^* \in X^* : \langle u^*, u\rangle = \sigma_C(u^*)\} \\ &= \{u^* \in X^* : \langle u^*, c-u\rangle \leqslant 0 \text{ for all } c \in C\}. \end{aligned}$$

Remark 3.113

Clearly $T_C(u) = T_{\overline{C}}(u)$. Also note that $h \in T_C(u)$ if and only if there exist sequences $\{\lambda_n\}_{n\geqslant 1} \subseteq (0,+\infty)$ and $\{h_n\}_{n\geqslant 1} \subseteq X$ such that $\lambda_n \searrow 0$ in $\mathbb{R}$, $h_n \longrightarrow h$ in X and $u+\lambda_n h_n \in \overline{C}$ for all $n \geqslant 1$. Note that $C \subseteq u + T_C(u)$. For the normal cone we see that

$$N_c(u) = T_C(u)^- = \{u^* \in X^* : \sigma_C(u^*) \leqslant 0\}$$

(the ***negative polar cone*** of $T_C(u)$). Recall that

$$\sigma_C(u^*) = \sup_{c\in C} \langle u^*, c\rangle$$

(the support function of C; see Definition 2.46). If

$$i_C(u) = \begin{cases} 0 & \text{if } u \in C, \\ +\infty & \text{if } u \notin C \end{cases}$$

(the ***indicator function*** of C), then

$$N_C(u) = \partial i_C(u).$$

Note that, if $C \subseteq X$ is a nonempty, closed, and convex set, then $i_C \in \Gamma_0(X)$.

Theorem 3.114
If H is a Hilbert space, $C \subseteq H$ is a nonempty, closed, and convex set, $a: H \times H \longrightarrow \mathbb{R}$ is a bilinear form which is continuous, i.e.,

$$|a(u,v)| \leqslant c_1\|u\|_H\|v\|_H \qquad \forall\, u,v \in H,$$

for some $c_1 > 0$, and coercive, i.e.,

$$a(u,u) \geqslant c_2\|u\|_H^2 \qquad \forall\, u \in H,$$

for some $c_2 > 0$,
then for every $h^* \in H^*$, there exists unique $u_0 \in H$ such that

$$u_0 \in C \quad \text{and} \quad a(u_0, v-u_0) \geqslant \langle h^*, v-u_0\rangle \qquad \forall\, v \in C. \tag{3.6}$$

Remark 3.115
A problem of type (3.6) is said to be a ***variational inequality***. If $C \subseteq H$ is a vector subspace, then by choosing $v = u_0 \pm h$ with $h \in C$, we obtain

$$a(u_0,h) \geqslant \langle h^*,h\rangle \quad \text{and} \quad a(u_0,-h) \geqslant \langle h^*,-h\rangle \qquad \forall\, h \in C,$$

so

$$a(u_0,h) = \langle h^*,h\rangle \qquad \forall\, h \in C.$$

This particular case of Theorem 3.114 is known as the ***Lax–Milgram Theorem***.

Proposition 3.116
Problem (3.6) (the variational inequality) is equivalent to the following problem:

$$\begin{cases} \text{Find } u_0 \in C, \text{ such that} \\ a(v, v-u_0) \geqslant \langle h^*, v-u_0\rangle \qquad \forall\, v \in C. \end{cases}$$

Let us state also an existence theorem for nonlinear variational inequalities.

Theorem 3.117
If X is a reflexive Banach space, $C \subseteq X$ is a nonempty, closed, and convex set, $A: C \longrightarrow X^*$ is a bounded (i.e., maps bounded sets to bounded sets), demicontinuous and pseudomonotone map (see Definitions 2.102, 2.110, and 2.129) and if C is also unbounded, then there exists $\widehat{u} \in C$ such that

$$\frac{\langle A(u), u-\widehat{u}\rangle}{\|u\|_X} \longrightarrow +\infty \quad \text{as } \|u\|_X \to +\infty,\ u \in C,$$

then given $h^* \in X^*$, there exists $u_0 \in C$ such that

$$\langle A(u_0) - h^*, v-u_0\rangle \geqslant 0 \qquad \forall\, v \in C. \tag{3.7}$$

Moreover, if A is monotone, then the set of $u_0 \in C$ satisfying (3.7) is closed, convex and it is also a singleton if A is strictly monotone.

3.1.3 Locally Lipschitz Functions

Throughout this section, X is a real Banach space with norm $\|\cdot\|_X$. By X^* we denote its dual and by $\langle\cdot,\cdot\rangle$ the duality brackets for the pair (X^*, X).

Definition 3.118
Let Y be a Banach space and let $\varphi\colon X \longrightarrow Y$ be a map. We say that φ is ***locally Lipschitz****, if for every $u \in X$, we can find a neighborhood U of u and $k > 0$ such that*

$$\|\varphi(v) - \varphi(h)\|_Y \leqslant k\|v - h\|_X \qquad \forall\, v, h \in U.$$

Remark 3.119
In most cases we will have $Y = \mathbb{R}$. If $\varphi\colon X \longrightarrow \mathbb{R}$ is a continuous and convex map, then from Proposition 3.63, we know that φ is locally Lipschitz.

The next result is an extension of Proposition 3.106 to the broader class of locally Lipschitz functions.

Theorem 3.120 *(Rademacher Theorem)*
If $\varphi\colon \mathbb{R}^N \longrightarrow \mathbb{R}^k$ is a locally Lipschitz map,
then φ is Fréchet differentiable almost everywhere.

There is also an extension of Theorem 3.108.

Theorem 3.121
If X is a separable Banach space, Y is a Banach space with the RNP (see Definition 1.45) and $\varphi\colon X \longrightarrow Y$ is a locally Lipschitz map,
then there exists a universally measurably set $D \subseteq X$ (see Definition I.4.47) such that $X \setminus D$ is Haar null and $\varphi|_D$ is Gâteaux differentiable.

In the sequel we focus exclusively on $\mathbb{R}$-valued locally Lipschitz functions (that is, $Y = \mathbb{R}$). For such functions, we can develop a coherent and complete subdifferential theory, which extends the one for continuous convex functions. The starting point is the following definition.

Definition 3.122
Given a locally Lipschitz function $\varphi\colon X \longrightarrow \mathbb{R}$, the ***generalized directional derivative*** *of φ at $u \in X$ in the direction $h \in X$ is given by*

$$\varphi^0(u;h) = \limsup_{\substack{v \to u \\ \lambda \searrow 0}} \tfrac{\varphi(v+\lambda h)-\varphi(v)}{\lambda} \in \mathbb{R}.$$

Proposition 3.123
If $\varphi\colon X \longrightarrow \mathbb{R}$ is a locally Lipschitz function,
then
(a) *$\varphi^0(u;\cdot)$ is sublinear (that is, subadditive and positively homogeneous) and Lipschitz continuous;*

(b) *$\varphi^0(u;-h) = (-\varphi)^0(u;h)$;*
(c) *the map $(u,h) \longmapsto \varphi^0(u;h)$ is upper semicontinuous.*

Definition 3.124
A locally Lipschitz function $\varphi\colon X \longrightarrow \mathbb{R}$ is said to be ***regular*** *at $u \in X$, if*
(a) *the directional derivative $\varphi'(u;h)$ (see Definition 3.65) exists for all $h \in X$;*
(b) *$\varphi^0(u;h) = \varphi'(u;h)$ for all $h \in X$.*

Remark 3.125
A continuous convex function $\varphi\colon X \longrightarrow \mathbb{R}$ is regular. Also, we say that $\varphi\colon X \longrightarrow \mathbb{R}$ is ***strictly differentiable***, if for all $u \in X$, there exists $\varphi'(u) \in X^*$ such that

$$\lim_{\substack{v \to u \\ \lambda \searrow 0}} \tfrac{\varphi(v+\lambda h)-\varphi(v)}{\lambda} = \langle \varphi'(u), h \rangle \qquad \forall\, h \in X$$

and the convergence is uniform for $h \in X$ in compact sets. If φ is Lipschitz continuous near zero, then this is automatically true. A strictly differentiable function is regular. A function $\varphi \in C^1(X)$ is strictly differentiable and locally Lipschitz.

Definition 3.126
Given a locally Lipschitz function $\varphi\colon X \longrightarrow \mathbb{R}$, the ***generalized*** *(or* ***Clarke****)* ***subdifferential*** *of φ at $u \in X$, is the set $\partial\varphi(u) \subseteq X^*$ defined by*

$$\partial\varphi(u) = \big\{u^* \in X^* : \langle u^*, h\rangle \leqslant \varphi^0(u;h) \text{ for all } h \in X\big\}.$$

When X is finite dimensional, then using the Rademacher theorem (see Theorem 3.120) we can have an alternative description of the generalized subdifferential, which is more intuitive and geometric.

Proposition 3.127
If $\varphi\colon \mathbb{R}^N \longrightarrow \mathbb{R}$ is locally Lipschitz and $E \subseteq \mathbb{R}^N$ is any Lebesgue-null set, then for every $u \in \mathbb{R}^N$, we have

$$\partial\varphi(u) = \operatorname{conv}\Big\{\lim_{n\to+\infty} \nabla\varphi(u_n) : \{u_n\}_{n\geqslant 1} \subseteq X,\ u_n \longrightarrow u,\ u_n \notin E \cup (X \setminus D_\varphi)\Big\},$$

where D_φ is the set of points of Fréchet differentiability of φ (from Theorem 3.120 we know that $\lambda^N(X \setminus D_\varphi) = 0$, with λ^N being the Lebesgue measure on $\mathbb{R}^N$). Also, we have

$$\varphi^0(u;h) = \limsup_{\substack{v \to u \\ v \notin (E \cup (X \setminus D_\varphi))}} \langle \nabla\varphi(v), h\rangle_{\mathbb{R}^N}.$$

The next proposition summarizes the main properties of the generalized subdifferential.

Proposition 3.128
If $\varphi\colon X \longrightarrow \mathbb{R}$ is a locally Lipschitz function, then
(a) for every $u \in X$, the set $\partial\varphi(u) \subseteq X^*$ is nonempty, convex, and w^*-compact;
(b) the multifunction $u \longmapsto \partial\varphi(u)$ is upper semicontinuous from X into $X^*_{w^*}$ (with $X^*_{w^*}$ being X^* with the w^*-topology);
(c) $\varphi^0(u;h) = \sigma_{\partial\varphi(u)}(h)$ for all $u, h \in X$;
(d) if φ is Gâteaux differentiable at $u \in X$, then $\varphi'_G(u) \in \partial\varphi(u)$;
(e) if $\varphi \in C^1(X)$, then $\partial\varphi(u) = \{\varphi'(u)\}$ for all $u \in X$;
(f) if φ is convex too, then the generalized subdifferential and the convex subdifferential (see Definition 3.70) coincide.

Proposition 3.129
If $\varphi\colon X \longrightarrow \mathbb{R}$ is a locally Lipschitz function, then

$$\partial(\lambda\varphi)(u) \;=\; \lambda\partial\varphi(u) \qquad \forall\, (\lambda, u) \in \mathbb{R} \times X,$$

hence $\partial(-\varphi)(u) = -\partial\varphi(u)$.

Proposition 3.130
If $\varphi, \psi\colon X \longrightarrow \mathbb{R}$ are both locally Lipschitz functions, then

$$\partial(\varphi + \psi)(u) \;\subseteq\; \partial\varphi(u) + \partial\psi(u) \qquad \forall\, u \in X.$$

Equality holds if both φ and ψ are regular at $u \in X$ or one of them is strictly differentiable.

Theorem 3.131 *(Mean Value Theorem)*
If $\varphi\colon X \longrightarrow \mathbb{R}$ is a locally Lipschitz map and $u, v \in X$,
then there exist $\widehat{\lambda} \in (0,1)$, $\widehat{u} = u + \widehat{\lambda}(v-u)$ and $\widehat{u}^* \in \partial\varphi(\widehat{u})$ such that

$$\varphi(v) - \varphi(u) \;=\; \langle \widehat{u}^*, v-u \rangle\,.$$

Proposition 3.132
If X and Y are two Banach spaces, $g \in C^1(X;Y)$ and $\varphi\colon Y \longrightarrow \mathbb{R}$ is a locally Lipschitz map,
then $\varphi \circ g\colon X \longrightarrow \mathbb{R}$ is locally Lipschitz and

$$\partial(\varphi \circ g)(u) \;\subseteq\; g'(u)^*\partial\varphi(g(u)) \qquad \forall\, u \in X. \tag{3.8}$$

Moreover, if φ is regular at $g(u)$, then $\varphi \circ g$ is regular at u and equality holds in (3.8). In addition, if g maps a neighborhood of $u \in X$ onto a dense subset of a neighborhood of $g(u)$ (for example, if $g'(u)$ is surjective), then we have equality in (3.8).

Let $C \subseteq X$ be a nonempty set and let $\operatorname{dist}(\cdot, C)$ be the ***distance function*** from C, i.e.,

$$\operatorname{dist}(u, C) \;=\; \inf_{c \in C} \|u - c\|_X.$$

Then $\operatorname{dist}(\cdot, C)$ is Lipschitz continuous and if C is convex, then $\operatorname{dist}(\cdot, C)$ is a convex function.

Definition 3.133
Let $C \subseteq X$ be a nonempty set and let $u \in \overline{C}$.
(a) The **Clarke tangent cone** *to C at u, is the set $T'_C(u) \subseteq X$ defined by*

$$T'_C(u) = \{h \in X : \operatorname{dist}^0(u, C; h) \leqslant 0\}.$$

(b) The **Clarke normal cone** *to C at u, is the set $N'_C(u) \subseteq X$ defined by*

$$N'_C(u) = \{u^* \in X^* : \langle u^*, h\rangle \leqslant 0 \text{ for all } h \in T'_C(u)\}.$$

Remark 3.134
Both cones are closed and convex sets and clearly $T'_C(u) = T'_{\overline{C}}(u)$ (so there is no loss of generality if we assume that C is closed). Evidently $N'_C(u)$ is the negative polar of $T'_C(u)$. If $u \in \operatorname{int} C$, then $T'_C(u) = X$.

The Clarke tangent cone is always convex and this is a nice and useful property. However, the price that we pay for this convenience is that the cone in some cases may be reduced to the trivial one $\{0\}$. For this reason we introduce the following notion.

Definition 3.135
Let $C \subseteq X$ and $u \in \overline{C}$. The **contingent** *(or* **Bouligand***)* **tangent cone** *to C at u is the set $T^b_C(u)$ defined by*

$$T^b_C(u) = \{h \in X : \liminf_{\lambda \searrow 0} \tfrac{1}{\lambda}\operatorname{dist}(u + \lambda h, C) = 0\}.$$

Remark 3.136
From Remark 2.89 we see that

$$T^b_C(u) = \limsup_{\lambda \searrow 0} \tfrac{1}{\lambda}(C - u).$$

Therefore $T^b_C(u)$ is a closed cone. Hence $h \in T^b_C(u)$ if and only if there exist sequences $\{\lambda_n\}_{n\geqslant 1} \subseteq (0, +\infty)$ and $\{h_n\}_{n\geqslant 1} \subseteq X$ such that

$$\lambda_n \searrow 0, \quad h_n \longrightarrow h \quad \text{in } X \quad \text{and} \quad u + \lambda_n h_n \in C \qquad \forall\, n \geqslant 1.$$

Hence (cf. Remark 3.134), we have that if $u \in \operatorname{int} C$, then $T^b_C(u) = X$.

3.1.4 Γ-Convergence and Relaxation

This section deals with two basic themes. First we define a convergence notion on the space of $\overline{\mathbb{R}}$-valued functions (where $\overline{\mathbb{R}} = \mathbb{R} \cup \{+\infty\}$), for which the maps

$$\varphi \longmapsto \inf_X \varphi \quad \text{and} \quad \varphi \longmapsto \operatorname{argmin}_X \varphi$$

are both continuous. This leads to the notion of Γ-convergence and related convergences. Subsequently, we deal with the notion of relaxation, which permits us to describe the minimizing sequences of functionals which are not lower semicontinuous, in terms of the minimizers of a suitable lower semicontinuous functional (the relaxed functional).

Definition 3.137
Let (X, d_X) be a metric space, let $\{\varphi_n : X \longrightarrow \overline{\mathbb{R}}\}_{n\geqslant 1}$ be a sequence of functions and let $\varphi : X \longrightarrow \overline{\mathbb{R}}$ be a function. We say that φ_n Γ-converges to φ, denoted by $\varphi_n \xrightarrow{\Gamma} \varphi$, if
(a) *for every sequence $\{u_n\}_{n\geqslant 1} \subseteq X$ such that $u_n \longrightarrow u$ in X we have*

$$\varphi(u) \leqslant \liminf_{n\to+\infty} \varphi_n(u_n);$$

(b) *there exists a sequence $\{v_n\}_{n\geqslant 1} \subseteq X$ such that $v_n \longrightarrow u$ in X and*

$$\limsup_{n\to+\infty} \varphi_n(v_n) \leqslant \varphi(u).$$

Remark 3.138
Evidently condition **(b)** in the above definition can be replaced by
(b)' there exists a sequence $\{v_n\}_{n\geqslant 1} \subseteq X$ such that $v_n \longrightarrow u$ in X such that

$$\lim_{n\to+\infty} \varphi_n(v_n) = \varphi(u).$$

If $\varphi_n = \varphi$ for all $n \geqslant 1$, then $\varphi_n \xrightarrow{\Gamma} \operatorname{cl}\varphi$ (see Definition 3.58). We can show that $\varphi_n \xrightarrow{\Gamma} \varphi$ if and only if $\operatorname{epi}\varphi_n \xrightarrow{K} \operatorname{epi}\varphi$ in $X \times \mathbb{R}$ (see Definition 2.88). For this reason Γ-convergence is also known in the literature as ***epiconvergence***.

Definition 3.139
Let (X, d_X) be a metric space and let $\{\varphi : X \longrightarrow \overline{\mathbb{R}}\}_{n\geqslant 1}$ be a sequence of functions. We define

$$\begin{aligned}
&\big(\Gamma\text{-}\liminf_{n\to+\infty} \varphi_n\big)(u) \\
&\quad = \min\Big\{\liminf_{n\to+\infty} \varphi_n(u_n) : \ \{u_n\}_{n\geqslant 1} \subseteq X,\ u_n \longrightarrow u \text{ in } X\Big\}, \\
&\big(\Gamma\text{-}\limsup_{n\to+\infty} \varphi_n\big)(u) \\
&\quad = \min\Big\{\limsup_{n\to+\infty} \varphi_n(u_n) : \ \{u_n\}_{n\geqslant 1} \subseteq X,\ u_n \longrightarrow u \text{ in } X\Big\}.
\end{aligned}$$

Remark 3.140
Evidently

$$\Gamma\text{-}\liminf_{n\to+\infty} \varphi_n \leqslant \Gamma\text{-}\limsup_{n\to+\infty} \varphi_n.$$

An alternative more topological description of these two $\overline{\mathbb{R}}$-valued functions is the following

$$\begin{aligned}
\big(\Gamma\text{-}\liminf_{n\to+\infty} \varphi_n\big)(u) &= \sup_{V\in\mathcal{N}(u)} \liminf_{n\to+\infty} \inf_{V} \varphi_n, \\
\big(\Gamma\text{-}\limsup_{n\to+\infty} \varphi_n\big)(u) &= \sup_{V\in\mathcal{N}(u)} \limsup_{n\to+\infty} \inf_{V} \varphi_n,
\end{aligned}$$

where $\mathcal{N}(u)$ denotes the filter of open neighborhoods of u. Note that these definitions make sense in every Hausdorff topological space, not necessarily a metric space (or even first countable; see Definition I.2.24 and Remark I.2.25). Of course $\mathcal{N}(u)$ can be replaced by a local basis $\mathcal{B}(u)$ at $u \in X$.

Proposition 3.141
If (X, d_X) is a metric space $\varphi\colon X \longrightarrow \overline{\mathbb{R}}$ is a function and $\{\varphi_n\colon \longrightarrow \overline{\mathbb{R}}\}_{n \geqslant 1}$ is a sequence of functions, then
(a) the functions $\Gamma\text{-}\liminf\limits_{n \to +\infty} \varphi_n$ and $\Gamma\text{-}\limsup\limits_{n \to +\infty} \varphi_n$ are both lower semicontinuous;
(b) $\varphi_n \xrightarrow{\Gamma} \varphi$ if and only if $\Gamma\text{-}\limsup\limits_{n \to +\infty} \varphi_n \leqslant \varphi \leqslant \Gamma\text{-}\liminf\limits_{n \to +\infty} \varphi_n$;
(c) if X is separable, then the sequence $\{\varphi_n\}_{n \geqslant 1}$ has a Γ-convergent subsequence.

Proposition 3.142
If (X, d_X) is a metric space, $\psi\colon X \longrightarrow \mathbb{R}$ is a function, $\{\varphi_n\colon X \longrightarrow \overline{\mathbb{R}}\}_{n \geqslant 1}$ is a sequence of functions, then
(a) $\Gamma\text{-}\liminf\limits_{n \to +\infty}(\varphi_n + \psi) = \Gamma\text{-}\liminf\limits_{n \to +\infty} \varphi_n + \psi$;
(b) $\Gamma\text{-}\limsup\limits_{n \to +\infty}(\varphi_n + \psi) = \Gamma\text{-}\limsup\limits_{n \to +\infty} \varphi_n + \psi$;
(c) if $\varphi_n \xrightarrow{\Gamma} \varphi$, then $\varphi_n + \psi \xrightarrow{\Gamma} \varphi + \psi$.

In general Γ-convergence and pointwise convergence are independent notions. However, in some special cases they are related.

Definition 3.143
Let (X, d_X) be a metric space and let $\{\varphi_n\colon X \longrightarrow \overline{\mathbb{R}}\}_{n \geqslant 1}$ be a sequence of functions. We say that the sequence $\{\varphi_n\}_{n \geqslant 1}$ is ***equi-lower semicontinuous*** *at $u \in X$, if for every $\varepsilon > 0$, we can find $U \in \mathcal{N}(u)$ such that*

$$\varphi_n(u) - \varepsilon \leqslant \varphi_n(h) \qquad \forall\, h \in U,\ n \geqslant 1.$$

We say that the sequence is ***equi-lower semicontinuous****, if it is equi-lower semicontinuous at every $u \in X$.*

Proposition 3.144
If (X, d_X) is a metric space and $\{\varphi_n\colon X \longrightarrow \overline{\mathbb{R}}\}_{n \geqslant 1}$ is a sequence which is equi-lower semicontinuous at $u \in X$, then

$$\begin{aligned}\big(\Gamma\text{-}\liminf_{n \to +\infty} \varphi_n\big)(u) &= \liminf_{n \to +\infty} \varphi_n(u) \quad \text{and} \quad \big(\Gamma\text{-}\limsup_{n \to +\infty} \varphi_n\big)(u) \\ &= \limsup_{n \to +\infty} \varphi_n(u).\end{aligned}$$

In particular, if $\{\varphi_n\}_{n \geqslant 1}$ is an equi-lower semicontinuous sequence, then $\varphi_n \xrightarrow{\Gamma} \varphi$ if and only if $\varphi_n \longrightarrow \varphi$ pointwise.

In fact the above result remains true, if X is a general topological space (see Remark 3.140).

Another instant where Γ and pointwise convergence coincide is the case of monotone sequences.

Proposition 3.145

(a) *If* (X, d_X) *is a metric space and* $\{\varphi_n : X \longrightarrow \overline{\mathbb{R}}\}_{n\geqslant 1}$ *is an increasing sequence of functions, then*

$$\Gamma\text{-}\lim_{n\to+\infty} \varphi_n = \lim_{n\to+\infty} \operatorname{cl}\varphi_n = \sup_{n\geqslant 1} \operatorname{cl}\varphi_n.$$

(b) *If* (X, d_X) *is a metric space,* $\{\varphi_n : X \longrightarrow \overline{\mathbb{R}}\}_{n\geqslant 1}$ *is a decreasing sequence of functions and* $\varphi_n \longrightarrow \varphi$ *pointwise, then* $\varphi_n \xrightarrow{\Gamma} \operatorname{cl}\varphi$.

The importance of the notion of Γ-convergence is a consequence of its variational character as it is illustrated by the following theorem.

Theorem 3.146

If (X, d_X) *is a metric space,* $\varphi : X \longrightarrow \overline{\mathbb{R}}$ *is a function,* $\{\varphi_n : X \longrightarrow \overline{\mathbb{R}}\}_{n\geqslant 1}$ *is a sequence of functions,* $\varphi_n \xrightarrow{\Gamma} \varphi$, $\{u_n\}_{n\geqslant 1} \subseteq X$ *is a relatively compact sequence and* $\{\varepsilon_n\}_{n\geqslant 1} \subseteq \mathbb{R}$ *is a sequence such that* $\varepsilon_n \searrow 0$ *and*

$$\varphi_n(u_n) \leqslant \inf_X \varphi_n + \varepsilon_n \qquad \forall\, n \geqslant 1,$$

then every limit point $u^* \in X$ *of the sequence* $\{u_n\}_{n\geqslant 1}$ *is a minimizer of* φ *and*

$$\varphi(u^*) = \lim_{n\to+\infty} \big(\inf_X \varphi_n\big).$$

Now we turn our attention to relaxation. Given $\varphi : X \longrightarrow \overline{\mathbb{R}}$, our goal is to apply the direct method of the calculus of variations on $\operatorname{cl}\varphi$ (see Definition 3.58). Directly from Definition 3.139 and Proposition 3.145, we have the following result.

Proposition 3.147

If (X, d_X) *is a metric space and* $\varphi : X \longrightarrow \overline{\mathbb{R}}$ *is a proper function, then*

$$\operatorname{cl}\varphi(u) = \inf\Big\{\liminf_{n\to+\infty} \varphi(u_n) : \{u_n\}_{n\geqslant 1} \subseteq X,\ u_n \longrightarrow u\Big\}$$

and so

(a) $\operatorname{cl}\varphi(u) \leqslant \liminf_{n\to+\infty} \varphi(u_n)$ *for all sequences* $u_n \to u$;

(b) $\limsup_{n\to+\infty} \varphi(v_n) \leqslant \varphi(u)$ *for some sequence* $v_n \longrightarrow u$.

The next theorem is the abstract "relaxation principle."

Theorem 3.148
If (X, d_X) *is a metric space,* $\varphi\colon X \longrightarrow \overline{\mathbb{R}}$ *is a map and* $\{u_n\}_{n\geqslant 1} \subseteq X$ *is a minimizing sequence for* φ *(that is,* $\varphi(u_n) \searrow \inf\limits_X \varphi$*) which is assumed to be compact,* *then* $\inf\limits_X \varphi = \min\limits_X \operatorname{cl} \varphi$ *and every limit point* u^* *of* $\{u_n\}_{n\geqslant 1}$ *satisfies*

$$\operatorname{cl} \varphi(u^*) \;=\; \min_X \operatorname{cl} \varphi.$$

The conclusion of the above theorem fails in the absence of compactness. Because of this and since in reflexive Banach spaces bounded sets are relatively w-compact, the following notion of convergence for sets and functions is used widely in the context of reflexive Banach spaces.

Definition 3.149
Let X *be a Banach space.*

(a) *Let* $\{C_n\}_{n\geqslant 1} \subseteq X$ *be a sequence of sets and let* $C \subseteq X$ *be a set. We say that the sequence* $\{C_n\}_{n\geqslant 1}$ *converges to* C *in the sense of Mosco, denoted by* $C_n \xrightarrow{M} C$*, if*

$$w\text{-}\limsup_{n\to+\infty} C_n \;\subseteq\; C \;\subseteq\; s\text{-}\liminf_{n\to+\infty} C_n,$$

where

$$\begin{aligned} s\text{-}\liminf_{n\to+\infty} C_n \;&=\; \Big\{u \in X :\; u = \lim_{n\to+\infty} u_n \text{ for some} \\ &\qquad \{u_n\}_{n\geqslant 1} \subseteq X \text{ with } u_n \in C_n \text{ for } n \geqslant 1\Big\}, \\ w\text{-}\limsup_{n\to+\infty} C_n \;&=\; \Big\{u \in X :\; u = w\text{-}\lim_{k\to+\infty} u_{n_k} \text{ for some} \\ &\qquad \{u_{n_k}\}_{k\geqslant 1} \subseteq X \text{ with } u_{n_k} \in C_{n_k} \text{ for } k \geqslant 1\Big\} \end{aligned}$$

(cf. Remark 2.89).

(b) *For a sequence* $\{\varphi_n\}_{n\geqslant 1} \subseteq \Gamma_0(X)$ *and a lower semicontinuous convex function* $\varphi\colon X \longrightarrow \overline{\mathbb{R}}$*, we say that the sequence* $\{\varphi_n\}_{n\geqslant 1}$ *converges to* φ *in the sense of Mosco, denoted by* $\varphi_n \xrightarrow{M} \varphi$*, if*

(b1) *for every* $u \in X$ *there exists a sequence* $\{v_n\}_{n\geqslant 1}$ *such that* $v_n \longrightarrow u$ *and*

$$\limsup_{n\to+\infty} \varphi_n(v_n) \;\leqslant\; \varphi(u);$$

(b2) *for every* $u \in X$ *and every sequence* $\{u_n\}_{n\geqslant 1} \subseteq X$ *such that* $u_n \xrightarrow{w} u$ *we have*

$$\varphi(u) \;\leqslant\; \liminf_{n\to+\infty} \varphi_n(u_n).$$

The importance of this mode of convergence comes from the fact that the operation of conjugation (see Definition 3.83) is bicontinuous for the Mosco convergence.

Theorem 3.150
<u>If</u> *X is a reflexive Banach space, $\{\varphi_n\}_{n\geqslant 1} \subseteq \Gamma_0(X)$ is a sequence and $\varphi \in \Gamma_0(X)$,*
<u>then</u> *$\varphi_n \xrightarrow{M} \varphi$ if and only if $\varphi_n^* \xrightarrow{M} \varphi^*$.*

Now we consider integral functionals of the form

$$I_\varphi(u) = \int_\Omega \varphi(z, u(z), \nabla u(z))\, dz. \tag{3.9}$$

To determine the semicontinuity properties of such functionals, first we consider integral functionals of the form

$$J_\varphi(u, v) = \int_\Omega \varphi(\omega, u(\omega), v(\omega))\, d\mu$$

with a finite measure space (Ω, Σ, μ) and a proper integrand $\varphi\colon \Omega \times \mathbb{R}^k \times \mathbb{R}^m \longrightarrow \overline{\mathbb{R}}$.

Proposition 3.151
<u>If</u> *φ is $\Sigma \times \mathcal{B}(\mathbb{R}^k) \times \mathcal{B}(\mathbb{R}^m)$-measurable, for μ-almost all $\omega \in \Omega$ the map $\varphi(\omega, \cdot, \cdot)$ is lower semicontinuous on $\mathbb{R}^k \times \mathbb{R}^m$ and for μ-almost all $z \in \Omega$ and all $u \in \mathbb{R}^k$ the map $\varphi(\omega, u, \cdot)$ is convex on $\mathbb{R}^m$,*
<u>then</u> *the functional $J_\varphi\colon L^1(\Omega; \mathbb{R}^k) \times L^1(\Omega; \mathbb{R}^m) \longrightarrow \overline{\mathbb{R}}$ is sequentially lower semicontinuous on $L^1(\Omega; \mathbb{R}^k) \times L^1(\Omega; \mathbb{R}^m)$, where $L^1(\Omega; \mathbb{R}^k)$ is equipped with the strong topology and $L^1(\Omega; \mathbb{R}^m)$ is equipped with the weak topology.*

Remark 3.152
In fact the result remains true if $\mathbb{R}^k$ is replaced by a separable Banach space Y and $\mathbb{R}^m$ is replaced by a separable and reflexive Banach space V. We can also assume that φ is $\overline{\mathbb{R}}$-valued and

$$\varphi(\omega, u, v) \geqslant \vartheta(\omega) - c(\|u\|_Y - \|v\|_V) \quad \text{for } \mu\text{-a.a. } \omega \in \Omega, \text{ all } u \in Y,\ v \in V,$$

with $\vartheta \in L^1(\Omega)$. Moreover, $L^1(\Omega; Y)$ and $L^1(\Omega; V)$ can be replaced by $L^p(\Omega; Y)$ and $L^q(\Omega; V)$ respectively (with $p, q \geqslant 1$).

Using Proposition 3.151 on the functional I_φ we obtain the following result.

Theorem 3.153
<u>If</u> *$\Omega \subseteq \mathbb{R}^N$ is an open set with a Lipschitz boundary $\partial\Omega$ and $\varphi\colon \Omega \times \mathbb{R}^k \times \mathbb{R}^{kN} \longrightarrow \overline{\mathbb{R}}_+$ is a $\mathcal{L} \times \mathcal{B}(\mathbb{R}^k) \times \mathcal{B}(\mathbb{R}^{kN})$-measurable integrand ($\mathcal{L}$ being the Lebesgue σ-field on Ω), such that for almost all $\omega \in \Omega$ the map $\mathbb{R}^k \times \mathbb{R}^{kN} \ni (u, v) \longmapsto \varphi(z, u, v)$ is lower semicontinuous and for almost all $z \in \Omega$ and all $u \in \mathbb{R}^k$ the map $\mathbb{R}^{kN} \ni v \longmapsto \varphi(z, u, v)$ is convex,*
<u>then</u> *the functional $I_\varphi\colon W^{1,1}(\Omega; \mathbb{R}^k) \longrightarrow \overline{\mathbb{R}}_+$ defined by (3.9) is sequentially weakly lower semicontinuous.*

In the scalar case (that is, $k = 1$), the convexity of $\varphi(z, u, \cdot)$ is also a necessary condition for the semicontinuity of I_φ.

Proposition 3.154

If $k = 1$ *and* $I_\varphi\colon W^{1,1}(\Omega) \longrightarrow \overline{\mathbb{R}} = [0, +\infty]$ *is a sequentially weakly lower semicontinuous functional for* $\varphi\colon \Omega \times \mathbb{R} \times \mathbb{R}^N \longrightarrow \overline{\mathbb{R}}_+$ *measurable and lower semicontinuous on* $\mathbb{R} \times \mathbb{R}^N$,
then for almost all $z \in \Omega$ *and all* $u \in \mathbb{R}$, *the map* $v \longmapsto \varphi(z, u, v)$ *is convex.*

In Problem 2.144, we saw that a sequence $\{u_n\}_{n\geqslant 1} \subseteq L^1(\Omega)$ which converges weakly but not strongly in $L^1(\Omega)$, oscillates violently around its weak limit. However, in the limit all this information about the fast oscillation is lost and only a mean value is registered. Clearly, this is not satisfactory. For example, suppose that $u_n \xrightarrow{w} u$ in $L^1(\Omega)$ and $h\colon \Omega \times \mathbb{R} \longrightarrow \mathbb{R}$ is a Carathéodory integrand. Let

$$N_h(u)(\cdot) \ = \ h(\cdot, u(\cdot)) \quad \text{and} \quad N_h(u_n)(\cdot) \ = \ h(\cdot, u_n(\cdot)) \qquad \forall\, n \geqslant 1.$$

Then, we cannot say that

$$N_h(u_n) \ \xrightarrow{w}\ N_h(u) \quad \text{in } L^1(\Omega)$$

unless $h(z, \cdot)$ is affine, a very restrictive requirement. The idea is that of relaxation. We embed the sequence $\{u_n\}_{n\geqslant 1}$ into a larger space and consider the limit there. This large space is a space of parametrized measures, known as "Young measures."

In what follows, Ω and T are locally compact and σ-compact metric spaces (hence separable), Σ is a σ-field on Ω which contains the Borel σ-field $\mathcal{B}(\Omega)$ and μ is a Radon measure on (Ω, Σ) such that Σ is μ-complete. Also, let

$$\begin{aligned} C_0(T) \ &= \ \{u\colon \longrightarrow \mathbb{R}:\ u \text{ is continuous and vanishes at infinity,} \\ &\qquad \text{i.e., for every } \varepsilon > 0, \text{ there exists compact set } K_\varepsilon \subseteq T, \\ &\qquad \text{such that } |u(t)| < \varepsilon \text{ for all } t \in T \setminus K_\varepsilon\}, \\ C_b(T) \ &= \ \{u\colon T \longrightarrow \mathbb{R}:\ u \text{ is continuous and bounded}\}. \end{aligned}$$

We know that

$$C_0(T)^* \ = \ M_b(T)$$

($M_b(T)$ being the space of Radon signed measures; see Theorem 1.88).

We set

$$\begin{aligned} M_b(T)_+ \ &= \ \{\lambda \in M(T):\ \lambda \geqslant 0\} \text{ the Radon measures on } T, \\ M_+^1(T) \ &= \ \{\lambda \in M(T)_+:\ \lambda(T) = 1\} \text{ the probability} \\ &\qquad \text{measures on } T, \\ SM_+^1(T) \ &= \ \{\lambda \in M(T)_+:\ \lambda(T) \leqslant 1\} \text{ the subprobability} \\ &\qquad \text{measures on } T. \end{aligned}$$

Definition 3.155

A ***transition probability*** *(respectively* ***transition subprobability****) on* Ω, *is a map* $\lambda\colon \Omega \longrightarrow M_+^1(T)$ *(respectively* $\lambda\colon \Omega \longrightarrow SM_+^1(T)$*) such that for every* $A \in \mathcal{B}(T)$ *(where* $\mathcal{B}(T)$ *is the Borel* σ*-field on* T*), we have that the map* $z \longmapsto \lambda(z)(A)$ *is* Σ*-measurable. By* $\widehat{R}(\Omega; T)$ *(respectively* $\widehat{SR}(\Omega; T)$*) we denote the space of all the transition probabilities (respectively transition subprobabilities) on* Ω.

Remark 3.156
We can endow $M_+^1(T)$ with the relative $w(M(T), C_b(T))$-topology, known as the ***narrow topology*** (see Definition 1.89). We denote this topology by w_n. From Theorem 1.96 we know that the space (M_+^1, w_n) is Polish.

Proposition 3.157
(a) *If* $\lambda\colon \Omega \longrightarrow M_+^1(T)$ *is* $(\Sigma, \mathcal{B}(M_+^1(T), w_n))$*-measurable,* *then* $\lambda \in \widehat{R}(\Omega; T)$.
(b) *If* T *is compact and* $\lambda \in \widehat{R}(\Omega; T)$*,* *then* λ *is* $(\Sigma, \mathcal{B}(M_+^1(T), w_n))$*-measurable.*

Let us recall the following definition.

Definition 3.158
Let $(S, \mathcal{Y})$ *be a measurable space,* V *a Hausdorff topological space,* $\vartheta\colon S \longrightarrow V$ *a* $(\mathcal{Y}, \mathcal{B}(V))$*-measurable map, and* $\lambda\colon \mathcal{Y} \longrightarrow \overline{\mathbb{R}}_+ = [0, +\infty]$ *a measure on* S*. The* ***image measure*** *of* λ *under* ϑ*, denoted by* $\lambda \circ \vartheta^{-1}$ *is defined by*

$$(\lambda \circ \vartheta^{-1})(A) = \lambda(\vartheta^{-1}(A)) \qquad \forall\, A \in \mathcal{B}(V).$$

Remark 3.159
Note that if $h\colon V \longrightarrow \mathbb{R}$ is a $\lambda \circ \vartheta^{-1}$-measurable map, then

$$\int_S h(\vartheta(s))\, d\lambda = \int_V h(v)\, d(\lambda \circ \vartheta^{-1})(v).$$

Definition 3.160
(a) *We say that* $\xi \in M_b(\Omega \times T)_+$ *is a* ***Young measure*** *with respect to* $\mu \in M_b(\Omega)_+$*, if*

$$\xi(A \times T) = \mu(A) \qquad \forall\, A \in \Sigma$$

(i.e., μ *is the* Ω*-marginal of* ξ*, that is, the image measure under the projection map on* proj_Ω*). By* $\mathcal{Y}(\Omega, T; \mu)$ *we denote the space of Young measures with respect to* μ*.*
(b) *We say that* $\xi \in M_b(\Omega \times T)_+$ *is a* ***Young submeasure*** *with respect to* $\mu \in M_b(\Omega)_+$*, if*

$$\xi(A \times T) \leqslant \mu(A) \qquad \forall\, A \in \Sigma.$$

By $S\mathcal{Y}(\Omega, T; \mu)$ *we denote the space of Young submeasures with respect to* μ*.*
(c) *If* $u\colon \Omega \longrightarrow T$*, then we say that the Young measure* $\xi \in \mathcal{Y}(\Omega, T; \mu)$ *is* ***associated with the function*** u*, if*

$$\int_{\Omega\times T} h(z, \tau)\, d\xi = \int_\Omega h(z, u(z))\, d\mu \qquad \forall\, h \in C_0(\Omega \times T).$$

Remark 3.161
Note that if $\xi \in \mathcal{Y}(\Omega, T; \mu)$, then $\mu = \xi \circ \mathrm{proj}_\Omega^{-1}$, where $\mathrm{proj}_\Omega\colon \Omega \times T \longrightarrow \Omega$ is the projection map. Moreover, $\xi = \mu \circ \beta^{-1}$ where $\beta\colon \Omega \longrightarrow \Omega \times T$ is defined by $\beta(z) = (z, u(z))$, in the case where ξ is associated with u (see Definition 3.160(c)).

Proposition 3.162
(a) We have

$$\begin{aligned} S\mathcal{Y}(\Omega, T; \mu) = \{ & \xi \in M_b(\Omega \times T)_+ : \xi(\Omega \times T) \leqslant \mu(\Omega) \text{ and} \\ & \text{for all } \eta \in C_0(\Omega),\ \eta \geqslant 0 \text{ and all } \gamma \in C(\Omega \times T)_+ \\ & \text{with } \gamma(z,\tau) \leqslant \eta(z) \text{ for all } (z,\tau) \in \Omega \times T, \\ & \text{we have } \int\limits_{\Omega \times T} \gamma(z,\tau)\, d\xi \leqslant \int\limits_{\Omega} \eta\, d\mu \}. \end{aligned}$$

(b) If T is compact, then

$$\begin{aligned} \mathcal{Y}(\Omega, T; \mu) = \{ & \xi \in S\mathcal{Y}(\Omega, T; \mu) : \text{ for all } \eta \in C_0(\Omega),\ \eta \geqslant 0, \\ & \text{we have } \int\limits_{\Omega} \eta\, d\mu = \int\limits_{\Omega \times T} \eta\, d\xi \}. \end{aligned}$$

From Theorem 1.88, we know that

$$M_b(\Omega \times T) = C_0(\Omega \times T)^*.$$

So, we can endow $\mathcal{Y}(\Omega, T; \mu)$ and $S\mathcal{Y}(\Omega, T; \mu)$ with the relative w^*-topology. Then we have the following result.

Theorem 3.163
(a) T is compact if and only if $\mathcal{Y}(\Omega, T; \mu)$ is w^*-compact.
(b) $S\mathcal{Y}(\Omega, T; \mu)$ with the relative w^*-topology is compact and metrizable.

Let $\lambda_1, \lambda_2 \in \widehat{SR}(\Omega; T)$. We introduce the following equivalence relation

$$\lambda_1 \sim \lambda_2 \iff \big[\lambda_1(z) = \lambda_2(z) \quad \text{for } \mu\text{-a.a. } z \in \Omega\big].$$

We set

$$R(\Omega; T) = \widehat{R}(\Omega; T)/_\sim \quad \text{and} \quad SR(\Omega; T) = \widehat{SR}(\Omega; T)/_\sim$$

(see Definition 3.155). We can relate Young measures with the transition probabilities.

Theorem 3.164
There is a bijection $\sigma\colon R(\Omega; T) \longrightarrow \mathcal{Y}(\Omega, T; \mu)$ defined by

$$\sigma(\lambda) = \xi,$$

where

$$\xi(C) = \int\limits_{\Omega} \int\limits_{T} \chi_C(z,\tau) \lambda(z)(d\tau)\, d\mu \qquad \forall\, C \in \Sigma \times \mathcal{B}(T).$$

Remark 3.165
Similarly we establish that σ is a bijection from $SR(\Omega;T)$ onto $S\mathcal{Y}(\Omega,T;\mu)$. Moreover, if ξ is the Young measure associated with a measurable function $u\colon\Omega\longrightarrow T$, then $\sigma(\delta_u)=\xi$. Here δ_u is the Dirac transition probability defined by

$$\delta_{u(z)}(D) \;=\; \begin{cases} 1 & \text{if} \quad u(z)\in D, \\ 0 & \text{if} \quad u(z)\notin D \end{cases} \qquad \forall\, D\in\mathcal{B}(T).$$

The identification established in Theorem 3.164 implies that w^*-topology on $\mathcal{Y}(\Omega,T;\mu)$ (resulting from the pair $(M_b(\Omega\times T),C_0(\Omega\times T))$) is equivalent to the w^*-topology on $R(\Omega;T)$ (resulting from the pair $(L^\infty(\Omega;M_b(T)),L^1(\Omega;C_0(T)))$). The fundamental idea behind the theory of Young measures is that, in some sense, Young measures form a completion of the space of ordinary functions. In this completion, the ordinary functions are the Dirac Young measures mentioned above.

The next approximation result is very useful in many situations.

Proposition 3.166
If $F\colon\Omega\longrightarrow 2^T\setminus\{\emptyset\}$ *is a graph measurable multifunction (i.e.,* $\operatorname{Gr}F\in\Sigma\times\mathcal{B}(T)$*) and*

$$R_F(\Omega;T) \;=\; \big(\lambda\in R(\Omega;T):\ \lambda(z)(F(z))=1 \text{ for } \mu\text{-a.a. } z\in\Omega\big),$$

then

$$R_F(\Omega;T) \;=\; \operatorname{cl}\big(\delta_u:\ u\colon\Omega\longrightarrow T \text{ is } \Sigma\text{-measurable and } u(z)\in F(z) \text{ for } \mu\text{-a.a. } z\in\Omega\big)$$

(the closure with respect to the w^**-topology on* $L^\infty(\Omega;M_b(T))$*).*

Definition 3.167
The ***narrow topology*** *on* $\mathcal{Y}(\Omega,T;\mu)$ *is the weakest topology which makes continuous the linear functionals of the form*

$$\xi \;\longmapsto\; \int\limits_{\Omega\times T}\varphi\,d\xi$$

for all measurable integrands $\varphi\colon\Omega\times T\longrightarrow\mathbb{R}$ *such that* $\varphi(z,\cdot)\in C_b(T)$ *for* μ*-almost all* $z\in\Omega$ *and* $z\longmapsto\|\varphi(z,\cdot)\|_{L^\infty(T)}$ *belongs in* $L^1(\Omega)$*. We say that the sequence* $\{\xi_n\}_{n\geqslant1}\subseteq\mathcal{Y}(\Omega,t;\mu)$ ***converges narrowly*** *to* ξ*,*

$$\lim_{n\to+\infty}\int\limits_{\Omega\times T}\varphi\,d\xi_n \;=\; \int\limits_{\Omega\times T}\varphi\,d\xi$$

for every integrand $\varphi(z,u)$ *as above. We write* $\xi_n\xrightarrow{n}\xi$*.*

We want to have a compactness criterion for the narrow topology. In this direction the following notion is important.

Definition 3.168
A set $\mathcal{D} \subseteq \mathcal{Y}(\Omega, T; \mu)$ *is said to be* ***uniformly tight****, if for every* $\varepsilon > 0$*, there exists a compact set* $K_\varepsilon \subseteq T$ *such that*

$$\sup_{\xi \in \mathcal{D}} \xi(\Omega \times (T \setminus K_\varepsilon)) \leqslant \varepsilon.$$

Remark 3.169
The uniform tightness of a set $\mathcal{D} \subseteq \mathcal{Y}(\Omega, T; \mu)$ is equivalent to saying that the set $C = \{\xi \circ \operatorname{proj}_T^{-1} : \xi \in \mathcal{D}\}$ is uniformly tight in $M_b(T)_+$.

Exploiting the identification established in Theorem 3.164, we can have the following alternative characterization of uniform tightness.

Proposition 3.170
A set $\mathcal{D} \subseteq \mathcal{Y}(\Omega, T; \mu)$ *is uniformly tight if and only if there exists a measurable function* $\psi \colon \Omega \times T \longrightarrow \overline{\mathbb{R}}_+$ *such that for* μ*-almost all* $z \in \Omega$ *and for every* $\eta \in \mathbb{R}$ *the set* $\{u \in T \colon \psi(z, u) \leqslant \eta\}$ *is compact and*

$$\sup_{\xi \in \mathcal{D}} \int_{\Omega \times T} \psi \, d\xi = \sup_{\xi \in \mathcal{D}} \int_\Omega \int_T \psi(z, u) \xi(z)(du) \, dt < +\infty.$$

Remark 3.171
A function $h \colon T \longrightarrow \overline{\mathbb{R}}$ with compact sublevel sets (as is the case for $\psi(z, \cdot)$ in the above proposition), is often called ***inf-compact***.

The notion of uniform tightness leads to a compactness criterion for the narrow topology (see Definition 3.167). The result is known as the Prohorov theorem for Young measures.

Theorem 3.172 *(Prohorov Theorem)*
A set $\mathcal{D} \subseteq \mathcal{Y}(\Omega, T; \mu)$ *is relatively compact for narrow topology if and only if* $\mathcal{D}$ *is uniformly tight.*

3.2 Problems

Problem 3.1**

Suppose that X and Y are two Banach spaces, $U \subseteq X$ is an open set and $f\colon U \longrightarrow Y$ is a compact and Fréchet differentiable map (see Definition 3.1). Show that for every $x \in U$, we have $f'(x) \in \mathcal{L}_c(X;Y)$.

Problem 3.2**

Let X be a Banach space with norm $\|\cdot\|_X$. Show that if $\|\cdot\|_X$ is differentiable at $u \neq 0$ and $t > 0$, then $\|\cdot\|_X$ is differentiable at tu and

$$\tfrac{d}{dt}\|tu\|_X \;=\; \tfrac{d}{dt}\|u\|_X.$$

Also show that $\left\|\frac{d}{du}\|u\|_X\right\|_* = 1$.

Problem 3.3**

Let $C \subseteq \mathbb{R}^N$ be a nonempty closed set and let us consider the function $\varphi_C(u) = \mathrm{dist}(u, C)$. Suppose that for some $u \in \mathbb{R}^N$, $\varphi_C'(u)$ exists and $\varphi_C'(u) \neq 0$. Show that $u \notin C$, $\mathrm{proj}_C(u) = \{c\}$ (where proj_C is the metric projection, that is, $\mathrm{proj}_C(u) = \{c \in C : \; |u - c| = \mathrm{dist}(u, C)\}$) and $\varphi_C'(u) = \frac{u-c}{|u-c|}$.

Problem 3.4*

Suppose that X and Y are two Banach spaces, $U \subseteq X$ is a nonempty, open and connected set and $\varphi\colon U \longrightarrow Y$ is a Fréchet differentiable function such that $\varphi'(u) = 0$ for all $u \in U$. Show that φ is constant.

Problem 3.5**

Suppose that (Ω, Σ, μ) is a finite measure space, $1 < p < +\infty$, $f\colon \Omega \times \mathbb{R} \longrightarrow \mathbb{R}$ is a Carathéodory function satisfying the following growth condition

$$|f(\omega, x)| \;\leqslant\; a(\omega) + c|x|^{p-1} \quad \text{for } \mu\text{-a.a. } \omega \in \Omega, \text{ all } x \in \mathbb{R},$$

where $a \in L^{p'}(\Omega)$ (with $\frac{1}{p} + \frac{1}{p'} = 1$), $c > 0$. Let

$$F(\omega, x) \;=\; \int_0^x f(\omega, s)\, ds$$

and let $N_f\colon L^p(\Omega) \longrightarrow L^{p'}(\Omega)$ be the Nemytskii operator corresponding to f (see Definition 1.145 and Problem 1.42). Let $\psi\colon L^p(\Omega) \longrightarrow \mathbb{R}$ be defined by

$$\psi(u) = \int_\Omega F(\omega, u(\omega)) \qquad \forall\, u \in L^p(\Omega).$$

Show that $\psi \in C^1(L^p(\Omega))$ and $\psi'(u) = N_f(u)$ for all $u \in L^p(\Omega)$.

Problem 3.6**

Suppose that (Ω, Σ, μ) is a finite measure space, $\varphi\colon \Omega \times \mathbb{R} \longrightarrow \mathbb{R}$ is a measurable functional such that $\varphi(\omega, \cdot) \in C^1(\mathbb{R})$ for μ-almost all $\omega \in \Omega$ and φ'_x satisfies the following growth condition

$$|\varphi'_x(\omega, x)| \leqslant a(\omega) + c|x|^m \quad \text{for } \mu\text{-a.a. } \omega \in \Omega, \text{ all } x \in \mathbb{R}$$

with $a \in L^n(\Omega)$ for some $n, m \geqslant 1$ and $c > 0$. Let $p = mn$ and $q = \frac{mn}{m+1}$ and let $N_\varphi\colon L^p(\Omega) \longrightarrow L^q(\Omega)$ be the Nemytskii operator corresponding to φ (see Definition 1.145 and Problem 1.42). Also, let $N_{\varphi'_x}\colon L^p(\Omega) \longrightarrow L^n(\Omega)$ be the Nemytskii operator corresponding to φ'_x. Show that $N_\varphi \in C^1(L^p(\Omega))$ and $N'_\varphi(u) = N_{\varphi'_x}(u)$ for all $u \in L^p(\Omega)$.

Problem 3.7***

Suppose that $\Omega \subseteq \mathbb{R}^N$ is a nonempty and bounded set, $p \in [1, +\infty)$, $\varphi\colon \Omega \times \mathbb{R} \longrightarrow \mathbb{R}$ is a measurable function such that $\varphi(z, \cdot) \in C^1(\mathbb{R})$ for almost all $z \in \Omega$ and

$$|\varphi'_x(z, x)| \leqslant M \quad \text{for a.a. } z \in \Omega, \text{ all } x \in \mathbb{R},$$

with $M > 0$. Then $N_\varphi\colon L^p(\Omega) \longrightarrow L^p(\Omega)$ and $N_{\varphi'_x}\colon L^p(\Omega) \longrightarrow L^\infty(\Omega)$. Suppose that N_φ is Fréchet differentiable. Show that

$$\varphi(z, x) = c(z)x + a(z) \quad \text{for a.a. } z \in \Omega, \text{ all } z \in \mathbb{R},$$

with $c \in L^\infty(\Omega)$, $a \in L^p(\Omega)$.

Problem 3.8*

Suppose that X and Y are two Banach spaces, $U \subseteq X$ is a nonempty and open set, $\varphi\colon U \longrightarrow Y$ is a C^1-map, and $\varphi'(u)$ is an isomorphism for every $u \in U$. Show that the set $\varphi(U) \subseteq Y$ is open.

Problem 3.9*

Suppose that X and Y are two Banach spaces, $U \subseteq X$ is a nonempty and open set, $\varphi \in C^1(U; Y)$ is an injective map and $\varphi'(u)$ is an isomorphism for every $u \in U$. Show that φ is a C^1-diffeomorphism from U onto $\varphi(U)$.

Problem 3.10**

Suppose that X, Y and Z are three Banach spaces, $U \subseteq X$ is a nonempty and open set, $\varphi\colon U \longrightarrow Y$ and $\psi\colon U \longrightarrow Z$ are two C^1-maps with $\varphi'(u)$ being an isomorphism for all $u \in U$. Suppose that there exists $\sigma\colon \varphi(U) \longrightarrow Z$ such that $\psi = \sigma \circ \varphi$. Show that σ is a C^1-map.

Problem 3.11*

Let X be a Banach space and let $\varphi\colon [a, b] \longrightarrow X$ be a continuous map which is right differentiable on $[a, b)$. Let

$$\vartheta(t) = \|\varphi(t)\|_X \qquad \forall\, t \in [a, b].$$

Show that ϑ is right differentiable on $[a, b)$ and

$$|\vartheta'_r(t)| \leqslant \|\varphi'_r(t)\|_X \qquad \forall\, t \in [a, b).$$

Problem 3.12**
Suppose that X is a Banach space, $C \subseteq X$ is a nonempty, closed, and convex set and $\varphi\colon [a, b] \longrightarrow X$ is a continuous map which is differentiable for almost all $t \in [a, b]$, $\varphi'(t) \in C$ and $\varphi' \in L^1((a, b); X)$. Show that $\frac{\varphi(b)-\varphi(a)}{b-a} \in C$.

Problem 3.13**
Let X be a Banach space and let $\varphi\colon [a, b] \longrightarrow X$ and $\vartheta\colon [a, b] \longrightarrow \mathbb{R}$ be two continuous functions which are both differentiable in (a, b). Suppose that

$$\|\varphi'(t)\|_X \leqslant \vartheta'(t) \qquad \forall\, t \in (a, b).$$

Show that

$$\|\varphi(b) - \varphi(a)\|_X \leqslant \vartheta(b) - \vartheta(a).$$

Problem 3.14*
Let X be a Banach space and let $A \in \mathcal{L}(X)$. Consider the map $\varphi\colon \mathbb{R} \longrightarrow \mathcal{L}(X)$ defined by

$$\varphi(t) = e^{tA} \qquad \forall\, t \in \mathbb{R}.$$

Show that $\varphi \in C^1(\mathbb{R}; \mathcal{L}(X))$.

Problem 3.15**
Let $X = C([0, 1])$ and let $\varphi\colon X \longrightarrow X$ be a map defined by

$$\varphi(u)(t) = \sin u(t) \qquad \forall\, t \in [0, 1].$$

Show that φ is Fréchet differentiable at every $u \in X$ and find $\varphi'(u)$.

Problem 3.16**
Let $p\colon l^\infty \longrightarrow \mathbb{R}$ be defined by

$$p(\widehat{u}) = \limsup_{n \to +\infty} |u| \qquad \forall\, \widehat{u} = \{u_n\}_{n \geqslant 1} \in l^\infty.$$

Show that the function p is continuous, but nowhere differentiable.

Problem 3.17*
Suppose that X and Y are two Banach spaces, $\varphi\colon X \longrightarrow Y$ is a map which is Fréchet differentiable at $u \in X$, and $M = \|\varphi'(u)\|_{\mathcal{L}}$. Show that for every $\varepsilon > 0$, there exists $\delta > 0$ such that

$$\|\varphi(u+h) - \varphi(u)\|_Y \leqslant (M + \varepsilon)\|h\|_X \qquad \forall\, h \in X, \text{ with } \|h\|_X \leqslant \delta.$$

Problem 3.18**
Suppose that X is a Banach space, $\varphi \in C^1(X; X)$, $\varphi'(u) \in \mathcal{L}(X; X)$ is an isomorphism for all $u \in X$ and there exists $c \in (0, 1)$ such that

$$\|\varphi(u) - \varphi(v) - (u - v)\|_X \leqslant c\|u - v\|_X \qquad \forall\, u, v \in X.$$

Show that φ is a C^1-diffeomorphism.

Problem 3.19*
Suppose that H is a Hilbert space, $A \in \mathcal{L}(H)$ and $\varphi\colon H \longrightarrow \mathbb{R}$ is a map defined by

$$\varphi(u) = (A(u), u)_H \qquad \forall\, u \in H.$$

Show that $\varphi \in C^1(H)$

Problem 3.20**
Let (Ω, Σ, μ) be a σ-finite measure space and consider the space $L^p(\Omega; \mathbb{R}^N)$ (with $1 < p < +\infty$). Show that the norm $\|\cdot\|_{L^p(\Omega;\mathbb{R}^N)}$ is Fréchet differentiable at every $u \neq 0$ and find the Fréchet derivative.

Problem 3.21***
Suppose that $\Omega \subseteq \mathbb{R}^N$ is a bounded open set with a Lipschitz boundary, $1 < p < +\infty$ and $\varphi\colon W_0^{1,p}(\Omega) \longrightarrow \mathbb{R}$ is the function defined by

$$\varphi(u) = \tfrac{1}{p}\|Du\|^p_{L^p(\Omega;\mathbb{R}^N)} \qquad \forall\, u \in L^p(\Omega;\mathbb{R}^N).$$

Show that $\varphi \in C^1(W_0^{1,p}(\Omega))$ and find $\varphi'(u)$.

Problem 3.22*
Suppose that X is a Banach space, $\varphi\colon X \longrightarrow \mathbb{R}$ is a Gâteaux differentiable map such that for some $u_0 \in X$ and some open neighborhood $U \subseteq X$ of u_0, we have

$$\varphi(u) - \varphi(u_0) - \langle \varphi'_G(u_0), u - u_0 \rangle \geqslant 0 \qquad \forall\, u \in U.$$

Show that φ is sequentially weakly lower semicontinuous at u_0. Also, show that, if the set U above is open convex, then $\varphi|_U$ is a convex map.

Problem 3.23*
Let X be a Banach space and let $\varphi\colon X \longrightarrow \mathbb{R}$ be a Gâteaux differentiable map such that

$$\langle \varphi'(u), u \rangle > 0 \qquad \forall\, u \in X, \text{ with } \|u\|_X = \varrho > 0$$

and $\inf\limits_{\|u\|_X \leqslant \varrho} \varphi(u)$ is realized at u_0. Show that $\|u_0\|_X < \varrho$.

Problem 3.24*
Let X and Y be two Banach spaces and let $\varphi\colon X \longrightarrow Y$ be a map which is Gâteaux differentiable at $u \in X$. Show that φ is hemicontinuous on X with the norm topology.

Problem 3.25**
Suppose that X and Y are two Banach spaces, $\varphi\colon X \longrightarrow Y$ is a Fréchet differentiable map such that φ and $\varphi'\colon X \longrightarrow \mathcal{L}(X;Y)$ are both compact. Show that φ is completely continuous.

Problem 3.26**
Let $X = C([0,1])$ and $\xi \in C^1(\mathbb{R})$. We introduce the function $\varphi\colon X \longrightarrow \mathbb{R}$ defined by

$$\varphi(u) = \int_0^1 \xi(u(t))\, dt \qquad \forall\, u \in X.$$

Show that $\varphi \in C^1(X)$ and find φ'.

Problem 3.27*
Suppose that X is a Banach space, $\varphi\colon X \longrightarrow \mathbb{R}$ is a map which is Fréchet differentiable at $u_0 + th$ and we set $\xi(\tau) = \varphi(u_0 + \tau h)$ for all $\tau \in \mathbb{R}$. Show that ξ is differentiable at t and

$$\xi'(t) = \langle \varphi'(u_0 + th), h \rangle .$$

Problem 3.28***
Let $M_n(\mathbb{R})$ be the space of all $n \times n$-real matrices and let $\xi\colon M_n(\mathbb{R}) \times M_n(\mathbb{R}) \longrightarrow M_n(\mathbb{R})$ be defined by $\xi(A, K) = AK$. Prove the following statements:

(a) $\xi \in C^1(M_n(\mathbb{R}) \times M_n(\mathbb{R}); M_n(\mathbb{R}))$ and find ξ'.

(b) The set $\mathrm{End}_n(\mathbb{R})$ of all invertible elements of $M_n(\mathbb{R})$ is open and the map $\beta\colon \mathrm{End}_n(\mathbb{R}) \longrightarrow \mathrm{End}_n(\mathbb{R})$ defined by $\beta(A) = A^{-1}$ is a C^1-diffeomorphism.

Problem 3.29*
Let X be a Banach space and let $\varphi\colon X \longrightarrow \mathbb{R}$ be a map. We say that φ is α-***homogeneous*** (with $\alpha \in \mathbb{R}$), if

$$\varphi(tu) = t^{\alpha}\alpha\varphi(u) \qquad \forall\, u \in X \setminus \{0\},\ t > 0.$$

Suppose that φ is Fréchet differentiable at every $u \neq 0$. Show that φ is α-homogeneous if and only if $\alpha\varphi(u) = \langle \varphi'(u), u \rangle$ for all $u \neq 0$.

Remark. The result of this problem is known in the literature as the ***Euler Formula***.

Problem 3.30*
Let $M_n(\mathbb{R})$ be the space of all $n \times n$ real matrices and let $\varphi\colon \mathbb{R} \times M_n(\mathbb{R}) \longrightarrow \mathbb{R}$ be defined by

$$\varphi(\lambda, A) = \det(\lambda I - A) \qquad \forall\, (\lambda, A) \in \mathbb{R} \times M_n(\mathbb{R}).$$

Let $\lambda \in \mathbb{R}$ be a simple eigenvalue of A. Show that there exists a neighborhood U of A and a C^{∞}-function $h\colon U \longrightarrow \mathbb{R}$ such that $h(A) = \lambda$ and for every $C \in U$, $h(C)$ is an eigenvalue of C.

Problem 3.31***
Let $\varphi \in C^1(\mathbb{R}^N; \mathbb{R}^N)$ be such that

$$\left(\varphi'(u)h, h\right)_{\mathbb{R}^N} \geqslant c|h|^2 \qquad \forall\, u, h \in \mathbb{R}^N,$$

for some $c > 0$. Show that φ is a C^1-diffeomorphism.

Problem 3.32**
Suppose that X is a Banach space, $\varphi \in C^1(X;X)$, $u \in X$ and $y = \varphi(u)$. Show that there exist an open neighborhood U of $(0,u)$ in $\mathbb{R} \times X$, an open neighborhood V of y in X and a C^1-map $h\colon U \longrightarrow X$ such that for all $(\lambda, v) \in U$, we have

$$\big[w \in V \text{ and } w = \varphi(\lambda w + v)\big] \iff w = h(\lambda, v).$$

Problem 3.33**
Suppose that X is a Banach space, $U \subseteq X$ is a nonempty, open, convex set, $0 \in U$ and $g \in C(U;X^*)$. Show that the following three statements are equivalent:
(a) There exists a function $\varphi \in C^1(U)$ such that $g = \varphi'$.
(b) $\int_\gamma \langle g(r), dr\rangle$ is independent of the path γ in U, with γ being simple and rectifiable.
(c) If $\xi(t) = tu + (1-t)v$ with $u, v \in U$, then

$$\int_0^1 \langle g(tu), u\rangle \, dt - \int_0^1 \langle g(tv), v\rangle \, dt = \int_0^1 \langle g(\xi(t)), u - v\rangle \, dt.$$

Also show that, if X is reflexive and $g \in C^1(U;X^*)$, then the above statements are equivalent to the following one:
(d) $g'(u) \in \mathcal{L}(X;X^*)$ is self-adjoint for all $u \in U$.

Problem 3.34**
Let X be a reflexive Banach space and let $\varphi \in C^1(X)$ be such that

$$\langle \varphi'(u), u\rangle \geqslant \xi(\|u\|_X) \qquad \forall\, u \in X,$$

where $\xi\colon (0,+\infty) \longrightarrow (0,+\infty)$ is a continuous function such that

$$\int_1^{+\infty} \tfrac{\xi(s)}{s} \, ds = +\infty.$$

Show that φ is coercive on X.

Problem 3.35*
Suppose that X is a Banach space, $U \subseteq X$ is nonempty, open set, $F\colon U \longrightarrow X$ is a vector field and $\varphi \in C^1(U)$. Show that φ is a first integral of F (see Definition 3.54(b)) if and only if $\langle \varphi'(u), F(u)\rangle = 0$ for all $u \in U$.

Problem 3.36**
Assume that X is a Banach space, $U \subseteq X$ is a nonempty open set, $F\colon U \longrightarrow X$ is a vector field, and $\varphi \in C^1(U)$ is a first integral of F such that $\varphi^{-1}(c) \subseteq X$ is a compact set for every $c \in \mathbb{R}$. Show that the vector field F is complete.

Problem 3.37**
Suppose that X is a finite dimensional Banach space, $U \subseteq X$ is a nonempty, open set and $F: U \longrightarrow X$ is a Lipschitz continuous vector field with Lipschitz constant $k > 0$. Show that F is complete.

Problem 3.38**
Let Y be a Banach space and let $\varphi: \mathbb{R}^m \longrightarrow Y$ be a map which is Gâteaux differentiable at $u_0 \in \mathbb{R}^m$ and Lipschitz continuous in an open neighborhood of u_0. Show that φ is Fréchet differentiable at u_0.

Problem 3.39***
Let X and Y be two Banach spaces and let $\varphi \in C^1(X;Y)$ be such that $\varphi'(u)$ is an isomorphism from X onto Y and $\sup\limits_{u \in X} \|\varphi'(u)^{-1}\|_{\mathcal{L}} < +\infty$ for all $u \in X$. Show that φ is a C^1-diffeomorphism.

Problem 3.40***
Let X be a reflexive Banach space and let $\varphi: [0,b] \longrightarrow X$ be a Lipschitz continuous map. Show that φ is differentiable at almost all $t \in [0,b]$.

Problem 3.41**
Let X be a reflexive Banach space with its dual X^* strictly convex and let $\sigma: X \longrightarrow \mathbb{R}$ be defined by $\sigma(u) = \|u\|_X$. Show that σ is Gâteaux differentiable on $X \setminus \{0\}$ and

$$\sigma'(u) = \frac{\mathcal{F}(u)}{\|u\|_X} \qquad \forall\, u \in X \setminus \{0\}$$

with $\mathcal{F}: X \longrightarrow X^*$ being the duality map of X.

Problem 3.42**
Let X be a reflexive Banach space with its dual X^* being locally uniformly convex and let $\sigma: X \longrightarrow \mathbb{R}$ be defined by $\sigma(u) = \|u\|_X$. Show that $\sigma \in C^1(X \setminus \{0\})$.

Problem 3.43**
Let X be a reflexive Banach space which is densely and continuously embedded in a Hilbert space H. Suppose that $\varphi \in \Gamma_0(X)$ (see Remark 3.56) and $c\|u\|_X^2 \leqslant \varphi(u)$ for all $u \in X$. We introduce the map $\widehat{\varphi}: H \longrightarrow \overline{\mathbb{R}} = \mathbb{R} \cup \{+\infty\}$ defined by

$$\widehat{\varphi}(u) = \begin{cases} \varphi(u) & \text{if } u \in X, \\ +\infty & \text{if } u \in H \setminus X. \end{cases}$$

Show that $\widehat{\varphi} \in \Gamma_0(H)$.

Problem 3.44*
Let X be a Banach space and let $\varphi: X \longrightarrow \overline{\mathbb{R}}$ be a convex function which is Gâteaux differentiable at $u_0 \in X$. Show that $\varphi'_G(u_0) \in \partial\varphi(u_0)$.

Problem 3.45***

Let X be a Banach space and let $\varphi \in \Gamma_0(X)$. Show that φ is Lipschitz continuous with Lipschitz constant k if and only if $\operatorname{dom}\varphi^* \subseteq k\overline{B}_1^*(0)$, where $\overline{B}_1^*(0) = (u^* \in X^* : \|u^*\|_*) \leqslant 1$.

Problem 3.46**

Let X be a reflexive Banach space and let $\varphi \in \Gamma_0(X)$. Show that the following two properties are equivalent:

(a) φ is strongly coercive (see Definition 3.105(b)).

(b) $\partial\varphi$ is surjective and $(\partial\varphi)^{-1}$ is bounded on bounded sets.

Problem 3.47**

Let X be a Banach space and let $\varphi\colon X \longrightarrow \overline{\mathbb{R}}$ be a proper, convex function which is lower semicontinuous at some point in $\operatorname{dom}\varphi$. Show that the following properties are equivalent:

(a) φ is coercive (see Definition 3.105(a)).

(b) There exist $c > 0$ and $b \in \mathbb{R}$ such that

$$\varphi(u) \geqslant c\|u\|_X + b \qquad \forall\, u \in X.$$

(c) $\liminf\limits_{\|u\|_X \to +\infty} \frac{\varphi(u)}{\|u\|_X} > 0$.

(d) for every $\lambda \in \mathbb{R}$, the set $\varphi^\lambda = \{u \in X : \varphi(u) \leqslant \lambda\}$ is bounded.

Problem 3.48**

Suppose that X is a Banach space, $u_0^* \in X^* \setminus \{0\}$ and $\varphi(u) = |\langle u_0^*, u\rangle|$ for all $u \in X$. Find $\partial\varphi(0)$.

Problem 3.49*

Let H be a Hilbert space and let $C \subseteq H$ be a nonempty and closed set. Let $\operatorname{dist}(\cdot, C)$ be the distance function from the set C, i.e.,

$$\operatorname{dist}(u, C) = \inf_{c \in C} \|u - c\|_H.$$

Let $\varphi = \frac{1}{2}\|\cdot\|_H^2 + i_C$. Show that

$$\operatorname{dist}(\cdot, C)^2 = \|\cdot\|_H^2 - 2\varphi^*.$$

Problem 3.50**

Suppose that X is a Banach space, $U \subseteq X$ is a nonempty, open, and convex set and $\varphi\colon U \longrightarrow \mathbb{R}$ is a Gâteaux differentiable function such that

$$\left\langle \varphi'_G(u) - \varphi'_G(v), u - v \right\rangle \geqslant 0 \qquad \forall\, u, v \in U.$$

Show that φ is convex.

Problem 3.51**
Suppose that X is a Banach space, $u^* \in X^* \setminus \{0\}$, $u_0 \in X$ and

$$D_+ = \{u \in X : \langle u^*, u \rangle \geqslant \langle u^*, u_0 \rangle\}.$$

Show that $\partial i_{D_+}(u_0) = \mathbb{R}_- u^*$ with $\mathbb{R}_- = (-\infty, 0]$.

Problem 3.52*
Suppose that X is a reflexive Banach space, $\varphi \in \Gamma_0(X)$ and $\varepsilon > 0$. Show that the following three conditions are equivalent:
(a) $u^* \in \partial_\varepsilon \varphi(u)$.
(b) $\varphi(u) + \varphi^*(u^*) - \langle u^*, u \rangle \leqslant \varepsilon$.
(c) $u \in \partial_\varepsilon \varphi^*(u^*)$.

Problem 3.53**
Let $\varphi\colon \mathbb{R} \longrightarrow \mathbb{R}$ be defined by $\varphi(u) = |u|$ and let $\varepsilon > 0$. Find $\partial_\varepsilon \varphi(u)$.

Problem 3.54**
Let X be a Banach space and let $\varphi(u) = \|u\|_X$ for all $u \in X$. Find $\partial\varphi(u)$.

Problem 3.55**
Find the points of Gâteaux differentiability for the norm of $L^1((0,1); \mathbb{R}^N)$.

Problem 3.56*
Suppose that X is a Banach space, $h \in \Gamma_0(\mathbb{R})$ is an even function and $\varphi(u) = h(\|u\|_X)$ for all $u \in X$. Show that

$$\varphi^*(u^*) = h^*(\|u^*\|_*) \qquad \forall\, u^* \in X^*.$$

Problem 3.57*
Let $a\colon \mathbb{R}_+ \longrightarrow \overline{\mathbb{R}} = \mathbb{R} \cup \{+\infty\}$ be a proper function (that is, $a \not\equiv +\infty$), $a(0) = 0$, $a(t) > 0$ for all $t > 0$ and

$$\frac{a(t_1)}{t_1} \leqslant \frac{a(t_2)}{t_2} \qquad \forall\, 0 < t_1 \leqslant t_2.$$

Show that $\lim\limits_{s \searrow 0} \frac{a^*(s)}{s} = 0$.

Problem 3.58***
Let $\varphi\colon \mathbb{R}^N \longrightarrow \overline{\mathbb{R}} = \mathbb{R} \cup \{+\infty\}$ be a proper function (that is, $\varphi \not\equiv +\infty$) and assume that $M(\varphi) = \{u \in \mathbb{R}^N : \varphi(u) = \inf\limits_{\mathbb{R}^N} \varphi\} = \{u_0\}$. Show that φ^* is differentiable at $u = 0$.

Problem 3.59**
Let X be a Banach space and let $\varphi\colon X \longrightarrow \overline{\mathbb{R}}$ be a proper convex function. Show that the following two properties are equivalent:

(a) φ is Lipschitz continuous on bounded subsets of X.
(b) $\partial\varphi$ maps bounded subsets of X into bounded subsets of X^*.

Problem 3.60**

Let X be a Banach space and let $\varphi\colon X \longrightarrow \overline{\mathbb{R}} = \mathbb{R} \cup \{+\infty\}$ be a convex function. Suppose that $D \subseteq X$ is a bounded, convex set and φ is bounded above on $C + \delta\overline{B}_1$ for some $\delta > 0$. Show that $\varphi|_D$ is Lipschitz continuous.

Problem 3.61**

Let X be a Banach space and let $\varphi\colon X \longrightarrow \mathbb{R}$ be a convex function which is bounded on bounded sets. Suppose that $\varphi_n \leqslant \varphi$ for all $n \geqslant 1$ and $\varphi_n^* \longrightarrow \varphi^*$ uniformly on bounded sets of $\operatorname{dom}\varphi^*$. Show that $\varphi_n \longrightarrow \varphi$ uniformly on bounded subsets of X.

Problem 3.62**

Suppose that X is a Banach space, $U \subseteq X$ is nonempty, open, and convex set and $\varphi\colon U \longrightarrow X$ is a twice differentiable function. Show that:
(a) φ is convex if and only if $\varphi''(u)(h,h) \geqslant 0$ for all $u \in U$ and all $h \in X$;
(b) if $\varphi''(u)(h,h) > 0$ for all $u \in U$ and all $h \in X \setminus \{0\}$, then φ is strictly convex;
(c) the converse of **(b)** is not true.

Problem 3.63***

Suppose that H is a Hilbert space, $C \subseteq H$ is a nonempty, closed, and convex set and $\varphi\colon H \longrightarrow \mathbb{R}_+$ is defined by

$$\varphi(u) = \tfrac{1}{2}\operatorname{dist}(u,C)^2 = \tfrac{1}{2}\inf_{c\in C}\|u-c\|_H^2 \qquad \forall\, u \in H.$$

Show that $\varphi \in C^1(H)$ and

$$\varphi'(u) = u - \operatorname{proj}_C(u) \qquad \forall\, u \in H,$$

where $\operatorname{proj}_C\colon H \longrightarrow C$ is the metric projection (i.e., $\operatorname{proj}_C(u)$ is the unique $c \in C$ such that $\operatorname{dist}(u,C) = \|u-c\|_H$).

Problem 3.64**

Suppose that H is a Hilbert space, $C \subseteq H$ is a nonempty, closed, and convex set, and $\varphi\colon K \longrightarrow \mathbb{R}_+$ is defined by

$$\varphi(u) = \tfrac{1}{2}\operatorname{dist}(u,C)^2 \qquad \forall\, u \in H$$

(see Problem 3.63). Show that φ is convex.

Problem 3.65**

Suppose that X is a Banach space, $\varphi \in \Gamma_0(X)$, $u_0 \in \operatorname{dom}\varphi$ and $\varepsilon > 0$. Show that there exists $u \in D(\partial\varphi)$ such that

$$|\varphi(u) - \varphi(u_0)| < \varepsilon \quad \text{and} \quad \|u-u_0\|_X < \varepsilon.$$

Problem 3.66***
Suppose that X is a Banach space, $\varphi\colon X \longrightarrow \overline{\mathbb{R}}$ is a lower semicontinuous (respectively sequentially weakly lower semicontinuous) function such that φ^{**} is proper and φ^* is Fréchet (respectively Gâteaux) differentiable at all $u^* \in D(\partial\varphi^*)$. Show that φ is convex.

Problem 3.67**
Let H be a Hilbert space and let $C \subseteq H$ be a nonempty and weakly closed set. Suppose that $\operatorname{dist}(\cdot, C)^2$ is Gâteaux differentiable. Show that C is convex.

Problem 3.68**
Suppose that H is a Hilbert space, $K \subseteq H$ is a nonempty, closed, and convex cone (i.e., $K \subseteq H$ is convex and $\lambda K \subseteq K$ for all $\lambda \geqslant 0$). Let $\operatorname{proj}_K\colon H \longrightarrow K$ be the metric projection map (see also Problem 3.63). Show the following properties:

(a) proj_K is characterized by

$$\big(u - \operatorname{proj}_K(u), c\big)_H \leqslant 0 \qquad \forall\, c \in C$$

and

$$\big(u - \operatorname{proj}_K(u), \operatorname{proj}_K(u)\big)_H = 0.$$

(b) proj_K satisfies $\operatorname{proj}_K(\lambda u) = \lambda \operatorname{proj}_K(u)$ for all $u \in H$ and all $\lambda > 0$, $\|u\|_H^2 = \|\operatorname{proj}_K(u)\|_H^2 + \|(I_H - \operatorname{proj}_K)(u)\|_H^2$ and $\|\operatorname{proj}_K(u)\|_H \leqslant \|u\|_H$ for all $u \in H$.

Problem 3.69**
Suppose that H is a Hilbert space, $K \subseteq H$ is a nonempty, closed, and convex cone (see Problem 3.68) and $K^0 \subseteq H$ is the polar of K, i.e,

$$K^0 = \{h \in H : \ (h, c)_H \leqslant 0 \qquad \forall\, c \in K\}.$$

Show that if $u = y + v$ with $y \in K$ and $v \in K^0$, $(y, v)_H = 0$, then $y = \operatorname{proj}_K(u)$ and $v = \operatorname{proj}_{K^0}(u)$. Moreover,

$$K = \operatorname{proj}_{K^0}^{-1}(0) \quad \text{and} \quad K^0 = \operatorname{proj}_K^{-1}(0).$$

Problem 3.70*
Suppose that X is a Banach space, $\varphi\colon X \longrightarrow \overline{\mathbb{R}}$ is a proper and convex function, $c > 0$ and $\eta \in \mathbb{R}$. Show that $\varphi \geqslant c\|\cdot\|_X + \eta$ if and only if $\varphi^* \leqslant -\eta$ on $c\overline{B}_1^*$.

Problem 3.71**
Let X be a Banach space and let $\varphi, \psi\colon X \longrightarrow \mathbb{R}$ be continuous and convex functions. Suppose that φ is not Fréchet differentiable at $u_0 \in X$. Can $\varphi + \psi$ be Fréchet differentiable at u_0? Justify your answer.

Problem 3.72*
Let X be a Banach space and let $\varphi, \psi\colon X \longrightarrow \mathbb{R}$ be two convex functions such that $\psi \leqslant \varphi$, φ is Fréchet differentiable at $u_0 \in X$ and $\varphi(u_0) = \psi(u_0)$. Show that ψ is Fréchet differentiable at u_0.

Problem 3.73**
Suppose that X is a Banach space, $K \subseteq X$ is a nonempty convex cone (see Problem 3.68) and $\varphi\colon K \longrightarrow \mathbb{R}$ is a positively homogeneous function (i.e., $\varphi(\lambda u) = \lambda\varphi(u)$ for all $u \in K$, $\lambda > 0$).

(a) Show that φ is convex if and only if $\varphi(u+v) \leqslant \varphi(u) + \varphi(v)$ for all $u, v \in K$.

(b) Show that if $\varphi \geqslant 0$ and the set $\varphi^1 = \{u \in K : \varphi(u) \leqslant 1\}$ is convex, then φ is convex.

Problem 3.74**
Suppose that $p \geqslant 1$, $K = \mathbb{R}^n_+ = \{\widehat{u} = (u_1, \ldots, u_n) : u_k \geqslant 0 \text{ for all } k \in \{1, \ldots, n\}\}$ and $\varphi\colon K \longrightarrow \mathbb{R}$ is defined by

$$\varphi(\widehat{u}) = \varphi(u_1, \ldots, u_n) = \Big(\sum_{k=1}^{n} u_k^p\Big)^{\frac{1}{p}} \qquad \forall\, \widehat{u} = (u_1, \ldots, u_n) \in K.$$

Show that φ is convex.

Problem 3.75**
Let $D = \{\widehat{u} = (u_1, \ldots, u_n) \in \mathbb{R}^n : u_k > 0 \text{ for all } k \in \{1, \ldots, n\}\}$ and let $\varphi\colon D \longrightarrow \mathbb{R}$ be defined by

$$\varphi(\widehat{u}) = -(u_1 \ldots u_n)^{\frac{1}{n}} \qquad \forall\, \widehat{u} = (u_1, \ldots, u_n) \in D.$$

Show that φ is convex.

Problem 3.76**
Let X be a Banach space and let $C \subseteq X$ be a nonempty, closed, and convex set. Show that for any $u \notin C$, we have $\|u^*\|_* = 1$ for all $u^* \in \partial\mathrm{dist}(u, C)$ (recall that $\mathrm{dist}(u, C) = \inf\limits_{c \in C} \|u - c\|_X$).

Problem 3.77**
Let X be a Banach space with a strictly convex dual X^* and let $C \subseteq X$ be a nonempty, closed, and convex set. Show that for every $u \notin C$, the set $\partial\mathrm{dist}(u, C)$ is a singleton.

Problem 3.78**

(a) Let $\varphi\colon \mathbb{R}^N \longrightarrow \mathbb{R}$ be a convex function. Show that $\partial\varphi$ maps bounded sets to bounded sets.

(b) Let $\{\varphi_n\colon \mathbb{R}^N \longrightarrow \mathbb{R}\}_{n \geqslant 1}$ be a sequence of convex functions such that $\varphi_n \longrightarrow \varphi$ pointwise, where $\varphi\colon \mathbb{R}^N \longrightarrow \mathbb{R}$ is a convex function. Also, let $\{u_n\}_{n \geqslant 1} \subseteq \mathbb{R}$ be a sequence such that $u_n \longrightarrow u \in \mathbb{R}$. Show that given $\varepsilon > 0$ we can find $n_0 = n_0(\varepsilon) \geqslant 1$ such that

$$\partial\varphi_n(u_n) \subseteq \partial\varphi(u) + \varepsilon B_1(0) \qquad \forall\, n \geqslant n_0.$$

Problem 3.79***

Let $\{\varphi_n: \mathbb{R}^N \longrightarrow \overline{\mathbb{R}}\}_{n\geqslant 1}$ be an increasing sequence of lower semicontinuous functions. Suppose that there exists a function $\xi: \mathbb{R}_+ \longrightarrow \mathbb{R}$ such that $\lim\limits_{t\to+\infty} \frac{\xi(t)}{t} = +\infty$ and $\varphi_n(u) \geqslant \xi(|u|)$ for all $n \geqslant 1$, all $u \in \mathbb{R}^N$. Show that

$$\sup_{n\geqslant 1} \varphi_n^{**} = \big(\sup_{n\geqslant 1} \varphi_n\big)^{**}.$$

Is the result true if we drop the boundedness below by ξ hypothesis? Justify your answer.

Problem 3.80***

Let X be a reflexive, locally uniformly convex Banach space with uniformly convex dual and let $\varphi \in \Gamma_0(X)$. For every $\lambda > 0$, we define

$$\varphi_\lambda(u) = \inf_{v\in X} \big\{\tfrac{1}{2\lambda}\|u-v\|_X^2 + \varphi(v)\big\}.$$

Show the following:

(a) For every $\lambda > 0$, the function φ_λ is convex, continuous, Gâteaux differentiable, and $(\partial\varphi)_\lambda = (\varphi_\lambda)'_G = \partial\varphi_\lambda$.

(b) $\varphi_\lambda(u) = \frac{1}{2\lambda}\|u - J_\lambda^{\partial\varphi}(u)\|_X^2 + \varphi(J_\lambda^{\partial\varphi}(u))$ for all $\lambda > 0$, $u \in X$ (here $J_\lambda^{\partial\varphi}$ is the resolvent of $\partial\varphi$; see Definition 2.122).

(c) $\lim\limits_{\lambda\searrow 0} \varphi_\lambda(u) = \varphi(u)$ for all $u \in X$.

(d) $\varphi(J_\lambda^{\partial\varphi}(u)) \leqslant \varphi_\lambda(u) \leqslant \varphi(u)$ for all $\lambda > 0$, $u \in X$.

Remark. The function φ_λ is known as the ***Moreau–Yosida regularization*** (or ***approximation***) of φ.

Problem 3.81***

(a) Suppose that H is a Hilbert space, $A: H \supseteq D(A) \longrightarrow H$ is a linear, self-adjoint and positive operator, $\varphi: H \longrightarrow \overline{\mathbb{R}}$ is defined by

$$\varphi(u) = \begin{cases} \frac{1}{2}\|A^{\frac{1}{2}}(u)\|_H & \text{if } u \in D(A^{\frac{1}{2}}), \\ +\infty & \text{otherwise} \end{cases}$$

(here $A^{\frac{1}{2}}$ denotes the square root of A). Show that $A = \partial\varphi$.

(b) Suppose that H is a Hilbert space, $A: H \supseteq D(A) \longrightarrow H$ is a linear, densely defined operator which is the subdifferential of a function $\psi \in \Gamma_0(H)$. Show that A is self-adjoint.

Problem 3.82**

Let H be a Hilbert space (not necessarily identified with its dual) and let $\varphi \in \Gamma_0(H)$. For every $\lambda > 0$, let φ_λ be the Moreau–Yosida approximation of φ (see Problem 3.80). Show that φ_λ is Fréchet differentiable on H.

Problem 3.83**
Let (Ω, Σ, μ) be a σ-finite measure space and let $f\colon L^1(\Omega) \longrightarrow L^\infty(\Omega)$ be the map defined by

$$f(u) = \{\|u\|_1 y : y \in L^\infty(\Omega),\ y(\omega) \in \operatorname{sign} u(\omega) \text{ for } \mu\text{-a.a. } \omega \in \Omega\},$$

where

$$\operatorname{sign} x = \begin{cases} \{-1\} & \text{if } x < 0, \\ [-1, 1] & \text{if } x = 0, \\ \{1\} & \text{if } x > 0. \end{cases}$$

Show that $f = \mathcal{F}$.

Problem 3.84**
Suppose that H is a Hilbert space, $\xi\colon \mathbb{R}_+ \longrightarrow \overline{\mathbb{R}}$ is an increasing function and $C \subseteq H$ is a nonempty set. Let $\varphi(u) = \xi(\operatorname{dist}(u, C))$ for all $u \in H$. Show that

$$\varphi^*(h) = \xi^*(\|h\|_H) + \sigma_C(h) \qquad \forall\, h \in H.$$

Problem 3.85**
Suppose that H is a Hilbert space, $C \subseteq H$ is a nonempty and convex set, $\xi\colon \mathbb{R}_+ \longrightarrow \overline{\mathbb{R}}$ is a proper function, and $\varphi(u) = \xi(\|u\|_H) + \sigma_C(u)$ for all $u \in H$. Show that

$$\varphi^*(h) = \xi^*(\operatorname{dist}(h, C)) \qquad \forall\, h \in H.$$

Problem 3.86**
Let X be a Banach space and let $\varphi\colon X \longrightarrow \mathbb{R}$ be a convex function such that

$$\varphi(u) \leqslant c\|u\|_X \qquad \forall\, u \in X,$$

for some $c > 0$. Show that φ is Lipschitz continuous with Lipschitz constant c.

Problem 3.87**
Show that Theorem 3.73 is no longer true if φ is only lower semicontinuous at $u \in X$ (instead of continuity at u).

Problem 3.88**
Let X be a Banach space and let $C \subseteq X$ be a nonempty, closed, and convex set. Show that

$$\operatorname{dist}(u, C) = \sup\{\langle u^*, u\rangle - \sigma_C(u^*) : \|u^*\|_* \leqslant 1\} \qquad \forall\, u \in X.$$

Problem 3.89**
Suppose that $\Omega \subseteq \mathbb{R}^N$ is a bounded open set with a C^1-boundary $\partial\Omega$, $1 < p < +\infty$ and $g\colon \mathbb{R} \longrightarrow \mathbb{R}$ is a continuous and convex function satisfying the following growth condition

$$|g(u)| \leqslant c(1 + |u|^p) \qquad \forall\, u \in \mathbb{R},$$

for some $c > 0$. Consider the integral operator $\varphi\colon W^{1,p}(\Omega) \longrightarrow \mathbb{R}$ defined by

$$\varphi(u) = \int_{\partial\Omega} g(\gamma(u)(z))\, d\sigma,$$

where γ is the trace map (see Definition 1.129) and σ denotes the $(N-1)$-dimensional Hausdorff (surface) measure. Show that φ is continuous, convex, and $u^* \in \partial\varphi(u)$ if and only if there exists $h \in L^{p'}(\partial\Omega)$ $(\frac{1}{p} + \frac{1}{p'} = 1)$ such that

$$h(z) \in \partial g(\gamma(u)(z)) \quad \text{for } \sigma\text{-a.a. } z \in \Omega$$

and

$$\langle u^*, v\rangle = \int_{\partial\Omega} h(z)\gamma(v)(z)\, d\sigma \qquad \forall\, v \in W^{1,p}(\Omega).$$

Problem 3.90*

Let $\Omega \subseteq \mathbb{R}^N$ be a bounded domain with a C^1-boundary $\partial\Omega$ and let $1 < p < +\infty$. Consider the functional $\varphi\colon L^p(\Omega) \longrightarrow \overline{\mathbb{R}} = \mathbb{R} \cup \{+\infty\}$ defined by

$$\varphi(u) = \begin{cases} \|Du\|_p^p & \text{if } u \in W^{1,p}(\Omega), \\ +\infty & \text{otherwise.} \end{cases}$$

Show that φ is lower semicontinuous.

Problem 3.91**

Suppose that H is a Hilbert space, $\varphi\colon H \longrightarrow \mathbb{R}$ is a lower semicontinuous and convex function and $u \in H$ is a minimizer of φ. Show that given $\varepsilon > 0$, we can find $\delta > 0$ such that

$$\varphi(u+h) \leqslant \varphi(u) + \varepsilon\|h\|_H \qquad \forall\, h \in N_{\partial\varphi(u)}(0),\ \|h\|_H \leqslant \delta.$$

Moreover, show that

$$\varphi(u+h) > \varphi(u) \qquad \forall\, h \in H \setminus N_{\partial\varphi(u)}(0).$$

Problem 3.92*

Let X be a Banach space and let $\varphi\colon X \longrightarrow \mathbb{R}$ be a continuous and convex function. Show that $0 \in \operatorname{int} \partial\varphi(u)$ if and only if there exists $\varepsilon > 0$ such that

$$\varphi(u+h) \geqslant \varphi(u) + \varepsilon\|h\|_X \qquad \forall\, h \in X.$$

Problem 3.93***

Let T be a compact metric space and let $\varphi_\lambda\colon \mathbb{R}^N \longrightarrow \mathbb{R}$ be convex functions for $\lambda \in T$, such that for each $u \in \mathbb{R}^N$, the function $\lambda \longmapsto \varphi_\lambda(u)$ is upper semicontinuous. We define

$$\varphi(u) = \sup_{\lambda\in T} \varphi_\lambda(u) \qquad \forall\, u \in \mathbb{R}^N$$

and

$$M(u) = \{\lambda \in T : \varphi_\lambda(u) = \varphi(u)\}.$$

Show that

$$\partial\varphi(u) = \operatorname{conv} \bigcup_{\lambda \in M(u)} \partial\varphi_\lambda(u) \qquad \forall\, u \in \mathbb{R}^N.$$

Problem 3.94**

Let $\{\varphi_n : \mathbb{R}^N \longrightarrow \mathbb{R}\}_{n\geqslant 1}$ be a sequence of differentiable and convex functions such that $\varphi_n \longrightarrow \varphi$ pointwise, for some convex and differentiable function $\varphi : \mathbb{R}^N \longrightarrow \mathbb{R}$. Show that $\varphi_n' \longrightarrow \varphi'$ uniformly on compact subsets of $\mathbb{R}^N$.

Problem 3.95**

Suppose that X is a Banach space, $C \subseteq X$ is a nonempty, closed, and convex set and $\varphi : X \longrightarrow \mathbb{R}$ is a continuous convex function. We consider the following minimization problem

$$(P) \qquad \inf_{u \in C} \varphi(u).$$

Show that the following statements are equivalent:

(a) $u_0 \in C$ is a solution of problem (P).

(b) $\varphi'(u; h) \geqslant 0$ for all $h \in T_C(u_0)$.

(c) $0 \in \partial\varphi(u_0) + N_C(u_0)$.

Problem 3.96*

Suppose that $I \subseteq \mathbb{R}$ is an open interval, $\varphi : I \longrightarrow \mathbb{R}$ is a convex function, and $s : I \longrightarrow \mathbb{R}$ is a function such that

$$s(u) \in \partial\varphi(u) \qquad \forall\, u \in I.$$

Show that

$$\varphi_-'(u) \leqslant s(u) \leqslant \varphi_+'(u) \qquad \forall\, u \in I,$$

where

$$\varphi_+'(u) = \lim_{h \searrow 0} \tfrac{\varphi(u+h)-\varphi(u)}{h} = \inf_{h>0} \tfrac{\varphi(u+h)-\varphi(u)}{h}$$

$$\varphi_-'(u) = \lim_{h \nearrow 0} \tfrac{\varphi(u+h)-\varphi(u)}{h} = \inf_{h<0} \tfrac{\varphi(u+h)-\varphi(u)}{h}$$

Problem 3.97**

Let $\varphi : [a, b] \longrightarrow \mathbb{R}$ be a continuous and convex function. Show that

$$\varphi\left(\tfrac{a+b}{2}\right) \leqslant \tfrac{1}{b-a} \int_a^b \varphi(t)\, dt \leqslant \tfrac{\varphi(a)+\varphi(b)}{2}.$$

Problem 3.98*
Suppose that X is a Banach space, $U \subseteq X$ is a nonempty, open, and convex set and $\varphi\colon U \longrightarrow \mathbb{R}$ is a convex function which attains a global maximum at an interior point of U. Show that φ is constant.

Problem 3.99*
Suppose that X is a Banach space, $K \subseteq X$ is a nonempty, convex, and compact set and $\varphi\colon K \longrightarrow \mathbb{R}$ is a continuous and convex function. Show that φ attains its global maximum at an extreme point of φ.

Problem 3.100**
Let X be a Banach space and let $\varphi \in \Gamma_0(X)$. Show that φ is continuous at $u_0 \in X$ if and only if φ^{**} is continuous at u_0.

Problem 3.101**
Let X be a Banach space and let $\varphi \in \Gamma_0(X)$. Show that φ is bounded on bounded sets of X if and only if φ^{**} is bounded on bounded sets of X^{**}.

Problem 3.102**
Let X be a Banach space and let $\varphi \in \Gamma_0(X)$. Show that φ is Fréchet differentiable at $u_0 \in X$ if and only if φ^{**} is Fréchet differentiable at $u_0 \in X$.

Problem 3.103*
Let X be a Banach space and let $\varphi\colon X \longrightarrow \overline{\mathbb{R}}$ be a proper convex function which is lower semicontinuous at some point of its domain. Show that φ is coercive if and only if φ^* is continuous at the origin.

Problem 3.104**
Let X be a Banach space and let $\varphi \in \Gamma_0(X)$. Show that φ is continuous at the origin if and only if φ^* has w^*-compact sublevel sets.

Problem 3.105**
Let X be a Banach space and let $\varphi \in \Gamma_0(X)$. Show the following statements:
(a) φ is strongly coercive if and only if φ^* is bounded on bounded sets;
(b) φ is bounded on bounded sets if and only if φ^* is strongly coercive.

Problem 3.106**
Suppose that X is a Banach space, $C \subseteq X$ is a nonempty, bounded, convex set and $\varphi\colon X \longrightarrow \overline{\mathbb{R}}$ is a convex function which is bounded above on $C_\delta = \{u \in X : \operatorname{dist}(u, C) \leqslant \delta\}$ for some $\delta > 0$. Show that φ is Lipschitz continuous on C.

Problem 3.107**
Let X be a Banach space and let $\varphi\colon X \longrightarrow \mathbb{R}$ be a convex function which is bounded on bounded sets. Show that

$$\varphi \boxdot n\|\cdot\|_X^2 \longrightarrow \varphi \quad \text{and} \quad \varphi \boxdot n\|\cdot\|_X \longrightarrow \varphi$$

uniformly on bounded subsets of X.

Problem 3.108*
Let X be a Banach space and let $C \subseteq X$ be a nonempty, closed, and convex set. Show that the support function $\sigma_C \colon X^* \longrightarrow \overline{\mathbb{R}}$ is proper, lower semicontinuous, convex (i.e., $\sigma_C \in \Gamma_0(X^*)$), and positively homogeneous of degree 1. Also, show that, if $C \subseteq X$ is in addition w-compact, then the function σ_C is continuous.

Problem 3.109*
Let X and Y be two Banach spaces and let $\xi \colon X \times Y \longrightarrow \overline{\mathbb{R}}$ be a proper and convex function. Consider the function $m \colon X \longrightarrow \mathbb{R} \cup \{\pm\infty\}$ defined by

$$m(u) = \inf_{y \in Y} \xi(u, y).$$

Show that the function m is convex.

Problem 3.110**
Suppose that X is a Banach space, $\varphi \in \Gamma_0(X)$, $u_0 \in \operatorname{dom}\varphi$, $u \in X$ and $\varepsilon > 0$. We set

$$\varphi^\infty(h) = \sup_{\lambda>0} q_\lambda(h), \quad \text{where } q_\lambda(h) = \tfrac{\varphi(u_0+\lambda h)-\varphi(u_0)}{\lambda}.$$

Show that $\varphi^\infty \in \Gamma_0(X)$, $\varphi^\infty(u) = \lim\limits_{\lambda \to +\infty} q_\lambda(u)$, φ^∞ does not depend on the choice of $u_0 \in \operatorname{dom}\varphi$ and φ^∞ is positively homogeneous of degree 1.

Problem 3.111**
(a) Let X be a Banach space and let $\varphi \in \Gamma_0(X)$ be such that $\inf\limits_X \varphi > -\infty$. Show that $\varphi^\infty \geqslant 0$ (where φ^∞ is defined in Problem 3.110).
(b) Suppose that X is a Banach space, $\varphi \in \Gamma_0(X)$, $\varphi \geqslant 0$, φ is positively homogeneous of degree $p > 1$, $\widehat{u}^* \in X^*$ and $\psi \colon X \longrightarrow \mathbb{R}$ is defined by

$$\psi(u) = -\langle \widehat{u}^*, u\rangle + \varphi(u) \qquad \forall\, u \in H.$$

Show that if the problem $\inf\limits_X \psi$ admits a solution, then

$$\langle \widehat{u}^*, u\rangle \leqslant 0 \qquad \forall\, u \in X,\ \varphi(u) = 0.$$

Problem 3.112***
Let X be a reflexive Banach space and let $\varphi \in \Gamma_0(X)$ be a function satisfying the following conditions:
(i) if $\{t_n\}_{n\geqslant 1} \subseteq \mathbb{R}$ and $\{u_n\}_{n\geqslant 1} \subseteq X$ are two sequences such that $t_n \longrightarrow +\infty$, $u_n \xrightarrow{w} u$ and $\{\varphi(t_n u_n)\}_{n\geqslant 1} \subseteq \mathbb{R}$ is bounded above, then $u_n \longrightarrow u$ in X;
(ii) $\varphi^\infty(h) \geqslant 0$ for all $h \in X$ (where φ^∞ is defined in Problem 3.110);
(iii) $\ker \varphi^\infty = \{u \in X : \varphi^\infty(u) = 0\}$ is a linear subspace of X.
Show that the problem $\inf\limits_{u \in X} \varphi(u)$ admits a solution.

Problem 3.113*
Let X be a Banach space and let $C \subseteq X$ be a nonempty, closed, and convex set. Show that the multifunction $u \longmapsto N_C(u)$ is closed.

Problem 3.114**
Let H be a finite dimensional Hilbert space and let $C \subseteq H$ be a nonempty, closed, and convex set. Show that the multifunction $u \longmapsto T_C(u)$ is lower semicontinuous.

Problem 3.115*
Let H be a Hilbert space and let $u \in \overline{B}_1$, where $\overline{B}_1 = \{u \in H : \|u\|_H \leqslant 1\}$. Show that

$$\begin{array}{rcll} T_{\overline{B}_1}(u) & = & H & \forall\, u \in \operatorname{int} \overline{B}_1 \\ T_{\overline{B}_1}(u) & = & \{h \in H : (h,u)_H \leqslant 0\} & \forall\, u \in H,\ \|u\|_H = 1 \\ N_{\overline{B}_1}(u) & = & \{0\} & \forall\, u \in \operatorname{int} \overline{B}_1 \\ N_{\overline{B}_1}(u) & = & \mathbb{R}_+ u & \forall\, u \in H,\ \|u\|_H = 1. \end{array}$$

Problem 3.116**
Suppose that X is a reflexive Banach space, $A \colon X \longrightarrow X^*$ is a monotone and demicontinuous map, $\varphi \in \Gamma_0(X)$ with $0 \in \operatorname{dom} \varphi$ and

$$\lim_{\|u\|_X \to +\infty} \frac{\langle A(u), u\rangle + \varphi(u)}{\|u\|_X} = +\infty.$$

Show that there exists $u \in X$ such that

$$\langle A(u), v-u\rangle + \varphi(v) - \varphi(u) \geqslant \langle u^*, v-u\rangle \qquad \forall\, v \in X.$$

Moreover, show that the set of solutions of the above problem is bounded, closed, and convex and if A is strictly monotone, then it is a singleton.

Problem 3.117*
Let $\Omega \subseteq \mathbb{R}^N$ be a bounded open set with C^1 boundary $\partial\Omega$ and let $g \in H^1(\Omega)$ be such that $g|_{\partial\Omega} \leqslant 0$. Let

$$C = \left\{u \in H^1_0(\Omega) : u(z) \geqslant g(z) \text{ for a.a. } z \in \Omega\right\}.$$

Show that there exists a unique $u \in C$ such that

$$\int_\Omega (Du, Dv - Du)_{\mathbb{R}^N}\, dz \geqslant 0 \qquad \forall\, v \in C.$$

Problem 3.118**
Suppose that X and Y are two Banach spaces, $u \in X$, $U \subseteq X$ is a nonempty open set containing u and $\varphi \colon U \longrightarrow Y$ is a strictly differentiable map (see Remark 3.125) with strict derivative $\varphi'_s(u) \in \mathcal{L}(X;Y)$. Show that φ is Lipschitz continuous near u.

Problem 3.119**

Suppose that X and Y are two Banach spaces, $u \in X$, $\varphi\colon X \longrightarrow Y$ is a function which is Lipschitz continuous near u and there exists $A \in \mathcal{L}(X;Y)$ such that

$$\lim_{\substack{v \to u \\ \lambda \searrow 0}} \frac{\varphi(v+\lambda h)-\varphi(v)}{\lambda} = A(h) \qquad \forall\, h \in X.$$

Show that φ is strictly differentiable at u and $\varphi_s'(u) = A$.

Problem 3.120***

Let X be a Banach space and let $\varphi\colon X \longrightarrow \mathbb{R}$ be a locally Lipschitz function. Must φ be locally Lipschitz on every bounded subset of X? Justify your answer.

Problem 3.121*

Let X be a Banach space and let $\varphi\colon X \longrightarrow \mathbb{R}$ be a locally Lipschitz function which is Gâteaux differentiable and regular at $u \in X$ (see Definition 3.124). Show that $\partial\varphi(u) = \{\varphi_G'(u)\}$.

Problem 3.122*

Suppose that X is a Banach space, $\varphi\colon X \longrightarrow \mathbb{R}$ is a locally Lipschitz function, $u \in X$ is a local extremum for φ (i.e., a local minimum or a local maximum). Show that $0 \in \partial\varphi(u)$.

Problem 3.123**

Let X be a Banach space and let $\varphi\colon X \longrightarrow \mathbb{R}$ be a locally Lipschitz function. We define

$$\mu(u) = \inf_{u^* \in \partial\varphi(u)} \|u^*\|_* \qquad \forall\, x \in X.$$

Show that the map $\mu\colon X \longmapsto \mathbb{R}$ is lower semicontinuous.

Problem 3.124**

Let X be a reflexive Banach space which is embedded continuously and densely into another Banach space Y and let $\varphi\colon Y \longrightarrow \mathbb{R}$ be a locally Lipschitz function. Show that

$$\partial(\varphi|_X)(u) = (\partial\varphi)|_X(u) \qquad \forall\, u \in X.$$

Problem 3.125**

Suppose that X is a Banach space, $\vartheta \in C^1([0,1];X)$ and $\varphi\colon X \longrightarrow \mathbb{R}$ is a locally Lipschitz function. Show that $\gamma = \varphi \circ \vartheta$ is differentiable almost everywhere and

$$\gamma'(t) \leqslant \max_{u^* \in \partial\varphi(\vartheta(t))} \langle u^*, \vartheta'(t)\rangle \quad \text{for a.a. } t \in T.$$

Problem 3.126***

Let X be a Banach space and let $\varphi\colon X \longrightarrow \mathbb{R}$ be a Gâteaux differentiable function such that the map $u \longmapsto \varphi_G'(u)$ is continuous at u_0. Show that

$$\varphi^0(u_0;h) = \langle \varphi_G'(u_0), h\rangle \qquad \forall\, h \in X.$$

Problem 3.127**
Suppose that X is a Banach space, $K \subseteq X$ is a nonempty, closed, and convex cone and $\varphi\colon X \longrightarrow \mathbb{R}$ is a locally Lipschitz function such that

$$\varphi(u) \leqslant \varphi(u+h) \qquad \forall\, u \in X,\ h \in K.$$

Show that for every $u \in X$ and $h \in K$, we have

$$\varphi^0(u;h) \leqslant \sigma_{K^*}(h) \quad \text{and} \quad \partial\varphi(u) \subseteq K^*,$$

where $K^* = \{h^* \in X^* : \langle h^*, h\rangle \geqslant 0 \text{ for all } h \in K\}$ (the ***dual cone***).

Problem 3.128**
Suppose that X is a Banach space, $C \subseteq X$ is a nonempty set, $\varphi\colon X \longrightarrow \mathbb{R}$ is a locally Lipschitz function and $u_0 \in C$ is such that

$$\varphi(u_0) = \inf_{u\in C} \varphi(u).$$

Show that $0 \in \partial\varphi(u_0) + N_C'(u_0)$.

Problem 3.129**
Suppose that X is a Banach space, $C \subseteq X$ is a nonempty, closed set and $u \in C$. Show that

$$\lim_{\substack{v \to u \\ v \in C}} \liminf_{\lambda \searrow 0} \frac{\operatorname{dist}(v+\lambda h, C)}{\lambda} = 0 \qquad \forall\, h \in T_C'(u).$$

Problem 3.130***
Suppose that H is a Hilbert space, $C \subseteq H$ is a nonempty, closed set and $u \in C$. Show that

$$(v-u, h)_H \leqslant 0 \qquad \forall\, h \in T_C'(u),\ v \in H,\ \operatorname{dist}(v, C) = \|v-u\|_H.$$

Problem 3.131**
Suppose that X is a Banach space, $C \subseteq X$ is a nonempty set, $\varphi\colon X \longrightarrow \mathbb{R}$ is a locally Lipschitz function and $u_0 \in C$ is such that

$$\varphi(u_0) = \inf_{u\in C} \varphi(u).$$

Show that

$$\varphi^0(u_0;h) \geqslant 0 \qquad \forall\, h \in T_C'(u_0).$$

Problem 3.132**
Suppose that $C \subseteq \mathbb{R}^N$ is a nonempty set and that $\operatorname{dist}'(u, C)$ exists and is nonzero. Show that $u \notin \overline{C}$, it admits a unique best approximation $c_0 \in \overline{C}$ and $\operatorname{dist}(u, C) = \frac{u-c_0}{|u-c_0|}$.

Problem 3.133**
Let X be a Banach space and let $\varphi \in \Gamma_0(X)$ be such that $\varphi(0) = 0$ and

$$\varphi(u) \geqslant -\|u\|_X \qquad \forall\, u \in X.$$

Show that there exists $u^* \in X^*$ with $\|u^*\|_* \leqslant 1$ such that $\langle u^*, u\rangle \leqslant \varphi(u)$ for all $u \in X$.

Problem 3.134*
Suppose that X is a Banach space, $\varphi\colon X \longrightarrow \mathbb{R}$ is a locally Lipschitz function, $\psi \in \Gamma_0(X)$ and $u \in X$. Show that the following two statements are equivalent:
(a) $0 \leqslant \varphi^0(u; v-u) + \psi(v) - \psi(u)$ for all $v \in X$.
(b) $u \in D(\partial\psi)$ and $0 \in \partial\varphi(u) + \partial\psi(u)$.
Here $\partial\varphi(u)$ is the generalized (Clarke) subdifferential of φ (see Definition 3.126) and $\partial\psi(u)$ is the convex subdifferential of ψ at u (see Definition 3.70).

Problem 3.135**
Let X be a Banach space and let $\varphi\colon X \longrightarrow \overline{\mathbb{R}}$ be a proper, lower semicontinuous function such that

$$\varphi(u) \geqslant -c_0(1 + \|u\|_X) \qquad \forall\, u \in X,$$

for some $c_0 > 0$. For every $n \geqslant 1$ we set

$$\varphi_n(u) = (\varphi \,\square\, n\|\cdot\|_X)(u) \qquad \forall\, u \in X.$$

Show the following:
(a) For every $n \geqslant c_0$ the function φ_n is n-Lipschitz.
(b) The sequence $\{\varphi_n\}_{n\geqslant 1}$ is increasing and $\varphi_n \nearrow \varphi$.
(c) If φ is convex, then so is φ_n for all $n \geqslant c_0$.

Problem 3.136**
Suppose that X is a Banach space, $u \in X$, $u^* \in X^* \setminus \{0\}$ and C is the half space given by

$$C = \{v \in X : \langle u^*, v-u\rangle \leqslant 0\}.$$

Show that $N_C(u) = \mathbb{R}_+ u^*$.

Problem 3.137***
Let X be a Banach space and let $\varphi\colon X \longrightarrow \mathbb{R}$ be a continuous and convex function. We set $C = \{u \in X : \varphi(u) \leqslant 0\}$ and we assume that there exists $u_0 \in X$ such that $\varphi(u_0) < 0$. Show that for every $u \in C$, we have

$$N_C(u) = \begin{cases} \{0\} & \text{if } \varphi(u) < 0,\\ \mathbb{R}_+\partial\varphi(u) & \text{if } \varphi(u) = 0.\end{cases}$$

Problem 3.138*
Suppose that X is a Banach space, $\psi_1, \ldots, \psi_n\colon X \longrightarrow \mathbb{R}$ are continuous and convex functions and $\varphi \in \Gamma_0(X)$. Suppose that there exists $u_0 \in \operatorname{dom}\varphi$ such that $\psi_k(u_0) < 0$ for all $k \in \{1, \ldots, n\}$. Show that the following statements are equivalent:

(a) $u \in X$ is a solution of

$$\min\left\{\varphi(v) : v \in X,\ \psi_k(v) \leqslant 0 \text{ for all } k \in \{1,\ldots,n\}\right\}.$$

(b) There exist numbers $\lambda_1,\ldots,\lambda_n \in \mathbb{R}_+$ such that

$$0 \in \partial\varphi(u) + \sum_{k=1}^{n} \lambda_k \partial\psi_k(u), \quad \psi_k(u) \leqslant 0,\ \lambda_k\psi_k(u) = 0$$
$$\forall\, k \in \{1,\ldots,n\}.$$

Problem 3.139**
Let (X, d_X) be a metric space and let $\{C_n\}_{n\geqslant 1}$ be a sequence of nonempty subsets of X. Show that $C_n \xrightarrow{K} C$ (see Definition 2.88) if and only if $i_{C_n} \xrightarrow{\Gamma} i_C$ (see Definition 3.137).

Problem 3.140*
Suppose that X is a Banach space, $C \subseteq X$ is a set and $\{C_n\}_{n\geqslant 1} \subseteq 2^X$ is a sequence of subsets of X. Show that $C_n \xrightarrow{M} C$ if and only if $i_{C_n} \xrightarrow{M} i_C$.

Problem 3.141*
Let (X, d_X) be a metric space.

(a) Suppose that $\varphi\colon X \longrightarrow \overline{\mathbb{R}}$ is a proper function and $\{\varphi_n\colon X \longrightarrow \overline{\mathbb{R}}\}_{n\geqslant 1}$ is a sequence of proper functions such that $\varphi_n(u) \longrightarrow \varphi(u)$ for all $u \in X$. Assume that the sequence $\{\varphi_n\}_{n\geqslant 1}$ is ***locally equicontinuous*** (i.e., for every $\varepsilon > 0$ and every $u \in X$, there exists a neighborhood U of u such that $|\varphi_n(u) - \varphi_n(v)| < \varepsilon$ for all $v \in U$, all $n \geqslant 1$). Show that $\varphi_n \xrightarrow{\Gamma} \varphi$.

(b) Suppose that $U \subseteq \mathbb{R}^N$ is an open and convex set, $\varphi\colon U \longrightarrow \overline{\mathbb{R}}$ is a proper function and $\{\varphi_n\colon U \longrightarrow \mathbb{R}\}_{n\geqslant 1}$ is a sequence of convex functions such that $\varphi_n(u) \longrightarrow \varphi(u)$ for all $u \in U$. Show that $\varphi_n \xrightarrow{\Gamma} \varphi$.

Problem 3.142***
Suppose that (X, d_X) is a metric space, $\varphi\colon X \longrightarrow \overline{\mathbb{R}}$ is a proper function, and $\{\varphi_n\colon X \longrightarrow \overline{\mathbb{R}}\}_{n\geqslant 1}$ is a sequence of proper functions such that $\varphi_n \xrightarrow{\Gamma} \varphi$. Show that:

(a) if $m_n = \inf\limits_X \varphi_n$ and $m = \inf\limits_X \varphi$, then $\limsup\limits_{n\to+\infty} m_n \leqslant m$;

(b) if for $\varepsilon \geqslant 0$ we set

$$S_n(\varepsilon) = \{u \in X : \varphi_n(u) \leqslant m_n + \varepsilon\} \quad \text{and}$$
$$S(\varepsilon) = \{u \in X : \varphi(u) \leqslant m + \varepsilon\},$$

then for $\varepsilon \geqslant 0$ sufficiently small we have

$$\limsup_{n\to+\infty} S_n(\varepsilon) \subseteq S(\varepsilon);$$

(c) if $\varepsilon_n \searrow 0$, then $\limsup\limits_{n\to+\infty} S_n(\varepsilon_n) \subseteq S(0)$;

(d) if $\limsup\limits_{n\to+\infty} S_n(\varepsilon_n) \neq \emptyset$, then $m_n \longrightarrow m$ and $S(0) \neq \emptyset$;

(e) if there exist two sequences $\{\varepsilon_n\}_{n\geqslant 1} \subseteq \mathbb{R}_+$ and $\{u_n\}_{n\geqslant 1} \subseteq X$ such that $\varepsilon_n \searrow 0$, $u_n \in S_n(\varepsilon_n)$ for $n \geqslant 1$ and the sequence $\{u_n\}_{n\geqslant 1}$ is relatively compact, then $m_n \longrightarrow m$.

Problem 3.143**

Suppose that (X, d_X) is a metric space, $\varphi\colon X \longrightarrow \overline{\mathbb{R}}$ is a proper function and $\{\varphi_n\colon X \longrightarrow \overline{\mathbb{R}}\}_{n\geqslant 1}$ is a sequence of proper functions such that $\varphi_n \xrightarrow{\Gamma} \varphi$, $m_n = \inf\limits_X \varphi_n$ for $n \geqslant 1$, $m = \inf\limits_X \varphi$,

$$S(0) = \{u \in X : \ m = \varphi(u)\} \quad \text{and} \quad S_n(\varepsilon) = \{u \in X : \ \varphi(u) \leqslant m_n + \varepsilon\}.$$

Show that the following statements are equivalent:

(a) $m_n \longrightarrow m$ and $S(0) \neq \emptyset$.

(b) For every $u_0 \in S(0)$, there exist two sequences $\{\varepsilon_n\}_{n\geqslant 1} \subseteq \mathbb{R}_+$ and $\{u_n\}_{n\geqslant 1} \subseteq X$ such that $\varepsilon_n \searrow 0$, $u_n \in S_n(\varepsilon_n)$ for $n \geqslant 1$ and $u_n \longrightarrow u_0$.

Problem 3.144***

Suppose that (X, d_X) is a metric space, $\varphi\colon X \longrightarrow \overline{\mathbb{R}}$ is a proper, lower semicontinuous function and $\{\varphi_n\colon X \longrightarrow \overline{\mathbb{R}}\}_{n\geqslant 1}$ is a sequence of proper functions such that for every compact set $K \subseteq X$, we have

$$\inf_K \varphi_n \longrightarrow \inf_K \varphi.$$

Show that $\varphi_n \xrightarrow{\Gamma} \varphi$.

Problem 3.145*

Suppose that (X, d_X) is a metric space, $\varphi\colon X \longrightarrow \overline{\mathbb{R}}$ is a proper, lower semicontinuous function and $\{\varphi_n\colon X \longrightarrow \overline{\mathbb{R}}\}_{n\geqslant 1}$ is a sequence of proper, lower semicontinuous functions such that $\varphi_n \longrightarrow \varphi$ uniformly on compact subsets of X. Show that $\varphi_n \xrightarrow{\Gamma} \varphi$.

Problem 3.146**

Let X be a Banach space and let $\{\varphi_n\colon X \longrightarrow \overline{\mathbb{R}}\}_{n\geqslant 1}$ be a sequence of proper and convex functions. Show that the function $\widehat{\varphi} = \Gamma\text{--}\limsup\limits_{n\to+\infty} \varphi_n$ is convex too but the function $\widetilde{\varphi} = \Gamma\text{--}\liminf\limits_{n\to+\infty} \varphi_n$ need not be convex.

Problem 3.147*

Let (X, d_X) be a metric space and let $\{\varphi_n\colon X \longrightarrow \overline{\mathbb{R}}\}_{n\geqslant 1}$ be a sequence of proper functions. We set $\widetilde{\varphi} = \Gamma\text{--}\liminf\limits_{n\to+\infty} \varphi_n$ and $\widehat{\varphi} = \Gamma\text{--}\limsup\limits_{n\to+\infty} \varphi_n$. Show that both functions $\widetilde{\varphi}$ and $\widehat{\varphi}$ are lower semicontinuous.

Problem 3.148**
Let (X, d_X) be a metric space and let $\{C_n\}_{n\geqslant 1}$ be a sequence of nonempty subsets of X. Show that sets $\liminf\limits_{n\to+\infty} C_n$ and $\limsup\limits_{n\to+\infty} C$ are both closed (possibly empty; see Definition 2.88).

Problem 3.149*
Suppose that X is a Hausdorff topological space, $\varphi\colon X \longrightarrow \overline{\mathbb{R}}$ is a proper function and $U \subseteq X$ is a nonempty open set. Show that

$$\inf_{u\in U} \varphi(u) = \inf_{u\in U} \operatorname{cl}\varphi(u),$$

where $\operatorname{cl}\varphi$ denotes the closure of φ (see Definition 3.58).

Problem 3.150**
Let X be a Hausdorff topological space and let $\left\{\varphi_n\colon X \longrightarrow \overline{\mathbb{R}}\right\}_{n\geqslant 1}$ be a sequence of proper functions. Show that

$$\Gamma\text{-}\liminf_{n\to+\infty} \varphi_n = \Gamma\text{-}\liminf_{n\to+\infty} \operatorname{cl}\varphi_n \quad \text{and} \quad \Gamma\text{-}\limsup_{n\to+\infty} \varphi_n = \Gamma\text{-}\limsup_{n\to+\infty} \operatorname{cl}\varphi_n.$$

Problem 3.151**
Let (X, d_X) be a metric space and let $\left\{\varphi_n\colon X \longrightarrow \overline{\mathbb{R}}\right\}_{n\geqslant 1}$ and $\left\{\psi_n\colon X \longrightarrow \overline{\mathbb{R}}\right\}_{n\geqslant 1}$ be two sequences of proper functions. Show that

$$\Gamma\text{-}\liminf_{n\to+\infty} \varphi_n + \Gamma\text{-}\liminf_{n\to+\infty} \psi_n \leqslant \Gamma\text{-}\liminf_{n\to+\infty} (\varphi_n + \psi_n)$$
$$\Gamma\text{-}\limsup_{n\to+\infty} \varphi_n + \Gamma\text{-}\liminf_{n\to+\infty} \psi_n \leqslant \Gamma\text{-}\limsup_{n\to+\infty} (\varphi_n + \psi_n)$$

and that these inequalities may be strict.

Problem 3.152**
Show that in a metric space Γ-convergence and pointwise convergence are in general distinct.

Problem 3.153**
Suppose that (X, d_X) is a metric space, $\varphi\colon X \longrightarrow \overline{\mathbb{R}}$ is a lower semicontinuous function, and $\left\{\varphi_n\colon X \longrightarrow \overline{\mathbb{R}}\right\}_{n\geqslant 1}$ is a sequence of lower semicontinuous functions. Assume that for every $\lambda \in \mathbb{R}$, there exists a sequence $\{\lambda_n\}_{n\geqslant 1} \subseteq \mathbb{R}$ such that $\lambda_n \longrightarrow \lambda$ and $\varphi_n^{\lambda_n} \xrightarrow{K} \varphi^{\lambda}$ (here $\varphi_n^{\lambda_n} = \{u \in X : \varphi_n(u) \leqslant \lambda_n\}$ and $\varphi^{\lambda} = \{u \in X : \varphi(u) \leqslant \lambda\}$). Show that $\varphi_n \xrightarrow{\Gamma} \varphi$.

Problem 3.154*
Suppose that (X, d_X) is a metric space, $\varphi\colon X \longrightarrow \overline{\mathbb{R}}$ is a lower semicontinuous function, and $\left\{\varphi_n\colon X \longrightarrow \overline{\mathbb{R}}\right\}_{n\geqslant 1}$ is a sequence of lower semicontinuous functions such that $\varphi_n \leqslant \varphi$ for all $n \geqslant 1$ and $\varphi_n \xrightarrow{\Gamma} \varphi$. Show that $\varphi_n \longrightarrow \varphi$ pointwise.

Problem 3.155*
Suppose that (X, d_X) is a metric space, $\varphi\colon X \longrightarrow \overline{\mathbb{R}}$ is a proper, lower semicontinuous function and $\{\varphi_n\colon X \longrightarrow \overline{\mathbb{R}}\}_{n\geqslant 1}$ is a sequence of proper, lower semicontinuous functions such that $\varphi_n \xrightarrow{\Gamma} \varphi$ and $\varphi_n \longrightarrow \varphi$ pointwise. Show that the sequence $\{\varphi_n\}_{n\geqslant 1}$ is equi-lower semicontinuous (see Definition 3.143).

Problem 3.156**
Let (X, d_X) be a metric space and let $\varphi\colon X \longrightarrow \overline{\mathbb{R}}$ be a proper function. Show that $\operatorname{cl}\varphi$ is characterized by the following two properties:
(a) for every sequence $\{u_n\}_{n\geqslant 1} \subseteq X$ such that $u_n \longrightarrow u$ in X we have

$$\operatorname{cl}\varphi(u) \leqslant \liminf_{n\to+\infty} \varphi(u_n).$$

(b) there exists a sequence $\{u_n\}_{n\geqslant 1} \subseteq X$ such that $u_n \longrightarrow u$ in X and

$$\limsup_{n\to+\infty} \varphi(u_n) \leqslant \operatorname{cl}\varphi(u).$$

Problem 3.157*
Suppose that $X = l^2$, $\{e_n\}_{n\geqslant 1}$ is the standard orthonormal basis of l^2, $\{\lambda_n\}_{n\geqslant 1} \subseteq \mathbb{R}_+$ is a sequence and $C_n = \operatorname{conv}\{0, \lambda_n e_n\}$ for $n \geqslant 1$. Determine whether the sequence $\{C_n\}_{n\geqslant 1}$ is Mosco convergent. Justify your answer.

Problem 3.158*
Suppose that X is a reflexive Banach space, $\varphi \in \Gamma_0(X)$ and $\{\varphi_n\}_{n\geqslant 1} \subseteq \Gamma_0(X)$ is a sequence such that $\varphi_n \xrightarrow{M} \varphi$. Let $m = \inf_X \varphi$, $m_n = \inf_X \varphi_n$, $S = \{u \in X : m = \varphi(u)\}$ and $S_n = \{u \in X : m_n = \varphi_n(u)\}$. Show that:
(a) $\limsup_{n\to+\infty} m_n \leqslant m$;
(b) if there exists a sequence $\{u_n\}_{n\geqslant 1} \subseteq X$ such that $u_n \in S_n$ for $n \geqslant 1$ and $u_n \xrightarrow{w} u$, then $u \in S$ and $m_n \longrightarrow m$.

Problem 3.159*
Find a sequence $\{C_n\}_{n\geqslant 1}$ of nonempty convex sets such that the Kuratowski limit in both the strong and weak topologies (see Definition 2.88) exist but are different. What can be said about the Mosco convergence of such a sequence? Justify your answer.

Problem 3.160*
Suppose that X is a Banach space, $C \subseteq X$ is a weakly closed subset of X and $\{C_n\}_{n\geqslant 1} \subseteq X$ is a sequence of weakly closed subsets of X. Show that $C_n \xrightarrow{M} C$ if and only if
(a) $C \subseteq s\text{-}\liminf_{n\to+\infty} C_n$; and

(b) for every w-compact set $K \subseteq X$, we have

$$w\text{–}\limsup_{n\to+\infty}(C_n \cap K) \subseteq C.$$

Problem 3.161**
Suppose that X is a Banach space, $u^* \in X^* \setminus \{0\}$ and $\{u_n^*\}_{n\geqslant 1} \subseteq X^* \setminus \{0\}$ is a sequence. Show that the following properties are equivalent:

(a) The sequence $\{u_n^*\}_{n\geqslant 1} \subseteq X^*$ is bounded and $u_n^* \xrightarrow{w^*} u^*$ in X^*.

(b) For every $\lambda \in \mathbb{R}$, we have $(u_n^*)^{-1}(\lambda) \xrightarrow{K} (u^*)^{-1}(\lambda)$.

(c) $(u_n^*)^{-1}(1) \xrightarrow{K} (u^*)^{-1}(1)$.

Problem 3.162**
Suppose that X is a Banach space, $u^* \in X^* \setminus \{0\}$ and $\{u_n^*\}_{n\geqslant 1} \subseteq X^* \setminus \{0\}$ is a sequence. Show that the following three properties are equivalent:

(a) $u_n^* \longrightarrow u^*$ uniformly on w-compact subsets of X.

(b) For every $\lambda \in \mathbb{R}$, we have $(u_n^*)^{-1}(\lambda) \xrightarrow{M} (u^*)^{-1}(\lambda)$.

(c) $(u_n^*)^{-1}(1) \xrightarrow{M} (u^*)^{-1}(1)$.

Problem 3.163*
Suppose that X is a reflexive Banach space, $\varphi \in \Gamma_0(X)$, $\{\varphi_n\}_{n\geqslant 1} \subseteq \Gamma_0(X)$ is a sequence such that $\varphi_n \xrightarrow{M} \varphi$. Show that for every $u^* \in X^*$ there exists a sequence $\{u_n^*\}_{n\geqslant 1} \subseteq X^*$ such that $u_n^* \longrightarrow u^*$ in X^* and

$$\inf_X (\varphi_n - u_n^*) \longrightarrow \inf_X (\varphi - u^*).$$

Problem 3.164**
Suppose that X is a reflexive Banach space, $C \in P_{fc}(X)$, $\{C_n\}_{n\geqslant 1} \subseteq P_{fc}(X)$ and there exists $W \in P_{wkc}(X)$ such that $C_n \subseteq W$ for all $n \geqslant 1$. Show that $C_n \xrightarrow{M} C$ if and only if $\sigma_{C_n}(u^*) \longrightarrow \sigma_C(u^*)$ for all $u^* \in X^*$.

Problem 3.165*
Suppose that X is a reflexive Banach space, $\varphi\colon X \longrightarrow \overline{\mathbb{R}}$ is a proper function, and $\{\varphi_n\colon X \longrightarrow \overline{\mathbb{R}}\}_{n\geqslant 1}$ is a sequence of proper functions. Show that $\varphi_n \xrightarrow{M} \varphi$ if and only if $\operatorname{epi}\varphi_n \xrightarrow{M} \operatorname{epi}\varphi$.

Problem 3.166**
Suppose that X is a reflexive Banach space, $\varphi \in \Gamma_0(X)$, $\{\varphi_n\}_{n\geqslant 1} \subseteq \Gamma_0(X)$ is a sequence and $m = \inf_X \varphi$. Show that:

(a) if $\varphi_n \xrightarrow{M} \varphi$ and $m < \lambda$, then $\varphi_n^\lambda \xrightarrow{M} \varphi^\lambda$;

(b) if $\varphi_n^\lambda \xrightarrow{M} \varphi^\lambda$ for every $\lambda \in \mathbb{R}$, then $\varphi_n \xrightarrow{M} \varphi$.

Problem 3.167*
Examine if in Problem 3.166(a) we can drop the hypothesis $m = \inf_X \varphi < \lambda$. Justify your answer.

Problem 3.168*
Is the result of Problem 3.166(b) true, if we have $\varphi_n^\lambda \xrightarrow{M} \varphi^\lambda$ only for $\lambda \geqslant \inf_X \varphi$? Justify your answer.

Problem 3.169*
Suppose that X is a reflexive Banach space, $C \in P_{fc}(X)$, $\{C_n\}_{n\geqslant 1} \subseteq P_{fc}(X)$ is a sequence such that $C_n \xrightarrow{M} C$ and $\varphi\colon X \longrightarrow \mathbb{R}$ is a continuous, convex and coercive function (i.e., $\varphi(u) \longrightarrow +\infty$ when $\|u\|_X \to +\infty$). Let $m_n = \inf_{C_n} \varphi$ for $n \geqslant 1$ and $m = \inf_C \varphi$. Show that $m_n \longrightarrow m$.

Problem 3.170**
Suppose that H is a Hilbert space, $a\colon H \times H \longrightarrow \mathbb{R}$ is a bilinear, symmetric, continuous and coercive form (see Theorem 3.114), $C \in P_{fc}(H)$, $\{C_n\}_{n\geqslant 1} \subseteq P_{fc}(H)$ is a sequence such that $C_n \xrightarrow{M} C$, $u^* \in H$, $\varphi\colon H \longrightarrow \mathbb{R}$ is a function defined by

$$\varphi(u) \;=\; \tfrac{1}{2}a(u,u) - (u^*,u)_H \qquad \forall\, u \in H$$

and

$$\begin{aligned} m_n &= \inf_{C_n} \varphi, \qquad m \;=\; \inf_C \varphi, \\ S_n &= \{u \in H : \; m_n = \varphi(u)\} \qquad \forall\, n \geqslant 1, \\ S &= \{u \in H : \; m = \varphi(u)\}. \end{aligned}$$

Show that S_n and S are singletons, that is, $S_n = \{u_n\}$ for $n \geqslant 1$, $S = \{u\}$ and $u_n \longrightarrow u$ in H.

Problem 3.171**
Let (X, d_X) be a metric space and let $\{\varphi_n\colon X \longrightarrow \mathbb{R}\}_{n\geqslant 1}$, $\{\psi_n\colon X \longrightarrow \mathbb{R}\}_{n\geqslant 1}$ be two sequences of functions such that

$$\begin{aligned} \varphi_n &\xrightarrow{\Gamma} \varphi \quad \text{and} \quad \varphi_n \longrightarrow \varphi \text{ pointwise}, \\ \psi_n &\xrightarrow{\Gamma} \psi \quad \text{and} \quad \psi_n \longrightarrow \psi \text{ pointwise}. \end{aligned}$$

Show that $\varphi_n + \psi_n \xrightarrow{\Gamma} \varphi + \psi$.

Problem 3.172**
Consider a lower semicontinuous function $\varphi\colon \mathbb{R}^N \longrightarrow \mathbb{R}$ and a sequence of functions $\{\varphi_n\colon \mathbb{R}^N \longrightarrow \mathbb{R}\}_{n\geqslant 1}$. Show that $\varphi \leqslant \Gamma\text{-}\liminf_{n\to+\infty} \varphi_n$ if and only if for every compact set $K \subseteq \mathbb{R}^N$, we have

$$\inf_K \varphi \;\leqslant\; \liminf_{n\to+\infty} (\inf_K \varphi_n).$$

Problem 3.173***

Let $\varphi\colon \mathbb{R}^N \longrightarrow \mathbb{R}$ be a function and let $\{\varphi_n\colon \mathbb{R}^N \longrightarrow \mathbb{R}\}_{n\geqslant 1}$ be a sequence of functions such that $\varphi_n \xrightarrow{\Gamma} \varphi$ and $\inf_{\mathbb{R}^N} \varphi \in \mathbb{R}$. Show that $\inf_{\mathbb{R}^N} \varphi_n \longrightarrow \inf_{\mathbb{R}^N} \varphi$ if and only if for every $\varepsilon > 0$ there exist a compact set $K \subseteq \mathbb{R}^N$ and $n_0 \geqslant 1$ such that

$$\inf_K \varphi_n \leqslant \inf_{\mathbb{R}^N} \varphi_n + \varepsilon \qquad \forall\, n \geqslant n_0.$$

Problem 3.174**

Let X be a Banach space and let $\{\varphi_n\colon X \longrightarrow \overline{\mathbb{R}}\}_{n\geqslant 1}$ be a sequence of proper and even functions. Let

$$\widetilde{\varphi} = \Gamma\text{-}\liminf_{n\to+\infty} \varphi_n, \quad \widehat{\varphi} = \Gamma\text{-}\limsup_{n\to+\infty} \varphi_n.$$

Show that both functions $\widetilde{\varphi}$ and $\widehat{\varphi}$ are even too. Also show that the result is no longer true for odd functions.

Problem 3.175**

Let X be a Banach space and let $\{C_n\}_{n\geqslant 1} \subseteq P_f(X)$. Show that there exists a subsequence $\{C_{n_k}\}_{k\geqslant 1}$ of $\{C_n\}_{n\geqslant 1}$ and $C \in P_f(X) \cup \{\emptyset\}$ such that $C_{n_k} \xrightarrow{K} C$.

Problem 3.176**

Suppose that X is a finite dimensional Banach space, $C \in P_f(X)$ and $\{C_n\}_{n\geqslant 1} \subseteq P_f(X)$ is a sequence of sets. Show that the following three conditions are equivalent:

(a) $C_n \xrightarrow{K} C$.
(b) $\operatorname{dist}(\cdot, C) \longrightarrow \operatorname{dist}(\cdot, C)$ pointwise.
(c) $\operatorname{dist}(\cdot, C) \xrightarrow{\Gamma} \operatorname{dist}(\cdot, C)$.

Problem 3.177**

Suppose that X is a reflexive Banach space, $\varphi \in \Gamma_0(X)$ and $\{\varphi_n\}_{n\geqslant 1} \subseteq \Gamma_0(X)$ is a sequence of functions. Show that the following statements are equivalent:

(a) $\varphi_n \xrightarrow{M} \varphi$.
(b) For every $u \in X$ there exists a sequence $\{u_n\}_{n\geqslant 1}$ such that $u_n \longrightarrow u$ in X and $\varphi_n(u_n) \longrightarrow \varphi(u)$ and for every $u^* \in X^*$ there exists a sequence $\{u_n^*\}_{n\geqslant 1} \subseteq X^*$ such that $u_n^* \longrightarrow u^*$ in X^* and $\varphi_n^*(u_n^*) \longrightarrow \varphi^*(u^*)$.

Problem 3.178**

Let X be a reflexive Banach space and let $\{\varphi_n\}_{n\geqslant 1} \subseteq \Gamma_0(X)$. Show that:

(a) if $\{\varphi_n\}_{n\geqslant 1}$ is an increasing sequence of functions, then $\varphi_n \xrightarrow{M} \varphi = \sup_{n\geqslant 1} \varphi_n$;

(b) if $\{\varphi_n\}_{n\geqslant 1}$ is a decreasing sequence of functions, then $\varphi_n \xrightarrow{M} \operatorname{cl}(\inf_{n\geqslant 1} \varphi_n)$.

Problem 3.179**
Let X be a reflexive Banach space and let $\{C_n\}_{n\geqslant 1} \subseteq P_{fc}(X)$ be a sequence of sets. Show the following statements:

(a) If the sequence $\{C_n\}_{n\geqslant 1}$ is increasing, then $C_n \xrightarrow{M} C = \overline{\bigcup_{n\geqslant 1} C_n}$.

(b) If the sequence $\{C_n\}_{n\geqslant 1}$ is decreasing, then $C_n \xrightarrow{M} C = \overline{\bigcap_{n\geqslant 1} C_n}$.

Problem 3.180*
Suppose that X is a Banach space, $C \subseteq X$ is a nonempty closed set and $\{C_n\}_{n\geqslant 1} \subseteq 2^X$ is a sequence of nonempty closed sets such that $\operatorname{dist}(\cdot, C_n) \xrightarrow{M} \operatorname{dist}(\cdot, C)$. Show that $C_n \xrightarrow{M} C$.

Problem 3.181***
Suppose that X is a reflexive Banach space, $C \subseteq X$ is a nonempty, closed, convex set and $\{C_n\}_{n\geqslant 1} \subseteq 2^X$ is a sequence of nonempty, closed, convex sets such that $C_n \xrightarrow{M} C$. Show that $\operatorname{dist}(\cdot, C_n) \xrightarrow{M} \operatorname{dist}(\cdot, C)$.

Problem 3.182**
Suppose that (Ω, Σ, μ) is a complete, σ-finite measure space, (X, d_X) is a locally compact, σ-compact metric space, $\varphi\colon \Omega \times X \longrightarrow \overline{\mathbb{R}}$ is a normal integrand (see Definition 3.109) and there exists a function $\beta\colon \Omega \longrightarrow \mathbb{R}$ (not necessarily measurable) such that

$$\beta(\omega) \leqslant \varphi(\omega, u) \quad \text{for } \mu\text{-a.a. } \omega \in \Omega, \text{ all } u \in \mathbb{R}.$$

Show that there exists a sequence of Carathéodory integrands $\{\varphi_n\colon \Omega \times X \longrightarrow \mathbb{R}\}_{n\geqslant 1}$ such that

$$\varphi_n(\omega, \cdot) \in C_0(X) \quad \text{for } \mu\text{-a.a. } \omega \in \Omega,$$

$$\beta(\omega) \leqslant \varphi_n(\omega, u) \leqslant n \quad \text{for } \mu\text{-a.a. } \omega \in \Omega, \text{ all } u \in X$$

and

$$\varphi_n(\omega, u) \nearrow \varphi(\omega, u) \quad \text{for } \mu\text{-a.a. } \omega \in \Omega, \text{ all } u \in X.$$

Problem 3.183**
Suppose that Ω, T are two locally compact, σ-compact metric spaces, Σ is a σ-field on Ω containing $\mathcal{B}(\Omega)$ and $\mu \in M_b(\Omega)_+$ is such that $\Sigma = \Sigma_\mu$ (i.e., Σ is μ-complete). Let $\varphi\colon \Omega \times T \longrightarrow \overline{\mathbb{R}}$ be a normal integrand such that $\varphi \geqslant 0$. Show that the map $\xi \longmapsto \int_{\Omega\times T} \varphi\, d\xi$ is w^*-lower semicontinuous on $\mathcal{SY}(\Omega, T; \mu)$ (see Definition 3.160). Is the result true if we drop the hypothesis $\varphi \geqslant 0$? Justify your answer.

Problem 3.184**
Let Ω, T, Σ, μ be as in Problem 3.183 and let $\widehat{K}_0(\Omega, \Sigma; T)$ be the space of all measurable integrands $\varphi\colon \Omega \times T \longrightarrow \mathbb{R}$ such that for μ-almost all $z \in \Omega$, we have

$\varphi(z,\cdot) \in C_0(T)$ and the map $z \longmapsto \|\varphi(z,\cdot)\|_{L^\infty(T)}$ belongs in $L^\infty(\Omega)$. On $\widehat{K}_0(\Omega,\Sigma;T)$ we introduce the following equivalence relation

$$\varphi \sim \psi \quad \Longleftrightarrow \quad \mu\big(\{z \in \Omega : \varphi(z,\cdot) \neq \psi(z,\cdot)\}\big) = 0.$$

Let $K_0(\Omega,\Sigma;T) = \widehat{K}_0(\Omega,\Sigma;T)/_\sim$. Show that $K_0(\Omega,\Sigma;T)$ is in bijection with $L^1(\Omega;C_0(T))$.

Problem 3.185**
Let Ω, T, Σ, μ be as in Problem 3.183. Show that the set $\mathcal{D} \subseteq \mathcal{Y}(\Omega,T;\mu)$ is uniformly tight if and only if for every $\varepsilon > 0$ there exists a measurable multifunction $F_\varepsilon \colon \Omega \longrightarrow P_k(T)$ such that

$$\sup_{\xi \in \mathcal{D}} \int_\Omega \xi(z)(T \setminus F_\varepsilon(z))\, d\mu \leqslant \varepsilon.$$

Problem 3.186**
Let Ω, T, Σ, μ be as in Problem 3.183 and let $\mathcal{D} \subseteq \mathcal{Y}(\Omega,T;\mu)$ be a uniformly tight set. Suppose that the set $\mathcal{D}$ is uniformly tight. Show that $\overline{\mathcal{D}}^n$ (i.e., the closure of $\mathcal{D}$ in the narrow topology) is uniformly tight too.

Problem 3.187**
Let Ω, Σ, μ be as in Problem 3.183 and let $T = \mathbb{R}^N$.

(a) Consider a bounded sequence $\{u_n\}_{n\geqslant 1} \subseteq L^1(\Omega;\mathbb{R}^N)$. Show that the sequence $\{\delta_{u_n}\}_{n\geqslant 1} \subseteq \mathcal{Y}(\Omega,T;\mu)$ (see Remark 3.165) is uniformly tight.

(b) Consider a measurable map $u\colon \Omega \longrightarrow \mathbb{R}^N$ and a sequence of measurable maps $\{u_n\colon \Omega \longrightarrow \mathbb{R}^N\}_{n\geqslant 1}$ such that $u_n \overset{\mu}{\longrightarrow} u$. Show that the sequence $\{\delta_{u_n}\}_{n\geqslant 1} \subseteq \mathcal{Y}(\Omega,T;\mu)$ is uniformly tight.

Problem 3.188*
Let Ω, T, Σ, μ be as in Problem 3.183. Let $\xi \in \mathcal{Y}(\Omega,T;\mu)$ and let $\{\xi_n\}_{n\geqslant 1} \subseteq \mathcal{Y}(\Omega,T;\mu)$ be a sequence. Show that $\xi_n \overset{n}{\longrightarrow} \xi$ in $\mathcal{Y}(\Omega,T;\mu)$ if and only if

$$\int_A \int_T \varphi(u)\xi_n(z)(du)\, d\mu \longrightarrow \int_A \int_T \varphi(u)\xi(z)(du)\, d\mu$$
$$\forall\, \varphi \in CU_b(T),\ A \in \Sigma,$$

where $CU_b(T)$ denotes the space of uniformly continuous and bounded functions on T.

Problem 3.189**
Let Ω, T, Σ, μ be as in Problem 3.183. Suppose that $\{u_m\colon\ \Omega \longrightarrow T\}_{m\geqslant 1}$ is a sequence of Σ-measurable maps such that

$$\xi_m = \delta_{u_m} \overset{n}{\longrightarrow} \xi \quad \text{in } \mathcal{Y}(\Omega,T;\mu)$$

and $\varphi\colon \Omega \times T \longrightarrow \overline{\mathbb{R}}$ is a measurable integrand such that for μ-almost all $z \in \Omega$, the map $\varphi(z,\cdot)$ is lower semicontinuous and the sequence $\{\, \varphi^-(\cdot, u_m(\cdot))\,\}_{m\geqslant 1}$ is uniformly integrable in $L^1(\Omega)$. Show that:

(a) if $\liminf\limits_{m\to+\infty}\int\limits_{\Omega}\varphi(z,u_m(z))\,d\mu<+\infty$, then $\int\limits_{\Omega\times T}\varphi^+(z,x)\,d\xi<+\infty$;

(b) $\int\limits_{\Omega\times T}\varphi(z,x)\,d\xi\leqslant\liminf\limits_{m\to+\infty}\int\limits_{\Omega}\varphi(z,u_m(z))\,d\mu$.

Problem 3.190*

Let Ω,Σ,μ and $\{u_m\colon\Omega\longrightarrow T\}_{m\geqslant1}$ be as in Problem 3.189. Suppose that $\widehat{\varphi}\colon\Omega\times T\longrightarrow\mathbb{R}$ is a Carathéodory integrand such that the sequence $\{\widehat{\varphi}(\cdot,u_n(\cdot))\}_{n\geqslant1}\subseteq L^1(\Omega)$ is uniformly integrable. Show that $\widehat{\varphi}$ is ξ-integrable and

$$\int\limits_{\Omega\times T}\widehat{\varphi}(z,x)\,d\xi = \lim_{m\to+\infty}\int\limits_{\Omega}\widehat{\varphi}(z,u_m(z))\,d\mu.$$

Problem 3.191**

Let Ω,Σ,μ be as in Problem 3.183 and let $T=\mathbb{R}^m$. Using Young measures, prove the Dunford–Pettis Theorem (see Theorem 1.28), namely show that if $\{u_n\}_{n\geqslant1}\subseteq L^1(\Omega;\mathbb{R}^m)$ is uniformly integrable, then there exist a subsequence $\{u_{n_k}\}_{k\geqslant1}$ of $\{u_n\}_{n\geqslant1}$ and $u\in L^1(\Omega;\mathbb{R}^m)$ such that $u_{n_k}\xrightarrow{w}u$ in $L^1(\Omega;\mathbb{R}^m)$.

Problem 3.192**

Let Ω,Σ,μ be as in Problem 3.183 and let $T=\mathbb{R}^m$. Suppose that $\{u_n\}_{n\geqslant1}\subseteq L^p(\Omega;\mathbb{R}^m)$ is a sequence such that $u_n\xrightarrow{w}u\in L^p(\Omega;\mathbb{R}^m)$ and $\xi_n=\delta_{u_n(\cdot)}\xrightarrow{n}\xi$. Show that for μ-almost all $z\in\Omega$, we have

$$u(z) = \int\limits_{\mathbb{R}^m}x\xi(z)(dx)$$

(***barycenter*** or ***expectation*** of $\xi(z)\in M_+^1(\mathbb{R}^m)$).

Problem 3.193**

Let Ω,T,Σ,μ be as in Problem 3.183. Let $u\colon\Omega\longrightarrow T$ be a Σ-measurable function and let $\{u_n\colon\Omega\longrightarrow T\}_{n\geqslant1}$ be a sequence of Σ-measurable functions. Show that the following properties are equivalent:

(a) $\delta_{u_n(\cdot)}\xrightarrow{n}\delta_{u(\cdot)}$ in $\mathcal{Y}(\Omega,T;\mu)$.

(b) $u_n\xrightarrow{\mu}u$.

Problem 3.194**

Let X be a Banach space and let $\varphi\colon X\longrightarrow\mathbb{R}$ be a locally Lipschitz function. Set

$$m(u) = \inf_{u^*\in\partial\varphi(u)}\|u^*\|_* \qquad \forall\, u\in X.$$

Show that $m\colon X\longrightarrow\mathbb{R}_+$ is lower semicontinuous.

Problem 3.195**

Let $C\subseteq\mathbb{R}^N$ be a nonempty set and let $\varphi\colon C\longrightarrow\mathbb{R}^m$ be a Lipschitz continuous map. Show that there exists a Lipschitz function $\widehat{\varphi}\colon\mathbb{R}^N\longrightarrow\mathbb{R}^m$ such that $\widehat{\varphi}|_C=\varphi$ and the Lipschitz constant $k_{\widehat{\varphi}}$ of $\widehat{\varphi}$ satisfies $k_{\widehat{\varphi}}\leqslant\sqrt{m}k_\varphi$, where k_φ is the Lipschitz constant of φ.

Problem 3.196**
Let $\Omega \subseteq \mathbb{R}^N$ be a bounded open domain with a Lipschitz boundary $\partial\Omega$. Show that $W^{1,\infty}(\Omega) = C^{0,1}(\Omega)$ (where $C^{0,1}(\Omega)$ denotes the space of all Lipschitz continuous functions on Ω).

Problem 3.197**
Suppose that H is a Hilbert space, X is a reflexive Banach space which is embedded compactly into H, $\varphi \in \Gamma_0(X)$ and for all $a, b \geqslant 0$ the sets

$$C_{a,b} = \{u \in X : \varphi(u) \leqslant a, \|u\|_H \leqslant b\}$$

are bounded. We set

$$\widehat{\varphi}(u) = \begin{cases} \varphi(u) & \text{if } u \in X, \\ +\infty & \text{if } u \in H \setminus X. \end{cases}$$

Show that $\widehat{\varphi} \in \Gamma_0(H)$ and if φ is Gâteaux differentiable at $u \in \operatorname{int} \operatorname{dom} \varphi$, then

$$\partial\widehat{\varphi}(u) = \{u^* \in H = H^* : (u^*, h)_H = \langle \varphi'(u), h \rangle \text{ for all } h \in X\}$$

(here by $(\cdot,\cdot)_H$ we denote the inner product of H and by $\langle \cdot,\cdot \rangle$ the duality brackets for the pair (X^*, X)).

Problem 3.198*
Let X be a Banach space. Show that a convex function $\varphi\colon X \longrightarrow \overline{\mathbb{R}} = \mathbb{R} \cup \{+\infty\}$ can be continuous at a point $u \in \overline{\operatorname{dom} \varphi} \setminus \operatorname{dom} \varphi$.

Problem 3.199***
Suppose that X is a Banach space, $C \subseteq X$ is a nonempty, closed, and convex set and $\varphi\colon C \longrightarrow \mathbb{R}$ is a continuous, convex function. Is it possible to have $\partial\varphi(u) = \emptyset$ for all u in a dense subset of C? Justify your answer.

Problem 3.200*
Give an example showing that the inclusion in Proposition 3.97 can be strict.

Problem 3.201***
Let H be a Hilbert space and let $C \subseteq H$ be a nonempty, closed, convex set. The ***metric projection*** map into C is the map $\operatorname{proj}_C : H \longrightarrow C$ defined by

$$\operatorname{proj}_C(u) = c_0 \in C, \quad \text{such that } \|u - c_0\|_H = \operatorname{dist}(u, C)$$

(that is, $c_0 \in C$ is the best approximation from C to u and it is unique). Show that proj_C is the Fréchet derivative of a convex function.

Problem 3.202*
Solve the following second order periodic problem

$$\begin{cases} -u''(t) + u'(t) + u(t) = 0 \text{ for a.a. } t \in [0,1], \\ u(0) = u(1), \; u'(0) = u'(1). \end{cases}$$

3.3 Solutions

Solution of Problem 3.1

We argue indirectly. So, suppose that for some $x_0 \in U$, we have $f'(x_0) \in \mathcal{L}(X;Y)$ is not compact. This means that $\overline{f'(x_0)(\partial B_1)} \subseteq Y$ is not compact (recall that $\partial B_1 = \{x \in X : \|x\|_X = 1\}$). So, we can find a sequence $\{v_n\}_{n\geqslant 1} \subseteq \partial B_1$ and $\varepsilon > 0$ such that

$$\|f'(x_0)(v_n - v_m)\|_Y \geqslant \varepsilon \qquad \forall\, n, m \geqslant 1,\ n \neq m.$$

On the other hand, from the Fréchet differentiability of f (see Definition 3.1), for $\lambda > 0$, we have

$$\begin{aligned}
& \|f(x_0 + \lambda h_n) - f(x_0 + \lambda h_m)\|_Y \\
\geqslant\ & \lambda \|f'(x_0)h_n - f'(x_0)h_m)\|_Y \\
& - \|f(x_0 + \lambda h_n) - f(x_0) - \lambda f'(x_0)h_n)\|_Y \\
& - \|f(x_0 + \lambda h_m) - f(x_0) - \lambda f'(x_0)h_m)\|_Y \\
\geqslant\ & \lambda\varepsilon - o(\lambda).
\end{aligned}$$

So, for $\lambda > 0$ small, we see that $\{f(x_0 + \lambda h_n)\}_{n\geqslant 1}$ cannot have a convergent subsequence, which contradicts the compactness of f (see Definition 2.1(a)).

Solution of Problem 3.2

For $h \in X \setminus \{0\}$ small we have

$$\begin{aligned}
\|tu + h\|_X &= t\|u + \tfrac{1}{t}h\|_X \\
&= t\big(\|u\|_X + \langle \tfrac{d}{du}\|u\|_X, \tfrac{h}{t}\rangle + e(\tfrac{h}{t})\big) \\
&= \|tu\|_X + \langle \tfrac{d}{du}\|u\|_X, h\rangle + e(h)
\end{aligned}$$

(see Definition 3.1), so $\frac{d}{du}\|tu\|_X$ exists at u and is equal to $\frac{d}{du}\|u\|_X$.

Next we show that $\big\|\frac{d}{du}\|u\|_X\big\|_* = 1$. We consider the function

$$\varphi(t) = \|tu\|_X \qquad \forall\, t > 0.$$

We have

$$\lim_{\tau \to 0} \tfrac{\|(t+\tau)u\|_X - \|tu\|_X}{\tau} = \lim_{\tau \to 0} \tfrac{\tau\|u\|_X}{\tau} = \|u\|_X,$$

so

$$\varphi'(t) = \|u\|_X \qquad \forall\, t > 0.$$

Note that $\varphi = \|\cdot\|_X \circ \xi$ where

$$\xi(t) = tu \qquad \forall\, t > 0.$$

Then by the chain rule (see Theorem 3.19), we have

$$\varphi'(t)\tau = \langle \tfrac{d}{du}\|tu\|_X, \tau u\rangle = \tau\langle \tfrac{d}{du}\|u\|_X, u\rangle,$$

so

$$\|u\|_X = \langle \tfrac{d}{du}\|u\|_X, u\rangle$$

and thus

$$\left\|\tfrac{d}{du}\|u\|_X\right\|_* \geqslant 1. \tag{3.10}$$

On the other hand, the Fréchet differentiability of $\|\cdot\|_X$ at u implies that

$$\|u+th\|_X - \|u\|_X - \left\langle \tfrac{d}{du}\|u\|_X, th\right\rangle = \|th\|_X\varepsilon(th),$$

with $\varepsilon(h) \longrightarrow 0$ as $h \to 0$. Then

$$\left|\left\langle \tfrac{d}{du}\|u\|_X, th\right\rangle\right| \leqslant \big|\|u+th\|_X - \|u\|_X\big| + \|th\|_X|\varepsilon(th)|,$$

so

$$\left|\left\langle \tfrac{d}{du}\|u\|_X, h\right\rangle\right| \leqslant \|h\|_X\big(1+|\varepsilon(th)|\big)$$

(by the triangle inequality) and thus

$$\left\|\tfrac{d}{du}\|u\|_X\right\|_* \leqslant 1. \tag{3.11}$$

From (3.10) and (3.11), we conclude that $\left\|\tfrac{d}{du}\|u\|_X\right\|_* = 1$.

Solution of Problem 3.3

If $u \in C$, then for any $h \in \mathbb{R}^N$, we have

$$\langle \varphi_C'(u), h\rangle = \lim_{t\to 0^+} \tfrac{\varphi_C(u+th)-\varphi_C(u)}{t} = \lim_{t\to 0^+} \tfrac{\varphi_C(u+th)}{t} \geqslant 0,$$

so $\varphi_C'(u) = 0$, a contradiction to our hypothesis. Therefore $u \notin C$. We know that $\operatorname{proj}_C(u) \neq \emptyset$. Let $c \in \operatorname{proj}_C(u)$. We will show that

$$\varphi_C'(u) = \tfrac{u-c}{|u-c|},$$

from which we infer that $\operatorname{proj}_C(u) = \{c\}$. For $t \in (0,1)$ the best approximation to $u + t(c-u)$ from the set C is still c and so

$$(1-t)|u-c| = \operatorname{dist}(u+t(c-u), C),$$

thus

$$\tfrac{\operatorname{dist}(u+t(c-u),C)-\operatorname{dist}(u,C)}{t} = -\operatorname{dist}(u,C) = -|u-c|$$

and hence

$$\langle \varphi_C'(u), c-u\rangle = -|u-c|. \tag{3.12}$$

Recall that $\varphi_C = \operatorname{dist}(\cdot, C)$ is nonexpansive (i.e., Lipschitz continuous with Lipschitz constant 1). Then

$$\|\varphi_C'(u)\| \leqslant 1$$

and so

$$\varphi_C'(u) = \tfrac{u-c}{|u-c|}$$

(see (3.12)).

Solution of Problem 3.4

We fix $u \in U$ and let $h \in U$ be another point. Since U is connected, we can connect u and h with a polygonal line. Let $\gamma \in C([0,1];X)$ be the piecewise linear path realizing this connection. We have $\gamma(0) = u$, $\gamma(1) = h$ and by hypothesis

$$\|\varphi(\gamma(t_{k+1})) - \varphi(\gamma(t_k))\|_Y = 0 \qquad \forall\, k \in \{0, \ldots, m-1\}$$

where $t_0 = 0$ and $t_m = 1$ (see Theorem 3.15). Therefore $\varphi(u) = \varphi(h)$. Since $h \in U$ is arbitrary, we conclude that φ is constant.

Solution of Problem 3.5

First note that the growth condition on f implies

$$|F(\omega, x)| \leqslant a_1(\omega) + c_1|x|^p \quad \text{for } \mu\text{-a.a. } \omega \in \Omega, \text{ and } x \in \mathbb{R},$$

with $a_1 \in L^1(\Omega)_+$, $c_1 > 0$. So, the functional ψ is well defined.

Let $u \in L^p(\Omega)$. For every $h \in L^p(\Omega)$, we define

$$e(h) = \int_\Omega F(\omega, u(\omega) + h(\omega))\, d\mu - \int_\Omega F(\omega, u(\omega))\, d\mu - \int_\Omega f(\omega, u(\omega))h(\omega)\, d\mu. \tag{3.13}$$

We have

$$F(\omega, u(\omega) + h(\omega)) - F(\omega, u(\omega)) = \int_0^1 \tfrac{d}{dt} F(\omega, u(\omega) + th(\omega))\, d\mu$$

$$= \int_0^1 f(\omega, u(\omega) + th(\omega))h(\omega)\, d\mu \tag{3.14}$$

(by the chain rule). Using (3.14) in (3.13), we obtain

$$e(h) = \int_\Omega \int_0^t \big(f(\omega, u(\omega) + th(\omega)) - f(\omega, u(\omega))\big)h(\omega)\, dt\, d\mu,$$

so

$$|e(h)| \leqslant \int_0^1 \|N_f(u + th) - N_f(u)\|_{p'}\, dt \|h\|_p.$$

Here we have used first the Fubini theorem and then the Hölder inequality (see Theorem 1.3 and Problem 1.27). Hence

$$\frac{|e(h)|}{\|h\|_p} \leqslant \int_0^1 \|N_f(u+th) - N_f(u)\|_{p'}\,dt,$$

so

$$\frac{|e(h)|}{\|h\|_p} \longrightarrow 0 \quad \text{as } \|h\|_p \to 0$$

(see Problem 1.42). So ψ is Fréchet differentiable and

$$\psi'(u) = N_f(u) \qquad \forall\, u \in L^p(\Omega).$$

But N_f is continuous from $L^p(\Omega)$ into $L^{p'}(\Omega)$ (see Problem 1.42). Therefore we conclude that $\psi \in C^1(L^p(\Omega))$.

Solution of Problem 3.6

Let $u, h \in L^p(\Omega)$. Then the growth condition on φ_x' implies that

$$\varphi_x'(\cdot, u(\cdot))h(\cdot) \in L^q(\Omega).$$

Then using the Hölder inequality (see Theorem 1.3 and Problem 1.27), we have

$$\int_\Omega |\varphi_x'(\omega, u(\omega))h(\omega)|^q\,d\mu \leqslant \Big(\int_\Omega |\varphi_x'(\omega, u(\omega))|^{\frac{pq}{p-q}}\,d\mu\Big)^{\frac{p-q}{p}} \|h\|_p^q.$$

Fix $u \in L^p(\Omega)$. Let

$$e(h) = N_\varphi(u+h) - N_\varphi(u) - N_{\varphi_x'}(u)h.$$

Similarly, as in the solution of Problem 3.5, for μ-almost all $\omega \in \Omega$, we can write

$$\begin{aligned}\varphi(\omega, u(\omega)+h(\omega)) - \varphi(\omega, u(\omega)) &= \int_0^1 \tfrac{d}{dt}\varphi(\omega, u(\omega)+th(\omega))\,dt \\ &= \int_0^1 \varphi_x'(\omega, u(\omega)+th(\omega))h(\omega)\,dt\end{aligned}$$

(by the chain rule), so

$$\|e(h)\|_q^q = \int_\Omega \Big|\int_0^1 \big(\varphi_x'(\omega, u(\omega)+th(\omega)) - \varphi_x'(\omega, u(\omega))\big)h(\omega)\,dt\Big|^q\,d\mu.$$

Using the Hölder inequality (see Theorem 1.3) and the Fubini theorem, we obtain

$$\|e(h)\|_q^q \;=\; \Big(\int_0^1\int_\Omega \big|\varphi_x'(\omega,u(\omega)+th(\omega))-\varphi_x'(\omega,u(\omega))\big|^n\,d\mu\,dt\Big)^{\frac{q}{n}}\|h\|_p^q,$$

thus

$$\frac{\|e(h)\|_q}{\|h\|_q}\longrightarrow 0 \quad \text{as } h\to 0 \quad \text{in } L^q(\Omega)$$

(from the continuity of $N_{\varphi_x'}$; see Problem 1.42) and hence

$$N_\varphi'(u)=N_{\varphi_x'}(u) \qquad \forall\, u\in L^p(\Omega).$$

The continuity of $N_{\varphi_x'}$ (see the growth assumption on φ_x') implies that $N_\varphi\in C^1(L^p(\Omega))$.

Solution of Problem 3.7

First let us compute the Gâteaux derivative of N_φ at $u\in L^p(\Omega)$ in the direction $h\in L^p(\Omega)$. To this end, let

$$\begin{aligned} w_t(z) &= \tfrac{1}{t}\big(\varphi(z,u(z)+th(z))-\varphi(z,u(z))\big)-\varphi_x'(z,u(z))h(z)\\ &= \int_0^1\big(\varphi_x'(z,u(z)+t\tau h(z))-\varphi_x'(z,u(z))\big)h(z)\,d\tau, \end{aligned}$$

so

$$\|w_t\|_p^p \leqslant \int_0^1\int_\Omega \big|\varphi_x'(z,u(z)+t\tau h(z))-\varphi_x'(z,u(z))\big|^p|h(z)|^p\,dz\,d\tau.$$

Note that

$$\big|\varphi_x'(z,u(z)+t\tau h(z))-\varphi_x'(z,u(z))\big| \longrightarrow 0 \quad \text{as } t\to 0 \quad \text{for a.a. } z\in\Omega$$

(as $\varphi(z,\cdot)\in C^1(\mathbb{R})$ for almost all $z\in\Omega$), so

$$\big|\varphi_x'(z,u(z)+t\tau h(z))-\varphi_x'(z,u(z))\big|^p \leqslant (2M)^p \quad \text{for a.a. } z\in\Omega$$

(see the assumption on the growth of φ_x'). Invoking the Lebesgue dominated convergence theorem, we infer that

$$\|w_t\|_p \longrightarrow 0 \quad \text{as } t\to 0,$$

so N_φ is Gâteaux differentiable and $(N_\varphi)_G'(u)=N_{\varphi_x'}(u)$. Since by hypothesis N_φ is Fréchet differentiable, it follows that

$$N_\varphi'(u) \;=\; N_{\varphi_x'}(u).$$

Without any loss of generality we may assume that $\varphi(z,0)=0$ for almost all $z \in \Omega$ (by replacing $\varphi(z,x)$ by $\varphi(z,x)-\varphi(z,0)$ if necessary). We have

$$\frac{1}{\|u\|_p} \int\limits_{\Omega} \left|\varphi(z,u(z)) - \varphi'_x(z,0)u(z)\right|^p dz \longrightarrow 0 \quad \text{as } \|u\|_p \to 0.$$

For fixed $\xi \in \mathbb{R}$ and a Lebesgue density point $z_0 \in \Omega$ of $\varphi(\cdot,u(\cdot)) - \varphi'_x(\cdot,0)u(\cdot)$ we set $u_n = \xi \chi_{B_{\frac{1}{n}}(z_0)}$. Then

$$\frac{1}{\xi^p \lambda^N(B_{\frac{1}{n}}(z_0))} \int\limits_{B_{\frac{1}{n}}(z_0)} \left|\varphi(z,\xi) - \varphi'_x(z,0)\xi\right|^p dz \longrightarrow 0 \quad \text{as } n \to +\infty,$$

so

$$\tfrac{1}{\xi^p}\left|\varphi(z_0,\xi) - \varphi'_x(z_0,0)\xi\right| = 0.$$

Therefore

$$\varphi(z,\xi) = \varphi'_x(z,0)\xi \quad \text{for a.a. } z \in \Omega$$

and so finally

$$\varphi(z,x) = c(z)x + a(z) \quad \text{for a.a. } z \in \Omega, \text{ all } z \in \mathbb{R},$$

with $c \in L^\infty(\Omega)$ and $a \in L^p(\Omega)$.

Solution of Problem 3.8

From the inverse function theorem (see Theorem 3.26), for every $u \in U$, we can find an open set V_u such that $u \in V_u \subseteq U$ and the set $\varphi(V_u) \subseteq Y$ is open. Then

$$\varphi(U) = \bigcup_{u \in U} \varphi(V_u)$$

and thus the set $\varphi(U)$ is open.

Solution of Problem 3.9

From Problem 3.8, we know that $\varphi(U) \subseteq Y$ is an open set. Since φ is injective, it is a bijection from U onto $\varphi(U)$. Since by the inverse function theorem (see Theorem 3.26), every $u \in U$ has an open neighborhood V_u such that $\varphi|_{V_u}$ is a C^1-

diffeomorphism, we conclude that globally φ is a C^1-diffeomorphism.

Solution of Problem 3.10

Let $y \in \varphi(U)$ and let $u \in U$ be such that $y = \varphi(u)$. Since $\varphi'(u)$ is an isomorphism, from the inverse function theorem (see Theorem 3.26), there exists an open neighborhood $V_u \subseteq U$ of u such that $\varphi\colon V_u \longrightarrow \varphi(V_u)$ is a C^1-diffeomorphism. Denote by φ_u^{-1} the inverse function of this diffeomorphism. Evidently $\varphi_u^{-1}\colon \varphi(V_u) \longrightarrow V_u$ is a C^1-map. By hypothesis we have that $\psi = \sigma \circ \varphi$. Hence $\sigma = \psi \circ \varphi_u^{-1}$ and so σ is a C^1-map on $\varphi(V_u)$. Since every $y \in \varphi(U)$ has an open neighborhood $\varphi(V_u)$ on which σ is C^1, we conclude that σ is C^1 on $\varphi(U)$.

Solution of Problem 3.11

Let $t \in [a,b)$ and let

$$\psi(s) = \varphi(t) + (s-t)\varphi_r'(t) \qquad \forall\, s \in [a,b].$$

Then

$$\lim_{s \searrow t} \tfrac{\psi(s)-\varphi(s)}{s-t} = 0,$$

so

$$\lim_{s \searrow t} \tfrac{\|\psi(s)\|_X - \|\varphi(s)\|_X}{s-t} = 0 \tag{3.15}$$

(by the triangle inequality). Let

$$\xi(s) = \|\psi(s)\|_X \qquad \forall\, s \in [a,b].$$

It is easy to see that ξ is convex and so it is right differentiable. Then for $s \neq t$, we have

$$\tfrac{\|\varphi(s)\|_X - \|\varphi(t)\|_X}{s-t} = \tfrac{\|\varphi(s)\|_X - \|\psi(s)\|_X}{s-t} + \tfrac{\|\psi(s)\|_X - \|\varphi(t)\|_X}{s-t},$$

so $\vartheta_r'(t) = \xi_r'(t)$ (see (3.15)). Because

$$\left|\tfrac{\|\varphi(s)\|_X - \|\varphi(t)\|_X}{s-t}\right| \leqslant \left\|\tfrac{\varphi(s)-\varphi(t)}{s-t}\right\|_X \qquad \forall\, s \neq t,$$

by passing to the limit as $s \searrow t$, we obtain $|\vartheta_r'(t)| \leqslant \|\varphi_r'(t)\|_X$ for all $t \in [a,b)$.

Solution of Problem 3.12

Let

$$\psi(t) \;=\; \int\limits_a^t \varphi'(s)\,ds \qquad \forall\, t\in[a,b].$$

Then $\psi(t)=\varphi(t)-\varphi(a)$ for all $t\in[a,b]$. From Problem 1.66, we have

$$\int\limits_a^b \varphi'(s)\,ds \;\in\; (b-a)C,$$

so

$$\varphi(b)-\varphi(a) \;\in\; (b-a)C$$

and thus

$$\tfrac{\varphi(b)-\varphi(a)}{b-a} \;\in\; C.$$

Solution of Problem 3.13

Let $\varepsilon>0$ and consider

$$\begin{aligned} S(\varepsilon) \;=\; \big\{t\in[a,b]:\ &\|\varphi(s)-\varphi(a)\|_X \leqslant \vartheta(s)-\vartheta(a)+\varepsilon(s-a)\\ &\text{for all } s\in[a,t]\big\}. \end{aligned} \tag{3.16}$$

Evidently $a\in S(\varepsilon)$. Also, if $t\in S(\varepsilon)$ and $t'\in[a,t]$, then $t'\in S(\varepsilon)$ and so the set $S(\varepsilon)$ is an interval. Let $\widehat{b}-\sup S(\varepsilon)$. We claim that $\widehat{b}\in S(\varepsilon)$. If $\widehat{b}=a$, then $\widehat{b}=a\subset S(t)$. So, suppose that $a<\widehat{b}$. If $t\in(a,\widehat{b})$, then

$$\|\varphi(t)-\varphi(a)\|_X \;\leqslant\; \vartheta(t)-\vartheta(a)+\varepsilon(t-a) \tag{3.17}$$

(see (3.16)). Exploiting the continuity of φ and ϑ, we see that (3.17) is also truc for $\widehat{b}$ and so $\widehat{b}\in S(\varepsilon)$.

Next we show that $\widehat{b}=b$. Arguing by contradiction, suppose that $\widehat{b}<b$. Then since by hypothesis $\|\varphi'(t)\|_X\leqslant\vartheta'(t)$ for all $t\in(a,b)$, we can find $\tau\in(0,b-\widehat{b})$ such that

$$\big\|\tfrac{\varphi(t)-\varphi(\widehat{b})}{t-\widehat{b}}\big\|_X \;\leqslant\; \tfrac{\vartheta(t)-\vartheta(\widehat{b})}{t-\widehat{b}}+\varepsilon \qquad \forall\, t\in(\widehat{b},\widehat{b}+\tau),$$

so

$$\|\varphi(t)-\varphi(\widehat{b})\|_X \;\leqslant\; \vartheta(t)-\vartheta(\widehat{b})+\varepsilon(t-\widehat{b}),$$

thus

$$\begin{aligned} \|\varphi(t)-\varphi(a)\|_X &\leqslant \|\varphi(t)-\varphi(\widehat{b})\|_X+\|\varphi(\widehat{b})-\varphi(a)\|_X\\ &\leqslant \big(\vartheta(t)-\vartheta(\widehat{b})+\varepsilon(t-\widehat{b})\big)+\big(\vartheta(\widehat{b})-\vartheta(a)+\varepsilon(\widehat{b}-a)\big)\\ &= \vartheta(t)-\vartheta(a)+\varepsilon(t-a) \end{aligned}$$

(see (3.17)) and hence $(\widehat{b}, \widehat{b}+\tau) \subseteq S(\varepsilon)$, which contradicts the fact that $\widehat{b} = \sup S(\varepsilon)$. Therefore $\widehat{b} = b$ and so $S(\varepsilon) = [a, b]$. Then we have

$$\|\varphi(b) - \varphi(a)\|_X \leqslant \vartheta(b) - \vartheta(a) + \varepsilon(b-a).$$

Let $\varepsilon \searrow 0$, to conclude that $\|\varphi(b) - \varphi(a)\|_X \leqslant \vartheta(b) - \vartheta(a)$.

Solution of Problem 3.14

Recall that the operators A, e^{tA} and e^{hA} (with $h \in \mathbb{R}$) commute. We have

$$\varphi(t+h) - \varphi(t) - hAe^{tA} = e^{tA}\left(e^{hA} - I_X - hA\right). \tag{3.18}$$

But note that

$$\begin{aligned} \|e^{hA} - I_X - hA\|_{\mathcal{L}} &= \Big\|\sum_{n\geqslant 2} \tfrac{h^n}{n!} A^n\Big\|_{\mathcal{L}} \\ \leqslant \sum_{n\geqslant 2} \tfrac{|h|^n}{n!} \|A\|_{\mathcal{L}}^n &= e^{|h|A} - 1 - |h|\|A\|_{\mathcal{L}}. \end{aligned} \tag{3.19}$$

Then from (3.18) and (3.19), we have

$$\varphi(t+h) - \varphi(t) - hAe^{tA} = e(|h|),$$

with $\frac{e(|h|)}{|h|} \longrightarrow 0$ as $h \to 0$. This implies that φ is differentiable and

$$\varphi'(t) = Ae^{tA} \qquad \forall\, t \in \mathbb{R}.$$

Therefore we conclude that $\varphi \in C^1(\mathbb{R}; \mathcal{L}(X))$.

Solution of Problem 3.15

Let $u, h \in X = C([0,1])$. Then for all $t \in [0,1]$, we have

$$\varphi(u+h)(t) - \varphi(u)(t) = \sin(u+h)(t) - \sin u(t). \tag{3.20}$$

From the Taylor formula, we know that there exists $\vartheta(t) \in (0,1)$ such that

$$\sin(u+h)(t) - \sin u(t) = h(t)\cos u(t) - \tfrac{h(t)^2}{2} \sin\left(u(t) + \vartheta(t)h(t)\right). \tag{3.21}$$

Let $\xi_u \colon X \longrightarrow X$ be defined by

$$\xi_u(h)(t) = h(t)\cos u(t) \qquad \forall\, h \in X,\ t \in [0,1].$$

Then ξ_u is linear and

$$\|\xi_u(h)\|_X = \sup_{t\in[0,1]} |h(t)\cos u(t)| \leqslant \sup_{t\in[0,1]} |h(t)| = \|h\|_X,$$

so $\xi_u \in \mathcal{L}(X)$. From (3.20) and (3.21), we have

$$\begin{aligned}&\|\varphi(u+h)-\varphi(u)-\xi_u(h)\|_X\\ =& \sup_{t\in[0,1]} \left|\tfrac{h(t)^2}{2}\sin\left(u(t)+\vartheta(t)h(t)\right)\right| \leqslant \tfrac{\|h\|_X^2}{2},\end{aligned}$$

so φ is Fréchet differentiable at $u \in X$ and $\varphi'(u)(h) = \xi_u(h)$ for all $u, h \in X$.

Solution of Problem 3.16

Note that

$$p(\widehat{u}+\widehat{v}) \leqslant p(\widehat{u})+p(\widehat{v}) \qquad \forall\, \widehat{u},\widehat{v} \in l^\infty$$

(i.e., p is subadditive) and

$$p(\lambda\widehat{u}) = |\lambda| p(\widehat{u}) \qquad \forall\, \lambda \in \mathbb{R},\ \widehat{u} \in l^\infty.$$

Also, clearly we have

$$p(\widehat{u}) \leqslant \|\widehat{u}\|_\infty \qquad \forall\, \widehat{u} \in l^\infty. \tag{3.22}$$

Suppose that $\widehat{u}_n \longrightarrow \widehat{u}$ in l^∞. Then

$$p(\widehat{u}_n) = p(\widehat{u}_n - \widehat{u} + \widehat{u}) \leqslant p(\widehat{u}_n - \widehat{u}) + p(\widehat{u}) \tag{3.23}$$

and

$$p(\widehat{u}) = p(\widehat{u} - \widehat{u}_n + \widehat{u}_n) \leqslant p(\widehat{u} - \widehat{u}_n) + p(\widehat{u}_n) \tag{3.24}$$

From (3.22), (3.23), and (3.24), it follows that

$$|p(\widehat{u}_n) - p(\widehat{u})| \leqslant \|\widehat{u}_n - \widehat{u}\|_\infty,$$

so p is continuous.

If $p(\widehat{u}) = 0$, then $u_n \longrightarrow 0$. Taking $\widehat{h} = (1,1,\ldots) \in l^\infty$, we have

$$\frac{p(\widehat{u}+\lambda\widehat{h})-p(\widehat{u})}{\lambda} = \frac{|\lambda|}{\lambda}$$

and this show that p is not Gâteaux differentiable at $\widehat{u}$ (as $\lim\limits_{\lambda\to 0}\frac{|\lambda|}{\lambda}$ does not exist).

If $p(\widehat{u}) > 0$, then without any loss of generality, we may assume that $p(\widehat{u}) = 1$. Then we can find a subsequence $\{u_{n_k}\}_{k\geqslant 1}$ of $\{u_n\}_{n\geqslant 1}$ such that $u_{n_k} \longrightarrow 1$. Moreover, without any loss of generality we may assume that $u_{n_k} > 0$ for all $k \geqslant 1$. Let

$$h_n = \begin{cases} 0 & \text{if } n \neq n_k \text{ or } n = n_k \text{ with } k \text{ odd},\\ 1 & \text{if } n = n_k \text{ with } k \text{ even},\end{cases}$$

for all $n \geqslant 1$. Let $\widehat{h} = \{h_n\}_{n \geqslant 1} \in l^\infty$. We have

$$\frac{p(\widehat{u}+\lambda\widehat{h})-p(\widehat{u})}{\lambda} = \begin{cases} 1 & \text{if } \lambda > 0 \\ 0 & \text{if } \lambda < 0. \end{cases}$$

Therefore p is nowhere Gâteaux differentiable.

Solution of Problem 3.17

Since φ is Fréchet differentiable at $u \in X$ (see Definition 3.1(b)), given $\varepsilon > 0$, we can find $\delta > 0$ such that

$$\|\varphi(u+h) - \varphi(u) - \varphi'(u)(h)\|_Y \leqslant \varepsilon\|h\|_X \qquad \forall\, h \in X, \text{ with } \|h\|_X \leqslant \delta. \tag{3.25}$$

Then we have

$$\begin{aligned} \|\varphi(u+h) - \varphi(u)\|_Y &\leqslant \|\varphi'(u)h\|_Y + \|\varphi(u+h) - \varphi(u) - \varphi'(u)h\|_Y \\ \leqslant M\|h\|_X + \varepsilon\|h\|_X &= (M+\varepsilon)\|h\|_X \qquad \forall\, h \in X, \text{ with } \|h\|_X \leqslant \delta \end{aligned}$$

(by the triangle inequality and using (3.25)).

Solution of Problem 3.18

First note that φ is injective. Indeed, if $\varphi(u) = \varphi(v)$, then

$$\|u - v\|_X \leqslant c\|u - v\|_X$$

and since $c \in (0,1)$, we infer that $u = v$.

Next we show that φ is surjective. Given any $h \in X$ we consider the equation $\varphi(u) = h$. Let $u_0 = h$ and inductively define $u_{n+1} = u_n + (h - \varphi(u_n))$ for all $n \geqslant 0$. We have

$$u_{n+1} - u_n = u_n - u_{n-1} + \varphi(u_{n-1}) - \varphi(u_n) \qquad \forall\, n \geqslant 1,$$

so

$$\|u_{n+1} - u_n\|_X = \|\varphi(u_{n-1}) - \varphi(u_n) - (u_{n-1} - u_n)\|_X \leqslant c\|u_{n-1} - u_n\|_X,$$

thus

$$\|u_{n+1} - u_n\|_X \leqslant c^n\|u_1 - u_0\|_X \qquad \forall\, n \geqslant 1,$$

hence

$$\sum_{n \geqslant 0} \|u_{n+1} - u_n\|_X \leqslant \Big(\sum_{n \geqslant 0} c^n\Big)\|u_1 - u_0\|_X \leqslant \tfrac{1}{1-c}\|u_1 - u_0\|_X.$$

This means that $\{u_n\}_{n\geqslant 0} \subseteq X$ is a Cauchy sequence and so we have $u_n \longrightarrow u$ in X. Hence $\varphi(u_n) \longrightarrow \varphi(u)$. Therefore, in the limit as $n \to +\infty$, we obtain $h = \varphi(u)$. Since $h \in X$ is arbitrary, we conclude that φ is surjective, hence φ is bijective. Since $\varphi'(u)$ is an isomorphism for every $u \in X$, from the inverse function theorem (see Theorem 3.26), we conclude that $\psi = \varphi^{-1} \in C^1(X;X)$ and so φ is a C^1-diffeomorphism.

Solution of Problem 3.19

Given $u \in H$, for every $h \in H$, we have

$$\varphi(u+h) = \varphi(u) + (A^*(u) + A(u), h)_H + (A(h), h)_H,$$

so φ is Fréchet differentiable at $u \in H$ and $\varphi'(u)(h) = (A^*(u) + A(u), h)_H$ for all $h \in H$. Thus $\varphi \in C^1(H)$.

Solution of Problem 3.20

First we consider the case $p \in (1,2)$. Let

$$\xi(t) = (|1+t|^p - 1 - pt)\tfrac{1}{|t|^p} \qquad \forall\, t \in \mathbb{R}.$$

Note that

$$\lim_{t\to\pm\infty} \xi(t) = 1 \quad \text{and} \quad \lim_{t\to 0} \xi(t) = 0.$$

So, ξ is bounded on $\mathbb{R}$ and we have

$$c_1|t|^p \leqslant |1+t|^p - 1 - pt \leqslant c_2|t|^p \qquad \forall\, t \in \mathbb{R} \tag{3.26}$$

for some $c_1, c_2 > 0$. Let $u, h \in L^p(\Omega;\mathbb{R}^N)$, $u(\omega) \neq 0$ for μ-almost all $\omega \in \Omega$. We set

$$y(\omega) = \tfrac{th(\omega)}{u(\omega)} \qquad \forall\, \omega \in \Omega.$$

From (3.26), it follows that

$$\begin{aligned} & c_1\|th\|^p_{L^p(\Omega;\mathbb{R}^N)} \\ \leqslant\ & \|u+th\|^p_{L^p(\Omega;\mathbb{R}^N)} - \|u\|^p_{L^p(\Omega;\mathbb{R}^N)} - pt\int_\Omega h(\omega)|u(\omega)|^{p-1}\operatorname{sign} u(\omega)\,d\mu \\ \leqslant\ & c_2\|th\|^p_{L^p(\Omega;\mathbb{R}^N)} \end{aligned} \tag{3.27}$$

Dividing (3.27) by t and letting $t \to 0$ we obtain

$$\left\langle (\|\cdot\|^p_{L^p(\Omega;\mathbb{R}^N)})'_G(u), h \right\rangle \;=\; p \int\limits_{\Omega} |u(\omega)|^{p-1} \operatorname{sign} u(\omega) h(\omega)\, d\mu.$$

So, the norm $\|\cdot\|_{L^p(\Omega;\mathbb{R}^N)}$ (with $1 < p < 2$) is Gâteaux differentiable at every $u \neq 0$ and

$$(\|\cdot\|^p_{L^p(\Omega;\mathbb{R}^N)})'_G(u) \;=\; p|u(\cdot)|^{p-1} \operatorname{sign} u(\cdot) \;=\; p|u(\cdot)|^{p-2} u(\cdot).$$

Since the map $u \longmapsto p|u(\cdot)|^{p-1}\operatorname{sign} u(\cdot)$ is continuous from $L^p(\Omega;\mathbb{R}^N)$ into $L^{p'}(\Omega;\mathbb{R}^N)$ (where $\frac{1}{p} + \frac{1}{p'} = 1$), from Proposition 3.3, it follows that the norm $\|\cdot\|_{L^p(\Omega;\mathbb{R}^N)}$ is Fréchet differentiable at every $u \neq 0$.

For $2 < p < +\infty$, we consider the function

$$\xi(t) \;=\; (|1+t|^p - 1 - pt)\frac{1}{|t|^p + t^2} \qquad \forall\, t > 0.$$

Reasoning as above we obtain

$$(\|\cdot\|^p_{L^p(\Omega;\mathbb{R}^N)})'(u) \;=\; p|u(\cdot)|^{p-1} \operatorname{sign} u(\cdot) \;=\; p|u(\cdot)|^{p-2} u(\cdot).$$

Finally for $p = 2$, $L^p(\Omega;\mathbb{R}^N)$ is a Hilbert space, hence the norm is Fréchet differentiable at every $u \neq 0$ and

$$(\|\cdot\|^2_{L^2(\Omega;\mathbb{R}^N)})'(u) \;=\; 2u(\cdot).$$

Let $\varphi(u) = \|u\|^p_{L^p(\Omega;\mathbb{R}^N)}$ and $\psi(u) = \|u\|_{L^p(\Omega;\mathbb{R}^N)}$ for all $u \in L^p(\Omega;\mathbb{R}^N)$. We have

$$\varphi(u) \;=\; \psi(u)^p \qquad \forall\, u \in L^p(\Omega;\mathbb{R}^N).$$

From the chain rule (see Theorem 3.19) we have

$$\varphi'(u) \;=\; p\psi(u)^{p-1}\psi'(u) \qquad \forall\, u \neq 0,$$

so

$$\psi'(u)(\cdot) \;=\; \frac{1}{\|u\|^{p-1}_{L^p(\Omega;\mathbb{R}^N)}} |u(\cdot)|^{p-2} u(\cdot) \qquad \forall\, u \in L^p(\Omega;\mathbb{R}^N),\ 1 < p < +\infty.$$

Solution of Problem 3.21

Consider the linear operator $T\colon W^{1,p}_0(\Omega) \longrightarrow L^p(\Omega;\mathbb{R}^N)$ defined by

$$T(u) \;=\; Du \qquad \forall\, u \in W^{1,p}_0(\Omega).$$

Evidently $T \in \mathcal{L}(W^{1,p}_0(\Omega); L^p(\Omega;\mathbb{R}^N))$. Also, let $\psi\colon L^p(\Omega;\mathbb{R}^N) \longrightarrow \mathbb{R}$ be the function defined by

$$\psi(y) = \tfrac{1}{p}\|y\|^p_{L^p(\Omega;\mathbb{R}^N)} \qquad \forall\, y \in L^p(\Omega;\mathbb{R}^N).$$

We see that

$$\varphi(u) = (\psi \circ T)(u) \qquad \forall\, u \in W_0^{1,p}(\Omega).$$

From the chain rule (see Theorem 3.19), we have

$$\varphi'(u) = T^* \psi'(Tu) = -\mathrm{div}\,(|Du|^{p-2} Du)$$

(see the solution of Problem 3.20 and recall that $T^* = -\mathrm{div}$). So, the derivative of φ is the negative p-Laplace differential operator denoted by $-\Delta_p$. We have

$$\langle \varphi'(u), h \rangle = \langle -\Delta_p u, h \rangle = \int_\Omega |Du|^{p-2} (Du, Dh)_{\mathbb{R}^N}\, dz$$
$$\forall\, u, h \in W_0^{1,p}(\Omega).$$

Next we show that the continuity of the map

$$W_0^{1,p}(\Omega) \ni u \longmapsto \varphi'(u) \in W^{-1,p'}(\Omega) = (W_0^{1,p}(\Omega))^*$$

(with $\frac{1}{p} + \frac{1}{p'} = 1$). First we deal with the case $1 < p \leqslant 2$. We recall the following elementary inequality:

$$\big| |x|^{p-2} x - |y|^{p-2} y \big| \leqslant c_1(p) |x - y|^{p-1} \qquad \forall\, x, y \in \mathbb{R}^N \tag{3.28}$$

where $1 < p \leqslant 2$. Then for all $u, v, h \in W_0^{1,p}(\Omega)$, we have

$$\begin{aligned} \big| \langle \varphi'(u) - \varphi'(v), h \rangle \big| &\leqslant \int_\Omega \big| |Du|^{p-2} Du - |Dv|^{p-2} Dv \big| |Dh|\, dz \\ &\leqslant c_1(p) \int_\Omega |Du - Dv|^{p-1} |Dh|\, dz \\ &\leqslant c_1(p) \|Du - Dv\|^{p-1}_{L^p(\Omega;\mathbb{R}^N)} \|Dh\|_{L^p(\Omega;\mathbb{R}^N)} \end{aligned} \tag{3.29}$$

(see (3.28) and use the Hölder inequality (see Theorem 1.3). By the Poincaré inequality (see Theorem 1.131), $\|Dh\|_{L^p(\Omega;\mathbb{R}^N)}$ is an equivalent norm on the Sobolev space $W_0^{1,p}(\Omega)$. So, from (3.29), we have

$$\|\varphi'(u) - \varphi'(v)\|_* \leqslant c_2(p) \|u - v\|^{p-1},$$

for some $c_2(p) > 0$ and thus the map $W_0^{1,p}(\Omega) \ni u \longmapsto \varphi'(u) \in W^{-1,p'}(\Omega) = (W_0^{1,p}(\Omega))^*$ is continuous when $1 < p \leqslant 2$.

Now assume that $p > 2$. Let $u_n \longrightarrow u$ in $W_0^{1,p}(\Omega)$. Then we have

$$Du_n \longrightarrow Du \quad \text{in } L^p(\Omega; \mathbb{R}^N).$$

By passing to a suitable subsequence if necessary, we may also assume that

$$|Du_n(z)| \leqslant \eta(z),\ |Du(z)| \leqslant \eta(z) \quad \text{for a.a. } z \in \Omega,\ \text{all } n \geqslant 1, \tag{3.30}$$

with $\eta \in L^p(\Omega)$. We recall another elementary inequality:

$$\big||x|^{p-2}x - |y|^{p-2}y\big| \leqslant c_3(p)|x-y|(|x|+|y|)^{p-2} \qquad \forall\, x,y \in \mathbb{R}^N \tag{3.31}$$

where $p > 2$. We have

$$\begin{aligned} \big|\langle \varphi'(u) - \varphi'(v), h\rangle\big| &\leqslant \int_\Omega \big||Du|^{p-2}Du - |Dv|^{p-2}Dv\big||Dh|\,dz \\ &\leqslant 2^{p-2}c_3(p)\int_\Omega \eta(z)^{p-2}|Du_n - Du||Dh|\,dz \end{aligned} \tag{3.32}$$

(see (3.30) and (3.31)). Note that $\eta^{p-2} \in L^{\frac{p}{p-2}}(\Omega)$ and $\frac{1}{p}+\frac{1}{p}+\frac{p-2}{p} = 1$. So, from (3.32) and the generalized Hölder inequality (see Theorem I.3.105), we have

$$\big|\langle \varphi'(u_n) - \varphi'(u), h\rangle\big| \leqslant c_4(p)\|Du_n - Du\|_{L^p(\Omega;\mathbb{R}^N)}\|Dh\|_{L^p(\Omega;\mathbb{R}^N)}. \tag{3.33}$$

As before using the Poincaré inequality, from (3.33), we obtain

$$\|\varphi'(u_n) - \varphi'(u)\|_* \leqslant c_5(p)\|u_n - u\|_{W_0^{1,p}(\Omega)} \qquad \forall\, n \geqslant 1,$$

for some $c_5 > 0$, thus

$$\varphi'(u_n) \longrightarrow \varphi'(u) \quad \text{in } W^{-1,p'}(\Omega)$$

(for the original sequence by the Urysohn criterion; see Problem I.1.3) and hence the map $W_0^{1,p}(\Omega) \ni u \longmapsto \varphi'(u) \in W^{-1,p'}(\Omega) = (W_0^{1,p}(\Omega))^*$ is continuous when $p > 2$.

Therefore we have proved that $\varphi \in C^1(W_0^{1,p}(\Omega))$.

Solution of Problem 3.22

Let $\{u_n\}_{n\geqslant 1} \subseteq U$ be a sequence and assume that $u_n \xrightarrow{w} u$ in X. From the hypothesis on φ, we have

$$\varphi(u_n) - \varphi(u_0) \geqslant \langle \varphi'_G(u_0), u_n - u_0\rangle,$$

so

$$\liminf_{n\to+\infty}(\varphi(u_n) - \varphi(u_0)) \geqslant 0,$$

thus

$$\varphi(u_0) \leqslant \liminf_{n\to+\infty}\varphi(u_n).$$

This shows the weak lower semicontinuity of φ.

Now, let us assume that U is open and convex. Let $u, v \in U$. Then

$$\varphi(v) \geqslant \varphi(u_0) + \langle \varphi'_G(u_0), v - u_0\rangle \tag{3.34}$$

and

$$\varphi(u) \geqslant \varphi(u_0) + \langle \varphi'_G(u_0), u - u_0 \rangle. \tag{3.35}$$

Multiplying (3.34) with t, (3.35) with $1-t$ (for $t \in (0,1)$), setting $u_0 = (1-t)u + tv$ and adding the two inequalities, we obtain

$$(1-t)\varphi(u) + t\varphi(v) \geqslant \varphi(u_0) + \langle \varphi'_G(u_0), (1-t)u + tv - u_0 \rangle = \varphi(u_0)$$

and thus $\varphi|_U$ is convex.

Solution of Problem 3.23

Arguing by contradiction, suppose that $\|u_0\|_X = \varrho$. Consider the function $\xi\colon [0,1] \longrightarrow \mathbb{R}$ defined by $\xi(t) = \varphi((1-t)u_0)$. Then

$$\xi(t) \geqslant \xi(0) \qquad \forall\, t \in [0,1]$$

and

$$0 \leqslant \xi'_+(0) = \lim_{t \to 0^+} \tfrac{\varphi((1-t)u_0) - \varphi(u_0)}{t} = \langle \varphi'(u_0), -u_0 \rangle < 0,$$

a contradiction.

Solution of Problem 3.24

Let $h \in X$ and let $\xi(t) = \varphi(u + th)$ for all $t \in \mathbb{R}$. Then the Gâteaux differentiability of φ at $u \in X$ implies the differentiability of ξ at $t = 0$ (see Definition 3.1(a)) and

$$\xi'(0) = \varphi'(u)h.$$

The differentiability of ξ at $t = 0$ implies the continuity of ξ at $t = 0$. Hence $\xi(t) \longrightarrow \xi(0)$ as $t \to 0$, which implies the hemicontinuity of φ at u for the strong topology on X (see Definition 2.110(b)).

Solution of Problem 3.25

First we claim that φ is sequentially continuous from X with the weak topology into Y with the weak topology. So, suppose that $\{u_n\}_{n\geqslant 1} \subseteq X$ is a sequence such that $u_n \xrightarrow{w} u$ in X. If our claim is not true, we can find $\varepsilon > 0$, a subsequence $\{u_{n_k}\}_{k\geqslant 1}$ of $\{u_n\}_{n\geqslant 1}$ and $y_0^* \in Y^*$ such that

$$\left| \langle y_0^*, \varphi(u_{n_k}) - \varphi(u) \rangle \right| \geqslant \varepsilon \qquad \forall\, k \geqslant 1. \tag{3.36}$$

The mean value theorem (see Theorem 3.16) implies that we can find $t_{n_k} \in (0,1)$ such that

$$\langle y_0^*, \varphi(u_{n_k}) - \varphi(u)\rangle = \langle y_0^*, \varphi'(u + t_{n_k}(u_{n_k} - u))(u_{n_k} - u)\rangle \qquad \forall\, k \geqslant 1. \tag{3.37}$$

By hypothesis $\varphi' \colon X \longrightarrow \mathcal{L}(X;Y)$ is compact (see Definition 2.1(a)). So, by passing to a subsequence if necessary, we may assume that

$$\varphi'(u + t_{n_k}(u_{n_k} - u)) \longrightarrow A \quad \text{in } \mathcal{L}(X;Y) \tag{3.38}$$

(with $A \in \mathcal{L}(X;Y)$). We have

$$\begin{aligned} &\big\langle y_0^*, \varphi'(u + t_{n_k}(u_{n_k} - u))(u_{n_k} - u)\big\rangle \\ &= \big\langle y_0^*, \big(\varphi'(u + t_{n_k}(u_n - u)) - A\big)(u_{n_k} - u)\big\rangle + \langle y_0^*, A(u_{n_k} - u)\rangle \longrightarrow 0 \end{aligned}$$

(see (3.38) and recall that $A \in \mathcal{L}(X;Y)$ is also weakly continuous), so

$$\langle y_0^*, \varphi(u_{n_k}) - \varphi(u)\rangle \longrightarrow 0$$

(see (3.37)) and this contradicts (3.36). This proves that $u \longmapsto \varphi(u)$ is sequentially continuous from X with the weak topology into Y with the weak topology.

Since φ is compact and $\{u_n\}_{n\geqslant 1} \subseteq X$ is bounded (being a weakly convergent sequence), we have that $\overline{\{\varphi(u_n)\}_{n\geqslant 1}} \subseteq Y$ is compact and so we can find a subsequence $\{\varphi(u_{n_k})\}_{k\geqslant 1}$ such that

$$\varphi(u_{n_k}) \longrightarrow y \quad \text{in } Y.$$

From the first part of the proof we have

$$\varphi(u_{n_k}) \xrightarrow{w} \varphi(u) \quad \text{in } Y.$$

Therefore $y = \varphi(u)$ and for the original sequence we have

$$\varphi(u_n) \longrightarrow \varphi(u) \quad \text{in } Y,$$

thus φ is completely continuous (see Definition 2.1(b)).

Solution of Problem 3.26

Let $u \in X$ and consider the function $h \longmapsto L(u)(h)$ defined by

$$L(u)(h) = \int_0^1 \xi'(u(t))h(t)\,dt \qquad \forall\, h \in X.$$

Clearly the function $L(u)(\cdot)$ is linear and

$$|L(u)(h)| \leqslant \|\xi' \circ u\|_X \|h\|_X,$$

so $L(u)(\cdot)$ is continuous, that is, $L(u) \in X^*$. We have

$$\begin{aligned} &\varphi(u+h) - \varphi(u) - L(u)(h) \\ &= \int_0^1 \big(\xi(u(t)+h(t)) - \xi(u(t)) - \xi'(u(t))h(t)\big)\,dt \\ &= \int_0^1 \big(\xi'(u(t)+\tau(t)h(t)) - \xi'(u(t))h(t)\big)\,dt, \end{aligned} \tag{3.39}$$

for some $\tau(t) \in (0,1)$ (by the mean value theorem; see Theorem 3.16). Without any loss of generality, we may assume that $\|h\|_X \leqslant 1$. So,

$$u(t)+\tau(t)h(t),\ u(t) \ \in\ \big[\|u\|_X - 1, \|u\|_X + 1\big] \qquad \forall\, t \in [0,1].$$

Since ξ' is continuous (recall that $\xi \in C^1(\mathbb{R})$), it is uniformly continuous on the interval $\big[\|u\|_X - 1, \|u\|_X + 1\big]$. Hence, given $\varepsilon > 0$, we can find $\delta \in (0,1)$ such that

$$\big|\xi'(u(t)+\tau(t)h(t)) - \xi'(u(t))\big| \ <\ \varepsilon \qquad \forall\, t \in [0,1],\ h \in X, \text{ with } \|h\|_X < \delta.$$

Then from (3.39), we have

$$|\varphi(u+h) - \varphi(h) - L(u)(h)| \ \leqslant\ \varepsilon\|h\|_X,$$

so φ is Fréchet differentiable at every $u \in X$ and

$$\langle \varphi'(u), h\rangle \ =\ \int_0^1 \xi'(u(t))h(t)\,dt \qquad \forall\, h \in X.$$

Because of the continuity of $X \ni u \longmapsto \varphi'(u) - L(u) \in X^*$, we conclude that $\varphi \in C^1(X)$.

Solution of Problem 3.27

If $h = 0$, then the result holds trivially.

So, suppose that $h \neq 0$. Then due to the Fréchet differentiability of φ at $u_0 + th$, we have

$$0 \ =\ \lim_{v\to 0} \frac{\varphi(u_0+th+v) - \varphi(u_0+th) - \langle \varphi'(u_0+th), v\rangle}{\|v\|_X}.$$

In particular, let $v = sh$. We have

$$\begin{aligned} 0 \ &=\ \lim_{s\to 0} \frac{\xi(t+s) - \xi(t) - s\langle \varphi'(u_0+th), h\rangle}{s} \\ &=\ \lim_{s\to 0} \frac{\xi(t+s) - \xi(t)}{s} - \big\langle \varphi'(u_0+th), h\big\rangle \end{aligned}$$

(since $h \neq 0$), so

$$\xi'(t) = \langle \varphi'(u_0 + th), h \rangle .$$

Solution of Problem 3.28

(a) The map ξ is bilinear and continuous (since $M_n(\mathbb{R})$ is a finite dimensional Banach space). So, it is C^1 and

$$\xi'(A, K)(L, T) = LK + AT. \tag{3.40}$$

(b) Let $A \in \mathrm{End}_n(\mathbb{R})$ and $M \in M_n(\mathbb{R})$. We set $T = A + M$. We write

$$T = (I + MA^{-1})A.$$

We have

$$\|MA^{-1}\|_{\mathcal{L}} \leqslant \|M\|_{\mathcal{L}} \|A^{-1}\|_{\mathcal{L}}.$$

So, if $\|M\|_{\mathcal{L}} < \frac{1}{\|A^{-1}\|_{\mathcal{L}}}$, then $\|MA^{-1}\|_{\mathcal{L}} < 1$ and so $I + MA^{-1} \in \mathrm{End}_n(\mathbb{R})$. Therefore, if $\varrho = \|A^{-1}\|_{\mathcal{L}}^{-1}$, then

$$B_\varrho(A) \subseteq \mathrm{End}_n(\mathbb{R})$$

and thus the set $\mathrm{End}_n(\mathbb{R}) \subseteq M_n(\mathbb{R})$ is open.
Let $A \in \mathrm{End}_n(\mathbb{R})$. Then for all $M \in M_n(\mathbb{R})$ we have

$$\xi_2'(A, A^{-1})(M) = AM$$

(see (3.40)), so $\xi_2'(A, A^{-1})$ is an isomorphism. So, we can apply the implicit function theorem (see Theorem 3.28) and find an open neighborhood U of A in $\mathrm{End}_n(\mathbb{R})$ and a C^1-map h defined on U such that

$$\xi(E, h(E)) = \xi(A, A^{-1}) = I \qquad \forall\, E \in U.$$

Note that $h = \beta|_U$ and so β is differentiable at $A \in \mathrm{End}_n(\mathbb{R})$. We have

$$\xi_1'(A, \beta(A)) + \xi_2'(A, \beta(A)) \circ \beta'(A) = 0$$

(see Proposition 3.12), so

$$MA^{-1} + A\beta'(A)(M) = 0 \qquad \forall\, M \in M_n(\mathbb{R}),$$

thus

$$\beta'(A)(M) = -A^{-1}MA^{-1} \qquad \forall\, M \in M_n(\mathbb{R}).$$

Evidently the map $A \longmapsto \beta'(A)$ is continuous and so we have $\beta \in C^1(\mathrm{End}_n(\mathbb{R}))$, β is injective and $\beta'(A)$ is an isomorphism. Therefore β is a C^1-diffeomorphism

(see Problem 3.9)

Solution of Problem 3.29

"$\Longrightarrow$": Suppose that φ is α-homogeneous. Let $\xi(t) = \varphi(tu)$, $u \neq 0$, $t > 0$. From Problem 3.27, we have

$$\xi'(1) = \langle \varphi'(u), u \rangle . \tag{3.41}$$

From the α-homogeneity of φ, we have

$$\xi(t) = \varphi(tu) = t^\alpha \varphi(u),$$

so

$$\xi'(1) = \alpha\varphi(u). \tag{3.42}$$

From (3.41) and (3.42), we conclude that $\alpha\varphi(u) = \langle \varphi'(u), u \rangle$ for all $u \neq 0$.

"$\Longleftarrow$": Suppose that $\alpha\varphi(u) = \langle \varphi'(u), u \rangle$ for all $u \neq 0$. As above let $\xi(t) = \varphi(tu)$, for $t > 0$. We consider the quotient $\frac{\xi(t)}{t^\alpha}$. We have

$$\frac{d}{dt}\left(\frac{\xi(t)}{t^\alpha}\right) = \frac{\xi'(t)t^\alpha - \alpha t^{\alpha-1}\xi(t)}{t^{2\alpha}} = \frac{\langle \varphi'(tu), tu\rangle - \alpha\varphi(tu)}{t^{\alpha+1}} = 0$$

(see Problem 3.27 and hypothesis). Therefore $\frac{\xi(t)}{t^\alpha} = c \in \mathbb{R}$ for all $t > 0$. Hence

$$\varphi(tu) = t^\alpha c \qquad \forall\, t > 0,$$

so $\varphi(u) = c$. Thus, we conclude that $\varphi(tu) = t^\alpha\varphi(u)$ for all $u \neq 0$, all $t > 0$, which means that φ is α-homogeneous.

Solution of Problem 3.30

Note that $\varphi(\lambda, A)$ is a polynomial of $n^2 + 1$ variables (the n^2 entries of the matrix A and λ). So, φ is C^∞-function.

Since $\lambda \in \mathbb{R}$ is a simple eigenvalue of A, we have $\frac{\partial\varphi(\lambda,A)}{\partial\lambda} \neq 0$ and so we can apply the implicit function theorem (see Theorem 3.28) and find an open neighborhood U of A and a C^∞-function $h\colon U \longrightarrow \mathbb{R}$ such that $h(A) = \lambda$ and

$$\varphi(h(C), C) = \varphi(\lambda, A) = 0 \qquad \forall\, C \in U,$$

so $h(C)$ is an eigenvalue of C.

Solution of Problem 3.31

Let $u, v \in \mathbb{R}^N$ and consider the function

$$\xi(t) = (\varphi(u + t(v-u)), v-u)_{\mathbb{R}^N} \qquad \forall\, t \in \mathbb{R}.$$

Then ξ is differentiable and by the chain rule, we have

$$\xi'(t) = (\varphi'(u + t(v-u))(v-u), v-u)_{\mathbb{R}^N} \geqslant c|v-u|^2$$

(see the hypothesis on φ). From Problem 3.13, we infer that

$$c|v-u|^2 \leqslant \xi(1) - \xi(0),$$

so

$$c|v-u|^2 \leqslant (\varphi(v) - \varphi(u), v-u)_{\mathbb{R}^N}. \tag{3.43}$$

Claim 1. φ is closed.

Let $C \subseteq \mathbb{R}^N$ be a closed set. We need to show that $\varphi(C)$ is closed. To this end, let $\{y_n\}_{n\geqslant 1} \subseteq \varphi(C)$ and assume that $y_n \longrightarrow y$ in $\mathbb{R}^N$. Let $\{u_n\}_{n\geqslant 1} \subseteq C$ be a sequence such that $y_n = \varphi(u_n)$ for all $n \geqslant 1$. From (3.43), we have

$$|u_n - u_m| \leqslant \tfrac{1}{c}|\varphi(u_n) - \varphi(u_m)| = \tfrac{1}{c}|y_n - y_m| \qquad \forall\, n, m \in \mathbb{N},$$

so $\{u_n\}_{n\geqslant 1} \subseteq C$ is a Cauchy sequence. Thus, we have that $u_n \longrightarrow u \in C$. Then $\varphi(u_n) \longrightarrow \varphi(u)$ and it follows that $\varphi(u) = y$. Thus $y \in \varphi(C)$ and so $\varphi(C) \subseteq \mathbb{R}^N$ is a close set.

Claim 2. φ is open.

Note that for all $u \in \mathbb{R}^N$, $\varphi'(u)$ is invertible. Indeed, it suffices to show that $\ker \varphi'(u) = \{0\}$. But this is an immediate consequence of the hypothesis of the problem. Then from Problem 3.8, we have that for every open set $U \subseteq \mathbb{R}^N$, the set $\varphi(U) \subseteq \mathbb{R}^N$ is open. This proves Claim 2.

From Claims 1 and 2, we have that $\varphi(\mathbb{R}^N)$ is both closed and open. Therefore $\varphi(\mathbb{R}^N) = \mathbb{R}^N$. Also, from (3.43), it is clear that φ is an injection. Therefore φ is a bijection. Problem 3.9 implies that φ is a C^1-diffeomorphism.

Solution of Problem 3.32

Consider the map $\psi\colon \mathbb{R} \times X \times X \longrightarrow X$ defined by

$$\psi(\lambda, v, w) = w - \varphi(\lambda w + v).$$

Evidently $\psi \in C^1(\mathbb{R} \times X \times X; X)$. We have

$$\psi'_w(\lambda, v, w) = I_X - \lambda\varphi'(\lambda w + v),$$

so $\psi'_w(0, u, y) = I_X$. Hence, $\psi'_w(0, u, y)$ is an isomorphism. Applying the implicit function theorem (see Theorem 3.28), we can find an open neighborhood $U \subseteq \mathbb{R} \times X$

of $(0, u)$, an open neighborhood $V \subseteq X$ of y, and a C^1-map $h\colon U \longrightarrow X$ such that for all $(\lambda, v) \in U$, we have

$$\big[w \in V \text{ and } w = \varphi(\lambda w + v)\big] \iff w = h(\lambda, v).$$

Solution of Problem 3.33

"(a) $\Longrightarrow$ (b)": Let $\{\xi(t)\}_{t\in[0,1]}$ be a parametrization of γ. We have

$$\begin{aligned}\int_\gamma \langle g(\xi(t)), d\xi(t)\rangle &= \int_\gamma \varphi'(\xi(t))\, d\xi(t) \\ &= \int_0^1 \tfrac{d}{dt}\varphi(\xi(t))\, dt = \varphi(\xi(1)) - \varphi(\xi(0)).\end{aligned}$$

"(b) $\Longrightarrow$ (c)": We can consider two paths γ_1 and γ_2 in U, connecting $u, v \in U$, namely let γ_1 be the line segment connecting u and v and let γ_2 be the line segment from 0 to u plus the line segment from 0 to v. We parameterize γ_1 by

$$\xi(t) = tu + (1-t)v \qquad \forall\, t \in [0,1]$$

and γ_2, by

$$\overline{\xi}_2(t) = tu \quad \text{and} \quad \overline{\overline{\xi}}_2(t) = tv \qquad \forall\, t \in [0,1]$$

(these are the two line segments constituting γ_2) and

$$\xi_2 = \overline{\xi}_2 \cup \overline{\overline{\xi}}_2.$$

We have

$$\int_{\gamma_1} \langle g(\xi(t)), d\xi(t)\rangle = \int_0^1 \tfrac{d}{dt}\varphi(\xi(t))\, dt = \int_0^1 \langle g(\xi(t)), u - v\rangle\ dt \tag{3.44}$$

(by the chain rule) and

$$\begin{aligned}\int_{\gamma_2} \langle g(\xi_2(t)), d\xi_2(t)\rangle &= \int_0^1 \tfrac{d}{dt}\varphi(\overline{\xi}_2(t))\, dt - \int_0^1 \tfrac{d}{dt}\varphi(\overline{\overline{\xi}}_2(t))\, dt \\ &= \int_0^1 \langle g(tu), u\rangle\ dt - \int_0^1 \langle g(tv), v\rangle\ dt \qquad (3.45)\end{aligned}$$

(again by the chain rule). But from **(b)**, we have

$$\int_{\gamma_1} \langle g(\xi(t)), d\xi(t)\rangle = \int_{\gamma_2} \langle g(\xi_2(t)), d\xi_2(t)\rangle ,$$

so

$$\int_0^1 \langle g(\xi(t)), u - v\rangle \ dt = \int_0^1 \langle g(tu), u\rangle \ dt - \int_0^1 \langle g(tv), v\rangle \ dt$$

(see (3.44) and (3.45)).

"**(c)** $\Longrightarrow$ **(a)**": Let

$$\varphi(u) = \int_0^1 \langle g(tu), u\rangle \ dt \qquad \forall\, u \in U.$$

Let $u \in U$ and $h \in X$. Then for all $\lambda \in \mathbb{R}$ with $|\lambda|$ small we have $u + \lambda h \in U$ and

$$\begin{aligned}
&\varphi(u + \lambda h) - \varphi(u) \\
&= \int_0^1 \langle g(t(u + \lambda h)), u + \lambda h\rangle \ dt - \int_0^1 \langle g(tu), u\rangle \ dt \\
&= \int_0^1 \langle g(u + t\lambda h), \lambda h\rangle \ dt
\end{aligned}$$

(see **(c)**), so

$$\tfrac{\varphi(u+\lambda h)-\varphi(u)}{\lambda} = \int_0^1 \langle g(u + t\lambda h), h\rangle \ dt.$$

Passing to the limit as $\lambda \to 0$ and recalling that g is continuous, we obtain that $\varphi \in C^1(U)$ and $\varphi'(u) = g(u)$ for all $u \in U$.

Now suppose that $g \in C^1(U; X^*)$.

"**(d)** $\Longrightarrow$ **(a)**": As before, let

$$\varphi(u) = \int_0^1 \langle g(tu), u\rangle \ dt \qquad \forall\, u \in U.$$

We have

$$\varphi(u + h) - \varphi(u) = \int_0^1 \langle g(t(u + h)), h\rangle \ dt + \int_0^1 \langle g(t(u + h)) - g(tu), u\rangle \ dt. \tag{3.46}$$

Since $g \in C^1(U; X^*)$, using **(d)** and the Fubini theorem, we have

$$\begin{aligned}
\int_0^1 \langle g(t(u+h)) - g(tu), u\rangle \, dt &= \int_0^1 \Big(\int_0^t \tfrac{d}{ds} \langle g(tu+sh), u\rangle \, ds\Big) \, dt \\
= \int_0^1 \int_0^t \big\langle g'(tu+sh)h, u\big\rangle \, ds \, dt &= \int_0^1 \int_0^t \big\langle g'(tu+sh)u, h\big\rangle \, ds \, dt \\
= \int_0^1 \int_s^1 \big\langle g'(tu+sh)u, h\big\rangle \, dt \, ds &= \int_0^1 \int_s^1 \tfrac{d}{dt} \langle g(tu+sh), h\rangle \, dt \, ds \\
= \int_0^1 \langle g(u+sh) - g(su+sh), h\rangle \, ds. & \qquad (3.47)
\end{aligned}$$

Returning to (3.46), we have

$$\varphi(u+h) - \varphi(u) = \int_0^1 \langle g(u+th), h\rangle \, dt. \tag{3.48}$$

Due to the fact that $g \in C^1(U; X^*)$, given $\varepsilon > 0$, we can find $\delta \in (0,1)$ such that

$$\|g(u+th) - g(u)\|_* \leqslant \varepsilon \qquad \forall\, t \in [0,1], \ \|h\|_X \leqslant \delta. \tag{3.49}$$

Then using (3.48) and (3.49), we have

$$\begin{aligned}
\big|\varphi(u+h) - \varphi(u) - \langle g(u), h\rangle \,\big| &= \Big|\int_0^1 \langle g(u+th) - g(u), h\rangle \, dt\Big| \\
&= \varepsilon\delta \leqslant \varepsilon \quad \forall\, h \in X, \text{ with } \|h\|_X \leqslant \delta,
\end{aligned}$$

so $\varphi \in C^1(U)$ and $\varphi'(u) = g(u)$ for all $u \in U$.

"**(a)** $\Longrightarrow$ **(d)**": This is an immediate consequence of Proposition 3.38.

Solution of Problem 3.34

Let $h \in X$ be such that $\|h\|_X = 1$ and let $t > 0$. Then

$$\varphi(th) - \varphi(rh) = \int_r^t \tfrac{d}{ds}\varphi(sh) \, ds = \int_r^t \langle \varphi'(sh), h\rangle \, ds \qquad \forall\, r \in (0,t).$$

By hypothesis, there exists $r > 0$ (independent of h) such that

$$\varphi(th) \geqslant \varphi(rh) + \int_r^t \tfrac{\xi(s)}{s} \, ds$$

(recall that $\|h\|_X = 1$), so

$$\liminf_{\|u\|_X \to +\infty} \varphi(u) \geqslant \int_r^{\infty} \frac{\xi(s)}{s}\, ds = +\infty,$$

thus φ is coercive.

Solution of Problem 3.35

"$\Longrightarrow$": Let σ be an integral curve for F (see Definition 3.43) with $\sigma(0) = u \in U$. By hypothesis

$$\varphi \circ \sigma|_{(-\varepsilon,\varepsilon)} \text{ is constant.}$$

So, we have

$$\begin{aligned} 0 &= \tfrac{d}{dt}(\varphi \circ \sigma)(t) = \big\langle \varphi'(\sigma(t)), \sigma'(t) \big\rangle \\ &= \big\langle \varphi'(\sigma(t)), F(\sigma(t)) \big\rangle \qquad \forall\, t \in (-\varepsilon, \varepsilon). \end{aligned}$$

(by the chain rule). For $t = 0$, we have

$$\big\langle \varphi'(u), F(u) \big\rangle = 0 \qquad \forall\, u \in U.$$

"$\Longleftarrow$": Let σ be an integral curve for F defined on an open interval I. Then

$$\tfrac{d}{dt}(\varphi \circ \sigma)(t) = \big\langle \varphi'(\sigma(t)), \sigma'(t) \big\rangle = 0 \qquad \forall\, t \in I,$$

so $\varphi \circ \sigma$ is constant on I.

Solution of Problem 3.36

Let σ be a maximal integral curve defined on an open interval $I = (a, b)$. Then there exists $c \in \mathbb{R}$ such that

$$(\varphi \circ \sigma)(t) = c \qquad \forall\, t \in I,$$

so

$$\sigma(t) \in \varphi^{-1}(c) \qquad \forall\, t \in I. \tag{3.50}$$

By hypothesis, the set $K = \varphi^{-1}(c) \subseteq X$ is compact. Suppose that $b < +\infty$. Let $t_n \nearrow b$. Then from (3.50), we have $\{\sigma(t_n)\}_{n \geqslant 1} \subseteq K$. We can find a subsequence $\{t_{n_k}\}_{k \geqslant 1}$ of $\{t_n\}_{n \geqslant 1}$ such that $\sigma(t_{n_k}) \longrightarrow \widehat{u} \in K$ in X. We can find $r > 0$ such that $B_r(\widehat{u}) = \{u \in X : \|u - \widehat{u}\|_X < r\} \subseteq U$ and an interval $\widehat{I} = (b-\delta, b+\delta)$ such that the maximal integral curve is defined on $\widehat{I}$ when the initial data are $(t_0, u) \in \widehat{I} \times B_r(\widehat{u})$. So, for $n \geqslant 1$ big, we can have

$$t_n > b - \delta \quad \text{and} \quad u(t_n) \in B_r(\widehat{u}).$$

Then the integral curve passing from $(t_n, u(t_n))$ is defined on $\widehat{I}$. But by Proposition 3.45 this integral curve is $\sigma|_{\widehat{I}}$, which is a contradiction since σ is not defined on

$[b, b+\delta)$. Therefore $b = +\infty$. Similarly we show that $a = -\infty$. So, $I = \mathbb{R}$ and this shows that F is complete (see Definition 3.54(a)).

Solution of Problem 3.37

Let σ be a maximal integral curve (see Definition 3.43) with $\sigma(t_0) = u_0$. Suppose that $I = (a, b)$ is the interval of definition of σ. If $b < +\infty$, then for $t \in [t_0, b)$, we have

$$\begin{aligned} \|\sigma(t) - u_0\|_X &= \Big\| \int_{t_0}^{t} F(\sigma(s))\, ds \Big\|_X \\ &\leqslant (t - t_0)\|F(u_0)\|_X + \int_{t_0}^{t} \|F(\sigma(s)) - F(u_0)\|_X\, ds \\ &\leqslant c + \int_{t_0}^{t} k\|\sigma(s) - u_0\|_X\, ds, \end{aligned}$$

with $c = (b - t_0)\|F(u_0)\|_X > 0$, so

$$\|\sigma(t) - u_0\|_X \leqslant ce^{k(b-t_0)} \qquad \forall\, t \in [t_0, b)$$

(by the Gronwall inequality). So, as t approaches b, the integral curve σ stays in the compact set $\overline{B}_{ce^{k(b-t_0)}}(u_0)$ (recall that X is finite dimensional). As in the solution of Problem 3.36, this leads to a contradiction. Therefore σ is defined on $[t_0, +\infty)$. A similar argument shows that σ is also defined on $(-\infty, t_0]$. Hence the vector field F is complete (see Definition 3.54(a)).

Solution of Problem 3.38

Arguing by contradiction, suppose that φ is not Fréchet differentiable at u_0. Then by virtue of Remark 3.2, we can find two sequences $\{t_n\}_{n\geqslant 1} \subseteq [-1, 1] \setminus \{0\}$, $\{h_n\}_{n\geqslant 1} \subseteq \overline{B}_1(0)$ (with $\overline{B}_1(0) = \{h \in \mathbb{R}^m : |h| \leqslant 1\}$) and $\varepsilon > 0$ such that $t_n \longrightarrow 0$ and

$$\Big| \frac{\varphi(u_0 + t_n h_n) - \varphi(u_0) - \langle \varphi'_G(u_0), t_n h_n \rangle}{t_n} \Big| \geqslant \varepsilon \qquad \forall\, n \geqslant 1. \tag{3.51}$$

Passing to a subsequence if necessary, we may assume that $h_n \longrightarrow h$ in $\mathbb{R}^m$. We have

$$\begin{aligned}
& \tfrac{1}{t_n}\big|\varphi(u_0+t_nh_n)-\varphi(u_0)-\langle\varphi'_G(u_0),t_nh_n\rangle\big| \\
\leqslant\ & \tfrac{1}{t_n}\big|\varphi(u_0+t_nh_n)-\varphi(u_0+t_nh)\big| \\
& +\tfrac{1}{t_n}\big|\varphi(u_0+t_nh)-\varphi(u_0)-\langle\varphi'_G(u_0),t_nh\rangle\big| \\
& +\big|\langle\varphi'_G(u_0),h_n-h\rangle\big| \\
\leqslant\ & k|h_n-h|+\tfrac{1}{t_n}\big|\varphi(u_0+t_nh)-\varphi(u_0)-\langle\varphi'_G(u_0),t_nh\rangle\big| \\
& +\big|\langle\varphi'_G(u_0),h_n-h\rangle\big|,
\end{aligned} \tag{3.52}$$

for all $n \geqslant 1$ big enough and with $k > 0$ being the Lipschitz constant of φ near u_0. Then passing to the limit as $n \to +\infty$ in (3.52), we obtain

$$\lim_{n\to+\infty} \frac{\big|\varphi(u_0+t_nh_n)-\varphi(u_0)-\langle\varphi'_G(u_0),t_nh_n\rangle\big|}{t_n} = 0. \tag{3.53}$$

Comparing (3.51) and (3.53) we reach a contradiction. This proves that φ is Fréchet differentiable at u_0.

Solution of Problem 3.39

First we show that φ is surjective. From Problem 3.8 we know that the set $\varphi(X) \subseteq Y$ is open. We show that it is also closed. From the inverse function theorem (see Theorem 3.26), for every $u \in X$, we can find convex open sets $U_u \subseteq X$ and $V_u \subseteq Y$ such that $u \in U_u$, $\varphi(u) \in V_u$ and φ is a C^1-diffeomorphism from U_u onto V_u. We have $\varphi(\overline{U}_u) \subseteq \overline{V}_u$. In fact the opposite inclusion also holds. Indeed, let $y_0 \in \overline{V}_u$. Then we can find a sequence $\{y_n\}_{n\geqslant 1} \subseteq V_u$ such that $y_n \longrightarrow y_0$ in Y. Let $u_n = \varphi^{-1}(y_n) \in U_u$ for $n \geqslant 1$. Since φ^{-1} is a C^1-map on V_u, from the mean value theorem (see Theorem 3.16), we have

$$\|u_n-u_m\|_X \leqslant \sup_{v\in[y_n,y_m]} \|(\varphi^{-1})'(v)\|_{\mathcal{L}}\|y_n-y_m\|_Y \qquad \forall\, n,m \geqslant 1, \tag{3.54}$$

where $[y_n,y_m] = \{y \in Y :\ y=(1-t)y_n+ty_m :\ t\in[0,1]\}$. By Theorem 3.26 and our hypothesis, we have

$$\sup_{v\in[y_n,y_m]} \|(\varphi^{-1})'(v)\|_{\mathcal{L}} = \sup_{v\in[y_n,y_m]} \|\varphi'(\varphi^{-1})(v))^{-1}\|_{\mathcal{L}} = M < +\infty.$$

So, from (3.54), we have

$$\|u_n-u_m\|_X \leqslant M\|y_n-y_m\|_Y \qquad \forall\, n,m \geqslant 1,$$

so $\{u_n\}_{n\geqslant 1} \subseteq X$ is a Cauchy sequence in X, thus $u_n \longrightarrow u_0$ for some $u_0 \in \overline{U}_u$ and $y_0 = \varphi(u_0) \in \varphi(\overline{U}_u)$. Therefore $\varphi(\overline{U}_u) = \overline{V}_u$. Evidently we may assume that the family $\{V_u\}_{u\in X}$ is locally finite. So, we have

$$\overline{\varphi(X)} \subseteq \overline{\bigcup_{u\in X} V_u} = \bigcup_{u\in X} \overline{V}_u = \bigcup_{u\in X} \varphi(\overline{U}_u) \subseteq \varphi(X),$$

thus $\overline{\varphi(X)} = \varphi(X)$ and hence $\varphi(X)$ is a closed set. Because $\varphi(X)$ is both open and closed, we conclude that $\varphi(X) = Y$, that is, φ is surjective.

Next we show that φ is injective. So, suppose that $\varphi(u_0)=\varphi(u_1)=y_0$. Let

$$u(t) = (1-t)u_0 + tu_1 \quad \text{with } t \in [0,1] \quad \text{and} \quad y(t) = \varphi(u(t)). \tag{3.55}$$

We define

$$H(s,t) = (1-s)y_0 + sy(t) \qquad \forall\, s,t \in [0,1]. \tag{3.56}$$

Note that $H(s,0) = y_0 = \varphi(u_0)$. From Theorem 3.26, we know that there are open neighborhoods U of u_0 and V of y_0 such that $\varphi\colon U \longrightarrow V$ is a homeomorphism. From the previous arguments, we have $\varphi(\overline{U}) = \overline{V}$. Note that

$$\|H(s,t) - y_0\|_Y = \|s(y(t) - y_0)\|_Y \leqslant \|y(t) - y_0\|_Y$$

(recall that $s \in [0,1]$). The continuity of $t \longmapsto y(t)$ implies that we can find $\delta \in (0,1]$ such that

$$H(s,t) \in \overline{V} \qquad \forall\, t \in [0,\delta].$$

Let $G(s,t) = \varphi^{-1}(H(s,t))$. Then $G(s,t) \in \overline{U}$ for all $s \in [0,1]$, all $t \in [0,\delta]$. Let δ_0 be the biggest such $\delta > 0$. If $\delta < 1$, then once again Theorem 3.26 implies that $G(s,\cdot)$ can be extended further than δ_0 keeping the relation $G(s,t) = \varphi^{-1}(H(s,t))$. So, $\delta_0 = 1$ and so $G(s,t)$ has been defined for all $s,t \in [0,1]$. Since H is differentiable (see (3.55) and (3.56)) and φ^{-1} is locally differentiable, we infer that G is differentiable too (see Theorem 3.19).

Since $H(s,1) = y_0$ for all $s \in [0,1]$ (see (3.56)), we have

$$0 = \tfrac{d}{ds}H(s,1) - \tfrac{d}{ds}\varphi(G(s,1)) = \varphi'(G(s,1))\tfrac{d}{ds}G(s,1)$$

(see Theorem 3.19), so

$$\tfrac{d}{ds}G(s,1) = 0,$$

and thus

$$G(0,1) = G(1,1). \tag{3.57}$$

Similarly, we have

$$\tfrac{d}{dt}G(0,t) = 0,$$

and thus

$$G(0,0) = G(0,1). \tag{3.58}$$

From (3.57) and (3.58), we have

$$G(0,0) = G(1,1),$$

so

$$u_0 = u_1$$

(see (3.56)). Therefore φ is a bijection. Invoking Problem 3.9, we conclude that φ is a C^1-diffeomorphism.

Solution of Problem 3.40

Since φ is continuous, $\overline{\text{span}}\,\varphi([0,b])$ is separable (see Problem I.1.38). So, without any loss of generality, we may assume that X is separable too. Let

$$\varphi_\varepsilon(t) = \tfrac{1}{\varepsilon}\left(\varphi(t+\varepsilon) - \varphi(t)\right) \qquad \forall\, \varepsilon \neq 0.$$

Since φ is Lipschitz continuous, we see that for every fixed $t \in [0,b]$, we have that

$$\{\varphi_\varepsilon(t)\}_{\varepsilon\neq 0} \subseteq X \quad \text{is bounded, hence relatively } w\text{-compact.} \tag{3.59}$$

Because X is separable and reflexive, so is its dual X^* (see Corollary I.5.84). Let $\{h_n^*\}_{n\geqslant 1} \subseteq X^*$ be a dense set in $\overline{B}_1^*(0) = \{u^* \in X^* : \|u^*\|_* \leqslant 1\}$. Then for every $n \geqslant 1$, the function $\xi_n \colon [0,b] \longrightarrow \mathbb{R}$ defined by

$$\xi_n(t) = \langle h_n^*, \varphi(t)\rangle$$

is Lipschitz continuous, thus differentiable for all $t \in T \setminus N_n$, with $\lambda(N_n) = 0$ (λ being the Lebesgue measure on $\mathbb{R}$; see Theorem 3.120). From (3.59) it follows that there exist $g \colon [0,b) \longrightarrow X$ such that

$$g(t) = w\text{-}\lim_{\varepsilon\to 0} \varphi_\varepsilon(t).$$

Then for all $t \in [0,b] \setminus N_n$, we have

$$\tfrac{d}{dt}\langle h_n^*, \varphi(t)\rangle = \langle h_n^*, g(t)\rangle\,. \tag{3.60}$$

Let $N = \bigcup\limits_{n\geqslant 1} N_n$. This is a Lebesgue-null set and for all $t \in [0,b] \setminus N$, we have

$$\|g(t)\|_X \leqslant c,$$

for some $c > 0$ (see (3.60)). Note that g is weakly measurable and so by the Pettis measurability theorem (see Theorem 1.36), it is measurable. So, $g \in L^1([0,b];X)$. We set

$$\psi(t) = \int_0^t g(s)\,ds + g(0),$$

so

$$\begin{aligned}
\langle h_n^*, \psi(t)\rangle &= \int_0^t \langle h_n^*, g(s)\rangle \, ds + \langle h_n^*, g(0)\rangle \\
&= \int_0^t \tfrac{d}{ds} \langle h_n^*, \varphi(s)\rangle \, ds + \langle h_n^*, \varphi(0)\rangle \\
&= \langle h_n^*, \varphi(t)\rangle \qquad \forall\, n \geqslant 1,\ t \in [0,b] \setminus N
\end{aligned}$$

(without any loss of generality, we have $g(0) = \varphi(0)$), thus

$$\psi(t) = \varphi(t) \quad \text{for a.a. } t \in [0,b]$$

and hence φ is differentiable at almost all $t \in T$.

Solution of Problem 3.41

Since X^* is strictly convex (see Definition I.5.168(a) and Remark I.5.169), the space X has Gâteaux differentiable norm (see Proposition I.5.170(a)). Hence σ is Gâteaux differentiable at every $u \in X \setminus \{0\}$.

For all $u, v \in X$, we have

$$\begin{aligned}
\langle \mathcal{F}(v), v-u\rangle &\geqslant \|v\|_X^2 - \|v\|_X\|u\|_X \geqslant \|v\|_X^2 - \tfrac{1}{2}\left(\|v\|_X^2 + \|u\|_X^2\right) \\
&= \tfrac{1}{2}\|v\|_X^2 - \tfrac{1}{2}\|u\|_X^2 \geqslant \|v\|_X\|u\|_X - \|u\|_X^2 \\
&\geqslant \langle \mathcal{F}(u), v-u\rangle . \qquad (3.61)
\end{aligned}$$

Let $v = u + th$ with $u \in X$ and $t \in \mathbb{R}$. Then from (3.61), we have

$$t\langle \mathcal{F}(u), h\rangle \leqslant \tfrac{1}{2}\left(\|u+th\|_X^2 - \|u\|_X^2\right) \leqslant t\langle \mathcal{F}(u+th), h\rangle ,$$

so

$$\lim_{t\to 0} \tfrac{1}{2t}\left(\|u+th\|_X^2 - \|u\|_X^2\right) = \langle \mathcal{F}(u), h\rangle \qquad \forall\, h \in X. \qquad (3.62)$$

So, if $\xi(u) = \frac{1}{2}\sigma(u)^2$, then

$$\left\langle \xi_G'(u), h\right\rangle = \langle \mathcal{F}(u), h\rangle \qquad \forall\, h \in X$$

(see (3.62)), thus

$$\xi_G'(u) = \mathcal{F}(u) \qquad \forall\, u \in X \setminus \{0\}. \qquad (3.63)$$

But from the chain rule (see Theorem 3.19), we have

$$\xi_G'(u) = \|u\|_X \sigma_G'(u) \qquad \forall\, u \in X \setminus \{0\},$$

so

$$\sigma_G'(u) = \tfrac{\mathcal{F}(u)}{\|u\|_X} \qquad \forall\, u \in X \setminus \{0\}$$

(see (3.63)).

Solution of Problem 3.42

Since X^* is locally uniformly convex (see Definition I.5.168(b) and Remark I.5.169), we have that σ is Fréchet differentiable on $X \setminus \{0\}$ (see Proposition I.5.170(b)). Moreover, from Problem 3.41, we have

$$\sigma'(u) = \frac{\mathcal{F}(u)}{\|u\|_X} \qquad \forall\, u \in X \setminus \{0\},$$

so the map $u \longmapsto \sigma'(u)$ is continuous on $X \setminus \{0\}$ (see Proposition 2.114(c)) and thus $\sigma \in C^1(X \setminus \{0\})$.

Solution of Problem 3.43

Evidently $\widehat{\varphi}$ is proper and convex. We need to show that $\widehat{\varphi}$ is also lower semicontinuous. To this end let $\lambda \in \mathbb{R}$ and let $\widehat{\varphi}^\lambda = (u \in H : \widehat{\varphi}(u) \leqslant \lambda)$. We need to show that $\widehat{\varphi}^\lambda$ is closed (see Proposition I.2.53). We consider a sequence $\{u_n\}_{n \geqslant 1} \subseteq \widehat{\varphi}^\lambda$ such that $u_n \longrightarrow u$ in H. Evidently the sequence $\{u_n\}_{n \geqslant 1} \subseteq X$ is bounded. Since X is a reflexive space, by passing to a suitable subsequence if necessary, we may assume that

$$u_n \xrightarrow{w} \widehat{u} \quad \text{in } X.$$

Since X is embedded continuously into H and $u_n \longrightarrow u$ in H, we have $\widehat{u} = u$ and so $u \in X$. Exploiting the lower semicontinuity and convexity of φ, we have

$$\lambda \geqslant \liminf_{n \to +\infty} \widehat{\varphi}(u_n) = \liminf_{n \to +\infty} \varphi(u_n) \geqslant \varphi(u) = \widehat{\varphi}(u)$$

(since $\{u_n\}_{n \geqslant 1} \subseteq X$ and $u \in X$; see Proposition I.2.48), so $u \in \widehat{\varphi}^\lambda$ and thus the set $\widehat{\varphi}^\lambda$ is closed. Because $\lambda \in \mathbb{R}$ is arbitrary, we conclude that $\widehat{\varphi} \in \Gamma_0(H)$.

Solution of Problem 3.44

Because of the convexity of φ, we have that for every $h \in X$, the map $t \longmapsto \frac{\varphi(u_0+th)-\varphi(u_0)}{t}$ is nondecreasing on $(0, +\infty)$. We have

$$\langle \varphi'_G(u_0), h \rangle = \lim_{t \searrow 0} \frac{\varphi(u_0+th)-\varphi(u_0)}{t}$$

(see Definition 3.1(a)) It follows that

$$\langle \varphi'_G(u_0), th \rangle \leqslant \varphi(u_0 + th) - \varphi(u_0) \qquad \forall\, t \geqslant 0,\ h \in X,\ \|h\|_X = 1,$$

so

$$\langle \varphi_G'(u_0), v\rangle \leqslant \varphi(u_0+v) - \varphi(u_0) \qquad \forall\, v \in X$$

and thus $\varphi_G'(u_0) \in \partial\varphi(u_0)$ (see Definition 3.70).

Solution of Problem 3.45

"$\Longrightarrow$": Let $\widehat{\varphi}(u) = \varphi(u) - \varphi(0)$. Then $\widehat{\varphi}$ is convex, k-Lipschitz and $\widehat{\varphi}(0) = 0$. Moreover, $\widehat{\varphi}^* = \varphi^* + \varphi(0)$ (see Remark 3.88). Therefore, without any loss of generality, we may assume that $\varphi(0) = 0$. So, we have

$$|\varphi(u)| \leqslant k\|u\|_X \qquad \forall\, u \in X. \tag{3.64}$$

Let $u^* \in X^*$ with $\|u^*\|_* > k$. Then

$$\langle u^*, u\rangle - \varphi(u) \geqslant \langle u^*, u\rangle - k\|u\|_X = \big(\langle u^*, \tfrac{u}{\|u\|_X}\rangle - k\big)\|u\|_X \qquad \forall\, u \in X \tag{3.65}$$

(see (3.64)). Note that

$$\sup_{u\in X} \big(\langle u^*, \tfrac{u}{\|u\|_X}\rangle - k\big) = \|u^*\|_* - k > 0,$$

so

$$\sup_{u\in X} \big(\langle u^*, u\rangle - \varphi(u)\big) = +\infty$$

(see (3.65)) and thus $u^* \notin \operatorname{dom}\varphi^*$ (see Definition 3.83). Therefore, we have $\operatorname{dom}\varphi^* \subseteq k\overline{B}_1^*(0)$.

"$\Longleftarrow$": Arguing by contradiction, suppose that φ is not k-Lipschitz. Then we can find $u, v \in X$ such that $\varphi(u) - \varphi(v) > \eta\|u - v\|_X$ with $\eta > k$ (we can have $\varphi(u) = +\infty$). Then

$$\varphi(u) > \varphi(v) + \eta\|u - v\|_X. \tag{3.66}$$

Let $\xi = \varphi(v) + \eta\|u - v\|_X$. Then from (3.66), we see that $(u, \xi) \notin \operatorname{epi}\varphi$. By the strong separation theorem (see Theorem I.5.29), we can find $z^* \in (X \times \mathbb{R})^*$ and $\delta > 0$ such that

$$\sup_{(y,\lambda)\in\operatorname{epi}\varphi} \langle z^*, (y,\lambda)\rangle_{X\times\mathbb{R}} \leqslant \langle z^*, (u,\xi)\rangle_{X\times\mathbb{R}} - \delta.$$

Identifying $(X \times \mathbb{R})^*$ with $X^* \times \mathbb{R}$, we have $z^* = (u^*, \beta) \in X^* \times \mathbb{R}$. Then

$$\sup_{(y,\lambda)\in\operatorname{epi}\varphi} \big(\langle u^*, y\rangle + \lambda\beta\big) \leqslant \langle u^*, u\rangle + \beta\xi - \delta. \tag{3.67}$$

Since $(u, \varphi(u) + n) \in \operatorname{epi}\varphi$ for all $n \geqslant 1$, using (3.66) we see that $\beta < 0$. Then from (3.67), we have

$$\big\langle \tfrac{u^*}{|\beta|}, h - u\big\rangle \leqslant \varphi(h) - \xi \qquad \forall\, h \in X, \tag{3.68}$$

so putting $\widehat{u}^* = \frac{u^*}{|\beta|}$, we have

$$\langle \widehat{u}^*, h\rangle - \varphi(h) \leqslant \langle \widehat{u}^*, u\rangle - \xi \qquad \forall\, h \in X,$$

thus $\widehat{u} \in \operatorname{dom}\varphi^*$ (see Definition 3.83) and $\|\widehat{u}^*\|_* \leqslant k$. Then we have

$$\langle \widehat{u}^*, v - u\rangle \leqslant -\eta\|u - v\|_X$$

(see (3.68) and recall that $\xi = \varphi(v) + \eta\|u - v\|_X$), so $\|\widehat{u}^*\|_* \geqslant \eta > k$, a contradiction to our hypothesis. Therefore, we conclude that φ is k-Lipschitz.

Solution of Problem 3.46

"(a) $\Longrightarrow$ (b)": Fix $u_0 \in D(\partial\varphi)$. We have

$$\langle u^*, u - u_0\rangle \geqslant \varphi(u) - \varphi(u_0) \qquad \forall\, u^* \in \partial\varphi(u)$$

(see Definition 3.70), so

$$\frac{\inf\{\langle u^*, u\rangle :\ u^* \in \partial\varphi(u)\}}{\|u\|_X} \longrightarrow +\infty$$

(see Definition 3.105). So, the map $u \longmapsto \partial\varphi(u)$ is strongly coercive (see Definition 2.98(f)). Also $\partial\varphi$ is maximal monotone (see Definition 2.100). Invoking Theorem 2.119 we infer that $R(\partial\varphi) = X^*$. Also, from the strong coercivity of $\partial\varphi$, it follows that $(\partial\varphi)^{-1}$ maps bounded sets to bounded sets.

"(b) $\Longrightarrow$ (a)": Since $\varphi \in \Gamma_0(X)$, it is bounded below by an affine function $u \longmapsto a(u)$. So, by replacing φ with $\widehat{\varphi} = \varphi - a$ if necessary, we see that without any loss of generality we may assume that $\varphi \geqslant 0$. Let $\varrho > 0$ and $u^* \in X^*$ with $\|u^*\|_* \leqslant \varrho$. By hypothesis we can find $u \in X$ and $r > 0$ such that

$$u^* \in \partial\varphi(u) \quad \text{and} \quad \|u\|_X \leqslant r.$$

From Definition 3.70, we have

$$\langle u^*, v - u\rangle \leqslant \varphi(v) - \varphi(u) \qquad \forall\, v \in X,$$

so

$$\langle u^*, v\rangle \leqslant \varphi(v) + r\varrho \qquad \forall\, v \in X,\ u^* \in \overline{B}^*_\varrho(0)$$

(recall that $\varphi \geqslant 0$), thus

$$\varrho\|v\|_X \leqslant \varphi(v) + r\varrho \qquad \forall\, v \in X$$

and hence

$$\varrho - \frac{r\varrho}{\|v\|_X} \leqslant \frac{\varphi(v)}{\|v\|_X} \qquad \forall\, v \in X.$$

Since $\varrho > 0$ is arbitrary, we conclude that

$$\lim_{\|v\|_X \to +\infty} \frac{\varphi(v)}{\|v\|_X} = +\infty.$$

Solution of Problem 3.47

"**(a)** $\Longrightarrow$ **(b)**": Let $u_0 \in \operatorname{dom}\varphi$ (recall that φ is proper, so $\operatorname{dom}\varphi \neq \emptyset$). By replacing φ with $\widehat{\varphi}(u) = \varphi(u_0 + u) - \varphi(u_0)$, we see that without any loss of generality, we may assume that $0 \in \operatorname{dom}\varphi$ and $\varphi(0) = 0$. The coercivity of φ implies that we can find $\varrho > 0$ such that

$$\varphi(u) \geqslant 1 \qquad \forall\, u \in X, \text{ with } \|u\|_X \geqslant \varrho. \tag{3.69}$$

Then from the convexity of φ, for all $u \in X$ with $\|u\|_X \geqslant \varrho$, we have

$$\varphi\big(\tfrac{\varrho}{\|u\|_X} u\big) \leqslant \tfrac{\|u\|_X - \varrho}{\|u\|_X}\varphi(0) + \tfrac{\varrho}{\|u\|_X}\varphi(u) = \tfrac{\varrho}{\|u\|_X}\varphi(u)$$

(recall that $\varphi(0) = 0$), so $\frac{\|u\|_X}{\varrho} \leqslant \varphi(u)$ (see (3.69)). The lower semicontinuity of φ at some point of its domain implies that there exists $\eta > 0$ such that

$$-\eta \leqslant \varphi(u) \qquad \forall\, u \in \varrho B_1(0),$$

so $\varphi \geqslant c\|\cdot\|_X + b$ with $c = \frac{1}{\varrho}$ and $b = -\eta$.

The implications "**(b)** $\Longrightarrow$ **(c)** $\Longrightarrow$ **(d)** $\longrightarrow$ **(a)**" are immediate consequences of the definitions.

Solution of Problem 3.48

Let $u^* \in \partial\varphi(0)$. Then $\langle u^*, h\rangle \leqslant \langle u_0^*, h\rangle$ for all $h \in X$. Let

$$V = \{u \in X : \langle u_0^*, u\rangle = 0\} = \ker u_0^*.$$

Then $u^*|_V \leqslant 0$ and since V is a linear subspace of X, we have $u^*|_V = 0$. Thus

$$\ker u_0^* \subseteq \ker u^*,$$

so

$$u^* = \lambda u_0^*,$$

for some $\lambda \in \mathbb{R}$ and hence

$$\lambda \langle u_0^*, h\rangle \leqslant |\langle u_0^*, h\rangle| \qquad \forall\, h \in X.$$

It follows that $\lambda \in [-1,1]$ and so

$$\partial\varphi(0) \subseteq [-1,1]u_0^*. \tag{3.70}$$

On the other hand, directly from Definition 3.70, we see that

$$[-1,1]u_0^* \subseteq \partial\varphi(0). \tag{3.71}$$

From (3.70) and (3.70), we conclude that

$$\partial\varphi(0) = [-1,1]u_0^*.$$

Solution of Problem 3.49
From Definitions 3.83 and 3.55, for all $u^* \in H$, we have

$$\begin{aligned}
\varphi^*(u^*) &= \sup_{u\in C}\left((u^*,u)_H - \tfrac{1}{2}\|u\|_H^2\right)\\
&= \sup_{u\in C}\left((u^*,u)_H - \tfrac{1}{2}(u,u)_H\right)\\
&= \sup_{u\in C}\left(\tfrac{1}{2}(u^*,u^*)_H + (u^*,u)_H - \tfrac{1}{2}(u^*,u^*)_H - \tfrac{1}{2}(u,u)_H\right)\\
&= \tfrac{1}{2}(u^*,u^*)_H + \sup_{u\in C}\left(-\tfrac{1}{2}(u^*,u^*)_H + (u^*,u)_H - \tfrac{1}{2}(u,u)_H\right)\\
&= \tfrac{1}{2}(u^*,u^*)_H - \tfrac{1}{2}\inf_{u\in C}\left((u^*,u^*)_H - 2(u^*,u)_H + (u,u)_H\right)\\
&= \tfrac{1}{2}(u^*,u^*)_H - \tfrac{1}{2}\inf_{u\in C}\|u^*-u\|_H^2\\
&= \tfrac{1}{2}\|u^*\|_H^2 - \tfrac{1}{2}\mathrm{dist}(u^*,C)^2,
\end{aligned}$$

so $\mathrm{dist}(\cdot,C)^2 = \|\cdot\|_H^2 - 2\varphi^*$.

Solution of Problem 3.50
Let $u_0 \in U$ and $h \in X$ with $\|h\|_X = 1$. We define

$$\xi(t) = \varphi(u_0 + th).$$

It suffices to show that ξ is convex. Let $\tau, t \in \mathbb{R}$ be such that $\xi(\tau)$ and $\xi(t)$ are defined. Then with $u = u_0 + \tau h$ and $v = u_0 + th$, we have

$$\begin{aligned}
\left(\xi'(\tau) - \xi'(t)\right)(\tau - t) &= (\tau - t)\left\langle \varphi_G'(u_0+\tau h) - \varphi_G'(u_0+th), h\right\rangle\\
&= \left\langle \varphi_G'(u_0+\tau h) - \varphi_G'(u_0+th), \tau h - th\right\rangle\\
&= \left\langle \varphi_G'(u) - \varphi_G'(v), u - v\right\rangle \geqslant 0
\end{aligned}$$

(see hypothesis and Theorem 3.19), so ξ' is nondecreasing and thus ξ is convex. Hence so is φ.

Solution of Problem 3.51

Let $v^* = \lambda u^*$ with $\lambda \leqslant 0$. Then

$$\langle v^*, u - u_0 \rangle = \lambda \langle u^*, u \rangle - \lambda \langle u^*, u_0 \rangle \leqslant i_{D_+}(u) - i_{D_+}(u_0) \qquad \forall\, u \in X.$$

Therefore $v^* \in \partial i_{D_+}(u_0)$ and so we have proved that

$$\mathbb{R}_- u^* \subseteq \partial i_{D_+}(u_0). \tag{3.72}$$

Next, let $v^* \in \partial i_{D_+}(u_0)$ and let $z \in \ker u^*$. Then $u_0 + z \in D_+$ and

$$\langle v^*, u_0 + z - u_0 \rangle \leqslant i_{D_+}(v_0 + z) - i_{D_+}(u_0) = 0,$$

hence

$$\langle v^*, z \rangle \leqslant 0 \qquad \forall\, z \in \ker u^*.$$

Since $\ker u^*$ is a closed linear subspace of X, we have

$$\langle v^*, z \rangle = 0 \qquad \forall\, z \in \ker u^*.$$

This means that $v^* = \lambda u^*$ for some $\lambda \in \mathbb{R}$. We have

$$0 \geqslant \langle v^*, u - u_0 \rangle = \lambda \langle u^*, u - u_0 \rangle \qquad \forall\, u \in D_+,$$

so $\lambda \leqslant 0$, thus $v^* = \lambda u^* \in \mathbb{R}_- u^*$ and hence

$$\partial i_{D_+}(u_0) \subseteq \mathbb{R}_- u^*. \tag{3.73}$$

From (3.72) and (3.73) we conclude that $\partial i_{D_+}(u_0) = \mathbb{R}_- u^*$.

Solution of Problem 3.52

From Theorem 3.85, we have

$$-\varphi(u) = \inf_{u^* \in X^*} \left(\varphi^*(u^*) - \langle u^*, u \rangle \right). \tag{3.74}$$

Let

$$\psi_u(u^*) = \varphi^*(u^*) - \langle u^*, u \rangle \qquad \forall\, u^* \in X^*.$$

We choose $u^* \in X^*$ such that

$$\varphi^*(u^*) - \langle u^*, u \rangle \leqslant \inf_{X^*} \psi_u + \varepsilon,$$

which is equivalent to

$$\varphi(u) + \varphi^*(u^*) - \langle u^*, u \rangle \leqslant \varepsilon$$

(see (3.74)), which is equivalent to

$$u^* \in \partial_\varepsilon \varphi(u)$$

(see Definition 3.99). By interchanging the roles of φ and φ^* in the above argument, we also show the equivalence with the relation $u \in \partial_\varepsilon \varphi^*(u^*)$.

Solution of Problem 3.53

Note that $\varphi^*(u^*) = i_{[-1,1]}(u^*)$. Then according to Problem 3.52, we have

$$\big[u^* \in \partial_\varepsilon \varphi(u)\big] \iff \big[|u| + i_{[-1,1]}(u^*) - u^* u \leqslant \varepsilon\big].$$

Therefore, we have

$$\partial_\varepsilon \varphi(u) = \begin{cases} [-1,1] & \text{if } u \in [-\frac{\varepsilon}{2}, \frac{\varepsilon}{2}], \\ \big[1 - \frac{\varepsilon}{u}, 1\big] & \text{if } u > \frac{\varepsilon}{2}, \\ \big[-1, -1 - \frac{\varepsilon}{u}\big] & \text{if } u < -\frac{\varepsilon}{2}. \end{cases}$$

Solution of Problem 3.54

Let $u \neq 0$. We will show that

$$\partial \varphi(u) = \{u^* \in X^* : \|u^*\|_* = 1, \ \langle u^*, u \rangle = \|u\|_X\} = C, \tag{3.75}$$

that is, $\partial \varphi(u) = \frac{1}{\|u\|_X} \mathcal{F}(u)$ with $\mathcal{F} \colon X \longrightarrow 2^{X^*} \setminus \{\emptyset\}$ being the duality map (see Definition 2.112). If $\|u^*\|_* = 1$ and $\langle u^*, u \rangle = \|u\|_X$, then $\langle u^*, h \rangle \leqslant \|h\|_X$ for all $h \in X$ and so

$$\langle u^*, h - u \rangle \leqslant \|h\|_X - \|u\|_X,$$

thus $u^* \in \partial \varphi(u)$ and hence

$$C \subseteq \partial \varphi(u) \tag{3.76}$$

(see (3.75)). On the other hand, if $u^* \in \partial \varphi(u)$, then

$$-\|u\|_X = -\varphi(u) = \varphi(0) - \varphi(u) \geqslant \langle u^*, 0 - u \rangle = -\langle u^*, u \rangle \tag{3.77}$$

(see Definition 3.70) and

$$\|u\|_X = 2\|u\|_X - \|u\|_X \geqslant \langle u^*, 2u - u \rangle = \langle u^*, u \rangle. \tag{3.78}$$

From (3.77) and (3.78) it follows that

$$\langle u^*, u\rangle = \|u\|_X. \tag{3.79}$$

For all $h \in X$ and all $t > 0$, we have

$$\|u + th\|_X - \|u\|_X \geqslant \langle u^*, th\rangle,$$

so

$$\langle u^*, h\rangle \leqslant \|\tfrac{u}{t} + h\|_X - \|\tfrac{u}{t}\|_X.$$

Let $t \to +\infty$. Then

$$\langle u^*, h\rangle \leqslant \|h\|_X,$$

so

$$\|u^*\|_* \leqslant 1. \tag{3.80}$$

From (3.79) we have $\|u^*\|_* \geqslant 1$ and thus combined with (3.80) gives $\|u^*\|_* = 1$. Therefore it follows that

$$\partial\varphi(u) \subseteq C \tag{3.81}$$

(see (3.75)). From (3.76) and (3.81) we conclude that

$$\partial\varphi(u) = C.$$

Now let $u = 0$. Then $u^* \in \partial\varphi(0)$ if and only if $\langle u^*, h\rangle \leqslant \|h\|_X$ for all $h \in X$ which is equivalent to saying that $\|u^*\|_* \leqslant 1$. Hence

$$\partial\varphi(0) = \overline{B}_1^*(0),$$

where $\overline{B}_1^*(0) = \{u^* \in X^* : \|u^*\|_* \leqslant 1\}$. So, finally we have

$$\partial\varphi(u) = \begin{cases} \frac{1}{\|u\|_X}\mathcal{F}(u) & \text{if } u \neq 0, \\ \overline{B}_1^*(0) & \text{if } u = 0. \end{cases}$$

Solution of Problem 3.55

From the Riesz representation theorem, we know that

$$L^1((0,1);\mathbb{R}^N)^* = L^\infty((0,1);\mathbb{R}^N).$$

Let $u \in L^1((0,1);\mathbb{R}^N)$, $u^* \in L^\infty((0,1);\mathbb{R}^N)$ and assume that $u^* \in \partial\|u\|_1$. Then from the solution of Problem 3.54, we know that, if $u \neq 0$, then

$$\|u^*\|_\infty = \operatorname*{ess\,sup}_{[0,1]} |u^*(t)| = 1$$

and

$$\langle u^*, u\rangle = \int_0^1 (u^*(t), u(t))_{\mathbb{R}^N}\, dt = \|u\|_1.$$

These equalities can hold if and only if $u^*(t) = \frac{u(t)}{|u(t)|}$ for almost all $t \in \{u \neq 0\}$.

If $u = 0$, then $u^* \in \overline{B}_1^*(0)$. So, finally

$$\partial\|u\|_1 = \begin{cases} \{u^* \in L^\infty((0,1);\mathbb{R}^N) : \ u^*(t) = \frac{u(t)}{|u(t)|} \text{ for a.a. } t \in \{u \neq 0\}\} \\ \qquad \text{if } \ u \neq 0, \\ \overline{B}_1^*(0) \quad \text{if } \ u = 0. \end{cases}$$

From this and Proposition 3.75, it follows that $\|\cdot\|_1$ is Gâteaux differentiable at those points $u \in L^1((0,1);\mathbb{R}^N)$ for which the set $\{t \in [0,1] : \ u(t) = 0\}$ is Lebesgue-null.

Solution of Problem 3.56

Evidently $\varphi \in \Gamma_0(X)$. We have

$$\begin{aligned} \varphi^*(u^*) &= \sup_{u\in X} \big(\langle u^*, u\rangle - \varphi(u)\big) = \sup_{t\geqslant 0} \sup_{\substack{u\in X \\ \|u\|_X = t}} \big(\langle u^*, u\rangle - h(\|u\|_X)\big) \\ &= \sup_{t\geqslant 0} \big(t\|u^*\|_* - h(t)\big) = \sup_{t\in\mathbb{R}} \big(t\|u^*\|_* - h(t)\big) = h^*(\|u^*\|_*) \end{aligned}$$

(since h is even).

Solution of Problem 3.57

Let $\varepsilon > 0$, $\delta = \frac{a(\varepsilon)}{\varepsilon}$ and $s \in (0,\delta)$. For $t \geqslant \varepsilon$, we have $t\frac{a(\varepsilon)}{\varepsilon} \leqslant a(t)$ and so

$$st - a(t) = t\big(s - \tfrac{a(t)}{t}\big) \leqslant t\big(s - \tfrac{a(\varepsilon)}{\varepsilon}\big) = t(s-\delta) < 0,$$

so

$$\sup_{t\geqslant\varepsilon} (st - a(t)) \leqslant 0.$$

For $t \in [0,\varepsilon)$, we have

$$st - a(t) \leqslant s\varepsilon - a(t) \leqslant s\varepsilon,$$

so

$$\sup_{t\in[0,\varepsilon)} (st - a(t)) \leqslant s\varepsilon.$$

Therefore, for $s \in (0, \delta)$, we have

$$\begin{aligned} 0 \leqslant a^*(s) &= \sup_{t \geqslant 0} (st - a(t)) \\ &= \max\Big\{ \sup_{t \geqslant \varepsilon}(st - a(t)),\ \sup_{t \in [0,\varepsilon)} (st - a(t))\Big\} \leqslant s\varepsilon, \end{aligned}$$

so $0 \leqslant \frac{a^*(s)}{s} \leqslant \varepsilon$, hence $\lim\limits_{s \searrow 0} \frac{a^*(s)}{s} = 0$.

Solution of Problem 3.58

In what follows for $t \geqslant 0$, let

$$\xi(t) = \inf\{\varphi(u) - \varphi(u_0) :\ |u - u_0| = t\}.$$

We have

$$-\varphi^*(0) = -\sup_{u \in \mathbb{R}^N} \big((0, u)_{\mathbb{R}^N} - \varphi(u)\big) = \inf_{\mathbb{R}^N} \varphi = \varphi(u_0) \tag{3.82}$$

(see Definition 3.83). So, for all $u^* \in \mathbb{R}^N$ we have

$$(u^*, u_0)_{\mathbb{R}^N} + \varphi^*(0) = (u^*, u_0)_{\mathbb{R}^N} - \varphi(u_0) \leqslant \varphi^*(u^*)$$

(see (3.82) and Definition 3.83). Therefore for all $u^* \in \mathbb{R}^N$, we can write

$$\begin{aligned} \varphi^*(u^*) &= \sup_{u \in \mathbb{R}^N} \big((u^*, u)_{\mathbb{R}^N} - \varphi(u)\big) \\ &\leqslant \sup_{u \in \mathbb{R}^N} \big((u^*, u)_{\mathbb{R}^N} - \xi(|u - u_0|) - \varphi(u_0)\big) \\ &= \sup_{v \in \mathbb{R}^N} \big((u^*, v + u_0)_{\mathbb{R}^N} - \xi(|v|) - \varphi(u_0)\big) \\ &\leqslant (u^*, u_0)_{\mathbb{R}^N} - \varphi(u_0) + \sup_{v \in \mathbb{R}^N} \big((u^*, v)_{\mathbb{R}^N} - \xi(|v|)\big) \\ &= (u^*, u_0)_{\mathbb{R}^N} + \varphi^*(0) + \xi^*(|u^*|) \end{aligned}$$

(see (3.82) and Problem 3.56), so

$$0 \leqslant \frac{\varphi^*(u^*) - \varphi^*(0) - (u^*, u_0)_{\mathbb{R}^N}}{|u^*|} \leqslant \frac{\xi^*(|u^*|)}{|u^*|}$$

(see (3.82)), thus

$$\lim_{u^* \to 0} \frac{|\varphi^*(u^*) - \varphi^*(0) - (u^*, u_0)_{\mathbb{R}^N}|}{|u^*|} = 0$$

(see Problem 3.57) and hence $(\varphi^*)'(0) = u_0$.

Solution of Problem 3.59

"**(a)** $\Longrightarrow$ **(b)**": Let $C \subseteq X$ be a nonempty and bounded set and let $k > 0$ be the Lipschitz constant of $\varphi|_C$. Let $u \in C$. From Theorem 3.71, we know that $\partial\varphi(u) \neq \emptyset$. Let $u^* \in \partial\varphi(u)$. We have

$$\langle u^*, h\rangle \leqslant \varphi(u+h) - \varphi(u) \leqslant k\|h\|_X \qquad \forall\, h \in \overline{B}_1(0),$$

so $\|u^*\|_* \leqslant k$ and thus $\partial\varphi(C) \subseteq k\overline{B}_1^*(0)$, where $\overline{B}_1^*(0) = \{u^* \in X^* : \|u^*\|_* \leqslant 1\}$.

"**(b)** $\Longrightarrow$ **(a)**": Arguing by contradiction, suppose that φ is not Lipschitz continuous on some nonempty and bounded set $C \subseteq X$. Then for each $n \geqslant 1$, we can find two sequences $\{u_n\}_{n\geqslant 1}, \{v_n\}_{n\geqslant 1} \subseteq C$ such that

$$|\varphi(v_n) - \varphi(u_n)| > n\|v_n - u_n\|_X \qquad \forall\, n \geqslant 1. \tag{3.83}$$

Let $v_n^* \in \partial\varphi(v_n)$ for all $n \geqslant 1$ (see Theorem 3.71). We have

$$\langle v_n^*, u_n - v_n\rangle \leqslant \varphi(u_n) - \varphi(v_n) < -n\|v_n - u_n\|_X \qquad \forall\, n \geqslant 1$$

(see (3.83)), so

$$\|v_n^*\|_* > n \qquad \forall\, n \geqslant 1,$$

which contradicts statements **(b)**.

Solution of Problem 3.60

Without any loss of generality we may assume that $0 \in D$. Let $\varrho > 0$ be such that $D \subseteq \varrho\overline{B}_1$ (recall that the set D is bounded) and let $M > 0$ be such that $\varphi(u) \leqslant M$ for all $u \in D + \delta\overline{B}_1$. Let $u \in D \setminus \{0\}$ and let $t = \|u\|_X$. From the convexity of φ, we have

$$\varphi(0) \leqslant \tfrac{t}{t+\delta}\varphi(-\tfrac{\delta}{t}u) + \tfrac{\delta}{t+\delta}\varphi(u),$$

so

$$\varphi(u) \geqslant \tfrac{t+\delta}{\delta}(\varphi(0) - M) \geqslant \tfrac{\varrho+\delta}{\delta}(\varphi(0) - M),$$

thus $\varphi|_D$ is bounded below. Let $\eta \in \mathbb{R}$ be such that $\varphi(u) \geqslant \eta$ for all $u \in C$. Suppose that $u, v \in C$, $u \neq v$ and let $t = \|u - v\|_X$. Then

$$y = \tfrac{\delta}{\delta+t}u + \tfrac{t}{\delta+t}(y + \tfrac{\delta}{t}(y - u)).$$

Because φ is convex, we have

$$\varphi(y) \leqslant \tfrac{\delta}{\delta+t}\varphi(u) + \tfrac{t}{\delta+t}M = \varphi(u) - \tfrac{t}{\delta+t}\varphi(u) + \tfrac{t}{\delta+t}M,$$

so

$$\varphi(y) - \varphi(u) \leqslant \tfrac{M-\eta}{\delta}\|y - u\|_X.$$

Reversing the roles of u and v, we conclude that $\varphi|_D$ is Lipschitz continuous.

Solution of Problem 3.61

Let $C \subseteq X$ be a nonempty and bounded set and let $\varepsilon > 0$. From Problem 3.59 we know that the set $\partial\varphi(C) \subseteq X^*$ is bounded (see Problem 3.60). Then by hypothesis we can find $n_0 = n_0(\varepsilon) \geqslant 1$ such that

$$\varphi_n^*(u^*) \leqslant \varphi^*(u^*) + \varepsilon \qquad \forall\, u^* \in \partial\varphi(C),\ n \geqslant n_0. \tag{3.84}$$

Let $u \in C$ and $u^* \in \partial\varphi(u)$ (see Theorem 3.71). For every $n \geqslant n_0$ we have

$$\langle u^*, u\rangle - \varphi(u) = \varphi^*(u^*) \geqslant \varphi_n^*(u^*) - \varepsilon \geqslant \langle u^*, u\rangle - \varphi_n(u) - \varepsilon$$

(see (3.84) and Proposition 3.84(b)), so

$$\varphi(u) - \varepsilon \leqslant \varphi_n(u) \leqslant \varphi(u) \qquad \forall\, u \in C,\ n \geqslant n_0$$

and thus $\varphi_n \longrightarrow \varphi$ uniformly on bounded subsets of X.

Solution of Problem 3.62

(a) Let $u, v \subset U$. From the Taylor theorem, we know that there exists $z \in (u, v)$ (where $(u, v) = \{h = (1-t)u + tv :\ t \in (0,1)\}$) such that

$$\varphi(v) - \varphi(u) - \langle \varphi'(u), v-u\rangle = \tfrac{1}{2}\varphi''(z)\,(v-u, v-u)\,. \tag{3.85}$$

If $\varphi''(u)(h,h) \geqslant 0$ for all $h \in X$, then from (3.85) we have

$$\varphi(v) - \varphi(u) \geqslant \langle \varphi'(u), v-u\rangle,$$

so $\varphi'(u)$ is monotone (interchange the roles of u and v in the above inequality and add) and thus φ is convex (see Problem 3.50).

Now suppose that φ is convex. Let $u \in U$ and let $\varrho > 0$ be small such that $B_\varrho(u) \subseteq U$. If $h \in B_\varrho(0)$ and $t \in (0,1]$, then from the Taylor formula, we have

$$\begin{aligned} 0 &\leqslant \varphi(u+th) - \varphi(u) - \langle \varphi'(u), th\rangle \\ &= \tfrac{1}{2}\varphi''(u)(th,th) + e(\|th\|_X^2), \end{aligned}$$

with $\frac{e(\|v\|_X)}{\|v\|_X} \to 0$ as $v \to 0$ (by the Taylor theorem), so

$$-\frac{e(\|th\|_X^2)}{t^2} \leqslant \tfrac{1}{2}\varphi''(u)(h,h).$$

Passing to the limit as $t \searrow 0$, we obtain

$$0 \leqslant \varphi''(u)(h,h).$$

If $h \in X \setminus B_\varrho(0)$, then there exists $\tau \neq 0$ such that $\tau h \in B_\varrho(0)$ and

$$\varphi''(u)(h,h) = \frac{1}{\tau^2}\varphi''(u)(\tau h, \tau h) \geqslant 0.$$

Therefore, we conclude that

$$\varphi''(u)(h,h) \geqslant 0 \qquad \forall\, h \in X.$$

(b) This is immediate from (3.85).
(c) Let $\varphi\colon \mathbb{R} \longrightarrow \mathbb{R}$ be defined by $\varphi(u) = u^4$. Then φ is strictly convex but $\varphi''(0) = 0$.

Solution of Problem 3.63
We have

$$\varphi(u) = \inf_{c \in C} \frac{1}{2}\|u-c\|_H^2 = \inf_{c \in H} \left(\frac{1}{2}\|u-c\|_H^2 + i_C(c)\right) = \left(\frac{1}{2}\|\cdot\|_H^2 \,\boxdot\, i_C\right)(u)$$

(see Definition 3.87). Then from Proposition 3.89(a), we have

$$\varphi^*(u^*) = \frac{1}{2}\|u^*\|_H^2 + \sigma_C(u^*) \qquad \forall\, u^* \in H$$

(see Definition 2.46). From Proposition 3.84(a), for all $u, u^* \in H$, we have

$$(u^*, u)_H - \frac{1}{2}\|u^*\|_H^2 - \sigma_C(u^*) \leqslant \varphi(u). \tag{3.86}$$

Fix $v \in H$ and let $u^* = v - \mathrm{proj}_C(v)$. From the properties of the metric projection, we have

$$0 \leqslant \left(v - \mathrm{proj}_C(v), \mathrm{proj}_C(v) - c\right)_H \qquad \forall\, c \in C,$$

so

$$\begin{aligned} \sigma_C(u^*) &= \sup_{c \in C}(u^*, c)_H = \sup_{c \in C}\left(v - \mathrm{proj}_C(v), c\right)_H \\ &= \left(v - \mathrm{proj}_C(v), \mathrm{proj}_C(v)\right)_H. \end{aligned} \tag{3.87}$$

Returning to (3.86) and using (3.87), we obtain

$$\left(v - \mathrm{proj}_C(v), u - \mathrm{proj}_C(v)\right)_H - \frac{1}{2}\|v - \mathrm{proj}_C(v)\|_H^2 \leqslant \varphi(u),$$

so

$$\left(v - \mathrm{proj}_C(v), u - \mathrm{proj}_C(v)\right)_H - \varphi(v) \leqslant \varphi(u),$$

hence

$$\begin{aligned}\varphi(u)-\varphi(v) &\geqslant \big(v-\operatorname{proj}_C(v), u-\operatorname{proj}_C(v)\big)_H - 2\varphi(u)\\ &\geqslant \big(v-\operatorname{proj}_C(v), u-\operatorname{proj}_C(v)\big)_H - \|v-\operatorname{proj}_C(v)\|_H^2\\ &= \big(v-\operatorname{proj}_C(v), u-v\big)_H\end{aligned}$$

and thus

$$0 \leqslant \varphi(u)-\varphi(v)-\big(v-\operatorname{proj}_C(v), u-v\big)_H. \tag{3.88}$$

Interchanging the roles of u and v in the above argument, we also have

$$0 \leqslant \varphi(v)-\varphi(u)-\big(u-\operatorname{proj}_C(u), v-u\big)_H,$$

so

$$\varphi(u)-\varphi(v) \leqslant \big(u-\operatorname{proj}_C(u), u-v\big)_H. \tag{3.89}$$

From (3.88) and (3.89), it follows that

$$\begin{aligned}0 &\leqslant \varphi(u)-\varphi(v)-\big(v-\operatorname{proj}_C(v), u-v\big)_H\\ &\leqslant \big(u-\operatorname{proj}_C(u)-(v-\operatorname{proj}_C(v)), u-v\big)_H\\ &\leqslant \|(I_H-\operatorname{proj}_C)(u-v)\|_H\|u-v\|_H\\ &\leqslant 2\|u-v\|_H^2\end{aligned}$$

(recall that proj_C is nonexpansive), so

$$0 \leqslant \frac{\varphi(u)-\varphi(v)-\big(v-\operatorname{proj}_C(v), u-v\big)_H}{\|u-v\|_H} \leqslant 2\|u-v\|_H.$$

Let $u \longrightarrow v$. Then we see that $\varphi'(v)$ exists and $\varphi'(v) = v-\operatorname{proj}_C(v)$. This proves that $\varphi \in C^1(H)$.

Solution of Problem 3.64

From Problem 3.63 we know that $\varphi \in C^1(H)$ and

$$\varphi'(u) = u-\operatorname{proj}_C(u) \qquad \forall\, u \in H.$$

Then we have

$$\begin{aligned}&\big(\varphi'(u+h)-\varphi'(u), h\big)_H\\ &\quad= \big(u+h-\operatorname{proj}_C(u+h)-u+\operatorname{proj}_C(u), h\big)_H\\ &\quad= \big(h-\operatorname{proj}_C(u+h)+\operatorname{proj}_C(u), h\big)_H\\ &\quad= \|h\|_H^2-\big(\operatorname{proj}_C(u+h)-\operatorname{proj}_C(u), h\big)_H\\ &\quad\geqslant \|h\|_H^2-\|\operatorname{proj}_C(u+h)-\operatorname{proj}_C(u)\|_H\|h\|_H \geqslant \|h\|_H^2-\|h\|_H^2 = 0\end{aligned}$$

(since proj_C is nonexpansive). Invoking Problem 3.50, we conclude that φ is convex.

Solution of Problem 3.65

Let $u_0^* \in \partial_{\frac{\varepsilon}{2}}\varphi(u_0)$ and $\lambda = \max\{\|u_0^*\|_*, 1\}$. From Theorem 3.102, we know that we can find $u \in \mathrm{dom}\,\varphi$ and $u^* \in X^*$ such that

$$u^* \in \partial\varphi(u), \quad \|u - u_0\|_X \leqslant \tfrac{\varepsilon}{2\lambda}, \quad \|u^* - u_0^*\|_X \leqslant \lambda. \tag{3.90}$$

Since $u_0^* \in \partial_{\frac{\varepsilon}{2}}\varphi(u_0)$, we have

$$\langle u_0^*, u - u_0\rangle \leqslant \varphi(u) - \varphi(u_0) + \tfrac{\varepsilon}{2}$$

(see Definition 3.99), so

$$-\|u_0^*\|_*\|u - u_0\|_X - \tfrac{\varepsilon}{2} \leqslant \varphi(u) - \varphi(u_0),$$

thus

$$-\tfrac{\varepsilon}{2\lambda}\lambda - \tfrac{\varepsilon}{2} \leqslant \varphi(u) - \varphi(u_0)$$

(see (3.90)) and hence

$$-\varepsilon \leqslant \varphi(u) - \varphi(u_0). \tag{3.91}$$

Note that $\|u^*\|_* \leqslant 2\lambda$ and since $u^* \in \partial\varphi(u)$, we have

$$\langle u^*, u_0 - u\rangle \leqslant \varphi(u_0) - \varphi(u),$$

so

$$-2\lambda\|u_0 - u\|_X \leqslant \varphi(u_0) - \varphi(u),$$

thus

$$-\varepsilon \leqslant \varphi(u_0) - \varphi(u) \tag{3.92}$$

(see (3.90)). From (3.91) and (3.92), we infer that

$$|\varphi(u) - \varphi(u_0)| \leqslant \varepsilon.$$

Clearly, from (3.90) we have $\|u - u_0\|_X \leqslant \varepsilon$.

Solution of Problem 3.66

From Proposition 3.84(d), we have that

$$\varphi^{**}|_X \leqslant \varphi \quad \text{and} \quad \varphi^{**}|_X \in \Gamma_0(X).$$

So, we need to show that $\varphi \leqslant \varphi^{**}|_X$. To simplify the notation, we set $\widehat{\varphi} = \varphi^{**}|_X$. If $u \notin \operatorname{dom} \widehat{\varphi}$, then $\widehat{\varphi} = +\infty$ and so $\varphi(u) \leqslant \widehat{\varphi}(u)$. Therefore we assume that $u \in \operatorname{dom} \widehat{\varphi}$. Suppose that $u \in D(\partial \widehat{\varphi})$ and let $u^* \in \partial \widehat{\varphi}(u)$. Since $\widehat{\varphi} \in \Gamma_0(X)$, we have $u \in \partial \varphi^*(u^*)$. Also, from Proposition 3.68, we have that φ^* is continuous at u^*. We choose $u_n \in X$ for $n \geqslant 1$ such that

$$\langle u^*, u_n \rangle - \varphi(u_n) \ \geqslant \ \varphi^*(u^*) - \varepsilon_n,$$

with $\varepsilon_n \searrow 0$ (see Definition 3.83), so

$$\langle u^*, u_n \rangle - \widehat{\varphi}(u_n) \ \geqslant \ \varphi^*(u^*) - \varepsilon_n \qquad \forall\, n \geqslant 1$$

and thus $u_n \in \partial_{\varepsilon_n} \varphi^*(u^*)$ for all $n \geqslant 1$ (see Definition 3.99). Then from Theorem 3.103(e) and Remark 3.104, we have $u_n \longrightarrow u$ in X (respectively $u_n \xrightarrow{w} u$ in X), so using the hypothesis on φ, we have

$$\liminf_{n \to +\infty} \varphi(u_n) \ \geqslant \ \varphi(u).$$

Hence, we have

$$\begin{aligned} \widehat{\varphi}(u) &= \langle u^*, u \rangle - \varphi^*(u^*) \ \geqslant \ \liminf_{n \to +\infty} \big(\langle u^*, u \rangle - (\langle u^*, u_n \rangle - \varphi(u_n) + \varepsilon_n) \big) \\ &= \liminf_{n \to +\infty} \varphi(u_n) \ \geqslant \ \varphi(u) \end{aligned}$$

(see Proposition 3.84(b)). So, we have proved that $\widehat{\varphi}(u) \geqslant \varphi(u)$ for all $u \in D(\partial \widehat{\varphi})$.

Next suppose that $u \notin D(\partial \widehat{\varphi})$. Problem 3.65 asserts that there exists a sequence $\{u_n\}_{n \geqslant 1} \subseteq D(\partial \widehat{\varphi})$ such that $u_n \longrightarrow u$ in X and $|\widehat{\varphi}(u_n) - \widehat{\varphi}(u)| \longrightarrow 0$. Then

$$\varphi(u) \ \leqslant \ \liminf_{n \to +\infty} \varphi(u_n) \ = \ \liminf_{n \to +\infty} \widehat{\varphi}(u_n) \ = \ \widehat{\varphi}(u)$$

(from the first part of the solution). So, we conclude that $\widehat{\varphi} = \varphi$, hence φ is convex (see Proposition 3.84(f)).

Solution of Problem 3.67

Let $\varphi = \frac{1}{2}\|\cdot\|_H^2 + i_C$. Then φ is sequentially weakly lower semicontinuous (recall that C is weakly closed). Also, $\varphi^* = \frac{1}{2}\left(\|\cdot\|_H^2 - \operatorname{dist}(\cdot, C)^2\right)$ (see Problem 3.49). So, φ^* is by hypothesis Gâteaux differentiable. Also, from Proposition 3.84(f), we have

$$\varphi^{**} \ = \ \overline{\operatorname{conv}}\, \varphi \ = \ \tfrac{1}{2}\|\cdot\|_H^2 + i_{\overline{\operatorname{conv}}\, C},$$

so φ^{**} is proper. So, we can use Problem 3.66 and infer that φ is convex. But then $\operatorname{dom} \varphi = C$ is convex.

Solution of Problem 3.68

(a) Suppose that $v \in K$ satisfies

$$(u-v,c)_H \leqslant 0 \quad \forall\, c \in C \quad \text{and} \quad (u-v,v)_H = 0. \tag{3.93}$$

From (3.93) it follows that

$$(u-v,c-v)_H \leqslant 0 \quad \forall\, c \in K,$$

so $v = \mathrm{proj}_K(u)$.
Conversely, from the properties of the metric projection, we have

$$\big(u-\mathrm{proj}_K(u), c-\mathrm{proj}_K(u)\big)_H \leqslant 0 \quad \forall\, c \in K, \tag{3.94}$$

so

$$\big(u-\mathrm{proj}_K(u), (\lambda-1)\mathrm{proj}_K(u)\big)_H \leqslant 0 \quad \forall\, \lambda > 0$$

and thus

$$(\lambda-1)\big(u-\mathrm{proj}_K(u), \mathrm{proj}_K(u)\big)_H \leqslant 0 \quad \forall\, \lambda > 0. \tag{3.95}$$

From (3.95) we see that

$$\big(u-\mathrm{proj}_K(u), \mathrm{proj}_K(u)\big)_H \geqslant 0 \quad \forall\, \lambda \in (0,1)$$

and

$$\big(u-\mathrm{proj}_K(u), \mathrm{proj}_K(u)\big)_H \leqslant 0 \quad \forall\, \lambda > 1.$$

Therefore

$$\big(u-\mathrm{proj}_K(u), \mathrm{proj}_K(u)\big)_H = 0,$$

so

$$\big(u-\mathrm{proj}_K(u), c\big)_H \leqslant 0 \quad \forall\, c \in K$$

(see (3.94)).

(b) Let $\lambda > 0$. Using part **(a)** we have

$$\begin{aligned}&\big(u-\tfrac{1}{\lambda}\mathrm{proj}_K(\lambda u), \tfrac{1}{\lambda}\mathrm{proj}_K(\lambda u)\big)_H \\ &= \tfrac{1}{\lambda^2}\big(\lambda u-\mathrm{proj}_K(\lambda u), \mathrm{proj}_K(\lambda u)\big)_H = 0\end{aligned}$$

and

$$\big(u-\tfrac{1}{\lambda}\mathrm{proj}_K(\lambda u), c\big)_H = \tfrac{1}{\lambda^2}\big(\lambda u-\mathrm{proj}_K(\lambda u), c\big)_H \leqslant 0 \quad \forall\, c \in K.$$

Then using once again part **(a)**, we have

$$\tfrac{1}{\lambda}\mathrm{proj}_K(\lambda u) = \mathrm{proj}_K(u),$$

so

$$\mathrm{proj}_K(\lambda u) = \lambda\mathrm{proj}_K(u).$$

Moreover, we have

$$\begin{aligned} \|u\|_H^2 &= \|u - \operatorname{proj}_K(u) + \operatorname{proj}_K(u)\|_H^2 \\ &= \|u - \operatorname{proj}_K(u)\|_H^2 + \|\operatorname{proj}_K(u)\|_H^2 \\ &\quad +2\left(u - \operatorname{proj}_K(u), \operatorname{proj}_K(u)\right)_H \\ &= \|u - \operatorname{proj}_K(u)\|_H^2 + \|\operatorname{proj}_K(u)\|_H^2 \end{aligned}$$

(see part **(a)**)

Solution of Problem 3.69

We have

$$(u-y,c)_H = (v,c)_H \leqslant 0 \qquad \forall\, c \in K$$

(since $v \in K^0$) and

$$(u-y,y)_H = (v,y)_H = 0$$

(by hypothesis). Then Problem 3.68(a) implies that $y = \operatorname{proj}_K(u)$. Similarly, we show that $v = \operatorname{proj}_{K^0}(u)$.

If $u \in K^0$, then

$$\operatorname{proj}_K(u) = u \quad \operatorname{proj}_{K^0}(u) = u - u = 0$$

and conversely, if $\operatorname{proj}_K(u) = 0$, then $u = \operatorname{proj}_{K^0}(u)$ and so $u \in K^0$. This proves that $K^0 = \operatorname{proj}_K^{-1}(0)$. Similarly, we show that $K = \operatorname{proj}_{K^0}^{-1}(0)$.

Solution of Problem 3.70

Note that

$$\varphi \geqslant c\|\cdot\|_X + \eta \quad \Longleftrightarrow \quad \varphi^* \leqslant (c\|\cdot\|_X + \eta)^*.$$

Also,

$$\varphi^* \leqslant (c\|\cdot\|_X + \eta)^* \quad \Longleftrightarrow \quad \varphi^* \leqslant i_{c\overline{B}_1^*} - \eta$$

(recall that $\|\cdot\|_X^* = i_{\overline{B}_1^*}$ and $(c\|\cdot\|_X)^*(u) = c i_{\overline{B}_1^*}(\frac{1}{c}u)$; see Remark 3.88).

Solution of Problem 3.71

No. To see this note that since φ is not Fréchet differentiable at $u_0 \in X$, according to Proposition 3.74, we can find $\varepsilon > 0$ such that

$$\varphi(u_0 + h) + \varphi(u_0 - h) - 2\varphi(u_0) > \varepsilon\|h\|_X \qquad \forall\, h \in X \setminus \{0\}. \tag{3.96}$$

Suppose that $\xi = \varphi + \psi$ is Fréchet differentiable at u_0. Then Proposition 3.74 implies that for $\varepsilon > 0$ as above we can find $\delta > 0$ such that

$$\xi(u_0 + h) + \xi(u_0 - h) - 2\xi(u_0) \leqslant \varepsilon\|h\|_X \qquad \forall\, h \in X,\ \|h\|_X \leqslant \delta. \tag{3.97}$$

From (3.96) and (3.97) it follows that

$$\psi(u_0 + h) + \psi(u_0 - h) < 2\psi(u_0) \qquad \forall\, h \in X,\ \|h\|_X \leqslant \delta. \tag{3.98}$$

From the convexity of ψ, for all $0 < \|h\|_X \leqslant \delta$, we have

$$\psi(u_0) \leqslant \tfrac{1}{2}\psi(u_0 + h) + \tfrac{1}{2}\psi(u_0 - h) < \psi(u_0)$$

(see (3.98)), a contradiction. This proves that $\xi = \varphi + \psi$ cannot be Fréchet differentiable at u_0.

Solution of Problem 3.72

From Proposition 3.74, we know that given $\varepsilon > 0$, we can find $\delta > 0$ such that

$$\varphi(u_0 + h) + \varphi(u_0 - h) - 2\varphi(u_0) \leqslant \varepsilon\|h\|_X \qquad \forall\, h \in X,\ \|h\|_X \leqslant \delta,$$

so

$$\psi(u_0 + h) + \psi(u_0 - h) - 2\psi(u_0) \leqslant \varepsilon\|h\|_X \qquad \forall\, h \in X,\ \|h\|_X \leqslant \delta,$$

and thus ψ is Fréchet differentiable at u_0 (see Proposition 3.74).

Solution of Problem 3.73

(a) Suppose that φ is convex. Let $u, v \in K$. Then

$$\tfrac{1}{2}\varphi(u + v) = \varphi(\tfrac{1}{2}(u + v)) \leqslant \tfrac{1}{2}\varphi(u) + \tfrac{1}{2}\varphi(v)$$

(by the positive homogeneity and convexity of φ), so

$$\varphi(u + v) \leqslant \varphi(u) + \varphi(v).$$

Conversely, suppose that $\varphi(u + v) \leqslant \varphi(u) + \varphi(v)$ for all $u, v \in K$. Then for $t \in [0, 1]$, we have

$$\varphi((1 - t)u + tv) \leqslant \varphi((1 - t)u) + \varphi(tv) = (1 - t)\varphi(u) + t\varphi(v)$$

(by hypothesis and by the positive homogeneity of φ), so φ is convex.

(b) Let $\lambda > \varphi(u)$ and $\mu > \varphi(v)$. Then $\lambda, \mu > 0$ (recall that $\varphi \geqslant 0$) and so

$$\varphi(\tfrac{u}{\lambda}) < 1 \quad \text{and} \quad \varphi(\tfrac{v}{\mu}) < 1$$

(by the positive homogeneity of φ), so $\frac{u}{\lambda}, \frac{v}{\mu} \in \varphi^1$. But by hypothesis φ^1 is convex, so

$$\tfrac{1}{\lambda+\mu}\varphi(u+v) = \varphi\big(\tfrac{u+v}{\lambda+\mu}\big) = \varphi\big(\tfrac{\lambda}{\lambda+\mu}\tfrac{u}{\lambda} + \tfrac{\mu}{\lambda+\mu}\tfrac{v}{\mu}\big) \leqslant 1,$$

thus

$$\varphi(u+v) \leqslant \lambda + \mu \qquad \forall\, \lambda > \varphi(u),\ \mu > \varphi(v),$$

hence

$$\varphi(u+v) \leqslant \varphi(u) + \varphi(v),$$

which means that φ is convex (see part **(a)**).

Solution of Problem 3.74

Evidently $\varphi \geqslant 0$ and it is positively homogeneous. Let $\widehat{p}_k$ be the projection of $\mathbb{R}^n$ on the k-th component. Then the map $K \ni \widehat{u} \longmapsto \widehat{p}_k(\widehat{u})^p = u_k^p$ is convex. Hence φ^p being the sum of convex functions is convex. Therefore the set $\{\widehat{u} \in K : \ \varphi^p(\widehat{u}) \leqslant 1\}$ is convex. But note that

$$\{\widehat{u} \in K : \ \varphi^p(\widehat{u}) \leqslant 1\} = \{\widehat{u} \in K : \ \varphi(\widehat{u}) \leqslant 1\},$$

so the set $\{\widehat{u} \in K : \ \varphi(\widehat{u}) \leqslant 1\}$ is convex and thus φ is convex (see Problem 3.73(b)).

Solution of Problem 3.75

For every $\widehat{u} \in D$, we have

$$\frac{\partial^2 \varphi}{\partial u_i \partial u_j}(\widehat{u}) = \frac{\varphi(\widehat{u})}{n^2 u_i u_j}(1 - n\delta_{ij}),$$

with δ_{ij} being the Kronecker symbol, i.e.,

$$\delta_{ij} = \begin{cases} 1 & \text{if} \quad i = j, \\ 0 & \text{if} \quad i \neq j, \end{cases} \qquad i, j \in \{1, \ldots, n\}.$$

Then for all $\widehat{h} = (h_1, \ldots, h_n) \in \mathbb{R}^N$, we have

$$\big(\varphi''(\widehat{u})\widehat{h}, \widehat{h}\big)_{\mathbb{R}^N} = \frac{\varphi(\widehat{u})}{n^2}\Big(\Big(\sum_{k=1}^n \frac{h_k}{u_k}\Big)^2 - n\sum_{k=1}^n \Big(\frac{h_k}{u_k}\Big)^2\Big). \tag{3.99}$$

Let $\|\cdot\|_1$ denote the l^1-norm of $\mathbb{R}^n$ and $\|\cdot\|_2$ the l^2-norm of $\mathbb{R}^n$. We know that

$$\|\cdot\|_1 \leqslant \sqrt{n}\|\cdot\|_2.$$

So the right-hand side of (3.99) is nonnegative (recall that $\varphi(\widehat{u}) \leqslant 0$). Then invoking Problem 3.62(a) we conclude that φ is convex.

Solution of Problem 3.76

Let $u \notin C$. Given $\varepsilon \in (0,1)$ we can find $c_\varepsilon \in C$ such that

$$0 < (1-\varepsilon)\|u - c_\varepsilon\|_X \leqslant \operatorname{dist}(u, C). \tag{3.100}$$

Then for any $u^* \in \partial\operatorname{dist}(u, C)$ we have

$$\langle u^*, c_\varepsilon - u\rangle \leqslant \operatorname{dist}(c_\varepsilon, C) - \operatorname{dist}(u, C) = -\operatorname{dist}(u, C),$$

so

$$0 < (1-\varepsilon)\|u - c_\varepsilon\|_X \leqslant \operatorname{dist}(u, C) \leqslant -\langle u^*, c_\varepsilon - u\rangle \leqslant \|u^*\|_*\|u - c_\varepsilon\|_X$$

(see (3.100)), thus

$$1 - \varepsilon \leqslant \|u^*\|_*.$$

Since $\varepsilon > 0$ is arbitrary, we let $\varepsilon \searrow 0$ to obtain

$$1 \leqslant \|u^*\|_*. \tag{3.101}$$

On the other hand, from Theorem 3.71, we have

$$\langle u^*, h\rangle \leqslant \operatorname{dist}'(u_0, C; h) = \lim_{\lambda \searrow 0} \tfrac{\operatorname{dist}(u+\lambda h, C) - \operatorname{dist}(u, C)}{\lambda} \leqslant \|h\|_X$$

(since $\operatorname{dist}(\cdot, C)$ is nonexpansive), so $\|u^*\|_* \leqslant 1$. Thus from (3.101), we conclude that $\|u^*\|_* = 1$.

Solution of Problem 3.77

Let $u \notin C$. We know that $\partial\operatorname{dist}(u, C) \subseteq X^*$ is convex and from Problem 3.76 we know that if $u^* \in \partial\operatorname{dist}(u, C)$, then $\|u^*\|_* = 1$. Since X^* is strictly convex (see Definition I.5.168 and Remark I.5.169), it follows that $\partial\operatorname{dist}(u, C)$ is a singleton.

Solution of Problem 3.78

(a) From Proposition 3.64, we know that φ is locally Lipschitz. Let $B \subseteq \mathbb{R}^N$ be a bounded set and let $\overline{B}_\varepsilon = \{u \in \mathbb{R}^N : \operatorname{dist}(u, B) \leqslant \varepsilon\}$, for $\varepsilon > 0$. Then $\overline{B}_\varepsilon \subseteq \mathbb{R}^N$ is compact and so $\varphi|_{\overline{B}_\varepsilon}$ is Lipschitz continuous with Lipschitz constant $k_\varepsilon > 0$. Let $(u, u^*) \in \operatorname{Gr} \partial\varphi$. Then for any $h \in \mathbb{R}^N$ with $|h| \leqslant \varepsilon$, we have

$$(u^*, h)_{\mathbb{R}^N} \leqslant \varphi(u+h) - \varphi(u) \leqslant k_\varepsilon |h|$$

(see Definition 3.70), so $|u^*| \leqslant k_\varepsilon$ and thus the set $\partial\varphi(B)$ is bounded in $\mathbb{R}^N$.

(b) From Proposition 3.64, we know that $\varphi_n \longrightarrow \varphi$ uniformly on bounded sets. Let $u_n^* \in \partial\varphi_n(u_n)$ for $n \geqslant 1$. Then for all $h \in \mathbb{R}^N$ with $|h| \leqslant 1$ we have

$$(u_n^*, h)_{\mathbb{R}^N} \leqslant \varphi_n(u_n + h) - \varphi_n(u_n) \qquad \forall\, n \geqslant 1.$$

The uniform convergence of φ_n to φ on compact sets, implies that given $\varepsilon > 0$, we can find $n_0 = n_0(\varepsilon) \geqslant 1$ such that

$$(u_n^*, h)_{\mathbb{R}^N} \leqslant \varphi(u+h) - \varphi(u) + \varepsilon \qquad \forall\, n \geqslant n_0. \tag{3.102}$$

Let $K = \{u\} \cup \{u_n : n \geqslant 1\} \subseteq \mathbb{R}^N$ and let $K_1 = \{u \in \mathbb{R}^N : \operatorname{dist}(u, K) \leqslant 1\}$. Both K and K_1 are compact. So, $\varphi|_{K_1}$ is Lipschitz continuous (see Proposition 3.64). Thus, from (3.102), we have

$$(u_n^*, h)_{\mathbb{R}^N} \leqslant k_1 |h| + \varepsilon \qquad \forall\, n \geqslant n_0,\ h \in \mathbb{R}^N,\ |h| \leqslant 1$$

for some $k_1 > 0$, so

$$|u_n^*| \leqslant k_1 + \varepsilon \qquad \forall\, n \geqslant n_0$$

and thus the sequence $\{u_n^*\}_{n \geqslant 1} \subseteq \mathbb{R}^N$ is bounded.

Arguing by contradiction, suppose that the result is not true. Then we can find $\varepsilon > 0$ and a subsequence (still denoted by the same index) such that $u_n^* \in \partial\varphi_n(u_n)$ and

$$u_n^* \notin \partial\varphi(u) + \varepsilon B_1(0) \qquad \forall\, k \geqslant 1. \tag{3.103}$$

From the boundedness of the sequence $\{u_n^*\}_{n \geqslant 1} \subseteq \mathbb{R}^N$, passing to a further subsequence if necessary, we may assume that $u_n^* \longrightarrow u^*$ in $\mathbb{R}^N$. Then from (3.103), it follows that

$$u^* \notin \partial\varphi(u) + \tfrac{\varepsilon}{2}\overline{B}_1(0). \tag{3.104}$$

On the other hand, for every $v \in \mathbb{R}^N$, we have

$$(u_n^*, v - u_n)_{\mathbb{R}^N} \leqslant \varphi_n(v) - \varphi_n(u_n) \qquad \forall\, n \geqslant 1,$$

so

$$(u_n^*, v - u)_{\mathbb{R}^N} \leqslant \varphi(v) - \varphi(u) \qquad \forall\, v \in \mathbb{R}^N$$

and hence $u^* \in \partial\varphi(u)$ (see Definition 3.70). Comparing this with (3.104), we reach a contradiction.

Solution of Problem 3.79

Recall that $\big(\sup\limits_{n\geqslant 1}\varphi_n\big)^{**}$ is the biggest lower semicontinuous and convex minorant of $\sup\limits_{n\geqslant 1}\varphi_n$. From Definition 3.83 and Proposition 3.84, we know that

$$\varphi_n^{**} \text{ is lower semicontinuous and convex} \qquad \forall\, n\geqslant 1 \tag{3.105}$$

and

$$\varphi_n^{**} \leqslant \varphi_n \qquad \forall\, n\geqslant 1. \tag{3.106}$$

From (3.105) and (3.106), it follows that $\sup\limits_{n\geqslant 1}\varphi_n^{**}$ is lower semicontinuous, convex and $\sup\limits_{n\geqslant 1}\varphi_n^{**} \leqslant \sup\limits_{n\geqslant 1}\varphi_n$. Thus

$$\sup_{n\geqslant 1}\varphi_n^{**} \leqslant \big(\sup_{n\geqslant 1}\varphi_n\big)^{**}. \tag{3.107}$$

Let $h(u) = (\widehat{u}^*, u)_{\mathbb{R}^N} + c$, with $\widehat{u}^* \in \mathbb{R}^N$ and $c\in\mathbb{R}$, be an affine function on $\mathbb{R}^N$ such that

$$h \leqslant \sup_{n\geqslant 1}\varphi_n = \varphi. \tag{3.108}$$

The superlinearity condition on ξ implies that we can find $M>0$ such that

$$h(u) \leqslant \xi(|u|) \qquad \forall\, u\in\mathbb{R}^N,\ |u|\geqslant M. \tag{3.109}$$

We claim that given $\varepsilon>0$, we can find $n_0=n_0(\varepsilon)\geqslant 1$ such that

$$\varphi_n(u) \geqslant h(u)-\varepsilon \qquad \forall\, n\geqslant n_0,\ u\in\mathbb{R}^N,\ |u|\leqslant M. \tag{3.110}$$

Indeed, if this is not the case, then we can find $\varepsilon_0>0$ and a subsequence (still denoted with the same index) such that

$$|u_n| \leqslant M \quad\text{and}\quad \varphi_n(u_n) < h(u_n)-\varepsilon_0 \qquad \forall\, n\geqslant 1.$$

Passing to a further subsequence if necessary, we may assume that $u_n \longrightarrow u$ in $\mathbb{R}^N$. Then from Proposition 3.64 and the continuity of h we have

$$\varphi(u) \leqslant h(u)-\varepsilon_0,$$

which contradicts (3.108). So, (3.110) holds and from (3.109) and (3.110), we infer that

$$\varphi_n^{**}(u) \geqslant h(u)-\varepsilon \qquad \forall\, n\geqslant n_0,\ u\in\mathbb{R}^N,$$

thus $\sup_{n\geqslant 1} \varphi_n^{**} \geqslant h$ (since $\varepsilon > 0$ is arbitrary) and hence

$$\sup_{n\geqslant 1} \varphi_n^{**} \;\geqslant\; \big(\sup_{n\geqslant 1} \varphi_n\big)^{**} \tag{3.111}$$

(see (3.108)). From (3.107) and (3.111), we conclude that

$$\sup_{n\geqslant 1} \varphi_n^{**} \;=\; \big(\sup_{n\geqslant 1} \varphi_n\big)^{**}.$$

The result fails if we drop the boundedness below by ξ hypothesis. To see this, consider the sequence

$$\varphi_n(u) \;=\; \begin{cases} 1 & \text{if } |u| < n, \\ 0 & \text{otherwise.} \end{cases}$$

Then, $\{\varphi_n\}_{n\geqslant 1}$ is an increasing sequence of lower semicontinuous function and $\sup_{n\geqslant 1} \varphi_n^{**} = 0$, but $\big(\sup_{n\geqslant 1} \varphi_n\big)^{**} = 1$.

Solution of Problem 3.80

For every $\lambda > 0$, let

$$\psi_\lambda(v) \;=\; \tfrac{1}{2\lambda}\|u-v\|_X^2 + \psi(v).$$

Then from Proposition 3.97 and (3.63) in the solution of Problem 3.41, we have

$$\partial\psi_\lambda(v) \;=\; \tfrac{1}{\lambda}\mathcal{F}(v-u) + \partial\varphi(v).$$

This means that every solution u_λ of the inclusion

$$0 \;\in\; \tfrac{1}{\lambda}\mathcal{F}(v-u) + \partial\varphi(v)$$

is the unique minimizer of the function

$$v \;\longmapsto\; \tfrac{1}{2\lambda}\|u-v\|_X^2 + \varphi(v).$$

But from Problem 2.161, we know that

$$J_\lambda^{\partial\varphi}(u) \;=\; u_\lambda,$$

so

$$\tfrac{1}{2\lambda}\|u - J_\lambda^{\partial\varphi}(u)\|_X^2 + \varphi(J_\lambda^{\partial\varphi}(u)) \;=\; \varphi_\lambda(u) \qquad \forall\, \lambda > 0.$$

This proves part (b) of the problem.

Directly from the definition of φ_λ, we see that

$$\varphi(J_\lambda^{\partial\varphi}(u)) \;\leqslant\; \varphi_\lambda(u) \;\leqslant\; \varphi(u) \qquad \forall\, \lambda > 0,\ u \in X \tag{3.112}$$

and this proves part **(d)** of the problem.

Next, let $u \in \operatorname{dom}\varphi$. Then from Problem 3.65 and Problem 2.164, we have

$$J_\lambda^{\partial\varphi}(u) \longrightarrow u \quad \text{in } X \quad \text{as } \lambda \searrow 0.$$

So, exploiting the lower semicontinuity of φ and using (3.112) we infer that

$$\varphi(u) \leqslant \liminf_{\lambda \searrow 0} \varphi(J_\lambda^{\partial\varphi}(u)) \leqslant \liminf_{\lambda \searrow 0} \varphi_\lambda(u) \leqslant \varphi(u),$$

so

$$\lim_{\lambda \searrow 0} \varphi_\lambda(u) = \varphi(u) \qquad \forall\, u \in \operatorname{dom}\varphi.$$

Now, suppose that $u \notin \operatorname{dom}\varphi$. Then $\varphi(u) = +\infty$. We show that $\varphi_\lambda(u) \longrightarrow +\infty$ as $\lambda \searrow 0$. Indeed, if this is not the case, we can find a sequence $\{\lambda_n\}_{n \geqslant 1}$ such that $\lambda_n \searrow 0$ and $M > 0$ such that

$$\varphi_{\lambda_n}(u) \leqslant M \qquad \forall\, n \geqslant 1. \tag{3.113}$$

Then from part **(b)** and (3.113), we see that

$$J_{\lambda_n}^{\partial\varphi}(u) \longrightarrow u \quad \text{in } X. \tag{3.114}$$

From (3.114) and part **(d)** it follows that

$$\varphi(u) \leqslant M$$

(recall that $\varphi \in \Gamma_0(X)$), a contradiction. This proves that

$$\varphi_\lambda(u) \longrightarrow +\infty \quad \text{as } \lambda \searrow 0 \qquad \forall\, u \notin \operatorname{dom}\varphi.$$

We conclude that part **(c)** of the problem is true.

From part **(b)** it is clear that $\varphi_\lambda \in \Gamma_0(X)$. In fact, if $u_0 \in \operatorname{dom}\varphi$, then

$$\varphi_\lambda(u) \leqslant \tfrac{1}{2\lambda}\|u - u_0\|_X^2 + \varphi(u_0),$$

so φ_λ is continuous (see Theorem 3.60).

Finally it remains to show that φ_λ is Gâteaux differentiable and

$$(\partial\varphi)_\lambda = (\varphi_\lambda)'_G = \partial\varphi_\lambda$$

(see Theorem 3.73). We have

$$\begin{aligned}
&\tfrac{1}{2\lambda}\big(\|v - J_\lambda^{\partial\varphi}(v)\|_X^2 - \|u - J_\lambda^{\partial\varphi}(u)\|_X^2\big)\\
\geqslant\ &\tfrac{1}{\lambda}\big\langle \mathcal{F}(u - J_\lambda^{\partial\varphi}(u)), v - u + J_\lambda^{\partial\varphi}(u) - J_\lambda^{\partial\varphi}(v)\big\rangle\\
=\ &\langle (\partial\varphi)_\lambda(u), v - u\rangle + \big\langle (\partial\varphi)_\lambda(u), J_\lambda^{\partial\varphi}(u) - J_\lambda^{\partial\varphi}(v)\big\rangle \qquad \forall\, v \in X.
\end{aligned} \tag{3.115}$$

Since $(\partial\varphi)_\lambda(u) \in \partial\varphi(J_\lambda^{\partial\varphi}(u))$ (see Problem 2.161), we have

$$\varphi(J_\lambda^{\partial\varphi}(v)) - \varphi(J_\lambda^{\partial\varphi}(u)) \geqslant \langle(\partial\varphi)_\lambda(u), J_\lambda^{\partial\varphi}(v) - J_\lambda^{\partial\varphi}(u)\rangle \qquad \forall\, v \in X. \tag{3.116}$$

Therefore, using part **(b)** and (3.115), (3.116), we have

$$\varphi_\lambda(v) - \varphi_\lambda(u) \geqslant \langle(\partial\varphi)_\lambda(u), v-u\rangle \qquad \forall\, v \in X, \tag{3.117}$$

so

$$(\partial\varphi)_\lambda(u) \in \partial\varphi_\lambda(u)$$

and thus

$$(\partial\varphi)_\lambda(u) = \partial\varphi_\lambda(u)$$

(since $(\partial\varphi)_\lambda$ is maximal monotone; see Theorem 3.92). From (3.117) by choosing $v = u + th$ with $t \in \mathbb{R}$, $h \in X$, we conclude that φ_λ is Gâteaux differentiable and $(\varphi_\lambda)'_G(u) = \partial\varphi_\lambda(u) = (\partial\varphi)_\lambda(u)$ for all $u \in X$. This proves part **(a)** of the problem.

Solution of Problem 3.81

(a) From Problem 2.197, we know that A is maximal monotone (see Definition 2.100). It is clear that $\varphi \in \Gamma_0(H)$. Also, for all $u \in D(A)$ and all $v \in D(A^{\frac{1}{2}})$, we have

$$\begin{aligned}
(A(u), v)_H &= (A^{\frac{1}{2}}(u), A^{\frac{1}{2}}(v))_H \\
&= \tfrac{1}{2}(A^{\frac{1}{2}}(u-v), A^{\frac{1}{2}}(v-u))_H + \tfrac{1}{2}\|A^{\frac{1}{2}}(u)\|_H^2 + \tfrac{1}{2}\|A^{\frac{1}{2}}(v)\|_H^2 \\
&\leqslant \tfrac{1}{2}\|A^{\frac{1}{2}}(v)\|_H^2 + \tfrac{1}{2}\|A^{\frac{1}{2}}(u)\|_H^2
\end{aligned} \tag{3.118}$$

(recall that $D(A)$ is dense in $D(A^{\frac{1}{2}})$, $A^{\frac{1}{2}}$ is self-adjoint (see Definition I.5.108(b)) and positive since A is self-adjoint and positive). Then, we have

$$\begin{aligned}
(A(u), v-u)_H &\leqslant \tfrac{1}{2}\|A^{\frac{1}{2}}(v)\|_H^2 + \tfrac{1}{2}\|A^{\frac{1}{2}}(u)\|_H - (A(u), u)_H \\
&= \tfrac{1}{2}\|A^{\frac{1}{2}}(v)\|_H^2 + \tfrac{1}{2}\|A^{\frac{1}{2}}(u)\|_H - \big(A^{\frac{1}{2}}(u), A^{\frac{1}{2}}(u)\big)_H \\
&= \tfrac{1}{2}\|A^{\frac{1}{2}}(v)\|_H^2 - \tfrac{1}{2}\|A^{\frac{1}{2}}(u)\|_H^2
\end{aligned}$$

(see (3.118)), so $A(u) \in \partial\varphi(u)$. But as we already said A is maximal monotone. Therefore $A = \partial\varphi$.

(b) Suppose that $A\colon H \supseteq D(A) \longrightarrow H$ is linear densely defined and $A = \partial\psi$ for some $\psi \in \Gamma_0(H)$. From Problem 3.80(a) we know that

$$A_\lambda = \psi'_\lambda \qquad \forall\, \lambda > 0$$

(see Definition 2.122), so

$$\tfrac{d}{dt}\psi_\lambda(tu) = t\,(A_\lambda(u), u)_H \qquad \forall\, u \in H,\ t \in [0,1],$$

thus

$$\psi_\lambda(u) = \frac{1}{2}\left(A_\lambda(u), u\right)_H \qquad \forall\, u \in H,\ \lambda > 0.$$

Similarly as in the solution of Problem 3.19, we get

$$\psi_\lambda'(u) \;=\; \tfrac{1}{2}(A_\lambda + A_\lambda^*),$$

so

$$A_\lambda \;=\; \tfrac{1}{2}(A_\lambda + A_\lambda^*),$$

hence $A_\lambda = A_\lambda^*$ for all $\lambda > 0$. Let $\lambda \searrow 0$, to conclude that $A = A^*$ (see Proposition 2.123(f)).

Solution of Problem 3.82

From Proposition 2.123(d), we know that the map $(\partial\varphi)_\lambda\colon H \longrightarrow H$ is Lipschitz continuous with Lipschitz constant $\frac{1}{\lambda}$. From Problem 3.80, we have

$$(\varphi_\lambda)_G'(u) \;=\; (\partial\varphi)_\lambda(u) \qquad \forall\, u \in H.$$

Invoking Proposition 3.3, we conclude that φ_λ is Fréchet differentiable on H.

Solution of Problem 3.83

Clearly $f(u) \subseteq \mathcal{F}(u)$ for all $u \in L^1(\Omega)$ (see Definition 2.112). On the other hand, from the solution of Problem 3.54, we have

$$\partial\|u\|_1 \;=\; \left\{y \in L^\infty(\Omega) :\ y(\omega) \in \operatorname{sign} u(\omega) \text{ for } \mu\text{-a.a. } \omega \in \Omega\right\}.$$

Then using Proposition 3.132, we have

$$\partial h(u) \;=\; f(u) \qquad \forall\, u \in L^1(\Omega), \tag{3.119}$$

where $h(u) = \frac{1}{2}\|u\|_1^2$. From Theorem 3.92, we know that the map ∂h is maximal monotone, so f is maximal monotone (see (3.119)). Recall that $f(u) \subseteq \mathcal{F}(u)$ for all $u \in L^1(\Omega)$ and $\mathcal{F}$ is maximal monotone (see Proposition 2.114(a)). Therefore $\mathcal{F} = f$.

Solution of Problem 3.84

Given $\varepsilon > 0$ and $u \in H$, let $c \in C$ be such that

$$\|u - c\|_H \leqslant \operatorname{dist}(u, C) + \varepsilon. \tag{3.120}$$

Recall that ξ is increasing. So, for all $h \in H$, we have

$$\begin{aligned} (h, u)_H - \varphi(u) &= (h, u)_H - \xi(\operatorname{dist}(u, C)) \\ \leqslant\ & (h, u - c)_H + (h, c)_H - \xi(\|u - c\|_H - \varepsilon) \\ \leqslant\ & \|h\|_H (\|u - c\|_H - \varepsilon) + (h, c)_H - \xi(\|u - c\|_H - \varepsilon) + \varepsilon\|h\|_H \\ \leqslant\ & \xi^*(\|h\|_H) + \sigma_C(h) + \varepsilon\|h\|_H \end{aligned}$$

(see (3.120), Proposition 3.84(a) and Definition 2.46). Taking the supremum over $u \in H$ and letting $\varepsilon \searrow 0$, we obtain

$$\varphi^*(h) \leqslant \xi^*(\|h\|_H) + \sigma_C(h) \qquad \forall\, h \in H. \tag{3.121}$$

On the other hand, let $h \in H$ and $\vartheta, \eta \in \mathbb{R}$ be such that

$$\vartheta < \sigma_C(h) \quad \text{and} \quad \eta < \xi^*(\|h\|_H).$$

We can find $c \in C$ such that

$$\vartheta < (h, c)_H \tag{3.122}$$

(see Definition 2.46). Also, from the solution of Problem 3.56, we know that we can find $t > 0$ such that

$$\eta < t\|h\|_H - \xi(t).$$

Let $u = c + t\frac{h}{\|h\|_H}$ if $h \neq 0$ and $u = c$ if $h = 0$. We have

$$\eta < (h, u - c)_H - \xi(\|u - c\|_H) \leqslant (h, u - c)_H - \xi(\operatorname{dist}(u, C)) \tag{3.123}$$

(since ξ is increasing). Adding (3.122) and (3.123), we obtain

$$\vartheta + \eta \leqslant (h, u)_H - \xi(\operatorname{dist}(u, C)) = (h, u)_H - \varphi(u) \leqslant \varphi^*(h)$$

(see Proposition 3.84(a)). Letting $\vartheta \nearrow \sigma_C(h)$ and $\eta \nearrow \xi^*(\|h\|_H)$, we have

$$\xi^*(\|h\|_H) + \sigma_C(h) \leqslant \varphi^*(h). \tag{3.124}$$

From (3.121) and (3.124), we conclude that

$$\varphi^*(h) = \xi^*(\|h\|_H) + \sigma_C(h) \qquad \forall\, h \in H.$$

Solution of Problem 3.85

For every $h \in H$, we have

$$\begin{aligned} \varphi^*(h) &= \sup_{u\in H} \big((h,u)_H - \varphi(u) \big) \\ &= \sup_{u\in H} \big((h,u)_H - \xi(\|u\|_H) - \sigma_C(u) \big) \\ &= \sup_{t\geqslant 0} \big(\big(\sup_{\|u\|_H=t} ((h,u)_H - \sigma_C(u)) \big) - \xi(t) \big). \end{aligned} \tag{3.125}$$

By homogeneity, we have

$$\begin{aligned} \sup_{\|u\|_H=t} \big((h,u)_H - \sigma_C(u) \big) &= \sup_{\|u\|_H\leqslant t} \big((h,u)_H - \sigma_C(u) \big) \\ &= \big(\sigma_C(\cdot) + i_{t\overline{B}_1}(\cdot) \big)^*(h). \end{aligned}$$

Note that

$$\sigma_C = i_C^* \quad \text{and} \quad i_{t\overline{B}_1} = (t\|\cdot\|_H)^*.$$

Invoking Proposition 3.89(a), we have

$$\big(\sigma_C + i_{t\overline{B}_1} \big)^* = \big(i_C \,\Box\, t\|\cdot\|_H \big)^{**} = \big(t\,\mathrm{dist}(\cdot, C) \big)^{**} \tag{3.126}$$

(see Remark 3.88). Since the set C is convex, the map $u \longmapsto \mathrm{dist}(u,C)$ is continuous and convex. Then Theorem 3.85 implies that

$$(t\,\mathrm{dist}(\cdot, C))^{**} = t\,\mathrm{dist}(\cdot, C),$$

so

$$\big(\sigma_C + i_{t\overline{B}_1} \big)^* = t\,\mathrm{dist}(\cdot, C)$$

(see (3.126)), thus

$$\varphi^*(h) = \sup_{t\geqslant 0} \big(t\,\mathrm{dist}(h,C) - \xi(t) \big) = \xi^*(\mathrm{dist}(h,C))$$

(see (3.125)).

Solution of Problem 3.86

Let $u \in X$ and $h \in X \setminus \{0\}$. From the convexity of φ, we have

$$\frac{\varphi(u+h)-\varphi(u)}{\|h\|_X} \leqslant \frac{\varphi(u+nh)-\varphi(u)}{n\|h\|_X} \qquad \forall\, n \geqslant 1. \tag{3.127}$$

Note that

$$\frac{\varphi(u+nh)-\varphi(u)}{n\|h\|_X} \leqslant c\Big\| \frac{u}{n\|h\|_X} + \frac{h}{\|h\|_X} \Big\|_X - \frac{\varphi(u)}{n\|h\|_X},$$

so

$$\limsup_{n\to+\infty} \frac{\varphi(u+nh)-\varphi(u)}{n\|h\|_X} \leqslant c$$

and thus

$$\varphi(u+h) - \varphi(u) \leqslant c\|h\|_X \qquad \forall\, u, h \in X \tag{3.128}$$

(see (3.127)). From (3.128) and Theorem 3.60 we infer that φ is c-Lipschitz continuous.

Solution of Problem 3.87

Let X be an infinite dimensional separable Banach space. Let $\{u_n\}_{n\geqslant 1} \subseteq \partial B_1(0)$ be norm dense in $\partial B_1(0)$. We define

$$C = \overline{\text{conv}}\left(\{\tfrac{1}{n}u_n\}_{n\geqslant 1} \cup \{0\}\right).$$

Let $\varphi = i_C$. Then $0 \in \partial\varphi(0)$. Suppose that $u^* \in \partial\varphi(0)$. Then

$$\left\langle u^*, \tfrac{1}{n}u_n\right\rangle \leqslant \varphi(\tfrac{1}{n}u_n) - \varphi(0) = 0 \qquad \forall\, n \geqslant 1,$$

so

$$\langle u^*, u_n\rangle \leqslant 0 \qquad \forall\, n \geqslant 1.$$

The density of $\{u_n\}_{n\geqslant 1}$ in $\partial B_1(0)$ implies that

$$\langle u^*, u\rangle \leqslant 0 \qquad \forall\, u \in X,$$

so $u^* = 0$. Therefore $\partial\varphi(0) = \{0\}$. The set C has empty interior (being compact in an infinite dimensional Banach space), so by Theorem 3.60, φ is not continuous at $u = 0$. However, it is lower semicontinuous, convex (since the set $C \subseteq X$ is closed and convex). Therefore, because of Proposition 3.68, φ cannot be Gâteaux differentiable at $u = 0$.

Solution of Problem 3.88

Recall that $\text{dist}(u, C) = \inf\limits_{c\in C} \|u - c\|_X$. We have

$$\text{dist}(u, C) = \inf_{c\in X} \left(\|u - c\|_X + i_C(c)\right) = \left(\|\cdot\|_X \boxdot i_C\right)(u)$$

(see Definition 3.87). From Proposition 3.89(a), we know that

$$\text{dist}(\cdot, C)^* = \left(\|\cdot\|_X \boxdot i_C\right)^* = \|\cdot\|_X^* + i_C^* = i_{\overline{B}_1(0)} + \sigma_C.$$

Then

$$\operatorname{dist}(\cdot, C)^{**}|_X(u) = \sup\{\langle u^*, u\rangle - \sigma_C(u^*) : \|u^*\|_* \leqslant 1\}$$

(see Definition 3.83). But since the set C is closed and convex, the function $u \longmapsto \operatorname{dist}(u, C)$ is continuous and convex. Therefore, Proposition 3.84(f) implies that

$$\operatorname{dist}(\cdot, C)^{**}|_X = \operatorname{dist}(\cdot, C),$$

so

$$\operatorname{dist}(u, C) = \sup\{\langle u^*, u\rangle - \sigma_C(u^*) : \|u^*\|_* \leqslant 1\} \qquad \forall\, u \in X.$$

Solution of Problem 3.89

Clearly φ is convex. From the growth condition on g, the continuity of the trace map γ (see Definition 1.129) and the Lebesgue dominated convergence theorem, we infer that φ is continuous. Let $\psi\colon L^p(\partial\Omega) \longrightarrow \mathbb{R}$ be defined by

$$\psi(y) = \int_{\partial\Omega} g(y(z))\, d\sigma \qquad \forall\, y \in L^p(\partial\Omega).$$

We have

$$\varphi(u) = (\psi \circ \gamma)(u) \qquad \forall\, u \in W^{1,p}(\Omega).$$

Since $\gamma \in \mathcal{L}(W^{1,p}(\Omega), L^p(\partial\Omega))$, from Proposition 3.132, we have

$$\partial\varphi(u) = \gamma^*\partial\psi(\gamma(u)) \qquad \forall\, u \in W^{1,p}(\Omega). \tag{3.129}$$

From Theorem 3.111, we know that

$$\partial\psi(\gamma(u)) = S^{p'}_{\partial g(\gamma(u)(\cdot))}. \tag{3.130}$$

Also, $\gamma^*\colon L^{p'}(\partial\Omega) \longrightarrow W^{1,p}(\Omega)^*$ is the adjoint of the trace map and so it is defined by

$$\langle\gamma^*(h), v\rangle = \int_{\partial\Omega} h(z)\gamma(v)(z)\, d\sigma \qquad \forall\, v \in W^{1,p}(\Omega),\ h \in L^{p'}(\partial\Omega).$$

Therefore, from (3.129) and (3.130), we have that: "$u^* \in \partial\varphi(u)$ if and only if there exits $h \in L^{p'}(\partial\Omega)$ with $h(z) \in \partial g(\gamma(u)(z))$ for σ-almost all $z \in \partial\Omega$ and

$$\langle u^*, v\rangle = \int_{\partial\Omega} h(z)\gamma(v)(z)\, d\sigma \qquad \forall\, v \in W^{1,p}(\Omega)."$$

Solution of Problem 3.90

We need to show that for every $\lambda \geqslant 0$, the set $\varphi^\lambda = \{u \in L^p(\Omega) : \varphi(u) \leqslant \lambda\}$ is closed (see Proposition I.2.53). To this end, let $\{u_n\}_{n\geqslant 1} \subseteq \varphi^\lambda$ be a sequence such that $u_n \longrightarrow u$ in $L^p(\Omega)$. Evidently $\{u_n\}_{n\geqslant 1} \subseteq W^{1,p}(\Omega)$ and it is bounded. So, we may assume that

$$u_n \xrightarrow{w} u \quad \text{in } W^{1,p}(\Omega),$$

so in particular

$$Du_n \xrightarrow{w} Du \quad \text{in } L^p(\Omega;\mathbb{R}^N)$$

and thus

$$\|Du\|_p^p \leqslant \liminf_{n\to+\infty} \|Du_n\|_p^p$$

(from the weak lower semicontinuity of the norm functional; see Proposition I.5.56(c)). Hence $\|Du\|_p^p \leqslant \lambda$, so $u \in \varphi^\lambda$ and thus the set φ^λ is closed. This proves the desired lower semicontinuity of φ.

Solution of Problem 3.91

From Definition 3.112 (see also Remark 3.113), we know that

$$T_{\partial\varphi(u)}(0)^- = N_{\partial\varphi(u)}(0).$$

Hence

$$h \in N_{\partial\varphi(u)}(0) \iff \left[(h,v)_H \leqslant 0 \ \ \forall v \in T_{\partial\varphi(u)}(0)\right],$$

so

$$h \in N_{\partial\varphi(u)}(0) \iff \left[(h,v)_H \leqslant 0 \ \ \forall v \in \partial\varphi(u)\right] \tag{3.131}$$

(see Definition 3.112). Since φ is continuous (see Proposition 3.63), from (3.131), Theorem 3.71 and since u is a minimizer of φ, we have

$$h \in N_{\partial\varphi(u)}(0) \iff \varphi'(u;h) = 0. \tag{3.132}$$

Recall that

$$\varphi(u+h) = \varphi(u) + \varphi'(u;h) + e(\|h\|_H), \tag{3.133}$$

with $\frac{e(\|h\|_H)}{\|h\|_H} \longrightarrow 0$ as $h \to 0$. Then from (3.132) and (3.133), it follows that given $\varepsilon > 0$, we can find $\delta > 0$ such that

$$\varphi(u+h) \leqslant \varphi(u) + \varepsilon\|h\|_H \qquad \forall\, h \in N_{\partial\varphi(u)}(0),\ \|h\|_H \leqslant \delta.$$

On the other hand, if $h \in H \setminus N_{\partial\varphi(u)}(0)$, then from (3.131) and Theorem 3.71, we have that $\varphi'(u;h) > 0$. Since $\varphi(u+h) - \varphi(u) \geqslant \varphi'(u;h)$, we conclude that $\varphi(u+h) > \varphi(u)$.

Solution of Problem 3.92

The condition $0 \in \operatorname{int} \partial\varphi(u)$ is equivalent to saying that there exists $\varepsilon > 0$ such that $B^*_\varepsilon(0) \subseteq \partial\varphi(u)$ (where $B^*_\varepsilon(0)=\{u^* \in X^* : \|u^*\|_*<\varepsilon\}$), so

$$\sigma_{B^*_\varepsilon(0)} \leqslant \sigma_{\partial\varphi(u)},$$

thus

$$\varepsilon\|\cdot\|_X \leqslant \varphi'(u;\cdot) \quad \text{on } X$$

(see Theorem 3.71) and hence

$$\varepsilon\|h\|_X \leqslant \varphi(u+h)-\varphi(u) \qquad \forall\, h \in X.$$

Solution of Problem 3.93

Evidently $M(u) \neq 0$ and it is compact in T for every $u \in \mathbb{R}^N$. Let

$$D = \operatorname{conv} \bigcup_{\lambda\in M(u)} \partial\varphi_\lambda(u).$$

First we show that

$$D \subseteq \partial\varphi(u) \tag{3.134}$$

Let $\lambda \in M(u)$ and $u^* \in \partial\varphi_\lambda(u)$. Then

$$(u^*, v-u)_{\mathbb{R}^N} \leqslant \varphi_\lambda(v)-\varphi_\lambda(u) \leqslant \varphi(v)-\varphi(u) \qquad \forall\, v \in \mathbb{R}^N,$$

so $u^* \in \partial\varphi(u)$ and thus

$$\partial\varphi_\lambda(u) \subseteq \partial\varphi(u) \qquad \forall\, \lambda \in M(u).$$

The set $\partial\varphi(u)$ is compact and convex (see Theorem 3.71 and Proposition 3.64) and so (3.134) follows.

Let

$$D_0 = \bigcup_{\lambda\in M(u)} \partial\varphi_\lambda(u).$$

Then $D = \operatorname{conv} D_0$. From (3.134) we see that the set D_0 is bounded. Let $\{u_n^*\}_{n\geqslant 1} \subseteq D_0$ be a sequence such that $u_n^* \longrightarrow u^*$ in $\mathbb{R}^N$. For each $n \geqslant 1$, there exists $\lambda_n \in M(u)$ such that $u_n^* \in \partial\varphi_{\lambda_n}(u)$ and

$$(u^*, v-u)_{\mathbb{R}^N} \leqslant \varphi_{\lambda_n}(v)-\varphi_{\lambda_n}(u) \qquad \forall\, v \in \mathbb{R}^N. \tag{3.135}$$

Because of the compactness of T, passing to a subsequence if necessary, we may assume that $\lambda_n \longrightarrow \lambda \in M(u)$. We have

$$\varphi_{\lambda_n}(u) = \varphi(u) = \varphi_\lambda(u) \tag{3.136}$$

and

$$\limsup_{n\to+\infty} \varphi_{\lambda_n}(v) \leqslant \varphi_\lambda(v) \tag{3.137}$$

(by hypothesis). So, if in (3.135) we pass to the limit as $n \to +\infty$ and use (3.136) and (3.137), we obtain

$$(u^*, v-u) \leqslant \varphi_\lambda(v) - \varphi_\lambda(u) \qquad \forall\, v \in \mathbb{R}^N,$$

so

$$u^* \in \partial\varphi_\lambda(u) \quad \text{with } \lambda \in M(u),$$

thus $u^* \in D_0$ and hence the set D_0 is closed. Therefore the set D_0 is compact and then so is $D = \operatorname{conv} D_0$.

For all $h \in \mathbb{R}^N$, we have

$$\varphi'(u;h) \leqslant \sigma_D(h) = \sup_{\lambda\in M(u)} \varphi'_\lambda(u;h). \tag{3.138}$$

Given $\varepsilon > 0$, we have

$$\tfrac{\varphi(u+th)-\varphi(u)}{t} > \varphi'(u;h) - \varepsilon \qquad \forall\, t > 0. \tag{3.139}$$

For $t > 0$, we define

$$T_t = \big\{\lambda \in T : \tfrac{\varphi_\lambda(u+th)-\varphi(u)}{\lambda} \geqslant \varphi'(u;h) - \varepsilon\big\}.$$

Evidently $T_t \neq \emptyset$ and it is compact. Note that

$$\tfrac{\varphi_\lambda(u+th)-\varphi(u)}{\lambda} = \tfrac{\varphi_\lambda(u+th)-\varphi_\lambda(u)}{\lambda} + \tfrac{\varphi_\lambda(u)-\varphi(u)}{\lambda},$$

so

$$T_{t_1} \subseteq T_{t_2} \quad \text{if } t_1 \leqslant t_2,$$

thus

$$\bigcap_{t>0} T_t \neq \emptyset.$$

Let $\lambda_* \in \bigcap\limits_{t>0} T_t \neq \emptyset$. We have

$$t\left(\varphi'(u;h) - \varepsilon\right) \leqslant \varphi_{\lambda_*}(u+th) - \varphi(u), \tag{3.140}$$

so $\lambda_* \in M(u)$ (just let $t \searrow 0$ and use the continuity of φ_{λ_*}). From (3.140), we obtain

$$\varphi'(u;h) - \varepsilon \leqslant \tfrac{\varphi_{\lambda_*}(u+th)-\varphi_{\lambda_*}(u)}{t} \qquad \forall\, t > 0,$$

so

$$\varphi'(u;h) - \varepsilon \leqslant \varphi'_{\lambda_*}(u;h) \qquad \forall\, h \in \mathbb{R}^N,\ \varepsilon > 0,$$

thus

$$\varphi'(u;h) \leqslant \varphi'_{\lambda_*}(u;h) \qquad \forall\, h \in \mathbb{R}^N,$$

hence

$$\sigma_{\partial\varphi(u)}(h) \leqslant \sigma_D(h) \qquad \forall\, h \in \mathbb{R}^N$$

and finally we get $\partial\varphi(u) \subseteq D$ (see Theorem 3.71). From this and (3.134), we conclude that $\partial\varphi(u) = D$.

Solution of Problem 3.94

Arguing by contradiction, suppose that there is a compact set $K \subseteq \mathbb{R}^N$, a sequence $\{u_n\}_{n\geqslant 1} \subseteq K$ and $\varepsilon > 0$ such that

$$\left|\varphi_n'(u_n) - \varphi'(u_n)\right| \geqslant \varepsilon \qquad \forall\, n \geqslant 1. \tag{3.141}$$

Because of the compactness of K, by passing to a suitable subsequence if necessary, we may assume that $u_n \longrightarrow u$ in $\mathbb{R}^N$. Since $\{\varphi_n'(u_n)\} = \partial\varphi_n(u_n)$ for all $n \geqslant 1$ and $\{\varphi'(u)\} = \partial\varphi(u)$ (see Theorem 3.73), from Problem 3.78(b), we have

$$\varphi_n'(u_n) \longrightarrow \varphi'(u) \quad \text{and} \quad \varphi'(u_n) \longrightarrow \varphi'(u) \quad \text{in } \mathbb{R}^N. \tag{3.142}$$

So, if in (3.141) we pass to the limit as $n \to +\infty$ and use (3.142), then we obtain $0 \geqslant \varepsilon$, a contradiction. Therefore, $\varphi_n' \longrightarrow \varphi'$ uniformly on compact subsets of $\mathbb{R}^N$.

Solution of Problem 3.95

"(a) $\Longrightarrow$ (b)": Let $v \in C$ and note that

$$u_0 + t(v - u_0) \in C \qquad \forall\, t \in [0,1].$$

We have

$$\varphi(u_0) \leqslant \varphi\left(u_0 + t(v - u_0)\right),$$

so

$$0 \leqslant \tfrac{\varphi(u_0+t(v-v_0))-\varphi(u_0)}{t} \qquad \forall\, t \in (0,1],$$

thus

$$0 \leqslant \varphi'(u_0, v - u_0) \qquad \forall\, v \in C.$$

Note that $v - u_0 \in T_C(u_0)$ (see Definition 3.112). From the positive homogeneity and from the continuity of $\varphi'(u_0;\cdot)$, we infer that

$$\varphi'(u_0;h) \geqslant 0 \qquad \forall\, h \in T_C(u_0).$$

"(b) $\Longrightarrow$ (a)": For any $v \in C$, we have $v - u_0 \in T_C(u_0)$ and so

$$0 \leqslant \varphi'(u_0; v - u_0) \leqslant \varphi(v) - \varphi(u_0)$$

(recall that the map $t \longmapsto \tfrac{\varphi(u_0+t(v-u_0))-\varphi(u_0)}{t}$ is increasing on $(0, +\infty)$), so

$$\varphi(u_0) \leqslant \varphi(v) \qquad \forall\, v \in C$$

and thus $u_0 \in C$ is a solution of problem (P).

"(b) $\Longleftrightarrow$ (c)": We have

$$\varphi'(u_0;h) + i_{T_C(u_0)}(h) \geqslant 0 \qquad \forall\, h \in X. \tag{3.143}$$

But since $T_C(u_0)$ is a closed, convex cone, we have

$$i_{T_C(u_0)}(h) = \sigma_{T_C(u_0)^-}(h) = \sigma_{N_C(u_0)}(h) \tag{3.144}$$

(see Definition 3.112). Also, from Theorem 3.71, we have

$$\varphi'(u_0;h) = \sigma_{\partial\varphi(u_0)}(h). \tag{3.145}$$

Returning to (3.143) and using (3.144) and (3.145), we obtain

$$\begin{aligned} 0 &\leqslant \sigma_{N_C(u_0)}(h) + \sigma_{\partial\varphi(u_0)}(h) \\ &= \sigma_{\partial\varphi(u_0)+N_C(u_0)}(h) \qquad \forall\, h \in X. \end{aligned} \tag{3.146}$$

But the set $\partial\varphi(u_0)$ is w^*-compact, convex (see Theorem 3.71) and the set $N_C(u_0) = T_C(u_0)^-$ is a closed convex cone (see Definition 3.112). So, $\partial\varphi(u_0) + N_C(u_0)$ is w^*-closed and convex. Then from (3.146) we conclude that $0 \in \partial\varphi(u_0) + N_C(u_0)$.

Solution of Problem 3.96

We set $\varphi|_{\mathbb{R}\setminus I} = +\infty$. Then for all $h \in \mathbb{R}$, we have

$$\tfrac{\varphi(u+th)-\varphi(u)}{t} \leqslant \varphi(u+h) - \varphi(u) \qquad \forall\, t \in (0,1],$$

so

$$\tfrac{\varphi(u+th)-\varphi(u)}{th} h \leqslant \varphi(u+h) - \varphi(u) \qquad \forall\, t \in (0,1],$$

thus

$$\varphi'_+(u)h \leqslant \varphi(u+h) - \varphi(u)$$

and hence $\varphi'_+(u) \in \partial\varphi(u)$. Similarly, for all $h \in \mathbb{R}$, we have

$$\tfrac{\varphi(u+th)-\varphi(u)}{t} \geqslant \varphi(u) - \varphi(u-h) \qquad \forall\, t \in [-1,0),$$

so

$$\varphi'_-(u)(-h) \leqslant \varphi(u-h) - \varphi(u)$$

thus $\varphi'_-(u) \in \partial\varphi(u)$. From Theorem 3.71, we have that

$$\partial\varphi(u) = \left[\varphi'_-(u), \varphi'_+(u)\right],$$

so $s(u) \in \left[\varphi'_-(u), \varphi'_+(u)\right]$ and s is nondecreasing.

Solution of Problem 3.97

The convexity of φ implies that

$$\varphi(u) \leqslant \varphi(a) + \tfrac{\varphi(b)-\varphi(a)}{b-a}(u-a) \qquad \forall\, u \in [a,b]$$

(use the three slopes inequality), so

$$\int_a^b \varphi(u)\,du \leqslant \varphi(a)(b-a) + \tfrac{1}{2}\left(\varphi(b)-\varphi(a)\right)(b-a),$$

thus

$$\tfrac{1}{b-a}\int_0^b \varphi(u)\,du \leqslant \tfrac{\varphi(a)+\varphi(b)}{2}.$$

Also, we have

$$\tfrac{1}{b-a}\int_0^b \varphi(u)\,du = \tfrac{1}{b-a}\Big(\int_a^{\frac{a+b}{2}} \varphi(u)\,du + \int_{\frac{a+b}{2}}^b \varphi(u)\,du\Big)$$

$$= \tfrac{1}{2}\int_0^1 \Big(\Big(\varphi\big(\tfrac{a+b-t(b-a)}{2}\big) + \varphi\big(\tfrac{a+b+t(b-a)}{2}\big)\Big)\Big)\,dt \geqslant \varphi\big(\tfrac{a+b}{2}\big)$$

(using the convexity of φ).

Solution of Problem 3.98

We argue by contradiction. So, suppose that φ is not constant. Let $u_0 \in \operatorname{int} U$ be the global maximizer of φ and choose $u \in U$ such that $\varphi(u) < \varphi(u_0)$. Take $t \in (0,1)$ small such that

$$h = u_0 + t(u_0 - u) \in U.$$

Then $u_0 = \frac{1}{1+t}(h + tu)$ and so from the convexity of φ we have

$$\varphi(u_0) \leqslant \tfrac{1}{1+t}\varphi(h) + \tfrac{t}{1+t}\varphi(u) < \tfrac{1}{1+t}\varphi(h) + \tfrac{t}{1+t}\varphi(h) = \varphi(h),$$

a contradiction.

Solution of Problem 3.99

Let $M(\varphi)$ be the set of global maximizers of φ. Evidently $M(\varphi) \subseteq K$ is nonempty, compact, and extremal (because of the convexity of φ). By the Krein–Milman theorem (see Theorem I.5.128), it has an extreme point, which is also an extreme point of K.

Solution of Problem 3.100

"$\Longrightarrow$": We assume that φ is continuous at u_0. Then according to Theorem 3.60, we can find $\varepsilon > 0$ and $M > 0$ such that

$$\varphi(u) \leqslant M \qquad \forall\, u \in \overline{B}_\varepsilon(u_0), \tag{3.147}$$

where $\overline{B}_\varepsilon(u_0) = \{h \in X : \ \|h - u_0\|_X \leqslant \varepsilon\}$.

Let $u^{**} \in \overline{B}_\varepsilon^{**}(0) = \{h^{**} \in X^{**} : \ \|h^{**} - u_0\|_{X^{**}} \leqslant \varepsilon\}$. By the Goldstine theorem (see Theorem I.5.70), we know that we can find a net $\{h_\alpha\}_{\alpha \in J} \subseteq \overline{B}_\varepsilon(0)$ such that $h_\alpha \xrightarrow{w^*} u^{**} - u_0$ in X^{**}. Recall that φ^{**} is w^*-lower semicontinuous. Therefore, we have

$$\varphi^{**}(u^{**}) \leqslant \liminf_{\alpha \in J} \varphi^{**}(h_\alpha) = \liminf_{\alpha \in J} \varphi(h_\alpha) \leqslant M$$

(recall that $\varphi^{**}|_X = \varphi$ (see Theorem 3.85) and see (3.147)), so $\varphi^{**}|_{\overline{B}_\varepsilon^{**}(u_0)}$ is bounded above and hence φ^{**} is continuous at u_0 (see Theorem 3.60).

"$\Longleftarrow$": Suppose that φ^{**} is continuous at $u_0 \in X$. From Theorem 3.85, we know that $\varphi^{**}|_X = \varphi$ and so φ is continuous at $u_0 \in X$.

Solution of Problem 3.101

"$\Longrightarrow$": Suppose that φ is bounded on bounded sets of X. So, we have

$$\varphi(u) \leqslant M \qquad \forall\, u \in \overline{B}_r(0),$$

where $\overline{B}_r(0) = \{h \in X : \ \|h\|_X \leqslant r\}$, so

$$\varphi^{**}(u) \leqslant M \qquad \forall\, u \in \overline{B}_r(0)$$

(see Theorem 3.85). Recall that φ^{**} is w^*-lower semicontinuous. So, if

$$(\varphi^{**})^M = \{h^{**} \in X^{**} : \varphi^{**}(h^{**}) \leqslant M\},$$

then $(\varphi^{**})^M$ is w^*-closed and $\overline{B}_r(0) \subseteq (\varphi^{**})^M$. By the Goldstine theorem (see Theorem I.5.70), we have that

$$\overline{B}_r^{**}(0) = \{h^{**} \in X^{**} : \|h^{**}\|_{X^{**}} \leqslant r\} \subseteq (\varphi^{**})^M,$$

so $\varphi^{**}|_{\overline{B}_r^{**}(0)} \leqslant M$.

"$\Longleftarrow$": Suppose that φ^{**} is bounded on bounded subsets of X^{**}. From Theorem 3.85, we have that $\varphi^{**}|_X = \varphi$. So, φ also bounded on bounded subsets of X.

Solution of Problem 3.102

"$\Longrightarrow$": Suppose that φ is Fréchet differentiable at $u_0 \in X$. From Proposition 3.74, we know that given $\varepsilon > 0$, we can find $\delta > 0$ such that

$$\varphi(u_0 + h) + \varphi(u_0 - h) - 2\varphi(u_0) \leqslant \varepsilon\|h\|_X \qquad \forall\ \|h\|_X \leqslant \delta. \tag{3.148}$$

Let $h^{**} \in X^{**}$ be such that $\|h^{**}\|_{X^{**}} \leqslant \delta$. By the Goldstine theorem (see Theorem I.5.70), we can find a net $\{h_\alpha\}_{\alpha\in J} \subseteq X$ such that $\|h_\alpha\|_X = \|h^{**}\|_{X^{**}} \leqslant \delta$ and $h_\alpha \xrightarrow{w^*} h^{**}$ in X^{**}. Since φ^{**} is w^*-lower semicontinuous and $\varphi^{**}|_X = \varphi$ (see Theorem 3.85), we have

$$\begin{aligned}
&\varphi^{**}(u_0 + h^{**}) + \varphi^{**}(u_0 - h^{**}) - 2\varphi^{**}(u_0) \\
\leqslant\ & \liminf_{\alpha\in J} \varphi^{**}(u_0 + h_\alpha) + \liminf_{\alpha\in J} \varphi^{**}(u_0 - h_\alpha) - 2\varphi^{**}(u_0) \\
=\ & \liminf_{\alpha\in J} \varphi(u_0 + h_\alpha) + \liminf_{\alpha\in J} \varphi(u_0 - h_\alpha) - 2\varphi(u_0) \\
\leqslant\ & \varepsilon\|h_\alpha\|_X = \varepsilon\|h^{**}\|_{X^{**}}
\end{aligned}$$

(see (3.148)), so φ^{**} is Fréchet differentiable at u_0 (see Proposition 3.74).

"$\Longleftarrow$": Suppose that φ^{**} is Fréchet differentiable at $u_0 \in X$. Since $\varphi^{**}|_X = \varphi$ (see Theorem 3.85), we infer that φ is also Fréchet differentiable at $u_0 \in X$.

Solution of Problem 3.103

According to Problem 3.47, the coercivity of φ (see Definition 3.105(a)) is equivalent to saying that there exist $c > 0$ and $b \in \mathbb{R}$ such that

$$\varphi(u) \geqslant c\|u\|_X + b \qquad \forall\ u \in X. \tag{3.149}$$

Problem 3.70 says that (3.149) is equivalent to

$$\varphi^*(u^*) \leqslant -b \qquad \forall\, u^* \in \overline{B}_c^*(0), \tag{3.150}$$

where $\overline{B}_c^*(0) = \{h^* \in X^* : \|h^*\|_* \leqslant c\}$. Finally, Theorem 3.60 says that (3.150) is equivalent to the continuity of φ^* at the origin.

Solution of Problem 3.104

From Problem 3.100 we know that φ is continuous at the origin if and only if φ^{**} is continuous at the origin. On the other hand, from Problem 3.103 we know that φ^{**} is continuous at the origin if and only if φ^* is coercive. From Problem 3.47 we know that φ^* is coercive if and only if the sublevel sets $(\varphi^*)^\lambda$ (for $\lambda \in \mathbb{R}$) are bounded (where $(\varphi^*)^\lambda = \{u^* \in X^* : \varphi^*(u^*) \leqslant \lambda\}$). Since $\varphi^* \colon X^* \longrightarrow \overline{\mathbb{R}}$ is proper and w^*-lower semicontinuous, the sets $(\varphi^*)^\lambda$ are w^*-closed (see Proposition I.2.53), hence by the Alaoglu theorem (see Theorem I.5.66), also w^*-compact.

Solution of Problem 3.105

(a) "$\Longrightarrow$": Suppose that φ is strongly coercive (see Definition 3.105(b)). Then given $\eta > 0$, we can find $M > 0$ such that

$$\varphi(u) \geqslant \eta\|u\|_X \qquad \forall\, \|u\|_X \geqslant M. \tag{3.151}$$

Also, we can find $\vartheta > 0$ such that

$$\varphi(u) \geqslant -\vartheta \qquad \forall\, \|u\|_X \leqslant M \tag{3.152}$$

(see Problem 3.47). From (3.151) and (3.152) it follows that

$$\varphi(u) \geqslant \eta\|u\|_X - \widehat{\vartheta} \qquad \forall\, u \in X, \tag{3.153}$$

for some $\widehat{\vartheta} > 0$. From (3.153) and Problem 3.70, it follows that $\varphi^*|_{\overline{B}_\eta^*(0)} \leqslant \widehat{\vartheta}$. Since $\eta > 0$ is arbitrary, we conclude that φ^* is bounded on bounded sets.

"$\Longleftarrow$": Suppose that φ^* is bounded on bounded sets. For any $r > 0$, there exists $M_r > 0$ such that

$$\varphi^*(u^*) \leqslant M_r \qquad \forall\, u^* \in \overline{B}_r^*(0),$$

so

$$\varphi(u) \geqslant r\|u\|_X + b \qquad \forall\, u \in X,$$

for some $b \in \mathbb{R}$ (see Problem 3.70), thus

$$\liminf_{\|u\|_X \to +\infty} \frac{\varphi(u)}{\|u\|_X} \geqslant r.$$

But $r > 0$ is arbitrary. Therefore we conclude that φ is strongly coercive.

(b) From part **(a)** we know that φ^* is strongly coercive if and only if φ^{**} is bounded on bounded sets (recall that $\varphi^* \in \Gamma_0(X^*)$; see Definition 3.83). But from Problem 3.101 we know that φ^{**} is bounded on bounded sets if and only if φ is bounded on bounded sets. So, finally we can say that φ^* is strongly coercive if and only if φ is bounded on bounded sets.

Solution of Problem 3.106

Translating C if necessary, we see that without any loss of generality, we may assume that $0 \in C$. Let $r, M > 0$ be such that

$$C \subseteq \overline{B}_r(0) \quad \text{and} \quad \varphi|_{C_\delta} \leqslant M \tag{3.154}$$

(recall that C is bounded and use the hypothesis). Let $u \in C \setminus \{0\}$ and let $\lambda = \|u\|_X$. Then the convexity of φ implies that

$$\varphi(0) \leqslant \tfrac{\lambda}{\lambda+\delta}\varphi\left(-\tfrac{\delta}{\lambda}u\right) + \tfrac{\delta}{\lambda+\delta}\varphi(u) \leqslant M + \tfrac{\delta}{\lambda+\delta}\varphi(u),$$

so

$$\tfrac{\lambda+\delta}{\delta}\left(\varphi(0) - M\right) \leqslant \varphi(u),$$

thus

$$\tfrac{r+\delta}{\delta}\left(\varphi(0) - M\right) \leqslant \varphi(u)$$

(since $\lambda = \|u\|_X \leqslant r$ and $\varphi(0) \leqslant M$), hence $\varphi|_C$ is bounded below and finally

$$\varphi(u) \geqslant \vartheta \qquad \forall\, u \in C, \tag{3.155}$$

for some $\vartheta \leqslant 0$. Let $u, v \in C$, $u \neq v$ and set $\lambda = \|v - u\|_X$. Then

$$v = \tfrac{\delta}{\lambda+\delta}u + \tfrac{\lambda}{\lambda+\delta}\left(v + \tfrac{\delta}{\lambda}(v-u)\right),$$

so

$$\varphi(v) \leqslant \tfrac{\delta}{\lambda+\delta}\varphi(u) + \tfrac{\lambda}{\lambda+\delta}M$$

(from the convexity of φ and (3.154)), thus

$$\varphi(v) - \varphi(u) \leqslant \tfrac{\lambda}{\lambda+\delta}(M - \varphi(u)) \leqslant \tfrac{\lambda}{\lambda+\delta}(M - \vartheta) \leqslant \tfrac{M-\vartheta}{\delta}\|v-u\|_X$$

(see (3.155) and recall that $\lambda = \|v - u\|_X$). If we reverse the roles of u and v in the above argument, we reach again the same inequality. Therefore we conclude that $\varphi|_C$ is Lipschitz continuous with Lipschitz constant $\frac{M-\vartheta}{\delta} > 0$.

Solution of Problem 3.107

From Proposition 3.89, we have

$$(\varphi \boxdot n\|\cdot\|_X^2)^* = \varphi^* + \tfrac{1}{2n}\|\cdot\|_*^2 \quad\text{and}\quad (\varphi \boxdot n\|\cdot\|_X)^* = \varphi^* + i_{\overline{B}_n^*(0)}$$

(see Definition 3.87). Hence

$$(\varphi \boxdot n\|\cdot\|_X^2)^* \longrightarrow \varphi^* \quad\text{and}\quad (\varphi \boxdot n\|\cdot\|_X)^* \longrightarrow \varphi^*$$

uniformly on bounded subsets of $\operatorname{dom}\varphi^*$. Moreover, note that

$$\varphi \boxdot n\|\cdot\|_X^2 \leqslant \varphi \quad\text{and}\quad \varphi \boxdot n\|\cdot\|_X \leqslant \varphi \qquad \forall\, n \geqslant 1.$$

Invoking Problem 3.61, we conclude that

$$\varphi \boxdot n\|\cdot\|_X^2 \longrightarrow \varphi \quad\text{and}\quad \varphi \boxdot n\|\cdot\|_X \longrightarrow \varphi$$

uniformly on bounded subsets of X.

Solution of Problem 3.108

By definition (see Definition 2.46), we have

$$\sigma_C(u^*) = \sup_{c\in C} \langle u^*, c\rangle \qquad \forall\, u^* \in X^*.$$

So, σ_C is the supremum of continuous convex (in fact linear) functions, hence it is lower semicontinuous, convex. We have $\sigma_C(0) = 0$ and so σ_C is proper. For every $u^* \in X^*$ and $t > 0$ we have

$$\sigma_C(tu^*) = \sup_{c\in C} t\,\langle u^*, c\rangle = t\sup_{c\in C}\langle u^*, c\rangle = t\sigma_C(u^*),$$

so σ_C is positively homogeneous of degree 1.

Now, suppose that the set $C \subseteq X$ is w-compact. Suppose that $\{u_n\}_{n\geqslant 1} \subseteq X^*$ and $\{c_n\}_{n\geqslant 1} \subseteq C$ are two sequences such that $u_n^* \longrightarrow u^*$ in X^* and $\sigma_C(u_n^*) = \langle u_n^*, c_n\rangle$ for all $n \geqslant 1$. Since the set C is w-compact and $\{c_n\}_{n\geqslant 1} \subseteq C$, then from the Eberlein–Smulian theorem (see Theorem I.5.78), we can find a subsequence $\{c_{n_k}\}_{k\geqslant 1}$ of $\{c_n\}_{n\geqslant 1}$ such that $c_{n_k} \xrightarrow{w} c \in C$. Then

$$\sigma_C(u_{n_k}^*) = \left\langle u_{n_k}^*, c_{n_k}\right\rangle \longrightarrow \langle u^*, c\rangle \leqslant \sigma_C(u^*). \tag{3.156}$$

On the other hand, from the lower semicontinuity of σ_C, we have

$$\sigma_C(u^*) \leqslant \liminf_{k\to+\infty} \sigma_C(u_{n_k}^*). \tag{3.157}$$

From (3.156) and (3.157), we have (for the original sequence)

$$\sigma_C(u_n^*) \longrightarrow \sigma_C(u^*),$$

so the function $u^* \longmapsto \sigma_C(u^*)$ is continuous.

Solution of Problem 3.109

Let $u, v \in X$ and $\lambda \in (0,1)$. Without any loss of generality, we may assume that $u, v \in \operatorname{dom} m$. Let $m(u) < \eta$ and $m(v) < \vartheta$. Then we can find $y_{u,\eta}, y_{v,\vartheta} \in Y$ such that

$$\xi(u, y_{u,\eta}) \leqslant \eta \quad \text{and} \quad \xi(v, y_{v,\vartheta}) \leqslant \vartheta. \tag{3.158}$$

Since ξ is convex, we have

$$\begin{aligned} & \xi((1+\lambda)u + \lambda v, (1-\lambda)y_{u,\eta} + \lambda y_{v,\vartheta}) \\ \leqslant\ & (1-\lambda)\xi(u, y_{u,\eta}) + \lambda\xi(v, y_{v,\vartheta}) \\ \leqslant\ & (1-\lambda)\eta + \lambda\vartheta \end{aligned}$$

(see (3.158)), so

$$m\left((1-\lambda)u + \lambda v\right) \leqslant (1-\lambda)\eta + \lambda\vartheta$$

(by the definition of m). Let $\eta \searrow m(u)$ and $\vartheta \searrow m(v)$, to obtain

$$m\left((1-\lambda)u + \lambda v\right) \leqslant (1-\lambda)m(u) + \lambda m(v),$$

so the function m is convex.

Solution of Problem 3.110

For every $\lambda > 0$, the map $h \longmapsto q_\lambda(h)$ is lower semicontinuous and convex. Note that

$$\operatorname{epi} \varphi^\infty = \bigcap_{\lambda>0} \operatorname{epi} q_\lambda,$$

so the set $\operatorname{epi} \varphi^\infty$ is closed and convex. Also, clearly $\varphi^\infty(0) = 0$. Therefore, we infer that $\varphi^\infty \in \Gamma_0(X)$ (see Proposition 3.57). Since the function $\lambda \longmapsto q_\lambda(u)$ is nondecreasing on $(0, +\infty)$, we infer that

$$\varphi^\infty(u) = \lim_{\lambda \to +\infty} q_\lambda(u). \tag{3.159}$$

We have

$$q_\lambda^*(h^*) = \tfrac{\varphi^*(h^*) + \varphi(u_0) - \langle h^*, u_0\rangle}{\lambda} \qquad \forall\, h^* \in X^*$$

(see Remark 3.88). If $h^* \notin \operatorname{dom} X^*$, then $q_\lambda^*(h^*) = +\infty$.

If $h^* \in \operatorname{dom} \varphi^*$, then

$$\varphi^*(h^*) + \varphi(u_0) - \langle h^*, u_0\rangle \geqslant 0$$

(see Proposition 3.84(a)) and so the map $\lambda \longmapsto q_\lambda^*(h^*)$ is decreasing on $(0,+\infty)$ and clearly $\lim_{\lambda\to+\infty} q_\lambda(h^*) = 0$.

So, it follows that

$$\inf_{\lambda>0} q_\lambda^*(h^*) = \lim_{\lambda\to+\infty} q_\lambda^*(h^*) = i_{\operatorname{dom}\varphi^*}(h^*). \tag{3.160}$$

Then for all $u \in X$, we have

$$\begin{aligned} \sup_{h^*\in X^*}\big(\langle h^*,u\rangle - i_{\operatorname{dom}\varphi^*(h^*)}\big) &= \sup_{h^*\in X^*}\big(\langle h^*,u\rangle - \inf_{\lambda>0} q_\lambda^*(h^*)\big) \\ = \sup_{\substack{h^*\in X^* \\ \lambda>0}}\big(\langle h^*,u\rangle - q_\lambda^*(h^*)\big) &= \sup_{\lambda>0} q_\lambda(u) = \varphi^\infty(u) \end{aligned}$$

(see (3.160), (3.159), Theorem 3.85 and since $q_\lambda \in \Gamma_0(X)$), so φ^∞ is independent of the choice of $u_0 \in \operatorname{dom}\varphi$. Finally we check the positive 1-homogeneity of φ^∞. So, for every $h \in X$ and $t > 0$, we have

$$\varphi^\infty(th) = \lim_{\lambda\to+\infty} \tfrac{\varphi(u_0+\lambda th)-\varphi(u_0)}{\lambda}.$$

Let $\vartheta = \lambda t$. Then

$$\varphi^\infty(th) = t \lim_{\vartheta\to+\infty} \tfrac{\varphi(u_0+\vartheta h)-\varphi(u_0)}{\vartheta} = t\varphi^\infty(h),$$

so φ^∞ is positively homogeneous of degree 1.

Solution of Problem 3.111

(a) Let $m = \inf_X \varphi > -\infty$ and let $u_0 \in \operatorname{dom}\varphi$. Then from Problem 3.110, we know that for every $h \in X$, we have

$$\varphi^\infty(h) = \lim_{\lambda\to+\infty} \tfrac{\varphi(u_0+\lambda h)-\varphi(u_0)}{\lambda} \geqslant \lim_{\lambda\to+\infty} \tfrac{m-\varphi(u_0)}{\lambda} = 0.$$

(b) Since φ is positively homogeneous, we have that $\varphi(0) = 0$. So, choosing $u_0 = 0$, we have

$$\psi^\infty(h) = \lim_{\lambda\to+\infty} \tfrac{\varphi(\lambda h)}{h} - \langle \widehat{u}^*, h\rangle \qquad \forall\, h \in X$$

(see Problem 3.110), so

$$\psi^\infty(h) = \lim_{\lambda\to+\infty} \lambda^{p-1}\varphi(h) - \langle \widehat{u}^*, h\rangle = \begin{cases} -\langle \widehat{u}^*, h\rangle & \text{if } \varphi(h) = 0, \\ +\infty & \text{otherwise.} \end{cases} \tag{3.161}$$

Then from part **(a)**, we have that, if the problem $\inf_X \psi$ admits a solution, then

$$\psi^\infty(h) \geqslant 0 \qquad \forall\, h \in X,$$

so

$$\langle \widehat{u}, h\rangle \leqslant 0 \qquad \forall\, h \in X,\ \varphi(h) = 0$$

(see (3.161)).

Solution of Problem 3.112

Let $\overline{B}_n = \{h \in X : \|h\|_X \leqslant n\}$, for $n \geqslant 1$. By reflexivity, each $\overline{B}_n$ is w-compact. So, by the Weierstrass theorem (see Theorem I.2.86), we can find $u_n \in \overline{B}_n$ such that

$$\varphi(u_n) = \inf_{\overline{B}_n} \varphi.$$

Suppose that for some $n \geqslant 1$, we have $\|u_n\|_X < n$. Then for each $h \in X$ and every $t \in (0,1)$, we have

$$\varphi(u_n + t(h - u_n)) \leqslant (1-t)\varphi(u_n) + t\varphi(h),$$

so

$$\tfrac{1}{t}\big(\varphi(u_n + t(h-u_n)) - \varphi(u_n)\big) = \varphi(h) - \varphi(u_n). \tag{3.162}$$

Note that for $t \in (0,1)$ small, we have

$$u_n + t(h - u_n) \in \overline{B}_n,$$

so

$$\varphi(u_n) \leqslant \varphi(u_n + t(h - u_n)),$$

thus

$$\varphi(u_n) \leqslant \varphi(h) \qquad \forall\, h \in X$$

and hence u_n is the desired solution.

Therefore, we need to consider the case $\|u_n\|_X = n$ for all $n \geqslant 1$. Let $y_n = \frac{u_n}{\|u_n\|_X}$ for $n \geqslant 1$. Then $\|y_n\|_X = 1$ for all $n \geqslant 1$ and so by the reflexivity of X and by passing to a suitable sequence if necessary, we may assume that $y_n \xrightarrow{w} y$ in X. Then

$$\varphi(ny_n) = \varphi(u_n) \leqslant \varphi(u_0) < +\infty, \tag{3.163}$$

for any $u_0 \in \operatorname{dom}\varphi$ and for $n \geqslant 1$ big. Hypothesis **(i)** implies that

$$y_n \longrightarrow y \quad \text{in } X.$$

Also, since $\varphi \in \Gamma_0(X)$, for every $t > 0$, we have

$$\begin{aligned} \varphi(ty + u_0) &\leqslant \liminf_{n\to+\infty} \varphi\left(ty_n + (1 - \tfrac{t}{n})u_0\right) \\ &\leqslant \liminf_{n\to+\infty} \left(\tfrac{t}{n}\varphi(ny_n) + (1-\tfrac{t}{n})\varphi(u_0)\right) \leqslant \varphi(u_0) \end{aligned}$$

(see (3.163)), so $\varphi^\infty(y) \leqslant 0$ and using hypothesis **(ii)**, we get

$$\varphi^\infty(y) = 0,$$

so $y \in \ker \varphi^\infty$. Using also hypothesis **(iii)**, we have $\varphi^\infty(-y) = 0$, so

$$\varphi(h - ty) \leqslant \varphi(h) \qquad \forall\, h \in X,\ t \geqslant 0 \tag{3.164}$$

(see Problem 3.110). Let $h = u_n$ and $t = n$. Since $\|y_n - y\|_X \longrightarrow 0$, we have

$$\|u_n - ny\|_X < n \qquad \forall\, n \text{ large},$$

so $u_n - ny$ is a solution of $\inf\limits_{\overline{B}_n} \varphi$ for $n \geqslant 1$ large (see (3.164)). Thus $u_n - ny$ for $n \geqslant 1$ large is the desired solution.

Solution of Problem 3.113

Let $\{(u_n, u_n^*)\}_{n\geqslant 1} \subseteq \operatorname{Gr} N_C$ be a sequence such that

$$u_n \longrightarrow u \quad \text{in } X \quad \text{and} \quad u_n^* \longrightarrow u^* \quad \text{in } X^*. \tag{3.165}$$

Then

$$\langle u_n^*, c\rangle \leqslant \langle u_n^*, u_n\rangle \qquad \forall\, c \in C,\ n \geqslant 1$$

(see Definition 3.112). Passing to the limit as $n \to +\infty$ and using (3.165), we obtain

$$\langle u^*, c\rangle \leqslant \langle u^*, u\rangle \qquad \forall\, c \in C,$$

so $u^* \in N_C(u)$.

Solution of Problem 3.114

Let $h \in T_C(u)$ (see Definition 3.112) and let $\{u_n\}_{n\geqslant 1} \subseteq H$ be a sequence such that $u_n \longrightarrow u$. We set

$$h_n = \operatorname{proj}_{T_C(u_n)}(h) \quad \text{and} \quad h_n^* = \operatorname{proj}_{N_C(u_n)}(h) \qquad \forall\, n \geqslant 1.$$

From Problem 3.68(b), we know that

$$\|h_n\|_H \leqslant \|h\|_H \quad \text{and} \quad \|h_n^*\|_H \leqslant \|h\|_H \qquad \forall\, n \geqslant 1.$$

So, by passing to a suitable subsequence, we may assume that

$$h_n \longrightarrow \widehat{h} \quad \text{and} \quad h_n^* \longrightarrow \widehat{h}^* \quad \text{in } H. \tag{3.166}$$

We have

$$(u_n, h_n^*) \in \operatorname{Gr} N_C \qquad \forall\, n \geqslant 1.$$

From Problem 3.113, we have $(u, \widehat{h}^*) \in \operatorname{Gr} N_C$. We have $h = h_n + h_n^*$ for all $n \geqslant 1$. Hence $h_n = h - h_n^* \longrightarrow h - \widehat{h}^* = \widehat{h}$. We know that

$$(h_n^*, h_n)_H = 0 \qquad \forall\, n \geqslant 1$$

(see Problem 3.68(a)), so $(\widehat{h}^*, h - \widehat{h}^*)_H = 0$, thus

$$\|\widehat{h}^*\|_H^2 = (\widehat{h}^*, h)_H \leqslant 0$$

(since $h \in T_C(u)$ and $\widehat{h}^* \in N_C(u)$; see Definition 3.112), hence $\widehat{h}^* = 0$. It follows that $h_n \longrightarrow h$ in H, with $h_n \in T_C(u_n)$ for all $n \geqslant 1$ and this means that the map $u \longmapsto T_C(u)$ is lower semicontinuous (see Proposition 2.38).

Solution of Problem 3.115

From Remark 3.113, it is clear that, if $u \in \operatorname{int} \overline{B}_1$ (that is, $\|u\|_H < 1$), then $T_{\overline{B}_1}(u) = H$ and by polarity, $N_{\overline{B}_1}(u) = \{0\}$. Now, suppose that $\|u\|_H = 1$. Then $h \in N_{\overline{B}_1(u)}$ if and only if $\|h\|_H = \sup\{(h,v)_H : \|v\|_H \leqslant 1\} = (h,u)_H$. Hence $h = \lambda u$ with $\lambda > 0$ and so $N_{\overline{B}_1} = \mathbb{R}_+ u$ which is equivalent (by polarity) to saying that $T_{\overline{B}_1} = \{h \in H : (h,u)_H \leqslant 0\}$.

Solution of Problem 3.116

From Theorem 2.127, we know that the map $u \longmapsto A(u) + \partial\varphi(u)$ is maximal monotone (see Definition 2.100). Also, for every $u^* \in \partial\varphi(u)$, we have

$$\frac{\langle A(u),u\rangle + \langle u^*,u\rangle}{\|u\|_X} \geqslant \frac{\langle A(u),u\rangle + \varphi(u) - \varphi(0)}{\|u\|_X} \longrightarrow +\infty \quad \text{as } \|u\|_X \to +\infty$$

(see the hypothesis), so the map $u \longmapsto A(u) + \partial\varphi(u)$ is strongly coercive (see Definition 2.98(f)). Invoking Theorem 2.119, we have that the map $u \longmapsto A(u) + \partial\varphi(u)$ is surjective. So, we can find $u \in X$ such that

$$u^* \in A(u) + \partial\varphi(u),$$

so

$$\langle u^* - A(u), v - u\rangle \leqslant \varphi(v) - \varphi(u) \qquad \forall\, v \in X,$$

hence

$$\langle u^*, v - u\rangle \leqslant \langle A(u), v - u\rangle + \varphi(v) - \varphi(u) \qquad \forall\, v \in X.$$

The set of solutions of the problem is equal to $(A + \partial\varphi)^{-1}(u^*)$. But we have seen earlier that the map $A + \partial\varphi$ is maximal monotone. Hence so is $(A + \partial\varphi)^{-1}$ and then Corollary 2.107, implies that the set $(A + \partial\varphi)^{-1}(u^*)$ is nonempty, closed, and convex (in fact w-compact). If A is strictly monotone, then so is $A + \partial\varphi$ and this means that the set $(A + \partial\varphi)^{-1}(u^*)$ is a singleton.

Solution of Problem 3.117

Let $H = H_0^1(\Omega)$ and let $a\colon H \times H \longrightarrow \mathbb{R}$ be the bilinear form defined by

$$a(u, v) = \int_\Omega (Du, Dv)_{\mathbb{R}^N}\, dz \qquad \forall\, u, v \in H_0^1(\Omega).$$

By the Cauchy–Schwarz inequality, we have

$$|a(u, v)| \leqslant \|Du\|_2 \|Dv\|_2 = \|u\|_H \|v\|_H$$

(by the Poincaré inequality (see Theorem 1.131), we can take $\|Du\|_2$ as the Sobolev norm for the space $H_0^1(\Omega)$). Also,

$$a(u, u) = \|Du\|_2^2 = \|u\|_H^2.$$

The set C is nonempty. Indeed, note that $g^+ \in H_0^1(\Omega)$ and $g^+ \geqslant g$. Also, the set C is convex. Finally, we check that the set C is also closed. So, let $\{u_n\}_{n\geqslant 1} \subseteq C$ be a sequence such that $u_n \longrightarrow u$ in $H_0^1(\Omega)$. Then $u_n \longrightarrow u$ in $L^2(\Omega)$ (from the continuity of the embedding $H_0^1(\Omega) \hookrightarrow L^2(\Omega)$) and so by passing to a suitable subsequence if necessary, we may also assume that

$$u_n(z) \longrightarrow u(z) \quad \text{for a.a. } z \in \Omega. \tag{3.167}$$

But we have

$$u_n(z) \geqslant g(z) \quad \text{for a.a. } z \in \Omega.$$

So, from (3.167), it follows that

$$u(z) \geqslant g(z) \quad \text{for a.a. } z \in \Omega.$$

So, $u \in C$ and we infer that the set $C \subseteq H_0^1(\Omega)$ is closed. Now, we can apply Theorem 3.114 (with $h^* = 0$) and produce the desired $u \in C$.

Solution of Problem 3.118

Arguing by contradiction, suppose that φ is not Lipschitz continuous near u. Then, we can find two sequences $\{u_n\}_{n\geqslant 1}, \{v_n\}_{n\geqslant 1} \subseteq X$ such that $u_n \longrightarrow x$, $v_n \longrightarrow x$ in X, $u_n, v_n \in B_{\frac{1}{n}}(u)$ and

$$\|\varphi(u_n) - \varphi(v_n)\|_H > n\|u_n - v_n\|_H \qquad \forall\, n \geqslant 1.$$

Let $\lambda_n > 0$ and $h_n \in X$ for $n \geqslant 1$ be such that

$$\|h_n\|_H = \tfrac{1}{\sqrt{n}} \quad \text{and} \quad v_n = u_n + \lambda_n h_n \qquad \forall\, n \geqslant 1.$$

Evidently $\lambda_n \longrightarrow 0$. Let $K = \{h_n\}_{n\geqslant 1} \cup \{0\}$. Then the set $K \subseteq X$ is compact and from the definition of the strict differentiability (see Remark 3.125), we know that given $\varepsilon > 0$, there exists $n_0 = n_0(\varepsilon) \geqslant 1$ such that

$$\left\|\tfrac{\varphi(u_n+\lambda_n h)-\varphi(u_n)}{\lambda_n} - \varphi_s'(u)h\right\|_Y < \varepsilon \qquad \forall\, n \geqslant n_0,\ h \in K. \tag{3.168}$$

On the other hand, if $h = h_n$, then

$$\left\|\tfrac{\varphi(u_n+\lambda_n h_n)-\varphi(u_n)}{\lambda_n}\right\|_Y > \tfrac{n\lambda_n\|h_n\|_X}{\lambda_n} = \sqrt{n}.$$

Then, we see that for $n \geqslant 1$ big we violate (3.168). Therefore φ is Lipschitz continuous near u.

Solution of Problem 3.119

Let $K \subseteq X$ be a compact set and let $\varepsilon > 0$. By hypothesis, for every $h \in X$, we can find $\delta = \delta(h) > 0$ such that

$$\left\|\tfrac{\varphi(v+\lambda h)-\varphi(v)}{\lambda} - A(h)\right\|_Y < \varepsilon \qquad \forall\, v \in B_\delta(u),\ \lambda \in (0,\delta).$$

Since φ is by hypothesis Lipschitz continuous near u, we see that for v close to u and for λ close to 0, we have

$$\left\|\tfrac{\varphi(v+\lambda h)-\varphi(v)}{\lambda} - \tfrac{\varphi(v+\lambda h')-\varphi(v)}{\lambda}\right\|_Y \leqslant k\|h-h'\|_X \qquad \forall\, h, h' \in X,$$

with $k > 0$ being the Lipschitz constant of φ near u. So, with a suitable redefinition of $\delta = \delta(h) > 0$, we have

$$\left\|\tfrac{\varphi(v+\lambda h')-\varphi(v)}{\lambda} - A(h')\right\|_Y < 3\varepsilon \qquad \forall\, v \in B_\delta(u),\ h' \in B_\delta(h),\ \lambda \in (0,\delta).$$

Then $\{B_\delta(h)\}_{h\in K}$ is an open cover of K. As the set K is compact, the cover $\{B_\delta(h)\}_{h\in K}$ admits a finite subcover $\{B_{\delta_k}(h_k)\}_{k=1}^m$ (with $\delta_k = \delta(h_k) > 0$). Let $\delta_0 = \min\{\delta_1, \ldots, \delta_m\} > 0$. We have

$$\left\|\tfrac{\varphi(v+\lambda h)-\varphi(v)}{\lambda} - A(h)\right\|_Y < 3\varepsilon \qquad \forall\, h \in K,\ v \in B_{\delta_0}(u),\ \lambda \in (0,\delta_0).$$

So, we conclude that φ is strictly differentiable at u with $\varphi_s'(u) = A \in \mathcal{L}(X;Y)$.

Solution of Problem 3.120

No. Let $X = l^2$ and let $\varphi\colon l^2 \longrightarrow \mathbb{R}$ be defined by

$$\varphi(\widehat{u}) = \sup_{n \geqslant 1} (2n|u_n| - n) \qquad \forall\, \widehat{u} = \{u_n\}_{n\geqslant 1} \in l^2.$$

Let

$$\varphi_n(\widehat{u}) = 2n|u_n| - n \qquad \forall\, \widehat{u} \in l^2,\ n \geqslant 1.$$

Evidently each φ_n is convex and so $\varphi = \sup\limits_{n\geqslant 1} \varphi_n$ is convex too. Also, we have

$$\varphi|_{\overline{B}_{\frac{1}{4}}(0)} = 0,$$

where $\overline{B}_{\frac{1}{4}}(0) = \{\widehat{u} \in l^2 : \ \|\widehat{u}\|_X \leqslant \frac{1}{4}\}$. Therefore, the convex function φ is bounded above on a nonempty open set. Invoking Theorem 3.60, we infer that φ is locally Lipschitz. On the other hand, let $\{e_n\}_{n\geqslant 1}$ be the standard orthonormal basis of l^2. Then $\varphi(e_n) = n$ and so $\varphi|_{\overline{B}_1}$ is not bounded. Hence $\varphi|_{\overline{B}_1}$ cannot be Lipschitz continuous.

Solution of Problem 3.121

Since φ is Gâteaux differentiable and regular at $u \in X$, we have

$$\langle \varphi_G'(u), h \rangle = \varphi'(u;h) = \varphi^0(u;h) \qquad \forall\, h \in X,$$

so $\varphi_G'(u) \in \partial\varphi(u)$ (see Definition 3.126). On the other hand, if $u^* \in \partial\varphi(u)$, then

$$\langle u^*, h \rangle \leqslant \varphi^0(u;h) = \varphi'(u;h) = \langle \varphi_G'(u), h \rangle \qquad \forall\, h \in X$$

(see Definition 3.126), so $u^* = \varphi_G'(u)$.

Solution of Problem 3.122

We assume that u is a local minimizer for φ (if u is a local maximizer for φ, then we employ the same argument with φ replaced by $-\varphi$). Then $0 \leqslant \varphi^0(u;h)$ for all

$h \in X$ (see Definition 3.122), so $0 \in \partial\varphi(u)$ (see Definition 3.126).

Solution of Problem 3.123

Let $\eta \geqslant 0$ and consider the sublevel set

$$\mu^\eta = \{u \in X : \mu(u) \leqslant \eta\}.$$

We need to show that the set $\mu^\eta \subseteq X$ is closed (see Proposition I.2.53). To this end, let $\{u_n\}_{n\geqslant 1} \subseteq \mu^\eta$ and assume that $u_n \longrightarrow u$ in X. We can find

$$u_n^* \in \partial\varphi(u_n) \qquad \forall\, n \geqslant 1,$$

such that $\|u_n^*\|_* \leqslant \eta + \frac{1}{n}$ for all $n \geqslant 1$. Hence the sequence $\{u_n^*\}_{n\geqslant 1} \subseteq X^*$ is bounded and so by the Alaoglu theorem (see Theorem I.5.66), we can find a subnet $\{u_\alpha^*\}_{\alpha\in J}$ such that $u_\alpha^* \xrightarrow{w^*} u^*$ in X^*. Exploiting the w^*-lower semicontinuity of the norm functional $\|\cdot\|_*$ (see Proposition I.5.65(c)), we obtain

$$\|u^*\|_* \leqslant \eta \quad \text{and} \quad u^* \in \partial\varphi(u)$$

(see Proposition 3.128(b)), so $\mu(u) \leqslant \eta$, thus $u \in \mu^\eta$ and hence μ^η is closed. This proves the lower semicontinuity of μ on X.

Solution of Problem 3.124

Let $i\colon X \longrightarrow Y$ be the continuous, linear embedding of X into Y. Then $\varphi|_X = \varphi \circ i$. From Proposition 3.132, we have

$$\partial(\varphi|_X)(u) = \partial(\varphi\circ i)(u) = i^*\partial\varphi(i(u)) \qquad \forall\, u \in X.$$

Recall that $i^*\colon Y^* \longrightarrow X^*$ is the continuous, linear and dense embedding of Y^* into X^*. So, finally

$$\partial(\varphi|_X)(u) = (\partial\varphi)|_X(u) \qquad \forall\, u \in X.$$

Solution of Problem 3.125

Clearly ϑ is Lipschitz continuous. Hence $\gamma = \varphi\circ\vartheta$ is locally Lipschitz and so it is differentiable almost everywhere. Let $t = t_0$ be a point of differentiability for γ. We have

$$\gamma'(t_0) = \lim_{s\to 0} \tfrac{\varphi(\vartheta(t_0+s)) - \varphi(\vartheta(t_0))}{s}$$

$$\begin{aligned}
&= \lim_{s\to 0} \tfrac{\varphi(\vartheta(t_0)+\vartheta'(t_0)s+e(s))-\varphi(\vartheta(t_0))}{s} \\
&= \lim_{s\to 0} \tfrac{\varphi(\vartheta(t_0)+\vartheta'(t_0)s)-\varphi(\vartheta(t_0))}{s} \\
&\leqslant \limsup_{\substack{h\to 0\\ s\searrow 0}} \tfrac{\varphi(\vartheta(t_0)+h+\vartheta'(t_0)s)-\varphi(\vartheta(t_0)+h)}{s} \\
&= \varphi^0(\vartheta(t_0);\vartheta'(t_0)) \;=\; \max_{u^*\in\partial\varphi(\vartheta(t_0))} \langle u^*, \vartheta'(t_0)\rangle
\end{aligned}$$

with $\frac{e(s)}{s} \longrightarrow 0$ as $s \to 0$ (since γ is locally Lipschitz and using Definition 3.122 and Proposition 3.128(c)).

Solution of Problem 3.126

From Remark 3.125 we know that φ is locally Lipschitz at u_0. The continuity of the Gâteaux derivative at u_0 implies that given $\varepsilon > 0$, we can find $\delta > 0$ such that

$$\text{if } \|u-u_0\|_X \;\leqslant\; \delta, \text{ then } \|\varphi'_G(u)-\varphi'_G(u_0)\|_* \;\leqslant\; \varepsilon. \tag{3.169}$$

Let $h \in X$, $v \in \overline{B}_{\frac{\delta}{2}}(u_0)$ and $t \in (0, \frac{\delta}{2\|h\|_X})$. We set $\vartheta(t) = \varphi(v+th)$. Then ϑ is differentiable and

$$\vartheta'(t) \;=\; \lim_{\lambda\to 0} \tfrac{\varphi(v+th+\lambda h)-\varphi(v+th)}{\lambda} \;=\; \left\langle \varphi'_G(v+\lambda h), h\right\rangle. \tag{3.170}$$

So, if $0 < \lambda \leqslant \frac{\delta}{2\|h\|_X}$, then

$$\begin{aligned}
\tfrac{\varphi(v+\lambda h)-\varphi(v)}{\lambda} - \left\langle \varphi'_G(u_0), h\right\rangle \;&=\; \tfrac{\vartheta(\lambda)-\vartheta(0)}{\lambda} - \left\langle \varphi'_G(u_0), h\right\rangle \\
= \; \tfrac{1}{\lambda}\int_0^\lambda \left\langle \varphi'_G(v+th)-\varphi'_G(u_0), h\right\rangle dt & \qquad\qquad (3.171)
\end{aligned}$$

(see (3.170)). Note that $\|v+th-u_0\|_X \leqslant \delta$. Then from (3.169) and (3.171), we have

$$\left| \tfrac{\varphi(v+\lambda h)-\varphi(v)}{\lambda} - \left\langle \varphi'_G(u_0), h\right\rangle \right| \;\leqslant\; \varepsilon\|h\|_X$$

for all $\lambda \in (0, \frac{\delta}{2\|h\|_X}]$ and $v \in \overline{B}_{\frac{\delta}{2}(u_0)}$. It follows that, if $\mu \in (0, \frac{\delta}{2}]$ and $s \in (0, \frac{\delta}{2\|h\|_X}]$, then

$$\sup_{\|v-u_0\|_X\leqslant\mu}\; \sup_{\lambda\leqslant s} \tfrac{\varphi(v+\lambda h)-\varphi(v)}{\lambda} \;\leqslant\; \left\langle \varphi'_G(u_0), h\right\rangle + \varepsilon\|h\|_X,$$

so

$$\varphi^0(u_0;h) \;\leqslant\; \left\langle \varphi'_G(u_0), h\right\rangle + \varepsilon\|h\|_X. \tag{3.172}$$

On the other hand, directly from Definitions 3.1 and 3.122, we have

$$\langle \varphi_G'(u_0), h \rangle \leqslant \varphi^0(u_0; h). \tag{3.173}$$

From (3.172) and (3.173) and since $\varepsilon > 0$ is arbitrary, we conclude that

$$\langle \varphi_G'(u_0), h \rangle = \varphi^0(u_0; h) \qquad \forall\, h \in X.$$

Solution of Problem 3.127

Fix $u \in X$. For every $v \in X$, $h \in K$ and $\lambda > 0$, we have

$$\varphi(v) = \varphi(v + \lambda h + \lambda(-h)) \leqslant \varphi(v + \lambda h)$$

(see the hypothesis), so

$$\tfrac{\varphi(v+\lambda h+\lambda(-h))-\varphi(v+\lambda h)}{\lambda} \leqslant 0,$$

thus

$$\varphi^0(u; -h) = \limsup_{\substack{v \to u \\ \lambda \searrow 0}} \tfrac{\varphi(v+\lambda h+\lambda(-h))-\varphi(v+\lambda h)}{\lambda} \leqslant 0 \qquad \forall\, h \in K. \tag{3.174}$$

On the other hand, we have

$$\sigma_{K^*}(h) = i_{(-K)}(h). \tag{3.175}$$

From (3.174) and (3.175), we infer that

$$\varphi^0(u; h) \leqslant \sigma_{K^*}(h) \qquad \forall\, h \in X,$$

so $\partial\varphi(u) \subseteq K^*$ (see Proposition 3.128(c)).

Solution of Problem 3.128

Since φ is locally Lipschitz, we can find $r, k > 0$ such that $\varphi|_{\overline{B}_r(u_0)}$ is Lipschitz continuous with Lipschitz constant $k > 0$. Let $\vartheta \in (0, \frac{r}{2})$ and $\varepsilon \in (0, \frac{r-2\vartheta}{2})$. Consider $u \in \overline{B}_\vartheta(u_0)$. We can find $c_\varepsilon \in C$ such that

$$\|u - c_\varepsilon\|_X \leqslant (1 + \varepsilon)\mathrm{dist}(u, C). \tag{3.176}$$

We have

$$\mathrm{dist}(u, C) \leqslant \|u - u_0\|_X \leqslant \vartheta,$$

so

$$\|u_0 - c_\varepsilon\|_X \leqslant \|u_0 - u\|_X + \|u - c_\varepsilon\|_X \leqslant (2+\varepsilon)\vartheta \leqslant r \tag{3.177}$$

(see (3.176) and recall the choice of ϑ and ε). Since $\varphi|_{\overline{B}_r(u_0)}$ is k-Lipschitz, from (3.177), we infer that

$$\varphi(c_\varepsilon) \leqslant \varphi(u) + k\|u - c_\varepsilon\|_X \leqslant \varphi(u) + k(1+\varepsilon)\mathrm{dist}(u, C)$$

(see (3.176)), so, for all $u \in \overline{B}_\vartheta(u_0)$, we have

$$\varphi(u_0) + k\,\mathrm{dist}(u_0, C) = \varphi(u_0) \leqslant \varphi(c_\varepsilon) \leqslant \varphi(u) + k(1+\varepsilon)\mathrm{dist}(u, C)$$

(see the hypothesis and recall that $u_0 \in C$). Let $\varepsilon \searrow 0$, to conclude that u_0 is a local minimizer of the map $u \longmapsto \varphi(u) + k\,\mathrm{dist}(u, C)$. Then according to Problem 3.122, we have

$$\begin{aligned} 0 \in \partial(\varphi + k\,\mathrm{dist}(\cdot, C))(u_0) &\subseteq \partial\varphi(u_0) + k\partial\mathrm{dist}(u_0, C) \\ &\subseteq \partial\varphi(u_0) + N_c'(u_0) \end{aligned}$$

(see Proposition 3.130 and Definition 3.133).

Solution of Problem 3.129

Clearly we have

$$\lim_{\substack{v \to u \\ v \in C}} \liminf_{\lambda \searrow 0} \frac{\mathrm{dist}(v+\lambda h, C)}{\lambda} \geqslant 0. \tag{3.178}$$

On the other hand, we have

$$\begin{aligned} \lim_{\substack{v \to u \\ v \in C}} \liminf_{\lambda \searrow 0} \frac{\mathrm{dist}(v+\lambda h, C)}{\lambda} &\leqslant \inf_{r>0} \sup_{\substack{\|v-u\|_X \leqslant r \\ v \in C}} \inf_{\eta>0} \sup_{\vartheta \leqslant \eta} \frac{\mathrm{dist}(v+\vartheta h, C)}{\vartheta} \\ &\leqslant \inf_{r,\eta>0} \sup_{\substack{\|v-u\|_X \leqslant r \\ v \in C}} \sup_{\vartheta \leqslant \eta} \frac{\mathrm{dist}(v+\vartheta h, C) - \mathrm{dist}(v, C)}{\vartheta} \leqslant \mathrm{dist}^0(u, C; h) \leqslant 0 \end{aligned} \tag{3.179}$$

(since $h \in T_C'(u)$; see Definition 3.133). From (3.178) and (3.179) we have that

$$\lim_{\substack{v \to u \\ v \in C}} \liminf_{\lambda \searrow 0} \frac{\mathrm{dist}(v+\lambda h, C)}{\lambda} = 0 \qquad \forall\, h \in T_C'(u).$$

Solution of Problem 3.130

Suppose that the statement is not true. Then we can find $h \in T'_C(u)$ (see Definition 3.133) and $v \in H$ such that $\operatorname{dist}(v, C) = \|v - u\|_H$ and $(v - u, h)_H = \delta > 0$. Let $D = H \setminus B_r(v)$, where $r = \operatorname{dist}(v, C)$. Then $C \subseteq D$ and so

$$\operatorname{dist}(u + \lambda h, D) \leqslant \operatorname{dist}(u + \lambda h, C). \tag{3.180}$$

Since D is the complement of a ball, we have

$$\operatorname{dist}(u + \lambda h, D) = \operatorname{dist}(v, C) - \|u + \lambda h - v\|_H. \tag{3.181}$$

We have

$$\begin{aligned} \|u + \lambda h - v\|_H^2 &= \|u - v\|_H^2 + 2\lambda (u - v, h)_H + \lambda^2 \|h\|_H^2 \\ &= \operatorname{dist}(v, C)^2 \big(1 - 2\lambda \tfrac{\delta}{\operatorname{dist}(v,C)^2} + \lambda^2 \tfrac{\|h\|_H^2}{\operatorname{dist}(v,C)^2}\big), \end{aligned}$$

so

$$\|u + \lambda h - v\|_H \leqslant \operatorname{dist}(v, C)\big(1 - \lambda \tfrac{\delta}{\operatorname{dist}(v,C)^2} + e(\lambda)\big),$$

with $\frac{e(\lambda)}{\lambda} \searrow 0$, thus

$$\lambda \tfrac{\delta}{\operatorname{dist}(v,C)} - e(\lambda)\operatorname{dist}(v, C) \leqslant \operatorname{dist}(u + \lambda h, D) \leqslant \operatorname{dist}(u + \lambda h, C)$$

(see (3.180) and (3.181)) and hence

$$\tfrac{\delta}{\operatorname{dist}(v,C)} \leqslant \liminf_{\lambda \searrow 0} \tfrac{\operatorname{dist}(u+\lambda h, C)}{\lambda}. \tag{3.182}$$

From Problem 3.129, we know that

$$\liminf_{\lambda \searrow 0} \tfrac{\operatorname{dist}(u+\lambda h, C)}{\lambda} = 0. \tag{3.183}$$

Comparing (3.182) and (3.183), we reach a contradiction.

Solution of Problem 3.131

From Problem 3.128, we know that

$$0 \in \partial\varphi(u_0) + N'_C(u_0).$$

So, there exists $u^* \in \partial\varphi(u_0)$ such that $-u^* \in N'_C(u_0)$. Hence

$$\langle u^*, h \rangle \geqslant 0 \qquad \forall\, h \in T'_C(u_0)$$

(see Definition 3.133), so

$$\varphi^0(u_0; h) \geqslant 0 \qquad \forall\, h \in T'_C(u_0)$$

(see Definition 3.126).

Solution of Problem 3.132

Suppose that $u \in \overline{C}$. Then for any $h \in \mathbb{R}^N$ and $\lambda > 0$, we have

$$0 \leqslant \frac{\operatorname{dist}(u+\lambda h, C)}{\lambda} = \frac{\operatorname{dist}(u+\lambda h, C)-\operatorname{dist}(u, C)}{\lambda}$$

(since $u \in \overline{C}$), so

$$0 \leqslant (\operatorname{dist}'(u, C), h)_{\mathbb{R}^N} \qquad \forall\, h \in \mathbb{R}^N,$$

thus $\operatorname{dist}'(u, C) = 0$, a contradiction to the hypothesis. So, $u \notin \overline{C}$. The local compactness of $\mathbb{R}^N$ implies that u admits at least one best approximation $c_0 \in \overline{C}$ (that is, $\operatorname{dist}(u, C) = |u - c_0|$).

We will show that $\operatorname{dist}'(u, C) = \frac{u-c_0}{|u-c_0|}$ from which we infer also the uniqueness of the best approximation c_0. So, let $\lambda \in (0, 1)$. Then the best approximation in $\overline{C}$ to $u + \lambda(c_0 - u)$ is still c_0. Therefore

$$\operatorname{dist}(u + \lambda(c_0 - u), C) = (1 - \lambda)|u - c_0| = (1 - \lambda)\operatorname{dist}(u, C),$$

so

$$\frac{\operatorname{dist}(u+\lambda(c_0-u), C)-\operatorname{dist}(u, C)}{\lambda} = -\operatorname{dist}(u, C) = -|u - c_0|,$$

thus

$$(\operatorname{dist}'(u, C), c_0 - u)_{\mathbb{R}^N} = -|u - c_0|. \tag{3.184}$$

Since $\operatorname{dist}(\cdot, C)$ is nonexpansive (i.e., Lipschitz continuous with Lipschitz constant 1), we have

$$|\operatorname{dist}'(u, C)| \leqslant 1.$$

Therefore from (3.184) we infer that

$$\operatorname{dist}'(u, C) = \frac{u-c_0}{|u-c_0|}.$$

Solution of Problem 3.133

Let $\vartheta(u) = \varphi(u) + \|u\|_X$. Evidently $\vartheta \in \Gamma_0(X)$, $\vartheta \geqslant 0$ and $\vartheta(0) = 0$. So, from Problem 3.122, we have

$$0 \in \partial\vartheta(0) = \partial\varphi(0) + \partial\|\cdot\|_X(0) \tag{3.185}$$

(see Proposition 3.97). We have $\partial\|\cdot\|_X(0) = \overline{B}_1^* = \{u^* \in X^* : \|u^*\|_* \leqslant 1\}$. Then from (3.185), we see that we can find $u^* \in X^*$ with $\|u^*\|_* \leqslant 1$ such that $u^* \in \partial\varphi(0)$, so

$\langle u^*, u \rangle \leqslant \varphi(u)$ for all $u \in X$.

Solution of Problem 3.134

"(a) $\Longrightarrow$ (b)": We assume that $0 \leqslant \varphi^0(u; v-u) + \psi(v) - \psi(u)$ for all $v \in X$, so

$$0 \leqslant \varphi^0(u;h) + \psi(u+h) - \psi(u) \qquad \forall\, h \in X.$$

Thus, the origin is a minimizer of the convex function

$$h \longmapsto \xi(h) = \varphi^0(u;h) + \psi(u+h) - \psi(u)$$

(see Proposition 3.123(a)). Hence $u \in D(\partial\psi)$ and

$$\begin{aligned} 0 \in \xi(0) &= \partial\big(\varphi^0(u;\cdot) + \psi(u+\cdot) - \psi(u)\big)(0) \\ &= \partial\big(\varphi^0(u;\cdot)\big)(0) + \partial\psi(u) \\ &= \partial\varphi(u) + \partial\psi(u) \end{aligned}$$

(see Proposition 3.97 and Definition 3.126).

"(b) $\Longrightarrow$ (a)": We assume that $u \in D(\partial\psi)$ and $0 \in \partial\varphi(u) + \partial\psi(u)$. Then we can find $u^* \in \partial\varphi(u)$ and $\widehat{u}^* \in \partial\psi(u)$ such that

$$u^* + \widehat{u}^* = 0,$$

so

$$\langle u^*, v-u \rangle + \langle \widehat{u}^*, v-u \rangle \leqslant \varphi^0(u; v-u) + \psi(v) - \psi(u) \qquad \forall\, v \in X$$

(see Definitions 3.70 and 3.122), thus

$$0 \leqslant \varphi^0(u; v-u) + \psi(v) - \psi(u) \qquad \forall\, v \in X.$$

Solution of Problem 3.135

(a) For every $n \geqslant c_0$, we have

$$\begin{aligned} \varphi_n(u) &= \inf_{v \in X} \big(\varphi(v) + n\|u-v\|_X\big) \\ &\geqslant \inf_{v \in X} \big(-c_0 - c_0\|v\|_X + n\|u-v\|_X\big) \\ &\geqslant \inf_{v \in X} \big(-c_0 - c_0\|v\|_X + n\|v\|_X - n\|u\|_X\big) \\ &\geqslant -c_0 - n\|u\|_X \;>\; -\infty \end{aligned}$$

(see Definition 3.87 and since $n \geqslant c_0$). Since φ is proper, there exists $u_0 \in \operatorname{dom} \varphi$ such that

$$\varphi_n(u) \leqslant \varphi(u_0) + n\|u - u_0\|_X < +\infty.$$

Therefore, for all $n \geqslant c_0$ and all $u \in X$, we have that $\varphi_n(u) \in \mathbb{R}$.
Next, let $u, v \in X$. Using the triangle inequality, we have

$$\varphi(y) + n\|v - y\|_X \leqslant \varphi(y) + n\|u - y\|_X + n\|v - u\|_X \qquad \forall\, y \in X,$$

thus

$$\varphi_n(v) \leqslant \varphi_n(u) + n\|v - u\|_X. \tag{3.186}$$

Interchanging the roles of v and u in the above argument, we obtain

$$\varphi_n(u) \leqslant \varphi_n(v) + n\|v - u\|_X. \tag{3.187}$$

From (3.186) and (3.187), we conclude that

$$|\varphi_n(u) - \varphi_n(v)| \leqslant n\|v - u\|_X,$$

so φ_n is n-Lipschitz for all $n \geqslant c_0$.

(b) In the definition of $\varphi_n(u)$, let $v = u$, to obtain

$$\varphi_n(u) \leqslant \varphi(u) \qquad \forall\, n \geqslant 1.$$

Also, from the definition of φ_n it is clear that $\{\varphi_n\}_{n \geqslant 1}$ is increasing. So, we have

$$\lim_{n \to +\infty} \varphi_n(u) \leqslant \varphi(u) \qquad \forall\, u \in X. \tag{3.188}$$

We will prove that the opposite inequality holds. If $\lim\limits_{n \to +\infty} \varphi_n(u) = +\infty$, then there is nothing to prove. So, we assume that $\lim\limits_{n \to +\infty} \varphi_n(u) < +\infty$. Let $\varepsilon_n \searrow 0$ and let $v_n \in X$ be such that

$$\varphi(v_n) + n\|u - v_n\|_X - \varepsilon_n \leqslant \varphi_n(u) \qquad \forall\, n \geqslant c_0, \tag{3.189}$$

so, by the hypothesis on φ, we have

$$-c_0(1 + \|v_n\|_X) + n\|u - v_n\|_X - \varepsilon_n \leqslant \sup_{n \geqslant 1} \varphi_n(u) < +\infty. \tag{3.190}$$

Since $n \geqslant c_0$, from (3.190) and the triangle inequality, it follows that

$$v_n \longrightarrow u \quad \text{in } X. \tag{3.191}$$

From (3.189), we have

$$\varphi(v_n) - \varepsilon_n \leqslant \varphi_n(u) \qquad \forall\, n \geqslant c_0,$$

so

$$\varphi(u) \leqslant \liminf_{n \to +\infty} \varphi(v_n) \leqslant \lim_{n \to +\infty} \varphi_n(u) \tag{3.192}$$

(see (3.191)). From (3.188) and (3.192), we conclude that

$$\varphi_n(u) \nearrow \varphi(u) \qquad \forall\, u \in X.$$

(c) Since φ and $n\|\cdot\|_X$ are both convex, then so is $\varphi_n = \varphi \boxdot n\|\cdot\|_X$ (see Remark 3.88).

Solution of Problem 3.136

Note that from the definition of C, we have

$$\langle u^*, v - u \rangle \leqslant 0 \qquad \forall\, v \in C,$$

so $u^* \in N_C(u)$ (see Definition 3.112) and thus

$$\mathbb{R}_+ u^* \subseteq N_C(u). \tag{3.193}$$

Next, let $v^* \in N_C(u) \setminus \{0\}$. From the definition of $N_C(u)$ (see Definition 3.112), we have

$$\langle v^*, v - u \rangle \leqslant 0 \qquad \forall\, v \in C. \tag{3.194}$$

Let $V = \{v \in X : \langle u^*, v - u \rangle = 0\}$ (an affine subspace of X). Note that $V = u + N$, where $N = \ker u^*$ and of course $V \subseteq C$. In (3.194), we choose $v \in V$ and have

$$\langle v^*, v \rangle \leqslant 0 \qquad \forall\, v \in N,$$

so

$$\langle v^*, v \rangle = 0 \qquad \forall\, v \in N,$$

thus $\ker u^* \subseteq \ker v^*$ and hence

$$v^* = \lambda u^*,$$

for some $\lambda \in \mathbb{R}$. From (3.194), we have

$$\lambda \langle u^*, v - u \rangle \leqslant 0 \qquad \forall\, v \in C,$$

so $\lambda \geqslant 0$. Thus, we conclude that

$$N_C(u) \subseteq \mathbb{R}_+ u^*. \tag{3.195}$$

From (3.193) and (3.195), we conclude that

$$N_C(u) = \mathbb{R}_+ u^*.$$

Solution of Problem 3.137

Let $u \in C$. If $\varphi(u) < 0$, then the continuity of φ implies that $u \in \operatorname{int} C$ and so $T_C(u) = X$, hence $N_C(u) = \{0\}$.

So, assume that $\varphi(u) = 0$. Let $u^* \in \partial\varphi(u)$. We have

$$\langle u^*, v - u \rangle \leqslant \varphi(v) - \varphi(u) \qquad \forall\, v \in X$$

(see Definition 3.70), so

$$\langle u^*, v - u \rangle \leqslant 0 \qquad \forall\, v \in C$$

(since $\varphi(v) \leqslant 0 = \varphi(u)$), thus $u^* \in N_C(u)$ and hence

$$\mathbb{R}_+ \partial\varphi(0) \subseteq N_C(u). \tag{3.196}$$

Next we prove the opposite inclusion. So, let $u^* \in N_C(u) \setminus \{0\}$. From Definition 3.112 we know that the map $v \longmapsto \langle u^*, v \rangle$ attains its maximum on C at u and

$$\langle u^*, v \rangle < \langle u^*, u \rangle \qquad \forall\, v \in \operatorname{int} C,$$

so

$$\langle u^*, v \rangle < \langle u^*, u \rangle \qquad \forall\, v \in X,\ \varphi(v) < 0,$$

thus

$$\varphi(v) \geqslant 0 \qquad \forall\, v \in X,\ \langle u^*, v - u \rangle \geqslant 0. \tag{3.197}$$

So, if we set $D = \{v \in X : \langle u^*, v - u \rangle \geqslant 0\}$, then $\varphi|_D \geqslant 0$. Note that $u \in D$ and recall that $\varphi(u) = 0$. Therefore

$$0 = \varphi(u) = \inf_D \varphi.$$

From Problem 3.128 (particular case of convex function and convex constraint set), we have

$$0 \in \partial\varphi(u) + N_D(u). \tag{3.198}$$

But from Problem 3.136, we have

$$N_D(u) = \mathbb{R}_+(-u^*). \tag{3.199}$$

So, from (3.198) and (3.199) we infer that there exists $\mu \leqslant 0$ such that

$$0 \in \partial\varphi(u) + \mu u^*.$$

If $\mu = 0$, then $0 \in \partial\varphi(u)$ and so $u \in X$ is a global minimizer of φ (see Proposition 3.82), a contradiction since by hypothesis $\varphi(u_0) < 0 = \varphi(u)$. Hence $\mu < 0$. Then $u^* \in -\frac{1}{\mu}\partial\varphi(u)$, so $u^* \in \mathbb{R}_+\partial\varphi(u)$ and hence

$$N_C(u) \subseteq \mathbb{R}_+\partial\varphi(u). \tag{3.200}$$

From (3.196) and (3.200), we conclude that $N_C(u) = \mathbb{R}_+\partial\varphi(u)$ for all $u \in X$, with $\varphi(u) = 0$.

Solution of Problem 3.138

Let $C = \{v \in X : \psi_k(x) \leqslant 0 \text{ for all } k \in \{1, \ldots, n\}\}$. Evidently this is a closed and convex set and $u_0 \in \operatorname{int} C$. Moreover, $u_0 \in \operatorname{dom} \varphi \cap \operatorname{int} C_+$ (see Theorem 5.129(e)). Therefore from Proposition 3.97 and Problem 3.128 (convex case), we have that u is a solution of the considered minimization problem if and only if

$$0 \in \partial\varphi(u) + N_C(u). \tag{3.201}$$

If $C_k = \{v \in X : \varphi_k(v) \leqslant 0\}$, then $C = \bigcap_{k=1}^{n} C_k$. Hence

$$i_C = \sum_{k=1}^{n} i_{C_k}.$$

Note that $u_0 \in \bigcap_{k=1}^{n} \operatorname{int} \operatorname{dom} i_{C_k}$, hence

$$\partial i_C = \sum_{k=1}^{n} \partial i_{C_k}$$

(see Proposition 3.97), so

$$N_C(u) = \sum_{k=1}^{n} N_{C_k}(u) \qquad \forall\, u \in C. \tag{3.202}$$

Using Problem 3.137, we can find $\lambda_1, \ldots, \lambda_n \subseteq \mathbb{R}_+$ such that

$$0 \in \partial\varphi(u) + \sum_{k=1}^{n} \lambda_k \partial\psi_k(u)$$

(see (3.201) and (3.202)), with $\lambda_k = 0$ if $\psi_k(u) < 0$ for all $k \in \{1, \ldots, n\}$. So, in all cases $\lambda_k\psi_k(u) = 0$ for all $k \in \{1, \ldots, n\}$.

Solution of Problem 3.139

Let $C' = K\text{-}\liminf\limits_{n\to+\infty} C_n$ and $\widehat{\varphi} = \Gamma\text{-}\limsup\limits_{n\to+\infty} i_{C_n}$. Evidently $\operatorname{im}\widehat{\varphi} = \{0,+\infty\}$. By Definition 2.88 we know that $u \in C'$ if and only if for all $U \in \mathcal{N}(u)$ (with $\mathcal{N}(u)$ being the filter of neighborhoods of u) there exists $n_0 \geqslant 1$ such that $U \cap C_n \neq \emptyset$ for all $n \geqslant n_0$. Note that

$$U \cap C_n \neq \emptyset \quad\Longleftrightarrow\quad \inf_{v\in U} i_{C_n}(v) \;=\; 0.$$

Therefore

$$u \;\in\; C' \quad\Longleftrightarrow\quad \limsup_{n\to+\infty} \inf_{v\in U} i_{C_n}(v) \;=\; 0 \qquad \forall\, U \in \mathcal{N}(u),$$

so

$$u \;\in\; C' \quad\Longleftrightarrow\quad \widehat{\varphi}(u) \;=\; 0.$$

So, we have

$$i_{C'} \;=\; \Gamma\text{-}\limsup_{n\to+\infty} i_{C_n} \;=\; \widehat{\varphi}. \tag{3.203}$$

Similarly, if $C'' = K\text{-}\limsup\limits_{n\to+\infty} C_n$ and $\widetilde{\varphi} = \Gamma\text{-}\liminf\limits_{n\to+\infty} \varphi_n$, then

$$i_{C''} \;=\; \Gamma\text{-}\limsup_{n\to+\infty} i_{C_n} \;=\; \widetilde{\varphi}. \tag{3.204}$$

From (3.127) and (3.128), we conclude that

$$C_n \xrightarrow{K} C \quad\Longleftrightarrow\quad i_{C_n} \xrightarrow{\Gamma} i_C.$$

Solution of Problem 3.140

"$\Longrightarrow$": Suppose that $C_n \xrightarrow{M} C$ (see Definition 3.149(a)) and let $u \in C$. Then we can find $u_n \in C_n$ for all $n \geqslant 1$, such that $u_n \longrightarrow u$ in X. Hence

$$0 \;=\; i_{C_n}(u) \;\longrightarrow\; i_C(u) \;=\; 0.$$

On the other hand, if $\{u_n\}_{n\geqslant 1} \subseteq X$ is a sequence such that $u_n \xrightarrow{w} u$ and for a subsequence $\{n_k\}_{k\geqslant 1}$ of $\{n\}_{n\geqslant 1}$ we have $u_{n_k} \in C_{n_k}$ for all $k \geqslant 1$, then $u \in w\text{-}\limsup\limits_{n\to+\infty} C_n = C$ and so

$$0 \;=\; i_C(u) \;\leqslant\; \liminf_{n\to+\infty} i_{C_n}(u_n).$$

From Definition 3.149(b), we conclude that $i_{C_n} \xrightarrow{M} i_C$.

"$\Longleftarrow$": Suppose that $i_{C_n} \xrightarrow{M} i_C$. Then for every $u \in X$, we can find a sequence $\{u_n\}_{n\geqslant 1} \subseteq X$ such that $u_n \longrightarrow u$ such that

$$\limsup_{n\to+\infty} i_{C_n}(u_n) \leqslant i_C(u).$$

So, if $u \in C$, then there exist $u_n \in C_n$ for all $n \geqslant 1$, such that $u_n \longrightarrow u$ in X.

Also, if $\{u_n\}_{n\geqslant 1} \subseteq X$ is a sequence such that $u_n \xrightarrow{w} u$ in X, then

$$i_C(u) \leqslant \liminf_{n\to+\infty} i_{C_n}(u_n).$$

So, if $\{n_k\}_{k\geqslant 1}$ is a subsequence of $\{n\}_{n\geqslant 1}$ and $u_{n_k} \in C_{n_k}$ for $k \geqslant 1$, then $u \in C$ and so

$$w\text{–}\limsup_{n\to+\infty} C_n \subseteq C,$$

thus $C_n \xrightarrow{M} C$.

Solution of Problem 3.141

(a) If $u_n \longrightarrow u$ in X and $\varepsilon > 0$, then we can find $n_0 \geqslant 1$ such that

$$\varphi_n(u_n) \geqslant \varphi_n(u) - \varepsilon \qquad \forall\, n \geqslant n_0,$$

so

$$\liminf_{n\to+\infty} \varphi_n(u_n) \geqslant \varphi(u) - \varepsilon.$$

Let $\varepsilon \searrow 0$ to obtain

$$\liminf_{n\to+\infty} \varphi_n(u_n) \geqslant \varphi(u).$$

Since $\varphi_n(u) \longrightarrow \varphi(u)$ for all $u \in X$, we conclude that $\varphi_n \xrightarrow{\Gamma} \varphi$.

(b) Note that by Proposition 3.64, the sequence $\{\varphi_n\}_{n\geqslant 1}$ is locally equi-Lipschitz, in particular then locally equicontinuous. So, from part **(a)**, we have that $\varphi_n \xrightarrow{\Gamma} \varphi$.

Solution of Problem 3.142

(a) First suppose that $m > -\infty$ and given any $\varepsilon > 0$, let $u \in X$ be such that

$$\varphi(u) \leqslant m + \varepsilon.$$

According to Definition 3.137, we can find a sequence $\{v_n\}_{n\geqslant 1}$ such that $v_n \longrightarrow u$ and

$$\limsup_{n\to+\infty} \varphi_n(v_n) \ \leqslant\ \varphi(u) \ \leqslant\ m+\varepsilon,$$

so

$$\limsup_{n\to+\infty} m_n \ \leqslant\ m+\varepsilon.$$

Since $\varepsilon > 0$ is arbitrary, we let $\varepsilon \searrow 0$, to conclude that

$$\limsup_{n\to+\infty} m_n \ \leqslant\ m.$$

If $m = -\infty$, then for every $\varepsilon > 0$, there exists $u \in X$ such that $\varphi(u) \leqslant -\frac{1}{\varepsilon}$ and as above we can find $v_n \longrightarrow u$ such that

$$\limsup_{n\to+\infty} \varphi_n(v_n) \ \leqslant\ \varphi(u) \ \leqslant\ -\tfrac{1}{\varepsilon},$$

so

$$\limsup_{n\to+\infty} m_n \ \leqslant\ -\tfrac{1}{\varepsilon}.$$

Let $\varepsilon \searrow 0$ to conclude that $\limsup\limits_{n\to+\infty} m_n = -\infty = m$.

(b) Let $u \in \limsup\limits_{n\to+\infty} S_n(\varepsilon)$. Then we can find a subsequence $\{n_k\}_{k\geqslant 1}$ of $\{n\}_{n\geqslant 1}$ and $u_{n_k} \in S_{n_k}(\varepsilon)$ for all $k \geqslant 1$ such that $u_{n_k} \longrightarrow u$ (see Definition 2.88). Suppose that $-\infty < \limsup\limits_{n\to+\infty} m_n$. Then from part **(a)** we have that $-\infty < m$. By passing to a further subsequence if necessary, we may assume that $m_{n_k} > -\infty$ for all $k \geqslant 1$. Then

$$\varphi_{n_k}(u_{n_k}) \ \leqslant\ m_{n_k} + \varepsilon,$$

so

$$\varphi(u) \ \leqslant\ \liminf_{k\to+\infty} \varphi_{n_k}(u_{n_k}) \ \leqslant\ \limsup_{k\to+\infty} m_{n_k} + \varepsilon \ \leqslant\ m+\varepsilon$$

(see part **(a)**), so $u \in S(\varepsilon)$ and thus $\limsup\limits_{n\to+\infty} S_n(\varepsilon) \subseteq S(\varepsilon)$.

Next suppose that $m_n \longrightarrow -\infty$. Then given $\varepsilon > 0$, we can find $k_0 \geqslant 1$ such that

$$m_{n_k} + \varepsilon \ \leqslant\ -\tfrac{1}{\varepsilon} \qquad \forall\, k \geqslant k_0,$$

so

$$\varphi_{n_k}(u_{n_k}) \ \leqslant\ -\tfrac{1}{\varepsilon},$$

thus

$$\varphi(u) \ \leqslant\ \liminf_{k\to+\infty} \varphi_{n_k}(u_{n_k}) \ \leqslant\ -\tfrac{1}{\varepsilon},$$

hence $u \in S(\varepsilon)$ and finally $\limsup\limits_{n\to+\infty} S_n(\varepsilon) \subseteq S(\varepsilon)$.

(c) We assume that $\limsup\limits_{n\to+\infty} S_n(\varepsilon_n) \neq \emptyset$ or there is nothing to prove. So, let $u \in \limsup\limits_{n\to+\infty} S_n(\varepsilon_n)$ and let $u_{n_k} \in S_{n_k}(\varepsilon_{n_k})$ for $k \geqslant 1$ be such that $u_{n_k} \longrightarrow u$. Given $\varepsilon > 0$, we have

$$u_{n_k} \in S_{n_k}(\varepsilon) \qquad \forall\, k \geqslant k_0.$$

From part **(b)**, we have

$$u \in \bigcap_{\varepsilon>0} S(\varepsilon) = S(0).$$

We have

$$m \leqslant \varphi(u) \leqslant \liminf_{k\to+\infty} \varphi_{n_k}(u_{n_k}) \leqslant \limsup_{k\to+\infty} \varphi_{n_k}(u_{n_k}) \leqslant \limsup_{k\to+\infty} m_{n_k}$$

(see Definition 3.137 and recall that $u_{n_k} \in S_{n_k}(\varepsilon_{n_k})$). Then the result follows from part **(a)**.

(d) Let $u \in \limsup\limits_{n\to+\infty} S_n(\varepsilon_n)$. Then from **(c)** we have $u \in S(0)$. We have $u_{n_k} \in S_{n_k}(\varepsilon_{n_k})$ such that $u_{n_k} \longrightarrow u$ in X, so

$$\liminf_{k\to+\infty} \varphi_{n_k}(u_{n_k}) \geqslant \varphi(u) = m$$

(see Definition 3.137), thus

$$\liminf_{k\to+\infty} m_{n_k} \geqslant m,$$

hence $m_m \longrightarrow m$ (see part **(a)**) and of course $S(0) \neq \emptyset$.

(e) By virtue of part **(a)**, it suffices to show that

$$m \leqslant \liminf_{n\to+\infty} m_n.$$

We assume that $m > -\infty$. Let $\{n_k\}_{k\geqslant 1}$ be a subsequence of $\{n\}_{n\geqslant 1}$ such that

$$\lim_{k\to+\infty} m_{n_k} = \liminf_{n\to+\infty} m_n.$$

We may also assume that $u_{n_k} \longrightarrow u$. We have

$$-\infty < m \leqslant \varphi(u) \leqslant \liminf_{k\to+\infty} \varphi_{n_k}(u_{n_k}),$$

so

$$\varphi_{n_k}(u_{n_k}) > -\infty \qquad \forall\, k \geqslant 1 \text{ sufficiently big},$$

thus

$$\varphi_{n_k}(u_{n_k}) \leqslant m_{n_k} + \varepsilon_{n_k} \qquad \forall\, k \geqslant 1 \text{ big},$$

hence

$$m \leqslant \lim_{k\to+\infty} m_{n_k} = \liminf_{n\to+\infty} m_n$$

and finally $m_n \longrightarrow m$ (by part **(a)**).

Solution of Problem 3.143

"**(a)** $\Longrightarrow$ **(b)**": Let $u_0 \in S(0)$. Then $m = \varphi(u_0)$. If $-\infty < m$, then $m_n > -\infty$ for all $n \geqslant 1$ big. From Definition 3.137, we know that we can find a sequence $\{u_n\}_{n\geqslant 1} \subseteq X$ such that $u_n \longrightarrow u_0$ and $\varphi_n(u_n) \longrightarrow \varphi(u_0)$. Then

$$\varphi_n(u_n) \;=\; \varphi_n(u_n) - \varphi(u_0) + \varphi(u_0) \;\leqslant\; |\varphi_n(u_n) - \varphi(u_0)| + m_n + \delta_n,$$

with $\delta_n \searrow 0$, so

$$u_n \;\in\; S_n(\varepsilon_n), \quad \text{with } \varepsilon_n = |\varphi_n(u_n) - \varphi(u_0)| + \delta_n > 0,$$

with $\varepsilon_n \searrow 0$. If $m = -\infty$, then there exist two sequences $\{\varepsilon_n\}_{n\geqslant 1} \subseteq \mathbb{R}_+$ and $\{u_n\}_{n\geqslant 1} \subseteq X$ such that $\varepsilon_n \searrow 0$, $u_n \longrightarrow u_0$ and

$$\varphi_n(u_n) \;\leqslant\; -\tfrac{1}{\varepsilon_n} \qquad \forall\, n \geqslant 1 \text{ big}$$

(we take $\varepsilon_n = -\frac{1}{\varphi_n(u_n)}$), so

$$u_n \;\in\; S_n(\varepsilon_n) \quad \text{and} \quad u_n \;\longrightarrow\; u_0.$$

"**(b)** $\Longrightarrow$ **(a)**": This follows from Problem 3.142(e).

Solution of Problem 3.144

By taking K to be a singleton, from the hypothesis, it follows that $\varphi_n \longrightarrow \varphi$ pointwise. So, part **(b)** of Definition 3.137 is true for the constant sequence. We need to check that part **(a)** of Definition 3.137 holds. Arguing by contradiction and by passing to a suitable subsequence if necessary, we can find $\eta > 0$ and a sequence $\{u_n\}_{n\geqslant 1}$ such that

$$\varphi_n(u_n) \;\leqslant\; \eta \;<\; \varphi(u) \qquad \forall\, n \geqslant 1. \tag{3.205}$$

We fix $n \geqslant 1$ and let $m > n$. Then

$$\inf_{k\geqslant n} \varphi_m(u_k) \;\leqslant\; \varphi_m(u_m) \;\leqslant\; \eta. \tag{3.206}$$

Let $K_n = \{u\} \cup \{u_k :\; k \geqslant n\}$. Then

$$\inf_{K_n} \varphi_m \;\leqslant\; \inf_{k\geqslant n} \varphi_m(u_k) \;\leqslant\; \varphi_m(u_m) \;\leqslant\; \eta$$

(see (3.206)). Passing to the limit as $m \to +\infty$ and using the hypothesis (with $K = K_n$), we obtain

$$\inf_{K_n} \varphi = \min \{\varphi(u), \inf_{k \geqslant n} \varphi(u_k)\} \leqslant \eta,$$

so

$$\inf_{k \geqslant n} \varphi(u_k) \leqslant \eta.$$

Let $n \to +\infty$. Since φ is lower semicontinuous, we obtain

$$\varphi(u) \leqslant \eta,$$

which contradicts the hypothesis. Hence

$$\varphi(u) \leqslant \liminf_{n \to +\infty} \varphi_n(u_n),$$

for every sequence $\{u_n\}_{n \geqslant 1}$ such that $u_n \longrightarrow u$, so $\varphi_n \xrightarrow{\Gamma} \varphi$.

Solution of Problem 3.145
By taking the compact set K to be singleton, we have

$$\varphi_n(u) \longrightarrow \varphi(u) \qquad \forall\, u \in X. \tag{3.207}$$

Next let $u_n \longrightarrow u$ in X and let $K = \{u\} \cup \{u_n :\ n \geqslant 1\} \subseteq X$ (a compact set). By hypothesis, given $\varepsilon > 0$, we can find $n_0 = n_0(\varepsilon) \geqslant 1$ such that

$$|\varphi_n(v) - \varphi(v)| \leqslant \varepsilon \qquad \forall\, n \geqslant n_0,\ v \in K,$$

so

$$\varphi_n(u_n) \geqslant \varphi(u_n) - \varepsilon \qquad \forall\, n \geqslant n_0,$$

thus

$$\liminf_{n \to +\infty} \varphi_n(u_n) \geqslant \liminf_{n \to +\infty} \varphi(u_n) - \varepsilon \geqslant \varphi(u) - \varepsilon$$

(since φ is lower semicontinuous). Since $\varepsilon > 0$ is arbitrary, we let $\varepsilon \searrow 0$ to conclude that

$$\liminf_{n \to +\infty} \varphi_n(u_n) \geqslant \varphi(u). \tag{3.208}$$

From (3.207) and (3.208) we conclude that $\varphi_n \xrightarrow{\Gamma} \varphi$ (see Definition 3.137).

Solution of Problem 3.146

Let $u, v \in X$ be such that $\widehat{\varphi}(u), \widehat{\varphi}(v) < +\infty$. For $t \in [0,1]$ we set $y = (1-t)u + tv$. Let $V \in \mathcal{N}(y)$ (where $\mathcal{N}(y)$ is the filter of neighborhoods of y). We can find $r > 0$ such that

$$W = \{(1-t)u' + tv' : u' \in B_r(u),\ v' \in B_r(v)\} \subseteq V.$$

Then for every $n \geqslant 1$, we have

$$\inf_V \varphi_n \leqslant \inf_W \varphi_n = \inf\{\varphi((1-t)u' + tv') : u' \in B_r(u),\ v' \in B_r(v)\}. \tag{3.209}$$

From Remark 3.140, we know that

$$\limsup_{n\to+\infty} \inf_{B_r(u)} \varphi_n \leqslant \widehat{\varphi}(u) < +\infty \quad \text{and} \quad \limsup_{n\to+\infty} \inf_{B_r(v)} \varphi_n \leqslant \widehat{\varphi}(v) < +\infty.$$

So, we can find $n_0 \geqslant 1$ such that

$$\inf_{B_r(u)} \varphi_n < +\infty, \quad \inf_{B_r(v)} \varphi_n < +\infty \qquad \forall\, n \geqslant n_0.$$

Then for all $n \geqslant n_0$, we have

$$\begin{aligned} &\inf_{u'\in B_r(u)} \inf_{v'\in B_r(v)} \varphi_n((1-t)u' + tv') \\ &\leqslant (1-t) \inf_{u'\in B_r(u)} \varphi_n(u') + t \inf_{v'\in B_r(v)} \varphi_n(v') \end{aligned}$$

(since φ_n is convex), so

$$\limsup_{n\to+\infty} \inf_V \varphi_n \leqslant (1-t)\widehat{\varphi}(u) + t\widehat{\varphi}(v)$$

(see (3.209)), thus

$$\widehat{\varphi}(y) \leqslant (1-t)\widehat{\varphi}(u) + t\widehat{\varphi}(v)$$

(see Remark 3.140).

To see that $\widetilde{\varphi} = \Gamma\text{-}\liminf\limits_{n\to+\infty} \varphi_n$ need not be convex, consider the following counterexample. Let $X = \mathbb{R}$ and consider the sequence

$$\varphi_n(u) = (u - (-1)^n)^2 \qquad \forall\, n \geqslant 1.$$

Evidently for each $n \geqslant 1$, the function φ_n is convex and then Proposition 3.144 implies that

$$\widetilde{\varphi}(u) = \min\{(u-1)^2,\ (u+1)^2\},$$

so $\widetilde{\varphi}$ is not convex.

Solution of Problem 3.147

For $\lambda \in \mathbb{R}$, let $L_\lambda = \{u \in X : \widetilde{\varphi}(u) \leqslant \lambda\}$. We need to show that the set L_λ is closed (see Proposition I.2.53). So, let $\{u_k\}_{k \geqslant 1} \subseteq L_\lambda$ be a sequence such that $u_k \longrightarrow u$. We have

$$\widetilde{\varphi}(u_k) \leqslant \lambda \qquad \forall\, k \geqslant 1.$$

We fix $k \geqslant 1$. Then from Definition 3.139, we see that given $\varepsilon > 0$, we can find a sequence $\{u_k^n\}_{n \geqslant 1}$ such that $u_k^n \longrightarrow u_k$ and

$$\liminf_{n \to +\infty} \varphi_n(u_k^n) \leqslant \widetilde{\varphi}(u_k) + \varepsilon \leqslant \lambda + \varepsilon. \tag{3.210}$$

We can find a mapping $n \longmapsto k(n)$, increasing (not necessarily strictly) to $+\infty$ such that

$$u_{k(n)}^n \longrightarrow u.$$

From Definition 3.139, we have

$$\widetilde{\varphi}(u) \leqslant \liminf_{n \to +\infty} \varphi_n(u_{k(n)}^n) \leqslant \lambda + \varepsilon$$

(see (3.210)). Since $\varepsilon > 0$ is arbitrary, we let $\varepsilon \searrow 0$ to conclude that

$$\widetilde{\varphi}(u) \leqslant \lambda,$$

so $u \in L_\lambda$, hence L_λ is closed. Therefore $\widetilde{\varphi}$ is lower semicontinuous. The above argument applies also to $\widehat{\varphi}$ and so we conclude that $\widehat{\varphi}$ is lower semicontinuous too.

Solution of Problem 3.148

Let us denote

$$\widetilde{C} = \liminf_{n \to +\infty} C_n, \quad \widehat{C} = \limsup_{n \to +\infty} C_n$$

and

$$\widetilde{i} = \Gamma\text{-}\liminf_{n \to +\infty} i_{C_n}, \quad \widehat{i} = \Gamma\text{-}\limsup_{n \to +\infty} i_{C_n}$$

From the solution of Problem 3.139, we know that

$$i_{\widetilde{C}} = \widehat{i} \quad \text{and} \quad i_{\widehat{C}} = \widetilde{i}. \tag{3.211}$$

From Problem 3.147 we know that $\widetilde{i}$ and $\widehat{i}$ are both lower semicontinuous. Hence from (3.211) it follows that both sets $\widetilde{C}, \widehat{C} \subseteq X$ are closed.

Solution of Problem 3.149

From Proposition 3.59 (stated there for the particular case of metric spaces), we have

$$\operatorname{cl} \varphi(u) = \sup_{U \in \mathcal{N}(u)} \inf_{v \in U} \varphi(v), \tag{3.212}$$

with $\mathcal{N}(u)$ being the filter of neighborhoods of u. From (3.212) we see that

$$\operatorname{cl}\varphi \leqslant \varphi,$$

so

$$\inf_U \operatorname{cl}\varphi \leqslant \inf_U \varphi. \tag{3.213}$$

Next, let $v \in U$. Then from (3.212) we have

$$\inf_U \varphi \leqslant \operatorname{cl}\varphi(v).$$

Since this is true for every $v \in U$, we have

$$\inf_U \varphi \leqslant \inf_U \operatorname{cl}\varphi. \tag{3.214}$$

From (3.213) and (3.214) we have the desired equality.

Solution of Problem 3.150

From Remark 3.140 and Problem 3.149, we have

$$\begin{aligned}\big(\Gamma \liminf_{n\to+\infty} \varphi_n\big)(u) &= \sup_{V\in\mathcal{N}(u)} \liminf_{n\to+\infty} \inf_V \varphi_n \\ = \sup_{V\in\mathcal{N}(u)} \liminf_{n\to+\infty} \inf_V \operatorname{cl}\varphi_n &= \big(\Gamma\text{-}\liminf_{n\to+\infty} \operatorname{cl}\varphi_n\big)(u).\end{aligned}$$

Similarly for $\Gamma\text{-}\limsup\limits_{n\to+\infty} \varphi_n$ using the fact that

$$\big(\Gamma\text{-}\limsup_{n\to+\infty} \varphi_n\big)(u) = \sup_{V\in\mathcal{N}(u)} \limsup_{n\to+\infty} \inf_V \varphi_n.$$

Solution of Problem 3.151

From Definition 3.139 we know that for $u \in X$ and for any given $\varepsilon > 0$, we can find a sequence $\{u_n\}_{n\geqslant 1} \subseteq X$ such that $u_n \longrightarrow u$ in X and

$$\liminf_{n\to+\infty} (\varphi_n + \psi_n)(u_n) \leqslant \big(\Gamma\text{-}\liminf_{n\to+\infty} (\varphi_n + \psi_n)\big)(u) + \varepsilon,$$

so

$$\liminf_{n\to+\infty} \varphi_n(u_n) + \liminf_{n\to+\infty} \psi_n(u_n) \leqslant \big(\Gamma\text{-}\liminf_{n\to+\infty} (\varphi_n + \psi_n)\big)(u) + \varepsilon,$$

thus

$$\big(\Gamma\!-\!\liminf_{n\to+\infty}\varphi_n\big)(u)+\big(\Gamma\!-\!\liminf_{n\to+\infty}\psi_n\big)(u) \;\leqslant\; \big(\Gamma\!-\!\liminf_{n\to+\infty}(\varphi_n+\psi_n)\big)(u)+\varepsilon$$

(see Definition 3.139). Let $\varepsilon \searrow 0$, to conclude that

$$\Gamma\!-\!\liminf_{n\to+\infty}\varphi_n+\Gamma\!-\!\liminf_{n\to+\infty}\psi_n \;\leqslant\; \Gamma\!-\!\liminf_{n\to+\infty}(\varphi_n+\psi_n).$$

Next let $u\in X$ and note that for any sequence $\{u_n\}_{n\geqslant 1}\subseteq X$ such that $u_n\longrightarrow u$ in X, we have

$$\limsup_{n\to+\infty}(\varphi_n+\psi_n)(u_n) \;\geqslant\; \limsup_{n\to+\infty}\varphi_n(u_n)+\liminf_{n\to+\infty}\psi_n(u_n),$$

so

$$\big(\Gamma\!-\!\limsup_{n\to+\infty}(\varphi_n+\psi_n)\big)(u) \;\geqslant\; \big(\Gamma\!-\!\limsup_{n\to+\infty}\varphi_n\big)(u)+\big(\Gamma\!-\!\liminf_{n\to+\infty}\psi_n\big)(u)$$

(see Definition 3.139).

To see that these two inequalities may be strict, let $X=\mathbb{R}$, $\varphi_n=\sin(nu)$ and $\psi_n(u)=-\sin(nu)$ for all $u\in\mathbb{R}$ and all $n\geqslant 1$. Then $\varphi_n\xrightarrow{\Gamma}\varphi\equiv -1$, $\psi_n\xrightarrow{\Gamma}\psi\equiv -1$, but $(\varphi_n+\psi_n)\equiv 0$ for all $n\geqslant 1$.

Solution of Problem 3.152

Let $X=\mathbb{R}$ and consider the sequence $\varphi_n(u)=nue^{-2n^2u^2}$ for $n\geqslant 1$. Evidently $\varphi_n\longrightarrow 0$ pointwise, but $\varphi_n\xrightarrow{\Gamma}\varphi$, where

$$\varphi(u) \;=\; \begin{cases} -\frac{1}{2}e^{-\frac{1}{2}} & \text{if } \; u=0,\\ 0 & \text{if } \; u\neq 0.\end{cases}$$

Next consider the sequence

$$\varphi_n(u) \;=\; \begin{cases} nue^{-2n^2u^2} & \text{if } \; \text{if } n \text{ is even},\\ 2nue^{-2n^2u^2} & \text{if } \; \text{if } n \text{ is odd}.\end{cases}$$

We have $\varphi_n\longrightarrow 0$ pointwise. On the other hand,

$$\big(\Gamma\!-\!\liminf_{n\to+\infty}\varphi_n\big)(u) \;=\; \begin{cases} -e^{-\frac{1}{2}} & \text{if } \; u=0,\\ 0 & \text{if } \; u\neq 0,\end{cases}$$

$$\big(\Gamma\!-\!\limsup_{n\to+\infty}\varphi_n\big)(u) \;=\; \begin{cases} -\frac{1}{2}e^{-\frac{1}{2}} & \text{if } \; u=0,\\ 0 & \text{if } \; u\neq 0.\end{cases}$$

So, we see that the sequence $\{\varphi_n\}_{n\geqslant 1}$ does not Γ-converge.

Finally consider the sequence $\varphi_n(u)=nue^{nu}$ for $n\geqslant 1$. Then $\varphi_n\xrightarrow{\Gamma}\varphi$, where

$$\varphi(u) \;=\; \begin{cases} 0 & \text{if } \; u<0,\\ -\frac{1}{e} & \text{if } \; u=0,\\ +\infty & \text{if } \; u>0.\end{cases}$$

On the other hand, $\varphi_n(u) \longrightarrow 0$ for all $u \leqslant 0$ and $\varphi_n(u) \longrightarrow +\infty$ for all $u > 0$.

Solution of Problem 3.153

Let $(u,\lambda) \in \operatorname{epi}\varphi$. By hypothesis we can find a sequence $\{\lambda_n\}_{n\geqslant 1} \subseteq \mathbb{R}$ such that $\lambda_n \longrightarrow \lambda$ and $\varphi^\lambda \subseteq \liminf\limits_{n\to+\infty} \varphi_n^{\lambda_n}$ (see Definition 2.88). We have $u \in \varphi^\lambda$. So, we can find $u_n \in \varphi_n^{\lambda_n}$ for $n \geqslant 1$ such that $u_n \longrightarrow u$. We have $(u_n,\lambda_n) \in \operatorname{epi}\varphi_n$ for $n \geqslant 1$ and $(u_n,\lambda_n) \longrightarrow (u,\lambda)$ and so

$$\operatorname{epi}\varphi \subseteq \liminf_{n\to+\infty} \operatorname{epi}\varphi_n. \tag{3.215}$$

We will show that we also have

$$\limsup_{n\to+\infty} \operatorname{epi}\varphi_n \subseteq \operatorname{epi}\varphi. \tag{3.216}$$

Arguing by contradiction, suppose that we can find $(u,\mu) \in \limsup\limits_{n\to+\infty} \operatorname{epi}\varphi_n$ and $(u,\mu) \notin \operatorname{epi}\varphi$. Then $\mu < \varphi(u)$. Let $(u_{n_k},\mu_{n_k}) \in \operatorname{epi}\varphi_{n_k}$ for $k \geqslant 1$, be such that

$$u_{n_k} \longrightarrow u \quad \text{in } X \quad \text{and} \quad \mu_{n_k} \longrightarrow \mu \quad \text{in } \mathbb{R}.$$

Let $\lambda \in (\mu,\varphi(u))$ and let $\{\lambda_n\}_{n\geqslant 1} \subseteq \mathbb{R}$ be a sequence such that $\lambda_n \longrightarrow \lambda$ and $\varphi^{\lambda_n} \xrightarrow{K} \varphi^\lambda$ (it exists by hypothesis). For $k \geqslant 1$ big, we will have $\mu_{n_k} < \lambda_{n_k}$ and so $u_{n_k} \in \varphi^{\lambda_{n_k}}$ for all $k \geqslant 1$ big. Since $\varphi^{\lambda_n} \xrightarrow{K} \varphi^\lambda$, we obtain that $u \in \varphi^\lambda$, which contradicts the choice of $\lambda \in (\mu,\varphi(u))$. This shows that (3.216) holds. From (3.215) and (3.216) it follows that

$$\operatorname{epi}\varphi_n \xrightarrow{K} \operatorname{epi}\varphi,$$

so $\varphi_n \xrightarrow{\Gamma} \varphi$ (see Remark 3.138).

Solution of Problem 3.154

From Definition 3.137, we have

$$\varphi(u) \leqslant \liminf_{n\to+\infty} \varphi_n(u) \leqslant \varphi(u) \qquad \forall\, u \in X,$$

so

$$\varphi(u) = \liminf_{n\to+\infty} \varphi_n(u) \qquad \forall\, u \in X. \tag{3.217}$$

On the other hand, by hypothesis, we have

$$\limsup_{n\to+\infty} \varphi_n(u) \leqslant \varphi(u) \qquad \forall\, u \in X. \tag{3.218}$$

From (3.217) and (3.218) we conclude that $\varphi_n \longrightarrow \varphi$ pointwise.

Solution of Problem 3.155

Let $u \in X$. Since by hypothesis $\varphi_n \xrightarrow{\Gamma} \varphi$, according to Remark 3.140, given $\varepsilon > 0$, we can find $V_0 \in \mathcal{N}(u)$ (with $\mathcal{N}(u)$ being the filter of neighborhoods of u) and $n_0 = n_0(\varepsilon) \geqslant 1$ such that

$$\big|\varphi(u) - \inf_{v \in V_0} \varphi_n(v)\big| \leqslant \tfrac{\varepsilon}{2} \qquad \forall\, n \geqslant n_0,$$

so

$$\varphi(u) - \tfrac{\varepsilon}{2} \leqslant \inf_{v \in V_0} \varphi_n(v) \leqslant \varphi_n(v) \qquad \forall\, v \in V_0,\ n \geqslant n_0. \tag{3.219}$$

Since by hypothesis $\varphi_n \longrightarrow \varphi$ pointwise, we can find $n_1 \geqslant n_0$ such that

$$|\varphi_n(u) - \varphi(u)| \leqslant \tfrac{\varepsilon}{2} \qquad \forall\, n \geqslant n_1,$$

so

$$\varphi_n(u) - \tfrac{\varepsilon}{2} \leqslant \varphi(u) \qquad \forall\, n \geqslant n_1. \tag{3.220}$$

From (3.219) and (3.220) it follows that

$$\varphi_n(u) - \varepsilon \leqslant \varphi_n(v) \qquad \forall\, v \in V_0 \qquad \forall\, n \geqslant n_1 \tag{3.221}$$

(recall that $n_1 \geqslant n_0$). Also, since by hypothesis each φ_n is lower semicontinuous, for each $n \in \{1, \ldots, n_1 - 1\}$, we can find $V_n \in \mathcal{N}(u)$ such that

$$\varphi_n(u) - \varepsilon \leqslant \varphi_n(v) \qquad \forall\, v \in V_n \qquad \forall\, n \in \{1, \ldots, n_1 - 1\}. \tag{3.222}$$

Let $V = \bigcap_{n=0}^{n_1} V_n \in \mathcal{N}(u)$. Then from (3.221) and (3.222), it follows that

$$\varphi_n(u) - \varepsilon \leqslant \varphi_n(v) \qquad \forall\, v \in V,\ n \geqslant 1,$$

so the sequence $\{\varphi_n\}_{n \geqslant 1}$ is equi-lower semicontinuous (see Definition 3.143).

Solution of Problem 3.156

(a) Recall that $\operatorname{cl}\varphi \leqslant \varphi$ and $\operatorname{cl}\varphi$ is lower semicontinuous (see Proposition 3.59). So, we have

$$\operatorname{cl}\varphi(u) \leqslant \liminf_{n \to +\infty} \operatorname{cl}\varphi(u_n) \leqslant \liminf_{n \to +\infty} \varphi(u_n).$$

This establishes property **(a)**.

(b) Evidently, we may assume that $\operatorname{cl}\varphi(u) < +\infty$. Let $\{\lambda_n\}_{n\geqslant 1} \subseteq \mathbb{R}$ be a sequence such that $\lambda_n \searrow \operatorname{cl}\varphi(u)$. Using Problem 3.149, we see that for every $n \geqslant 1$, we can find $u_n \in B_{\frac{1}{n}}(u)$ for $n \geqslant 1$ such that

$$\lambda_n > \varphi(u_n) \qquad \forall\, n \geqslant 1. \tag{3.223}$$

Note that $u_n \longrightarrow u$ in X and

$$\limsup_{n\to+\infty} \varphi(u_n) \leqslant \lim_{n\to+\infty} \lambda_n = \operatorname{cl}\varphi(u)$$

(see (3.223)). This proves property **(b)**.

Solution of Problem 3.157

Yes. Note that $0 \in C_n$ for all $n \geqslant 1$. Also, suppose that $u_{n_k} \in C_{n_k}$ for all $k \geqslant 1$ and assume that $u_{n_k} \xrightarrow{w} u$ in X. Then the subsequence $\{u_{n_k}\}_{k\geqslant 1}$ is norm bounded and so we can find $c > 0$ such that

$$\|u_{n_k}\|_X \leqslant c \qquad \forall\, k \geqslant 1.$$

We have $u_{n_k} = \lambda'_{n_k} e_{n_k}$ for some $\lambda'_{n_k} \geqslant 0$. Hence $|\lambda'_{n_k}| \leqslant c$ for all $k \geqslant 1$. Also, we know that $e_{n_k} \xrightarrow{w} 0$. Hence $u_{n_k} \xrightarrow{w} 0$ in X and so we conclude that

$$C_n \xrightarrow{M} \{0\}$$

(see Definition 3.149).

Solution of Problem 3.158

(a) Let $u \in X$. Since $\varphi_n \xrightarrow{M} \varphi$, according to Definition 3.149, we can find a sequence $\{u_n\}_{n\geqslant 1} \subseteq X$ such that $u_n \longrightarrow u$ in X such that

$$\limsup_{n\to+\infty} \varphi_n(u_n) \leqslant \varphi(u).$$

Then we have

$$\limsup_{n\to+\infty} m_n \leqslant \limsup_{n\to+\infty} \varphi_n(u_n) \leqslant \varphi(u) \qquad \forall\, u \in X,$$

so

$$\limsup_{n\to+\infty} m_n \leqslant m$$

(b) Let $\{u_n\}_{n\geqslant 1} \subseteq X$ be a sequence such that $u_n \in S_n$ for $n \geqslant 1$ and $u_n \xrightarrow{w} u$. From Definition 3.149, we have

$$\varphi(u) \leqslant \liminf_{n\to+\infty} \varphi_n(u_n) = \liminf_{n\to+\infty} m_n \leqslant \limsup_{n\to+\infty} m_n \leqslant m$$

(see part **(a)**), so $\varphi(u) = m$ and thus $u \in S$ and $m_n \longrightarrow m$.

Solution of Problem 3.159

Suppose that $X = l^2$, $\{e_n\}_{n\geqslant 1}$ is the canonical orthonormal basis of X and $C_n = \operatorname{conv}\{0, e_\tau + e_{n+\tau}\}$ for $n \geqslant 1$. So, if s denotes the strong topology on X and w denotes the weak topology on X, then we have

$$C_n \xrightarrow{K_s} \{0\} \quad \text{and} \quad C_n \xrightarrow{K_w} \operatorname{conv}\{0, e_\tau\}.$$

It follows that

$$s\text{–}\liminf_{n\to+\infty} C_n = \{0\} \subsetneq \operatorname{conv}\{0, e_\tau\} = w\text{–}\limsup_{n\to+\infty} C_n$$

and so we cannot have Mosco convergence for the sequence $\{C_n\}_{n\geqslant 1}$.

Solution of Problem 3.160

Suppose that $C_n \xrightarrow{M} C$. Then **(a)** holds (see Definition 3.149) and since

$$C_n \cap K \subseteq C_n \qquad \forall\, n \geqslant 1,$$

we have

$$w\text{–}\limsup_{n\to+\infty} (C_n \cap K) \subseteq w\text{–}\limsup_{n\to+\infty} C_n \subseteq C,$$

which shows that part **(b)** holds.

Next suppose that **(a)** and **(b)** hold. We need to show that

$$w\text{–}\limsup_{n\to+\infty} C_n \subseteq C \tag{3.224}$$

(see Definition 3.149). So, let $u \in w\text{–}\limsup\limits_{n\to+\infty} C_n$. Then we can find a subsequence $\{n_k\}_{k\geqslant 1}$ of $\{n\}_{n\geqslant 1}$ and $u_{n_k} \subseteq C_{n_k}$ such that $u_{n_k} \xrightarrow{w} u$ in X. Let $K = \{u\} \cup \{u_{n_k} : k \geqslant 1\}$. Then K is w-compact and $u_{n_k} \in C_{n_k} \cap K$. So, according to **(b)**, we have that $u \in C$, hence (3.224) holds and we can conclude that $C_n \xrightarrow{M} C$.

Solution of Problem 3.161

"**(a)** $\Longrightarrow$ **(b)**": Since $u^* \in X^* \setminus \{0\}$ and $u_n^* \in X^* \setminus \{0\}$ for all $n \geqslant 1$, we can find $\delta > 0$ such that

$$\|u_n^*\|_* \geqslant \delta \qquad \forall\, n \geqslant 1.$$

Let $u \in (u^*)^{-1}(\lambda)$. We have $\langle u_n^*, u\rangle \longrightarrow \langle u^*, u\rangle = \lambda$. So, given $\varepsilon > 0$, we can find $n_0 = n_0(\varepsilon) \geqslant 1$ such that

$$|\langle u_n^*, u\rangle - \lambda| \leqslant \varepsilon \qquad \forall\, n \geqslant n_0. \tag{3.225}$$

Note that the choice of $\delta > 0$ implies that

$$\big[\langle u_n^*, u\rangle - \delta\varepsilon, \langle u_n^*, u\rangle + \delta\varepsilon\big] \subseteq \operatorname{im}\big(u_n^*|_{u+\varepsilon\overline{B}_1}\big).$$

Hence from (3.225) we infer that

$$(u_n^*)^{-1}(\lambda) \cap (u + \varepsilon\overline{B}_1) \neq \emptyset \qquad \forall\, n \geqslant n_0,$$

so

$$(u^*)^{-1}(\lambda) \subseteq \liminf_{n\to+\infty} (u_n^*)^{-1}(\lambda). \tag{3.226}$$

Next, let $u \in \limsup\limits_{n\to+\infty} (u_n^*)^{-1}(\lambda)$. We can find a subsequence $\{n_k\}_{k\geqslant 1}$ of $\{n\}_{n\geqslant 1}$ and $u_{n_k} \in (u_{n_k}^*)^{-1}(\lambda)$ such that $u_{n_k} \longrightarrow u$. We have

$$\begin{aligned}
\big|\big\langle u_{n_k}^*, u_{n_k}\big\rangle - \langle u^*, u\rangle\big| &\leqslant \big|\big\langle u_{n_k}^*, u_{n_k} - u\big\rangle + \big\langle u_{n_k}^* - u^*, u\big\rangle\big| \\
&\leqslant \|u_{n_k}^*\|_* \|u_{n_k} - u\|_X + \Big|\big\langle u_{n_k}^* - u^*, u\big\rangle\Big| \longrightarrow 0 \quad \text{as } k \to +\infty.
\end{aligned}$$

But $\big\langle u_{n_k}^*, u_{n_k}\big\rangle = \lambda$ for all $k \geqslant 1$ and so $\langle u^*, u\rangle = \lambda$, that is, $u \in (u^*)^{-1}(\lambda)$. Therefore

$$\limsup_{n\to+\infty} (u_n^*)^{-1}(\lambda) \subseteq (u^*)^{-1}(\lambda). \tag{3.227}$$

From (3.226) and (3.227) we conclude that

$$(u_n^*)^{-1}(\lambda) \xrightarrow{K} (u^*)^{-1}(\lambda).$$

"**(b)** $\Longrightarrow$ **(c)**": It is obvious.

"**(c)** $\Longrightarrow$ **(a)**": We claim that the sequence $\{u_n^*\}_{n\geqslant 1} \subseteq X^*$ is bounded. Indeed, if this is not the case, then we can find a subsequence $\{u_{n_k}^*\}_{k\geqslant 1}$ of $\{u_n^*\}_{n\geqslant 1}$ such that

$$\|u_{n_k}^*\|_* \geqslant k \qquad \forall\, k \geqslant 1.$$

So, we can find $u_k \in \frac{1}{k}\overline{B}_1(0)$ (where $\overline{B}_1(0) = \{u \in X : \|u\|_X \leqslant 1\}$) for $k \geqslant 1$, such that $u_k \in (u^*_{n_k})^{-1}(1)$ and so

$$0 \in \limsup_{n\to+\infty}(u^*_n)^{-1}(1) = (u^*)^{-1}(1),$$

a contradiction. So, the sequence $\{u^*_n\}_{n\geqslant 1} \subseteq X^*$ is bounded. Given $u \in (u^*)^{-1}(1)$, let $u_n \in (u^*_n)^{-1}(1)$ for $n \geqslant 1$ be such that $u_n \longrightarrow u$ (see **(c)**). Then

$$\begin{aligned} &|\langle u^*_n, u\rangle - \langle u^*, u\rangle| \\ \leqslant\ &|\langle u^*_n, u - u_n\rangle| + |\langle u^*_n - u^*, u_n\rangle| + |\langle u^*, u_n - u\rangle| \longrightarrow 0. \end{aligned}$$

So, the sequence $\{u^*_n\}_{n\geq 1}$ converges pointwise to u^* on $(u^*)^{-1}(1)$. Exploiting the linearity of u^*_n, we have pointwise convergence of $\{u^*_n\}_{n\geq 1}$ to u^* on $(u^*)^{-1}(\lambda)$ for all $\lambda \in \mathbb{R}\setminus\{0\}$. Since the sequence $\{u^*_n\}_{n\geqslant 1} \subseteq X^*$ is bounded and converges pointwise to u^* on a norm dense subset of X (the complement of $\ker u^*$), we must have pointwise convergence on all X. Therefore $u^*_n \xrightarrow{w^*} u^*$ in X^*.

Solution of Problem 3.162

"**(a)** $\Longrightarrow$ **(b)**": Evidently the sequence $\{u^*_n\}_{n\geqslant 1} \subseteq X^*$ is bounded and $u^*_n \xrightarrow{w^*} u^*$. So, from Problem 3.161, we have

$$(u^*_n)^{-1}(\lambda) \xrightarrow{K} (u^*)^{-1}(\lambda),$$

so

$$(u^*)^{-1}(\lambda) \subseteq s\text{-}\liminf_{n\to+\infty}(u^*_n)^{-1}(\lambda). \tag{3.228}$$

On the other hand, let $u \in w\text{-}\limsup\limits_{n\to+\infty}(u^*_n)^{-1}(\lambda)$. Then we can find a subsequence $\{n_k\}_{k\geqslant 1}$ of $\{n\}_{n\geqslant 1}$ and $u_{n_k} \in (u^*_{n_k})^{-1}(\lambda)$ for all $k \geqslant 1$ such that $u_{n_k} \xrightarrow{w} u$ in X. Since by hypothesis $u^*_{n_k} \longrightarrow u^*$ uniformly on w-compact sets of X, we have

$$\lambda = \left\langle u^*_{n_k}, u_{n_k}\right\rangle \longrightarrow \langle u^*, u\rangle,$$

so $\lambda = \langle u^*, u\rangle$, that is, $u \in (u^*)^{-1}(\lambda)$. Hence

$$w\text{-}\limsup_{n\to+\infty}(u^*_n)^{-1}(\lambda) \subseteq (u^*)^{-1}(\lambda). \tag{3.229}$$

From (3.228) and (3.229), we infer that

$$(u^*_n)^{-1}(\lambda) \xrightarrow{M} (u^*)^{-1}(\lambda).$$

"**(b)** $\Longrightarrow$ **(c)**": This is obvious.

"**(c) $\Longrightarrow$ (a)**": Since $(u_n^*)^{-1}(\lambda) \xrightarrow{M} (u^*)^{-1}(\lambda)$, we have $(u_n^*)^{-1}(\lambda) \xrightarrow{K_s} (u^*)^{-1}(\lambda)$ and so Problem 3.161 implies that the sequence $\{u_n^*\}_{n\geqslant 1} \subseteq X^*$ is bounded and $u_n^* \xrightarrow{w^*} u^*$ in X^*. Suppose that $u_n \xrightarrow{w} u$ in X. We need to show that, if $\eta = \lim\limits_{n\to+\infty} \langle u_n^*, u_n\rangle$, then $\eta = \langle u^*, u\rangle$.

Let $v \in X$ be such that $\langle u^*, v\rangle = 1 - \eta$. From Problem 3.161, we know that we can find $v_n \in (u_n^*)^{-1}(1-\eta)$ for $n \geqslant 1$, such that $v_n \longrightarrow v$ in X. Then

$$\langle u_n^*, u_n + v_n\rangle \longrightarrow \eta + 1 - \eta = 1.$$

So, we may assume that $\langle u_n^*, u_n + v_n\rangle \neq 0$ for all $n \geqslant 1$ and therefore we can define

$$\eta_n = \tfrac{1}{\langle u_n^*, u_n + v_n\rangle} \qquad \forall\, n \geqslant 1.$$

Then

$$\eta_n(u_n + v_n) \xrightarrow{w} u + v \quad \text{in } X.$$

Since $\eta_n(u_n + v_n) \in (u_n^*)^{-1}(1)$ for all $n \geqslant 1$, from the hypothesis, we have $u + v \in (u^*)^{-1}(1)$, so $\langle u^*, u + v\rangle = 1$ and thus

$$\langle u^*, u\rangle = 1 - \langle u^*, v\rangle = 1 - 1 + \eta = \eta,$$

which is what we wanted.

Solution of Problem 3.163

From Theorem 3.150, we know that

$$\varphi_n^* \xrightarrow{M} \varphi^*.$$

Hence, given $u^* \in X^*$, we can find a sequence $u_n^* \longrightarrow u^*$ in X^* such that

$$\varphi_n^*(u_n^*) \longrightarrow \varphi^*(u^*),$$

so

$$\inf_X (\varphi_n - u_n^*) = -\varphi_n^*(u_n^*) \longrightarrow -\varphi^*(u^*) = \inf_X (\varphi - u^*)$$

(see Definition 3.83).

Solution of Problem 3.164

From Problem 3.140 we know that

$$C_n \xrightarrow{M} C \iff i_{C_n} \xrightarrow{M} i_C. \tag{3.230}$$

Since $i^*_{C_n} = \sigma_{C_n}(\cdot)$ for $n \geqslant 1$ and $i^*_C = \sigma_C$, from Theorem 3.150, we have

$$i_{C_n} \xrightarrow{M} i_C \iff \sigma_{C_n}(\cdot) \xrightarrow{M} \sigma_C. \tag{3.231}$$

But note that for every $n \geqslant 1$ and every $u^*, v^* \in X^*$ we have

$$\left|\sigma_{C_n}(u^*) - \sigma_{C_n}(v^*)\right| \leqslant \|u^* - v^*\|_* \sup_{y \in W} \|y\|_X \qquad \forall\, n \geqslant 1,$$

so the sequence $\{\sigma_{C_n}(\cdot)\}_{n\geqslant 1}$ is equicontinuous (see Definition I.1.83). Thus, from Proposition 3.144 we have

$$\sigma_{C_n}(\cdot) \xrightarrow{M} \sigma_C \iff \sigma_{C_n}(\cdot) \longrightarrow \sigma_C \text{ pointwise.} \tag{3.232}$$

From (3.230), (3.231) and (3.232) we conclude that

$$C_n \xrightarrow{M} C \iff \sigma_{C_n}(u^*) \longrightarrow \sigma_C(u^*) \qquad \forall\, u^* \in X^*.$$

Solution of Problem 3.165

"$\Longrightarrow$": Suppose that $\varphi_n \xrightarrow{M} \varphi$. Let $(u, \lambda) \in w\text{-}\limsup\limits_{n\to+\infty} \operatorname{epi} \varphi_n$. Then we can find a subsequence $\{n_k\}_{k\geqslant 1}$ of $\{n\}_{n\geqslant 1}$ and $(u_{n_k}, \lambda_{n_k}) \in \operatorname{epi} \varphi_{n_k}$ for $k \geqslant 1$ such that

$$(u_{n_k}, \lambda_{n_k}) \longrightarrow (u, \lambda) \quad \text{in } X_w \times \mathbb{R},$$

with X_w being the Banach space X furnished with the weak topology. Since $\varphi_n \xrightarrow{M} \varphi$, we have

$$\varphi(u) \leqslant \liminf_{k\to+\infty} \varphi_{n_k}(u_{n_k}) \leqslant \lambda,$$

so $(u, \lambda) \in \operatorname{epi} \varphi$ and thus

$$w\text{-}\limsup_{n\to+\infty} \operatorname{epi} \varphi_n \subseteq \operatorname{epi} \varphi. \tag{3.233}$$

Let $(u, \lambda) \in \operatorname{epi} \varphi$. Then $\varphi(u) \leqslant \lambda$. Since $\varphi_n \xrightarrow{M} \varphi$, we can find a sequence $\{u_n\}_{n\geqslant 1} \subseteq X$ such that $u_n \longrightarrow u$ and

$$\varphi_n(u_n) \longrightarrow \varphi(u)$$

(see Definition 3.149). Let $\lambda_n = \varphi_n(u_n) + (\lambda - \varphi(u))$. Evidently $(u_n, \lambda_n) \in \operatorname{epi} \varphi_n$ for $n \geqslant 1$. Then $(u_n, \lambda_n) \longrightarrow (u, \lambda)$ and thus

$$\operatorname{epi} \varphi \subseteq s\text{-}\liminf_{n\to+\infty} \operatorname{epi} \varphi_n. \tag{3.234}$$

From (3.233) and (3.234) we infer that $\operatorname{epi} \varphi_n \xrightarrow{M} \operatorname{epi} \varphi$.

"$\Longleftarrow$": Suppose that $\operatorname{epi}\varphi_n \xrightarrow{M} \operatorname{epi}\varphi$. Let $\{u_n\}_{n\geqslant 1} \subseteq X$ be a sequence such that $u_n \xrightarrow{w} u$ in X. Then $(u_n, \varphi_n(u_n)) \in \operatorname{epi}\varphi_n$. We assume that $\liminf\limits_{n\to+\infty} \varphi_n(u_n) < +\infty$ and consider a subsequence $\{u_{n_k}\}_{k\geqslant 1}$ of the sequence $\{u_n\}_{n\geqslant 1}$ such that

$$\liminf_{n\to+\infty} \varphi_n(u_n) \;=\; \lim_{k\to+\infty} \varphi_{n_k}(u_{n_k}) \;=\; \lambda.$$

Then $(u_{n_k}, \varphi_{n_k}(u_{n_k})) \longrightarrow (u, \lambda) \in X_w \times \mathbb{R}$ and so $(u, \lambda) \in \operatorname{epi}\varphi$, hence

$$\varphi(u) \;\leqslant\; \lambda \;=\; \liminf_{n\to+\infty} \varphi(u_n). \tag{3.235}$$

On the other hand, let $u \in X$. Suppose that $u \in \operatorname{dom}\varphi$ (or otherwise there is nothing to prove). Then $(u, \varphi(u)) \in \operatorname{epi}\varphi$ and so we can find $(u_n, \lambda_n) \in \operatorname{epi}\varphi_n$ for $n \geqslant 1$ such that $(u_n, \lambda_n) \longrightarrow (u, \varphi(u))$ in $X \times \mathbb{R}$. We have

$$\varphi_n(u_n) \;\leqslant\; \lambda_n \qquad \forall\, n \geqslant 1,$$

so

$$\limsup_{n\to+\infty} \varphi_n(u_n) \;\leqslant\; \varphi(u). \tag{3.236}$$

From (3.235) and (3.236), we conclude that $\varphi_n \xrightarrow{M} \varphi$ (see Definition 3.149(b)).

Solution of Problem 3.166

(a) Let $u \in w\text{–}\limsup\limits_{n\to+\infty} \varphi_n^\lambda$. Then we can find a subsequence $\{n_k\}_{k\geqslant 1}$ of $\{n\}_{n\geqslant 1}$ and $u_{n_k} \in \varphi_{n_k}^\lambda$ for $k \geqslant 1$ such that $u_{n_k} \xrightarrow{w} u$. Since $\varphi_n \xrightarrow{M} \varphi$, from Definition 3.149, we have

$$\varphi(u) \;\leqslant\; \liminf_{k\to+\infty} \varphi_{n_k}(u_{n_k}) \;\leqslant\; \lambda,$$

so $u \in \varphi^\lambda$, thus

$$w\text{–}\limsup_{n\to+\infty} \varphi_n^\lambda \;\subseteq\; \varphi^\lambda. \tag{3.237}$$

Next, let $u \in \varphi^\lambda$. First we assume that $\varphi(u) < \lambda$. Since $\varphi_n \xrightarrow{M} \varphi$, from Definition 3.149, we know that we can find $u_n \longrightarrow u$ in X such that $\varphi_n(u_n) \longrightarrow \varphi(u) < \lambda$, hence $u_n \in s\text{–}\liminf\limits_{n\to+\infty} \varphi_n^\lambda$. Therefore

$$\varphi^\lambda \;\subseteq\; s\text{–}\liminf_{n\to+\infty} \varphi_n^\lambda. \tag{3.238}$$

Now assume that $\varphi(u) = \lambda$. By hypothesis $m = \inf\limits_X \varphi < \lambda$. So, we can find $v \in X$ such that $\varphi(v) < \varphi(u) = \lambda$. Let $y_k = (1 - \frac{1}{k})u + \frac{1}{k}v$ for all $k \geqslant 1$. The convexity of φ implies that

$$\varphi(y_k) \;\leqslant\; (1 - \tfrac{1}{k})\varphi(u) + \tfrac{1}{k}\varphi(v) \;<\; \lambda \qquad \forall\, k \geqslant 1.$$

So, from the first part of the solution we know that we can find $y_k^n \in \varphi_n^\lambda$ such that $y_k^n \longrightarrow y_k$ as $n \to +\infty$. Also, we have $y_k \longrightarrow u$. Therefore, we can find a map $n \longmapsto k(n)$ increasing (not necessarily strictly) to $+\infty$ such that $y_{k(n)}^n \longrightarrow u$ in X. Let $\widehat{y}_n = y_{k(n)}^n \in \varphi_n^\lambda$ for $n \geqslant 1$. Then $\widehat{y}_n \longrightarrow u \in s\text{-}\liminf\limits_{n\to+\infty} \varphi_n^\lambda$ and so again we have (3.238). From (3.237) and (3.238), we conclude that

$$\varphi_n^\lambda \xrightarrow{M} \varphi^\lambda.$$

(b) Let $u \in \operatorname{dom}\varphi$ and let $\lambda = \varphi(u) < +\infty$. By hypothesis, we can find $u_n \in \varphi_n^\lambda$ for $n \geqslant 1$ such that $u_n \longrightarrow u$ in X. We have $\varphi_n(u_n) \leqslant \lambda$ for all $n \geqslant 1$ and so

$$\limsup_{n\to+\infty} \varphi_n(u_n) \leqslant \varphi(u) = \lambda.$$

Next, let $(u_n, \lambda_n) \in \operatorname{epi}\varphi_n$ for $n \geqslant 1$ and assume that $(u_n, \lambda_n) \longrightarrow (u, \lambda)$ in $X_w \times \mathbb{R}$ (here X_w denotes the space X furnished with the weak topology). Then, given $\varepsilon > 0$, we can find $n_0 = n_0(\varepsilon) \geqslant 1$ such that $\lambda_n \leqslant \lambda + \varepsilon$ for all $n \geqslant n_0$. So, $u_n \in \varphi_n^{\lambda+\varepsilon}$ for all $n \geqslant n_0$. Since $u_n \xrightarrow{w} u$ in X, we have $u \in w\text{-}\limsup\limits_{n\to+\infty} \varphi_n^{\lambda+\varepsilon} = \varphi^{\lambda+\varepsilon}$ (by hypothesis). Hence

$$\varphi(u) \leqslant \lambda + \varepsilon.$$

Since $\varepsilon > 0$ is arbitrary, we let $\varepsilon \searrow 0$, to conclude that $\varphi(u) \leqslant \lambda$, that is, $(u, \lambda) \in \operatorname{epi}\varphi$. Therefore

$$w\text{-}\limsup_{n\to+\infty} \operatorname{epi}\varphi_n \subseteq \operatorname{epi}\varphi,$$

so $\varphi_n \xrightarrow{M} \varphi$ (see Problem 3.165).

Solution of Problem 3.167

No. We cannot drop the hypothesis that $m = \inf\limits_X \varphi < \lambda$. To see this consider the following example. Let $X = \mathbb{R}$ and consider the functions

$$\varphi_n(u) = \begin{cases} \frac{1}{n} & \text{if } u = 0, \\ +\infty & \text{if } n \neq 0 \end{cases} \qquad \forall\, n \geqslant 1$$

and $\varphi(u) = \delta_{\{0\}}(u)$. Evidently $\varphi_n \xrightarrow{M} \varphi$, but $\varphi_n^0 = \emptyset$ for all $n \geqslant 1$, while $\varphi^0 = \{0\}$.

Solution of Problem 3.168

No. The result of Problem 3.166(b) fails if we have $\varphi_n^\lambda \xrightarrow{M} \varphi^\lambda$ only for $\lambda \geqslant \inf_X \varphi$. To see this, consider the following example. Let $X = \mathbb{R}$ and consider the functions

$$\varphi_n(u) = \begin{cases} -1+\frac{1}{n} & \text{if } u=0, \\ +\infty & \text{if } u \neq 0, \end{cases} \qquad \forall\, n \geqslant 1$$
$$\varphi(u) = \begin{cases} 0 & \text{if } u=0, \\ +\infty & \text{if } u \neq 0. \end{cases}$$

For all $\lambda \geqslant 0$, we have $\varphi_n^\lambda = \varphi^\lambda = \{0\}$ for $n \geqslant 1$, but φ_n does not converge to φ in Mosco sense.

Solution of Problem 3.169

From Problem 3.140 we know that $i_{C_n} \xrightarrow{M} i_C$. We show that

$$\varphi + i_{C_n} \xrightarrow{M} \varphi + i_C. \tag{3.239}$$

To this end, let $u \in X$ and let $u_n \longrightarrow u$ in X be such that

$$\limsup_{n\to+\infty} i_{C_n}(u_n) \leqslant i_C(u),$$

so

$$\limsup_{n\to+\infty} \big(\varphi(u_n) + i_{C_n}(u_n)\big) \leqslant \varphi(u) + i_C(u) \tag{3.240}$$

(recall that φ is continuous). Next, suppose that $u_n \xrightarrow{w} u$ in X. The convexity of φ implies that φ is weakly lower semicontinuous. Hence we have

$$\begin{aligned} \liminf_{n\to+\infty} \left(\varphi + i_{C_n}\right)(u_n) &\geqslant \liminf_{n\to+\infty} \varphi(u_n) + \liminf_{n\to+\infty} i_{C_n}(u_n) \\ &\geqslant \varphi(u) + i_C(u) \end{aligned} \tag{3.241}$$

(since $i_{C_n} \xrightarrow{M} i_C$; see Definition 3.149). From (3.240) and (3.241), we conclude that (3.239) holds.

Let

$$\begin{aligned} S_n &= \{u \in C_n :\ m_n = \varphi(u)\} \qquad \forall\, n \geqslant 1, \\ S &= \{u \in C :\ m = \varphi(u)\}. \end{aligned}$$

By the Weierstrass theorem (see Theorem I.1.75), we have that $S_n \neq \emptyset$ for all $n \geqslant 1$ and $S \neq \emptyset$.

Let $u \in C$ and let $\{u_n\}_{n\geqslant 1} \subseteq X$ be a sequence such that $u_n \longrightarrow u$ in X and

$$\limsup_{n\to+\infty} (\varphi + i_{C_n})(u_n) \leqslant (\varphi + i_C)(u) < +\infty$$

(see (3.239)), so

$$(\varphi + i_{C_n})(u_n) \leqslant c_0 \qquad \forall\, n \geqslant 1, \tag{3.242}$$

for some $c_0 > 0$. Let $v_n \in S_n$ for $n \geqslant 1$. Then

$$(\varphi + i_{C_n})(v_n) \leqslant c_0 \qquad \forall\, n \geqslant 1,$$

so

$$\varphi(v_n) \leqslant c_0 \qquad \forall\, n \geqslant 1,$$

thus the sequence $\{v_n\}_{n \geqslant 1} \subseteq X$ is bounded (since φ is coercive). So, by passing to a suitable subsequence if necessary, we may assume that $v_n \xrightarrow{w} v$ in X (by the Alaoglu and Eberlein–Smulian theorems; see Theorems I.5.66 and I.5.78). From Problem 3.158(b), we have that $v \in S$ and $m_n \longrightarrow m$.

Solution of Problem 3.170

We have

$$\varphi(u) \geqslant c_0\|u\|_H^2 - \|u^*\|_H\|u\|_H \qquad \forall\, u \in H,$$

for some $c_0 > 0$, so φ is coercive (that is, $\varphi(u) \longrightarrow +\infty$ as $\|u\|_H \to +\infty$). Also, φ is strictly convex and continuous. Then we know that S and S_n (for $n \geqslant 1$) are nonempty and because of the strict convexity of φ, they are singletons, let us say $S = \{u\}$ and $S_n = \{u_n\}$ for $n \geqslant 1$. From the solution of Problem 3.169, we have

$$u_n \xrightarrow{w} u \quad \text{in } H. \tag{3.243}$$

We have

$$\tfrac{1}{2}a(u_n, u_n) = \tfrac{1}{2}a(u, u) + a(u_n - u, u) + \tfrac{1}{2}a(u_n - u, u_n - u),$$

so

$$\varphi(u_n) \geqslant \varphi(u) - (u^*, u_n - u)_H + a(u_n - u, u) + \tfrac{c}{2}\|u_n - u\|_H^2,$$

for some $c > 0$ (because a is coercive), so

$$m_n - m + (u^*, u_n - u)_H - a(u_n - u, u) \geqslant \tfrac{c}{2}\|u_n - u\|_H^2 \qquad \forall\, n \geqslant 1. \tag{3.244}$$

From Problem 3.169 we have $m_n \longrightarrow m$. Also, because of (3.243), we have

$$(u^*, u_n - u)_H \longrightarrow 0 \quad \text{and} \quad a(u_n - u, u) \longrightarrow 0.$$

Therefore from (3.244) we conclude that $u_n \longrightarrow u$ in H.

Solution of Problem 3.171

From Problem 3.151, we have

$$\Gamma\text{-}\liminf_{n\to+\infty}\varphi_n+\Gamma\text{-}\liminf_{n\to+\infty}\psi_n \leqslant \Gamma\text{-}\liminf_{n\to+\infty}(\varphi_n+\psi_n), \tag{3.245}$$

$$\Gamma\text{-}\limsup_{n\to+\infty}(\varphi_n+\psi_n) \leqslant \Gamma\text{-}\limsup_{n\to+\infty}\varphi_n+\Gamma\text{-}\limsup_{n\to+\infty}\psi_n. \tag{3.246}$$

From the pointwise convergence of the sequence, we have

$$\begin{aligned}\lim_{n\to+\infty}(\varphi_n+\psi_n) &= \varphi+\psi\\ &= \Gamma\text{-}\limsup_{n\to+\infty}\varphi_n+\Gamma\text{-}\limsup_{n\to+\infty}\psi_n \geqslant \Gamma\text{-}\limsup_{n\to+\infty}(\varphi_n+\psi_n)\end{aligned} \tag{3.247}$$

(see (3.246)). Also, we have

$$\varphi+\psi = \Gamma\text{-}\liminf_{n\to+\infty}\varphi_n+\Gamma\text{-}\liminf_{n\to+\infty}\psi_n \leqslant \Gamma\text{-}\liminf_{n\to+\infty}(\varphi_n+\psi_n) \tag{3.248}$$

(see (3.245)). From (3.247), (3.248) and Proposition 3.141(b), we conclude that $\varphi_n+\psi_n \xrightarrow{\Gamma} \varphi+\psi$.

Solution of Problem 3.172

"$\Longrightarrow$": Suppose that $\varphi \leqslant \Gamma\text{-}\liminf\limits_{n\to+\infty}\varphi_n$ and $K\subseteq\mathbb{R}^N$ is a compact set. Given $\varepsilon>0$, we can find $y_n\in K$ for $n\geqslant 1$, such that

$$\varphi_n(y_n) \leqslant \inf_K \varphi_n+\varepsilon \qquad \forall\, n\geqslant 1. \tag{3.249}$$

Since the set K is compact, passing to a subsequence if necessary, we may assume that $y_n\longrightarrow y\in K$. We have

$$\big(\Gamma\text{-}\liminf_{n\to+\infty}\varphi_n\big)(y) \leqslant \liminf_{n\to+\infty}\varphi_n(y_n) \leqslant \liminf_{n\to+\infty}(\inf_K\varphi_n)+\varepsilon$$

(see Definition 3.139 and (3.249)), so

$$\inf_K\varphi \leqslant \varphi(y) \leqslant \liminf_{n\to+\infty}(\inf_K\varphi_n)+\varepsilon.$$

Because $\varepsilon>0$ is arbitrary, we let $\varepsilon\searrow 0$ to conclude that

$$\inf_K\varphi \leqslant \liminf_{n\to+\infty}(\inf_K\varphi_n).$$

"$\Longleftarrow$": Now, suppose that for every compact set $K\subseteq\mathbb{R}^N$, we have

$$\inf_K\varphi \leqslant \liminf_{n\to+\infty}(\inf_K\varphi_n).$$

According to Remark 3.140, we have

$$\big(\Gamma\text{-}\liminf_{n\to+\infty}\varphi_n\big)(u) \;=\; \sup_{r>0}\liminf_{n\to+\infty}\big(\inf_{\overline{B}_r(u)}\varphi_n\big). \tag{3.250}$$

But $\overline{B}_r(u)$ is compact in $\mathbb{R}^N$. So, by hypothesis, we have

$$\inf_{\overline{B}_r(u)}\varphi \;\leqslant\; \liminf_{n\to+\infty}\big(\inf_{\overline{B}_r(u)}\varphi_n\big),$$

thus

$$\sup_{r>0}\inf_{\overline{B}_r(u)}\varphi \;\leqslant\; \big(\Gamma\text{-}\liminf_{n\to+\infty}\varphi_n\big)(u)$$

(see (3.250)) and hence

$$\varphi(u) \;\leqslant\; \big(\Gamma\text{-}\liminf_{n\to+\infty}\varphi_n\big)(u)$$

(since φ is lower semicontinuous; see Remark I.2.47).

Solution of Problem 3.173

"$\Longrightarrow$": Suppose that $\inf\limits_{\mathbb{R}^N}\varphi_n \longrightarrow \inf\limits_{\mathbb{R}^N}\varphi$. Let $\varepsilon>0$ be given and choose $u\in\mathbb{R}^N$ such that

$$\varphi(u) \;\leqslant\; \inf_{\mathbb{R}^N}\varphi+\tfrac{\varepsilon}{3}. \tag{3.251}$$

Since by hypothesis $\varphi_n \xrightarrow{\Gamma} \varphi$, we can find a sequence $\{u_n\}_{n\geqslant 1}\subseteq\mathbb{R}^N$ such that $u_n\longrightarrow u$ and

$$\limsup_{n\to+\infty}\varphi_n(u_n) \;\leqslant\; \varphi(u) \tag{3.252}$$

(see Definition 3.137). Let $K=\{u\}\cup\{u_n:\ n\geqslant 1\}$. This is a compact set in $\mathbb{R}^N$. We have

$$\inf_K\varphi_n \;\leqslant\; \varphi_n(u_n) \qquad \forall\, n\geqslant 1,$$

so, we can find $n_1\geqslant 1$ such that

$$\inf_K\varphi_n \;\leqslant\; \varphi(u)+\tfrac{\varepsilon}{3} \qquad \forall\, n\geqslant n_1 \tag{3.253}$$

(see (3.252)). Also, since by hypothesis $\inf\limits_{\mathbb{R}^N}\varphi_n \longrightarrow \inf\limits_{\mathbb{R}^N}\varphi$, we can find $n_2\geqslant n_1$ such that

$$\inf_{\mathbb{R}^N}\varphi \;\leqslant\; \inf_{\mathbb{R}^N}\varphi_n+\tfrac{\varepsilon}{3} \qquad \forall\, n\geqslant n_2. \tag{3.254}$$

Then from (3.251), (3.253) and (3.254) we have

$$\inf_K\varphi_n \;\leqslant\; \inf_{\mathbb{R}^N}\varphi_n+\varepsilon \qquad \forall\, n\geqslant n_2.$$

"$\Longleftarrow$": From Problem 3.142 (see also Problem 3.158), we have

$$\limsup_{n\to+\infty}(\inf_{\mathbb{R}^N}\varphi_n) \leqslant \inf_{\mathbb{R}^N}\varphi. \tag{3.255}$$

By hypothesis, given $\varepsilon > 0$, we can find a compact set $K \subseteq \mathbb{R}^N$ and $n_0 \geqslant 1$ such that

$$\inf_K \varphi_n \leqslant \inf_{\mathbb{R}^N}\varphi_n + \varepsilon \qquad \forall\, n \geqslant n_0. \tag{3.256}$$

Note that φ is lower semicontinuous (see Problem 3.147). So, we can use Problem 3.172 and have

$$\inf_{\mathbb{R}^N}\varphi \leqslant \inf_K \varphi \leqslant \liminf_{n\to+\infty}(\inf_K \varphi_n) \leqslant \liminf_{n\to+\infty}(\inf_{\mathbb{R}^N}\varphi_n) + \varepsilon$$

(see (3.256)). Since $\varepsilon > 0$ is arbitrary, we let $\varepsilon \searrow 0$ and have

$$\inf_{\mathbb{R}^N}\varphi \leqslant \liminf_{n\to+\infty}(\inf_{\mathbb{R}^N}\varphi_n). \tag{3.257}$$

From (3.255) and (3.257) we conclude that $\inf\limits_{\mathbb{R}^N}\varphi_n \longrightarrow \inf\limits_{\mathbb{R}^N}\varphi$.

Solution of Problem 3.174

Using Remark 3.140 we have

$$\widetilde{\varphi} = \sup_{r>0}\liminf_{n\to+\infty}\inf_{B_r(u)}\varphi_n = \sup_{r>0}\liminf_{n\to+\infty}\inf_{B_r(-u)}\varphi_n = \widetilde{\varphi}(-u)$$

(where $B_r(u) = \{v \in X : \|v-u\|_X < r\}$ and since φ_n are even), so $\widetilde{\varphi}$ is even.

Similarly for $\widehat{\varphi}$, since

$$\widehat{\varphi} = \sup_{r>0}\limsup_{n\to+\infty}\inf_{B_r(u)}\varphi_n.$$

The result fails for odd functions. To see this, let $X = \mathbb{R}$ and let

$$\varphi_n(u) = u\cos(nu) \qquad \forall\, n \geqslant 1.$$

Clearly φ_n is odd for all $n \geqslant 1$. Let

$$\varphi(u) = -|u| \qquad \forall\, u \in \mathbb{R}.$$

Directly from Definition 3.139 we have $\varphi_n \xrightarrow{\Gamma} \varphi$ and φ is not odd.

Solution of Problem 3.175

Consider the sequence $\{i_{C_n}\}_{n\geqslant 1}$. According to Proposition 3.141(c), we can find a subsequence $\{i_{C_{n_k}}\}_{n\geqslant 1}$ of $\{i_{C_n}\}_{n\geqslant 1}$ and a lower semicontinuous function $\varphi\colon X \longrightarrow \overline{\mathbb{R}}$ (see Problem 3.147) such that

$$i_{C_{n_k}} \stackrel{\Gamma}{\longrightarrow} \varphi.$$

Evidently $\operatorname{im}\varphi \subseteq \{0,+\infty\}$. Hence we can find a set $C \in P_f(X) \cup \{\emptyset\}$ such that $\varphi = i_C$. Then

$$i_{C_{n_k}} \longrightarrow i_C,$$

so $C_{n_k} \stackrel{K}{\longrightarrow} C$ (see Problem 3.139).

Solution of Problem 3.176

From Remark 2.89 and Problems 2.134 and 2.135 (recall that X is finite dimensional), we obtain that conditions **(a)** and **(b)** are equivalent. But for each $n \geqslant 1$, the function $\operatorname{dist}(\cdot, C_n)$ is nonexpansive (that is, Lipschitz continuous with Lipschitz constant 1). Hence $\{\operatorname{dist}(\cdot, C_n)\}_{n\geqslant 1}$ is a sequence of equicontinuous functions (see Definition I.1.83). Invoking Proposition 3.144, we obtain that conditions **(b)** and **(c)** are equivalent.

Solution of Problem 3.177

"**(a)** $\Longrightarrow$ **(b)**": Assume that $\varphi_n \stackrel{M}{\longrightarrow} \varphi$. From Theorem 3.150 we know that

$$\varphi_n^* \stackrel{M}{\longrightarrow} \varphi^*.$$

Then according to Definition 3.149, we can find subsequences of $\{u_n\}_{n\geqslant 1}$ and $\{u_n^*\}_{n\geqslant 1}$ such that

$$u_n \longrightarrow u \quad \text{in } X, \quad u_n^* \longrightarrow u^* \quad \text{in } X^*,$$

$$\varphi_n(u_n) \longrightarrow \varphi(u) \quad \text{and} \quad \varphi_n^*(u_n^*) \longrightarrow \varphi^*(u^*).$$

"**(b)** $\Longrightarrow$ **(a)**": Suppose that $u_n \stackrel{w}{\longrightarrow} u$ in X. By hypothesis for every $u^* \in X^*$, we can find a sequence $\{u_n^*\}_{n\geqslant 1} \subseteq X^*$ such that $u_n^* \longrightarrow u^*$ in X^* and $\varphi_n^*(u_n^*) \longrightarrow \varphi^*(u^*)$. We have

$$\varphi_n = (\varphi_n^*)^* \qquad \forall\, n \geqslant a$$

(see Theorem 3.85) and so

$$\langle u_n^*, u_n\rangle - \varphi_n^*(u_n^*) \leqslant \varphi_n(u_n) \qquad \forall\, n \geqslant 1$$

(see Definition 3.83). Hence we have

$$\langle u^*, u\rangle - \varphi^*(u^*) \leqslant \liminf_{n\to+\infty} \varphi_n(u_n). \tag{3.258}$$

Since $u^* \in X^*$ is arbitrary, from (3.258) and Theorem 3.85 (recall that $\varphi \in \Gamma_0(X)$), we have

$$\varphi(u) \leqslant \liminf_{n\to+\infty} \varphi_n(u_n). \tag{3.259}$$

From (3.259) and hypothesis **(b)**, we infer that $\varphi_n \xrightarrow{M} \varphi$ (see Definition 3.149).

Solution of Problem 3.178

(a) From Proposition 3.145(a), we know that $\varphi_n \xrightarrow{\Gamma} \varphi = \sup\limits_{n\geqslant 1} \varphi_n$ for the strong topology. Therefore for every $u \in X$, we can find a sequence $\{u_n\}_{n\geqslant 1} \subseteq X$ such that $u_n \longrightarrow u$ in X and

$$\varphi_n(u_n) \longrightarrow \varphi(u). \tag{3.260}$$

From Proposition 3.84(c), we see that the sequence $\{\varphi_n^*\}_{n\geqslant 1} \subseteq \Gamma_0(X^*)$ is decreasing and so Proposition 3.145(b) implies that $\varphi_n^* \xrightarrow{\Gamma} \varphi^* = \mathrm{cl}\,(\inf\limits_{n\geqslant 1} \varphi_n^*)$ for the strong topology on X^*. So, given $u^* \in X^*$, we can find a sequence $\{u_n^*\}_{n\geqslant 1} \subseteq X^*$ such that $u_n^* \longrightarrow u^*$ in X^* and

$$\varphi_n^*(u_n^*) \longrightarrow \mathrm{cl}\,(\inf_{n\geqslant 1} \varphi_n^*)(u^*). \tag{3.261}$$

From (3.260), (3.261) and Problem 3.177, we infer that $\varphi_n \xrightarrow{M} \sup\limits_{n\geqslant 1} \varphi_n$.

(b) This is done similarly as part **(a)**.

Solution of Problem 3.179

(a) Let the sequence $\{C_n\}_{n\geqslant 1}$ be increasing (i.e., $C_n \subseteq C_{n+1}$ for all $n \geqslant 1$). Evidently the sequence $\{i_{C_n}\}_{n\geqslant 1}$ is decreasing. So, from Problem 3.178(b), we have

$$i_{C_n} \xrightarrow{M} \varphi,$$

with $\varphi = \mathrm{cl}\,(\inf\limits_{n\geqslant 1} i_{C_n}) = i_C$ with $C = \overline{\bigcup\limits_{n\geqslant 1} C_n}$, so

$$i_{C_n} \xrightarrow{M} i_C$$

and thus

$$C_n \xrightarrow{M} C = \overline{\bigcup_{n\geqslant 1} C_n}$$

(see Problem 3.140).

(b) This is done similarly as part **(a)**, using this time Problem 3.178(a).

Solution of Problem 3.180

Let $u \in C$. Then $\operatorname{dist}(u, C) = 0$. Since $\operatorname{dist}(\cdot, C_n) \xrightarrow{M} \operatorname{dist}(\cdot, C)$, we can find a sequence $\{u_n\}_{n\geqslant 1} \subseteq X$ such that $u_n \longrightarrow u$ in X and

$$\operatorname{dist}(u_n, C) \longrightarrow \operatorname{dist}(u, C) = 0 \tag{3.262}$$

(see Definition 3.149). Let $c_n \in C_n$ for $n \geqslant 1$ be such that

$$\|u_n - c_n\|_X \leqslant \operatorname{dist}(u_n, C_n) + \tfrac{1}{n} \qquad \forall\, n \geqslant 1.$$

Then

$$\|u_n - c_n\|_X \longrightarrow 0$$

(see (3.262)), so

$$c_n \longrightarrow u \quad \text{in } X, \quad \text{with } c_n \in C_n,$$

thus

$$u \in s\text{-}\liminf_{n\to+\infty} C_n. \tag{3.263}$$

On the other hand, let $u \in w\text{-}\limsup\limits_{n\to+\infty} C_n$. Then we can find a subsequence $\{n_k\}_{k\geqslant 1}$ of $\{n\}_{n\geqslant 1}$ and $u_{n_k} \in C_{n_k}$ for $k \geqslant 1$ such that $u_{n_k} \xrightarrow{w} u$ in X. Since $\operatorname{dist}(\cdot, C_{n_k}) \xrightarrow{M} \operatorname{dist}(\cdot, C)$, we have

$$\operatorname{dist}(u, C) \leqslant \liminf_{n\to+\infty} \operatorname{dist}(u_{n_k}, C_{n_k}),$$

so $\operatorname{dist}(u, C) = 0$, hence $u \in C$ and thus

$$w\text{-}\limsup_{n\to+\infty} C_n \subseteq C. \tag{3.264}$$

From (3.263) and (3.264), we conclude that $C_n \xrightarrow{M} C$.

Solution of Problem 3.181
Since $C_n \xrightarrow{M} C$, we have $i_{C_n} \xrightarrow{M} i_C$ (see Problem 3.140) and so

$$\sigma_{C_n}(\cdot) = i^*_{C_n}(\cdot) \xrightarrow{M} i^*_C = \sigma_C$$

(see Theorem 3.150). Note that for every $n \geqslant 1$, we have

$$\mathrm{dist}(\cdot, C_n) = (i_{C_n} \boxdot \|\cdot\|_X)$$

(see Definition 3.87), so

$$\mathrm{dist}(\cdot, C_n)^* = i^*_{C_n} + \|\cdot\|^* = \sigma_{C_n} + i_{\overline{B}^*_1} \tag{3.265}$$

(see Proposition 3.89(a); here $\overline{B}^*_1 = \{u^* \in X^* : \|u^*\|_* \leqslant 1\}$). If $u^*_n \xrightarrow{w} u^*$ in X^*, then

$$\begin{aligned} \liminf_{n\to+\infty} \mathrm{dist}(\cdot, C_n)^*(u^*_n) &\geqslant \liminf_{n\to+\infty} \sigma_{C_n}(u^*_n) + \liminf_{n\to+\infty} i_{\overline{B}^*_1}(u^*_n) \\ &\geqslant \sigma_C(u^*) + i_{\overline{B}^*_1}(u^*) \end{aligned} \tag{3.266}$$

(since $\sigma_{C_n}(\cdot) \xrightarrow{M} \sigma_C$; see Definition 3.149 and $i_{\overline{B}^*_1} \in \Gamma_0(X^*)$).

On the other hand, because $\sigma_{C_n} \xrightarrow{M} \sigma_C$, given $u^* \in X^*$, we can find a sequence $\{u^*_n\}_{n\geqslant 1} \subseteq X^*$ such that $u^*_n \longrightarrow u^*$ in X^* and

$$\sigma_{C_n}(u^*_n) \longrightarrow \sigma_C(u^*)$$

(see Definition 3.149), so

$$\|u^*_n\|_* \sigma_{C_n}\Big(\frac{u^*_n}{\|u^*_n\|_*}\Big) \longrightarrow \|u^*\|_* \sigma_C\Big(\frac{u^*}{\|u^*\|_*}\Big)$$

(exploiting the positive homogeneity of the support function), thus

$$\sigma_{C_n}\Big(\frac{u^*_n}{\|u^*_n\|_*}\Big) \longrightarrow \sigma_C\Big(\frac{u^*}{\|u^*\|_*}\Big). \tag{3.267}$$

Then since $\Big\|\frac{u^*_n}{\|u^*_n\|_*}\Big\|_* = 1$ for $n \geqslant 1$ and $\Big\|\frac{u^*}{\|u^*\|_*}\Big\|_* = 1$, we have

$$\mathrm{dist}(\cdot, C_n)^*\Big(\frac{u^*_n}{\|u^*_n\|_*}\Big) = \sigma_{C_n}\Big(\frac{u^*_n}{\|u^*_n\|_*}\Big) \longrightarrow \sigma_C\Big(\frac{u^*}{\|u^*\|_*}\Big) = \mathrm{dist}\Big(\frac{u^*}{\|u^*\|_*}, C\Big)^* \tag{3.268}$$

(see (3.267)). From (3.266) and (3.268), it follows that

$$\mathrm{dist}(\cdot, C_n)^* \xrightarrow{M} \mathrm{dist}(\cdot, C)^*. \tag{3.269}$$

But the functions $\mathrm{dist}(\cdot, C)$ and $\mathrm{dist}(\cdot, C_n)$ for $n \geqslant 1$ are continuous and convex. Then from (3.269) and Theorem 3.150 we conclude that $\mathrm{dist}(\cdot, C_n) \xrightarrow{M} \mathrm{dist}(\cdot, C)$.

Solution of Problem 3.182
For every $n \geqslant 1$, we set

$$\widehat{\varphi}_n(\omega, u) = \inf_{v \in X} \big(\varphi(\omega, v) + n d_X(v, u)\big).$$

Clearly, we have

$$\beta(\omega) \leqslant \widehat{\varphi}_1(\omega, u) \leqslant \ldots \leqslant \widehat{\varphi}_n(\omega, u) \leqslant \widehat{\varphi}_{n+1}(\omega, u) \leqslant \ldots \leqslant \varphi(\omega, u)$$

for μ-almost all $\omega \in \Omega$ and all $u \in \mathbb{R}$. Fix $n \geqslant 1$, $u \in X$ and $\lambda \in \mathbb{R}$. Then

$$\begin{aligned} & \{\omega \in \Omega : \widehat{\varphi}_n(\omega, u) < \lambda\} \\ = \ & \operatorname{proj}_\Omega \{(\omega, v) \in \Omega \times X : \varphi(\omega, v) + n d_X(v, u) < \lambda\}, \end{aligned} \tag{3.270}$$

with $\operatorname{proj}_\Omega$ denoting the projection of $\Omega \times X$ on Ω. Let

$$E_\lambda = \{(\omega, v) \in \Omega \times X : \varphi(\omega, v) + n d_X(v, u) < \lambda\}.$$

Evidently $E_\lambda \in \Sigma \times \mathcal{B}(X)$ and the Yankov–von Neumann–Aumann projection theorem (see Theorem I.4.65) implies that $\operatorname{proj}_\Omega E_\lambda \in \Sigma$ (recall that μ is complete), so

$$\{\omega \in \Omega : \widehat{\varphi}_n(\omega, u) < \lambda\} \in \Sigma \qquad \forall\, n \geqslant 1$$

(see (3.270)). So, for every $n \geqslant 1$ and every $u \in X$, the function $\omega \longmapsto \widehat{\varphi}_n(\omega, u)$ is measurable.

As in the solution of Problem 3.135, we obtain that

$$|\widehat{\varphi}_n(\omega, u) - \widehat{\varphi}(\omega, u)| \leqslant n d_X(u, v) \quad \text{for a.a. } \omega \in \Omega,\ u, v \in X$$

and

$$\widehat{\varphi}_n(\omega, u) \nearrow \varphi(\omega, u) \quad \text{for } \mu\text{-a.a. } \omega \in \Omega, \text{ all } u \in X.$$

In particular then each $\widehat{\varphi}_n$ is $\Sigma \times \mathcal{B}(X)$-measurable (being a Carathéodory integrand). Let $\{\vartheta_n\}_{n \geqslant 1}$ be an increasing sequence in $C_0(X)$ which converges pointwise to 1. We set

$$\varphi_n(\omega, u) = \inf\{n\vartheta_n(u), \widehat{\varphi}_n(\omega, u)\}.$$

Evidently the sequence $\{\varphi_n\}_{n \geqslant 1}$ is the desired sequence of integrands.

Solution of Problem 3.183
From Problem 3.182, we know that there exists a sequence $\{\varphi_n\}_{n \geqslant 1}$ of Carathéodory integrands on $\Omega \times T$ such that

$$0 \leqslant \varphi_n(z, u) \leqslant n \quad \text{for } \mu\text{-a.a. } z \in \Omega, \text{ all } u \in T,\ n \geqslant 1, \tag{3.271}$$

$$\varphi_n(z, \cdot) \in C_0(T) \qquad \forall\, z \in \Omega \quad \text{and} \quad \varphi_n \nearrow \varphi. \tag{3.272}$$

From the monotone convergence theorem, we have

$$\int_{\Omega\times T} \varphi\, d\xi = \lim_{n\to+\infty} \int_{\Omega\times T} \varphi_n\, d\xi.$$

Since $\varphi_n \in L^1(\Omega; C_0(T))$ for all $n \geqslant 1$ (see (3.271) and (3.272)), we see that the functions

$$\xi \longmapsto \int_{\Omega\times T} \varphi_n\, d\xi \text{ are } w^*\text{-continuous on } S\mathcal{Y}(\Omega, T; \mu)$$

(see Remark 3.165), so the function $\xi \longmapsto \int_{\Omega\times T} \varphi\, d\xi$ is w^*-lower continuous on $S\mathcal{Y}(\Omega, T; \mu)$.

If we drop the hypothesis that $\varphi \geqslant 0$, the result fails. To see this, let $T = \mathbb{R}$, $\varphi(t) = -1$ and $t_n = n$. Then $\delta_{t_n} \longrightarrow \lambda = 0$, but $\int_T \varphi\, d\lambda = 0 > \lim_{n\to+\infty} \int_T \varphi\, d\delta_{t_n} = -1$.

Solution of Problem 3.184

Let $\varphi \in K_0(\Omega, \Sigma; T)$ and let $\sigma(\varphi)(z) = \varphi(z, \cdot)$. By definition, for all $z \in \Omega$, $\sigma(\varphi)(z) \in C_0(T)$. Consider the map $\Omega \ni z \longmapsto \sigma(\varphi)(z) \in C_0(T)$. We claim that it is measurable. To this end, let $\{u_n\}_{n\geqslant 1} \subseteq T$ be a dense sequence. Then for every $h \in C_0(T)$, we have

$$\|\sigma(\varphi)(z) - h\|_{C_0(T)} = \sup_{n\geqslant 1} |\varphi(z, u_n) - h(u_n)|,$$

so the map $z \longmapsto \|\sigma(\varphi)(z) - h\|_{C_0(T)}$ is Σ-measurable for all $C_0(T)$. The space $C_0(T)$ is separable. So, it follows that the map $\Omega \ni z \longmapsto \sigma(\varphi)(z) \in C_0(T)$ is strongly measurable (see Definition 1.35). Moreover, we have

$$\int_{\Omega} \|\sigma(\varphi)(z)\|_{C_0(T)}\, d\mu < +\infty,$$

so $\sigma(\varphi) \in L^1(\Omega; C_0(T))$. Hence the map $\sigma\colon K_0(\Omega, \Sigma; T) \longrightarrow L^1(\Omega; C_0(T))$ is injective.

Let $f \in L^1(\Omega; C_0(T))$. Then $f = \sigma(\varphi)$ with $\varphi(z, u) = f(z)(u)$. Therefore σ is in fact a bijection.

Solution of Problem 3.185

"$\Longrightarrow$": Suppose that the set $\mathcal{D} \subseteq \mathcal{Y}(\Omega, T; \mu)$ is uniformly tight (see Definition 3.168). Then according to Proposition 3.170 there exists an inf-compact (see Remark 3.171) measurable integrand $\psi \geqslant 0$ such that

$$0 \leqslant \beta = \sup_{\xi \in \mathcal{D}} \int_\Omega \int_T \psi(z,u)\xi(z)(du)\, d\mu < +\infty.$$

We define

$$F_\varepsilon(z) = \left\{u \in T : \psi(z,u) \leqslant \tfrac{\beta}{\varepsilon}\right\}.$$

Then since ψ is a measurable inf-compact integrand, it follows that F_ε is a measurable multifunction with compact values. Also for every $\xi \in \mathcal{D}$ we have

$$\frac{\beta}{\varepsilon} \int_\Omega \xi(z)(T \setminus F_\varepsilon(z))\, d\mu \leqslant \int_\Omega \int_T \psi(z,u)\xi(z)(du)\, d\mu \leqslant \beta.$$

"$\Longleftarrow$": Let $\varepsilon_n = \frac{1}{3^n}$ for $n \geqslant 1$. We can always assume that the sequence $\{F_{\varepsilon_n} = F_n\}_{n \geqslant 1}$ is increasing. Set $F_0 = \emptyset$ and define

$$\psi(z,u) = \begin{cases} 2^n & \text{if } u \in F_n(z) \setminus F_{n-1}(z), \qquad \forall\, n \geqslant 1, \\ +\infty & \text{if } u \notin \bigcup\limits_{n \geqslant 1} F_n(z). \end{cases}$$

Then clearly ψ is a measurable inf-compact integrand and

$$\sup_{\xi \in \mathcal{D}} \int_\Omega \int_T \psi(z,u)\xi(z)(du)\, d\mu < +\infty,$$

so the set $\mathcal{D} \subseteq \mathcal{Y}(\Omega, T; \mu)$ is uniformly tight (see Proposition 3.170).

Solution of Problem 3.186

According to Definition 3.168, given $\varepsilon > 0$ we can find a compact set $K_\varepsilon \subseteq T$ such that

$$\sup_{\xi \in \mathcal{D}} \xi(\Omega \times (T \setminus K_\varepsilon)) \leqslant \varepsilon.$$

We set $\varphi = \chi_{T \setminus K_\varepsilon}$. Then φ is lower semicontinuous (since the set $T \setminus K_\varepsilon$ is open) and $\varphi \geqslant 0$. Also, if $\xi \in \mathcal{Y}(\Omega, T; \mu)$, then

$$\int_{\Omega \times T} \varphi(u)\, d\xi \leqslant \varepsilon \iff \xi(\Omega \times (T \setminus K_\varepsilon)) \leqslant \varepsilon.$$

Therefore by Problem 3.183 we see that, if $\xi \in \overline{\mathcal{D}}^n$, then

$$\xi(\Omega \times (T \setminus K_\varepsilon)) \leqslant \varepsilon,$$

so $\overline{\mathcal{D}}^n$ is uniformly tight.

Solution of Problem 3.187

(a) Let $\psi(z,u) = |u|$ for all $(z,u) \in \Omega \times \mathbb{R}^N$. Evidently this is a measurable, inf-compact (see Remark 3.171) integrand and

$$\sup_{n \geqslant 1} \int_\Omega \psi(z, u_n(z))\, d\mu \;=\; \sup_{n \geqslant 1} \|u_n\|_{L^1(\Omega;\mathbb{R}^N)} \;<\; +\infty,$$

so the sequence $\{\delta_{u_n}\}_{n\geqslant 1} \subseteq \mathcal{Y}(\Omega, T;\mu)$ is uniformly tight (see Proposition 3.170).

(b) Since by hypothesis $u_n \xrightarrow{\mu} u$ (see Definition 1.11), given $\varepsilon > 0$, we can find $n_0 = n_0(\varepsilon) \geqslant 1$ such that

$$\mu(\{|u_n - u| \geqslant 1\}) \;\leqslant\; \varepsilon \qquad \forall\, n \geqslant n_0. \tag{3.273}$$

Choose $\eta \geqslant 1$ big such that

$$\max_{1\leqslant n \leqslant n_0} \mu(\{|u_n - u| \geqslant \eta\}) \;\leqslant\; \varepsilon. \tag{3.274}$$

Then from (3.273) and (3.274), we infer that

$$\sup_{n\geqslant 1} \mu(\{|u_n - u| \geqslant \eta\}) \;\leqslant\; \varepsilon. \tag{3.275}$$

Let $F_\varepsilon(z) = \{v \in \mathbb{R}^N : |v - u(z)| \leqslant \eta\}$. Evidently F_ε is measurable with compact values and we have

$$\sup_{n\geqslant 1} \int_\Omega \delta_{u_n(z)}(T \setminus F_\varepsilon(z))\, d\mu \;=\; \sup_{n\geqslant 1} \mu\left(\{|u_n(z) - u(z)| \geqslant \eta\}\right) \;\leqslant\; \varepsilon$$

(see (3.275)), so the sequence $\{\delta_{u_n}\}_{n\geqslant 1} \subseteq \mathcal{Y}(\Omega, T;\mu)$ is uniformly tight (see Problem 3.185).

Solution of Problem 3.188

"$\Longrightarrow$": This implication is immediate, since $\chi_A\varphi$ is a measurable integrand on $\Omega\times T$, for all $z \in \Omega$, $\chi_Z(z)\varphi(\cdot) \in C_b(T)$ and the map $z \longmapsto \|\chi_A(z)\varphi(\cdot)\|_{C_b(T)}$ is in $L^1(\Omega)$. So, the implication follows from Definition 3.167.

"$\Longleftarrow$": Exploiting the density of $CU_b(T)$ in $C_b(T)$, we see that the map $\xi \longmapsto \int_{\Omega\times T} s\, d\xi$ is continuous for every simple function $s \in L^1(\Omega; C_b(T))$. The result follows from Definition 1.38.

Solution of Problem 3.189

(a) We fix $d \geqslant 0$ and let $\varphi_d = \max\{\varphi, -d\}$. Since $\varphi_d \geqslant -d$, we have

$$\int_{\Omega\times T} (\varphi_d(z,x) + d)\, d\xi \;\leqslant\; \liminf_{m\to+\infty} \int_{\Omega} (\varphi_d(z,u_m(z)) + d)\, d\mu,$$

so

$$\int_{\Omega\times T} \varphi_d(z,x)\, d\xi \;\leqslant\; \liminf_{m\to+\infty} \int_{\Omega} \varphi_d(z,u_m(z))\, d\mu. \tag{3.276}$$

If we let $d = 0$, then from (3.276) and the uniform integrability of the sequence $\{\varphi^-(\cdot,u_m(\cdot))\}_{m\geqslant 1} \subseteq L^1(\Omega)$, we conclude that the sequence $\{\varphi^+(\cdot,u_m(\cdot))\}_{m\geqslant 1} \subseteq L^1(\Omega)$ is bounded and so

$$\int_{\Omega\times T} \varphi^+(z,x)\, d\xi \;<\; +\infty.$$

(b) Let $D_{m,d} = \{z \in \Omega :\ \varphi(z,u_m(z)) < -d\}$. We have

$$\int_{D_{m,d}} \varphi(z,u_m(z)) \;\leqslant\; 0 \qquad \forall\, m \geqslant 1,$$

so

$$\int_{D_{m,d}} \varphi^+(z,u_m(z))\, d\mu - \int_{D_{m,d}} \varphi^-(z,u_m(z))\, d\mu \;\leqslant\; 0 \qquad \forall\, m \geqslant 1.$$

Exploiting part **(a)** and the hypothesis, given $\varepsilon > 0$, we can find $d > 0$ big, such that

$$-\,\varepsilon \;\leqslant\; \int_{D_{m,d}} \varphi(z,u_m(z))\, d\mu \;\leqslant\; 0 \qquad \forall\, m \geqslant 1. \tag{3.277}$$

Note that

$$\varphi_d(z,u_m(z)) \;=\; \begin{cases} \varphi(z,u_m(z)) & \text{for } \ z \in \Omega \setminus D_{m,d}, \\ -d & \text{for } \ z \in D_{m,d}. \end{cases}$$

Therefore

$$\begin{aligned}
\int_{\Omega} \varphi(z,u_m(z))\, d\mu &= \int_{D_{m,d}} \varphi(z,u_m(z))\, d\mu + \int_{\Omega\setminus D_{m,d}} \varphi(z,u_m(z))\, d\mu \\
&= \int_{D_{m,d}} \varphi(z,u_m(z))\, d\mu + \int_{\Omega\setminus D_{m,d}} \varphi_d(z,u_m(z))\, d\mu \\
&= \int_{D_{m,d}} \varphi(z,u_m(z))\, d\mu + \int_{\Omega} \varphi_d(z,u_m(z))\, d\mu - \int_{D_{m,d}} \varphi_d(z,u_m(z))\, d\mu
\end{aligned}$$

$$\geqslant \int_{D_{m,d}} \varphi(z, u_m(z))\, d\mu + \int_\Omega \varphi_d(z, u_m(z))\, d\mu$$

$$\geqslant \int_\Omega \varphi_d(z, u_m(z))\, d\mu - \varepsilon \qquad \forall\, m \geqslant 1$$

(see (3.277)), so

$$\liminf_{m\to+\infty} \int_\Omega \varphi(z, u_m(z))\, d\mu \;\geqslant\; \liminf_{m\to+\infty} \int_\Omega \varphi_d(z, u_m(z))\, d\mu - \varepsilon$$

$$\geqslant \int_{\Omega\times T} \varphi_d(z, x)\, d\xi - \varepsilon \;\geqslant\; \int_{\Omega\times T} \varphi(z, x)\, d\xi - \varepsilon$$

(since $d_{u_m} \xrightarrow{n} \xi$ and $\varphi \leqslant \varphi_d$). Let $\varepsilon \searrow 0$ to conclude that

$$\int_{\Omega\times T} \varphi(z, x)\, d\xi \;\leqslant\; \liminf_{m\to+\infty} \int_\Omega \varphi(z, u_m(z))\, d\mu$$

Solution of Problem 3.190

Apply Problem 3.189 to $\varphi = \widehat{\varphi}$ and to $\varphi = -\widehat{\varphi}$.

Solution of Problem 3.191

The uniform integrability of the sequence $\{u_n\}_{n\geqslant 1}$ implies that it is $L^1(\Omega; \mathbb{R}^m)$-bounded (see Problem 1.7). Then Problem 3.187(a) implies that the sequence $\{\xi_n = \delta_{u_n(\cdot)}\}_{n\geqslant 1}$ of Young measures is uniformly tight. So, by Theorem 3.172 and by passing to a suitable subsequence if necessary, we may assume that

$$\xi_n \xrightarrow{n} \xi$$

(see Definition 3.167). Let $h \in L^\infty(\Omega)$ and set $\widehat{\varphi}(z, x) = h(z)x$. This is a Carathéodory integrand and the sequence $\{\widehat{\varphi}(\cdot, u_n(\cdot))\}_{n\geqslant 1} \subseteq L^1(\Omega)$ is uniformly integrable (see Definition 1.18). So, from Problem 3.190 we have

$$\lim_{n\to+\infty} \int_\Omega h(z)u_n(z)\, d\mu \;=\; \int_\Omega h(z)\Big(\int_{\mathbb{R}^m} x\xi(z)(dx)\Big)\, d\mu.$$

Then the barycenter $u(z) = \int_{\mathbb{R}^m} x\xi(z)(dx)$ is the w-limit in $L^1(\Omega;\mathbb{R}^m)$ of $\{u_n\}_{n\geqslant 1}$ (that is, $u_n \xrightarrow{w} u$ in $L^1(\Omega;\mathbb{R}^m)$).

Solution of Problem 3.192

Arguing componentwise if necessary, we may assume without any loss of generality that $m = 1$. We consider $\widehat{\varphi}(z,x) = \vartheta(z)x$ with $\vartheta \in C_c(\Omega)$. Since $u_n \xrightarrow{w} u$ in $L^p(\Omega)$ we have that the sequence $\{\widehat{\varphi}(\cdot, u_n(\cdot))\}_{n\geqslant 1} \subseteq L^1(\Omega)$ is uniformly integrable (see Theorem 1.28). So, by Problem 3.190, we have

$$\int_{\Omega\times\mathbb{R}} \vartheta(z)x\,d\xi = \lim_{n\to+\infty} \int_\Omega \vartheta(z)u_n(z)\,d\mu = \int_\Omega \vartheta(z)u(z)\,d\mu,$$

so

$$\int_\Omega \vartheta(z)\Big(\int_{\mathbb{R}} x\xi(z)(dx)\Big)\,d\mu = \int_\Omega \vartheta(z)u(z)\,d\mu,$$

thus

$$\int_{\mathbb{R}} x\xi(z)(dx) = u(z) \quad \text{for } \mu\text{-a.a. } z \in \Omega$$

(since $\vartheta \in C_c(\Omega)$ is arbitrary).

Solution of Problem 3.193

"**(a)** $\Longrightarrow$ **(b)**": Let $\varepsilon > 0$ be arbitrary and define

$$\widehat{\varphi}(z,x) = \begin{cases} -1 & \text{if } d_T(x,u(z)) \geqslant \varepsilon, \\ 0 & \text{otherwise,} \end{cases}$$

with d_T being the metric on T. Evidently $\widehat{\varphi}$ is a measurable integrand such that for μ-almost all $z \in \Omega$, $\widehat{\varphi}(z,\cdot)$ is lower semicontinuous and it is bounded below. Since $\delta_{u_n(\cdot)} \xrightarrow{n} \delta_{u(\cdot)}$ (see Definition 3.167), from Problem 3.189 we have

$$0 \leqslant \int_\Omega \widehat{\varphi}(z,u(z))\,d\mu \leqslant \liminf_{n\to+\infty} \int_\Omega \widehat{\varphi}(z,u_n(z))\,d\mu,$$

so

$$\lim_{n\to+\infty} \mu\big\{\big(z\in\Omega:\ d_T(u_n(z),u(z)) \geqslant \varepsilon\big)\big\} = 0,$$

hence $u_n \xrightarrow{\mu} u$ (see Definition 1.11).

"(b) $\Longrightarrow$ (a)": Let $D \in \Sigma$ and $\psi \in UC(T)$ (where $UC(T)$ denotes the space of uniformly continuous functions on T). Given $\eta > 0$, we can find $\varepsilon > 0$ such that

$$|\psi(x) - \psi(x')| \leqslant \eta, \quad \text{when } d_T(x, x') \leqslant \varepsilon.$$

Then we have

$$\begin{aligned} &\int_D |\psi(u_n(z)) - \psi(u(z))| \, d\mu \\ &\leqslant \eta\mu(\Omega) + 2\mu(\{z \in \Omega : d_T(u_n(z), u(z)) \geqslant \varepsilon\}) \sup_{z \in T} |\psi(x)|, \end{aligned}$$

so $\delta_{u_n(\cdot)} \xrightarrow{n} \delta_{u(\cdot)}$.

Solution of Problem 3.194

We need to show that for every $\lambda \geqslant 0$, the set

$$L_\lambda = \{u \in X : m(u) \leqslant \lambda\}$$

is closed. So, let $\{u_n\}_{n \geqslant 1} \subseteq L_\lambda$ be a sequence such that $u_n \longrightarrow u$ in X. Because the set $\partial\varphi(u_n) \subseteq X^*$ is w^*-compact and the norm is w^*-lower semicontinuous, by the Weierstrass theorem (see Theorem I.1.75), we can find $u_n^* \in \partial\varphi(u_n)$ such that

$$m(u_n) = \|u_n^*\|_* \qquad \forall\, n \geqslant 1.$$

Recall that $\partial\varphi\colon X \longrightarrow 2^{X^*} \setminus \{\emptyset\}$ is upper semicontinuous from X with the norm topology into X^* with the w^*-topology and has w^*-compact, convex values (see Proposition 3.128). So, from Problem 2.56, we infer that the sequence $\{u_n^*\}_{n \geqslant 1} \subseteq X^*$ is relatively w^*-compact. Thus, we can find a subnet $\{u_\beta^*\}_{\beta \in I}$ of the sequence $\{u_n^*\}_{n \geqslant 1}$ such that

$$u_\beta^* \xrightarrow{w^*} u^* \quad \text{in } X^*.$$

Then $u^* \in \partial\varphi(u)$ (see Proposition 2.43) and $\|u^*\|_* \leqslant \lambda$. Hence $m(u) \leqslant \lambda$ and so $u \in L_\lambda$. This proves the closedness of L_λ, hence the lower semicontinuity of m.

Solution of Problem 3.195

First let $m = 1$. We set

$$\widehat{\varphi}(u) = \inf_{v \in C} \big(\varphi(v) + k_\varphi |v - u|\big). \tag{3.278}$$

Note that if $h \in C$, then $\widehat{\varphi}(h) \leqslant \varphi(h)$. On the other hand, from (3.278), we see that

$$\varphi(h) \leqslant \varphi(v) + k_\varphi |v-h| \qquad \forall\, v \in C$$

and so $\varphi(h) \leqslant \widehat{\varphi}(h)$. Therefore $\widehat{\varphi}|_C = \varphi$.

If $u, y \in \mathbb{R}^N$, then

$$\widehat{\varphi}(u) \leqslant \inf_{v \in C} \big(\varphi(v) + k_\varphi(|v-y| + |y-u|)\big) = \widehat{\varphi}(y) + k_\varphi |y-u|. \tag{3.279}$$

Similarly, we show that

$$\widehat{\varphi}(y) \leqslant \widehat{\varphi}(u) + k_\varphi |u-y|. \tag{3.280}$$

From (3.279) and (3.280), we infer that

$$|\widehat{\varphi}(u) - \widehat{\varphi}(y)| \leqslant k_\varphi |u-y|,$$

that is, $\widehat{\varphi}$ is Lipschitz continuous.

In the general case $m > 1$, we have $\varphi = (\varphi_1, \ldots, \varphi_m)$ with $\varphi_k \colon C \longrightarrow \mathbb{R}$ being Lipschitz continuous for $k \in \{1, \ldots, m\}$. Using the first part of the solution, each φ_k admits a Lipschitz continuous extension $\widehat{\varphi}_k \colon \mathbb{R}^N \longrightarrow \mathbb{R}$, for $k \in \{1, \ldots, m\}$. We set $\widehat{\varphi} = (\widehat{\varphi}_1, \ldots, \widehat{\varphi}_m)$. Then $\widehat{\varphi} \colon \mathbb{R}^N \longrightarrow \mathbb{R}^m$ is a Lipschitz continuous extension of φ. Moreover, for every $u, y \in \mathbb{R}^N$, we have

$$|\widehat{\varphi}(u) - \widehat{\varphi}(y)|^2 = \sum_{k=1}^m |\widehat{\varphi}_k(u) - \widehat{\varphi}_k(y)|^2 \leqslant m k_\varphi^2 |u-y|^2$$

(from the first part of the solution), so

$$|\widehat{\varphi}(u) - \widehat{\varphi}(y)| \leqslant \sqrt{m} k_\varphi |u-y|,$$

thus $k_{\widehat{\varphi}} \leqslant \sqrt{m} k_\varphi$.

Remark. In fact we can have a Lipschitz extension $\widehat{\varphi}$ of φ such that $k_{\widehat{\varphi}} = k_\varphi$. The result is known as the Kirszbraun theorem and can be found in Federer [10, p. 201].

Solution of Problem 3.196

We already know that $W^{1,\infty}(\mathbb{R}^N) \subseteq C^{0,1}(\mathbb{R}^N)$ (see Remark 1.124). From Theorem 1.125, we know that every function $u \in W^{1,\infty}(\Omega)$ admits an extension to a function in $W^{1,\infty}(\mathbb{R}^N)$. Also, from Problem 3.195, we know that every Lipschitz continuous function on Ω admits a Lipschitz continuous extension on all of $\mathbb{R}^N$. Finally note that the restriction of an element of $W^{1,\infty}(\mathbb{R}^N)$ (respectively of $C^{0,1}(\mathbb{R}^N)$) on Ω is an element of $W^{1,\infty}(\Omega)$ (respectively of $C^{0,1}(\Omega)$). This completes the solution of the problem.

Solution of Problem 3.197

Clearly $\widehat{\varphi}$ is convex and $\operatorname{dom}\widehat{\varphi} \neq \emptyset$. If we show that $\widehat{\varphi}$ is also lower semicontinuous, then $\widehat{\varphi} \in \Gamma_0(H)$. So, let $\{u_n\}_{n\geqslant 1} \subseteq H$ be a sequence such that $u_n \longrightarrow u$ in H and without any loss of generality, we may assume that $\lim\limits_{n\to+\infty} \widehat{\varphi}(u_n) = m$ exists and $m < +\infty$. Let $a > m$ and $b > \|u_n\|_H$ for all $n \geqslant 1$. Then there exists $n_0 \geqslant 1$ such that $u_n \in C_{a,b}$ for all $n \geqslant n_0$. By hypothesis the set $C_{a,b}$ is bounded in X and so we have $u_n \xrightarrow{w} u$ in X (recall $u_n \longrightarrow u$ in H). Since $\varphi \in \Gamma_0(X)$, the set $L_a = \{y \in X : \varphi(y) \leqslant a\}$ is closed and convex and so $u \in L_a$. Since $a > \lim\limits_{n\to+\infty} \varphi(u_n)$ is arbitrary, it follows that

$$\widehat{\varphi}(u) = \varphi(u) \leqslant \lim_{n\to+\infty} \varphi(u_n) = \liminf_{n\to+\infty} \widehat{\varphi}(u_n),$$

so $\widehat{\varphi} \in \Gamma_0(H)$.

Now assume that φ is Gâteaux differentiable at $u \in \operatorname{int}\operatorname{dom}\varphi \subseteq X$. Let $u^* \in \partial\widehat{\varphi}(u)$. Then by definition

$$(u^*, v-u)_H \leqslant \widehat{\varphi}(v) - \widehat{\varphi}(u) \qquad \forall\, v \in H.$$

Let $v = u \pm th$, $t > 0$, $h \in X$. Then

$$\pm\,(u^*, h)_H \leqslant \pm\tfrac{\varphi(u\pm th)-\varphi(u)}{t} \qquad \forall\, t > 0,\ h \in X,$$

so

$$(u^*, h)_H = \langle \varphi'(u), h\rangle \qquad \forall\, h \in X,$$

thus

$$\partial\widehat{\varphi}(u) \subseteq \{u^* \in H = H^* : (u^*, h)_H = \langle \varphi'(u), h\rangle \text{ for all } h \in X\}.$$

We need to show that the opposite inclusion also holds. So, let $u^* \in H = H^*$ be such that

$$(u^*, h)_H = \langle \varphi'(u), h\rangle \qquad \forall\, h \in X. \tag{3.281}$$

From the convexity of φ, we have

$$\begin{aligned} \varphi(u+t(h-u)) &= \varphi(th+(1-t)u) \leqslant t\varphi(h) + (1-t)\varphi(u) \\ &= \varphi(u) + t(\varphi(h) - \varphi(u)), \end{aligned}$$

so

$$\tfrac{\varphi(u+t(h-u))-\varphi(u)}{t} \leqslant \varphi(h) - \varphi(u) \qquad \forall\, t \in (0,1],\ h \in X,$$

thus

$$(u^*, h-u)_H = \langle \varphi'(u), h-u\rangle \leqslant \varphi(h) - \varphi(u) \qquad \forall\, h \in X$$

(see (3.281)), hence $u^* \in \partial\widehat{\varphi}(u)$ and finally

$$\partial\widehat{\varphi}(u) = \big\{u^* \in H = H^* : \ (u^*, h)_H = \langle \varphi'(u), h\rangle \text{ for all } h \in X\big\}.$$

Solution of Problem 3.198

Recall that the neighborhoods of $+\infty$ in $\overline{\mathbb{R}}$ are all sets of the form $(c, +\infty]$ with $c \in \mathbb{R}$. So, if we define

$$\varphi(u) = \begin{cases} \frac{1}{u} & \text{if } u > 0, \\ +\infty & \text{if } u \leqslant 0, \end{cases}$$

then φ is convex and continuous at $0 \in \overline{\operatorname{dom}\varphi} \setminus \operatorname{dom}\varphi$.

Solution of Problem 3.199

Yes. Let $X = l^2$ and let $C \subseteq l^2$ be the nonempty, closed (in fact compact), and convex set, defined by

$$C = \big\{\widehat{u} = \{u_n\}_{n\geqslant 1} \in l^2 : \ |u_n| \leqslant \tfrac{1}{2^n} \text{ for all } n \geqslant 1\big\}.$$

Let $\varphi\colon C \longrightarrow \mathbb{R}$ be defined by

$$\varphi(\widehat{u}) = \sum_{n\geqslant 1} -(u_n + \tfrac{1}{2^n})^{\frac{1}{2}}. \tag{3.282}$$

Note that for every $n \geqslant 1$, the function

$$\varphi_n(u) = -(u + \tfrac{1}{2^n})^{\frac{1}{2}} \qquad \forall\, u \in (-\tfrac{1}{2^n}, \tfrac{1}{2^n})$$

is continuous, convex and

$$|\varphi_n(u)| \leqslant \frac{1}{2^{\frac{n-1}{2}}} \qquad \forall\, u \in (-\tfrac{1}{2^n}, \tfrac{1}{2^n}).$$

Since $\sum\limits_{n\geqslant 1} \frac{1}{2^{\frac{n-1}{2}}} < +\infty$, it follows from (3.282) that the series converges uniformly and so φ is continuous and convex. We claim that $\partial\varphi(\widehat{u}) = \emptyset$ for each $\widehat{u} = \{u_n\}_{n\geqslant 1} \in C$ with $u_n > -\frac{1}{2^n}$ for infinitely many $n \geqslant 1$. To see that, let e_n be the n-th element of the standard orthonormal basis of l^2. If $\widehat{u}^* \in \partial\varphi(\widehat{u})$ with $\widehat{u} = \{u_n\}_{n\geqslant 1} \subseteq C$ such that $\widehat{u} = \{u_n\}_{n\geqslant 1} \in C$ with $u_n > -\frac{1}{2^n}$ for infinitely many $n \geqslant 1$, then

$$\begin{aligned} -\|\widehat{u}^*\| \ \leqslant\ (\widehat{u}^*, e_n) \ \leqslant\ \varphi'_+(\widehat{u}; e_n) &= \lim_{t\to 0^+} \tfrac{\varphi(\widehat{u}+te_n)-\varphi(\widehat{u})}{t} \\ &= -\tfrac{1}{2}(u_n + \frac{1}{2^n})^{-\frac{1}{2}} \qquad \forall\, n \geqslant 1, \end{aligned}$$

a contradiction. Therefore $\partial\varphi(\widehat{u}) = \emptyset$.

Solution of Problem 3.200

Let $X = \mathbb{R}^2$, $q(u) = u^2$, $C = \operatorname{epi} q$, $L = \{(x,0) \in \mathbb{R}^2 : x \in \mathbb{R}\}$ (the x-axis) and define $\varphi = \delta_C$ and $\psi = \delta_L$. Note that $C \cap L = \{0\}$. Also, we have $\partial\varphi(0) = \mathbb{R}_- e$ and $\partial\psi(0) = \mathbb{R}e$ with $\mathbb{R}_- = (-\infty, 0]$ and $e = \begin{pmatrix} 1 \\ 0 \end{pmatrix} \in \mathbb{R}^2$. So, we have

$$\partial\varphi(0) + \partial\psi(0) \subsetneq \partial(\varphi + \psi)(0) = \mathbb{R}^2.$$

Solution of Problem 3.201

Consider the convex function $\varphi \colon H \longrightarrow \mathbb{R}$ defined by

$$\varphi(u) = \tfrac{1}{2}\|u\|_H^2 - \tfrac{1}{2}\|u - \operatorname{proj}_C(u)\|_H^2 \qquad \forall\, u \in H.$$

From the definition of the metric projection, we have

$$\|u + h - \operatorname{proj}_C(u+h)\|_H \leqslant \|u + h - \operatorname{proj}_C(u)\|_H \qquad \forall\, h \in H. \tag{3.283}$$

So, we have

$$\begin{aligned} \varphi(u+h) &= \tfrac{1}{2}\|u+h\|_H^2 - \tfrac{1}{2}\|(u+h) - \operatorname{proj}_C(u+h)\|_H^2 \\ &\geqslant \tfrac{1}{2}\|u+h\|_H^2 - \tfrac{1}{2}\|(u+h) - \operatorname{proj}_C(u)\|_H^2 \end{aligned}$$

(see (3.283)), so

$$\begin{aligned} &\varphi(u+h) - \varphi(u) \geqslant \tfrac{1}{2}\|u+h\|_H^2 - \tfrac{1}{2}\|u\|_H^2 \\ &-\left(\tfrac{1}{2}\|(u+h) - \operatorname{proj}_C(u)\|_H^2 - \tfrac{1}{2}\|u - \operatorname{proj}_C(u)\|_H^2\right) \\ \geqslant\ &(u,h)_H - \left(u - \operatorname{proj}_C u, h\right)_H = \left(\operatorname{proj}_C(u), h\right)_H \end{aligned}$$

(here $(\cdot,\cdot)_H$ denotes the inner product of H), thus

$$\varphi(u+h) - \varphi(u) - \left(\operatorname{proj}_C(u), h\right)_H \geqslant 0. \tag{3.284}$$

Also, note that

$$\|u - \operatorname{proj}_C(u)\|_H \leqslant \|u - \operatorname{proj}_C(u+h)\|_H. \tag{3.285}$$

Hence, we have

$$\varphi(u+h) - \varphi(u) \leqslant \tfrac{1}{2}\|u+h\|_H^2 - \tfrac{1}{2}\|u\|_H^2$$

$$+\left(\tfrac{1}{2}\|u-\operatorname{proj}_C(u+h)\|_H^2-\tfrac{1}{2}\|(u+h)-\operatorname{proj}_C(u+h)\|_H^2\right)$$
$$\leqslant\ (u+th,h)_H+\left(\operatorname{proj}_C(u+th)-u,h\right)_H, \tag{3.286}$$

for some $t\in(0,1)$ (see Theorem 3.16). Using (3.284) and (3.286) and recalling that proj_C is nonexpansive, we have

$$\begin{aligned}0 &\leqslant \varphi(u+h)-\varphi(u)-\left(\operatorname{proj}_C(u),h\right)_H\\ &\leqslant ((u+th)-u,h)_H+\left(\operatorname{proj}_C(u+h)-\operatorname{proj}_C(u),h\right)_H\\ &\leqslant t\|h\|_H^2+\|h\|_H^2\ =\ (1+t)\|h\|_H^2,\end{aligned}$$

so

$$\lim_{h\to 0}\frac{1}{\|h\|_H}\left(\varphi(u+h)-\varphi(u)-\left(\operatorname{proj}_C(u),h\right)_H\right)\ =\ 0,$$

thus $\varphi'(u)=\operatorname{proj}_C(u)$ for all $u\in H$.

Solution of Problem 3.202

Let $W^{1,2}_{\mathrm{per}}(0,1)=\{u\in W^{1,2}(0,1):\ u(0)=u(1)\}$. This is a closed linear subspace of $W^{1,2}(0,1)$ which is well defined (see Theorem 1.100). Therefore $W^{1,2}_{\mathrm{per}}(0,1)$ is a Hilbert space furnished with the inner product and consequently the norm of $W^{1,2}(0,1)$. Consider the bilinear form $a\colon W^{1,2}_{\mathrm{per}}(0,1)\times W^{1,2}_{\mathrm{per}}(0,1)\longrightarrow\mathbb{R}$ defined by

$$a(u,v)\ =\ \int_0^1(u'v'+u'v+uv)\,dt\qquad\forall\,u,v\in W^{1,2}_{\mathrm{per}}(0,1). \tag{3.287}$$

Using the Cauchy–Schwarz inequality, we see that

$$|a(u,v)|\ \leqslant\ c\|u\|\,\|v\|\qquad\forall\,u,v\in W^{1,2}_{\mathrm{per}}(0,1),$$

with $c>0$. Also, since

$$\int_0^1 u'u\,dt\ =\ \tfrac{1}{2}\left(u(1)^2-u(0)^2\right)\ =\ 0\qquad\forall\,u\in W^{1,2}_{\mathrm{per}}(0,1)$$

(by the integration by parts), we have

$$a(u,u)\ \geqslant\ \|u\|^2\qquad\forall\,u\in W^{1,2}_{\mathrm{per}}(0,1).$$

So, we can use the Lax–Milgram theorem (see Remark 3.115) and find unique $u_0\in W^{1,2}_{\mathrm{per}}(0,1)$ such that

$$a(u_0,v)\ =\ 0\qquad\forall\,v\in W^{1,2}_{\mathrm{per}}(0,1). \tag{3.288}$$

Since $W_0^{1,2}(0,1) \subseteq W_{\text{per}}^{1,2}(0,1)$ and $\int_0^1 u_0' v\, dt = 0$ for all $v \in W_0^{1,2}(0,1)$ (by the integration by parts), we have

$$\int_0^1 (u_0' v' + u_0 v)\, dt = 0 \qquad \forall\, v \in W_0^{1,2}(0,1).$$

Recalling that $u_0'' \in W^{-1,2}(0,1) = W_0^{1,2}(0,1)^*$ (see Theorem 1.140), we have

$$\langle -u_0'' + u_0, v\rangle = 0 \qquad \forall\, v \in W_0^{1,2}(0,1),$$

so

$$-u_0'' + u_0 = 0. \tag{3.289}$$

Then from (3.287), (3.288), and (3.289), we have

$$\int_0^1 u_0' v\, dt = 0 \qquad \forall\, v \in W_{\text{per}}^{1,2}(0,1),$$

so $u_0' = 0$. Hence $u_0 \equiv c \in \mathbb{R}$ and from the equation we have $u_0(t) = 0$ for all $t \in (0,1)$. Therefore the periodic problem has only the trivial solution.

Bibliography

[1] Attouch, H.: Variational Convergence for Functions and Operators. Pitman, Boston (1984)

[2] Borwein, J.M., Vanderwerff, J.D.: Convex Functions: Constructions, Characterizations and Counterexamples. Cambridge University Press, Cambridge (2010)

[3] Buttazzo, G.: Semicontinuity, Relaxation and Integral Representation in the Calculus of Variations. Longman Scientific & Technical, New York (1989)

[4] Cartan, H.: Formes Différentielles. Applications Élḿentaires au Calcul des Variations et a la Théorie des Courbes et des Surfaces. Hermann, Paris (1967)

[5] Clarke, F.H.: Optimization and Nonsmooth Analysis. SIAM, Philadelphia (1990)

[6] Coleman, R.: Calculus on Normed Vector Spaces. Springer, New York (2012)

[7] Dal Maso, G.: An Introduction to Γ-Convergence. Birkhäuser, Boston (1993)

[8] Denkowski, Z., Migórski, S., Papageorgiou, N.S.: An Introduction to Nonlinear Analysis: Applications. Kluwer Academic, Boston (2003)

[9] Ekeland, I., Témam, R.: Convex Analysis and Variational Problems. SIAM, Philadelphia (1999)

[10] Federer, H.: Geometric Measure Theory. Springer, New York (1969)

[11] Gasiński, L., Papageorgiou, N.S.: Nonlinear Analysis. Chapman & Hall/CRC, Boca Raton (2006)

[12] Giles, J.R.: Convex Analysis with Application in the Differentiation of Convex Functions. Pitman, Boston (1982)

[13] Ioffe, A.D., Tihomirov, V.M.: Theory of Extremal Problems. North-Holland, Amsterdam (1979)

[14] Kinderlehrer, D., Stampacchia, G.: An Introduction to Variational Inequalities and Their Applications. Academic, New York (1980)

[15] Papageorgiou, N.S., Kyritsi-Yiallourou, S.Th.: Handbook of Applied Analysis. Springer, New York (2009)

[16] Phelps, R.R.: Convex Functions, Monotone Operators and Differentiability. Springer, Berlin (1993)

[17] Rockafellar, R.T., Wets R.: Variational Analysis. Springer, Berlin (1998)

[18] Roubiček, T.: Relaxation in Optimization Theory and Variational Calculus. Walter de Gruyter & Co., Berlin (1997)

Chapter 4

Degree Theory and Fixed Point Theory

4.1 Introduction

4.1.1 Degree Theory

Degree theory deals with equations of the form $\varphi(u) = h$ on a space X (finite of infinite dimensional). It addresses the questions of existence, uniqueness, or multiplicity of solutions and their distribution in the space. Moreover, it examines how sensitive are these properties to variations of φ and h.

We start with an outline of degree theory for continuous functions defined on subset of $\mathbb{R}^N$ and more generally for any finite dimensional normed space.

In what follows $\Omega \subseteq \mathbb{R}^N$ is a bounded open set.

Definition 4.1
*Let $\varphi \in C^1(\Omega;\mathbb{R}^N) \cap C(\overline{\Omega};\mathbb{R}^N)$ and $u_0 \in \Omega$. We say that u_0 is a **critical point of** φ, if $J_\varphi(u_0) = \det \varphi'(u_0) = 0$ ($J_\varphi(u_0)$ is called the **Jacobian** of φ at u_0). Let*

$$S_\varphi = \{u \in \Omega : J_\varphi(u) = 0\} \quad (\text{the } \textbf{critical set} \text{ of } \varphi).$$

*Also $\varphi(S_\varphi)$ is the set of critical values of φ (or the **crease** of φ). If $h \notin \varphi(S_\varphi)$, then we say that h is a **regular value** of φ.*

Next we recall the following important result concerning the critical values of φ.

Theorem 4.2 *(Sard Theorem)*
<u>If</u> $\Omega \subseteq \mathbb{R}^N$ is an open set and $\varphi \in C^1(\Omega;\mathbb{R}^N)$,
<u>then</u> $\lambda^N(\varphi(S_\varphi)) = 0$, where λ^N denotes the Lebesgue measure on $\mathbb{R}^N$.

L. Gasiński, N.S. Papageorgiou, *Exercises in Analysis: Part 2*, Problem Books in Mathematics, DOI 10.1007/978-3-319-27817-9_4

Remark 4.3
In fact there is a powerful generalization of the above theorem to Sobolev functions. Namely, if $p > N$ and $u \in W^{1,p}(\Omega;\mathbb{R}^N)$, then for every measurable set $D \subseteq \Omega$, we have that the set $u(D) \subseteq \mathbb{R}^N$ is measurable and

$$\lambda^N(u(D)) \leqslant \int_D |J_u(z)|\,dz.$$

Definition 4.4
If $\varphi \in C^1(\Omega;\mathbb{R}^N) \cap C(\overline{\Omega};\mathbb{R}^N)$ and $h \in \mathbb{R}^N \setminus \varphi(\partial\Omega \cup S_\varphi)$, then the ***degree*** *of φ at h with respect to Ω is defined by*

$$d(\varphi,\Omega,h) = \sum_{u\in\varphi^{-1}(h)} \operatorname{sgn} J_\varphi(u),$$

where the ***sign function*** sgn *is defined by*

$$\operatorname{sgn}(t) = \begin{cases} 1 & \textit{if } t>0,\\ 0 & \textit{if } t=0,\\ -1 & \textit{if } t<0. \end{cases}$$

Remark 4.5
From Problem 4.1 we know that since $h \notin \varphi(S_\varphi)$, then $\varphi^{-1}(h)$ is finite.

Our first goal is to remove the restriction that $h \notin \varphi(S_\varphi)$. This can be achieved with the help of Theorem 4.2 and of the next proposition in which the sum in Definition 4.4 is replaced by a suitable integral.

Proposition 4.6
If $\varphi \in C^1(\Omega;\mathbb{R}^N) \cap C(\overline{\Omega};\mathbb{R}^N)$, $h \in \mathbb{R}^N \setminus \varphi(\partial\Omega \cup S_\varphi)$ and $\{\xi_\varepsilon\}_{\varepsilon>0}$ is a family of mollifiers (see Definition 1.117),
then there exists $\varepsilon_0 = \varepsilon_0(\varphi,h) > 0$ such that

$$d(\varphi,\Omega,h) = \int_\Omega \xi_\varepsilon(\varphi(z)-h)J_\varphi(z)\,dz \qquad \forall\, \varepsilon \in (0,\varepsilon_0].$$

Using this expression, we can show that $d(\varphi,\Omega,\cdot)$ is constant for all regular values in a connected component of $\mathbb{R}^N \setminus \varphi(\partial\Omega)$. We can use this fact to get rid of the restriction $h \notin \varphi(S_\varphi)$. Indeed by Theorem 4.2, for every $\varepsilon > 0$, we have

$$B_\varepsilon(h) \cap (\mathbb{R}^N \setminus \varphi(S_\varphi)) \neq \emptyset$$

(here $B_\varepsilon(h) = \{y \in \mathbb{R}^N : |y-h| < \varepsilon\}$). Let $y_\varepsilon \in B_\varepsilon(h) \cap (\mathbb{R}^N \setminus \varphi(S_\varphi))$. Moreover, since $\varphi(\partial\Omega) \subseteq \mathbb{R}^N$ is compact and $h \notin \varphi(\partial\Omega)$, by choosing $\varepsilon > 0$ small, we can have $B_\varepsilon(h) \cap \varphi(\partial\Omega) = \emptyset$. So, there exists $y \in \mathbb{R}^N \setminus \varphi(\partial\Omega \cup S_\varphi)$ such that $|y-h| < \operatorname{dist}(h,\varphi(\partial\Omega))$. If $|y_k - h| < \operatorname{dist}(h,\varphi(\partial\Omega))$ for $k = 1,2$ are such that $y_k \in \mathbb{R}^N \setminus \varphi(\partial\Omega \cup S_\varphi)$ for $k = 1,2$, then $y_k \in B_{\operatorname{dist}(h,\varphi(\partial\Omega))}(h) \subseteq \mathbb{R}^N \setminus \varphi(\partial\Omega)$ for $k = 1,2$. Since the set $B_{\operatorname{dist}(h,\varphi(\partial\Omega))}(h) \subseteq \mathbb{R}^N$ is connected, it is included in a connected component of $\mathbb{R}^N \setminus \varphi(\partial\Omega \cup S_\varphi)$ and so by the initial remark we have $d(\varphi,\Omega,y_1) = d(\varphi,\Omega,y_2)$. Therefore we have justified the following definition.

Definition 4.7
Suppose that $\varphi \in C^1(\Omega;\mathbb{R}^N) \cap C(\overline{\Omega};\mathbb{R}^N)$ *and* $h \notin \varphi(\partial\Omega)$. *We define* $d(\varphi,\Omega,h)$, *the* ***degree*** *of* φ *at* h *with respect to* Ω, *to be the number* $d(\varphi,\Omega,y)$ *where* $y \in \mathbb{R}^N \setminus \varphi(\partial\Omega \cup S_\varphi)$ *such that* $|y-h| < \operatorname{dist}(h,\varphi(\partial\Omega))$.

Now we are ready for the step in which we extend the definition of degree to all continuous functions on $\overline{\Omega}$. So, let $\varphi \in C(\overline{\Omega};\mathbb{R}^N)$. Using the Tietze extension theorem (see Theorem I.2.138), we can extend φ to a continuous function $\widehat{\varphi}$ on all of $\mathbb{R}^N$. Subsequently using mollifications, we obtain the following result.

Proposition 4.8
<u>If</u> $\varphi \in C(\overline{\Omega};\mathbb{R}^N)$ *and* $\varepsilon > 0$,
<u>then</u> there exists $\varphi_\varepsilon \in C^\infty(\mathbb{R}^N;\mathbb{R}^N)$ *such that*

$$|\varphi_\varepsilon(z) - \varphi(z)| \leqslant \varepsilon \qquad \forall\, z \in \overline{\Omega}.$$

This approximation result together with the fact that the degree map from Definition 4.7 is invariant with respect to C^1-homotopies (that is, maps $h\colon [0,1]\times\overline{\Omega} \longrightarrow \mathbb{R}^N$ such that $h(t,\cdot) \in C^1(\overline{\Omega};\mathbb{R}^N)$ and $h(t,\cdot) \longrightarrow h(t_0,\cdot)$ in $C^1(\overline{\Omega};\mathbb{R}^N)$ as $t \to t_0$), lead to the following final definition of the degree.

Definition 4.9
Suppose that $\varphi \in C(\overline{\Omega};\mathbb{R}^N)$ *and* $h \notin \varphi(\partial\Omega)$. *We define* $d(\varphi,\Omega,h)$, *the* ***degree*** *of* φ *at* h *with respect to* Ω, *to be the number* $d(\psi,\Omega,h)$ *for any* $\psi \in C^1(\overline{\Omega};\mathbb{R}^N)$ *such that*

$$\big|\psi(z) - \varphi(z)\big| < \operatorname{dist}(h,\varphi(\partial\Omega)) \qquad \forall\, z \in \overline{\Omega}.$$

Remark 4.10
In fact this definition is still valid if we replace $\mathbb{R}^N$ by a finite dimensional normed space X with a fixed basis. Problem 4.2 shows that this definition does not depend on the choice of the fixed basis. This degree is known as the ***Brouwer degree***.

In the next theorem, we have gathered the main properties of the degree function we have just defined.

Theorem 4.11
<u>If</u> $\mathcal{Y} = \big\{(\varphi,\Omega,h) : \Omega \subseteq \mathbb{R}^N$ *is open and bounded,* $\varphi \in C(\overline{\Omega};\mathbb{R}^N),\ \ h \notin \varphi(\partial\Omega)\big\}$ *and* $d\colon \mathcal{Y} \longrightarrow \mathbb{Z}$ *is the degree function defined in Definition 4.9,*
<u>then</u> the following properties hold:

(a) *Normalization:* $d(I_{\mathbb{R}^N},\Omega,h) = 1$ *for* $h \in \Omega$ $(d(I_{\mathbb{R}^N},\Omega,h) = 0$ *for* $h \notin \Omega)$.

(b) *Domain Decomposition: If* Ω_1,Ω_2 *are disjoint open subsets of* Ω *and* $h \notin \varphi(\overline{\Omega}\setminus(\Omega_1\cup\Omega_2))$, *then*

$$d(\varphi,\Omega,h) = d(\varphi,\Omega_1,h) + d(\varphi,\Omega_2,h).$$

(c) *Homotopy Invariance: If* $\widehat{h}\colon [0,1]\times\overline{\Omega} \longrightarrow \mathbb{R}^N$ *and* $y\colon [0,1] \longrightarrow \mathbb{R}^N$ *are continuous with* $y(t) \notin \widehat{h}(t,\partial\Omega)$ *for all* $t \in [0,1]$, *then* $d(\widehat{h}(t,\cdot),\Omega,y(t))$ *is independent of* $t \in [0,1]$;

(d) *Existence: If* $d(\varphi, \Omega, h) \neq 0$*, then there exists* $u \in \Omega$ *such that* $\varphi(u) = h$.
(e) *Excision: If* $K \subseteq \overline{\Omega}$ *is compact and* $h \notin \varphi(K)$*, then*

$$d(\varphi, \Omega, h) = d(\varphi, \Omega \setminus K, h).$$

(f) *Dependence on* φ *and* h*: If* $\|\psi - \varphi\|_{C(\overline{\Omega})} < \operatorname{dist}(h, \varphi(\partial\Omega))$*, then*

$$d(\varphi, \Omega, h) = d(\psi, \Omega, h)$$

and if h_1, h_2 *belong in the same connected component of* $\mathbb{R}^N \setminus \varphi(\partial\Omega)$*, then*

$$d(\varphi, \Omega, h_1) = d(\varphi, \Omega, h_2).$$

Definition 4.12

(a) *Given* $\varphi \in C(\overline{\Omega}; \mathbb{R}^N)$ *and* $h \in \mathbb{R}^N$*, we say that* $u \in \overline{\Omega}$ *is an* ***h-point*** *of* φ *if* $\varphi(u) = h$.
(b) *Suppose that* $\varphi \in C(\overline{\Omega}; \mathbb{R}^N)$*,* $h \notin \varphi(\partial\Omega)$ *and* u_0 *is an isolated* h*-point of* φ*, then the index of* φ *with respect to* (u_0, h) *is defined by*

$$i(\varphi, u_0, h) = d(\varphi, U, h),$$

where $U \in \mathcal{D}$ *(*$\mathcal{D}$ *denotes the collection of all open sets* $V \subseteq \mathbb{R}^N$ *such that* $u_0 \in V$ *and* V *contains no other* h*-points).*
(c) *Suppose that* $\varphi \in C(\mathbb{R}^N; \mathbb{R}^N)$ *and* $h \in \mathbb{R}^N$ *are such that there exists* $R_0 > 0$ *for which we have*

$$\varphi(u) \neq h \qquad \forall\, u \in \mathbb{R}^N,\ |u| \geqslant R_0.$$

Then the ***index of*** φ ***at infinity*** *with respect to* h *is defined by*

$$i_\infty(\varphi, h) = d(\varphi, B_R(0), h),$$

for any $R \geqslant R_0$*. By the excision property of the Brouwer degree (see Theorem 4.11(e)), this definition does not depend on* $R \geqslant R_0$*.*

Remark 4.13

For $U \in \mathcal{D}$, we have $h \notin \varphi(\partial U)$ and so $d(\varphi, U, h)$ is well defined. Moreover, if $U_1, U_2 \in \mathcal{D}$, then $U_1 \cup U_2 \in \mathcal{D}$ and by the excision property of the Brouwer degree (see Theorem 4.11(e)), we have

$$d(\varphi, U_1, h) = d(\varphi, U_1 \cup U_2, h) = d(\varphi, U_2, h).$$

Therefore $d(\varphi, U, h)$ is the same for all $U \in \mathcal{D}$ and thus $i(\varphi, u_0, h)$ is well defined.

Now note that from Theorem 4.11(f), if D is a connected component of $\mathbb{R}^N \setminus \varphi(\partial\Omega)$, then $d(\varphi, \Omega, h)$ is the same for all $h \in D$ and so we can use the notation $d(\varphi, \Omega, D)$ to denote this common value.

Theorem 4.14 *(Multiplication Theorem)*
If $\varphi \in C(\overline{\Omega};\mathbb{R}^N)$, *$V$ is a bounded open set containing $\varphi(\overline{\Omega})$, $E = V \setminus \varphi(\partial\Omega)$, $\{D_i\}_{i \geqslant 1}$ are the connected components of E, $\psi \in C(\overline{V};\mathbb{R}^N)$ and $h \notin (\psi \circ \varphi)(\partial E) \cup \psi(\partial V)$, then*

$$d(\psi \circ \varphi, \Omega, h) = \sum_{i \geqslant 1} d(\psi, D_i, h) d(\varphi, E, D_i).$$

In Theorem 4.11(c), we have seen that homotopic maps have the same degree. In fact the converse is also true for spheres.

Theorem 4.15 *(Hopf Theorem)*
If $\Omega = B_r(0) \subseteq \mathbb{R}^N$ $(N \geqslant 2)$, $\varphi, \psi \in C(\overline{\Omega};\mathbb{R}^N)$ and $d(\varphi,\Omega,0) = d(\psi,\Omega,0)$, then φ and ψ are homotopic, i.e., there exists a continuous map $\widehat{h}\colon [0,1] \times \overline{\Omega} \longrightarrow \mathbb{R}^N$ such that $0 \notin \widehat{h}(t,\partial\Omega)$ for all $t \in [0,1]$ and $\widehat{h}(0,\cdot) = \varphi$, $\widehat{h}(1,\cdot) = \psi$.

Continuing with functions defined on spheres, we can define the (Brouwer) degree for such functions. In what follows $\overline{B}_1^N$ is the closed unit ball in $\mathbb{R}^N$ and S^{N-1} is the unit sphere (i.e., $S^{N-1} = \{z \in \mathbb{R}^N : |z| = 1\}$).

Definition 4.16
Let $\psi \in C(S^{N-1};S^{N-1})$ and $\varphi\colon \overline{B}_1^N \longrightarrow \mathbb{R}^N$ be a continuous extension of ψ (it exists by the Dugundji extension theorem; see Theorem 2.8). We define

$$d_1(\psi) = d(\varphi, \overline{B}_1^N, 0)$$

*(the **(Brouwer) degree** of ψ).*

Remark 4.17
Note that

$$\varphi(\partial B_1^N) = \varphi(S^{N-1}) = \psi(S^{N-1}) \subseteq S^{N-1}$$

and so $d(\varphi, \overline{B}_1^N, 0)$ is well defined. Moreover, from Problem 4.3 we see that the definition of $d_1(\psi)$ is independent of the particular extension φ of ψ.

According to Theorem 4.11, if we want to show that the equation $\varphi(u) = h$ has a solution, it suffices to show that $d(\varphi,\Omega,h) \neq 0$. The next result is a useful tool in this direction.

Theorem 4.18 *(Borsuk Theorem)*
If $\Omega \subseteq \mathbb{R}^N$ is a bounded, open, symmetric set with $0 \in \Omega$, $\varphi \in C(\overline{\Omega};\mathbb{R}^N)$, $0 \notin \varphi(\partial\Omega)$ and

$$\frac{\varphi(u)}{|\varphi(u)|} \neq \frac{\varphi(-u)}{|\varphi(-u)|} \qquad \forall\, u \in \partial\Omega, \tag{4.1}$$

then $d(\varphi,\Omega,0)$ is odd, in particular $d(\varphi,\Omega,0) \neq 0$.

Remark 4.19
If $\varphi \in C(\overline{\Omega};\mathbb{R}^N)$ is an odd map with $0 \notin \varphi(\partial\Omega)$, then (4.1) is automatically true. Condition (4.1) means that on the boundary $\partial\Omega$, vectors $\varphi(u)$ and $\varphi(-u)$ do not point in the same direction.

As a consequence of the Borsuk theorem (see Theorem 4.18), we obtain the following result.

Theorem 4.20 *(Borsuk–Ulam Theorem)*
<u>If</u> $\Omega \subseteq \mathbb{R}^N$ is a bounded, open, symmetric set with $0 \in \Omega$, $m < N$, $\varphi\colon \partial\Omega \longrightarrow \mathbb{R}^m$ is a continuous function
<u>then</u> there exists $u \in \partial\Omega$ such that $\varphi(u) = \varphi(-u)$.

The next result is useful in extending degree theory to infinite dimensional spaces.

Proposition 4.21
<u>If</u> X_n, X_m are two finite dimensional topological vector spaces, X_m is a subspace of X_n, $\dim X_m = m < n = \dim X_n$, $\Omega \subseteq X_n$ is a bounded open set, $\varphi \in C(\overline{\Omega};X_m)$ and $h \in X_m \setminus (I_{X_n} - \varphi)(\partial\Omega)$, <u>then</u>

$$d(I_{X_n} - \varphi, \Omega, h) = d((I_{X_n} - \varphi)|_{\overline{\Omega \cap X_m}}, \Omega \cap X_m, h).$$

We conclude our discussion of the finite dimensional degree theory with a known classical result, which can be proved using degree theory (see Theorem 4.14).

Theorem 4.22 *(Jordan Separation Theorem)*
<u>If</u> $\Omega_1, \Omega_2 \subseteq \mathbb{R}^N$ are homeomorphic compact sets,
<u>then</u> the sets $\mathbb{R}^N \setminus \Omega_1$ and $\mathbb{R}^N \setminus \Omega_2$ have the same number of connected components.

Remark 4.23
We mention that there exists a unique function $d\colon \mathcal{Y} \longrightarrow \mathbb{Z}$ (see Theorem 4.11) exhibiting the normalization, domain decomposition and homotopy invariance properties. Moreover, these properties imply that, if $0 \in \Omega$ and A is an invertible $N \times N$-matrix, then $d(A, \Omega, 0) = \operatorname{sgn} \det A$ (see Problem 4.5). Also, if $\varphi \in C(\overline{\Omega})$, $u_0 \in \Omega$ and φ is C^1 in a neighborhood of u_0 with $J_\varphi(u_0) \neq 0$, then

$$i(\varphi, u_0, \varphi(u_0)) = \operatorname{sgn} J_\varphi(u_0).$$

The next logical step is to try to extend the Brouwer degree to infinite dimensional Banach spaces. In finite dimensional spaces, the degree was defined to all continuous functions. In infinite dimensional Banach spaces, this is no longer true as the following example illustrates.

Example 4.24
Let $X = C([0,1])$ and let $u_0 \in X$ be the constant function $u_0(t) = \frac{1}{2}$ for all $t \in [0,1]$. We set $\Omega = \{u \in X : \|u - u_0\|_X < \frac{1}{2}\}$. Let $g \in X$ be such that $g(0) = 0$, $g(1) = 1$, $0 \leqslant g(t) \leqslant 1$ for all $t \in [0,1]$ and consider the map $\varphi\colon \overline{\Omega} \longrightarrow X$ defined by

$\varphi(u) = g \circ u$ for all $u \in \overline{\Omega}$. Evidently $\varphi(\overline{\Omega}) \subseteq \overline{\Omega}$. We consider the affine homotopy $\widehat{h}\colon [0,1] \times \overline{\Omega} \longrightarrow X$ defined by

$$\widehat{h}(s,u) = s\varphi(u) + (1-s)u. \tag{4.2}$$

Let $v \in \partial\Omega$. Then $\|v - u_0\|_X = \|v - \frac{1}{2}\|_X = \frac{1}{2}$, hence $0 \leqslant v(t) \leqslant 1$ for all $t \in [0,1]$ and for some $t_0 \in [0,1]$, we have $v(t_0) = 0$ or $v(t_0) = 1$. If $v(t_0) = 0$, then $\widehat{h}(s,v)(t_0) = 0$, while, if $v(t_0) = 1$, then $\widehat{h}(s,v)(t_0) = 1$. Moreover, since $0 \leqslant g \leqslant 1$, we see that $0 \leqslant \widehat{h}(s,v)(t)$ for all $s,t \in [0,1]$. Therefore, we have $\widehat{h}(s,\partial\Omega) \subseteq \partial\Omega$. Suppose that in analogy with the finite dimensional theory, there exists a degree d on $C(\overline{\Omega};X)$, exhibiting the normalization, existence, and homotopy invariance properties. Then from the admissible homotopy (4.2), for any $h \in \Omega$, we have

$$d(\varphi,\Omega,h) = d(I_X,\Omega,h) = 1.$$

So, by the existence property, there exists $u \in \Omega$ such that $\varphi(u) = h$. On the other hand, let

$$g(t) = \begin{cases} t & \text{if } t \in [0,\frac{1}{2}], \\ 1-t & \text{if } t \in (\frac{1}{2},\frac{5}{8}], \\ \frac{5}{3}(t-1)+1 & \text{if } t \in (\frac{5}{8},1], \end{cases}$$

$$h(t) = \tfrac{1}{2}t + \tfrac{1}{4}.$$

For these choices the equation $\varphi(u) = h$ has no solution, a contradiction.

This example illustrates that some restriction has to be placed on the map φ. So, we are going to consider maps of the form $I_X - f$, with f being compact and we will exploit the fact that compact maps can be uniformly approximated by finite rank maps (see Theorem 2.6).

So, suppose that X is a Banach space, $\Omega \subseteq X$ is a bounded open set, $\varphi\colon I_X - f\colon \overline{\Omega} \longrightarrow X$ is a function with $f\colon \overline{\Omega} \longrightarrow X$ being compact (see Definition 2.1) and $h \notin \varphi(\partial\Omega)$. From Proposition 2.16 we know that $\varphi = I_X - f$ is proper, that is, for every compact set $C \subseteq X$, the set $\varphi^{-1}(C) \subseteq X$ is compact (see Definition 2.13). Let $r = \operatorname{dist}(h,\varphi(\partial\Omega)) > 0$ and let $\varepsilon \in (0,r)$. Using Theorem 2.6 we can find a finite rank map $f_\varepsilon\colon \overline{\Omega} \longrightarrow X$ such that

$$\|f(u) - f_\varepsilon(u)\|_X < \varepsilon \qquad \forall\, u \in \overline{\Omega}.$$

Now choose a finite dimensional subspace $X_\varepsilon \subseteq X$ such that

$$h \in X_\varepsilon \quad \text{and} \quad f_\varepsilon(\overline{\Omega}) \subseteq X_\varepsilon.$$

We can define $d(I_X - f_\varepsilon, \Omega \cap X, h)$. It turns out that this degree is independent of $\varepsilon \in (0,r)$. So, we are led to the following definition.

Definition 4.25
Suppose that X is a Banach space, $\Omega \subseteq X$ is a bounded open set, $\varphi\colon I_X - f\colon \overline{\Omega} \longrightarrow X$ is a function with $f \in K(\overline{\Omega}; X)$ (see Definition 2.1) and $h \notin \varphi(\partial\Omega)$. Let $r = \operatorname{dist}(h, \varphi(\partial\Omega)) > 0$ and let $\widehat{f}\colon \overline{\Omega} \longrightarrow X$ be a finite rank map such that

$$\|f(u) - \widehat{f}(u)\|_X \; < \; r \qquad \forall\, u \in \overline{\Omega}.$$

Choose a finite dimensional subspace X_0 of X such that $h \in X_0$ and $\widehat{f}(\Omega) \subseteq X_0$. Let $\Omega_0 = \Omega \cap X_0$. Setting $\widehat{\varphi} = I_X - \widehat{f}$, we define

$$d_{LS}(\varphi, \Omega, h) \; = \; d(\widehat{\varphi}, \Omega_0, h).$$

Remark 4.26
The degree map that we have just defined is known as the ***Leray–Schauder degree*** (this also explains the notation d_{LS}). A closer examination of Definition 4.25 reveals that Ω need not be bounded. It suffices that Ω meets every finite dimensional subspace of X in a bounded set (such sets are called ***finitely bounded***). So, in the theory the word *bounded* may be replaced by *finitely bounded.*

As we did in Theorem 4.11 for the Brouwer degree, in the next theorem we have gathered the main properties of the Leray–Schauder degree. In the light of Definition 4.25, it is not a surprise that these properties are similar to those of the Brouwer degree.

Theorem 4.27
If X is a Banach space,

$$\begin{aligned} \mathcal{Y}_0 \; = \; & \{(\varphi = I_X - f, \Omega, h) : \; \Omega \subseteq X \text{ is bounded open}, \\ & f \in K(\overline{\Omega}; X), \; h \notin \varphi(\partial\Omega)\} \end{aligned}$$

and $d_{LS}\colon \mathcal{Y}_0 \longrightarrow \mathbb{Z}$ is the degree map defined in Definition 4.25,
then the following properties hold:

(a) *Normalization: $d_{LS}(I_X, \Omega, h) = 1$ for $h \in \Omega$ ($d_{LS}(I_X, \Omega, h) = 0$ for $h \notin \Omega$).*

(b) *Domain Decomposition: If Ω_1, Ω_2 are disjoint open subsets of Ω and $h \notin \varphi(\overline{\Omega} \setminus (\Omega_1 \cup \Omega_2))$, then*

$$d_{LS}(\varphi, \Omega, h) \; = \; d_{LS}(\varphi, \Omega_1, h) + d_{LS}(\varphi, \Omega_2, h).$$

(c) *Homotopy Invariance: If $\widehat{h}\colon [0,1] \times \overline{\Omega} \longrightarrow X$ is compact and $y\colon [0,1] \longrightarrow X$ is continuous with $y(t) \notin (I_X - \widehat{h}(t,\cdot))(\partial\Omega)$ for all $t \in [0,1]$, then $d_{LS}(I_X - \widehat{h}(t,\cdot), \Omega, y(t))$ is independent of $t \in [0,1]$;*

(d) *Existence: If $d_{LS}(\varphi, \Omega, h) \neq 0$, then there exists $u \in \Omega$ such that $\varphi(u) = h$.*

(e) *Excision: If $K \subseteq \overline{\Omega}$ is closed and $h \notin \varphi(K)$, then*

$$d_{LS}(\varphi, \Omega, h) \; = \; d_{LS}(\varphi, \Omega \setminus K, h).$$

(f) *Dependence on φ and h: If $\|g-f\|_{C(\overline{\Omega})} < r = \operatorname{dist}(h, \varphi(\partial\Omega))$, $g \in K(\overline{\Omega}; X)$ then*

$$d_{LS}(I_X - g, \Omega, h) = d(I_X - f, \Omega, h)$$

and if h_1, h_2 belong in the same connected component of $X \setminus (I_X - f)(\partial\Omega)$, then

$$d_{LS}(I_X - f, \Omega, h_1) = d(I_X - f, \Omega, h_2).$$

Let $(\varphi = I_X - f, \Omega, h) \in \mathcal{Y}_0$ and let u_0 be an isolated solution of $\varphi(u) = h$ (i.e., an isolated h-point; see Definition 4.12(a)). So, there exists $r_0 > 0$ such that no other solution of the equation exists in $\overline{B}_{r_0}(u_0) = \{u \in X : \|u - u_0\|_X \leqslant r_0\}$. Then the excision property of d_{LS} (see Theorem 4.27(e)) implies that

$$d_{LS}(\varphi, B_r(u_0), h) = d_{LS}(\varphi, B_{r_0}(u_0), h) \qquad \forall\, r \in (0, r_0]. \tag{4.3}$$

So, the following definition is legitimate.

Definition 4.28
Given $(\varphi = I_X - f, \Omega, h) \in \mathcal{Y}_0$ and an isolated solution $u_0 \in \Omega$ of the equation $\varphi(u) = h$, the ***index*** *of φ with respect to (u_0, h) is defined by*

$$i(\varphi, u_0, h) = d_{LS}(\varphi, B_r(u_0), h) \qquad \forall\, r \in (0, r_0]$$

(see (4.3)).

When $\varphi = I_X - L$ with $L \in \mathcal{L}_c(X)$, we can have a relation between $d_{LS}(I_X - L, \Omega, h)$ and the algebraic multiplicity of the eigenvalues of L. Recall that for every $\lambda \in \sigma(L) \setminus \{0\}$ there exists a least natural number $k = k(\lambda)$ such that

$$\ker\left((L - \lambda I_X)^n\right) = \ker\left((L - \lambda I_X)^k\right) \qquad \forall\, n \geqslant k.$$

Then $\dim\ker\left((L - \lambda I_X)^k\right)$ is called the ***algebraic multiplicity*** of $\lambda \in \sigma(L) \setminus \{0\}$.

Theorem 4.29
<u>If</u> X is a Banach space, $\varphi = I_X - L$ with $L \in \mathcal{L}_c(X)$, $\Omega \subseteq X$ is a bounded open set with $0 \in \Omega$, $\lambda \neq 0$ and $\frac{1}{\lambda}$ is not an eigenvalue of L,
<u>then</u> $d_{LS}(I_X - \lambda L, \Omega, 0) = (-1)^{m(\lambda)}$ with $m(\lambda)$ being the sum of the algebraic multiplicities of the real eigenvalues ϑ such that $\vartheta\lambda > 0$ and $m(\lambda) = 0$ when L has no such eigenvalues.

Again Theorem 4.27(f) leads to a multiplication formula (cf., Theorem 4.14).

Theorem 4.30 *(Multiplication Theorem)*
<u>If</u> $\varphi = I_X - f\colon \overline{\Omega} \longrightarrow X$ is a function with $f \in K(\overline{\Omega}; X)$, V is a bounded open set containing $\varphi(\overline{\Omega})$, $E = V \setminus \varphi(\partial\Omega)$, $\{D_i\}_{i \geqslant 1}$ are the connected components of E, $\psi = I_X - g\colon \overline{V} \longrightarrow X$ is a function with $g \in K(\overline{V}; X)$ and $h \notin \psi(\varphi(\partial\Omega)) \cup \psi(\partial V)$,
<u>then</u>

$$d_{LS}(\psi \circ \varphi, \Omega, h) = \sum_{i \geqslant 1} d_{LS}(\psi, D_i, h) d_{LS}(\varphi, E, D_i).$$

Also, there is an infinite dimensional version of the Borsuk theorem (see Theorem 4.18).

Theorem 4.31
If X *is a Banach space,* $\Omega \subseteq X$ *is a bounded, open, symmetric set with* $0 \in \Omega$ *and* $\varphi = I_X - f$ *with* $f \in K(\overline{\Omega}; X)$ *satisfying*

$$\frac{\varphi(u)}{\|\varphi(u)\|_X} \neq \frac{\varphi(-u)}{\|\varphi(-u)\|_X} \qquad \forall\, u \in \partial\Omega,$$

then $d_{LS}(\varphi, \Omega, 0)$ *is odd; in particular* $d_{LS}(\varphi, \Omega, 0) \neq 0$.

Using Theorem 4.30 one can prove the following version of the Jordan separation theorem (see Theorem 4.22).

Theorem 4.32
If X *is a Banach space,* $C, D \subseteq X$ *are two closed bounded sets and there exists a homeomorphism* $\varphi = I_X - f$ *from* C *onto* D *with* $f \in K(\overline{\Omega}; X)$,
then $X \setminus C$ *and* $X \setminus D$ *have the same number of connected components.*

Remark 4.33
By virtue of Definition 4.25, the uniqueness of the Brouwer degree (see Remark 4.23) implies that $d_{LS} \colon \mathcal{Y}_0 \longrightarrow \mathbb{Z}$ is the unique function satisfying the normalization, domain decomposition and homotopy invariance properties.

We can extend the Leray–Schauder degree to a broader class of maps. So, let X be a Banach space and let γ denote the Kuratowski measure of noncompactness α or the Hausdorff measure of noncompactness β (see Definition 2.31). Let $\Omega \subseteq X$ be a bounded open set. Recall that (see Definition 2.31)

$$\begin{aligned} SC_k(\overline{\Omega}; X) &= \big\{ f \in C(\overline{\Omega}; X) : \ f \text{ is a } k\text{-set contraction} \big\}, \\ S(\overline{\Omega}; X) &= \big\{ f \in C(\overline{\Omega}; X) : \ f \text{ is condensing} \big\} \end{aligned}$$

(recall that $f \in C(\overline{\Omega}; X)$ is a k-set contraction (with $0 \leqslant k < 1$) if for every bounded set $C \subseteq X$, we have $\gamma(f(C)) \leqslant k\gamma(C)$ and f is condensing if for every bounded set $C \subseteq X$ with $\gamma(C) > 0$, we have $\gamma(f(C)) < \gamma(C)$). Note that

$$K(\overline{\Omega}; X) \subseteq SC_k(\overline{\Omega}; X) \subseteq S(\overline{\Omega}; X).$$

Moreover, from Problem 2.10, we have that every $f \in S(\overline{\Omega}; X)$ is bounded and proper.

Now, suppose that $f \in SC_k(\overline{\Omega}; X)$.We set

$$D_1 = \overline{\operatorname{conv}}\, f(\overline{\Omega})$$

and inductively define

$$D_n = \overline{\operatorname{conv}}\, f(D_{n-1} \cap \overline{\Omega}) \qquad \forall\, n \geqslant 2.$$

Let $D = \bigcap_{n\geqslant 1} D_n$. Evidently the set D is closed and convex. Also the sequence $\{D_n\}_{n\geqslant 1}$ is decreasing. Indeed, clearly $D_1 \supseteq D_2$ and assume that $D_{n-1} \supseteq D_n$ (induction hypothesis). Then

$$D_n = \overline{\operatorname{conv}} f(D_{n-1} \cap \overline{\Omega}) \supseteq \overline{\operatorname{conv}} f(D_n \cap \overline{\Omega}) = D_{n+1}.$$

Moreover, we have

$$\gamma(D_n) = \gamma(f(D_{n-1} \cap \overline{\Omega})) \leqslant \gamma(f(D_{n-1})) \leqslant k\gamma(D_{n-1})$$

(see Proposition 2.32), so

$$\gamma(D_n) \leqslant k^n \gamma(\overline{\Omega}) \longrightarrow 0$$

(since $k \in [0,1)$), thus $D \neq \emptyset$ and it is compact and convex (see Problem 2.12).

Invoking the Dugundji extension theorem (see Theorem 2.8), we can find a continuous map $\widehat{f} \colon \overline{\Omega} \longrightarrow D$ such that $\widehat{f}|_{D\cap\overline{\Omega}} = f|_{D\cap\overline{\Omega}}$. Note that if for some $u \in \overline{\Omega}$, we have $f(\widehat{u}) = u$, then $u \in D$ and so $\widehat{f}(u) = f(u)$. Hence, if $0 \notin (I_X - f)(\partial\Omega)$, then $0 \notin (I_X - \widehat{f})(\partial\Omega)$. Therefore, recalling that the range of $\widehat{f}$ is compact (the set D), we see that $d_{LS}(I_X - \widehat{f}, \Omega, 0)$ is well defined. We also claim that this degree is the same for all "Dugundji extensions" $\widehat{f}$. To see this, let $\widetilde{f} \colon \overline{\Omega} \longrightarrow D$ be another such extension. We consider the homotopy

$$\widehat{h}(t,u) = t\widehat{f}(u) + (1-t)\widetilde{f}(u) \qquad \forall\, (t,u) \in [0,1] \times \overline{\Omega}.$$

Since both $\widehat{f}$ and $\widetilde{f}$ maps into D, we see that if $f(u) = u$, then $u = \widehat{h}(t,u)$ and so $0 \notin (I_X - h(t,\cdot))(\partial\Omega)$ for all $t \in [0,1]$. Hence the homotopy invariance of the Leray–Schauder degree (see Theorem 4.27(c)) implies that

$$d_{LS}(I_X - \widehat{f}, \Omega, 0) = d_{LS}(I_X - \widetilde{f}, \Omega, 0).$$

This leads to the following definition.

Definition 4.34
Let $f \in SC_k(\overline{\Omega}; X)$ and assume that $0 \notin (I_X - f)(\partial\Omega)$. Then the degree $d_{SC}(I_X - f, \Omega, 0)$ is defined by

$$d_{SC}(I_X - f, \Omega, 0) = d_{LS}(I_X - \widehat{f}, \Omega, 0),$$

where $\widehat{f} \colon \partial\Omega \longrightarrow D$ is any "Dugundji extension" of $f|_{\overline{\Omega}\cap D}$. For $h \notin (I_X - f)(\partial\Omega)$, we define

$$d_{SC}(I_X - f, \Omega, h) = d_{SC}(I_X - f - h, \Omega, 0),$$

with $(I_X - f - h)(u) = u - f(u) - h$ for all $u \in \overline{\Omega}$.

Remark 4.35
In the above case the admissible homotopies are of the form

$$I_X - \widehat{h}(t,\cdot),$$

with $\widehat{h} \in C([0,1] \times \overline{\Omega}; X)$ and $\gamma(\widehat{h}(t,C)) \leqslant k\gamma(C)$ for all $t \in [0,1]$ and all $C \subseteq \overline{\Omega}$.

Theorem 4.36
If X is a Banach space,

$$\begin{aligned}\mathcal{Y}_{SC} &= \{(\varphi = I_X - f, \Omega, h) : \Omega \subseteq X \text{ is bounded open,} \\ &\quad f \in SC_k(\overline{\Omega}; X),\ h \notin \varphi(\partial\Omega)\}\end{aligned}$$

and $d_{SC} : \mathcal{Y}_{SC} \longrightarrow \mathbb{Z}$ is the degree map defined in Definition 4.34, then d_{SC} is the unique degree map on $\mathcal{Y}_{SC}$ satisfying the following properties (see Theorem 4.27): normalization, domain decomposition, homotopy invariance (for the admissible homotopies given in Remark 4.35), existence, excision, and dependence on φ and h.

Next we extend the definition of the degree to maps $\varphi = I_X - f$ with $f \in S(\overline{\Omega}; X)$. This can be done by approximating $f \in S(\overline{\Omega}; X)$ with elements from $SC_k(\overline{\Omega}; X)$. First recall that φ is closed (see Problem 2.10 and Proposition 2.24). Hence $\varphi(\partial\Omega)$ is closed. Since by hypothesis $0 \notin \varphi(\partial\Omega)$, we have that $\vartheta = \operatorname{dist}(0, \varphi(\partial\Omega)) > 0$. Let $f_0 \in SC_k(\overline{\Omega}; X)$ for some $k \in (0,1)$ such that

$$\|f(u) - f_0(u)\|_X < \tfrac{\vartheta}{3} \qquad \forall\, u \in \overline{\Omega}.$$

Such a choice is always possible, since we can take $f_0 = \lambda f$ with $\lambda \in (0,1)$ and $1 - \lambda$ sufficiently small. Let $\varphi_0 = I_X - f_0$. Then $\varphi_0(\partial\Omega)$ is closed (see Problem 2.10 and Proposition 2.24) and $\operatorname{dist}(0, \varphi_0(\partial\Omega)) \geqslant \frac{2}{3}\vartheta$. Let $f_1 \in SC_l(\overline{\Omega}; X)$ ($l \in (0,1)$) be another set-contraction such that

$$\|f(u) - f_1(u)\|_X < \tfrac{\vartheta}{3} \qquad \forall\, u \in \overline{\Omega}.$$

Let $m = \max\{k, l\}$ and let $f_0, f_1 \in SC_m(\overline{\Omega}; X)$. We consider the homotopy

$$\widehat{h}(t,u) = t f_0(u) + (1-t) f_1(u) \qquad \forall\, (t,u) \in [0,1] \times \overline{\Omega}.$$

If $\widehat{h}(t,u) = u$ for some $(t,u) \in [0,1] \times \partial\Omega$, then since $\|f_1(u) - f_2(u)\|_X < \frac{2\vartheta}{3}$ for all $u \in \overline{\Omega}$, we have

$$\|u - f_1(u)\|_X < \tfrac{2\vartheta}{3},$$

a contradiction to the fact that $\|u - f_1(u)\|_X \geqslant \frac{2\vartheta}{3}$. So, $u \neq \widehat{h}(t,u)$ for all $(t,u) \in [0,1] \times \partial\Omega$. Invoking the homotopy invariance of the degree d_{SC} (see Theorem 4.36), we have

$$d_{SC}(I_X - f_0, \Omega, 0) = d_{SC}(I_X - f_1, \Omega, 0).$$

This leads to the following definition.

Definition 4.37
Let $\varphi = I_X - f$ with $f \in S(\overline{\Omega}; X)$ and assume that $0 \notin (I_X - f)(\partial\Omega)$. Then the degree $d_C(I_X - f, \Omega, 0)$ is defined by

$$d_C(I_X - f, \Omega, 0) = d_{SC}(I_X - f_0, \Omega, 0),$$

where $f_0 \in SC_k(\overline{\Omega}; X)$ *for some* $k \in (0,1)$ *is such that*

$$\|f_0(u) - f(u)\|_X < \delta \qquad \forall\, u \in \overline{\Omega}$$

for some $\delta > 0$ *small. For* $h \notin (I_X - f)(\partial\Omega)$, *we define*

$$d_C(I_X - f, \Omega, h) = d_C(I_X - f - h, \Omega, 0),$$

with $(I_X - f - h)(u) = u - f(u) - h$ *for all* $u \in \overline{\Omega}$.

Remark 4.38
In the above definition, we may take $\delta = \frac{1}{3}\mathrm{dist}(0, \varphi(\partial\Omega))$. The admissible homotopies are of the form

$$I_X - \widehat{h}(t, \cdot) \qquad \forall\, t \in [0,1],$$

where $\widehat{h} \in C([0,1] \times \overline{\Omega}; X)$ and $\gamma(\widehat{h}(t, C)) < \gamma(C)$ for all $t \in [0,1]$ and all noncompact sets $C \subseteq \overline{\Omega}$.

Theorem 4.39
If X *is a Banach space,*

$$\begin{aligned} \mathcal{Y}_c = \{ & (\varphi = I_X - f, \Omega, h) : \Omega \subseteq X \text{ is bounded open}, \\ & f \in S(\overline{\Omega}; X),\ h \notin \varphi(\partial\Omega) \} \end{aligned}$$

and $d_C : \mathcal{Y}_C \longrightarrow \mathbb{Z}$ *is the degree map defined in Definition 4.37,*
then d_C *is the unique degree map on* $\mathcal{Y}_C$ *satisfying the following properties (see Theorem 4.36): normalization, domain decomposition, homotopy invariance (for the admissible homotopies given in Remark 4.38), existence, excision and dependence on* φ *and* h.

Next we present a degree for operators of type $(S)_+$ defined on a reflexive Banach space X with values in the dual X^*.

So, let X be a reflexive Banach space with topological dual X^* and duality brackets $\langle \cdot, \cdot \rangle$. Let $\Omega \subseteq X$ be a bounded open set and let $f : \overline{\Omega} \longrightarrow X^*$ be a demicontinuous map of type $(S)_+$ (see Definitions 2.110(a) and 2.137). Let $\{X_\alpha\}_{\alpha \in J}$ be the family of all finite dimensional subspaces of X such that $\Omega_\alpha = \Omega \cap X_\alpha \neq \emptyset$. We order J by inclusion (that is, $\alpha_1 \leqslant \alpha_2$ if and only if $X_{\alpha_1} \subseteq X_{\alpha_2}$). By i_α we denote the injection of X_α into X and by i_α^* we denote the adjoint of i_α which is the projection of X^* onto X_α^*. To simplify our presentation, we choose as reference point $h^* \in X^*$ the origin. There is no loss of generality in doing this, since we can always replace f by $f - h^*$ which remains demicontinuous and of type $(S)_+$. Let $f_\alpha : \overline{\Omega} \cap X_\alpha \longrightarrow X_\alpha^*$ be the Galerkin approximation of f with respect to X_α defined by

$$\langle f_\alpha(u), h \rangle_\alpha = \langle f(u), h \rangle \qquad \forall\, u \in \overline{\Omega} \cap X_\alpha,\ h \in X_\alpha.$$

Here by $\langle \cdot, \cdot \rangle_\alpha$ we denote the duality brackets for the pair (X_α^*, X_α).

Proposition 4.40
<u>If</u> Y is a finite dimensional subspace of X,
<u>then</u> a degree map $d_{(S)_+}$ is defined on the demicontinuous, $(S)_+$-map $f\colon \overline{\Omega} \longrightarrow X^*$ with a bounded open set $\Omega \subseteq X$ which is invariant under affine homotopies, that is, homotopies $\widehat{h}(t,u)$ of the form

$$\widehat{h}(t,u) = tf_0(u) + (1-t)f_1(u) \qquad \forall\, (t,u) \in [0,1] \times \overline{\Omega},$$

with demicontinuous, $(S)_+$-maps $f_0, f_1\colon \overline{\Omega} \longrightarrow X^*$ and it is normalized by a duality map $\mathcal{F}$ of type $(S)_+$. Moreover, if $f\colon \overline{\Omega} \longrightarrow X^*$ is such a demicontinuous $(S)_+$-map, f_Y is the Galerkin approximation of f with respect to Y and $d_{(S)_+}(f,\Omega,0)$ is not defined or $d(f_Y,\Omega\cap Y,0)$ is not defined or both are well defined but $d_{(S)_+}(f,\Omega,0) \neq d(f_Y,\Omega\cap Y,0)$, then there exists $u \in \partial\Omega$ such that $\langle f(u),u\rangle \leqslant 0$ and $\langle f(u),y\rangle = 0$ for all $y \in Y$.

Remark 4.41
From the Troyanski renorming theorem (see Theorem I.5.192 or Remark 2.115), we can equivalently renorm X so that both X and X^* are locally uniformly convex. Then from Problem 2.185, we know that the duality map $\mathcal{F}\colon X \longrightarrow X^*$ is of type $(S)_+$.

Proposition 4.42
<u>If</u> f,Ω,X are as above and $0 \notin f(\partial\Omega)$,
<u>then</u> there exists $\alpha_0 \in J$ such that for all $\alpha \geqslant \alpha_0$ (that is, $X_\alpha \supseteq X_{\alpha_0}$), we have $0 \notin f_\alpha(\partial\Omega_\alpha)$ ($\Omega_\alpha = \Omega \cap X_\alpha$) and $d(f_\alpha,\Omega_\alpha,0)$ is independent of $\alpha \geqslant \alpha_0$.

This proposition suggests a degree for demicontinuous $(S)_+$-maps. So, for a reflexive Banach space X, we introduce the family

$$\begin{aligned} \mathcal{Y}_{(S_+)} = \big\{(f,\Omega,h^*) :\ & \Omega \subseteq X \text{ is bounded open}, \\ & f\colon \overline{\Omega} \longrightarrow X^* \text{ is demicontinuous and of type } (S)_+, \\ & h^* \notin f(\partial\Omega)\big\}. \end{aligned}$$

Definition 4.43
We can define a degree $d_{(S)_+}$ on $\mathcal{Y}_{(S_+)}$ be setting

$$d_{(S)_+}(f,\Omega,h^*) = d(f_\alpha,\Omega\cap X_\alpha,h^*),$$

for X_α sufficiently big.

Remark 4.44
In the above definition, for the sake of generality we have used a general reference point $h^* \in X^*$. For this degree possible admissible homotopies are the affine homotopies defined by

$$\widehat{h}(t,u) = tf_0(u) + (1-t)f_1(u) \qquad \forall\, (t,u) \in [0,1] \times \overline{\Omega},$$

with $f_0, f_1 : \overline{\Omega} \longrightarrow X^*$ being demicontinuous and of type $(S)_+$. However, this family is too narrow for some important applications. For this reason we broaden the class of admissible homotopies as follows.

Definition 4.45
Let X be a reflexive Banach space and let $\Omega \subseteq X$ be a bounded open set. A map $\widehat{h} : [0,1] \times \overline{\Omega} \longrightarrow X^*$ is said to be a ***homotopy of type*** *$(S)_+$, if for any sequences $\{t_n\}_{n\geqslant 1} \subseteq [0,1]$ and $\{u_n\}_{n\geqslant 1} \subseteq \overline{\Omega}$ such that $t_n \longrightarrow t$ in $[0,1]$, $u_n \xrightarrow{w} u$ in X and*

$$\limsup_{n\to+\infty} \langle h(t_n, u_n), u_n - u\rangle \leqslant 0,$$

we have $u_n \longrightarrow u$ in X and $h(t_n, u_n) \xrightarrow{w} h(t,u)$ in X^*.

Theorem 4.46
The degree map $d_{(S)_+} : \mathcal{Y}_{(S)+} \longrightarrow \mathbb{Z}$ defined in Definition 4.43 is the unique degree having the following three fundamental properties:

(a) *Normalization: $d_{(S)_+}(\mathcal{F}, \Omega, h^*) = 1$, where $\mathcal{F} : X \longrightarrow X^*$ is the duality map corresponding to the norm of X which makes both X and X^* locally uniformly convex (this is always possible by the Troyanski renorming theorem; see Theorem I.5.192 or Remark 2.115) and $h^* \in \mathcal{F}(\Omega)$ (zero otherwise).*

(b) *Domain Decomposition: If Ω_1, Ω_2 are disjoint open subsets of Ω and $h^* \notin f(\overline{\Omega} \setminus (\Omega_1 \cup \Omega_2))$, then*

$$d_{(S)_+}(f, \Omega, h^*) = d_{(S)_+}(f, \Omega_1, h^*) + d_{(S)_+}(f, \Omega_2, h^*).$$

(c) *Homotopy Invariance: If $\widehat{h} : [0,1] \times \overline{\Omega} \longrightarrow X^*$ is a homotopy of type $(S)_+$ and $y^* : [0,1] \longrightarrow X^*$ is a continuous map such that $y^*(t) \notin \widehat{h}(t, \partial\Omega)$ for all $t \in [0,1]$, then $d_{(S)_+}(\widehat{h}(t,\cdot), \overline{\Omega}, y^*(t))$ is independent of $t \in [0,1]$.*

Remark 4.47
If $X = H$ is a Hilbert space with $H = H^*$ (pivot Hilbert space), then $d_{(S)_+}$ contains as a special case the Leray–Schauder degree. Indeed, note that the Kadec–Klee property of Hilbert spaces (see Corollary 1.26) implies that $u \longmapsto u - f(u)$ is of type $(S)_+$ if $f : \overline{\Omega} \longrightarrow H$ is a compact map. Of course, the degree map $d_{(S)_+}$ has the other important properties such as

- Excision: If $C \subseteq \overline{\Omega}$ is a closed set and $h^* \notin f(C \cup \partial\Omega)$, then
$$d_{(S)_+}(f, \Omega, h^*) = d_{(S)_+}(f, \Omega \setminus C, h^*).$$
- Existence: If $d_{(S)_+}(f, \Omega, h^*) \neq 0$, then there exists $\widehat{u} \in \Omega$ such that $f(\widehat{u}) = h^*$.
- Dependence on boundary value: If $f, g : \overline{\Omega} \longrightarrow X^*$ are demicontinuous $(S)_+$ maps such that
$$f|_{\partial\Omega} = g|_{\partial\Omega} \quad \text{and} \quad h^* \notin f(\partial\Omega),$$
then $d_{(S)_+}(f, \Omega, h^*) = d_{(S)_+}(g, \Omega, h^*)$.

The following theorem is useful in applications on partial differential equations.

Theorem 4.48
If X is a reflexive Banach space, $\Omega \subseteq X$ is an open set, $\varphi\colon \Omega \longrightarrow \mathbb{R}$ is a Gâteaux differentiable function with $\varphi'_G\colon \Omega \longrightarrow X^*$ being demicontinuous and of type $(S)_+$ and there exist $a, b \in \mathbb{R}$, $a < b$ and $u_0 \in X$ such that:
(i) the set $V = \{\varphi < b\}$ is bounded and $\overline{V} \subseteq \Omega$;
(ii) if $u \in \{\varphi \leqslant a\}$, then $tu + (1-t)u_0 \in V$ for all $t \in [0,1]$;
(iii) $\varphi'_G(u) \neq 0$ for all $u \in \{a \leqslant \varphi \leqslant b\}$,
then $d_{(S)_+}(\varphi'_G, V, 0) = 1$.

Now we will extend the degree to maps of the form $f + A$, with f being bounded, demicontinuous and of type $(S)_+$ and A being maximal monotone such that $(0,0) \in \operatorname{Gr} A$. We assume that X is a reflexive Banach space which by virtue of the Troyanski renorming theorem (see Theorem I.5.192 or Remark 2.115) is furnished with a norm so that X itself and its dual X^* are both locally uniformly convex. Suppose that $\Omega \subseteq X$ is a bounded open set, $f\colon \overline{\Omega} \longrightarrow X^*$ is a bounded, demicontinuous map of type $(S)_+$ and $A\colon X \supseteq D \longrightarrow 2^{X^*}$ is a maximal monotone map such that $(0,0) \in \operatorname{Gr} A$. We assume that $h^* \notin (f+A)(\partial\Omega)$. For every $\lambda > 0$, let A_λ be the corresponding Yosida approximation of A (see Problem 2.161 and the Remark following it). Then $f + A_\lambda$ is a bounded, demicontinuous map of type $(S)_+$ (see Problem 2.163). Then it can be shown that

$$h^* \notin (f + A_\lambda)(\partial\Omega) \tag{4.4}$$

and

$$d_{(S)_+}(f + A_\lambda, \Omega, h^*) \quad \text{stabilize for } \lambda > 0 \text{ small.} \tag{4.5}$$

So, if X is a reflexive Banach space normed in such a way so that both X and X^* are locally uniformly convex and we introduce the family

$$\begin{aligned} \mathcal{Y}_M = \big\{&(f + A, \Omega, h^*) : \ \Omega \subseteq X \text{ is bounded open,} \\ &f\colon \overline{\Omega} \longrightarrow X^* \text{ is bounded, demicontinuous, and of type } (S)_+, \\ &A\colon X \supseteq D \longrightarrow 2^{x^*} \text{ is maximal monotone with } (0,0) \in \operatorname{Gr} A \\ &\text{and } h^* \notin (f + A)(\partial\Omega)\big\}, \end{aligned}$$

then (4.4) and (4.5) lead to the following definition of a degree map on $\mathcal{Y}_M$.

Definition 4.49
We can define a degree d_M on $\mathcal{Y}_M$ by setting

$$d_M(f + A, \Omega, h^*) = d_{(S)_+}(f + A_\lambda, \Omega, h^*) \qquad \forall\, \lambda > 0 \text{ small.}$$

For this degree map, the admissible homotopies $\widehat{h}$ are defined by

$$\widehat{h}(t,u) = f_t(u) + A^t(u) \qquad \forall\, (t,u) \in [0,1] \times \overline{\Omega},$$

with f_t being a homotopy of type $(S)_+$ (see Definition 4.45) and A^t being a pseudomonotone homotopy defined in what follows.

Definition 4.50

Let $\{A^t\colon X \longrightarrow X^*\}_{t\in[0,1]}$ be a family of maximal monotone maps such that $(0,0) \in \operatorname{Gr} A^t$ for all $t \in [0,1]$. We say that $\{A^t\}_{t\in[0,1]}$ is a ***pseudomonotone homotopy of maximal monotone maps****, if it satisfies one of the following equivalent conditions:*

(a) If $\{t_n\}_{n\geqslant 1} \subseteq \mathbb{R}$, $\{u_n\}_{n\geqslant 1} \subseteq X$ and $\{u_n^*\}_{n\geqslant 1} \subseteq X^*$ are three sequences such that $t_n \longrightarrow t$, $(u_n, u_n^*) \in \operatorname{Gr} A^{t_n}$ for all $n \geqslant 1$, $u_n \xrightarrow{w} u$ in X, $u_n^* \xrightarrow{w} u^*$ in X^* and

$$\limsup_{n\to+\infty} \langle u_n^*, u_n - u\rangle \leqslant 0,$$

then $(u,u^*) \in \operatorname{Gr} A^t$ and $\langle u_n^*, u_n\rangle \longrightarrow \langle u^*, u\rangle$.

(b) For every $\lambda > 0$ (equivalently for some $\lambda > 0$), the map $[0,1] \times X \ni (t,u) \longmapsto J_\lambda^{A^t}(u) \in X$ is continuous.

(c) For every $u \in X$ and every $\lambda > 0$ (equivalently for some $\lambda > 0$), the map $[0,1] \ni t \longmapsto J_\lambda^{A^t}(u) \in X$ is continuous.

(d) If $\{t_n\}_{n\geqslant 1} \subseteq [0,1]$ is a sequence such that $t_n \longrightarrow t$ in $[0,1]$ and $(u,u^*) \in \operatorname{Gr} A^t$, then we can find two sequences $\{u_n\}_{n\geqslant 1} \subseteq X$ and $\{u_n^*\}_{n\geqslant 1} \subseteq X^*$ such that $(u_n, u_n^*) \in \operatorname{Gr} A^{t_n}$ for all $n \geqslant 1$, $u_n \longrightarrow u$ in X and $u_n^* \xrightarrow{w} u^*$ in X^*.

Remark 4.51

Note that Definition 4.50(d) says that, if $t_n \longrightarrow t$ in $[0,1]$, then

$$\operatorname{Gr} A^t \subseteq (s \times w)\text{–}\liminf_{n\to+\infty} \operatorname{Gr} A^{t_n}$$

(see Definition 2.88).

In general affine homotopies need not be pseudomonotone homotopies. More precisely, if A^1, A^2 are maximal monotone maps with $(0,0) \in \operatorname{Gr} A^k$ for $k = 1,2$ and $\overline{D(A^1)} \neq \overline{D(A^2)}$, then $A^t = tA^1 + (1-t)A^2$ need not be a pseudomonotone homotopy. In fact it may happen that for some $t \in (0,1)$, A^t may not be maximal monotone. But even if A^t is maximal monotone for all $t \in [0,1]$, the condition $\overline{D(A^1)} = \overline{D(A^2)}$ is only necessary for $\{A^t\}_{t\in[0,1]}$ to be a pseudomonotone homotopy. If $D(A^1) = X$ and $A^1\colon X \longrightarrow X^*$ is continuous monotone, then the family $\{A^t\}_{t\in[0,1]}$ is a pseudomonotone homotopy.

Theorem 4.52

The degree $d_M : \mathcal{Y}_M \longrightarrow \mathbb{Z}$ defined in Definition 4.49 is the unique degree having the following three fundamental properties:

(a) Normalization: $d_M(\mathcal{F}, \Omega, h^*) = 1$, where $\mathcal{F}\colon X \longrightarrow X^*$ is the duality map corresponding to the norm of X which makes both X and X^* locally uniformly convex (this is always possible by the Troyanski renorming theorem; see Theorem I.5.192 or Remark 2.115) and $h^* \in \mathcal{F}(\Omega)$.

(b) Domain Decomposition: If Ω_1, Ω_2 are disjoint open subsets of Ω and $h^* \notin (f+A)(\overline{\Omega} \setminus (\Omega_1 \cup \Omega_2))$, then

$$d_M(f+A, \Omega, h^*) = d_M(f+A, \Omega_1, h^*) + d_M(f+A, \Omega_2, h^*).$$

(c) Homotopy Invariance: If $\{f_t\}_{t\in[0,1]}$ is a homotopy of type $(S)_+$ with f_t being bounded, $\{A^t\}_{t\in[0,1]}$ is a pseudomonotone homotopy of maximal monotone maps, $y^*\colon [0,1] \longrightarrow X^*$ is a continuous map such that

$$y^*(t) \notin (f_t + A^t)(\partial\Omega) \quad \forall\, t \in [0,1],$$

then $d_M(f_t + A^t, \overline{\Omega}, y^*(t))$ is independent of $t \in [0,1]$.

Remark 4.53
Of course both $d_{(S)_+}$ and d_M have other useful properties such as excision and existence which can be formulated in the obvious way (see Remark 4.47).

We can also have an extension of the Leray–Schauder degree to multifunctions. So, let X be a Banach space and let $\Omega \subseteq X$ be a bounded open set.

Definition 4.54
A multifunction $F\colon \overline{\Omega} \longrightarrow 2^X\setminus\{\emptyset\}$ is said to be **compact**, *if it is upper semicontinuous and for every compact set $C \subseteq \overline{\Omega}$, the set $F(C) = \bigcup\limits_{u\in C} F(u) \subseteq X$ is relatively compact.*

Suppose that $F\colon \overline{\Omega} \longrightarrow P_{fc}(X)$ is a compact multifunction. Then according to Problem 2.83, given $\varepsilon > 0$ we can find a continuous map $f_\varepsilon\colon \overline{\Omega} \longrightarrow X$ such that

$$h^*\left(\operatorname{Gr} f_\varepsilon, \operatorname{Gr} F\right) < \varepsilon \quad \text{and} \quad f_\varepsilon(u) \subseteq \overline{\operatorname{conv}}\, F(\overline{\Omega}) \qquad \forall\, u \in \overline{\Omega}. \tag{4.6}$$

If $h \notin (I_X - F)(\partial\Omega)$, then we can show that for $\varepsilon > 0$ small we have

$$h \notin (I_X - f_\varepsilon(\partial\Omega) \quad \text{and} \quad d_{LS}(I_X - f_\varepsilon, \Omega, h) \quad \text{stabilizes.} \tag{4.7}$$

So, for a Banach space X, let

$$\begin{aligned} \mathcal{Y}_{SV} = \{(F,\Omega,h) :\ & \Omega \subseteq X \text{ is bounded open,} \\ & F\colon \overline{\Omega} \longrightarrow P_{fc}(X) \text{ is compact multifunction} \\ & \text{and } h \notin (I_X - F)(\partial\Omega)\}, \end{aligned}$$

Then (4.7) leads to the following degree map on $\mathcal{Y}_{SV}$.

Definition 4.55
We can define a degree d_{SV} on $\mathcal{Y}_{SV}$ be setting

$$d_{SV}(I_X - F, \Omega, h) = d_{LS}(I_X - f_\varepsilon, \Omega, h) \qquad \forall\, \varepsilon > 0 \text{ small,}$$

where $f_\varepsilon\colon \overline{\Omega} \longrightarrow X$ is any continuous approximate selector of F (in the sense of (4.6)).

Remark 4.56
For this degree map, the admissible homotopies are of the form

$$I_X - \widehat{h}(t,\cdot) \qquad \forall\, t \in [0,1],$$

where $\widehat{h}\colon [0,1] \times \overline{\Omega} \longrightarrow P_{fc}(X)$ is compact.

Theorem 4.57
The degree map $d_{SV}\colon \mathcal{Y}_{SV} \longrightarrow \mathbb{Z}$ defined in Definition 4.55 is the unique degree having the following three fundamental properties:
(a) *Normalization: $d_{SV}(I_X,\Omega,h)=1$ for $h\in\Omega$ ($d_{SV}(I_X,\Omega,h)=0$ for $h\notin\Omega$).*
(b) *Domain Decomposition: If Ω_1,Ω_2 are disjoint open subsets of Ω and $h\notin(I_X-F)(\overline{\Omega}\setminus(\Omega_1\cup\Omega_2))$, then*

$$d_{SV}(I_X-F,\Omega,h) = d_{SV}(I_X-F,\Omega_1,h)+d_{SV}(I_X-F,\Omega_2,h).$$

(c) *Homotopy Invariance: If $\widehat{h}\colon[0,1]\times\overline{\Omega}\longrightarrow P_{fc}(X)$ is a compact map and $h\notin(I_X-h(t,\cdot))(\partial\Omega)$ for all $t\in[0,1]$ then $d_{SV}(I_X-\widehat{h}(t,\cdot),\Omega,h)$ is independent of $t\in[0,1]$.*

4.1.2 Metric Fixed Point Theory

There is an informal classification of fixed point theorems to metric fixed points, topological fixed points, and order fixed points. In metric fixed point theory, the results are always formulated in a metric space setting (usually in a Banach space setting) and the methods involved are based on the metric structure and geometry of the space and on the metric properties of the maps.

Undoubtedly, the major result in this class is the celebrated Banach fixed point theorem. First a definition fixing our terminology.

Definition 4.58
Let (X,d_X) be a metric space, let $\varphi\colon X\supseteq C\longrightarrow X$ be a map and let $k\in[0,1)$.
(a) *We say that φ is a **k-contraction**, if*

$$d_X(\varphi(u),\varphi(v)) \leqslant kd_X(u,v) \qquad \forall\, u,v\in C.$$

(b) *We say that φ is **nonexpansive**, if*

$$d_X(\varphi(u),\varphi(v)) \leqslant d_X(u,v) \qquad \forall\, u,v\in C.$$

(c) *We say that φ is **contractive**, if*

$$d_X(\varphi(u),\varphi(v)) < d_X(u,v) \qquad \forall\, u,v\in C,\ u\neq v.$$

Theorem 4.59 *(Banach Fixed Point Theorem)*
If (X,d_X) is a complete metric space and $\varphi\colon X\longrightarrow X$ is a k-contraction, then φ has a unique fixed point $\widehat{u}$ and for every $u\in X$, we have $\varphi^{(n)}(u)\longrightarrow\widehat{u}$.

Remark 4.60
The proof of the above theorem is constructive and produces interesting information about the iteration process. More precisely, we have

$$d_X\big(\varphi^{(n)}(u),\varphi^{(n+m)}(u)\big) \leqslant \tfrac{k^n}{1-k}d_X\,(u,\varphi(u)) \qquad \forall\, m\geqslant 1$$

and

$$\begin{aligned} d_X\left(\varphi^{(n)}(u), \widehat{u}\right) &= \lim_{m \to +\infty} d_X\left(\varphi^{(n)}(u), \varphi^{(n+m)}(u)\right) \\ &\leqslant \tfrac{k^n}{1-k} d_X\left(u, \varphi(u)\right) \qquad \forall\, n \geqslant 1. \end{aligned} \tag{4.8}$$

Also we have

$$d_X(u, \widehat{u}) \leqslant \tfrac{1}{1-k} d_X(u, \varphi(u)) \qquad \forall\, u \in X.$$

Evidently (4.8) gives an *a priori* error estimate at the n-th step of the iteration (approximation) process starting from $u \in X$.

There is a parametric version of Theorem 4.59.

Theorem 4.61
If (X, d_X) *is a complete metric space,* (T, ϱ) *is another metric space (the parameter space),* $\{\varphi_t \colon X \longrightarrow X\}_{t \in T}$ *is a family of* k*-contractions with* $k \in [0,1)$ *independent of* $t \in T$ *and for each* $u \in X$*, the map* $t \longmapsto \varphi_t(u)$ *is continuous,*
then for every $t \in T$ *there exists a unique* $u_t \in X$ *such that* $\varphi_t(u_t) = u_t$ *and the map* $t \longmapsto u_t$ *is continuous.*

There are some generalizations of Theorem 4.59.

Proposition 4.62
If (X, d_X) *is a complete metric space and* $\varphi \colon X \longrightarrow X$ *is a map such that for some* $n \geqslant 1$, $\varphi^{(n)}$ *is a* k*-contraction,*
then φ *has a unique fixed point.*

Remark 4.63
A map $\varphi \colon X \longrightarrow X$ as in the above proposition need not be continuous. For example

$$\varphi(u) = \begin{cases} 1 & \text{if } u \in \mathbb{Q}, \\ 0 & \text{if } u \in \mathbb{R} \setminus \mathbb{Q}. \end{cases}$$

Evidently φ is not continuous, but $\varphi^{(2)} \equiv 1$.

Another generalization of Theorem 4.59 is provided by the next proposition.

Proposition 4.64
If (X, d_X) *is a metric space and* $\varphi \colon X \longrightarrow X$ *is a map satisfying*

$$d_X\left(\varphi(u), \varphi(v)\right) \leqslant \xi\left(d_X(u, v)\right),$$

where $\xi \colon \mathbb{R}_+ \longrightarrow \mathbb{R}_+$ *is a nondecreasing (not necessarily continuous) function such that* $\xi^{(n)}(t) \longrightarrow 0$ *for every* $t > 0$,
then φ *has a unique fixed point* $\widehat{u} \in X$ *and for every* $u \in X$*, we have* $\varphi^{(n)}(u) \longrightarrow \widehat{u}$.

The next theorem is an interesting consequence of Theorem 4.59.

Theorem 4.65
If X is Banach space, $U \subseteq X$ is an open set, $\varphi\colon U \longrightarrow X$ is a k-contraction and

$$f(u) = u - \varphi(u) \qquad \forall\, u \in U,$$

then
(a) *f is an open map (in particular the set $f(U)$ is open);*
(b) *$f\colon U \longrightarrow f(U)$ is a homeomorphism.*

A far reaching generalization of Theorem 4.59 is the following fixed point theorem.

Theorem 4.66 *(Caristi Fixed Point Theorem)*
If (X, d_X) is a complete metric space, $\xi\colon X \longrightarrow \overline{\mathbb{R}} = \mathbb{R} \cup \{+\infty\}$ is a lower semicontinuous function which is not identically $+\infty$ and is bounded below and $F\colon X \longrightarrow 2^X \setminus \{\emptyset\}$ is a multifunction such that

$$\xi(y) \leqslant \xi(u) - d_X(u, y) \qquad \forall\, u \in X,\ y \in F(u) \tag{4.9}$$

then there exists $\widehat{u} \in X$ such that $\widehat{u} \in F(\widehat{u})$.

Remark 4.67
The remarkable feature of the above result is that no continuity assumption is made on F. Suppose that $\varphi\colon X \longrightarrow X$ is a k-contraction, then by setting $\xi(u) = \frac{1}{1-k} d_X(u, \varphi(u))$, we satisfy (4.9) with $F = \varphi$. So, indeed Theorem 4.66 is a generalization of Theorem 4.59.

We can also have a set-valued version of Theorem 4.59

Theorem 4.68
If (X, d_X) is a complete metric space, $F\colon X \longrightarrow P_{bf}(X)$ is a multifunction such that

$$h(F(u), F(v)) \leqslant k d_X(u, v) \qquad \forall\, u, v \in X,$$

with $k \in [0, 1)$, where h is the Hausdorff metric on $P_{bf}(X)$ (see Definition 2.49), then F has a fixed point, that is, there exists $\widehat{u} \in X$ such that $\widehat{u} \in F(\widehat{u})$.

Remark 4.69
Of course in this case we do not expect the fixed point to be unique. Indeed, to see this consider a constant multifunction.

For nonexpansive maps, the situation is more involved. Such maps in general do not need to have fixed points. Indeed, let X be a Banach space and let $\varphi\colon X \longrightarrow X$ be defined by

$$\varphi(u) = u + h \qquad \forall\, u \in X,$$

with $h \in X \setminus \{0\}$ fixed. Evidently φ is nonexpansive and fixed point free. Also, if a nonexpansive map has a fixed point, then this need not be unique. Indeed, think of the identity map on X.

Theorem 4.70
If X is a uniformly convex Banach space, $C \subseteq X$ is a nonempty, bounded, closed, and convex set and $F\colon C \longrightarrow P_k(C)$ satisfies

$$h(F(u), F(v)) \leqslant \|u - v\|_X \qquad \forall\, u, v \in C,$$

then F has a fixed point, that is, there exists $\widehat{u} \in C$ such that $\widehat{u} \in F(\widehat{u})$.

Remark 4.71
Using the parallelogram law (see Remark I.5.94), we see that Hilbert spaces are uniformly convex.

To go beyond uniformly convex Banach spaces, we need to introduce the following geometric notion.

Definition 4.72
Let X be a Banach space and let $C \subseteq X$ be a nonempty set.
(a) A point $u \in C$ is said to be a ***diametral point*** *of C, if*

$$\sup_{c \in C} \|u - c\|_X = \operatorname{diam} C.$$

(b) A convex set C is said to have ***normal structure****, if every bounded convex set $D \subseteq C$ with $\operatorname{diam} D > 0$ contains a nondiametral point.*
(c) A convex set C in a dual Banach space X^* is said to have ***w^*-normal structure****, if every bounded, w^*-closed, convex set $D \subseteq C$ with $\operatorname{diam} D > 0$, contains a nondiametral point.*

Theorem 4.73
If X is a Banach space, $C \subseteq X$ is a nonempty, w-compact, convex set with normal structure and $\varphi\colon C \longrightarrow C$ is a nonexpansive map,
then φ has a fixed point.

Theorem 4.74
If X is a Banach space, $C \subseteq X^*$ is a nonempty, w^*-compact, convex set with w^*-normal structure and $\varphi\colon C \longrightarrow C$ is a nonexpansive map,
then φ has a fixed point.

So far we have considered maps that take the set C into itself. Next we will deal with maps that need not take C into itself.

Theorem 4.75
If X is a uniformly convex space, $C \subseteq X$ is a nonempty, bounded, closed, and convex set and $\varphi\colon C \longrightarrow X$ is a nonexpansive map and $\varphi(u) - u \in T_C(u)$ for all $u \in C$ with $T_C(u)$ being the tangent cone to C at u (see Definition 3.112),
then φ has a unique fixed point.

Remark 4.76
A map $\varphi\colon C \longrightarrow X$ which satisfies condition $\varphi(u) - u \in T_C(u)$ for all $u \in C$ is called ***weakly inward***.

We mention a basic structural property of nonexpansive maps. First a definition.

Definition 4.77
Let X be a Banach space and let $\xi\colon X \supseteq D(\xi) \longrightarrow X$ be a map. We say that ξ is ***demiclosed*** *on $D(\xi)$ if for every sequence $\{u_n\}_{n\geqslant 1} \subseteq D(\xi)$ such that $u_n \xrightarrow{w} u$ and $\xi(u_n) \longrightarrow y$, we have that $u \in D(\xi)$ and $\xi(u) = y$.*

Proposition 4.78
If X is a uniformly convex Banach space, $C \subseteq X$ is a bounded, closed, convex set and $\varphi\colon C \longrightarrow X$ is a nonexpansive map,
then $I_X - \varphi$ is demiclosed.

We conclude our survey of the basic metric fixed point theory with a nonlinear alternative theorem for k-contractions.

Theorem 4.79
If X is a Banach space, $C \subseteq X$ is a nonempty, closed, convex set, U is a relatively open subset of C with $0 \in U$ and $\varphi\colon \overline{U} \longrightarrow X$ is a bounded k-contraction (where $0 \leqslant k < 1$),
then at least one of the following properties holds:
(a) *φ has a unique fixed point;*
(b) *there exist $u_0 \in \partial U$ and $t \in (0,1)$ such that $u_0 = t\varphi(u_0)$.*

There is an analogous result for nonexpansive maps.

Theorem 4.80
If X is a uniformly convex Banach space, $U \subseteq X$ is a bounded, open and convex set with $0 \in U$ and $\varphi\colon \overline{U} \longrightarrow X$ is nonexpansive,
then at least one of the following properties holds:
(a) *φ has a unique fixed point;*
(b) *there exist $u_0 \in \partial\overline{U}$ and $t \in (0,1)$ such that $u_0 = t\varphi(u_0)$.*

4.1.3 Topological Fixed Point Theory

In this section, we present some of the main results from the topological fixed point theory. Now in the analysis the main tools are the topological properties of the spaces and/or of the maps involved. The notion of compactness is central in this theory.

Topological fixed point theory was born with the celebrated Brouwer fixed point theorem.

Theorem 4.81 *(Brouwer Fixed Point Theorem)*
If $C \subseteq \mathbb{R}^N$ is homeomorphic to a compact, convex set and $\varphi\colon C \longrightarrow C$ is continuous,
then φ has a fixed point.

Remark 4.82
A nonempty, bounded, closed, and convex subset of $\mathbb{R}^N$ with nonempty interior is homeomorphic to the closed unit ball $\overline{B}_1$ in $\mathbb{R}^N$. So, if $U \subseteq \mathbb{R}^N$ is a bounded, open and convex set and $\varphi: \overline{U} \longrightarrow \overline{U}$ is continuous, then φ has a fixed point. Recall that a convex set $C \subseteq \mathbb{R}^N$ may have empty interior. However, we always have $\operatorname{rint} C \neq \emptyset$, where $\operatorname{rint} C$ denotes the relative interior of C with respect to $\operatorname{span} C$.

Next we present two fixed point theorems for open sets in $\mathbb{R}^N$.

Theorem 4.83
If $U \subseteq \mathbb{R}^N$ is a nonempty, bounded and open set, $\varphi \in C(\overline{U}; \mathbb{R}^N)$ and there exists $y \in U$ such that

$$\varphi(u) - y \neq \lambda(u - y) \qquad \forall\, u \in \partial U,\ \lambda > 1, \tag{4.10}$$

then φ has a fixed point in $\overline{U}$.

Remark 4.84
Condition (4.10) is known in the literature as the ***Leray–Schauder boundary condition***. It says that for some point $y \in U$, for no $u \in \partial U$ does $\varphi(u)$ lie in the continuation of the segment $[y, u]$ beyond u.

The second fixed point theorem for open sets is the following.

Theorem 4.85 *(Borsuk Fixed Point Theorem)*
If $U \subseteq \mathbb{R}^N$ is a nonempty, bounded, open and symmetric set with $0 \in U$ and $\varphi \in C(\overline{U}; \mathbb{R}^N)$ satisfies

$$-\varphi(u) = \varphi(-u) \qquad \forall\, u \in \partial U,$$

then φ has a fixed point.

It is natural to ask if there is an infinite dimensional version of the Brouwer fixed point theorem (see Theorem 4.81). The next example illustrates that without additional restrictions on φ Theorem 4.81 fails in infinite dimensional spaces.

Example 4.86
Let $H = l^2$ and let $\varphi: \overline{B}_1 \longrightarrow \overline{B}_1$ (where $\overline{B}_1 = \{\widehat{u} \in l^2 : \|\widehat{u}\|_{l^2} \leqslant 1\}$) be defined by

$$\varphi(\widehat{u}) = \left((1 - \|u\|_{l^2}^2)^{\frac{1}{2}}, u_1, u_2, \ldots\right) \qquad \forall\, \widehat{u} = \{u_n\}_{n \geqslant 1} \in \overline{B}_1.$$

Note that for all $\widehat{u} \in \overline{B}_1$, we have

$$\|\varphi(\widehat{u})\|_{l^2} = \left((1 - \|u\|_{l^2}^2) + \|u\|_{l^2}^2\right)^{\frac{1}{2}} = 1,$$

so

$$\varphi(\overline{B}_1) \subseteq \partial B_1 \tag{4.11}$$

and φ is continuous. The map φ is fixed point free. Indeed, if $\widehat{u} = \varphi(\widehat{u})$ with $\widehat{u} = \{u_n\}_{n\geqslant 1} \in l^2$, then from (4.11), we have

$$u_1 = 0 \quad \text{and} \quad u_1 = u_2 = u_3 = \ldots,$$

so

$$u_n = 0 \qquad \forall\, n \geqslant 1$$

and thus $\widehat{u} = 0$, a contradiction (see (4.11)).

In fact we have the following result illustrating in an emphatic way the failure of Theorem 4.81 in infinite dimensions.

Proposition 4.87
If X is a Banach space,
then every continuous map $\varphi\colon \overline{B}_1 \longrightarrow \overline{B}_1$ *has a fixed point if and only if* $\dim X < +\infty$.

Remark 4.88
The problem is that in infinite dimensional Banach spaces, bounded and closed sets need not be compact. For this reason as we did in Sect. 4.1.1 in the definition of the Leray–Schauder degree, we need to introduce some compactness structure.

The next result is the infinite dimensional analog of Theorem 4.81.

Theorem 4.89 *(Schauder Fixed Point Theorem)*
If X *is a Banach space,* $C \subseteq X$ *is a nonempty, bounded, closed, and convex set and* $\varphi\colon C \longrightarrow C$ *is compact,*
then φ *has a fixed point.*

We can also have an infinite dimensional analog of the Borsuk fixed point theorem (see Theorem 4.85).

Theorem 4.90
If X *is a Banach space,* $U \subseteq X$ *is a nonempty, open, convex, and symmetric set with* $0 \in U$ *and* $\varphi\colon \overline{U} \longrightarrow X$ *is compact and satisfies*

$$-\varphi(u) = \varphi(-u) \qquad \forall\, u \in \partial U,$$

then φ *has a fixed point.*

There is also a nonlinear alternative theorem, in the spirit of Theorem 4.79.

Theorem 4.91
If X *is a Banach space,* $C \subseteq X$ *is a nonempty, bounded, convex set,* $U \subseteq C$ *is a relatively open subset of* C *with* $0 \in U$ *and* $\varphi\colon \overline{U} \longrightarrow X$ *is a compact map,*
then at least one of the following properties holds:

(a) *φ has a fixed point in $\overline{U}$;*
(b) *there exist $u_0 \in \partial U$ and $t \in (0,1)$ such that $u_0 = t\varphi(u_0)$.*

Remark 4.92
The above theorem suggests that we have to impose conditions on φ which will prevent the realization of **(b)**. Then necessarily φ has a fixed point.

Theorem 4.91 leads to the following theorem.

Theorem 4.93 *(Leray–Schauder Alternative Theorem)*
If X is a Banach space, $C \subseteq X$ is a nonempty, convex set with $0 \in C$, $\varphi\colon C \longrightarrow C$ is compact and $E(\varphi) = \{u \in C: \ u = \lambda\varphi(u) \text{ for some } \lambda \in (0,1)\}$,
then either $E(\varphi)$ is unbounded or φ has a fixed point.

Remark 4.94
In the context of boundary value problems, roughly speaking, the above theorem says that *a priori* estimates lead to existence results.

Next we consider the parametric equation $u = \lambda\varphi(u)$ (that is, the existence of an invariant direction for the map φ, an eigenelement (λ, u)). We have the following result.

Theorem 4.95 *(Birkhoff–Kellogg Invariant Direction Theorem)*
If X is an infinite dimensional Banach space, $U \subseteq X$ is a bounded open set with $0 \in U$ and $\varphi\colon \partial U \longrightarrow X$ is a compact map such that

$$\inf_{u \in \partial U} \|\varphi(u)\|_X \ = \ m \ > \ 0,$$

then there exist $\lambda > 0$ and $u \in \partial U$ such that $u = \lambda\varphi(u)$.

Remark 4.96
The above theorem fails in finite dimensional Banach spaces. In fact the eigenvalue problem $u = \lambda\varphi(u)$, $u \in \partial U$ has a solution in $\mathbb{R}^N$ when N is odd. Recall the classical "hedgehog theorem" of Poincaré–Brouwer, which says that the surface of a ball in $\mathbb{R}^N$ ($N \geqslant 2$) admits a nowhere vanishing continuous tangential vector field if and only if N is even.

Definition 4.97
Let X be a Banach space and let $\varphi\colon X \longrightarrow X$ be a map.
(a) *We say that φ is **quasibounded**,*

$$|\varphi|_\infty \ = \ \limsup_{\|u\|_X \to +\infty} \frac{\|\varphi(u)\|_X}{\|u\|_X} \ = \ \inf_{r>0} \sup_{\|u\|_X \geqslant r} \frac{\|\varphi(u)\|_X}{\|u\|_X} \ < \ +\infty.$$

*The number $|\varphi|_\infty$ is called the **quasinorm** of φ.*

(b) We say that φ is ***asymptotically linear****, if there exists $A \in \mathcal{L}(X)$ such that*

$$\lim_{\|u\|_X \to +\infty} \frac{\|A(u) - \varphi(u)\|_X}{\|u\|_X} = 0.$$

The operator $A \in \mathcal{L}(X)$ is uniquely determined, it is called ***asymptotic derivative*** *of φ and is denoted by $\varphi'(\infty)$.*

Remark 4.98
Evidently every $A \in \mathcal{L}(X)$ is quasibounded and $|A|_\infty = \|A\|_{\mathcal{L}}$.

Theorem 4.99
If X is a Banach space and $\varphi\colon X \longrightarrow X$ is quasibounded and compact, then for every $h \in X$ and every $\lambda \in \mathbb{R}$ with $|\lambda| \leqslant \frac{1}{|\varphi|_\infty}$, the equation

$$h = u - \lambda\varphi(u)$$

has at least one solution. In particular then the compact map $\lambda\varphi$ has a fixed point.

The next result extends Theorem 4.75 to infinite dimensional spaces.

Theorem 4.100
If X is a Banach space, $C \subseteq X$ is a nonempty, bounded, closed, and convex set and $\varphi\colon C \longrightarrow X$ is γ-condensing (with γ being the Kuratowski or Hausdorff measure of noncompactness; see Definition 2.31) and

$$\varphi(u) - u \in T_C(u) \qquad \forall\, u \in C,$$

with $T_C(u)$ being the tangent cone to C at u (see Definition 3.112), then φ has a fixed point.

The main results of the topological fixed point theory extend to locally convex spaces (see Definition I.5.3). We mention the extension of the Schauder fixed point theorem (see Theorem 4.89).

Theorem 4.101 *(Tichonov Fixed Point Theorem)*
If X is a locally convex space, $C \subseteq X$ is a nonempty, bounded, closed, and convex set and $\varphi\colon C \longrightarrow C$ is a compact map, then φ has a fixed point.

There is a version of this result for angelic locally convex spaces, which is useful in applications. First a definition.

Definition 4.102
A Hausdorff topological space X is said to be ***angelic*** *(or that it has* ***countably determined compactness****), if for every relatively countably compact set $C \subseteq X$ (see Definition I.2.88), the following hold:*

(a) C is relatively compact;
(b) for every $u \in \overline{C}$ we can find a sequence $\{u_n\}_{n\geqslant 1} \subseteq C$ such that $u_n \longrightarrow u$.

Remark 4.103
Clearly for a set C in an angelic space, we have

$$\begin{aligned} C \text{ is compact} &\iff C \text{ is countably compact} \\ &\iff C \text{ is sequentially compact.} \end{aligned}$$

All metrizable locally convex spaces (in particular all normed spaces) are angelic when furnished with their weak topologies. If X is a normed space, then X_w^* (the space X^* furnished with the weak topology) is angelic, but $X_{w^*}^*$ (the space X^* furnished with the w^*-topology) need not be angelic.

Proposition 4.104
If X is a locally convex space which is angelic when furnished with the weak topology, $C \subseteq X$ is a nonempty, w-compact, and convex set and $\varphi\colon C \longrightarrow C$ is a sequentially weakly continuous map,
then φ has a fixed point.

Next we present some fixed point theorems for multifunctions. The first is a set valued version of Theorem 4.100.

Theorem 4.105
If X is a Banach space, $C \subseteq X$ is a nonempty, bounded, closed, and convex set and $F\colon C \longrightarrow P_{fc}(X)$ is upper semicontinuous and γ-condensing (i.e., for all sets $D \subseteq X$ with $\gamma(D) > 0$ we have $\gamma(F(D)) < \gamma(D)$), where γ is either the Kuratowski or Hausdorff measure of noncompactness (see Definition 2.31) and

$$(F(u) - u) \cap T_C(u) \neq \emptyset \qquad \forall\, u \in C,$$

with $T_C(u)$ being the tangent cone to C at u (see Definition 3.112),
then F has a fixed point, i.e., there exists $u \in C$ such that $u \in F(u)$.

Remark 4.106
If $C \subseteq X$ is a nonempty, compact, and convex set, then we can drop the hypothesis that F is γ-condensing. In fact in this case the result is true for general locally convex spaces.

As a particular case of this theorem, we obtain a multivalued version of the Schauder fixed point theory (see Theorem 4.89).

Theorem 4.107 *(Kakutani Fixed Point Theorem)*
If X is a Banach space, $C \subseteq X$ is a nonempty, bounded, closed, and convex set and $F\colon C \longrightarrow P_{fc}(C)$ is a compact multifunction (see Definition 4.54),
then F has a fixed point, i.e., there exists $u \in C$ such that $u \in F(u)$.

Remark 4.108
The result is also true for locally convex spaces. More precisely, let X be a locally convex space, let $C \subseteq X$ be a nonempty, compact, and convex set and let $F: C \longrightarrow P_{fc}(C)$ be an upper semicontinuous multifunction. Then F has a fixed point. In this form, the result is known as the ***Ky Fan fixed point theorem***.

We have the following multivalued version of the Leray–Schauder alternative principle (see Theorem 4.93).

Theorem 4.109
If X and Y are two Banach spaces, $F: X \longrightarrow P_{wkc}(Y)$ is an upper semicontinuous multifunction from X into Y_w (where Y_w is the Banach space Y furnished with the weak topology), $G: Y \longrightarrow X$ is a completely continuous map and $S = G \circ F$ maps bounded sets to relatively compact sets,
then the set

$$E(S) = \{u \in X : u \in \lambda S(u) \text{ for some } \lambda \in (0,1)\}$$

is unbounded or otherwise S has a fixed point.

The next finite dimensional theorem, provides the main tool for many important fixed point theorems.

Theorem 4.110 *(Knaster–Kuratowski–Mazurkiewicz Theorem)*
If $u_k \in \mathbb{R}^n$ and $C_k \in P_f(\mathbb{R}^n)$ for $k \in \{1,\dots,m\}$ are such that for every subset $D \subseteq \{1,\dots,m\}$ we have $\operatorname{conv}\{u_k : k \in D\} \subseteq \bigcup_{k\in D} C_k$, then

$$\Big(\bigcap_{k=1}^{m} C_k\Big) \cap \operatorname{conv}\{u_1,\dots,u_m\} \in P_k(\mathbb{R}^n).$$

This result suggests the following definition.

Definition 4.111
Let X be a vector space and let $C \subseteq X$ be an arbitrary set. A multifunction $F: C \longrightarrow 2^X \setminus \{0\}$ is said to be a ***KKM-multifunction****, if*

$$\operatorname{conv}\{u_1,\dots,u_n\} \subseteq \bigcup_{k=1}^{n} F(u_k)$$

for every finite subset $\{u_1,\dots,u_n\} \subseteq X$.

We conclude this section by introducing some topological notions which will be also used in Chap. 5.

Definition 4.112
Let X be a Hausdorff topological space.
(a) A continuous map $h: [0,1] \times X \longrightarrow X$ is a ***deformation*** *of X, if $h(0,\cdot) = I_X$. Moreover, if $h(1,X) \subseteq C \subseteq X$, then we say that X is* ***deformable into*** *C.*

(b) *A closed set $C \subseteq X$ is a* ***deformation retract*** *(respectively a* ***strong deformation retract****) of X, if there exists a deformation $h(t,u)$ of X into C such that*

$$h(1,\cdot)|_C = I_C$$

$$(\text{respectively } h(t,\cdot)|_C = I_C \text{ for all } t \in [0,1]).$$

Definition 4.113
Let X and Y be two Hausdorff topological spaces and let $C \subseteq X$.
(a) *A continuous map $\varphi\colon C \longrightarrow Y$ is said to be* ***extendable*** *to X (respectively* ***compactly extendable*** *to X), if there exists a continuous (respectively compact) map $\widehat{\varphi}\colon X \longrightarrow Y$ such that $\widehat{\varphi}|_C = \varphi$.*
(b) *Two maps $\varphi, \psi\colon X \longrightarrow Y$ are said to be* ***homotopic****, if there exists a continuous map $\widehat{h}\colon [0,1] \times X \longrightarrow Y$ such that*

$$\widehat{h}(0,\cdot) = \varphi(\cdot) \quad \text{and} \quad \widehat{h}(1,\cdot) = \psi(\cdot).$$

The map $\widehat{h}$ is called a ***homotopy*** *(or* ***continuous homotopy****) of φ to ψ. When φ and ψ are homotopic, we write $\varphi \simeq \psi$. If $\varphi, \psi\colon X \longrightarrow Y$ are compact maps and the homotopy $\widehat{h}$ is compact, then φ and ψ are said to be* ***compactly homotopic*** *and we write $\varphi \simeq \psi$ in $K(X;Y)$. In this case $\widehat{h}$ is called a* ***compact homotopy****.*
(c) *A map $\varphi\colon X \longrightarrow Y$ is said to be* ***null-homotopic*** *(respectively* ***compactly null-homotopic****), if it is homotopic (respectively compactly homotopic) to a constant map. We write $\varphi \simeq 0$ (respectively $\varphi \simeq 0$ in $K(X;Y)$)*
(d) *We say that the space X is* ***contractible****, if $I_X \simeq 0$ (i.e., the identity map on X is null-homotopic).*

Remark 4.114
Evidently $\simeq$ is an equivalence relation. So, it decomposes $C(X;Y)$ into pairwise disjoint classes known as ***homotopy classes***.

Definition 4.115
A Hausdorff topological space is said to have the ***fixed point property*** *(or is a* ***fixed point space****), if every continuous map $\varphi\colon X \longrightarrow X$ has a fixed point.*

4.1.4 Order Fixed Point Theory

In this section we present some basic results of fixed point theory which are based on the order structure of the underlying spaces.

Definition 4.116
Let X be a set. A ***partial order*** *on X is any relation $\preceq$ in $X \times X$ which satisfies the following conditions:*
(a) *$u \preceq u$ (reflexivity);*
(b) *if $u \preceq v$ and $v \preceq u$, then $u = v$ (antisymmetry);*

(c) if $u \preceq v$ and $v \preceq w$, then $u \preceq w$ (transitivity).
We say that relation $\preceq$ is ***linear*** *(or* ***total****) if in addition to the above conditions it also satisfies:*
(d) for all $u, v \in X$, we have $u \preceq v$ or $v \preceq u$.
If $u \preceq v$ and $u \neq v$, the we write $u \prec v$.
If $\preceq$ is a partial order on X, then a pair $(X, \preceq)$ is called a ***partially ordered set.***
If $\preceq$ is a linear order on X, then a pair $(X, \preceq)$ is called a ***linearly ordered set*** *(or a* ***totally ordered set****).*

Remark 4.117
Some authors drop the property of reflexivity from the definition of partial order. We will call such a relation ***strict partial order***. Evidently $\prec$ is a strict partial order.

Definition 4.118
Let $(X, \preceq)$ be a partially ordered set and let $C \subseteq X$ be a nonempty set. An ***upper bound*** *for C is an element $u \in X$ such that $c \preceq u$ for all $c \in C$. A* ***maximal element*** *in X is an element $u \in X$ such that if $u \preceq v$, then $u = v$. A* ***lower bound*** *of C and a* ***minimal element*** *of X are defined by reversing the inequalities in the above relations. We say that the set C is* ***bounded above*** *(respectively* ***bounded below****), if there exists an upper bound (respectively a lower bound) for the set C. If C is both bounded above and bounded below, then we say that C is* ***order bounded.*** *An upper bound $u \in X$ for C is said to be a* ***least upper bound*** *(or* ***supremum****), if for every upper bound $v \in X$, we have $u \preceq v$. Similarly we define the* ***greatest lower bound*** *(or* ***infimum****) of C. The supremum and infimum of C are denoted by $\sup C$ and $\inf C$ respectively. When they exist they are necessarily unique. We say that the partially ordered set X is a* ***lattice****, if for all $u, v \in X$, both $\sup\{u, v\}$ and $\inf\{u, v\}$ exist. We say that X is a* ***complete lattice****, if for every nonempty set $C \subseteq X$, both $\sup C$ and $\inf C$ exist. A subset C of a partially ordered set $(X, \preceq)$ is a* ***chain****, if the relation $\preceq|_{C \times C}$ is total.*

Remark 4.119
Maximal elements need not be unique. For such elements, the basic result is the one that follows.

Theorem 4.120 *(Kuratowski–Zorn Lemma)*
If $(X, \preceq)$ is a partially ordered set and every chain has an upper bound, then X has at least one maximal element.

Definition 4.121
Let $(X, \preceq)$ be a partially ordered set. Given $u \in X$, we define

$$S_+(u) = \{v \in X : u \preceq v\} \quad \text{(the \textbf{upper} (or \textbf{right}) \textbf{section} of } u\text{)},$$
$$S_-(u) = \{v \in X : v \preceq u\} \quad \text{(the \textbf{lower} (or \textbf{left}) \textbf{section} of } u\text{)}.$$

Then, given $u, v \in X$, the ***order interval*** *$[u, v]$ determined by u and v is defined by*

$$[u, v] = S_+(u) \cap S_-(v) = \{y \in X : u \preceq y \preceq v\}.$$

Suppose that $(X, \preceq_X)$ *and* $(Y, \preceq_Y)$ *are two partially ordered sets and* $\varphi\colon X \longrightarrow Y$ *is a map. We say that* φ *is* ***increasing*** *(respectively* ***decreasing****), if*

$$u \preceq_X v \quad \Longrightarrow \quad \varphi(u) \preceq_Y \varphi(v)$$

$$(\text{respectively } u \preceq_X v \quad \Longrightarrow \quad \varphi(v) \preceq_Y \varphi(u)).$$

We say that φ *is* ***strictly increasing*** *(respectively* ***strictly decreasing****), if*

$$u \prec_X v \quad \Longrightarrow \quad \varphi(u) \prec_Y \varphi(v)$$

$$(\text{respectively } u \prec_X v \quad \Longrightarrow \quad \varphi(v) \prec_Y \varphi(u)).$$

Theorem 4.122 *(Bourbaki–Kneser Fixed Point Theorem)*
If $(X, \preceq)$ *is a partially ordered set in which every chain has an upper bound (respectively a lower bound) and* $\varphi\colon X \longrightarrow X$ *satisfies*

$$u \preceq \varphi(u) \qquad \forall\, u \in X$$

$$(\text{respectively } \varphi(u) \preceq u \qquad \forall\, u \in X),$$

then φ *admits a fixed point.*

In applications, the partial order is defined on a Banach space via the use of a cone.

Definition 4.123
Let X *be a Banach space and let* $K \subseteq X$ *be a nonempty set. We say that* K *is an* ***order cone****, if* K *is closed, convex,*

$$\lambda K \subseteq K \qquad \forall\, \lambda \geqslant 0 \quad \text{and} \quad K \cap (-K) = \{0\}.$$

We say that the order cone K *is* ***solid****, if* $\operatorname{int} K \neq \emptyset$.

Remark 4.124
Given an order cone K in a Banach space, for every $u, v \in K$ we define

$$u \leqslant v \quad \Longleftrightarrow \quad v - u \in K.$$

Then it is easy to see that $\leqslant$ is a partial order on X (see Definition 4.116). If the cone K is solid, then we write

$$u \ll v \quad \Longleftrightarrow \quad v - u \in \operatorname{int} K.$$

The Banach space together with the partial order $\leqslant$ introduced by a cone K is said to be an ***ordered Banach space*** (an ***OBS*** for short) and the elements of K are said to be ***positive*** in X. Finally recall that in general a set $C \subseteq X$ is a ***cone*** if

$$\lambda u \in C \qquad \forall\, u \in C,\ \lambda > 0.$$

Evidently an order cone is a cone, but the converse is not true in general.

Some properties of order cones which are intuitive and maybe obvious in finite dimensional spaces, need not hold in infinite dimensions. For this reason, we introduce the following definitions.

Definition 4.125
Let X be an OBS with order cone K.
*(a) We say that K is **generating** (or **reproducing**), if $X = K - K$. We say that K is **total**, if $X = \overline{K - K}$.*
*(b) We say that K is **normal**, if*

$$\inf_{u,v \in K \cap \partial B_1} \|u + v\|_X > 0$$

(here $\partial B_1 = \{z \in X : \|z\|_X = 1\}$).
*(c) We say that K is **regular**, if every increasing sequence $\{u_n\}_{n\geqslant 1} \subseteq X$ which is order bounded, is strongly convergent in X.*
*(d) We say that K is **fully regular**, if every increasing sequence $\{u_n\}_{n\geqslant 1} \subseteq X$ which is norm bounded, is strongly convergent in X.*
*(e) We say that K is **minihedral**, if for every $u, v \in X$, $\sup\{u, v\}$ exists.*
*(f) We say that K is **strongly minihedral**, if for every set $C \subseteq X$ which is bounded from above, $\sup C$ exists.*

Remark 4.126
If K is generating, then every $u \in X$ can be written as

$$u = v_1 - v_2, \quad \text{with } v_1, v_2 \in K.$$

Geometrically, normality means that the cone cannot be too large, since the angle between two unit norm positive elements is bounded away from π. Also, regularity and full regularity can be equivalently defined using decreasing sequences. Similarly, K is minihedral if and only if for all $u, v \in X$, $\inf\{u, v\}$ exists, while K is strongly minihedral if and only if every set $C \subseteq X$ which is bounded below has an infimum. Evidently, K is minihedral if and only if for every finite set C, $\sup C$ and $\inf C$ exist.

Definition 4.127
*Let X be an OBS with order cone K. The **dual cone** $K^* \subseteq X^*$ is defined by*

$$K^* = \{u^* \in X^* : \langle u^*, u\rangle \geqslant 0 \text{ for all } u \in K\}.$$

Remark 4.128
The dual cone $K^* \subseteq X^*$ need not be an order cone, since the condition $K^* \cap (-K^*) = \{0\}$ may be violated (see Definition 4.123). However, it is easy to see that K^* is an order cone if and only if K is total. Note that $u \in K$ if and only if $\langle u^*, u\rangle \geqslant 0$ for all $u^* \in K^*$ and if $u \in K \setminus \{0\}$, then there exists $u^* \in K^*$ such that $\langle u^*, u\rangle > 0$.

The next result relates K and the dual cone K^*.

Theorem 4.129 *(Krein Theorem)*
If X is an OBS with order cone K and $K^* \subseteq X^*$ is the dual cone,
then
(a) K is generating if and only if K^* is normal.
(b) K is normal if and only if K^* is generating.

With additional structure on X, we can relate the notions introduced in Definition 4.125.

Proposition 4.130
If X is a reflexive OBS with order cone K,
then the following conditions are equivalent:
(a) K is normal.
(b) K is regular.
(c) K is fully regular.

Remark 4.131
In general, we have that the following implications hold:

$$\begin{gathered} K \text{ is fully regular} \\ \Downarrow \\ K \text{ is regular} \\ \Downarrow \\ K \text{ is normal} \end{gathered}$$

To be able to produce more fixed point theorems based on the order structure, we need to exploit some tools available from degree theory.

We start by recalling the following notions from topology.

Definition 4.132
Let X be a Hausdorff topological space.
*(a) A set $C \subseteq X$ is a **retract**, if there exists a continuous map $r: X \longrightarrow C$ such that $r|_C = I_C$. The map r is a **retraction** of X onto C.*
*(b) We say that X is an **absolute retract** (an **AR** for short), if X is metrizable and for any metrizable space V and any closed set $D \subseteq V$ each continuous map $\varphi: D \longrightarrow X$ is extendable on V.*
*(c) We say that X is an **absolute neighborhood retract** (an **ANR** for short), if X is metrizable and for any metrizable space V and any closed set $D \subseteq V$ each continuous map $\varphi: D \longrightarrow X$ is extendable on some neighborhood U of D.*

Remark 4.133
A retract is always closed. By virtue of the Dugundji extension theorem (see Theorem 2.8), a closed convex set of a locally convex space is a retract and in fact AR. If D is a retract of C and C is a retract of X, then D is a retract of X. Every AR is path-connected (see Definition I.1.97) and every ANR is locally path-connected (see Definition I.1.103).

Definition 4.134
Let X be a Banach space and consider the family:

$$\begin{aligned}\mathcal{Y}_i &= \{(\varphi,\Omega,K): K\subseteq X \text{ is a retract, } \Omega\subseteq K \text{ is bounded,}\\ &\quad \text{relatively open}\\ &\quad \varphi\colon \overline{\Omega}\longrightarrow X \text{ is compact and } \mathrm{Fix}(\varphi)\cap\partial\Omega=\emptyset\},\end{aligned}$$

where $\mathrm{Fix}(\varphi)=\{u\in\overline{\Omega}: u=\varphi(u)\}$. *Let $r\colon X\longrightarrow K$ be a retraction of X onto K. Then for every $(\varphi,\Omega,K)\in\mathcal{Y}_i$ we define the **fixed point index** $i\colon\mathcal{Y}_i\longrightarrow\mathbb{Z}$ of φ over Ω with respect to K, by*

$$i(\varphi,\Omega,K) = d_{LS}(I_X-\varphi\circ r, r^{-1}(\Omega),0).$$

Remark 4.135
Using the properties of the Leray–Schauder degree (see Theorem 4.27), in particular the homotopy invariance and the excision properties, we can show that the above definition is independent of the retraction r used. When $K=X$, we see that we recover the Leray–Schauder degree.

Directly from Theorem 4.27, we infer the following fundamental theorem for the fixed point index.

Theorem 4.136
*There is a unique map $i\colon\mathcal{Y}_i\longrightarrow\mathbb{Z}$ called the **fixed point index** of φ on Ω with respect to K, which satisfies:*
(a) *Normalization: If $\varphi(u)=y_0\in\Omega$ for all $u\in\overline{\Omega}$, then $i(\varphi,\Omega,K)=1$.*
(b) *Domain Decomposition: If Ω_1,Ω_2 are disjoint open subsets of Ω and*

$$\mathrm{Fix}(\varphi)\cap(\overline{\Omega}\setminus(\Omega_1\cup\Omega_2)) = \emptyset,$$

then

$$i(\varphi,\Omega,K) = i(\varphi,\Omega_1,K)+i(\varphi,\Omega_2,K).$$

(c) *Homotopy Invariance: If $\widehat{h}\colon[0,1]\times\overline{\Omega}\longrightarrow K$ is compact and*

$$\widehat{h}(t,u) \neq u \qquad \forall\, t\in[0,1],\ u\in\partial\Omega,$$

then $i(\widehat{h}(t,\cdot),\Omega,K)$ is independent of $t\in[0,1]$;
(d) *Existence: If $i(\varphi,\Omega,K)\neq 0$, then there exists $u\in\Omega$ such that $\varphi(u)=u$.*
(e) *Excision: If Ω_0 is an open subset of Ω and $\mathrm{Fix}(\varphi)\cap(\overline{\Omega}\setminus\Omega_0)=\emptyset$, then*

$$i(\varphi,\Omega,K) = d(\varphi,\Omega_0,K).$$

(f) *Reduction: If K_1 is a retract of K and $\varphi(\overline{\Omega})\subseteq K_1$, then*

$$i(\varphi,\Omega,K) = i(\varphi,\Omega\cap K_1,K_1).$$

(g) *Dependence on boundary value: If $\varphi|_{\partial\Omega}=\psi|_{\partial\Omega}$, then $i(\varphi,\Omega,K)=i(\psi,\Omega,K)$.*

Remark 4.137
Of course the definition of the fixed point index can be extended to γ-condensing maps, that is, in Definition 4.134 replace d_{LS} by d_C (see Theorem 4.39). Also, we can define a fixed point index for compact (or γ-condensing) multifunctions, using the degree map d_{SV} (see Theorem 4.57).

Theorem 4.138
If X is an OBS with order cone K, $\Omega_1, \Omega_2 \subseteq X$ are two nonempty, bounded, open sets such that

$$0 \in \Omega_1 \quad \text{and} \quad \overline{\Omega}_1 \subseteq \Omega_2,$$

$\varphi\colon K \cap (\overline{\Omega}_2 \setminus \Omega_1) \longrightarrow K$ is a compact map and one of the following conditions holds:
(i) $\varphi(u) \ngeq u$ for all $u \in K \cap \partial\Omega_1$ and $\varphi(u) \nleq u$ for all $u \in K \cap \partial\Omega_2$; or
(ii) $\varphi(u) \nleq u$ for all $u \in K \cap \partial\Omega_1$ and $\varphi(u) \ngeq u$ for all $u \in K \cap \partial\Omega_2$,
then φ has at least one fixed point in $K \cap (\Omega_2 \setminus \overline{\Omega}_1)$.

Theorem 4.139
If X is an OBS with order cone K, $\Omega_1, \Omega_2 \subseteq X$ are two nonempty, bounded, open sets such that

$$0 \in \Omega_1 \quad \text{and} \quad \overline{\Omega}_1 \subseteq \Omega_2,$$

$\varphi\colon K \cap (\overline{\Omega}_2 \setminus \Omega_1) \longrightarrow K$ is a compact map and one of the following conditions holds:
(i) $\|\varphi(u)\|_X \leqslant \|u\|_X$ for all $u \in K \cap \partial\Omega_1$ and $\|\varphi(u)\|_X \geqslant \|u\|_X$ for all $u \in K \cap \partial\Omega_2$; or
(ii) $\|\varphi(u)\|_X \geqslant \|u\|_X$ for all $u \in K \cap \partial\Omega_1$ and $\|\varphi(u)\|_X \leqslant \|u\|_X$ for all $u \in K \cap \partial\Omega_2$,
then φ has at least one fixed point in $K \cap (\Omega_2 \setminus \overline{\Omega}_1)$.

Remark 4.140
Both the above existence theorems employ conditions of cone expansion and compression.

We can also have a multiplicity theorem.

Theorem 4.141
If X is an OBS with order cone K which is normal and solid, $y_1, u_1, y_2, u_2 \in X$ satisfy

$$y_1 < u_1 < y_2 < u_2$$

and $\varphi\colon [y_1, y_2] \longrightarrow X$ is compact and strongly increasing (i.e., if $u < v$, then $\varphi(u) \ll \varphi(v)$) and in addition we have

$$y_1 \leqslant \varphi(y_1), \quad \varphi(u_1) < u_1, \quad y_2 < \varphi(y_2) \quad \text{and} \quad \varphi(u_2) \leqslant u_2,$$

then φ has at least three fixed points $\widehat{u}_1, \widehat{u}_2, \widehat{u}_3 \in [y_1, y_2]$ such that

$$y_1 \leqslant \widehat{u}_1 \ll u_1, \quad y_2 \ll \widehat{u}_2 \leqslant u_2 \quad \text{and} \quad y_2 \nleq \widehat{u}_3 \nleq u_1.$$

We conclude with two useful auxiliary results.

Theorem 4.142 *(Whyburn Lemma (Separation Lemma))*
If K is a compact metric space and K_1, K_2 are nonempty, closed disjoint subsets of K,
then one of the following holds:

(a) *there exists a connected component C of K such that $C \cap K_1 \neq \emptyset$, $C \cap K_2 \neq \emptyset$;*
or

(b) *there exist disjoint compact sets $\widetilde{K}_1$ and $\widetilde{K}_2$ such that $K = \widetilde{K}_1 \cup \widetilde{K}_2$ and $K_1 \subseteq \widetilde{K}_1$ and $K_2 \subseteq \widetilde{K}_2$*

Theorem 4.143 *(Arens–Eells Theorem)*
Every metric space (X, d_X) can be isometrically embedded as a closed subset of a normed space.

4.2 Problems

Problem 4.1**
Suppose that $\Omega \subseteq \mathbb{R}^N$ is a bounded open set, $\varphi \in C^1(\overline{\Omega}; \mathbb{R}^N)$ and $h \notin \varphi(S_\varphi)$ (see Definition 4.1). Show that the set $\varphi^{-1}(h)$ is finite.

Problem 4.2**
Let X be a finite dimensional normed space. Show that the Brouwer degree (see Definition 4.9) can be extended to triples (Ω, φ, h) for bounded open set $\Omega \subseteq X$, $\varphi \in C(\overline{\Omega}; X)$ and $h \notin \varphi(\partial\Omega)$.

Problem 4.3*
Suppose that $\Omega \subseteq \mathbb{R}^N$ is a bounded open set, $\varphi, \psi \in C(\overline{\Omega}; \mathbb{R}^N)$ are such that $\varphi|_{\partial\Omega} = \psi|_{\partial\Omega}$ and $h \notin \varphi(\partial\Omega)$. Show that $d(\varphi, \Omega, h) = d(\psi, \Omega, h)$.

Problem 4.4*
Let $(\varphi, \Omega, h) \in \mathcal{Y}$ (see Theorem 4.11). Show that for all $y \in \mathbb{R}^N$, we have

$$d(\varphi, \Omega, h) = d(\varphi - y, \Omega, h - y).$$

Problem 4.5**
Let $(\varphi, \Omega, h) \in \mathcal{Y}$ (see Theorem 4.11) and suppose that in Ω the equation $\varphi(u) = h$ has finite number of solutions $u_1, \ldots, u_n \in \Omega$. Suppose that φ is a C^1-function in a neighborhood of each u_k and $J_\varphi(u_k) \neq 0$ for $k \in \{1, \ldots, n\}$. Show that

$$d(\varphi, \Omega, h) = \sum_{k=1}^{n} \operatorname{sgn} J_\varphi(u_k).$$

Problem 4.6**
Suppose that $\Omega \subseteq \mathbb{R}^N$ is a bounded open set, $\varphi \in C^1(\overline{\Omega}; \mathbb{R}^N)$, $u_0 \in \varphi^{-1}(h)$ and $J_\varphi(u_0) \neq 0$. Show that $i(\varphi, u_0, h) = (-1)^\mu$ where μ is the number of real negative eigenvalues of $\varphi'(u_0)$ counting algebraic multiplicity.

Problem 4.7*
Suppose that $\Omega \subseteq \mathbb{R}^N$ is a bounded, open, symmetric set with $0 \in \Omega$, $\varphi \in C(\overline{\Omega}; \mathbb{R}^N)$, $0 \notin \varphi(\partial\Omega)$ and $\varphi(-u) \neq \lambda\varphi(u)$ for all $u \in \partial\Omega$ and all $\lambda \geqslant 1$. Show that $d(\varphi, \Omega, 0)$ is odd.

Problem 4.8*
Let $(\varphi, \Omega, h) \in \mathcal{Y}$ (see Theorem 4.11) and assume that there exists $u \in \Omega$ such that $\varphi(u) = h$. Is it true that $d(\varphi, \Omega, h) \neq 0$? Justify your answer.

Problem 4.9***
Let $\Omega \subseteq \mathbb{R}^N$ be a nonempty open set and let $\varphi \colon \Omega \longrightarrow \mathbb{R}^N$ be a continuous and locally injective map. Show that the set $\varphi(\Omega)$ is open.

Problem 4.10**

Suppose that $D \subseteq \mathbb{R}^N$ is a set with nonempty interior and $\varphi \in C(D;\mathbb{R}^N)$ is an injection. Show that $\operatorname{int}\varphi(D) = \varphi(\operatorname{int} D)$.

Problem 4.11**

Let $\varphi\colon \mathbb{R}^N \longrightarrow \mathbb{R}^N$ be a continuous injection such that $|\varphi(u)| \longrightarrow +\infty$ as $|u| \to +\infty$. Show that φ is a surjection.

Problem 4.12*

Let $\Omega \subseteq \mathbb{R}^N$ be a bounded, open, symmetric set with $0 \in \Omega$ and let $\varphi \in C(\overline{\Omega};\mathbb{R}^N)$ be a function which is odd on $\partial\Omega$. Show that:

(a) there exists $u \in \overline{\Omega}$ such that $\varphi(u) = 0$;

(b) there exists $v \in \overline{\Omega}$ such that $\varphi(v) = v$.

Problem 4.13*

Let $(\varphi, \Omega, h) \in \mathcal{Y}$ (see Theorem 4.11) be such that $d(\varphi, \Omega, h) \neq 0$. Show that $\varphi(\Omega)$ is a neighborhood of h (i.e., $\varphi(\Omega)$ contains an open set U with $h \in U$).

Problem 4.14***

Let X be an infinite dimensional Banach space. Show that $\partial B_r(0)$ is a retract of $\overline{B}_r(0)$, where $\partial B_r(0) = \{u \in X : \|u\|_X = r\}$ and $\overline{B}_r(0) = \{u \in X : \|u\|_X \leqslant r\}$.

Problem 4.15**

Show that for every $r > 0$, $\partial B_r(0)$ is not a retract of $\overline{B}_r$ (see Definition 4.132(a)), where $\partial B_r(0) = \{u \in \mathbb{R}^N : |u| = r\}$ and $\overline{B}_r(0) = \{u \in \mathbb{R}^N : |u| \leqslant r\}$. (Compare with Problem 4.14.)

Problem 4.16***

(a) In $\mathbb{R}^N$, show that $\partial B_r(0)$ (with $r > 0$) is not contractible (see Definition 4.113(d)), where $\partial B_r(0) = \{u \in \mathbb{R}^N : |u| = r\}$.

(b) In an infinite dimensional Banach space X, show that $\partial B_r(0)$ (with $r > 0$) is contractible, where $\partial B_r(0) = \{u \in X : \|u\|_X = r\}$.

Problem 4.17***

Suppose that $a, b \in \mathbb{R}$, $a < b$, $V \subseteq [a, b] \times \mathbb{R}^N$ is a bounded open set, $\widehat{h}\colon \overline{V} \longrightarrow \mathbb{R}^N$ is a continuous map and $h \notin \widehat{h}\big(\partial_{[a,b]\times\mathbb{R}^N} V\big)$. For every $t \in [a, b]$ we define $V_t = \{u \in \mathbb{R}^N : (t, u) \in V\}$. Show that the map $[a, b] \ni t \longmapsto d(\widehat{h}(t,\cdot), V_t, h)$ is constant (here we use the convention that $d(\widehat{h}(t,\cdot), \emptyset, h) = 0$).

Problem 4.18*

Let $\Omega \subseteq \mathbb{R}^N$ be a bounded open set and let $\varphi, \psi \in C(\overline{\Omega};\mathbb{R}^N)$ be such that

$$|\psi(u)| < |\varphi(u)| \qquad \forall\, u \in \partial\Omega.$$

Show that

$$d(\varphi + \psi, \Omega, 0) = d(\varphi, \Omega, 0).$$

Problem 4.19**
Let $\Omega \subseteq \mathbb{R}^N$ be a bounded open set and let $\varphi \in C(\overline{\Omega};\mathbb{R}^N)$ be such that $\varphi(\overline{\Omega}) \subseteq \overline{\Omega}$ and $\varphi(u) = u$ for all $u \in \partial\Omega$. Show that $\varphi(\overline{\Omega}) = \overline{\Omega}$.

Problem 4.20**
Let $\Omega \subseteq \mathbb{R}^N$ be a bounded open set with $0 \in \Omega$ and let $\varphi \in C(\overline{\Omega};\mathbb{R}^N)$ be a map such that

$$(\varphi(u), u)_{\mathbb{R}^N} \geqslant 0 \qquad \forall\, u \in \partial\Omega.$$

Show that there exists $u_0 \in \overline{\Omega}$ such that $\varphi(u_0) = 0$.

Problem 4.21*
Let $m, n \in \mathbb{N}$ be such that $m < n$. Show that there is no continuous injection $\varphi\colon \mathbb{R}^n \longrightarrow \mathbb{R}^m$.

Problem 4.22**
Let $m, n \in \mathbb{N}$ be such that $n < m$ and assume that $\varphi\colon \mathbb{R}^n \longrightarrow \mathbb{R}^m$ is a continuous injection. Show that the set $\mathbb{R}^m \setminus \varphi(\mathbb{R}^n)$ is dense in $\mathbb{R}^m$.

Problem 4.23**
Suppose that N is an odd positive integer, $\Omega \subseteq \mathbb{R}^N$ is a bounded open set with $0 \in \Omega$ and $\varphi \in C(\overline{\Omega};\mathbb{R}^N)$ with $0 \notin \varphi(\partial\Omega)$. Show that there exist $u \in \partial\Omega$ and $\lambda \neq 0$ such that $\varphi(u) = \lambda u$.

Problem 4.24**
Suppose that N is an even positive integer, $S^N = \partial B_1^{N+1} = \{u \in \mathbb{R}^{N+1} : |u| = 1\}$ and $\varphi\colon S^N \longrightarrow S^N$ is a continuous map such that $\varphi(u) \neq -u$ for all $u \in S^N$. Show that there exists $\widehat{u} \in S^N$ such that $\varphi(\widehat{u}) = \widehat{u}$.

Problem 4.25***
Let $\varphi\colon S^N \longrightarrow S^N$ be a continuous map. Show that the following statements are equivalent:

(a) φ is not homotopic to a constant.

(b) Every continuous extension $\widehat{\varphi}\colon \overline{B}_1^{N+1} \longrightarrow \mathbb{R}^{N+1}$ has a zero, that is, there exists $u_0 \in B_1^{N+1}$ with $\widehat{\varphi}(u_0) = 0$ (here $\overline{B}_1^{N+1} = \{u \in \mathbb{R}^{N+1} : |u| \leqslant 1\}$).

(c) Every continuous extension $\widehat{\varphi}\colon \overline{B}_1^{N+1} \longrightarrow \mathbb{R}^{N+1}$ satisfies $d(\widehat{\varphi}, B_1^{N+1}, 0) \neq 0$.

Problem 4.26**
Let $\varphi\colon S^N \longrightarrow S^N$ be a continuous map. Show that:

(a) if $d_1(\varphi) \neq (-1)^{N+1}$, then φ has a fixed point (see Definition 4.16);

(b) if $d_1(\varphi) \neq 1$, then there exists $u_0 \in S^N$ such that $\varphi(u_0) = -u_0$.

Problem 4.27**

(a) Let $\Omega \subseteq \mathbb{R}^N$ be a bounded open symmetric set such that $0 \in \Omega$ and let $\varphi \in C(\partial\Omega;\mathbb{R}^N)$ be an odd map such that $\varphi(\partial\Omega)$ is contained in a proper subspace of $\mathbb{R}^N$. Show that there exists $u_0 \in \partial\Omega$ such that $\varphi(u_0) = 0$.

(b) Let $m, n \in \mathbb{N}$ be such that $n > m$. Show that there is no odd map $\varphi \in C(S^n; S^m)$.

Problem 4.28***
Let $\Omega \subseteq \mathbb{R}^N$ be a bounded open set and let $\varphi \in C(\overline{\Omega}; \mathbb{R}^N)$ be an injection. Suppose that $h \in \varphi(\Omega)$. Show that $d(\varphi, \Omega, h) = \pm 1$.

Problem 4.29**
Suppose that $\Omega \subseteq \mathbb{R}^N$ is a bounded open set, $\varphi = I_{\mathbb{R}^N} + f$ with $f \colon \overline{\Omega} \longrightarrow \mathbb{R}^N$ nonexpansive, (that is, $|f(u) - f(v)| < |u - v|$ for all $u, v \in \overline{\Omega}$, $u \neq v$). Show that φ is an injection and $d(\varphi, \Omega, h) = \pm 1$ for every $h \in \varphi(\Omega)$.

Problem 4.30*
Let $\Omega \subseteq \mathbb{R}^N$ be a bounded open set, $\varphi, \psi \in C(\overline{\Omega}; \mathbb{R}^N)$ and

$$h \notin [\varphi(u), \psi(u)] \;=\; \{v = (1-t)\varphi(u) + t\psi(u) : t \in [0,1]\} \qquad \forall\, u \in \partial\Omega.$$

Show that

$$d(\varphi, \Omega, h) \;=\; d(\psi, \Omega, h).$$

Problem 4.31**
Suppose that $\Omega = (a, b)$ (with $a < b$), $\varphi \in C([a, b])$ and $h \notin \{\varphi(a), \varphi(b)\}$. Show that $d(\varphi, \Omega, h) \in \{\pm 1, 0\}$

Problem 4.32*
Let $N \in \mathbb{N}$ be odd. Show that $-I_{S^N}, I_{S^N} : S^N \longrightarrow S^N$ are homotopic.

Problem 4.33*
Let $(\varphi, \Omega, 0) \in \mathcal{Y}$ (see Theorem 4.11) and suppose that there exists $y \neq 0$ such that

$$\varphi(u) \;\neq\; \lambda y \qquad \forall\, u \in \partial\Omega,\ \lambda > 0.$$

Show that $d(\varphi, \Omega, 0) = 0$.

Problem 4.34*
Suppose that X is a Banach space, $\Omega \subseteq X$ is a bounded open set and $\widehat{h}_0 \colon [0,1] \times \overline{\Omega} \longrightarrow X$ is a compact map. We set

$$\widehat{h}(t, u) = u - \widehat{h}_0(t, u) \qquad \forall\, (t, u) \in [0,1] \times \overline{\Omega}$$

and assume that $0 \notin \widehat{h}(t, \partial\Omega)$ for all $t \in [0,1]$. Show that

$$\inf\left\{\|\widehat{h}(t,u)\|_X :\ t \in [0,1],\ u \in \partial\Omega\right\} \;>\; 0.$$

Problem 4.35**
Let $(\varphi, \Omega, h) \in \mathcal{Y}_0$ (see Theorem 4.27). Show that for all $y \in X$ we have

$$d_{LS}(\varphi, \Omega, h) \;=\; d_{LS}(\varphi - y, \Omega, h - y).$$

(Compare with Problem 4.4.)

Problem 4.36*
Suppose that X is a Banach space, $\Omega \subseteq X$ is a bounded open set and $y \in X$. Show that for all $h \notin \partial\Omega - y$, we have

$$d_{LS}(I_X - y, \Omega, h) = \begin{cases} 1 & \text{if } h + y \in \Omega, \\ 0 & \text{if } h + y \notin \overline{\Omega}. \end{cases}$$

Problem 4.37**
Suppose that X is a Banach space, $\Omega \subseteq X$ is a bounded open set and $\varphi: I_X - f: \overline{\Omega} \longrightarrow X$ with $f \in K(\overline{\Omega}; X)$ being injective. Show that $\varphi^{-1}: \varphi(\overline{\Omega}) \longrightarrow X$ is also a compact perturbation of the identity and the set $\varphi(\Omega) \subseteq X$ is open.

Problem 4.38**
Suppose that X is a Banach space, $U \subseteq X$ is an open set and $\varphi = I_X - f$ is an injection with $f \in K(\overline{U}; X)$. Show that φ is an open map and φ is a homeomorphism of U onto $\varphi(U)$.

Problem 4.39*
Suppose that X is a Banach space, $\varphi = I_X - f$ with $f \in K(X; X)$ is such that

$$\|\varphi(u) - \varphi(v)\|_X \geqslant c\|u - v\|_X \qquad \forall\, u, v \in X$$

for some $c > 0$. Show that φ is a homeomorphism of X onto itself.

Problem 4.40*
Suppose that $(\varphi, \Omega, h), (\psi, \Omega, h) \in \mathcal{Y}_0$ (see Theorem 4.27) and assume that $\varphi|_{\partial\Omega} = \psi|_{\partial\Omega}$. Show that $d_{LS}(\varphi, \Omega, h) = d_{LS}(\psi, \Omega, h)$. (Compare with Problem 4.3.)

Problem 4.41**
In Problem 4.14 we saw that in an infinite dimensional Banach space X, $\partial B_r(0)$ is a retract of $\overline{B}_r(0)$ (with $r > 0$). Is it possible to find a retraction $\xi: \overline{B}_r(0) \longrightarrow \partial B_r(0)$ of the form $\xi = I_X - f$ with $f \in K(\overline{B}_r(0); X)$? Justify your answer.

Problem 4.42***
Suppose that X is a Banach space, $\Omega \subseteq X$ is an open set with $0 \in \Omega$, $\varphi: \overline{\Omega} \longrightarrow X$ is Fréchet differentiable at $u = 0$, $\varphi = I_X - f$ with $f \in K(\overline{\Omega}; X)$ compact, $f(0) = 0$, $\mu = 1$ is not an eigenvalue of $f'(0)$. Show that $u = 0$ is an isolated solution of $\varphi(u) = 0$ and $i(\varphi, 0, 0) = (-1)^m$, with m being the sum of the algebraic multiplicities of the eigenvalues μ of $f'(0)$ such that $\mu > 1$.

Problem 4.43**
Suppose that X is a Banach space with $X = \bigoplus_{k=1}^{m} V_k$ where $V_k \subseteq X$ are closed subspaces of X, $L \in \mathcal{L}(X)$ is injective and $L = I_X - A$ with $A \in \mathcal{L}_c(X)$. Show that

$$d_{LS}(L, B_1(0), 0) = \prod_{k=1}^{m} d_{LS}(L|_{V_k}, B_1(0) \cap V_k, 0).$$

Problem 4.44**
Suppose that X is a Banach space, $a, b \in \mathbb{R}$, $a < b$, $V \subseteq [a, b] \times X$ is a bounded open set, $k\colon \overline{V} \longrightarrow X$ is a compact map, $\widehat{h}(t, u) = u - k(t, u)$ and $h \notin \widehat{h}(\partial_{[a,b]\times X} V)$. Set

$$V_t = \{u \in X : (t, u) \in V\} \qquad \forall\, t \in [a, b].$$

Show that the map $t \longmapsto d_{LS}(\widehat{h}(t, \cdot), V_t, h)$ is constant on $[a, b]$ (here we use the convention that $d(\widehat{h}(t, \cdot), \emptyset, h) = 0$). (Compare with Problem 4.17.)

Problem 4.45**
Suppose that X is a Banach space, $\Omega \subseteq X$ is a bounded open set and $f \in K(\overline{\Omega}; X)$ satisfies

$$(I_X - f)(\overline{\Omega}) \subseteq \overline{\Omega} \quad \text{and} \quad f(u) = 0 \qquad \forall\, u \in \partial\Omega.$$

Show that $(I_X - f)(\overline{\Omega}) = \overline{\Omega}$. (Compare with Problem 4.19.)

Problem 4.46*
Suppose that X and Y are two Banach spaces, $A \in \mathcal{L}(X; Y)$ is an isomorphism, $\Omega \subseteq X$ is a bounded open set, $h \in A(\Omega)$, $k\colon [0, 1] \times \overline{\Omega} \longrightarrow Y$ is a compact map such that $k(0, u) = 0$ for all $u \in \overline{\Omega}$ and

$$A(u) - k(t, u) \neq h \qquad \forall\, (t, u) \in (0, 1] \times \partial\Omega.$$

Show that there exists $u \in \overline{\Omega}$ such that $A(u) - k(1, u) = h$.

Problem 4.47**
Suppose that X and Y are two Banach spaces, $A \in \mathcal{L}(X; Y)$ is an isomorphism, $k\colon [0, 1] \times X \longrightarrow Y$ is a compact map such that $k(0, u) = 0$ for all $u \in X$ and there exists $R > \|A^{-1}(y)\|_X$ such that for all $(t, u) \in (0, 1) \times X$, we have

$$\text{if } A(u) - k(t, u) = y, \text{ then } \|u\|_X \leqslant R$$

(*a priori* bound). Show that there exists $u \in \overline{B}_R(0)$ such that $A(u) - k(1, u) = h$.

Problem 4.48**
Suppose that X and Y are two Banach spaces, $A \in \mathcal{L}(X; Y)$ is an isomorphism and $\varphi\colon X \longrightarrow Y$ is a compact map such that

$$\lim_{\|u\|_X \to +\infty} \frac{\varphi(u)}{\|u\|_X} = 0.$$

Show that for every $y \in Y$ there exists $u \in X$ such that $A(u) - \varphi(u) = y$.

Problem 4.49*
Suppose that X is a Banach space and $(\varphi, \Omega, 0), (\psi, \Omega, 0) \in \mathcal{Y}_0$ (see Theorem 4.27) are such that

$$d_{LS}(\varphi, \Omega, 0) \neq d_{LS}(\psi, \Omega, 0).$$

Show that there exist $\lambda < 0$ and $\widehat{u} \in \partial\Omega$ such that $\varphi(\widehat{u}) = \lambda\psi(\widehat{u})$.

Problem 4.50**
Suppose that X is a Banach space, $\Omega \subseteq X$ is a bounded open set and $f\colon \overline{\Omega} \longrightarrow X$ is a compact map such that $f(u) \neq u$ for all $u \in \partial\Omega$. Let $\vartheta = d_{LS}(I_X - f, \Omega, 0)$ and consider the eigenvalue problem

$$f(u) \;=\; \mu u \quad \text{with } u \in \partial\Omega,\ \mu \in \mathbb{R}.$$

Show that this problem has a solution $(\mu, \widehat{u}) \in (1, +\infty) \times \partial\Omega$, if
(a) $0 \in \Omega$ and $\vartheta \neq 1$; or
(b) $0 \notin \Omega$ and $\vartheta \neq 0$.

Problem 4.51*
Suppose that X is a Banach space, $(\varphi, \Omega, h) \in \mathcal{Y}_0$ (see Theorem 4.27) and $\psi\colon \overline{\Omega} \longrightarrow X$ is a map such that $\psi = I_X - g$ with $g \in K(\overline{\Omega}; X)$. Suppose that

$$\|\varphi(u) - \psi(u)\|_X \;<\; \|\psi(u) - h\|_X \qquad \forall\, u \in \partial\Omega.$$

Show that $(\psi, \Omega, h) \in \mathcal{Y}_0$ and $d_{LS}(\varphi, \Omega, h) = d_{LS}(\psi, \Omega, h)$.

Problem 4.52**
Suppose that X is a Banach space, $(\varphi, \Omega, h) \in \mathcal{Y}_0$ (see Theorem 4.27) and there exists a ray $R_h^y = \{u \in X : \ u = (1-t)h + ty,\ t \in [0, +\infty)\}$ emanating from h in the direction $y \neq h$ such that $R_h^y \cap \varphi(\partial\Omega) = \emptyset$. Show that $d_{LS}(\varphi, \Omega, h) = 0$.

Problem 4.53***
Suppose that X is a Banach space, $(\varphi, \Omega, 0) \in \mathcal{Y}_0$ (see Theorem 4.27) with $\varphi = I_X - f$, $d_{LS}(\varphi, \Omega, 0) \neq 0$ and $f \in K(\overline{\Omega}; X)$ satisfies the following condition:

"for every $\varepsilon > 0$, there exists $f_\varepsilon \in K(\overline{\Omega}; X)$ such that

$$\|f_\varepsilon(u) - f(u)\|_X \;\leqslant\; \varepsilon \qquad \forall\, u \in \overline{\Omega}$$

and for every $h \in X$ with $\|h\|_X \leqslant \varepsilon$, the equation

$$u - f_\varepsilon(u) = h$$

has at most one solution in $\overline{\Omega}$."

Show that the set $S(\varphi) = \{u \in \Omega : \ \varphi(u) = 0\}$ is connected.

Problem 4.54**
Suppose that X is a Banach space, $\Omega \subseteq X$ is a bounded, open, symmetric set with $0 \in \Omega$, $\varphi = I_X - f\colon \partial\Omega \longrightarrow X$ is odd with $f \in K(\partial\Omega; X)$ and $\varphi(\partial\Omega) \subseteq V$, where V is a closed linear subspace of X of codimension 1. Show that we can find $\widehat{u} \in \partial\Omega$ such that $\varphi(\widehat{u}) = 0$.

Problem 4.55**
Suppose that X is a Banach space, $\Omega \subseteq X$ is a bounded, open, symmetric set with $0 \in \Omega$, $\varphi\colon I_X - f\colon \partial\Omega \longrightarrow X$ with $f \in K(\partial\Omega; X)$ and $\varphi(\partial\Omega) \subseteq V$ where V is a closed linear subspace of X of codimension 1. Show that there exists $\widehat{u} \in \partial\Omega$ such that $\varphi(\widehat{u}) = \varphi(-\widehat{u})$.

Problem 4.56***
Suppose that X is a Banach space, $\varphi = I_X - f \colon X \longrightarrow X$ with $f \in K(X;X)$ which is asymptotically linear with $f'(\infty) \in \mathcal{L}_c(X)$ (see Definition 4.97(b)). Show that for a sufficiently large bounded open set $\Omega \subseteq X$ containing the origin, we have

$$d_{LS}(\varphi, \Omega, 0) = (-1)^m,$$

with m being the sum of the algebraic multiplicities of the real eigenvalues of $\varphi'(\infty)$ in $(1, +\infty)$.

Problem 4.57**
Suppose that X is a Banach space, $(\varphi, \Omega, h) \in \mathcal{Y}_0$ (see Theorem 4.27) and assume that $\varphi(\overline{\Omega}) \subseteq V$ with V being a proper linear subspace of X. Show that $d_{LS}(\varphi, \Omega, h) = 0$.

Problem 4.58**
Suppose that X is a Banach space, $\varphi = I_X - f \colon X \longrightarrow X$ with $f \in K(X;X)$ and there is $h \in X$, such that $d_{LS}(\varphi, B_r(0), h) \neq 0$ for all open balls $B_r(0) \supseteq \varphi^{-1}(h)$. Show that φ is surjective.

Problem 4.59**
Suppose that X is an infinite dimensional Banach space, $\Omega \subseteq X$ is a bounded open set with $0 \notin \partial\Omega$, $f \in K(\overline{\Omega};X)$ with $0 \notin \overline{f(\partial\Omega)}$ and $f(u) \neq \lambda u$ for all $(\lambda, u) \in [0,1] \times \partial\Omega$. Show that $d_{LS}(I_X - f, \Omega, 0) = 0$.

Problem 4.60**
Suppose that X is a Banach space and $\varphi = I_X - f \colon X \longrightarrow X$ with $f \in K(X;X)$ being odd. Show that φ is surjective.

Problem 4.61**
Suppose that X is a Banach space, $(\varphi, \Omega, 0) \in \mathcal{Y}_c$ (see Theorem 4.39) and $d_C(\varphi, \Omega, 0) \neq 0$ (see Definition 4.37). Show that for every $r > 0$ such that $\varphi(\partial\Omega) \cap B_r(0) = \emptyset$, we have $B_r(0) \subseteq \varphi(\Omega)$.

Problem 4.62***
Suppose that X is a Banach space, $\Omega \subseteq X$ is a bounded, convex, symmetric, and open set with $0 \in \Omega$, $\varphi \colon I_X - f \colon \overline{\Omega} \longrightarrow X$ with f condensing (that is, $f \in S(\overline{\Omega};X)$), $0 \notin \varphi(\partial\Omega)$ and $\varphi|_{\partial\Omega}$ is odd. Show that $d_C(\varphi, \Omega, 0)$ is odd.

Problem 4.63**
Suppose that X is a Banach space, $\Omega \subseteq X$ is a bounded, convex, symmetric, and open set with $0 \in \Omega$, $\varphi = I_X - f \colon \overline{\Omega} \longrightarrow X$ with f being condensing (that is, $f \in S(\overline{\Omega};X)$) and

$$\varphi(u) \neq t\varphi(-u) \qquad \forall\, (t,u) \in [0,1] \times \partial\Omega.$$

Show that $d_C(\varphi, \Omega, 0)$ is odd.

Problem 4.64**
Suppose that X is a Banach space, $(\varphi, \Omega, 0) \in \mathcal{Y}_c$ (see Theorem 4.39) with $0 \in \Omega$ and for every $v \in \partial\Omega$, let $R_0^v = \{u \in X : u = tv \text{ for some } t \geqslant 0\}$. Suppose that $d_C(\varphi, \Omega, 0) \neq 0$. Show that

$$R_0^v \cap \varphi(\partial\Omega) \neq \emptyset \qquad \forall\, v \in \partial\Omega.$$

Problem 4.65**
Suppose that X is a Banach space, $V \subseteq X$ is a linear subspace of codimension 1. Suppose that $\varphi = I_V - f\colon \overline{B}_1(0) \longrightarrow V$ with f being condensing. Show that there exists $\widehat{u} \in \partial B_1(0)$ such that $\varphi(\widehat{u}) = \varphi(-\widehat{u})$.

Problem 4.66**
Let $\varphi \in C(\mathbb{R}^N; \mathbb{R}^N)$ be a strongly coercive function. Show that φ is surjective.

Problem 4.67**
Suppose X is an infinite dimensional Banach space, $\Omega \subseteq X$ is a bounded open set with $0 \in \Omega$, $f\colon \overline{\Omega} \longrightarrow X$ is a condensing map (that is, $f \in S(\overline{\Omega}; X)$), $g\colon \partial\Omega \longrightarrow X$ is a compact map (that is, $g \in K(\partial\Omega; X)$) and

$$(1-t)f(u) + tg(u) \neq u \qquad \forall\, (t, u) \in [0,1] \times \partial\Omega$$

and

$$\|u\|_X \leqslant \|g(u)\|_X \qquad \forall\, u \in \partial\Omega.$$

Show that $d_C(I_X - f, \Omega, 0) = 0$. (Compare with Problem 4.59.)

Problem 4.68**
Suppose that X is an infinite dimensional Banach space, $\Omega \subseteq X$ is a bounded open set with $0 \in \Omega$, $h \in X$ is such that

$$\|h\|_X > \sup_{u \in \partial\Omega} \|u\|_X$$

and $f\colon \overline{\Omega} \longrightarrow X$ is a condensing map (that is, $f \in S(\overline{\Omega}; X)$). Suppose that

$$f(u) \neq \lambda u + (\lambda - 1)h \qquad \forall\, \lambda > 1,\ u \in \partial\Omega$$

and

$$\|f(u)\|_X > \|u\|_X \qquad \forall\, u \in \partial\Omega.$$

Show that $d_C(I_X - f, \Omega, 0) = 0$.

Problem 4.69**
Suppose that $\Omega \subseteq \mathbb{R}^N$ is an open set with $u_0 \in \Omega$ and $\varphi \in C(\Omega; \mathbb{R}^N)$ is such that φ is differentiable at u_0 and $J_\varphi(u_0) \neq 0$. Let $h = \varphi(u_0)$. Show that

$$i(\varphi, u_0, h) = \operatorname{sgn} J_\varphi(u_0)$$

(see Definition 4.12(b)).

Problem 4.70**

Suppose that $\Omega \subseteq \mathbb{R}^N$ is an open set with $0 \in \Omega$, $\varphi \in C(\Omega; \mathbb{R}^N)$ and $h \in C(\mathbb{R}^N; \mathbb{R}^N)$ is α-homogeneous (see Problem 3.29) with $h^{-1}(0) = \{0\}$ (that is, h is ***nondegenerate***). Suppose that

$$\lim_{u \to 0} \tfrac{\varphi(u) - h(u)}{|u|^\alpha} = 0.$$

Show that $i(\varphi, 0, 0) = i(h, 0, 0)$ (see Definition 4.12(b)).

Problem 4.71**

Suppose that $\varphi, h \in C(\mathbb{R}^N; \mathbb{R}^N)$, h is α-homogeneous and nondegenerate (see Problem 4.70). Suppose that

$$\lim_{|u| \to +\infty} \tfrac{\varphi(u) - h(u)}{|u|^\alpha} = 0.$$

Show that for every $y \in \mathbb{R}^N$, we have

$$i_\infty(\varphi, y) = i_\infty(h, 0)$$

(see Definition 4.12(c)).

Remark. Hence, if $\varphi \in C(\mathbb{R}^N; \mathbb{R}^N)$ is asymptotically linear with $\varphi'(\infty)$ invertible (see Definition 4.97(b)), we have $i_\infty(\varphi, y) = \operatorname{sgn} \det \varphi'(\infty)$ for every $y \in \mathbb{R}^N$.

Problem 4.72***

Suppose that $a, b \in \mathbb{R}$, $V \subseteq [a, b] \times \mathbb{R}^N$ is a bounded and open (for the relative topology from $\mathbb{R}^{N+1}$) set, $\varphi \in C(\overline{V}; \mathbb{R}^N)$, $h \in \mathbb{R}^N$ with $h \notin \varphi(\partial_{[a,b] \times \mathbb{R}^N} V)$, $S = \{(t, u) \in V : \varphi(t, u) = h\}$ and for each $t \in [a, b]$ we introduce the section

$$S_t = \{u \in \mathbb{R}^N : (t, u) \in V\}.$$

Suppose that there exists $\widehat{t} \in [a, b]$ such that $d(\varphi(\widehat{t}, \cdot), V_{\widehat{t}}, h) \neq 0$. Show that there exists a connected component C of S such that

$$C \cap (\{a\} \times S_a) \neq \emptyset \quad \text{and} \quad C \cap (\{b\} \times S_b) \neq \emptyset.$$

Problem 4.73***

Suppose that $\Omega \subseteq \mathbb{R}^N$ is an open set and $\varphi \in C^{2-0}(\Omega; \mathbb{R}^N)$ (that is, $\varphi \in C^1(\Omega; \mathbb{R}^N)$ and φ' is locally Lipschitz). Suppose that there exists $\eta \in \mathbb{R}$ such that the set $\overline{\varphi^\eta} = \{u \in \Omega : \varphi(u) \leqslant \eta\}$ is compact and also there exist $\vartheta < \eta$, $r > 0$ and $u_0 \in \Omega$ such that $\overline{\varphi^\vartheta} = \{u \in \Omega : \varphi(u) \leqslant \vartheta\} \subseteq \overline{B}_r(u_0) \subseteq \varphi^\eta = \{u \in \Omega : \varphi(u) < \eta\}$ and

$$\varphi'(u) \neq 0 \qquad \forall\, u \in \overline{\varphi^\eta} \setminus \varphi^\vartheta.$$

Show that

$$d(\nabla\varphi, \varphi^\eta, 0) = 1$$

(here $\nabla\varphi$ is the gradient of φ, that is, $\nabla\varphi(u) \in \mathbb{R}^N$ and $\langle \varphi'(u), h \rangle = (\nabla\varphi(u), h)_{\mathbb{R}^N}$ for all $h \in \mathbb{R}^N$).

Problem 4.74**
Let $\varphi \in C^{2-0}(\mathbb{R}^N;\mathbb{R}^N)$ (see Problem 4.73) be such that

$$\varphi'(u) \neq 0 \qquad \forall\, u \in \mathbb{R}^N,\ |u| \geqslant r_0 > 0$$

and φ is coercive (that is, $\varphi(u) \longrightarrow +\infty$ and $|u| \to +\infty$). Show that

$$d(\nabla\varphi, B_r(0), 0) \;=\; 1 \qquad \forall\, r \geqslant r_0.$$

Problem 4.75**
Suppose that $\Omega \subseteq \mathbb{R}^N$ is an open set, $u_0 \in \Omega$ and $\varphi \in C^{2-0}(\Omega;\mathbb{R}^N)$ (see Problem 4.73). Suppose that u_0 is an isolated critical point of φ which is a local minimizer of φ. Show that for all $\varrho > 0$ small, we have $d(\nabla\varphi, B_\varrho(u_0), 0) = 1$.

Problem 4.76***
Suppose that $\Omega \subseteq \mathbb{R}^N$ is a bounded, open, and connected set with $\operatorname{int}\overline{\Omega} = \Omega$, $\varphi_0 \in C(\overline{\Omega};\mathbb{R}^N)$ is an injection, $\varphi \in C^1(\Omega;\mathbb{R}^N) \cap C(\overline{\Omega};\mathbb{R}^N)$ with $J_\varphi(u) > 0$ for all $u \in \Omega$ and such that $\varphi_0|_{\partial\Omega} = \varphi|_{\partial\Omega}$. Show that

$$\varphi_0(\Omega) \;=\; \varphi(\Omega), \quad \varphi_0(\overline{\Omega}) \;=\; \varphi(\overline{\Omega})$$

and φ is a homeomorphism on $\overline{\Omega}$ and a C^1-diffeomorphism on Ω.

Problem 4.77**
Let X be a reflexive Banach space and let $\varphi\colon X \longrightarrow \mathbb{R}$ be a continuous and Gâteaux differentiable function such that $\varphi'_G\colon X \longrightarrow X^*$ is demicontinuous and $(S)_+$, $\varphi(u) \longrightarrow +\infty$ and $\|u\|_X \to +\infty$ (that is, φ is coercive) and there exists $r_0 > 0$ such that $\varphi'_G(u) \neq 0$ for all $\|u\|_X \geqslant r_0$. Show that there exists $r_1 \geqslant r_0$ such that

$$d_{(S)_+}(\varphi'_G, B_r(0), 0) \;=\; 1 \qquad \forall\, r \geqslant r_1.$$

Problem 4.78***
Suppose that X is a reflexive Banach space, $U \subseteq X$ is a nonempty open set and $\varphi \in C^1(U;\mathbb{R})$ is such that $\varphi'\colon U \longrightarrow X^*$ is an $(S)_+$-map. Suppose that $u_0 \in U$ is a local minimizer and an isolated critical point of φ. Show that there exists $r_1 > 0$ such that

$$d_{(S)_+}(\varphi', B_r(u_0), 0) \;=\; 1 \qquad \forall\, r \in (0, r_1].$$

Problem 4.79***
Suppose that X is a reflexive Banach space furnished with a norm such that both X and X^* are locally uniformly convex (by the Troyanski renorming theorem this is always possible; see Theorem I.5.192 or Remark 2.115), $U \subseteq X$ is a nonempty, bounded, open set, $A\colon X \supseteq D(A) \longrightarrow 2^{X^*}$ is a maximal monotone map with $(0,0) \in \operatorname{Gr} A$ and $h^* \in (\mathcal{F} + A)(U)$ with $\mathcal{F}\colon X \longrightarrow X^*$ being the duality map. Show that $d_M(\mathcal{F} + A, U, h^*) = 1$.

Problem 4.80***
Suppose that X is a reflexive Banach space furnished with a norm such that both X and X^* are locally uniformly convex (by the Troyanski renorming theorem this is always possible; see Theorem I.5.192 or Remark 2.115), $U \subseteq X$ is a nonempty, bounded, open set, $f \colon \overline{U} \longrightarrow X^*$ is a bounded, demicontinuous, monotone and $(S)_+$-map, $A \colon X \supseteq D(A) \longrightarrow 2^{X^*}$ is a maximal monotone map with $(0,0) \in \operatorname{Gr} A$ and $h^* \in (f+A)(U) \setminus (f+A)(\partial U)$. Show that $d_M(f+A, U, h^*) = 1$.

Problem 4.81***
Suppose that X is a reflexive Banach space furnished with a norm such that both X and X^* are locally uniformly convex (by the Troyanski renorming theorem this is always possible; see Theorem I.5.192 or Remark 2.115), $\{\varphi^t\}_{t\in[0,1]} \subseteq \Gamma_0(X)$ (see Remark 3.56) is such that

$$\varphi^t(u) \geqslant \varphi^t(0) \qquad \forall\, t \in [0,1],\ u \in X$$

and assume that $\varphi^{t_n} \xrightarrow{M} \varphi^t$ as $t_n \to t$ in $[0,1]$ (see Definition 3.149(b)). Show that $\{\partial\varphi^t\}_{t\in[0,1]}$ is a pseudomonotone homotopy.

Problem 4.82***
Suppose that X is a reflexive Banach space furnished with a norm such that both X and X^* are locally uniformly convex (by the Troyanski renorming theorem this is always possible; see Theorem I.5.192 or Remark 2.115), $\varphi^0, \varphi^1 \in \Gamma_0(X)$ (see Remark 3.56) with $\varphi^0(0) \leqslant \varphi^0(u)$, $\varphi^1(0) \leqslant \varphi^1(u)$ for all $u \in X$ and $\overline{\operatorname{dom}\varphi^0} = \overline{\operatorname{dom}\varphi^1}$. Suppose that there are functions $c_k \colon (0,1] \longrightarrow [0,+\infty)$ (for $k \in \{1,2\}$) such that $\lim\limits_{\lambda\searrow 0} c_k(\lambda) = 0$ and

$$\begin{aligned}
\varphi^0(J_\lambda^{\partial\varphi^1}(u)) &\leqslant (1+c_1(\lambda))\varphi^0(u) + c_2(\lambda) && \forall\, \lambda > 0,\ u \in \operatorname{dom}\varphi^0,\\
\varphi^1(J_\lambda^{\partial\varphi^0}(u)) &\leqslant (1+c_1(\lambda))\varphi^1(u) + c_2(\lambda) && \forall\, \lambda > 0,\ u \in \operatorname{dom}\varphi^1.
\end{aligned}$$

Let $\varphi^t = (1-t)\varphi^0 + t\varphi^1$ for all $t \in [0,1]$. Show that $\{\partial\varphi^t\}_{t\in[0,1]}$ is a pseudomonotone homotopy.

Problem 4.83**
Let (X, d_X) be a complete metric space and let $\varphi \colon X \longrightarrow X$ be a function. Suppose that there exists a right continuous function $\vartheta \colon [0,+\infty) \longrightarrow [0,+\infty)$ such that $\vartheta(r) < r$ if $r > 0$ and

$$d_X(\varphi(u), \varphi(v)) \leqslant \vartheta(d_X(u,v)) \qquad \forall\, u, v \in X$$

Show that φ has a unique fixed point $\widehat{u} \in X$ and for any $u_0 \in X$, we have $\varphi^{(n)}(u_0) \longrightarrow \widehat{u}$ in X.

Problem 4.84**
Let (X, d_X) be a complete metric space and let $\varphi \colon X \longrightarrow X$ be a continuous map. Suppose that there exists a function $\vartheta \colon X \longrightarrow [0,+\infty)$ such that

$$d_X(u, \varphi(u)) \leqslant \vartheta(u) - \vartheta(\varphi(u)) \qquad \forall\, u \in X.$$

Show that φ has a fixed point and for any $u_0 \in X$, the sequence $\{\varphi^{(n)}(u_0)\}_{n\geqslant 1}$ converges to a fixed point of φ.

Problem 4.85**
Let (X, d_X) be a compact metric space and let $\varphi\colon X \longrightarrow X$ be contractive (see Definition 4.58(c)). Show that φ has a unique fixed point $\widehat{u} \in X$ and for any $u_0 \in X$, we have $\varphi^{(n)}(u_0) \longrightarrow \widehat{u}$.

Problem 4.86**
Let (X, d_X) be a locally compact complete metric space, let $\varphi_0\colon X \longrightarrow X$ be a map and let $\{\varphi_n\colon X \longrightarrow X\}_{n\geqslant 1}$ be a sequence of map such that for each $n \geqslant 0$ there exists $m_n \geqslant 1$ such that $\varphi^{(m_n)}$ is a k_n-contraction. Assume that $\varphi_n(u) \longrightarrow \varphi_0(u)$ for all $u \in X$ and that the sequence $\{\varphi_n\}_{n\geqslant 1}$ is equicontinuous. Show that

$$\widehat{u}_n = \varphi_n(\widehat{u}_n) \longrightarrow \varphi_0(\widehat{u}_0) = \widehat{u}_0,$$

where $\widehat{u}_n$ is the unique fixed point of φ_n for all $n \geqslant 0$ (see Proposition 4.62).

Problem 4.87***
Let (X, d_X) be a complete metric space and let $\{\varphi_n\colon X \longrightarrow X\}_{n\geqslant 1}$ be a sequence of maps such that each φ_n has at least one fixed point $\widehat{u}_n \in X$ for all $n \geqslant 1$. Suppose that $\varphi_n \longrightarrow \varphi$ uniformly on X with $\varphi\colon X \longrightarrow X$ a uniformly continuous map such that $\varphi^{(m)}$ is a k-contraction for some $m \geqslant 1$. Show that φ has a unique fixed point $\widehat{u} \in X$ and

$$\varphi_n(\widehat{u}_n) = \widehat{u}_n \longrightarrow \widehat{u} = \varphi(\widehat{u}).$$

Problem 4.88*
Suppose that (X, d_X) is a complete metric space, $B_r(u) = \{v \in X : d_X(v, u) < r\}$ and $\varphi\colon B_r(u) \longrightarrow X$ is a k-contraction such that $d_X(\varphi(u), u) < (1-k)r$. Show that φ has a fixed point.

Problem 4.89*
Suppose that X is a Banach space, $\varphi\colon X \longrightarrow X$ is a k-contraction and $f = I_X - \varphi$. Show that f is a homeomorphism of X onto itself.

Problem 4.90*
Suppose that X is a Banach space, $C \subseteq X$ is a nonempty, closed, and convex set, U is a relatively open subset of C, $0 \in U$ and $\varphi\colon \overline{U} \longrightarrow C$ is a bounded k-contraction such that for all $u \in \partial U$ at least one of the following conditions holds:
(a) $\|\varphi(u)\|_X \leqslant \|u\|_X$;
(b) $\|\varphi(u)\|_X \leqslant \|u - \varphi(u)\|_X$;
(c) $\|\varphi(u)\|_X^2 \leqslant \|u\|_X^2 + \|u - \varphi(u)\|_X^2$;
(d) X is a Hilbert space with the inner product $(\cdot,\cdot)_X$ and $(\varphi(u), u)_X \leqslant \|u\|_X^2$.
Show that φ has a unique fixed point.

Problem 4.91*
Suppose that X is a Banach space, $U \subseteq X$ is an open, symmetric set with $0 \in U$ and $\varphi\colon \overline{U} \longrightarrow X$ is a bounded k-contraction such that $\varphi|_{\partial U}$ is odd. Show that φ has a unique fixed point.

Problem 4.92*
Suppose that X is a Banach space, $U \subseteq X$ is a bounded open set with $0 \in U$ and $\varphi, \psi\colon \overline{U} \longrightarrow X$ are two k-contractions such that $\varphi|_{\partial U} = \psi|_{\partial U}$. Show that φ has a fixed point if and only if ψ has a fixed point.

Problem 4.93**
Suppose that H is a Hilbert space, $U \subseteq H$ is an open set and $\varphi\colon U \longrightarrow H$ is a Lipschitz map such that

$$(\varphi(u) - \varphi(v), u - v)_H \geqslant c\|u - v\|_H^2 \qquad \forall\, u, v \in H,$$

with $c > 0$. Show that φ is an open map and a homeomorphism of U onto $\varphi(U)$.

Problem 4.94**
Suppose that (X, d_X) is a complete metric space, $\{\varphi_n\colon X \longrightarrow X\}_{n\geqslant 1}$ is a sequence of k-contractions, $\varphi\colon X \longrightarrow X$ is a k-contraction all with the same $k \in (0,1)$,

$$\varphi_n(u) \longrightarrow \varphi(u) \qquad \forall\, u \in X$$

and let $\widehat{u}_n = \varphi_n(\widehat{u}_n)$ for $n \geqslant 1$ and $\widehat{u} = \varphi(\widehat{u})$ (see the Banach fixed point theorem; Theorem 4.59). Show that $\widehat{u}_n \longrightarrow \widehat{u}$ in X and this convergence is no longer true if the contraction constant $k \in (0,1)$ is not the same for all φ_n.

Problem 4.95**
Suppose that X is a uniformly convex Banach space, $C \subseteq X$ is a nonempty, bounded, closed, and convex set and $\varphi\colon C \longrightarrow C$ is a nonexpansive map. Show that the set $\mathrm{Fix}(\varphi) = \{u \in C : u = \varphi(u)\}$ is nonempty, w-compact, and convex.

Problem 4.96*
Suppose that X is a Banach space, $C \subseteq X$ is a nonempty, closed, and convex set and $\varphi\colon C \longrightarrow C$ is a nonexpansive map with $\overline{\varphi(C)}$ being compact. Show that φ has a fixed point.

Problem 4.97*
Suppose that X is a Banach space, $C \subseteq X$ is a bounded, closed, and convex set and $\varphi\colon C \longrightarrow C$ is a nonexpansive map. Show that for every $\varepsilon > 0$ there exists $\widehat{u}_\varepsilon \in C$ such that $\|\varphi(\widehat{u}_\varepsilon) - \widehat{u}_\varepsilon\|_X < \varepsilon$

Problem 4.98***
Let (X, d_X) be a complete metric space such that for any points $u, v \in X$ and for some $\varepsilon > 0$, we can find points $y_0, y_1 \ldots, y_n \in X$ with $y_0 = u$, $y_n = v$ such that

$d_X(y_{k-1}, y_k) < \varepsilon$ for all $k \in \{1, \ldots, n\}$ (such a metric space is usually called ε-***chainable*** and the collection $y_0, y_1 \ldots, y_n$ is an ε-***chain*** for (u, v)). Suppose that $F\colon X \longrightarrow P_{bf}(X)$ is a multifunction such that

$$\text{if } d_X(u,v) < \varepsilon, \text{ then } h(F(u), F(v)) \leqslant \xi d_X(u,v)$$

with $\xi \in [0,1)$ (here h denotes the Hausdorff metric on $P_{bf}(X)$; see Definition 2.49). Show that F has a fixed point.

Problem 4.99**

Suppose that X is a uniformly convex Banach space, $C \subseteq X$ is a bounded, closed, and convex set and $\varphi\colon C \longrightarrow X$ is a nonexpansive map such that

$$\inf_{u \in C} \|u - \varphi(u)\|_X = 0.$$

Show that φ has a fixed point.

Problem 4.100*

Suppose that (X, d_X) is a metric space, $K \subseteq X$ is a compact set, $\varphi\colon X \longrightarrow K$ is a continuous map and for every $\varepsilon > 0$ we can find $u_\varepsilon \in X$ such that $d_X(\varphi(u_\varepsilon), u_\varepsilon) < \varepsilon$. Show that φ has a fixed point.

Problem 4.101**

Is it true that the set of fixed points of a nonexpansive map when it is nonempty must be connected? Justify your answer.

Problem 4.102**

Let H be a Hilbert space and let $\overline{B}_\varrho = \{u \in H : \|u\|_H \leqslant \varrho\}$ (with $\varrho > 0$). Consider the radial retraction map

$$r(u) = \begin{cases} u & \text{if } \|u\|_H \leqslant \varrho, \\ \varrho \frac{u}{\|u\|_H} & \text{if } \|u\|_H \geqslant \varrho. \end{cases}$$

Show that $r\colon H \longrightarrow \overline{B}_\varrho$ is nonexpansive.

Problem 4.103**

Suppose that (X, d_X) is a complete metric space, $u_0 \in X$, $r > 0$, $\overline{B}_r(u_0) = \{u \in X : d_X(u, u_0) \leqslant r\}$, $F\colon \overline{B}_r(u_0) \longrightarrow P_{bf}(X)$ is a multivalued k-contraction, i.e.,

$$h(F(u), F(v)) \leqslant k d_X(u,v) \qquad \forall\, u, v \in X,$$

with $k \in (0,1)$, where h is the Hausdorff metric corresponding to d_X (see Definition 2.49) and suppose that

$$\operatorname{dist}(u_0, F(u_0)) < (1-k)r.$$

Show that F has a fixed point (this is the multivalued analog of Problem 4.88).

Problem 4.104***
Suppose that (X, d_X) is a complete metric space, $U \subseteq X$ is an open set, $F, G\colon \overline{U} \longrightarrow P_{bf}(X)$ are two multifunctions and $H\colon [0,1] \times \overline{U} \longrightarrow P_{bf}(X)$ is a multifunctions such that:

(i) $H(0, \cdot) = F$ and $H(1, \cdot) = G$;
(ii) $u \notin H(t, \partial U)$ for all $t \in [0,1]$, all $u \in \partial U$;
(iii) $h(H(t,u), H(t,v)) \leqslant k d_X(u,v)$ for all $t \in [0,1]$, all $u, v \in \overline{U}$, with $k \in [0,1)$;
(iv) there exists a continuous, nondecreasing function $\vartheta\colon [0,1] \longrightarrow \mathbb{R}$ such that

$$h(H(t,u), H(s,u)) \leqslant |\vartheta(t) - \vartheta(s)| \qquad \forall\, t, s \in [0,1],\ u \in \overline{U}.$$

Suppose that F has a fixed point in U. Show that G has a fixed point in U too.

Problem 4.105**
Suppose that X is a Banach space, $U \subseteq X$ is an open set with $0 \in U$ and $G\colon \overline{U} \longrightarrow P_f(X)$ is a multivalued k-contraction such that the set $G(\overline{U}) \subseteq X$ is bounded. Show that at least one of the following properties holds:

(a) G has a fixed point in $\overline{U}$.
(b) There exist $t \in (0,1)$ and $u \in \partial U$ such that $u \in tG(u)$.

Problem 4.106**
Suppose that X is a Banach space, $U \subseteq X$ is an open set with $0 \in U$ and $F\colon \overline{U} \longrightarrow P_f(X)$ is a multivalued k-contraction such that the set $F(\overline{U}) \subseteq X$ is bounded and $m_F(u) = h(\{0\}, F(u))$. Suppose that one of the following conditions holds:

(i) $m_F(u) \leqslant \|u\|_X$ for all $u \in \overline{U}$;
(ii) $m_F(u) \leqslant \operatorname{dist}(u, F(u))$ for all $u \in \overline{U}$;
(iii) $m_F(u) \leqslant \left((\operatorname{dist}(u, F(u)))^2 + \|u\|_X^2\right)^{\frac{1}{2}}$ for all $u \in \overline{U}$;
(iv) $m_F(u) \leqslant \max\left\{\|u\|_X, \operatorname{dist}(u, F(u))\right\}$ for all $u \in \overline{U}$.

Show that F has a fixed point in $\overline{U}$.

Problem 4.107**
Let X be a Banach space and let $F\colon X \longrightarrow P_{bf}(X)$ be a multifunction such that for every $r > 0$, $F|_{\overline{B}_r(0)}$ is a multivalued k-contraction. Let

$$E(F) = \{u \in X : \ u \in tF(u) \text{ for some } t \in [0,1]\}.$$

Show that either $E(F)$ is unbounded or F has a fixed point.

Problem 4.108**
Suppose that X is a uniformly convex Banach space, $C \subseteq X$ is a nonempty, bounded, closed, and convex set and $\{\varphi_t\colon C \longrightarrow C\}_{t \in T}$ is a family of commuting nonexpansive maps (i.e., $\varphi_t \circ \varphi_s = \varphi_s \circ \varphi_t$ for all $t, s \in T$). Show that the family $\{\varphi_t\}_{t\in T}$ has a common fixed point.

Problem 4.109**

Suppose that X and Y are two Banach spaces, $u_0 \in X$, $r > 0$, $B_r(u_0) = \{u \in X : \|u - u_0\|_X < r\}$ and $\varphi\colon B_r(u_0) \longrightarrow Y$ is a Fréchet differentiable map. Suppose that

(i) $\varphi'(u_0)$ is invertible;

(ii) $\|\varphi'(u) - \varphi'(u_0)\|_{\mathcal{L}} \leqslant \eta\|u - u_0\|_X$ for all $u \in B_r(u_0)$ and some $\eta > 0$;

(iii) $\mu = 4\eta\|\varphi'(u_0)^{-1}\|_{\mathcal{L}}^2\|\varphi(u_0)\|_Y \leqslant 1$;

(iv) $\varrho = 2\|\varphi'(u_0)^{-1}\|_{\mathcal{L}}\|\varphi(u_0)\|_Y < r$.

Show that there exists unique $\widehat{u} \in \overline{B}_\varrho(u_0)$ such that $\varphi(\widehat{u}) = 0$.

Problem 4.110*

Using a fixed point argument, show that the equation $\cos u = u$ has a solution in $[0, \frac{\pi}{2}]$.

Problem 4.111*

Let X be any nonempty set and let $\varphi, \psi\colon X \longrightarrow X$ be two maps such that $\varphi \circ \psi = \psi \circ \varphi$. Suppose that φ has a unique fixed point $\widehat{u} \in X$. Show that $\widehat{u}$ is also a fixed point of ψ.

Problem 4.112**

Let X be a complete metric space and let $\{\varphi_t\colon X \longrightarrow X\}_{t\in T}$ be a commuting family of maps (i.e., $\varphi_t \circ \varphi_s = \varphi_s \circ \varphi_t$ for all $t, s \in X$). Suppose that for some $t_0 \in T$, the map φ_{t_0} is a k-contraction. Show that the family $\{\varphi_t\}_{t\in T}$ has a common fixed point.

Problem 4.113*

Suppose that X is a Banach space, $r > 0$, $B_r(0) = \{u \in X : \|u\|_X < r\}$ and $\varphi\colon \overline{B}_r(0) \longrightarrow X$ is a compact map such that

$$\varphi(u) \neq tu \qquad \forall\, u \in \partial B_r(0),\ t > 0.$$

Show that there exists $\widehat{u} \in \overline{B}_r(0)$ such that $\varphi(\widehat{u}) = 0$.

Problem 4.114*

(a) Suppose that X is a Banach space, $r > 0$, $B_r(0) = \{u \in X : \|u\|_X < r\}$ and $\varphi\colon \overline{B}_r(0) \longrightarrow X$ is a compact map. Suppose that for every $u \in \partial B_r(0)$ there exists $u^* = u^*(u) \in X^*$ such that

$$\langle u^*, u\rangle = \|u\|_X \quad \text{and} \quad \langle u^*, \varphi(u)\rangle \geqslant 0.$$

Show that there exists $\widehat{u} \in \overline{B}_r(0)$ such that $\varphi(\widehat{u}) = 0$.

(b) Let $\varphi\colon \mathbb{R}^N \longrightarrow \mathbb{R}^N$ be a continuous map such that

$$\lim_{\|u\|_X \to +\infty} \frac{(\varphi(u), u)_{\mathbb{R}^N}}{|u|} = +\infty.$$

Show that φ is surjective (see also Problem 4.66).

Problem 4.115**

Suppose that X is a Banach space, $C \subseteq X$ is a nonempty, closed, and convex set and $\varphi, \psi \colon C \longrightarrow X$ are two maps such that:

(i) φ is compact;
(ii) ψ is a k-contraction;
(iii) $\varphi(C) + \psi(C) \subseteq C$.

Show that there exists $\widehat{u} \in C$ such that $\varphi(\widehat{u}) + \psi(\widehat{u}) = \widehat{u}$.

Remark. In the literature, this result is known as the ***Krasnoselskii fixed point theorem***.

Problem 4.116**

Let X and V be two Banach spaces with V being reflexive and assume that $V \hookrightarrow X \hookrightarrow V^*$ with both embeddings being compact and dense. Also, let $A \in \mathcal{L}(V;V^*)$ and let $\varphi \colon X \longrightarrow V^*$ be a continuous and bounded (i.e., maps bounded sets to bounded sets) map. Suppose that there exist $c_0 > 0$, $c \geqslant 0$ and $\tau \in [0,2)$ such that

$$\langle A(u), u \rangle \geqslant c_0 \|u\|_V^2 \quad \text{and} \quad \langle \varphi(u), u \rangle \geqslant -c_1(1 + \|u\|_V^\tau) \qquad \forall\, u \in V,$$

with $\langle \cdot, \cdot \rangle$ being the duality brackets for the pair (V^*, V). Show that $A+\varphi$ is surjective.

Problem 4.117**

Suppose that X is a Banach space, $C \subseteq X$ is a closed, convex set, $U \subseteq C$ is a relatively open set, $u_0 \in U$ and $\varphi \colon \overline{U} \longrightarrow C$ is a compact map such that

$$u \neq \lambda\varphi(u) + (1-\lambda)u_0 \qquad \forall\, u \in \partial U,\ \lambda \in (0,1).$$

Show that φ has a fixed point in $\overline{U}$.

Problem 4.118*

Suppose that (X, d_X) is a metric space, $\{F_n \colon X \longrightarrow P_{bf}(X)\}_{n \geqslant 1}$ is a sequence of multivalued contractions with fixed points $\{\widehat{u}_n\}_{n \geqslant 1}$ respectively and $F_0 \colon X \longrightarrow P_{bf}(X)$ is another multivalued contraction such that

$$F_n(u) \xrightarrow{h} F_0(u) \qquad \forall\, u \subset X$$

and the sequence $\{\widehat{u}_n\}_{n \geqslant 1}$ admits a subsequence $\{\widehat{u}_{n_k}\}_{k \geqslant 1}$ such that $\widehat{u}_{n_k} \longrightarrow \widehat{u}$. Show that $\widehat{u} \in F_0(\widehat{u})$.

Problem 4.119***

Suppose that (X, d_X) is a complete metric space, $\{F_n \colon X \longrightarrow P_k(X)\}_{n \geqslant 1}$ is a sequence of multivalued contractions, $\widehat{u}_n \in F_n(\widehat{u}_n)$ for all $n \geqslant 1$ (see Theorem 4.68) and let $F_0 \colon X \longrightarrow P_k(X)$ be another multivalued contraction. Suppose that one of the following conditions holds:

(i) All multifunctions F_n (for $n \geqslant 1$) have the same contraction constant $k \in (0,1)$ and $F_n(u) \xrightarrow{h} F(u)$ for all $u \in X$.

(ii) $\sup_{u \in X} h(F_n(u), F(u)) \longrightarrow 0.$

(iii) (X, d_X) is a locally compact metric space and $F_n(u) \xrightarrow{h} F(u)$ for all $u \in X$.

Show that there exists a subsequence $\{\widehat{u}_{n_k}\}_{k \geqslant 1}$ of $\{\widehat{u}\}_{n \geqslant 1}$ such that $\widehat{u}_{n_k} \longrightarrow \widehat{u}$, with $\widehat{u} \in F_0(\widehat{u})$.

Problem 4.120***

Suppose that X is a locally convex space, $C \subseteq X$ is a nonempty, compact, convex set and $F: C \longrightarrow P_{fc}(C)$ is an upper semicontinuous multifunction. Show that there exists $\widehat{u} \in C$ such that $\widehat{u} \in F(\widehat{u})$.

Problem 4.121**

Suppose that X is a locally convex space, $C \subseteq X$ is a nonempty, compact, and convex set, and $F: C \longrightarrow 2^C \setminus \{\emptyset\}$ is a multifunction with convex values and closed graph. Show that F has a fixed point.

Problem 4.122**

Suppose that X is a locally convex space, $C \subseteq X$ is a nonempty, compact, and convex set and $F: C \longrightarrow P_{fc}(X)$ is an upper semicontinuous multifunction such that $F(u) \cap C \neq \emptyset$ for all $u \in C$. Show that F has a fixed point (that is, there exists $\widehat{u} \in C$ such that $\widehat{u} \in F(\widehat{u})$).

Problem 4.123**

Let X be a Banach space and let $C \subseteq X$ be a nonempty, compact set. Suppose that there exists a family $\{h_u: [0,1] \longrightarrow C\}_{u \in C}$ of functions such that

$$\begin{aligned} h_u(1) &= u \qquad \forall\, u \in C, \\ \|h_u(t) - h_v(t)\|_X &\leqslant \xi(t)\|u - v\|_X \qquad \forall\, u, v \in C, \end{aligned}$$

with $\xi: (0,1) \longrightarrow (0,1)$ and if $t_n \longrightarrow t$ in $[0,1]$, $u_n \longrightarrow u$ in C, then $h_{u_n}(t_n) \longrightarrow h_u(t)$ in C. Show that every nonexpansive map $\varphi: C \longrightarrow C$ has a fixed point.

Problem 4.124**

Suppose that X is a Banach space, $C \subseteq X$ is a compact, star-shaped set with star center y_0 and $\varphi: C \longrightarrow C$ is a nonexpansive map. Show that φ has a fixed point.

Problem 4.125*

Suppose that X is a Hausdorff topological space with the fixed point property (see Definition 4.115) and $C \subseteq X$ is a retract of X. Show that the set C also has the fixed point property.

Problem 4.126***

Let $\mathcal{H}$ be the Hilbert cube, i.e.,

$$\mathcal{H} = \Big\{ h = \{u_k\}_{k \geqslant 1} \in l^2 : |u_k| \leqslant \tfrac{1}{k} \text{ for all } k \geqslant 1 \Big\}.$$

Show that:

(a) $\mathcal{H}$ has the fixed point property (see Definition 4.115);
(b) if X is a Banach space and $C \subseteq X$ is a compact and convex set, then C is isomorphic to a compact, convex subset of $\mathcal{H}$;
(c) any closed, convex subset of $\mathcal{H}$ has the fixed point property.

Problem 4.127*
Suppose that X is a reflexive Banach space, $C \subseteq X$ is a nonempty closed convex set and $\varphi\colon C \longrightarrow C$ is a sequentially weakly continuous map such that $\varphi(C)$ is bounded. Show that φ has a fixed point.

Problem 4.128***
Let X be a Banach space and let $\varphi\colon X \longrightarrow X$ be a sequentially weakly continuous map which is weakly compact (i.e., maps bounded sets to relatively w-compact sets) and $\limsup\limits_{\|u\|_X \to +\infty} \frac{\|\varphi(u)\|_X}{\|u\|_X} < 1$. Show that $I_X + \varphi$ is surjective.

Problem 4.129**
Suppose that X is a Banach space, $\overline{B}_r = \{u \in X : \|u\|_X \leqslant r\}$ and $\varphi\colon \overline{B}_r \longrightarrow X$ is a continuous map such that $\varphi(\overline{B}_r)$ is relatively compact in X.
(a) Show that if for every $u \in X$ with $\|u\|_X = r$, we have

$$\|\varphi(u) - u\|_X^2 \geqslant \|\varphi(u)\|_X^2 - \|u\|_X^2,$$

then φ has a fixed point.
(b) Show that, if $\varphi(\partial B_r) \subseteq \overline{B}_r$, then φ has a fixed point.

Problem 4.130**
Suppose that X is a Banach space, $C \subseteq X$ is a nonempty, closed, convex set, and $\varphi\colon C \longrightarrow X$ is a compact map such that $\varphi(\partial C) \subseteq C$. Show that φ has a fixed point.

Problem 4.131***
Suppose that X is a uniformly convex Banach space, $C \subseteq X$ is a nonempty, bounded, closed, and convex set and $\varphi\colon C \longrightarrow C$ is a nonexpansive, compact map. Given $u_0 \in C$ we define

$$u_{n+1} = \tfrac{1}{2}\big(u_n + \varphi(u_n)\big) \qquad \forall\, n \geqslant 0.$$

Show that there exists $\widehat{u} \in X$ such that $u_n \longrightarrow \widehat{u}$ in X and $\widehat{u} = \varphi(\widehat{u})$.

Problem 4.132**
Suppose that X is a Banach space, $y \in X$, $r > 0$, $\overline{B}_r(y) = \{u \in X : \|u - y\|_X \leqslant r\}$ and $\varphi\colon \overline{B}_r(y) \longrightarrow X$ is a k-Lipschitz map ($k > 0$). Show that for every $\lambda \in \mathbb{R}$ with $|\lambda|k < 1$, the map $I_X - \lambda\varphi$ is injective,

$$\overline{B}_\varrho(y - \lambda\varphi(y)) \subseteq R(I_X - \lambda\varphi),$$

where $\varrho = 1 - |\lambda|kr > 0$ and

$$\|(I_X - \lambda\varphi)^{-1}(u) - (I_X - \lambda\varphi)(v)\|_X \leqslant (1 - |\lambda|k)^{-1}\|u - v\|_X \qquad \forall\, u, v \in R(I_X - \lambda\varphi).$$

Problem 4.133***
Show that a contractive map (see Definition 4.58(c)) on a complete metric space into itself need not have a fixed point.

Problem 4.134*
Suppose that X is a Banach space, $C \subseteq X$ is a nonempty, bounded, closed, and convex set and $\varphi\colon C \longrightarrow C$ is a continuous map which is γ-condensing (with γ being the Kuratowski or Hausdorff measure of noncompactness). Show that the set

$$\operatorname{Fix}(\varphi) = \{u \in C : \varphi(u) = u\}$$

is nonempty and compact.

Problem 4.135**
Suppose that X and Y are two Banach spaces, $C \subseteq X$ is a nonempty, closed, and convex set, $\varphi\colon C \longrightarrow Y$ is a compact map and $\xi\colon C \times \varphi(C) \longrightarrow C$ is a map such that:
(i) the map $u \longmapsto \xi(u, y)$ is a k-contraction for every $y \in \varphi(C)$; and
(ii) the map $y \longmapsto \xi(u, y)$ is continuous on $\varphi(C)$ uniformly with respect to $u \in C$.
Show that there exists $\widehat{u} \in C$ such that $\widehat{u} = \xi(\widehat{u}, \varphi(\widehat{u}))$.

Problem 4.136**
Suppose that X and Y are two Banach spaces, $C \subseteq X$ is a nonempty, closed set, $D \subseteq Y$ is a nonempty set and $\xi\colon C \times D \longrightarrow X$ is a map such that:
(i) for every $y \in D$, the map $u \longmapsto \xi(u, y)$ is continuous and there exists $y_0 \in D$ such that ξ is continuous at every (u, y_0), with $u \in C$;
(ii) there exists a unique $u(y_0) \in C$ such that $u(y_0) = \xi(u(y_0), y_0)$;
(iii) for every subset $C_0 \subseteq C$ with $\gamma(C_0) > 0$ (with γ being the Kuratowski or Hausdorff measure of noncompactness), three exists an open neighborhood V of y_0 such that for any relatively compact set $D_0 \subseteq D \cap V$ we have

$$\gamma(\xi(C_0, D_0)) < \gamma(C_0);$$

(iv) for every $y \in D \cap V$, the equation $u = \xi(u, y)$ has a solution $u(y) \in C$.
Show that $u(y) \longrightarrow u(y_0)$ in X as $y \to y_0$ in Y.

Problem 4.137**
Suppose that X and Y are two Banach spaces, $C \subseteq X$ is a nonempty, closed set, $D \subseteq Y$ is a nonempty set and $\xi\colon C \times D \longrightarrow X$ is a map such that:
(i) for every $y \in D$, the map $u \longmapsto \xi(u, y)$ is continuous, γ-condensing (with γ being the Kuratowski or Hausdorff measure of noncompactness) and there exists $y_0 \in D$ such that the map $y \longmapsto \xi(u, y)$ is continuous at y_0 uniformly for all $u \in C$;
(ii) there exists a unique $u(y_0) \in C$ such that $u(y_0) = \xi(u(y_0), y_0)$;
(iii) there exists an open neighborhood V of y_0 such that for all $y \in D \cap V$, the equation $u = \xi(u, y)$ has a solution $u(y) \in C$.
Show that $u(y) \longrightarrow u(y_0)$ in X as $y \to y_0$ in Y.

Problem 4.138**
Let $A = (a_{ij})_{i,j=1}^n$ be a $n \times n$ matrix with $a_{ij} > 0$ for $i, j \in \{1, \dots, n\}$. Show that A has a strictly positive eigenvalue.

Problem 4.139*
Suppose that X is a vector space, $C \subseteq X$ is a nonempty, convex set and $F\colon C \longrightarrow 2^C$ is a multifunction such that $u \in F(u)$ for all $u \in C$ and the sets $X \setminus F^{-1}(u)$ are convex for all $u \in C$ (here $F^{-1}(u) = \{v \in C\colon\ u \in F(v)\}$). Show that F is a KKM-multifunction (see Definition 4.111).

Problem 4.140*
Suppose that X is a vector space, $C \subseteq X$ is a nonempty convex set and $F\colon C \longrightarrow 2^C \setminus \{\emptyset\}$ is a multifunction such that $u \longmapsto C \setminus F^{-1}(u)$ is not a KKM-multifunction. Show that there exists $\widehat{u} \in C$ such that $\widehat{u} \in \operatorname{conv} F(\widehat{u})$.

Problem 4.141**
Suppose that X is a locally convex space, $C \subseteq X$ is a nonempty, compact, and convex set and $F\colon C \longrightarrow 2^C$ is a multifunction with convex values such that for every $u \in C$, the set $F^{-1}(u)$ is open or F has open values and for every $u \in C$, the set $F^{-1}(u)$ is nonempty convex. Show that F has a fixed point.

Problem 4.142**
Suppose that X is a locally convex space, $C \subseteq X$ is a nonempty, compact, convex set and $\varphi\colon C \times C \longrightarrow \mathbb{R}$ is a function such that:
(i) for all $y \in C$, the function $u \longmapsto \varphi(u, y)$ is lower semicontinuous;
(ii) for all $u \in C$, the function $y \longmapsto \varphi(u, y)$ is concave.
Show that for any $\lambda \in \mathbb{R}$ one of the following properties holds:
(a) There exists $\widehat{u} \in C$ such that $\sup\limits_{y \in C} \varphi(\widehat{u}, y) \leqslant \lambda$;
(b) There exists $\widetilde{y} \in C$ such that $\lambda < \varphi(\widetilde{y}, \widetilde{y})$.

Problem 4.143**
Suppose that X is a locally convex space, $C \subseteq X$ is a nonempty, compact, and convex set and $\varphi\colon C \times C \longrightarrow \mathbb{R}$ is a function such that:
(i) for all $y \in C$, the function $u \longmapsto \varphi(u, y)$ is lower semicontinuous;
(ii) for all $u \in C$, the function $y \longmapsto \varphi(u, y)$ is concave.
Show that there exists $\widehat{u} \in C$ such that

$$\sup_{y \in C} \varphi(\widehat{u}, y) \ \leqslant\ \sup_{y \in C} \varphi(y, y).$$

Problem 4.144**
Suppose that X is a uniformly convex Banach space, $C \subseteq X$ is a nonempty, bounded, closed, and convex set and $\varphi, \psi\colon C \longrightarrow C$ are two maps such that:
(i) $(\varphi + \psi)(C) \subseteq C$;
(ii) φ is nonexpansive;
(iii) ψ is completely continuous.

Show that there exists $\widehat{u} \in C$ such that $\varphi(\widehat{u}) + \psi(\widehat{u}) = \widehat{u}$. (Compare with Problem 4.115.)

Problem 4.145***
Is it possible to extend Theorem 4.70 to k-set-Lipschitz maps (see Definition 2.31) with $k = 1$? Justify your answer. (Compare with Problem 4.133.)

Problem 4.146***
Suppose that (X, d_X) is a complete metric pace and $F, G \colon X \longrightarrow P_{bf}(X)$ are two k-Lipschitz multifunctions with $k \in [0,1)$ (that is, $h(F(u), F(v)) \leqslant k d_X(u,v)$ and $h(G(u), G(v)) \leqslant k d_X(u,v)$ for all $u, v \in X$). Let

$$\operatorname{Fix}(F) = \{u \in X : u \in F(u)\}, \quad \operatorname{Fix}(G) = \{u \in X : u \in G(u)\}.$$

From Theorem 4.68, we know that both sets are nonempty. Show that

$$h(\operatorname{Fix}(F), \operatorname{Fix}(G)) \leqslant \tfrac{1}{1-k} \sup_{u \in X} h(F(u), G(u)).$$

Problem 4.147**
Suppose that (X, d_X) is a complete metric space, $F \colon X \longrightarrow P_{bf}(X)$ is a k-Lipschitz multifunction and $\{F_n \colon X \longrightarrow P_{bf}(X)\}_{n \geqslant 1} \subseteq P_{bf}(X)$ is a sequence of k-Lipschitz multifunctions (with $k \in [0,1)$) such that

$$\sup_{u \in X} h(F_n(u), F(u)) \longrightarrow 0.$$

Also let

$$\operatorname{Fix}(F_n) = \big\{u \in X : u \in F_n(u)\big\} \qquad \forall\, n \geqslant 1$$

and

$$\operatorname{Fix}(F) = \big\{u \in X : u \in F(u)\big\}.$$

Show that $\operatorname{Fix}(F_n) \xrightarrow{h} \operatorname{Fix}(F)$. (Compare with Problem 4.119)

Problem 4.148**
Suppose that H is a Hilbert space, $A \colon H \supseteq D(A) \longrightarrow 2^H$ is a maximal monotone map and $G \colon H \supseteq D(G) \longrightarrow H$ is a monotone map such that the set $D(G) \subseteq H$ is closed and

$$\|G(u) - G(v)\|_H \leqslant k\|u - v\|_H \qquad \forall\, u, v \in D(G),$$

with $k \in (0,1)$. Show that $A + G$ is maximal monotone.

Problem 4.149*
Let $X = c_0 = \big\{\widehat{u} = \{u_k\}_{k \geqslant 1} : u_k \longrightarrow 0\big\}$ and $\overline{B}_1 = \big\{\widehat{u} = \{u_n\}_{n \geqslant 1} \in X : \|\widehat{u}\|_X \leqslant 1\big\}$. Produce an isometry on $\overline{B}_1$ into itself which is fixed point free.

Problem 4.150*
Give an example of a continuous function φ on a complete metric space which is not a k-contraction (with $k \in (0,1)$), but $\varphi^{(n)}$ is a k-contraction for some $n \geqslant 1$.

Problem 4.151*
Suppose that X is a Banach space, $C \subseteq X$ is a nonempty, w-compact, convex set and $\varphi\colon C \longrightarrow C$ is a function. Show that there exists a nonempty, closed, and convex set $\widehat{C} \subseteq C$ such that $\varphi(\widehat{C}) \subseteq \widehat{C}$ (that is, $\widehat{C}$ is φ-invariant) and $\widehat{C}$ is minimal with respect to inclusion.
Remark. If $\widehat{C}$ is a singleton, then φ has a fixed point.

Problem 4.152**
Let X be a compact Hausdorff topological space and let $\varphi\colon X \longrightarrow X$ be a continuous map. Show that there exists $\widehat{\mu} \in M_1^+(X)$ (see Proposition 1.95) such that $\widehat{\mu} = \widehat{\mu}\varphi^{-1}$ (that is, $\widehat{\mu}$ coincides with its image measure under φ; such a measure is usually called ***invariant*** under φ).

Problem 4.153*
Let X be a Hausdorff topological space which is an AR and let $C \subseteq X$ be a retract of X. Show that the set C is an AR too.

Problem 4.154**
Let X be an infinite dimensional Banach space and let $\partial B_r = \{u \in X : \ \|u\|_X = r\}$, with $r > 0$. Show that ∂B_r is an AR.

Problem 4.155*
Let X be a nonempty set and let $B(X)$ be the complete metric space of all bounded functions $\varphi\colon X \longrightarrow \mathbb{R}$ furnished with the supremum metric $d_{B(X)}$ defined by

$$d_{B(X)}(\varphi,\psi) \;=\; \sup_{u\in X} |\varphi(u) - \psi(u)|.$$

Let V be a closed, linear subspace of $B(X)$ which contains the constant functions. Suppose that $S\colon V \longrightarrow V$ is a map such that:
(i) if $\varphi,\psi \in V$ and $\varphi \leqslant \psi$, then $S(\varphi) \leqslant S(\psi)$ (monotonicity of S);
(ii) there exists $\eta \in [0,1)$ such that $S(\varphi + c) \leqslant S(\varphi) + \eta c$ for each $c \in \mathbb{R}$.
Show that S has a unique fixed point.

Problem 4.156**
Let K be a compact metric space and let $\varphi\colon K \longrightarrow K$ be a continuous map such that there exist an AR X and continuous maps $f\colon K \longrightarrow X$ and $g\colon X \longrightarrow K$ such that $\varphi = g \circ f$. Show that φ has a fixed point.

Problem 4.157**
Let X be an AR and let $\varphi\colon X \longrightarrow X$ be a continuous map such that the set $\varphi(X) \subseteq X$ is relatively compact. Show that φ has a fixed point.

Problem 4.158*
Let X be an ANR. Show that any open subset of X is an ANR too.

Problem 4.159**
Let X be a Hausdorff topological space. Show that X is AR if and only if it is a contractible ANR.

Problem 4.160*
Suppose that Z and X are two metric spaces, X is an AR, and $\varphi, \psi \colon Z \longrightarrow X$ are continuous maps. Show that φ and ψ are homotopic.

Problem 4.161*
Suppose that X is a vector space, $D \subseteq X$ is a nonempty set, and $F \colon D \longrightarrow 2^X \setminus \{\emptyset\}$ is a KKM-multifunction. Suppose that C is a convex subset of D. Show that the multifunction $u \longmapsto G(u) = F(u) \cap C$ is also a KKM-multifunction.

Problem 4.162**
Suppose that X is a topological vector space, $D \subseteq X$ is a nonempty subset and $F \colon D \longrightarrow P_f(X)$ is a KKM-multifunction such that $F(u_0) \in P_k(X)$ for some $u_0 \in D$. Show that $\bigcap\limits_{u \in D} F(u) \in P_k(X)$.

Problem 4.163**
Suppose that X is a locally convex space, $C \subseteq X$ is a nonempty, compact, and convex set and $\prec$ is a binary relation on C which has the following properties:
(i) For every $u \in C$, we have $u \notin \operatorname{conv}\{y \in C : u \prec y\}$;
(ii) For every $u \in C$, the set $\{y \in C : u \prec y\}$ is open in C.
Show that the set of the elements of C which are $\prec$-maximal is nonempty and compact.

Problem 4.164**
Suppose that X is a locally convex space, $C \subseteq X$ is a nonempty, compact, and convex set and $\xi^* \colon C \longrightarrow X^*$ is a continuous map. Show that there exists $u_0 \in C$ such that

$$\langle \xi^*(u_0), u - u_0 \rangle \geqslant 0 \qquad \forall\, u \in C.$$

Problem 4.165**
Suppose that X is a locally convex space, $C \subseteq X$ is a nonempty, compact, and convex set and $F \colon C \longrightarrow P_{fc}(X)$ is an upper semicontinuous multifunction such that

$$(F(u) - u) \cap T_C(u) \neq \emptyset \qquad \forall\, u \in C,$$

with $T_C(u)$ being the tangent cone to C at $u \in C$ (see Definition 3.112). Show that F has a fixed point.

Problem 4.166**
Suppose that X is a locally convex space, $C \subseteq X$ is a nonempty, compact, and convex set and $F \colon C \longrightarrow P_{fc}(X)$ is an upper semicontinuous multifunction such that for every $u \in C$, there exist $v \in C$ and $\lambda \geqslant 0$ such that $u + \lambda(v - u) \in F(u)$. Show that F has a fixed point.

Problem 4.167**
Suppose that X is a locally convex space, $C \subseteq X$ is a nonempty, compact, and convex set and $F, G\colon C \longrightarrow P_{fc}(X)$ are two upper semicontinuous multifunctions such that for every $u \in C$ at least one of sets $F(u)$ or $G(u)$ is compact and for every $u \in C$, we can find $y \in F(u)$, $h \in G(u)$ and $\lambda > 0$ such that $u + \lambda(y - h) \in C$. Show that there exists $u \in C$ such that $F(u) \cap G(u) \neq \emptyset$.

Problem 4.168**
Suppose that X is a locally convex space, $C \subseteq X$ is a nonempty, compact, and convex set and $F\colon C \longrightarrow P_{fc}(X)$ is an upper semicontinuous multifunctions such that for every $u \in C$ we can find $y \in F(u)$ and $\lambda > 0$ such that $u + \lambda y \in C$. Show that there exists $u \in C$ such that $0 \in F(u)$.

Problem 4.169**
Suppose that X and Y are two metric spaces, Y is an ANR, $C \subseteq X$ is a nonempty, closed set. Show that every $\varphi \in K(C;Y)$ is compactly extendable to an open set $U \supseteq C$.

Problem 4.170*
Suppose that X and Y are two metric spaces, Y is an ANR, $C \subseteq X$ is a nonempty, closed set and $\varphi, \psi \in K(C;Y)$ are compactly homotopic (see Definition 4.113(b)). Suppose that φ is compactly extendable to $\widehat{\varphi} \in K(X;Y)$ (see Definition 4.113(a)). Show that ψ is also compactly extendable to a $\widehat{\psi} \in K(X;Y)$ and $\widehat{\psi}$ can be chosen in such a way so that the given homotopy of φ to ψ extends to a compact homotopy of $\widehat{\varphi}$ to $\widehat{\psi}$.

Problem 4.171**
Let X be an ANR and let $\varphi \in K(X;X)$ be a compactly nullhomotopic map. Show that φ has a fixed point.

Problem 4.172**
Suppose that X is an ANR, $U \subseteq X$ is a nonempty open set and $\varphi \in K(\overline{U};X)$ is compactly nullhomotopic to a constant map in U and fixed point free on ∂U. Show that φ has a fixed point.

Problem 4.173**
Let X be a Hausdorff topological space and let $C \subseteq X$ be a nonempty closed set. Show that C is a deformation retract of X if and only if C is a retract of X and X is deformable into C.

Problem 4.174*
Let X be an AR and let $C \subseteq X$ be a nonempty closed set. Show that:
(a) if C is an AR, then C is a strong deformation retract;
(b) if C is a deformation retract, then C is an AR.

Problem 4.175**
Is the Hilbert cube $\mathcal{H} = \left\{\widehat{u} = \{u_k\}_{k\geqslant 1} \in l^2 : \ |u_k| \leqslant \frac{1}{k} \text{ for all } k \geqslant 1\right\}$ an AR? Justify your answer.

Problem 4.176**
Let $(X, \preceq)$ be a partially ordered space (see Definition 4.116) and let $\varphi: X \longrightarrow \mathbb{R}$ be a map such that:

(i) if $u \preceq v$ and $u \neq v$, then $\varphi(u) < \varphi(v)$;

(ii) for any increasing sequence $\{u_n\}_{n\geqslant 1} \subseteq X$ such that $\varphi(u_n) \leqslant M < +\infty$ for all $n \geqslant 1$ and some $M > 0$, there exists $\widetilde{u} \in X$ such that $u_n \preceq \widetilde{u}$ for all $n \geqslant 1$;

(iii) for every $u \in X$, the set $\varphi(S(u))$ is bounded above, where $S(u) = \{v \in X : \ u \preceq v\}$.

Show that for every $u \in X$, there exists $\widehat{u} \in S(u)$ such that $S(\widehat{u}) = \{\widehat{u}\}$ (that is, $\widehat{u}$ is $\preceq$-maximal).

Problem 4.177**
Use Problem 4.176 to prove Theorem 4.66 (the Caristi fixed point theorem).

Problem 4.178*
Let $(X, \preceq)$ be a partially ordered set. Show that the following statements are equivalent:

(a) X has a maximal element.

(b) Every multifunction $F: X \longrightarrow 2^X \setminus \{\emptyset\}$ which has the property:

"for every $u \in X$, there exists $y \in F(u)$ such that $u \preceq y$"

admits a fixed point.

Problem 4.179**
Suppose that (X, d_X) is a complete metric space, $\vartheta: \mathbb{R}_+ \longrightarrow \mathbb{R}_+$ is a nondecreasing continuous function such that there exist $c > 0$ and $\delta > 0$ for which we have

$$\vartheta(t) \geqslant ct \qquad \forall\, t \in [0, \delta],$$

$\xi: X \longrightarrow \mathbb{R}$ is lower semicontinuous function which is bounded below and $\varphi: X \longrightarrow X$ is a map which satisfies

$$\vartheta(d_X(u, \varphi(u))) \leqslant \xi(u) - \xi(\varphi(u)) \qquad \forall\, u \in X.$$

Show that φ has a fixed point.

Problem 4.180*
Let X be an OBS with ordered cone K (see Definition 4.124). Suppose that $u_0 \in \operatorname{int} K$. Show that for every $u \in K$, we can find $\lambda_u > 0$ such that $\lambda_u u_0 - u \in K$.

Problem 4.181***
Let X be an OBS with order cone K. Show that the following properties are equivalent:

(a) K is normal.
(b) There exists $c > 0$ such that $\|u+v\|_X \geqslant c\max\{\|u\|_X, \|v\|_X\}$ for all $u, v \in K$.
(c) There exists $c_1 > 0$ such that $\|u\|_X \leqslant c_1\|v\|_X$ for all $u, v \in K$ with $u \leqslant v$.
(d) There exists an equivalent norm $|\cdot|$ on X such that $|u| \leqslant |v|$ for all $u, v \in K$ with $u \leqslant v$.
(e) If $u_n \leqslant y_n \leqslant v_n$ for all $n \geqslant 1$ and $u_n \longrightarrow u$, $v_n \longrightarrow u$ in X, then $y_n \longrightarrow u$ in X.
(f) If $\overline{B}_1 = \{u \in X : \|u\|_X \leqslant 1\}$, then the set $(\overline{B}_1 + K) \cap (\overline{B}_1 - K)$ is bounded.
(g) Every order interval $[u, v] = \{y \in X : u \leqslant y \leqslant v\}$ is bounded.

Problem 4.182***
Let X be an OBS with order cone K. Show that the following implications hold:

$$K \text{ is fully regular} \quad\Longrightarrow\quad K \text{ is regular} \quad\Longrightarrow\quad K \text{ is normal}.$$

Also, show that the reverse implications are not in general true.

Problem 4.183*
Let X be an OBS with order cone K and let $v \in X \setminus K$. Show that there exists $v^* \in K^*$ (see Definition 4.127) such that $\langle v^*, v\rangle < 0$.

Problem 4.184**
Let X be an OBS with order cone K. Suppose that $\operatorname{int} K \neq \emptyset$. Show that $u \in \operatorname{int} K$ if and only if $\langle u^*, u\rangle > 0$ for all $u^* \in K^* \setminus \{0\}$.

Problem 4.185**
Let X be a separable OBS with order cone K. Show that there exists $u^* \in K^*$ such that

$$\langle u^*, u\rangle > 0 \qquad \forall\, u \in K \setminus \{0\}.$$

Problem 4.186*
Let X be an OBS with order cone K. Suppose that $\operatorname{int} K \neq \emptyset$. Show that K is generating.

Problem 4.187*
Let X be an OBS with order cone K and let $\varphi\colon [u_0, v_0] \longrightarrow [u_0, v_0]$ be an increasing map (that is, if $u, v \in [u_0, v_0]$ with $u \leqslant v$, then $\varphi(u) \leqslant \varphi(v)$). Suppose that one of the following conditions is satisfied:
(i) K is regular and φ is continuous;
(ii) K is normal and φ is γ-condensing (with γ being the Kuratowski of Hausdorff measure of noncompactness).
Show that φ has a fixed point.

Problem 4.188**
Let X be an OBS with order cone K and let $\varphi\colon [u_0, v_0] \longrightarrow [u_0, v_0]$ be an increasing map (see Problem 4.187). Suppose that the set $\varphi([u_0, v_0]) \subseteq X$ is relatively compact. Show that φ has a fixed point.

Problem 4.189**
Let X be an OBS with order cone K and let $\varphi\colon [u_0, v_0] \longrightarrow [u_0, v_0]$ be an increasing map (see Problem 4.187). Suppose that K is minihedral and the set $\varphi([u_0, v_0]) \subseteq X$ is relatively compact. Show that φ has extremal fixed points in $[u_0, v_0]$, that is, a maximal fixed point $\widehat{u} \in [u_0, v_0]$ and a minimal fixed point $\widehat{v} \in [u_0, v_0]$.

Problem 4.190*
Suppose that X is an OBS with order cone K, $\Omega \subseteq X$ is a bounded open set with $0 \in \Omega$ and $\varphi\colon \Omega \cap K \longrightarrow K$ is a γ-condensing map (with γ being the Kuratowski or Hausdorff measure of noncompactness) such that

$$\varphi(u) \neq \lambda u \qquad \forall\, u \in \partial(\Omega \cap K) = \partial\Omega \cap K,\ \lambda \geqslant 1.$$

Show that $i(\varphi, \Omega \cap K, K) = 1$.

Problem 4.191***
Suppose that X is an OBS with order cone K, $\Omega \subseteq X$ is a bounded open set and $\varphi\colon \overline{\Omega} \cap K \longrightarrow K$ and $g\colon \partial\Omega \cap K \longrightarrow K$ are two compact maps. Suppose that
(i) $\inf\limits_{u \in \partial\Omega \cap K} \|g(u)\|_X > 0$;
(ii) $u - \varphi(u) \neq \lambda g(u)$ for all $\lambda \geqslant 0$, all $u \in \partial\Omega \cap K = \partial(\Omega \cap K)$.
Show that $i(\varphi, \Omega \cap K, K) = 0$.

Problem 4.192**
Suppose that X is an OBS with order cone K, $\Omega \subseteq X$ is a bounded open set and $\varphi\colon \overline{\Omega} \cap K \longrightarrow K$ is a compact map such that:
(i) $\inf\limits_{u \in \partial\Omega \cap K} \|\varphi(u)\|_X > 0$;
(ii) $\varphi(u) \neq \lambda u$ for all $\lambda \in (0, 1]$, all $u \in \partial\Omega \cap K$.
Show that $i(\varphi, \Omega \cap K, K) = 0$.

Problem 4.193*
Let $(\varphi, \Omega, K) \in \mathcal{Y}_i$ (see Definition 4.134) and suppose that there exists $v \in \Omega$ such that

$$\varphi(u) - u \neq \lambda(u - v) \qquad \forall\, \lambda > 0,\ u \in \partial\Omega.$$

Show that $i(\varphi, \Omega, K) = 1$.

Problem 4.194*
Let $(\varphi, \Omega, K) \in \mathcal{Y}_i$ (see Definition 4.134) and suppose that there exists $v \in K \setminus \overline{\Omega}$ such that

$$\varphi(u) - u \neq \lambda(u - v) \qquad \forall\, \lambda > 0,\ u \in \partial\Omega.$$

Show that $i(\varphi, \Omega, K) = 0$.

Problem 4.195**
Suppose that X is an OBS with order cone K, $C \subseteq X$ is a nonempty, bounded, closed, and convex set, $\Omega_1, \Omega_2 \subseteq C$ are two relatively open sets with $\overline{\Omega}_1 \cap \overline{\Omega}_2 = \emptyset$ and $\varphi\colon C \longrightarrow C$ is a compact map such that:

(i) there exists $v_1 \in \Omega_1$ such that

$$\varphi(u) - u \neq \lambda(u - v_1) \qquad \forall\, \lambda \geqslant 0,\ u \in \partial\Omega_1;$$

(ii) there exists $v_2 \in \Omega_2$ such that

$$\varphi(u) - u \neq \lambda(u - v_2) \qquad \forall\, \lambda \geqslant 0,\ u \in \partial\Omega_2.$$

Show that φ has at least three fixed points $u_1, u_2, u_3 \in C$ such that

$$u_1 \in \Omega_1, \quad u_2 \in \Omega_2 \quad \text{and} \quad u_3 \in C \setminus \overline{\Omega_1 \cup \Omega_2}.$$

Problem 4.196**
Suppose that X is an OBS with order cone K, $\varphi\colon K \longrightarrow K$ is a compact map such that $\varphi(0) = 0$ and there exists $L \in \mathcal{L}(X;X)$ such that

$$\lim_{\substack{h \to 0 \\ h \in K}} \tfrac{\|\varphi(h) - L(h)\|_X}{\|h\|_X} = 0,$$

1 is not an eigenvalue of L corresponding to a positive eigenvector and L has no positive eigenvector for eigenvalues $\lambda > 1$. Show that there exists $r_0 > 0$ such that

$$i(\varphi, K_r, K) = 1 \qquad \forall\, r \in (0, r_0],$$

where $K_r = B_r \cap K$ with $B_r = \{u \in X : \|u\|_X < r\}$.

Problem 4.197**
Let X, K, $\varphi\colon K \longrightarrow K$ and $L \in \mathcal{L}(X,X)$ be as in Problem 4.196. Suppose that 1 is not an eigenvalue of L corresponding to a positive eigenvector and L has a positive eigenvector corresponding to an eigenvalue $\lambda > 1$. Show that there exists $r_0 > 0$ such that

$$i(\varphi, K_r, K) = 0 \qquad \forall\, r \in (0, r_0].$$

Problem 4.198**
Let X be a separable OBS with order cone K. Show that if K is regular and minihedral, then it is strongly minihedral.

Problem 4.199**
Let X be an OBS with order cone K which is strongly minihedral and let $\varphi\colon [u_0, v_0] \longrightarrow [u_0, v_0]$ be an increasing map (see Problem 4.187). Show that φ has a greatest fixed point $\widehat{u}$ and a smallest fixed point $\widehat{v}$ in $[u_0, v_0]$.

Problem 4.200**
Let X be an OBS with order cone K and let $\varphi\colon [u_0, v_0] \longrightarrow [u_0, v_0]$ be an increasing map (see Problem 4.187). Suppose that K is minihedral and the set $\varphi([u_0, v_0]) \subseteq X$ is relatively compact. Show that φ has a greatest fixed point $\widehat{u}$ and a smallest fixed point $\widehat{v}$ in $[u_0, v_0]$.

Problem 4.201**
Suppose that X is an OBS with order cone K, $\Omega \subseteq X$ is a bounded open set and $\varphi\colon \overline{\Omega} \cap K \longrightarrow K$ is a compact map such that:
(i) $\varphi(u) \neq \lambda u$ for all $\lambda \in [0,1]$, all $\partial\Omega \cap K = \partial(\Omega \cap K)$;
(ii) the set $\left\{\frac{\varphi(u)}{\|\varphi(u)\|_X} : u \in \partial\Omega \cap K\right\}$ is relatively compact.
Show that $i(\varphi, \Omega \cap K, K) = 0$.

Problem 4.202**
Suppose that X is an OBS with order cone K which is normal and $\varphi\colon [u_0, v_0] \longrightarrow [u_0, v_0]$ is an increasing map (see Problem 4.187) which is γ-condensing (with γ being the Kuratowski or Hausdorff measure of noncompactness). Show that φ has a greatest fixed point $\widehat{u}$ and a smallest fixed point $\widehat{v}$ in $[u_0, v_0]$ and

$$\widehat{u} \;=\; \lim_{n\to+\infty} v_n \quad \text{and} \quad \widehat{v} \;=\; \lim_{n\to+\infty} u_n$$

where $v_n = \varphi(v_{n-1})$ and $u_n = \varphi(u_{n-1})$ for all $n \geqslant 1$.

Problem 4.203**
Suppose that X is a Banach space, $C \subseteq X$ is a nonempty, bounded, closed, and convex set and $\varphi\colon C \longrightarrow C$ is a weakly continuous map which is ξ-condensing (i.e., $\xi(\varphi(D)) < \xi(D)$ for all $D \subseteq C$ with $\xi(D) > 0$, where ξ denotes the weak measure of noncompactness; see Definition 2.33). Show that the set

$$\operatorname{Fix}(\varphi) \;=\; \{u \in C : \varphi(u) = u\}$$

is nonempty and w-compact.

Problem 4.204**
Suppose that X is a Banach space, $C \subseteq X$ is a nonempty, closed, and convex set and $\varphi\colon C \longrightarrow C$ is a continuous map with the following property:

(W): "For any weakly convergent sequence $\{u_n\}_{n\geq 1} \subseteq C$, the sequence $\{\varphi(u_n)\}_{n\geqslant 1} \subseteq C$ has a strongly convergent subsequence."

Suppose that φ is ξ-condensing (see Problem 4.203) and $\varphi(C) \subseteq C$ is bounded. Show that φ has a fixed point.
Remark. Mappings that satisfy condition (W) need not be weakly continuous.

Problem 4.205***
Suppose that X is a Banach space, $C \subseteq X$ is a nonempty, bounded, closed, and convex set, $\varphi\colon C \longrightarrow X$ and $\psi\colon X \longrightarrow X$ are two maps such that:
(i) φ satisfies condition (W) (see Problem 4.204), it is continuous and

$$\xi(\varphi(D)) \;\leqslant\; k\xi(D) \qquad \forall\, D \subseteq C,$$

with $k \in [0,1)$, where ξ is as in Problem 4.203;
(ii) $\|\psi(u) - \psi(v)\|_X \leqslant \eta(\|u-v\|_X)$ for all $u, v \in X$, with $\eta\colon \mathbb{R}_+ \longrightarrow \mathbb{R}_+$ being continuous, nondecreasing and such that $\eta(t) < (1-k)t$ for all $t > 0$ and if $\{u_n\}_{n\geqslant 1} \subseteq X$ is a weakly convergent sequence, then the sequence $\{\psi(u_n)\}_{n\geqslant 1}$ has a weakly convergent subsequence;

(iii) if $u = \varphi(v) + \psi(u)$ for some $v \in C$, then $u \in C$.
Show that there exists $\widehat{u} \in C$ such that $\widehat{u} = \varphi(\widehat{u}) + \psi(\widehat{u})$.

Problem 4.206**

Let X be a Banach space and let $\varphi \colon X \longrightarrow X$ be a continuous map which is ξ-condensing (see Problem 4.203). Suppose that φ satisfies condition (W) (see Problem 4.204). Show that at least one of the following two properties holds:
(a) φ has a fixed point; or
(b) for some $t \in (0,1)$ the set $\{u \in X : \ u = t\varphi(u)\}$ is unbounded.

Problem 4.207**

Suppose that X is a Banach space, $C \subseteq X$ is a nonempty, bounded, closed, and convex set and $\varphi \colon C \longrightarrow X$ and $\psi \colon X \longrightarrow X$ are two maps such that:
(i) φ is completely continuous and the set $\varphi(C) \subseteq X$ is relatively w-compact;
(ii) ψ is nonexpansive and ξ-condensing (see Problem 4.203);
(iii) $I_X - \psi$ is demiclosed (see Definition 4.77);
(iv) if $t \in (0,1)$ and $u = t\psi(u) + \varphi(v)$ for some $v \in C$, then $u \in C$.
Show that there exists $\widehat{u} \in C$ such that $\widehat{u} = \varphi(\widehat{u}) + \psi(\widehat{u})$.

4.3 Solutions

Solution of Problem 4.1

The set $\overline{\Omega} \subseteq \mathbb{R}^N$ is compact. So, if we show that each h-point (see Definition 4.12) is isolated, then we will have that the set $\varphi^{-1}(h)$ is finite. Arguing by contradiction suppose that we can find a sequence $\{u_n\}_{n\geqslant 1} \subseteq \overline{\Omega}$ such that $u_n \longrightarrow u$ and $\varphi(u_n) = h$ for all $n \geqslant 1$. We have that $u \in \overline{\Omega}$ and $\varphi(u) = h$. Hence

$$0 \;=\; \varphi(u_n) - \varphi(u) \;=\; \big(\varphi'(u), u_n - u\big)_{\mathbb{R}^N} + r(|u_n - u|), \tag{4.12}$$

with $\frac{r(t)}{t} \longrightarrow 0$ as $t \searrow 0$. Since u is a regular point, $\varphi'(u)$ is invertible and so

$$|\varphi'(u)v| \;\geqslant\; c|v| \qquad \forall\, v \in \mathbb{R}^N, \tag{4.13}$$

for some $c > 0$. Form (4.12) we see that for $n \geqslant 1$ large, we have

$$\big|\, \big(\varphi'(u), u_n - u\big)_{\mathbb{R}^N} \,\big| \;\leqslant\; \tfrac{c}{2}|u_n - u|. \tag{4.14}$$

Comparing (4.13) and (4.14) we reach a contradiction. So, each h-point is isolated which in turn implies that the set $\varphi^{-1}(h)$ is finite.

Solution of Problem 4.2

Let $N = \dim X$. Then X is homeomorphic to $\mathbb{R}^N$. We fix a basis on X and consider the linear homeomorphism $\xi\colon X \longrightarrow \mathbb{R}^N$ corresponding to this basis. Let $\widehat{\Omega} = \xi(\Omega) \subseteq \mathbb{R}^N$, $\psi = \xi \circ \varphi \circ \xi^{-1}$ and $\widehat{h} = \xi(h) \in \mathbb{R}^N$. Then $\widehat{\Omega}$ is a bounded open set in $\mathbb{R}^N$ and we define the Brouwer degree on X by setting

$$d(\varphi, \Omega, h) \;=\; d(\psi, \widehat{\Omega}, \widehat{h}). \tag{4.15}$$

In order for this to be a well-defined notion, we need to show that this is independent of the choice of the basis on X. To this end, we consider another basis on X and let $\widetilde{\xi}\colon X \longrightarrow \mathbb{R}^N$ be the linear homeomorphism corresponding to this new basis. We know that there exists an $N \times N$ matrix A with $\det A \neq 0$ such that $\widetilde{\xi} = A\xi$. We set

$$\widetilde{\Omega} \;=\; \widetilde{\xi}(\Omega), \quad \widetilde{\psi} \;=\; \widetilde{\xi} \circ \varphi \circ \widetilde{\xi}^{-1} \quad \text{and} \quad \widetilde{u} \;=\; A(u).$$

If $\varphi \in C^1(\overline{\Omega}; \mathbb{R}^N)$, then

$$J_{\widetilde{\psi}}(\widetilde{u}) \;=\; (\det A^{-1}) J_{\psi}(u) (\det A) \;=\; J_{\psi}(u).$$

So, the degree remains the same in both bases (see (4.15)), provided $\varphi \in C^1(\overline{\Omega}; X)$ and $h \notin \varphi(S_\varphi)$ (see Definition 4.4). By suitable approximation of φ and h, we extend this definition to all $\varphi \in C(\overline{\Omega}; X)$ and $h \notin \varphi(\partial\Omega)$.

Remark. We can be more general and consider two N-dimensional normed spaces X and Y. Let $\Omega \subseteq X$ be a bounded open set, $\varphi \in C(\overline{\Omega}; Y)$ and $y \notin \varphi(\partial\Omega)$. Then we can define $d(\varphi, \Omega, y)$ provided that X and Y are oriented. We call a finite dimensional normed space oriented, provided we consider as admissible only those bases on X for which the transition matrix has a positive determinant.

Solution of Problem 4.3

Let

$$\widehat{h}(t,u) = (1-t)\varphi(u) + t\psi(u) \qquad \forall\, (t,u) \in [0,1] \times \mathbb{R}^N.$$

Then

$$\widehat{h}(t,u) = \varphi(u) = \psi(u) \qquad \forall\, (t,u) \in [0,1] \times \partial\Omega,$$

so $h \notin \widehat{h}([0,1] \times \partial\Omega)$. Thus, from the homotopy invariance of the Brouwer degree (see Theorem 4.11(c)), we have $d(\varphi, \Omega, h) = d(\psi, \Omega, h)$.

Solution of Problem 4.4

For every $t \in [0,1]$ consider the triple $(\varphi - ty, \Omega, h - ty)$. Evidently

$$h - ty \notin (\varphi - ty)(\partial\Omega).$$

Hence

$$(\varphi - ty, \Omega, h - ty) \in \mathcal{Y} \qquad \forall\, t \in [0,1].$$

Thus we can define the map $\vartheta \colon [0,1] \longrightarrow \mathbb{Z}$ be setting

$$\vartheta(t) = d(\varphi - ty, \Omega, h - ty) \qquad \forall\, t \in [0,1].$$

From Theorem 4.11(f), we know that ϑ is continuous, hence it must be constant (being $\mathbb{Z}$-valued). Therefore

$$d(\varphi, \Omega, h) = d(\varphi - y, \Omega, h - y)$$

(as $\vartheta(0) = \vartheta(1)$).

Solution of Problem 4.5

For each $k \in \{1, \ldots, n\}$ we can find $r_k > 0$ such that $\varphi|_{B_{r_k}(u_k)\cap\Omega}$ is a C^1-function, where $B_{r_k}(u_k) = \{u \in \mathbb{R}^N : |u - u_k| < r_k\}$. Let

$$\widehat{\Omega} = \bigcup_{k=1}^{n} (B_{r_k}(u_k) \cap \Omega).$$

This is an open set in Ω which contains all the solutions of the equation $\varphi(u) = h$. Hence the excision property of the Brouwer degree (see Theorem 4.11(e)) implies that

$$d(\varphi, \Omega, h) = d(\varphi, \widehat{\Omega}, h). \tag{4.16}$$

Evidently we can always choose $r_k > 0$ small such that the sets $\{B_{r_k}(u) \cap \Omega\}_{k=1}^{n}$ are pairwise disjoint. Hence by the domain decomposition property of the Brouwer degree (see Theorem 4.11(b)), we have

$$d(\varphi, \widehat{\Omega}, h) = \sum_{k=1}^{n} d(\varphi, B_{r_k}(u_k) \cap \Omega, h). \tag{4.17}$$

From Remark 4.23, we have

$$d(\varphi, B_{r_k}(u_k) \cap \Omega, h) = i(\varphi, u_k, h) = \operatorname{sgn} J_\varphi(u_k) \qquad \forall\, k \in \{1, \ldots, n\}.$$

So, from (4.16) and (4.17), we conclude that

$$d(\varphi, \Omega, h) = \sum_{k=1}^{n} \operatorname{sgn} J_\varphi(u_k).$$

Solution of Problem 4.6

From Problem 4.1 we know that u_0 is an isolated h-point. Let $U \subseteq \Omega$ be an open set such that $u_0 \in U$, $\varphi(u) \neq h$ for all $u \in U \setminus \{u_0\}$. From Definition 4.12(b), we have

$$i(\varphi, u_0, h) = d(\varphi, U, h) = \operatorname{sgn} J_\varphi(u_0) \tag{4.18}$$

(see Definition 4.4). Let $\lambda_1, \ldots, \lambda_N$ be the eigenvalues of $\varphi'(u_0)$. We know that

$$J_\varphi(u_0) = \lambda_1 \ldots \lambda_N.$$

Complex eigenvalues appear in conjugate pairs λ and $\overline{\lambda}$ and so $\lambda\overline{\lambda} > 0$. Therefore

$$J_\varphi(u_0) = (-1)^\mu$$

and so $i(\varphi, u_0, h) = (-1)^\mu$ (see (4.18)).

Solution of Problem 4.7

Let $\widehat{h}(t,u) = \varphi(u) - t\varphi(-u)$ for all $t \in [0,1]$ and all $u \in \Omega$. Then we see that

$$\widehat{h}(t,u) \neq 0 \qquad \forall\ (t,u) \in [0,1] \times \partial\Omega.$$

So, by the homotopy invariance property of the Brouwer degree (see Theorem 4.11(c)), we have

$$d(\varphi, \Omega, 0) = d(\widehat{h}(1,\cdot), \Omega, 0). \tag{4.19}$$

But $\widehat{h}(1,u) = \varphi(u) - \varphi(-u)$, hence $\widehat{h}(1,\cdot)$ is odd. Invoking the Borsuk theorem (see Theorem 4.18), we have that $d(\widehat{h}(1,\cdot), \Omega, 0)$ is odd and thus $d(\varphi, \Omega, 0)$ is odd (see (4.19)).

Solution of Problem 4.8

No. Consider $\Omega = (-2,2) \subseteq \mathbb{R}$, $\varphi \colon \overline{\Omega} \longrightarrow \mathbb{R}$ defined by $\varphi(u) = u^2 - 1$ and $h = 0$. The equation $\varphi(u) = 0$ has two solutions ± 1 in Ω and $h = 0$ is a regular value of φ. Note that

$$\varphi'(1) > 0 \quad \text{and} \quad \varphi'(-1) < 0.$$

So, from Definition 4.4 we have

$$d(\varphi, \Omega, 0) = \operatorname{sgn} \varphi'(1) + \operatorname{sgn} \varphi'(-1) = 1 - 1 = 0.$$

Solution of Problem 4.9

We need to show that given any $u_0 \in \Omega$, we can find $r > 0$ small enough such that

$$B_r(\varphi(u_0)) \subseteq \varphi(\Omega)$$

(where $B_r(\varphi(u_0)) = \{y \in \mathbb{R}^N :\ |y - \varphi(u_0)| < r\}$). If we replace Ω by $\widehat{\Omega} = \Omega - u_0$ and φ by $\widehat{\varphi}(u) = \varphi(u + u_0) - \varphi(u_0)$ for all $u \in \widehat{\Omega}$, we see that without any loss of generality, we may assume that $u_0 = 0$ and $\varphi(0) = 0$.

Since by hypothesis φ is locally injective, we can find $\varrho > 0$ such that $\overline{B}_\varrho(0) \subseteq \Omega$ and $\varphi|_{\overline{B}_\varrho(0)}$ is injective. We consider the homotopy $\widehat{h}(t,u)$ defined by

$$\widehat{h}(t,u) = \varphi(\tfrac{1}{1+t}u) - \varphi(-\tfrac{t}{1+t}u) \qquad \forall\ (t,u) \in [0,1] \times \overline{B}_\varrho(0).$$

Clearly $\widehat{h}$ is continuous and

$$\widehat{h}(0,\cdot) = \varphi(\cdot), \quad \widehat{h}(1,u) = \varphi(\tfrac{1}{2}u) - \varphi(-\tfrac{1}{2}u) \qquad \forall\, u \in \overline{B}_\varrho(0)$$

and so $h(1,\cdot)$ is odd. Suppose that $\widehat{h}(t,u) = 0$ for some $u \in \partial B_\varrho(0)$ and all $t \in [0,1]$. Since φ is an injection on $\overline{B}_\varrho(0)$, we must have

$$\tfrac{1}{1+t}u = \tfrac{t}{1+t}u \qquad \forall\, t \in [0,1],$$

so $u = 0$, a contradiction (since $u \in \partial B_\varrho(0)$). Thus, $\widehat{h}$ is an admissible homotopy and by the homotopy invariance property of the Brouwer degree (see Theorem 4.11(c)), we have

$$d(\varphi, B_\varrho(0), 0) = d(\widehat{h}(1,\cdot), B_\varrho(0), 0). \tag{4.20}$$

Since $\widehat{h}(1,\cdot)$ is odd, from the Borsuk theorem (see Theorem 4.18), we have

$$d(\widehat{h}(1,\cdot), B_\varrho(0), 0) \neq 0,$$

so

$$d(\varphi, B_\varrho(0), 0) \neq 0$$

(see (4.20)) and thus

$$d(\varphi, B_\varrho(0), h) \neq 0 \qquad \forall\, h \in B_\varepsilon(0),$$

with $\varepsilon > 0$ small (see Theorem 4.11(f)). Then the existence property of the Brouwer degree (see Theorem 4.11(d)) implies that

$$B_\varepsilon(0) \subseteq \varphi(B_\varrho(0)) \subseteq \varphi(\Omega),$$

so the set $\varphi(\Omega)$ is open.

Solution of Problem 4.10

From Problem 4.9 we have that

$$\varphi(\operatorname{int} D) \subseteq \operatorname{int} \varphi(D). \tag{4.21}$$

Let $h \in \operatorname{int}\varphi(D)$ and $r > 0$ be such that $B_r(h) \subseteq \varphi(D)$. Let $u \in D$ be such that $h = \varphi(u) \in B_r(h) \subseteq \varphi(D)$. Then $u \in \varphi^{-1}(B_r(h)) \subseteq \varphi^{-1}(\varphi(D)) = D$ (since φ is an injection). The continuity of φ implies that the set $\varphi^{-1}(B_r(h))$ is open and we infer that $u \in \operatorname{int} D$, hence $h \in \varphi(\operatorname{int} D)$. Therefore

$$\operatorname{int}\varphi(D) \subseteq \varphi(\operatorname{int} D),$$

so $\operatorname{int}\varphi(D) = \varphi(\operatorname{int} D)$ (see (4.21)).

Solution of Problem 4.11

From Problem 4.9, we know that $\varphi(\mathbb{R}^N)$ is open. We will show that $\varphi(\mathbb{R}^N)$ is also closed, hence $\varphi(\mathbb{R}^N) = \mathbb{R}^N$. So, let $\{h_n\}_{n\geqslant 1} \subseteq \varphi(\mathbb{R}^N)$ be a sequence such that $h_n \longrightarrow h$. We have $h_n = \varphi(u_n)$ for all $n \geqslant 1$ for some sequence $\{u_n\}_{n\geqslant 1} \subseteq \mathbb{R}^N$. Our hypothesis on φ implies that the sequence $\{u_n\}_{n\geqslant 1} \subseteq \mathbb{R}^N$ is bounded. Hence, passing to a subsequence if necessary, we may assume that $u_n \longrightarrow u$ in $\mathbb{R}^N$ and so $h = \varphi(u)$, that is, $h \in \varphi(\mathbb{R}^N)$ and so $\varphi(\mathbb{R}^N)$ is closed. Therefore $\varphi(\mathbb{R}^N) = \mathbb{R}^N$.

Solution of Problem 4.12

(a) If $0 \in \varphi(\partial\Omega)$, then there exists $u \in \partial\Omega \subseteq \overline{\Omega}$ such that $\varphi(u) = 0$. Otherwise the degree $d(\varphi, \Omega, 0)$ is well defined and nonzero (from the Borsuk theorem; see Theorem 4.18). So, the existence property of the Brouwer degree (see Theorem 4.11(d)) implies that there exists $u \in \Omega$ such that $\varphi(u) = 0$.

(b) Let $\psi(u) = u - \varphi(u)$ for all $u \in \overline{\Omega}$. Evidently $\psi \in C(\overline{\Omega}; \mathbb{R}^N)$ and $\psi|_{\partial\Omega}$ is odd. Therefore by part **(a)** we can find $v \subset \overline{\Omega}$ such that $\psi(v) = 0$. Then $\varphi(v) - v$.

Solution of Problem 4.13

From the existence property of the Brouwer degree (see Theorem 4.11(d)), we know that there exists $u_0 \in \Omega$ such that $\varphi(u_0) = h$. Let V be the connected component of $\mathbb{R}^N \setminus \varphi(\partial\Omega)$ containing h. From Theorem 4.11(f) we know that

$$0 \neq d(\varphi, \Omega, h) = d(\varphi, \Omega, h') \qquad \forall\, h' \in V,$$

so $V \subseteq \varphi(\Omega)$. Since V is an open set, we conclude that $\varphi(\Omega)$ is a neighborhood of h.

Solution of Problem 4.14

Let $V \subsetneq X$ be a dense subspace of X (for example, the kernel of an unbounded linear functional on X). Then $D_V = \partial B_r(0) \cap V$ is dense in $\partial B_r(0)$ and by the Dugundji extension theorem (see Theorem 2.8), $I_{\partial B_r(0)}$ has a continuous extension

$\varphi\colon \overline{B}_r(0) \longrightarrow \partial B_r(0) \cup \overline{\operatorname{conv}} D_V \subseteq \partial B_r(0) \cup (\overline{B}_r(0) \cap V) \subseteq \overline{B}_r(0)$. Choose any $u_0 \in \overline{B}_{\frac{r}{3}}(0) \setminus \varphi(\overline{B}_r(0))$ and let $\varrho = \frac{r}{2}$. We introduce $\xi_0\colon \overline{B}_r(0) \setminus \{u_0\} \longrightarrow \partial B_r(0)$ defined by

$$\xi_0(u) = \begin{cases} \frac{\varrho(u-u_0)+|u-u_0|u_0}{\|\varrho(u-u_0)+|u-u_0|u_0\|_X} & \text{if } 0 < \|u-u_0\|_X < \varrho, \\ \frac{u}{\|u\|_X} & \text{if } \|u-u_0\|_X \geqslant \varrho. \end{cases}$$

Evidently ξ_0 is continuous. Hence $\xi_0 \circ \varphi\colon \overline{B}_r(0) \longrightarrow \partial B_r(0)$ is continuous and $\xi_0 \circ \varphi|_{\partial B_r(0)} = I_{\partial B_r(0)}$. So, $\partial B_r(0)$ is a retract of $\overline{B}_r(0)$.

Solution of Problem 4.15

Arguing by contradiction, suppose that $\partial B_r(0)$ is a retract of $\overline{B}_r(0)$ and let $\xi\colon \overline{B}_r(0) \longrightarrow \partial B_r(0)$ be the retraction (i.e., ξ is continuous and $\xi|_{\partial B_r(0)} = I_{\partial B_r(0)}$). Let $\varphi = -\xi$. Then $\varphi\colon \overline{B}_r(0) \longrightarrow \overline{B}_r(0)$ is a continuous map which has no fixed point, contradicting the Brouwer fixed point theorem (see Theorem 4.81).

Solution of Problem 4.16

(a) Arguing by contradiction, suppose that $\partial B_r(0)$ is contractible. This means that there exists $\widetilde{h} \in C([0,1] \times \partial B_r(0), \partial B_r(0))$ such that

$$\widetilde{h}(0,u) = u \quad \forall\, u \in \partial B_r(0) \quad \text{and} \quad \widetilde{h}(1,u) = u_0, \tag{4.22}$$

for some $u_0 \in \partial B_r(0)$ fixed. Invoking the Tietze extension theorem (see Theorem I.2.138), we can find a continuous extension $\widehat{h}\colon [0,1] \times \overline{B}_r(0) \longrightarrow \mathbb{R}^N$ of $\widetilde{h}$. We set

$$\widehat{\varphi}(\cdot) = \widehat{h}(0,\cdot) \quad \text{and} \quad \widehat{\psi}(\cdot) = \widehat{h}(1,\cdot). \tag{4.23}$$

From the homotopy invariance property of the Brouwer degree (see Theorem 4.11(c)), we have

$$d(\widehat{\varphi}, B_r(0), 0) = d(\widehat{\psi}, B_r(0), 0),$$

so

$$d(I_{\mathbb{R}^N}, B_r(0), 0) = d(u_0, B_r(0), 0)$$

(see (4.22), (4.23) and Problem 4.3). Here by u_0, we mean the constant function $\psi(u) = u_0$ for all $u \in \partial B_r(0)$. Note that $d(I_{\mathbb{R}^N}, B_r(0), 0) \neq 0$ (see the Borsuk theorem; Theorem 4.18), while $d(u_0, \overline{B}_r(0), 0) = 0$ (recall that $u_0 \in \partial B_r(0)$). This leads to a contradiction which proves that $\partial B_r(0)$ is not contractible.

(b) From Problem 4.14 we know that $\partial B_r(0)$ is a retract of $\overline{B}_r(0)$. Let $\xi\colon \overline{B}_r(0) \longrightarrow \partial B_r(0)$ be such a retraction. Then

$$\widehat{h}(t,u) = \xi((1-t)u) \quad \forall\, (t,u) \in [0,1] \times \partial B_r(0)$$

is a deformation which deforms $\partial B_r(0)$ into $\{\xi(0)\}$. This proves that $\partial B_r(0)$ is contractible.

Solution of Problem 4.17

We have $\{t\} \times \partial V_t \subseteq \partial_{[a,b]\times\mathbb{R}^N} V$. Therefore $d(\widehat{h}(t,\cdot), V_t, h)$ is well defined for all $t \in [a,b]$. Let $K = \{(t,u) \in \overline{V} : \widehat{h}(t,u) = h\}$. This is a compact subset of V. So, we can find an open set $W \subseteq [a,b] \times \mathbb{R}^N$ such that $K \subseteq W \subseteq \overline{W} \subseteq V$. Given $t_0 \in [a,b]$, we can find a subinterval $[\gamma,\delta]$ of $[a,b]$ containing t_0 such that $[\gamma,\delta] \times W_{t_0} \subseteq V$ and for all $(t,u) \in \overline{V}$ with $t \in [\gamma,\delta]$, $u \notin W_{t_0}$, we have $\widehat{h}(t,u) = h$. From the excision property of the Brouwer degree (see Theorem 4.11(e)), we have

$$d(\widehat{h}(t,\cdot), V_t, h) = d(\widehat{h}(t,\cdot), W_{t_0}, h). \tag{4.24}$$

Moreover, from the homotopy invariance property of the degree (see Theorem 4.11(c)), we have that the function $t \longmapsto d(\widehat{h}(t,\cdot), W_{t_0}, h)$ is constant on $[\gamma,\delta]$. From (4.24) it follows that the map $t \longmapsto d(\widehat{h}(t,\cdot), V_t, h)$ is locally constant, $\mathbb{Z}$-valued on $[a,b]$, hence it is constant.

Solution of Problem 4.18

Let $\widehat{h}\colon [0,1] \times \overline{\Omega} \longrightarrow \mathbb{R}^N$ be defined by

$$\widehat{h}(t,u) = \varphi(u) + t\psi(u) \qquad \forall\, (t,u) \in [0,1] \times \overline{\Omega}.$$

Evidently $\widehat{h}(\cdot,\cdot)$ is continuous. Suppose that we can find $(t,u) \in [0,1] \times \partial\Omega$ such that

$$\widehat{h}(t,u) = 0,$$

so

$$-t\psi(u) = \varphi(u)$$

and thus

$$t|\psi(u)| = |\varphi(u)| > |\psi(u)|,$$

a contradiction (since $t \in [0,1]$). So, it follows that $0 \notin \widehat{h}(t,\partial\Omega)$ for all $t \in [0,1]$. Then the homotopy invariance of the Brouwer degree (see Theorem 4.11(c)) implies that $d(\varphi + \psi, \Omega, 0) = d(\varphi, \Omega, 0)$.

Solution of Problem 4.19

Let $v \in \Omega$ and consider the map $\widehat{h}\colon [0,1] \times \overline{\Omega} \longrightarrow \mathbb{R}^N$ defined by

$$\widehat{h}(t,u) \;=\; (1-t)(\varphi(u)-v)+t(u-v) \qquad \forall\, (t,u) \in [0,1] \times \overline{\Omega}.$$

Evidently $\widehat{h}(\cdot,\cdot)$ is continuous and for all $(t,u) \in [0,1] \times \partial\Omega$, we have

$$\widehat{h}(t,u) \;=\; (1-t)(u-v)+t(u-v) \;=\; u-v \;\neq\; 0.$$

So, the homotopy invariance of the Brouwer degree (see Theorem 4.11(c)) implies that

$$d(\varphi - v, \Omega, 0) \;=\; d(I_{\mathbb{R}^N} - v, \Omega, 0). \tag{4.25}$$

From Problem 4.4 and the normalization property of the Brouwer degree (see Theorem 4.11(a)), we have

$$d(I_{\mathbb{R}^N} - v, \Omega, 0) \;=\; d(I_{\mathbb{R}^N}, \Omega, v) \;=\; 1,$$

so

$$d(\varphi - v, \Omega, 0) \;\neq\; 0$$

(see (4.25)). The existence property of the Brouwer degree (see Theorem 4.11(d)) implies that we can find $u \in \Omega$ such that $\varphi(u) = v$. Since $v \in \Omega$ is arbitrary and $\varphi(\partial\Omega) = \partial\Omega$, we conclude that $\varphi(\overline{\Omega}) = \overline{\Omega}$.

Solution of Problem 4.20

We assume that $0 \notin \varphi(\partial\Omega)$ or otherwise there is nothing to show. Since by hypothesis $(\varphi(u), u)_{\mathbb{R}^N} \geqslant 0$ for all $u \in \partial\Omega$, we see that for all $u \in \partial\Omega$, $\varphi(u)$ never points opposite to u, that is,

$$\varphi(u) + \lambda u \;\neq\; 0 \qquad \forall\, u \in \partial\Omega,\ \lambda \geqslant 0. \tag{4.26}$$

We consider the continuous map $\widehat{h}\colon [0,1] \times \overline{\Omega} \longrightarrow \mathbb{R}^N$ defined by

$$\widehat{h}(t,u) \;=\; (1-t)\varphi(u) + tu \qquad \forall\, (t,u) \in [0,1] \times \overline{\Omega}.$$

Because of (4.26) we see that $0 \notin \widehat{h}(t, \partial\Omega)$ for all $t \in [0,1]$. Hence from the homotopy invariance of the Brouwer degree (see Theorem 4.11(c)), we have

$$d(\varphi, \Omega, 0) \;=\; d(I_{\mathbb{R}^N}, \Omega, 0). \tag{4.27}$$

The normalization property of the Brouwer degree (see Theorem 4.11(a)) implies that

$$d(I_{\mathbb{R}^N}, \Omega, 0) \;=\; 1,$$

so

$$d(\varphi, \Omega, 0) \;\neq\; 0$$

(see (4.27)). So, the existence property of the Brouwer degree (see Theorem 4.11(d)) implies that there exists $u_0 \in \Omega$ such that $\varphi(u_0) = 0$.

Solution of Problem 4.21

Arguing by contradiction, suppose that there exists a continuous injection $\varphi\colon \mathbb{R}^n \longrightarrow \mathbb{R}^m$. Let $\psi\colon \mathbb{R}^n \longrightarrow \mathbb{R}^n$ be defined by

$$\psi(u) = (\varphi(u), 0) \qquad \forall\, u \in \mathbb{R}^n, \text{ with } 0 \in \mathbb{R}^{n-m}.$$

Clearly ψ is a continuous injection and so from Problem 4.9 we have that the set

$$\psi(\mathbb{R}^n) = \varphi(\mathbb{R}^n) \times \{0\}$$

is open in $\mathbb{R}^n$, a contradiction.

Solution of Problem 4.22

First we show that if for every $k \in \mathbb{N}$, $I_k = [-k, k]$, then the set $\mathbb{R}^m \setminus \varphi(I_k^n)$ is dense in $\mathbb{R}^m$. To show this, we argue by contradiction. So, suppose that we can find $v_0 \in \mathbb{R}^m$ and $r > 0$ such that

$$\overline{B}_r(v_0) \subseteq \varphi(I_k^n),$$

where $\overline{B}_r(v_0) = \{v \in \mathbb{R}^m : |v - v_0| \leqslant r\}$. Let $K = \varphi^{-1}(\overline{B}_r(v_0))$. Then the set $K \subseteq \mathbb{R}^n$ is compact and since φ is injective, $f = \varphi|_K : K \longrightarrow \overline{B}_r(v_0)$ is a homeomorphism. Then $f^{-1}\colon \overline{B}_r(v_0) \longrightarrow K$ is a homeomorphism, which contradicts the Borsuk–Ulam theorem (see Theorem 4.20). This proves that the set $\mathbb{R}^m \setminus \varphi(I_k^n)$ is dense in $\mathbb{R}^m$ for all $k \in \mathbb{N}$.

Next note that $\mathbb{R}^n = \bigcup\limits_{k\geqslant 1} I_k^n$. Hence

$$\varphi(\mathbb{R}^n) = \bigcup_{k\geqslant 1} \varphi(I_k^n),$$

so

$$\mathbb{R}^m \setminus \varphi(\mathbb{R}^n) = \bigcap_{k\geqslant 1} (\mathbb{R}^m \setminus \varphi(I_k^n)).$$

From the first part of the solution, we know that for every $k \geqslant 1$, the set $\mathbb{R}^m \setminus \varphi(I_k^n)$ is dense and open in $\mathbb{R}^m$, so the set

$$\mathbb{R}^m \setminus \varphi(\mathbb{R}^n) = \bigcap_{k\geqslant 1} (\mathbb{R}^m \setminus \varphi(I_k^n))$$

is dense (by the Baire category theorem; see Theorem I.1.26).

Solution of Problem 4.23

Arguing by contradiction suppose that we have

$$\varphi(u) \neq \lambda u \qquad \forall\, u \in \partial\Omega,\ \lambda \in \mathbb{R}. \tag{4.28}$$

We consider the following two homotopies

$$\begin{aligned} \widehat{h}_1(t,u) &= (1-t)\varphi(u) + tu && \forall\, t \in [0,1],\ u \in \overline{\Omega}, \\ \widehat{h}_2(t,u) &= (1-t)\varphi(u) - tu && \forall\, t \in [0,1],\ u \in \overline{\Omega}. \end{aligned}$$

From (4.28) we see that

$$0 \notin \widehat{h}_k(t,\partial\Omega) \qquad \forall\, t \in [0,1],\ k \in \{1,2\}.$$

So, the homotopy invariance of the Brouwer degree (see Theorem 4.11(c)) implies that

$$d(\varphi,\Omega,0) = d(I_{\mathbb{R}^N},\Omega,0) = 1 \tag{4.29}$$

(see the normalization property of the Brouwer degree; Theorem 4.11(a)) and

$$d(\varphi,\Omega,0) = d(-I_{\mathbb{R}^N},\Omega,0) = (-1)^N = -1 \tag{4.30}$$

(see Remark 4.23 and recall that N is odd). Comparing (4.29) and (4.30), we reach a contradiction. So, (4.28) cannot be true and we can find $u \in \partial\Omega$ and $\lambda \neq 0$ such that $\varphi(u) = \lambda u$.

Solution of Problem 4.24

By the Tietze extension theorem (see Theorem I.2.138), we can find a continuous map $\widehat{\varphi}\colon \mathbb{R}^{N+1} \longrightarrow \mathbb{R}^{N+1}$ such that

$$\widehat{\varphi}|_{S^N} = \varphi. \tag{4.31}$$

Note that $N+1$ is odd and $\widehat{\varphi}(S^N) = \varphi(S^N) \subseteq S^N$, hence $0 \notin \widehat{\varphi}(S^N)$. So, we can use Problem 4.23 and find $\widehat{u} \in S^N$ and $\lambda \neq 0$ such that

$$\widehat{\varphi}(\widehat{u}) = \lambda\widehat{u},$$

so

$$\varphi(\widehat{u}) = \lambda\widehat{u}$$

(see (4.31)). We have $1 = |\varphi(\widehat{u})| = |\lambda|$. Since $\varphi(\widehat{u}) \neq -\widehat{u}$, we must have $\lambda = 1$ and so we conclude that $\varphi(\widehat{u}) = \widehat{u}$.

Solution of Problem 4.25

"(a) $\Longrightarrow$ (b)": Note that for a constant map c, we have

$$d_1(c) = 0 \tag{4.32}$$

(see Definition 4.16). So, if φ is not homotopic to a constant map, then

$$d(\widehat{\varphi}, B_1^{N+1}, 0) \neq 0$$

(see (4.32) and Definition 4.16). By the existence property of the Brouwer degree (see Theorem 4.11(d)), we see that we can find $u_0 \in B_1^{N+1}$ such that $\widehat{\varphi}(u_0) = 0$.

"(b) $\Longrightarrow$ (c)": We prove the contrapositive. So, suppose that for some continuous extension $\widehat{\varphi}\colon \overline{B}_1^{N+1} \longrightarrow \mathbb{R}^{N+1}$ we have $d(\widehat{\varphi}, B_1^{N+1}, 0) = 0$. Then by the Hopf theorem (see Theorem 4.15), $\widehat{\varphi}$ is homotopic to a constant map $c \in S^N$. So, we can find a continuous map $\widehat{h}\colon [0,1] \times S^N \longrightarrow S^N$ such that

$$\widehat{h}(0,\cdot) = c \quad \text{and} \quad \widehat{h}(1,\cdot) = \varphi(\cdot). \tag{4.33}$$

Let $\widehat{\varphi}_0\colon \overline{B}_1^{N+1} \longrightarrow S^N$ be defined by

$$\widehat{\varphi}_0(u) = \widehat{h}\big(|u|, \tfrac{u}{|u|}\big) \qquad \forall\, u \in \overline{B}_1^{N+1}. \tag{4.34}$$

Using (4.33), we can easily see that $\widehat{\varphi}_0$ is continuous. We have

$$\widehat{\varphi}_0\big|_{S^N} = \varphi$$

(see (4.33)) and

$$\widehat{\varphi}_0(u) \neq 0 \qquad \forall\, u \in \overline{B}_1^{N+1}.$$

"(c) $\Longrightarrow$ (a)": Again we prove the contrapositive. So, suppose that φ is homotopic to a constant $c \in S^N$. Let $\widehat{h}(t,u)$ be the corresponding homotopy as above (see (4.33)). Again we introduce $\widehat{\varphi}_0$. This is a continuous extension of φ which is homotopic to the constant $c \in S^N$. Hence by Theorem 4.15, we have $d(\widehat{\varphi}_0, B_1^{N+1}, 0) = 0$.

Solution of Problem 4.26

(a) Recall that $d(-I_{\mathbb{R}^{N+1}}, B_1^{N+1}, 0) = (-1)^{N+1}$. So, the hypothesis and the Hopf theorem (see Theorem 4.15) imply that, if

$$\widehat{h}(t,u) = (1-t)\varphi(u) - tu \qquad \forall\, (t,u) \in [0,1] \times S^N,$$

then we can find $t_0 \in [0,1]$ and $u_0 \in S^N$ such that

$$\widehat{h}(t_0, u_0) = 0,$$

so

$$\varphi(u_0) = \tfrac{t_0}{1-t_0} u_0$$

thus

$$1 = \tfrac{t_0}{1-t_0}$$

and hence $t_0 = \frac{1}{2}$. Therefore, we have $\varphi(u_0) = u_0$.

(b) Arguing by contradiction, suppose that $\varphi(u) \neq -u$ for all $u \in S^N$. This means that $-\varphi$ has no fixed point and so by part **(a)** we have $d_1(-\varphi) = (-1)^{N+1}$. But from Definition 4.16, we have $d_1(-\varphi) = (-1)^{N+1} d_1(\varphi)$. Therefore

$$(-1)^{N+1} = (-1)^{N+1} d_1(\varphi),$$

so $d_1(\varphi) = 1$, a contradiction to the hypothesis.

Solution of Problem 4.27

(a) Arguing by contradiction, suppose that $0 \notin \varphi(\partial\Omega)$. By hypothesis $\varphi(\partial\Omega) \subseteq \mathbb{R}^{N-1}$. Using the Tietze extension theorem (see Theorem I.2.138), we can find $\widehat{\varphi} \in C(\overline{\Omega}; \mathbb{R}^{N-1})$ such that $\widehat{\varphi}|_{\partial\Omega} = \varphi$ and $\widehat{\varphi}(\overline{\Omega}) \subseteq \mathbb{R}^{N-1}$. From the Borsuk theorem (see Theorem 4.18) and Problem 4.13, we know that $\widehat{\varphi}(\Omega)$ contains a neighborhood of the origin in $\mathbb{R}^N$, contradicting the fact that $\widehat{\varphi}(\overline{\Omega}) \subseteq \mathbb{R}^{N-1}$. Therefore $0 \in \varphi(\partial\Omega)$ and so we can find $u_0 \in \partial\Omega$ such that $\varphi(u_0) = 0$.

(b) Reasoning indirectly, suppose that there exists an odd map $\varphi \in C(S^n; S^m)$. Then from part **(a)**, we can find $u_0 \in S^n$ such that $\varphi(u_0) = 0$, a contradiction (since φ has values in S^m).

Solution of Problem 4.28

From Problem 4.9 we know that the set $\varphi(\Omega) \subseteq \mathbb{R}^N$ is open and $\varphi\colon \Omega \longrightarrow \varphi(\Omega)$ is a homeomorphism onto $\varphi(\Omega)$. Let $r > 0$ be such that $B_r(h) \subseteq \varphi(\Omega)$ (where $B_r(h) = \{u \in \mathbb{R}^N : |u-h| < r\}$). Then the set $\varphi^{-1}(B_r(h))$ is connected. We consider the composition $\varphi \circ \varphi^{-1}\colon B_r(h) \longrightarrow \varphi(\Omega)$ and apply the multiplication theorem (see Theorem 4.14). Note that by the Jordan separation theorem (see Theorem 4.22), the set $\mathbb{R}^N \setminus \varphi(\partial B_r(h))$ has two connected components Δ and $\widetilde{\Delta}$ with Δ bounded and $\widetilde{\Delta}$ unbounded. So, we have

$$1 = d(\varphi \circ \varphi^{-1}, B_r(h), h) = d(\varphi, \Delta, h) d(\varphi^{-1}, B_r(h), \Delta) \tag{4.35}$$

(see the normalization property of the Brouwer degree; Theorem 4.11(a)). But by the excision property of the Brouwer degree (see Theorem 4.11(e)) and the injectivity of φ, we have

$$d(\varphi, \Omega, h) \;=\; d(\varphi, \Delta, h),$$

so

$$1 \;=\; d(\varphi, \Omega, h) d(\varphi^{-1}, B_r(h), \Delta)$$

(see (4.35)) and thus $d(\varphi, \Omega, h) = \pm 1$.

Solution of Problem 4.29
Let $u, v \in \overline{\Omega}$ with $u \neq v$. We have

$$|\varphi(u) - \varphi(v)| \;=\; |u - v + f(u) - f(v)| \;\geqslant\; |u - v| - |f(u) - f(v)| \;>\; 0,$$

so $\varphi(u) \neq \varphi(v)$ and hence φ is an injection. Then invoking Problem 4.28 we infer that $d(\varphi, \Omega, h) = \pm 1$ for every $h \in \varphi(\Omega)$.

Solution of Problem 4.30
Clearly, we can define $d(\varphi, \Omega, h)$ and $d(\psi, \Omega, h)$. We consider the continuous map $\widehat{h}\colon [0,1] \times \overline{\Omega} \longrightarrow \mathbb{R}^N$ defined by

$$\widehat{h}(t, u) \;=\; (1-t)\varphi(u) + t\psi(u) \qquad \forall\, (t, u) \in [0,1] \times \overline{\Omega}.$$

By hypothesis $h \notin \widehat{h}(t, \partial\Omega)$ for all $t \in [0,1]$. So, the homotopy invariance of the Brouwer degree (see Theorem 4.11(c)) implies that $d(\varphi, \Omega, h) = d(\psi, \Omega, h)$.

Solution of Problem 4.31
Evidently $d(\varphi, \Omega, h)$ exists. Let

$$\psi(u) = \tfrac{\varphi(b) - \varphi(a)}{b - a}(u - a) + \varphi(a) \qquad \forall\, u \in [a, b].$$

If $\varphi(a) = \varphi(b)$, then ψ is constant ($\psi(u) = \varphi(a)$ for all $u \in [a, b]$) and so

$$d(\psi, \Omega, h) \;=\; 0$$

(recall that $h \neq \varphi(a)$). If $\varphi(a) \neq \varphi(b)$, then

$$d(\psi, \Omega, h) = \begin{cases} \operatorname{sgn} \frac{\varphi(b)-\varphi(a)}{b-a} & \text{if } h \in [\varphi(a), \varphi(b)], \\ 0 & \text{if } h \notin [\varphi(a), \varphi(b)] \end{cases} \tag{4.36}$$

(see Definition 4.4). Note that $\psi(a) = \varphi(a)$ and $\psi(b) = \varphi(b)$. Then, from Problem 4.3, we have

$$d(\varphi, \Omega, h) = d(\psi, \Omega, h),$$

so $d(\varphi, \Omega, h) \in \{\pm 1, 0\}$ (see (4.36)).

Solution of Problem 4.32
From Definition 4.16 we have

$$d_1(-I_{S^N}) = (-1)^{N+1} = 1$$

(since N is odd), so

$$d_1(-I_{S^N}) = d_1(I_{S^N})$$

and thus $-I_{S^N}$ and I_{S^N} are homotopic (by the Hopf theorem; see Theorem 4.15).

Solution of Problem 4.33
Let

$$\widehat{h}(t, u) = (1-t)\varphi(u) - ty \qquad \forall\, t \in [0, 1].$$

This is a continuous homotopy and by hypothesis

$$\widehat{h}(t, u) \neq 0 \qquad \forall\, t \in [0, 1],\ u \in \partial\Omega.$$

Then from the homotopy invariance of the Brouwer degree (see Theorem 4.11(c)), we have

$$d(\varphi, \Omega, 0) = d(y, \Omega, 0) = 0$$

(since $y \neq 0$).

Solution of Problem 4.34
Arguing by contradiction, suppose that

$$\inf \left\{ \|\widehat{h}(t,u)\|_X : \ t \in [0,1], \ u \in \partial\Omega \right\} \ = \ 0.$$

Then we can find two sequences $\{t_n\}_{n\geqslant 1} \subseteq [0,1]$ and $\{u_n\}_{n\geqslant 1} \subseteq \partial\Omega$ such that

$$\widehat{h}(t_n,u_n) \ \longrightarrow 0 \quad \text{in } X.$$

Because of the compactness of $\widehat{h}_0(\cdot,\cdot)$ and by passing to a subsequence if necessary, we may assume that

$$\widehat{h}_0(t_n,u_n) \ \longrightarrow \ v_0 \quad \text{in } X.$$

Evidently, we may assume that $t_n \longrightarrow t_0 \in [0,1]$. Then

$$(t_n,u_n) \ \longrightarrow \ (t_0,v_0) \in [0,1] \times \partial\Omega,$$

so

$$\widehat{h}(t_0,v_0) \ = \ \lim_{n\to+\infty} \widehat{h}(t_n,u_n) \ = \ 0,$$

which contradicts our hypothesis that $0 \notin \widehat{h}(t,\partial\Omega)$ for all $t \in [0,1]$. Therefore

$$\inf \left\{ \|\widehat{h}(t,u)\|_X : \ t \in [0,1], \ u \in \partial\Omega \right\} \ > \ 0.$$

Solution of Problem 4.35
Let $\widehat{\varphi}\colon \overline{\Omega} \longrightarrow X$ be such that $\widehat{f} = I_X - \widehat{\varphi} \in K(\overline{\Omega};X)$ with $h, y, \widehat{f}(\overline{\Omega})$ contained in a finite dimensional subspace $X_0 \subseteq X$ and

$$\|\widehat{\varphi}(u) - \varphi(u)\|_X \ < \ \operatorname{dist}(h,\varphi(\partial\Omega)) \qquad \forall\, u \in \overline{\Omega}.$$

From Definition 4.25, we know that

$$d_{LS}(\varphi,\Omega,h) \ = \ d(\widehat{\varphi},\Omega_0,h), \tag{4.37}$$

where $\Omega_0 = \Omega \cap X_0$. Similarly, we have

$$d_{LS}(\varphi - y,\Omega,h-y) \ = \ d(\widehat{\varphi} - y,\Omega_0,h-y). \tag{4.38}$$

But from Problem 4.4 and Remark 4.10, we have

$$d(\widehat{\varphi},\Omega,h) \ = \ d(\widehat{\varphi} - y,\Omega_0,h-y). \tag{4.39}$$

Combining (4.37), (4.38), and (4.39), we have

$$d_{LS}(\varphi, \Omega, h) = d_{LS}(\varphi - y, \Omega, h - y).$$

Solution of Problem 4.36

Let $X_0 = \operatorname{span}\{h, y\}$. Then from Definition 4.25 and Proposition 4.21, we have

$$d_{LS}(I_X - y, \Omega, h) = d((I_X - y)|_{\Omega \cap X_0}, \Omega \cap X_0, h). \tag{4.40}$$

From the normalization property of the Brouwer degree (see Theorem 4.11(a)), we have

$$d((I_X - y)|_{\Omega \cap X_0}, \Omega \cap X_0, h) = \begin{cases} 1 & \text{if } h + y \in \Omega, \\ 0 & \text{if } h + y \notin \overline{\Omega}, \end{cases}$$

so

$$d_{LS}(I_X - y, \Omega, h) = \begin{cases} 1 & \text{if } h + y \in \Omega, \\ 0 & \text{if } h + y \notin \overline{\Omega} \end{cases}$$

(see (4.40)).

Solution of Problem 4.37

Note that

$$(I_X + f \circ \varphi^{-1}) \circ \varphi = \varphi + f = I_X,$$

so $\varphi^{-1} = I_X + f \circ \varphi^{-1}$. We claim that $f \circ \varphi^{-1} \in K(\varphi(\overline{\Omega}); X)$. First we show the continuity of $f \circ \varphi^{-1}$. To this end let $\{y_n\}_{n \geqslant 1} \subseteq \varphi(\overline{\Omega})$ be a sequence such that $y_n \longrightarrow y \in \varphi(\overline{\Omega})$ in X. We can find a sequence $\{u_n\}_{n \geqslant 1} \subseteq \overline{\Omega}$ and $u \in \overline{\Omega}$ such that

$$y_n = \varphi(u_n) \quad \forall\, n \geqslant 1 \quad \text{and} \quad y = \varphi(u).$$

Since the sequence $\{y_n\}_{n \geqslant 1} \subseteq X$ is relatively compact and using Proposition 2.16, we have that the sequence $\{u_n = \varphi^{-1}(y_n)\}_{n \geqslant 1} \subseteq X$ is relatively compact. So, by passing to a suitable subsequence if necessary, we may assume that

$$u_n \longrightarrow \widehat{u} \in \overline{\Omega},$$

so

$$y_n = \varphi(u_n) \longrightarrow \varphi(\widehat{u}) = y = \varphi(u).$$

Since φ is injective, we get $\widehat{u} = u$, so

$$u_n \longrightarrow u \quad \text{in } X,$$

for the original sequence and hence φ^{-1} is continuous.

Also, let $D \subseteq \varphi(\overline{\Omega})$ be a bounded set. Then

$$\varphi^{-1}(D) \subseteq \varphi^{-1}(\varphi(\overline{\Omega})) = \overline{\Omega}$$

(since φ is injective) is bounded. Since f is compact, we have

$$(f \circ \varphi^{-1})(D) \;=\; f(\varphi^{-1}(D)) \;\subseteq\; X$$

is relatively compact. Therefore $\varphi^{-1} \in K(\varphi(\overline{\Omega}); X)$.

Finally note that $\varphi(\Omega) = (\varphi^{-1})^{-1}(\Omega)$ and so from the continuity of φ^{-1}, it follows that the set $\varphi(\Omega) \subseteq X$ is open.

Solution of Problem 4.38

Let $V \subseteq U$ be an open set and let $h \in \varphi(V)$. Let B be a ball such that $\varphi^{-1}(h) \subseteq B \subseteq V$. Then $h \in \varphi(B) \subseteq \varphi(V)$ and from Problem 4.37, we know that the set $\varphi(B) \subseteq X$ is open. Since $h \in \varphi(V)$ is arbitrary, it follows that the set $\varphi(V) \subseteq X$ is open and so φ is an open map. Finally from the solution of Problem 4.37, we know that φ^{-1} is continuous and so we conclude that φ is a homeomorphism from U onto $\varphi(U)$.

Solution of Problem 4.39

From the hypothesis on φ, it follows that given $\varepsilon > 0$, we can find $\delta > 0$ such that

$$\operatorname{diam} \varphi^{-1}(B_\delta(u)) \;<\; \varepsilon \qquad \forall\, u \in X.$$

Then we have

$$\text{if } \|(u-v) - (f(u) - f(v))\|_X < \delta, \text{ then } \|u-v\|_X < \varepsilon. \tag{4.41}$$

Let

$$g(u) \;=\; f(u) - f(0) \qquad \forall\, u \in X$$

and let $\overline{B}_\varepsilon(0) = \{u \in X : \ \|u\|_X \leqslant \varepsilon\}$.

We consider the homotopy $h\colon [0,1] \times \overline{B}_\varepsilon(0) \longrightarrow X$ defined by

$$h(t,u) \;=\; f(\tfrac{u}{1+t}) - f(\tfrac{-tu}{1+t}). \tag{4.42}$$

Suppose that $h(t,u) = u$ for $t \in [0,1]$. Then

$$f(\tfrac{u}{1+t}) - f(\tfrac{-tu}{1+t}) \;=\; u \;=\; \tfrac{u}{1+t} - (\tfrac{-tu}{1+t}),$$

so $\|u\|_X < \varepsilon$ (see (4.41)) and thus

$$0 \notin (I_X - h(t,\cdot))(\partial B_\varepsilon(0)).$$

So, by the homotopy invariance property of the Leray–Schauder degree (see Theorem 4.27(c)), we have

$$d_{LS}(I_X - g, B_\varepsilon(0), 0) \;=\; d_{LS}(I_X - h(1,\cdot), B_\varepsilon(0), 0). \tag{4.43}$$

But from (4.42), it is clear that $I_X - h(1,\cdot)$ is odd. Hence Theorem 4.31 implies that

$$d_{LS}(I_X - h(1,\cdot), B_\varepsilon(0), 0) \;\neq\; 0,$$

so

$$d_{LS}(I_X - g, B_\varepsilon(0), 0) \;\neq\; 0 \tag{4.44}$$

(see (4.43)).

Next, let $h \in X$ with $\|h\|_X < \delta$ and consider the compact homotopy

$$\widehat{h}(t,u) \;=\; g(u) + th \qquad \forall\, (t,u) \in [0,1] \times \overline{B}_\varepsilon(0).$$

Suppose that $\widehat{h}(t,u) = u$. Then

$$\|f(u) - f(0) - u\|_X \;\leqslant\; t\|h\|_X \;<\; \delta,$$

so $\|u\|_X < \varepsilon$ (see (4.41)), thus

$$0 \;\notin\; (I_X - h(t,\cdot))(\partial B_\varepsilon(0)) \qquad \forall\, t \in [0,1].$$

Once again the homotopy invariance property implies that

$$0 \;\neq\; d_{LS}(I_X - g, B_\varepsilon(0), 0) \;=\; d_{LS}(I_X - g + h, B_\varepsilon(0), 0),$$

so $B_\delta(\varphi(0)) \subseteq \varphi(B_\varepsilon(0))$ (by the existence property; see Theorem 4.27(d)). So, for each $h \in \varphi(X)$, we have $B_\delta(h) \subseteq \varphi(X)$ and so φ is surjective.

Solution of Problem 4.40

We have $\varphi = I_X - f$ and $\psi = I_X - g$ with $f, g \in K(\overline{\Omega}; X)$. Let

$$\widehat{h}(t,u) \;=\; (1-t)f(u) + tg(u) \qquad \forall\, (t,u) \in [0,1] \times \overline{\Omega}.$$

This is a compact map and since $\varphi|_{\partial\Omega} = \psi|_{\partial\Omega}$, we see that

$$h \;\notin\; (I_X - \widehat{h}(t,\cdot))(\partial\Omega) \qquad \forall\, t \in [0,1].$$

Hence we can use the homotopy invariance property of the Leray–Schauder degree (see Theorem 4.27(c)) and infer that $d_{LS}(\varphi, \Omega, h) = d_{LS}(\psi, \Omega, h)$.

Solution of Problem 4.41

No. Arguing by contradiction, suppose that we could find a retraction $\xi \colon \overline{B}_r(0) \longrightarrow \partial B_r(0)$ of the form $\xi = I_X - f$ with $f \in K(\overline{B}_r(0); X)$. Since $\xi|_{\partial B_r(0)} = I_{\partial B_r(0)}$, from Problem 4.40, we have

$$d_{LS}(\xi, B_r(0), 0) = d_{LS}(I_X, B_r(0), 0) = 1$$

(see Theorem 4.27(a)). From the existence property of the Leray–Schauder degree (see Theorem 4.27(c)), we see that we can find $u \in B_r(0)$ such that

$$\xi(u) = 0,$$

so $r = \|\xi(u)\|_X = 0$, a contradiction.

Solution of Problem 4.42

The Fréchet differentiability of φ at $u = 0$ implies that

$$\varphi(u) = u - f'(0)u + r(u),$$

with $\frac{r(u)}{\|u\|_X} \longrightarrow 0$ in X as $u \to 0$. Let

$$\widetilde{h}(t, u) = \widetilde{h}_t(u) = u - f'(0)u + tr(u) \qquad \forall\, (t, u) \in [0, 1] \times \overline{\Omega}.$$

Then we have

$$\widetilde{h}_t(u) = 0,$$

so

$$u = (I_X - f'(0))^{-1}(-tr(u)). \tag{4.45}$$

Since by hypothesis $\mu = 1$ is not an eigenvalue of $f'(0)$, we have $(I_X - f'(0))^{-1} \in \mathcal{L}(X)$. Since $r(u) = o(\|u\|_X)$, from (4.45), we see that we can find $\varepsilon > 0$ such that

$$\widetilde{h}_t(u) \neq 0 \qquad \forall\, u \in \overline{B}_\varepsilon(0) \setminus \{0\},\ t \in [0, 1],$$

so

$$i(\varphi, 0, 0) = d_{LS}(\varphi, B_\varepsilon(0), 0) \tag{4.46}$$

(see Definition 4.12 and Remark 4.26). From Problem 3.1, we know that $f'(0) \in \mathcal{L}_c(X)$ and so it follows that $r \in K(\overline{\Omega}; X)$. Therefore $\widetilde{h}_t(u)$ is an admissible homotopy (see Theorem 4.27(c)). Hence, we have

$$d_{LS}(\varphi, B_\varepsilon(0), 0) = d_{LS}(I_X - f'(0), B_\varepsilon(0), 0). \tag{4.47}$$

But from Theorem 4.29 we have

$$d_{LS}(I_X - f'(0), B_\varepsilon(0), 0) = (-1)^m,$$

so $i(\varphi, 0, 0) = (-1)^m$ (see (4.46) and (4.47)).

Solution of Problem 4.43

It suffices to assume that $m = 2$ and the general case follows by induction. Let $\text{proj}_k \colon X \longrightarrow V_k$ (for $k \in \{1, 2\}$) be projection operators. We set

$$S_k = I_X - A \circ \text{proj}_k \qquad \forall\, k \in \{1, 2\}.$$

Then S_1, S_2 are injective and $A \circ \text{proj}_1, A \circ \text{proj}_2 \in \mathcal{L}_c(X)$. We have

$$L = S_1 \circ S_2.$$

From the multiplication theorem (see Theorem 4.30), we have

$$d_{LS}(L, B_1(0), 0) = d_{LS}(S_1, B_1(0), 0) d_{LS}(S_2, B_1(0), 0). \tag{4.48}$$

From the reduction property of the Leray–Schauder degree (see Remark 4.26), we have

$$d_{LS}(S_k, B_1(0), 0) = d_{LS}(S_k|_{\Omega \cap V_k}, B_1(0) \cap V_k, 0) \qquad \forall\, k \in \{1, 2\},$$

so

$$d_{LS}(L, B_1(0), 0) = \prod_{k=1}^{2} d_{LS}(L|_{V_k}, B_1(0) \cap V_k, 0).$$

Solution of Problem 4.44

As in the solution of Problem 4.17, we have that $d_{LS}(\widehat{h}(t, \cdot), V_t, h)$ is defined for all $t \in [a, b]$. Let $\varepsilon > 0$ be such that

$$B_\varepsilon(h) \cap \widehat{h}(\partial_{[a,b]\times X} V) = \emptyset$$

and consider a finite rank map $k_1 \colon \overline{V} \longrightarrow X$ such that $\|k - k_1\|_\infty < \varepsilon$ (see Theorem 2.6). Let Y be a finite dimensional subspace of X such that $k_1(\overline{V}) \subseteq Y$ and $h \in Y$. We set

$$\widehat{h}_1(t, u) = u - k_1(t, u) \qquad \forall\, (t, u) \in \overline{V}.$$

From Problem 4.17, we know that the map

$$t \longmapsto d(\widehat{h}_1(t,\cdot), V_t \cap Y, h) \;=\; d_{LS}(\widehat{h}_1(t,\cdot), V_t \cap Y, h)$$

(see Definition 4.25) is constant on $[a,b]$.

Solution of Problem 4.45

Let $v \in \Omega$ and define

$$\widehat{h}(t,u) \;=\; tf(u) - v \qquad \forall\, t \in [0,1],\ u \in \overline{\Omega}.$$

Evidently this is a compact map. Suppose that for some $(t,u) \in [0,1] \times \partial\Omega$, we have

$$\widehat{h}(t,u) \;=\; u,$$

so $-v = u$ (since $f(u) = 0$), a contradiction (since $v \in \Omega$, $u \in \partial\Omega$). Therefore $\widehat{h}(\cdot,\cdot)$ is an admissible compact homotopy. Hence by the homotopy invariance property of the Leray–Schauder degree (see Theorem 4.27(c)), we have

$$d_{LS}(I_X - f - v, \Omega, 0) \;=\; d_{LS}(I_X - v, \Omega, 0). \tag{4.49}$$

From Problem 4.36, we have

$$d_{LS}(I_X - v, \Omega, 0) \;=\; 1,$$

so $d_{LS}(I_X - f - v, \Omega, 0) = 1$ (see (4.49)). Then from the existence property of the Leray–Schauder degree (see Theorem 4.27(d)), we know that there exists $u \in \Omega$ such that

$$(I_X - f)(u) \;=\; v.$$

Since $(I_X - f)|_{\partial\Omega} = I_{\partial\Omega}$ and $v \in \Omega$ is arbitrary, we conclude that $(I_X - f)(\overline{\Omega}) = \overline{\Omega}$.

Solution of Problem 4.46

Note that $A^{-1} \circ k \colon [0,1] \times \overline{\Omega} \longrightarrow X$ is a compact map. So, if we replace k with $A^{-1} \circ k$ and k with $A^{-1}(h)$, without any loss of generality, we may assume that $X = Y$ and $A = I_X$.

If we can find $u \in \partial\Omega$ such that $u - k(1,u) - h$, then we are done. So, because of our hypothesis, we may assume that

$$u - k(t,u) \;\neq\; h \qquad \forall\, (t,u) \in [0,1] \times \partial\Omega.$$

Invoking the homotopy invariance property of the Leray–Schauder degree (see Theorem 4.27(c)), we have

$$1 = d(I_X, \Omega, h) = d(I_X - k(1,\cdot), \Omega, h)$$

(see Theorem 4.27(a)). So, the existence property (see Theorem 4.27(d)) implies the existence of $u \in \Omega$ such that $u - k(1,u) = h$.

Solution of Problem 4.47

Let $\{R_n\}_{n\geqslant 1} \subseteq \mathbb{R}$ be a sequence such that $R_n > R$ for all $n \geqslant 1$ and $R_n \searrow R$. Using Problem 4.46 with $\Omega = B_{R_n}(0)$, we obtain $u_n \in \overline{B}_{R_n}(0)$ such that

$$A(u_n) - k(1, u_n) = h \qquad \forall\, n \geqslant 1. \tag{4.50}$$

The compactness of k implies that by passing to a subsequence if necessary, we may assume that

$$k(1, u_n) \longrightarrow v \quad \text{in } Y.$$

Then from (4.50) we see that

$$A(u_n) \longrightarrow h + v \quad \text{in } Y,$$

so

$$u_n \longrightarrow A^{-1}(h+v) = u$$

(recall that A is an isomorphism). Evidently $u \in \overline{B}_R(0)$ and from the continuity of A and $k(1,\cdot)$ and from (4.50), we have

$$A(u) - k(1,u) = h.$$

Solution of Problem 4.48

Since A is an isomorphism, we can find $c > 0$ such that

$$c\|u\|_X \leqslant \|A(u)\|_Y \qquad \forall\, u \in X. \tag{4.51}$$

Let $k\colon [0,1]\times X \longrightarrow Y$ be the compact map defined by $k(t,u) = t\varphi(u)$. So, $k(0,u) = 0$ for all $u \in X$. Let $y \in Y$. Using the hypothesis on φ, we see that we can find $R > \|A^{-1}(y)\|_X$ such that

$$\tfrac{\|\varphi(u)\|_Y + \|y\|_Y}{\|u\|_X} < c \qquad \forall\, u \in X,\ \|u\|_X > R. \tag{4.52}$$

If $(t, u) \in (0,1) \times X$ is such that $A(u) - t\varphi(u) = y$, then

$$c\|u\|_X \leqslant \|A(u)\|_Y \leqslant \|\varphi(u)\|_Y + \|y\|_Y,$$

so

$$c \leqslant \frac{\|\varphi(u)\|_Y + \|y\|_Y}{\|u\|_X} \qquad \forall\, u \neq 0,$$

thus $\|u\|_X \leqslant R$ (see (4.52)). So, we can use Problem 4.47 and find $u \in \overline{B}_R(0)$ such that $A(u) - \varphi(u) = y$.

Solution of Problem 4.49

Let

$$\widehat{h}(t, u) = (1-t)\varphi(u) + t\psi(u) \qquad \forall\, (t, u) \in [0,1] \times \overline{\Omega}.$$

This cannot be an admissible homotopy, or otherwise the homotopy invariance property of the Leray–Schauder degree (see Theorem 4.27(c)) would imply that

$$d_{LS}(\varphi, \Omega, 0) = d_{LS}(\psi, \Omega, 0),$$

a contradiction to our hypothesis. Since $\widehat{h}(t, u)$ is not admissible, there exist $(\widehat{t}, \widehat{u}) \in (0,1) \times \partial\Omega$ (we cannot have $\widehat{t} = 0$ or $\widehat{t} = 1$, since $0 \notin \varphi(\partial\Omega)$, $0 \notin \psi(\partial\Omega)$) such that

$$\widehat{h}(\widehat{t}, \widehat{u}) = 0,$$

so $\varphi(\widehat{u}) = \lambda\psi(\widehat{u})$ with $\lambda = -\frac{\widehat{t}}{1-\widehat{t}} < 0$.

Solution of Problem 4.50

Note that in both cases, we have

$$\vartheta \neq d(I_X, \Omega, 0).$$

So, according to Problem 4.49, we can find $\lambda < 0$ and $\widehat{u} \in \partial\Omega$ such that

$$\widehat{u} - f(\widehat{u}) = \lambda\widehat{u},$$

so

$$f(u) = \mu\widehat{u} \quad \text{with } \mu = 1 - \lambda > 1.$$

Solution of Problem 4.51

Let $\varphi = I_X - f$ with $f \in K(\overline{\Omega}; X)$ and consider the compact homotopy $\widehat{h}\colon [0,1] \times \overline{\Omega} \longrightarrow X$ defined by

$$\widehat{h}(t,u) = (1-t)f(u) + tg(u) \qquad \forall\,(t,u) \in [0,1] \times \overline{\Omega}.$$

Then we have

$$u - \widehat{h}(t,u) - h = \psi(u) - h + (1-t)(\varphi(u) - \psi(u)),$$

so

$$\begin{aligned} \|u - \widehat{h}(t,u) - h\|_X &\geqslant \|\psi(u) - h\|_X - \|\varphi(u) - \psi(u)\|_X \\ &> 0 \qquad \forall\, t \in [0,1],\ u \in \partial\Omega, \end{aligned}$$

thus

$$h \notin (I_X - \widehat{h}(t,\cdot))(\partial\Omega) \qquad \forall\, t \in [0,1],$$

hence $(\psi, \Omega, h) \in \mathcal{Y}_0$. So, $\widehat{h}(t,u)$ is admissible and the homotopy invariance property of the Leray–Schauder degree (see Theorem 4.27(c)) implies that

$$d_{LS}(\varphi, \Omega, h) = d_{LS}(\psi, \Omega, h).$$

Solution of Problem 4.52

The set $\varphi(\overline{\Omega})$ is bounded. So, we can find $\varrho > 0$ such that $v \notin \varphi(\overline{\Omega})$ for all $\|v\|_X \geqslant \varrho$. Let $h^* \in R_h^y$ with $\|h^*\|_X \geqslant \varrho$ be such that $d(\varphi, \Omega, h^*) = 0$ (see Theorem 4.27(d)). We define

$$\widehat{h}(t,u) = \varphi(u) - ((1-t)h + th^*) \qquad \forall\,(t,u) \in [0,1] \times \overline{\Omega}.$$

Note that

$$[h, h^*] \subseteq R_h^y,$$

where $[h, h^*] = \{u \in X : u = (1-t)h + th^*,\ t \in [0,1]\}$ and by hypothesis $R_h^y \cap \varphi(\partial\Omega) = \emptyset$. We infer that

$$\widehat{h}(t,u) \neq 0 \qquad \forall\,(t,u) \in [0,1] \times \partial\Omega.$$

therefore $\widehat{h}$ is an admissible homotopy and by the homotopy invariance property (see Theorem 4.27(c)), we have

$$d_{LS}(\varphi - h, \Omega, 0) = d_{LS}(\varphi - h^*, \Omega, 0),$$

so

$$d_{LS}(\varphi, \Omega, h) = d_{LS}(\varphi, \Omega, h^*) = 0$$

(see Problem 4.35).

Solution of Problem 4.53

Since by hypothesis $d_{LS}(\varphi, \Omega, 0) \neq 0$, we see that $S(\varphi) \neq \emptyset$. Also from Proposition 2.16, we know that the set $S(\varphi) \subseteq X$ is compact. Arguing by contradiction, suppose that the set $S(\varphi)$ is not connected. Then we have

$$S(\varphi) = C_1 \cup C_2,$$

with $C_1, C_2 \subseteq X$ nonempty, compact, and disjoint. We can find open sets $\Omega_1, \Omega_2 \subseteq \Omega$ with $\overline{\Omega}_1 \cap \overline{\Omega}_2 = \emptyset$ such that

$$C_1 \subseteq \Omega_1 \quad \text{and} \quad C_2 \subseteq \Omega_2.$$

From the excision and the domain decomposition properties of the Leray–Schauder degree (see Theorem 4.27(e) and (d)), we have

$$d_{LS}(\varphi, \Omega, 0) = d_{LS}(\varphi, \Omega_1, 0) + d_{LS}(\varphi, \Omega_2, 0). \tag{4.53}$$

Suppose that $\varphi(u_1) = u_1 \in C_1$. Given $\varepsilon > 0$, let f_ε be the compact ε-approximation of f postulated by our hypothesis. We define

$$\psi_\varepsilon(u) = u - f_\varepsilon(u) - (u_1 - f_\varepsilon(u_1)) \qquad \forall\, u \in \overline{\Omega} \tag{4.54}$$

and consider the homotopy

$$\widehat{h}(t,u) = (1-t)\varphi(u) + t\psi_\varepsilon(u) \qquad \forall\, (t,u) \in [0,1] \times \overline{\Omega}.$$

Let $m = \big\{\|\varphi(u)\|_X :\ u \in \overline{\Omega} \setminus (\Omega_1 \cup \Omega_2)\big\} > 0$ and let $\varepsilon \in (0, \frac{m}{4})$. We have

$$\begin{aligned} \|\widehat{h}(t,u)\|_X &= \|\varphi(u) + t(f(u) - f_\varepsilon(u)) - t(u_1 - f_\varepsilon(u_1))\|_X \\ &\geqslant \|\varphi(u)\|_X - t\|f(u) - f_\varepsilon(u)\|_X - t\|u_1 - f_\varepsilon(u_1)\|_X \\ &\geqslant \|\varphi(u)\|_X - 2\varepsilon \geqslant m - 2\varepsilon > 0 \qquad \forall\, u \in \overline{\Omega} \setminus (\Omega_1 \cup \Omega_2), \end{aligned}$$

so $\widehat{h}$ is an admissible homotopy. Thus, the homotopy invariance property (see Theorem 4.27(c)) implies that

$$0 \neq d_{LS}(\varphi, \Omega, 0) = d_{LS}(\psi_\varepsilon, \Omega, 0).$$

From (4.54) and our hypothesis, we infer that ψ_ε has only one zero $\widehat{u}$. Since $u_1 \in C_1$, we see that $\widehat{u} \in \Omega_1$ and so it follows that ψ_ε has no zeros on Ω_2. Therefore

$$d_{LS}(\varphi, \Omega_2, 0) = d_{LS}(\psi_\varepsilon, \Omega_2, 0) = 0.$$

If $u_1 \in C_2$, then a similar argument shows that

$$d_{LS}(\varphi, \Omega_1, 0) = d_{LS}(\psi_\varepsilon, \Omega_1, 0) = 0.$$

From (4.53) we conclude that $d_{LS}(\varphi, \Omega, 0) = 0$, a contradiction to our hypothesis.

Solution of Problem 4.54

Arguing by contradiction, suppose that $\varphi(u) \neq 0$ for all $u \in \partial\Omega$. Using the Dugundji extension theorem (see Theorem 2.8), we can find $\widehat{f} \in K(\overline{\Omega}; X)$ such that $\widehat{f}|_{\partial\Omega} = f$. We define

$$\widehat{\varphi} = I_X - \widehat{f} \colon \overline{\Omega} \longrightarrow X.$$

Since $\widehat{\varphi}|_{\partial\Omega} = \varphi$, we have $0 \notin \widehat{\varphi}(\partial\Omega)$ and then by Theorem 4.31, we have that

$$d_{LS}(\widehat{\varphi}, \Omega, 0) \neq 0. \tag{4.55}$$

On the other hand, since $\varphi(\partial\Omega) \subseteq V$, we can find a ray R_0^y (with $y \in X \setminus V$) such that $R_0^y \cap \widehat{\varphi}(\partial\Omega) = \emptyset$. So, invoking Problem 4.52, we have

$$d_{LS}(\widehat{\varphi}, \Omega, 0) = 0. \tag{4.56}$$

Comparing (4.55) and (4.56), we reach a contradiction. This means that there exists $\widehat{u} \in \partial\Omega$ such that $\varphi(\widehat{u}) = 0$.

Solution of Problem 4.55

We introduce

$$\psi(u) = \tfrac{\varphi(u)-\varphi(-u)}{2} = u - \tfrac{f(u)-f(-u)}{2} \qquad \forall\, u \in \partial\Omega.$$

Then ψ satisfies the conditions of Problem 4.54. So, we can find $\widehat{u} \in \partial\Omega$ such that

$$\psi(\widehat{u}) = 0,$$

thus

$$\varphi(\widehat{u}) = \varphi(-\widehat{u}).$$

Solution of Problem 4.56

Let $A = f'(\infty) \in \mathcal{L}_c(X)$. From Definition 4.97(b), we have $\varphi = I_X - A + e$ with

$$\tfrac{\|e(u)\|_X}{\|u\|_X} \longrightarrow 0 \quad \text{as } \|u\|_X \to +\infty. \tag{4.57}$$

Since $A \in \mathcal{L}_c(X)$, for $\Omega \subseteq X$ large, bounded, open with $0 \in \Omega$, we have

$$0 \notin \varphi(\partial\Omega).$$

Let

$$\widehat{h}(t,u) = A(u) + te(u).$$

Because $A(\overline{\Omega}) \subseteq X$ is compact and using (4.57), we see that by taking Ω even bigger if necessary, we have that $\widehat{h}(t,u)$ is a compact homotopy. We have

$$0 \notin (I_X - \widehat{h}(t,\cdot))(\partial\Omega) \qquad \forall\, t \in [0,1].$$

Indeed, suppose that

$$u = \widehat{h}(t,u),$$

for some $t \in [0,1]$ and $u \in \partial\Omega$. So

$$u = A(u) + te(u)$$

and hence

$$c\|u\|_X \leqslant \|(I_X - A)(u)\|_X = t\|e(u)\|_X \leqslant \|e(u)\|_X$$

for some $c > 0$ (since $I_X - A$ is invertible). This contradicts (4.57). So, we can use the homotopy invariance property of the Leray–Schauder degree (see Theorem 4.27(c)) and have

$$d_{LS}(\varphi,\Omega,0) = d_{LS}(I_X - A,\Omega,0). \tag{4.58}$$

But from Theorem 4.29, we have

$$d_{LS}(I_X - A,\Omega,0) = (-1)^m$$

(recall that $A \in \mathcal{L}_c(X)$), so $d_{LS}(\varphi,\Omega,0) = (-1)^m$.

Solution of Problem 4.57

Let D_h be the connected component of $X \setminus \varphi(\partial\Omega)$ containing h. Since by hypothesis $\varphi(\overline{\Omega}) \subseteq V$, we can find $y \in D_h$ such that $y \notin \varphi(\overline{\Omega})$ or otherwise $D_h \subseteq \varphi(\overline{\Omega})$, hence $\operatorname{int} \varphi(\overline{\Omega}) \neq \emptyset$, a contradiction to the hypothesis that V is a proper linear subspace of X. From Theorem 4.27(f), we have

$$d_{LS}(\varphi,\Omega,h) = d_{LS}(\varphi,\Omega,y) = 0$$

(see Theorem 4.27(d)).

Solution of Problem 4.58

Let $y \in X$ and let $I = [y, h] = \{v \in X : v = (1-t)y + th, \ t \in [0,1]\}$. Evidently $I \subseteq X$ is compact and so $\varphi^{-1}(I) \subseteq X$ is compact (see Proposition 2.16), hence it is bounded. Let $r > 0$ be large enough so that $B_r(0) \supseteq \varphi^{-1}(I)$. Then by hypothesis

$$d_{LS}(\varphi, B_r(0), h) \neq 0. \tag{4.59}$$

Let $\xi(t) = (1-t)y + th$ for $t \in [0,1]$. Hence $I = \xi([0,1])$. From Theorem 4.27(c), we have

$$d(\varphi, B_r(0), y) \ = \ d(\varphi, B_r(0), h) \neq 0$$

(see (4.59)). So, from Theorem 4.27(d), we know that there exists $u \in B_r(0)$ such that

$$\varphi(u) \ = \ y.$$

Since $y \in X$ is arbitrary, we conclude that φ is surjective.

Solution of Problem 4.59

We start by showing that there exists $\varepsilon_0 > 0$ such that

$$f_\varepsilon(u) \neq \lambda u \qquad \forall\, (\lambda, u) \in [0,1] \times \partial\Omega, \ \varepsilon \in (0, \varepsilon_0), \tag{4.60}$$

with f_ε being the uniform finite rank ε-approximation of f (that is, $\|f_\varepsilon(u) - f(u)\|_X < \varepsilon$ for all $u \in \overline{\Omega}$; see Theorem 2.6). Suppose that (4.60) is not true. This means that we can find three sequences $\{\varepsilon_n\}_{n\geqslant 1} \subseteq \mathbb{R}$, $\{u_n\}_{n\geqslant 1} \subseteq \partial\Omega$ and $\{\lambda_n\}_{n\geqslant 1} \subseteq [0,1]$ such that $\varepsilon_n \searrow 0$, $\lambda_n \longrightarrow \widehat{\lambda}$ and

$$f_{\varepsilon_n}(u_n) \ = \ \lambda_n u_n \qquad \forall\, n \geqslant 1,$$

so

$$f(u_n) - \lambda_n u_n \ \longrightarrow \ 0 \quad \text{in } X. \tag{4.61}$$

Since by hypothesis $0 \notin \overline{f(\partial\Omega)}$, from (4.61) we infer that $\widehat{\lambda} \neq 0$. Also, from (4.61), it follows that $\{u_n\}_{n\geqslant 1} \subseteq X$ is relatively compact and so we may assume that $u_n \longrightarrow \widehat{u}$ in X. Thus in the limit as $n \to +\infty$, we have

$$f(\widehat{u}) \ = \ \widehat{\lambda}\widehat{u},$$

which contradicts our hypothesis. So, (4.60) holds. By taking $\varepsilon \in (0, \varepsilon_0)$ small from Definition 4.25, we know that

$$d_{LS}(I_X - f, \Omega, 0) \ = \ d(I_X - f_\varepsilon, \Omega \cap V, 0), \tag{4.62}$$

with V being finite dimensional subspace of X such that span $f_\varepsilon(\overline{\Omega}) \subseteq V$. We consider $\widehat{h}_\varepsilon \in C([0,1] \times (\Omega \cap V); V)$ defined by

$$\widehat{h}_\varepsilon(t, u) \ = \ tu - f_\varepsilon(u) \qquad \forall\, (t, u) \in [0,1] \times (\Omega \cap V).$$

Our hypothesis implies that $0 \notin \widehat{h}_\varepsilon(t, \partial(\Omega \cap V))$ for all $t \in [0,1]$. So, by the homotopy invariance of the Brouwer degree (see Theorem 4.11(c)), we have

$$d(I_X - f_\varepsilon, \Omega \cap V, 0) \;=\; d(-f_\varepsilon, \Omega \cap V, 0). \tag{4.63}$$

We can always assume that span $f_\varepsilon(\overline{\Omega})$ is a strict subspace of V. So, for $h \in V$ close to 0, we have

$$d(-f_\varepsilon, \Omega \cap V, 0) \;=\; d(-f_\varepsilon, \Omega \cap V, h) \;=\; 0,$$

therefore

$$d(I_X - f, \Omega, 0) \;=\; 0$$

(see (4.62) and (4.63)).

Solution of Problem 4.60

The set $\varphi^{-1}(0)$ is bounded (see Proposition 2.16). Hence for $r > 0$ big, we have

$$d_{LS}(\varphi, B_r(0), 0) \;\neq\; 0.$$

We can apply Problem 4.58 to conclude that φ is surjective.

Solution of Problem 4.61

We have $\varphi = I_X - f$, with $f\colon \overline{\Omega} \longrightarrow X$ condensing (see Definition 2.31(d)). Let $h \in B_r(0)$ and consider the condensing map $f_0\colon \overline{\Omega} \longrightarrow X$ defined by $f_0(u) = f(u) + h$ and $\varphi_0 = I_X - f_0$. We consider the homotopy $\widehat{h}\colon \overline{\Omega} \times X \longrightarrow X$ defined by

$$\widehat{h}(t,u) \;=\; (1-t)f(u) + tf_0(u) \qquad \forall\, (t,u) \in [0,1] \times \overline{\Omega}.$$

Evidently this is a condensing homotopy. If for some $(t,u) \in [0,1] \times \partial\Omega$ we have that $0 = \widehat{h}(t,u)$, then

$$\|u - f(u)\|_X \;=\; t\|h\|_X \;<\; r \;<\; \inf_{u \in \partial\Omega} \|u - f(u)\|_X$$

(since $\varphi(\partial\Omega) \cap B_r(0) = \emptyset$), a contradiction. So, $\widehat{h}(t,u)$ is an admissible homotopy and by Theorem 4.39, we have

$$d_C(\varphi_0, \Omega, 0) \;=\; d_C(\varphi, \Omega, 0) \;\neq\; 0.$$

Hence there exists $\widehat{u} \in \Omega$ such that $\varphi_0(\widehat{u}) = 0$. So

$$\widehat{u} - f(\widehat{u}) \;=\; h,$$

with $\widehat{u} \in \Omega$, thus $B_r(0) \subseteq \varphi(\Omega)$.

Solution of Problem 4.62

First we show that

$$\varepsilon = \inf_{u \in \partial\Omega} \|u - f(u)\|_X > 0. \tag{4.64}$$

Arguing by contradiction, suppose that we can find a sequence $\{u_n\}_{n\geqslant 1} \subseteq \partial\Omega$ such that

$$\|u_n - f(u_n)\|_X \searrow 0.$$

So, given $\varepsilon > 0$ we can find $n_0 = n_0(\varepsilon) \geqslant 1$ such that

$$u_n \in f(u_n) + \varepsilon B_1(0) \quad \text{and} \quad f(u_n) \in u_n + \varepsilon B_1(0) \qquad \forall\, n \geqslant n_0. \tag{4.65}$$

Then we have

$$\begin{aligned} \gamma(\{u_n\}_{n\geqslant 1}) &= \gamma(\{u_n\}_{n\geqslant n_0}) \leqslant \gamma(\{f(u_n)\}_{n\geqslant n_0}) + \vartheta\varepsilon \\ &= \gamma(\{f(u_n)\}_{n\geqslant 1}) + \vartheta\varepsilon, \end{aligned} \tag{4.66}$$

$$\begin{aligned} \gamma(\{f(u_n)\}_{n\geqslant 1}) &= \gamma(\{f(u_n)\}_{n\geqslant n_0}) \leqslant \gamma(\{u_n\}_{n\geqslant n_0}) + \vartheta\varepsilon \\ &= \gamma(\{u_n\}_{n\geqslant 1}) + \vartheta\varepsilon, \end{aligned} \tag{4.67}$$

with $\vartheta = 1$ or $\vartheta = 2$ (see (4.65)). Since $\varepsilon > 0$ is arbitrary, from (4.66) and (4.67) we have

$$\gamma(\{u_n\}_{n\geqslant 1}) = \gamma(\{f(u_n)\}_{n\geqslant 1}),$$

so

$$\gamma(\{u_n\}_{n\geqslant 1}) = 0$$

(since f is condensing) and hence the sequence $\{u_n\}_{n\geqslant 1} \subseteq \partial\Omega$ is relatively compact. So, by passing to a subsequence if necessary, we may assume that $u_n \longrightarrow u \in \partial\Omega$. Then

$$\|u - f(u)\|_X = 0,$$

so $0 \in \varphi(\partial\Omega)$, a contradiction. Thus, $\varepsilon > 0$ (see (4.64)). Let $m = \sup_{u\in\Omega} \|f(u)\|_X$ and choose $\eta \in (1 - \frac{\varepsilon}{m}, 1)$. We have

$$\|\eta f(u) - f(u)\|_X < \varepsilon \qquad \forall\, u \in \Omega$$

and ηf is a strict γ-contraction. Moreover, via the condensing homotopy

$$\widehat{h}(t,u) = (1-t)f(u) + t\eta f(u) \qquad \forall\, (t,u) \in [0,1] \times \overline{\Omega},$$

we see that f and ηf are homotopic. Exploiting the homotopy invariance property of the degree d_C (see Theorem 4.39), we see that without any loss of generality, we may assume that $f \in SC_\eta(\overline{\Omega}; X)$.

Inductively, we define

$$C_1 = \overline{\text{conv}}\,\big(f(\overline{\Omega}) \cup (-f(\overline{\Omega}))\big)$$

and

$$C_k = \overline{\text{conv}}\,\big(f(\overline{\Omega} \cap C_{k-1}) \cup (-f(\overline{\Omega} \cap C_{k-1}))\big) \qquad \forall\, k \geqslant 2.$$

Then $C = \bigcap\limits_{k \geqslant 1} C_k$ is a nonempty, compact, convex, and symmetric subset of X. So by the Dugundji extension theorem (see Theorem 2.8), C is a retract of X. Let $r\colon X \longrightarrow C$ be a retraction of X onto C. We set

$$r_0(u) = \tfrac{1}{2}(r(u) - r(-u)).$$

Then this is an odd retraction of X onto C. As above, we can check that f and $r_0 \circ f$ are homotopic in $SC_\eta(\overline{\Omega}; X)$ (see Remark 4.35). But $r_0 \circ f$ is compact. Hence

$$d_C(\varphi, \Omega, 0) = d_{LS}(\varphi_0, \Omega, 0),$$

with $\varphi_0 = I_X - r_0 \circ f$. We apply Theorem 4.31, to conclude that $d_C(\varphi, \Omega, 0)$ is odd.

Solution of Problem 4.63

Let $\widehat{h}\colon [0,1] \times \overline{\Omega} \longrightarrow X$ be the homotopy defined by

$$\widehat{h}(t,u) = \tfrac{1}{1+t} f(u) - \tfrac{t}{1+t} f(-u) \qquad \forall\, (t,u) \in [0,1] \times \overline{\Omega}.$$

From the hypothesis on φ, we have

$$0 \notin (I_X - \widehat{h}(t,\cdot))(\partial\Omega) \qquad \forall\, t \in [0,1].$$

Also, for every $D \subseteq [0,1] \times \overline{\Omega}$, we have

$$\widehat{h}(D) \subseteq \overline{\text{conv}}\,\big(f(D) \cup (-f(-D))\big),$$

so $\widehat{h}(t,u)$ is an admissible condensing homotopy (see Remark 4.38). Therefore

$$d_C(\varphi, \Omega, 0) = d_C(I_X - \widehat{h}(1,\cdot), \Omega, 0) \tag{4.68}$$

(see Theorem 4.39). Note that

$$\widehat{h}(1,u) = \tfrac{1}{2}\big(f(u) - f(-u)\big) \qquad \forall\, u \in \overline{\Omega},$$

so $\widehat{h}(1,\cdot)$ is odd, thus $d_C(I_X - \widehat{h}(1,\cdot),\Omega,0)$ is odd (see Problem 4.62) and hence $d_C(\varphi,\Omega,0)$ is odd (see (4.68)).

Solution of Problem 4.64

Arguing by contradiction, suppose that we can find $v \in \partial\Omega$ such that

$$u \neq f(u) + tv \qquad \forall\, u \in \partial\Omega,\ t > 0, \tag{4.69}$$

where $\varphi = I_X - f$, with $f \in S(\overline{\Omega}; X)$. The set $\varphi(\overline{\Omega}) \subseteq X$ is bounded. So, there exists $\vartheta > 0$ such that

$$m = \sup_{u\in\overline{\Omega}} \|\varphi(u)\|_X < \vartheta\|v\|_X.$$

Then we define the homotopy

$$\widehat{h}(t,u) = f(u) + t\vartheta v \qquad \forall\, (t,u) \in [0,1]\times\overline{\Omega}.$$

Evidently $\widehat{h}$ is condensing and $\widehat{h}(t,u) \neq u$ for all $t \in [0,1]$, all $u \in \partial\Omega$ (see (4.69)). So, $\widehat{h}$ is an admissible condensing homotopy (see Remark 4.38) and by Theorem 4.39, we have

$$0 \neq d_C(\varphi,\Omega,0) = d_C(I_X - \widehat{h}(1,\cdot),\Omega,0). \tag{4.70}$$

But $\widehat{h}(1,u) \neq u$ for all $u \in \Omega$. To see this, note that if for some $u \in \Omega$, we have $u = \widehat{h}(1,u)$, then $\varphi(u) = \vartheta u$ and so $\vartheta\|v\|_X \leqslant m$, a contradiction. So, we have

$$d_C(I_X - \widehat{h}(1,\cdot),\Omega,0) = 0. \tag{4.71}$$

Comparing (4.70) and (4.71), we reach a contradiction. This shows that

$$R_0^v \cap \varphi(\partial\Omega) \neq \emptyset \qquad \forall\, v \in \partial\Omega.$$

Solution of Problem 4.65

Arguing by contradiction, suppose that $\varphi(u) \neq \varphi(-u)$ for all $u \in \partial B_1(0)$. Let

$$\psi(u) = \tfrac{1}{2}\big(\varphi(u) - \varphi(-u)\big) \qquad \forall\, u \in \overline{B}_1(0).$$

Then ψ is odd and $0 \notin \psi(\partial B_1(0))$. So, $(\psi, B_1(0), 0) \in \mathcal{Y}_c$ (see Theorem 4.39). From Problem 4.63, we have $d_C(\psi, B_1(0), 0) \neq 0$ and this by Problem 4.64 implies that

$$R_0^v \cap \varphi(\partial\Omega) \neq \emptyset \qquad \forall\, v \in \partial\Omega. \tag{4.72}$$

But, if $v \in (X \setminus V) \cap \partial B_1(0)$, then

$$R_0^v \cap \varphi(\partial\Omega) = \emptyset. \tag{4.73}$$

Comparing (4.72) and (4.73), we reach a contradiction. So, we conclude that there exists $\widehat{u} \in \partial B_1(0)$ such that $\varphi(\widehat{u}) = \varphi(-\widehat{u})$.

Solution of Problem 4.66

Let $h \in \mathbb{R}^N$. From the strong coercivity of φ (see Definition 3.105(b)), we see that we can find $M > 0$ such that

$$(\varphi(u) - h, u)_{\mathbb{R}^N} \geqslant (\varphi(u), u)_{\mathbb{R}^N} - |h||u| \geqslant 0 \qquad \forall\, |u| \geqslant M. \tag{4.74}$$

Let $\varphi_0(u) = \varphi(u) - h$ for $u \in \mathbb{R}^N$. Then from (4.74) and Problem 4.20, we know that there exist $u_0 \in \overline{B}_M(0)$ such that $\varphi_0(u_0) = 0$, hence $\varphi(u_0) = h$. Since $h \in \mathbb{R}^N$ is arbitrary, we conclude that φ is surjective.

Solution of Problem 4.67

Using the Dugundji extension theorem (see Theorem 2.8), we can find a compact map $\widehat{g}\colon \overline{\Omega} \longrightarrow X$ (that is, $\widehat{g} \in K(\overline{\Omega}; X)$) such that $\widehat{g}|_{\partial\Omega} = g$. Let $\widehat{h}\colon [0,1] \times \overline{\Omega} \longrightarrow X$ be the homotopy defined by

$$\widehat{h}(t,u) = (1-t)f(u) + t\widehat{g}(u) \qquad \forall\, (t,u) \in [0,1] \times \overline{\Omega}.$$

This is a condensing homotopy and by hypothesis

$$0 \notin (I_X - \widehat{h}(t,\cdot))(\partial\Omega) \qquad \forall\, t \in [0,1].$$

So, it is admissible in the sense of Remark 4.38. Hence according to Theorem 4.39, we have

$$d_C(I_X - f, \Omega, 0) = d_{LS}(I_X - \widehat{g}, \Omega, 0). \tag{4.75}$$

But the hypothesis on $g = \widehat{g}|_{\partial\Omega}$ implies that we can use Problem 4.59 and have

$$d_{LS}(I_X - \widehat{g}, \Omega, 0) = 0,$$

so

$$d_{LS}(I_X - f, \Omega, 0) = 0.$$

Solution of Problem 4.68

Consider the map $g\colon \partial\Omega \longrightarrow X$ defined by $g(u) = -h$ for all $u \in \partial\Omega$. Evidently $g \in K(\partial\Omega; X)$. Suppose that

$$(1-t)f(u) + tg(u) = u$$

for some $t \in [0,1]$ and some $u \in \partial\Omega$. Evidently the hypotheses on $f \in S(\overline{\Omega}; X)$ and $h \in X$ imply that $t \in (0,1)$. We have

$$f(u) = \tfrac{1}{1-t}u + th = \lambda u - (1-\lambda)h, \quad \text{with } \lambda = \tfrac{1}{1-t} > 1,$$

which contradicts our hypothesis on f. Therefore

$$(1-t)f(u) + tg(u) \neq u \qquad \forall\, (t,u) \in [0,1] \times \partial\Omega.$$

So, we can use Problem 4.67 and conclude that $d_C(I_X - f, \Omega, 0) = 0$.

Solution of Problem 4.69

By hypothesis $\varphi'(u_0) \in \mathcal{L}(\mathbb{R}^N; \mathbb{R}^N)$ is invertible. So, there exists $c > 0$ such that

$$c|u| \leqslant |\varphi'(u_0)u| \qquad \forall\, u \in \mathbb{R}^N. \tag{4.76}$$

Also, there exists $\vartheta_0 > 0$ such that

$$\text{if } |u-v| \leqslant \vartheta_0, \text{ then } |f(u) - f(v) - f'(u_0)(u-v)| \leqslant \tfrac{c}{2}|u-v|. \tag{4.77}$$

So, for all $\vartheta \in (0, \vartheta_0]$, all $t \in [0,1]$ and all $u \in \partial B_\vartheta(u_0)$, we have

$$\begin{aligned} & \big|(1-t)(\varphi(u) - \varphi(u_0)) + t\varphi'(u_0)(u-u_0)\big| \\ = \; & \big|\varphi'(u_0)(u-u_0) + (1-t)\big(\varphi(u) - \varphi(u_0) - \varphi'(u_0)(u-u_0)\big)\big| \\ \geqslant \; & |\varphi'(u_0)(u-u_0)| - |\varphi(u) - \varphi(u_0) - \varphi'(u_0)(u-u_0)| \\ \geqslant \; & \tfrac{c}{2}|u-u_0| = \tfrac{c}{2}\vartheta > 0 \end{aligned}$$

(see (4.76) and (4.77)). Then we have

$$\begin{aligned} & d(\varphi, B_\vartheta(u_0), h) = d(\varphi - \varphi(u_0), B_\vartheta(u_0), 0) \\ = \; & d(\varphi'(u_0)(\cdot - u_0), B_\vartheta(u_0), 0) = d(\varphi'(u_0), B_\vartheta(0), 0) \end{aligned}$$

(see Problem 4.4 and use the homotopy invariance of the Brouwer degree; see Theorem 4.11(c)), so $i(\varphi, u_0, h) = \operatorname{sgn} J_\varphi(u_0)$ (see Definition 4.12(b) and Remark 4.33).

Solution of Problem 4.70

The nondegeneracy hypothesis on h implies that there exists $c > 0$ such that

$$|h(u)| \geqslant c \qquad \forall\, u \in \partial B_1(0). \tag{4.78}$$

From the hypothesis on φ and h we see that there exists $\varepsilon_0 \in (0,1)$ such that

$$|\varphi(u) - h(u)| \leqslant \tfrac{c}{2}|u|^\alpha \qquad \forall\, u \in \overline{B}_{\varepsilon_0}(0). \tag{4.79}$$

Let $\varepsilon \in (0, \varepsilon_0)$, $u \in \overline{B}_\varepsilon(0) \setminus \{0\}$ and $t \in [0,1]$. We have

$$\begin{aligned} |(1-t)\varphi(u) + th(u)| &= |u|^\alpha \big|h(\tfrac{u}{|u|}) - (1-t)\tfrac{h(u)-\varphi(u)}{|u|^\alpha}\big| \\ \geqslant |u|^\alpha\big(|h(\tfrac{u}{|u|})| - \tfrac{|h(u)-\varphi(u)|}{|u|^\alpha}\big) &\geqslant \tfrac{c}{2}|u|^\alpha > 0 \end{aligned}$$

(see (4.78) and (4.79)). Therefore by the homotopy invariance property of the Brouwer degree (see Theorem 4.11(c)), we have

$$d(\varphi, B_\varepsilon(0), 0) = d(h, B_\varepsilon(0), 0),$$

so $i(\varphi, 0, 0) = i(h, 0, 0)$ (see Definition 4.12(b)).

Solution of Problem 4.71

Since h is nondegenerate, we can find $c > 0$ such that

$$|h(u)| \geqslant c \qquad \forall\, u \in \partial B_1(0). \tag{4.80}$$

Let $y \in \mathbb{R}^N$. From the assumption on φ and h, we see that we can find $R_0 \geqslant 1$ such that

$$|\varphi(u) - y - h(u)| \leqslant \tfrac{c}{2}|u|^\alpha \qquad \forall\, |u| \geqslant R_0. \tag{4.81}$$

Then for all $u \in \mathbb{R}^N$ with $|u| \geqslant R_0$ and all $t \in [0,1]$, we have

$$\begin{aligned} \big|(1-t)(\varphi(u) - y) + th(u)\big| &= |u|^\alpha \big|h(\tfrac{u}{|u|}) - (1-t)\big(\tfrac{h(h)-\varphi(u)}{|u|^\alpha}\big)\big| \\ \geqslant |u|^\alpha\big(|h(\tfrac{u}{|u|})| - \tfrac{|h(u)-\varphi(u)|}{|u|^\alpha}\big) &\geqslant \tfrac{c}{2}|u|^\alpha \geqslant \tfrac{c}{2}R_0 > 0 \end{aligned}$$

(because of the homogeneity of h and using (4.80) and (4.81)). So, for all $R \geqslant R_0$, we have

$$d(\varphi, B_R(0), y) = d(\varphi - y, B_R(0), 0) = d(h, B_R(0), 0)$$

(see Problem 4.4 and use the homotopy invariance property of the Brouwer degree (see Theorem 4.11(c)), so

$$i_\infty(\varphi, y) = i_\infty(h, 0)$$

(see Definition 4.12(c)).

Solution of Problem 4.72

From Problem 4.17, we know that the map $t \longmapsto d(\varphi(t,\cdot), V_t, h)$ is constant, hence it is different from zero for all $t \in [a,b]$. Evidently the set S is compact and for every $t \in [a,b]$, the set $\{t\} \times S_t \subseteq S$ is closed. It follows that the sets

$$K_1 = (\{a\} \times V_a) \cap S \quad \text{and} \quad K_2 = (\{b\} \times V_b) \cap S$$

are nonempty, disjoint, closed subsets of S. Arguing by contradiction, suppose that no component of S intersects both $\{a\} \times S_a$ and $\{b\} \times S_b$. Then invoking the Whyburn separation lemma (see Theorem 4.142), we can find compact sets $\widetilde{K}_1$ and $\widetilde{K}_2$ such that

$$K_1 \subseteq \widetilde{K}_1, \quad K_2 \subseteq \widetilde{K}_2, \quad S = \widetilde{K}_1 \cup \widetilde{K}_2 \quad \text{and} \quad \widetilde{K}_1 \cap \widetilde{K}_2 = \emptyset. \tag{4.82}$$

Let $\varepsilon = \frac{1}{2} \min \{ |v_1 - v_2| : v_1 \in \widetilde{K}_1, \ v_2 \in \widetilde{K}_2 \} > 0$ and

$$U = \{ (t,u) \in V : \operatorname{dist}((t,u), \widetilde{K}_1) < \varepsilon \}.$$

The set U is bounded open and from the choice of $\varepsilon > 0$ and (4.82), we see that $U \cap K_2 = \emptyset$ and $\partial U \cap S = \emptyset$. Then from Problem 4.17, we have

$$0 \neq d(\varphi(a,\cdot), U_a, h) = d(\varphi(b,\cdot), U_b, h). \tag{4.83}$$

But note that $U_b = \emptyset$. So, $d(\varphi(b,\cdot), U_b, h) = 0$, a contradiction (see (4.83)).

Solution of Problem 4.73

We consider the gradient system

$$v'(t) = -\nabla \varphi(v(t)). \tag{4.84}$$

For every $\lambda \leqslant \eta$, the set $\overline{\varphi^\lambda} = \{u \in \Omega : \varphi(u) \leqslant \lambda\}$ is compact and positively invariant for the flow of (4.84). The compactness is evident from the inclusion $\overline{\varphi^\lambda} \subseteq \overline{\varphi^\eta}$. For the positive invariance let $v(\cdot, x)$ be the unique local solution of (4.84) emanating from x (that is, $v(0, x) = x$). We have

$$\begin{aligned} \tfrac{d}{dt}\varphi(v(t,x)) &= \big(\nabla\varphi(v(t,x)), v_t'(t,x)\big)_{\mathbb{R}^N} \\ &= -|\nabla\varphi(v(t,x))|^2 \leqslant 0 \end{aligned} \tag{4.85}$$

(see (4.84)). Hence, if $x \in \overline{\varphi^\lambda}$, then from (4.85), we see that

$$\varphi(v(t,x)) \leqslant \varphi(v(0,x)) = \varphi(x) \leqslant \lambda \qquad \forall\, t \in [0, \tau_+(x)),$$

with $\tau_+(x)$ being the maximal time of existence of the local solution v. We have

$$v(t,x) \subseteq \overline{\varphi^\lambda} \qquad \forall\, t \in [0, \tau_+(x))$$

and since $\overline{\varphi^\lambda}$ is compact, we infer that $\tau_+(x) = +\infty$ (that is, the solution is global). If $x \in \partial\varphi^\eta$ (that is, $\varphi(x) = \eta$), then

$$\tfrac{d}{dt}\varphi(v(t,x))\Big|_{t=0} = -|\nabla\varphi(x)|^2 < 0$$

(see (4.85) and the hypotheses), so

$$v(t,x) < v(x) \qquad \forall\, t > 0.$$

Therefore

$$d(I_{\mathbb{R}^N} - v(t,\cdot), \varphi^\eta, 0) = (-1)^N d(-\nabla\varphi, \varphi^\eta, 0) = d(\nabla\varphi, \varphi^\eta, 0) \tag{4.86}$$

(see the solution of Problem 4.69). We introduce

$$m = \min\big\{|\nabla\varphi(x)|^2 : x \in \overline{\varphi^\eta} \setminus \varphi^\vartheta\big\} > 0$$

(see the hypotheses).

If $x \in \partial\varphi^\eta$, then from the positive invariance of $\overline{\varphi^\eta}$, we have $v(t,x) \in \overline{\varphi^\eta}$ for all $t \geqslant 0$. If $v(t,x) \in \overline{\varphi^\eta} \setminus \varphi^\vartheta$ for all $t \in [0, \tau]$, then

$$\begin{aligned} \eta - \vartheta \geqslant \varphi(x) - \varphi(v(\tau,x)) &= \int_\tau^0 \tfrac{d}{dt}\varphi(v(t,x))\,dt \\ &= \int_0^\tau |\nabla\varphi(v(t,x))|^2\,dt \geqslant m\tau, \end{aligned}$$

so $\tau \leqslant \frac{\eta-\vartheta}{m}$. Thus, for $t > \frac{\eta-\vartheta}{m}$, we have $v(t,x) \in \varphi^\vartheta$ and so by hypothesis

$$v(t,x) \in B_r(u_0).$$

Therefore, for every $s \in [0,1]$ and every $x \in \partial\varphi^\eta$, we have

$$\begin{aligned} \big|(1-s)(x-u_0) + s(x - v(\tau,x))\big| &= \big|x - u_0 - t(v(\tau,x) - u_0)\big| \\ &\geqslant |x - v_0| - |v(\tau,x) - u_0| > 0 \end{aligned}$$

(for $x \in \partial\varphi^\eta$, we have $|x - u_0| > r$). By the homotopy invariance property of the Brouwer degree (see Theorem 4.11(c)), we have

$$d(I_{\mathbb{R}^N} - v(\tau, \cdot), \varphi^\eta, 0) \;=\; d(I_{\mathbb{R}^N} - u_0, \varphi^\eta, 0) \;=\; d(I_{\mathbb{R}^N}, \varphi^\eta, u_0) \;=\; 1$$

(see Problem 4.4 and the homotopy invariance of the Brouwer degree; Theorem 4.11(c)), so $d(\nabla\varphi, \varphi^\eta, 0) = 1$ (see (4.86)).

Solution of Problem 4.74

Let $\vartheta = \max\{\varphi(u) : |u| \leqslant r_0\}$. The coercivity of φ implies that φ^ϑ is bounded and so we can find $r > 0$ such that

$$\overline{\varphi^\vartheta} \;\subseteq\; \overline{B}_r(0).$$

Let $\eta > \max\{\varphi(u) : |u| \leqslant r\}$. Then with $u_0 = 0$, we can use Problem 4.73 and have that

$$d(\nabla\varphi, \varphi^\eta, 0) \;=\; 1. \tag{4.87}$$

Since by hypothesis $\nabla\varphi(u) \neq 0$ for all $|u| \geqslant r_0$, from the excision property of the Brouwer degree (see Theorem 4.11(e)) and (4.87), we have

$$d(\nabla\varphi, B_r(0), 0) = 1 \qquad \forall\, r \geqslant r_0.$$

Solution of Problem 4.75

By replacing φ by $\widehat{\varphi}(u) = \varphi(u + u_0) - \varphi(u_0)$, we may assume without any loss of generality that $u_0 = 0$ and $\varphi(0) = 0$. Also, we can take $\Omega = B_r(0)$ such that $\varphi(u) > 0$ for all $u \in B_\varrho(0) \setminus \{0\}$ (recall that $u_0 = 0$ is a strict local minimizer of φ). Let $r_1, r_2 \in (0, r)$ with $r_1 < r_2$ and let

$$\eta \;=\; \min\{\varphi(u) : r_1 \leqslant |u| \leqslant r_2\} \;>\; 0.$$

The level set φ^η is open and $0 \in \varphi^\eta$. So, we can find $\varrho \in (0, r_1)$ such that $\overline{B}_\varrho(0) \subseteq \varphi^\eta$. Let

$$\vartheta \;=\; \tfrac{1}{2}\min\{\varphi(u) : \varrho \leqslant |u| \leqslant r_2\} \;>\; 0.$$

We have $\vartheta < \eta$ and $\overline{\varphi^\vartheta} \subseteq \overline{B}_\varrho(0) \subseteq \varphi^\eta$ with $\overline{\varphi^\eta}$ compact. So, we can use Problem 4.73 and have

$$d(\nabla\varphi, \varphi^\eta, 0) \;=\; 1. \tag{4.88}$$

Since $\nabla\varphi(u) \neq 0$ for all $u \in \overline{\varphi^\eta} \setminus B_\varrho(0)$, from the excision property of the Brouwer degree (see Theorem 4.11(e)) and (4.88), we have

$$d(\nabla\varphi, B_\varrho(u_0), 0) = 1 \qquad \forall\, \varrho > 0 \text{ small.}$$

Solution of Problem 4.76

First we show that $\varphi_0(\Omega) \subseteq \varphi(\Omega)$ and $\varphi^{-1}(h)$ is a singleton for all $h \in \varphi_0(\Omega)$. Let $h \in \varphi_0(\Omega)$. Using Problems 4.3 and 4.28 we have

$$d(\varphi, \Omega, h) = d(\varphi_0, \Omega, h) = \pm 1, \tag{4.89}$$

so $\varphi^{-1}(h) \neq \emptyset$ (see the existence property of the Brouwer degree; Theorem 4.11(d)). From Definition 4.4 and the hypothesis on φ, we have

$$d(\varphi, \Omega, h) = \sum_{u \in \varphi^{-1}(h)} \operatorname{sgn} J_\varphi(u) = \operatorname{card} \varphi^{-1}(h). \tag{4.90}$$

From (4.89) and (4.90) we infer that $\operatorname{card} \varphi^{-1}(h) = 1$ and so $\varphi|_{\varphi_0(\Omega)}$ is an injection and $\varphi_0(\Omega) \subseteq \varphi(\Omega)$.

Next we show that

$$\varphi(\overline{\Omega}) \subseteq \varphi_0(\overline{\Omega}). \tag{4.91}$$

Let $h \in \mathbb{R}^N \setminus \varphi_0(\overline{\Omega})$. Then from the existence property of the Brouwer degree (see Theorem 4.11(d)) and Problem 4.3, we have

$$0 = d(\varphi_0, \Omega, h) = d(\varphi, \Omega, h). \tag{4.92}$$

From (4.90) and (4.92), it follows that $h \notin \varphi(\Omega)$. Since $h \notin \varphi_0(\partial\Omega)$ and $\varphi_0|_{\partial\Omega} = \varphi|_{\partial\Omega}$, it follows that $h \notin \varphi(\partial\Omega)$. Therefore, $h \in \mathbb{R}^N \setminus \varphi(\overline{\Omega})$ and so we have proved (4.91).

We have that the set $\varphi_0(\overline{\Omega}) \subseteq \mathbb{R}^N$ is compact and so

$$\overline{\varphi_0(\Omega)} \subseteq \varphi_0(\overline{\Omega}). \tag{4.93}$$

Suppose that $h \in \varphi_0(\overline{\Omega})$. Then $h = \varphi_0(u)$ with $u \in \overline{\Omega}$. Let $\{u_n\}_{n\geqslant 1} \subseteq \Omega$ be a sequence such that $u_n \longrightarrow u$ and $h_n = \varphi(u_n) \longrightarrow \varphi(u) = h$. Since $h_n \in \varphi(\Omega)$ for all $n \geqslant 1$, we have $h \in \overline{\varphi_0(\Omega)}$ and so

$$\overline{\varphi_0(\Omega)} = \varphi_0(\overline{\Omega}) \tag{4.94}$$

(see (4.93)). From (4.89), (4.91), and (4.94), we infer that

$$\varphi_0(\overline{\Omega}) = \varphi(\overline{\Omega}). \tag{4.95}$$

From Problem 4.10 and the hypothesis on Ω, we have

$$\operatorname{int} \varphi_0(\overline{\Omega}) = \varphi_0(\operatorname{int} \overline{\Omega}) = \varphi_0(\Omega). \tag{4.96}$$

Moreover, since $J_\varphi(u) > 0$ for all $u \in \Omega$, from the inverse mapping theorem (see Theorem 3.26), we have that φ is locally an injection and so using Problem 4.9, we have

$$\operatorname{int}\varphi(\overline{\Omega}) \supseteq \varphi(\operatorname{int}\overline{\Omega}) = \varphi(\Omega). \tag{4.97}$$

From (4.95), (4.96), and (4.97), it follows that

$$\varphi_0(\Omega) = \varphi(\Omega). \tag{4.98}$$

From (4.89) and (4.98) it follows that $\varphi|_\Omega$ is an injection. Also, because $\varphi_0|_{\partial\Omega} = \varphi|_{\partial\Omega}$, we have that $\varphi|_{\partial\Omega}$ is an injection. Finally suppose that $u \in \Omega$, $v \in \partial\Omega$. From (4.98) we see that, we can find $z \in \Omega$ such that

$$\varphi(u) = \varphi_0(z). \tag{4.99}$$

Also, since $\varphi_0|_{\partial\Omega} = \varphi|_{\partial\Omega}$, we have

$$\varphi(v) = \varphi_0(v). \tag{4.100}$$

Since $\partial\Omega \ni v \neq z \in \Omega$ and φ_0 is an injection, we have that $\varphi_0(z) \neq \varphi_0(v)$, hence $\varphi(u) \neq \varphi(v)$. This means that φ is an injection and so φ is a homeomorphism and $\varphi|_\Omega$ is a C^1-diffeomorphism.

Solution of Problem 4.77

Let $\eta > \sup\limits_{u \in B_{r_0}(0)} \varphi(u)$ and $r_1 = \sup\limits_{u \in \{\varphi \leqslant \eta\}} \|u\|$. For $r \geqslant r_1$, let

$$\vartheta > \max\big\{\eta, \sup_{u \in B_r(0)} \varphi(u)\big\}.$$

Apply Theorem 4.48 with $u_0 = 0$, $\Omega = X$, $b = \vartheta$, $a = \eta$ and $V = \{\varphi < \vartheta\}$. We obtain

$$d_{(S)_+}(\varphi'_G, V, 0) = 1. \tag{4.101}$$

Finally the excision property with $C = \overline{V} \setminus B_r(0)$ implies that $d_{(S)_+}(\varphi'_G, B_r(0), 0) = 1$ for all $r \geqslant r_1$.

Solution of Problem 4.78

Let $\widehat{\varphi}(u) = \varphi(u) - \varphi(u_0)$ for all $u \in U$ and let $\widehat{U} \subseteq U$ be a convex open neighborhood of u_0. We will show that $\widehat{\varphi}|_{\widehat{U}}$ is sequentially weakly lower semicontinuous. Arguing by contradiction, suppose that we can find a sequence $\{u_n\}_{n\geqslant 1} \subseteq \widehat{U}$ such that

$$u_n \longrightarrow u \in \widehat{U} \quad \text{and} \quad \lim_{n\to+\infty} \widehat{\varphi}(u_n) < \widehat{\varphi}(u). \tag{4.102}$$

By the mean value theorem (see Theorem 3.16), we can find $t_n \in (0,1)$ such that

$$\widehat{\varphi}(u_n) - \widehat{\varphi}(u) = \langle \widehat{\varphi}'(u + t_n(u_n - u)), u_n - u \rangle, \tag{4.103}$$

so

$$\limsup_{n \to +\infty} \langle \widehat{\varphi}'(u + t_n(u_n - u)), u_n - u \rangle \leqslant 0$$

(see (4.102)), hence

$$u + t_n(u_n - u) \longrightarrow u \quad \text{in } X$$

(since φ' is an $(S)_+$-map; see Definition 2.137) and thus

$$\widehat{\varphi}(u_n) \longrightarrow \widehat{\varphi}(u)$$

(see (4.103) and recall that φ' is continuous). This contradicts (4.102). Therefore the map $\widehat{\varphi}|_{\widehat{U}}$ is sequentially weakly lower semicontinuous.

Since, by hypothesis $u_0 \in \widehat{U} \subseteq U$ is a local minimizer and an isolated critical point of $\widehat{\varphi}$, we can find $r_0 > 0$ such that $\overline{B}_{r_0}(u_0) \subseteq \widehat{U}$ and

$$0 = \widehat{\varphi}(u_0) < \widehat{\varphi}(u) \quad \text{and} \quad \widehat{\varphi}'(u) \neq 0 \qquad \forall\, u \in \overline{B}_{r_0}(u_0) \setminus \{u_0\}. \tag{4.104}$$

We claim that

$$\widehat{m}_r = \inf_{u \in \overline{B}_{r_0}(u_0) \setminus B_r(u_0)} \widehat{\varphi}(u) > 0 \qquad \forall\, r \in (0, r_0]. \tag{4.105}$$

Again we argue indirectly. So, suppose that we can find $r \in (0, r_0]$ and a sequence $\{u_n\}_{n \geqslant 1} \subseteq \overline{B}_{r_0}(u_0) \setminus B_r(u_0)$ such that

$$\lim_{n \to +\infty} \widehat{\varphi}(u_n) = \widehat{\varphi}(u_0) = 0. \tag{4.106}$$

Since $\{u_n\}_{n \geqslant 1} \subseteq \overline{B}_{r_0}(u_0)$ and the space X is reflexive, by passing to a suitable subsequence if necessary, we may assume that $u_n \xrightarrow{w} u \in \overline{B}_{r_0}(u_0) \subseteq \widehat{U}$. Recall that the map $\widehat{\varphi}|_{\widehat{U}}$ is sequentially weakly lower semicontinuous. So, we have

$$\widehat{\varphi}(u) \leqslant \liminf_{n \to +\infty} \widehat{\varphi}(u_n) = \widehat{\varphi}(u_0) = 0$$

(see (4.106)), so $u = u_0$ (by the hypothesis on $u_0 \in U$). The mean value theorem (see Theorem 3.16) implies the existence of $t_n \in (0,1)$ such that

$$\widehat{\varphi}(u_n) - \widehat{\varphi}(\tfrac{u_n+u_0}{2}) = \langle \varphi'(y_n), \tfrac{u_n-u_0}{2} \rangle \qquad \forall\, n \geqslant 1, \tag{4.107}$$

where $y_n = t_n u_n + (1-t_n)\frac{u_n+u_0}{2}$. Then using the fact that $\frac{u_n+u_0}{2} \xrightarrow{w} u_0$ in X and that $\widehat{\varphi}|_{\widehat{U}}$ is sequentially weakly lower semicontinuous, from (4.106) and (4.107), we obtain

$$\limsup_{n \to +\infty} \langle \varphi'(y_n), \tfrac{u_n-u_0}{2} \rangle \leqslant 0,$$

so

$$\limsup_{n\to+\infty}(1+t_n)\left\langle \varphi'(y_n), \tfrac{u_n-u_0}{2}\right\rangle \leqslant 0,$$

thus

$$y_n \longrightarrow u_0 \quad \text{in } X \tag{4.108}$$

(since φ' is an $(S)_+$-map; see Definition 2.137). But note that

$$\|y_n - u_0\|_X = \tfrac{1+t_n}{2}\|u_n - u_0\|_X \geqslant \tfrac{r}{2} \qquad \forall\, n \geqslant 1. \tag{4.109}$$

Comparing (4.108) and (4.109) we reach a contradiction. This proves (4.105).

Let

$$V = \left\{u \in B_{r_0}(u_0) : \widehat{\varphi}(u) < \widehat{m}_{\frac{r_0}{2}}\right\}.$$

Clearly the set V is nonempty, open and $V \subseteq B_{\frac{r_0}{2}}(u_0)$. Fix $r_1 \in (0, \frac{r_0}{2})$ such that $\overline{B}_{r_1}(u_0) \subseteq V$ and choose $\lambda \in (0, \widehat{m}_r)$. We have

$$\{u \in B_{r_0}(u_0) : \widehat{\varphi}(u) \leqslant \lambda\} \subseteq B_r(u_0) \subseteq \overline{B}_r(u_0) \subseteq V \qquad \forall\, r \in (0, r_1]. \tag{4.110}$$

Because of (4.104) and (4.110), we can apply Theorem 4.48 to the functional $\widehat{\varphi}|_{B_{r_0}(u_0)}$ and the numbers $\lambda < \widehat{m}_{\frac{r_0}{2}}$ and have

$$d_{(S)_+}(\varphi', V, 0) = 1.$$

Since $0 \notin \varphi'(C)$ with $C = \overline{V} \setminus B_r(u_0)$, from the excision property (see Remark 4.47), we have

$$d_{(S)_+}(\varphi', B_r(u_0), 0) = 1 \qquad \forall\, r \in (0, r_0].$$

Solution of Problem 4.79

Let $u_0, u_\lambda \in X$ be the unique solutions of

$$h^* \in \mathcal{F}(u_0) + A(u_0) \quad \text{and} \quad h^* = \mathcal{F}(u_\lambda) + A_\lambda(u_\lambda),\ \lambda > 0 \tag{4.111}$$

(see Theorem 2.116 and Problem 2.163). Since $h^* \in (\mathcal{F} + A)(U)$, it follows that $u_0 \in U$ and that $h^* \notin (\mathcal{F} + A)(\partial U)$. Therefore $d_M(\mathcal{F} + A, U, h^*)$ is well defined (see Theorem 4.52).

Claim 1: $u_\lambda \longrightarrow u_0$ in X as $\lambda \searrow 0$.

For every $\lambda > 0$, we have

$$\langle h^*, u_\lambda\rangle = \langle \mathcal{F}(u_\lambda), u_\lambda\rangle + \langle A_\lambda(u_\lambda), u_\lambda\rangle$$

(see (4.111)), so

$$\|u_\lambda\|_X^2 \leqslant \|h^*\|_*\|u_\lambda\|_X$$

(recall that $A_\lambda(0) = 0$), thus the family $\{u_\lambda\}_{\lambda>0} \subseteq X$ is bounded and hence the family $\{A_\lambda(u_\lambda)\}_{\lambda>0} \subseteq X^*$ is bounded (see Problem 2.163).

Consider a sequence $\{\lambda_n\}_{n\geqslant 1} \subseteq (0,+\infty)$ such that

$$\begin{aligned} &\lambda_n \searrow 0, \quad u_n = u_{\lambda_n} \xrightarrow{w} u \text{ in } X^* \quad \text{and} \\ &A_{\lambda_n}(u_{\lambda_n}) \xrightarrow{w} u^* \text{ in } X^* \text{ as } n \to +\infty. \end{aligned} \tag{4.112}$$

We have

$$h^* = \mathcal{F}(u_n) + A_{\lambda_n}(u_n) \qquad \forall\, n \geqslant 1. \tag{4.113}$$

Note that

$$\liminf_{n\to+\infty} \langle \mathcal{F}(u_n), u_n - u\rangle \geqslant 0. \tag{4.114}$$

Indeed, if (4.114) is not true, then by passing to a suitable subsequence, we would have

$$\lim_{n\to+\infty} \langle \mathcal{F}(u_n), u_n - u\rangle < 0. \tag{4.115}$$

But $\mathcal{F}$ is an $(S)_+$-map (see Problem 2.185). So, from (4.114) it follows that

$$u_n \longrightarrow u \quad \text{in } X,$$

so

$$\langle \mathcal{F}(u_n), u_n - u\rangle \longrightarrow 0,$$

a contradiction to (4.115). Therefore (4.114) holds.

From the definition of A_λ and J_λ^A (see Problem 2.160), we have

$$\|J_\lambda^A(u_n) - u_n\|_X = \lambda_n \|A_{\lambda_n}(u_n)\|_* \longrightarrow 0 \quad \text{as } n \to +\infty. \tag{4.116}$$

So, we have

$$\begin{aligned} &\limsup_{n\to+\infty} \left\langle A_{\lambda_n}(u_n) - u^*, J_{\lambda_n}^A(u_n) - u_n\right\rangle \\ = &\limsup_{n\to+\infty} \langle A_{\lambda_n}(u_n) - u^*, u_n - u\rangle \\ = &\limsup_{n\to+\infty} \langle A_{\lambda_n}(u_n), u_n - u\rangle \\ = &\lim_{n\to+\infty} \langle \mathcal{F}(u_n) + A_{\lambda_n}(u_n), u_n - u\rangle - \liminf_{n\to+\infty} \langle \mathcal{F}(u_n), u_n - u\rangle \leqslant 0 \end{aligned}$$

(see (4.114)), thus

$$\limsup_{n\to+\infty} \left\langle A_{\lambda_n}(u_n), J_{\lambda_n}^A(u_n)\right\rangle \leqslant \langle u^*, u\rangle .$$

Recalling that $(J_{\lambda_n}^A(u_n), A_{\lambda_n}(u_n)) \in \operatorname{Gr} A$ for all $n \geqslant 1$ (see Problem 2.163) and using Problem 2.162, we have that

$$(u, u^*) \in \operatorname{Gr} A \quad \text{and} \quad \langle A_{\lambda_n}(u_n), u_n\rangle \longrightarrow \langle u^*, u\rangle , \tag{4.117}$$

so

$$\lim_{n\to+\infty} \langle \mathcal{F}(u_n), u_n - u\rangle = 0,$$

thus $u_n \longrightarrow u$ in X (since $\mathcal{F}$ is an $(S)_+$-map; see Problem 2.185). So, finally, we have

$$h^* = \mathcal{F}(u) + u^* \in \mathcal{F}(u) + A(u)$$

(see (4.117)), so $u = u_0$ (see (4.111)) and hence $u_\lambda \longrightarrow u_0$ in X as $\lambda \searrow 0$. This proves Claim 1.

Since $u_0 \in U$, from Claim 1 and Definition 4.49, for small $\lambda > 0$, we have

$$u_\lambda \in U \quad \text{and} \quad d_M(\mathcal{F} + A, U, h^*) = d_{(S)_+}(\mathcal{F} + A_\lambda, U, h^*). \tag{4.118}$$

Because the set U is bounded, we can find $r > 0$ such that $U \subseteq B_r(0)$. From (4.111), we have

$$h^* \notin (\mathcal{F} + A_\lambda)(\overline{B}_r(0) \setminus U),$$

so

$$d_{(S)_+}(\mathcal{F} + A_\lambda, U, h^*) = d_{(S)_+}(\mathcal{F} + A_\lambda, B_r(0), h^*) \qquad \forall\, \lambda > 0 \text{ small} \tag{4.119}$$

(by the excision property).

Claim 2: $d_{(S)_+}(\mathcal{F} + A_\lambda, B_r(0), h^*) = 1$ for $\lambda > 0$ small.

Let $h(t, u) = \mathcal{F}(u) + tA_\lambda(u)$ for all $(t, u) \in [0, 1] \times \overline{B}_r(0)$. Recall that $\mathcal{F}$ is of type $(S)_+$ (see Problem 2.185) and A_λ is maximal monotone (see Problem 2.163). Moreover, both are demicontinuous. Therefore h is a homotopy of type $(S)_+$ (see Definition 4.45). We will show that

$$th^* \notin h(t, \partial B_r(0)) \qquad \forall\, t \in [0, 1]. \tag{4.120}$$

Arguing by contradiction, suppose that we can find $t \in [0, 1]$ and $u \in \partial B_r(0)$ such that

$$th^* = \mathcal{F}(u) + tA_\lambda(u). \tag{4.121}$$

Since $h^* \notin (\mathcal{F} + A_\lambda)(\overline{B}_r(0) \setminus U)$, we must have $t \in [0, 1)$. From (4.111) and (4.121), we have

$$t(\mathcal{F}(u_\lambda) + A_\lambda(u_\lambda)) - t(\mathcal{F}(u) + A_\lambda(u)) = (1 - t)\mathcal{F}(u),$$

so

$$0 \leqslant (1 - t)\,\langle \mathcal{F}(u), u_\lambda - u \rangle$$

(since $\mathcal{F} + A_\lambda$ is monotone), thus

$$0 \leqslant (1 - t)(r\|u_\lambda\|_X - r^2) < 0$$

(since $u_\lambda \in U \subseteq B_r(0)$ for $\lambda > 0$ small), a contradiction. This proves Claim 2.

Using the homotopy invariance property of $d_{(S)_+}$ (see Theorem 4.46), we have

$$d_{(S)_+}(\mathcal{F} + A_\lambda, B_r(0), h^*) = d_{(S)_+}(\mathcal{F}, B_r(0), h^*) \qquad \forall\, \lambda > 0 \text{ small},$$

so $d_M(\mathcal{F}+A, U, h^*) = 1$ (see (4.118), (4.119) and Theorem 4.46(a)).

Solution of Problem 4.80

Let $u_0 \in U$ be such that $h^* = f(u_0) + u_0^*$ with $u_0^* \in A(u_0)$. Let $y^* \in \mathcal{F}(u_0) + u_0^*$. We introduce the map

$$h(t, u) = (1-t)f(u) + t\mathcal{F}(u) + A(u) \qquad \forall\, (t, u) \in [0,1] \times \overline{U}.$$

We claim that

$$(1-t)h^* + ty^* \notin h(t, \partial U) \qquad \forall\, t \in [0,1]. \tag{4.122}$$

Arguing indirectly, suppose that we can find $(t, u) \in [0,1] \times \partial U$ (evidently $u \neq u_0$) such that

$$(1-t)h^* + ty^* = (1-t)f(u) + t\mathcal{F}(u) + u^* \quad \text{with } u^* \in A(u). \tag{4.123}$$

If $t = 0$, then $h^* = f(u) + u^* \in (f+A)(\partial U)$, a contradiction. Hence $t \in (0,1]$. We have

$$(1-t)f(u_0) + t\mathcal{F}(u_0) + u_0^* - (1-t)f(u) - t\mathcal{F}(u) - u^* = 0 \tag{4.124}$$

(see (4.123)). On (4.124) we act with $u_0 - u$ and using the monotonicity of f and A and the strict monotonicity of $\mathcal{F}$, we reach a contradiction. Therefore (4.122) holds. Also $u \longmapsto t\mathcal{F}(u) + A(u)$ is a pseudomonotone homotopy (see Definition 4.50(a)). Then the homotopy invariance property of d_M (see Theorem 4.52(c)) implies that

$$d_M(f+A, U, h^*) = d(\mathcal{F}+A, U, y^*),$$

so

$$d_M(f+A, U, h^*) = 1$$

(see Problem 4.79)

Solution of Problem 4.81

We will verify Definition 4.50(a). So, suppose that $t_n \longrightarrow t$ in $[0,1]$, $(u_n u_n^*) \in \operatorname{Gr} \partial\varphi^{t_n}$ for all $n \geqslant 1$, $u_n \xrightarrow{w} u$ in X, $u_n^* \xrightarrow{w} u^*$ in X^* and

$$\limsup_{n\to+\infty} \langle u_n^*, u_n \rangle \leqslant \langle u^*, u \rangle.$$

Let $v \in \operatorname{dom} \varphi^t$. From Definition 3.149(b), we know that we can find a sequence $\{v_n\}_{n \geqslant 1} \subseteq X$ such that $v_n \longrightarrow v$ in X and $\varphi^{t_n}(v_n) \longrightarrow \varphi^t(v)$. Since $(u_n, u_n^*) \in \operatorname{Gr} \partial \varphi^{t_n}$ for all $n \geqslant 1$, we have

$$\langle u_n^*, v_n - u_n \rangle \leqslant \varphi^{t_n}(v_n) - \varphi^{t_n}(u_n) \qquad \forall\, n \geqslant 1, \tag{4.125}$$

so

$$\langle u^*, v - u \rangle \leqslant \varphi^t(v) - \varphi^t(u) \tag{4.126}$$

(see Definition 3.149(b)). Since $v \in \operatorname{dom} \varphi^t$ is arbitrary, from (4.126) we infer that

$$(u, u^*) \in \operatorname{Gr} \partial \varphi^t.$$

If we take $v = u$ (hence $v_n \longrightarrow u$ in X), then from (4.125) we have that

$$\langle u^*, u_n \rangle \longrightarrow \langle u^*, u \rangle,$$

so $\{ \partial \varphi^t \}_{t \in [0,1]}$ is a pseudomonotone homotopy (see Definition 4.50(a)).

Solution of Problem 4.82

Evidently $\varphi^t \in \Gamma_0(X)$ for all $t \in [0,1]$. Suppose that $t_n \longrightarrow t$ in $[0,1]$. We claim that

$$\varphi^{t_n} \xrightarrow{M} \varphi^t \tag{4.127}$$

(see Definition 3.149(b)). Since $\varphi^0, \varphi^1 \in \Gamma_0(X)$, if $u_n \xrightarrow{w} u$ in X, then

$$\varphi^t(u) \leqslant \liminf_{n \to +\infty} \varphi^{t_n}(u_n). \tag{4.128}$$

Next, let $u \in \operatorname{dom} \varphi^t$. If $t \in (0,1)$, then $\operatorname{dom} \varphi^t = \operatorname{dom} \varphi^0 \cap \operatorname{dom} \varphi^1$ and so the sequence $\{u_n = u\}_{n \geqslant 1}$ satisfies $\varphi^{t_n}(u_n) \longrightarrow \varphi^t(u)$ and we conclude that (4.127) holds (see (4.128) and Definition 3.149(b)).

Suppose that $t = 0$. Then we define

$$u_n = \begin{cases} J_{t_n}^{\partial \varphi^1}(u) & \text{if } t_n > 0 \\ u & \text{if } t_n = 0 \end{cases} \qquad \forall\, n \geqslant 1.$$

By hypothesis, we have $u \in \operatorname{dom} \varphi^0 \subseteq \overline{\operatorname{dom} \varphi^1} = \overline{D(\partial \varphi^1)}$. So, from Problem 2.164, we have $u_n \longrightarrow u$ in X. By hypothesis

$$\varphi^0(u_n) \leqslant (1 + c_1(t_n))\varphi^0(u) + c_2(t_n),$$

when $t_n > 0$, so

$$\limsup_{n \to +\infty} \varphi^0(u_n) \leqslant \varphi^0(u)$$

and hence

$$\varphi^0(u_n) \longrightarrow \varphi^0(u) \tag{4.129}$$

(since $\varphi^0 \in \Gamma_0(X)$). On the other hand, from Problem 2.161 and the hypothesis, we have

$$\varphi^1(0) \leqslant \varphi^1(u_n) \leqslant \varphi^1(u) \qquad \forall\, n \geqslant 1,$$

so the sequence $\{\varphi^1(u_n)\}_{n\geqslant 1} \subseteq \mathbb{R}$ is bounded. This fact and (4.129) imply that

$$\varphi^{t_n}(u_n) \longrightarrow \varphi^0(u) \tag{4.130}$$

(recall $t = 0$). From (4.128) and (4.130), we infer that (4.127) holds.

Similarly, if $t = 1$, by interchanging the roles of φ^0 and φ^1 in the above argument.

So, we can use Problem 4.81, to conclude that $\{\partial\varphi^t\}_{t\in[0,1]}$ is a pseudomonotone homotopy.

Solution of Problem 4.83

Suppose that $\widehat{u}, \widehat{v} \in X$ are fixed points of φ. Then

$$d_X(\widehat{u}, \widehat{v}) = d_X(\varphi(\widehat{u}), \varphi(\widehat{v})) \leqslant \vartheta d_X(\widehat{u}, \widehat{v}),$$

so $\widehat{u} = \widehat{v}$. This proves the uniqueness of the fixed point.

Next we prove the existence. So, let $u_0 \in X$ and set $u_1 = \varphi(u_0)$ and $u_{n+1} = \varphi(u_n)$ for all $n \geqslant 1$. We claim that $\{u_n\}_{n\geqslant 1}$ is a Cauchy sequence in (X, d_X).

Let $\xi_n = d_X(u_n, u_{n+1})$ for all $n \geqslant 1$. Then

$$\xi_{n+1} \leqslant \vartheta(\xi_n) \leqslant \xi_n \qquad \forall\, n \geqslant 1, \tag{4.131}$$

so the sequence $\{\xi_n\}_{n\geqslant 1}$ is nonincreasing. Let $\xi_n \searrow \xi \geqslant 0$. Exploiting the right continuity of ϑ, from (4.131) we have $\xi = \vartheta(\xi)$, so $\xi = 0$. Arguing by contradiction, suppose that $\{u_n\}_{n\geqslant 1}$ is not d_X-Cauchy. Then we can find $\varepsilon > 0$ and for every $k \geqslant 1$ we can find integers $m_k > n_k \geqslant k$ such that

$$\beta_k = d_X(u_{m_k}, u_{n_k}) \geqslant \varepsilon \qquad \forall\, k \geqslant 1.$$

Moreover, by choosing the smallest m_k, we may assume that

$$d_X(u_{m_{k-1}}, u_{n_k}) < \varepsilon \qquad \forall\, k \geqslant 1 \text{ big} \tag{4.132}$$

(recall that $\xi_n \longrightarrow 0$). So, for $k \geqslant 1$ big, we have

$$\varepsilon \leqslant \beta_k \leqslant d_X(u_{m_k}, u_{m_{k-1}}) + d_X(u_{m_{k-1}}, u_{n_k}) < \xi_{m_k} + \varepsilon$$

(by the triangle inequality and (4.132)), so $\beta_k \longrightarrow \varepsilon$ (recall that $\xi_{m_k} \searrow 0$).

Also, we have

$$\beta_k \leqslant \beta_{k+1} + \xi_{m_{k+1}} + \xi_{n_{k+1}} \leqslant \vartheta(\beta_k) + \xi_{m_{k+1}} + \xi_{n_{k+1}},$$

so $\varepsilon \leqslant \vartheta(\varepsilon)$, a contradiction.

So, $\{\varphi^{(n)}(u_0)\}_{n \geqslant 1}$ is a Cauchy sequence. The completeness of (X, d_X) implies that

$$\varphi(u_{n-1}) = \varphi^{(n)}(u_0) \longrightarrow \widehat{u} \quad \text{in } (X, d_X),$$

so $\varphi(\widehat{u}) = \lim_{n \to +\infty} \varphi(u_n) = \widehat{u}$.

Solution of Problem 4.84

Let $u_n = \varphi(u_{n-1})$ for all $n \geqslant 1$ (that is, $u_n = \varphi^{(n)}(u_0)$ for all $n \geqslant 1$). From the assumption, we see that the sequence $\{\vartheta(u_n)\}_{n \geqslant 0}$ is decreasing, hence convergent. Then for $n, m \geqslant 1$, we have

$$\begin{aligned} d_X(u_{n+m}, u_n) &\leqslant \sum_{k=0}^{m-1} d_X(u_{n+k+1}, u_{n+k}) \\ &\leqslant \sum_{k=0}^{m-1} (\vartheta(u_{n+k}) - \vartheta(u_{n+k+1})) = \vartheta(u_n) - \vartheta(u_{n+m}), \end{aligned}$$

so $\{u_n\}_{n \geqslant 0}$ is a Cauchy sequence. The completeness of (X, d_X) implies that $u_n \longrightarrow \widehat{u}$. Then from the continuity of φ, we have

$$\begin{aligned} \varphi(\widehat{u}) &= \lim_{n \to +\infty} \varphi^{(n)}(u_0) = \lim_{n \to +\infty} \varphi(\varphi^{(n-1)}(u_0)) \\ &= \lim_{n \to +\infty} \varphi(u_{n-1}) = \lim_{n \to +\infty} u_n = \widehat{u}, \end{aligned}$$

so $\widehat{u} \in X$ is a fixed point of φ and $\varphi^{(n)}(u_0) \longrightarrow \widehat{u}$ in (X, d_X).

Solution of Problem 4.85

The uniqueness of the fixed point follows at once from the contractivity property of φ. We prove the existence of a fixed point. Let $\xi \colon X \longrightarrow \mathbb{R}_+$ be defined by

$$\xi(u) = d_X(u, \varphi(u)) \qquad \forall\, u \in X.$$

Evidently ξ is continuous. The compactness of X implies the existence of $\widehat{u} \in X$ such that

$$\xi(\widehat{u}) = \inf_{u \in X} \xi(u).$$

If $\widehat{u} \neq \varphi(\widehat{u})$, then

$$d_X(\widehat{u}, \varphi(\widehat{u})) = \min_{u \in X} d_X(u, \varphi(u)) \leqslant d_X(\varphi(\widehat{u}), \varphi(\varphi(\widehat{u}))) < d_X(\widehat{u}, \varphi(\widehat{u})),$$

a contradiction. So, $\widehat{u} = \varphi(\widehat{u})$, that is, $\widehat{u} \in X$ is the unique fixed point of φ. Next, let $u_0 \neq \widehat{u}$ be given and let $\beta_n = d_X(\varphi^{(n)}(u_0), \widehat{u})$. We have

$$\begin{aligned} \beta_{n+1} &= d_X(\varphi^{(n+1)}(u_0), \widehat{u}) = d_X(\varphi^{(n+1)}(u_0), \varphi(\widehat{u})) \\ &< d_X(\varphi^{(n)}(u_0), \widehat{u}) = \beta_n, \end{aligned}$$

so the sequence $\{\beta_n\}_{n\geqslant 1}$ is strictly decreasing and so $\beta_n \searrow \beta^* \geqslant 0$.

Let $\{\varphi^{n_k}(u_0)\}_{k\geqslant 1}$ be a subsequence of $\{\varphi^{(n)}(u_0)\}_{n\geqslant 0}$ such that $\varphi^{(n_k)}(u_0) \longrightarrow \widetilde{u}$. Suppose that $\widetilde{u} \neq \widehat{u}$. Then

$$\begin{aligned} \beta^* &= d_X(\widetilde{u}, \widehat{u}) = \lim_{k\to+\infty} \beta_{n_k} = \lim_{k\to+\infty} \beta_{n_{k+1}} \\ &= \lim_{k\to+\infty} d_X(\varphi(\varphi^{(n_k)}(u_0)), \widehat{u}) = d_X(\varphi(\widetilde{u}), \widehat{u}) \\ &= d_X(\varphi(\widetilde{u}), \varphi(\widehat{u})) < d_X(\widetilde{u}, \widehat{u}), \end{aligned}$$

a contradiction. So, $\widetilde{u} = \widehat{u}$ and the compactness of X implies that $\varphi^{(n)}(u_0) \longrightarrow \widehat{u}$.

Solution of Problem 4.86

Proposition 4.62 implies that for every $n \geqslant 0$, φ_n has a unique fixed point $\widehat{u}_n$. Since the space X is locally compact, we can find $r > 0$ such that $\overline{B}_r(\widehat{u}_0)$ is compact (where $\overline{B}_r(\widehat{u}_0) = \{u \in X : d_X(u, \widehat{u}_0) \leqslant r\}$). Then, since the sequence $\{\varphi_n\}_{n\geqslant 1}$ is equicontinuous (see Definition I.1.83) and converges pointwise to φ_0, by the Arzela–Ascoli theorem (see Theorem I.2.181), we have that $\varphi_n \longrightarrow \varphi_0$ uniformly on $\overline{B}_r(\widehat{u}_0)$. Hence, given $\varepsilon \in (0, r]$, we can find $n_0 = n_0(\varepsilon) \geqslant 1$ such that

$$d_X(\varphi_n^{(m_0)}(u), \varphi_0^{(m_0)}(u)) \leqslant \varepsilon(1 - k_0) \qquad \forall\, u \in \overline{B}_\varepsilon(\widehat{u}_0),\ n \geqslant n_0. \tag{4.133}$$

Then for all $u \in \overline{B}_\varepsilon(\widehat{u}_0)$ and all $n \geqslant n_0$, we have

$$\begin{aligned} d_X(\varphi_n^{(m_0)}(u), \widehat{u}_0) &= d_X(\varphi_n^{(m_0)}(u), \varphi_0^{(m_0)}(\widehat{u}_0)) \\ &\leqslant d_X(\varphi_n^{(m_0)}(u), \varphi_0^{(m_0)}(u)) + d_X(\varphi_0^{(m_0)}(u), \varphi_0^{(m_0)}(\widehat{u}_0)) \\ &\leqslant \varepsilon(1 - k_0) + k_0 d_X(u, \widehat{u}_0) \leqslant \varepsilon \end{aligned}$$

(see (4.133)). So, we have proved that

$$\varphi_n^{(m_0)}(\overline{B}_\varepsilon(\widehat{u}_0)) \subseteq \overline{B}_\varepsilon(\widehat{u}_0) \qquad \forall\, n \geqslant n_0,$$

thus

$$d_X(\widehat{u}_n, \widehat{u}_0) \leqslant \varepsilon \qquad \forall\, n \geqslant n_0$$

and hence $\widehat{u}_n \longrightarrow \widehat{u}_0$ in X.

Solution of Problem 4.87

The existence of a unique fixed point $\widehat{u} \in X$ for φ, follows from Proposition 4.62.

First suppose that $m = 1$ (that is, φ is a k-contraction). Since $\varphi_n \longrightarrow \varphi$ uniformly on X, given $\varepsilon > 0$, we can find $n_0 = n_0(\varepsilon) \geqslant 1$ such that

$$d_X(\varphi_n(u), \varphi(u)) \leqslant \varepsilon(1-k) \qquad \forall\, u \in X,\ n \geqslant n_0. \tag{4.134}$$

Then for all $n \geqslant n_0$, we have

$$\begin{aligned} d_X(\widehat{u}_n, \widehat{u}) &= d_X(\varphi_n(\widehat{u}_n), \varphi(\widehat{u})) \\ &\leqslant d_X(\varphi_n(\widehat{u}_n), \varphi(\widehat{u}_n)) + d_X(\varphi(\widehat{u}_n), \varphi(\widehat{u})) \\ &\leqslant \varepsilon(1-k) + k d_X(\widehat{u}_n, \widehat{u}) \end{aligned}$$

(see (4.134)), so

$$d_X(\widehat{u}_n, \widehat{u}) \leqslant \varepsilon \qquad \forall\, n \geqslant n_0,$$

thus

$$\widehat{u}_n \longrightarrow \widehat{u} \quad \text{in } X.$$

Now, we consider the general case $m \geqslant 1$. Then

$$d_X(\varphi^{(m)}(u), \varphi^{(m)}(v)) \leqslant k^m d_X(u,v) \qquad \forall\, u, v \in X \tag{4.135}$$

with some $k \in (0,1)$. We consider the following metric on X equivalent to d_X:

$$\widehat{d}_X(u,v) = \sum_{i=0}^{m-1} \tfrac{1}{k^i} d_X(\varphi^{(i)}(u), \varphi^{(i)}(v)) \qquad \forall\, u, v \in X.$$

Since φ is uniformly continuous, we see that $\varphi_n \longrightarrow \varphi$ uniformly on $(X, \widehat{d}_X)$. Also, for all $u, v \in X$, we have

$$\begin{aligned} \widehat{d}_X(\varphi(u), \varphi(v)) &= \sum_{i=0}^{m-1} \tfrac{1}{k^i} d_X(\varphi^{(i+1)}(u), \varphi^{(i+1)}(v)) \\ &= k \sum_{i=1}^{m-1} \tfrac{1}{k^i} d_X(\varphi^{(i)}(u), \varphi^{(i)}(v)) + \tfrac{1}{k^{m-1}} d_X(\varphi^{(m)}(u), \varphi^{(m)}(v)) \\ &\leqslant k \sum_{i=1}^{m-1} \tfrac{1}{k^i} d_X(\varphi^{(i)}(u), \varphi^{(i)}(v)) + k d_X(u,v) \\ &= k \sum_{i=0}^{m-1} \tfrac{1}{k^i} d_X(\varphi^{(i)}(u), \varphi^{(i)}(v)) = k \widehat{d}_X(u,v) \end{aligned}$$

(see (4.135)), so φ is a k-contraction for the new metric $\widehat{d}_X$. Therefore we are back to the previous case.

Solution of Problem 4.88
Let $\varepsilon \in (0, r)$ be such that

$$d_X(\varphi(u), u) \leqslant (1-k)\varepsilon < (1-k)r$$

and let $\overline{B}_\varepsilon(u) = \{v \in X : d_X(v, u) \leqslant \varepsilon\}$. For $v \in \overline{B}_\varepsilon(u)$, we have

$$\begin{aligned} d_X(\varphi(v), u) &\leqslant d_X(\varphi(v), \varphi(u)) + d_X(\varphi(u), u) \\ &\leqslant k d_X(v, u) + (1-k)\varepsilon \leqslant \varepsilon. \end{aligned}$$

So, $\varphi\colon \overline{B}_\varepsilon(u) \longrightarrow \overline{B}_\varepsilon(u)$ and is a k-contraction. Invoking the Banach fixed point theorem (see Theorem 4.59), we infer that φ has a fixed point.

Solution of Problem 4.89
According to Theorem 4.65, we need to show that f is a surjection. To this end, given any $h_0 \in X$, let $\psi\colon X \longrightarrow X$ be defined by

$$\psi(u) = h_0 + \varphi(u) \qquad \forall\, u \in X.$$

Evidently ψ is a k-contraction. So, the Banach fixed point theorem (see Theorem 4.59) implies the existence of a (unique) $\widehat{u} \in X$ such that

$$\psi(\widehat{u}) = h_0 + \varphi(\widehat{u}) = \widehat{u},$$

so $h_0 = f(\widehat{u})$. Since $h_0 \in X$ is arbitrary, we conclude that f is surjective.

Solution of Problem 4.90
We do the solution under condition **(c)**. The other cases are treated similarly.

Arguing indirectly, suppose that φ has no fixed point. Then according to Theorem 4.79, we can find $u_0 \in \partial U$ and $t \in (0, 1)$ such that

$$u_0 = t\varphi(u_0). \tag{4.136}$$

Since **(c)** holds, we have

$$\begin{aligned} \|\varphi(u_0)\|_X^2 &\leqslant \|u_0\|_X^2 + \|u_0 - \varphi(u_0)\|_X^2 \\ &= t^2\|\varphi(u_0)\|_X^2 + (1-t)^2\|\varphi(u_0)\|_X^2 \end{aligned}$$

(see (4.136)), so

$$1 \leqslant t^2 + (1-t)^2 < t + (1-t) = 1,$$

a contradiction. Therefore, φ has a unique fixed point (see Theorem 4.79). Similarly for the other conditions.

Solution of Problem 4.91

Note that for all $u \in \partial U$, we have

$$\|\varphi(u) - \varphi(-u)\|_X \leqslant 2k\|u\|_X,$$

so

$$\|\varphi(u)\|_X \leqslant k\|u\|_X$$

(since $\varphi|_{\partial U}$ is odd). So, we can apply Problem 4.90(a) and conclude that φ has a unique fixed point.

Solution of Problem 4.92

Suppose that φ has no fixed points. Then Theorem 4.79 implies that there exist $u_0 \in \partial U$ and $t \in (0,1)$ such that

$$u_0 = t\varphi(u_0),$$

so

$$u_0 = t\psi(u_0)$$

(since $\varphi|_{\partial U} = \psi|_{\partial U}$) and hence ψ has no fixed points (see Theorem 4.79).

Solution of Problem 4.93

For every $u, v \in U$ and $t > 0$, we have

$$\begin{aligned} & \|(I_H - t\varphi)(u) - (I_H - t\varphi)(v)\|_H^2 \\ = \; & \|u-v\|_H^2 - 2t\,(\varphi(u) - \varphi(v), u-v)_H + t^2\|\varphi(u) - \varphi(v)\|_H^2 \\ \leqslant \; & \|u-v\|_H^2 - 2tc\|u-v\|_H^2 + t^2\xi^2\|u-v\|_H^2 \\ = \; & (1 + t^2\xi^2 - 2tc)\|u-v\|_H^2. \end{aligned}$$

for some $\xi > 0$. For $t \in (0, \frac{2c}{\xi^2})$ we have that $k = 1 + t^2\xi^2 - 2tc < 1$. So, $I_H - t\varphi$ is a k-contraction. Using Theorem 4.65, we conclude that φ is an open map, $\varphi(U)$ is open and φ is a homeomorphism onto $\varphi(U)$.

Solution of Problem 4.94
From Remark 4.60 (see (4.8)), we know that for every $n \geqslant 1$, we have

$$d_X(\varphi_n^{(m)}(u), \widehat{u}_n) \leqslant \tfrac{k^m}{1-k} d_X(u, \varphi_n(u)) \qquad \forall\, u \in X,\ m \geqslant 0.$$

Let $m = 0$ and $u = \widehat{u}$. Then

$$d_X(\widehat{u}, \widehat{u}_n) \leqslant \tfrac{1}{1-k} d_X(\widehat{u}, \varphi_n(\widehat{u})) = \tfrac{1}{1-k} d_X(\varphi(\widehat{u}), \varphi_n(\widehat{u})),$$

so

$$\lim_{n \to +\infty} d_X(\widehat{u}, \widehat{u}_n) = 0.$$

The result fails if the contraction constant $k \in (0,1)$ is not the same. To see this let $X = l^2$ and consider maps $\varphi_n \colon l^2 \longrightarrow l^2$ defined by

$$\varphi_n(\{u_k\}_{k \geqslant 1}) = \left(0, \ldots, 0, (1 - \tfrac{1}{n})u_n + \tfrac{1}{n}, 0, \ldots\right) \qquad \forall\, \{u_k\}_{k \geqslant 1} \in l^2.$$

Clearly each φ_n is a k_n-contraction with $k_n = 1 - \frac{1}{n}$ and $\varphi_n \longrightarrow \varphi \equiv 0$ pointwise. But $\|\widehat{u}_n\|_X = 1$ for all $n \geqslant 1$ and so $\widehat{u}_n \not\longrightarrow \widehat{u} = 0$.

Solution of Problem 4.95
From Theorem 4.70 we know that $\operatorname{Fix}(\varphi) \neq \emptyset$ and clearly it is closed and bounded. Let $u, v \in \operatorname{Fix}(\varphi)$ with $u \neq v$ and define

$$y_\lambda = (1 - \lambda)u + \lambda v \quad \text{with } \lambda \in (0,1).$$

We have

$$\begin{aligned} & \|u - \varphi(y_\lambda)\|_X + \|\varphi(y_\lambda) - v\|_X \\ = \ & \|\varphi(u) - \varphi(y_\lambda)\|_X + \|\varphi(y_\lambda) - \varphi(v)\|_X \\ \leqslant \ & \|u - y_\lambda\|_X + \|y_\lambda - v\|_X = \|u - v\|_X \\ \leqslant \ & \|u - \varphi(y_\lambda)\|_X + \|\varphi(y_\lambda) - v\|_X \end{aligned}$$

(since $u, v \in \operatorname{Fix}(\varphi)$ and φ is nonexpansive, that is, Lipschitz continuous with Lipschitz constant 1), so $u, \varphi(y_\lambda), v$ are colinear and

$$\|u - y_\lambda\|_X = \|u - \varphi(y_\lambda)\|_X, \quad \|v - y_\lambda\|_X = \|v - \varphi(y_\lambda)\|_X.$$

The uniform convexity of X implies that $y_\lambda = \varphi(y_\lambda)$ which means that the set $\mathrm{Fix}\,(\varphi)$ is convex. Therefore $\mathrm{Fix}\,(\varphi)$ is also w-compact.

Solution of Problem 4.96
Let $u_0 \in C$ and let

$$\varphi_n(u) = (1-\tfrac{1}{n})\varphi(u) + \tfrac{1}{n}u_0 \qquad \forall\, n \geqslant 2.$$

Then since the set C is convex, we have that $\varphi_n \colon C \longrightarrow C$ and it is a k-contraction. So, we can apply the Banach fixed point theorem (see Theorem 4.59) and find $\widehat{u}_n \in C$ such that

$$\widehat{u}_n = \varphi_n(\widehat{u}_n) = (1-\tfrac{1}{n})\varphi(\widehat{u}_n) + \tfrac{1}{n}u_0. \tag{4.137}$$

Since $\varphi_n(\widehat{u}_n) \in \varphi(C)$ for every $n \geqslant 2$, by passing to a suitable subsequence if necessary, we may assume that

$$\varphi(\widehat{u}_n) \longrightarrow \widehat{u} \quad \text{in } X,$$

so

$$\widehat{u}_n \longrightarrow \widehat{u} \quad \text{in } X \tag{4.138}$$

(see (4.137)). The continuity of φ implies that

$$\varphi(\widehat{u}_n) \longrightarrow \varphi(\widehat{u}),$$

so

$$\widehat{u}_n \longrightarrow \varphi(\widehat{u}) \quad \text{in } X \tag{4.139}$$

(see (4.137)). From (4.138) and (4.139) it follows that $\widehat{u} = \varphi(\widehat{u})$.

Solution of Problem 4.97
It suffices to show that

$$\inf_{u\in C} \|\varphi(u) - u\|_X = 0.$$

To this end fix $u_0 \in C$ and $\varepsilon \in (0,1)$ and consider the map $\varphi_\varepsilon \colon C \longrightarrow C$ defined by

$$\varphi_\varepsilon(u) = (1-\varepsilon)\varphi(u) + \varepsilon u_0 \qquad \forall\, u \in C. \tag{4.140}$$

Note that for all $u, v \in C$ we have

$$\|\varphi_\varepsilon(u) - \varphi_\varepsilon(v)\|_X = (1-\varepsilon)\|\varphi(u) - \varphi(v)\|_X \leqslant (1-\varepsilon)\|u - v\|_X,$$

so φ_ε is a k-contraction with $k = 1 - \varepsilon < 1$. Using the Banach fixed point theorem (see Theorem 4.59), we can find $u_\varepsilon \in C$ such that

$$u_\varepsilon = \varphi_\varepsilon(u_\varepsilon). \tag{4.141}$$

Then we have

$$\begin{aligned} \|u_\varepsilon - \varphi(u_\varepsilon)\|_X &= \|(1-\varepsilon)\varphi(u_\varepsilon) + \varepsilon u_0 - \varphi(u_\varepsilon)\|_X \\ \leqslant \varepsilon\|u_0 - \varphi(u_\varepsilon)\|_X &\leqslant \varepsilon \operatorname{diam} C \longrightarrow 0 \quad \text{as } \varepsilon \searrow 0. \end{aligned}$$

(see (4.140) and (4.141)).

Solution of Problem 4.98

We introduce a new metric on X, by

$$\widehat{d}_X(u,v) = \inf\Big\{\sum_{k=1}^n d_X(y_{k-1},y_k) : y_0,\dots,y_n \text{ is an } \varepsilon\text{-chain for } (u,v)\Big\}.$$

It is easy to see that $\widehat{d}_X$ is indeed a metric on X. Note that

$$d_X(u,v) \leqslant \widehat{d}_X(u,v) \qquad \forall\, u,v \in X$$

and

$$\text{if } d_X(u,v) < \varepsilon, \text{ then } d_X(u,v) = \widehat{d}_X(u,v). \tag{4.142}$$

So, $(X, \widehat{d}_X)$ is also a complete metric space. Let $\widehat{h}$ be the Hausdorff metric on $P_{bf}(X)$ generated by the metric $\widehat{d}_X$.

Let $u, v \in X$ and let $y_0,\dots,y_n$ be an ε-chain for (u,v). We have

$$h(F(y_{k-1}),F(y_k)) \leqslant \xi d_X(y_{k-1},y_k) \qquad \forall\, k \in \{1,\dots,n\}$$

(see the hypothesis on F), so by the triangle inequality, also

$$\begin{aligned} \widehat{h}(F(u),F(v)) &\leqslant \sum_{k=1}^n \widehat{h}(F(y_{k-1}),F(y_k)) \\ = \sum_{k=1}^n h(F(y_{k-1}),F(y_k)) &= \xi\sum_{k=1}^n d_X(y_{k-1},y_k) \end{aligned}$$

(see (4.142)), so

$$\widehat{h}(F(u),F(v)) \leqslant \xi\widehat{d}_X(u,v) \qquad \forall\, u,v \in X.$$

So, we can apply Theorem 4.68 and obtain $\widehat{u} \in X$ such that $\widehat{u} \in F(\widehat{u})$.

Solution of Problem 4.99

Let $\{u_n\}_{n\geqslant 1} \subseteq C$ be a sequence such that

$$(I_X - \varphi)(u_n) \longrightarrow 0 \quad \text{in } X. \tag{4.143}$$

Since X is uniformly convex, from the Milman–Pettis theorem (see Theorem I.5.89), we have that X is reflexive. Hence C is w-compact. So, by the Eberlein–Smulian theorem (see Theorem I.5.78) and by passing to a suitable subsequence if necessary, we may assume that

$$u_n \xrightarrow{w} \widehat{u} \quad \text{in } X. \tag{4.144}$$

From (4.143), (4.144) and Proposition 4.78, we have $0 = (I_X - \varphi)(\widehat{u})$, so $\widehat{u} = \varphi(\widehat{u})$.

Solution of Problem 4.100

Let $\{u_n\}_{n\geqslant 1} \subseteq X$ be a sequence such that $d_X(\varphi(u_n), u_n) \searrow 0$. We have that $\{\, \varphi(u_n)\,\}_{n\geqslant 1} \subseteq K$ and so by passing to a suitable subsequence if necessary, we may assume that

$$d_X(\varphi(u_n), y) \longrightarrow 0, \tag{4.145}$$

with $y \in K$, so

$$d_X(u_n, y) \leqslant d_X(u_n, \varphi(u_n)) + d_X(\varphi(u_n), y) \longrightarrow 0$$

and by the continuity of φ, we get

$$d_X(\varphi(u_n), \varphi(y)) \longrightarrow 0$$

thus $\varphi(y) = y$ (see (4.145)).

Solution of Problem 4.101

No. Suppose that $X = c_0 = \{\widehat{u} = \{u_k\}_{k\geqslant 1} : u_k \longrightarrow 0\}$, $\overline{B}_1 = \{u \in X : \|u\|_X \leqslant 1\}$ and $\varphi\colon \overline{B}_1 \longrightarrow \overline{B}_1$ is defined by

$$\varphi(u_1, u_2, \ldots) = (u_1, 1 - |u_1|, u_2, u_3, \ldots).$$

Note that φ is nonexpansive. Suppose that $\widehat{u} = \varphi(\widehat{u})$ with $\widehat{u} = \{u_k\}_{k\geqslant 1} \in c_0$. Then

$$u_k = 1 - |u_1| \qquad \forall\, k \geqslant 2.$$

But since $u_k \longrightarrow 0$ as $k \to +\infty$, we must have

$$|u_1| = 1 \quad \text{and} \quad u_k = 0 \qquad \forall\, k \geqslant 2.$$

So, the set of fixed points of φ is $\widehat{e}_1, -\widehat{e}_1$ with $\widehat{e}_1 = (1, 0, 0, \ldots)$, which is disconnected.

Solution of Problem 4.102

Note that r is the metric projection map on $\overline{B}_\varrho$ (the best approximation map). So, for any $u, v \neq 0$, we have

$$(u - r(u), r(v) - r(u))_H \leqslant 0. \tag{4.146}$$

Then

$$u - v = r(u) - r(v) + u - r(u) + r(v) - v = r(u) - r(v) + h,$$

with $h = u - r(u) + r(v) - v$. Hence

$$\|u - v\|_H^2 = \|r(u) - r(v)\|_H^2 + 2\,(h, r(u) - r(y))_H + \|h\|_H^2.$$

We have

$$(h, r(u) - r(v))_H = -\,(u - r(u), r(v) - r(u))_H - (v - r(v), r(u) - r(v))_H \geqslant 0$$

(see (4.146)), so

$$\|u - v\|_H^2 \geqslant \|r(u) - r(v)\|_H^2$$

and hence r is nonexpansive.

Solution of Problem 4.103

Because F is a multivalued k-contraction, we can find $u_1 \in F(u_0)$ such that

$$d_X(u_1, u_0) < (1 - k)r.$$

Now suppose that we have generated

$$u_n \in F(u_{n-1}) \quad \text{such that } d_X(u_n, u_{n-1}) < k^{n-1}(1 - k)r \qquad \forall\, n > 1. \tag{4.147}$$

From the k-contractivity of F and (4.147) we have

$$h(F(u_n), F(u_{n-1})) \leqslant k d_X(u_n, u_{n-1}) < k^{n-1}(1 - k)r.$$

So, from Definition 2.49 we see that we can find $u_{n+1} \in F(u_n)$ such that

$$d_X(u_{n+1}, u_n) < k^{n-1}(1 - k)r.$$

Therefore by induction we have produced a sequence $\{u_n\}_{n\geqslant 1} \subseteq \overline{B}_r(u_0)$ which satisfies (4.147). Note that for $m > n$ we have

$$d_X(x_m, x_n) \;<\; (1 + k + \ldots + k^{m-n})k^n(1-k)r$$

and so we infer that $\{u_n\}_{n\geqslant 1} \subseteq X$ is a Cauchy sequence. The completeness of X implies that $u_n \longrightarrow \widehat{u} \in \overline{B}_r(u_0)$. We have

$$d_X(u_{n+1}, F(\widehat{u})) \;\leqslant\; h(F(u_n), F(\widehat{u})) \;\leqslant\; kd_X(u_n, \widehat{u}),$$

so $d_X(\widehat{u}, F(\widehat{u})) = 0$, hence $\widehat{u} \in F(\widehat{u})$ (since F is closed valued).

Solution of Problem 4.104

Let

$$S \;=\; \{(t,u) \in [0,1] \times U : \; u \in H(t,u)\}.$$

By hypothesis $S \neq \emptyset$. On S we introduce the following partial order

$$(t,u) \preceq (s,v) \quad\Longleftrightarrow\quad t \leqslant s \text{ and } d_X(u,v) \leqslant \tfrac{2(\vartheta(s)-\vartheta(t))}{1-k}. \tag{4.148}$$

It is easy to see that $\preceq$ is a partial order (see Definition 4.116). Let T be a chain in S (see Definition 4.118) and define

$$t_0 \;=\; \sup_{(t,u)\in T} t.$$

We consider a $\preceq$-increasing sequence $\{(t_n, u_n)\}_{n\geqslant 1} \subseteq T$ such that $t_n \longrightarrow \widehat{t}$. We have

$$d_X(u_m, u_n) \;\leqslant\; \tfrac{2(\vartheta(t_m)-\vartheta(t_n))}{1-k} \qquad \forall\, m > n$$

(see (4.148)), so $\{u_n\}_{n\geqslant 1} \subseteq X$ is a Cauchy sequence.

The completeness of X implies that $u_n \longrightarrow u_0 \in \overline{U}$ in X. From hypotheses **(iii)** and **(iv)** and the triangle inequality for the Hausdorff metric, we see that the map $(t,u) \longmapsto H(t,u)$ is h-continuous. Since $(t_n, u_n) \in S$ for all $n \geqslant 1$, we have $u_n \in H(t_n, u_n)$ for all $n \geqslant 1$ and so $\widehat{u} \in H(\widehat{t}, \widehat{u})$. Also

$$(t,u) \;\preceq\; (t_0, u_0) \qquad \forall\, (t,u) \in T,$$

so (t_0, u_0) is an upper bound for T.

Since T is an arbitrary chain in S, from the Kuratowski–Zorn lemma (see Theorem 4.120), the set S admits a maximal element $(\widehat{t}, \widehat{u}) \in S$. We need to show that $\widehat{t} = 1$. We proceed by contradiction. So, suppose that $\widehat{t} < 1$. We can choose $r > 0$ and $t \in (\widehat{t}, 1]$ such that

$$\overline{B}_r(\widehat{u}) \;\subseteq\; U \quad\text{and}\quad r \;=\; \tfrac{2(\vartheta(t)-\vartheta(\widehat{t}))}{1-k}.$$

We have

$$\begin{aligned} d_X(\widehat{u}, H(t,\widehat{u})) &\leqslant d_X(\widehat{u}, H(\widehat{t},\widehat{u})) + h(H(t,\widehat{u}), H(\widehat{t},\widehat{u})) \\ = \ h(H(t,\widehat{u}), H(\widehat{t},\widehat{u})) &\leqslant \vartheta(t) - \vartheta(\widehat{t}) \ < \ (1-k)r \end{aligned}$$

(where we used Proposition 2.51(a), the fact that $(\widehat{t},\widehat{u}) \in S$ and hypothesis **(iv)**). Invoking Problem 4.103 we can find $\widetilde{u} \in \overline{B}_r(\widehat{u}) \subseteq U$ such that $\widetilde{u} \in H(t,\widetilde{u})$, so

$$(t,\widetilde{u}) \in S \quad \text{and} \quad (\widehat{t},\widehat{u}) \preceq (t,\widetilde{u}), \quad (\widehat{t},\widehat{u}) \neq (t,\widetilde{u}),$$

which contradicts the maximality of $(\widehat{t},\widehat{u})$. Therefore $\widehat{t} = 1$ and so G has a fixed point in U.

Solution of Problem 4.105

Suppose that **(b)** does not hold and $G|_{\partial U}$ has no fixed point. Hence $u \notin tG(u)$ for all $t \in [0,1]$, all $u \in \partial U$. Let $H\colon [0,1] \times \overline{U} \longrightarrow P_{bf}(X)$ be defined by

$$H(t,u) \ = \ tG(u).$$

Let F be the trivial multifunction, i.e., $F(u) = \{0\}$ for all $u \in \overline{U}$. Then we can apply Problem 4.104 and conclude that G has a fixed point in U.

Solution of Problem 4.106

We do the proof when condition **(iii)** holds. The other cases can be treated similarly. Suppose that we can find $t \in (0,1)$ and $u \in \partial U$ such that

$$u \ \in \ tF(u). \tag{4.149}$$

Then using Proposition 2.51(a) and (4.149) we have

$$\tfrac{1}{t}\|u\|_X \ \leqslant \ m_F(u) \quad \text{and} \quad \operatorname{dist}(u, F(u)) \ \leqslant \ (\tfrac{1}{t} - 1)\|u\|_X. \tag{4.150}$$

Because of **(iii)** and (4.150), we have

$$\tfrac{1}{t}\|u\|_X \ \leqslant \ \left((\tfrac{1}{t} - 1)^2\|u\|_X^2 + \|u\|_X^2\right)^{\frac{1}{2}},$$

so

$$\tfrac{1}{t^2}\|u\|_X^2 \ \leqslant \ \tfrac{1}{t^2}\|u\|_X^2 - \tfrac{2}{t}\|u\|_X^2 + 2\|u\|_X^2,$$

thus $\frac{1}{t} \leqslant 1$, a contradiction. Therefore (4.149) cannot hold for $t \in (0,1)$ and $u \in \partial U$. Using Problem 4.105, we conclude that F has a fixed point in $\overline{U}$. Similarly for the other cases.

Solution of Problem 4.107

Suppose that the set $E(F)$ is bounded. Then we can find $r > 0$ such that

$$E(F) \subseteq B_r(0) = \{u \in X : \|u\|_X < r\}. \tag{4.151}$$

By hypothesis $F|_{\overline{B}_r(0)}$ is a multivalued k-contraction. Also, for every $u \in \overline{B}_r(0)$ and every $y \in F(u)$, we have

$$d_X(y, F(0)) \leqslant h(F(u), F(0)) \leqslant k\|u\|_X \leqslant r,$$

so

$$\|y\|_X \leqslant r + |F(0)|,$$

where $|F(0)| = \sup_{v \in F(0)} \|v\|_X < +\infty$ (since $F(0)$ is bounded), thus the set $F(\overline{B}_r(0))$ is bounded.

Also note that (4.151) implies that for all $u \in \partial B_r(0)$ we have $u \notin tF(u)$ with $t \in [0,1]$. Therefore invoking Problem 4.105 we conclude that F has a fixed point.

Solution of Problem 4.108

For every $t \in T$ let

$$\mathrm{Fix}(\varphi_t) = \{u \in C : u = \varphi_t(u)\}.$$

From Theorem 4.70 and Problem 4.95 we know that $\mathrm{Fix}(\varphi_t)$ is nonempty, w-compact, and convex. Also, since the family $\{\varphi_t\}_{t \in T}$ is commuting, the family of sets $\{\mathrm{Fix}(\varphi_t)\}_{t \in T}$ has the finite intersection property. Since the set C is w-compact, we infer that

$$\bigcap_{t \in T} \mathrm{Fix}(\varphi_t) \neq \emptyset,$$

so the family $\{\varphi_t\}_{t \in T}$ has a common fixed point.

Solution of Problem 4.109

Let $\psi \colon \overline{B}_\varrho(u_0) \longrightarrow X$ be defined by

$$\psi(u) = u - \varphi'(u_0)^{-1}(\varphi(u)) \qquad \forall\, u \in \overline{B}_\varrho(u_0).$$

Evidently ψ is Fréchet differentiable and

$$\psi'(u) = I_X - \varphi'(u_0)^{-1} \circ \varphi'(u) \qquad \forall\, u \in \overline{B}_\varrho(u_0),$$

so

$$\begin{aligned}\|\psi'(u)\|_{\mathcal{L}} &\leqslant \|\varphi'(u_0)^{-1}\|_{\mathcal{L}}\|\varphi'(u_0) - \varphi'(u)\|_{\mathcal{L}} \leqslant \eta\varrho\|\varphi'(u_0)^{-1}\|_{\mathcal{L}} \\ &= 2\eta\|\varphi'(u_0)^{-1}\|_{\mathcal{L}}^2\|\varphi(u_0)\|_Y = \tfrac{\mu}{2}\end{aligned}$$

(use hypotheses **(ii)**, **(iv)**, **(iii)** and recall that $u \in \overline{B}_\varrho(u_0)$). So, by the mean value theorem it follows that ψ is Lipschitz continuous with Lipschitz constant $\xi \leqslant \frac{\mu}{2} \leqslant \frac{1}{2}$ (see **(iii)**). Also, we have

$$\|\psi(u_0) - u_0\|_X \leqslant \|\varphi'(u_0)^{-1}\|_{\mathcal{L}}\|\varphi(u_0)\|_Y = \tfrac{\varrho}{2},$$

thus

$$\begin{aligned}\|\psi(u) - u_0\|_X &\leqslant \|\psi(u) - \psi(u_0)\|_X + \|\psi(u_0) - u_0\|_X \\ &\leqslant \tfrac{\mu}{2}\|u - u_0\|_X + \tfrac{\varrho}{2} \leqslant \varrho \qquad \forall\, u \in \overline{B}_\varrho(u_0).\end{aligned}$$

Hence $\psi\colon \overline{B}_\varrho(u_0) \longrightarrow \overline{B}_\varrho(u_0)$ and it is a ξ-contraction. So, using the Banach fixed point theorem (see Theorem 4.59), we can find a unique $\widehat{u} \in \overline{B}_\varrho(u_0)$ such that

$$\psi(\widehat{u}) = \widehat{u},$$

so $\varphi(\widehat{u}) = 0$.

Solution of Problem 4.110

Note that the function $u \longmapsto \cos u$ is not a k-contraction on $[0, \frac{\pi}{2}]$. So, we cannot apply the Banach fixed point theorem (see Theorem 4.59). On the other hand, $\cos^{(2)}(u)$ is a k-contraction since

$$\tfrac{d}{du}(\cos^{(2)}(u)) = \sin(\cos u)\sin u,$$

so

$$\left|\tfrac{d}{du}\cos^{(2)}(u)\right| \leqslant \sin 1 \in (0,1).$$

So, by the mean value theorem, we have that $u \longmapsto \cos^{(2)}(u)$ is a k-contraction on $[0, \frac{\pi}{2}]$, with $k \leqslant \sin 1$. Then we can apply Proposition 4.62 and have $\widehat{u} \in [0, \frac{\pi}{2}]$ such that $\cos \widehat{u} = \widehat{u}$.

Solution of Problem 4.111

Since by hypothesis φ and ψ commute, we have

$$\psi(\widehat{u}) = \psi(\varphi(\widehat{u})) = \varphi(\psi(\widehat{u})),$$

so $\psi(\widehat{u}) = \widehat{u}$ (because $\widehat{u} \in X$ is the unique fixed point of φ).

Solution of Problem 4.112

By the Banach fixed point theorem (see Theorem 4.59), the map φ_{t_0} has a unique fixed point $\widehat{u} \in X$. Using Problem 4.111 we conclude that $\widehat{u} \in X$ is a fixed point for all φ_t with $t \in T$.

Solution of Problem 4.113

Arguing by contradiction, suppose that φ has no zeros in $\overline{B}_r(0)$. Then $\psi\colon \overline{B}_r(0) \longrightarrow \overline{B}_r(0)$ defined by

$$\psi(u) = \tfrac{r\varphi(u)}{\|\varphi(u)\|_X} \qquad \forall\, u \in \overline{B}_r(0)$$

is continuous and the set $\overline{\varphi(\overline{B}_r(0))}$ is compact. So, we can apply the Schauder fixed point theorem (see Theorem 4.89) and obtain $\overline{u} \in \overline{B}_r(0)$ such that $\psi(\overline{u}) = \overline{u}$. Thus

$$\varphi(\overline{u}) = \tfrac{\|\varphi(\overline{u})\|_X}{r}\overline{u},$$

so $\|\overline{u}\|_X = r$ and this contradicts our hypothesis on φ.

Solution of Problem 4.114

(a) We argue indirectly. So, suppose that φ has no zeros in $\overline{B}_r(0)$. Hence we can define the map $\psi_0\colon \overline{B}_r(0) \longrightarrow \overline{B}_r(0)$ by

$$\psi_0(u) = -\tfrac{r\varphi(u)}{\|\varphi(u)\|_X} \qquad \forall\, u \in \overline{B}_r(0).$$

The Schauder fixed point theorem (see Theorem 4.89) implies that there exists $\overline{u} \in \overline{B}_r(0)$ such that $\psi_0(\overline{u}) = \overline{u}$, so

$$\varphi(\overline{u}) = -\tfrac{\|\varphi(\overline{u})\|_X}{r}\overline{u},$$

thus $\|\overline{u}\|_X = r$.

Let $\overline{u}^* \in X^*$ be such that $\langle \overline{u}^*, \overline{u}\rangle = \|\overline{u}\|_X$ (it exists by the Hahn–Banach theorem; see Theorem I.5.24). Then

$$\langle \overline{u}^*, \varphi(\overline{u})\rangle = -\tfrac{\|\varphi(\overline{u})\|_X}{r}\|\overline{u}\|_X = -\|\varphi(\overline{u})\|_X < 0,$$

which contradicts our hypothesis on φ.

(b) We fix $h_0 \in \mathbb{R}^N$ and consider the map

$$\psi(u) = \varphi(u) - h_0 \qquad \forall\, u \in \mathbb{R}^N.$$

From our hypothesis on φ, we see that for $r > 0$ big, we have

$$\frac{(\psi(u),u)_{\mathbb{R}^N}}{|u|} \geqslant 0 \qquad \forall\, u \in \partial B_r(0). \tag{4.152}$$

For every $u \in \partial B_r(0)$, let $u^* = u^*(u) = \frac{u}{|u|} \in \mathbb{R}^N$. Then $(u^*, u)_{\mathbb{R}^N} = |u|$ and from (4.152), we have $(u^*, \psi(u))_{\mathbb{R}^N} \geqslant 0$. Using part **(a)** of the problem, we can find $\widehat{u} \in \mathbb{R}^N$ such that $\psi(\widehat{u}) = 0$, so $\varphi(\widehat{u}) = h_0$. Since $h_0 \in \mathbb{R}^N$ is arbitrary, we conclude that φ is surjective.

Solution of Problem 4.115

We have that $I_X - \psi$ is continuous and for all $u, v \in C$ we have

$$\begin{aligned}\|(I_X - \psi)(u) - (I_X - \psi)(v)\|_X &\geqslant \|u - v\|_X - \|\psi(u) - \psi(v)\|_X \\ &\geqslant (1-k)\|u - v\|_X,\end{aligned}$$

so $(I_X - \psi)^{-1}$ is continuous. Hence $I_X - \psi$ is a homeomorphism of C onto $(I_X - \psi)(C)$.

For fixed $u \in C$, the map $v \longmapsto \varphi(u) + \psi(v)$ is a k-contraction on C and so by the Banach fixed point theorem (see Theorem 4.59) (see hypothesis **(iii)**), there exists a unique $\widehat{v} = \widehat{v}(u) \in C$ such that

$$\widehat{v} = \varphi(u) + \psi(\widehat{v}),$$

so

$$\widehat{v} = (I_X - \psi)^{-1}(\varphi(u)) \in C.$$

Hence the map $(I_X - \psi)^{-1} \circ \varphi \colon C \longrightarrow C$ is compact (being the composition of a compact and a continuous map). So, the Schauder fixed point theorem (see Theorem 4.89) implies that there exists $\widehat{u} \in C$ such that

$$(I_X - \psi)^{-1}(\varphi(\widehat{u})) = \widehat{u},$$

so $\varphi(\widehat{u}) + \psi(\widehat{u}) = \widehat{u}$.

Remark. See also Remark after the solution of Problem 4.135.

Solution of Problem 4.116

From the hypothesis, we see that A^{-1} exists and $A^{-1} \in \mathcal{L}(V^*; V)$. Moreover, from Theorem 2.119, we have that A is surjective. So, given $h^* \in V^*$ and $x \in X$, we can find a unique $v \in V$ such that

$$A(v) = h^* - \varphi(x),$$

hence

$$v = A^{-1}(h^* - \varphi(x)).$$

Let $g\colon X \longrightarrow X$ be defined by

$$g(x) = A^{-1}(h^* - \varphi(x)) \qquad \forall\, x \in X.$$

Exploiting the compactness of the embedding of V into X and the continuity of φ, we see that g is compact.

Let $\lambda \in (0,1)$ and suppose that $v_\lambda = \lambda g(v_\lambda)$. Then we have

$$A(v_\lambda) + \lambda\varphi(v_\lambda) = \lambda h^*. \tag{4.153}$$

Acting on (4.153) with v_λ and using the assumptions on A and φ, we obtain

$$c_0\|v_\lambda\|_V^2 \leqslant c_1(1 + \|u_\lambda\|_V^\tau) + \|h^*\|_{V^*}\|v_\lambda\|_V \qquad \forall\, \lambda \in (0,1). \tag{4.154}$$

Since $\tau \in [0,2)$, from (4.154) we infer that there exists $M > 0$ such that

$$\|v_\lambda\|_V \leqslant M \qquad \forall\, \lambda \in (0,1).$$

Applying the Leray–Schauder alternative theorem (see Theorem 4.93), we obtain $\widehat{v} \in V$ such that $\widehat{v} = g(\widehat{v})$, so

$$A(\widehat{v}) + \varphi(\widehat{v}) = h^*.$$

Since $h^* \in V^*$ is arbitrary, we conclude that $A + \varphi$ is surjective.

Solution of Problem 4.117

We assume that $\varphi|_{\partial U}$ is fixed point free (or otherwise there is nothing to prove). So, we have

$$u \neq \lambda\varphi(u) + (1-\lambda)u_0 \qquad \forall\, u \in \partial U,\ \lambda \in [0,1]. \tag{4.155}$$

Let

$$D = \big\{u \in \overline{U}:\ u = \lambda\varphi(u) + (1-\lambda)u_0 \text{ for some } \lambda \in [0,1]\big\}.$$

Evidently $u_0 \in D$ and so $D \neq \emptyset$. Also, D is closed and $D \cap \partial U = \emptyset$ (see (4.155)). Invoking the Urysohn lemma (see Theorem I.2.136), we can find a continuous function $\vartheta\colon \overline{U} \longrightarrow [0,1]$ such that $\vartheta|_D = 1$ and $\vartheta|_{\partial U} = 0$. Let $g\colon C \longrightarrow C$ be defined by

$$g(u) = \begin{cases} \vartheta(u)\varphi(u) + (1-\vartheta(u))u_0 & \text{if } u \in \overline{U}, \\ u_0 & \text{if } u \in C \setminus \overline{U}. \end{cases} \tag{4.156}$$

Evidently g is compact. So, the Schauder fixed point theorem (see Theorem 4.89) implies that we can find $\widehat{u} \in C$ such that $g(\widehat{u}) = \widehat{u}$. Since $u_0 \in U$, from (4.156), we see that $\widehat{u} \in \overline{U}$. Hence

$$\vartheta(\widehat{u})\varphi(\widehat{u}) + (1 - \vartheta(\widehat{u}))u_0 = \widehat{u},$$

so $\widehat{u} \in D$ and thus $\varphi(\widehat{u}) = \widehat{u}$ (since $\vartheta|_D = 1$).

Solution of Problem 4.118

Given $\varepsilon > 0$, we can find $n_0 = n_0(\varepsilon) \geqslant 1$ such that

$$d_X(\widehat{u}_{n_k}, \widehat{u}) \leqslant \tfrac{\varepsilon}{2} \quad \text{and} \quad h(F_{n_k}(\widehat{u}), F_0(\widehat{u})) \leqslant \tfrac{\varepsilon}{2} \qquad \forall\, n_k \geqslant n_0. \tag{4.157}$$

Then for $n_k \geqslant n_0$ we have

$$\begin{aligned} h(F_{n_k}(\widehat{u}_{n_k}), F_0(\widehat{u})) &\leqslant h(F_{n_k}(\widehat{u}_{n_k}), F_{n_k}(\widehat{u})) + h(F_{n_k}(\widehat{u}), F_0(\widehat{u})) \\ &\leqslant d_X(\widehat{u}_{n_k}, \widehat{u}) + h(F_{n_k}(\widehat{u}), F_0(\widehat{u})) \leqslant \varepsilon \end{aligned}$$

(see (4.157)), so

$$F_{n_k}(\widehat{u}_{n_k}) \stackrel{h}{\longrightarrow} F_0(\widehat{u}).$$

Since $\widehat{u}_{n_k} \in F_{n_k}(\widehat{u}_{n_k})$, we have

$$\operatorname{dist}(\widehat{u}_{n_k}, F_0(\widehat{u})) \longrightarrow 0,$$

so $\operatorname{dist}(\widehat{u}, F_0(\widehat{u})) = 0$, hence $\widehat{u} \in F_0(\widehat{u})$.

Solution of Problem 4.119

Let $R_n \colon P_k(X) \longrightarrow P_k(X)$ (for $n \geqslant 0$) be as in Problem 2.70. We know that R_n is a k_n-contraction on the complete metric space $(P_k(X), h)$ (here $k_n \in (0,1)$ denotes the contraction constant of the multifunction F_n). By the Banach fixed point theorem (see Theorem 4.59), R_n admits a unique fixed point $C_n \in P_k(X)$ (for $n \geqslant 0$). If hypothesis **(i)** holds, then

$$F_n(u) \stackrel{h}{\longrightarrow} F(u) \quad \text{uniformly for } u \in K \in P_k(X),$$

so

$$R_n(K) \longrightarrow R_0(K) \qquad \forall\, K \in P_k(X).$$

If hypothesis **(ii)** holds, then

$$\sup_{K \in P_k(X)} h(R_n(K), R_0(K)) \longrightarrow 0 \quad \text{as } n \to +\infty,$$

so, from Problem 4.87, we get

$$C_n \xrightarrow{h} C_0.$$

If hypothesis **(iii)** holds, then $(P_k(X), h)$ is locally compact and from Problem 4.86, we get

$$C_n \xrightarrow{h} C_0.$$

Next we show that $D = \bigcup\limits_{n \geqslant 0} C_n \in P_k(X)$. To this end, let $\{\widehat{u}_n\}_{n \geqslant 1} \subseteq D$. Then either one C_k contains infinitely many $\widehat{u}_n$'s or otherwise each C_k contains only a finite number of $\widehat{u}_n$'s. In the first case, we can extract a subsequence converging to an element of C_k. In the second case, every limit point of $\{\widehat{u}_n\}_{n \geqslant 1}$ belongs in C_0 (as $C_n \xrightarrow{h} C_0$). Therefore the set D is indeed compact.

From the Banach fixed point theorem (see Theorem 4.59), we know that $R_n^{(m)}(\widehat{u}_n) \xrightarrow{h} C_n$ as $m \to +\infty$ and since $\widehat{u}_n \in R_n^{(m)}(\widehat{u}_n)$ for all $n \geqslant 1$, it follows that $\widehat{u}_n \in C_n$. So, $\{\widehat{u}_n\}_{n \geqslant 1} \subseteq D$. Since the set D is compact, we can find a subsequence $\{\widehat{u}_{n_k}\}_{k \geqslant 1}$ of $\{\widehat{u}_n\}_{n \geqslant 1}$ such that $\widehat{u}_{n_k} \longrightarrow \widehat{u}$ with $\widehat{u} \in F_0(\widehat{u})$ (see Problem 4.118).

Solution of Problem 4.120

Let $\mathcal{B}$ be a local basis at the origin consisting of open, balanced (i.e., for every $U \in \mathcal{B}$, we have $\lambda U \subseteq U$ when $|\lambda| \leqslant 1$) and convex sets. Such a local basis exists since X is locally convex. For every $U \in \mathcal{B}$, we define

$$D_U = \{u \in C : \ u \in F(u) + \overline{U}\}.$$

We claim that the set D_U is closed. So, let $u_0 \notin D_U$. Then

$$u_0 \in C \quad \text{and} \quad u_0 \notin F(u_0) + \overline{U}.$$

So, there exists $V \in \mathcal{N}(0)$ (with $\mathcal{N}(0)$ being the filter of neighborhoods of the origin) such that

$$(u_0 + V) \cap (F(u_0) + \overline{U} + V) = \emptyset. \tag{4.158}$$

Since by hypothesis F is upper semicontinuous (see Definition 2.36(a)), we can find $W(u_0)$, a neighborhood of u_0 in C such that

$$F(u) \subseteq F(u_0) + V \qquad \forall\, u \in W(u_0).$$

Hence, we have

$$(u_0 + V) \cap (F(u) + \overline{U}) = \emptyset \qquad \forall\, u \in W(u_0)$$

(see (4.158)), so the set D_U is closed (since its complement is open).

Next we show that the set D_U is nonempty. Since the set $C \subseteq X$ is compact, we can find $c_1, \dots c_m$ such that $\{c_1 + U, \dots, c_m + U\}$ is an open cover of C. Let $K = \operatorname{conv}\{c_1, \dots, c_m\} \subseteq C$ and define

$$F_U(u) \;=\; (F(u) + \overline{U}\,) \cap K.$$

Since $F(u) \subseteq C$ and $U = -U$, we see that $F_U(x) \in P_{fc}(C)$. The set K lies in a finite dimensional subspace of X which can be identified with $\mathbb{R}^m$. The multifunction $F_U \colon K \longrightarrow P_{fc}(K)$ is upper semicontinuous (having closed graph) and so by the Ky Fan fixed point theorem (see Remark 4.108), we can find $\widehat{u} \in F_U(\widehat{u})$. Hence $D_U \neq \emptyset$.

Evidently $\{D_U\}_{U \in \mathcal{B}}$ has the finite intersection property. Then the compactness of C implies that $\bigcap\limits_{U \in \mathcal{B}} D_U \neq \emptyset$. Let $\widehat{u} \in \bigcap\limits_{U \in \mathcal{B}} D_U$. Then $\widehat{u} \in F(\widehat{u})$.

Solution of Problem 4.121

Since the multifunction F has closed graph, it has closed values. Also, the compactness of C and Proposition 2.45 imply that F is upper semicontinuous. So, we can use Problem 4.120 and conclude that F has a fixed point.

Solution of Problem 4.122

Let $G(u) = F(u) \cap C$ for all $u \in C$. Then $G \colon C \longrightarrow P_{fc}(C)$ and by Proposition 2.59(b), it is upper semicontinuous. So, we can use Problem 4.120 and find $\widehat{u} \in C$ such that $\widehat{u} \in G(\widehat{u}) \subseteq F(\widehat{u})$.

Solution of Problem 4.123

Let $t_n = \frac{n}{n+1}$ for $n \geqslant 1$ and let $\varphi_n \colon C \longrightarrow C$ be defined by

$$\varphi_n(u) \;=\; h_{\varphi(u)}(t_n) \qquad \forall\, u \in C,\; n \geqslant 1.$$

For every $n \geqslant 1$ and every $u, v \in X$, we have

$$\|\varphi_n(u) - \varphi_n(v)\|_X \;=\; \|h_{\varphi(u)}(t_n) - h_{\varphi(v)}(t_n)\| \;\leqslant\; \xi(t_n)\|u - v\|_X,$$

so φ_n is a $\xi(t_n)$-contraction. The Banach fixed point theorem (see Theorem 4.59) implies that φ_n has a unique fixed point $\widehat{u}_n \in C$ (for $n \geqslant 1$). Due to the compactness of C and by passing to a subsequence if necessary, we may assume that $\widehat{u}_n \longrightarrow \widehat{u} \in C$. We have

$$\varphi_n(\widehat{u}_n) \;=\; \widehat{u}_n \qquad \forall\, n \geqslant 1,$$

so

$$\varphi_n(\widehat{u}_n) \longrightarrow \widehat{u} \quad \text{in } X. \tag{4.159}$$

Also, from the definition of φ_n, we have

$$\varphi_n(\widehat{u}_n) = h_{\varphi(\widehat{u}_n)}(t_n) \longrightarrow h_{\varphi(\widehat{u})}(1) = \varphi(\widehat{u}). \tag{4.160}$$

From (4.159) and (4.160), we conclude that $\varphi(\widehat{u}) = \widehat{u}$.

Solution of Problem 4.124

Since the set C is star-shaped with star center $y_0 \in C$, we have

$$ty_0 + (1-t)u \in C \qquad \forall\, u \in C,\ t \in [0,1].$$

Then for every $u \in C$, let $h_u\colon [0,1] \longrightarrow C$ be defined by

$$h_u(t) = (1-t)y_0 + tu.$$

We have

$$\|h_u(t) - h_v(t)\|_X = t\|u-v\|_X.$$

So, we can use Problem 4.123, to conclude that φ has a fixed point.

Solution of Problem 4.125

Let $r\colon X \longrightarrow C$ be a retraction and let $\varphi\colon C \longrightarrow C$ be a continuous map. Then $\varphi \circ r\colon X \longrightarrow C$ is a continuous map and so there exists $\widehat{u} \in X$ such that

$$(\varphi \circ r)(\widehat{u}) = \widehat{u} \in C.$$

So, $r(\widehat{u}) = \widehat{u}$ and it follows that $\varphi(\widehat{u}) = \widehat{u}$. Therefore the set C has the fixed point property.

Solution of Problem 4.126

(a) Let $P_n\colon \mathcal{H} \longrightarrow \mathcal{H}$ be the continuous map defined by

$$P_n(h) = (u_1, \ldots, u_n, 0, \ldots) \qquad \forall\, h = \{u_k\}_{k \geqslant 1} \in \mathcal{H}.$$

Then given $\varepsilon > 0$, we can find $n \geqslant 1$ big enough such that

$$\|P_n(h) - h\|_{\mathcal{H}} \leqslant \Big(\sum_{k \geqslant n+1} \frac{1}{k^2} \Big)^{\frac{1}{2}} < \varepsilon \qquad \forall\, h \in \mathcal{H}.$$

Since $P_n(\mathcal{H})$ is compact, it follows that $\mathcal{H}$ is compact. Let $C_n = P_n(\mathcal{H})$ for $n \geqslant 1$. Evidently, we can view C_n as a compact, convex set in $\mathbb{R}^n$. If $\varphi\colon \mathcal{H} \longrightarrow \mathcal{H}$ is continuous, then $P_n \circ \varphi\colon C_n \longrightarrow C_n$ is continuous and so by the Brouwer fixed point theorem (see Theorem 4.81), we can find $\widehat{h}_n \in C_n$ such that

$$\|\widehat{h}_n - \varphi(\widehat{h}_n)\|_{\mathcal{H}} \leqslant \Big(\sum_{k\geqslant n+1} \tfrac{1}{k^2}\Big)^{\frac{1}{2}}. \tag{4.161}$$

Since $\mathcal{H}$ is compact and $\{\widehat{h}_n\}_{n\geqslant 1} \subseteq \mathcal{H}$, we may assume that $\widehat{h}_n \longrightarrow \widehat{h}$, thus $\widehat{h} = \varphi(\widehat{h})$ (see (4.161)).
Since $\varphi\colon \mathcal{H} \longrightarrow \mathcal{H}$ is an arbitrary continuous map, we conclude that $\mathcal{H}$ has the fixed point property.

(b) Without any loss of generality, we may assume that $C \subseteq \overline{B}_1$ (where $\overline{B}_1 = \{u \in X : \|u\|_X \leqslant 1\}$). Let $\{u_n\}_{n\geqslant 1} \subseteq X$ and $\{u_n^*\}_{n\geqslant 1} \subseteq X^*$ be two sequences such that $\{u_n\}_{n\geqslant 1}$ is dense in $\overline{\operatorname{span}}\, C$ and

$$\langle u_n^*, u_n\rangle = \frac{\|u_n\|_X}{n} \quad\text{and}\quad \|u_n^*\|_* = \frac{1}{n} \qquad \forall\, n \geqslant 1.$$

Let $L\colon X \longrightarrow l^2$ be the linear map defined by

$$L(u) = \{\langle u_n^*, u\rangle\}_{n\geqslant 1} \qquad \forall\, u \in X.$$

Then $L(C) \subseteq \mathcal{H}$ and $L \in \mathcal{L}(X; l^2)$. Moreover, if $u, v \in \overline{\operatorname{span}}\, C$ and $u \neq v$, then

$$\begin{aligned} \big|\,\langle u_n^*, u\rangle - \langle u_n^*, v\rangle\,\big| &\geqslant \big|\,\langle u_n^*, u_n\rangle\,\big| - \big|\,\langle u_n^*, u - v - u_n\rangle\,\big| \\ &\geqslant \frac{\|u_n\|_X}{n} - \frac{1}{n}\|(u-v) - u_n\|_X > 0 \end{aligned}$$

(choosing u_n sufficiently close to $u - v$), so $L|_{\overline{\operatorname{span}}\, C}$ is an injection. Therefore L is a homeomorphism of C onto $L(C) \subseteq \mathcal{H}$ and $L(C)$ is a compact and convex set.

(c) Let $C \subseteq \mathcal{H}$ be a closed and convex set. From the Dugundji extension theorem (see Theorem 2.8), D is a retract of $\mathcal{H}$. By part **(a)** the cube $\mathcal{H}$ has the fixed point property. So, we can use Problem 4.125 to conclude that D has the fixed point property.

Solution of Problem 4.127

By the Mazur theorem (see Theorem I.5.58), the set C is also weakly closed. The reflexivity of X implies that $\varphi(C)$ is a relatively w-compact subset of X, hence the set $\overline{\operatorname{conv}}\,\varphi(C) \subseteq C$ is w-compact and convex (by the Krein–Smulian theorem; see Theorem I.5.86). Invoking the Tichonov fixed point theorem (see Theorem 4.101) for the locally convex space X_w (where X_w is the Banach space X furnished with the weak topology), we produce a fixed point for φ.

Solution of Problem 4.128

Let $h \in X$ and consider the map

$$\psi(u) = h - \varphi(u) \qquad \forall\, u \in X.$$

We need to show that ψ has a fixed point. Let $r > 0$ and let $\overline{B}_r = \{u \in X : \|u\|_X \leqslant r\}$ be equipped with the relatively weak topology. Then by hypothesis, $\psi|_{\overline{B}_r}$ is continuous into X_w (where X_w is the Banach space X furnished with the weak topology) and $\overline{\psi(B_r)}^w$ is w-compact in X. Let

$$m_r = \sup_{u \in \overline{B}_r} \|\varphi(u)\|_X.$$

Then from our hypothesis on φ, we have

$$\limsup_{n \to +\infty} \tfrac{m_r}{r} < 1. \tag{4.162}$$

Also, if $\|u\|_X \leqslant r$, then $\|\psi(u)\|_X \leqslant \|h\|_X + m_r$. So, from (4.162), we see that if $r > 0$ is chosen big, then $\psi \colon \overline{B}_r \longrightarrow \overline{B}_r$ and we can apply the Tichonov fixed point theorem (see Theorem 4.101), to conclude that there exists $\widehat{u} \in \overline{B}_r$ such that $\psi(\widehat{u}) = \widehat{u}$. Then $\widehat{u} + \varphi(\widehat{u}) = h$. Since $h \in X$ is arbitrary, we conclude that $I_X + \varphi$ is surjective.

Solution of Problem 4.129

(a) Arguing by contradiction, suppose that φ has no fixed point. Then Theorem 4.91 implies that there exist $u \in \overline{B}_r$ and $t \in (0, 1)$ such that

$$\|u\|_X = r \quad \text{and} \quad u = t\varphi(u). \tag{4.163}$$

Then by hypothesis, we have

$$\|\varphi(u) - u\|_X^2 \geqslant \|\varphi(u)\|_X^2 - \|u\|_X^2,$$

so $t \geqslant 1$ (see (4.163)), a contradiction (since $t \in (0, 1)$). So, φ must have a fixed point.

(b) For all $u \in \partial B_r$, we have

$$\begin{aligned}
\|\varphi(u) - u\|_X^2 &\geqslant \big(\|\varphi(u)\|_X - \|u\|_X\big)^2 \\
&= \|\varphi(u)\|_X^2 - 2\|u\|_X \|\varphi(u)\|_X + \|u\|_X^2 \\
&\geqslant \|\varphi(u)\|_X^2 - 2\|u\|_X^2 + \|u\|_X^2 = \|\varphi(u)\|_X^2 - \|u\|_X^2
\end{aligned}$$

(since $\varphi(\partial B_r) \subseteq \overline{B}_r$). So, we can apply part **(a)** and produce a fixed point for φ.

Solution of Problem 4.130
The result is trivial if $\operatorname{int} C = \emptyset$. So, we assume that $\operatorname{int} C \neq \emptyset$ and by translating things if necessary, without any loss of generality we may assume that $0 \in \operatorname{int} C$. Let $r\colon X \longrightarrow C$ be defined by

$$r(u) = \frac{u}{\max\{1, \mathrm{p}_C(u)\}} \qquad \forall\, u \in X.$$

Here p_C is the Minkowski functional of C defined by

$$\mathrm{p}_C(u) = \inf_{\substack{\lambda > 0 \\ u \in \lambda C}} \lambda.$$

Note that $r\colon X \longrightarrow C$ is continuous, surjective and

$$\text{if } r(u) \in \operatorname{int} C, \text{ then } r(u) = u$$

and

$$\text{if } u \notin C, \text{ then } r(u) \in \partial C.$$

Let $\psi = r \circ \varphi$. Then $\psi\colon C \longrightarrow C$ and it is compact (since φ is compact). Using the Schauder fixed point theorem (see Theorem 4.89), we can find $\widehat{u} \in C$ such that $\psi(\widehat{u}) = \widehat{u}$.

If $\widehat{u} \in \partial C$, then by hypothesis $\varphi(\widehat{u}) \in C$ and so $\widehat{u} = r(\varphi(\widehat{u})) = \varphi(\widehat{u})$.

If $u \in \operatorname{int} C$, then $(r \circ \varphi)(\widehat{u}) \in \operatorname{int} C$ and so

$$\widehat{u} = r(\varphi(\widehat{u})) = \varphi(\widehat{u}).$$

Solution of Problem 4.131
Let

$$\operatorname{Fix}(\varphi) = \{v \in C : v = \varphi(v)\}$$

(the set of fixed points of the map φ). From the Schauder fixed point theorem (see Theorem 4.89), we know that $\operatorname{Fix}(\varphi) \neq \emptyset$. We claim that

$$\|u_{n+1} - v\|_X \leqslant \|u_n - v\|_X \qquad \forall\, v \in \operatorname{Fix}(\varphi),\ n \geqslant 0. \tag{4.164}$$

To this end, we have

$$\begin{aligned} \|u_{n+1} - v\|_X &= \left\|\tfrac{1}{2}(u_n + \varphi(u_n)) - \tfrac{1}{2}(v + \varphi(v))\right\|_X \\ &= \left\|\tfrac{1}{2}(u_n - v) + \tfrac{1}{2}(\varphi(u_n) - \varphi(v))\right\|_X \\ &\leqslant \tfrac{1}{2}\|u_n - v\|_X + \tfrac{1}{2}\|u_n - v\|_X = \|u_n - v\|_X \end{aligned}$$

(since $v = \varphi(v)$ and φ is nonexpansive). So, (4.164) holds.

Suppose that we can find $\varepsilon > 0$ and $n_0 \geqslant 1$ such that

$$\|u_n - \varphi(u_n)\|_X \geqslant \varepsilon \qquad \forall\, n \geqslant n_0. \tag{4.165}$$

Then

$$\|u_n - v - (\varphi(u_n) - \varphi(v))\|_X \geqslant \varepsilon \qquad \forall\, n \geqslant n_0,\ v \in \operatorname{Fix}(\varphi). \tag{4.166}$$

Also, since φ is nonexpansive, we have

$$\|\varphi(u_n) - \varphi(v)\|_X \leqslant \|u_n - v\|_X \leqslant \|u_0 - v\|_X \tag{4.167}$$

(see (4.164)). Because X is uniformly convex (see Definition I.5.88), from (4.166) we see that there exists $\delta = \delta(\varepsilon) \in (0,1)$ such that

$$\begin{aligned} \|u_{n+1} - v\|_X &= \left\|\tfrac{1}{2}(u_n - v) + \tfrac{1}{2}(\varphi(u_n) - \varphi(v))\right\|_X \\ &\leqslant \delta \max\left\{\|u_n - v\|_X, \|\varphi(u_n) - \varphi(v)\|_X\right\} \\ &\leqslant \delta\|u_n - v\|_X \qquad \forall\, n \geqslant n_0 \end{aligned}$$

(since $v \in \operatorname{Fix}(\varphi)$ and φ is nonexpansive), so $u_n \longrightarrow v$ (see (4.164) and recall that $\delta < 1$).

If there are no $\varepsilon > 0$ and $n_0 \geqslant 1$ for which (4.165) is true, then by passing to a subsequence if necessary, we may assume that

$$u_n - \varphi(u_n) \longrightarrow 0 \quad \text{in } X$$

and

$$\varphi(u_n) \longrightarrow \widehat{u} \quad \text{in } X$$

(recall that φ is compact). So, we have

$$u_n \longrightarrow \widehat{u} \quad \text{in } X \quad \text{and} \quad \varphi(\widehat{u}) = \widehat{u}.$$

Then, from (4.164) with $v = \widehat{u} \in \operatorname{Fix}(\varphi)$, we conclude that the original sequence $\{u_n\}_{n\geqslant 1} \subseteq C$ converges to $\widehat{u}$.

Solution of Problem 4.132

For $u, v \in \overline{B}_r(y)$, we have

$$\begin{aligned} \|(I_X - \lambda\varphi)(u) - (I_X - \lambda\varphi)(v)\|_X &\geqslant \|u - v\|_X - |\lambda|\|\varphi(u) - \varphi(v)\|_X \\ &\geqslant (1 - |\lambda|k)\|u - v\|_X \end{aligned}$$

(since φ is k-Lipschitz), so $I_X - \lambda\varphi$ is an injection and $(1-|\lambda|k)^{-1}$ is a Lipschitz constant of $(I_X - \lambda\varphi)^{-1}$.

Let $h \in \overline{B}_\varrho(y - \lambda\varphi(y))$ and consider the map $\psi\colon \overline{B}_r(y) \longrightarrow X$ defined by

$$\psi(u) \;=\; \lambda\varphi(u) + h \qquad \forall\, u \in \overline{B}_r(y).$$

Then for $u \in \overline{B}_r(y)$, we have

$$\begin{aligned} \|\psi(u) - y\|_X \;&=\; \|\lambda\varphi(u) + h - y\|_X \\ &\leqslant\; \|\lambda\varphi(u) - \lambda\varphi(y)\|_X + \|h - (y - \lambda\varphi(y))\|_X \\ &\leqslant\; |\lambda|k\|u - y\|_X + (1 - \lambda|k|)r \;\leqslant\; r \end{aligned}$$

(since φ is nonexpansive, $h \in \overline{B}_\varrho(y - \lambda\varphi(y))$ and $u \in \overline{B}_r(y)$). This means that $\psi\colon \overline{B}_r(y) \longrightarrow \overline{B}_r$ and

$$\|\psi(u) - \psi(v)\|_X \;=\; |\lambda|\|\varphi(u) - \varphi(v)\|_X \;\leqslant\; |\lambda|k\|u - v\|_X \qquad \forall\, u, v \in \overline{B}_r(y),$$

so ψ is a $|\lambda|k$-contraction.

So, we can apply the Banach fixed point theorem (see Theorem 4.59) and produce $\widehat{u} \in \overline{B}_r(y)$ such that $\psi(\widehat{u}) = \widehat{u}$, so

$$\lambda\varphi(\widehat{u}) + h \;=\; \widehat{u},$$

thus

$$h \;=\; \widehat{u} - \lambda\varphi(\widehat{u})$$

and hence $\overline{B}_\varrho(y - \lambda\varphi(y)) \subseteq R(I_X - \lambda\varphi)$.

Solution of Problem 4.133

Let $c_0 = \Big\{\widetilde{u} = \{u_n\}_{n\geqslant 1} : \, u_n \longrightarrow 0\Big\}$. We know that c_0 furnished with the norm

$$\|\widetilde{u}\|_{c_0} \;=\; \max\{|u_n| : \, n \geqslant 1\} \qquad \forall\, \widetilde{u} = \{u_n\}_{n\geqslant 1} \in c_0$$

is a Banach space. Let $\overline{B}_1 = \{\widetilde{u} \in c_0 : \, \|\widetilde{u}\|_{c_0} \leqslant 1\}$ and consider the map $\varphi\colon \overline{B}_1 \longrightarrow X$ defined by

$$\varphi(\widetilde{u}) \;=\; \widetilde{v},$$

where $\widetilde{v} = \{v_n\}_{n\geqslant 1} \in c_0$ is defined by

$$v_1 \;=\; \tfrac{1}{2}(1 + \|\widetilde{u}\|_{c_0}) \quad \text{and} \quad v_n = (1 - \tfrac{1}{2^{n+1}})u_{n-1} \qquad \forall\, n \geqslant 2.$$

Since $|v_1| \leqslant 1$ and $|v_n| \leqslant |u_n| \leqslant 1$ it follows that $\varphi(\overline{B}_1) \subseteq \overline{B}_1$.

Let $\widetilde{u}, \widetilde{v} \in \overline{B}_1$, $\widetilde{u} \neq \widetilde{v}$. We have

$$\begin{aligned} \|\varphi(\widetilde{u}) - \varphi(\widetilde{v})\|_{c_0} \;&=\; \max\Big\{\tfrac{1}{2}\big|\|\widetilde{u}\|_{c_0} - \|\widetilde{v}\|_{c_0}\big|, \, \max_{n\geqslant 2}(1 - \tfrac{1}{2^{n+1}})|u_{n-1} - v_{n-1}|\Big\} \\ &<\; \|\widetilde{u} - \widetilde{v}\|_{c_0}, \end{aligned}$$

so φ is contractive. Suppose that there exists $\widetilde{u} \in \overline{B}_1$ such that $\varphi(\widetilde{u}) = \widetilde{u} = \{u_n\}_{n\geqslant 1}$. Then

$$u_1 \;=\; \tfrac{1}{2}(1+\|\widetilde{u}\|_{c_0}) \;>\; 0$$

and

$$\begin{aligned} |u_n| \;&=\; \big|(1-\tfrac{1}{2^{n+1}})u_{n-1}\big| \;=\; \big|(1-\tfrac{1}{2^{n+1}})(1-\tfrac{1}{2^{n}})u_{n-2}\big| \\ =\; \ldots \;&=\; \Big|\prod_{i=0}^{n-2}(1-\tfrac{1}{2^{n+1-i}})u_i\Big| \\ \geqslant\; &(1-\sum_{i=1}^{n-2}\tfrac{1}{2^{n+1-i}})|u_1| = (1-\sum_{i=3}^{n+1}\tfrac{1}{2^i})|u_1| \geqslant \tfrac{3}{4}|u_1| \;>\; 0 \qquad \forall\, n\geqslant 2, \end{aligned}$$

a contradiction to the fact that $u_n \longrightarrow 0$. So, φ has no fixed points.

Solution of Problem 4.134

Since $C \subseteq u + T_C(u)$ for all $u \in C$ (see Definition 3.112), the nonemptiness of $\operatorname{Fix}(\varphi)$ follows from Theorem 4.100. Clearly the set $\operatorname{Fix}(\varphi)$ is closed. Suppose that $\gamma(\operatorname{Fix}(\varphi)) > 0$. Then since φ is γ-condensing (see Definition 2.31), we have

$$\gamma(\operatorname{Fix}(\varphi)) \;=\; \gamma(\varphi(\operatorname{Fix}(\varphi))) \;<\; \gamma(\operatorname{Fix}(\varphi)),$$

a contradiction. So, $\gamma(\operatorname{Fix}(\varphi)) = 0$ and so $\operatorname{Fix}(\varphi)$ is compact.

Solution of Problem 4.135

Let $y \in \varphi(C)$ and consider the map $\xi_y(\cdot) = \xi(\cdot, y) \colon C \longrightarrow C$. By hypothesis **(i)**, the map ξ_y is a k-contraction and so the Banach fixed point theorem (see Theorem 4.59) implies that there exists a unique $u(y) \in C$ such that

$$u(y) \;=\; \xi(u(y), y). \tag{4.168}$$

Hypothesis **(ii)** implies that the map $\vartheta \colon \varphi(C) \ni y \longmapsto u(y)$ is continuous with the relative topology on $\varphi(C)$. Then the map $\vartheta \circ \varphi \colon C \longrightarrow C$ is compact and so the Schauder fixed point theorem (see Theorem 4.89) implies that there exists $\widehat{u} \in C$ such that

$$\widehat{u} \;=\; (\vartheta\circ\varphi)(\widehat{u}) \;=\; \xi\big((\vartheta\circ\varphi)(\widehat{u}), \varphi(\widehat{u})\big) \;=\; \xi\big(\widehat{u}, \varphi(\widehat{u})\big)$$

(see (4.168)).

Remark. In fact this problem can be used to provide an alternative solution to Problem 4.115. Indeed, using the notation of Problem 4.115, we set $\xi(u, y) = \psi(u) + y$

and φ is the same. Then it is easy to see that the conditions of Problem 4.115 are satisfied and so we can find $\widehat{u} \in C$ such that

$$\widehat{u} = \xi(\widehat{u}, \varphi(\widehat{u})) = \varphi(\widehat{u}) + \psi(\widehat{u}).$$

Solution of Problem 4.136

Let $\{y_n\}_{n\geqslant 1} \subseteq D \cap V$ be a sequence such that $y_n \longrightarrow y_0$ in Y. Let $u_n = \xi(u_n, y_n)$ for $n \geqslant 1$ (it exists by hypothesis **(iv)**). Let $C_0 = \{u(y_n) = u_n : n \geqslant 1\}$ and clearly we may assume that $D_0 = \{y_n : n \geqslant 1\} \subseteq D \cap V$. Suppose that $\gamma(C_0) > 0$. Then using hypothesis **(iii)**, we have

$$\gamma(C_0) = \gamma(\{\xi(u_n, y_n) : n \geqslant 1\}) \leqslant \gamma(\xi(C_0, D_0)) < \gamma(C_0),$$

a contradiction. So, $\gamma(C_0) = 0$, which mans that the set C_0 is relatively compact. So, we can find a subsequence $\{u(y_{n_k})\}_{k\geqslant 1}$ of $\{u(y_n)\}_{n\geqslant 1}$ such that

$$u(y_{n_k}) \longrightarrow h \in C \quad \text{in } X.$$

Using hypothesis **(i)**, we have $h = \xi(h, y_0)$ and so $h = u(y_0)$ (see hypothesis **(ii)**).

Moreover, the uniqueness of $u(y_0) \in C$ implies that for the original sequence, we have

$$u(y_n) \longrightarrow u(y_0) \quad \text{in } X,$$

so

$$u(y) \longrightarrow u(y_0) \quad \text{in } X \quad \text{as } y \to y_0 \quad \text{in } Y.$$

Solution of Problem 4.137

Hypothesis **(i)** implies that given $\varepsilon > 0$, we can find $\delta = \delta(\varepsilon) > 0$ such that

$$\text{if } \|y - y_0\|_Y < \delta, \text{ then } \|\xi(u, y) - \xi(u, y_0)\|_X < \varepsilon \qquad \forall\, u \in C. \tag{4.169}$$

Let $V = B_\delta(y_0) = \{y \in Y : \|y - y_0\|_Y < \delta\}$ and $C_0 \subseteq C$ with $\gamma(C_0) > 0$ (see Definition 2.31). Hypothesis **(i)** implies that

$$\gamma(\xi(C_0, y_0)) \leqslant k\gamma(C_0), \tag{4.170}$$

with $k = k(C_0, y_0) \in [0, 1)$. For $\varepsilon > 0$, let $2\varepsilon = \delta\gamma(C_0)$ and choose $\varepsilon > 0$ small such that $k + \delta < 1$. We have

$$\xi(u, y) = \xi(u, y_0) + \big(\xi(u, y) - \xi(u, y_0)\big),$$

so

$$\begin{aligned}\gamma(\xi(C_0, D_0)) &\leqslant \gamma(\xi(C_0, y_0)) + 2\varepsilon \leqslant k\gamma(C_0) + 2\varepsilon \\ &\leqslant (k+\delta)\gamma(C_0) < \gamma(C_0)\end{aligned}$$

(see (4.169)). So, we can use Problem 4.136 and conclude that $u(y) \longrightarrow u(y_0)$ in X as $y \to y_0$ in Y.

Solution of Problem 4.138

Let $K \subseteq \mathbb{R}^n$ be the compact, convex set defined by

$$K = \Big\{u=(u_1, \ldots, u_n) \in \mathbb{R}^n : \sum_{k=1}^{n} u_k=1,\ u_k \geqslant 0 \text{ for all } k \in \{1, \ldots, n\}\Big\}.$$

Consider the function $\varphi\colon K \longrightarrow \mathbb{R}^n$ defined by

$$\varphi(u) = \tfrac{Au}{\|u\|_1},$$

with $\|u\|_1 = \sum\limits_{k=1}^{n} |u_k|$ for all $u = (u_1, \ldots, u_n) \in \mathbb{R}^n$. If $u \in K$, then all the components of u are nonnegative and at least one is positive. So, all the components of Au are positive. Hence $\varphi(K) \subseteq K$ and of course it is continuous. Applying the Brouwer fixed point theorem (see Theorem 4.81), we can find $\widehat{u} \in K$ such that $\|\widehat{u}\|_1\widehat{u} = A(\widehat{u})$.

Solution of Problem 4.139

Let $u_1, \ldots, u_n \in C$ and let $h \in \operatorname{conv}\{u_1, \ldots, u_n\}$. By hypothesis $h \in F(h)$ and so $h \notin X \setminus F^{-1}(h)$. Hence $\operatorname{conv}\{u_1, \ldots, u_n\}$ is not contained in $X \setminus F^{-1}(h)$. But by hypothesis $X \setminus F^{-1}(h)$ is convex. So, there exists $k_0 \in \{1, \ldots, n\}$ such that $u_{k_0} \notin X \setminus F^{-1}(h)$, so $h \in F(u_{k_0})$. Therefore

$$\operatorname{conv}\{u_1, \ldots, u_n\} \subseteq \bigcup_{k=1}^{n} F(u_n),$$

so F is a KKM-multifunction (see Definition 4.111).

Solution of Problem 4.140

Since by hypothesis the $u \longmapsto C \setminus F^{-1}(u)$ is not a KKM-multifunction (see Definition 4.111), we can find $u_1, \ldots, u_n \in C$ and $\widehat{u} \in \operatorname{conv}\{u_1, \ldots, u_n\}$ such that

$$\widehat{u} \in C \setminus \bigcup_{k=1}^{n} (C \setminus F^{-1}(u_k)) = \bigcap_{k=1}^{n} F^{-1}(u_k),$$

so

$$u_k \in F(\widehat{u}) \qquad \forall\, k \in \{1, \ldots, n\}$$

and thus $\widehat{u} \in \operatorname{conv} F(\widehat{u})$.

Solution of Problem 4.141

We consider the first case (that is, F has convex values and $F^{-1}(u)$ is open for all $u \in C$). The other case is similar. For every $u \in C$, the set $C \setminus F^{-1}(u)$ is compact. The surjectivity of F^{-1} implies that

$$\emptyset = C \setminus \bigcup_{u \in C} F^{-1}(u) = \bigcap_{u \in C} (C \setminus F^{-1}(u)),$$

so $u \longmapsto C \setminus F^{-1}(u)$ is not a KKM-multifunction (see Definition 4.111). So, using Problem 4.140 and since F has convex values, we can find $\widehat{u} \in C$ such that $\widehat{u} \in F(\widehat{u})$.

Solution of Problem 4.142

Given $\lambda \in \mathbb{R}$, we consider the multifunction $F \colon C \longrightarrow 2^C$ defined by

$$F(y) = \{u \in C : \varphi(u, y) > \lambda\} \qquad \forall\, y \in C.$$

From hypothesis **(i)** and **(ii)** it follows that F has open values and for every $u \in C$, the set $F^{-1}(u)$ is convex. If F is not surjective, then we can find $\widehat{u} \in C$ such that $F^{-1}(\widehat{u}) = \emptyset$, so $\varphi(\widehat{u}, y) \leqslant \lambda$ for all $y \in C$, thus property **(a)** holds.

If F is surjective, then we can use Problem 4.141 and find $\widetilde{y} \in C$ such that $\widetilde{y} \in F(\widetilde{y})$, so $\lambda < \varphi(\widetilde{y}, \widetilde{y})$.

Solution of Problem 4.143

Clearly we may assume that $\lambda = \sup_{y \in C} \varphi(y,y) < +\infty$. Then according to Problem 4.142 we can find $\widehat{u} \in C$ such that

$$\sup_{y \in C} \varphi(\widehat{u}, y) \leqslant \lambda = \sup_{y \in C} \varphi(y,y).$$

Solution of Problem 4.144

By translating things if necessary, without any loss of generality, we may assume that $0 \in C$. For every $n \geqslant 1$, let $\xi_n : C \longrightarrow C$ be defined by

$$\xi_n(u) = (1 - \tfrac{1}{n})(\varphi + \psi)(u) \qquad \forall\, u \in C$$

(by hypothesis **(i)**). Since X is uniformly convex, it is reflexive (see the Milman–Pettis theorem; Theorem I.5.89). So, Proposition 2.4 implies that ψ is compact. Then for any subset $D \subseteq C$ we have

$$\gamma(\xi_n(D)) \leqslant (1 - \tfrac{1}{n})\gamma(\varphi(D)) \leqslant (1 - \tfrac{1}{n})\gamma(D)$$

(since φ is nonexpansive), so ξ_n is a γ-contraction, in particular γ-condensing (see Definition 2.31). Thus, Theorem 4.105 implies that we can find $u_n \in C$ such that

$$\xi_n(u_n) = u_n \qquad \forall\, n \geqslant 1. \tag{4.171}$$

Since $\{u_n\}_{n \geqslant 1} \subseteq C$, the set $C \subseteq X$ is bounded and X is a reflexive space, passing to a subsequence if necessary, we may assume that $u_n \xrightarrow{w} \widehat{u} \in C$ in X (note that C being convex is weakly closed). Then, we have

$$\big\|\xi_n(u_n) - (\varphi + \psi)(u_n)\big\|_X \leqslant \tfrac{1}{n}\big(\|\varphi(u_n)\|_X + \|\psi(u_n)\|_X\big) \longrightarrow 0 \tag{4.172}$$

(since C is bounded). Since ψ is completely continuous, we have

$$\psi(u_n) \longrightarrow \psi(\widehat{u}) \quad \text{in } X. \tag{4.173}$$

Then from (4.171), (4.172), and (4.173), it follows that

$$u_n - \varphi(u_n) \longrightarrow \psi(\widehat{u}) \quad \text{in } X.$$

From Proposition 4.78, we know that $I_X - \varphi$ is demiclosed. So, we have

$$\widehat{u} - \varphi(\widehat{u}) = \psi(\widehat{u}),$$

thus $\varphi(\widehat{u}) + \psi(\widehat{u}) = \widehat{u}$.

Solution of Problem 4.145

No. Consider the uniformly convex Banach space l^p (with $1 < p < +\infty$). Let $\varphi\colon X \longrightarrow X$ be defined by $\varphi(\widehat{u}) = \widehat{y}$ for all $\widehat{u} = \{u_n\}_{n\geqslant 1} \in X$ with $\widehat{y} = \{y_n\}_{n\geqslant 1}$ satisfying

$$y_1 \;=\; 0 \quad\text{and}\quad y_n \;=\; u_{n-1} \qquad \forall\, n \geqslant 2.$$

Also, let $\psi\colon X \longrightarrow X$ be defined by $\psi(\widehat{u}) = \widehat{y}$ for all $\widehat{u} = \{u_n\}_{n\geqslant 1} \in X$ with $\widehat{y} = \{y_n\}_{n\geqslant 1}$ satisfying

$$y_1 \;=\; 1 - \|\widehat{u}\|_X \quad\text{and}\quad y_n = 0 \qquad \forall\, n \geqslant 2.$$

We have

$$\|\varphi(\widehat{u}) - \varphi(\widehat{v})\|_X \;=\; \|\widehat{u} - \widehat{v}\|_X \qquad \forall\, \widehat{u}, \widehat{v} \in X$$

(that is, φ is nonexpansive) and ψ is compact.

Also, with α being the Kuratowski measure of noncompactness (see Definition 2.31(a)), we have

$$\alpha((\varphi + \psi)(D)) \;\leqslant\; \alpha(\varphi(D)) \qquad \forall\, D \subseteq X \text{ bounded},$$

so $\varphi + \psi$ is k-set-Lipschitz with $k = 1$ (see Definition 2.31).

If $\widehat{u} \in X$ with $\|\widehat{u}\|_X \leqslant 1$, then

$$\|(\varphi + \psi)(\widehat{u})\|_X^p \;=\; (1 - \|\widehat{u}\|_X)^p + \|\widehat{u}\|_X^p \;\leqslant\; 1,$$

so $\varphi + \psi\colon \overline{B}_1 \longrightarrow \overline{B}_1$ where $\overline{B}_1 = \{u \in X : \; \|u\|_X \leqslant 1\}$.

We claim that $\varphi + \psi$ is fixed point free on $\overline{B}_1$. Indeed, suppose that we can find $\widehat{u} \in \overline{B}_1$ such that $(\varphi + \psi)(\widehat{u}) = \widehat{u}$. Then

$$u_1 \;=\; 1 - \|\widehat{u}\|_X \quad\text{and}\quad u_n \;=\; u_{n-1} \qquad \forall\, n \geqslant 2,$$

so $u_n = 1 - \|\widehat{u}\|_X$ for all $n \geqslant 1$. But recall that $u_n \longrightarrow 0$. So, $1 - \|\widehat{u}\|_X = 0$ and $\|\widehat{u}\|_X = 1$, a contradiction.

Solution of Problem 4.146

Given $\varepsilon > 0$, we choose $c > 0$ such that

$$c \sum_{n\geqslant 1} n k^n \;<\; 1$$

and we set $\varepsilon_1 = \frac{\varepsilon c}{1-k} > 0$. Let $u_0 \in \operatorname{Fix}(F)$ and choose $u_1 \in G(u_0)$ such that

$$d_X(u_0, u_1) \;\leqslant\; h(F(u_0), G(u_0)) + \varepsilon. \tag{4.174}$$

Since G is k-Lipschitz, we can find $u_2 \in G(u_1)$ such that

$$d_X(u_2, u_1) \;\leqslant\; k d_X(u_1, u_0) + k\varepsilon_1.$$

Then inductively we produce a sequence $\{u_n\}_{n\geqslant 1} \subseteq X$ such that

$$u_{n+1} \in G(u_n) \quad \text{and} \quad d_X(u_{n+1}, u_n) \leqslant k d_X(u_n, u_{n-1}) + k^n \varepsilon_1 \qquad \forall\, n \geqslant 1,$$

thus

$$d_X(u_{n+1}, u_n) \leqslant k^n d_X(u_1, u_0) + n k^n \varepsilon_1 \qquad \forall\, n \geqslant 1.$$

So, we have

$$\sum_{n\geqslant m} d_X(u_{n+1}, u_n) \leqslant \tfrac{k^m}{1-k} d_X(u_1, u_0) + \sum_{n\geqslant m} n k^n \varepsilon_1 \longrightarrow 0 \quad \text{as } m \to +\infty,$$

thus $\{u_n\}_{n\geqslant 1} \subseteq X$ is a Cauchy sequence.

The completeness of X implies that $u_n \longrightarrow u$ in X. We have

$$G(u_n) \xrightarrow{h} G(u),$$

so $u \in G(u)$, that is, $u \in \mathrm{Fix}\,(G)$.

Moreover, we have

$$d_X(u_0, u) \leqslant \sum_{n\geqslant 0} d_X(u_{n+1}, u_n) \leqslant \tfrac{1}{1-k} d_X(u_1, u_0) + \sum_{n\geqslant 1} n k^n \varepsilon_1$$
$$\leqslant \tfrac{1}{1-k}\big(d_X(u_1, u_0) + \varepsilon\big) \leqslant \tfrac{1}{1-k}\big(h(F(u_0), G(u_0)) + 2\varepsilon\big) \tag{4.175}$$

(see (4.174)). Reversing the roles of F and G in the above argument, we see that for every $v_0 \in \mathrm{Fix}\,(G)$, we can find $v \in \mathrm{Fix}\,(F)$ such that

$$d_X(v_0, u) \leqslant \tfrac{1}{1-k}\big(h(F(v_0), G(v_0)) + 2\varepsilon\big). \tag{4.176}$$

From (4.175) and (4.176) and by letting $\varepsilon \searrow 0$, we obtain

$$h(\mathrm{Fix}\,(F), \mathrm{Fix}\,(G)) \leqslant \tfrac{1}{1-k} \sup_{u\in X} h(F(u), G(u)).$$

Solution of Problem 4.147

From Theorem 4.68 we know that sets $\mathrm{Fix}\,(F)$ and $\mathrm{Fix}\,(F_n)$ (for $n \geqslant 1$) are nonempty. Given $\varepsilon > 0$, we can find $n_0 = n_0(\varepsilon) \geqslant 1$ such that

$$\sup_{u\in X} h(F_n(u), F(u)) \leqslant (1-k)\varepsilon \qquad \forall\, n \geqslant n_0.$$

Using Problem 4.146 we see that

$$h(\mathrm{Fix}\,(F_n), \mathrm{Fix}\,(F)) \leqslant \varepsilon \qquad \forall\, n \geqslant n_0,$$

so $\operatorname{Fix}(F_n) \xrightarrow{h} \operatorname{Fix}(F)$.

Solution of Problem 4.148

According to Theorem 2.116, it suffices to show that $R(A+G+I_X) = H$ (recall that in this case, since $H = H^*$, we have $\mathcal{F} = I_H$, with $\mathcal{F}$ being the duality map). Let $h \in H$ and let $V(u) = (A + I_H)^{-1}(h - G(u))$ for $u \in D(G)$. Since $(A + I_H)^{-1}$ is nonexpansive (see Proposition 2.123(b)), we have

$$\|V(u) - V(y)\|_H \leqslant \|G(u) - G(v)\|_H \leqslant k\|u - v\|_H \qquad \forall\, u, v \in D(G).$$

Since by hypothesis the set $D(G) \subseteq H$ is closed, we can apply the Banach fixed point theorem (see Theorem 4.59) and find $\widehat{u} \in D(G)$ such that

$$V(\widehat{u}) = \widehat{u},$$

so

$$h = A(\widehat{u}) + G(\widehat{u}),$$

thus

$$R(A + G + I_H) = H$$

(since $h \in H$ is arbitrary) and hence $A + G$ is maximal monotone.

Solution of Problem 4.149

Let $\varphi \colon \overline{B}_1 \longrightarrow \partial B_1$ be defined by

$$\varphi(\widehat{u}) = (1, u_1, u_2, \ldots) \qquad \forall\, \widehat{u} = \{u_n\}_{n\geqslant 1} \in \overline{B}_1.$$

Evidently

$$\|\varphi(\widehat{u}) - \varphi(\widehat{v})\|_X = \|\widehat{u} - \widehat{v}\|_X \qquad \forall\, \widehat{u}, \widehat{v} \in \overline{B}_1$$

(that is, φ is an isometry). If $\widehat{u} = \varphi(\widehat{u})$ with $\widehat{u} = \{u_n\}_{n\geqslant 1} \in \overline{B}_1$, then $u_n = 1$ for all $n \geqslant 1$, but $(1, 1, \ldots) \notin X$.

Solution of Problem 4.150

Let $T = [0, b]$, $X = C(T)$ and consider the map $\varphi \colon X \longrightarrow X$ defined by

$$\varphi(u)(t) = \int_0^t u(s)\, ds \qquad \forall\, u \in X,\ t \in T.$$

Evidently, if $b > 1$, then φ is not a k-contraction. On the other hand, for $n > 1$, we have

$$\varphi^{(n)}(u)(t) = \tfrac{1}{(n-1)!} \int_0^t (t-s)^{n-1} u(s)\, ds \qquad \forall\, u \in X,\ t \in T.$$

So, we have

$$\|\varphi^{(n)}(u) - \varphi^{(n)}(v)\|_X \leqslant \tfrac{b^{n-1}}{(n-1)!} \|u - v\|_\infty \qquad \forall\, u, v \in X,$$

so $\varphi^{(n)}$ is a k-contraction for some big $n \geqslant 1$.

Solution of Problem 4.151

Let

$$\mathcal{Y} = \big\{ D \subseteq C :\ D \text{ is nonempty, closed, convex, and } \varphi\text{-invariant} \big\}.$$

We order $\mathcal{Y}$ by reverse inclusion, that is,

$$D_1 \preccurlyeq D_2 \iff D_2 \subseteq D_1.$$

The finite intersection property for the weak topology implies that every chain in $\mathcal{Y}$ has an upper bound (namely the intersection of the elements in the chain). So, by the Kuratowski–Zorn lemma (see Theorem 4.120), we know that $\mathcal{Y}$ has a maximal element $\widehat{C} \in \mathcal{Y}$. Then $\widehat{C}$ is φ-invariant and minimal with respect to inclusion.

Solution of Problem 4.152

Let $\varphi^* \colon M_1^+(X) \longrightarrow M_1^+(X)$ be defined by

$$\varphi^*(\mu) = \mu\varphi^{-1}$$

(recall that the image measure $\mu\varphi^{-1}$ is defined by $\mu\varphi^{-1}(D) = \mu(\varphi^{-1}(D))$ for all $D \in \mathcal{B}(X)$; see Definition 3.158). Recall that $C(X)^* = M_b(X)$ (the signed Radon measures on X; see Theorem 1.88). We furnish $M_1^+(X)$ with the relative w^*-topology. We claim that for this topology on $M_1^+(X)$, φ^* is continuous. To this end, let $\{\mu_\alpha\}_{\alpha \in J} \subseteq M_1^+(X)$ be a net such that $\mu_\alpha \xrightarrow{w^*} \mu \in M_1^+(X)$. Then for all $f \in C(X)$, we have

$$\int_X f\, d(\mu_\alpha\varphi^{-1}) = \int_X f \circ \varphi\, d\mu_\alpha \longrightarrow \int_X f \circ \varphi\, d\mu = \int_X f\, d(\mu\varphi^{-1}),$$

so

$$\varphi^*(\mu_\alpha) \xrightarrow{w^*} \varphi^*(\mu),$$

thus φ^* is continuous. But $M_1^+(X)$ is w^*-closed in $\overline{B}_1^* = \{\mu \in M_b(X) : \|\mu\|_{M_b(X)}\} \leqslant 1$ which is w^*-compact (by the Alaoglu theorem; see Theorem I.5.66). Hence $M_1^+(X)$ is w^*-compact and clearly convex. Invoking the Tichonov fixed point theorem (see Theorem 4.101), we can find $\widehat{\mu} \in M_1^+(X)$ such that $\varphi^*(\widehat{\mu}) = \widehat{\mu}$, so $\widehat{\mu} = \widehat{\mu}\varphi^{-1}$.

Solution of Problem 4.153

Suppose that V is a metrizable space, $D \subseteq V$ is a closed set and $\varphi\colon D \longrightarrow C$ is a continuous map. Let $r\colon X \longrightarrow C$ be a retraction of X onto C. Since by hypothesis X is an AR (see Definition 4.132), φ admits a continuous extension $\widehat{\varphi}\colon V \longrightarrow X$. Let $\widehat{\psi} = r \circ \widehat{\varphi}\colon V \longrightarrow C$. Then $\widehat{\psi}$ is the desired continuous extension of φ, which shows that the set C is an AR.

Solution of Problem 4.154

Let $\overline{B}_r = \{u \in X : \|u\|_X \leqslant r\}$. From the Dugundji extension theorem (see Theorem 2.8), we know that $\overline{B}_r$ is an AR (see Definition 4.132(b)). Also, from Problem 4.14, we know that ∂B_r is a retract of $\overline{B}_r$. Hence using Problem 4.153, we conclude that ∂B_r is an AR.

Solution of Problem 4.155

Let $\varphi, \psi \in V$ and let $c = d_{B(X)}(\varphi, \psi)$. Then from the definition of the supremum metric we have

$$\varphi \leqslant \psi + c \quad \text{and} \quad \psi \leqslant \varphi + c,$$

so

$$S(\varphi) \leqslant S(\psi + c) \leqslant S(\psi) + \eta c \tag{4.177}$$

and

$$S(\psi) \leqslant S(\varphi + c) \leqslant S(\varphi) + \eta c \tag{4.178}$$

(see hypothesis **(i)** and **(ii)**). From (4.177) and (4.178), it follows that

$$\left|S(\varphi)(u) - S(\psi)(u)\right| \leqslant \eta c \qquad \forall\, u \in X,$$

so

$$d_{B(X)}\big(S(\varphi), S(\psi)\big) \leqslant \eta c = \eta d_{B(X)}(\varphi, \psi).$$

hence S is a contraction. The space V is a closed subspace of the complete metric space $(B(X), d_{B(X)})$. So, $(V, d_{B(X)})$ is a complete metric space itself and we can use the Banach fixed point theorem (see Theorem 4.59), to conclude that S has a unique fixed point.

Solution of Problem 4.156

We know that K is homeomorphic to a compact subset K_0 of the Hilbert cube $\mathcal{H}$. Let $h\colon K \longrightarrow K_0$ be this homeomorphism and consider the map $f\circ h^{-1}\colon K_0 \longrightarrow X$. Since X is an AR (see Definition 4.132(b)), the map $f\circ h^{-1}$ admits a continuous extension $\xi\colon \mathcal{H} \longrightarrow X$. Let $i\colon K_0 \longrightarrow \mathcal{H}$ be the inclusion map and consider the continuous maps

$$i\circ h\colon K \longrightarrow \mathcal{H} \quad \text{and} \quad g\circ\xi\colon \mathcal{H} \longrightarrow K.$$

We have

$$(g\circ\xi)\circ(i\circ h) \;=\; g\circ(\xi\circ i)\circ h \;=\; g\circ(f\circ h^{-1})\circ h \;=\; g\circ f \;=\; \varphi.$$

Note that $(i\circ h)\circ(g\circ\xi)\colon \mathcal{H} \longrightarrow \mathcal{H}$ is a continuous map and from Problem 4.126(a), we know that $\mathcal{H}$ has the fixed point property. So, we can find $\widehat{u}_0 \in \mathcal{H}$ such that

$$(i\circ h)\circ(g\circ\xi)(\widehat{u}_0) \;=\; \widehat{u}_0,$$

hence

$$(g\circ\xi)\circ(i\circ h)\big((g\circ\xi)(\widehat{u}_0)\big) \;=\; (g\circ\xi)(\widehat{u}_0),$$

thus $\varphi(\widetilde{u}_0) = \widetilde{u}_0$, with $\widetilde{u}_0 = (g\circ\xi)(\widehat{u}_0)$.

Solution of Problem 4.157

Let $K = \overline{\varphi(X)}$. Then K is a compact metric space. Let $i\colon K \longrightarrow X$ be the inclusion map. Let $\psi = \varphi\circ i$. Then Problem 4.156 implies that there exists $\widehat{u} \in K$ such that $\psi(\widehat{u}) = \widehat{u}$, so $\varphi(\widehat{u}) = \widehat{u}$.

Solution of Problem 4.158

Suppose that $U \subseteq X$ is a nonempty open set, V is a metrizable space, $C \subseteq V$ is a closed set and $\varphi\colon C \longrightarrow U$ is a continuous map. Since X is an ANR (see Definition 4.132(c)), there exists a neighborhood W of C and a continuous map $\widehat{\varphi}\colon W \longrightarrow X$ extending φ (see Definition 4.132(c)). We have that the set $\widehat{\varphi}^{-1}(U) \subseteq V$

is open and $C \subseteq \widehat{\varphi}^{-1}(U)$. So, $\widehat{\varphi}|_{\widehat{\varphi}^{-1}(U)} : \widehat{\varphi}^{-1}(U) \longrightarrow U$ is the desired extension of φ.

Solution of Problem 4.159

First assume that X is an AR (see Definition 4.132(b)). Then from Definition 4.132(b), it is clear that X is an ANR (see Definition 4.132(c)). Let $V = [0,1]\times X$ and let $C = (\{0\}\times X)\cup(\{1\}\times X) \subseteq V$ be a closed set. Consider the map $h\colon C \longrightarrow X$ defined by

$$h(0,u) \;=\; u \quad\text{and}\quad h(1,u) \;=\; u_0 \,\in\, X \qquad \forall\, u \in X.$$

Because X is an AR, the map h admits a continuous extension $\widehat{h}\colon [0,1]\times X \longrightarrow X$. Evidently $\widehat{h}$ is a homotopy, $\widehat{h}(0,\cdot) = I_X$ and $\widehat{h}(1,\cdot) = u_0$. Therefore $i_X \simeq 0$ and so X is contractible (see Definition 4.113(d)).

Next assume that X is a contractible ANR. From Definition 4.113(d), we know that there exists a homotopy $\widehat{h}\colon [0,1]\times X \longrightarrow X$ such that

$$\widehat{h}(0,u) \;=\; u \quad\text{and}\quad \widehat{h}(1,u) \;=\; u_0 \,\in\, X \qquad \forall\, u \in X.$$

Suppose that V is a metrizable space, $C \subseteq V$ is a nonempty, closed set and $\varphi\colon C \longrightarrow X$ is a continuous map. Since X is an ANR, we can find a continuous extension $\widehat{\varphi}\colon W \longrightarrow X$ of φ with $W \subseteq V$ being an open set containing C. We choose an open set $E \subseteq W$ such that

$$C \;\subseteq\; E \;\subseteq\; \overline{E} \;\subseteq\; W.$$

Invoking the Urysohn lemma (see Theorem I.2.136), we find a continuous function $\vartheta\colon V \longrightarrow [0,1]$ such that

$$\vartheta|_C \;=\; 0 \quad\text{and}\quad \vartheta|_{X\setminus E} \;=\; 1.$$

Then we introduce the map $\psi\colon V \longrightarrow X$ defined by

$$\psi(u) \;=\; \begin{cases} \widehat{h}(\vartheta(v),\varphi(v)) & \text{for } v \in \overline{E} \\ u_0 & \text{for } v \in V\setminus E. \end{cases}$$

Evidently ψ is continuous. We have

$$\psi(v) \;=\; \widehat{h}(\vartheta(v),\varphi(v)) \;=\; \widehat{h}(0,\varphi(v)) \;=\; \varphi(v) \qquad \forall\, v \in C,$$

so ψ is the desired extension of φ to all of X. Thus X is an AR.

Solution of Problem 4.160

Let $V = [0,1]\times Z$ and let $C = (\{0\}\times Z)\cup(\{1\}\times Z) \subseteq V$ be a closed set. We consider the continuous map $h\colon C \longrightarrow X$ defined by

$$h(0,u) \;=\; \varphi(u) \quad\text{and}\quad h(1,u) \;=\; \psi(u) \qquad \forall\, u \in Z.$$

Since X is an AR (see Definition 4.132(b)), h admits a continuous extension

$\widehat{h}: V = [0,1] \times Z \longrightarrow X$. This is the desired homotopy from φ to ψ. Therefore φ and ψ are homotopic.

Solution of Problem 4.161

Let $\{u_1, \ldots, u_n\} \subseteq C$. The convexity of C implies that $\operatorname{conv}\{u_1, \ldots, u_n\} \subseteq C$, while the fact that F is a KKM-multifunction (see Definition 4.111) implies $\operatorname{conv}\{u_1, \ldots, u_n\} \subseteq \bigcup_{k=1}^{n} F(u_k)$. Therefore

$$\begin{aligned} \operatorname{conv}\{u_1, \ldots, u_n\} &= \operatorname{conv}\{u_1, \ldots, u_n\} \cap C \subseteq \Big(\bigcup_{k=1}^{n} F(u_k)\Big) \cap C \\ &= \bigcup_{k=1}^{n} F(u_k \cap C) = \bigcup_{k=1}^{n} G(u_k), \end{aligned}$$

so $u \longmapsto G(u) = F(u) \cap C$ is also a KKM-multifunction.

Solution of Problem 4.162

Since $F(u_0) \in P_k(X)$, it suffices to show that the family $\{F(u) \colon u \in D\}$ exhibits the finite intersection property. Let $\{u_1, \ldots, u_n\} \subseteq D$ and let $V = \operatorname{span}\{u_1, \ldots, u_n\}$. Then $C_k = F(u_k) \cap \operatorname{conv}\{u_1, \ldots, u_n\} \in P_f(V)$. Invoking the Knaster–Kuratowski–Mazurkiewicz theorem (see Theorem 4.110), we have

$$\emptyset \neq \bigcap_{k=1}^{n} C_k \subseteq \bigcap_{k=1}^{n} F(u_k),$$

so $\bigcap_{u \in D} F(u) \neq \emptyset$ and of course it is compact (since $F(u_0)$ is compact).

Solution of Problem 4.163

We consider the multifunction $F \colon C \longrightarrow 2^C$ defined by

$$F(u) = \{y \in C : u \prec y\}.$$

Hypothesis **(i)** and Problem 4.140 imply that

$$u \longmapsto G(u) = X \setminus F^{-1}(u) = \{y \in X : u \not\prec y\}$$

is a KKM-multifunction (see Definition 4.111). Then Problem 4.161 implies that $u \longmapsto G_0(u) = G(u) \cap C = \big(X \setminus F^{-1}(u)\big) \cap C = C \setminus F^-(u)$ is a KKM-multifunction with values in C. Hypothesis **(ii)** implies that G_0 is compact valued. Then Problem 4.162 implies that $\bigcap\limits_{u \in C} G_0(u) \in P_k(X)$. Note that

$$y \in G_0(u) \quad \Longleftrightarrow \quad u \not\prec y,$$

so $\bigcap\limits_{u \in C} G_0(u)$ is the set of $\prec$-maximal elements.

Solution of Problem 4.164

Arguing indirectly, suppose that the result is not true. Then for every $u_0 \in C$, we can find $y = y(u_0) \in C$ such that

$$\langle \xi^*(u_0), y - u_0 \rangle \; < \; 0.$$

Given $y \in C$, let

$$V_y \;=\; \{u \in C : \; \langle \xi^*(u), y - u \rangle < 0\}.$$

Then $\{V_y\}_{y \in C}$ is an open cover of C. As the set C is compact, we can find a finite subcover $\{V_{y_1}, \ldots, V_{y_n}\}$ and a corresponding continuous partition of unity $\{\vartheta_1, \ldots, \vartheta_n\}$ subordinate to this subcover. Let $\xi\colon C \longrightarrow X$ be defined by

$$\xi(u) \;=\; \sum_{k=1}^{n} \vartheta_k(u) y_k.$$

Evidently ξ is continuous and since the set C is convex and $\{y_1, \ldots, y_n\} \subseteq C$, it follows that $\xi(C) \subseteq C$. In fact, if $C_0 = C \cap \operatorname{span}\{y_1, \ldots, y_n\}$, then $\xi(C_0) \subseteq C_0$ and so we can apply the Brouwer fixed point theorem (see Theorem 4.81) and produce $\widehat{u} \in C_0$ such that $\xi(\widehat{u}) = \widehat{u}$.

For each $u \in C$, we have

$$\langle \xi^*(u), \xi(u) - u \rangle \;=\; \sum_{k=1}^{n} \vartheta_k(u) \, \langle \xi^*(u), y_k - u \rangle \; < \; 0,$$

so

$$0 \;=\; \langle \xi^*(\widehat{u}), \xi(\widehat{u}) - \widehat{u} \rangle \; < \; 0,$$

a contradiction. So, we conclude that there exists $u_0 \in C$ such that

$$\langle \xi^*(u_0), u - u_0 \rangle \; \geqslant \; 0 \qquad \forall \, u \in C.$$

Alternative solution.

On C we define the following binary relation

$$u \prec y \quad \Longleftrightarrow \quad \langle \xi^*(u), y \rangle < \langle \xi^*(u), u \rangle \,. \tag{4.179}$$

Note that for every $u \in C$, we have

$$u \notin \text{conv}\,\{y \in C:\ u \prec y\} \;=\; \{y \in C:\ u \prec y\}.$$

Also, because of the continuity of ξ^*, we have that the set

$$\{u \in C:\ u \prec y\} \;=\; \{u \in C:\ \langle \xi^*(u), y - u\rangle < 0\}$$

is open. So, we can use Problem 4.163 and infer that there exists $u_0 \in C$ which is $\prec$-maximal. According to (4.179) this means that

$$\langle \xi^*(u_0), u - u_0\rangle \;\geqslant\; 0 \qquad \forall\, u \in C.$$

Solution of Problem 4.165

We argue by contradiction. So, suppose that F has no fixed point. Then $u \notin F(u)$ for all $u \in C$. The strong separation theorem (see Theorem I.5.29) implies that we can hand $y_u^* \in X^* \setminus \{0\}$ and $\eta_u \in \mathbb{R}$ such that

$$\sigma_{F(u)}(y_u^*) \;<\; \eta_u \;<\; \langle y_u^*, u\rangle \tag{4.180}$$

(see Definition 2.46). Let

$$V_u \;=\; F^+\left(\{v \in X:\ \langle y_u^*, v\rangle < \eta_u\}\right) \cap \{v \in X:\ \langle y_u^*, v\rangle > \eta_u\}$$

(see Definition 2.35(b)). The upper semicontinuity of F implies that the set V_u is open (see Definition 2.36). Also, from (4.180), we see that $u \in V_u$. The family $\{V_u\}_{u \in C}$ is an open cover of C. As the set C is compact, we can find a finite subcover $\{V_{u_1}, \ldots, V_{u_n}\}$ and a subordinate continuous partition of unity $\{\vartheta_1, \ldots, \vartheta_n\}$. Then, if $\vartheta_k(u) > 0$, we have

$$\left\langle y_{u_k}^*, v\right\rangle \;<\; \eta_{u_k} \qquad \forall\, v \in F(u_k) \quad \text{and} \quad \left\langle y_{u_k}^*, u\right\rangle \;>\; \eta_{u_k}.$$

Let $\xi^*\colon C \longrightarrow X^*$ be defined by

$$\xi^*(u) \;=\; \sum_{k=1}^n \vartheta_k(u) y_{u_k}^*.$$

Clearly ξ^* is continuous and so we can use Problem 4.164 and find $u_0 \in C$ such that

$$\langle \xi^*(u_0), u - u_0\rangle \;\geqslant\; 0 \qquad \forall\, u \in C. \tag{4.181}$$

For every $u \in C$ and every $y \in F(u)$, we have

$$\langle \xi^*(u), u\rangle \;=\; \sum_{k=1}^n \vartheta_k(u)\left\langle y_{u_k}^*, u\right\rangle \;>\; \sum_{k=1}^n \vartheta_k(u)\eta_{u_k} \;>\; \langle \xi^*(u), y\rangle. \tag{4.182}$$

From the tangentiality hypothesis, we can find $v \in F(u_0)$ and $\lambda > 0$ such that $u_0 + \lambda(v - u_0) \in C$. In (4.181), we let $u = u_0 + \lambda(v - u_0)$ and obtain

$$\langle \xi^*(u_0), u_0\rangle \;\leqslant\; \langle \xi^*(u_0), v\rangle,$$

which contradicts (4.182) with $y = v \in F(u_0)$. So, F must have a fixed point.

Solution of Problem 4.166

From Definition 3.112, we have

$$\lambda(v-u) \in T_C(u).$$

So, by hypothesis, we have

$$(F(u)-u) \cap T_C(u) \neq \emptyset \qquad \forall\, u \in C.$$

Since F is upper semicontinuous, from Problem 4.165, we infer that F has a fixed point.

Solution of Problem 4.167

We argue by contradiction. So, suppose that $F(u) \cap G(u) = \emptyset$ for all $u \in C$. Since the sets $F(u)$ and $G(u)$ are disjoint, closed, convex, and at least one of them is compact, we can apply the strong separation theorem (see Theorem I.5.29). Starting from the strong separation and reasoning as in the solution of Problem 4.165, we can find a continuous map $\xi^* \colon C \longrightarrow X^*$ such that

$$\langle \xi^*(u), y \rangle < \langle \xi^*(u), h \rangle \qquad \forall\, y \in F(u),\ h \in G(u). \tag{4.183}$$

From Problem 4.164, we know that there exists $u_0 \in C$ such that

$$\langle \xi^*(u_0), u-u_0 \rangle \geqslant 0 \qquad \forall\, u \in C. \tag{4.184}$$

By hypothesis we can find $y \in F(u_0)$, $h \in G(u_0)$ and $\lambda > 0$ such that

$$u_0 + \lambda(y-h) \in C.$$

Then in (4.184) we choose $u = u_0 + \lambda(y-h) \in C$ and obtain

$$\langle \xi^*(u_0), y \rangle \geqslant \langle \xi^*(u_0), h \rangle ,$$

which contradicts (4.183) for $u = u_0$. Therefore we conclude that there is $u \in C$ such that $F(u) \cap G(u) \neq \emptyset$.

Solution of Problem 4.168

Just use Problem 4.167 with $G(u) = \{0\}$ for all $u \in C$. All the conditions of that problem are satisfied and so we can find $u \in C$ such that $F(u) \cap G(u) \neq \emptyset$, hence $0 \in F(u)$.

Solution of Problem 4.169

Let $K = \overline{\varphi(C)} \subseteq Y$. Then K is a compact metric space and so it is homeomorphic to a closed subset K_0 of the compact metric space $[0,1]^{\mathbb{N}}$ (the Tichonov cube). Let $h\colon K \longrightarrow K_0$ be this homeomorphism and let $i\colon K \longrightarrow Y$ be the inclusion map. We define

$$f = h \circ \varphi \quad \text{and} \quad g = i \circ h^{-1}.$$

Then $f\colon C \longrightarrow K_0$ and $g\colon K_0 \longrightarrow Y$ are continuous maps and we have $\varphi = g \circ f$. Using the Tietze extension theorem (see Theorem I.2.138), we can find a continuous map $\widehat{f}\colon X \longrightarrow [0,1]^{\mathbb{N}}$ such that $\widehat{f}|_C = f$. Also, since by hypothesis Y is an ANR (see Definition 4.132(c)), we can find an open set $W \subseteq [0,1]^{\mathbb{N}}$ such that $W \supseteq K_0$ and a continuous map $\widehat{g}\colon W \longrightarrow Y$ such that $\widehat{g}|_{K_0} = g$. We can find an open set $V \subseteq [0,1]^{\mathbb{N}}$ such that $K_0 \subseteq V \subseteq \overline{V} \subseteq W$ and let $U = \widehat{f}^{-1}(V)$. Then $\widehat{\varphi} = \widehat{g} \circ \widehat{f}|_U : U \longrightarrow Y$ is a continuous extension of φ to an open neighborhood U of C. Note that

$$\widehat{\varphi}(U) = (\widehat{g} \circ \widehat{f})(\widehat{f}^{-1}(V)) \subseteq \widehat{g}(V) \subseteq \widehat{g}(\overline{V})$$

and the latter is compact. So $\widehat{\varphi} \in K(U;Y)$

Solution of Problem 4.170

Let $D = ([0,1] \times C) \cup (\{0\} \times X)$ and let $\widehat{h}\colon [0,1] \times C \longrightarrow Y$ be the compact homotopy of φ to ψ. Let $\widehat{h}_0\colon D \longrightarrow Y$ be defined by

$$\widehat{h}_0(t,u) = \begin{cases} \widehat{h}(t,u) & \text{if } (t,u) \in [0,1] \times C, \\ \widehat{\varphi}(u) & \text{if } u \in X. \end{cases}$$

Evidently $\widehat{h}_0 \in K(D;Y)$. Since Y is an ANR (see Definition 4.132(c)) and $D \subseteq [0,1] \times X$ is a closed set, we can use Problem 4.169 and find an open set $V \supseteq D$ and $\widetilde{h}_0 \in K(V;Y)$ such that $\widetilde{h}_0|_D = \widehat{h}_0$. We can find an open set $U \supseteq C$ such that $[0,1] \times U \subseteq V$. The sets C and $X \setminus U$ are disjoint, closed and so by the Urysohn lemma (see Theorem I.2.136), we can find a continuous function $\tau\colon X \longrightarrow [0,1]$ such that

$$\tau|_C = 1 \quad \text{and} \quad \tau|_{X \setminus U} = 0.$$

Let $\eta(t,u) = (\tau(u)t, u)$ for all $(t,u) \in [0,1] \times X$. If $u \in U$, then $(\tau(u)t, u) \in V$ and if $u \in X \setminus U$, then $\eta(t,u) = (0,u) \in D \subseteq V$. Therefore $\eta|_D = I_D$.

Now let $h^*\colon [0,1]\times X \longrightarrow Y$ be defined by $h^* = \widetilde{h}_0 \circ \eta$. Then $h^* \in K([0,1]\times X;Y)$ and for all $(t,u) \in D$, we have

$$h^*(t,u) \;=\; \widetilde{h}_0(\eta(t,u)) \;=\; \widetilde{h}_0(t,u) \;=\; \widehat{h}_0(t,u),$$

so $u \longmapsto h^*(t,u) = \widehat{\psi}(u)$ is a compact extension of ψ to X. Moreover, the homotopy h^* of $\widehat{\varphi}$ to $\widehat{\psi}$ is compact.

Solution of Problem 4.171

Using the Arens–Eells theorem (see Theorem 4.143), we can view X as a closed subset of a normed space Y. By hypothesis $\varphi \in K(X;X)$ is compactly homotopic to a constant map $c_0\colon X \longrightarrow Y$ (see Definition 4.113(c)). Evidently c_0 is compactly extendable to Y. Hence Problem 4.169 says that we can find $\widehat{\varphi}\colon K(Y;X)$ such that $\widehat{\varphi}|_X = \varphi$. Using Theorem 4.101, we can find $\widehat{u} \in Y$ such that $\widehat{\varphi}(\widehat{u}) = \widehat{u}$. Since $\widehat{\varphi}(Y) \subseteq X$, we have that $\widehat{u} \in X$ and so $\varphi(\widehat{u}) = \widehat{u}$ (recall that $\widehat{\varphi}|_X = \varphi$).

Solution of Problem 4.172

Since by hypothesis φ is compactly nullhomotopic to a constant map $c_0 \in U$, we can find $\widehat{h} \in K([0,1] \times \overline{U};X)$ such that

$$\widehat{h}(0,\cdot) \;=\; \varphi(\cdot) \quad\text{and}\quad \widehat{h}(1,\cdot) \;=\; c_0 \;\in\; U.$$

Using Problem 4.170, we can find $\widetilde{h} \in K([0,1] \times X;X)$ such that $\widetilde{h}|_{[0,1]\times\overline{U}} = \widehat{h}$.

Let

$$F \;=\; \big\{u \in X \setminus U : \; \widetilde{h}(t,u) = u \text{ for some } t \in [0,1]\big\}.$$

We assume that $F \neq \emptyset$ or otherwise by Problem 4.171, φ has a fixed point in U and so we are done. Since by hypothesis $\varphi|_{\partial U}$ is fixed point free, the sets F and $\overline{U}$ are closed and disjoint and so by the Urysohn lemma (see Theorem I.2.136) there exists a continuous function $\tau\colon X \longrightarrow [0,1]$ such that

$$\tau|_F \;=\; 1 \quad\text{and}\quad \tau|_{\overline{U}} \;=\; 0. \tag{4.185}$$

Let $\widehat{\varphi}\colon X \longrightarrow X$ be defined by

$$\widehat{\varphi}(u) \;=\; \widetilde{h}(\tau(u),u) \qquad \forall\, u \in X.$$

Evidently $\widehat{\varphi} \in K(X;X)$. Suppose that we can find $u_0 \in X \setminus U$ such that

$$u_0 \;=\; \widehat{\varphi}(u_0) \;=\; \widetilde{h}(\tau(u_0),u_0),$$

so $u_0 \in F$ and thus $u_0 = \widetilde{h}(1, u_0) = c_0 \in U$ (see (4.185)), a contradiction.

Therefore $\widehat{\varphi}|_{X \setminus U}$ is fixed point free. Then Problem 4.171 implies that $\widehat{\varphi}$ has a fixed point in U. But note that

$$\widehat{\varphi}|_U = \widetilde{h}(\tau(\cdot), \cdot)|_U = \widetilde{h}(0, \cdot)|_U = \widehat{h}(0, \cdot) = \varphi.$$

So, φ has a fixed point.

Solution of Problem 4.173

"$\Longrightarrow$": This implication is clear from Definition 4.112(b).

"$\Longleftarrow$": Let $r\colon X \longrightarrow C$ be a retraction of X onto C (see Definitions 4.132) and let $\widehat{h}\colon [0,1] \times X \longrightarrow X$ be a deformation of X into C (see Definition 4.112(a)). We define

$$\widetilde{h}(t, u) = \begin{cases} \widehat{h}(2t, u) & \text{if } t \in [0, \frac{1}{2}], \\ (r \circ \widehat{h})(2 - 2t, u) & \text{if } t \in (\frac{1}{2}, 1]. \end{cases}$$

Evidently $\widetilde{h} \in C([0,1] \times X; X)$, $\widetilde{h}(0, \cdot) = I_X$ and $\widetilde{h}(1, \cdot)|_C = r|_C = I_C$. Therefore C is a deformation retract of X.

Solution of Problem 4.174

(a) Let $\widetilde{h}\colon [0,1] \times C \longrightarrow C$ be defined by

$$\widetilde{h}(t, u) = u \qquad \forall\ (t, u) \in [0,1] \times C.$$

Since $C \subseteq X$ is closed and by hypothesis C is an AR (see Definition 4.132(b)), we can find a continuous map $\widehat{h}\colon [0,1] \times X \longrightarrow C$ such that $\widehat{h}|_{[0,1] \times C} = \widetilde{h}$. So, we have

$$\widehat{h}(t, u) = \widetilde{h}(t, u) = u \qquad \forall\ (t, u) \in [0,1] \times C,$$

so C is a strong deformation retract of X (see Definition 4.112(b)).

(b) Suppose that Y is a metric space, $D \subseteq Y$ is a nonempty closed set and $\varphi\colon D \longrightarrow C$ is a continuous map. Since by hypothesis C is a deformation retract of X, there exists a deformation $\widehat{h}\colon [0,1] \times X \longrightarrow X$ of X into C (that is, $\widehat{h}(1, X) \subseteq C$) such that $\widehat{h}(1, \cdot)|_C = I_C$. Since by hypothesis X is an AR, we can find a continuous map $\widehat{\varphi}\colon Y \longrightarrow X$ such that $\widehat{\varphi}|_D = \varphi$. We set

$$\widetilde{\varphi}(v) = \widehat{h}(1, \widehat{\varphi}(v)) \qquad \forall\ v \in Y.$$

Evidently $\widetilde{\varphi}\colon Y \longrightarrow C$ is continuous and for all $v \in D$, we have

$$\widetilde{\varphi}(v) = \widehat{h}(1, \widehat{\varphi}(v)) = \widehat{h}(1, \varphi(v)) = \varphi(v),$$

so C is an AR (see Definition 4.132(b)).

Solution of Problem 4.175

Yes. Note that $\mathcal{H} \subseteq l^2$ is a compact convex set. So, by the Dugundji extension theorem (see Theorem 2.8), $\mathcal{H}$ is a retract of l^2. Also, from Problem 4.174(b), it is clear that l^2 is an AR (see Definition 4.132(b)). So, we can use Problem 4.153 to conclude that $\mathcal{H}$ is an AR.

Solution of Problem 4.176

For every $u \in X$, let

$$m(u) = \sup_{v \in S(u)} \varphi(v).$$

Arguing by contradiction, suppose that the claim of the problem is not true for some $u \in X$. Then inductively we can produce a sequence $\{u_n\}_{n \geqslant 1} \subseteq X$ such that

$$u_1 = u, \quad u_{n+1} \in S(u_n) \quad \text{and} \quad m(u_n) \leqslant \varphi(u_{n+1}) + \tfrac{1}{n} \qquad \forall\, n \geqslant 1. \tag{4.186}$$

Then $\{u_n\}_{n \geqslant 1} \subseteq X$ is increasing and because $\varphi(u_{n+1}) \leqslant m(u) < +\infty$ (see hypothesis **(iii)**), from hypothesis **(ii)**, we know that there exists $\widetilde{u} \in X$ such that $u_n \preceq \widetilde{u}$ for all $n \geqslant 1$. From the assumption concerning u, we see that there exists $v \in X$ such that $\widetilde{u} \preceq v$, $\widetilde{u} \neq v$. Then hypothesis **(i)** implies that

$$\varphi(\widetilde{u}) < \varphi(v). \tag{4.187}$$

By transitivity, we have $u_n \preceq v$ for all $n \geqslant 1$ and so

$$\varphi(v) \leqslant m(u_n) \leqslant \varphi(u_{n+1}) + \tfrac{1}{n} \leqslant \varphi(\widetilde{u}) + \tfrac{1}{n} \qquad \forall\, n \geqslant 1$$

(see (4.186)), so $\varphi(v) \leqslant \varphi(\widetilde{u})$, which contradicts (4.187). Therefore for every $u \in X$, there exists $\widehat{u} \in S(u)$ such that $S(\widehat{u}) = \{\widehat{u}\}$.

Solution of Problem 4.177

On the complete metric space (X, d_X), we introduce the partial order $\preceq$ by

$$u \preceq y \iff d_X(u, y) \leqslant \xi(u) - \xi(y). \tag{4.188}$$

Then from the hypothesis of Theorem 4.66, we have

$$u \preceq y \qquad \forall\, u \in X, y \in F(u). \tag{4.189}$$

Let $\varphi = -\xi$. We will verify that φ satisfies hypotheses **(i)**, **(ii)**, **(iii)** of Problem 4.176.

Hypothesis **(i)** is evident from (4.188).

Suppose that the sequence $\{u_n\}_{n\geqslant 1}$ is increasing. Then from (4.188) it is clear that the sequence $\{\xi(u_n)\}_{n\geqslant 1}$ is decreasing and bounded below. So, $\xi(u_n) \longrightarrow \eta$ in $\mathbb{R}$. Then it follows that $\{u_n\}_{n\geqslant 1} \subseteq X$ is a Cauchy sequence and the completeness of X implies that $u_n \longrightarrow \widetilde{u}$ in X. Then lower semicontinuity of ξ (see Theorem 4.66) implies that

$$\xi(\widetilde{u}) \leqslant \eta,$$

so

$$d_X(u_n, \widetilde{u}) \leqslant \xi(u_n) - \eta \leqslant \xi(u_n) - \xi(\widetilde{u}),$$

thus

$$u_n \preceq \widetilde{u} \qquad \forall\, n \geqslant 1$$

(see (4.188)). Hence hypothesis **(ii)** is satisfied.

Finally hypothesis **(iii)** is satisfied since ξ is bounded below (see Theorem 4.66). Therefore, we can use Problem 4.176 and infer that for all $u \in X$, we can find $\widehat{u} \in S(u)$ such that $S(\widehat{u}) = \{\widehat{u}\}$. Thus $\widehat{u} \in F(\widehat{u})$ (see (4.189)).

Solution of Problem 4.178

"**(a)** $\Longrightarrow$ **(b)**": Evidently every maximal element $\widehat{u} \in X$ of X is a fixed point of F.

"**(b)** $\Longrightarrow$ **(a)**": Arguing by contradiction, suppose that X has no maximal elements. Let $F\colon X \longrightarrow 2^X$ be defined by

$$F(u) = \{y \in X :\ u \preceq y,\ u \neq y\} \qquad \forall\, u \in X.$$

By hypothesis $F(u) \neq \emptyset$ for all $u \in X$ and there exists $\widehat{u} \in X$ such that $\widehat{u} \in F(\widehat{u})$ which contradicts the definition of F.

Solution of Problem 4.179

Let $m_0 = \inf\limits_{u\in X} \xi(u) > -\infty$. For every $\varepsilon > 0$ we define

$$S_\varepsilon = \{u \in X :\ \xi(u) \leqslant m_0 + \varepsilon\}.$$

The lower semicontinuity of ξ implies that the set $S_\varepsilon \subseteq X$ is nonempty and closed. On S_ε, we introduce the binary relation $\preceq$ defined by

$$u \preceq y \iff \vartheta(d_X(u,y)) \leqslant \xi(u) - \xi(y). \tag{4.190}$$

Then clearly $\preceq$ is reflexive and antisymmetric, but not transitive in general (unless ϑ is subadditive).

If $u, y \in S_\varepsilon$ and $u \preceq y$, then from (4.190) we have

$$m_0 \leqslant \xi(y) \leqslant \xi(u) \leqslant m_0 + \varepsilon,$$

so

$$\vartheta(d_X(u, y)) \leqslant \xi(u) - \xi(y) \leqslant \xi(u) - m_0 \leqslant \varepsilon.$$

The continuity of ϑ implies that we can find $\widehat{\varepsilon} > 0$ such that

$$\vartheta^{-1}([0, \widehat{\varepsilon}]) \subseteq [0, \delta].$$

Then from the hypothesis, we have

$$c_2 d_X(u, y) \leqslant \xi(u) - \xi(y) \qquad \forall\, u, y \in S_{\widehat{\varepsilon}}, \tag{4.191}$$

with $u \preceq y$. So, based on (4.190) on the set $S_{\widehat{\varepsilon}}$ we introduce the binary relation $\preceq^*$ defined by

$$u \preceq^* y \iff d_X(u, y) \leqslant \tfrac{1}{c}\xi(u) - \tfrac{1}{c}\xi(y).$$

Then $(S_{\widehat{\varepsilon}}, \preceq^*)$ is a partially ordered set and it has maximal element (see Problem 4.176). Let $\widehat{u}$ be such a maximal element of $(S_{\widehat{\varepsilon}}, \preceq^*)$. We claim that $\widehat{u}$ is maximal also on $(X, \preceq)$. To this end, let $u \in X$ be such that $\widehat{u} \preceq u$. Then from the hypothesis, we have

$$\vartheta(d_X(u, \widehat{u})) \leqslant \xi(\widehat{u}) - \xi(u),$$

so

$$\xi(u) \leqslant \xi(\widehat{u}) \leqslant m_0 + \widehat{\varepsilon},$$

thus $u \in S_{\widehat{\varepsilon}}$. As above, we have

$$\vartheta(d_X(u, \widehat{u})) \leqslant \widehat{\varepsilon},$$

so

$$c d_X(u, \widehat{u}) \leqslant \xi(\widehat{u}) - \xi(u)$$

(see the hypothesis), thus $\widehat{u} \preceq^* u$ (see (4.191)) and hence $\widehat{u} = u$ (due to the $\preceq^*$-maximality of $\widehat{u}$). Finally from the hypothesis, we see that $u \preceq \varphi(u)$ for all $u \in X$ and so $\widehat{u} = \varphi(\widehat{u})$.

Solution of Problem 4.180

Let $\varrho > 0$ be such that $B_\varrho(u_0) \subseteq K$ (where $B_\varrho(u_0) = \{u \in X : \|u - u_0\|_X < \varrho\}$). Let $\lambda_u > \frac{\|u\|_X}{\varrho}$. Then we have

$$u_0 - \tfrac{1}{\lambda_u} u \in B_\varrho(u_0) \subseteq K,$$

so $\lambda_u u_0 - u \in K$.

Solution of Problem 4.181

"(a) $\Longrightarrow$ (b)": Let K be a normal cone (see Definition 4.125(b)). Let $u, v \in K$ and without any loss of generality, we may assume that $\|u\|_X = 1$ and $\|v\|_X \leqslant 1$. We have

$$1 \;=\; \|u\|_X \;\leqslant\; \|u+v\|_X + \|v\|_X \tag{4.192}$$

and

$$\begin{aligned}\|u+v\|_X \;&=\; \big\|u + \tfrac{v}{\|v\|_X} + (1 - \tfrac{1}{\|v\|_X})v\big\|_X \\ \geqslant\; &\big\|u + \tfrac{v}{\|v\|_X}\big\|_X - 1 + \|v\|_X \;\geqslant\; m - 1 + \|v\|_X,\end{aligned} \tag{4.193}$$

where $m = \inf\limits_{y,h \in \partial B_1} \|y+h\|_X > 0$. From (4.192) and (4.193), it follows that

$$\|u+v\|_X \;\geqslant\; m - \|u+v\|_X,$$

so

$$\|u+v\|_X \;\geqslant\; \tfrac{m}{2} \;>\; 0.$$

"(b) $\Longrightarrow$ (c)": Arguing by contradiction, suppose that we can find two sequences $\{u_n\}_{n\geqslant 1}, \{v_n\}_{n\geqslant 1} \subseteq K$ such that

$$u_n \;\leqslant\; v_n \quad\text{and}\quad \|u_n\|_X \;>\; n\|v_n\|_X \qquad \forall\, n \geqslant 1. \tag{4.194}$$

Let $y_n = \frac{u_n}{\|u_n\|_X} + \frac{v_n}{n\|v_n\|_X}$ and $h_n = -\frac{u_n}{\|u_n\|_X} + \frac{v_n}{n\|v_n\|_X}$ for all $n \geqslant 1$. From (4.194), we have

$$y_n, h_n \;\in\; K \quad\text{and}\quad \|y_n\|_X \;\geqslant\; 1 - \tfrac{1}{n}, \quad \|h_n\|_X \;\geqslant\; 1 - \tfrac{1}{n} \qquad \forall\, n \geqslant 1. \tag{4.195}$$

From property **(b)**, we have

$$\tfrac{2}{n} \;=\; \|y_n + h_n\|_X \;\geqslant\; c\big(1 - \tfrac{1}{n}\big) \qquad \forall\, n \geqslant 1$$

(see (4.195)). Passing to the limit as $n \to +\infty$, we obtain $0 \geqslant c > 0$, a contradiction.

"(c) $\Longrightarrow$ (d)": We define

$$|u| \;=\; \inf_{y \leqslant u} \|y\|_X + \inf_{h \geqslant u} \|h\|_X. \tag{4.196}$$

First we show that this is in fact a norm on X.

Suppose that $|u| = 0$. Then given $\varepsilon > 0$, we can find $y, h \in X$ such that

$$y \;\leqslant\; u \;\leqslant\; h \quad\text{and}\quad \|y\|_X \;\leqslant\; \varepsilon, \quad \|h\|_X \;\leqslant\; \varepsilon \tag{4.197}$$

(see (4.196)). From property **(c)**, we have

$$\|u\|_X \leqslant \|u-y\|_X + \|y\|_X \leqslant c_1\|h-y\|_X + \varepsilon \leqslant (2c_1+1)\varepsilon$$

(see (4.197)). Letting $\varepsilon \searrow 0$, we have $\|u\|_X = 0$, hence $u = 0$.

From (4.196) it is clear that $|\lambda u| = |\lambda||u|$ for all $\lambda \in \mathbb{R}$.

Finally let $u, v \in X$. Given $\varepsilon > 0$, we can find $y, h, \widehat{y}, \widehat{h} \in X$ such that

$$\|y\|_X + \|h\|_X \leqslant |u| + \varepsilon, \quad \|\widehat{y}\|_X + \|\widehat{h}\|_X \leqslant |v| + \varepsilon. \tag{4.198}$$

and

$$y \leqslant u \leqslant h \quad \text{and} \quad \widehat{y} \leqslant v \leqslant \widehat{h} \tag{4.199}$$

Note that $y + \widehat{y} \leqslant u + v \leqslant h + \widehat{h}$ and so from (4.196), we have

$$|u+v| \leqslant \|y\|_X + \|\widehat{y}\|_X + \|h\|_X + \|\widehat{h}\|_X \leqslant |u| + |v| + 2\varepsilon$$

(see (4.198)–(4.199)). Letting $\varepsilon \searrow 0$, we have

$$|u+v| \leqslant |u| + |v|,$$

so $|\cdot|$ is a norm on X.

If $u, v \in K$ with $u \leqslant v$, then

$$\inf_{y \leqslant u} \|y\|_X = \inf_{y \leqslant v} \|y\|_X = 0$$

and so from (4.196) we have

$$|u| = \inf_{u \leqslant h} \|h\|_X \leqslant \inf_{v \leqslant h} \|h\|_X = |v|,$$

so $|\cdot|$ is monotone.

Finally we need to show that the norms $\|\cdot\|_X$ and $|\cdot|$ are equivalent. From (4.196) it is clear that we have

$$|u| \leqslant 2\|u\|_X \qquad \forall\, u \in X. \tag{4.200}$$

Also, if $y \leqslant u \leqslant h$, then using property **(c)**, we have

$$\begin{aligned} \|u\|_X &\leqslant \|u-y\|_X + \|y\|_X \leqslant c_1\|h-y\|_X + \|y\|_X \\ &\leqslant (c_1+1)(\|h\|_X + \|y\|_X), \end{aligned}$$

so

$$\|u\|_X \leqslant (c_1+1)|u| \tag{4.201}$$

(see (4.196)). From (4.200) and (4.201), we conclude that $\|\cdot\|_X$ and $|\cdot|$ are equivalent norms on X.

"**(d)** $\Longrightarrow$ **(e)**": We have

$$y_n - u_n,\ v_n - u_n \in K \quad \text{and} \quad y_n - u_n \leqslant v_n - u_n \qquad \forall\, n \geqslant 1,$$

so

$$|y_n - u_n| \leqslant |v_n - u_n| \qquad \forall\, n \geqslant 1 \tag{4.202}$$

(see property **(d)**). Exploiting the equivalence of the norms, we have

$$c_2\|y_n - u_n\|_X \leqslant |y_n - u_n| \leqslant |v_n - u_n| \leqslant c_3\|v_n - u_n\| \qquad \forall\, n \geqslant 1,$$

for some $c_2, c_3 > 0$ (see (4.202)), so $\|y_n - u_n\|_X \longrightarrow 0$, thus

$$\|y_n - u\|_X \leqslant \|y_n - u_n\|_X + \|u_n - u\|_X \longrightarrow 0$$

and hence $y_n \longrightarrow u$ in X.

"**(e)** $\Longrightarrow$ **(f)**": Arguing by contradiction, suppose that the set $(\overline{B}_1 + K) \cap (\overline{B}_1 - K)$ is unbounded. So, we can find a sequence $\{w_n\}_{n\geqslant 1} \subseteq (\overline{B}_1 + K) \cap (\overline{B}_1 - K)$ such that $\|w_n\|_X \longrightarrow +\infty$. Let $\{e_n\}_{n\geqslant 1}, \{z_n\}_{n\geqslant 1} \subseteq \overline{B}_1$ be two sequences such that

$$e_n \leqslant w_n \leqslant z_n \qquad \forall\, n \geqslant 1. \tag{4.203}$$

Let

$$u_n = \tfrac{e_n}{\|w_n\|_X}, \quad y_n = \tfrac{w_n}{\|w_n\|_X}, \quad v_n = \tfrac{z_n}{\|w_n\|_X} \qquad \forall\, n \geqslant 1.$$

Then from (4.203), we have

$$u_n \leqslant y_n \leqslant v_n \qquad \forall\, n \geqslant 1.$$

Note that $u_n \longrightarrow 0$ and $v_n \longrightarrow 0$ in X, but $\|y_n\|_X = 1$ for all $n \geqslant 1$. This contradicts property **(e)**.

"**(f)** $\Longrightarrow$ **(g)**": Let $r > 0$ be such that $(\overline{B}_1 + K) \cap (\overline{B}_1 - K) \subseteq r\overline{B}_1$. Let $y \in [u, v]$ and let $m = \max\{\|u\|_X, \|v\|_X\}$. Then $\frac{1}{m}y \in (\overline{B}_1 + K) \cap (\overline{B}_1 - K)$ and so $[u, v] \subseteq mr\overline{B}_1$.
"**(g)** $\Longrightarrow$ **(a)**": Arguing by contradiction, suppose that K is not normal. Then we can find two sequences $\{u_n\}_{n\geqslant 1}, \{v_n\}_{n\geqslant 1} \subseteq K \cap \overline{B}_1$ such that

$$\|u_n + v_n\|_X \leqslant \tfrac{1}{n^3} \qquad \forall\, n \geqslant 1.$$

Let

$$y_n = \frac{u_n}{\|u_n + v_n\|_X^{\frac{1}{2}}} \quad \text{and} \quad h_n = \frac{u_n + v_n}{\|u_n + v_n\|_X^{\frac{1}{2}}} \qquad \forall\, n \geqslant 1.$$

We have

$$0 \leqslant y_n \leqslant h_n \qquad \forall\, n \geqslant 1 \quad \text{and} \quad \sum_{n\geqslant 1} \|h_n\|_X \leqslant \sum_{n\geqslant 1} \tfrac{1}{n^{\frac{3}{2}}} < +\infty. \tag{4.204}$$

Let $h = \sum\limits_{n\geqslant 1} h_n$ (see (4.204)). Then $0 \leqslant y_n \leqslant h_n \leqslant h$ for all $n \geqslant 1$ (see (4.204)) and

$$\|y_n\|_X = \frac{1}{\|u_n + v_n\|_X^{\frac{1}{2}}} > n^{\frac{3}{2}} \qquad \forall\, n \geqslant 1,$$

so $[0, v]$ is unbounded, a contradiction.

Solution of Problem 4.182

First we show that if K is not normal, then K is not fully regular and it is not regular. Since K is not normal, according to Problem 4.181(c), we can find two sequences $\{u_n\}_{n\geqslant 1}, \{v_n\}_{n\geqslant 1} \subseteq K$ such that

$$u_n \leqslant v_n \quad \text{and} \quad \|u_n\|_X > n^2\|v_n\|_X \qquad \forall\, n \geqslant 1. \tag{4.205}$$

Let $y_n = \frac{u_n}{\|u_n\|_X}$ and $h_n = \frac{v_n}{n^2\|v_n\|_X}$ for all $n \geqslant 1$. We have that $\{y_n\}_{n\geqslant 1}, \{h_n\}_{n\geqslant 1} \subseteq K$ and

$$y_n \leqslant \frac{u_n}{n^2\|v_n\|_X} \leqslant \frac{v_n}{n^2\|v_n\|_X} = h_n \qquad \forall\, n \geqslant 1 \tag{4.206}$$

(see (4.205)) and

$$\sum_{n\geqslant 1} \|h_n\|_X = \sum_{n\geqslant 1} \frac{1}{n^2} < +\infty. \tag{4.207}$$

Let $h = \sum\limits_{n\geqslant 1} h_n$ (see (4.207)) and define

$$z_n = \begin{cases} \sum\limits_{k=1}^{n} h_k & \text{if} \quad n = 2m,\ m \geqslant 1, \\ \sum\limits_{k=1}^{n} h_k + y_{2m+1} & \text{if} \quad n = 2m+1,\ m \geqslant 1. \end{cases} \tag{4.208}$$

From (4.206), we see that the sequence $\{z_n\}_{n\geqslant 1} \subseteq K$ is increasing and order bounded by h. But $\{z_n\}_{n\geqslant 1}$ does not converge strongly since

$$\|z_{2m} - z_{2m+1}\|_X = \|y_{2m+1}\|_X = 1 \qquad \forall\, m \geqslant 1.$$

This shows that K is not regular and also it is not fully regular since $\|u_n\|_X \leqslant c_1\|h\|_X$ (see Problem 4.181(c)).

Next we show that full regularity implies regularity. So, if K is fully regular, then from the first part of the solution, we have that K is normal. Let $\{u_n\}_{n\geqslant 1} \subseteq X$ be an increasing sequence which is order bounded by $h \in K$ (that is, $u_1 \leqslant u_2 \leqslant \ldots \leqslant u_n \leqslant \ldots \leqslant h$). Then from Problem 4.181(c), we have

$$\|h - u_n\|_X \leqslant c_1\|h - u_1\|_X \qquad \forall\, n \geqslant 1,$$

so the sequence $\{u_n\}_{n\geqslant 1}$ is norm bounded, thus $u_n \longrightarrow u$ in X (since K is fully regular) and hence K is regular (see Definition 4.125(c)).

Now we show that reverse implications are not in general true. According to Proposition 4.130, in order to produce counterexample, we need to consider nonreflexive Banach spaces. Let X=C([0,1]) with order cone

$$K = \{u \in X : u(t) \geqslant 0 \text{ for all } t \in [0,1]\}.$$

Clearly $\|\cdot\|_X$ is monotone and so from Problem 4.181(d) we have that K is normal. However, K is not regular. To see this, let $u_n(t) = 1 - t^n$ for all $t \in [0,1]$, all $n \geqslant 1$. Evidently the sequence $\{u_n\}_{n\geqslant 1} \subseteq K$ is increasing and order bounded by $h \equiv 1$. But the sequence $\{u_n\}_{n\geqslant 1} \subseteq X$ is not strongly convergent in X.

Let $X = c_0 = \{\widehat{u} = \{u_k\}_{k\geqslant 1} : u_k \longrightarrow 0\}$, $\|\widehat{u}\|_X = \sup\limits_{k\geqslant 1} |u_k|$ and

$$K = \Big\{\widehat{u} = \{u_k\}_{k\geqslant 1} \in X : u_k \geqslant 0 \text{ for all } k \geqslant 1\Big\}.$$

Suppose that the sequence $\{\widehat{u}^n\}_{n\geqslant 1} \subseteq X$ is increasing and order bounded. Then we have

$$\widehat{u}^n = \{u_k^n\}_{k\geqslant 1} \longrightarrow \widehat{u} = \{u_k\}_{k\geqslant 1} \quad \text{in } X \quad \text{as } n \to +\infty,$$

with $u_k = \lim\limits_{k\to+\infty} u_k^n$ (increasing bounded sequences in $\mathbb{R}$ converge). So, K is regular. On the other hand, let $\widehat{u}^n = \{u_k^n\}_{k\geqslant 1}$ with

$$u_k^n = \begin{cases} 1 & \text{if } k \leqslant n, \\ 0 & \text{if } k > n. \end{cases}$$

Then the sequence $\{\widehat{u}^n\}_{n\geqslant 1} \subseteq K$ is increasing and $\|\widehat{u}^n\|_X = 1$. But the sequence $\{\widehat{u}^n\}_{n\geqslant 1}$ does not converge in X and so K is not fully regular.

Solution of Problem 4.183

Since K is closed and convex (see Definition 4.123) and $v \notin K$, by the strong separation theorem (see Theorem I.5.29), we can find $v^* \in X^* \setminus \{0\}$ and $\varepsilon > 0$ such that

$$\langle v^*, v\rangle + \varepsilon \leqslant \langle v^*, u\rangle \qquad \forall\, u \in K, \tag{4.209}$$

so

$$\langle v^*, v\rangle \leqslant -\varepsilon < 0$$

(since $u = 0 \in K$; see Definition 4.123).

Let $u \in K$. Then $nu \in K$ for all $n \geqslant 1$ (see Definition 4.123) and so from (4.209), we have

$$\big\langle v^*, \tfrac{v}{n}\big\rangle_X + \tfrac{\varepsilon}{n} \leqslant \langle v^*, u\rangle \qquad \forall\, n \geqslant 1,$$

so

$$0 \leqslant \langle v^*, u\rangle_X. \tag{4.210}$$

Since $u \in K$ is arbitrary, from (4.210) we conclude that $v^* \in K^*$ (see Definition 4.127).

Solution of Problem 4.184
"$\Longrightarrow$": Since $u \in \operatorname{int} K$, we can find $\varrho > 0$ such that $\overline{B}_\varrho(u) \subseteq K$ (where $\overline{B}_\varrho(u) = \{v \in X : \|v-u\|_X \leqslant \varrho\}$). Then for all $u^* \in K^* \setminus \{0\}$, we have

$$\langle u^*, u - \varrho h\rangle \geqslant 0 \qquad \forall\, h \in \overline{B}_1$$

(where $\overline{B}_1 = \{v \in X : \|v\|_X \leqslant 1\}$), so

$$\langle u^*, u\rangle \geqslant \varrho \langle u^*, h\rangle \qquad \forall\, h \in \overline{B}_1$$

and thus

$$\langle u^*, u\rangle \geqslant \varrho \|u^*\|_{X^*} > 0.$$

"$\Longleftarrow$": Arguing by contradiction, suppose that $u \notin \operatorname{int} K$. Since $\operatorname{int} K$ is convex, by the weak separation theorem (see Theorem I.5.28) we can find $\widehat{u}^* \in X^* \setminus \{0\}$ such that

$$\langle \widehat{u}^*, u\rangle \leqslant \langle \widehat{u}^*, v\rangle \qquad \forall\, v \in \operatorname{int} K,$$

so

$$\langle \widehat{u}^*, u\rangle \leqslant \langle \widehat{u}^*, v\rangle \qquad \forall\, v \in K \tag{4.211}$$

(since $\overline{\operatorname{int} K} = K$).

If $\widehat{u}^* \notin K^* \setminus \{0\}$, then there exists $\widehat{v} \in K$ such that

$$\langle \widehat{u}^*, \widehat{v}\rangle < 0. \tag{4.212}$$

Because $\lambda \widehat{v} \in K$ for all $\lambda > 0$ from (4.211) and (4.212) we have a contradiction. Hence $\widehat{u}^* \in K^* \setminus \{0\}$ and from (4.211), we have

$$\langle \widehat{u}^*, u\rangle \leqslant 0$$

(by taking $v = 0 \in K$), which contradicts the hypothesis. Therefore $u \in \operatorname{int} K$.

Solution of Problem 4.185
Since X is separable, the dual unit ball $\overline{B}_1^* = \{u^* \in X^* : \|u^*\|_* \leqslant 1\}$ furnished with the relative weak topology is compact and metrizable. So, the set $D^* = K^* \cap \overline{B}_1^*$ is a compact and metrizable space too, hence separable. Suppose that $\{u_n^*\}_{n\geqslant 1} \subseteq D^*$ is w^*-dense in D^*. Let

$$\widetilde{u}^* = \sum_{n\geqslant 1} \frac{u_n^*}{2^n} \in D^* \subseteq K^*.$$

If for some $\overline{u} \in K \setminus \{0\}$ we have $\langle \widetilde{u}^*, \overline{u}\rangle = 0$, then

$$\langle u_n^*, \overline{u}\rangle = 0 \qquad \forall\, n \geqslant 1,$$

so

$$\langle u^*, \overline{u} \rangle = 0 \qquad \forall\, u^* \in D^*,$$

thus

$$\langle u^*, \overline{u} \rangle = 0 \qquad \forall\, u^* \in K^*,$$

which contradicts Remark 4.128.

Solution of Problem 4.186

Let $u_0 \in \operatorname{int} K$ and let $\varrho > 0$ be such that $\overline{B}_\varrho(u_0) \subseteq K$ (where $\overline{B}_\varrho(u_0) = \{u \in X : \|u - u_0\|_X \leqslant \varrho\}$). For any $u \in X$, we have

$$u_0 \pm \tfrac{\varrho}{\|u\|_X} u \in K.$$

Let $v_1 = \frac{\|u\|_X}{2\varrho}\big(u_0 + \frac{\varrho}{\|u\|_X} u\big) \in K$ and $v_2 = \frac{\|u\|_X}{2\varrho}\big(u_0 - \frac{\varrho}{\|u\|_X} u\big) \in K$. We have

$$u = v_1 - v_2,$$

so K is generating (see Definition 4.125(a)).

Solution of Problem 4.187

Since φ maps $[u_0, v_0]$ into itself, we have

$$u_0 \leqslant \varphi(u_0) \quad \text{and} \quad \varphi(v_0) \leqslant v_0.$$

Suppose that condition **(i)** holds. Then the sequence $\{\varphi^{(n)}(u_0)\}_{n \geqslant 1} \subseteq [u_0, v_0]$ is increasing. The regularity of K implies that

$$u_n = \varphi^{(n)}(u_0) \longrightarrow y \in [u_0, v_0] \quad \text{in } X,$$

so

$$\varphi(u_n) \longrightarrow \varphi(y)$$

(since φ is continuous). But $\varphi(u_n) = \varphi^{(n+1)}(u_0) \longrightarrow y$. So, $y = \varphi(y)$.

Next suppose that condition **(ii)** holds. Then the set $[u_0, v_0]$ is bounded (see Problem 4.181(g) and recall that K is normal), closed, and convex. So, invoking Theorem 4.105, we conclude that φ has a fixed point.

Solution of Problem 4.188

Let

$$S = \{u \in \varphi([u_0, v_0]) : u \leqslant \varphi(u)\}.$$

Note that $u_0 \leqslant \varphi(u_0)$, hence $\varphi(u_0) \leqslant \varphi(\varphi(u_0))$ and so $\varphi(u_0) \in S$, which means that $S \neq \emptyset$. Let $C \subseteq S$ be a chain. We have

$$C \subseteq S \subseteq \varphi([u_0, v_0]),$$

so $C \subseteq X$ is relatively compact, hence separable.

Let $D = \{u_n\}_{n \geqslant 1} \subseteq C$ be a dense sequence and let

$$y_n = \max\{u_1, \dots, u_n\} \in C \qquad \forall\, n \geqslant 1$$

(recall that $C \subseteq S$ is a chain). So, passing to a suitable subsequence if necessary, we may assume that $y_n \longrightarrow u^*$ in X, so

$$u^* \in \overline{C} \subseteq \overline{S} \subseteq \overline{\varphi([u_0, v_0])} \subseteq [u_0, v_0].$$

Since $u_n \leqslant y_n \leqslant u^*$, we have $u \leqslant u^*$ for all $u \in C$ and so

$$u \leqslant \varphi(u) \leqslant \varphi(u^*) \qquad \forall\, u \in C, \tag{4.213}$$

so $\varphi(u^*)$ is an upper bound for C. From (4.213), we have

$$y_n \leqslant \varphi(u^*) \qquad \forall\, n \geqslant 1,$$

so

$$u^* \leqslant \varphi(u^*)$$

and thus

$$\varphi(u^*) \leqslant \varphi(\varphi(u^*))$$

and so $\varphi(u^*) \in S$. Therefore $\varphi(u^*)$ is an upper bound of C in S. Invoking the Kuratowski–Zorn theorem (see Theorem 4.120), we infer that S has a maximal element $\widehat{u} \in S$. We have

$$\widehat{u} \leqslant \varphi(\widehat{u}), \tag{4.214}$$

so

$$\varphi(\widehat{u}) \leqslant \varphi(\varphi(\widehat{u}))$$

(since φ is increasing), thus $\varphi(\widehat{u}) \in S$. From the maximality of $\widehat{u}$ and (4.214) we conclude that $\varphi(\widehat{u}) = \widehat{u}$.

Solution of Problem 4.189

Let $S = \{u \in \varphi([u_n, v_0]) : u \leqslant \varphi(u)\}$. From the solution of Problem 4.188, we know that S has a maximal element $\widehat{u} \in S$ such that $\varphi(\widehat{u}) = \widehat{u}$. We claim that $\widehat{u} \in S$ is the maximal fixed point of φ in $[u_0, v_0]$. To this end let $\widetilde{u} \in [u_0, v_0]$ be a fixed point of φ. Since by hypothesis K is minihedral (see Definition 4.125(e)), we have that $h = \{\widehat{u}, \widetilde{u}\}$ exists and we have

$$\widehat{u} \leqslant h, \quad \widetilde{u} \leqslant h,$$

so

$$\widehat{u} = \varphi(\widehat{u}) \leqslant \varphi(h), \quad \widetilde{u} = \varphi(\widetilde{u}) \leqslant \varphi(h), \tag{4.215}$$

thus

$$h \leqslant \varphi(h),$$

hence $\varphi(h) \leqslant \varphi(\varphi(h))$ and also $\varphi(h) \in S$. The maximality of $\widehat{u}$ in S implies that $\varphi(h) = \widehat{u}$ and so $\widetilde{u} \leqslant \widehat{u}$ (see (4.215)). This shows that $\widehat{u} \in [u_0, v_0]$ is the maximal fixed point of φ in $[u_0, v_0]$.

In a similar fashion, let $T = \{v \in \varphi([u_0, v_0]) : \varphi(v) \leqslant v\}$ and use the Kuratowski–Zorn theorem (see Theorem 4.120), to produce a minimal element $\widehat{v} \in T$ of T such that $\varphi(\widehat{v}) = \widehat{v}$. Then reasoning as above we show that $\widehat{v} \in [u_0, v_0]$ is the minimal fixed point of φ in $[u_0, v_0]$.

Solution of Problem 4.190

Let $\widehat{h}\colon [0,1] \times (\overline{\Omega} \cap K) \longrightarrow K$ be defined by

$$\widehat{h}(t,u) = t\varphi(u) \qquad \forall\, (t,u) \in [0,1] \times (\overline{\Omega} \cap K) = [0,1] \times \overline{\Omega \cap K}.$$

Evidently $\widehat{h}$ is continuous and $\widehat{h}(t,\cdot)$ is γ-condensing for every $t \in [0,1]$ (see Definition 2.31). By hypothesis, we have

$$\widehat{h}(t,u) \neq u \qquad \forall\, (t,u) \in [0,1] \times (\partial\Omega \cap K).$$

So, by the homotopy invariance property of the fixed point index (see Theorem 4.136(c)), we have

$$i(\varphi, \Omega \cap K, K) = i(0, \Omega \cap K, K) = 1$$

(since $0 \in \Omega \cap K$; see Theorem 4.136(a)).

Solution of Problem 4.191

From the Dugundji extension theorem (see Theorem 2.8), there exists a compact map $\widehat{g}\colon \overline{\Omega}\cap K \longrightarrow K$ such that

$$\widehat{g}|_{\partial\Omega\cap K} = g \quad \text{and} \quad \widehat{g}(\overline{\Omega}\cap K) \subseteq \overline{\operatorname{conv}}\, g(\partial\Omega\cap K).$$

We claim that

$$\inf_{u\in\overline{\operatorname{conv}}\, g(\partial\Omega\cap K)} \|u\|_X > 0. \tag{4.216}$$

Let $Y = \overline{\operatorname{span}}\, g(\partial\Omega\cap K)$. Since $g(\partial\Omega\cap K)$ is relatively compact, the space $Y \subseteq X$ is separable. Let $K_Y = K\cap Y$. Then $\overline{\operatorname{conv}}\, g(\partial\Omega\cap K) \subseteq K_Y$. Invoking Problem 4.185, we can find $\overline{y}^* \in Y^*$ such that

$$\langle \overline{y}^*, y\rangle > 0 \qquad \forall\, y\in K_Y\setminus\{0\}.$$

Let $m = \inf\limits_{y\in g(\partial\Omega\cap K)} \langle \overline{y}^*, y\rangle$. Suppose that $m=0$. Then we can find a sequence $\{y_n\}_{n\geqslant 1} \subseteq g(\partial\Omega\cap K)$ such that $\langle \overline{y}^*, y_n\rangle \longrightarrow 0$. Since the set $\overline{g(\partial\Omega\cap K)}$ is compact, passing to a subsequence if necessary, we may assume that $y_n \longrightarrow \overline{y}\in K_Y$ and so

$$\langle \overline{y}^*, y_n\rangle \longrightarrow \langle \overline{y}^*, \overline{y}\rangle = 0,$$

so $\overline{y}=0$ and $y_n\longrightarrow 0$, a contradiction to hypothesis **(i)**. Therefore $m>0$. We have

$$\langle \overline{y}^*, y\rangle \geqslant m > 0 \qquad \forall\, y\in \overline{\operatorname{conv}}\, g(\partial\Omega\cap K). \tag{4.217}$$

The set $\overline{\operatorname{conv}}\, g(\partial\Omega\cap K)\subseteq X$ is compact (by the Mazur theorem; see Theorem I.5.58) and so, we can find $\overline{u}\in\overline{\operatorname{conv}}\, g(\partial\Omega\cap K)$ such that

$$\inf_{u\in\overline{\operatorname{conv}}\, g(\partial\Omega\cap K)} \|u\|_X = \|\overline{u}\|_X. \tag{4.218}$$

Then we have

$$\langle \overline{y}^*, \overline{u}\rangle \geqslant m > 0$$

(see (4.217)), so

$$\|\overline{u}\|_X > 0.$$

From (4.218) we infer that (4.216) holds. Hence

$$\inf_{u\in\overline{\Omega}\cap K} \|g(u)\|_X = \widehat{m} > 0. \tag{4.219}$$

If $i(\varphi,\Omega\cap K,K)\neq 0$, then using hypothesis **(ii)** and the homotopy invariance property of the fixed point index (see Theorem 4.136(c)), we have

$$i(\varphi+tg,\Omega\cap K,K) = i(\varphi,\Omega\cap K,K) \neq 0 \qquad \forall\, t>0. \tag{4.220}$$

Let $\beta = \sup\limits_{u\in\overline{\Omega}\cap K}\|u\|_X$, $\gamma = \sup\limits_{u\in\overline{\Omega}\cap K}\|\varphi(u)\|_X$ and $\widehat{t} > \frac{\beta+\gamma}{\widehat{m}} > 0$ (see (4.219)). Then

$$i(\varphi+\widehat{t}g,\Omega\cap K,K) \neq 0$$

(see (4.220)). Then by the solution property of the fixed point index (see Theorem 4.136(d)), we know that there exists $\widehat{u} \in \Omega \cap K$ such that

$$\varphi(\widehat{u}) + \widehat{t}g(\widehat{u}) = \widehat{u},$$

so

$$\widehat{t} = \frac{\|\widehat{u}-\varphi(\widehat{u})\|_X}{\|g(\widehat{u})\|_X} \leqslant \frac{\beta+\gamma}{\widehat{m}},$$

a contradiction. So, we conclude that $i(\varphi, \Omega \cap K, K) = 0$.

Solution of Problem 4.192

Suppose that we can find $\lambda_0 \geqslant 0$ and $u_0 \in \partial\Omega \cap K$ such that

$$u_0 - \varphi(u_0) = \lambda_0 \varphi(u_0),$$

so

$$\widehat{\lambda}_0 u_0 = \varphi(u_0),$$

with $\widehat{\lambda}_0 = \frac{1}{1+\lambda_0} \in (0,1]$, which contradicts hypothesis **(ii)**. Therefore

$$u - \varphi(u) \neq \lambda\varphi(u) \qquad \forall\, \lambda \geqslant 0,\ u \in \partial\Omega \cap K.$$

This fact and hypothesis **(i)** permit the use of Problem 4.191. So, we have

$$i(\varphi, \Omega \cap K, K) = 0.$$

Solution of Problem 4.193

Let $\widehat{h}\colon [0,1] \times \overline{\Omega} \longrightarrow X$ be defined by

$$\widehat{h}(t,u) = (1-t)\varphi(u) + tu \qquad \forall\, (t,u) \in [0,1] \times \overline{\Omega}.$$

Suppose that we can find $t_0 \in [0,1]$ and $u_0 \in \partial\Omega$ such that

$$(1-t_0)\varphi(u_0) + t_0 v = u_0,$$

so $t_0 \neq 1$ (recall that $v \in \Omega$), thus

$$\varphi(u_0) - u_0 = \frac{t_0}{1-t_0}(u_0 - v) \quad \text{and} \quad \frac{t_0}{1-t_0} > 0,$$

which contradicts the hypothesis. Hence

$$\widehat{h}(t,u) \neq u \qquad \forall\, t \in [0,1],\ u \in \partial\Omega.$$

Using the homotopy invariance property of the fixed point index (see Theorem 4.136(c)), we have

$$i(\varphi, \Omega, K) = i(v, \Omega, K) = 1$$

(see Theorem 4.136(a)).

Solution of Problem 4.194
Consider the homotopy $\widehat{h}\colon [0,1] \times \overline{\Omega} \longrightarrow X$ defined by

$$\widehat{h}(t,u) = (1-t)\varphi(u) + tv \qquad \forall\ (t,u) \in [0,1] \times \overline{\Omega}.$$

By hypothesis, we have

$$\widehat{h}(t,u) \neq u \qquad \forall\ (t,u) \in [0,1] \times \Omega.$$

So from the homotopy invariance property of the fixed point index (see Theorem 4.136(c)), we have

$$i(\varphi, \Omega, K) = i(v, \Omega, K) = 0$$

(since $v \in K \setminus \overline{\Omega}$).

Solution of Problem 4.195
We have

$$i(\varphi, C, C) = 1. \tag{4.221}$$

Also, from Problem 4.193, we have

$$i(\varphi, \Omega_1, C) = i(\varphi, \Omega_2, C) = 1. \tag{4.222}$$

Then from (4.221), (4.222) and the domain decomposition property of the fixed point index (see Theorem 4.136(c)), we have

$$i(\varphi, C \setminus \overline{\Omega_1 \cup \Omega_2}, C) = i(\varphi, C, C) - i(\varphi, \Omega_1, C) - i(\varphi, \Omega_2, C) = -1. \tag{4.223}$$

From (4.222), (4.223) and the existence property of the fixed point index (see Theorem 4.136(d)), we see that φ has at least three fixed points

$$u_1 \in \Omega_1, \quad u_2 \in \Omega_2 \quad \text{and} \quad u_3 \in C \setminus \overline{\Omega_1 \cup \Omega_2}.$$

Solution of Problem 4.196
We claim that

$$\inf_{u\in\partial B_1\cap K} \|u - Lu\|_X = m > 0. \tag{4.224}$$

Arguing by contradiction, suppose that we could find a sequence $\{u_n\}_{n\geqslant 1} \subseteq \partial B_1 \cap K$ such that

$$\|u_n - L(u_n)\|_X \longrightarrow 0. \tag{4.225}$$

As in Problem 3.1 we can check that L maps bounded sets of K into relatively compact sets. So, we may assume that

$$L(u_n) \longrightarrow y \quad \text{in } X,$$

so

$$u_n \longrightarrow y \quad \text{in } X$$

(see (4.225)), thus $y \in \partial B_1 \cap K$ and $y = L(y)$, a contradiction to the hypotheses.

So, we see that (4.224) is true. From (4.224), we have

$$\|u - Lu\|_X \geqslant m\|u\|_X \qquad \forall\, u \in K. \tag{4.226}$$

Let $r_0 > 0$ be such that

$$\|\varphi(u) - L(u)\|_X \leqslant \tfrac{m}{2}\|u\|_X \qquad \forall\, u \in \overline{B}_r \cap K. \tag{4.227}$$

Then we have

$$\begin{aligned} &\|u-(t\varphi(u)+(1-t)L(u))\|_X \geqslant \|u-L(u)\|_X - t\|\varphi(u)-L(u)\|_X \\ \geqslant\ & m\|u\|_X - \tfrac{m}{2}\|u\|_X = \tfrac{m}{2}r > 0 \quad \forall\,(t,u) \in [0,1]\times\partial K_r,\ r\in(0,r_0] \end{aligned}$$

(see (4.226) and (4.227)). Using the homotopy invariance property of the fixed point index (see Theorem 4.136(c)), we have

$$i(\varphi, K_r, K) = i(L, K_r, K) \qquad \forall\, r \in (0, r_0]. \tag{4.228}$$

But from the hypotheses and Problem 4.190, we have

$$i(L, K_r, K) = 1 \qquad \forall\, r \in (0, r_0],$$

so

$$i(\varphi, K_r, K) = 1 \qquad \forall\, r \in (0, r_0]$$

(see (4.228)).

Solution of Problem 4.197

By hypothesis we can find $\lambda > 1$ and $y \in K \setminus \{0\}$ such that

$$L(y) = \lambda y. \tag{4.229}$$

We claim that

$$u - L(y) = ty \qquad \forall\, u \in K \setminus \{0\},\ t \geqslant 0. \tag{4.230}$$

First note that if (4.230) is true for some $u \in K \setminus \{0\}$ and $t \geqslant 0$, then $t > 0$. Also, from (4.229), we have $u \geqslant ty$. Let $t_0 = \sup\limits_{u \geqslant ty} t$. We have

$$u = L(u) + ty \geqslant L(t_0 y) + ty = (\lambda t_0 + t)y$$

(see (4.229)). Since $\lambda t_0 + t > t_0$, we contradict the definition of t_0. Therefore (4.230) holds. Then Problem 4.191 implies that

$$i(L, K_r, K) = 0 \qquad \forall\, r > 0 \tag{4.231}$$

On the other hand, as in the solution of Problem 4.196, we can show that there exists $r_0 > 0$ such that

$$i(\varphi, K_r, K) = i(L, K_r, K) \qquad \forall\, r \in (0, r_0] \tag{4.232}$$

(see (4.228)). From (4.231) and (4.232), we conclude that

$$i(\varphi, K_r, K) = 0 \qquad \forall\, r \in (0, r_0].$$

Solution of Problem 4.198

Let $C \subseteq X$ be a bounded above set and let $\{u_n\}_{n \geqslant 1} \subseteq C$ be a sequence dense in C (recall that X is separable). Set $y_n = \sup\{u_1, \ldots, u_n\}$ for all $n \geqslant 1$ (these suprema exist since by hypothesis K is minihedral; see Definition 4.125(e)). We have

$$y_1 \leqslant y_2 \leqslant \ldots \leqslant y_n \leqslant \ldots \leqslant h,$$

for some $h \in X$. The regularity of K (see Definition 4.125(c)) implies that $y_n \longrightarrow y$ in X. We claim that $y = \sup C$. Clearly, we have

$$u_n \leqslant y_n \leqslant y \qquad \forall\, n \geqslant 1. \tag{4.233}$$

Let $v \in C$. Then, we can find a subsequence $\{u_{n_k}\}_{k \geqslant 1}$ of $\{u_n\}_{n \geqslant 1}$ such that

$$u_{n_k} \longrightarrow v \quad \text{in } X,$$

so $v \leqslant y$ (see (4.233)) and thus y is an upper bound of C.

If $w \in X$ is another upper bound of C, then

$$u_n \leqslant w \qquad \forall\, n \geqslant 1,$$

so $v \leqslant w$ and thus $v = \sup C \in X$. Therefore C is strongly minihedral (see Definition 4.125(f)).

Solution of Problem 4.199

Let $S = \{u \in [u_0, v_0] : u \leqslant \varphi(u)\}$. Evidently $u_0 \in S$ and v_0 is an upper bound for the set S. Since by hypothesis K strongly minihedral (see Definition 4.125(f)), $\sup S = \widehat{u} \in X$. We claim that $\widehat{u}$ is a maximal fixed point of φ in $[u_0, v_0]$. For any $u \in S$, we have

$$u_0 \leqslant u \leqslant \widehat{u} \leqslant v,$$

so

$$u_0 \leqslant \varphi(u_0) \leqslant \varphi(u) \leqslant \varphi(\widehat{u}) \leqslant \varphi(v) \leqslant v$$

(since φ is increasing), thus $u \leqslant \varphi(\widehat{u})$. As $\widehat{u} = \sup S$, we have

$$\widehat{u} \leqslant \varphi(\widehat{u}),$$

so

$$\varphi(\widehat{u}) \leqslant \varphi(\varphi(\widehat{u}))$$

(since φ is increasing), thus $\varphi(\widehat{u}) \in S$ and hence $\widehat{u} = \varphi(\widehat{u})$ (since $\widehat{u} = \sup S$).

Let $y \in [u_0, v_0]$ be any other fixed point of φ. Then $y \in S$ and $y \leqslant \widehat{u}$. Therefore $\widehat{u}$ is the greatest fixed point of φ in $[u_0, v_0]$.

Similarly, if $T = \{v \in [u_0, v_0] : \varphi(v) \leqslant v\}$, then working as above we show that $\widehat{v} = \inf T$ exists and is the smallest fixed point of φ in $[u_0, v_0]$.

Solution of Problem 4.200

From Problem 4.199, we know that φ has a maximal fixed point $\widehat{u}$ and a minimal fixed point $\widehat{v}$ in $[u_0, v_0]$. Moreover, if $S = \{u \in [u_0, v_0] : u \leqslant \varphi(u)\}$, then $\widehat{u}$ is a maximal point in S (see the solution of Problem 4.199). Let $\widetilde{u} \in [u_0, v_0]$ be a fixed point of φ. Since K is minihedral (see Definition 4.125(e)), $\sup\{\widetilde{u}, \widehat{u}\} = h$ exists. We have

$$\widetilde{u} = \varphi(\widetilde{u}) \leqslant \varphi(h) \quad \text{and} \quad \widehat{u} = \varphi(\widehat{u}) \leqslant \varphi(h)$$

(since φ is increasing), so $h \leqslant \varphi(h)$ and thus $h \in S$. Since K is minihedral, as above $w = \{\widehat{u}, h\}$ exists and belong in S. The maximality of $\widehat{u}$ implies that $\widehat{u} = w$, so $h \leqslant \widehat{u}$, thus $\widetilde{u} \leqslant \widehat{u}$, which means that $\widehat{u}$ is the greatest fixed point of φ in $[u_0, v_0]$.

Similarly, using the set $T = \{v \in [u_0, v_0] : \varphi(v) \leqslant v\}$, we show that the minimal fixed point $\widehat{v} \in [u_0, v_0]$ is in fact the smallest fixed point of φ in $[u_0, v_0]$.

Solution of Problem 4.201

Let $\eta = \sup\limits_{u \in \partial\Omega \cap K} \|\varphi(u)\|_X > 0$ and consider the map $\psi\colon \partial\Omega \cap K \longrightarrow K$ defined by

$$\psi(u) \;=\; \eta\frac{\varphi(u)}{\|\varphi(u)\|_X} \qquad \forall\, u \in \partial\Omega \cap K.$$

By hypothesis **(ii)**, ψ is compact. Let $\widehat{\psi}\colon \overline{\Omega} \cap K = \overline{\Omega \cap K} \longrightarrow K$ be the compact extension of ψ postulated by the Dugundji extension theorem (see Theorem 2.8). From Problem 4.191, we have

$$i(\widehat{\psi}, \Omega \cap K, K) \;=\; 0. \tag{4.234}$$

We consider the compact homotopy $\widehat{h}\colon [0,1] \times (\overline{\Omega} \cap K) \longrightarrow K$ defined by

$$\widehat{h}(t,u) \;=\; t\varphi(u) + (1-t)\widehat{\psi}(u) \qquad \forall\, (t,u) \in [0,1] \times (\overline{\Omega} \cap K) = [0,1] \times \overline{\Omega \cap K}.$$

Hypothesis **(i)** implies that

$$u \;\neq\; \widehat{h}(t,u) \qquad \forall\, (t,u) \in [0,1] \times (\partial\Omega \cap K).$$

So, the homotopy invariance property of the fixed point index (see Theorem 4.136(c)) implies that

$$i(\varphi, \Omega \cap K, K) \;=\; i(\widehat{\psi}, \Omega \cap K, K),$$

so $i(\varphi, \Omega \cap K, K) = 0$ (see (4.234)).

Solution of Problem 4.202

Since φ is increasing, we have

$$u_0 \leqslant u_1 \leqslant \ldots \leqslant u_n \leqslant \ldots \leqslant v_n \leqslant \ldots \leqslant v_1 \leqslant v_0.$$

The normality of K implies that $[u_0, v_0] \subseteq X$ is bounded (see Problem 4.181(g)) and so the set $D = \{u_n : n \geqslant 1\} \subseteq X$ is bounded and $D = \varphi(D) \cup \{u_0\}$. So, we have

$$\gamma(D) \;=\; \gamma(\varphi(D)),$$

thus $\gamma(D) = 0$ (since φ is γ-condensing; see Definition 2.31)) and the set D is relatively compact.

Therefore, we can find a subsequence $\{u_{n_k}\}_{k\geqslant 1}$ of $\{u_n\}_{n\geqslant 1}$ for which we have that

$$u_{n_k} \longrightarrow \widehat{v} \quad \text{in } X, \tag{4.235}$$

so

$$u_n \leqslant \widehat{v} \leqslant v_n \qquad \forall\, n \geqslant 1. \tag{4.236}$$

For $n \geqslant n_k$, we have

$$0 \leqslant \widehat{v} - u_n \leqslant \widehat{v} - u_{n_k},$$

so

$$\|\widehat{v} - u_n\|_X \leqslant c_1\|\widehat{v} - u_{n_k}\|_X$$

(since K is normal; see Definition 4.125(b)), thus

$$u_n \longrightarrow \widehat{v} \quad \text{in } X$$

(see (4.235)). Recall that $u_n = \varphi(u_{n-1})$ for all $n \geqslant 1$, so

$$\widehat{v} = \varphi(\widehat{v})$$

(since φ is γ-condensing; see Definition 2.31).

Similarly, we have

$$v_n \longrightarrow \widehat{u} \quad \text{in } X \quad \text{and} \quad \widehat{u} = \varphi(\widehat{u}).$$

Let $\widetilde{u} \in [u_0, v_0]$ be a fixed point of φ. Then

$$u_n \leqslant \widetilde{u} \leqslant v_n \qquad \forall\, n \geqslant 1$$

(since φ is increasing), so $\widehat{v} \leqslant \widetilde{u} \leqslant \widehat{u}$.

Solution of Problem 4.203

Evidently $\operatorname{Fix}(\varphi)$ is weakly closed and we have $\operatorname{Fix}(\varphi) = \varphi(\operatorname{Fix}(\varphi))$. Therefore, we have $\xi(\operatorname{Fix}(\varphi)) = 0$ and so $\operatorname{Fix}(\varphi)$ is w-compact.

Let $u_0 \in C$ and let

$$\mathcal{Y}_0 = \{D \in P_{fc}(X) : \ u_0 \in D \text{ and } \varphi(D) \subseteq D\}.$$

Note that $\mathcal{Y}_0 \neq \emptyset$ since $C \in \mathcal{Y}_0$. Let $\widehat{D} = \bigcap_{D \in \mathcal{Y}_0} D$. Then the set $\widehat{D} \subseteq X$ is closed, convex, and $u_0 \in \widehat{D}$. If $u \in \widehat{D}$, then $u \in D$ for all $D \in \mathcal{Y}_0$ and so $\varphi(u) \in D$ for all $D \in \mathcal{Y}_0$ (recall the definition of $\mathcal{Y}_0$). Hence $\varphi(\widehat{D}) \subseteq \widehat{D}$. Note that $\widehat{D} \subseteq C$ is w-compact. Indeed, if $\xi(\widehat{D}) > 0$, then $\xi(\varphi(\widehat{D})) < \xi(\widehat{D})$. We set

$$\widetilde{D} = \overline{\operatorname{conv}}\,(\varphi(\widehat{D}) \cup \{u_0\}).$$

Then $\widetilde{D} \subseteq \widehat{D}$ and so $\varphi(\widetilde{D}) \subseteq \varphi(\widehat{D}) \subseteq \widetilde{D}$. Therefore $\widetilde{D} \in \mathcal{Y}_0$ and so $\widehat{D} \subseteq \widetilde{D}$. We conclude that $\widehat{D} = \widetilde{D}$. But from Proposition 2.34, we have

$$\xi(\widetilde{D}) = \xi(\overline{\operatorname{conv}}\,(\varphi(\widehat{D}) \cup \{u_0\})) = \xi(\varphi(\widehat{D})) < \xi(\widehat{D}) = \xi(\widetilde{D}),$$

a contradiction. This proves that $\widehat{D} \in P_{wkc}(X)$. Invoking the Tichonov fixed point theorem (see Theorem 4.101), we conclude that $\varphi\colon \widehat{D} \longrightarrow \widehat{D}$ has a fixed point and so $\operatorname{Fix}(\varphi) \neq \emptyset$.

Solution of Problem 4.204

We argue as in the solution of Problem 4.203. Let $u_0 \in C$ and let

$$\mathcal{Y}_0 = \{D \in P_{fc}(X) : u_0 \in D \text{ and } \varphi(D) \subseteq D\}.$$

Since $C \in \mathcal{Y}_0$, we have that $\mathcal{Y}_0 \neq \emptyset$. Let $\widehat{D} = \bigcap_{D \in \mathcal{Y}_0} D$. As in the solution of Problem 4.203, we introduce

$$\widetilde{D} = \overline{\operatorname{conv}}\,\big(\varphi(\widehat{D}) \cup \{u_0\}\big).$$

Using the hypotheses on φ, as in the solution of Problem 4.203, we obtain

$$\widetilde{D} = \widehat{D} \in P_{wkc}(X). \tag{4.237}$$

We have that $\varphi(\widehat{D}) \subseteq \widehat{D}$ and we claim that the set $\varphi(\widehat{D}) \subseteq X$ is relatively compact. To see this, let $\{y_n\}_{n\geqslant 1} \subseteq \varphi(\widehat{D})$. We have

$$y_n = \varphi(u_n) \quad \text{with } u_n \in \widehat{D} \qquad \forall\, n \geqslant 1.$$

From (4.237) and the Eberlein–Smulian theorem (see Theorem I.5.78), by passing to a suitable subsequence if necessary, we may assume that $u_n \xrightarrow{w} u$. Then condition (W) of φ implies that the sequence $\{y_n\}_{n\geqslant 1}$ admits a strongly convergent subsequence. Hence $\varphi(\widehat{D})$ is relatively compact in X. Since φ is continuous, invoking the Schauder fixed point theorem (see Theorem 4.89), we conclude that φ has a fixed point.

Solution of Problem 4.205

First we show that $u \longmapsto (I_X - \psi)(u)$ is a homeomorphism of X onto X. To this end, for any $h \in X$, let $\vartheta_h\colon X \longrightarrow X$ be defined by

$$\vartheta_h(u) = \psi(u) + h \qquad \forall\, u \in X.$$

We have

$$\|\vartheta_h(u) - \vartheta_h(v)\|_X \leqslant \eta(\|u - v\|_X) \qquad \forall\, u, v \in X$$

(see hypothesis **(ii)**). Problem 4.83 implies that ϑ_h has a unique fixed point $u \in X$, that is,

$$u = \psi(u) + h,$$

so

$$h = (I_X - \psi)(u),$$

thus $I_X - \psi$ is surjective on X.

If $u, v \in X$ and $u \neq v$, then

$$\begin{aligned} \|(u - \psi(u)) - (v - \psi(v))\|_X &\geqslant \|u - v\|_X - \|\psi(u) - \psi(v)\|_X \\ &\geqslant \|u - v\|_X - \eta(\|u - v\|_X) > 0, \end{aligned}$$

so $I_X - \psi$ is an injection and thus we can define $(I_X - \psi)^{-1}$.

Finally we show the continuity of $(I_X - \psi)^{-1}$. So, let

$$(I_X - \psi)(u_n) \longrightarrow (I_X - \psi)(u) \quad \text{in } X.$$

Suppose that $\limsup\limits_{n \to +\infty} \|u_n - u\|_X = t$. Then

$$\|(I_X - \psi)(u_n) - (I_X - \psi)(u)\|_X \geqslant \|u_n - u\|_X - \eta(\|u_n - u\|_X) \qquad \forall\, n \geqslant 1,$$

so $0 \geqslant t - \eta(t)$, thus $t = 0$ and hence $u_n \longrightarrow u$ in X. This proves the continuity of $(I_X - \psi)^{-1}$ and we conclude that the map $u \longmapsto (I_X - \psi)(u)$ is a homeomorphism.

We set $f(u) = (I_X - \psi)^{-1} \circ \varphi(u)$ for all $u \in C$. For any $h \in C$, from the first part of the solution we know that there exists a unique $u \in X$ such that

$$\varphi(h) = u - \psi(u),$$

so $u \in C$ (see hypothesis **(iii)**) and thus $f(C) \subseteq C$. Evidently f is continuous. Also, if $\{u_n\}_{n \geqslant 1} \subseteq C$ is a sequence such that $u_n \xrightarrow{w} u$, then by hypothesis **(i)** (in particular condition (W); see Problem 4.204), the sequence $\{\varphi(u_n)\}_{n \geqslant 1}$ admits a strongly convergent subsequence. The continuity of $(I_X - \psi)^{-1}$ implies that the sequence $\{(I_X - \psi)^{-1}(\varphi(u_n))\}_{n \geqslant 1}$ is strongly convergent. Therefore the function

$$u \longmapsto f(u) = (I_X - \psi)^{-1} \circ \varphi(u)$$

satisfies condition (W) (see Problem 4.204).

We have

$$f = (I_X - \psi)^{-1} \circ \varphi = \varphi + \psi \circ (I_X - \psi)^{-1} \circ \varphi. \tag{4.238}$$

So, if $D \subseteq C$ is such that $\xi(D) > 0$, then

$$f(D) \subseteq \varphi(D) + \psi(f(D))$$

(see (4.238)), thus

$$\xi(f(D)) \leqslant \xi(\varphi(D)) + \xi(\psi(f(D))) \leqslant k\xi(D) + \eta(\xi(f(D))) \tag{4.239}$$

(see Proposition 2.34 and hypotheses **(i)** and **(ii)**).

If $k = 0$, then $\xi(f(D)) = 0$.

If $k \in (0, 1)$, then from (4.239) and hypothesis **(ii)**, we have

$$\xi(f(D)) < k\xi(D) + (1 - k)\xi(f(D)),$$

so $\xi(f(D)) < \xi(D)$.

Therefore f is ξ-condensing. Hence we can use Problem 4.204 and conclude that

f has a fixed point $\widehat{u} \in C$, that is, $\widehat{u} = f(\widehat{u})$, hence $\widehat{u} = \varphi(\widehat{u}) + \psi(\widehat{u})$.

Solution of Problem 4.206

Let $R > 0$ and let $r\colon X \longrightarrow X$ be defined by

$$r(u) = \begin{cases} u & \text{if } \|u\|_X \leqslant R, \\ \frac{R}{\|u\|_X} u & \text{if } \|u\|_X > R \end{cases}$$

(that is, r is the R-radial retraction). Then r is continuous and $r(X) \subseteq \overline{B}_R$ (where $\overline{B}_R = \{u \in X : \ \|u\|_X \leqslant R\}$). Let $\varphi_r\colon \overline{B}_R \longrightarrow \overline{B}_R$ be defined by $\varphi_r = (r \circ \varphi)|_{\overline{B}_R}$. Clearly φ_r is continuous and satisfies condition (W), since φ does.

Let $D \subseteq \overline{B}_R$ be a set such that $\xi(D) > 0$. For $u \in D$, we have two possibilities:

(1) $\|\varphi(u)\|_X \leqslant R$ in which case

$$r(\varphi(u)) = \varphi(u) \in \varphi(D) \subseteq \operatorname{conv}(\varphi(D) \cup \{0\})$$

(2) $\|\varphi(u)\|_X > R$ in which case

$$r(\varphi(u)) = \frac{R}{\|\varphi(u)\|_X}\varphi(u) \in \varphi(D) \subseteq \operatorname{conv}(\varphi(D) \cup \{0\}).$$

Therefore, we have established that

$$\varphi_r(D) \subseteq \operatorname{conv}(\varphi(D) \cup \{0\}),$$

so

$$\xi(\varphi_r(D)) \leqslant \xi(\varphi(D)) < \xi(D)$$

(see Proposition 2.34), thus φ_r is ξ-condensing. Now, we can use Problem 4.204 on the map $\varphi_r\colon \overline{B}_R \longrightarrow \overline{B}_R$ and obtain $\widehat{u} \in \overline{B}_R$ such that $\widehat{u} = \varphi_r(\widehat{u})$.

If $\varphi(\widehat{u}) \in \overline{B}_R$, then

$$\widehat{u} = \varphi_r(\widehat{u}) = r(\varphi(\widehat{u})) = \varphi(\widehat{u}).$$

If $\varphi(\widehat{u}) \notin \overline{B}_R$, then

$$\widehat{u} = \varphi_r(\widehat{u}) = \frac{R}{\|\varphi(\widehat{u})\|_X}\varphi(\widehat{u})$$

and so $\widehat{u}$ is a solution of

$$u = t\varphi(u),$$

with $t = \frac{R}{\|\varphi(\widehat{u})\|_X} \in (0,1)$, $R = \|\widehat{u}\|_X$.

So, if there is no $u \in \overline{B}_R$ such that $u = \varphi(u)$ for every $R > 0$, then from the above argument we see that for some $t \in (0,1)$ the set $\{u \in X : \ u = t\varphi(u)\}$ is unbounded.

Solution of Problem 4.207

Let $t \in (0,1)$ and consider the maps $\varphi\colon C \longrightarrow X$ and $t\psi\colon X \longrightarrow X$. Then for this pair, we can use Problem 4.205 and produce $\widehat{u}_t \in C$ such that

$$\widehat{u}_t = \varphi(\widehat{u}_t) + t\psi(\widehat{u}_t).$$

Let $\{t_n\}_{n\geqslant 1} \subseteq (0,1)$ be such that $t_n \longrightarrow 1$ and let $\widehat{u}_n = \widehat{u}_{t_n} \in C$ for all $n \geqslant 1$. We have

$$\widehat{u}_n = \varphi(\widehat{u}_n) + t_n\psi(\widehat{u}_n) \qquad \forall\, n \geqslant 1. \tag{4.240}$$

Let $D = \{\widehat{u}_n :\ n \geqslant 1\}$ and suppose that $\xi(D) > 0$. Then we have

$$\begin{aligned} \xi(D) &= \xi(\{\varphi(\widehat{u}_n) + t_n\psi(\widehat{u}_n) :\ n \geqslant 1\}) \\ &\leqslant \xi(\varphi(D)) + t_n\xi(\psi(D)) = t_n\xi(\psi(D)) \\ &\leqslant \xi(\psi(D)) < \xi(D) \end{aligned}$$

(see Proposition 2.34, (4.240), hypotheses **(i)** and **(ii)** and recall that $t_n < 1$), a contradiction. Therefore, $\xi(D) = 0$ and so the set D is relatively w-compact. Thus, by the Eberlein–Smulian theorem (see Theorem I.5.78) and by passing to a suitable subsequence if necessary, we may assume that

$$\widehat{u}_n \xrightarrow{w} \widehat{u} \in C \quad \text{in } X. \tag{4.241}$$

We have

$$\varphi(\widehat{u}_n) \longrightarrow \varphi(\widehat{u}) \quad \text{in } X$$

(see hypothesis **(i)**), so

$$(I_X - t_n\psi)(\widehat{u}_n) \longrightarrow \varphi(\widehat{u}) \quad \text{in } X. \tag{4.242}$$

Note that since $C \subseteq X$ is bounded and ψ is nonexpansive, we have that the sequence $\{\psi(\widehat{u}_n)\}_{n\geqslant 1} \subseteq X$ is bounded. We have

$$(1-t_n)\|\psi(\widehat{u}_n)\|_X \longrightarrow 0. \tag{4.243}$$

Hence

$$\begin{aligned} &\|(I_X - \psi)(\widehat{u}_n) - \varphi(\widehat{u})\|_X \\ &\leqslant \|(I_X - \psi)(\widehat{u}_n) - (I_X - t_n\psi)(\widehat{u}_n)\|_X + \|(I_X - t_n\psi)(\widehat{u}_n) - \varphi(\widehat{u})\|_X \\ &= (1-t_n)\|\psi(\widehat{u}_n)\|_X + \|(I_X - t_n\psi)(\widehat{u}_n) - \varphi(\widehat{u})\|_X \longrightarrow 0 \end{aligned}$$

(see (4.242) and (4.242)), so

$$(I_X - \psi)(\widehat{u}_n) \longrightarrow \varphi(\widehat{u}) \quad \text{in } X. \tag{4.244}$$

Since $I_X - \psi$ is demiclosed (see hypothesis **(iii)**), from (4.241) and (4.244), we infer that

$$(I_X - \psi)(\widehat{u}) = \varphi(\widehat{u})$$

(see Definition 4.77) and thus $\widehat{u} = \varphi(\widehat{u}) + \psi(\widehat{u})$.

Bibliography

[1] Cronin, J.: Fixed Points and Topological Degree in Nonlinear Analysis. American Mathematical Society, Providence (1964)

[2] Deimling, K.: Nonlinear Functional Analysis. Springer, Berlin (1985)

[3] Denkowski, Z., Migórski, S., Papageorgiou, N.S.: An Introduction to Nonlinear Analysis: Applications. Kluwer Academic, Boston (2003)

[4] Fonseca, I., Gangbo, W.: Degree Theory in Analysis and Applications. Oxford University Press, New York (1995)

[5] Goebel, K., Kirk, W.A.: Topics in Metric Fixed Point Theory. Cambridge University Press, Cambridge (1990)

[6] Górniewicz, L.: Topological Fixed Point Theory of Multivalued Mappings. Springer, Dordrecht (2006)

[7] Granas, A., Dugundji, J.: Fixed Point Theory. Springer, New York (2003)

[8] Krasnosel'skii, M.A.: Positive Solutions of Operator Equations. Noordhoff Ltd., Groningen (1964)

[9] Krawcewicz, W., Wu, J.: Theory of Degrees with Applications to Bifurcations and Differential Equations. Wiley, New York (1997)

[10] Lloyd, N.G.: Degree Theory. Cambridge University Press, Cambridge (1978)

[11] Papageorgiou, N.S., Kyritsi-Yiallourou, S.Th.: Handbook of Applied Analysis. Springer, New York (2009)

[12] Rothe, E.H.: Introduction to Various Aspects of Degree Theory in Banach Spaces. American Mathematical Society, Providence (1986)

[13] Smart, D.R.: Fixed Point Theorems. Cambridge University Press, London (1974)

[14] Zeidler, E.: Nonlinear Functional Analysis and its Applications, vol. I. Springer, New York (1990)

Chapter 5

Variational and Topological Methods

5.1 Introduction

5.1.1 Minimization Methods

Let X be a set and let $\varphi\colon X \longrightarrow \overline{\mathbb{R}} = \mathbb{R} \cup \{+\infty\}$ be a map. One of the main problems of the Calculus of Variations is to find the minimum value

$$m(\varphi) \;=\; \inf_{u \in X} \varphi(u)$$

and also investigate the set of minimizers, that is,

$$S(\varphi) \;=\; \{u \in X : \; m(\varphi) = \varphi(u)\}.$$

The classical way to approach this problem is the so called ***direct method*** of Tonelli. This method consists of determining a suitable topology on X for which the function φ is lower semicontinuous and coercive.

We have already encountered the notion of lower semicontinuity (see Definition 3.55). For completeness we recall it here.

Let (X, τ) be a topological space (τ denotes the topology). Recall that all topologies considered in this book are Hausdorff. Given $u \in X$, by $\mathcal{N}(u)$ we denote the filter of open neighborhoods of u. Since we will be dealing with $\overline{\mathbb{R}} = \mathbb{R} \cup \{+\infty\}$-valued functions φ, we always assume that φ is not identically $+\infty$. Finally $\operatorname{dom} \varphi = \{u \in X : \; \varphi(u) < +\infty\}$ (the effective domain of φ)

Definition 5.1

(a) *We say that $\varphi = X \longrightarrow \overline{\mathbb{R}}$ is **τ-lower semicontinuous** at $u \in X$, if for every $\lambda \in \mathbb{R}$ with $\lambda < \varphi(u)$, we can find $U \in \mathcal{N}(u)$ such that $\lambda < \varphi(u)$ for all $v \in U$. We say that φ is **τ-lower semicontinuous**, if it is τ-lower semicontinuous at every $u \in X$.*

L. Gasiński, N.S. Papageorgiou, *Exercises in Analysis: Part 2*, Problem Books in Mathematics, DOI 10.1007/978-3-319-27817-9_5

(b) *We say that* $\varphi = X \longrightarrow \overline{\mathbb{R}}$ *is* ***sequentially*** τ***-lower semicontinuous*** *at* $u \in X$, *if for every sequence* $\{u_n\}_{n\geqslant 1} \subseteq X$ *such that* $u_n \xrightarrow{\tau} u$ *in* X, *we have*

$$\varphi(u) \leqslant \liminf_{n\to+\infty} \varphi(u_n).$$

We say that φ *is* ***sequentially*** τ***-lower semicontinuous***, *if it is sequentially* τ*-lower semicontinuous at every* $u \in X$.

Remark 5.2

Evidently τ-lower semicontinuity of φ is equivalent to saying that for every $\lambda \in \mathbb{R}$, the sublevel set $L_\lambda = \{u \in X : \varphi(u) \leqslant \lambda\}$ is τ-closed. Also $\varphi\colon X \longrightarrow \overline{\mathbb{R}}$ is τ-lower semicontinuous at $u \in X$ if and only if

$$\varphi(u) = \sup_{U\in\mathcal{N}(u)} \inf_{v\in U} \varphi(u).$$

We say that φ is τ-upper semicontinuous, if $-\varphi$ is τ-lower semicontinuous.

Definition 5.3

Let τ_{seq} *be the topology on* X, *for which closed sets are the sequentially* τ*-closed sets.*

Proposition 5.4

(a) τ_{seq} *is the strongest topology on* X *for which convergent sequences are the* τ*-convergent sequences.*

(b) $\tau \subseteq \tau_{seq}$ *and the two are equivalent if and only if* τ *is first countable (see Definition I.2.24).*

(c) $\varphi\colon X \longrightarrow \overline{\mathbb{R}}$ *is sequentially* τ*-lower semicontinuous if and only if it is* τ_{seq}*-lower semicontinuous.*

Definition 5.5

(a) *We say that* $\varphi = X \longrightarrow \overline{\mathbb{R}}$ *is* τ***-coercive*** *(or* τ*-*inf***-compact****), if for every* $\lambda \in \mathbb{R}$, *the set* $L_\lambda = \{u \in X : \varphi(u) \leqslant \lambda\}$ *is relatively* τ*-compact.*

(b) *We say that* $\varphi = X \longrightarrow \overline{\mathbb{R}}$ *is* ***sequentially*** τ***-coercive***, *if for every* $\lambda \in \mathbb{R}$, *the set* $L_\lambda = \{u \in X : \varphi(u) \leqslant \lambda\}$ *is relatively sequentially* τ*-compact (that is, relatively* τ_{seq}*-compact).*

The ***direct method*** is summarized by the following theorem.

Theorem 5.6

If $\varphi\colon X \longrightarrow \overline{\mathbb{R}}$ *is* τ*-lower semicontinuous and* τ*-coercive (respectively sequentially* τ*-lower semicontinuous and sequentially* τ*-coercive),*
then there exists $u_0 \in X$ *such that*

$$\varphi(u_0) = m(\varphi) = \inf_{u\in X} \varphi(u)$$

(that is, $S(\varphi) \neq \emptyset$*).*

So far we have used only a topological structure of X and correspondingly on the function φ. In what follows, we will introduce more structure on both the space X and the function φ. First we recall the following notion (see Definition 3.21).

Definition 5.7
*Suppose that X is a Banach space, $U \subseteq X$ is a nonempty open set and $\varphi\colon U \longrightarrow \mathbb{R}$ is a function. We say that $u_0 \in U$ is a **local minimum** (respectively **maximum**) of φ if there exists $r > 0$ such that*

$$\varphi(u) \geqslant \varphi(u_0) \quad (\textit{respectively } \varphi(u) \leqslant \varphi(u_0)) \qquad \forall\, u \in B_r(u_0),$$

*where $B_r(u_0) = \{u \in X : \|u - u_0\|_X < r\}$. We say that $u \in X$ is a **local extremum** of φ if it is a local minimum or local maximum of φ.*

*If the above inequalities are strict for all $u \in B_r(u_0) \setminus \{u_0\}$, then we say that we have a **strict local minimum** (respectively **strict local maximum**) and **strict local extremum** in both cases. If the inequalities are true for all $u \in X$, then we say that we have a **global minimum** (respectively **global maximum**) and **global extremum** in both cases.*

Remark 5.8
It is always possible to pass from a maximization problem to a minimization one by simply replacing φ with $-\varphi$. In what follows we focus exclusively on minimization problems.

Definition 5.9
Suppose that X and Y are two Banach spaces, $U \subseteq X$ and $V \subseteq Y$ are nonempty open sets and $\varphi\colon U \longrightarrow V$ is a Fréchet differentiable map.

*(a) We say that $u_0 \in U$ is a **critical point** of φ if $\varphi'(u_0) \in \mathcal{L}(X;Y)$ is not surjective. Then $\varphi(u_0) \in V$ is a **critical value** of φ.*

*(b) We say that $u_0 \in U$ is a **regular point** of φ if $\varphi'(u_0) \in \mathcal{L}(X;Y)$ is surjective. Then $\varphi(u_0) \in V$ is a **regular value** of φ.*

Remark 5.10
Note that, if $Y = \mathbb{R}$, then $u_0 \in U$ is a critical (respectively regular) point of the function $\varphi\colon U \longrightarrow \mathbb{R}$ if and only if $\varphi'(u_0) = 0$ (respectively $\varphi'(u_0) \neq 0$) (see Definition 3.42(b)).

The next result is the classical Fermat condition for local extremal points. So, for differentiable functions, the local extremal points are necessarily contained in the set of critical points of φ (see also Proposition 3.25).

Proposition 5.11
<u>If</u> X is a Banach space, $U \subseteq X$ is a nonempty open set and $\varphi\colon U \longrightarrow \mathbb{R}$ is a Fréchet differentiable function,
<u>then</u> every local extremal point of φ is a critical point of φ.

If we improve the structure of φ, then we can say more (see also Proposition 3.40).

Proposition 5.12
If X is a Banach space, $U \subseteq X$ is a nonempty open set, $\varphi \in C^2(U;\mathbb{R})$ and $u_0 \in U$ is a local minimum point of φ, then
(a) *$\varphi'(u_0) = 0$ (that is, u_0 is a critical point of φ; see Proposition 5.11).*
(b) *$\varphi''(u_0) \in \mathcal{L}(X \times X;\mathbb{R}) = \mathcal{L}(X;X^*)$ is nonnegative definite, i.e.,*

$$\varphi''(u_0)(h,h) \;=\; \langle \varphi''(u_0)h, h\rangle \;\geqslant\; 0 \qquad \forall\, h \in X.$$

In fact this result almost has an inverse.

Proposition 5.13
If X is a Banach space, $U \subseteq X$ is a nonempty open set, $\varphi \in C^2(U;\mathbb{R})$ and for $u_0 \in U$, we have
(i) *u_0 is a critical point of φ (i.e., $\varphi'(u_0) = 0$);*
(ii) *$\varphi''(u_0)$ is positive definite, i.e.,*

$$\inf_{\substack{h \in X \\ \|h\|_X = 1}} \langle \varphi''(u_0)h, h\rangle \;>\; 0,$$

then φ has a strict local minimum at $u_0 \in U$.

For convex functions, stronger results hold.

Proposition 5.14
If X is a Banach space, $U \subseteq X$ is a nonempty, open, convex set, $\varphi \in C^1(U;\mathbb{R})$ is a convex function,
then every critical point of φ is a global minimum.

Remark 5.15
For strictly convex functions, the global minimum is unique.

In many situations in the Calculus of Variations, we must find critical points of a functional φ in the presence of constraints, that is, critical points of φ restricted to a set of constraints $M \subseteq X$. This requires for M to have a differentiable structure. This leads to the notion of differentiable manifolds modeled on a Banach space.

Definition 5.16
*Let X be a Banach space. A topological space M is a C^k-**Banach manifold** modeled on X, if there is an open cover $\{U_i\}_{i\in I}$ of M and a family of maps $\{\psi_i \colon U_i \longrightarrow X\}_{i\in I}$ such that:*
(a) *$\psi_i \colon U_i \longrightarrow V_i = \psi_i(U_i)$ is a homeomorphism;*
(b) *$\psi_j \circ \psi_i^{-1} \colon \psi_i(U_i \cap U_j) \longrightarrow \psi_j(U_i \cap U_j)$ is of class C^k.*
*The pairs (U_i,ψ_i), for $i \in I$, are called **charts** and the maps $\psi_j \circ \psi_i^{-1}$ are **changes of charts**. The pair (V_i,ψ_i^{-1}) is a **local parametrization** (or **local coordinates**) of the elements of M. If $X = \mathbb{R}^n$, then M is an n-dimensional C^k-manifold.*

Remark 5.17
In fact, we can be more general and allow each map ψ_i to map U_i to a possibly different Banach space X_i. The basic idea behind the above definition, is to introduce a local object which will support the operation of differentiation and then patch these local objects together smoothly.

An important example of a differential manifold is the tangent bundle of a C^1-manifold. There are several alternative ways to define this. Here we use the so-called ***tangent curves approach***.

Definition 5.18
Let M be a C^1-Banach manifold and let $u \in M$. A **curve** *at u is a C^1-map $\sigma\colon I \longrightarrow M$ (where I is an open interval in $\mathbb{R}$ with $0 \in I$) such that $\sigma(0) = u$. Let σ_1, σ_2 be two curves at u. We say that σ_1 and σ_2 are* **tangent at** *$u \in M$ if there is a chart (U, ψ) with $u \in U$ such that*

$$(\psi \circ \sigma_1)'(0) = (\psi \circ \sigma_2)'(0).$$

Remark 5.19
In fact it is easy to see that σ_1 and σ_2 are actually tangent with respect to any chart (U, ψ) with $u \in U$. The notion of tangency at $u \in M$ is an equivalence relation among curves at u. An equivalence class of such curves, is denoted by $[\sigma]_u$ with σ being a representative of the class.

Definition 5.20
Let M be a C^1-Banach manifold and let $u \in M$. The **tangent space** *of M at u is the set of all equivalence classes of curves at u*

$$T_u M = \{[\sigma]_u : \sigma \text{ is a curve at } u\}.$$

The **tangent bundle** *TM of M is defined by*

$$TM = \bigcup_{u \in M} T_u M.$$

The map $\tau_M\colon TM \longrightarrow M$ defined by

$$\tau_M([\sigma]_u) = u \in M$$

is the **tangent bundle projection** *of M.*

Remark 5.21
From the above definition, it follows that $T_u M$ is a Banach space homeomorphic to X.

Definition 5.22
Let M and N be two C^k-Banach manifolds and let $f\colon M \longrightarrow N$ be a map. We say that f is m-times differentiable ($m \leqslant k$), if $\psi_j \circ f \circ \psi_i^{-1}$ is C^m whenever it is defined (here, as usual. (U_i, ψ_i) and (U_j, ψ_j) are charts).

Remark 5.23
Using the chain rule, we can see that his definition is independent of the special choice of charts. Note that the above definition establishes the differentiability of f, but does not produce the derivatives of f. This is done below using the notion of tangent space introduced in Definition 5.20.

Definition 5.24
*Let M and N be two C^1-Banach manifolds and let $f\colon M \longrightarrow N$ be a C^1-map. The **differential** of f at $u \in M$ is defined by*

$$df(u)([\sigma]_u) = [f \circ \sigma]_{f(u)}.$$

Evidently $df(u) \in \mathcal{L}(T_uM, T_{f(u)}N)$ for all $u \in M$.

Proposition 5.25
If M, N, and E are three C^1-Banach manifolds and $f\colon M \longrightarrow N$, $g\colon N \longrightarrow E$ are C^1-maps, then
(a) $g \circ f\colon M \longrightarrow E$ is a C^1-map;
(b) for every $u \in M$, we have

$$d(g \circ f)(u) = dg(f(u)) \circ df(u)$$

(chain rule);
(c) if $h\colon M \longrightarrow M$ is the identity map, then for every $u \in M$, $df(u)$ is the identity map on T_uM;
(d) if $f\colon M \longrightarrow N$ is a diffeomorphism, then for every $u \in M$, $df(u)\colon T_uM \longrightarrow T_{f(u)}N$ is an isomorphism and

$$df(u)^{-1} = df^{-1}(u).$$

Next we present an important example that we encounter in applications.

Example 5.26
Let X be a Banach space and let $M \subseteq X$ be a C^1-Banach manifold. Let $\varphi \in C^1(X)$ and $\varphi_M = \varphi|_M$. Then for every $u \in M$, $d\varphi_M(u) \in \mathcal{L}(T_uM; \mathbb{R})$ is the restriction of $\varphi'(u) \in \mathcal{L}(X; \mathbb{R}) = X^*$ to T_uM. If X is a Hilbert space, then $d\varphi_M(u)$ defines the ***constrained gradient*** of φ at u be setting

$$(\nabla\varphi_M(u), h) = d\varphi_M(u)(h) \qquad \forall\, h \in T_uM,$$

with $(\cdot,\cdot)$ denoting the inner product of X. Hence the relation between the constrained gradient $\nabla\varphi_M(u)$ and the free gradient $\nabla\varphi(u)$ is given by

$$(\nabla\varphi_M, h) = (\nabla\varphi(u), h) \qquad \forall\, h \in T_uM,$$

which means that $\nabla\varphi_M(u)$ is the orthogonal projection of $\nabla\varphi(u)$ on T_uM.
A particular case of interest is when

$$M = \{u \in X : \; g(u) = 0\},$$

where $g \in C^k(X;\mathbb{R})$ with $g'(u) \neq 0$ for all $u \in M$. Then M is a C^k-Banach manifold of codimension 1 (that is, it is modeled on a subspace of codimension 1 in X) and T_uM is given by

$$T_uM = \ker g'(u) = \{v \in X : \langle g'(u), v\rangle = 0\}.$$

The next theorem provides a generalization of this particular example. First a definition.

Definition 5.27
Suppose that X and Y are two Banach spaces, $D \subseteq X$ is a nonempty open set, $g\colon D \longrightarrow Y$ is a map and $u_0 \in D$. We say that g is ***submersion*** *at u_0 (respectively* ***immersion*** *at u_0) if the following conditions hold:*
(a) g is continuously differentiable in an open neighborhood of u_0;
(b) $g'(u_0)$ is surjective (respectively $g'(u_0)$ is injective);
(c) $\ker g'(u_0)$ is complemented in X, i.e., there exists a closed linear subspace $V \subseteq X$ such that $X = \ker g'(u_0) \oplus V$ (respectively $R(g'(u_0))$ is complemented in Y, i.e., there exists a closed linear subspace $Z \subseteq Y$ such that $Y = R(g'(u_0)) \oplus Z$).

Remark 5.28
Condition **(c)** means that there exists a continuous projection operator from X onto $\ker \varphi'(u_0)$ such that

$$X = \ker \varphi'(u_0) \oplus (I_X - P)(X).$$

Similarly for $R(\varphi'(u_0))$ in the case of an immersion. We recall that a closed linear subspace of a Banach space is ***complemented***, if it is finite dimensional or it has finite codimension. Also, if X is a Hilbert space, then $\ker \varphi'(u_0)$ is complemented since it is always a closed subspace.

We can now state the extension of the situation described in Example 5.26.

Theorem 5.29 *(Ljusternik Theorem)*
If X and Y are two Banach spaces, $D \subseteq X$ is a nonempty, open set, $g\colon D \longrightarrow Y$ is a function, $M = \{u \in D : g(u) = 0\}$ and g is a submersion at every $u \in M$, then
(a) M is a C^1-Banach manifold;
(b) $T_uM = \ker g'(u)$ for all $u \in M$;
(c) if $\ker g'(u_0) = \{0\}$, then u_0 is an isolated solution of the equation $g(u) = 0$.

Next we give the precise definition of a critical point of $\varphi_M = \varphi|_M$ for a function $\varphi\colon X \longrightarrow \mathbb{R}$.

Definition 5.30
Suppose that X is a Banach space, $M \subseteq X$ is a C^1-Banach manifold and $\varphi\colon X \longrightarrow \mathbb{R}$ is a map. We say that $u_0 \in M$ is a ***critical point*** *of $\varphi_M = \varphi|_M$, if $\frac{d}{dt}\varphi(\sigma(t))|_{t=0} = 0$ for every C^1-curve $\sigma\colon (-\varepsilon, \varepsilon) \longrightarrow M$ such that $\sigma(0) = u_0$.*

Existence of critical points for a constrained functional, is closely related to the existence of Lagrange multipliers.

Proposition 5.31
If X and Y are two Banach spaces, $\widehat{D}, D \subseteq X$ are two nonempty open subsets of X, $\varphi\colon \widehat{D} \longrightarrow \mathbb{R}$, $g\colon D \longrightarrow Y$ are two maps, $M = \{u \in D:\ g(u) = 0\}$, $u_0 \in M$, φ is Fréchet differentiable at u_0 and g is a submersion at u_0,
then u_0 is a critical point of $\varphi_M = \varphi|_M$ if and only if there exists $y^* \in Y^*$ such that

$$\varphi'(u_0) - y^* \circ g'(u_0) \;=\; 0.$$

An alternative version of this result under weaker conditions is the following one.

Proposition 5.32
If X and Y are two Banach spaces, $\widehat{D}, D \subseteq X$ are two nonempty open subsets of X, $\varphi\colon \widehat{D} \longrightarrow \mathbb{R}$, $g\colon D \longrightarrow Y$ are two maps, $M = \{u \in D:\ g(u) = 0\}$, $u_0 \in M$, φ is Fréchet differentiable at u_0 and g is Fréchet differentiable in an open neighborhood of u_0 with $y \longmapsto g'(u)$ continuous at u_0, $R(g'(u_0)) \subseteq Y$ is closed and u_0 is a critical point of $\varphi_M = \varphi|_M$
then there exist $\lambda_0 \in \mathbb{R}$ and $y^* \in Y^*$, not both equal to zero, such that

$$\lambda_0\varphi'(u_0) - y^* \circ g'(u_0) \;=\; 0.$$

Moreover, in the nondegenerate case $R(g'(u_0)) = Y$, we can have $\lambda_0 = 1$.

Now, let X be a Banach space, $M \subseteq X$ a C^1-Banach manifold and $\varphi \in C^1(X)$. If $u_0 \in M$ is a critical point of the unconstrained functional φ, then $u_0 \in M$ is also a critical point of $\varphi_M = \varphi|_M$. This is deduced from Definition 5.30 (use the chain rule). The converse is not in general true.

Definition 5.33
Suppose that X is a Banach space, $M \subseteq X$ is a C^1-Banach manifold and $\varphi \in C^1(X)$. We say that M is a **natural constraint** *for φ if there exists $\widetilde{\varphi} \in C^1(X)$ such that*

$$\widetilde{\varphi}'_M(u) \;=\; 0 \quad\Longleftrightarrow\quad \varphi'(u) \;=\; 0.$$

Remark 5.34
In the light of Proposition 5.32, this means that $y^* = 0$.

The simple example of the exponential function shows that a smooth function which is bounded below, need not attain its infimum. A much better understanding of this situation was achieved with the celebrated Ekeland variational principle, which in the setting of C^1-functionals guarantees the existence of minimizing sequence, which consists of almost critical points.

We start by giving the general form of the Ekeland variational principle, which is valid in a general complete metric space.

Theorem 5.35 *(Ekeland Variational Principle)*
If (X, d_X) *is a complete metric space,* $\varphi \colon X \longrightarrow \overline{\mathbb{R}}$ *is a lower semicontinuous function which is bounded below and* $\widehat{u} \in \operatorname{dom} \varphi$,
then there exists $\widetilde{u} \in \operatorname{dom} \varphi$ *such that*

$$\varphi(\widetilde{u}) + d_X(\widetilde{u}, \widehat{u}) \leqslant \varphi(\widehat{u})$$

and

$$\varphi(\widetilde{u}) < \varphi(u) + d_X(u, \widetilde{u}) \qquad \forall\, u \in X,\ u \neq \widetilde{u}.$$

From this formulation, we can easily derive the following more familiar one.

Theorem 5.36 *(Ekeland Variational Principle)*
If (X, d_X) *is a complete metric space,* $\varphi \colon X \longrightarrow \overline{\mathbb{R}}$ *is a lower semicontinuous function which is bounded below and* $\varepsilon > 0$, $\widehat{u} \in X$ *satisfy*

$$\varphi(\widehat{u}) \leqslant \inf_X \varphi + \varepsilon,$$

then for every $\lambda > 0$, *there exists* $u_\lambda \in X$ *such that*

$$\varphi(u_\lambda) \leqslant \varphi(\widehat{u}), \quad d_X(u_\lambda, \widehat{u}) \leqslant \lambda$$

and

$$\varphi(u_\lambda) \leqslant \varphi(u) + \tfrac{\varepsilon}{\lambda} d_X(u, u_\lambda) \qquad \forall\, u \in X$$

Remark 5.37
In the above theorem, the conclusions

$$d(u_\lambda, \widehat{u}) \leqslant \lambda \quad \text{and} \quad \varphi(u_\lambda) \leqslant \varphi(u) + \tfrac{\varepsilon}{\lambda} d_X(u, u_\lambda) \qquad \forall\, u \in X$$

are in a sense complementary. Indeed, the choice of $\lambda > 0$ determines a balance between the two. So, if $\lambda > 0$ is large, the inequality $d_X(u_\lambda, \widehat{u}) \leqslant \lambda$ gives little information on the whereabouts of u_λ, while the inequality $\varphi(u_\lambda) \leqslant \varphi(u) + \frac{\varepsilon}{\lambda} d_X(u, u_\lambda)$ for $u \in X$, becomes sharper and says that u_λ is close to being a global minimizer of φ. The situation is reversed when $\lambda > 0$ is small. If we choose $\lambda = 1$, it means that we are not interested on the whereabouts of u_λ, while if we choose $\lambda = \sqrt{\varepsilon}$, it means that we need information from both inequalities.

Corollary 5.38 *(Ekeland Variational Principle)*
If (X, d_X) *is a complete metric space and* $\varphi \colon X \longrightarrow \overline{\mathbb{R}}$ *is a lower semicontinuous function which is bounded below,*
then for every $\varepsilon > 0$ *we can find* $u_\varepsilon \in X$ *such that*

$$\varphi(u_\varepsilon) \leqslant \inf_X \varphi + \varepsilon \quad \textit{and} \quad \varphi(u_\varepsilon) \leqslant \varphi(u) + \varepsilon d_X(u, u_\varepsilon) \qquad \forall\, u \in X.$$

The inequality $\varphi(u_\varepsilon) \leqslant \varphi(u) + \varepsilon d_X(u, u_\varepsilon)$ for all $u \in X$, has a clear interpretation when φ is defined on a Banach space and it is Gâteaux differentiable.

Proposition 5.39
<u>If</u> X *is a Banach space,* $\varphi\colon X \longrightarrow \mathbb{R}$ *is a lower semicontinuous function, bounded below and Gâteaux differentiable,*
<u>then</u> *for every* $\varepsilon > 0$, *we can find* $u_\varepsilon \in X$ *such that*

$$\varphi(u_\varepsilon) \leqslant \inf_X \varphi + \varepsilon \quad \text{and} \quad \|\varphi'_G(u_\varepsilon)\|_* \leqslant \varepsilon.$$

Proposition 5.40
<u>If</u> X *is a Banach space,* $\varphi\colon X \longrightarrow \mathbb{R}$ *is a lower semicontinuous, bounded below and Gâteaux differentiable function,* $\varepsilon > 0$ *and* $u_\varepsilon \in X$ *satisfies*

$$\varphi(u_\varepsilon) \leqslant \inf_X \varphi + \varepsilon,$$

<u>then</u> *we can find* $v_\varepsilon \in X$ *such that*

$$\varphi(v_\varepsilon) \leqslant \varphi(u_\varepsilon), \quad \|v_\varepsilon - u_\varepsilon\|_X \leqslant \sqrt{\varepsilon} \quad \text{and} \quad \|\varphi'_G(v_\varepsilon)\|_* \leqslant \sqrt{\varepsilon}.$$

There is a "weighted" extension of the Ekeland variational principle (see Theorem 5.36).

Theorem 5.41
<u>If</u> (X, d_X) *is a complete metric space,* $u_0 \in X$ *if fixed,* $\varphi\colon X \longrightarrow \overline{\mathbb{R}}$ *is a lower semicontinuous and bounded below function,* $\xi\colon \mathbb{R}_+ \longrightarrow \mathbb{R}_+$ *is a continuous nondecreasing function such that* $\int\limits_0^{+\infty} \frac{1}{1+\xi(s)}\, ds = +\infty$, $\lambda > 0$, $\varepsilon > 0$ *and* $u \in X$ *satisfy*

$$\varphi(u) \leqslant \inf_X \varphi + \varepsilon,$$

$r_0 = d_X(u, u_0)$ *and we fix* $\overline{r} > 0$ *such that* $\int\limits_{r_0}^{r_0+\overline{r}} \frac{1}{1+\xi(s)}\, ds \geqslant \lambda$,
<u>then</u> *there exists* $v \in X$ *such that*

$$\varphi(v) \leqslant \varphi(u), \quad d_X(v, u_0) \leqslant r_0 + \overline{r}$$

and

$$\varphi(v) \leqslant \varphi(y) + \frac{\varepsilon}{\lambda(1+\xi(d_X(v,u_0)))} d_X(v, y) \qquad \forall\, y \in X.$$

Remark 5.42
<u>If</u> $\xi \equiv 0$, $u_0 = u$ and $\overline{r} = \lambda$, then Theorem 5.41 reduces to Theorem 5.36.

Corollary 5.43
<u>If</u> X *is a Banach space,* $\varphi\colon X \longrightarrow \mathbb{R}$ *is a lower semicontinuous, bounded below, Gâteaux differentiable function and* $\xi\colon \mathbb{R}_+ \longrightarrow \mathbb{R}_+$ *is a continuous nondecreasing function such that* $\int\limits_0^{+\infty} \frac{1}{1+\xi(s)}\, ds = +\infty$,
<u>then</u> *given* $\varepsilon > 0$ *and* $u \in X$ *satisfying*

$$\varphi(u) \leqslant \inf_X \varphi + \varepsilon$$

and $\lambda > 0$, $\overline{r} > 0$ *such that* $\int_0^{\overline{r}} \frac{1}{1+\xi(s)}\, ds \geqslant \lambda$, *we can find* $v \in X$ *such that*

$$\varphi(v) \leqslant \varphi(u), \quad \|v-u\|_X \leqslant \overline{r} \quad \text{and} \quad \|\varphi'_G(v)\|_* \leqslant \frac{\varepsilon}{\lambda(1+\xi(\|v\|_X))}.$$

Corollary 5.44
If X *is a Banach space,* $\varphi\colon X \longrightarrow \mathbb{R}$ *is a lower semicontinuous, bounded below, Gâteaux differentiable function and* $\xi\colon \mathbb{R}_+ \longrightarrow \mathbb{R}_+$ *is a continuous nondecreasing function such that* $\int_0^{+\infty} \frac{1}{1+\xi(s)}\, ds = +\infty$,
then for every $\varepsilon > 0$, *we can find* $u_\varepsilon \in X$, *such that*

$$\varphi(u_\varepsilon) \leqslant \inf_X \varphi + \varepsilon \quad \text{and} \quad \|\varphi'_G(u_\varepsilon)\|_* \leqslant \frac{\varepsilon}{1+\xi(\|u_\varepsilon\|_X)}.$$

5.1.2 Minimax Methods for Critical Points

We have seen in the beginning of this chapter the importance of compactness in variational problems. In infinite dimensional spaces norm compactness is difficult to realize (recall that if X is an infinite dimensional Banach space and $C \subseteq X$ is a norm compact, then $\operatorname{int} C = \emptyset$). To overcome this difficulty, following Palais–Smale [12], we transfer the compactness requirement from the space X to the function φ itself. This reminds us of the situation with compact operators for the spectral theory (see Chap. 2) and for the degree theory (see Chap. 4). This approach leads to the following definition.

Definition 5.45
Let X *be a Banach space and let* $\varphi \in C^1(X)$.

(a) We say that φ *satisfies the* ***Palais–Smale condition*** *at level* $c \in \mathbb{R}$ *(****PS***$_c$***-condition*** *for short), if every sequence* $\{u_n\}_{n\geqslant 1} \subseteq X$ *such that* $\varphi(u_n) \longrightarrow c$ *and* $\varphi'(u_n) \longrightarrow 0$ *in* X^* *admits a strongly convergent subsequence. We say that* φ *satisfies the* ***Palais–Smale condition (PS-condition*** *for short), if it satisfies the PS*$_c$*-condition for every* $c \in \mathbb{R}$.

(b) We say that φ *satisfies the* ***Cerami condition*** *at level* $c \in \mathbb{R}$ *(****C***$_c$***-condition*** *for short), if every sequence* $\{u_n\}_{n\geqslant 1} \subseteq X$ *such that* $\varphi(u_n) \longrightarrow c$ *and* $(1+\xi(\|u_n\|_X))\varphi'(u_n) \longrightarrow 0$ *in* X^* *admits a strongly convergent subsequence, with* $\xi\colon \mathbb{R}_+ \longrightarrow \mathbb{R}_+$ *being a continuous nondecreasing function such that* $\int_0^{+\infty} \frac{1}{1+\xi(s)}\, ds = +\infty$. *We say that* φ *satisfies the* ***Cerami condition (C-condition*** *for short), if it satisfies the C*$_c$*-condition for every* $c \in \mathbb{R}$.

Remark 5.46
Usually the C-condition is defined with $\xi \equiv I_x$. The compactness condition originally used by Palais–Smale [12] was none of the above. It was the following CPS-condition (Palais–Smale [12] called it originally C-condition and it was formulated for C^2-functions on a Hilbert space; here we state a more general version):

> Let X be a Banach space and let $\varphi \in C^1(X)$. We say that φ satisfies the CPS-condition, if for any subset $S \subseteq X$ such that $\varphi|_S$ is bounded but $\|\varphi'(\cdot)\|_*|_S$ is not bounded away from zero, the function φ has a critical point in $\overline{S}$.

Evidently the PS-condition implies the CPS-condition, but the converse is not in general true (consider any constant function).

There are nonsmooth versions of the above compactness conditions.

Definition 5.47

(a) *Let X be a Banach space and let $\varphi\colon X \longrightarrow \mathbb{R}$ be a locally Lipschitz function. We say that φ satisfies the* ***nonsmooth Cerami condition*** *at level $c \in \mathbb{R}$ (**nonsmooth C_c-condition** for short), if every sequence $\{u_n\}_{n\geqslant 1} \subseteq X$ such that $\varphi(u_n) \longrightarrow c$ and $(1+\|u_n\|_X)m(u_n) \longrightarrow 0$, with $m(u_n) = \min\limits_{u^*\in\partial\varphi(u_n)} \|u^*\|_*$ for $n \geqslant 1$, admits a strongly convergent subsequence. We say that φ satisfies the* ***nonsmooth Cerami condition (nonsmooth C-condition*** *for short), if it satisfies the nonsmooth C_c-condition for every $c \in \mathbb{R}$.*

(b) *Let X be a Banach space and let $\varphi \in C^1(X)$ and $\psi \in \Gamma_0(X)$ (see Remark 3.56). We say that $\varphi+\psi$ satisfies the* ***SC_c-condition*** *(with $c\in\mathbb{R}$), if every sequence $\{u_n\}_{n\geqslant 1} \subseteq X$ such that $\varphi(u_n)\longrightarrow x$ and*

$$\langle \varphi'(u_n), v-u_n\rangle + \psi(v) - \psi(u_n) \geqslant -\varepsilon_n\|v-u_n\| \qquad \forall\, v \in X,\ n \geqslant 1,$$

with $\varepsilon_n \searrow 0$, admits a strongly convergent subsequence. We say that φ satisfies the ***SC-condition****, it if satisfies the SC_c-condition for every $c \in \mathbb{R}$.*

There is a strict relation between the set of solutions of a boundary value problem and the critical points of a certain functional defined on a Banach space. This leads to the investigation of the existence of critical points. In general it is not easy to see if a given functional defined on an infinite dimensional Banach space has a critical point. One effective approach is based on the investigation of the sublevel sets of the functional itself. More precisely, differences in the topological properties between sublevel sets will imply the presence of critical points. To detect changes in the topological structure of the sublevel sets, one uses flows of differential equations, in particular the steepest descent flow. The first basic result in the so-called first deformation lemma. First we introduce some notation.

Let X be a Banach space, $\varphi \in C^1(X)$ and $a, c \in \mathbb{R}$ with $a < c$. We set

$$\begin{aligned}
K_\varphi &= \{u \in X:\ \varphi'(u) = 0\} \text{ (the critical set of } \varphi)\\
K_\varphi^c &= \{u \in K_\varphi:\ \varphi(u) = c\} \text{ (the critical set of } \varphi \text{ at level } c)\\
\varphi^c &= \{u \in X:\ \varphi(u) \leqslant c\} \text{ (the sublevel of } \varphi \text{ at } c)\\
\varphi_a &= \{u \in X:\ a \leqslant \varphi(u)\} \text{ (the superlevel of } \varphi \text{ at } a)\\
\varphi_a^c &= \{u \in X:\ a \leqslant \varphi(u) \leqslant c\}.
\end{aligned}$$

Theorem 5.48 *(First Deformation Lemma)*
If X is a Banach space, $\varphi \in C^1(X)$, $-\infty < a \leqslant c \leqslant +\infty$ (we can have $c = +\infty$ if $a \in \mathbb{R}$) and

$$\inf_{\substack{u \in X \\ a \leqslant \varphi(u) \leqslant c}} \|\varphi'(u)\|_* > 0, \tag{5.1}$$

then φ^a is a strong deformation retract of φ^c (see Definition 4.112(b)).

Remark 5.49
The above proposition says that if in the interval $[a, c]$ there are no critical values of $\varphi \in C^1(X)$, then the sublevel sets corresponding to the values in this interval, all have the same topological structure. If X is a Hilbert space and $\varphi \in C^{1,1}_{loc}(X; \mathbb{R})$ (that is, φ is C^1 and its derivative is locally Lipschitz), following the plan outlined earlier, we follow the flow generated by the steepest descent. So, let $u \in \varphi^c$ and consider the abstract Cauchy problem

$$\left\{ \begin{array}{rcl} \sigma'(t) & = & -\frac{\nabla\varphi(\sigma(t))}{\|\nabla\varphi(\sigma(t))\|_X^2}, \quad t \in \mathbb{R} \\ \sigma(0) = u. & & \end{array} \right. \tag{5.2}$$

Here $\nabla\varphi$ denotes the gradient of φ (so, $(\nabla\varphi(u), h)_X = \langle \varphi'(u), h \rangle$ for all $u, h \in X$, with $(\cdot, \cdot)_X$ being the inner product of X and $\langle \cdot, \cdot \rangle$ are the duality brackets for the pair (X^*, X)). Since by hypothesis $\nabla\varphi$ is locally Lipschitz, problem (5.2) has a unique local flow $\sigma\colon (-\delta, \delta) \longrightarrow X$. We look for the biggest domain of definition of σ in the future (this is, for $t > 0$). We have

$$\tfrac{d}{dt}\varphi(\sigma(t)) = \big(\nabla\varphi(\sigma(t)), \sigma'(t)\big)_X = -1. \tag{5.3}$$

We show that σ exists on the interval $[0, \varphi(u) - a]$. Indeed, let $[0, b)$ be the maximal forward interval of the existence of σ, with $b \leqslant \varphi(u) - a$. From (5.3) we have

$$\varphi(\sigma(t)) = \varphi(u) - t > a \qquad \forall\, t \in [0, b).$$

But from (5.1), we know that

$$\|\nabla\varphi(u)\| \geqslant \varepsilon > 0 \qquad \forall\, u \in \varphi_a^c.$$

Hence, from (5.2), we have

$$\|\sigma'(t)\| \leqslant \tfrac{1}{\varepsilon} \qquad \forall\, t \in [0, b),$$

so σ is Lipschitz continuous on X

Hence $\lim\limits_{t \to b^-} \sigma(t)$ exists, contradicting the maximality of $[0, b)$. So, the flow σ exists on $[0, \varphi(u) - a]$ and we define

$$h(t, u) = \left\{ \begin{array}{ll} \sigma(t(\varphi(u) - a), u) & \text{if} \quad u \in \varphi_a^c, \\ 0 & \text{if} \quad u \in \varphi^a \end{array} \right.$$

(here we have denoted the flow σ as a function of the time $t > 0$ and of the initial condition u). This is the desired strong deformation (see Definition 4.112(b)).

We have presented the argument of the above particular case (X being a Hilbert space and $\varphi \in C^{1,1}_{loc}(X;\mathbb{R})$), in order to underline the difficulties we face when we are in the more general setting of Theorem 5.48. In a general Banach space X, the notion of gradient does not exist and φ is only C^1. We overcome these difficulties with the use of the so-called pseudogradient vector field.

Definition 5.50
Suppose that E is a metric space, X is a Banach space and $\psi\colon E \longrightarrow X^* \setminus \{0\}$ is a continuous map. A ***pseudogradient vector field*** *for ψ on E is a locally Lipschitz map $g\colon E \longrightarrow X$ such that*

$$\|g(v)\|_X \leqslant 2\|\psi(v)\|_* \quad \text{and} \quad \langle \psi(v), g(v)\rangle \geqslant \|\psi(v)\|_*^2 \qquad \forall\, v \in E.$$

Remark 5.51
In our case $E = \{u \in X : \varphi'(u) \neq 0\} = X \setminus K_\varphi$ and $\psi = \varphi'$.

Proposition 5.52
If E is a metric space, X is a Banach space and $\psi\colon E \longrightarrow X^* \setminus \{0\}$ is a continuous map,
then a pseudogradient vector field for ψ exists.

Remark 5.53
Such a map exists, but is not unique. Any convex combination of pseudogradient vector fields is again a pseudogradient vector field. If X is a Hilbert space and $\varphi \in C^{1,1}_{loc}(X;\mathbb{R})$, then $\nabla\varphi$ is a pseudogradient vector field for φ. If φ is even, then it admits an odd pseudogradient vector field.

Using a pseudogradient vector field and the flow generated by it, we can extend the first deformation lemma (see Theorem 5.48) to handle the presence of critical points in the strip φ_a^c. This leads to the classical deformation lemma.

Theorem 5.54 *(Deformation Lemma)*
If X is a Banach space, $\varphi \in C^1(X)$, $c \in \mathbb{R}$ satisfies the C_c-condition, $\varepsilon_0, \vartheta > 0$ and U is an open neighborhood of K_φ^c,
then there exist $\varepsilon \in (0,\varepsilon_0)$ and a continuous map $h\colon [0,1]\times X \longrightarrow X$ such that for all $(t,u) \in [0,1]\times X$, we have
(a) *$\|h(t,u)-u\|_X \leqslant \vartheta(1+\|u\|_X)t$;*
(b) *$\varphi(h(t,u)) \leqslant \varphi(u)$;*
(c) *if $h(t,u) \neq u$, then $\varphi(h(t,u)) < \varphi(u)$;*
(d) *if $|\varphi(u)-c| > \varepsilon_0$, then $h(t,u) = u$;*
(e) *$h(1,\varphi^{c+\varepsilon}) \subseteq \varphi^{c-\varepsilon} \cup U$ and $h(1,\varphi^{c+\varepsilon}\setminus U) \subseteq \varphi^{c-\varepsilon}$;*
(f) *h has the semigroup property, that is, $h(t,\cdot)\circ h(s,\cdot) = h(t+s,\cdot)$ for all $t,s \geqslant 0$.*

Remark 5.55
If $K_\varphi^c = \emptyset$, then we can take $U = \emptyset$ and so from **(e)** we have $h(1,\varphi^{c+\varepsilon}) \subseteq \varphi^{c-\varepsilon}$. Conclusion **(e)** says that starting a little bit above a critical level c, you will either

bypass the critical neighborhood U and reach a harmless level $c-\varepsilon$ or else you will end up in U, the only place near $\{u \in X:\ \varphi(u)=c\}$ where topologically interesting things can happen. The same deformation result holds if $\varphi\colon X \longrightarrow \mathbb{R}$ is only locally Lipschitz and satisfies the nonsmooth C_c-condition (see Definition 5.47(a)). In this case too we can show the existence of a pseudogradient vector field (see Gasiński–Papageorgiou [6, p. 130].

We can extend Theorem 5.48 and allow for the presence of critical points at the level a (so, condition (5.1) is no longer true). The result is known as the second deformation lemma. In the statement we allow $c=+\infty$, in which case $\varphi^c \setminus K_\varphi^c = X$.

Theorem 5.56 *(Second Deformation Lemma)*
If X is a Banach space, $\varphi \in C^1(X)$, $a \in \mathbb{R}$, $c \in (a,+\infty]$, φ satisfies the C_d-condition for every $d \in [a,c)$, φ has no critical values in (a,c) and $\varphi^{-1}(a)$ contains at most a finite number of critical points of φ,
then there exists a deformation $h\colon [0,1]\times(\varphi^c \setminus K_\varphi^c) \longrightarrow \varphi^c \setminus K_\varphi^c$ of $\varphi^c \setminus K_\varphi^c$ into φ^a such that:
(a) if $u \in \varphi^a$, then $h(t,u)=u$ for all $t \in [0,1]$ (hence φ^a is a strong deformation retract of $\varphi^c \setminus K_\varphi^c$);
(b) $\varphi(h(t,u)) \leqslant \varphi(h(s,u))$ for all $(t,u),(s,u) \in [0,1]\times(\varphi^c \setminus K_\varphi^c)$, $s \leqslant t$.

Corollary 5.57
If X is a Banach space, $\varphi \in C^1(X)$, $a \in \mathbb{R}$, $a<c\leqslant+\infty$, φ satisfies the C_d-condition for every $d \in [a,c)$ and φ has no critical values in $[a,c]$,
then φ^a is a strong deformation retract of φ^a.

There is another version of the deformation lemma, known as ***quantitative deformation lemma***. The advantage of this result is that no *a priori* compactness condition on φ is needed.

For $S \subseteq X$ and $\delta>0$, we use the notation

$$S_\delta \;=\; \{u \in X:\ d_X(u,s) \leqslant \delta\}.$$

Proposition 5.58
If X is a Banach space, $\varphi \in C^1(X)$, $S \subseteq X$, $c \in \mathbb{R}$, $\varepsilon,\delta>0$ and

$$\|\varphi'(v)\|_* \;\geqslant\; \tfrac{8\varepsilon}{\delta} \qquad \forall\, v \in \varphi_{c-2\varepsilon}^{c+2\varepsilon} \cap S_{2\delta},$$

then there exists a continuous map $h\colon [0,1]\times X \longrightarrow X$ such that:
(a) $h(t,u)=u$ if $t=0$ or if $u \notin \varphi_{c-2\varepsilon}^{c+2\varepsilon} \cap S_{2\delta}$;
(b) $h(1,\varphi^{c+\varepsilon} \cap S) \subseteq \varphi^{c-\varepsilon}$;
(c) $h(1,\cdot)$ is a homeomorphism of X for all $t \in [0,1]$;
(d) $\|h(t,u)-u\|_X \leqslant \delta$ for all $t \in [0,1]$ and all $u \in X$;
(e) the map $t \longmapsto \varphi(h(t,u))$ is nonincreasing for all $u \in X$;
(f) $\varphi(h(t,u))<c$ for all $t \in [0,1]$ and all $u \in \varphi^c \cap S_\delta$.

Now we can pass to the minimax theory of the critical values of φ. We start with a notion which in different forms is central in critical point theory.

Definition 5.59

*Suppose that V is a Hausdorff topological space, $E_0 \subseteq E$ and D are nonempty subsets of V and $\gamma_* \in C(E_0; V)$. We say that the pair $\{E_0, E\}$ **links** D in V via γ_*, if the following holds:*

(a) $\gamma_*(E_0) \cap D = \emptyset$;

(b) *if* $\gamma \in C(E; V)$ *and* $\gamma|_{E_0} = \gamma_*$, *then* $\gamma(E) \cap D \neq \emptyset$.

Remark 5.60

We also say that the sets $\{E_0, E, D\}$ are ***linking sets*** in V via γ_*. If $\gamma_* = I_{E_0}$ (which is the usual case), then we say that the sets $\{E_0, E, D\}$ are ***linking sets*** (in V).

Example 5.61

We present some characteristic cases of linking sets. Let X be a Banach space.

(a) $E_0 = \{u_0, u_1\} \subseteq X$, $E = \{(1-t)u_0 + tu_1 : \ t \in [0,1]\}$ and $D = \partial U$ where $U \subseteq X$ is open, $u_0 \in U$ and $u_1 \notin \overline{U}$. Then $\{E_0, E, D\}$ are linking sets.

(b) Suppose that $X = Y \oplus V$ with $\dim Y < +\infty$. Let $r > 0$ and define

$$\begin{aligned} E_0 &= \{u \in Y : \ \|u\|_X = r\} = \partial B_r \cap Y, \\ E &= \{u \in Y : \ \|u\|_X \leqslant r\} = \overline{B}_r \cap Y, \quad D = V. \end{aligned}$$

Then $\{E_0, E, D\}$ are linking sets.

(c) Suppose that $X = Y \oplus V$ with $\dim Y < +\infty$. Let $r_1 > \varrho > 0$, $r_2 > 0$ and $v_0 \in V$ with $\|v_0\|_X = 1$. We define

$$\begin{aligned} E_0 &= \{y + \lambda v_0 : \ y \in Y, \ [0 < \lambda < r_1, \ \|y\|_X = r_2] \\ &\qquad \text{or } [\lambda \in \{0, r_1\}, \ \|y\|_X \leqslant r_2]\}, \\ E &= \{y + \lambda v_0 : \ y \in Y, \ 0 \leqslant \lambda \leqslant r_1, \ \|y\|_X \leqslant r_2\}, \\ D &= \partial B_\varrho \cap V. \end{aligned}$$

Evidently E is a cylinder with basis $\overline{B}_{r_2} \cap Y$ and height r_1, while E_0 is the boundary of this cylinder (lateral surface union with the two bases, the bottom one ($\lambda = 0$) and the top one ($\lambda = r_1$)). Then $\{E_0, E, D\}$ are linking sets.

(d) Suppose that $X = Y \oplus V$ with $\dim Y < +\infty$. Let $r > \varrho > 0$ and $v_0 \in V$ with $\|v_0\|_X = 1$. We define

$$\begin{aligned} E_0 &= \{u = y + \lambda v_0 : \ y \in Y, \ [\lambda \geqslant 0, \ \|u\|_X = r] \\ &\qquad \text{or } [\lambda = 0, \ \|y\|_X \leqslant r]\}, \\ E &= \{u = y + \lambda v_0 : \ y \in Y, \ \lambda \geqslant 0, \ \|u\|_X \leqslant r\}, \\ D &= \partial B_\varrho \cap V. \end{aligned}$$

Evidently E is the upper-half ball of radius $r > 0$ and E_0 is its boundary, that is, the "northern" hemisphere union the "equator" disc. Then $\{E_0, E, D\}$ are linking sets.

Using the notion of linking sets, we can have a general minimax principle. In what follows for $A, C \subseteq X$ we set

$$\operatorname{dist}(A, C) = \inf\{\|a - c\|_X : a \in A,\ c \in C\}.$$

Theorem 5.62
If X is a Banach space, $\{E_0, E, D\}$ are linking sets in X via γ_*, the set $\gamma_*(E_0)$ is bounded, the set D is closed, $\operatorname{dist}(\gamma_*(E_0), D) > 0$, $\Gamma = \{\gamma \in C(E; X) : \gamma|_{E_0} = \gamma_*\}$, $\varphi \in C^1(X)$ with

$$a = \sup_{\gamma_*(E_0)} \varphi \leqslant \inf_D \varphi = b, \qquad c = \inf_{\gamma \in \Gamma} \sup_{u \in E} \varphi(\gamma(u))$$

and assume that φ satisfies the C_c-condition,
then $c \geqslant b$ and c is a critical value of φ (i.e., $K_\varphi^c \neq \emptyset$).
Moreover, if $c = b$, then $K_\varphi^c \cap D \neq \emptyset$.

With suitable choices of the linking sets, we can have the classical minimax theorems of the critical point theory.

If the linking sets $\{E_0, E, D\}$ in Theorem 5.62 are those from Example 5.61(a), then we obtain the following theorem.

Theorem 5.63 *(Mountain Pass Theorem)*
If X is a Banach space, $\varphi \in C^1(X)$, $u_0, u_1 \in X$, $\|u_1 - u_0\|_X > r > 0$,

$$\max\{\varphi(u_0), \varphi(u_1)\} \leqslant \inf\{\varphi(u) : \|u - u_0\|_X = r\} = m_r,$$

$$c = \inf_{\gamma \in \Gamma} \max_{0 \leqslant t \leqslant 1} \varphi(\gamma(t)),$$

with $\Gamma = \{\gamma \in C([0,1]; X) : \gamma(0) = u_0,\ \gamma(1) = u_1\}$ and φ satisfies the C_c-condition, then $c \geqslant m_r$, c is a critical value of φ and if $c = m_r$, then $K_\varphi^c \cap \partial B_r(u_0) \neq \emptyset$.

If the linking sets $\{E_0, E, D\}$ in Theorem 5.62 are those from Example 5.61(b), then we obtain the following theorem.

Theorem 5.64 *(Saddle Point Theorem)*
If X is a Banach space, $X = Y \oplus V$ with $\dim Y < +\infty$, $\varphi \in C^1(X)$, there exists $r > 0$ such that

$$\max_{u \in \partial B_r \cap Y} \varphi(u) \leqslant \inf_{v \in V} \varphi(u) = b$$

$$c = \inf_{\gamma \in \Gamma} \max_{u \in \overline{B}_r \cap Y} \varphi(\gamma(u)),$$

with $\Gamma = \left\{\gamma \in C(\overline{B}_r \cap Y; X) : \gamma|_{\partial B_r \cap Y} = I_{\partial B_r \cap Y}\right\}$ and φ satisfies the C_c-condition, then $c \geqslant b$, c is a critical value of φ and if $c = b$, then $K_\varphi^c \cap V \neq \emptyset$.

If the linking sets $\{E_0, E, D\}$ in Theorem 5.62 are those from Example 5.61(c) or (d), then we obtain the following theorem.

Theorem 5.65 *(Linking Theorem (Generalized Mountain Pass Theorem))*
If X is a Banach space, $X = Y \oplus V$ with $\dim Y < +\infty$, $\{E_0, E, D\}$ are the linking sets from Example 5.61(c) or (d), $\varphi \in C^1(X)$,

$$\sup_{E_0} \varphi \leqslant \inf_D \varphi = b$$

$$c = \inf_{\gamma\in\Gamma} \max_{u\in E} \varphi(\gamma(u)),$$

with $\Gamma = \{\gamma \in C(E;X) : \gamma|_{E_0} = I_{E_0}\}$ and φ satisfies the C_c-condition, then $c \geqslant b$, c is a critical value of φ and if $c = b$, then $K_\varphi^c \cap V \neq \emptyset$.

The mountain pass geometry (see Theorem 5.63) implies the existence of a C-sequence. First a couple of definition.

Definition 5.66
Suppose that X is a Banach space, $C \subseteq X$ is a nonempty closed set and $u_0, u_1 \in X$, $u_0 \neq u_1$. We say that C ***separates*** *u_0 and u_1 in X, if u_0 and u_1 belong to different connected components of $X \setminus C$.*

Remark 5.67
Since $X \setminus C$ is locally connected, the components of $X \setminus C$ are open subsets of X. So, if U_0 is the component of $X \setminus C$ containing u_0 and U_1 is the union of all the other components of $X \setminus C$, then $\{U_0, U_1\}$ is an open partition of $X \setminus C$ and $u_k \in U_k$ for $k \in \{0, 1\}$.

Definition 5.68
For a curve $\sigma \in C^1([0,1]; X)$, we define the ***geodesic length*** *$l(\sigma)$ of σ by*

$$l(\sigma) = \int_0^1 \frac{\|\sigma'(t)\|_X}{1+\|\sigma(t)\|_X}\, dt.$$

Then the ***geodesic distance*** *$\delta(u_0, u_1)$ between two points $u_0, u_1 \in X$ is defined by*

$$\delta(u_0, u_1) = \inf\{l(\sigma) : \sigma \in C^1([0,1]; X),\ \sigma(0) = u_0,\ \sigma(1) = u_1\}.$$

Remark 5.69
Clearly we have $\delta(u_0, u_1) \leqslant \|u_0 - u_1\|$. Conversely, for every norm bounded set $D \subseteq X$ we can find $\widehat{c} > 0$ such that

$$\widehat{c}\|u_0 - u_1\|_X \leqslant \delta(u_0, u_1) \qquad \forall\, u_0, u_1 \in D.$$

Note that $\delta(0, u) = \ln(1 + \|u\|_X)$.

Theorem 5.70
If X is a Banach space, $\varphi\colon X \longrightarrow \mathbb{R}$ is continuous, Gâteaux differentiable with $\varphi'_G\colon X \longrightarrow X^*$ norm-to-weak-continuous (see Definition 3.1(a)), $u_0, u_1 \in X$,*

$$c = \inf_{\gamma\in\Gamma} \max_{0\leqslant t\leqslant 1} \varphi(\gamma(t)),$$

with $\Gamma = \{\gamma \in C([0,1];X) : \gamma(0) = u_0, \ \gamma(1) = u_1\}$ and $C \subseteq X$ is a nonempty closed set such that

$$C \cap \varphi_c \text{ separates } u_0 \text{ and } u_1 \text{ in } X$$

then there exists a sequence $\{u_n\}_{n\geqslant 1} \subseteq X$ such that

$$\delta(u_n, c) \longrightarrow 0, \quad \varphi(u_n) \longrightarrow c$$

and

$$(1 + \|u_n\|_X)\varphi'_G(u_n) \longrightarrow 0 \quad \text{in } X^*.$$

Definition 5.71
Let X be a Banach space, $\varphi \in C^1(X)$ and $\widehat{u} \in K_\varphi$. We say that $\widehat{u}$ is a point of ***mountain pass type****, if for every open neighborhood U of $\widehat{u}$ in X, the set*

$$U \cap \{u \in X : \varphi(u) < \varphi(\widehat{u})\}$$

is nonempty and disconnected.

It may happen that the critical points found via the use of the mountain pass theorem are not points of mountain pass type. However, we have the following result.

Theorem 5.72
If X is a Banach space, $\varphi \in C^1(X)$, $u_0, u_1 \in X$, $\|u_1 - u_0\|_X > r > 0$,

$$\max\{\varphi(u_0), \varphi(u_1)\} < \inf\{\varphi(u) : \|u - u_0\|_X = r\} = m_r,$$

$$c = \inf_{\gamma\in\Gamma} \max_{0\leqslant t\leqslant 1} \varphi(\gamma(t)),$$

with $\Gamma = \{\gamma \in C([0,1];X) : \gamma(0) = u_0, \ \gamma(1) = u_1\}$ and φ satisfies the C_c-condition, then K_φ^c is nonempty and contains either a point of mountain pass type or a local minimizer.
Moreover, if the set K_φ^c is discrete, then it contains a point of mountain pass type.

We can have nonsmooth version of these minimax principles. Indicatively, we state the mountain pass theorem for:

- $\varphi\colon X \longrightarrow \mathbb{R}$ being locally Lipschitz;
- $\Phi = \varphi + \psi$ with $\varphi \in C^1(X)$ and $\psi \in \Gamma_0(X)$ (see Remark 3.56).

First we define what we mean by a critical point for the above nonsmooth cases.

Definition 5.73
Let X be a Banach space.
*(a) Given a locally Lipschitz function $\varphi\colon X \longrightarrow \mathbb{R}$, we say that $u_0 \in X$ is a **critical point** of φ, if $0 \in \partial\varphi(u_0)$. We set*

$$\begin{aligned} K_\varphi &= \{u \in X : \ 0 \in \partial\varphi(u)\} \quad \text{(the critical set of } \varphi\text{)},\\ K_\varphi^c &= \{u \in K_\varphi : \ \varphi(u) = c\} \quad \text{(with } c \in \mathbb{R}\text{)}. \end{aligned}$$

*(b) Given a function $\Phi\colon X \longrightarrow \overline{\mathbb{R}}$ with $\Phi = \varphi + \psi$ where $\varphi \in C^1(X)$ and $\psi \in \Gamma_0(X)$ (see Remark 3.56), we say that $u_0 \in X$ is a **critical point** of Φ, if $0 \in \varphi'(u_0) + \partial\psi(u_0)$. We set*

$$\begin{aligned} K_\Phi &= \big\{u \in X : \ -\varphi'(u) \in \partial\psi(u)\big\} \quad \text{(the critical set of } \Psi\text{)},\\ K_\Phi^c &= \{u \in K_\Phi : \ \Phi(u) = c\} \quad \text{(with } c \in \mathbb{R}\text{)}. \end{aligned}$$

Remark 5.74
Clearly, if $\varphi \in C^1(X)$, then φ is locally Lipschitz and since $\partial\varphi(u) = \{\varphi'(u)\}$, from Definition 5.73(a) we recover the usual notion of critical points. Also, note that $u \in K_\Phi$ (see Definition 5.73(b)) if and only if u is a solution of the variational inequality

$$\varphi'(u; h-u) + \psi(h) - \psi(u) \geqslant 0 \qquad \forall\, h \in X. \tag{5.4}$$

In particular, if $\psi = i_C$ (the indicator function of $C \in P_{fc}(X)$; see Definition 3.55), then (5.4) becomes

$$\varphi'(u; h-u) \geqslant 0 \qquad \forall\, h \in C,$$

with $u \in C$.

Theorem 5.75 *(Nonsmooth Mountain Pass Theorem)*
<u>If</u> X is a Banach space, $\varphi\colon X \longrightarrow \mathbb{R}$ is a locally Lipschitz function, $u_0, u_1 \in X$, $\|u_0 - u_1\|_X > r > 0$,

$$\begin{aligned} \max\{\,\varphi(u_0), \varphi(u_1)\,\} \leqslant \inf\{\,\varphi(u) : \ \|u-u_0\|_X = r\,\} &= m_r,\\ c &= \inf_{\gamma\in\Gamma}\max_{0\leqslant t\leqslant 1}\varphi(\gamma(t)), \end{aligned}$$

with $\Gamma = \{\gamma \in C([0,1];X) : \ \gamma(0) = u_0, \ \gamma(1) = u_1\}$ and φ satisfies the nonsmooth C_c-condition,
<u>then</u> $c \geqslant m_r$, $K_\varphi^c \neq \emptyset$ and if $c = m_r$, then $K_\varphi^c \cap \partial B_r(u_0) \neq \emptyset$.

Theorem 5.76 *(Nonsmooth Mountain Pass Theorem)*
<u>If</u> X is a Banach space, $\Phi = \varphi + \psi\colon X \longrightarrow \mathbb{R}$ with $\varphi \in C^1(X)$, $\psi \in \Gamma_0(X)$ (see Remark 3.56), $u_0, u_1 \in X$, $\|u_0 - u_1\|_X > r > 0$,

$$\begin{aligned} \max\{\Phi(u_0), \Phi(u_1)\} \leqslant \inf\{\Phi(u) : \ \|u-u_0\|_X = r\} &= m_r,\\ c &= \inf_{\gamma\in\Gamma}\max_{0\leqslant t\leqslant 1}\Phi(\gamma(t)), \end{aligned}$$

with $\Gamma = \{\gamma \in C([0,1];X) : \ \gamma(0) = u_0, \ \gamma(1) = u_1\}$,
<u>then</u> $c \geqslant m_r$, $K_\Phi^c \neq \emptyset$.

Now we turn our attention to the existence of multiple critical points. First we present multiplicity results in which no symmetry assumption is imposed on φ.

Definition 5.77
*Let X be a Banach space, $X = Y \oplus V$ and $\varphi = C^1(X)$. We say that φ has a **local linking** at 0 (with respect to the direct sum pair (Y, V)), if there exists $r > 0$ such that*

$$\begin{aligned} \varphi(u) &\leqslant 0 \quad \text{if} \quad u \in Y, \ \|u\|_X \leqslant r, \\ \varphi(u) &\geqslant 0 \quad \text{if} \quad u \in V, \ \|u\|_X \leqslant r. \end{aligned}$$

Remark 5.78
Note that $\varphi'(0) = 0$ in X^*. So, $u = 0$ is a critical point of φ. If $X = H$ is a Hilbert space, $\varphi \in C^2(H;\mathbb{R})$, $\varphi(0) = 0$ and $\varphi''(0)$ is invertible (that is, $u = 0$ is a nondegenerate critical point), then φ has a local linking at 0.

Theorem 5.79
If X is a Banach space, $X = Y \oplus V$ with $\dim Y < +\infty$, $\varphi \in C^1(X)$, $\varphi(0) = 0$, φ is bounded below, $\inf\limits_X \varphi < 0$, φ satisfies the PS-condition and φ has a local linking at 0, then φ has at least two nontrivial critical points.

In this multiplicity theorem, φ is bounded below. It is natural to ask what happens if φ is indefinite. The next theorem treats this case. First we need to introduce an extension of the PS-condition.

Let X be a Banach space with a direct sum decomposition $X = Y \oplus V$. Consider two sequences of subspaces

$$Y_1 \subseteq Y_2 \subseteq \ldots \subseteq Y_n \subseteq \ldots \subseteq Y$$

and

$$V_1 \subseteq V_2 \subseteq \ldots \subseteq V_n \subseteq \ldots \subseteq V,$$

which satisfy

$$Y = \overline{\bigcup_{n \geqslant 1} Y_n} \quad \text{and} \quad V = \overline{\bigcup_{n \geqslant 1} V_n}.$$

For every multiindex $\alpha = (i, j) \in \mathbb{N}^2$, let $X_\alpha = Y_i \oplus V_j$. A sequence $\{\alpha_n\}_{n \geqslant 1} \subseteq \mathbb{N}^2$ of multiindices is ***admissible***, if for every $\alpha \in \mathbb{N}^2$, there is $n_0 \in \mathbb{N}$ such that

$$\alpha_n \geqslant \alpha \qquad \forall\, n \geqslant n_0$$

(that is, $\alpha_n = (i_n, j_n)$ with $i_n, j_n \to +\infty$).

Definition 5.80
Let X be a Banach space with a direct sum decomposition $X = Y \oplus V$ and let $\varphi \in C^1(X)$.

(a) We say that φ satisfies the PS_c^*-condition (with $c \in \mathbb{R}$), if every sequence $\{u_{\alpha_n}\}_{n \geqslant 1} \subseteq X$ with $\{\alpha_n\}_{n \geqslant 1} \subseteq \mathbb{N}^2$ admissible such that

$$u_{\alpha_n} \in X_{\alpha_n}, \quad \varphi(u_{\alpha_n}) \longrightarrow c \quad \text{and} \quad \varphi'_{\alpha_n}(u_{\alpha_n}) \longrightarrow 0 \quad \text{in } X^*,$$

admits a strongly convergent subsequence.

(b) We say that φ satisfies the PS^*-condition, if it satisfies the PS_c^*-condition for every $c \in \mathbb{R}$.

Remark 5.81
Note that in the above definition, we can replace $\mathbb{N}^2$ by any directed set. If $Y_n = Y$ and $V_n = \{0\}$ for all $n \geqslant 1$, then the PS_c^*-condition coincides with the PS_c-condition (see Definition 5.45(a)).

Theorem 5.82
If X is a Banach space which admits a direct sum decomposition $X = Y \oplus V$ with $V \neq \{0\}$, $\varphi \in C^1(X)$ and

(i) φ has a local linking at 0 with respect to the direct sum pair (Y, V);
(ii) φ satisfies the PS^*-condition;
(iii) φ maps bounded sets to bounded sets;
(iv) for every $n \geqslant 1$, $\varphi(u) \longrightarrow -\infty$ as $\|u\|_X \longrightarrow +\infty$ with $u \in Y \oplus V_n$,

then φ has at least one nontrivial critical point.

We can use the extension of the PS-condition introduced in Definition 5.80 in order to have a version of Theorem 5.79 in which we drop the requirement that $\dim Y < +\infty$ (see Theorem 5.79). In what follows, if $\varphi \in C^1(X)$ has a local linking at 0 and satisfies the PS^*-condition, then we always assume that the same direst sum decomposition of X holds for both properties and that $\dim Y_n < +\infty$, $\dim V_n < +\infty$ for all $n \geqslant 1$.

Theorem 5.83
If X is a Banach space which admits a direct sum decomposition $X = Y \oplus V$, $\varphi \in C^1(X)$ and

(i) φ has a local linking at 0 with respect to the direct sum pair (Y, V);
(ii) φ satisfies the PS^*-condition;
(iii) φ maps bounded sets to bounded sets;
(iv) φ is bounded below and $\inf\limits_X \varphi < 0$,

then φ has at least two nontrivial critical points.

Stronger multiplicity theorems can be obtained in the presence of symmetries, that is, when there exists some topological group G acting continuously on X and the functional φ is invariant under this group action.

Let X be a Banach space and let G be a topological group.

Definition 5.84
A ***representation*** *of G over X is a family $\{L(g)\}_{g\in G} \subseteq \mathcal{L}(X)$ such that:*
(a) *$L(e) = I_X$ with $e \in G$ being the identity element of the group;*
(b) *$L(g_1g_2) = L(g_1) \circ L(g_2)$ for all $g_1, g_2 \in G$;*
(c) *the map $G \times X \ni (g,u) \longmapsto L(g)(u) \in X$ is continuous.*
We say that the representation is ***isometric****, if*

$$\|L(g)(u)\|_X = \|u\|_X \qquad \forall\, g \in G,\ u \in X$$

(this means that for every $g \in G$, $L(g) \in \mathcal{L}(X)$ is an isometry). A set $C \subseteq X$ is ***invariant*** *(or* ***G-invariant****), if $L(g)(C) \subseteq C$ for all $g \in G$. Similarly, a function $\varphi\colon X \longrightarrow \mathbb{R}$ is* ***invariant****, if $\varphi \circ L(g) = \varphi$ for all $g \in G$. A map $\xi\colon X \longrightarrow X$ is* ***equivariant****, if $L(g) \circ \xi = \xi \circ L(g)$ for all $g \in G$. Finally the set of* ***invariant*** *(or* ***fixed****) points of X is the set*

$$\mathrm{Fix}_G(X) = \{u \in X :\ L(g)(u) = u \text{ for all } g \in G\}.$$

Remark 5.85
Often we identify $L(g)$ with g and speak of the ***linear action*** of G on X. For notational economy we will adopt this custom here.

In what follows X is a Banach space and we have a linear action on a topological group G on X. We start with the equivariant version of the deformation lemma (see Theorem 5.54).

Proposition 5.86
If X is a Banach space, G is a compact topological group with isometric action on X, $\varphi \in C^1(X)$, $c \in \mathbb{R}$ satisfies the C_c-condition, $\varepsilon_0, \vartheta > 0$ and U is an invariant open neighborhood of K_φ^c,
then there exist $\varepsilon \in (0, \varepsilon_0)$ and a continuous map $h\colon [0,1] \times X \longrightarrow X$ such that for all $(t,u) \in [0,1] \times X$ conditions ***(a)–(f)*** *of Theorem 5.54 hold and*
(g) *$h(t,\cdot)$ is equivariant.*

Theorem 5.87 *(Ljusternik–Schnirelmann Theorem)*
If $\varphi \in C(\mathbb{R}^N)$ is even,
then $\varphi|_{S^{N-1}}$ admits at least N-couples $(u,-u)$ of critical points (recall that $S^{N-1} = \{u \in \mathbb{R}^N :\ |u| = 1\}$).

We would like to extend this result to infinite dimensional spaces and to other linear group actions different from $G = \mathbb{Z}_2 = \{I_X, -I_X\}$. In infinite dimensions the compactness property of the unit sphere is lost and so an additional condition is needed. As before, the C-condition on φ is a suitable additional assumption.

Let $\mathcal{Y} = \{D \subseteq X :\ D \text{ is closed and } G\text{-invariant}\}$.

Definition 5.88
*An **index** (or G-**index**) on X is a map* $i\colon \mathcal{Y} \longrightarrow \mathbb{N} \cup \{+\infty\}$ *such that:*
(a) $i(D) = 0$ *if and only if* $D = \emptyset$;
(b) *if* $h\colon D \longrightarrow E$ *is equivariant and continuous* ($D, E \in \mathcal{Y}$), *then* $i(D) \leqslant i(E)$ *(monotonicity);*
(c) $i(D \cup E) \leqslant i(D) + i(E)$ *for all* $D, E \in \mathcal{Y}$ *(subadditivity);*
(d) *if* $D \in \mathcal{Y}$ *is compact, then there exists an invariant open neighborhood U of D such that* $i(\overline{U}) = i(D)$ *(continuity).*

Remark 5.89
Property **(a)** implies that, if $D, E \in \mathcal{Y}$ and $D \subseteq E$, then $i(D) \leqslant i(E)$. Having a G-index $i\colon \mathcal{Y} \longrightarrow \mathbb{N} \cup \{+\infty\}$, we define

$$\mathcal{Y}_k = \{D \subseteq X : D \text{ is compact, invariant and } i(D) \geqslant k\} \qquad \forall\, k \geqslant 1.$$

Given $\varphi \in C^1(X)$, we define the following values

$$c_k = \inf_{D \in \mathcal{Y}_k} \max_{u \in D} \varphi(u). \tag{5.5}$$

Clearly $-\infty \leqslant c_1 \leqslant c_2 \leqslant \ldots$.

Theorem 5.90
If X is a Banach space, G is a compact topological group with isometric linear action on X, $\varphi \in C^1(X)$ *is invariant and satisfies the C-condition and for some* $k \geqslant 1$, *we have* $c_k > -\infty$ *(see (5.5)),*
then given $m \geqslant 1$, $m \leqslant k$ *such that* $c_m = c_k = c > -\infty$ *we have* $i(K_\varphi^c) \geqslant k - m + 1$ *(in particular* c_k *is a critical value of* φ*).*

Of special interest is the group $G = \mathbb{Z}_2 = \{I_X, -I_X\}$. Then

$$\mathcal{Y} = \{D \subseteq X : D \text{ is closed and symmetric (i.e., } D = -D)\}. \tag{5.6}$$

For this collection, a particular index is the ***Krasnoselskii genus***.

Definition 5.91
For $D \in \mathcal{Y}$ *(see (5.6)), the **Krasnoselskii genus*** $\gamma(D)$ *of D is defined by*

$$\begin{aligned}\gamma(\emptyset) &= 0,\\ \gamma(D) &= \inf\left\{m \geqslant 1 : \textit{there exists } h \in C(D; \mathbb{R}^m \setminus \{0\}) \textit{ odd}\right\} \quad \forall\, D \neq \emptyset.\end{aligned}$$

(we use the convention that $\inf \emptyset = +\infty$*).*

Using γ we can have the following generalization of the Ljusternik–Schnirelmann theorem (see Theorem 5.87).

Theorem 5.92
If X is a Banach space $\varphi \in C^1(X)$ *is even and for some* $r > 0$, $\varphi|_{\partial B_r}$ *is bounded below and satisfies the PS-condition,*
then φ *has infinitely many distinct pairs of critical points.*

Another extension of Theorem 5.87 is proved by the next theorem.

Theorem 5.93
If X is a Banach space, $\varphi \in C^1(X)$ is even, bounded below and satisfies the C-condition and there exists a compact, symmetric set $D \subseteq X$ such that

$$\gamma(D) = m \quad \text{and} \quad \sup_D \varphi < \varphi(0),$$

then φ has at least m distinct pairs $(u_1, -u_1), \ldots, (u_m, -u_m)$ of critical points with

$$\varphi(u_k) = \varphi(-u_k) < \varphi(0) \qquad \forall\, k \in \{1, \ldots, m\}.$$

A related multiplicity result is the following.

Theorem 5.94 *(Symmetric Mountain Pass Theorem)*
If X is an infinite dimensional Banach space, $\varphi \in C^1(X)$, $\varphi(0) = 0$ φ is even and satisfies the PS-condition and the following conditions hold:

(i) *there exists linear subspace V of X of finite codimension and numbers $\beta, r > 0$ such that $\varphi|_{\partial B_r \cap V} \geqslant \beta$;*

(ii) *there exists a finite dimensional linear subspace Y of X with $\dim Y > \operatorname{codim} V$ such that $\varphi|_Y$ is anticoercive (i.e., $\varphi(y) \longrightarrow -\infty$ as $\|y\|_X \longrightarrow +\infty$, with $y \in Y$),*

then φ has at least $k = \dim Y - \operatorname{codim} V$ distinct pairs of nontrivial critical points.

*Moreover, if **(ii)** is replaced by*

(ii)' *for every $k \geqslant 1$ there is a k-dimensional linear subspace Y of X such that $\varphi|_Y$ is anticoercive,*

then φ has infinitely many distinct pairs of nontrivial critical points.

Another important index (historically, the first one to be defined in 1934), is the Ljusternik–Schnirelmann category.

Definition 5.95
*Let Y be a Hausdorff topological space and let $D \subseteq Y$ be a closed subset. The **Ljusternik–Schnirelmann category** of D in Y, denoted by $cat_Y(D)$ is defined by*

$$cat_Y(D) = \Big\{m \geqslant 1 : D \subseteq \bigcup_{k=1}^m D_k,\ D_k \text{ is closed and contractible for all } k \in \{1, \ldots, m\}\Big\}$$

(as before $\inf \emptyset = +\infty$).

Remark 5.96
Since closedness and contractibility are topological properties (that is, are preserved under homeomorphisms), we get the same values for homeomorphic Y or homeomorphic D. However, in $cat_Y(D)$, the subscript Y is essential, since if we have continuous embedding $Y \hookrightarrow V$, then $cat_V(D) \leqslant cat_Y(D)$ (sets may be contractible in the big space V but not in the smaller space Y). If Y is a Banach space, then cat_Y is an

"index" in the sense of Definition 5.88 for $G = \{I_X\}$ and $\mathcal{Y} = P_f(X) \cup \{\emptyset\}$. In fact more generally, cat_Y has the properties of Definition 5.88 if Y is a Hausdorff topological space which is path-connected and an ANR (see Definition 4.132(b)).

Theorem 5.97

If X is a Banach space, $\varphi \in C^1(X)$ is bounded below and satisfies the PS-condition, $\Sigma_k = \{D \subseteq X : cat_X(D) \geqslant k\}$ and $c_k = \inf\limits_{D\in\Sigma_k} \sup\limits_{u\in D} \varphi(u)$ for $\Sigma_k \neq \emptyset$, then

(a) $\operatorname{card}(K_\varphi) \geqslant cat_X(X)$ *and for every finite c_k, we have $c_k \in \varphi(K_\varphi)$;*

(b) *if $\varphi(K_\varphi)$ is bounded and there exists $k_0 \geqslant 1$, such that $c_k < +\infty$ only if $k \leqslant k_0$, then $k_0 = cat_X(X)$;*

(c) *if $c_k = c_{k+1} = \eta < +\infty$, then* $\operatorname{card}(K_\varphi \cap \varphi^{-1}(\eta)) = +\infty$;

(d) *if $c_k = \ldots = c_{k+m} = \eta < +\infty$, then $cat_X(K_\varphi \cap \varphi^{-1}(\eta)) \geqslant m+1$.*

Under suitable conditions, we can guarantee infinitely many critical values for an invariant functional. The following notion is important in this direction.

Definition 5.98

Let V be a finite dimensional Banach space and let G be a compact topological group which acts diagonally on V^k $(k \geqslant 2)$, that is,

$$g(v_1, \ldots, v_k) = (gv_1, \ldots, gv_k).$$

We say that this action is ***admissible****, if for every integer $k \geqslant 2$ and every invariant, bounded, open neighborhood of the origin $U \subseteq V^k$, we have that every continuous, equivariant map $h: \partial U \longrightarrow V^{k-1}$ has a zero.*

Remark 5.99

By the Borsuk–Ulam theorem (see Theorem 4.20), the linear action $G = \mathbb{Z}_2 = \{I_X, -I_X\}$ on $V = \mathbb{R}$ is admissible.

Now, suppose that X is a Banach space, V is a finite dimensional Banach space and G is a compact topological group with an admissible linear action on V. Suppose that

$$X = \overline{\bigoplus_{k\geqslant 1} X_k},$$

where for each integer $k \geqslant 1$, the space X_k is invariant and there is an equivalent isomorphism from V onto X_k. We set

$$Z_m = \bigoplus_{k=1}^{m} X_k \quad \text{and} \quad Y_m = \overline{\bigoplus_{k\geqslant m} X_k}.$$

We can have the following multiplicity theorems.

Theorem 5.100 *(Fountain Theorem)*
If the setting is as above, $\varphi \in C^1(X)$ is invariant and satisfies the PS-condition and for every $m \geqslant 1$ there are $\varrho_m > r_m > 0$ such that:
(i) $\max\{\varphi(u) : u \in Z_m, \|u\|_X = \varrho_m\} \leqslant 0$,
(ii) $\inf\{\varphi(u) : u \in Y_m, \|u\|_X = r_m\} \longrightarrow +\infty$ *as* $m \to +\infty$,
then φ has an unbounded sequence of critical values.

Theorem 5.101 *(Symmetric Criticality Theorem)*
If H is a Hilbert space, G is a topological group with isometric action on H, $\varphi \in C^1(H)$ is invariant and $u_0 \in K_{\varphi|_{\mathrm{Fix}_G(H)}}$ (see Definition 5.84),
then $u_0 \in K_\varphi$.

5.1.3 Morse Theory: Critical Groups

Morse theory and critical groups are useful in distinguishing between different types of critical points and in generating multiple critical points for a functional.

In what follows, by H_* we denote the singular homology with coefficients in $\mathbb{Z}$ (for a compact review of the singular homology theory; see Gasiński–Papageorgiou, Part 1).

Definition 5.102
*Let X be a Banach space, $\varphi \in C^1(X)$ and $u \in K_\varphi$ is an isolated critical point. The **critical groups** of φ at u are defined by*

$$C_k(\varphi, u) = H_k(\varphi^c \cap U, \varphi^c \cap U \setminus \{0\}) \qquad \forall\, k \geqslant 0,$$

where $c = \varphi(u)$ and U is a neighborhood of u such that $K_\varphi \cap \varphi^c \cap U = \{u\}$. For integers $k < 0$, we set $C_k(\varphi, u) = 0$.

Remark 5.103
The excision property of singular homology implies that the above definition is independent of the choice of the neighborhood U. The critical groups $C_k(\varphi, u)$ depend only on the behavior of φ near u. In fact, that makes sense even if φ is defined only in a neighborhood of u.

The strongest results concerning critical groups can be obtained if $X = H$ is a Hilbert space and φ is a C^2-function. Recall that in this case $\varphi''(u) \in \mathcal{L}(H)$ is a self-adjoint and we have the orthogonal direct sum decomposition

$$H = \ker \varphi''(u) \oplus \operatorname{im} \varphi''(u). \tag{5.7}$$

Definition 5.104
Suppose that H is a Hilbert space, $U \subseteq H$ is a nonempty open set, $\varphi \in C^2(U;\mathbb{R})$ and $u \in K_\varphi$.

(a) *The* ***Morse index*** *of u is defined to be the supremum of the dimensions of the vector subspaces of H on which $\varphi''(u)$ is negative definite.*
(b) *The* ***nullity*** *of u'' is the dimension of* $\ker \varphi''(u)$.
(c) *We say that $u \in K_\varphi$ is* ***nondegenerate****, if $\varphi''(u)$ is invertible.*

Remark 5.105
The inverse mapping theorem (see Theorem 3.26) implies that a nondegenerate critical point is always isolated. From (5.7) it is clear that if the nullity of u is finite, then $\varphi''(u)$ is a Fredholm operator (see Definition 2.29) of index zero (i.e., $\dim\ker\varphi''(u) = \operatorname{codim}\varphi''(u) < +\infty$).

Theorem 5.106
If H is a Hilbert space, $U \subseteq H$ is a nonempty open set, $\varphi \in C^2(U;\mathbb{R})$ and $u \in K_\varphi$ is nondegenerate with Morse index m (possibly $+\infty$),
then $C_k(\varphi, U) = \delta_{k,m}\mathbb{Z}$ for all $k \geqslant 0$, where $\delta_{k,m}$ is the Kronecker symbol, i.e.,

$$\delta_{k,m} = \begin{cases} 1 & \text{if } k = m, \\ 0 & \text{if } k \neq m, \end{cases} \qquad k, m \geqslant 0.$$

For the degenerate case, there is the so-called ***shifting theorem***, which says that for a degenerate critical point, the critical groups depend on the Morse index and on the "degenerate part" of the function. For this we need the following result, which describes the local behavior of φ near the critical point u.

Theorem 5.107 *(Generalized Morse Lemma)*
If H is a Hilbert space, $U \subseteq H$ is a nonempty open set, $\varphi \in C^2(U;\mathbb{R})$ and $u \in K_\varphi$ is isolated and of finite nullity,
then there exist a neighborhood of the origin $V \subseteq H$, a homeomorphism $h\colon V \longrightarrow H$, a neighborhood W of the origin in $\ker\varphi''(u)$ and $\widehat{\varphi} \in C^2(W;\mathbb{R})$ such that:
- $h(0) = u$, $\widehat{\varphi}(0) = 0$, $\widehat{\varphi}'(0) = 0$, $\widehat{\varphi}''(0) = 0$,
- $\varphi(h(y)) = \varphi(u) + \frac{1}{2}(\varphi''(u)y, y)_H + \widehat{\varphi}(z)$ *for all $y \in V$, where $y = z + v$ with $z \in \ker\varphi''(u)$, $v \in \operatorname{im}\varphi''(u)$ (see (5.7)).*

Remark 5.108
If $z \in V \cap W$ is a critical point of $\widehat{\varphi}$, then $h(z)$ is a critical point of φ. Moreover, since $u = h(0)$ is an isolated critical point of φ, then $z = 0$ is an isolated critical point of $\widehat{\varphi}$ and so the critical groups $C_k(\widehat{\varphi}, 0)$, for $k \geqslant 0$, are well defined (see Definition 5.104). If $u \in K_\varphi$ is nondegenerate, then $\widehat{\varphi} \equiv 0$ (since $\ker\varphi''(u) = \{0\}$) and using this version of Theorem 5.107 one can prove Theorem 5.106.

Theorem 5.109 *(Shifting Theorem)*
If H is a Hilbert space, $U \subseteq H$ is a nonempty open set, $\varphi \in C^2(U;\mathbb{R})$ and $u \in K_\varphi$ is an isolated critical point with finite nullity and finite Morse index m,
then $C_k(\varphi, u) = C_{k-m}(\widehat{\varphi}, 0)$ for all $k \geqslant 0$ with $\widehat{\varphi}$ as in Theorem 5.107.

Remark 5.110
When $u \in K_\varphi$ is nondegenerate, then $\widehat{\varphi} \equiv 0$ and so we recover Theorem 5.106.

An alternative formulation of the shifting theorem is the following.

Theorem 5.111
If H is a Hilbert space, $U \subseteq H$ is a nonempty open set, $\varphi \in C^2(U;\mathbb{R})$ and $u \in K_\varphi$ is an isolated critical point with finite nullity ν and finite Morse index m,
then one of the following holds:
(a) $C_k(\varphi, u) = 0$ for all $k \leqslant m$ and all $k \geqslant m+\nu$;
(b) $C_k(\varphi, u) = \delta_{k,m}\mathbb{Z}$ for all $k \geqslant 0$;
(c) $C_k(\varphi, u) = \delta_{k,m+\nu}\mathbb{Z}$ for all $k \geqslant 1$.

Remark 5.112
In fact the above result as well as Theorem 5.109 remain valid even if $\varphi^{2-0}(U;\mathbb{R})$ (that is, φ is continuously differentiable and the map $u \longmapsto \varphi'(u)$ is locally Lipschitz).

Definition 5.113
*Let X be a Banach space, $\varphi \in C^1(X)$ satisfy the C-condition and $\inf \varphi(K_\varphi) > -\infty$. The **critical groups** of φ at infinity are defined by*

$$C_k(\varphi, \infty) = H_k(X, \varphi^c),$$

for any $c < \inf \varphi(K_\varphi)$.

Remark 5.114
From the second deformation lemma (see Theorem 5.56), we know that if $b < c < \inf \varphi(K_\varphi)$, then φ^b is a strong deformation retract of φ^a. Hence

$$C_k(X, \varphi^c) = C_k(X, \varphi^b) \qquad \forall\, k \geqslant 0.$$

Therefore Definition 5.113 is independent of the choice of the level $c < \inf \varphi(K_\varphi)$.

Definition 5.115
*Let X be a Banach space and let $\varphi \in C^1(X)$ satisfy the C-condition and let K_φ be finite. The **Morse polynomials** of φ are defined by*

$$M(t,u) = \sum_{k\geqslant 0} \operatorname{rank} C_k(\varphi, u) t^k \qquad \forall\, t \in \mathbb{R},\ u \in K_\varphi,$$

$$P(t,\infty) = \sum_{k\geqslant 0} \operatorname{rank} C_k(\varphi, \infty) t^k \qquad \forall\, t \in \mathbb{R}.$$

Theorem 5.116 *(Morse Inequality)*
There exists a polynomial $Q(t)$ with nonnegative integer coefficients such that

$$\sum_{u\in K_\varphi} M(t,u) = P(t,\infty) + (1+t)Q(t) \qquad \forall\, t \in \mathbb{R}.$$

Next we state a result for the critical groups at infinity, analogous to Theorem 5.111. Let H be a Hilbert space with inner product $(\cdot,\cdot)_H$ and a function $\varphi \in C^1(H)$ which satisfies the following condition.

> $\underline{(A_\infty)}$: $\varphi(u) = \frac{1}{2}(A(u),u)_H + \psi(u)$ for all $u \in H$, with a self-adjoint $A \in \mathcal{L}(H)$ such that 0 is isolated in the spectrum of A, $\psi \in C^1(H)$ and $\lim\limits_{\|u\|_H \to +\infty} \frac{\psi(u)}{\|u\|_H^2} = 0$. Moreover, we assume that ψ and ψ' are bounded (that is, map bounded sets to bounded sets), $\inf \varphi(K_\varphi) > -\infty$ and it satisfies the C-condition.

We set $Y = \ker A$ and $Z = Y^\perp$. We know that $Z = Z_+ \oplus Z_-$ with both subspaces A-invariant and

$$A|_{Z_+} > 0, \qquad A|_{Z_-} < 0.$$

So, there exists $c_0 > 0$ such that

$$\pm(A(u),u)_H \geqslant c_0\|u\|_H^2 \qquad \forall\, u \in Z_\pm.$$

We set $m = \dim Z_-$ (Morse index of φ at infinity) and $\nu = \dim Y$ (nullity of φ at infinity).

Theorem 5.117

If H is a Hilbert space, $\varphi \in C^1(H)$ satisfies condition (A_∞) and m, ν are finite, then

(a) *$C_k(\varphi,\infty) = 0$ for all $k \notin \{m, m+1, \ldots, m+\nu\}$;*

(b) *$C_k(\varphi,\infty) = \delta_{k,m}\mathbb{Z}$ for all $k \geqslant 0$ provided the following angle condition holds:*

> $\underline{(A_\infty^+)}$: *there exists $M > 0$ and $\vartheta \in (0,1)$ such that*
>
> $$\langle \varphi'(u), y\rangle_H \geqslant 0$$
>
> *for $u = y + z$ with $y \in Y$, $z \in Z$, $\|u\|_X \geqslant M$ and $\|z\|_X \leqslant \vartheta\|u\|_X$;*

(c) *$C_k(\varphi,\infty) = \delta_{k,m+\nu}\mathbb{Z}$ for all $k \geqslant 0$, provided the following angle condition holds:*

> $\underline{(A_\infty^-)}$: *there exists $M > 0$ and $\vartheta \in (0,1)$ such that*
>
> $$-\langle \varphi'(u), y\rangle_H \geqslant 0$$
>
> *for $u = y + z$ with $y \in Y$, $z \in Z$, $\|u\|_X \geqslant M$ and $\|z\|_X \leqslant \vartheta\|u\|_X$.*

In Definition 5.59 we introduced the notion or linking sets which as we saw is crucial in the minimax theory of critical values. That notion is often referred to "homotopical linking" in order to distinguish it from the notion that we are about to introduce and which is known as "homological linking."

Definition 5.118

Let X be a Banach space and let $E_0 \subseteq E$ and D be nonempty subsets of X such that $E_0 \cap D = \emptyset$. We say that the pair $\{E_0, E\}$ ***homologically links*** *D in dimension m, if the homomorphism $i_* \colon H_m(E,E_0) \longrightarrow H_m(X, X\setminus D)$ induced by the corresponding inclusion map $i\colon (E,E_0) \longrightarrow (X, X\setminus D)$ is nontrivial.*

Remark 5.119
The sets from Example 5.61 are also homologically linking. In general we can say that homological linking implies (homotopical) linking.

Proposition 5.120
If X is a Banach space, the pair of sets $\{E_0, E\}$ homologically links D in dimension m, $\varphi \in C^1(X)$ and $a < b \leqslant +\infty$ are such that

$$\varphi|_{E_0} \leqslant a < \varphi|_D \quad \text{and} \quad \sup_E \varphi \leqslant b,$$

then
(a) $H_m(\varphi^b, \varphi^a) \neq 0$;
(b) if φ satisfies the C-condition, a, b are regular values of φ and $\varphi^{-1}(a,b) \cap K_\varphi$ is finite, then there exists $u \in \varphi^{-1}(a,b) \cap K_\varphi$ such that $C_m(\varphi, u) \neq 0$.

In definition 5.77 we introduced the notion of "local linking" which is useful in obtaining multiple critical points (see Theorem 5.79 and 5.83). We can have a homological version of this notion.

Definition 5.121
Suppose that X is a Banach space, $\varphi \in C^1(X)$, $\varphi(0) = 0$, $0 \in K_\varphi$ is isolated and $m, n \geqslant 1$. We say that φ has **local** *(m,n)***-linking** *near the origin, if there are a neighborhood U of 0 and nonempty sets $E_0 \subseteq E \subseteq U$ and $D \subseteq X$ such that $0 \notin E_0$, $E_0 \cap D = \emptyset$ and*
(a) $\varphi^0 \cap U \cap K_\varphi = \{0\}$;
(b) $\operatorname{rank} \operatorname{im} i_{m-1} - \operatorname{rank} \operatorname{im} j_{m-1} \geqslant n$, *where* $i_{m-1} \colon H_{m-1}(E_0) \longrightarrow H_{m-1}(X \setminus D)$ *and* $j_{m-1} \colon H_{m-1}(E_0) \longrightarrow H_{m-1}(E)$ *are the homomorphisms induced by the corresponding inclusion maps* $i \colon E_0 \longrightarrow X \setminus D$ *and* $j \colon E_0 \longrightarrow E$;
(c) $\varphi|_E \leqslant 0 \leqslant \varphi|_{D \cap U \setminus \{0\}}$.

In the next proposition we relate this "homological" notion with the one introduced in Definition 5.77.

Proposition 5.122
If X is a Banach space, $X = Y \oplus V$ with $\dim Y < +\infty$, $\varphi \in C^1(X)$, φ has a local linking at the origin with respect to the pair (Y, V) (see Definition 5.77), $0 \in K_\varphi$ is isolated and one of the following conditions holds:
(i) 0 is a strict local minimizer of $\varphi|_V$; or
(ii) $X = H$ is a Hilbert space and φ' is Lipschitz continuous near 0,
then φ has a local $(m,1)$-linking at 0.

The next proposition estimates the critical groups at the origin if the functional has a homological local linking 0.

Proposition 5.123
If X is a Banach space, $\varphi \in C^1(X)$, $0 \in K_\varphi$ is isolated and φ has a local (m,n)-linking near origin (see Definition 5.121),
then $\operatorname{rank} C_m(\varphi, 0) \geqslant n$.

Proposition 5.124
If H is a Hilbert space, $\varphi \in C^2(H;\mathbb{R})$, $u_0 \in K_\varphi$ is an isolated critical point, $\varphi''(u_0) \in \mathcal{L}(H)$ is a Fredholm operator and $\dim\ker\varphi''(u_0) \leqslant 1$ *if $\sigma(\varphi''(u_0)) \subseteq [0,+\infty)$ ($\sigma(\varphi''(u_0))$ being the spectrum of $\varphi''(u_0)$),*
then the following statement are equivalent:
(a) $u_0 \in K_\varphi$ is of mountain pass type (see Definition 5.71);
(b) $C_1(\varphi, u_0) \neq 0$;
(c) $C_k(\varphi, u_0) = \delta_{k,1}\mathbb{Z}$ for all $k \geqslant 0$.

The next two theorems are useful in the computation of critical groups. The first shows that as in degree theory, critical groups exhibit homotopy invariance.

Theorem 5.125
If X is a Banach space, $\{h_t\colon X \longrightarrow \mathbb{R}\}_{t\in[0,1]}$ is a family of C^1-functions, $U \subseteq X$ is a nonempty open set, $u_t \in U$ is the only critical point of h_t on $\overline{U}$ for all $t \in [0,1]$ and
(i) for every $t_0 \in [0,1]$, $\|h_t|_U - h_{t_0}|_U\|_{C^1(\overline{U})} \longrightarrow 0$ as $t \to t_0$;
(ii) for every $t \in [0,1]$, φ_t satisfies the C-condition on $\overline{U}$ (see Definition 5.45),
then for all $k \geqslant 0$, $C_k(h_t, u_t)$ is independent of $t \in [0,1]$.

The second theorem says that critical groups are continuous in the C^1-norm.

Theorem 5.126
If X is a Banach space, $\varphi \in C^1(X)$, $U \subseteq X$ is a nonempty open set, $u_0 \in U$ is the only critical point of φ in $\overline{U}$ and φ satisfies the C-condition in $\overline{U}$,
then there exists $\varepsilon > 0$ such that for every $\psi \in C^1(X)$ having a unique critical point $v_0 \in U$, satisfying C-condition in $\overline{U}$ and such that $\|\psi|_U - \varphi|_U\|_{C^1(\overline{U})} \leqslant \varepsilon$, we have

$$C_k(\varphi, u_0) = C_k(\psi, v_0) \qquad \forall\, k \geqslant 0.$$

5.1.4 Dirichlet Elliptic Problems

Let $\Omega \subseteq \mathbb{R}^N$ be a bounded domain (that is, a bounded, open and connected subset of $\mathbb{R}^N$) with a C^2-boundary $\partial\Omega$ and let $m \in L^s(\Omega)$ with $s \in (\frac{N}{2}, +\infty)$. We consider the following weighted linear eigenvalue problem

$$\begin{cases} -\Delta u(z) = \widehat{\lambda} m(z) u(z) & \text{in } \Omega, \\ u|_{\partial\Omega} = 0. \end{cases} \tag{5.8}$$

Definition 5.127
We say that $(\widehat{\lambda}(m), \widehat{u}) \in \mathbb{R} \times (H_0^1(\Omega) \setminus \{0\})$ *is an* ***eigenpair*** *for problem (5.8), if*

$$\int_\Omega (D\widehat{u}, Dh)_{\mathbb{R}^N}\, dz \;=\; \widehat{\lambda}(m) \int_\Omega m\widehat{u}h\, dz \qquad \forall\, h \in H_0^1(\Omega).$$

Then $\widehat{\lambda}(m)$ *is an* ***eigenvalue*** *of* $(-\Delta, H_0^1(\Omega), m)$ *and* $\widehat{u} \in H_0^1(\Omega) \setminus \{0\}$ *the corresponding* ***eigenfunction***.

Remark 5.128
In fact $\widehat{u} \in C^1(\overline{\Omega})$ (regularity theory) and the equation in (5.8) is satisfied pointwise for almost all $z \in \Omega$. If $m \equiv 1$, then we write $\widehat{\lambda}$ instead of $\widehat{\lambda}(1)$.

Theorem 5.129
If $\Omega \subseteq \mathbb{R}^N$ *is a bounded open set with a* C^2*-boundary* $\partial\Omega$ *and* $m \in L^s(\Omega)$ *with* $s \in (\frac{N}{2}, \infty)$, *then*

(a) *in general problem (5.8) has a double sequence of distinct eigenvalues*

$$\begin{aligned} \ldots \;<\; \widehat{\lambda}_{-n}(m) \;\ldots \;<\; \widehat{\lambda}_{-1}(m) \;<\; 0 \;<\; \widehat{\lambda}_1(m) \;<\\ \ldots \;<\; \widehat{\lambda}_n(m) \;<\; \ldots \end{aligned}$$

$$\widehat{\lambda}_n(m) \longrightarrow +\infty \quad \text{and} \quad \widehat{\lambda}_{-n}(m) \longrightarrow -\infty.$$

(b) *if* $m^+ = 0$ *(respectively* $m^- = 0$*), then there are no positive (respectively negative) eigenvalues;*

(c) *if* $m^+ \neq 0$, *then*

$$\widehat{\lambda}_n(m) \;=\; \inf_{V \in \mathcal{Y}_n} \sup_{u \in V} \Big\{ \|Du\|_2^2 : \int_\Omega m(z)u^2\, dz = 1 \Big\},$$

where

$$\begin{aligned} \mathcal{Y}_n \;=\; \{&V \subseteq H :\, V \text{ is a finite dimensional linear subspace of}\\ &H \text{ with } \dim V = n\}. \end{aligned}$$

(d) *if* $m^- \neq 0$, *then*

$$\widehat{\lambda}_{-n}(m) \;=\; \inf_{V \in \mathcal{Y}_n} \sup_{u \in V} \Big\{ \|Du\|_2^2 : \int_\Omega m(z)u^2\, dz = -1 \Big\};$$

(e) $\widehat{\lambda}_1(m)$ *and* $\widehat{\lambda}_{-1}(m)$ *if they exist are simple and if* $E(\widehat{\lambda}_1(m))$ *and* $E(\widehat{\lambda}_{-1}(m))$ *are the corresponding eigenspaces, then*

$$E(\widehat{\lambda}_1(m)) \;=\; \mathbb{R}\widehat{u}_1 \quad \text{and} \quad E(\widehat{\lambda}_{-1}(m)) \;=\; \mathbb{R}\widehat{u}_{-1},$$

with $\widehat{u}_1, \widehat{u}_{-1} \in \operatorname{int} C_+$, *where*

$$C_+ \;=\; \big\{ u \in C_0^1(\overline{\Omega}) :\; u(z) \geqslant 0 \text{ for all } z \in \overline{\Omega} \big\}$$

and

$$\operatorname{int} C_+ \;=\; \left\{u \in C_0^1(\overline{\Omega}) : \, u(z) > 0 \text{ for all } z \in \Omega, \, \tfrac{\partial u}{\partial n} < 0 \text{ on } \partial\Omega\right\}$$

and $n(\cdot)$ denotes the outward unit normal on $\partial\Omega$;

(f) *for every $n \in \mathbb{Z}$, $E(\widehat{\lambda}_n(m)) \subseteq C^1(\Omega)$ is finite dimensional and has the* ***unique continuation property****, i.e., if $u \in E(\widehat{\lambda}_n(m))$ and u vanishes on a set of positive measure, then $u \equiv 0$.*

Remark 5.130

We also have

$$\begin{aligned}
\widehat{\lambda}_1(m) &= \inf\Big\{\|Du\|_2^2 : \int_\Omega m(z)u^2\,dz = 1, \; u \in H^1(\Omega)\Big\},\\
\widehat{\lambda}_{-1}(m) &= \inf\Big\{\|Du\|_2^2 : \int_\Omega m(z)u^2\,dz = -1, \; u \in H^1(\Omega)\Big\},\\
\widehat{\lambda}_n(m) &= \inf\Big\{\|Du\|_2^2 : \int_\Omega m(z)u^2\,dz=1, \; u \in \widehat{H}_n=\overline{\bigoplus_{k\geqslant n} E(\widehat{\lambda}_k(m))}\Big\}\\
&= \sup\Big\{\|Du\|_2^2 : \int_\Omega m(z)u^2\,dz=1, \; u \in \overline{H}_n=\bigoplus_{k=1}^{n} E(\widehat{\lambda}_k(m))\Big\},\\
\widehat{\lambda}_{-n}(m) &= \inf\Big\{\|Du\|_2^2 : \int_\Omega m(z)u^2\,dz = -1, \; u \in \widehat{H}_{-n}\\
&= \overline{\bigoplus_{k\geqslant n} E(\widehat{\lambda}_{-k}(m))}\Big\}\\
&= \sup\Big\{\|Du\|_2^2 : \int_\Omega m(z)u^2\,dz = -1, \; u \in \overline{H}_{-n}\\
&= \bigoplus_{k=1}^{n} E(\widehat{\lambda}_{-k}(m))\Big\},
\end{aligned}$$

for $n \geqslant 2$. In these expressions the infima and suprema are realized on the corresponding eigenspaces.

Proposition 5.131

The functions $m \longmapsto \widehat{\lambda}_k(m)$ and $m \longmapsto \widehat{\lambda}_{-k}(m)$ are continuous on $L^s(\Omega)$ and if $m(z) \leqslant \widetilde{m}(z)$ for almost all $z \in \Omega$ and the inequality is strict on a set of positive measure, then

$$\widehat{\lambda}_k(\widetilde{m}) \;<\; \widehat{\lambda}_k(m) \quad \text{and} \quad \widehat{\lambda}_{-k}(m) \;<\; \widehat{\lambda}_{-k}(\widetilde{m}) \qquad \forall\, k \geqslant 1.$$

We consider also the corresponding nonlinear eigenvalue problem for the p-Laplacian:

$$\begin{cases} -\Delta_p u(z) \;=\; \widehat{\lambda}|u(z)|^{p-2}u(z) & \text{in } \Omega,\\ u|_{\partial\Omega} \;=\; 0. \end{cases} \tag{5.9}$$

Again $(\widehat{\lambda}, \widehat{u}) \in \mathbb{R} \times (W_0^{1,p}(\Omega) \setminus \{0\})$ is an ***eigenpair*** for problem (5.9), if

$$\int_\Omega |D\widehat{u}|^{p-2} (D\widehat{u}, Dh)_{\mathbb{R}^N}\, dz = \widehat{\lambda} \int_\Omega |\widehat{u}|^{p-2}\widehat{u}h\, dz \qquad \forall\, h \in W_0^{1,p}(\Omega).$$

Then $\widehat{\lambda}$ is an ***eigenvalue*** of $(-\Delta_p, W_0^{1,p}(\Omega))$ and $\widehat{u} \in W_0^{1,p}(\Omega)\setminus\{0\}$ the corresponding ***eigenfunction***.

Proposition 5.132
There is a smallest eigenvalue $\widehat{\lambda}_1 > 0$ *given by*

$$\widehat{\lambda}_1 = \inf_{\substack{u \in W_0^{1,p}(\Omega)\\ u \neq 0}} \frac{\|Du\|_p^p}{\|u\|_p^p} \tag{5.10}$$

and $\widehat{\lambda}_1$ *is isolated and simple.*

Remark 5.133
From (5.10) it is clear that the corresponding eigenfunction does not change sign. Let $\widehat{u}_1$ be the L^p-normalized (i.e., $\|\widehat{u}_1\|_p = 1$), positive eigenfunction. Then $\widehat{u}_1 \in \operatorname{int} C_+$. Also, the infimum in (5.10) is realized on $\mathbb{R}\widehat{u}_1$.

Let $\sigma_0(p)$ denote the set of eigenvalues of $(-\Delta_p, W_0^{1,p}(\Omega))$. The set is closed and Proposition 5.132 implies that

$$\widehat{\lambda}_2 = \inf\{\widehat{\lambda} :\ \widehat{\lambda} \in \sigma_0(p),\ \widehat{\lambda} > \widehat{\lambda}_1\}$$

is well defined and is the second eigenvalue of $(-\Delta_p, W_0^{1,p}(\Omega))$.

We define

$$\begin{aligned}
\mathcal{Y} &= \{A \subseteq W_0^{1,p}(\Omega) :\ A \text{ is compact and } A = -A\}\\
\mathcal{Y}_k &= \{A \subseteq \mathcal{Y} :\ 0 \notin A,\ \gamma(A) \geqslant k\}\\
\widehat{\lambda}_k^{LS} &= \inf_{A \in \mathcal{Y}_k} \sup_{u \in A} \frac{\|Du\|_p^p}{\|u\|_p^p} \qquad \forall\, k \geqslant 1,
\end{aligned}$$

with γ being the Krasnoselskii genus (see Definition 5.91).

Proposition 5.134
$\{\widehat{\lambda}_k^{LS}\}_{k\geqslant 1}$ *are eigenvalues of (5.9),* $\widehat{\lambda}_1^{LS} = \widehat{\lambda}_1$, $\widehat{\lambda}_2^{LS} = \widehat{\lambda}_2$ *and* $\lambda_k^{LS} \longrightarrow +\infty$.

Remark 5.135
We do not know if the sequence exhausts $\sigma_0(p)$. This is the case if $N = 1$ or $p = 2$. All the eigenfunctions belong in $C_0^1(\overline{\Omega})$ (nonlinear regularity).

There is an alternative minimax characterization of $\widehat{\lambda}_2$. Let

$$\begin{aligned}
\partial B_1^{L^p} &= \{u \in L^p(\Omega) :\ \|u\|_p = 1\}\\
M &= W_0^{1,p}(\Omega) \cap \partial B_1^{L^p}.
\end{aligned}$$

Proposition 5.136
$\widehat{\lambda}_2 = \inf_{\widehat{\gamma}\in\widehat{\Gamma}} \max_{-1\leqslant t\leqslant 1} \|D\widehat{\gamma}(t)\|_p^p$, *where*

$$\widehat{\Gamma} = \big\{\widehat{\gamma} \in C([-1,1];M) : \widehat{\gamma}(-1) = -\widehat{u}_1,\ \widehat{\gamma}(1) = \widehat{u}_1\big\}.$$

Let $f\colon \Omega \times \mathbb{R} \longrightarrow \mathbb{R}$ be a Carathéodory function (that is, for all $x \in \mathbb{R}$, the map $z \longmapsto f(z,x)$ is measurable and for almost all $z \in \Omega$, the map $x \longmapsto f(z,x)$ is continuous) and consider the following Dirichlet problem

$$\begin{cases} -\Delta_p u(z) = f(z,u(z)) & \text{in } \Omega, \\ u|_{\partial\Omega} = 0, \end{cases} \tag{5.11}$$

with $1 < p < +\infty$. Assume that

$$|f(z,x)| \leqslant a(z)(1+|x|^{r-1}) \quad \text{for a.a. } z \in \Omega, \text{ all } x \in \mathbb{R}, \tag{5.12}$$

with $a \in L^\infty(\Omega)_+$ and $p \leqslant r \leqslant p^*$, where

$$p^* = \begin{cases} \frac{Np}{N-p} & \text{if } p < N, \\ +\infty & \text{if } p \geqslant N. \end{cases}$$

Recall that by a ***weak solution*** of (5.11), we mean a function $u_0 \in W_0^{1,p}(\Omega)$ such that

$$\int_\Omega |Du_0|^{p-2}(Du_0, Dh)_{\mathbb{R}^N}\, dz = \int_\Omega f(z,u_0)h\, dz \qquad \forall\, h \in C_c^\infty(\Omega).$$

The regularity theory of weak solutions, is performed in two steps. First we show that $u_0 \in L^\infty(\Omega)$ (this is done by the so-called Moser iteration technique) and then we show the global regularity of u_0 (that is, the continuity of u_0 and Du_0 on $\overline{\Omega}$).

Proposition 5.137
If $u_0 \in W_0^{1,p}(\Omega)$ *is a weak solution of (5.11),*
then $u_0 \in L^\infty(\Omega)$ *and for any* $\vartheta \in (r,+\infty)$ *(see (5.12)),* $\vartheta \leqslant p^*$, *we have*

$$\|u\|_\infty \leqslant M(1+\|u\|_\vartheta)^{\frac{\vartheta-p}{\vartheta-r}},$$

with $M = M(a,\Omega,N,p,\vartheta) > 0$.

Proposition 5.138
If $u_0 \in W_0^{1,p}(\Omega)$ *is a weak solution of (5.11),*
then $u_0 \in C_0^{1,\alpha}(\overline{\Omega})$ *with* $\alpha \in (0,1)$ *and*

$$\|u_0\|_{C_0^{1,\alpha}(\overline{\Omega})} \leqslant \widehat{M} = \widehat{M}(a,\Omega,N,p,\|u_0\|_\infty) > 0.$$

The following nonlinear version of the maximum principle, is a basic tool in obtaining solutions of constant sign.

Theorem 5.139 *(Nonlinear Maximum Principle)*
If $z_1 \in \partial\Omega$, $\widehat{u} \in C^1(\Omega \cup \{z_1\}) \setminus \{0\}$, $\widehat{u} \geqslant 0$ *and for some* $c > 0$, $\widehat{u}$ *satisfies*

$$\Delta_p \widehat{u} \leqslant c\widehat{u}^{p-1}$$

in the distributional sense, i.e.,

$$-\int_\Omega |D\widehat{u}|^{p-2} (D\widehat{u}, Dh)_{\mathbb{R}^N}\, dz \leqslant c \int_\Omega \widehat{u} h\, dz \qquad \forall\, h \in C_c^\infty(\Omega),\ h \geqslant 0,$$

then $\widehat{u}(z) > 0$ *for all* $z \in \Omega$ *and if* $\widehat{u}(z_1) = 0$, *then*

$$\tfrac{\partial u}{\partial z}(z_1) = (Du(z_1), n(z_1))_{\mathbb{R}^N} < 0,$$

with $n(\cdot)$ *being the outward unit normal on* $\partial\Omega$.

Remark 5.140
In the semilinear case (i.e., p=2), more can be said. So, suppose that (5.12) holds with $a \in L^{\frac{N}{2}}(\Omega)$ and $2 = p = r$. Suppose that $u_0 \in H_0^1(\Omega)$ is a weak solution of (5.11) (with $p = 2$). We have that $u_0 \in L^q(\Omega)$ for all $q \in [1, +\infty)$. Then the Calderon–Zygmund inequality (see Gilbarg–Trudinger [9, p. 230], implies that $u_0 \in W^{2,q}(\Omega) \cap H_0^1(\Omega)$, $1 \leqslant q < +\infty$ and so using the Sobolev embedding theorem (see Theorem 1.109), we have $u_0 \in C_0^{1,\alpha}(\overline{\Omega})$ with $\alpha \in (0, 1)$.

These regularity results lead to a useful result relating Hölder and Sobolev local minimizers for certain C^1-functionals. So, suppose that (5.12) holds,

$$F(z, x) = \int_0^x f(x, s)\, ds \qquad \forall\, (z, x) \in \Omega \times \mathbb{R}$$

and let $\varphi\colon W_0^{1,p}(\Omega) \longrightarrow \mathbb{R}$ be the C^1-functional defined by

$$\varphi(u) = \tfrac{1}{p}\|Du\|_p^p - \int_\Omega F(z, u(z))\, dz \qquad \forall\, u \in W_0^{1,p}(\Omega).$$

Proposition 5.141
If $u_0 \in W_0^{1,p}(\Omega)$ *is a local* $C_0^1(\overline{\Omega})$*-minimizer of* φ, *i.e., there exists* $\varrho_0 > 0$ *such that*

$$\varphi(u_0) \leqslant \varphi(u_0 + h) \qquad \forall\, h \in C_0^1(\overline{\Omega}),\ \|u\|_{C_0^1(\overline{\Omega})} \leqslant \varrho_0,$$

then $u_0 \in C_0^{1,\alpha}(\overline{\Omega})$ *for some* $\alpha \in (0, 1)$ *and* u_0 *is a local* $W_0^{1,p}(\Omega)$*-minimizers of* φ, *i.e., there exists* $\varrho_1 > 0$ *such that*

$$\varphi(u_0) \leqslant \varphi(u_0 + h) \qquad \forall\, h \in W_0^{1,p}(\Omega),\ \|u\|_{W^{1,p}(\Omega)} \leqslant \varrho_1.$$

The following identities are a useful tool in many occasions in the study of boundary value problems.

Theorem 5.142 *(Picone Identity)*
If $\Omega \subseteq \mathbb{R}^N$ is a bounded domain with a C^2-boundary, $u, v\colon \Omega \longrightarrow \mathbb{R}$ are differentiable functionals such that $v(z) > 0$ and $u(z) \geqslant 0$ for all $z \in \Omega$ and we define

$$\begin{aligned} L(u,v) &= |Du|^p + (p-1)\left(\tfrac{u}{v}\right)^p |Dv|^p - p\left(\tfrac{u}{v}\right)^{p-1} \left(|Dv|^{p-2}Dv, Du\right)_{\mathbb{R}^N}, \\ R(u,v) &= |Du|^p - \left(|Dv|^{p-2}Dv, D\left(\tfrac{u^p}{v^{p-1}}\right)\right)_{\mathbb{R}^N}, \quad 1<p<+\infty, \end{aligned}$$

then $L(u,v)(z) = R(u,v)(z) \geqslant 0$ for almost all $z \in \Omega$. Moreover $L(u,v)(z) = 0$ for almost all $z \in \Omega$ if and only if $u = \xi v$ for some $\xi \in [0,+\infty)$.

Theorem 5.143 *(Pohozaev Identity)*
If $\Omega \subseteq \mathbb{R}^N$ is a bounded domain with a Lipschitz boundary $\partial\Omega$ and $u \in C^2(\Omega) \cap C^1(\overline{\Omega})$ is a solution of the Dirichlet problem:

$$\begin{cases} -\Delta u(z) = f(u(z)) & \text{in } \Omega, \\ u|_{\partial\Omega} = 0, \end{cases}$$

where $f\colon \mathbb{R} \longrightarrow \mathbb{R}$ is a continuous function,
then

$$N \int_\Omega F(u)\,dz - \tfrac{N-2}{2} \int_\Omega u f(u)\,dz = \tfrac{1}{2} \int_{\partial\Omega} \left|\tfrac{\partial u}{\partial n}\right|^2 (z,n)_{\mathbb{R}^N}\, d\sigma,$$

where $F(x) = \int_0^x f(s)\,ds$.

We conclude by recalling some elementary facts from the singular homology theory. In what follows all the spaces are Hausdorff topological spaces.

Proposition 5.144

(a) *If the topological pairs (X,A) and (Y,B) are homotopy equivalent (see Definition I.2.191),*
then $H_k(X,A) = H_k(Y,B)$ for all $k \geqslant 0$ (i.e., the two abelian groups are isomorphic).

(b) *If $A \subseteq X$ is a deformation retract (see Definition 4.112(b)),*
then $H_k(X,A) = 0$ for all $k \geqslant 0$.
In particular $H_k(X,X) = 0$ for all $k \geqslant 0$.

(c) *If $C \subseteq A \subseteq X$ and C is a deformation retract of A,*
then $H_k(X,A) = H_k(X,C)$ for all $k \geqslant 0$.

(d) *If $C \subseteq A \subseteq X$ and A is a deformation retract of X,*
then $H_k(X,C) = H_k(A,C)$ for all $k \geqslant 0$.

(e) *If $A \subseteq X$ is a retract,*
then $H_k(X) = H_k(A) \oplus H_k(X,A)$ for all $k \geqslant 0$.

(f) <u>*If*</u> *X is contractible and $\star \in X$,*
<u>*then*</u> *$H_k(X, \star) = 0$ for all $k \geqslant 0$.*

(g) <u>*If*</u> *$A \subseteq X$ is contractible and $\star \in A$,*
<u>*then*</u> *$H_k(X, A) = H_k(X, \star)$ for all $k \geqslant 0$.*

We also mention the following topological result (see Problem I.2.184).

Proposition 5.145
<u>*If*</u> *$X_1 \subseteq X_2 \subseteq X_3 \subseteq X_4$ are Hausdorff topological spaces,* <u>*then*</u> *for all $k \geqslant 0$, we have*

$$\begin{aligned} \operatorname{rank} H_k(X_3, X_2) &\leqslant \operatorname{rank} H_{k-1}(X_2, X_1) + \operatorname{rank} H_{k+1}(X_4, X_3) \\ &+ \operatorname{rank} H_k(X_4, X_1). \end{aligned}$$

From minimax theory we will also need the following result.

Theorem 5.146 *(von Neumann Minimax Theorem)*
<u>*If*</u> *X and Y are two topological spaces and $\varphi\colon X \times Y \longrightarrow \mathbb{R}$ satisfies*

(i) *for all $y \in Y$, the map $x \longmapsto \varphi(x, y)$ is convex and lower semicontinuous;*

(ii) *there exists $y_0 \in Y$ such that the map $x \longmapsto \varphi(x, y_0)$ is* inf-*compact (that is, $\varphi(\cdot, y_0)$ has compact sublevel sets; see Definition 5.5);*

(iii) *for all $x \in X$, the map $y \longmapsto \varphi(x, y)$ is concave and upper semicontinuous;*

(iv) *there exits $x_0 \in X$ such that the map $y \longmapsto \varphi(x_0, y)$ is* sup-*compact (that is, $\varphi(x_0, \cdot)$ has compact superlevel sets),*

<u>*then*</u> *there exists a saddle point $(\widehat{x}, \widehat{y}) \in X \times Y$.*

5.2 Problems

Problem 5.1*
Let X^* be the dual of a Banach space furnished with the w^*-topology. Show that the function $\varphi\colon X^* \longrightarrow \overline{\mathbb{R}} = \mathbb{R} \cup \{+\infty\}$ is w^*-coercive (see Definition 5.5(a)) if and only if $\lim\limits_{\|u^*\|_* \to +\infty} \varphi(u^*) = +\infty$.

Problem 5.2**
Let X be a reflexive Banach space, and let $\xi\colon X \times X \longrightarrow \mathbb{R}$ be a function such that:
(i) for all $v \in X$, the map $u \longmapsto \xi(u, v)$ is continuous and convex on X;
(ii) for every bounded set $D \subseteq X$, the family $\{\xi(u, \cdot)\}_{u \in D}$ is weakly (sequentially) equicontinuous on X.

Show that the function $u \longmapsto \varphi(u) = \xi(u, u)$ is weakly (sequentially) lower semicontinuous.

Problem 5.3**
Let X be a reflexive Banach space and let $\xi\colon X \times X \longrightarrow \mathbb{R}$ be a function which satisfies conditions **(i)** and **(ii)** of Problem 5.2 for the weak topology. Suppose that $C \subseteq X$ is a nonempty, bounded and weakly closed set and set $\varphi(u) = \xi(u, u)$ for all $u \in X$. Show that there exists $\widehat{u} \in C$ such that $\varphi(\widehat{u}) = \inf\limits_C \varphi$.

Problem 5.4**
Let X be a reflexive Banach space and let $\xi\colon X \times X \longrightarrow \mathbb{R}$ be a function which satisfies conditions **(i)** and **(ii)** of Problem 5.2 for the weak topology. Let $\varphi(u) = \xi(u, u)$ for all $u \in X$ and assume that φ is weakly coercive. Show that there exists $\widehat{u} \in X$ such that $\varphi(\widehat{u}) = \inf\limits_X \varphi$.

Problem 5.5*
Suppose that X is a reflexive Banach space, $C \subseteq X$ is a nonempty, weakly closed set, and $\varphi\colon C \longrightarrow X$ is a weakly lower semicontinuous function. Show that φ realizes its infimum on C if and only if it has a bounded minimizing sequence.

Problem 5.6**
Suppose that X is a reflexive Banach space, $\varphi\colon X \longrightarrow \mathbb{R}$ is a weakly lower semicontinuous and weakly coercive function, $f\colon X \longrightarrow \mathbb{R}$ is a function which is weakly continuous, $c \in \mathbb{R}$ and $M = f^{-1}(c) = \{u \in X : f(u) = c\} \neq \emptyset$. Show that there exists $\widehat{u} \in M$ such that $\varphi(\widehat{u}) = \inf\limits_M \varphi$.

Problem 5.7**
Let $\Omega \subseteq \mathbb{R}^N$ be a bounded open set with a Lipschitz boundary and let $G\colon \mathbb{R}^N \longrightarrow \mathbb{R}$ be a convex map such that

$$|G(y)| \leqslant c_1(1 + |y|^p) \quad \text{and} \quad G(y) \geqslant c_2|y|^p - c_3 \qquad \forall\, y \in \mathbb{R}^N,$$

where $c_1, c_2, c_3 > 0$ and $1 < p < +\infty$. Given $f \in L^{p'}(\Omega)$ (where $\frac{1}{p} + \frac{1}{p'} = 1$), we consider the functional $\varphi \colon W_0^{1,p}(\Omega) \longrightarrow \mathbb{R}$ defined by

$$\varphi(u) = \int_\Omega G(Du(z))\,dz - \int_\Omega f(z)u(z)\,dz \qquad \forall\, u \in W_0^{1,p}(\Omega).$$

Show that there exists $\widehat{u} \in W_0^{1,p}(\Omega)$ such that $\varphi(\widehat{u}) = \inf\limits_{u \in W_0^{1,p}(\Omega)} \varphi(u)$.

Problem 5.8**

Let $\Omega \subseteq \mathbb{R}^N$ be a bounded open set with a Lipschitz boundary and let $G \in C^1(\mathbb{R}^N)$ be a convex function satisfying

$$|\nabla G(y)| \leqslant c_1(1 + |y|^{p-1}) \quad \text{and} \quad G(y) \geqslant c_2|y|^p - c_3 \qquad \forall\, y \in \mathbb{R}^N,$$

where $c_1, c_2, c_3 > 0$. Given $f \in L^{p'}(\Omega)$ (where $\frac{1}{p} + \frac{1}{p'} = 1$), show that there exists $\widehat{u} \in W_0^{1,p}(\Omega)$ such that

$$\int_\Omega (\nabla G(D\widehat{u}), Dh)_{\mathbb{R}^N}\,dz = \int_\Omega fh\,dz \qquad \forall\, h \in W_0^{1,p}(\Omega).$$

Problem 5.9*

Let $\Omega \subseteq \mathbb{R}^N$ be a bounded open set with a Lipschitz boundary and let $G \in C^1(\mathbb{R}^N)$ be a convex map satisfying the hypotheses of Problem 5.8. Suppose that for some $\widehat{u} \in W_0^{1,p}(\Omega)$, we have

$$\int_\Omega (\nabla G(D\widehat{u}), Dh)_{\mathbb{R}^N}\,dz = \int_\Omega fh\,dz \qquad \forall\, h \in W_0^{1,p}(\Omega).$$

Show that $\widehat{u}$ is a minimizer of the functional $\varphi \colon W_0^{1,p}(\Omega) \longrightarrow \mathbb{R}$ defined by

$$\varphi(u) = \int_\Omega G(Du(z))\,dz - \int_\Omega f(z)u(z)\,dz \qquad \forall\, u \in W_0^{1,p}(\Omega).$$

Problem 5.10**

Let $\Omega \subseteq \mathbb{R}^N$ be a nonempty open set, $1 < p < +\infty$ and let $\varphi \colon L^p(\Omega) \longrightarrow \overline{\mathbb{R}}$ be a function satisfying:

(i) for every $\lambda \in \mathbb{R}$, we have $\{\varphi \leqslant \lambda\} \subseteq W^{1,p}(\Omega)$ and $\|Du\|_p$ is bounded on $\{\varphi \leqslant \lambda\}$ (here $\{\varphi \leqslant \lambda\} = \{u \in L^p(\Omega) : \varphi(u) \leqslant \lambda\}$);

(ii) $\varphi|_{W^{1,p}(\Omega)}$ is sequentially weakly lower semicontinuous on $W^{1,p}(\Omega)$.

Show that φ is lower semicontinuous for the strong topology on $L^p(\Omega)$.

Problem 5.11*

Let $\Omega \subseteq \mathbb{R}^N$ be a nonempty, bounded and open set with a Lipschitz boundary, $1 \leqslant p < +\infty$ and let $\varphi \colon L^p(\Omega) \longrightarrow \overline{\mathbb{R}}$ be a function such that for every $\lambda \in \mathbb{R}$ the set $L_\lambda = \{u \in L^p(\Omega) : \varphi(u) \leqslant \lambda\} \subseteq W^{1,p}(\Omega)$ and it is bounded in $W^{1,p}(\Omega)$. Show that φ is coercive for the strong topology on $L^p(\Omega)$ (see Definition 5.5(a)).

Problem 5.12**
Suppose that $\Omega \subseteq \mathbb{R}^N$ (with $N \geqslant 3$) is a nonempty, bounded and open set with a Lipschitz boundary, $\xi \in L^\infty(\Omega)$ with $\xi(z) \geqslant 0$ for almost all $z \in \Omega$ and $f\colon \mathbb{R} \longrightarrow \mathbb{R}$ is a continuous function such that:
(i) $|f(x)| \leqslant c_1(1 + |x|^{p^*-1})$ for all $x \in \mathbb{R}$, with $c_1 > 0$;
(ii) $f(x)x \leqslant 0$ and $(f(x) - f(y))(x - y) \leqslant 0$ for all $x, y \in \mathbb{R}$.
Show that for every $g \in L^2(\Omega)$, the Dirichlet problem:

$$\begin{cases} -\Delta u(z) + \xi(z)u(z) = f(u(z)) + g(z) & \text{in } \Omega, \\ u|_{\partial\Omega} = 0, \end{cases}$$

has a unique weak solution, i.e., there exists unique $\widehat{u} \in H_0^1(\Omega)$ such that

$$\int_\Omega (D\widehat{u}, Dh)_{\mathbb{R}^N}\, dz + \int_\Omega \xi(z)\widehat{u}h\, dz = \int_\Omega f(\widehat{u})h\, dz + \int_\Omega gh\, dz \quad \forall\, h \in H_0^1(\Omega).$$

Problem 5.13*
Let X be a reflexive Banach space and let $\varphi \in C^2(X; \mathbb{R})$ be such that

$$\varphi''(u)(h, h) \geqslant 0 \qquad \forall\, u \in \overline{B}_r,\ h \in X,$$

where $\overline{B}_r = \{v \in X : \|v\|_X \leqslant r\}$. Show that there exists $\widehat{u} \in \overline{B}_r$ such that $\varphi(\widehat{u}) = \inf\limits_{u \in \overline{B}_r} \varphi(u)$.

Problem 5.14*
Suppose that X is a Banach space, $\varphi \in C^1(X)$ is such that $\varphi|_{\partial B_{t_0}}$ is bounded below (here $\partial B_{t_0} = \{u \in X : \|u\|_X = t_0\}$), $\vartheta\colon (0, +\infty) \longrightarrow \mathbb{R}$ is a continuous function such that

$$\int_{t_0}^{+\infty} \frac{\vartheta(s)}{s}\, ds = +\infty$$

for some $t_0 > 0$ and

$$\langle \varphi'(u), u \rangle \geqslant \vartheta(\|u\|_X) \qquad \forall\, u \in X.$$

Show that $\varphi(u) \longrightarrow +\infty$ as $\|u\|_X \longrightarrow +\infty$.

Problem 5.15*
Suppose that X is a reflexive Banach space, $\varphi\colon X \longrightarrow \mathbb{R}$ is a weakly lower semicontinuous function, $C \subseteq X$ is a nonempty, bounded, open set, $\partial_w C = \overline{C}^w \setminus C$ and

$$\varphi(v) \geqslant \varphi(\widetilde{u}) \qquad \forall\, v \in \partial_w C,$$

for some $\widetilde{u} \in C$. Show that $K_\varphi \cap C \neq \emptyset$

Problem 5.16*
Suppose that H is a Hilbert space, $M = \partial B_r = \{u \in H : \|u\|_H = 1\}$, $A \in \mathcal{L}(H)$ is a symmetric operator and $\varphi(u) = \frac{1}{2}(A(u), u)_H$ for all $u \in H$. Find the critical points of $\varphi|_M$.

Problem 5.17***
Suppose that $\Omega \subseteq \mathbb{R}^N$ (with $N \geqslant 3$) is a nonempty open set with a Lipschitz boundary and $2 < p < 2^* = \frac{2N}{N-2}$, $\varphi \colon H_0^1(\Omega) \longrightarrow \mathbb{R}$ is the C^1-functional defined by

$$\varphi(u) \;=\; \tfrac{1}{2}\|Du\|_2^2 - \tfrac{1}{p}\|u\|_p^p \qquad \forall\, u \in H_0^1(\Omega)$$

and $M = \big\{u \in H_0^1(\Omega) : \langle \varphi'(u), u\rangle = 0,\ u \neq 0\big\}$. Show that $M \neq \emptyset$ and it is a natural constraint for φ.

Remark. The set M is known as the ***Nehari manifold*** for φ and by the definition contains all the nontrivial critical points of φ. The next problem generalizes Problem 5.17.

Problem 5.18**
Suppose that X is a Banach space, $\varphi \in C^2(X;\mathbb{R})$, $M = \{u \in X : \langle \varphi'(u), u\rangle = 0,\ u \neq 0\}$, $M \neq \emptyset$,

$$\varphi''(u)(u,u) \;=\; \big\langle \varphi''(u)u, u\big\rangle \;\neq\; 0 \qquad \forall\, u \in M$$

and there exists $r > 0$ such that $B_r \cap M = \emptyset$ (where $B_r = \{u \in X : \|u\|_X < r\}$). Show that M is a C^1-Banach manifold of codimension 1 and that it is a natural constraint for φ.

Problem 5.19**
Let $\Omega \subset \mathbb{R}^N$ be a nonempty, bounded, and open set with a Lipschitz boundary $\partial\Omega$ and let $f \colon \mathbb{R} \longrightarrow \mathbb{R}$ be a C^1-function which is strictly increasing, $f(0) = 0$ and

$$0 \;<\; f'(x) \;\leqslant\; c_1 \qquad \forall\, x \in \mathbb{R},$$

for some $c_1 > 0$. We set $F(x) = \int\limits_0^x f(s)\,ds$ and consider the C^1-functional $\varphi \colon H^1(\Omega) \longrightarrow \mathbb{R}$ defined by

$$\varphi(u) \;=\; \tfrac{1}{2}\|Du\|_2^2 - \int\limits_\Omega F(u(z))\,dz \qquad \forall\, u \in H^1(\Omega).$$

Consider the set $M \subseteq H^1(\Omega)$ defined by

$$M \;=\; \Big\{u \in H^1(\Omega) : \int\limits_\Omega f(u)\,dz = 0\Big\}.$$

Show that M is a C^1-Banach manifold of codimension 1 and that M is a natural constraint for φ.

Problem 5.20**

Suppose that $\Omega \subseteq \mathbb{R}^N$ (with $N \geqslant 1$) is a nonempty, bounded, open set, $2 < p < 2^*$ with

$$2^* = \begin{cases} \frac{2N}{N-2} & \text{if } N \geqslant 3, \\ +\infty & \text{if } N = 1, 2, \end{cases}$$

$\lambda > 0$ and $\varphi_\lambda\colon H_0^1(\Omega) \longrightarrow \mathbb{R}$ is the C^2-functional defined by

$$\varphi_\lambda(u) = \tfrac{1}{2}\|Du\|_2^2 - \tfrac{\lambda}{2}\|u\|_2^2 + \tfrac{1}{p}\|u\|_p^p \qquad \forall\, u \in H_0^1(\Omega).$$

Define $M_\lambda = \{u \in H_0^1(\Omega) : \langle \varphi_\lambda'(u), u\rangle = 0,\ u \neq 0\}$ (the Nehari manifold for φ_λ; see the Remark after Problem 5.17).

Show that:

(a) $M_\lambda = \emptyset$ if and only if $\lambda \leqslant \widehat{\lambda}_1$ (see Remark 5.128);

(b) for all $\lambda > \widehat{\lambda}_1$, M_λ is a natural constraint for φ_λ.

Problem 5.21**

Suppose that $\Omega \subseteq \mathbb{R}^N$ (with $N \geqslant 1$) is a nonempty, bounded, open set, $2 < p < 2^*$ with

$$2^* = \begin{cases} \frac{2N}{N-2} & \text{if } N \geqslant 3, \\ +\infty & \text{if } N = 1, 2 \end{cases}$$

and $f \in C^1(\mathbb{R})$ satisfies

(i) $|f'(u)| \leqslant c(1 + |u|^{p-2})$ for all $u \in \mathbb{R}$ and some $c > 0$;

(ii) $\frac{f(u)}{u} < f'(u)$ for all $u \neq 0$.

Let $F(x) = \int\limits_0^x f(s)\,ds$ and consider the functional $\varphi\colon H^1(\Omega) \longrightarrow \mathbb{R}$ defined by

$$\varphi(u) = \tfrac{1}{p}\|u\|_p^p - \int\limits_\Omega F(u(z))\,dz \qquad \forall\, u \in H_0^1(\Omega).$$

We introduce the set

$$M = \Big\{u \in H_0^1(\Omega) : \|Du\|_2^2 = \int\limits_\Omega f(u)u\,dz,\ u \neq 0\Big\}.$$

Show that M is a natural constraint for the functional φ.

Problem 5.22**

Let X be a Banach space and let $\varphi\colon X \longrightarrow \mathbb{R}$ be a lower semicontinuous function which is bounded below, Gâteaux differentiable and

$$\varphi(u) \geqslant c_1\|u\|_X - c_2 \qquad \forall\, u \in X,$$

for some $c_1, c_2 > 0$. Show that the set $\varphi_G'(X)$ is dense in $c_1\overline{B}_1^*$ (here $\overline{B}_1^* = \{u^* \in X^* : \|u^*\|_* \leqslant 1\}$).

Problem 5.23****

Let X be a Banach space and let $\varphi\colon X \longrightarrow \mathbb{R}$ be a lower semicontinuous and Gâteaux differentiable function such that

$$\varphi(u) \geqslant \vartheta(\|u\|_X) \qquad \forall\, u \in X,$$

where $\vartheta\colon \mathbb{R}_+ \longrightarrow \mathbb{R}$ is a continuous functions satisfying $\frac{\vartheta(r)}{r} \longrightarrow +\infty$ as $r \to +\infty$. Show that the set $\varphi'_G(X)$ is dense in X^*.

Problem 5.24***

Suppose that X is a Banach space, $\varphi\colon X \longrightarrow \mathbb{R}$ is a lower semicontinuous, bounded below, and Gâteaux differentiable function, $\varepsilon > 0$, $y \in X$ are such that

$$\varphi(y) \leqslant \inf_X \varphi + \varepsilon,$$

$\lambda > 0$ and $\xi\colon \mathbb{R}_+ \longrightarrow \mathbb{R}_+$ is a continuous nondecreasing function such that

$$\int_0^{+\infty} \tfrac{1}{1+\xi(s)}\, ds = +\infty.$$

Show that there exists $u_\lambda \in X$ such that

$$\varphi(u_\lambda) \leqslant \varphi(y), \quad \|u_\lambda - y\|_X \leqslant \overline{r}, \quad \|\varphi'_G(u_\lambda)\|_* \leqslant \tfrac{\varepsilon}{\lambda(1+\xi(\|u_\lambda\|_X))},$$

where $\overline{r} > 0$ is such that $\int_{\|y\|_X}^{\|y\|_X+\overline{r}} \frac{1}{1+\xi(s)}\, ds \geqslant \lambda$.

Problem 5.25****

Suppose that X is a Banach space, $\varphi \in C^1(X)$ is bounded below and satisfies the C_m-condition (see Definition 5.47(a)), where $m = \inf_X \varphi$. Show that there exists $u_0 \in X$ such that $\varphi(u_0) = \inf_X \varphi$.

Problem 5.26***

Show that Theorem 4.66 (the Caristi fixed point theorem) and Corollary 5.38 (the Ekeland variational principle) are equivalent.

Problem 5.27****

Suppose that X is a Banach space, $\varphi \in C^1(X)$ satisfies the C-condition and $u_0 \in X$ is a local minimum point of φ. Show that given $r' \in (0, r]$, one of the following conditions holds:

(a) there exists $\widehat{r} \in (0, r')$ such that

$$\varphi(u_0) < \inf\{\varphi(u) :\ \|u - u_0\|_X = \widehat{r}\} = m_{\widehat{r}};$$

or

(b) for every $\varrho \in (0, r')$, φ has a local minimum at a point $u_\varrho \in X$ with

$$\|u_\varrho - u_0\|_X = \varrho \quad \text{and} \quad \varphi(u_0) = \varphi(u_\varrho).$$

Problem 5.28**

Suppose that X is a Banach space, $\varphi \in C^1(X)$ satisfies the C-condition and $u_0, u_1 \in X$ are two distinct local minimizers of φ. Show that φ has a third critical point.

Problem 5.29**

Let X be a complete metric space and let $\varphi\colon X \longrightarrow \mathbb{R}$ be a lower semicontinuous, bounded below function satisfying the following compactness type condition:

$\underline{\text{NC}_c^+}$: "If $\{u_n\}_{n\geqslant 1} \subseteq X$, $\{\varepsilon_n\}_{n\geqslant 1}, \{\delta_n\}_{n\geqslant 1} \subseteq (0, +\infty)$ are sequences such $\underline{\text{that}}$

$$\varphi(u_n) \longrightarrow c \quad \text{and} \quad \varphi(u_n) \leqslant \varphi(v) + \varepsilon_n d_X(v, u_n)$$
$$\forall\, v \in X,\ d_X(v, u_n) \leqslant \delta_n$$

then there exists a strongly convergent subsequence of $\{u_n\}_{n\geqslant 1}$."

Show that, if $c = \inf\limits_X \varphi$, then we can find $u_0 \in X$ such that $\varphi(u_0) = c$.

Problem 5.30**

Let X be a Banach space and let $\varphi\colon X \longrightarrow \mathbb{R}$ be a lower semicontinuous and bounded below function. Let $\widehat{m} = \inf\limits_X \varphi$ and assume that φ satisfies the nonsmooth $\text{C}_{\widehat{m}}$-condition. Show that there exists $u_0 \in X$ such that $\varphi(u_0) = \widehat{m}$.

Problem 5.31*

Let $\varphi \in C^1(\mathbb{R}^N)$ and assume that

$$|\varphi(u)| + |\varphi'(u)| \longrightarrow +\infty \quad \text{as } |u| \to +\infty.$$

Show that φ satisfies the PS-condition.

Problem 5.32*

Let X be a Banach space and let $\varphi \in C^1(X)$ satisfy

$$\varphi'(u) = A(u) + g(u) \qquad \forall\, u \in X,$$

where $A \in \mathcal{L}(X; X^*)$ is invertible and $g\colon X \longrightarrow X^*$ is compact. Suppose that every Palais–Smale sequence is bounded. Show that φ satisfies the PS-condition.

Problem 5.33**

Suppose that X is a Banach space, $C = \{u \in X : 0 \leqslant a \leqslant \|u\|_X \leqslant b\}$ and $\varphi\colon C \longrightarrow \mathbb{R}$ is a C^1-function which satisfies the following compactness type condition:

Every sequence $\{u_n\}_{n\geqslant 1} \subseteq X$ such that $a + \delta \leqslant \|u_n\|_X \leqslant b - \delta$ for some $\delta > 0$, $\{\,\varphi(u_n)\,\}_{n\geqslant 1} \subseteq \mathbb{R}$ is bounded, $\varphi'(u_n) \longrightarrow 0$ in X^*, has a strongly convergent subsequence.

Suppose that $K_\varphi \cap C = \emptyset$ and set

$$m(r) = \inf\{\varphi(u) : \|u\|_X = r\} \qquad \forall\, r \in [a,b].$$

Show that, if $a < r_1 < r < r_2 < b$, then $m(r) > \min\{m(r_1), m(r_2)\}$.

Problem 5.34**
Suppose that X is a Banach space, $\varphi\colon X \longrightarrow \overline{\mathbb{R}}$ is a lower semicontinuous, convex, and bounded below function, $u_0 \in \operatorname{dom}\varphi$, $\varepsilon > 0$. Show that there exist $u_\varepsilon \in \operatorname{dom}\varphi$ and $u_\varepsilon^* \in \partial\varphi(u_\varepsilon)$ such that

$$\|u_\varepsilon - u_0\|_X \leqslant \tfrac{1}{\varepsilon}\big(\varphi(u_0) - \varphi(u_\varepsilon)\big) \quad \text{and} \quad \|u_\varepsilon^*\|_* \leqslant \varepsilon.$$

Problem 5.35***
Show that if the Ekeland variational principle in the form of Corollary 5.38 holds on a metric space (X, d_X), then (X, d_X) is complete.

Problem 5.36**
Suppose that X is a Banach space, $\varphi \in C^1(X)$ and $c \in \mathbb{R}$ is such that φ^λ in unbounded for $\lambda > c$ and bounded for $\lambda < c$. Show that there exists a sequence $\{u_n\}_{n\geqslant 1} \subseteq X$ such that

$$\varphi(u_n) \longrightarrow c, \quad (1 + \|u_n\|_X)\varphi'(u_n) \longrightarrow 0 \quad \text{in } X^*$$
$$\text{and} \quad \|u_n\|_X \longrightarrow +\infty.$$

Problem 5.37**
Let X be a Banach space and let $\varphi \in C^1(X)$ be a bounded below function satisfying the C-condition. Show that $\|u\|_X \longrightarrow +\infty$ implies $\varphi(u) \longrightarrow +\infty$.

Problem 5.38**
Let X be a Banach space and let $\varphi \in C^1(X)$ be a function bounded below. Show that φ satisfies the PS-condition if and only if it satisfies the C-condition.

Problem 5.39**
Let H be a Hilbert space and let $\varphi \in C^2(H;\mathbb{R})$ be a bounded below function satisfying the CPS-condition (see Remark 5.46). Given $u \in H$, let $\sigma_u(t)$ (for $t \geqslant 0$) be the negative gradient flow of φ starting from $u \in H$, i.e., the solution of the Cauchy problem

$$\begin{cases} \sigma'(t) = -\nabla\varphi(\sigma(t)) & t \geqslant 0, \\ \sigma(0) = u, \end{cases}$$

where $\nabla\varphi(y)$ is the gradient of φ at $y \in H$, i.e.,

$$\langle \varphi'(y), h\rangle = (\nabla\varphi(y), h)_H \qquad \forall\, h \in H,$$

with $\langle\cdot,\cdot\rangle$ being the duality brackets for the pair (H^*, H) and $(\cdot,\cdot)_H$ being the inner product of H. Show that the flow σ is global (i.e., $\sigma_u(t)$ exists for all $t \geqslant 0$) and $c = \lim\limits_{t\to+\infty} \varphi(\sigma_u(t))$ is a critical value of φ.

Problem 5.40**

Suppose that X is a Banach space, $\xi = \varphi + \psi$ with $\varphi \in C^1(X)$, $\psi \in \Gamma_0(X)$ (see Remark 3.56) and ξ is bounded below and satisfies the SC_m-condition with $m = \inf\limits_X \xi$ (see Definition 5.47). Show that there exists $\widehat{u} \in X$ such that $m = \xi(\widehat{u})$ and $-\varphi'(\widehat{u}) \in \partial\psi(\widehat{u})$.

Problem 5.41**

Let X be a reflexive Banach space and let $\varphi\colon X \longrightarrow \mathbb{R}$ be a Gâteaux differentiable, lower semicontinuous and convex function such that $\varphi(u) \longrightarrow +\infty$ when $\|u\|_X \to +\infty$. Show that φ satisfies the following condition:

WC: "If $\{u_n\}_{n\geqslant 1} \subseteq X$ is a sequence such that:

(i) the sequence $\{\varphi(u_n)\}_{n\geqslant 1} \subseteq \mathbb{R}$ is bounded,

(ii) $\varphi'_G(u_n) \neq 0$ for all $n \geqslant 1$,

(iii) $\varphi'_G(u_n) \longrightarrow 0$ in X^*,

then there exists $\overline{u} \in X$ such that

$$\liminf_{n\to+\infty} \varphi(u_n) \leqslant \varphi(\overline{u}) \leqslant \limsup_{n\to+\infty} \varphi(u_n), \quad \text{and} \quad \varphi'_G(\overline{u}) = 0.\text{"}$$

Problem 5.42***

Suppose that X is reflexive Banach space, $A \in \mathcal{L}_c(X; X^*)$ (see Definition 2.1), $A = A^*$ and $\psi \in C^1(X)$ is a convex functions satisfying:

(i) $\psi(u) \geqslant c_1\|u\|_X - c_2$ for all $u \in X$, with $c_1, c_2 > 0$;

(ii) $\langle \psi'(u), u\rangle \leqslant c_3\|u\|_X + c_4$ for all $u \in X$, with $c_3 \in [0, 2)$, $c_4 > 0$.

Let $\varphi(u) = \frac{1}{2}\langle A(u), u\rangle + \psi(u)$ for all $u \in X$. Show that φ satisfies the WC-condition from Problem 5.41.

Problem 5.43***

Suppose that X is a Banach space, $X = Y \oplus V$ with $\dim Y < +\infty$, $\varphi\colon X \longrightarrow \mathbb{R}$ is a continuous, Gâteaux differentiable function satisfying the WC-condition (see Problem 5.41), $\varphi'_G\colon X \longrightarrow X^*$ is norm-to-weak* continuous and there exists $r > 0$ such that:

(i) $\varphi(y, 0) < 0$ for all $y \in Y \cap \partial B_r$ (where $\partial B_r = \{u \in X : \|u\|_X = r\}$);

(ii) $\varphi(0, v) \geqslant 0$ for all $v \in V$.

Show that there exists $\widehat{u} \in X$ such that $\varphi'_G(\widehat{u}) = 0$ and $\varphi(\widehat{u}) \geqslant 0$.

Problem 5.44**

Suppose that X is a Banach space, $\varphi \in C^1(X)$ satisfies the C-condition and there exists $\eta \in \mathbb{R}$ such that the set $\varphi^{-1}(\eta)$ is bounded. Show that $|\varphi|(u) = |\varphi(u)| \longrightarrow +\infty$ as $\|u\|_X \longrightarrow +\infty$.

Problem 5.45**
Suppose that X is a Banach space, $\varphi \in C^1(X)$ is a bounded below function which satisfies the C_m-condition for $m = \inf_X \varphi$. Show that $\varphi^{m+\lambda}$ is bounded for some $\lambda > 0$.

Problem 5.46*
Suppose that X is a Banach space, $\varphi \in C^1(X)$ is a bounded below function satisfying the PS_m-condition where $m = \inf_X \varphi$. Show that every minimizing sequence φ admits a convergent subsequence whose limit is a global minimizer of φ.

Problem 5.47**
Suppose that X is a Banach space, X^* is its topological dual, $\varphi\colon X^* \longrightarrow \overline{\mathbb{R}}$ is w^*-lower semicontinuous function which is bounded below. Show that there exists $\widetilde{u}^* \in X^*$ such that if a net $\{u_\alpha^*\}_{\alpha\in J} \subseteq X^*$ satisfies

$$\varphi(u_\alpha^*) + \|\widetilde{u}^* - u_\alpha^*\|_* \longrightarrow \varphi(\widetilde{u}^*),$$

then $u_\alpha^* \xrightarrow{w^*} \widetilde{u}^*$ in X^*.

Problem 5.48**
Let X be a reflexive Banach space and let $\varphi\colon X \longrightarrow \overline{\mathbb{R}}$ be a lower semicontinuous, convex, and bounded below function. Show that there exists $\widetilde{u} \in X$ such that if a sequence $\{u_n\}_{n\geqslant 1} \subseteq X$ satisfies

$$\varphi(u_n) + \|\widetilde{u} - u_n\|_X \longrightarrow \varphi(\widetilde{u}),$$

then $u_n \longrightarrow \widetilde{u}$ in X.

Problem 5.49*
Suppose that X is a Banach space and $\xi = \varphi+\psi$ is a function such that $\varphi\colon X \longrightarrow \mathbb{R}$ is locally Lipschitz and $\psi \in \Gamma_0(X)$ (see Remark 3.56). We say that $\widehat{u} \in X$ is a ***critical point*** of ξ if

$$0 \leqslant \varphi^0(\widehat{u}; h) + \psi(\widehat{u}+h) - \psi(\widehat{u}) \qquad \forall\, h \in X.$$

Suppose that $\{u_n\}_{n\geqslant 1} \subseteq X$ is a sequence such that

$$u_n \longrightarrow \widehat{u} \ \text{ in } X, \quad \varphi(u_n) \longrightarrow c$$

and

$$-\varepsilon_n\|u - u_n\| \leqslant \varphi^0(u_n; u - u_n) + \psi(u) - \psi(u_n) \qquad \forall\, u \in X,$$

with $\varepsilon_n \searrow 0$. Show that $\widehat{u} \in X$ is a critical point of ξ with critical value c.

Problem 5.50*
Suppose that X is a Banach space, $\xi = \varphi + \psi$ is a functions such that $\varphi\colon X \longrightarrow \mathbb{R}$ is locally Lipschitz and $\psi \in \Gamma_0(X)$ (see Remark 3.56). Suppose that ξ is bounded below and satisfies the following compactness-type condition:

GNPS$_m$: "Every sequence $\{u_n\}_{n\geqslant 1} \subseteq X$ such that

$$\xi(u_n) \longrightarrow m = \inf_X \xi$$

and

$$-\varepsilon_n\|u - u_n\|_X \leqslant \varphi^0(u_n; u - n_n) + \psi(u) - \psi(u_n) \qquad \forall\, u \in X,$$

with $\varepsilon \searrow 0$, admits a strongly convergent subsequence."
Show that ξ has a critical point $\widehat{u} \in X$ (in the sense of Problem 5.49) and $\xi(\widehat{u}) = m$.

Problem 5.51***
Suppose that X is a Banach space, $\varphi\colon X \longrightarrow \mathbb{R}$ is a locally Lipschitz function, E is a compact metric space, $E_0 \subseteq E$ is a nonempty closed subset and $\gamma_0\colon E_0 \longrightarrow X$ is a continuous map. Define

$$\Gamma = \left\{\gamma \in C(E;X) : \gamma|_{E_0} = \gamma_0\right\} \quad \text{and} \quad c = \inf_{\gamma\in\Gamma}\max_{u\in E}\varphi(\gamma(u)).$$

Suppose that $c > \max_{u\in E_0}\varphi(\gamma_0(u))$. Show that there exists a sequence $\{v_n\}_{n\geqslant 1} \subseteq X$ such that

$$\varphi(v_n) \longrightarrow c \quad \text{and} \quad m(v_n) \longrightarrow 0$$

(see Definition 5.47(a)).

Problem 5.52**
Let X be a Banach space and let $\varphi\colon X \longrightarrow \mathbb{R}$ be a locally Lipschitz function such that $\varphi(0) = 0$ and there exists $u_1 \in X \setminus \{0\}$ such that $\varphi(u_1) \leqslant 0$. Define

$$\Gamma = \{\gamma \in C([0,1];X) : \gamma(0) = 0,\ \gamma(1) = u_1\}$$

$$\text{and} \quad c = \inf_{\gamma\in\Gamma}\max_{0\leqslant t\leqslant 1}\varphi(\gamma(t)).$$

Suppose that $c > 0$ and φ satisfies the nonsmooth PS$_c$-condition. Show that c is a critical value of φ.

Problem 5.53**
Suppose that X is a Banach space, $\varphi\colon X \longrightarrow \mathbb{R}$ is a Gâteaux differentiable function such that $\varphi'_G\colon X \longrightarrow X^*$ is continuous from X with norm topology into X^* with the w^*-topology, E is a compact metric space, $E_0 \subseteq E$ is a nonempty closed subset and $\gamma_0\colon E_0 \longrightarrow X$ is a continuous map. Define

$$\Gamma = \left\{\gamma \in C(E;X) : \gamma|_{E_0} = \gamma_0\right\} \quad \text{and} \quad c = \inf_{\gamma\in\Gamma}\max_{u\in E}\varphi(\gamma(u)).$$

Suppose that $c > \max_{u\in E_0}\varphi(\gamma_0(u))$. Show that there exists a sequence $\{u_n\}_{n\geqslant 1} \subseteq X$ such that

$$\varphi(u_n) \longrightarrow c \quad \text{and} \quad \|\varphi'(u_n)\|_* \longrightarrow 0$$

and if φ satisfies the PS$_c$-condition, then we can find $\widehat{u} \in X$ such that

$$\varphi(\widehat{u}) = c \quad \text{and} \quad \varphi'(\widehat{u}) = 0.$$

Problem 5.54**

Let X be a Banach space and let $\varphi\colon X \longrightarrow \mathbb{R}$ be a locally Lipschitz function which is bounded below. Show that there exists a sequence $\{u_n\}_{n\geqslant 1} \subseteq X$ such that

$$\varphi(u_n) \longrightarrow \inf_X \varphi \quad \text{and} \quad (1+\|u\|_X)m(u_n) \longrightarrow 0$$

(see Definition 5.47(a)).

Problem 5.55**

Let X be a Banach space and let $\varphi\colon X \longrightarrow \mathbb{R}$ be a locally Lipschitz function. We consider the following two compactness type conditions on φ:

<u>NC_c</u>: "If $\{u_n\}_{n\geqslant 1} \subseteq X$, $\{\varepsilon_n\}_{n\geqslant 1}, \{\delta_n\}_{n\geqslant 1} \subseteq (0,+\infty)$ are sequences such that

$$\varepsilon_n \longrightarrow 0, \quad \delta_b \longrightarrow 0, \quad \varphi(u_n) \longrightarrow c$$

and

$$|\varphi(u_n) - \varphi(u)| \leqslant \varepsilon\|u_n - u\|_X \qquad \forall\, u \in X,\ n \geqslant 1,\ \|u_n - u\|_X \leqslant \delta_n,$$

then there exists a strongly convergent subsequence of $\{u_n\}_{n\geqslant 1}$"
(see Problem 5.29 for a unilateral version of this condition) and

<u>NPS_c</u>: "If $\{u_n\}_{n\geqslant 1} \subseteq X$ is a sequence such that

$$\varphi(u_n) \longrightarrow c \quad \text{and} \quad m(u_n) \longrightarrow 0$$

(see Definition 5.47(a)), then there exists a strongly convergent subsequence of $\{u_n\}_{n\geqslant 1}$"
(see Definition 5.47(a) for the nonsmooth Cerami analog of this notion). Show that both above condition are equivalent.

Problem 5.56**

Suppose that H is a Hilbert space, Y is a reflexive Banach space, $H \hookrightarrow Y$ compactly and densely and $\psi\colon Y \longrightarrow \mathbb{R}$ is a locally Lipschitz function. Let $\widehat{\psi} = \psi|_H$ and define

$$\varphi(u) = \tfrac{1}{2}\|u\|_H^2 - \widehat{\psi}(u) \qquad \forall\, u \in H.$$

Suppose that φ satisfies the following compactness type condition:

<u>WNPS_c</u>: "If $\{u_n\}_{n\geqslant 1} \subseteq X$ is a sequence such that

$$\varphi(u_n) \longrightarrow c \quad \text{and} \quad m(u_n) \longrightarrow 0$$

(see Definition 5.47(a)), then there exists a weakly convergent subsequence of $\{u_n\}_{n\geqslant 1}$."

Show that φ satisfies the NPS_c-condition (see Problem 5.55).

Problem 5.57***

Let X ba a Banach space and let $\varphi \in \Gamma_0(X)$ (see Remark 3.56). Show that given any $\widehat{u} \in \operatorname{dom}\varphi$, we can find a sequence $\{u_n\}_{n\geqslant 1} \subseteq X$ such that

$$\varphi(u_n) \longrightarrow \varphi(\widehat{u}), \quad \|u_n - \widehat{u}\|_X \leqslant \tfrac{1}{n} \quad \text{and} \quad \partial\varphi(u_n) \neq 0 \qquad \forall\, n \geqslant 1.$$

Remark. In particular this problem shows that $D(\partial\varphi)$ is dense in $\operatorname{dom}\varphi$ for $\varphi \in \Gamma_0(X)$.

Problem 5.58**
In the definition of the PS_c-condition, we have two requirements, namely

$$\varphi(u_n) \longrightarrow c \quad \text{and} \quad \varphi'(u_n) \longrightarrow 0 \quad \text{in } X^*$$

(see Definition 5.45(a)). Show that they are independent.

Problem 5.59**
Let $\Omega \subseteq \mathbb{R}^N$ be a nonempty, bounded, open set with a Lipschitz boundary and let $\varphi_\lambda\colon H_0^1(\Omega) \longrightarrow \mathbb{R}$ be the C^1-functional defined by

$$\varphi_\lambda(u) \;=\; \tfrac{1}{2}\|Du\|_2^2 - \tfrac{\lambda}{2}\|u\|_2^2 \qquad \forall\, u \in H_0^1(\Omega).$$

Show that φ satisfies the PS-condition if and only is $\lambda \neq \widehat{\lambda}_n$ for all $n \geqslant 1$ (here $\{\widehat{\lambda}_n\}_{n\geqslant 1}$ are the eigenvalues of $(-\Delta, H_0^1(\Omega))$).

Problem 5.60*
Suppose that H is a Hilbert space, $\varphi \in C^1(H)$ is a bounded below function, $g \in C^1(H)$, $M = g^{-1}(0)$, $g'(u) \neq 0$ for all $u \in M$, $m = \inf\limits_M \varphi > -\infty$ and assume that the following condition holds:

$\underline{\mathrm{PS}_m^M}$: "Every sequence $\{u_n\}_{n\geqslant 1} \subseteq M$ such that

$$\varphi(u_n) \longrightarrow m \quad \text{and} \quad \varphi_M'(u_n) \longrightarrow 0$$

admits a strongly convergent subsequent."

Show that there exists $\widehat{u} \in M$ and $\lambda \in \mathbb{R}$ such that

$$\varphi(\widehat{u}) \;=\; m \quad \text{and} \quad \varphi'(\widehat{u}) \;=\; \lambda g'(\widehat{u}).$$

Problem 5.61**
Let $\Omega \subseteq \mathbb{R}^N$ be a nonempty, bounded, open set with a Lipschitz boundary and let $p \in (2, p^*)$, where $p^* = \frac{2N}{N-2}$ if $N \geqslant 3$ and $p^* = +\infty$ is $N \in \{1,2\}$. Consider the C^1-functional $\varphi\colon H_0^1(\Omega) \longrightarrow \mathbb{R}$ defined by

$$\varphi(u) \;=\; \tfrac{1}{2}\|Du\|_2^2 - \tfrac{1}{p}\|u\|_p^p \qquad \forall\, u \in H_0^1(\Omega).$$

Show that the functional φ satisfies the C-condition.

Problem 5.62*
Let X be a Banach space and $\xi = \varphi + \psi$ is a function such that $\varphi\colon X \longrightarrow \mathbb{R}$ is locally Lipschitz and $\psi \in \Gamma_0(X)$ (see Remark 3.56). Show that if $\widehat{u} \in X$ is a local minimizer of ξ, then $\widehat{u}$ is a critical point of ξ (see Problem 5.49).

Problem 5.63**
Suppose that X and Y are two Banach spaces with X being reflexive, $X \hookrightarrow Y$ compactly and densely, $\widehat{\varphi}\colon Y \longrightarrow \mathbb{R}$ is a locally Lipschitz function, $\psi\colon X \longrightarrow \mathbb{R}$ is a convex and Gâteaux differentiable function, $\varphi = \widehat{\varphi}|_X$, $\xi = \varphi + \psi$ and $\widehat{u} \in X$ is a critical point of ξ (see Problem 5.49). Show that $-\psi'_G(\widehat{u}) \in \partial\widehat{\varphi}(\widehat{u})$.

Problem 5.64**
Suppose that X and Y are two Banach spaces with X being reflexive, $X \hookrightarrow Y$ compactly and densely, $\widehat{\varphi}\colon Y \longrightarrow \mathbb{R}$ is a locally Lipschitz functions, $\psi\colon X \longrightarrow \mathbb{R}$ is a convex and Gâteaux differentiable function, $\varphi = \widehat{\varphi}|_X$ and $\xi = \varphi + \psi$ is such that

$$\xi(u) \longrightarrow +\infty \quad \text{as } \|u\|_X \to +\infty.$$

Show that there exists $\widehat{u} \in X$ such that

$$\xi(\widehat{u}) = \inf_X \xi \quad \text{and} \quad -\psi'_G(\widehat{u}) \in \partial\widehat{\varphi}(\widehat{u}).$$

Problem 5.65*
Let X be a reflexive Banach space and let $\varphi \in C^1(X)$ satisfy

$$\varphi(u) \longrightarrow +\infty \quad \text{as } \|u\|_X \to +\infty.$$

Suppose that $\varphi' = A + K$ with $A\colon X \longrightarrow X^*$ of type $(S)_+$ (see Definition 2.137) and $K\colon X \longrightarrow X^*$ being completely continuous (see Definition 2.1). Show that φ satisfies the C-condition.

Problem 5.66**
Let $\Omega \subseteq \mathbb{R}^N$ be a nonempty, bounded, open set with a Lipschitz boundary, $1 < p < +\infty$ and let $f\colon \Omega \times \mathbb{R} \longrightarrow \mathbb{R}$ be a Carathéodory function such that

$$|f(z,x)| \leqslant a(z)(1+|x|^{q-1}) \quad \text{for a.a. } z \in \Omega, \text{ all } x \in \mathbb{R},$$

with $a \in L^\infty(\Omega)_+$, $1 < q < p$. We set $F(z,x) = \int\limits_0^x f(z,s)\,ds$ and consider the C^1-functional defined by

$$\varphi(u) = \tfrac{1}{p}\|Du\|_p^p - \int_\Omega F(z,u(z))\,dz \qquad \forall\, u \in W_0^{1,p}(\Omega).$$

Show that φ satisfies the C-condition and the PS-condition.

Problem 5.67**
Let $\Omega \subseteq \mathbb{R}^N$ be a nonempty, bounded, open set with a C^2-boundary and let $\vartheta \in L^1(\Omega)$ be such that $\vartheta(z) \leqslant \widehat{\lambda}_1$ for almost all $z \in \Omega$ with strict inequality on a set of positive Lebesgue measure (here $\widehat{\lambda}_1 > 0$ is the first eigenvalue of $(-\Delta_p, W_0^{1,p}(\Omega))$). Show that there exists $c_0 > 0$ such that

$$\|Du\|_p^p - \int_\Omega \vartheta(z)|u(z)|^p\,dz \geqslant c_0\|u\|^p \qquad \forall\, u \in W_0^{1,p}(\Omega).$$

Problem 5.68***

Let $\Omega \subseteq \mathbb{R}^N$ be a nonempty, bounded, open set with a C^2-boundary, $1 < p < +\infty$ and let $f\colon \Omega \times \mathbb{R} \longrightarrow \mathbb{R}$ be a Carathéodory function such that:

(i) $|f(z,x)| \leqslant a(z)(1+|x|^{p-1})$ for almost all $z \in \Omega$, all $x \in \mathbb{R}$ with some $a \in L^\infty(\Omega)_+$;

(ii) $\limsup\limits_{x \to \pm\infty} \frac{f(z,x)}{|x|^{p-2}x} \leqslant \widehat{\lambda}_1$ uniformly for almost all $z \in \Omega$ (here $\widehat{\lambda}_1 > 0$ is the first eigenvalue of $(-\Delta_p, W_0^{1,p}(\Omega))$);

(iii) if $F(z,x) = \int\limits_0^x f(z,s)\,ds$, then

$$\lim_{x \to \pm\infty} \big(f(z,x)x - pF(z,x)\big) \;=\; +\infty$$

uniformly for almost all $z \in \Omega$.

Consider the C^1-functional $\varphi\colon W_0^{1,p}(\Omega) \longrightarrow \mathbb{R}$ defined by

$$\varphi(u) \;=\; \tfrac{1}{p}\|Du\|_p^p - \int\limits_\Omega F(z,u(z))\,dz \qquad \forall\, u \in W_0^{1,p}(\Omega).$$

Show that φ satisfies the C-condition and the PS-condition.

Problem 5.69**

Suppose that X is a Banach space, $\xi = \varphi + \psi$ with $\varphi\colon X \longrightarrow \mathbb{R}$ locally Lipschitz and $\psi \in \Gamma_0(X)$ (see Remark 3.56) and consider the following refinement of Definition 5.47(a):

<u>SC_c</u>: "Every sequence $\{u_n\}_{n\geqslant 1} \subseteq X$ such that $\xi(u_n) \longrightarrow c$ and

$$(1+\|u_n\|_X)\big(\varphi^0(u_n;h) + \psi(u_n+h) - \psi(u_n)\big) \;\geqslant\; -\varepsilon_n\|h\|_X$$

for every $n \geqslant 1$ with $\varepsilon_n \searrow 0$, admits a strongly convergent subsequence."

Show that when $\psi \equiv 0$, then the SC_c-condition coincides with the nonsmooth C_c-condition.

Problem 5.70*

Suppose that X is a Banach space, $\varphi \in C^1(X)$, c is not a critical value of φ and φ satisfies the C_c-condition. Show that there exists $\delta > 0$ such that

$$\text{if } |\varphi(u) - c| \;\leqslant\; \delta, \text{ then } (1+\|u\|_X)\|\varphi'(u)\|_* \;\geqslant\; \delta.$$

Problem 5.71*

Let X be a finite dimensional Banach space and let $\varphi \in C^1(X)$. Show that φ is bounded below and satisfies the C-condition if and only if

$$\varphi(u) \;\longrightarrow\; +\infty \quad \text{as } \|u\|_X \;\longrightarrow\; +\infty.$$

Problem 5.72**
Let X be a Banach space and let $\varphi \in C^1(X)$ be a bounded below function such that

$$\varphi(u) \longrightarrow +\infty \quad \text{as } \|u\|_X \longrightarrow +\infty.$$

Is it true that φ satisfies the C-condition? Justify your answer.

Problem 5.73*
Suppose that X is a Banach space and let $\varphi \in C^1(X)$ be a function satisfying the C_c-condition for some $c \in \mathbb{R}$. Show that the set K_φ^c is compact.

Problem 5.74*
Find a nonconstant function which satisfies the CPS-condition (see Remark 5.46) but not the PS-condition (see Definition 5.45(a)).

Problem 5.75*
Find a C^1-function φ which is bounded below, $c = \inf\limits_X \varphi$, but φ does not satisfy the PS_c-condition.

Problem 5.76**
Let H be a Hilbert space with inner product $(\cdot,\cdot)_H$, $h \in H$ and let $\varphi \in C^1(H)$ be defined by

$$\varphi(u) = \tfrac{1}{2}\|u\|^2 - \tfrac{1}{2}(u,h)_H^2 \qquad \forall\, u \in H.$$

Show that φ satisfies the PS-condition if and only if $\|h\|_H \neq 1$

Problem 5.77***
Let $\Omega \subseteq \mathbb{R}^N$ be a nonempty bounded open set with a Lipschitz boundary and let $f\colon \Omega \times \mathbb{R} \longrightarrow \mathbb{R}$ be a Carathéodory function such that:

(i) $|f(z,x)| \leqslant a(z)$ for almost all $z \in \Omega$, all $x \in \mathbb{R}$, with $a \in L^\infty(\Omega)_+$;

(ii) there exist $f_+, f_- \in L^\infty(\Omega)$ such that

$$f_\pm(z) = \lim_{x \to \pm\infty} f(z,x) \quad \text{for a.a. } z \in \Omega;$$

(iii) we have

$$\int_\Omega f_+(z)\widehat{u}_1\,dz < 0 < \int_\Omega f_-(z)\widehat{u}_1\,dz$$

$$\text{or} \quad \int_\Omega f_-(z)\widehat{u}_1\,dz < 0 < \int_\Omega f_+(z)\widehat{u}_1\,dz,$$

where $\widehat{u}_1$ is the L^p-normalized, positive eigenfunction corresponding to the first eigenvalue $\widehat{\lambda}_1 > 0$ of $(-\Delta, H_0^1(\Omega))$.

Let $F(z,x) = \int\limits_0^x f(z,s)\,ds$ and consider $\varphi \in C^1(H_0^1(\Omega);\mathbb{R})$ defined by

$$\varphi(u) = \tfrac{1}{2}\|Du\|_2^2 - \tfrac{\widehat{\lambda}_1}{2}\|u\|_2^2 - \int\limits_\Omega F(z,u(z))\,dz \qquad \forall\, h \in H_0^1(\Omega).$$

Show that φ satisfies the PS-condition.

Problem 5.78*

Let $\Omega \subseteq \mathbb{R}^N$ be a nonempty, bounded, open set with a Lipschitz boundary and let $f\colon \Omega \times \mathbb{R} \longrightarrow \mathbb{R}$ be a Carathéodory function such that

$$|f(z,x)| \leqslant a(z) \quad \text{for a.a. } z \in \Omega, \text{ all } x \in \mathbb{R},$$

with $a \in L^\infty(\Omega)_+$. Suppose that there exist functions $f_+, f_- \in L^\infty(\Omega)$ such that

$$f_\pm(z) = \lim_{x \to \pm\infty} f(z,x) \quad \text{for a.a. } z \in \Omega.$$

Assume that

$$f_-(z) \leqslant f(z,x) \leqslant f_+(z) \quad \text{for a.a. } z \in \Omega, \text{ all } x \in \mathbb{R},$$

with strict inequalities on sets of positive Lebesgue measure. Let $u_0 \in H_0^1(\Omega)$ be a weak solution of the following Dirichlet boundary value problem:

$$\begin{cases} -\Delta u(z) = \widehat{\lambda}_1 u(z) + f(z,u(z)) \text{ in } \Omega, \\ u|_{\partial\Omega} = 0. \end{cases}$$

Show that

$$\int\limits_\Omega f_-(z)\widehat{u}_1\,dz < 0 < \int\limits_\Omega f_+(z)\widehat{u}_1\,dz.$$

Problem 5.79***

Suppose that X is a Banach space,

$$\begin{aligned}\mathcal{Y} = \Big\{g\colon X \longrightarrow \mathbb{R} :\ & g(u) = \tfrac{1}{2}\sum_{n \geqslant 1} \lambda_n \|u_n - u\|_X^2, \text{ with} \\ & \{u_n\}_{n\geqslant 1} \subseteq X \text{ convergent}, \ \lambda_n \geqslant 0, \ \sum_{n\geqslant 1}\lambda_n = 1\Big\},\end{aligned}$$

$\varphi\colon X \longrightarrow \mathbb{R}$ is a lower semicontinuous function which is bounded below, $\varepsilon > 0$ and $u_0 \in X$ are such that

$$\varphi(u_0) \leqslant \inf_X \varphi + \varepsilon.$$

Show that there exist $u_\varepsilon \in X$ and $g \in \mathcal{Y}$ such that

$$\varphi(u_\varepsilon) \leqslant \inf_X \varphi + \varepsilon, \quad \|u_\varepsilon - u_0\|_X \leqslant 1$$

and

$$\varphi(u_\varepsilon) + 2\varepsilon(g(u_\varepsilon) - g(u)) < \varphi(u) \qquad \forall\, u \neq u_\varepsilon.$$

Problem 5.80***

Suppose that Problem 5.79 holds in the Banach space X with

$$\mathcal{Y} = \{g\colon X \longrightarrow \mathbb{R} : g(u) = \tfrac{1}{2}\|u-h\|_X^2 \text{ for all } u \in X, \text{ with } h \in X\}.$$

Show that X is reflexive.

Problem 5.81*

Let H be a Hilbert space and let $\varphi \in C^2(H;\mathbb{R})$ be a bounded below function. Show that there exists a sequence $\{u_n\}_{n\geqslant 1} \subseteq H$ such that

$$\varphi(u_n) \longrightarrow \inf_H \varphi, \quad \varphi'(u_n) \longrightarrow 0 \quad \text{in } H^*$$

and

$$\langle \varphi''(u_n)y, y\rangle \geqslant 0 \qquad \forall\, y \in H.$$

Problem 5.82**

Let X be a Banach space, $\varphi \in C^1(X)$ and $0 < \lambda < \mu$. Show that there exists a locally Lipschitz map $\widehat{g}\colon X \setminus K_\varphi \longrightarrow X$ such that

$$\|\widehat{g}(u)\|_X \leqslant \mu\|\varphi'(u)\|_*$$

and

$$\lambda\|\varphi'(u)\|_*^2 \leqslant \langle \varphi'(u), \widehat{g}(u)\rangle \qquad \forall\, u \in X \setminus K_\varphi.$$

Problem 5.83**

Let X be a Banach space, $\varphi \in C^1(X)$ and $0 < \lambda < \mu$. Show that there exist locally Lipschitz maps $\widehat{g}, \widetilde{g}\colon X \setminus K_\varphi \longrightarrow X$ such that:

(a) $\|\widehat{g}(u)\|_X \leqslant \mu$ and $\lambda\|\varphi'(u)\|_* \leqslant \langle \varphi'(u), \widehat{g}(u)\rangle$ for all $u \in X \setminus K_\varphi$;

(b) $\lambda \leqslant \langle \varphi'(u), \widetilde{g}(u)\rangle \leqslant \|\varphi'(u)\|_*\|\widetilde{g}(u)\|_X \leqslant \mu$ for all $u \in X \setminus K_\varphi$.

Problem 5.84***

Suppose that X is a Banach space, $\varphi \in C^1(X)$, $c \in \mathbb{R}$ and φ satisfies the following compactness type condition at level c:

<u>PS_c^θ</u>: "There exists a nonincreasing function $\vartheta\colon (0,+\infty) \longrightarrow (0,+\infty)$ such that $\int_1^{+\infty} \vartheta(s)\,ds = +\infty$ for which every sequence $\{u_n\}_{n\geqslant 1} \subseteq X$ satisfying

$$\varphi(u_n) \longrightarrow c \quad \text{and} \quad \frac{1}{\vartheta(\|u_n\|_X)}\varphi'(u_n) \longrightarrow 0 \quad \text{in } X^*,$$

admits a strongly convergent subsequence."

Assume that $K_\varphi^c = \emptyset$ and let $r > 0$. Show that there exist $\varepsilon > 0$ and a continuous map $h\colon [0,1] \times X \longrightarrow X$ such that:

(a) $h(0,u) = u$ for all $u \in X$, for every $t \in [0,1]$, $h(t,\cdot)\colon X \longrightarrow X$ is a homeomorphism and

$$\varphi(h(t,u)) \leqslant \varphi(h(s,u)) \qquad \forall\, t,s \in [0,1],\ t \leqslant s,\ u \in X;$$

(b) if $|\varphi(u) - c| \geqslant 2\varepsilon$, then $h(t,u) = u$ for all $t \in [0,1]$;

(c) $h(1, \varphi_{c-\varepsilon} \cap \overline{B}_r) \subseteq \varphi_{c+\varepsilon}$, where $\overline{B}_r = \{u \in X : \|u\|_X \leqslant r\}$.

Remark. Note that if ϑ is a constant, then PS_c^θ coincides with the usual PS_c-condition (see Definition 5.45(a)). On the other hand, if $\vartheta = \frac{1}{1+s}$, then the PS_c^θ-condition coincides with the C_c-condition (see Definition 5.45(b)).

Problem 5.85*★*

Suppose that X is a Banach space, $\varphi \in C^1(X)$, $c \in \mathbb{R}$, φ satisfies the C_c-condition, $\eta > 0$ and U is an open neighborhood of the set K_φ^c. Show that there exists $\varepsilon_0 > 0$ such that

$$\inf_{u \in \varphi_{c-\varepsilon}^{c+\varepsilon} \setminus U} (1 + \|u\|_X)\|\varphi'(u)\|_* \geqslant \eta\varepsilon \qquad \forall\, \varepsilon \in (0,\varepsilon_0).$$

Problem 5.86★★

Suppose that X is a Banach space, $\varphi \in C^1(X)$, $u_0 \in X \setminus \{0\}$,

$$\mathcal{D} = \{U \subseteq X : U \text{ is a bounded, open neighborhood of } 0,\ u_0 \notin \overline{U}\}$$

and we define

$$c = \sup_{U \in \mathcal{D}} \inf_{u \in \partial U} \varphi(u).$$

Suppose that φ satisfies the PS_c^θ-condition (see Problem 5.84) and

$$\varphi(0), \varphi(u_0) \leqslant 0, \quad \varphi(u) \geqslant m > 0 \qquad \forall\, u \in \partial U_0,$$

for some $U_0 \in \mathcal{D}$. Show that c is a critical value of φ.

Problem 5.87★★

Suppose that X is a Banach space, $\varphi \in C^1(X)$, φ satisfies the PS_c^θ-condition for every $c \in \mathbb{R}$ (see Problem 5.84), $u = 0$ is a local minimizer of φ and there exists $u_0 \in X \setminus \{0\}$ such that $\varphi(u_0) \leqslant \varphi(0) = 0$. Show that we can find $\widehat{u} \in K_\varphi$, $\widehat{u} \notin \{0, u_0\}$.

Problem 5.88***
Let X be a Banach space and let $\varphi \in C^1(X)$ satisfy the following conditions:
(i) there exists $r > 0$ such that

$$\varphi(0) < \inf\{\varphi(u) : \|u\|_X = r\} = m_r;$$

(ii) there exists $u_0 \in X$ such that $\|u_0\|_X > r$ and $\varphi(u_0) < m_r$;
(iii) φ satisfies the WC-condition (see Problem 5.41).
Show that there exists $\widehat{u} \in K_\varphi$ with $m_r \leqslant \varphi(\widehat{u})$.

Problem 5.89**
Suppose that H is a Hilbert space, $\varphi \in C^1(H)$ has a local linking at the origin with respect to the direct sum pair (Y, V) (see Definition 5.77) and φ' is Lipschitz continuous in a neighborhood of the origin. Show that there exist a closed ball $\overline{B}$ centered at the origin and a homeomorphism $h\colon H \longrightarrow H$ such that:
(a) $K_\varphi \cap h(\overline{B}) = \{0\}$;
(b) $h|_{\overline{B}\cap Y} = I_{\overline{B}\cap Y}$;
(c) $\varphi(v) > 0$ for all $v \in h(\overline{B} \cap V \setminus \{0\})$.

Problem 5.90**
Consider the function $\varphi\colon \mathbb{R}^2 \longrightarrow \mathbb{R}$ defined by

$$\varphi(u,v) = u^2(1+v)^3 + v^2 \qquad \forall\, (u,v) \in \mathbb{R}^2.$$

Show that φ has the mountain pass geometry (see Theorem 5.63) but fails to satisfy the PS-condition.

Problem 5.91**
Let X be a Banach space and let $\varphi\colon X \longrightarrow \mathbb{R}$ be a locally Lipschitz function. Consider the following two compactness type conditions:

NPS$_c$: "Every sequence $\{u_n\}_{n\geqslant 1} \subseteq X$ such that

$$\varphi(u_n) \longrightarrow c \quad \text{and} \quad m(u_n) = \inf_{u^* \in \partial\varphi(u_n)} \|u^*\|_* \longrightarrow 0$$

admits a strongly convergent subsequence."

NPS$'_c$: "Every sequence $\{u_n\}_{n\geqslant 1} \subseteq X$ such that

$$\varphi(u_n) \longrightarrow c$$

and for every $\varepsilon > 0$, there exists $n_0 = n_0(\varepsilon) \geqslant 1$ such that

$$0 \leqslant \varphi^0(u_n; h) + \varepsilon\|h\|_X \qquad \forall\, n \geqslant n_0,\ h \in X,$$

admits a strongly convergent subsequence."

Show that both above conditions are equivalent.

Problem 5.92*
Suppose that X is a Banach space, $\varphi\colon X \longrightarrow \mathbb{R}$ is a locally Lipschitz function which is bounded below and $\widehat{u} \in X$ and $\varepsilon > 0$ are given. Show that there exists $\widetilde{u} \in X$ such that

$$\varphi(\widetilde{u}) + \varepsilon\|\widetilde{u} - \widehat{u}\|_X \leqslant \varphi(\widehat{u}) \quad \text{and} \quad 0 \leqslant \varphi^0(\widetilde{u}; h) + \varepsilon\|h\|_X \qquad \forall\, h \in X.$$

Problem 5.93**
Suppose that X is a Banach space, $\varphi\colon X \longrightarrow \mathbb{R}$ is a locally Lipschitz function which is bounded below and satisfies the NPS$_c$-condition for every $c \in \mathbb{R}$ (see Problem 5.91). Show that $\varphi(u) \longrightarrow +\infty$ as $\|u\|_X \to +\infty$.

Problem 5.94**
Suppose that X is a Banach space, $\varphi\colon X \longrightarrow \mathbb{R}$ is a locally Lipschitz function which satisfies the NPS$_c$-condition for every $c \in \mathbb{R}$ (see Problem 5.91) and has two local minimizers $u_0, u_1 \in X$. Show that φ has a third critical point (that is, there exists $\widehat{u} \in X \setminus \{u_0, u_1\}$ such that $0 \in \partial\varphi(\widehat{u})$; see Definition 5.73(a)).

Problem 5.95***
Let $\Omega \subseteq \mathbb{R}^N$ be a nonempty bounded open set with a Lipschitz boundary and let $f\colon \Omega \times \mathbb{R} \longrightarrow \mathbb{R}$ be a Carathéodory function such that:

(i) for every $\varrho > 0$, there exists $a_\varrho \in L^q(\Omega)$ with $q > \frac{2N}{N+2}$ if $N \geqslant 3$, $q > 1$ if $N = 2$ and $q \geqslant 1$ if $N = 1$ such that

$$|f(z, x| \leqslant a_\varrho(z) \quad \text{for a.a. } z \in \Omega,\ |x| \leqslant \varrho;$$

(ii) $\lim\limits_{x \to \pm\infty} f(z, x) = 0$ uniformly for almost all $z \in \Omega$;

(iii) $\lim\limits_{x \to \pm\infty} F(z, x) = F_\pm(z)$ uniformly for almost all $z \in \Omega$ with $F_\pm \in L^1(\Omega)$, where $F(z, x) = \int\limits_0^x f(z, s)\, ds$.

Consider the C^1-functional $\varphi\colon H_0^1(\Omega) \longrightarrow \mathbb{R}$ defined by

$$\varphi(u) = \tfrac{1}{2}\|Du\|_2^2 - \tfrac{\widehat{\lambda}_k}{2}\|u\|_2^2 - \int\limits_\Omega F(z, u(z))\, dz \qquad \forall\, u \in H_0^1(\Omega),$$

where $\widehat{\lambda}_k > 0$ for $k \geqslant 1$ is an eigenvalue of $(-\Delta, H_0^1(\Omega))$. Let $E(\widehat{\lambda}_k)$ be the corresponding eigenspace and let

$$D = \Big\{c = -\int\limits_{\{\widehat{u}>0\}} F_+\, dz - \int\limits_{\{\widehat{u}<0\}} F_-\, dz : \ \widehat{u} \in E(\widehat{\lambda}_k),\ \|\widehat{u}\| = 1\Big\}$$

(here $\|\widehat{u}\| = \|D\widehat{u}\|_2$ by the Poincaré inequality (see Theorem 1.131)). Show that φ satisfies the PS$_c$-condition if and only if $c \in \mathbb{R} \setminus D$.

Problem 5.96**
Let X be a Banach space and let $\xi = \varphi + \psi$ be a function such that $\varphi\colon X \longrightarrow \mathbb{R}$ is locally Lipschitz and $\psi \in \Gamma_0(X)$ (see Remark 3.56). Suppose that ξ is bounded below and $c = \liminf\limits_{\|u\|_X \to +\infty} \xi(u)$ is finite. Show that there exists a sequence $\{u_n\}_{n\geqslant 1} \subseteq X$ such that

$$\|u_n\|_X \longrightarrow +\infty, \quad \xi(u_n) \longrightarrow c$$

and

$$-\varepsilon_n\|u-u_n\|_X \leqslant \varphi^0(u_n; u-u_n) + \psi(u) + \psi(u_n) \qquad \forall\, u \in X,$$

with $\varepsilon_n \searrow 0$.

Problem 5.97**
Let X be a Banach space and let $\xi = \varphi + \psi$ be a function such that $\varphi\colon X \longrightarrow \mathbb{R}$ is locally Lipschitz and $\psi \in \Gamma_0(X)$ (see Remark 3.56). Suppose that ξ is bounded below and satisfies the SPS-condition (see Definition 5.47). Show that $\xi(u) \longrightarrow +\infty$ as $\|u\|_X \longrightarrow +\infty$.

Problem 5.98***
Suppose that X and Y are two reflexive Banach spaces, X and X^* are both uniformly convex, $X \hookrightarrow Y$ compactly and densely, $\widehat{\psi}\colon Y \longrightarrow \mathbb{R}$ is locally Lipschitz, $\psi = \widehat{\psi}|_X$ and $\varphi\colon X \longrightarrow \mathbb{R}$ is the locally Lipschitz function defined by

$$\varphi(u) = \tfrac{1}{p}\|u\|_X^p - \psi(u) \qquad \forall\, u \in X,$$

with $1 < p < +\infty$. Suppose that φ satisfies the following compactness type condition:

<u>NPS_w</u>: "Every sequence $\{u_n\}_{n\geqslant 1} \subseteq X$ such that $\{\varphi(u_n)\}_{n\geqslant 1} \subseteq \mathbb{R}$ is bounded and

$$m(u_n) \longrightarrow 0$$

(see Definition 5.47(a)) admits a weakly convergent subsequence."

Show that φ satisfies the NPS_c-condition for every $c \in \mathbb{R}$ (see Problem 5.55).

Problem 5.99*
Suppose that (X, d_X) is a complete metric space, $C \subseteq X$ is a nonempty, closed set and $\varphi\colon X \times X \longrightarrow \mathbb{R}$ is a function such that:

(i) $\varphi(u,u) = 0$ for all $u \in C$;
(ii) $\varphi(u,y) \leqslant \varphi(u,v) + \varphi(v,y)$ for all $u, y, v \in C$;
(iii) for every $u \in X$, $\varphi(u,\cdot)$ is lower semicontinuous and bounded below.

Show that for every $\varepsilon > 0$ and every $\widehat{u} \in C$, we can find $\widetilde{u} \in C$ such that

$$\varphi(\widehat{u}, \widetilde{u}) + \varepsilon d_X(\widetilde{u}, \widehat{u}) \leqslant 0 \ \text{ and } \ 0 < \varphi(\widetilde{u}, u) + \varepsilon d_X(u, \widetilde{u}) \qquad \forall\, u \in C,\ u \neq \widetilde{u}.$$

Problem 5.100**
Suppose that X is a Banach space, $C \subseteq X$ is a nonempty and w-compact set and $\varphi\colon C \times C \longrightarrow \mathbb{R}$ is a function such that:

(i) $\varphi(u,u) = 0$ for all $u \in C$;
(ii) $\varphi(u,y) \leqslant \varphi(u,v) + \varphi(v,y)$ for all $u, y, v \in C$;

(iii) for every $u \in X$, $\varphi(u,\cdot)$ is lower semicontinuous and bounded below;
(iv) for every $y \in X$, $\varphi(\cdot,y)$ is sequentially weakly upper semicontinuous.
Show that there exists $\widehat{u} \in C$ such that

$$0 \leqslant \varphi(\widehat{u},y) \qquad \forall\, y \in C.$$

Problem 5.101**

Suppose that X is a reflexive Banach space, $C \subseteq X$ is a nonempty and w-closed set and $\varphi\colon C \times C \longrightarrow \mathbb{R}$ is a function such that:

(i) $\varphi(u,u) = 0$ for all $u \in C$;
(ii) $\varphi(u,y) \leqslant \varphi(u,v) + \varphi(v,y)$ for all $u,y,v \in C$;
(iii) there exists a nonempty w-compact set $K \subseteq C$ such that for every $u \in C \setminus K$, there exists $y \in C$ such that $\|y\|_X < \|u\|_X$ and $\varphi(u,y) \leqslant 0$;
(iv) for every $u \in C$, $\varphi(u,\cdot)|_K$ is sequentially weakly lower semicontinuous;
(v) for every $u \in K$, $\varphi(u,\cdot)|_K$ is bounded below;
(vi) for every $y \in K$, $\varphi(\cdot,y)|_K$ is sequentially weakly upper semicontinuous. Show that there exists $\widehat{u} \in C$ such that

$$0 \leqslant \varphi(\widehat{u},y) \qquad \forall\, y \in C.$$

Problem 5.102***

Let $\Omega \subseteq \mathbb{R}^N$ be a bounded domain with a C^2-boundary $\partial\Omega$ and let $f\colon \Omega \times \mathbb{R} \longrightarrow \mathbb{R}$ be a Carathéodory function such that $f(z,0) = 0$ for almost all $z \in \Omega$ and

(i) $|f(z,x)| \leqslant a(z)(1+x)$ for almost all $z \in \Omega$, all $x \geqslant 0$, with $a \in L^\infty(\Omega)$;
(ii) $\limsup\limits_{x\to+\infty} \frac{f(z,x)}{x} \leqslant \vartheta(z)$ uniformly for almost all $z \in \Omega$ with $\vartheta \in L^\infty(\Omega)$ and $\vartheta(z) \leqslant \widehat{\lambda}_1$ for almost all $z \in \Omega$ with strict inequality on a set of positive Lebesgue measure;
(iii) $\liminf\limits_{x\to 0^+} \frac{f(z,x)}{x} \geqslant \eta(z)$ uniformly for almost all $z \in \Omega$ with $\eta \in L^\infty(\Omega)$ and $\eta(z) \geqslant \widehat{\lambda}_1$ for almost all $z \in \Omega$ with strict inequality on a set of positive Lebesgue measure;
(iv) for almost all $z \in \Omega$, the map $x \longmapsto \frac{f(z,x)}{x}$ is decreasing on $(0,+\infty)$;
(v) for every $\varrho > 0$, there exists $\xi_\varrho > 0$ such that for almost all $z \in \Omega$, the map $x \longmapsto f(z,x) + \xi_\varrho x$ is nondecreasing on $[0,\varrho]$.

Consider the Dirichlet problem

$$\begin{cases} -\Delta u(z) = f(z,u(z)) & \text{in } \Omega, \\ u|_{\partial\Omega} = 0. \end{cases}$$

Show that it has a unique positive solution $u_0 \in \operatorname{int} C_+$ (see Theorem 5.129(e)).

Problem 5.103**

Let $\Omega \subseteq \mathbb{R}^N$ be a bounded domain with a C^2-boundary $\partial\Omega$, $1 < p < +\infty$ and let λ be an eigenvalue of $(-\Delta_p, W_0^{1,p}(\Omega))$ such that $\lambda \neq \widehat{\lambda}_1$. Show that every eigenfunction $u \in W_0^{1,p}(\Omega)$ corresponding to λ is nodal (that is, sign-changing).

Problem 5.104**

Suppose that $\Omega \subseteq \mathbb{R}^N$ is a bounded domain with a C^2-boundary $\partial\Omega$, $1 < p < +\infty$, $\sigma_0(p)$ is the set of eigenvalues of $(-\Delta_p, W_0^{1,p}(\Omega))$, $E(\lambda)$ is the set of normalized eigenfunctions corresponding to $\lambda \in \sigma_0(p)$. Show that $\sigma_0(p)$ is closed and $E(\lambda) \subseteq W_0^{1,p}(\Omega)$ is compact.

Problem 5.105**

Suppose that $\Omega \subseteq \mathbb{R}^N$ is a bounded domain with a C^2-boundary $\partial\Omega$, $1 < p < +\infty$, $u \in W_0^{1,p}(\Omega)$ is an eigenfunction of $(-\Delta_p, W_0^{1,p}(\Omega))$ corresponding to an eigenvalue $\lambda > \widehat{\lambda}_1$ and

$$\Omega_+ = \{z \in \Omega : u(z) > 0\}, \quad \Omega_- = \{z \in \Omega : u(z) < 0\}$$

and $|\cdot|_N$ denotes the Lebesgue measure on $\mathbb{R}^N$. Show that

$$\min\{|\Omega_+|_N, |\Omega_-|_N\} \geqslant (\lambda c^p)^r,$$

with $c > 0$ depending only on p, N and $r = \frac{N}{p}$ if $p < N$, $r = 2$ if $N \leqslant p$.

Problem 5.106*

Suppose that $\Omega \subseteq \mathbb{R}^N$ is a bounded domain with a C^2-boundary $\partial\Omega$, $1 < p < +\infty$, $\xi, g \in L^\infty(\Omega)$, $g(z) \geqslant 0$ for almost all $z \in \Omega$, $g \neq 0$ and $u \in W_0^{1,p}(\Omega)$ is a solution of the following problem

$$\begin{cases} -\Delta_p u(z) + \xi(z)|u(z)|^{p-2}u(z) = g(z) & \text{for a.a. } z \in \Omega, \\ u \geqslant 0. \end{cases}$$

Show that $u \in \operatorname{int} C_+$ (see Theorem 5.129(e)).

Problem 5.107***

Suppose that $\Omega \subseteq \mathbb{R}^N$ is a bounded domain with a C^2-boundary $\partial\Omega$, $1 < p < +\infty$, $\xi \in L^\infty(\Omega)$ and consider the following nonlinear eigenvalue problem:

$$\begin{cases} -\Delta_p u(z) + \xi(z)|u(z)|^{p-2}u(z) = \lambda|u(z)|^{p-2}u(z) & \text{in } \Omega, \\ u|_{\partial\Omega} = 0, \ 1 < p < +\infty. \end{cases}$$

Show that the above problem has a first (smallest) eigenvalue $\widehat{\lambda}_1(\xi)$ which is simple, isolated, expressed by

$$\widehat{\lambda}_1(\xi) - \inf_{u \in W_0^{1,p}(\Omega)\setminus\{0\}} \frac{\|Du\|_p^p + \int_\Omega \xi(x)|u|^p\,dz}{\|u\|_p^p}$$

and the corresponding eigenfunctions have constant sign and belong in $C_0^1(\overline{\Omega})$.

Problem 5.108**
Suppose that $\Omega \subseteq \mathbb{R}^N$ is a bounded domain with a C^2-boundary $\partial\Omega$, $1 < p < +\infty$, $\xi \in L^\infty(\Omega)$ and for every $g \in L^\infty(\Omega)$, $g(z) \geqslant 0$ for almost all $z \in \Omega$, $g \neq 0$, every solution $u \in W_0^{1,p}(\Omega)$ of the problem:

$$-\Delta_p u(z) + \xi(z)|u(z)|^{p-2}u(z) \;=\; g(z) \quad \text{in } \Omega,$$

satisfies $u(z) > 0$ for almost all $z \in \Omega$. Show that $\widehat{\lambda}_1(\xi) > 0$ (see Problem 5.107).

Problem 5.109**
Suppose that $\Omega \subseteq \mathbb{R}^N$ is a bounded domain with a C^2-boundary $\partial\Omega$, $1 < p < +\infty$, $\xi, g \in L^\infty(\Omega)$ are such that $g(z) \geqslant 0$ for almost all $z \in \Omega$, $g \neq 0$ and $v \in W_0^{1,p}(\Omega)$ is a positive solution of the following problem:

$$-\Delta_p v(z) + \xi(z)v(z)^{p-1} \;=\; g(z) \quad \text{in } \Omega.$$

Show that $\widehat{\lambda}_1(\xi) > 0$ (see Problem 5.107).

Problem 5.110*
Suppose that $\Omega \subseteq \mathbb{R}^N$ is a bounded domain with a C^2-boundary $\partial\Omega$, $1 < p < +\infty$, $\xi \in L^\infty(\Omega)$ and $\widehat{\lambda}_1(\xi) > 0$ (see Problem 5.107). Show that for every $g \in L^\infty(\Omega)$ with $g(z) \geqslant 0$ for almost all $z \in \Omega$, every solution $u \in W_0^{1,p}(\Omega)$ of the Dirichlet problem:

$$\begin{cases} -\Delta_p u(z) + \xi(z)|u(z)|^{p-2}u(z) \;=\; g(z) & \text{in } \Omega, \\ u|_{\partial\Omega} \;=\; 0, \end{cases}$$

satisfies $u \in C_+$ (that is, $u \in C_0^1(\overline{\Omega})$, $u(z) \geqslant 0$ for all $z \in \Omega$).

Problem 5.111**
Suppose that $\Omega \subseteq \mathbb{R}^N$ is a bounded domain with a C^2-boundary $\partial\Omega$, $1 < p < +\infty$, $\xi \in L^\infty(\Omega)$ and $\widehat{\lambda}_1(\xi) > 0$ (see Problem 5.107). Show that for every $g \in L^\infty(\Omega)$ with $g(z) \geqslant 0$ for almost all $z \in \Omega$, the Dirichlet problem from Problem 5.110 has a unique nonnegative solution $u_0 \in C_+$.

Problem 5.112**
Suppose that $\Omega \subseteq \mathbb{R}^N$ is a bounded domain with a Lipschitz boundary $\partial\Omega$, $\psi\colon \Omega \longrightarrow \mathbb{R} \cup \{\pm\infty\}$ is a function and $g \in W^{1,p}(\Omega)$ (with $1 < p < +\infty$). We introduce the following set

$$K_{\psi,g} \;=\; \big\{u \in W^{1,p}(\Omega) : \; u(z) \geqslant \psi(z) \text{ for a.a. } z \in \Omega, \; u - g \in W_0^{1,p}(\Omega)\big\}.$$

Suppose that $u_0 \in K_{\psi,g}$ satisfies

$$\int_\Omega |Du_0|^{p-2}\,(Du_0, Du - Du_0)_{\mathbb{R}^N}\, dz \;\geqslant\; 0 \qquad \forall\, u \in K_{\psi,g}.$$

Also $\overline{u} \in W^{1,p}(\Omega)$ satisfies

$$\int_\Omega |D\overline{u}|^{p-2} (D\overline{u}, Dh)_{\mathbb{R}^N}\, dz \geqslant 0 \qquad \forall\, h \in C_c^\infty(\Omega),\ h \geqslant 0.$$

Assume that $\min\{u_0, \overline{u}\} \in K_{\psi,g}$. Show that $\overline{u}(z) \geqslant u_0(z)$ for almost all $z \in \Omega$.

Problem 5.113**

Suppose that $\Omega \subseteq \mathbb{R}^N$ is a bounded domain with a Lipschitz boundary $\partial\Omega$, $f\colon \Omega \times \mathbb{R} \longrightarrow \mathbb{R}$ is a Carathéodory function such that

$$|f(z,x)| \leqslant a(z)(1 + |x|^{p-1}) \quad \text{for a.a. } z \in \Omega, \text{ all } x \in \mathbb{R},$$

with $a \in L^\infty(\Omega)$, $1 < p < +\infty$. Let $\xi_\pm \in L^\infty(\Omega)$ be such that

$$\xi_\pm(z) \leqslant \liminf_{x \to \pm\infty} \tfrac{f(z,x)}{|x|^{p-2}x}$$

uniformly for almost all $z \in \Omega$ and assume that $\widehat{\lambda}_1(\xi_\pm) > 0$ (see Problem 5.107). Show that for each $g \in W^{-1,p'}(\Omega) = W_0^{1,p}(\Omega)^*$ (with $\frac{1}{p} + \frac{1}{p'} = 1$), the Dirichlet problem:

$$\begin{cases} -\Delta_p u(z) + f(z, u(z)) = g(z) & \text{in } \Omega, \\ u|_{\partial\Omega} = 0 \end{cases}$$

has at least one weak solution.

Problem 5.114***

Let $\Omega \subseteq \mathbb{R}^N$ be bounded domain with a Lipschitz boundary $\partial\Omega$ and consider the following Dirichlet problem:

$$\begin{cases} -\Delta u(z) = \lambda u(z) + u(z)^{p-1} & \text{in } \Omega, \\ u|_{\partial\Omega} = 0,\ u \geqslant 0,\ u \neq 0, \end{cases}$$

with $2 < p < 2^*$ where $2^* = \frac{2N}{N-2}$ if $N \geqslant 3$ and $2^* = +\infty$ if $N \in \{1,2\}$. Show that the considered Dirichlet problem has a weak solution if and only if $\lambda < \widehat{\lambda}_1$ with $\widehat{\lambda}_1$ being the first eigenvalue of $(-\Delta, H_0^1(\Omega))$.

Problem 5.115**

Let $\Omega \subseteq \mathbb{R}^N$ be a bounded domain with a Lipschitz boundary $\partial\Omega$ and let $f\colon \Omega \times \mathbb{R} \longrightarrow \mathbb{R}$ be a measurable function such that for almost all $z \in \Omega$ we have $f(z,\cdot) \in C^1(\mathbb{R})$ and

(i) for every $\varrho > 0$, there exists $a_\varrho \in L^\infty(\Omega)$ such that

$$|f'_x(z,x)| \leqslant a_\varrho(z) \quad \text{for a.a. } z \in \Omega, \text{ all } |x| \leqslant \varrho;$$

(ii) $f'_x(z,x) \leqslant \widehat{\lambda}_2$ for almost all $z \in \Omega$, all $x \in \mathbb{R}$ and the inequality is strict on a set of positive measure (here $\widehat{\lambda}_2$ is the second eigenvalue of $(-\Delta, H_0^1(\Omega))$).

Suppose that $u_1, u_2 \in H_0^1(\Omega) \cap L^\infty(\Omega)$ are two weak solutions of the following Dirichlet problem:

$$\begin{cases} -\Delta u(z) = f(z, u(z)) & \text{in } \Omega, \\ u|_{\partial\Omega} = 0. \end{cases}$$

Show that $u_1 \leqslant u_2$ or $u_2 \leqslant u_1$.

Problem 5.116**

Let $\Omega \subseteq \mathbb{R}^N$ $(N \geqslant 3)$ be a bounded domain with a Lipschitz boundary $\partial\Omega$ and let $f\colon \Omega \times \mathbb{R} \longrightarrow \mathbb{R}$ be a Carathéodory function such that:

(i) for every $\varrho > 0$, there exists $a_\varrho \in L^\infty(\Omega)$ such that

$$|f(z, x)| \leqslant a_\varrho(z) \quad \text{for a.a. } z \in \Omega, \text{ all } |x| \leqslant \varrho;$$

(ii) there exist functions $\xi_\pm \in L^\infty(\Omega)$ such that

$$\xi_\pm(z) = \lim_{x \to \pm\infty} \tfrac{f(z,x)}{x} \quad \text{uniformly for a.a. } z \in \Omega.$$

Suppose that the following Dirichlet problem:

$$\begin{cases} -\Delta u(z) = \xi_+(z)u(z)^+ - \xi_-(z)u(z)^- & \text{in } \Omega, \\ u|_{\partial\Omega} = 0, \end{cases}$$

admits only the trivial solution. Let $F(z, x) = \int\limits_0^x f(z, s)\, ds$ and consider the C^1-functional $\varphi\colon H_0^1(\Omega) \longrightarrow \mathbb{R}$ defined by

$$\varphi(u) = \tfrac{1}{2}\|Du\|_2^2 - \int\limits_\Omega F(z, u(z)) \qquad \forall\, u \in H_0^1(\Omega).$$

Show that φ satisfies the PS-condition.

Problem 5.117**

Let $\Omega \subseteq \mathbb{R}^N$ be a bounded domain with a Lipschitz boundary $\partial\Omega$ and let $\{\widehat{\lambda}_k\}_{n \geqslant 1}$ be the sequence of distinct eigenvalues of $(-\Delta, H_0^1(\Omega))$. Let $m \geqslant 1$ and set

$$\overline{H}_m = \bigoplus_{k=1}^m E(\widehat{\lambda}_k), \quad \widehat{H}_m = \overline{\bigoplus_{k \geqslant m} E(\widehat{\lambda}_k)}.$$

Show that:

(a) if $\eta \in L^\infty(\Omega)$ satisfies $\eta(z) \leqslant \widehat{\lambda}_m$ for almost all $z \in \Omega$, with strict inequality on a set of positive Lebesgue measure, then there exists $c_1 > 0$ such that

$$\|Du\|_2^2 - \int\limits_\Omega \eta(z)u^2\, dz \geqslant c_1\|u\|^2 \qquad \forall\, u \in \widehat{H}_m;$$

(b) if $\vartheta \in L^\infty(\Omega)$ satisfies $\vartheta(z) \geqslant \widehat{\lambda}_m$ for almost all $z \in \Omega$, with strict inequality on a set of positive Lebesgue measure, then there exists $c_2 > 0$ such that

$$\|Du\|_2^2 - \int_\Omega \vartheta(z) u^2 \, dz \leqslant -c_2 \|u\|^2 \qquad \forall\, u \in \overline{H}_m.$$

Problem 5.118[**]

Let $\Omega \subseteq \mathbb{R}^N$ be a bounded domain with a Lipschitz boundary $\partial\Omega$ and let $f\colon \Omega \times \mathbb{R} \longrightarrow \mathbb{R}$ be a Carathéodory function such that for almost all $z \in \Omega$ we have $f(z,0) = 0$ and

(i) there exists $a \in L^\infty(\Omega)$ such that

$$|f(z,x)| \leqslant a(z)(1 + |x|) \quad \text{for a.a. } z \in \Omega, \text{ all } x \in \mathbb{R};$$

(ii) $\limsup\limits_{x \to \pm\infty} \frac{f(z,x)}{x} \leqslant \widehat{\lambda}_1$ uniformly for almost all $z \in \Omega$;

(iii) for $F(z,x) = \int\limits_0^x f(z,s)\, ds$, we have

$$\lim_{x \to \pm\infty} \big(f(z,x)x - 2F(z,x)\big) = +\infty$$

uniformly for almost all $z \in \Omega$;

(iv) there exist integer $m \geqslant 1$, function $\vartheta \in L^\infty(\Omega)$ and $\delta_0 > 0$ such that

$$\vartheta(z) \leqslant \widehat{\lambda}_{m+1} \quad \text{for a.a. } z \in \Omega,$$

the above inequality is strict on a set of positive Lebesgue measure,

$$\limsup_{x \to 0} \frac{F(z,x)}{x^2} \leqslant \vartheta(z)$$

uniformly for almost all $z \in \Omega$,

$$F(z,x) \geqslant \frac{\widehat{\lambda}_m}{2} |x|^2 \quad \text{for a.a. } z \in \Omega, \text{ all } |x| \leqslant \delta_0.$$

Show that the following Dirichlet problem

$$\begin{cases} -\Delta u(z) = f(z, u(z)) & \text{in } \Omega, \\ u|_{\partial\Omega} = 0. \end{cases}$$

has at least two nontrivial weak solutions $u_0, \widehat{u} \in H_0^1(\Omega)$.

Problem 5.119[**]

Suppose that $\Omega \subseteq \mathbb{R}^N$ is a bounded domain with a C^2-boundary $\partial\Omega$, $(\widehat{\lambda}_1, \widehat{u}_1)$ is the principal eigenpair of $(-\Delta_p, W_0^{1,p}(\Omega))$ with $\widehat{u}_1 \in \operatorname{int} C_+$ (see Proposition 5.132 and Remark 5.133) and consider the vector subspace

$$V = \Big\{ u \in W_0^{1,p}(\Omega) : \int_\Omega u \widehat{u}_1^{p-1} \, dz = 0 \Big\}.$$

Define

$$\widehat{\lambda}_V = \inf_{\substack{u \in V \\ u \neq 0}} \frac{\|Du\|_p^p}{\|u\|_p^p}.$$

Show that $\widehat{\lambda}_1 < \widehat{\lambda}_V \leqslant \widehat{\lambda}_2$ with $\widehat{\lambda}_2$ being the second eigenvalue of $(-\Delta_p, W_0^{1,p}(\Omega))$.

Problem 5.120**

Show that the infimum

$$\inf\Big\{\int_0^1 tu'(t)^2\,dt:\ u \in C^1([0,1]),\ u(0)=1,\ u(1)=0\Big\} = m$$

is not attained and $m = 0$.

Problem 5.121*

Let $\Omega \subseteq \mathbb{R}^N$ be a bounded domain with a Lipschitz boundary $\partial\Omega$ and let $f\colon \Omega \times \mathbb{R} \longrightarrow \mathbb{R}$ be a Carathéodory function such that for almost all $z \in \Omega$ we have $f(z,0) = 0$ and $f(z,\cdot)$ is convex and for every $\varrho > 0$, there exists $a_\varrho \in L^2(\Omega)$ such that

$$|f(z,x)| \leqslant a_\varrho(z) \quad \text{for a.a. } z \in \Omega, \text{ all } |x| \leqslant \varrho.$$

Suppose that $u_0 \in H_0^1(\Omega) \cap L^\infty(\Omega)$ is a weak solution of the following Dirichlet problem:

$$\begin{cases} -\Delta u(z) = f(z,u(z)) & \text{in } \Omega, \\ u|_{\partial\Omega} = 0. \end{cases}$$

Show that:

(a) if $\overline{u} = \lambda u_0$ with $\lambda > 1$, then

$$\int_\Omega (-\Delta\overline{u})h\,dz \leqslant \int_\Omega f(z,\overline{u})h\,dz \qquad \forall\, h \in H_0^1(\Omega),\ h \geqslant 0;$$

(b) if $\underline{u} = \lambda u_0$ with $0 < \lambda < 1$, then

$$\int_\Omega (-\Delta\underline{u})h\,dz \geqslant \int_\Omega f(z,\underline{u})h\,dz \qquad \forall\, h \in H_0^1(\Omega),\ h \geqslant 0.$$

Problem 5.122*

Let X be a Banach space and let $\varphi \in C^1(X)$ be a convex function. Show that every critical point of φ is a global minimizer of φ.

Problem 5.123***

Let $\Omega \subseteq \mathbb{R}^N$ be a bounded domain with a Lipschitz boundary $\partial\Omega$ and let $f\colon \Omega \times \mathbb{R} \longrightarrow \mathbb{R}$ be a Carathéodory function such that:

(i) for every $\varrho > 0$, there exists $a_\varrho \in L^\infty(\Omega)_+$ such that

$$|f(z,x)| \leqslant a_\varrho \quad \text{for a.a. } z \in \Omega, \text{ all } |x| \leqslant \varrho;$$

(ii)$_1$ there exist integer $m \geqslant 1$ and functions $\eta, \widehat{\eta} \in L^\infty(\Omega)$ such that

$$\begin{aligned} &\widehat{\lambda}_m \leqslant \eta(z) \leqslant \widehat{\eta}(z) \leqslant \widehat{\lambda}_{m+1} \quad \text{for a.a. } z \in \Omega, \\ &\eta \not\equiv \widehat{\lambda}_m, \ \widehat{\eta} \not\equiv \widehat{\lambda}_{m+1}, \\ &\eta(z) \leqslant \liminf_{z \to \pm\infty} \frac{f(z,x)}{x} \leqslant \limsup_{z \to \pm\infty} \frac{f(z,x)}{x} \leqslant \widehat{\eta}(z) \end{aligned}$$

uniformly for almost all $z \in \Omega$;

or

(ii)$_2$ there exist functions $\eta_0, \widehat{\eta}_0 \in L^\infty(\Omega)$ such that

$$\begin{aligned} &\widehat{\eta}_0(z) \leqslant \widehat{\lambda}_1 \quad \text{for a.a. } z \in \Omega, \ \widehat{\eta}_0 \not\equiv \widehat{\lambda}_1, \\ &\eta_0(z) \leqslant \liminf_{z \to \pm\infty} \frac{f(z,x)}{x} \leqslant \limsup_{z \to \pm\infty} \frac{f(z,x)}{x} \leqslant \widehat{\eta}_0(z) \end{aligned}$$

uniformly for almost all $z \in \Omega$.

Consider the following Dirichlet problem:

$$\begin{cases} -\Delta u(z) = f(z,u(z)) + g(z) & \text{in } \Omega, \\ u|_{\partial\Omega} = 0, \end{cases}$$

with $g \in L^2(\Omega)$. Show that the above Dirichlet problem has at least one weak solution $u_0 \in H_0^1(\Omega)$.

Problem 5.124**

Let $\Omega \subseteq \mathbb{R}^N$ be a bounded domain with a Lipschitz boundary $\partial\Omega$ and let $f\colon \Omega \times \mathbb{R} \longrightarrow \mathbb{R}$ be a measurable function such that for almost all $z \in \Omega$, $f(z,\cdot) \in C^1(\mathbb{R})$ and

(i) there exists $a \in L^\infty(\Omega)$ such that

$$|f'_x(z,x)| \leqslant a(z) \quad \text{for a.a. } z \in \Omega, \text{ all } x \in \mathbb{R};$$

(ii) $f'_x(z,x) \leqslant \widehat{\lambda}_1$ for almost all $z \in \Omega$ and all $x \in \mathbb{R}$ and there exists a set $D \subseteq \Omega$ with $|D|_N > 0$ such that $f'_x(z,x) < \widehat{\lambda}_1$ for all $z \in D$, all $x \in \mathbb{R}$.

Show that for every $g \in L^2(\Omega)$ the Dirichlet problem from Problem 5.123 has a unique solution $u_0 \in H_0^1(\Omega)$.

Problem 5.125*
Let $\Omega \subseteq \mathbb{R}^N$ (with $N \geqslant 3$) be a bounded domain with a Lipschitz boundary $\partial\Omega$. Assume that Ω is star-shaped with respect to the origin $0 \in \mathbb{R}^N$. Let $u \in C^2(\Omega) \cap C^1(\overline{\Omega})$ be a solution of the following Dirichlet problem:

$$\begin{cases} -\Delta u(z) = |u(z)|^{p-2}u(z) & \text{in } \Omega, \\ u|_{\partial\Omega} = 0. \end{cases}$$

Show that $p < 2^* = \frac{2N}{N-2}$.

Problem 5.126**
Let $\Omega \subseteq \mathbb{R}^N$ (with $N \geqslant 3$) be a bounded domain with a C^2-boundary $\partial\Omega$. Assume that Ω is star-shaped with respect to the origin $0 \in \mathbb{R}^N$. Consider the following Dirichlet problem:

$$\begin{cases} -\Delta u(z) = \lambda u(z) + |u(z)|^{2^*-2}u(z) & \text{in } \Omega, \\ u|_{\partial\Omega} = 0, \end{cases}$$

Show that, if $\lambda \leqslant 0$, then the Dirichlet problem has only the trivial solution.

Problem 5.127*
Let $\Omega \subseteq \mathbb{R}^N$ be a bounded domain with a C^2-boundary $\partial\Omega$. Show that every eigenpair $(\widehat{\lambda}, \widehat{u}) \in (0, \infty) \times H_0^1(\Omega)$ of $(-\Delta, H_0^1(\Omega))$ satisfies

$$\widehat{\lambda}\|\widehat{u}\|_2^2 = \frac{1}{2}\int_{\partial\Omega} \left(\frac{\partial \widehat{u}}{\partial n}\right)^2 (z, n)_{\mathbb{R}^N}\, d\sigma.$$

Problem 5.128**
Let X be a Banach space and let G be a group acting on X (see Definition 5.84). We also consider the action of G on X^* defined naturally by

$$\langle gu^*, u \rangle = \langle u^*, g^{-1}u \rangle \qquad \forall\, g \in G,\ u \in X,\ u^* \in X^*.$$

We introduce the following sets

$$\begin{aligned} \mathrm{Fix}_G(X) &= \{u \in X :\ gu = u\}, \\ \mathrm{Fix}_G(X^*) &= \{u^* \in X^* :\ gu^* = u^*\}. \end{aligned}$$

Show that for every $\varphi \in C^1(X)$ which is G-invariant, we have that

$$\left[K_{\varphi|_{\mathrm{Fix}_G(X)}} \subseteq K_\varphi\right] \iff \left[\mathrm{Fix}_G(X^*) \cap \mathrm{Fix}_G(X)^\perp = \{0\}\right],$$

where $\mathrm{Fix}_G(X)^\perp = \{u^* \in X^* :\ \langle u^*, u \rangle = 0 \text{ for all } u \in \mathrm{Fix}_G(X)\}$.

Problem 5.129*
Let X be a Banach space and let G be a group acting on X, with the action being isometric (see Definition 5.84). Consider the action of G on X^* defined in Problem 5.128. Show that this action is also isometric.

Problem 5.130**

Let X be a reflexive and strictly convex Banach space and let G be a group acting on X with the action being isometric. Let $\mathcal{F}\colon X \longrightarrow 2^{X^*} \setminus \{\emptyset\}$ be the duality map (see Definition 2.112). Show that $\mathcal{F}^{-1}(\mathrm{Fix}_G(X^*)) \subseteq \mathrm{Fix}_G(X)$ (see Problem 5.128).

Problem 5.131**

Let X be a Banach space and let G be a compact topological group acting on X. Show that for every G-invariant $\varphi \in C^1(X)$, we have $K_{\varphi|_{\mathrm{Fix}_G(X)}} \subseteq K_\varphi$.

Problem 5.132**

Let $\Omega \subseteq \mathbb{R}^N$ be a bounded domain with a Lipschitz boundary $\partial\Omega$ and let $f\colon \Omega \times \mathbb{R} \longrightarrow \mathbb{R}$ be a Carathéodory function such that:

(i) $|f(z,x))| \leqslant c(1+|x|^{r-1})$ for almost all $z \in \Omega$, all $x \in \mathbb{R}$ with $1 < p < r < p^*$ where $p^* = \frac{Np}{N-p}$ if $p < N$ and $p^* = +\infty$ if $N \leqslant p$;

(ii) there exist $q > p$ and $M > 0$ such that

$$0 < qF(z,x) \leqslant f(z,x)x \quad \text{for a.a. } z \in \Omega, \text{ all } |x| \geqslant M,$$

where $F(z,x) = \int\limits_0^x f(z,s)\,ds$.

Consider the C^1-functional $\varphi\colon W_0^{1,p}(\Omega) \longrightarrow \mathbb{R}$ defined by

$$\varphi(u) = \frac{1}{p}\|Du\|_p^p - \int\limits_\Omega F(z,u(z))\,dz \qquad \forall\, u \in W_0^{1,p}(\Omega).$$

Show that φ satisfies the PS-condition.

Problem 5.133**

Let $(T,\Sigma\,\mu)$ be a measurable space, $1<p<+\infty$ and let $f\colon T \times \mathbb{R} \longrightarrow \mathbb{R}$ be a Carathéodory function which satisfies

(i) there exist $q > p$ and $M > 0$ such that $0 < qF(t,x) \leqslant f(t,x)x$ for μ-almost all $t \in T$ and all $|x| \geqslant M$, where $F(t,x) = \int\limits_0^x f(t,s)\,ds$;

(ii) $0 < \operatorname*{ess\,inf}\limits_T F(\cdot, \pm M)$.

Show that $\lim\limits_{x\to\pm\infty} \frac{f(t,x)}{|x|^{p-2}x} = +\infty$ uniformly for μ-almost all $t \in T$. Is the converse true? Justify your answer.

Problem 5.134***

Let $\Omega \subseteq \mathbb{R}^N$ be a bounded domain with a Lipschitz boundary $\partial\Omega$ and let $f\colon \Omega \times \mathbb{R} \longrightarrow \mathbb{R}$ be a Carathéodory function such that:

(i) for almost all $z \in \Omega$, the function $f(z,\cdot)$ is odd;

(ii) $|f(z,x)| \leqslant c(1+|x|^{r-1})$ for almost all $z \in \Omega$, all $x \in \mathbb{R}$ with $2 < r < 2^*$, where $2^* = \frac{2N}{N-2}$ when $N \geqslant 3$ and $2 < r < +\infty$ when $N = 1,2$;

(iii) there exist $q > 2$ and $M > 0$ such that

$$0 < qF(z,x) \leqslant f(z,x)x \quad \text{for a.a. } z \in \Omega, \text{ all } |x| \geqslant M$$

and

$$0 < \operatorname*{ess\,inf}_{\Omega} F(\cdot, \pm M),$$

where $F(z,x) = \int_0^x f(z,s)\,ds$.

Consider the following Dirichlet problem:

$$\begin{cases} -\Delta u(z) &= f(z,u(z)) \quad \text{in } \Omega, \\ u|_{\partial\Omega} &= 0. \end{cases}$$

Show that the above problem admits an unbounded sequence $\{\pm u_n\}_{n\geqslant 1}$ of pairs of weak solutions.

Problem 5.135***
Let $\Omega \subseteq \mathbb{R}^N$ be a bounded domain with a C^2-boundary $\partial\Omega$ and $g \in L^\infty(\Omega)$ with $g(z) \geqslant 0$ for almost all $z \in \Omega$, $g \not\equiv 0$. Consider the following parametric Dirichlet problem:

$$\begin{cases} -\Delta u(z) &= \lambda u(z) + g(z) \quad \text{in } \Omega, \\ u|_{\partial\Omega} &= 0, \ \lambda > 0. \end{cases}$$

Show that there exists $\delta = \delta(g) > 0$ such that if $\lambda \in (\widehat{\lambda}_1, \widehat{\lambda}_1 + \delta)$ any weak solution $u \in H_0^1(\Omega)$ of the above problem satisfies $u \in -\operatorname{int} C_+$ (see Theorem 5.129(e)).

Problem 5.136**
Let $\Omega \subseteq \mathbb{R}^N$ be an open, bounded, symmetric neighborhood of the origin. Show that $\gamma(\partial\Omega) = N$, where γ is the Krasnoselskii genus.

Problem 5.137**
Let X be an infinite dimensional separable Banach space and let $\partial B_1 = \{u \in X : \|u\| = 1\}$. Show that $\gamma(\partial B_1) = +\infty$.

Problem 5.138*
Let X be a Banach space and let $C \subseteq X$ be a nonempty closed set such that $C \cap (-C) = \emptyset$. Show that $\gamma(C \cup (-C)) = 1$.

Problem 5.139*
Let $\Omega \subseteq \mathbb{R}^N$ be an open, bounded, symmetric neighborhood of the origin and let $h \in C(\partial\Omega; \mathbb{R}^m)$ be odd, where $m \leqslant N - 1$. Consider the set

$$D = \{u \in \partial\Omega : h(u) = 0\}.$$

Show that $\gamma(D) \geqslant N - m$.

Problem 5.140**
Suppose that X is a Banach space, G is a compact topological group acting on X, $V \subseteq X$ is a closed, G-invariant subspace which admits a topological complement. Show that V has a G-invariant topological complement.

Problem 5.141*
Let G be a compact topological group with a representation $\{L(g)\}_{g\in G}$ over $\mathbb{R}^N$. Show that there exists an invertible matrix A such that $AL(g)A^{-1}$ is an isometric representation of G over $\mathbb{R}^N$.

Problem 5.142*
Suppose that X is a Banach space, G is a topological group acting on X and $\varphi \in C^1(X)$ is G-invariant. Show that:
(a) $\langle \varphi'(gu), v\rangle = \big\langle \varphi'(u), g^{-1}v\big\rangle$ for all $g \in G$, $u, v \in X$;
(b) if the action of G on X is isometric, then $\|\varphi'(gu)\|_* = \|\varphi'(u)\|_*$ for all $g \in G$ and all $u \in X$.

Problem 5.143**
Suppose that X is a Banach space, G is a compact topological group acting on X isometrically and $\varphi \in C^1(X)$ is G-invariant. Show that there exists an equivariant pseudogradient vector field $\widehat{v}\colon X \setminus K_\varphi \longrightarrow X$ for φ (see Definition 5.50).

Problem 5.144*
Suppose that X is a Banach space, G is a topological group acting on X and $\varphi \subset \Gamma_0(X)$ (see Remark 3.56) is G-invariant. Show that the subdifferential $\partial\varphi$ is equivariant (that is, $\partial\varphi(gu) = g\partial\varphi(u)$ for all $g \in G$, $u \in X$).

Problem 5.145***
Let $\Omega \subseteq \mathbb{R}^N$ (with $N \geqslant 3$) be a bounded domain with a Lipschitz boundary $\partial\Omega$, $\lambda \in (0, \lambda_1)$ and consider the C^1-functional $\varphi_\lambda\colon H_0^1(\Omega) \longrightarrow \mathbb{R}$ defined by

$$\varphi_\lambda(u) \;=\; \tfrac{1}{2}\|Du\|_2^2 - \tfrac{\lambda}{2}\|u\|_2^2 - \tfrac{1}{2^*}\|u\|_{2^*}^{2^*},$$

with $2^* = \frac{2N}{N-2}$. Show that φ satisfies the PS_c-condition for all $c < \frac{S^{\frac{N}{2}}}{N}$, with S being the optimal constant for the Gagliardo–Nirenberg–Sobolev inequality (see Theorem 1.133), that is, $S = \inf\big\{\|Du\|_2^2 : \ u \in H_0^1(\Omega),\ \|u\|_{2^*} = 1\big\}$.

Problem 5.146**
Suppose that $X_1 \subseteq \ldots \subseteq X_{m+1}$ are Hausdorff topological spaces. Show that

$$\operatorname{rank} H_k(X_{m+1}, X_1) \;\leqslant\; \sum_{n=1}^{m} \operatorname{rank} H_k(X_{n+1}, X_n) \qquad \forall\, k \geqslant 0.$$

Problem 5.147**

Suppose that X is a Banach space, $\varphi \in C^1(X)$, $-\infty < a < c < b \leqslant +\infty$, K_φ^c is finite, $K_\varphi \cap [a,b] = \{c\}$, φ satisfies the C$_d$-condition for every $d \in [a,b] \cap \mathbb{R}$. Show that

$$H_k(\varphi^b, \varphi^a) = \bigoplus_{u \in K_\varphi^c} C_k(\varphi, u) \qquad \forall\, k \geqslant 0.$$

Problem 5.148**

Suppose that X is a Banach space, $\varphi \in C^1(X)$, $-\infty < a < b \leqslant +\infty$ are regular values of φ, $\varphi_a^b \cap K_\varphi$ is a finite set and φ satisfies the C$_c$-condition for every $c \in [a,b] \cap \mathbb{R}$. Show that

$$\operatorname{rank} H_k(\varphi^b, \varphi^a) \leqslant \sum_{u \in K_\varphi[a,b]} \operatorname{rank} C_k(\varphi, u),$$

where $K_\varphi[a,b] = \{u \in K_\varphi : \ a \leqslant \varphi(u) \leqslant b\}$.

Problem 5.149**

Suppose that X is a Banach space, $\varphi \in C^1(X)$, φ satisfies the C-condition, K_φ is finite, $\inf \varphi(K_\varphi) > -\infty$ and $C_k(\varphi, \infty) \neq 0$ for some $k \geqslant 0$. Show that there exists $u \in K_\varphi$ such that $C_k(\varphi, u) \neq 0$.

Problem 5.150*

Suppose that X is a Banach space, $\varphi \in C^1(X)$ and $u \in X$ is a local minimizer of φ which is an isolated critical point of φ. Show that $C_k(\varphi, u) = \delta_{k,0}\mathbb{Z}$.

Problem 5.151**

Suppose that X is a finite dimensional Banach space, $\varphi \in C^1(X)$ and $u \in X$ is a local maximizer of φ which is an isolated critical point of φ. Show that $C_k(\varphi, u) = \delta_{k,d}\mathbb{Z}$ for all $k \geqslant 0$ with $d = \dim X$. What can be said if X is infinite dimensional? Justify your answer.

Problem 5.152**

Let $\varphi \in C^1(\mathbb{R})$ and let $u \in K_\varphi$ be isolated. Determine all the critical groups $C_k(\varphi, u)$ for $k \geqslant 0$.

Problem 5.153**

Suppose that X is a Banach space, $\varphi \in C^1(X)$ satisfies the C-condition and is bounded below and K_φ is finite. Show that $C_k(\varphi, \infty) = \delta_{k,0}\mathbb{Z}$ for all $k \geqslant 0$ and φ has a global minimizer $u_0 \in X$ such that $C_k(\varphi, u_0) = \delta_{k,0}\mathbb{Z}$ for all $k \geqslant 0$.

Problem 5.154**

Suppose that X is a Banach space, $\varphi \in C^1(X)$ satisfies the C-condition and $K_\varphi = \{0\}$. Show that $C_k(\varphi, 0) = C_k(\varphi, \infty)$ for all $k \geqslant 0$.

Problem 5.155**
Suppose that X is a Banach space, $\varphi \in C^1(X)$ satisfies the C-condition, $\inf \varphi(K_\varphi) > -\infty$, $0 \in K_\varphi$ is isolated and $C_{k_0}(\varphi, 0) \neq C_{k_0}(\varphi, \infty)$ for some $k_0 \geqslant 0$. Show that $K_\varphi \cap (X \setminus \{0\}) \neq \emptyset$.

Problem 5.156**
Suppose that X is a Banach space, $\varphi \in C^1(X)$ satisfies the C-condition, $0 \in K_\varphi$ is isolated and $-\infty < \inf \varphi(K_\varphi)$. Show that:

(a) if $C_k(\varphi, 0) = 0$ and $C_k(\varphi, \infty) \neq 0$ for some $k \geqslant 0$, then we can find $u \in K_\varphi \setminus \{0\}$ and $C_k(\varphi, u) \neq 0$;

(b) if $C_k(\varphi, 0) \neq 0$ and $C_k(\varphi, \infty) = 0$ for some $k \geqslant 0$, then we can find $u \in K_\varphi \setminus \{0\}$ such that $\varphi(u) < 0$ and $C_{k-1}(\varphi, u) \neq 0$ or $\varphi(u) > 0$ and $C_{k+1}(\varphi, u) \neq 0$.

Problem 5.157**
Suppose that X is a Banach space, $\varphi \in C^1(X)$ satisfies the C-condition, $0 \in K_\varphi$ is isolated, $-\infty < \inf \varphi(K_\varphi)$, $C_0(\varphi, 0) \neq 0$ and $C_0(\varphi, \infty) = 0$. Show that φ has a critical point u of mountain pass type (see Definition 5.71) such that $\varphi(u) > 0$.

Problem 5.158**
Suppose that X is a Banach space, $\varphi \in C^1(X)$ satisfies the C-condition and is bounded below, $0 \in K_\varphi$ is isolated and $C_k(\varphi, 0) \neq 0$ for some $k \geqslant 1$. Show that φ has a critical point $u_0 \neq 0$. Moreover, if $k \geqslant 2$, then there is a second nontrivial critical point $\widehat{u}$ of φ.

Problem 5.159***
Suppose that $\Omega \subseteq \mathbb{R}^N$ be a bounded domain with a C^2-boundary $\partial\Omega$ and let $f, g\colon \Omega \times \mathbb{R} \longrightarrow \mathbb{R}$ be Carathéodory functions such that

$$|f(z,x)|,\ |g(z,x)| \ \leqslant\ c(1 + |x|^{r-1}) \quad \text{for a.a. } z \in \Omega, \text{ all } x \in \mathbb{R},$$

with $c > 0$ and $r \in [2, 2^*)$, where

$$2^* \ =\ \begin{cases} \frac{2N}{N-2} & \text{if } N \geqslant 3, \\ +\infty & \text{if } N = 1, 2. \end{cases}$$

We set $F(z,x) = \int\limits_0^x f(z,s)\,ds$, $G(z,x) = \int\limits_0^x g(z,s)\,ds$ and consider the C^1-functionals $\varphi, \widehat{\varphi}\colon H_0^1(\Omega) \longrightarrow \mathbb{R}$ defined by

$$\begin{aligned} \varphi(u) &= \tfrac{1}{2}\|Du\|_2^2 - \int\limits_\Omega F(z,u)\,dz \qquad \forall\, u \in H_0^1(\Omega), \\ \widehat{\varphi}(u) &= \tfrac{1}{2}\|Du\|_2^2 - \int\limits_\Omega G(z,u)\,dz \qquad \forall\, u \in H_0^1(\Omega). \end{aligned}$$

Suppose that u_0 is an isolated critical point of φ and there exists $\delta > 0$ such that

$$f(z, u_0(z) + x) \;=\; g(z, u_0(z) + x) \quad \text{for a.a. } z \in \Omega, \text{ all } |x| \leqslant \delta.$$

Show that u_0 is also an isolated critical point of $\widehat{\varphi}$ and $C_k(\varphi, u_0) = C_k(\widehat{\varphi}, u_0)$ for all $k \geqslant 0$.

Problem 5.160**

Suppose that $\Omega \subseteq \mathbb{R}^N$ is a bounded domain with a Lipschitz boundary $\partial\Omega$, $\sigma(2)$ denotes the set of eigenvalues of $(-\Delta, H_0^1(\Omega))$ (i.e., $\sigma(2) = \{\widehat{\lambda}_k : k \geqslant 1\}$), $\lambda \in \mathbb{R} \setminus \sigma(2)$ and $\varphi_\lambda \colon H_0^1(\Omega) \longrightarrow \mathbb{R}$ is defined by

$$\varphi_\lambda(u) \;=\; \tfrac{1}{2}\|Du\|_2^2 - \tfrac{\lambda}{2}\|u\|_2^2 \qquad \forall\, u \in H_0^1(\Omega).$$

Show that:

(a) if $\lambda < \widehat{\lambda}_1$, then $C_k(\varphi_\lambda, 0) = \delta_{k,0}\mathbb{Z}$ for all $k \geqslant 0$;

(b) if $\lambda \in (\widehat{\lambda}_m, \widehat{\lambda}_{m+1})$ for some $m \geqslant 1$, then $C_k(\varphi_\lambda, 0) = \delta_{k,d_m}$ for all $k \geqslant 0$, with $d_m = \dim \bigoplus\limits_{i=1}^{m} E(\widehat{\lambda}_i)$.

Problem 5.161***

Let $\Omega \subseteq \mathbb{R}^N$ be a bounded domain with a C^2-boundary $\partial\Omega$ and let $f \colon \Omega \times \mathbb{R} \longrightarrow \mathbb{R}$ be a Carathéodory function such that

$$|f(z,x)| \;\leqslant\; c(1 + |x|^{r-1}) \quad \text{for a.a. } z \in \Omega, \text{ all } x \in \mathbb{R},$$

with $c > 0$, $r \in [2, 2^*)$, where

$$2^* \;=\; \begin{cases} \frac{2N}{N-2} & \text{if } N \geqslant 3, \\ +\infty & \text{if } N = 1, 2. \end{cases}$$

Let $F(z,x) = \int\limits_0^x f(z,s)\,ds$ and consider the C^1-functional $\varphi \colon H_0^1(\Omega) \longrightarrow \mathbb{R}$ defined by

$$\varphi(u) \;=\; \tfrac{1}{2}\|Du\|_2^2 - \int_\Omega F(z,u)\,dz \qquad \forall\, u \in H_0^1(\Omega).$$

Suppose that $u = 0$ is an isolated critical point of φ. Show that:

(a) if there exists $\delta > 0$ such that

$$\tfrac{f(z,x)}{x} \;\leqslant\; \widehat{\lambda}_1 \quad \text{for a.a. } z \in \Omega, \text{ all } 0 < |x| \leqslant \delta,$$

then $C_k(\varphi, 0) = \delta_{k,0}\mathbb{Z}$ for all $k \geqslant 0$;

(b) if there exist $\delta > 0$ and $m \geqslant 1$ such that

$$\widehat{\lambda}_m \;\leqslant\; \tfrac{f(z,x)}{x} \;\leqslant\; \widehat{\lambda}_{m+1} \quad \text{for a.a. } z \in \Omega, \text{ all } 0 < |x| \leqslant \delta,$$

then $C_k(\varphi, 0) = \delta_{k,d_m}\mathbb{Z}$ for all $k \geqslant 0$, with $d_m = \dim \overline{H}_m$, where $H_m = \bigoplus\limits_{i=0}^{m} E(\widehat{\lambda}_i)$.

Problem 5.162**

Let $\Omega \subseteq \mathbb{R}^N$ be a bounded domain with a C^2-boundary $\partial\Omega$ and let $f: \Omega \times \mathbb{R} \longrightarrow \mathbb{R}$ be a Carathéodory function such that

$$|f(z,x)| \leqslant c(1+|x|^{r-1}) \quad \text{for a.a. } z \in \Omega, \text{ all } x \in \mathbb{R},$$

with $c > 0$, $r \in [2, 2^*)$, where

$$2^* = \begin{cases} \frac{2N}{N-2} & \text{if } N \geqslant 3, \\ +\infty & \text{if } N = 1, 2. \end{cases}$$

Suppose that there exists $\delta > 0$ such that

$$f(z,x)x \leqslant 0 \quad \text{for a.a. } z \in \Omega, \text{ all } |x| \leqslant \delta.$$

Let $F(z,x) = \int\limits_0^x f(x,s)\,ds$ and consider the C^1-functional $\varphi\colon H_0^1(\Omega) \longrightarrow \mathbb{R}$ defined by

$$\varphi(u) = \tfrac{1}{2}\|Du\|_2^2 - \int\limits_\Omega F(z,u)\,dz \qquad \forall\, u \in H_0^1(\Omega).$$

Show that $C_k(\varphi,0) = \delta_{k,0}\mathbb{Z}$ for all $k \geqslant 0$.

Problem 5.163**

Suppose that $\varphi \in C^2(\mathbb{R}^N)$, $U \subseteq \mathbb{R}^N$ is a nonempty, bounded open set and $K \subseteq U$ is a nonempty compact set such that $K_\varphi \cap \overline{U \setminus K} = \emptyset$. Show that given any $\varepsilon > 0$, we can find $\widehat{\varphi} \in C^2(\mathbb{R}^N)$ such that:

(a) $|\varphi(u) - \widehat{\varphi}(u)| + |\varphi'(u) - \widehat{\varphi}'(u)| \leqslant \varepsilon$ for all $u \in \mathbb{R}^N$.

(b) $\varphi(u) = \widehat{\varphi}(u)$ for all $u \notin U$.

(c) the set $K_{\widehat{\varphi}} \cap \overline{U}$ is finite and its elements are nondegenerate.

Problem 5.164***

Let $\Omega \subseteq \mathbb{R}^N$ be a bounded domain with a C^2-boundary $\partial\Omega$ and let $f: \Omega \times \mathbb{R} \longrightarrow \mathbb{R}$ be a Carathéodory function such that:

(i) $|f(z,x)| \leqslant c(1+|x|^{r-1})$ for almost all $z \in \Omega$, $x \in \mathbb{R}$, with $c > 0$, $r \in [2, 2^*)$, where

$$2^* = \begin{cases} \frac{2N}{N-2} & \text{if } N \geqslant 3, \\ +\infty & \text{if } N = 1, 2. \end{cases}$$

(ii) there exist functions $\eta, \widehat{\eta} \in L^\infty(\Omega)$ such that either

(I) $\widehat{\eta}(z) \leqslant \widehat{\lambda}_1$ for almost all $z \in \Omega$ with strict inequality on a set of positive measure; or

(II) $\widehat{\lambda}_m \leqslant \eta(z) \leqslant \widehat{\eta}(z) \leqslant \widehat{\lambda}_{m+1}$ for some $m \geqslant 1$ and for almost all $z \in \Omega$ with the first and the third inequalities strict on sets of positive measures and

$$\eta(z) \leqslant \liminf_{x \to 0} \frac{f(z,x)}{x} \leqslant \limsup_{x \to 0} \frac{f(z,x)}{x} \leqslant \widehat{\eta}(z)$$

uniformly for a.a. $z \in \Omega$.

Set $F(z,x)=\int_0^x f(z,s)\,ds$ and consider the C^1-functional $\varphi\colon H_0^1(\Omega)\longrightarrow\mathbb{R}$ defined by

$$\varphi(u) \;=\; \tfrac{1}{2}\|Du\|_2^2-\int_\Omega F(z,u)\,dz \qquad \forall\, u\in H_0^1(\Omega).$$

Show that

$$C_k(\varphi,0) \;=\; \delta_{k,0}\mathbb{Z} \qquad \forall\, k\geqslant 0$$

if **(ii)(I)** holds and

$$C_k(\varphi,0) \;=\; \delta_{k,d_m}\mathbb{Z} \qquad \forall\, k\geqslant 0,$$

with $d_m=\dim\bigoplus_{i=1}^m E(\widehat{\lambda}_i)$ if **(ii)(II)** holds.

Problem 5.165**
Let $\Omega\subseteq\mathbb{R}^N$ be a bounded domain with a Lipschitz boundary $\partial\Omega$ and let $\sigma(2)$ be the set of eigenvalues of $(-\Delta,H_0^1(\Omega))$ (that is, $\sigma(2)=\{\widehat{\lambda}_k:\ k\geqslant 1\}$). Suppose that $\lambda\in\mathbb{R}\setminus\sigma(2)$ and consider the C^1-functional $\varphi_\lambda\colon H_0^1(\Omega)\longrightarrow\mathbb{R}$ defined by

$$\varphi_\lambda(u) \;=\; \tfrac{1}{2}\|Du\|_2^2-\tfrac{\lambda}{2}\|u\|_2^2 \qquad \forall\, u\in H_0^1(\Omega).$$

Compute $C_k(\varphi_\lambda,0)$ for all $k\geqslant 0$.

Problem 5.166***
Let $\Omega\subseteq\mathbb{R}^N$ be a bounded domain with a Lipschitz boundary $\partial\Omega$ and let $f\colon\Omega\times\mathbb{R}\longrightarrow\mathbb{R}$ be a Carathéodory function such that:

(i) there exists $c>0$ such that

$$|f(x,z)| \;\leqslant\; c(1+|x|^{p-1}) \quad \text{for a.a. } z\in\Omega, \text{ all } x\in\mathbb{R};$$

(ii) if $F(z,x)=\int_0^x f(z,s)\,ds$, then

$$\limsup_{x\to\pm\infty}\tfrac{pF(z,x)}{|x|^p} \;\leqslant\; \widehat{\lambda}_1 \quad \text{uniformly for a.a. } z\in\Omega;$$

(iii) $\lim_{x\to\pm\infty}\big(f(z,x)x-pF(z,x)\big)=+\infty$ uniformly for almost all $z\in\Omega$.

Let $\varphi\colon W_0^{1,p}(\Omega)\longrightarrow\mathbb{R}$ be the C^1-functional defined by

$$\varphi(u) \;=\; \tfrac{1}{p}\|Du\|_p^p-\int_\Omega F(z,u)\,dz \qquad \forall\, u\in W_0^{1,p}(\Omega).$$

Assuming that K_φ is finite, compute the critical groups of φ at infinity.

Problem 5.167*
Suppose that X is a Banach space, $E_0\subseteq E$ and D are nonempty subsets of X such that $E_0\cap D=\emptyset$, $\{E_0,E\}$ and D homologically link in dimension m (see Definition 5.118) and $h\colon X\longrightarrow X$ is a homeomorphism. Show that $\{h(E_0),h(E)\}$ and $h(D)$ homologically link in dimension m.

Problem 5.168**
Show that the notion of homological linking (see Definition 5.118) is essentially independent of the choice of $E \supseteq E_0$ contractible.

Problem 5.169**
Suppose that X is a Banach space, $u_0, u_1 \in X$, U is an open neighborhood of u_0 such that $u_1 \notin U$, $E_0 = \{u_0, u_1\}$, $E = \{tu_0 + (1-t)u_1 : t \in [0,1]\}$ and $D = \partial U$. Show that $\{E_0, E\}$ and D homologically link in dimension 1 (see Definition 5.118).

Problem 5.170**
Suppose that X is a Banach space, $X = Y \oplus V$ with $\dim Y < +\infty$, $\varphi \in C^1(X)$ satisfies the C-condition, $-\infty < \inf \varphi(K_\varphi)$ and
(i) $\varphi|_V$ is bounded below;
(ii) $\varphi(y) \longrightarrow -\infty$ as $\|y\| \to +\infty$, $y \in Y$.
Show that $C_d(\varphi, \infty) \neq 0$, where $d = \dim Y$.

Problem 5.171*
Suppose that X is a Banach space, $\varphi \in C^1(X)$ satisfies the C-condition and $a < \inf \varphi(K_\varphi) \leqslant \sup \varphi(K_\varphi) < b$. Show that $C_k(\varphi, \infty) = H_k(\varphi^b, \varphi^a)$ for all $k \geqslant 0$.

Problem 5.172*
Suppose that X is a Banach space, $\varphi \in C^1(X)$ satisfies the C-condition, and $K_\varphi = \emptyset$. Show that $C_k(\varphi, \infty) = 0$ for all $k \geqslant 0$.

Problem 5.173*
Suppose that X is a Banach space, $\varphi \in C^1(X)$ satisfies the C-condition, $\inf \varphi(K_\varphi) > -\infty$ and $C_0(\varphi, \infty) \neq 0$. Show that $C_k(\varphi, \infty) = \delta_{k,0}\mathbb{Z}$ for all $k \geqslant 0$.

Problem 5.174**
Let $\Omega \subseteq \mathbb{R}^N$ be a bounded domain with a Lipschitz boundary $\partial\Omega$ and let $f \colon \Omega \times \mathbb{R} \longrightarrow \mathbb{R}$ be a Carathéodory function such that:
(i) $|f(z,x)| \leqslant c(1+|x|^{r-1})$ for almost all $z \in \Omega$, all $x \in \mathbb{R}$, with $c > 0$, $1 < p < r < p^*$, where

$$p^* = \begin{cases} \frac{Np}{N-p} & \text{if } p < N, \\ +\infty & \text{if } N \leqslant p; \end{cases}$$

(ii) there exists $\vartheta \in L^\infty(\Omega)$ such that $\vartheta(x) \leqslant \widehat{\lambda}_1$ for almost all $z \in \Omega$ with strict inequality on a set of positive Lebesgue measure ($\widehat{\lambda}_1$ being the first eigenvalue of $(-\Delta_p, W_0^{1,p}(\Omega))$; see Proposition 5.132) and if $F(z,x) = \int\limits_0^x f(z,s)\,ds$, then

$$\limsup_{x \to \pm\infty} \frac{pF(z,x)}{|x|^p} \leqslant \vartheta(z) \quad \text{uniformly for a.a. } z \in \Omega.$$

We consider the following Dirichlet boundary value problem:

$$\begin{cases} -\Delta_p u(z) = f(z, u(z)) & \text{in } \Omega, \\ u|_{\partial\Omega} = 0. \end{cases}$$

Let $\varphi\colon W_0^{1,p}(\Omega) \longrightarrow \mathbb{R}$ be the C^1-energy functional for the above problem, defined by

$$\varphi(u) = \tfrac{1}{p}\|Du\|_p^p - \int_\Omega F(z,u)\,dz \qquad \forall\, u \in W_0^{1,p}(\Omega).$$

Assume that K_φ is discrete. Show that the Dirichlet problem has a solution $u_0 \in W_0^{1,p}(\Omega)$ such that $C_k(\varphi, u_0) = \delta_{k,0}\mathbb{Z}$ for all $k \geqslant 0$.

Problem 5.175*

Suppose that X is a Banach space, $\varphi \in C^1(X)$ satisfies the C-condition and $\inf \varphi(K_\varphi) > -\infty$. Suppose that φ is not bounded below and pick $a < \inf \varphi(K_\varphi)$. Show that

$$C_k(\varphi, \infty) = H_{k-1}(\varphi^a, \star) \qquad \forall\, k \geqslant 0,$$

with $\star \in \varphi^a$.

Problem 5.176***

Let $\Omega \subseteq \mathbb{R}^N$ be a bounded domain with a C^2-boundary $\partial\Omega$ and let $f\colon \Omega \times \mathbb{R} \longrightarrow \mathbb{R}$ be a Carathéodory function such that $f(z,0) = 0$ for almost all $z \in \Omega$ and

(i) $|f(z,x)| \leqslant a(z)(1+|x|^{r-1})$ for almost all $z \in \Omega$, all $x \in \mathbb{R}$, with $a \in L^\infty(\Omega)$ and $2 < r < 2^*$;

(ii) $\lim\limits_{x\to\pm\infty} \frac{f(z,x)}{x} = +\infty$ uniformly for almost all $z \in \Omega$;

(iii) $\lim\limits_{x\to 0} \frac{f(z,x)}{x} = 0$ uniformly for almost all $z \in \Omega$.

Consider the following parametric Dirichlet problem:

$$\begin{cases} -\Delta u(z) = \lambda u(z) - f(z, u(z)) & \text{in } \Omega, \\ u|_{\partial\Omega} = 0, \ \lambda > 0. \end{cases}$$

Show that for every $\lambda > \widehat{\lambda}_2$, the above Dirichlet problem has at least three nontrivial solutions $u_0 \in \operatorname{int} C_+$ (see Theorem 5.129(e)), $v_0 \in -\operatorname{int} C_+$ and $y_0 \in C_0^1(\overline{\Omega})$.

5.3 Solutions

Solution of Problem 5.1

"$\Longrightarrow$": Arguing by contradiction, suppose that we can find $\lambda \in \mathbb{R}$ and a sequence $\{u_n^*\}_{n\geqslant 1} \subseteq X^*$ such that

$$\varphi(u_n^*) \leqslant \lambda \quad \text{and} \quad \|u_n^*\|_* > n \qquad \forall\, n \geqslant 1.$$

But this contradicts Definition 5.5(a). So $\lim\limits_{\|u^*\|_* \to +\infty} \varphi(u^*) = +\infty$.

"$\Longleftarrow$": The limit condition implies that for every $\lambda \in \mathbb{R}$, we can find $M_\lambda > 0$ such that

$$\text{if } \|u^*\|_* > M_\lambda, \text{ then } \varphi(u^*) > \lambda. \tag{5.13}$$

By the Alaoglu theorem (see Theorem I.5.66), we have that the set

$$C_\lambda = \big\{u^* \in X^* : \|u^*\|_* \leqslant M_\lambda\big\}$$

is w^*-compact. Since $L_\lambda = \big\{u^* \in X^* : \varphi(u^*) \leqslant \lambda\big\} \subseteq C_\lambda$ (see (5.13)), it follows that L_λ is a relatively w^*-compact set. Because $\lambda \in \mathbb{R}$ is arbitrary, we conclude that φ is w^*-coercive.

Solution of Problem 5.2

We do the solution for the weak topology, the situation for the w_{seq}-topology being similar, with the nets replaced by sequences.

Note that hypothesis **(i)** implies that for all $v \in X$, the map $u \longmapsto \xi(u,v)$ is weakly lower semicontinuous (by the Mazur theorem; see Theorem I.5.58). Let $\lambda \in \mathbb{R}$. We need to show that the set $L_\lambda = \{u \in X : \varphi(u) \leqslant \lambda\}$ is weakly closed (see Remark 5.2). So, let $\{u_\alpha\}_{\alpha\in J} \subseteq L_\lambda$ be a net such that $u_\alpha \xrightarrow{w} u$ in X. We have

$$\varphi(u_\alpha) = \xi(u_\alpha, u_\alpha) = \xi(u_\alpha, u) + \xi(u_\alpha, u_\alpha) - \xi(u_\alpha, u). \tag{5.14}$$

We know that the sequence $\{u_\alpha\}_{\alpha\in J} \subseteq X$ is bounded and so by hypothesis **(ii)** the family $\{\psi_\alpha(\cdot) = \xi(u_\alpha, \cdot)\}_{\alpha\in J}$ is weakly equicontinuous on X. We see that for every $v \in X$, $\psi_\alpha(v) \longrightarrow \psi_u(v) = \xi(u,v)$. The set $\overline{\{u_\alpha\}}^{w}_{\alpha\in J}$ is weakly compact. Since by the weak equicontinuity, the topology of pointwise convergence coincides with the topology of uniform convergence on compacta (for the weak topology), we have

$$\xi(u_\alpha, u_\alpha) = \psi_\alpha(u_\alpha) \longrightarrow \psi_u(u) = \xi(u,u),$$

so

$$\begin{aligned} &\big|\xi(u_\alpha, u_\alpha) - \xi(u_\alpha, u)\big| \\ \leqslant\ & \big|\xi(u_\alpha, u) - \xi(u,u)\big| + \big|\xi(u,u) - \xi(u_\alpha, u)\big| \overset{\alpha\in J}{\longrightarrow} 0. \end{aligned} \tag{5.15}$$

Returning to (5.14) and using (5.15), we obtain

$$\lambda \geqslant \liminf_{\alpha \in J} \varphi(u_\alpha) = \liminf_{\alpha \in J} \xi(u_\alpha, u) \geqslant \xi(u, u) = \varphi(u),$$

so $u \in L_\lambda$ and thus L_λ is weakly closed.

This means that φ is weakly lower semicontinuous. Similarly for the "sequential" case.

Solution of Problem 5.3

The reflexivity of X implies that $C \subseteq X$ is weakly compact by the Alaoglu theorem (see Theorem I.5.66). Also, from Problem 5.2, we know that φ is weakly lower semicontinuous. Hence we can use Theorem 5.6, to conclude that there exists $\widehat{u} \in C$ such that $\varphi(\widehat{u}) = \inf\limits_{C} \varphi$.

Solution of Problem 5.4

Because of the reflexivity of X, for every $r > 0$, the set $\overline{B}_r = \{u \in X : \|u\|_X \leqslant r\}$ is weakly compact (by the Alaoglu theorem; see Theorem I.5.66). Also, by Problem 5.2, φ is weakly lower semicontinuous. So, we can find $\widehat{u} \in \overline{B}_r$ such that

$$\varphi(\widehat{u}) = \inf_{u \in \overline{B}_r} \varphi(u). \tag{5.16}$$

On the other hand, the weak coercivity hypothesis on φ and Problem 5.1 imply that we can find $r > 0$ such that

$$\text{if } \|u\|_X > r, \text{ then } \varphi(u) > \varphi(0) \geqslant \inf_{\overline{B}_r} \varphi. \tag{5.17}$$

Then from (5.16) and (5.17), we conclude that $\varphi(\widehat{u}) = \inf\limits_{X} \varphi$.

Solution of Problem 5.5

"$\Longrightarrow$": Suppose that there exists $\widehat{u} \in C$ such that

$$\varphi(\widehat{u}) = \inf_{C} \varphi.$$

Then for $r > 0$, we have

$$\varphi(\widehat{u}) = \inf_C \varphi = \inf_{C \cap \overline{B}_r(\widehat{u})} \varphi,$$

where $\overline{B}_r(\widehat{u}) = \{u \in X : \|u - \widehat{u}\|_X \leqslant r\}$. Hence every minimizing sequence in $C \cap \overline{B}_r(\widehat{u})$ is bounded.

"$\Longleftarrow$": Let $\{u_n\}_{n \geqslant 1} \subseteq C$ be the bounded minimizing sequence. Because of the reflexivity of X, the set $\{u_n : n \geqslant 1\} \subseteq C$ is relatively weakly compact. Then by the Eberlein–Smulian theorem (see Theorem I.5.78) and by passing to a subsequence if necessary, we may assume that $u_n \xrightarrow{w} \widehat{u}$ in X. Since the set $C \subseteq X$ is weakly closed, we have $\widehat{u} \in C$. Also, the weak lower semicontinuity of φ implies that

$$\varphi(\widehat{u}) \leqslant \liminf_{n \to +\infty} \varphi(u_n) = \inf_C \varphi,$$

so $\varphi(\widehat{u}) = \inf_C \varphi$ (since $\widehat{u} \in C$).

Solution of Problem 5.6

Since by hypothesis φ is weakly coercive, it is bounded below. Also, using Problem 5.1, we can find $r > 0$ such that

$$\text{if } \|u\|_X > r, \text{ then } \varphi(u) > \inf_M \varphi.$$

Then for $M_r = M \cap \overline{B}_r$, where $\overline{B}_r = \{u \in X : \|u\|_X \leqslant r\}$, we have

$$\inf_M \varphi = \inf_{M_r} \varphi. \tag{5.18}$$

Since by hypothesis f is weakly continuous, the set M_r is bounded and weakly closed, hence weakly compact (recall that X is reflexive). So, by Theorem 5.6, we can find $\widehat{u} \in M_r$ such that $\varphi(\widehat{u}) = \inf_M \varphi$ (see (5.18)).

Solution of Problem 5.7

Evidently the convexity of G implies that φ is convex. We show that φ is strongly continuous on $W_0^{1,p}(\Omega)$. So, suppose that $u_n \longrightarrow u$ in $W_0^{1,p}(\Omega)$. Then by passing to a suitable subsequence if necessary, we may assume that

$$Du_n(z) \longrightarrow Du(z) \quad \text{for a.a. } z \in \Omega \tag{5.19}$$

and

$$|Du_n(z)| \leqslant \vartheta(z) \quad \text{for a.a. } z \in \Omega, \text{ all } n \geqslant 1, \tag{5.20}$$

with $\vartheta \in L^p(\Omega)$. The function G is convex and locally bounded (by the hypothesis). So, from Theorem 3.60, we have that G is continuous. Hence, from (5.19), we have

$$G(Du_n(z)) \longrightarrow G(Du(z)) \quad \text{for a.a. } z \in \Omega. \tag{5.21}$$

For almost all $z \in \Omega$ and all $n \geqslant 1$, we have

$$|G(Du_n(z))| \leqslant c_1(1 + |Du_n|^p) \leqslant c_1(1 + \vartheta(z)^p) \tag{5.22}$$

(by hypothesis). Then from (5.21), (5.22) and the Lebesgue dominated convergence theorem (see Theorem I.3.94), we obtain

$$\int_\Omega G(Du_n)\,dz \longrightarrow \int_\Omega G(Du)\,dz. \tag{5.23}$$

Also, since $u_n \longrightarrow u$ in $L^p(\Omega)$, we have

$$\int_\Omega f u_n\,dz \longrightarrow \int_\Omega f u\,dz. \tag{5.24}$$

From (5.23) and (5.24) we infer that φ is strongly continuous. In particular then φ is lower semicontinuous and the convexity implies that φ is weakly lower semicontinuous on $W_0^{1,p}(\Omega)$. From the hypothesis on G and the Hölder inequality (see Theorem 1.3), we have

$$\varphi(u) \geqslant c_2\|Du_n\|_p^p - c_3|\Omega|_N - \|u\|_p\|h\|_{p'} \geqslant c_2\|u\|^p - c_3|\Omega|_N - c_4\|u\|, \tag{5.25}$$

for some $c_4 > 0$, where $|\cdot|_N$ denotes the Lebesgue measure on $\mathbb{R}^N$ and $\|\cdot\|$ the norm of the Sobolev space $W_0^{1,p}(\Omega)$. By the Poincaré inequality (see Theorem 1.131), we have that $\|u\| = \|Du\|_p$ for all $u \in W_0^{1,p}(\Omega)$. From (5.25) and since $p > 1$, we infer that φ is coercive (see Problem 5.1). So, we can apply Theorem 5.6 and conclude that there exists $\widehat{u} \in W_0^{1,p}(\Omega)$ such that $\varphi(\widehat{u}) = \inf\limits_{u \in W_0^{1,p}(\Omega)} \varphi(u)$.

Solution of Problem 5.8

Note that by the mean value theorem (see Theorem 3.16), we see that the hypotheses on G imply that the hypotheses of Problem 5.7 hold. Using Problem 5.7, we infer that there exists $\widehat{u} \in W_0^{1,p}(\Omega)$ such that

$$\varphi(\widehat{u}) = \inf_{u \in W_0^{1,p}(\Omega)} \varphi(u), \tag{5.26}$$

with $\varphi\colon W_0^{1,p}(\Omega) \longrightarrow \mathbb{R}$ being defined by

$$\varphi(u) = \int_\Omega G(Du(z))\,dz - \int_\Omega f(z)u(z)\,dz \qquad \forall\, u \in W_0^{1,p}(\Omega)$$

(see Problem 5.7). For every $t \in \mathbb{R}$ and every $h \in W_0^{1,p}(\Omega)$, we have that $\widehat{u} + th \in W_0^{1,p}(\Omega)$. So, we can define $\vartheta(t) = \varphi(\widehat{u} + th)$ for all $t \in [-1,1]$. From (5.26) we see that ϑ attains its minimum at $t = 0$.

We will show that $\vartheta \in C^1([-1,1])$. From the definition of ϑ we have

$$\vartheta(t) = \int_\Omega G(D\widehat{u} + tDh)\,dz - \int_\Omega f\widehat{u}\,dz - t\int_\Omega fh\,dz.$$

Let $H(z,t) = G(Du(z) + tDh(z))$. Clearly $H(z,\cdot)$ is continuously differentiable and by the chain rule (see Theorem 3.19), we have

$$\tfrac{\partial H(z,t)}{\partial t} = (\nabla G(D\widehat{u}(z) + tDh(z)), Dh(z))_{\mathbb{R}^N},$$

so

$$\begin{aligned} \left|\tfrac{\partial H(z,t)}{\partial t}\right| &\leqslant \left|\nabla G(D\widehat{u}(z) + tDh(z))\right||Dh(z)| \\ &\leqslant c_1\big(1 + \left|D\widehat{u}(z) + tDh(z)\right|^{p-1}\big)|Dh(z)| \\ &\leqslant c_4\big(1 + |D\widehat{u}(z)|^{p-1} + |Dh(z)|^{p-1}\big)|Dh(z)| \quad \text{for a.a. } z \in \Omega, \end{aligned} \tag{5.27}$$

for some $c_4 > 0$ (by the hypotheses on G). Note that $|D\widehat{u}(\cdot)|^{p-1} \in L^{p'}(\Omega)$ and $|Dh(\cdot)| \in L^p(\Omega)$. So, by the Hölder inequality (see Theorem 1.3), we have $|D\widehat{u}(\cdot)|^{p-1}|Dh(\cdot)| \in L^1(\Omega)$. Clearly $|Dh(\cdot)|^p \in L^1(\Omega)$ (since $h \in W_0^{1,p}(\Omega)$). Therefore, from (5.27), we see that $\left|\frac{\partial H(z,t)}{\partial t}\right|$ is bounded by an L^1-function for all $t \in [-1,1]$. Hence the map $t \longmapsto \int_\Omega H(z,t)\,dz$ is C^1 and its derivative is given by $t \longmapsto \int_\Omega \frac{\partial H(z,t)}{\partial t}\,dz$. Therefore, we have

$$\vartheta'(t) = \int_\Omega (\nabla G(\widehat{u} + th), Dh)_{\mathbb{R}^N}\,dz - \int_\Omega fh\,dz. \tag{5.28}$$

Since $t = 0$ is a minimizer of ϑ, we have that $\vartheta'(0) = 0$ and so

$$\int_\Omega (\nabla G(\widehat{u}), Dh)_{\mathbb{R}^N}\,dz = \int_\Omega fh\,dz \qquad \forall\, h \in W_0^{1,p}(\Omega).$$

Solution of Problem 5.9

Since G is convex and C^1, we have

$$(\nabla G(y), v-y)_{\mathbb{R}^N} \leqslant G(v) - G(y) \qquad \forall\, v, y \in \mathbb{R}^N \tag{5.29}$$

(see Definition 3.70 and Theorem 3.73). Then we have

$$\begin{aligned}\int\limits_\Omega G(Dh)\,dz - \int\limits_\Omega G(D\widehat{u})\,dz &\geqslant \int\limits_\Omega (\nabla G(D\widehat{u}), Dh - D\widehat{u})_{\mathbb{R}^N}\,dz \\ &= \int\limits_\Omega f(h-\widehat{u})\,dz \qquad \forall\, h \in W_0^{1,p}(\Omega)\end{aligned}$$

(by hypothesis), so

$$\int\limits_\Omega G(Dh)\,dz - \int\limits_\Omega fh\,dz \geqslant \int\limits_\Omega G(D\widehat{u})\,dz - \int\limits_\Omega f\widehat{u}\,dz \qquad \forall\, h \in W_0^{1,p}(\Omega),$$

thus $\varphi(h) \geqslant \varphi(\widehat{u})$ for all $h \in W_0^{1,p}(\Omega)$.

Solution of Problem 5.10

We need to show that for every $\lambda \in \mathbb{R}$, the set $L_\lambda = \{\varphi \leqslant \lambda\} \subseteq L^p(\Omega)$ is strongly closed. So, let $\{u_n\}_{n\geqslant 1} \subseteq L_\lambda$ be a sequence such that $u_n \longrightarrow u$ in $L^p(\Omega)$. From hypothesis **(i)** we have that $\{u_n\}_{n\geqslant 1} \subseteq W^{1,p}(\Omega)$ and $\{Du_n\}_{n\geqslant 1} \subseteq L^p(\Omega;\mathbb{R}^N)$ is bounded. It follows that the sequence $\{u_n\}_{n\geqslant 1} \subseteq W^{1,p}(\Omega)$ is bounded and so by passing to a suitable subsequence if necessary, we may assume that

$$u_n \overset{w}{\longrightarrow} u \quad \text{in } W^{1,p}(\Omega),$$

so

$$\varphi(u) \leqslant \liminf_{n\to+\infty} \varphi(u_n) \leqslant \lambda$$

(see hypothesis **(ii)**), thus $u \in L_\lambda$ and hence the set L_λ is closed. Therefore φ is lower semicontinuous on $L^p(\Omega)$ with the strong topology.

Solution of Problem 5.11

Let $\lambda \in \mathbb{R}$. By hypothesis the set $L_\lambda \subseteq W^{1,p}(\Omega)$ is bounded in $W^{1,p}(\Omega)$. Then from the Rellich–Kondrachov embedding theorem (see Theoremt.1.133), we have that the set L_λ is relatively compact in $L^p(\Omega)$, which implies the coercivity of φ for the strong topology on $L^p(\Omega)$.

Solution of Problem 5.12

Let $F(x) = \int\limits_0^x f(s)\,ds$ and let $\varphi\colon H_0^1(\Omega) \longrightarrow \mathbb{R}$ be the functional defined by

$$\varphi(u) = \tfrac{1}{2}\|Du\|_2^2 + \tfrac{1}{2}\int\limits_\Omega \xi(z)u^2\,dz - \int\limits_\Omega F(u)\,dz - \int\limits_\Omega gu\,dz \qquad \forall\, u \in H_0^1(\Omega).$$

Then $\varphi \in C^1(H_0^1(\Omega))$ and for all $u, h \in H_0^1(\Omega)$, we have

$$\langle \varphi'(u), h\rangle = \int\limits_\Omega (Du, Dh)_{\mathbb{R}^N}\,dz + \int\limits_\Omega \xi(z)uh\,dz - \int\limits_\Omega f(u)h\,dz - \int\limits_\Omega gh\,dz \tag{5.30}$$

(see Problems 3.5 and 3.21). From (5.30), for all $u, v \in H_0^1(\Omega)$, we have

$$\begin{aligned}
&\langle \varphi'(u) - \varphi'(v), u - v\rangle \\
&= \|D(u-v)\|_2^2 + \int\limits_\Omega \xi(z)(u-v)^2\,dz - \int\limits_\Omega (f(u) - f(v))(u-v)\,dz \\
&\geqslant \|D(u-v)\|_2^2 = \|u-v\|^2
\end{aligned}$$

(see hypothesis **(ii)**, the fact that $\xi \geqslant 0$ and using Poincaré inequality; see Theorem 1.131). So, φ' is strongly monotone (see Definition 2.98) and this implies that φ is strictly convex.

Note that hypothesis **(ii)** implies that $F(x) \leqslant 0$ for all $x \in \mathbb{R}$. Hence

$$\varphi(u) \geqslant \tfrac{1}{2}\|u\|^2 - \|g\|_2\|u\|_2 \geqslant \tfrac{1}{2}\|u\|^2 - c_2\|u\| \qquad \forall\, u \in H_0^1(\Omega),$$

for some $c_2 > 0$, thus φ is coercive.

Also, φ being convex and locally bounded, it is continuous (see Theorem 3.60). So, by Theorem 5.6, there exists $\widehat{u} \in H_0^1(\Omega)$ such that $\varphi'(\widehat{u}) = 0$ and so $\widehat{u} \in H_0^1(\Omega)$ is a weak solution of the Dirichlet problem.

Moreover, the strict convexity of φ implies that $\widehat{u}$ is the unique critical point of φ, hence the unique weak solution of the Dirichlet problem (see Proposition 5.14 and Remark 5.15).

Solution of Problem 5.13

From the mean value theorem (see Theorem 3.16), we can find $\lambda, \lambda' \in [0,1]$ such that

$$\begin{aligned}
\varphi(v) - \varphi(u) &= \langle \varphi'(u + \lambda(v-u)), v - u\rangle \\
&= \langle \varphi'(u), v-u\rangle + \langle \varphi'(u + \lambda(v-u)) - \varphi'(u), v-u\rangle \\
&= \langle \varphi'(u), v-u\rangle + \lambda'\varphi''(u + \lambda(v-u))(v-u, v-u)
\end{aligned} \tag{5.31}$$

Suppose that $\{u_\alpha\}_{\alpha\in J} \subseteq \overline{B}_r(u)$ is a net such that $u_\alpha \xrightarrow{w} u$. Then $u_\alpha - u \in \overline{B}_r$ and from the hypothesis on φ'' and (5.31) we have

$$\varphi(u_\alpha) - \varphi(u) \geqslant \langle \varphi'(u), u_\alpha - u \rangle \qquad \forall\, \alpha \in J,$$

so

$$\liminf_{\alpha\in J} \varphi(u_\alpha) \geqslant \varphi(u),$$

thus φ is weakly lower semicontinuous.

Since X is reflexive, $\overline{B}_r$ is w-compact. So, Theorem 5.6 implies that there exists $\widehat{u} \in \overline{B}_r$ such that $\varphi(\widehat{u}) = \inf\limits_{u\in\overline{B}_r} \varphi(u)$.

Solution of Problem 5.14

For all $e \in X$ with $\|e\|_X = 1$ and all $t > t_0$, we have

$$\varphi(te) - \varphi(t_0 e) = \int_{t_0}^{t} \tfrac{d}{ds}\varphi(se)\,ds = \int_{t_0}^{t} \langle \varphi'(se), e\rangle\, ds \geqslant \int_{t_0}^{t} \tfrac{\vartheta(s)}{s}\, ds$$

(by the chain rule; see Theorem 3.19, the hypotheses and since $\|e\|_X = 1$). Then for $u \in X$, we have

$$\varphi(u) \geqslant \varphi(t_0 \tfrac{u}{\|u\|_X}) + \int_{t_0}^{\|u\|_X} \tfrac{\vartheta(s)}{s}\, ds \longrightarrow +\infty \quad \text{as } \|u\|_X \longrightarrow +\infty.$$

Solution of Problem 5.15

Note that $\overline{C}^w$ is bounded, w-closed, hence w-compact (due to the reflexivity of X). Since by hypothesis φ is weakly lower semicontinuous, by Theorem 5.6, we can find $\widehat{u} \in \overline{C}^w$ such that

$$\varphi(\widehat{u}) = \inf_{u\in\overline{C}^w} \varphi(u).$$

Because of the hypothesis on φ, we can have $\widehat{u} \in C$. Then $\widehat{u} \in K_\varphi \cap C$.

Solution of Problem 5.16

From Proposition 5.31, we know that $u \in K_{\varphi|_M}$ if and only if there exists $\lambda \in \mathbb{R}$ such that $\varphi'(u) = \lambda u$, hence $A(u) = \lambda u$. Evidently $\lambda = (A(u), u)_H$. So, the critical points of $\varphi|_M$ are the normalized eigenvectors of A.

Solution of Problem 5.17

Note that

$$M = \left\{ u \in H_0^1(\Omega) : \|Du\|_2^2 = \|u\|_p^p, \ u \neq 0 \right\}.$$

Then from Theorem 5.29, we know that M is a C^1-Banach manifold.

For every $u \in H_0^1(\Omega) \setminus \{\emptyset\}$ and $t > 0$ consider the equation

$$\|D(tu)\|_2^2 = \|tu\|_p^p,$$

so

$$t = \Big(\frac{\|Du\|_2^2}{\|u\|_p^p}\Big)^{\frac{1}{p-2}} = \Big(\frac{\|u\|^2}{\|u\|_p^p}\Big)^{\frac{1}{p-2}} > 0$$

(see the Poincaré inequality; Theorem 1.131). Thus, for this $t > 0$, $tu \in M$ and it follows that $M \neq \emptyset$.

From the definition of M, we have

$$\|u\|^2 = \|Du\|_2^2 = \|u\|_p^p \leqslant c_1 \|u\|^p \qquad \forall\, u \in M,$$

for some $c_1 > 0$, so

$$\Big(\frac{1}{c_1}\Big)^{\frac{1}{p-2}} \leqslant \|u\| \qquad \forall\, u \in M \tag{5.32}$$

(recall that $u \neq 0$). Let $m = \inf\limits_M \varphi$. Note that $\varphi|_M(u) = (\frac{1}{2} - \frac{1}{p})\|u\|^2$ and so

$$m = \big(\tfrac{1}{2} - \tfrac{1}{p}\big) \inf_M \|u\|^2 \geqslant \big(\tfrac{1}{2} - \tfrac{1}{p}\big)\Big(\frac{1}{c_1}\Big)^{\frac{2}{p-2}} > 0.$$

Let $\{u_n\}_{n \geqslant 1} \subseteq M$ be a minimizing sequence. Since $\varphi(u_n) = \varphi(|u_n|)$ for all $n \geqslant 1$, we may assume that $u_n \geqslant 0$ for all $n \geqslant 1$. Because $\varphi|_M(u) = (\frac{1}{2} - \frac{1}{p})\|u\|^2$ and $p > 2$, we see that the sequence $\{u_n\}_{n \geqslant 1} \subseteq H_0^1(\Omega)$ is bounded. So, by passing to a suitable subsequence if necessary, we may assume that

$$u_n \xrightarrow{w} u \quad \text{in } H_0^1(\Omega) \quad \text{and} \quad u_n \longrightarrow u \quad \text{in } L^p(\Omega) \tag{5.33}$$

(here we have used the Rellich–Kondrachov embedding theorem since $p < 2^*$; see Theorem 1.135). Evidently $u \geqslant 0$ and by the weak lower semicontinuity of the norm functional in a Banach space, we have

$$\begin{aligned} \varphi(u) &= \tfrac{1}{2}\|u\|^2 - \tfrac{1}{p}\|u\|_p^p \leqslant \liminf_{n \to +\infty} \big(\tfrac{1}{2}\|u_n\|^2 - \tfrac{1}{p}\|u_n\|_p^p\big) \\ &= \liminf_{n \to +\infty} \varphi(u_n) = m. \end{aligned}$$

For every $n \geqslant 1$, we have

$$\|u_n\|^2 = \|u_n\|_p^p \tag{5.34}$$

(since $\{u_n\}_{n\geqslant 1} \subseteq M$), so

$$\|u\|^2 \leqslant \|u\|_p^p \tag{5.35}$$

(see (5.34)). Note that since $u_n \in M$ for all $n \geqslant 1$, we have

$$\|u_n\| \geqslant \left(\tfrac{1}{c_1}\right)^{\frac{1}{p-2}} \qquad \forall\, n \geqslant 1$$

(see (5.32)), so

$$\|u_n\|_p^p \geqslant \left(\tfrac{1}{c_1}\right)^{\frac{2}{p-2}} \qquad \forall\, n \geqslant 1,$$

thus

$$\|u\|_p^p \geqslant \left(\tfrac{1}{c_1}\right)^{\frac{2}{p-2}}$$

and hence $u \neq 0$.

If the inequality in (5.35) is strict, then $u \notin M$, $u \neq 0$. But then from the first part of the solution, we know that for $t = \left(\frac{\|u\|^2}{\|u\|_p^p}\right)^{\frac{1}{p-2}} \in (0,1)$ (see (5.35) with strict inequality), we have $tu \in M$. Hence

$$\begin{aligned} 0 < m \leqslant \varphi(tu) &= t^2\left(\tfrac{1}{2} - \tfrac{1}{p}\right)\|u\|^2 \\ \leqslant t^2 \liminf_{n\to+\infty}\left(\tfrac{1}{2} - \tfrac{1}{p}\right)\|u_n\|^2 &= t^2 \liminf_{n\to+\infty} \varphi(u_n) = t^2 m < m, \end{aligned}$$

a contradiction. So, in (5.35) we have equality, which means that $u \in M$ and

$$\varphi(u) = \inf_M \varphi.$$

Let $h \in H_0^1(\Omega)$. We can find $\varepsilon > 0$ small, such that $u + sh$ is not identically zero for $s \in (-\varepsilon, \varepsilon)$. We set

$$t(s) \in \left(\frac{\|u+sh\|^2}{\|u+sh\|_p^p}\right)^{\frac{1}{p-2}}, \quad s \in (-\varepsilon, \varepsilon).$$

From the previous arguments, we know that $s \longmapsto t(s)$ is a curve in M, it is differentiable (being the composition of differentiable maps) and $t(0) = 1$.

Let $\xi\colon (-\varepsilon, \varepsilon) \longrightarrow \mathbb{R}$ be defined by

$$\xi(s) = \varphi(t(s)(u + sh)).$$

We know that $s = 0$ is a minimizer of ξ. Then

$$\begin{aligned} 0 &= \xi'(0) = \langle \varphi'(u), t'(0)u + h \rangle \\ &= t'(0)\langle \varphi'(u), u \rangle + \langle \varphi'(u), h \rangle = \langle \varphi'(u), h \rangle \end{aligned}$$

(by the chain rule (see Theorem 3.19) and since $u \in M$), so $\varphi'(u) = 0$, that is, $u \in M$ is a free critical point of φ. This proves that M is a natural constraint for φ (see Definition 5.33).

Solution of Problem 5.18
Let $\vartheta(u) = \langle \varphi'(u), u \rangle$. Then $\vartheta \in C^1(X)$ and we have

$$M = \{u \in X : \vartheta(u) = 0, \ u \neq 0\}. \tag{5.36}$$

Note that

$$\langle \vartheta'(u), u \rangle = \langle \varphi''(u)u, u \rangle + \langle \varphi'(u), u \rangle = \langle \varphi''(u)u, u \rangle \neq 0 \qquad \forall\, u \in M. \tag{5.37}$$

So, $K_\vartheta \cap M = \emptyset$. Then from Theorem 5.29, we have that M is a C^1-Banach manifold of codimension 1.

Next let $u \in K_{\varphi|_M}$. Then we have

$$\langle \varphi'(u), h \rangle = 0 \qquad \forall\, h \in T_u M = \ker \varphi'(u)$$

(see Theorem 5.29, and (5.36), (5.37)), so

$$\ker \vartheta'(u) \subseteq \ker \varphi'(u),$$

thus

$$\varphi'(u) = \lambda \vartheta'(u), \tag{5.38}$$

for some $\lambda \in \mathbb{R}$. Since $M \subseteq \vartheta^{-1}(0)$, we have

$$0 = \vartheta(u) = \langle \varphi'(u), u \rangle = \lambda \langle \vartheta'(u), u \rangle \qquad \forall\, u \in M$$

(see (5.38)), so $\lambda = 0$ (see (5.37)), thus $\varphi'(u) = 0$ (see (5.38)) and hence $K_{\varphi|_M} \subseteq K_\varphi$. This shows that M is a natural constraint for φ (see Definition 5.33).

Remark A critical point $u \in K_\varphi$ which is a global minimizer of φ on M is called ***ground state*** (or ***least energy***) critical point of φ.

Solution of Problem 5.19
Note that $M \neq \emptyset$ since $0 \in M$.

Let $g\colon H^1(\Omega) \longrightarrow \mathbb{R}$ be the functional defined by

$$g(u) = \int_\Omega f(u)\, dz \qquad \forall\, u \in H^1(\Omega).$$

By the hypothesis on f', we have that $g \in C^1(H^1(\Omega);\mathbb{R})$ and

$$\langle g'(u), h\rangle = \int_\Omega f'(u)h\,dz \qquad \forall\, u,h \in H^1(\Omega)$$

(see Problem 3.26), so

$$g'(u) \neq 0.$$

So, we can use Theorem 5.29 and infer that M is a C^1-Banach manifold of codimension 1.

Next, let $u \in K_{\varphi|_M}$. Proposition 5.31 says that there exists $\lambda \in \mathbb{R}$ such that

$$\int_\Omega (Du, Dh)_{\mathbb{R}^N}\,dz - \int_\Omega f(u)h\,dz = \lambda \int_\Omega f'(u)h\,dz \qquad \forall\, h \in H^1(\Omega) \tag{5.39}$$

(see Problem 3.26). In (5.39) we choose $h = 1$. Then since $u \in M$, we have

$$0 = \lambda \int_\Omega f'(u)\,dz,$$

so $\lambda = 0$ (see the hypothesis on f) and thus $\varphi'(u) = 0$ (see (5.39)). Hence $K_{\varphi|_M} \subseteq K_\varphi$ and this proves that M is a natural constraint for φ (see Definition 5.33).

Solution of Problem 5.20

(a) "$\Longrightarrow$": We will show the contrapositive. So, suppose that $\lambda > \widehat{\lambda}_1$. Let $\widehat{u}_1$ be the L^2-normalized positive eigenfunction corresponding to $\widehat{\lambda}_1 > 0$. We define

$$\xi(t) = \langle \varphi'_\lambda(t\widehat{u}_1), t\widehat{u}_1\rangle \qquad \forall\, t \geqslant 0.$$

Evidently the map $t \longmapsto \xi(t)$ is continuous. We have

$$\begin{aligned}\xi(t) &= t^2\|D\widehat{u}_1\|_2^2 - \lambda t^2\|\widehat{u}_1\|_2^2 + t^p\|\widehat{u}_1\|_p^p \\ &= t^2(\widehat{\lambda}_1 - \lambda) + t^p\|\widehat{u}_1\|_p^p.\end{aligned}$$

Since $\widehat{\lambda}_1 < \lambda$ and $p > 2$ we see that

$$\begin{aligned}\xi(t) &< 0 \qquad \forall\, t \in (0,1) \text{ small},\\ \xi(t) &> 0 \qquad \forall\, t > 1 \text{ big}.\end{aligned}$$

So, by the Bolzano theorem (see Theorem I.1.90), we can find $t_0 > 0$ such that

$$\xi(t_0) = 0,$$

so

$$\langle \varphi_\lambda'(t_0\widehat{u}_1), t_0\widehat{u}_1 \rangle = 0, \quad t_0\widehat{u}_1 \neq 0$$

and hence $t_0\widehat{u}_1 \in M_\lambda$. Therefore, if $M_\lambda = 0$, then we must have $\lambda \leqslant \widehat{\lambda}_1$.

"$\Longleftarrow$": For every $u \in H_0^1(\Omega) \setminus \{0\}$, we have

$$\langle \varphi_\lambda'(u), u \rangle = \|Du\|_2^2 - \lambda\|u\|_2^2 + \|u\|_p^p \geqslant \|u\|_p^p > 0,$$

since $\lambda \leqslant \widehat{\lambda}_1$ (see Remark 5.130). So, $M_\lambda = \emptyset$.

(b) Let $g_\lambda(u) = \langle \varphi_\lambda'(u), u \rangle$. Then $g_\lambda \in C^2(H_0^1(\Omega))$ and we have

$$M_\lambda = \left\{ u \in H_0^1(\Omega) : \ g_\lambda(u) = 0, \ u \neq 0 \right\}.$$

Note that for $u \in M_\lambda$, we have

$$\langle g_\lambda'(u), u \rangle = \langle \varphi_\lambda''(u)u, u \rangle + \langle \varphi_\lambda'(u), u \rangle = \langle \varphi_\lambda''(u)u, u \rangle,$$

so

$$\begin{aligned} \langle g_\lambda'(u), u \rangle &= \|Du\|_2^2 - \lambda\|u\|_2^2 + (p-1)\|u\|_2^2 \\ &= -\|u\|_p^p + (p-1)\|u\|_p^p = (p-2)\|u\|_p^p > 0 \end{aligned}$$

(since $p > 2$, $u \in M_\lambda$), thus

$$g_\lambda'(u) \neq 0 \qquad \forall\, u \in M_\lambda.$$

Theorem 5.29 implies that M_λ is a C^1-Banach manifold of codimension 1. Suppose that $u \in K_{\varphi_\lambda|_{M_\lambda}}$. Then Proposition 5.31 implies that there exists $\mu \in \mathbb{R}$ such that

$$A(u) \quad \lambda u + |u|^{p-2}u - \mu\varphi''(u)u \quad \text{in } H^{-1}(\Omega), \tag{5.40}$$

where $A \in \mathcal{L}(H_0^1(\Omega); H^{-1}(\Omega))$ is defined by

$$\langle A(u), h \rangle = \int_\Omega (Du, Dh)_{\mathbb{R}^N}\, dz \qquad \forall\, u, h \in H_0^1(\Omega)$$

(recall that $H^{-1}(\Omega) = H_0^1(\Omega)^*$; see Theorem 1.140). Acting on (5.40) with u we obtain

$$\|Du\|_2^2 - \lambda\|u\|_2^2 + \|u\|_p^p = \mu(p-2)\|u\|_p^p,$$

so

$$\mu(p-2)\|u\|_p^p = 0$$

(since $u \in M_\lambda$), thus $\mu = 0$ (since $u \in M_\lambda$ and $p > 2$), hence $\varphi_\lambda'(u) = 0$. This proves that M_λ is a natural constraint for φ_λ (see Definition 5.33).

Solution of Problem 5.21

From hypothesis **(i)** and the Rellich–Kondrachov embedding theorem (see Theoremt.1.133), we have that $\varphi \in C^2(H_0^1(\Omega))$. Let

$$g(u) = \|Du\|_2^2 - \int_\Omega f(u)u\,dz \qquad \forall\, u \in H_0^1(\Omega).$$

Clearly $g \in C^2(H_0^1(\Omega))$ and $M = \{u \in H_0^1(\Omega) : \ g(u) = 0, \ u \neq 0\}$. Note that

$$\langle g'(u), h\rangle = \int_\Omega (Du, Dh)_{\mathbb{R}^N} - \int_\Omega (f(u) + f'(u)u)h\,dz \qquad \forall\, u, h \in H_0^1(\Omega). \tag{5.41}$$

In (5.41) we choose $h = u \in M$. Then

$$\begin{aligned}\langle g'(u), u\rangle &= \|Du\|_2^2 - \int_\Omega f(u)u\,dz - \int_\Omega f'(u)u^2\,dz \\ &= -\int_\Omega f'(u)u^2\,dz < -\int_\Omega f(u)u\,dz = -\|Du\|_2^2 < 0 \end{aligned}\tag{5.42}$$

(since $u \in M$ and using hypothesis **(ii)**), so $g'(u) \neq 0$ for all $u \in M$. Invoking Theorem 5.29 we infer that M is a C^1-Banach manifold of codimension 1.

Suppose that $u \in K_{\varphi|_M}$. Proposition 5.31 implies that we can find $\lambda \in \mathbb{R}$ such that

$$\varphi'(u) = \lambda g'(u),$$

so

$$\|Du\|_p^p - \int_\Omega f(u)u\,dz = \lambda \langle g'(u), u\rangle,$$

thus

$$\lambda \langle g'(u), u\rangle = 0$$

(since $u \in M$) and hence $\lambda = 0$ (see (5.42)). So, $\varphi'(u) = 0$ and this proves that M_λ is a natural constraint for φ (see Definition 5.33).

Solution of Problem 5.22

Let $u^* \in c_1\overline{B}_1^*$ and consider the function $\psi\colon X \longrightarrow \mathbb{R}$ defined by

$$\psi(u) = \varphi(u) - \langle u^*, u\rangle \qquad \forall\, u \in X.$$

Clearly ψ is lower semicontinuous, bounded below (see the hypothesis on φ) and Gâteaux differentiable. So, according to Proposition 5.39, given any $\varepsilon > 0$, we can find $u_\varepsilon \in X$ such that

$$\|\psi_G'(u_\varepsilon)\|_* \leqslant \varepsilon,$$

so

$$\|\varphi'_G(u_\varepsilon) - u^*\|_* \leqslant \varepsilon$$

and thus the set $\varphi'_G(X)$ is dense in $c_1\overline{B}_1^*$ (since $\varepsilon > 0$ is arbitrary).

Solution of Problem 5.23

The hypothesis on ϑ implies that given any $c_1 > 0$ we can find $r_0 > 0$ such that

$$\vartheta(r) \geqslant c_1 r \qquad \forall\, r \geqslant r_0.$$

Let $c_2 = \inf\limits_{r\in[0,r_0]} \vartheta(r)$. Then

$$\vartheta(r) \geqslant c_1 r - c_2 \qquad \forall\, r \geqslant 0,$$

so

$$\varphi(u) \geqslant c_1\|u\|_X - c_2 \qquad \forall\, u \in X$$

(see the hypothesis on φ). So, from Problem 5.22, we know that the set $\varphi'_G(X)$ is dense in $c_1\overline{B}_1^*$. Because $c_1 > 0$ is arbitrary, we conclude that the set $\varphi'_G(X)$ is dense in X^*.

Solution of Problem 5.24

Let $u_0 = 0$. Using Theorem 5.41, we can find $u_\lambda \in X$ such that

$$\varphi(u_\lambda) \leqslant \varphi(y), \quad \|u_\lambda - y\|_X \leqslant \overline{r}$$

and for all $h \in X$ with $\|h\|_X = 1$ and all $t > 0$ we have

$$-\tfrac{\varepsilon}{\lambda(1+\xi(\|u_\lambda\|_X))} \leqslant \frac{\varphi(u_\lambda + th) - \varphi(u_\lambda)}{t}.$$

We let $t \searrow 0$ and obtain

$$-\tfrac{\varepsilon}{\lambda(1+\xi(\|u_\lambda\|_X))} \leqslant \langle \varphi'_G(u_\lambda), h\rangle \qquad \forall\, h \in X,\ \|h\|_X = 1,$$

so

$$\|\varphi'_G(u_\lambda)\|_* \leqslant \tfrac{\varepsilon}{\lambda(1+\xi(\|u_\lambda\|_X))},$$

Solution of Problem 5.25
From Problem 5.24 with $\xi(t) = t$ for all $t \geqslant 0$, we see that we can find a sequence $\{u_n\}_{n\geqslant 1} \subseteq X$ such that

$$\varphi(u_n) \longrightarrow m = \inf_X \varphi \quad \text{and} \quad (1 + \|u_n\|_X)\varphi'(u_n) \longrightarrow 0 \quad \text{in } X^*.$$

Since by hypothesis φ satisfies the C_m-condition (see Definition 5.47(a)), by passing to a suitable subsequence if necessary, we may assume that $u_n \longrightarrow u \in X$ in X. Then clearly $\varphi(u) = m$.

Solution of Problem 5.26
"Theorem 4.66 $\Longrightarrow$ Corollary 5.38":
Arguing by contradiction, suppose that the implication is not true. Then we cannot find $u_\varepsilon \in X$ such that

$$\varphi(u_\varepsilon) \;<\; \varphi(u) + \varepsilon d_X(u, u_\varepsilon) \qquad \forall\, u \neq u_\varepsilon.$$

We introduce

$$F(u) \;=\; \{y \in X :\; \varphi(u) \geqslant \varphi(y) + \varepsilon d_X(u, y),\; y \neq u\}.$$

Then $F(u) \neq \emptyset$ for all $u \in X$ and using Theorem 4.66 (with the metric $d_\varepsilon = \varepsilon d_X$), we can find $u_0 \in F(u_0)$, a contradiction.

"Corollary 5.38 $\Longrightarrow$ Theorem 4.66":
For $\varepsilon = 1$, we can find $u_0 \in X$ such that

$$\varphi(u_0) \;<\; \varphi(u) + d_X(u, u_0) \qquad \forall\, u \neq u_0. \tag{5.43}$$

We claim that $u_0 \in F(u_0)$. If this is not true, then for every $u \in F(u_0)$, we have

$$\varphi(u) \;\leqslant\; \varphi(u_0) - d_X(y, u_0), \quad u \neq u_0. \tag{5.44}$$

Comparing (5.43) and (5.44) we reach a contradiction.

Solution of Problem 5.27
Let $r' \in (0, r]$ be given and suppose that statement **(a)** does not hold. Then for every fixed $\varrho \in (0, r']$ we have

$$m_\varrho \;=\; \varphi(u_0). \tag{5.45}$$

Choose $\delta > 0$ such that

$$0 < \varrho - \delta < \varrho + \delta < r'$$

and let $R = \{u \in X : \varrho - \delta \leqslant \|u - u_0\|_X \leqslant \varrho + \delta\}$. We can find a sequence $\{u_n\}_{n \geqslant 1} \subseteq X$ such that

$$\|u_n - u_0\|_X = \varrho \quad \text{and} \quad \varphi(u_n) \leqslant \varphi(u_0) + \tfrac{1}{n} \qquad \forall\, n \geqslant 1 \tag{5.46}$$

(see (5.45)). Using Theorem 5.41, we can find a sequence $\{v_n\}_{n \geqslant 1} \subseteq \mathbb{R}$ such that

$$\varphi(v_n) \leqslant \varphi(u_n), \quad \|v_n - u_n\|_X \leqslant \tfrac{1}{n} \tag{5.47}$$

and

$$\varphi(v_n) \leqslant \varphi(u) + \tfrac{1}{n(1+\|v_n\|_X)}\|u - v_n\| \qquad \forall\, u \in \mathbb{R}. \tag{5.48}$$

From (5.46), we see that $v_n \in \operatorname{int} R$ for all $n \geqslant n_0$. Then with $u = v_n + te$, where $e \in X$, $\|e\|_X = 1$ and $|t| > 0$ small (so that $u = v_n + th \in R$), we have

$$(1 + \|v_n\|_X)\|\varphi'(v_n)\|_* \leqslant \tfrac{1}{n} \qquad \forall\, n \geqslant n_0,$$

so

$$(1 + \|v_n\|_X)\varphi'(v_n) \longrightarrow 0 \quad \text{in } X^*. \tag{5.49}$$

Since φ satisfies the C-condition (see Definition 5.45(b)), from (5.47), (5.48), and (5.49), we infer that at least for a subsequence, we have

$$v_n \longrightarrow \widehat{v} \quad \text{in } X,$$

so

$$u_n \longrightarrow \widehat{v}$$

(see (5.47), (5.48)), thus

$$\|\widehat{v} - u_0\|_X = \varrho \quad \text{and} \quad \varphi(\widehat{v}) = \varphi(u_0)$$

(see (5.46)).

Solution of Problem 5.28

Without any loss of generality, we may assume that $\varphi(u_0) \leqslant \varphi(u_1)$. Since $u_1 \in X$ is by hypothesis a local minimizer of φ, we can find $r > 0$ such that

$$\varphi(u_1) \leqslant \varphi(u) \qquad \forall\, u \in \overline{B}_r(u_0)$$

(where $\overline{B}_r(u_0) = \{v \in X : \|v - u_0\|_X \leqslant r\}$). Suppose that for some $\varrho \in (0, r)$, we have

$$\varphi(u_0) \leqslant \varphi(u_1) < \inf\{\varphi(u) : \|u - u_1\|_X = \varrho\} = m_\varrho. \tag{5.50}$$

Because of (5.50) and since by hypothesis φ satisfies the C-condition (see Definition 5.45(b)), we see that we can apply the mountain pass theorem (see Theorem 5.63) and find $y \in X$ such that

$$y \in K_\varphi \quad \text{and} \quad m_\varrho \leqslant \varphi(y). \tag{5.51}$$

From (5.50) and (5.51), it follows that $y \in K_\varphi \setminus \{u_0, u_1\}$.

If no $\varrho \in (0, r)$ can be found so that (5.50) holds, then statement **(b)** in Problem 5.27 is true and we have a whole sequence of distinct critical points of φ, different from u_0 and u_1.

Solution of Problem 5.29

Let $\{u_n\}_{n\geqslant 1} \subseteq X$ be a minimizing sequence for φ, that is, $\varphi(u_n) \searrow c$. Let $\delta_n = \varphi(u_n) - c > 0$ and $\varepsilon_n = \delta_n > 0$ for all $n \geqslant 1$. Using the Ekeland variational principle (see Theorem 5.36), we can find a sequence $\{v_n\}_{n\geqslant 1} \subseteq X$ such that

$$\varphi(v_n) \leqslant \varphi(u_n), \quad d_X(v_n, u_n) \leqslant \delta_n^{\frac{1}{2}} \tag{5.52}$$

and

$$\varphi(v_n) \leqslant \varphi(u) + \delta_n^{\frac{1}{2}} d_X(u, v_n) \qquad \forall\, u \in X. \tag{5.53}$$

Evidently $\varphi(v_n) \longrightarrow c$. Since φ satisfies the NC_c^+-condition, by passing to a subsequence if necessary, we may assume that

$$v_n \longrightarrow u_0 \quad \text{in } X,$$

so

$$\varphi(u_0) \leqslant \varphi(u) \qquad \forall\, u \in X$$

(see (5.52), (5.53)), thus $\varphi(u_0) = c$.

Solution of Problem 5.30

By Theorem 5.41, we can find a sequence $\{u_n\}_{n\geqslant 1} \subseteq X$ such that

$$\varphi(u_n) \searrow \widehat{m} \quad \text{and} \quad \varphi(u_n) \leqslant \varphi(u) + \tfrac{1}{n(1+\|u_n\|_X)} \|u - u_n\|_X \qquad \forall\, u \in X. \tag{5.54}$$

Let $u = u_n + th$ with $t > 0$ and $h \in X$. From (5.54) we have

$$-\tfrac{1}{n(1+\|u_n\|_X)} \|h\|_X \leqslant \tfrac{\varphi(u_n+th)-\varphi(u_n)}{t} \qquad \forall\, n \geqslant 1,$$

so

$$-\tfrac{1}{n(1+\|u_n\|_X)} \|h\|_X \leqslant \varphi^0(u_n; h) \qquad \forall\, n \geqslant 1$$

(see Definition 3.122), thus

$$-\|h\|_X \leqslant n(1+\|u_n\|_X)\varphi^0(u_n;h) \triangleq \psi_n(h) \qquad \forall\, n \geqslant 1.$$

Note that ψ_n is sublinear and continuous (see Proposition 3.123(a)). So, we can use Problem 3.133 and find $u_n^* \in X^*$ with $\|u_n^*\|_* \leqslant 1$ such that

$$\langle u_n^*, h\rangle \leqslant \psi_n(h) \qquad \forall\, h \in X,\ n \geqslant 1.$$

If $v_n^* = \frac{u_n^*}{n(1+\|u_n\|_X)}$, then $v_n^* \in \partial\varphi(u_n)$ for all $n \geqslant 1$ and $\|v_n^*\|_* \leqslant \frac{1}{n(1+\|u_n\|_X)}$ for all $n \geqslant 1$. We have

$$(1+\|u_n\|_X)m(u_n) \leqslant (1+\|u_n\|_X)\|v_n^*\|_* \leqslant \tfrac{1}{n}\|u_n^*\|_X \leqslant \tfrac{1}{n} \longrightarrow 0$$

(see Definition 5.47(a)). Because φ satisfies the nonsmooth C_m-condition (see Definition 5.47), passing to a subsequence, we may assume that

$$u_n \longrightarrow u_0 \quad \text{in } X,$$

so $\varphi(u_0) = \widehat{m}$.

Solution of Problem 5.31

Let $\{u_n\}_{n\geqslant 1} \subseteq \mathbb{R}^N$ be a Palais–Smale sequence, i.e., the sequence $\{\varphi(u_n)\}_{n\geqslant 1} \subseteq \mathbb{R}$ is bounded and $\varphi'(u_n) \longrightarrow 0$ in $\mathbb{R}^N$. Then from the hypothesis on φ, we infer that the sequence $\{u_n\}_{n\geqslant 1} \subseteq \mathbb{R}$ is bounded and so it admits a convergent subsequence. Then φ satisfies the PS-condition (see Definition 5.45(a)).

Solution of Problem 5.32

Let $\{u_n\}_{n\geqslant 1} \subseteq X$ be a Palais–Smale sequence i.e., the sequence $\{\varphi(u_n)\}_{n\geqslant 1} \subseteq \mathbb{R}$ is bounded and $\varphi'(u_n) \longrightarrow 0$ in X^*. We have

$$A(u_n) + g(u_n) \longrightarrow 0 \quad \text{in } X^*,$$

so

$$u_n + A^{-1}(g(u_n)) \longrightarrow 0 \quad \text{in } X. \tag{5.55}$$

By hypothesis, the sequence $\{u_n\}_{n\geqslant 1} \subseteq X$ is bounded. Since g is compact, so is $A^{-1} \circ g$ and it follows that the sequence $\{A^{-1}(g(u_n))\}_{n\geqslant 1} \subseteq X$ is relatively compact. So, it admits a strongly convergent subsequence. Because of (5.55) so does the sequence $\{u_n\}_{n\geqslant 1} \subseteq X$, which means that φ satisfies the PS-condition (see Definition 5.45(a)).

Solution of Problem 5.33

We argue by contradiction. So, suppose that for the triple $r_1 < r < r_2$, the claim of the problem is not rue. Let $\{u_n\}_{n\geqslant 1} \subseteq X$ be a sequence such that

$$\|u_n\|_X = r \quad \text{and} \quad \varphi(u_n) < m(r) + \tfrac{1}{n^2} \qquad \forall\, n \geqslant 1. \tag{5.56}$$

Let $D = \{u \in X : r_1 \leqslant \|u\|_X \leqslant r_2\}$ and use the Ekeland variational principle (see Theorem 5.35) to find $v_n \in D$ such that

$$\varphi(v_n) \leqslant \varphi(u_n) - \tfrac{1}{n}\|u_n - v_n\|_X \tag{5.57}$$

and

$$\varphi(v_n) \leqslant \varphi(u) + \tfrac{1}{n}\|u - v_n\| \qquad \forall\, u \in D. \tag{5.58}$$

If $v_n \in \partial D$, say $\|v_n\|_X = r_1$, then

$$\begin{aligned} m(r_1) &\leqslant \varphi(v_n) \leqslant \varphi(u_n) - \tfrac{1}{n}\|u_n - v_n\| \\ &\leqslant m(r) + \tfrac{1}{n}\big(\tfrac{1}{n} - (r - r_1)\big) \end{aligned} \tag{5.59}$$

(see (5.56), (5.57), and (5.58)). From (5.59), for all $n \geqslant 1$ big, we have

$$m(r_1) < m(r),$$

contradicting our hypothesis in the beginning of the solution.

Therefore, $v_n \notin \partial D$ for all $n \geqslant 1$. From (5.57) and choosing $u = v_n + th$ for $t > 0$ small and $h \in X$, we have

$$\tfrac{1}{t}\big(\varphi(v_n) - \varphi(v_n + th)\big) \leqslant \tfrac{1}{n}\|h\|_X,$$

so

$$-\big\langle \varphi'(v_n), h \big\rangle \leqslant \tfrac{1}{n}\|h\|_X,$$

thus

$$\|\varphi'(v_n)\|_* \leqslant \tfrac{1}{n} \qquad \forall\, n \geqslant 1$$

and hence $\varphi'(v_n) \longrightarrow 0$ in X^*. The compactness type hypothesis on φ implies that at least for a subsequence, we have

$$v_n \longrightarrow v \in D \quad \text{in } X.$$

Then $\varphi'(v) = 0$, a contradiction to the hypothesis of the problem.

Solution of Problem 5.34

From the Ekeland variational principle (see Theorem 5.36), we know that there exists $u_\varepsilon \in X$ such that

$$\varepsilon\|u_\varepsilon - u_0\|_X \leqslant \varphi(u_0) - \varphi(u_\varepsilon)$$

and

$$\varphi(u_\varepsilon) \leqslant \varphi(u) + \varepsilon\|u - u_\varepsilon\|_X \qquad \forall\, u \in X. \tag{5.60}$$

The function $u \longmapsto \psi(u) = \varphi(u) + \varepsilon\|u - u_\varepsilon\|_X$ is lower semicontinuous, convex, and attains its infimum at u_ε (see (5.60)). So, we have

$$0 \in \partial\psi(u_\varepsilon)$$

(see Proposition 3.82), thus

$$0 \in \partial\varphi(u_\varepsilon) + \varepsilon\overline{B}_1^* \tag{5.61}$$

(see Proposition 3.97), where $\overline{B}_1^* = \{u^* \in X^* : \ \|u^*\|_* \leqslant 1\}$. Then we can find $u^* \in \overline{B}_1^*$ such that

$$-\varepsilon u^* \in \partial\varphi(u_\varepsilon)$$

(see (5.61)). If we set $u_\varepsilon^* = -\varepsilon u^*$, then we are finished.

Solution of Problem 5.35

Let $\{u_n\}_{n\geqslant 1} \subseteq X$ be a Cauchy sequence and let $\varphi(v) = \lim\limits_{n\rightarrow+\infty} d_X(u_n, v)$ for all $x \in X$. Then φ is lower semicontinuous and nonnegative. Given $\varepsilon \in (0,1)$ and using Corollary 5.38, we can find $y \in X$ such that

$$\varphi(y) \leqslant \varepsilon \quad \text{and} \quad \varphi(y) \leqslant \varphi(v) + \varepsilon d_X(v, y) \qquad \forall\, v \in X.$$

Let $v = u_n$. Then

$$\varphi(y) \leqslant \varphi(u_n) + \varepsilon d_X(u_n, y) \qquad \forall\, n \geqslant 1.$$

Passing to the limit as $n \rightarrow +\infty$ and since $\lim\limits_{k,n\rightarrow+\infty} d_X(u_n, u_k) = 0$, we obtain

$$\varphi(y) \leqslant \varepsilon\varphi(y),$$

so

$$\varphi(y) = 0$$

(since $\varepsilon \in (0,1)$), thus $u_n \longrightarrow y \in X$ in X. This proves that (X, d_X) is complete.

Solution of Problem 5.36

By hypothesis for every $n \geqslant 1$, we can find $r_n \geqslant n$ such that

$$\varphi^{c-\frac{1}{n}} \subseteq \overline{B}_{r_n},$$

where $\overline{B}_{r_n} = \{u \in X : \|u\|_X \leqslant r_n\}$. Let $D_n = X \setminus \overline{B}_{r_n}$ for all $n \geqslant 1$. We have

$$c - \tfrac{1}{n} \leqslant \inf_{D_n} \varphi = c_n. \tag{5.62}$$

By hypothesis $\varphi^{c+\frac{1}{n}} \subseteq X$ is unbounded. So, we can find $v_n \in X$ such that

$$\varphi(v_n) \leqslant c + \tfrac{1}{n} \quad \text{and} \quad \|v_n\|_X \geqslant r_n + 1 + \tfrac{1}{\sqrt{n}}. \tag{5.63}$$

From (5.62) and (5.63), we have

$$\varphi(v_n) \leqslant c + \tfrac{1}{n} \leqslant c_n + \tfrac{2}{n}.$$

We apply Theorem 5.41 on $\varphi|_{D_n}$ with $\varepsilon = \frac{2}{n}$, $\xi(s) = s$, $u_0 = v_n$ and $\lambda = \frac{1}{\sqrt{n}}$ and produce a sequence $\{u_n\}_{n \geqslant 1} \subseteq X$ such that $u_n \in D_n$ for all $n \geqslant 1$ and

$$c - \tfrac{1}{n} \leqslant c_n \leqslant \varphi(u_n) \leqslant \varphi(v_n) \leqslant c + \tfrac{1}{n} \leqslant c_n + \tfrac{2}{n} \qquad \forall\, n \geqslant 1, \tag{5.64}$$

$$(1 + \|u_n\|_X)\|\varphi'(u_n)\|_* \leqslant \tfrac{2}{\sqrt{n}} \qquad \forall\, n \geqslant 1, \tag{5.65}$$

$$\|u_n - v_n\|_X \leqslant \overline{r}_n \qquad \forall\, n \geqslant 1, \tag{5.66}$$

where $\overline{r}_n > 0$ is such that $\ln(1 + \overline{r}_n) = \frac{1}{\sqrt{n}}$. From (5.66) and (5.63), we have

$$\|u_n\|_X \geqslant r_n + \tfrac{1}{\sqrt{n}} - e^{\frac{1}{\sqrt{n}}} \qquad \forall\, n \geqslant 1. \tag{5.67}$$

Therefore for the sequence $\{u_n\}_{n \geqslant 1} \subseteq X$ we have

$$\varphi(u_n) \longrightarrow c$$

(see (5.64)),

$$(1 + \|u_n\|_X)\varphi'(u_n) \longrightarrow 0 \quad \text{in } X^*$$

(see (5.65)),

$$\|u_n\|_X \longrightarrow +\infty$$

(see (5.67)).

Solution of Problem 5.37
We argue indirectly. So, suppose that the claim of the problem is not true. Let

$$m = \inf_X \varphi \quad \text{and} \quad C = \{\lambda \in \mathbb{R} : \varphi^\lambda \text{ is bounded}\}.$$

Recalling that the empty set is bounded, we see that $(-\infty, m) \subseteq C$, so $C \neq \emptyset$. Let $c_0 = \sup C$. Because we have assumed that the claim of the problem is not true, we have that $c_0 < +\infty$. Hence for all $\lambda > c_0$, the set φ^λ is unbounded, while for $\lambda < c_0$, the set φ^λ is bounded. Problem 5.36 implies that φ does not satisfy the C_{c_0}-condition (see Definition 5.45(b)), a contradiction. This proves that the conclusion of the problem holds.

Solution of Problem 5.38
According to Definition 5.45, we only need to show that the C-condition implies the PS-condition. So, suppose that $\{u_n\}_{n\geqslant 1} \subseteq X$ is a Palais–Smale sequence, that is, the sequence $\{\varphi(u_n)\}_{n\geqslant 1} \subseteq \mathbb{R}$ is bounded and $\varphi'(u_n) \longrightarrow 0$ in X^*. From Problem 5.37 it follows that the sequence $\{u_n\}_{n\geqslant 1} \subseteq X$ is bounded. Therefore

$$(1 + \|u_n\|_X)\varphi'(u_n) \longrightarrow 0 \quad \text{in } X^*.$$

But φ satisfies the C-condition. So, the sequence $\{u_n\}_{n\geqslant 1} \subseteq X$ admits a strongly convergent subsequence.

Solution of Problem 5.39
Let $[0, t^*)$ be the forward maximal interval of existence for the flow $t \longmapsto \sigma_u(t)$. From the Cauchy problem, we have that for all $0 \leqslant t_1 \leqslant t_2 < t^*$,

$$\|\sigma_u(t_2) - \sigma_u(t_1)\|_H^2 \leqslant (t_2 - t_1) \int_{t_1}^{t_2} \|\nabla\varphi(\sigma_u(t))\|_H^2 \, dt. \tag{5.68}$$

Note that

$$\tfrac{d}{dt}\varphi(\sigma_u(t)) = \langle \varphi'(\sigma_u(t)), \sigma_u'(t)\rangle = -\|\nabla\varphi(\sigma_u(t))\|_H^2 \qquad \forall\, t \in [0, t^*)$$

(by the chain rule (see Theorem 3.19) and from the Cauchy problem), so the map $t \longmapsto \varphi(\sigma_u(t))$ is nonincreasing. Thus, we have

$$\varphi(y) - \varphi(\sigma_u(t)) = \int_0^t \|\nabla\varphi(\sigma_u(s))\|_H^2 \, ds \qquad \forall\, t \in [0, t^*). \tag{5.69}$$

Let $m = \inf_X \varphi > -\infty$ (recall that by hypothesis φ is bounded below). Then from (5.69), we have

$$\int_0^t \|\nabla\varphi(\sigma_u(s))\|_H^2\, ds \;\leqslant\; \varphi(u) - m \;=\; \widehat{m} \qquad \forall\, t \in [0, t^*). \tag{5.70}$$

Using (5.70) in (5.69), we obtain

$$\|\sigma_u(t_2) - \sigma_u(t_1)\|_H \;\leqslant\; |t_2 - t_1|\widehat{m},$$

so $\sigma_u(t) \longrightarrow h \in H$ as $t \to t^*$. This means that $t^* = +\infty$ and so the forward (positive) flow $t \longmapsto \sigma_u(t)$ is global.

Let $c = \lim_{t\to+\infty} \varphi(\sigma_u(t))$. Then from (5.69), we have

$$\varphi(u) - c \;=\; \int_0^{+\infty} \|\nabla\varphi(\sigma_u(s))\|_H^2\, ds.$$

Therefore, we can find a sequence $\{t_n\}_{n\geqslant 1} \subseteq \mathbb{R}_+$ such that $t_n \to +\infty$ and

$$\nabla\varphi(\sigma_u(t)) \;\longrightarrow\; 0 \quad \text{in } X.$$

Let $S = \{\sigma_u(t_n) : n \geqslant 1\}$. Then since φ satisfies the CPS-condition (see Remark 5.46), we have that $\overline{S}$ contains a critical point of φ, hence c is a critical value of φ.

Solution of Problem 5.40

Using the Ekeland variational principle (see Corollary 5.38), we can find a minimizing sequence $\{u_n\}_{n\geqslant 1} \subseteq X$ for ξ such that

$$-\tfrac{1}{n}\|v - u_n\|_X \;\leqslant\; \xi(v) - \xi(u_n) \qquad \forall\, v \in X,\ n \geqslant 1. \tag{5.71}$$

Let $v = (1-t)u_n + th$ with $t \in (0,1)$ and $h \in X$. From (5.71) and since ψ is convex, we have

$$\begin{aligned} &\varphi(u_n + t(h - u_n)) - \varphi(u_n) + t(\psi(h) - \psi(u_n)) \\ &\geqslant\; \xi(v) - \xi(u_n) \;\geqslant\; -\tfrac{1}{n}\|v - u_n\|_X \;=\; -\tfrac{1}{n}t\|h - u_n\| \qquad \forall\, n \geqslant 1, \end{aligned}$$

so

$$\langle \varphi'(u_n), h - u_n \rangle + \psi(h) - \psi(u_n) \;\geqslant\; -\tfrac{1}{n}\|h - u_n\|. \tag{5.72}$$

Since by hypothesis ξ satisfies the SC_m-condition (see Definition 5.47(b) and recall that $\xi(u_n) \searrow m$), we may assume that

$$u_n \longrightarrow \widehat{u} \quad \text{in } X.$$

Then exploiting the lower semicontinuity of ξ, we have

$$\xi(\widehat{u}) \leqslant \liminf_{n\to+\infty} \xi(u_n) = m,$$

so $\xi(\widehat{u}) = m$. Also from (5.72), we obtain

$$\langle \varphi'(\widehat{u}), h - \widehat{u}\rangle + \psi(h) - \psi(\widehat{u}) \geqslant 0 \qquad \forall\, h \in X,$$

so

$$-\varphi'(\widehat{u}) \in \partial\psi(\widehat{u})$$

(see Definition 3.70).

Solution of Problem 5.41

Consider a sequence $\{u_n\}_{n\geqslant 1} \subseteq X$ which satisfies **(i)**, **(ii)** and **(iii)** in the WC-condition. Let $\{u_{n_k}\}_{k\geqslant 1}$ be a subsequence of $\{u_n\}_{n\geqslant 1}$ such that

$$\lim_{k\to+\infty} \varphi(u_{n_k}) = \limsup_{n\to+\infty} \varphi(u_n).$$

Because of the hypothesis on φ and **(i)**, we see that the subsequence $\{u_{n_k}\}_{k\geqslant 1} \subseteq X$ is bounded. Since X is reflexive, by passing to a further subsequence if necessary we may assume that

$$u_{n_k} \xrightarrow{w} \overline{u} \quad \text{in } X.$$

The function φ is weakly lower semicontinuous (being convex). So, we have

$$\varphi(\overline{u}) \leqslant \lim_{k\to+\infty} \varphi(u_{n_k}) = \limsup_{k\to+\infty} \varphi(u_{n_k}). \tag{5.73}$$

The convexity of φ implies that

$$\langle \varphi'_G(u_{n_k}), h - u_{n_k}\rangle \leqslant \varphi(h) - \varphi(u_{n_k}) \qquad \forall\, h \in X,$$

so

$$\varphi(\overline{u}) \leqslant \varphi(h) \qquad \forall\, h \in X$$

(see **(i)** and (5.73)). Therefore $\overline{u} \in X$ is a minimizer of φ and $\varphi'_G(\overline{u}) = 0$. Evidently $\varphi(u_n) \longrightarrow \varphi(\overline{u})$ and so we conclude that the WC-condition holds.

Solution of Problem 5.42

Let $\{u_n\}_{n\geqslant 1} \subseteq X$ be a sequence such that

$$-M \leqslant \varphi(u_n) = \tfrac{1}{2}\langle A(u_n), u_n\rangle + \psi(u_n) \leqslant M \qquad \forall\, n \geqslant 1, \tag{5.74}$$

for some $M > 0$, so

$$0 \neq \varphi'(u_n) = A(u_n) + \psi'(u_n) = v_n^* \longrightarrow 0 \quad \text{in } X^*. \tag{5.75}$$

We rewrite (5.74) as follows

$$-M \leqslant \tfrac{1}{2}\langle v_n^* - \psi'(u_n), u_n\rangle + \psi(u_n) \leqslant M \qquad \forall\, n \geqslant 1,$$

so

$$\tfrac{1}{2}\langle v_n^*, u_n\rangle + (1 - \tfrac{c_3}{2})\psi(u_n) - c_4 \leqslant M \qquad \forall\, n \geqslant 1$$

(see hypothesis **(ii)**), thus

$$\tfrac{1}{2}\langle v_n^*, u_n\rangle + (1 - \tfrac{c_3}{2})(c_1\|u_n\|_X^2 - c_0) - c_4 \leqslant M \qquad \forall\, n \geqslant 1$$

(see hypothesis **(i)**), hence

$$-\tfrac{1}{2}\|v_n^*\|_*\|u_n\|_X + (1 - \tfrac{c_3}{2})c_1\|u_n\|_X^2 \leqslant M_1 \qquad \forall\, n \geqslant 1, \tag{5.76}$$

for some $M_1 > 0$. Since $\|v_n^*\|_* \leqslant M_2$ for all $n \geqslant 1$ and some $M_2 > 0$ (see (5.75)), from (5.76) we infer that the sequence $\{u_n\}_{n\geqslant 1} \subseteq X$ is bounded. The reflexivity of X implies that by passing to a suitable subsequence if necessary, we may assume that

$$u_n \xrightarrow{w} \overline{u} \quad \text{in } X. \tag{5.77}$$

Since $A \in \mathcal{L}_c(X; X^*)$, we have

$$A(u_n) \longrightarrow A(\overline{u}) \quad \text{in } X^*$$

(see Proposition 2.3), thus

$$\psi'(u_n) = v_n^* - A(u_n) \longrightarrow -A(\overline{u}) \quad \text{in } X^*. \tag{5.78}$$

The function ψ is continuous, convex; hence it is weakly lower semicontinuous. So, we have

$$\psi(\overline{u}) \leqslant \liminf_{n\to+\infty} \psi(u_n) \tag{5.79}$$

(see (5.77)). Also, the convexity of φ implies that

$$\langle \psi'(u_n), v - u_n\rangle \leqslant \psi(v) - \psi(u_n) \qquad \forall\, n \geqslant 1,\ v \in X,$$

so

$$\langle -A(\overline{u}), v - \overline{u}\rangle \leqslant \psi(v) - \psi(\overline{u}) \qquad \forall\, v \in X$$

(see (5.78) and (5.79)), thus

$$-A(\overline{u}) = \psi'(\overline{u})$$

(see Definition 3.70 and Theorem 3.73), hence

$$\varphi'(\overline{u}) = 0. \tag{5.80}$$

We have

$$\langle \psi'(u_n), \overline{u} - u_n \rangle \leqslant \psi(\overline{u}) - \psi(u_n) \qquad \forall\, n \geqslant 1,$$

so

$$\limsup_{n \to +\infty} \psi(u_n) \leqslant \psi(\overline{u})$$

(see (5.77) and (5.78)), thus

$$\psi(u_n) \longrightarrow \psi(\overline{u}) \tag{5.81}$$

(see (5.79)). The compactness of A implies

$$\langle A(u_n), u_n \rangle \longrightarrow \langle A(\overline{u}), \overline{u} \rangle\,. \tag{5.82}$$

Hence we infer that

$$\varphi(u_n) \longrightarrow \varphi(\overline{u}) \tag{5.83}$$

(see (5.81) and (5.82)). Then (5.80) and (5.83) imply that the WC-condition (see Problem 5.41) is satisfied.

Solution of Problem 5.43

Let $\overline{B}_r = \{u \in X : \|u\|_X \leqslant r\}$ and define

$$\overline{B}_r^Y = \overline{B}_r \cap Y, \quad \partial B_r^V = \partial B_r \cap V$$

$$\text{and} \quad \Gamma = \big\{\gamma \in C(\overline{B}_r^Y; V) : \gamma(\partial B_r^Y) = 0\big\}.$$

We equip Γ with the supremum metric and this makes Γ a complete metric space. Let $\xi \colon \Gamma \longrightarrow \mathbb{R}$ be defined by

$$\xi(\gamma) = \max_{y \in \overline{B}_r^Y} \varphi(y, \gamma(y)) \tag{5.84}$$

(recall that $\overline{B}_r^Y \subseteq Y$ is compact). The maximum cannot be realized on ∂B_r^Y because of hypotheses **(i)** and **(ii)**. Indeed, we have

$$\varphi(y, \gamma(y)) = \varphi(y, 0) < 0 \leqslant \varphi(0, \gamma(0)) \qquad \forall\, y \in \partial B_r^Y.$$

Claim 1. ξ is lower semicontinuous and bounded below.

To show the lower semicontinuity of ξ, we need to show that for every $\lambda \in \mathbb{R}$, the set $L_\lambda = \{\gamma \in \Gamma : \xi(\gamma) \leqslant \lambda\}$ is closed (see Remark 5.2). So, let $\{\gamma_n\}_{n\geqslant 1} \subseteq L_\lambda$ be a sequence such that $\gamma_n \longrightarrow \gamma$ in Γ. Thus

$$\varphi(y, \gamma_n(y)) \longrightarrow \varphi(y, \gamma(y)) \qquad \forall\, y \in \overline{B}_r^Y. \tag{5.85}$$

We have

$$\varphi(y, \gamma_n(y)) \leqslant \xi(\gamma_n) \leqslant \lambda \qquad \forall\, y \in \overline{B}_r^Y,\ n \geqslant 1,$$

so

$$\varphi(y, \gamma(y)) \leqslant \lambda \qquad \forall\, y \in \overline{B}_r^Y,$$

thus $\xi(\gamma) \leqslant \lambda$ (see (5.84)) and hence $\gamma \in L_\lambda$. This proves lower semicontinuity of ξ.

Also, from (5.84), we have

$$\xi(\gamma) \geqslant \varphi(0, \gamma(0)) \geqslant 0,$$

so ξ is bounded below by zero. This proves Claim 1.

Because of Claim 1, we can use the Ekeland variational principle (see Corollary 5.38) and given $\varepsilon > 0$, we can find $\gamma_\varepsilon \in \Gamma$ such that

$$\xi(\gamma_\varepsilon) \leqslant \inf_\Gamma \xi + \varepsilon \quad \text{and} \quad \xi(\gamma_\varepsilon) \leqslant \xi(\gamma) + \varepsilon\|\gamma - \gamma_\varepsilon\|_\infty \qquad \forall\, \gamma \in \Gamma. \tag{5.86}$$

From (5.86), we have

$$-\,\varepsilon t\|\gamma\|_\infty \leqslant \xi(y_\varepsilon + t\gamma) - \xi(\gamma_\varepsilon) \qquad \forall\, \gamma \in \Gamma,\ t \geqslant 0. \tag{5.87}$$

Claim 2. There exists $\widehat{y}_\varepsilon \in \overline{B}_r^Y$ such that

$$\varphi(\widehat{y}_\varepsilon, \gamma_\varepsilon(\widehat{y}_\varepsilon)) = \xi(\gamma_\varepsilon) \quad \text{and} \quad \|\varphi'_G(\widehat{y}_\varepsilon, \gamma_\varepsilon(\widehat{y}_\varepsilon))\|_* \leqslant \varepsilon.$$

Note that

$$\begin{aligned}
&\xi(\gamma_\varepsilon + t\gamma) - \xi(\gamma_\varepsilon)\\
&= \max_{y\in\overline{B}_r^Y} \varphi(y, (\gamma_\varepsilon + t\gamma)(y)) - \max_{y\in\overline{B}_r^Y} \varphi(y, \gamma_\varepsilon(y))\\
&= \max_{y\in\overline{B}_r^Y} \big(\varphi(y, \gamma_\varepsilon(y)) + t\,\langle \varphi'_G(y, \gamma_\varepsilon(y)), \gamma(y)\rangle\,\big) + o(t)\\
&\quad - \max_{y\in\overline{B}_r^Y} \varphi(y, \gamma_\varepsilon(y)), \quad \text{with } \tfrac{o(t)}{t} \longrightarrow 0, \text{ as } t \searrow 0.
\end{aligned}$$

Let $f = \varphi(y, \gamma_\varepsilon(y))$ and $g = \langle \varphi'_G(y, \gamma_\varepsilon(y)), \gamma(y)\rangle$. Evidently $f, g \in C(\overline{B}_r^Y)$. Let $\eta \colon C(\overline{B}_r^Y) \longrightarrow \mathbb{R}$ be defined by

$$\eta(\vartheta) = \max_{y\in\overline{B}_r^Y} |\vartheta(y)| \qquad \forall\, \vartheta \in C(\overline{B}_r^Y).$$

This function is continuous, convex, hence subdifferentiable everywhere (see Theorem 3.71). Recall that $C(\overline{B}_r^Y)^* = M_b(\overline{B}_r^Y)$ (with $M_b(\overline{B}_r^Y)$ being the space of Radon measures on $\overline{B}_r^Y$; see Theorem 1.88). We can easily establish that

$$\partial\eta(0) = \{\mu \geqslant 0 : \mu(\overline{B}_r^Y) = 1, \ \operatorname{supp}\mu \subseteq M(\vartheta)\},$$

where $M(\vartheta) = \{y \in \overline{B}_r^Y : \vartheta(y) = \eta(\vartheta)\}$. Then we have

$$\begin{aligned} -\varepsilon\|\gamma\|_\infty &\leqslant \lim_{t\to 0} \tfrac{\xi(\gamma_\varepsilon+t\gamma)-\xi(\gamma_\varepsilon)}{t} \\ &= \lim_{t\to 0} \tfrac{\eta(f+tg)-\eta(t)}{t} = \max_{\mu\in\partial\eta(f)} \langle g,\mu\rangle_{C(\overline{B}_r^Y)} \\ &= \max_{\mu\in\partial\eta(f)} \int_{\overline{B}_r^Y} \big\langle \varphi'_G(y,\gamma_\varepsilon(y)),\gamma(y)\big\rangle\, d\mu \end{aligned} \tag{5.88}$$

(see (5.87); by $\langle\cdot,\cdot\rangle_{C(\overline{B}_r^Y)}$ we denote the duality brackets for the pair $(M_b(\overline{B}_r^Y), C(\overline{B}_r^Y))$). In (5.88) we take the infimum over all $\gamma \in \Gamma$ of both sides. Then

$$-\varepsilon \leqslant \max_{\mu\in\partial\eta(f)} \Big(-\int_{\overline{B}_r^Y} \|\varphi'(\gamma_\varepsilon(y))\|_*\, d\mu\Big) = -\min_{y\in M(\varphi\circ\gamma_\varepsilon)} \|\varphi'(\gamma_\varepsilon(y))\|_*.$$

So, there exists some $\widehat{y}_\varepsilon \in \overline{B}_r^Y$ such that

$$\varphi(\widehat{y}_\varepsilon,\gamma_\varepsilon(\widehat{y}_\varepsilon)) = \max_{y\in\overline{B}_r^Y} \varphi(y,\gamma_\varepsilon(y)) \quad \text{and} \quad \|\varphi'_G(\widehat{y}_\varepsilon,\gamma_\varepsilon(\widehat{y}_\varepsilon))\|_* \leqslant \varepsilon.$$

This proves Claim 2.

If in Claim 2, we choose $\varepsilon = \frac{1}{n}$ for $n \geqslant 1$, we see that the result of the problem is a consequence of the fact that φ satisfies the WC-condition (see Problem 5.41).

Solution of Problem 5.44

Let $u_0 \in \varphi^{-1}(\eta)$ and set $\widehat{\varphi}(u) = \varphi(u) - \varphi(u_0)$ for all $u \in X$. Evidently $\widehat{\varphi}$ satisfies the C-condition (since by hypothesis φ satisfies the C-condition; see Definition 5.45(b)) and

$$\widehat{\varphi}^{-1}(0) = \varphi^{-1}(\eta).$$

So, we see that without any loss of generality, we may assume that $\eta = 0$. Since by hypothesis the set $\varphi^{-1}(\eta) = \varphi^{-1}(0)$ is bounded, we can find $n_0 \geqslant 1$ such that

$$\varphi(u) \neq 0 \qquad \forall\, u \in X,\ \|u\|_X \geqslant e^{n_0} - 1. \tag{5.89}$$

Arguing by contradiction, suppose that

$$c = \liminf_{\|u\|_X \to +\infty} |\varphi|(u) \in \mathbb{R}.$$

So, for every $n \geqslant n_0$, we can find $u_n \in X$ such that

$$\|u_n\|_X \geqslant 2(e^n - 1) \quad \text{and} \quad |\varphi|(u_n) \leqslant c + \tfrac{1}{n}. \tag{5.90}$$

Using Problem 5.24 with $\xi(s) = s$, $\varepsilon = c + \frac{1}{n} - \inf\limits_X |\varphi|$, $\lambda = n$ and $\overline{r} = e^n - 1$, we produce a sequence $\{v_n\}_{n\geqslant 1} \subseteq X$ such that

$$|\varphi|(v_n) \leqslant |\varphi|(u_n), \quad \|v_n - u_n\|_X \leqslant e^n - 1 \tag{5.91}$$

and

$$\|\varphi'(v_n)\|_* \leqslant \tfrac{\varepsilon}{n(1+\|v_n\|_X)}. \tag{5.92}$$

Here we have used the fact that $\varphi(y) \neq 0$ in a neighborhood of v_n (see (5.89)). Note that

$$\|v_n\|_X \geqslant \|u_n\|_X - \|v_n - u_n\|_X \geqslant e^{n-1}$$

(see (5.90), (5.91), and (5.92)). Therefore, we have

$$\|v_n\|_X \longrightarrow +\infty, \quad |\varphi(v_n)| \longrightarrow c \quad \text{and} \quad (1 + \|v_n\|_X)\|\varphi'(v_n)\|_* \longrightarrow 0,$$

a contradiction to the hypothesis that φ satisfies the C-condition.

Solution of Problem 5.45

We argue indirectly. Suppose that $\varphi^{m+\lambda}$ is unbounded for all $\lambda > 0$. So, we can find a sequence $\{u_n\}_{n\geqslant 1} \subseteq X$ such that

$$m \leqslant \varphi(u_n) \leqslant m + \tfrac{1}{n} \quad \text{and} \quad \|u_n\|_X \geqslant 2(e^n - 1) \qquad \forall\, n \geqslant 1. \tag{5.93}$$

Using Problem 5.24 with $\xi(s) = s$, $\varepsilon = \frac{1}{n}$, $\lambda = n$ and $\overline{r} = e^n - 1$, we can find a sequence $\{v_n\}_{n\geqslant 1} \subseteq X$ such that

$$m \leqslant \varphi(v_n) \leqslant \varphi(u_n) \leqslant m + \tfrac{1}{n}, \quad \|v_n - u_n\|_X \leqslant e^n - 1 \tag{5.94}$$

and

$$\|\varphi'(v_n)\|_* \leqslant \tfrac{1}{n^2(1+\|v_n\|_X)} \qquad \forall\, n \geqslant 1. \tag{5.95}$$

From (5.93), (5.94), and (5.95), we see that

$$\varphi(v_n) \longrightarrow m, \quad \|v_n\|_X \geqslant \|u_n\|_X - \|v_n - u_n\|_X \geqslant e^n - 1$$

and

$$(1 + \|v_n\|_X)\|\varphi'(v_n)\|_* \longrightarrow 0,$$

which contradicts the assumption that φ satisfies the C_m-condition (see Definition 5.45(b)).

Solution of Problem 5.46

Let $\{u_n\}_{n\geqslant 1} \subseteq X$ be a minimizing sequence for φ. By passing to a subsequence if necessary, we may assume that

$$\varphi(u_n) \leqslant m + \tfrac{1}{n^2} \qquad \forall\, n \geqslant 1.$$

Using the Ekeland variational principle (see Corollary 5.38), we can find a sequence $\{v_n\}_{n\geqslant 1} \subseteq X$ such that

$$\varphi(v_n) \longrightarrow m, \quad \|v_n - u_n\| \longrightarrow 0 \quad \text{and} \quad \varphi'(v_n) \longrightarrow 0 \quad \text{in } X^*. \tag{5.96}$$

But by hypothesis φ satisfies the PS_m-condition (see Definition 5.45(a)). So, from (5.96), we infer that the sequence $\{v_n\}_{n\geqslant 1} \subseteq X$ admits a strongly convergent subsequence. Since $\|v_n - u_n\|_X \longrightarrow 0$, it follows that the corresponding subsequence of $\{u_n\}_{n\geqslant 1}$ is also strongly convergent to the same limit. The continuity of φ implies that this limit is a global minimizer of φ.

Solution of Problem 5.47

Let $\widehat{u}^* \in \operatorname{dom}\varphi$. From the Ekeland variational principle (see Theorem 5.35), we know that we can find $\widetilde{u}^* \in X^*$ such that

$$\varphi(\widetilde{u}^*) + \|\widetilde{u}^* - \widehat{u}^*\| \leqslant \varphi(\widehat{u}^*) \tag{5.97}$$

and

$$\varphi(\widetilde{u}^*) < \varphi(u^*) + \|u^* - \widetilde{u}^*\|_* \qquad \forall\, u^* \in X^*,\ u^* \neq \widetilde{u}^*. \tag{5.98}$$

From (5.97), we see that $\widetilde{u}^* \in \operatorname{dom}\varphi$. Let $\{u^*_\alpha\}_{\alpha\in J} \subseteq X^*$ be a net such that

$$\varphi(u^*_\alpha) + \|\widetilde{u}^* - u^*_\alpha\|_* \longrightarrow \varphi(\widetilde{u}^*). \tag{5.99}$$

We have

$$\begin{aligned}
\limsup_{\alpha\in J} \|u^*_\alpha\|_* &\leqslant \|\widetilde{u}^*\|_* + \limsup_{\alpha\in J} \|\widetilde{u}^* - u^*_\alpha\|_* \\
&\leqslant \|\widetilde{u}^*\|_* + \limsup_{\alpha\in J} \big(\varphi(u^*_\alpha) + \|\widetilde{u}^* - u^*_\alpha\|_*\big) - \inf_X \varphi \\
&= \|\widetilde{u}^*\|_* + \varphi(\widetilde{u}^*) - \inf_X \varphi < +\infty
\end{aligned}$$

(see (5.99) and recall that $\widetilde{u}^* \in \operatorname{dom}\varphi$). So, $\{u_\alpha^*\}_{\alpha\in J} \subseteq X^*$ is bounded. Thus, it admits a subnet $\{u_\beta^*\}_{\beta\in I}$ such that

$$u_\beta^* \xrightarrow{w^*} \overline{u}^* \quad \text{in } X^*.$$

Suppose that $\overline{u}^* \neq \widetilde{u}^*$. Recalling that the norm of X^* is w^*-lower semicontinuous, we have

$$\varphi(\overline{u}^*) + \|\widetilde{u}^* - \overline{u}^*\|_* \leqslant \varliminf_{\beta\in I} \left(\varphi(u_\beta^*) + \|\widetilde{u}^* - u_\beta^*\|_*\right) = \varphi(\widetilde{u}^*)$$

(see (5.99)). But this contradicts (5.98). So, we conclude that for the original net $\{u_\alpha^*\}_{\alpha\in J} \subseteq X^*$, we have $u_\alpha^* \xrightarrow{w^*} \widetilde{u}^*$ in X^*.

Solution of Problem 5.48

Note that the convexity of φ implies that φ is weakly lower semicontinuous. Also, due to the reflexivity of X, the weak and weak* topologies on X coincide. Then using Problem 5.47, we can find $\widetilde{u} \in \operatorname{dom}\varphi$, satisfying the claim of that problem. We will show that this is the desired $\widetilde{u} \in X$. So, let $\{u_n\}_{n\geqslant 1} \subseteq X$ be a sequence such that

$$\varphi(u_n) + \|\widetilde{u} - u_n\|_X \longrightarrow \varphi(\widetilde{u}). \tag{5.100}$$

According to Problem 5.47, we have

$$u_n \xrightarrow{w} \widetilde{u} \quad \text{in } X.$$

Suppose that this convergence is not true in the strong topology. Then, we can find $\varepsilon > 0$ and a subsequence $\{u_{n_k}\}_{k\geqslant 1}$ of $\{u_n\}_{n\geqslant 1}$ such that

$$\|u_{n_k} - \widetilde{u}\|_X > 2\varepsilon \quad \text{and} \quad \varphi(u_{n_k}) + \|\widetilde{u} - u_{n_k}\|_X < \varphi(\widetilde{u}) + \varepsilon \tag{5.101}$$

(see (5.100)). From (5.101), we have

$$\varphi(u_{n_k}) < \varphi(\widetilde{u}) - \varepsilon \qquad \forall\, k \geqslant 1,$$

so

$$\varphi(\widetilde{u}) \leqslant \varphi(\widetilde{u}) - \varepsilon$$

(from the weak lower semicontinuity of φ), a contradiction. Therefore, we conclude that $u_n \longrightarrow \widetilde{u}$ in X.

Solution of Problem 5.49

Recall that the map $(u,h) \longmapsto \varphi^0(u;h)$ is upper semicontinuous (see Proposition 3.123(b)). Then since $u_n \longrightarrow \widehat{u}$ in X and $\psi \in \Gamma_0(X)$, we have

$$\begin{aligned} 0 &\leqslant \varphi^0(\widehat{u}; u-\widehat{u}) + \psi(u) - \liminf_{n\to+\infty} \psi(u_n) \\ &\leqslant \varphi^0(\widehat{u}; u-\widehat{u}) + \psi(u) - \psi(\widehat{u}) \qquad \forall\, u \in X, \end{aligned} \tag{5.102}$$

so $\widehat{u}$ is a critical point of ξ.

The inequality $\psi(\widehat{u}) \leqslant \liminf\limits_{n\to+\infty} \psi(u_n)$ cannot be strict. Indeed, if the inequality is strict, then from (5.102) with $y = \widehat{u}$, we have

$$0 \leqslant \psi(\widehat{u}) - \liminf_{n\to+\infty} \psi(u_n) < 0,$$

a contradiction. So, $\varphi(\widehat{u}) = \liminf\limits_{n\to+\infty} \varphi(u_n) = c$.

Solution of Problem 5.50

Using the Ekeland variational principle (see Corollary 5.38), we can find a sequence $\{u_n\}_{n\geqslant 1} \subseteq X$ such that, for all $u \in X$ and $n \geqslant 1$, we have

$$\xi(u_n) \longrightarrow m \quad \text{and} \quad -\tfrac{1}{n}\|u-u_n\|_X \leqslant \xi(u) - \xi(u_n). \tag{5.103}$$

Let $u = (1-t)u_n + th$ with $t \in [0,1]$ and $h \in X$. From (5.103) and the convexity of ψ, we have

$$-\tfrac{1}{n}\|h-u_n\| \leqslant \tfrac{1}{t}\big(\varphi(u_n + t(h-u_n)) - \varphi(u_n)\big) + \psi(h) - \psi(u_n) \qquad \forall\, n \geqslant 1,$$

so

$$-\tfrac{1}{n}\|h-u_n\| \leqslant \varphi^0(u_n; h-u_n) + \psi(h) - \psi(u_n) \qquad \forall\, n \geqslant 1. \tag{5.104}$$

Since $\xi(u_n) \longrightarrow m$ (see (5.103)) and since ξ satisfies the GNPS_m-condition, we may assume that $u_n \longrightarrow \widehat{u}$ in X. Then

$$\xi(\widehat{u}) = m \quad \text{and} \quad 0 \leqslant \varphi^0(\widehat{u}; h-u) + \psi(h) - \psi(\widehat{u})$$

(see (5.103), Proposition 3.123(b) and recall that $\psi \in \Gamma_0(X)$), so $\widehat{u} \in X$ is a critical point of ξ (see Problem 5.49) with critical level m.

Solution of Problem 5.51

First note that by the Dugundji extension theorem (see Theorem 2.8), we have that $\Gamma \neq \emptyset$ and $c \geqslant \max_{u \in E_0} \varphi(\gamma_0(u))$. On Γ we consider the supremum metric d_∞ defined by

$$d_\infty(\gamma, \widehat{\gamma}) = \max_{u \in E} \|\gamma(u) - \widehat{\gamma}(u)\|_X \qquad \forall \gamma, \widehat{\gamma} \in \Gamma.$$

Evidently (Γ, d_∞) is a complete metric space. We consider the function $\vartheta \colon \Gamma \longrightarrow \mathbb{R}$ defined by

$$\vartheta(\gamma) = \max_{u \in E} \varphi(\gamma(u)). \tag{5.105}$$

Claim. ϑ is continuous and bounded below.

Suppose that $\{\gamma_n\}_{n \geqslant 1} \subseteq \Gamma$ is a sequence such that $\gamma_n \longrightarrow \gamma$ in (Γ, d_∞). Since the set E is compact and $\varphi \circ \gamma_n$ is continuous for every $n \geqslant 1$, we can find a sequence $\{u_n\}_{n \geqslant 1} \subseteq E$ such that

$$\vartheta(\gamma_n) = \varphi(\gamma_n(u_n)) \qquad \forall n \geqslant 1.$$

We may assume that $u_n \longrightarrow \widehat{u}$ in E. Then $\gamma_n(u_n) \longrightarrow \gamma(\widehat{u})$ in X (recall that $\gamma_n \xrightarrow{d_\infty} \gamma$) and so $\vartheta(\gamma_n) = \varphi(\gamma_n(u_n)) \longrightarrow \varphi(\gamma(\widehat{u}))$. We have

$$\varphi(\gamma_n(u)) \leqslant \vartheta(\gamma_n) \qquad \forall u \in X,\ n \geqslant 1,$$

so

$$\varphi(\gamma(u)) \leqslant \varphi(\gamma(\widehat{u})) \qquad \forall u \in X,$$

thus

$$\vartheta(\gamma) = \varphi(\gamma(\widehat{u}))$$

and hence ϑ is continuous.

Also, for every $\gamma \in \Gamma$, we have

$$\vartheta(\gamma) \geqslant \max_{u \in E_0} \varphi(\gamma_0(u)),$$

so ϑ is bounded below. This proves the Claim.

We have $c = \inf_{\gamma \in \Gamma} \vartheta(\gamma)$ (see the definitions of c and ϑ). Because of the Claim, we can use the Ekeland variational principle (see Corollary 5.38). So, given $\varepsilon > 0$, we can find $\widehat{\gamma} \in \Gamma$ such that

$$c \leqslant \vartheta(\widehat{\gamma}) \leqslant c + \varepsilon \quad \text{and} \quad \vartheta(\widehat{\gamma}) \leqslant \vartheta(\gamma) + \varepsilon d_\infty(\gamma, \widehat{\gamma}) \qquad \forall \gamma \in \Gamma. \tag{5.106}$$

Let $S(\widehat{\gamma}) = \{u \in E : \vartheta(\widehat{\gamma}) = \varphi(\widehat{\gamma}(u))\}$. Invoking Problem 2.113, we can find a continuous map $\xi \colon S(\widehat{\gamma}) \longrightarrow X$ such that for every $u \in S(\widehat{\gamma})$ and every $u^* \in \partial\varphi(\widehat{\gamma}(u))$, we have

$$\|\xi(u)\|_X \leqslant 1 \quad \text{and} \quad \langle u^*, \xi(u) \rangle \geqslant \eta - \varepsilon, \tag{5.107}$$

where $\eta = \inf\limits_{u\in S(\widehat{\gamma})} m(\widehat{\gamma}(u))$. Then for every $u \in S(\widehat{\gamma})$, we have

$$\begin{aligned}\varphi^0(\widehat{\gamma}(u); -\xi(u)) &= \max_{u^*\in\partial\varphi(\widehat{\gamma}(u))} \langle u^*, -\xi(u)\rangle \\ &= -\min_{u^*\in\partial\varphi(\widehat{\gamma}(u))} \langle u^*, \xi(u)\rangle \leqslant -\eta+\varepsilon \end{aligned} \tag{5.108}$$

(see Proposition 3.128(c) and (5.107)). By hypothesis, we have $S(\widehat{\gamma})\cap E_0 = \emptyset$. So, we can find a continuous extension $\widehat{\xi}\colon E \longrightarrow X$ of ξ such that $\widehat{\xi}|_{E_0} = 0$ and $\|\widehat{\xi}(u)\|_X \leqslant 1$ for all $u \in E$.

Let $\gamma_t(u) = \widehat{\gamma}(u) - t\widehat{\xi}(u)$ for all $u \in E$ and $t > 0$. From (5.106), we have

$$-\varepsilon \leqslant -\varepsilon\|\widehat{\xi}\|_\infty \leqslant \tfrac{\vartheta(\gamma_t)-\vartheta(\widehat{\gamma})}{t}. \tag{5.109}$$

Let $t_n \searrow 0$ and let $\{u_n\}_{n\geqslant 1} \subseteq E$ be a sequence such that

$$\varphi(\gamma_n(u_n)) = \vartheta(\gamma_n)$$

(where $\gamma_n = \gamma_{t_n}$). We may assume that $u_n \longrightarrow \widehat{u}$ in E. From the proof of the Claim, we know that $\widehat{u} \in S(\widehat{\gamma})$. We have

$$\tfrac{\vartheta(\gamma_n)-\vartheta(\widehat{\gamma})}{t_n} = \tfrac{\vartheta(\widehat{\gamma}-t_n\widehat{\xi})-\vartheta(\widehat{\gamma})}{t_n} \leqslant \tfrac{\varphi(\widehat{\gamma}(u_n)-t_n\widehat{\xi}(u_n))-\varphi(\widehat{\gamma}(u_n))}{t_n},$$

so

$$\begin{aligned}-\varepsilon &\leqslant \tfrac{\varphi(\widehat{\gamma}(u_n)-t_n\widehat{\xi}(u_n))-\varphi(\widehat{\gamma}(u_n))}{t_n} \\ &\leqslant \tfrac{\varphi(\widehat{\gamma}(u_n)-t_n\widehat{\xi}(u_n))-\varphi(\widehat{\gamma}(u_n)-t_n\widehat{\xi}(\widehat{u}))}{t_n} + \tfrac{\varphi(\widehat{\gamma}(u_n)-t_n\widehat{\xi}(\widehat{u}))-\varphi(\widehat{\gamma}(u_n))}{t_n}.\end{aligned} \tag{5.110}$$

Since φ is locally Lipschitz and $u_n \longrightarrow \widehat{u}$ in E, we see that

$$\tfrac{\varphi(\widehat{\gamma}(u_n)-t_n\widehat{\xi}(u_n))-\varphi(\widehat{\gamma}(u_n)-t_n\widehat{\xi}(\widehat{u}))}{t_n} \longrightarrow 0. \tag{5.111}$$

So, if in (5.110) we pass to the limit as $n \to +\infty$ and use (5.111), then

$$-\varepsilon \leqslant \limsup_{n\to+\infty} \tfrac{\varphi(\widehat{\gamma}(\widehat{u})+w_n-t_n\widehat{\xi}(\widehat{u}))-\varphi(\widehat{\gamma}(\widehat{u})+w_n)}{t_n}, \tag{5.112}$$

with $w_n = \widehat{\gamma}(u_n) - \widehat{\gamma}(\widehat{u}) \longrightarrow 0$. So, from (5.112), we have

$$-\varepsilon \leqslant \varphi^0(\widehat{\gamma}(\widehat{u}); -\widehat{\xi}(\widehat{u})) = \varphi^0(\widehat{\gamma}(\widehat{u}); -\xi(\widehat{u})) \leqslant -\eta+\varepsilon$$

(see (5.108) and recall that $0 \in S(\widehat{\gamma})$), so $\eta \leqslant 2\varepsilon$. But m is lower semicontinuous and E is compact. So, we can find $\widetilde{u}_\varepsilon \in E$ such that

$$m(\widehat{\gamma}(\widetilde{u}_\varepsilon)) \leqslant 2\varepsilon.$$

Finally, let $\varepsilon_n = \frac{1}{n}$ and $v_n = \widehat{\gamma}(\widetilde{u}_n)$ (with $\widetilde{u}_n = \widetilde{u}_{\varepsilon_n}$ for all $n \geqslant 1$). We have

$$\varphi(v_n) \longrightarrow c \quad \text{and} \quad m(v_n) \longrightarrow 0.$$

Solution of Problem 5.52

Let $E = [0,1]$, $E_0 = \{0,1\}$ and $\gamma_0 \colon E_0 \longrightarrow X$ be defined by $\gamma_0(0) = 0$, $\gamma(1) = u_1$. Since by hypothesis $c > 0$, we have

$$c > \max_{t \in E_0} \gamma_0(t).$$

So, we can use Problem 5.51 and find a sequence $\{u_n\}_{n \geqslant 1} \subseteq X$ such that

$$\varphi(u_n) \longrightarrow c \quad \text{and} \quad m(u_n) \longrightarrow 0. \tag{5.113}$$

Since φ satisfies the nonsmooth PS_c-condition (see Definition 5.47(a)), we may assume that $u_n \longrightarrow \widehat{u}$ in X. Because m is lower semicontinuous (see Problem 3.194), we have

$$\varphi(\widehat{u}) = c \quad \text{and} \quad m(\widehat{u}) = 0$$

(see (5.113)), so $0 \in \partial\varphi(\widehat{u})$ and $\varphi(\widehat{u}) = c$. So, $\widehat{u}$ is a critical point of φ with critical value c.

Solution of Problem 5.53

The continuity hypothesis on φ' implies that the map $u \longmapsto \varphi'(u)$ is locally bounded. So, for $t > 0$ small and $v \in X$ with $\|v\|_X$ small, we have

$$\big|\varphi(u+v+th) - \varphi(u+v)\big| = \big|t\left\langle \varphi'_G(u+v+t\eta h), h\right\rangle\big| \leqslant c\|th\|_X,$$

for some $\eta \in (0,1)$ and $c > 0$ (see Theorem 3.16), so φ is locally Lipschitz. Moreover, from the continuity hypothesis on φ' we have

$$\varphi^0(u;h) = \left\langle \varphi'_G(u), h\right\rangle \qquad \forall\, u, h \in X,$$

so $\partial\varphi(u) = \{\, \varphi'_G(u) \,\}$. Then the conclusion of the problem follows from Problem 5.51.

Solution of Problem 5.54

We use Theorem 5.41 with $u_0 = 0$, $\varepsilon = \frac{1}{n}$ for all $n \geqslant 1$, $\xi(s) = s$ and $\lambda = 1$. So, we can find a sequence $\{u_n\}_{n \geqslant 1} \subseteq X$ such that

$$\varphi(u_n) \searrow \inf_X \varphi \tag{5.114}$$

and

$$\varphi(u_n) \leqslant \varphi(u) + \tfrac{1}{n(1+\|u_n\|_X)} \qquad \forall\, u \in X,\ n \geqslant 1. \tag{5.115}$$

From (5.115) with $u = u_n + th$, $t > 0$, $h \in X$, we have

$$-\tfrac{\|h\|_X}{n(1+\|u_n\|_X)} \leqslant \tfrac{\varphi(u_n+th)-\varphi(u_n)}{t} \qquad \forall\, t > 0,\ n \geqslant 1,$$

so

$$-\tfrac{\|h\|_X}{n(1+\|u_n\|_X)} \leqslant \varphi^0(u_n;h) \qquad \forall\, h \in X,\ n \geqslant 1 \tag{5.116}$$

(see Definition 3.122). Let $\psi_n = n(1+\|u_n\|_X)\varphi^0(u_n;h)$. Then ψ_n is sublinear, continuous (see Proposition 3.123) and $\psi_n(0) = 0$. From (5.116), we have

$$-\|h\|_X \leqslant \psi_h(h) \qquad \forall\, h \in X,\ n \geqslant 1.$$

Using Problem 3.133, we can find $v_n^* \in X^*$ such that

$$\|v_n^*\|_* \leqslant 1 \quad \text{and} \quad \langle v_n^*, h\rangle \leqslant \psi_n(h) \qquad \forall\, h \in X,\ n \geqslant 1.$$

Let $u_n^* = \frac{1}{n(1+\|u_n\|_X)} v_n^*$. Then

$$\langle u_n^*, h\rangle \leqslant \varphi^0(u_n;h) \qquad \forall\, h \in X,\ n \geqslant 1. \tag{5.117}$$

From (5.117), we see that

$$u_n^* \in \partial\varphi(u_n) \qquad \forall\, n \geqslant 1,$$

so

$$(1+\|u_n\|_X)m(u_n) \leqslant (1+\|u_n\|_X)\|u_n^*\|_* \leqslant \tfrac{1}{n} \qquad \forall\, n \geqslant 1,$$

thus $(1+\|u_n\|_X)m(u_n) \longrightarrow 0$. Using also (5.114), we conclude that $\{u_n\}_{n\geqslant 1} \subseteq X$ is the desired sequence.

Solution of Problem 5.55

"$\text{NC}_c \Longrightarrow \text{NPS}_c$":

Let $\{u_n\}_{n\geqslant 1} \subseteq X$ be a sequence such that

$$\varphi(u_n) \longrightarrow c \quad \text{and} \quad m(u_n) \longrightarrow 0.$$

Let $u_n^* \in \partial\varphi(u_n)$ be such that $m(u_n) = \|u_n^*\|_*$ for all $n \geqslant 1$. It exists since the norm functional on X^* is w^*-lower semicontinuous and the multifunction $u \longmapsto \partial\varphi(u)$ has nonempty, w^*-compact, and convex values in X^* (see Proposition 3.128). We have

$$|\langle u_n^*, h\rangle| \leqslant m(u_n) = \varepsilon_n \qquad \forall\, h \in X,\ \|h\|_X = 1,\ n \geqslant 1,$$

so

$$|\langle u_n^*, h\rangle| \leqslant |\varphi^0(u_n;h)| \leqslant \varepsilon_n \qquad \forall\, h \in X,\ \|h\|_X = 1,\ n \geqslant 1$$

(see Proposition 3.128(c)), thus

$$\tfrac{1}{t}\big(\varphi(u_n+th)-\varphi(u_n)\big) \leqslant \langle u_n^*, h\rangle + \varepsilon_n \qquad \forall\, 0<t\leqslant \delta_n = \delta_n(\varepsilon_n),\ n \geqslant 1$$

and hence

$$\varphi(u_n+th) \leqslant \varphi(u_n) + \delta\varepsilon_n \qquad \forall\, n \geqslant 1.$$

Since NC_c holds, we conclude that the sequence $\{u_n\}_{n\geqslant 1}$ admits a strongly convergent subsequence. Therefore NPS_c holds.

"$\mathrm{NPS}_c \Longrightarrow \mathrm{NC}_c$":
Let $\{u_n\}_{n\geqslant 1} \subseteq X$ and $\{\varepsilon_n\}_{n\geqslant 1}, \{\delta_n\}_{n\geqslant 1} \subseteq (0,+\infty)$ be sequences such that

$$\varepsilon_n \longrightarrow 0, \quad \delta_b \longrightarrow 0, \quad \varphi(u_n) \longrightarrow c$$

and

$$|\varphi(u_n)-\varphi(u)| \leqslant \varepsilon\|u_n-u\|_X \qquad \forall\, u \in X,\ n \geqslant 1,\ \|u_n-u\|_X \leqslant \delta_n.$$

Let $u = u_n + th$ with $|t| \in (0,\delta_n]$, $\|h\|_X = 1$. Then

$$-\varepsilon_n|t| \leqslant \varphi(u_n+th)-\varphi(u_n) \leqslant \varepsilon_n|t|,$$

so

$$\varphi^0(u_n;-h),\ \varphi^0(u_n;h) \leqslant \varepsilon_n \qquad \forall\, h \in X,\ \|h\|_X = 1,\ n \geqslant 1. \tag{5.118}$$

Let $u_n^* \in \partial\varphi(u_n)$ for $n \geqslant 1$ be such that

$$\langle u_n^*, h\rangle = \varphi^0(u_n;h) \qquad \forall\, h \in X,\ n \geqslant 1$$

(see Proposition 3.128). Then from (5.118), we have

$$|\langle u_n^*, h\rangle| \leqslant \varepsilon_n \qquad \forall\, h \in X,\ \|h\|_X = 1,\ n \geqslant 1,$$

so

$$\|u_n^*\|_* \leqslant \varepsilon_n \qquad \forall\, n \geqslant 1,$$

thus $m(u_n) \longrightarrow 0$. Since the NPS_c holds, we conclude that the sequence $\{u_n\}_{n\geqslant 1}$ admits a strongly convergent subsequence.

Solution of Problem 5.56

Let $\{u_n\}_{n\geqslant 1} \subseteq H$ be a sequence such that

$$\varphi(u_n) \longrightarrow c \quad \text{and} \quad m(u_n) \longrightarrow 0. \tag{5.119}$$

Because the WNPS$_c$-condition holds, passing to a subsequence if necessary, we may assume that

$$u_n \xrightarrow{w} u \quad \text{in } H. \tag{5.120}$$

Let $u_n^* \in \partial\varphi(u_n)$ be such that $m(u_n) = \|u_n^*\|_*$ for all $n \geqslant 1$. We have

$$u_n^* = A(u_n) - v_n^* \qquad \forall\, n \geqslant 1, \tag{5.121}$$

where $A \in \mathcal{L}(H;H^*)$ is the canonical isomorphism (since H is a Hilbert space) and $v_n^* \in \partial\widehat{\psi}(u_n) = \partial\psi(u_n)|_H$ for all $n \geqslant 1$ (see Problem 3.124). From (5.120) and since $H \hookrightarrow Y$ compactly, we have

$$u_n \longrightarrow u \quad \text{in } Y,$$

so the set $\bigcup\limits_{n\geqslant 1} \partial\psi(u_n) \subseteq Y^*$ is relatively w-compact (see Problem 5.57(a) and Proposition 3.128(a)).

But the embedding $Y^* \hookrightarrow H^*$ is compact and dense (see Problem 1.71). Therefore by passing to a subsequence if necessary, we may assume that

$$v_n^* \longrightarrow v^* \quad \text{in } H^*. \tag{5.122}$$

From (5.121), we have

$$u_n = A^{-1}(u_n^* + v_n^*) \longrightarrow A^{-1}(v^*)$$

(see (5.119) and (5.122)), so φ satisfies the NPS$_c$-condition.

Solution of Problem 5.57

Since $\varphi \in \Gamma_0(X)$, it is minorized by a continuous affine function (see Proposition 3.84). So, we can find $\widehat{u}^* \in X^*$ and $c \in \mathbb{R}$ such that

$$\psi(u) = \varphi(u) - \langle \widehat{u}^*, u\rangle - c > 0 \qquad \forall\, u \in X.$$

Evidently ψ is lower semicontinuous and bounded below. So, we may apply the Ekeland variational principle (see Theorem 5.36) with $\varepsilon = 2(\psi(\widehat{u}) - \inf\limits_X \psi) > 0$ and $\lambda = \frac{1}{n}$ for $n \geqslant 1$. So, we can find a sequence $\{u_n\}_{n\geqslant 1} \subseteq X$ such that

$$\psi(u_n) \leqslant \psi(\widehat{u}), \tag{5.123}$$

$$\|u_n - \widehat{u}\|_X \leqslant \tfrac{1}{n} \tag{5.124}$$

and

$$\psi(u_n) < \psi(u) + \varepsilon n \|u - u_n\|_X \qquad \forall\, u \in X,\ u \neq u_n. \tag{5.125}$$

Consider the functional $\xi\colon X \longrightarrow \overline{\mathbb{R}}$ defined by

$$\xi_n(u) = \psi(u) + \varepsilon n \|u - u_n\|_X \qquad \forall\, n \geqslant 1. \tag{5.126}$$

Clearly $\xi_n \in \Gamma_0(X)$ and from (5.125) we see that $u = u_n$ is the unique global minimizer of ξ_n. Therefore

$$0 \in \partial \xi_n(u_n) \qquad \forall\, n \geqslant 1 \tag{5.127}$$

(see Proposition 3.82). Let $d_n \in \Gamma_0(X)$ be defined by

$$d_n(u) = \|u - u_n\|_X \qquad \forall\, u \in X,\ n \geqslant 1.$$

From (5.126), (5.127) and Proposition 3.97, we have

$$0 = w_n^* + \varepsilon_n v_n^* \qquad \forall\, n \geqslant 1$$

with $w_n^* \in \partial \psi(u_n)$, $v_n^* \in \partial d_n(u_n)$. Note that

$$w_n^* = u_n^* - \widehat{u}^* \qquad \forall\, n \geqslant 1,$$

with $u_n^* \in \partial \varphi(u_n)$, so

$$\partial \varphi(u_n) \neq \emptyset \qquad \forall\, n \geqslant 1.$$

Also, from (5.123), we have

$$\varphi(u_n) \leqslant \varphi(\widehat{u}) + \langle \widehat{u}^*, u_n - \widehat{u} \rangle \qquad \forall\, n \geqslant 1,$$

so

$$\limsup_{n \to +\infty} \varphi(u_n) \leqslant \varphi(\widehat{u}) \tag{5.128}$$

(see (5.124)). On the other hand, from the lower semicontinuity of φ, we have

$$\varphi(\widehat{u}) \leqslant \liminf_{n \to +\infty} \varphi(u_n). \tag{5.129}$$

From (5.128) and (5.129) we conclude that $\varphi(u_n) \longrightarrow \varphi(\widehat{u})$.

Solution of Problem 5.58

Let $X = \mathbb{R}$ and let $\varphi\colon \mathbb{R} \longrightarrow \mathbb{R}$ be defined by

$$\varphi(u) = \sin u^2 \qquad \forall\, u \in \mathbb{R}.$$

Let $c = \inf\limits_{\mathbb{R}} \varphi = -1$. Consider the sequence $u_n = \left(\frac{3\pi}{2} + 2\pi n + \frac{1}{\sqrt{2\pi n}}\right)^{\frac{1}{2}}$ for $n \geqslant 1$. Then

$$\varphi(u_n) \longrightarrow c = -1, \quad \text{but } \varphi'(u_n) \longrightarrow 2.$$

So, this sequence although minimizing is not a Palais–Smale sequence.

Solution of Problem 5.59

Note that for all $u \in H_0^1(\Omega)$ we have

$$\varphi_\lambda'(u) = A(u) - \lambda u,$$

where $A \in \mathcal{L}(H_0^1(\Omega); H^{-1}(\Omega))$ (recall that $H^{-1}(\Omega) = H_0^1(\Omega)^*$) is defined by

$$\langle A(u), v\rangle = \int\limits_\Omega (Du, Dv)_{\mathbb{R}^N}\, dz \qquad \forall\, u, v \in H_0^1(\Omega)$$

(the canonical isomorphism). If $\lambda \neq \widehat{\lambda}_n$ for all $n \geqslant 1$, then $A - \lambda I_{\mathbb{R}^N}$ is invertible, while by the Rellich–Kondrachov embedding theorem (see Theorem 1.135), the linear map $u \longmapsto \int\limits_\Omega hu\, dz$ is compact. So, using Problem 5.32, we infer that φ satisfies the PS-condition (see Definition 5.45(a)).

Now suppose that $\lambda = \widehat{\lambda}_k$ for some $k \geqslant 1$. Then there exists $\widehat{u}_k \in H_0^1(\Omega) \setminus \{0\}$ such that

$$A(\widehat{u}_k) = \widehat{\lambda}_k \widehat{u}_k, \tag{5.130}$$

so

$$\|D\widehat{u}_k\|_2^2 = \widehat{\lambda}_k \|\widehat{u}_k\|_2^2,$$

thus

$$\varphi_\lambda(\widehat{u}_k) = 0. \tag{5.131}$$

Let $\widehat{v}_n = n\widehat{u}_k$ for $n \geqslant 1$. Then

$$\varphi_\lambda(\widehat{v}_n) = n\varphi_\lambda(\widehat{u}_k) = 0 \qquad \forall\, n \geqslant 1$$

(see (5.131)) and

$$\varphi_\lambda'(\widehat{v}_n) = 0 \qquad \forall\, n \geqslant 1$$

(see (5.130)). Therefore $\{\widehat{v}_n\}_{n\geqslant 1}$ is a Palais–Smale sequence with no convergence subsequence. Hence $\varphi_{\widehat{\lambda}_k}$ (with $k \geqslant 1$) does not satisfy the PS-condition.

Solution of Problem 5.60

According to the Ekeland variational principle (see Corollary 5.38), we can find a sequence $\{u_n\}_{n\geqslant 1} \subseteq M$ such that

$$\varphi(u_n) \searrow m \quad \text{and} \quad \varphi(u_n) < \varphi(u) + \tfrac{1}{n}\|u - u_n\|_X \qquad \forall\, u \in M,\ u \neq u_n. \tag{5.132}$$

For each $h \in T_{u_n}M$, there exists a curve $\sigma\colon (-1,1) \longrightarrow M$ such that $\sigma(0) = u_n$ and $\sigma'(0) = h$. Then from (5.132) with $u = \sigma(t)$, we have

$$-\tfrac{1}{n}\|\sigma(t) - u_n\|_X \leqslant \varphi(\sigma(t)) - \varphi(u_n).$$

Dividing by $t > 0$ and letting $t \searrow 0$, we obtain

$$-\tfrac{1}{n}\|\sigma'(0)\|_X \leqslant \left\langle \varphi'(u_n), \sigma'(0)\right\rangle,$$

so

$$-\tfrac{1}{n}\|h\|_X \leqslant \left\langle \varphi'(u_n), h\right\rangle \qquad \forall\, h \in T_{u_n}M,$$

thus

$$\|\varphi'_M(u_n)\|_* \leqslant \tfrac{1}{n} \qquad \forall\, n \geqslant 1$$

and hence $\varphi'_M(u_n) \longrightarrow 0$. Because of the PS_m^M-condition, passing to a subsequence if necessary, we may assume that $u_n \longrightarrow \widehat{u}$ in X. Then $\widehat{u} \in M$, $\varphi(\widehat{u}) = 0$ and $\varphi'_M(\widehat{u}) = 0$. Invoking Proposition 5.31, we see that we can find $\lambda \in \mathbb{R}$ such that $\varphi'(\widehat{u}) = \lambda g(\widehat{u})$.

Solution of Problem 5.61

Let $\{u_n\}_{n\geqslant 1} \subseteq H_0^1(\Omega)$ be a sequence such that the sequence $\{\,\varphi(u_n)\,\}_{n\geqslant 1} \subseteq \mathbb{R}$ is bounded and

$$(1 + \|u_n\|_X)\varphi'(u_n) \longrightarrow 0 \quad \text{in } H^{-1}(\Omega) = H_0^1(\Omega)^*. \tag{5.133}$$

From the boundedness of the sequence $\{\,\varphi(u_n)\,\}_{n\geqslant 1} \subseteq \mathbb{R}$, we have

$$\tfrac{p}{2}\|Du\|_2^2 - \|u\|_p^p \leqslant M_1 \qquad \forall\, n \geqslant 1, \tag{5.134}$$

for some $M_1 > 0$, while from (5.133), we have

$$\Big|\left\langle A(u_n), h\right\rangle - \int_\Omega |u_n|^{p-2}u_n h\, dz\Big| \leqslant \tfrac{\varepsilon_n\|h\|_X}{1+\|u_n\|_X} \qquad \forall\, h \in H_0^1(\Omega), \tag{5.135}$$

with $\varepsilon_n \searrow 0$. Recall that $A \in \mathcal{L}(H_0^1(\Omega); H^{-1}(\Omega))$ is defined by

$$\left\langle A(u), h\right\rangle = \int_\Omega (Du, Dh)_{\mathbb{R}^N}\, dz \qquad \forall\, u, h \in H_0^1(\Omega).$$

In (5.135) we choose $h = u_n \in H_0^1(\Omega)$ and obtain

$$-\|Du_n\|_2^2 + \|u_n\|_p^p \leqslant \varepsilon_n \qquad \forall\, n \geqslant 1. \tag{5.136}$$

Adding (5.134) and (5.136), we have

$$(\tfrac{p}{2}-1)\|Du_n\|_2^2 \leqslant M_2 \qquad \forall\, n \geqslant 1,$$

for some $M_2 > 0$, so the sequence $\{u_n\}_{n\geqslant 1} \subseteq H_0^1(\Omega)$ is bounded (by the Poincaré inequality; see Theorem 1.131). So, by passing to a subsequence if necessary and using the Rellich–Kondrachov embedding theorem (see Theorem 1.135), we can say that

$$u_n \xrightarrow{w} u \quad \text{in } H_0^1(\Omega) \quad \text{and} \quad u_n \longrightarrow u \quad \text{in } L^p(\Omega). \tag{5.137}$$

Note that the sequence $\left\{|u_n|^{p-2}u_n\right\}_{n\geqslant 1} \subseteq L^{p'}(\Omega)$ (with $\frac{1}{p}+\frac{1}{p'}=1$) is bounded. Therefore by the Hölder inequality (see Theorem 1.3) and (5.137), we have

$$\int\limits_{\Omega} |u_n|^{p-2}u_n(u_n-u)\,dz \longrightarrow 0. \tag{5.138}$$

So, if in (5.135) we choose $h = u_n - u \in H_0^1(\Omega)$, pass to the limit as $n \to +\infty$ and use (5.138), then

$$\lim_{n\to+\infty} \langle A(u_n), u_n - u\rangle = 0,$$

so

$$\|Du\|_2 \longrightarrow \|Du\|_2$$

(note that $A(u_n) \xrightarrow{w} A(u)$ in $H^{-1}(\Omega)$; see (5.137)). Hence, we have

$$u_n \xrightarrow{w} u \quad \text{in } H_0^1(\Omega) \quad \text{and} \quad \|u_n\| \longrightarrow \|u\|. \tag{5.139}$$

But a Hilbert space has the Kadec–Klee property (see Corollary 1.26). So, from (5.139), it follows that

$$u_n \longrightarrow u \quad \text{in } H_0^1(\Omega),$$

thus φ satisfies the C-condition.

Solution of Problem 5.62

Let $h \in \operatorname{dom}\psi$. By hypothesis, we can find $\delta \in (0,1)$ such that

$$\xi(\widehat{u}+th) \geqslant \xi(\widehat{u}) \qquad \forall\, t \in (0,\delta). \tag{5.140}$$

The convexity of ψ implies that

$$\begin{aligned}\psi(\widehat{u}+th) &= \psi((1-t)\widehat{u}+t(\widehat{u}+h)) \leqslant (1-t)\psi(\widehat{u})+t\psi(\widehat{u}+h)\\ &= t(\psi(\widehat{u}+h)-\psi(\widehat{u}))+\psi(\widehat{u}).\end{aligned} \tag{5.141}$$

So, we have

$$\tfrac{\varphi(\widehat{u}+th)-\varphi(\widehat{u})}{t} + \psi(\widehat{u}+h)-\psi(\widehat{u}) \geqslant \tfrac{\xi(\widehat{u}+th)-\xi(\widehat{u})}{t} \geqslant 0 \qquad \forall\, t \in (0,\delta)$$

(see (5.140) and (5.141)), thus $\varphi^0(\widehat{u};h)+\psi(\widehat{u}+h)-\psi(\widehat{u}) \geqslant 0$ for all $h \in X$ (see Definition 3.122), hence $\widehat{u} \in X$ is a critical point of ξ.

Solution of Problem 5.63

Since $\widehat{u} \in X$ is a critical point of ξ, we have

$$\varphi^0(\widehat{u}; th) + \psi(\widehat{u} + th) - \psi(\widehat{u}) \geqslant 0 \qquad \forall\, t > 0,\ h \in X$$

(see Problem 5.62), so

$$\varphi^0(u; h) + \tfrac{1}{t}\big(\psi(\widehat{u} + th) - \psi(\widehat{u})\big) \geqslant 0,$$

thus

$$-\psi'_G(\widehat{u}) \in \partial\varphi(\widehat{u})$$

(see Proposition 3.128) and hence $-\psi'_G(\widehat{u}) \in \partial\widehat{\varphi}(\widehat{u})$ (see Problem 3.124).

Solution of Problem 5.64

The compactness of the embedding of $X \hookrightarrow Y$ implies that φ is sequentially weakly continuous, in particular then sequentially weakly lower semicontinuous. So, because of coercivity hypothesis on ξ and the reflexivity of X, we can use Theorem 5.6 and obtain $\widehat{u} \in X$ such that

$$\xi(\widehat{u}) = \inf_X \xi.$$

Then from Problem 5.62, we have that $\widehat{u} \in X$ is a critical point of ξ. Using Problem 5.63, we conclude that $-\psi'_G(\widehat{u}) \in \partial\widehat{\varphi}(\widehat{u})$.

Solution of Problem 5.65

Let $\{u_n\}_{n\geqslant 1} \subseteq X$ be a sequence such that

$$|\varphi(u_n)| \leqslant M \qquad \forall\, n \geqslant 1, \tag{5.142}$$

for some $M > 0$ and

$$(1 + \|u_n\|_X)\varphi'(u_n) \longrightarrow 0 \quad \text{in } X^*. \tag{5.143}$$

From coercivity hypothesis on φ and (5.142), it follows that the sequence $\{u_n\}_{n\geqslant 1} \subseteq X$ is bounded. The reflexivity of X and the Eberlein–Smulian theorem (see Theorem I.5.78) imply that, passing to a subsequence if necessary, we may assume that

$$u_n \xrightarrow{w} u \quad \text{in } X. \tag{5.144}$$

The complete continuity of K implies that

$$K(u_n) \longrightarrow K(u) \quad \text{in } X^*. \tag{5.145}$$

From (5.143), we have

$$\left|\langle A(u_n),h\rangle+\langle K(u_n),h\rangle\right| \leqslant \tfrac{\varepsilon_n\|h\|_X}{1+\|u_n\|_X} \qquad \forall\, h\in X, \tag{5.146}$$

with $\varepsilon_n \searrow 0$. In (5.146) we choose $h=u_n-u\in X$, pass to the limit as $n\to+\infty$ and using (5.145), we obtain

$$\lim_{n\to+\infty}\langle A(u_n),u_n-u\rangle = 0,$$

so using (5.144) and exploiting the fact that A is of $(S)_+$-type, we obtain

$$u_n \longrightarrow u \quad \text{in } X.$$

This proves that φ satisfies the C-condition (see Definition 5.45(b)).

Solution of Problem 5.66

From the growth condition on f, we have

$$|F(z,x)| \leqslant c(1+|x|^q) \quad \text{for a.a. } z\in\Omega,\ \ x\in\mathbb{R}, \tag{5.147}$$

with $c>0$. Then for all $u\in W_0^{1,p}(\Omega)$, we have

$$\begin{aligned}\varphi(u) &\geqslant \tfrac{1}{p}\|Du\|_p^p-c\|u\|_q^q-c|\Omega|_N\\ &\geqslant \tfrac{1}{p}\|u\|^p-\widehat{c}\|u\|^q-c|\Omega|_N,\end{aligned} \tag{5.148}$$

for some $\widehat{c}>0$ (from the Poincaré inequality (see Theorem 1.131) and the Rellich–Kondrachov embedding theorem (Theorem 1.135)). From (5.148) and since $q<p$, we infer that

$$\varphi(u) \longrightarrow +\infty \quad \text{as } \|u\| \to +\infty. \tag{5.149}$$

Note that for all $u\in W_0^{1,p}(\Omega)$, we have

$$\varphi'(u) = A(u)-N_f(u),$$

with $A\colon W_0^{1,p}(\Omega)\longrightarrow W^{-1,p'}(\Omega)=W_0^{1,p}(\Omega)^*$ (where $\frac{1}{p}+\frac{1}{p'}=1$) defined by

$$\langle A(u),h\rangle = \int_\Omega |Du|^{p-2}(Du,Dh)\,dz \qquad \forall\, u,h\in W_0^{1,p}(\Omega)$$

and

$$N_f(u)(\cdot) = f(\cdot,u(\cdot))$$

(see Problems 3.21 and 3.5). From Problem 3.191 we know that A is of type $(S)_+$, while from the compactness of the embedding $W_0^{1,p}(\Omega)\hookrightarrow L^p(\Omega)$ (see the Rellich–Kondrachov embedding theorem; Theorem 1.135) and Problem 1.45, we see that the map $u\longmapsto N_f(u)$ is completely continuous. So, using Problem 5.65 we conclude that φ satisfies the C-condition (see Definition 5.45(b)). Because of (5.149) and Problem 5.38, it also satisfies the PS-condition (see Definition 5.45(a)).

Solution of Problem 5.67

Let $\psi\colon W_0^{1,p}(\Omega) \longrightarrow \mathbb{R}$ be the C^1-functional defined by

$$\psi(u) \;=\; \|Du\|_p^p - \int_\Omega \vartheta(z)|u|^p\,dz \qquad \forall\, u \in W_0^{1,p}(\Omega).$$

From Proposition 5.132 we know that $\psi \geqslant 0$. Suppose that the claim of the problem is not true. Exploiting the p-homogeneity of ψ, we can find a sequence $\{u_n\}_{n\geqslant 1} \subseteq W_0^{1,p}(\Omega)$ such that

$$\|u_n\|_p \;=\; 1 \qquad \forall\, n \geqslant 1 \quad \text{and} \quad \psi(u_n) \searrow 0. \tag{5.150}$$

From (5.150), we see that the sequence $\{u_n\}_{n\geqslant 1}W_0^{1,p}(\Omega)$ is bounded. Therefore, by passing to a subsequence if necessary, we may assume that

$$u_n \xrightarrow{\;w\;} u \quad \text{in } W_0^{1,p}(\Omega), \quad \text{and} \quad u_n \longrightarrow u \quad \text{in } L^p(\Omega) \tag{5.151}$$

(see the Rellich–Kondrachov embedding theorem; Theorem 1.135). Because of (5.151) and since the norm functional in a Banach space is weakly lower semicontinuous, we have

$$\psi(u) \;\leqslant\; 0,$$

so

$$\|Du\|_p^p \;\leqslant\; \int_\Omega \vartheta(z)|u|^p\,dz \;\leqslant\; \widehat{\lambda}_1\|u\|_p^p, \tag{5.152}$$

thus

$$\|Du\|_p^p \;=\; \widehat{\lambda}_1\|u\|_p^p,$$

hence $u = \xi\widehat{u}_1$ for some $\xi \in \mathbb{R}$ (see Remark 5.133). Since $\|u\|_p = 1$ (see (5.150) and (5.151)), we see that $\xi \neq 0$. Then from (5.152) and since $\widehat{u}_1 \in \operatorname{int} C_+$ (see Theorem 5.129(e) and Remark 5.133), we have

$$\|Du\|_p^p \;<\; \widehat{\lambda}\|u\|_p^p,$$

which contradicts Proposition 5.132. So, we can find $c_0 > 0$ such that

$$\|Du\|_p^p - \int_\Omega \vartheta(z)|u(z)|^p\,dz \;\geqslant\; c_0\|u\|^p \qquad \forall\, u \in W_0^{1,p}(\Omega).$$

Solution of Problem 5.68

First we show that φ satisfies

$$\varphi(u) \;\longrightarrow\; +\infty \quad \text{as } \|u\| \;\to\; +\infty. \tag{5.153}$$

Suppose that (5.153) is not true. Then we can find a sequence $\{u_n\}_{n\geqslant 1} \subseteq W_0^{1,p}(\Omega)$ and $M_1 > 0$ such that

$$\|u_n\| \longrightarrow +\infty \quad \text{and} \quad \varphi(u_n) \leqslant M_1 \qquad \forall\, n \geqslant 1. \tag{5.154}$$

Let $y_n = \frac{u_n}{\|u_n\|}$ for $n \geqslant 1$. Then $\|y_n\| = 1$ for all $n \geqslant 1$ and so by passing to a subsequence if necessary, we may assume that

$$y_n \xrightarrow{w} y \quad \text{in } W_0^{1,p}(\Omega) \quad \text{and} \quad y_n \longrightarrow y \quad \text{in } L^p(\Omega) \tag{5.155}$$

(see the Rellich–Kondrachov embedding theorem; Theorem 1.135). From (5.154), we have

$$\tfrac{1}{p}\|Dy_n\|_p^p - \int\limits_\Omega \tfrac{F(z,u_n)}{\|u_n\|^p}\,dz \leqslant \frac{M_1}{\|u_n\|^p} \qquad \forall\, n \geqslant 1. \tag{5.156}$$

Hypotheses **(i)** and **(ii)** imply that

$$|F(z,x)| \leqslant c(1+|x|^p) \quad \text{for a.a. } z \in \Omega, \text{ all } x \in \mathbb{R}, \tag{5.157}$$

for some $c > 0$ and

$$\limsup_{x\to\pm\infty} \tfrac{pF(z,x)}{|x|^p} \leqslant \widehat{\lambda}_1 \tag{5.158}$$

uniformly for almost all $z \in \Omega$. From (5.157), we see that the sequence $\{\frac{F(\cdot,u_n(\cdot))}{\|u_n\|^p}\}_{n\geqslant 1} \subseteq L^1(\Omega)$ is uniformly integrable (Definition 1.18). So, by the Dunford–Pettis theorem (see Theorem 1.28), we may assume that

$$\frac{F(\cdot,u_n(\cdot))}{\|u_n\|^p} \xrightarrow{w} g \quad \text{in } L^1(\Omega). \tag{5.159}$$

Moreover, from hypothesis **(i)** and (5.155) (note that we can always assume that in addition $y_n(z) \longrightarrow y(z)$ for almost all $z \in \Omega$ and $|y_n(z)| \leqslant \eta(z)$ for almost all $z \in \Omega$, all $n \geqslant 1$ for some $\eta \in L^1(\Omega)$), we see that there exists $\widehat{\eta} \in L^1(\Omega)_+$ such that

$$-\widehat{\eta}(z) \leqslant \tfrac{F(z,u_n(z))}{\|u_n\|^p} \quad \text{for a.a. } z \in \Omega, \text{ all } n \geqslant 1. \tag{5.160}$$

Given $\varepsilon > 0$, for every $n \geqslant 1$ we introduce the set

$$D_n^\varepsilon = \big\{z \in \Omega:\ u_n(z) \neq 0,\ -(\widehat{\eta}(z)+\varepsilon) \leqslant \tfrac{F(z,u_n(z))}{|u_n(z)|^p} \leqslant \tfrac{1}{p}(\widehat{\lambda}_1+\varepsilon)\big\}.$$

From (5.158) and (5.160) and since $|u_n(z)| \longrightarrow +\infty$ for almost all $z \in \{y \neq 0\}$, we see that

$$\chi_{D_n^\varepsilon}(z) \longrightarrow 1 \quad \text{for a.a. } z \in \{y \neq 0\}.$$

Note that

$$\big\|(1-\chi_{D_n^\varepsilon})(\cdot)\tfrac{F(\cdot,u_n(\cdot))}{\|u_n\|^{p-1}}\big\|_{L^1(\{y\neq 0\})} \longrightarrow 0,$$

so

$$\chi_{D_n^\varepsilon}(\cdot)\tfrac{F(\cdot,u_n(\cdot))}{\|u_n\|^{p-1}} \xrightarrow{w} g \quad \text{in } L^1(\{y \neq 0\}).$$

From the definition of D_n^ε we have

$$\begin{aligned}-(\widehat{\eta}(z)+\varepsilon)|y_n(z)|^p &\leqslant \chi_{D_n^\varepsilon}(z)\tfrac{F(z,u_n(z))}{|u_n(z)|^p}|y_n(z)|^p \\ &\leqslant \tfrac{1}{p}(\widehat{\lambda}_1+\varepsilon)|y_n(z)|^p \quad \text{for a.a. } z\in\Omega,\end{aligned}$$

so

$$-(\widehat{\eta}(z)+\varepsilon)|y(z)|^p \leqslant g(z) \leqslant \tfrac{1}{p}\widehat{\lambda}_1|y(z)|^p \quad \text{for a.a. } z\in\{y\neq 0\}.$$

Let $\varepsilon\searrow 0$, to obtain

$$-\widehat{\eta}(z)|y(z)|^p \leqslant g(z) \leqslant \tfrac{1}{p}\widehat{\lambda}_1|y(z)|^p \quad \text{for a.a. } z\in\{y\neq 0\}.$$

On the other hand, from (5.157), we see that

$$g(z) = 0 \quad \text{for a.a. } z\in\{y=0\}.$$

So, finally we have

$$-\widehat{\eta}(z)|y(z)|^p \leqslant g(z) \leqslant \tfrac{1}{p}\widehat{\lambda}_1|y(z)|^p \quad \text{for a.a. } z\in\Omega,$$

so

$$g(z) = \vartheta(z)|y(z)|^p \quad \text{for a.a. } z\in\Omega, \tag{5.161}$$

with $-\widehat{\eta}\leqslant\vartheta\leqslant\frac{1}{p}\widehat{\lambda}_1$. Therefore, if in (5.156) we pass to the limit as $n\to+\infty$ and use (5.155), (5.159), and (5.161), then

$$\tfrac{1}{p}\|Dy\|_p^p \leqslant \int_\Omega \vartheta(z)|y|^p\,dz. \tag{5.162}$$

If $\vartheta\not\equiv\widehat{\lambda}_1$, then using Problem 5.67, we have

$$c_0\|y\|^p \leqslant 0,$$

so $y=0$ and thus

$$y_n \longrightarrow 0 \quad \text{in } W_0^{1,p}(\Omega)$$

(see (5.156)), a contradiction to the fact that $\|y_n\|=1$ for all $n\geqslant 1$.

If $\vartheta(z)=\widehat{\lambda}_1$ for almost all $z\in\Omega$, then from (5.162) and Proposition 5.132, we have

$$\|Dy\|_p^p = \widehat{\lambda}_1\|y\|_p^p,$$

so $y=\xi\widehat{u}_1$ for some $\xi\in\mathbb{R}$.

If $\xi=0$, then $y=0$ and as above, we have a contradiction to the fact that $\|y_n\|=1$ for all $n\geqslant 1$.

If $\xi\neq 0$, then $y\in \operatorname{int} C_+$ if $\xi>0$ or $y\in -\operatorname{int} C_+$ if $\xi<0$ (see Theorem 5.129(e) and Remark 5.133). Hence $|y(z)|>0$ for all $z\in\Omega$ and so

$$|u_n(z)| \longrightarrow +\infty \quad \text{for a.a. } z\in\Omega. \tag{5.163}$$

Hypothesis **(iii)** implies that given $\gamma > 0$, we can find $M_2 = M_2(\gamma) > 0$ such that

$$f(z,x)x - pF(z,x) \geqslant \gamma \quad \text{for a.a. } z \in \Omega, \text{ all } |x| \geqslant M_2. \tag{5.164}$$

Then we have

$$\begin{aligned} \frac{d}{dx}\frac{F(z,x)}{x^p} &= \frac{f(z,x)x^p - pF(z,x)x^{p-1}}{x^{2p}} = \frac{f(z,x)x - pF(z,x)}{x^{p+1}} \\ &\geqslant \frac{\gamma}{x^{p-1}} \quad \text{for a.a. } z \in \Omega, \text{ all } x \geqslant M_2 \end{aligned}$$

(see (5.164)), so

$$\frac{F(z,u)}{u^p} - \frac{F(z,v)}{v^p} \geqslant -\frac{\gamma}{p}\left(\frac{1}{u^p} - \frac{1}{v^p}\right) \quad \text{for a.a. } z \in \Omega, \text{ all } u \geqslant v \geqslant M_2,$$

thus

$$\frac{1}{p}\widehat{\lambda}_1 - \frac{F(z,v)}{v^p} \geqslant \frac{\gamma}{pv^p} \quad \text{for a.a. } z \in \Omega, \text{ all } v \geqslant M_2$$

(let $u \to +\infty$ and use (5.158)), hence

$$\frac{1}{p}\widehat{\lambda}_1 v^p - F(z,v) \geqslant \frac{\gamma}{p} \quad \text{for a.a. } z \in \Omega, \text{ all } v \geqslant M_2.$$

Since $\gamma > 0$ is arbitrary, we have that

$$\lim_{x \to +\infty} \left(\frac{\widehat{\lambda}_1}{p}x^p - F(z,x)\right) = +\infty \quad \text{uniformly for a.a. } z \in \Omega.$$

Similarly, we show that

$$\lim_{x \to -\infty} \left(\frac{\widehat{\lambda}_1}{p}|x|^p - F(z,x)\right) = +\infty \quad \text{uniformly for a.a. } z \in \Omega.$$

So, we have

$$\lim_{x \to \pm\infty} \left(\frac{\widehat{\lambda}_1}{p}|x|^p - F(z,x)\right) = +\infty \quad \text{uniformly for a.a. } z \in \Omega. \tag{5.165}$$

From (5.163), (5.165) and the Fatou lemma (see Theorem I.3.95), we have

$$\int_\Omega \left(\frac{\widehat{\lambda}_1}{p}|u_n|^p - F(z,u_n)\right) dz \longrightarrow +\infty. \tag{5.166}$$

On the other hand, from (5.154), we have

$$\frac{1}{p}\|Du_n\|_p^p - \int_\Omega F(z,u_n)\, dz \leqslant M_1 \qquad \forall\, n \geqslant 1,$$

so

$$\frac{1}{p}\widehat{\lambda}_1\|u_n\|_p^p - \int_\Omega F(z,u_n)\, dz \leqslant M_1 \qquad \forall\, n \geqslant 1$$

(see Proposition 5.132) and thus

$$\int_{\Omega} \left(\tfrac{\widehat{\lambda}_1}{p}|u_n|^p - F(z,u_n)\right) dz \leqslant M_1. \tag{5.167}$$

Comparing (5.166) and (5.167), we reach a contradiction. Therefore (5.153) holds.

We have

$$\varphi'(u) = A(u) - N_f(u)$$

(see the solution of Problem 5.67) and A is of type $(S)_+$ (see Definition 2.137), while the map $u \longmapsto N_f(u)$ is completely continuous. From Problem 5.65, we conclude that φ satisfies the C-condition (see Definition 5.45(b)). Then (5.153) and Problem 5.38 imply that φ satisfies the PS-condition too (see Definition 5.45(a)).

Solution of Problem 5.69

Suppose that $\{u_n\}_{n\geqslant 1} \subseteq X$ is a sequence such that

$$\varphi(u_n) \longrightarrow c \quad \text{and} \quad (1+\|u_n\|_X)m(u_n) \longrightarrow 0 \tag{5.168}$$

(see Definition 5.47(a)). Let $u_n^* \in \partial\varphi(u_n)$ be such that $m(u_n) = \|u_n^*\|_*$ for every $n \geqslant 1$. Then for all $h \in X$, we have

$$\begin{aligned}(1+\|u_n\|_X)\varphi^0(u_n;h) &\geqslant (1+\|u_n\|_X)\,\langle u_n^*, h\rangle \\ &\geqslant -(1+\|u_n\|_X)\|u_n^*\|_*\|h\|_X \qquad \forall\, n \geqslant 1,\end{aligned}$$

so

$$(1+\|u_n\|_X)\varphi^0(u_n;h) \geqslant -\varepsilon_n\|h\|_X \qquad \forall\, n \geqslant 1,$$

where $\varepsilon_n = (1+\|u_n\|_X)m(u_n) \searrow 0$ (see (5.168)). Thus the SC_c-condition holds.

Conversely, suppose that the SC_c-condition holds. Let

$$\sigma_n(h) = \tfrac{1+\|u_n\|_X}{\varepsilon_n}\varphi^0(u_n;h) \qquad \forall\, h \in X,\ n \geqslant 1.$$

Note that σ_n is continuous, sublinear and $\sigma_n(0) = 0$. Also, we have

$$\sigma_n(h) \geqslant -\varepsilon_n\|h\|_X \qquad \forall\, h \in X,\ n \geqslant 1.$$

From Problem 3.133, we know that there exists $\widehat{u}_n^* \in X^*$ with $\|\widehat{u}_n^*\|_* \leqslant 1$ such that

$$\sigma_n(h) \geqslant \langle \widehat{u}_n^*, h\rangle \qquad \forall\, h \in X,\ n \geqslant 1.$$

Then $u_n^* = \frac{\varepsilon_n}{1+\|u_n\|_X}\widehat{u}_n^* \in \partial\varphi(u_n)$ and

$$(1+\|u_n\|_X)m(u_n) \leqslant (1+\|u_n\|_X)\|u_n^*\|_* \leqslant \varepsilon_n \qquad \forall\, n \geqslant 1$$

(recall that $\|\widehat{u}_n^*\|_* \leqslant 1$). Hence the nonsmooth C_c-condition holds.

Solution of Problem 5.70

Arguing by contradiction, suppose that the conclusion of the problem is not true. Then we can find a sequence $\{u_n\}_{n\geqslant 1} \subseteq X$ such that

$$\varphi(u_n) \longrightarrow c \quad \text{and} \quad (1+\|u_n\|_X)\varphi'(u_n) \longrightarrow 0 \quad \text{in } X^*. \tag{5.169}$$

Since by hypothesis φ satisfies the C_c-condition (see Definition 5.45(b)), because of (5.169) and by passing to a suitable subsequence if necessary, we may assume that

$$u_n \longrightarrow u \quad \text{in } X,$$

so

$$\varphi'(u) = 0 \quad \text{and} \quad \varphi(u) = c,$$

a contradiction to the hypothesis that c is not a critical value of φ.

Solution of Problem 5.71

"$\Longrightarrow$": This implication follows from Problem 5.37.
"$\Longleftarrow$": Since X is finite dimensional, from the coercivity assumption on φ, it follows that φ is bounded below. Let $\{u_n\}_{n\geqslant 1} \subseteq X$ be a C-sequence. Since the sequence $\{\varphi(u_n)\}_{n\geqslant 1} \subseteq \mathbb{R}$ is bounded, from our hypothesis we get that the sequence $\{u_n\}_{n\geqslant 1} \subseteq X$ is bounded and so there is a convergent subsequence. Therefore φ satisfies the C-condition (see Definition 5.45(b)).

Solution of Problem 5.72

No. In the light of Problem 5.71, to produce a counterexample, we need to consider an infinite dimensional Banach space. So, we assume that X is an infinite dimensional Banach space and let $\xi\colon \mathbb{R}_+ \longrightarrow \mathbb{R}$ be a smooth function such that

$$\xi(t) = \begin{cases} 0 & \text{for } t = [0,2], \\ t & \text{for } t \geqslant 3. \end{cases}$$

Consider the functional $\varphi(u) = \xi(\|u\|_X)$ for all $u \in X$. Evidently φ is bounded below and satisfies the required coercivity condition. On the other hand, if $\{u_n\}_{n\geqslant 1} \subseteq X$ is such that $\|u_n\|_X = 1$ for all $n \geqslant 1$, then $\varphi(u_n) = 0$ and $(1+\|u_n\|_X)\varphi'(u_n) = 0$ for all $n \geqslant 1$, but since X is infinite dimensional, there need not be any strongly convergent subsequence of $\{u_n\}_{n\geqslant 1} \subseteq X$.

Solution of Problem 5.73

Let $\{u_n\}_{n\geqslant 1} \subseteq K_\varphi^c$ be a sequence. Then

$$\varphi(u_n) = c \quad \text{and} \quad \varphi'(u_n) = 0 \qquad \forall\, n \geqslant 1.$$

Since by hypothesis φ satisfies the C_c-condition, the sequence $\{u_n\}_{n\geqslant 1} \subseteq K_\varphi^c$ admits a strongly convergent subsequence. Therefore the set K_φ^c is compact.

Solution of Problem 5.74

Let $X = \mathbb{R}$ and $\varphi(u) = \sin u$ for all $u \in \mathbb{R}$. Consider a sequence $\{u_n\}_{n\geqslant 1} \subseteq \mathbb{R}$ such that

$$\varphi(u_n) = \sin u_n \longrightarrow c \quad \text{and} \quad \varphi'(u_n) = \cos u_n \longrightarrow 0.$$

We write

$$u_n = 2\lambda_n\pi + h_n \qquad \forall\, n \geqslant 1,$$

with $\lambda_n \in \mathbb{N}$ and $h_n \in [0, 2\pi]$. Then we have

$$\sin h_n \longrightarrow c \quad \text{and} \quad \cos h_n \longrightarrow 0.$$

The sequence $\{h_n\}_{n\geqslant 1}$ has a convergent subsequence with limit $h \in [0, 2\pi]$. We have

$$\sin h = c \quad \text{and} \quad \cos h = 0,$$

so $h \in K_\varphi^c$. So, φ satisfies the CPS-condition, but does not satisfy the PS_c-condition, for $c = 1$ or $c = -1$.

Solution of Problem 5.75

Let $X = \mathbb{R}$ and $\varphi = e^{-\frac{1}{u^2}}$ for all $u \in \mathbb{R}$. Then $c = \inf\limits_{\mathbb{R}} \varphi = 0$. We have $\varphi'(u) = \frac{2}{|u|^2 u} e^{-\frac{1}{u^2}}$. Hence

$$\varphi(n) \longrightarrow 0 \quad \text{and} \quad \varphi'(n) \longrightarrow 0,$$

so the PS_0-condition does not hold (see Definition 5.45(a)).

Solution of Problem 5.76

Suppose that $\|h\|_H \neq 1$ and let $\{u_n\}_{n\geqslant 1} \subseteq H$ be a sequence such that the sequence $\{\varphi(u_n)\}_{n\geqslant 1} \subseteq \mathbb{R}$ is bounded and

$$\varphi'(u_n) \longrightarrow 0 \quad \text{in } H^*. \tag{5.170}$$

By the Riesz representation theorem, let $\nabla\varphi(u) \in H$ be the gradient of φ at $u \in X$, i.e.,

$$\langle \varphi'(u), v \rangle = (\nabla\varphi(u), v)_H \qquad \forall\, u, v \in H.$$

We have

$$\nabla\varphi(u) = u - (u, h)_H\, h \qquad \forall\, u \in H. \tag{5.171}$$

From (5.170) and (5.171), we have

$$\big| (u_n, v)_H - (u_n, h)_H\, (h, v)_H \big| \leqslant \varepsilon_n \|v\|_H \qquad \forall\, v \in H, \tag{5.172}$$

with $\varepsilon_n \searrow 0$. In (5.172), we choose $v = h$ and have

$$|(u_n, h)_H|\, \big|1 - \|h\|^2\big| \leqslant \varepsilon_n \|h\|_H \qquad \forall\, n \geqslant 1,$$

so

$$(u_n, h)_H \longrightarrow 0 \tag{5.173}$$

(since $\|h\|_H \neq 1$). Because

$$|\varphi(u_n)| \leqslant M \qquad \forall\, n \geqslant 1,$$

for some $M > 0$ (see (5.170)), it follows that the sequence $\{u_n\}_{n\geqslant 1} \subseteq H$ is bounded. So, by passing to a suitable subsequence if necessary, we may assume that

$$u_n \xrightarrow{w} u \quad \text{in } H. \tag{5.174}$$

From (5.172) with $v = u_n - u$, we have

$$\big| (u_n, u_n - u)_H - (u_n, h)_H\, (h, u_n - u)_H \big| \leqslant \varepsilon_n \|u_n - u\|_H \qquad \forall\, n \geqslant 1,$$

so

$$\big| \|u_n\|^2 - (u_n, u)_H \big| \longrightarrow 0.$$

But $(u_n, u)_H \longrightarrow \|u\|^2$ (see (5.174)). Therefore

$$\|u_n\|_H \longrightarrow \|u\|_H. \tag{5.175}$$

From (5.174) and (5.175) and the Kadec–Klee property of Hilbert spaces (see Corollary 1.26), we have that

$$u_n \longrightarrow u \quad \text{in } H,$$

so φ satisfies the PS condition (see Definition 5.45(a)).

Now suppose that $\|h\|_H = 1$. Let $u_n = nh \in H$. Then

$$\varphi(u_n) = \tfrac{n^2}{2} - \tfrac{n^2}{2} = 0 \qquad \forall\, n \geqslant 1,$$

so

$$\varphi'(u_n) = nh - nh = 0 \qquad \forall\, n \geqslant 1$$

(recall that $\|h\|_H = 1$). So, $\{u_n = nh\}_{n\geqslant 1}$ is a PS-sequence with no convergent subsequence. Therefore φ does not satisfy the PS-condition.

Solution of Problem 5.77

Let $\{u_n\}_{n\geqslant 1} \subseteq H_0^1(\Omega)$ be a sequence such that

$$|\varphi(u_n)| \leqslant M \qquad \forall\, n \geqslant 1, \tag{5.176}$$

for some $M > 0$ and

$$\varphi'(u_n) \longrightarrow 0 \quad \text{in } H^{-1}(\Omega) = H_0^1(\Omega)^*. \tag{5.177}$$

From (5.177), we have

$$\big|\langle A(u_n), h\rangle - \widehat{\lambda}_1 \int\limits_\Omega u_n h\, dz - \int\limits_\Omega f(z, u_n) h\, dz\big| \leqslant \varepsilon_n \|h\| \qquad \forall\, h \in H_0^1(\Omega), \tag{5.178}$$

with $\varepsilon_n \searrow 0$. Suppose that the sequence $\{u_n\}_{n\geqslant 1} \subseteq H_0^1(\Omega)$ is not bounded. Then, passing to a subsequence if necessary, we may assume that

$$\|u_n\| \longrightarrow +\infty. \tag{5.179}$$

Let $y_n = \frac{u_n}{\|u_n\|}$ for all $n \geqslant 1$. Then $\|y_n\| = 1$ for all $n \geqslant 1$ and so we may assume that

$$y_n \xrightarrow{w} y \quad \text{in } H_0^1(\Omega) \quad \text{and} \quad y_n \longrightarrow y \quad \text{in } L^2(\Omega). \tag{5.180}$$

From (5.176), we have

$$\tfrac{1}{2}\|Dy_n\|_2^2 - \tfrac{\widehat{\lambda}_1}{2}\|y_n\|_2^2 - \int\limits_\Omega \tfrac{F(z,u_n)}{\|u_n\|^2}\, dz \leqslant \tfrac{M}{\|u_n\|^2} \qquad \forall\, n \geqslant 1. \tag{5.181}$$

Since $|F(z,x)| \leqslant a(z)|x|$ for almost all $z \in \Omega$, all $x \in \mathbb{R}$, we see that

$$\int\limits_\Omega \tfrac{F(z,u_n)}{\|u_n\|^2}\, dz \longrightarrow 0.v \tag{5.182}$$

So, if in (5.181) we pass to the limit as $n \to +\infty$ and use (5.180) and (5.182), then

$$\|Dy\|_2^2 \leqslant \widehat{\lambda}_1 \|y\|_2^2,$$

so

$$\|Dy\|_2^2 = \widehat{\lambda}_1 \|y\|_2^2,$$

(see Remark 5.130), thus $y = \xi \widehat{u}_1$ for some $\xi \in \mathbb{R}$ (see Theorem 5.129(e)).

Note that, if $\xi = 0$, then $y = 0$ and so from (5.181) we see that

$$y_n \longrightarrow 0 \quad \text{in } H_0^1(\Omega),$$

a contradiction to the fact that $\|y_n\| = 1$ for all $n \geqslant 1$. Therefore $\xi \neq 0$ and we assume that $\xi > 0$ (the reasoning is similar if we assume that $\xi < 0$). Then $y(z) > 0$ for all $z \in \Omega$ and so

$$u_n(z) \longrightarrow +\infty \quad \text{for a.a. } z \in \Omega. \tag{5.183}$$

From (5.176), we have

$$-2M \leqslant \|Du_n\|_2^2 - \widehat{\lambda}_1 \|u_n\|_2^2 - \int_\Omega 2F(z, u_n)\, dz \leqslant 2M \qquad \forall\, n \geqslant 1. \tag{5.184}$$

Also, from (5.178) with $h = u_n \in H_0^1(\Omega)$, we have

$$\begin{aligned} -\varepsilon_n \|u_n\| &\leqslant -\|Du_n\|_2^2 \\ &+ \widehat{\lambda}_1 \|u_n\|_2^2 + \int_\Omega f(z, u_n) u_n\, dz \leqslant \varepsilon_n \|u_n\| \qquad \forall\, n \geqslant 1. \end{aligned} \tag{5.185}$$

Adding (5.184) and (5.185) and dividing by $\|u_n\|$, we obtain

$$\Big| \int_\Omega \big(f(z, u_n) y_n - 2\tfrac{F(z,u_n)}{\|u_n\|}\big)\, dz \Big| \leqslant \varepsilon_n' \qquad \forall\, n \geqslant 1, \tag{5.186}$$

with $\varepsilon_n' \searrow 0$. Let

$$g(z, x) = \begin{cases} \frac{F(z,x)}{x} & \text{if } x \neq 0, \\ f(z, 0) & \text{if } x = 0. \end{cases}$$

This is a Carathéodory function and from (5.186) we have

$$\Big| \int_\Omega \big(f(z, u_n) y_n - 2g(z, u_n) y_n\big)\, dz \Big| \leqslant \varepsilon_n' \qquad \forall\, n \geqslant 1. \tag{5.187}$$

From hypothesis **(ii)** and (5.183), we have

$$\begin{aligned} f(z, u_n(z)) &\longrightarrow f_+(z) \quad \text{for a.a. } z \in \Omega, \\ g(z, u_n(z)) &\longrightarrow f_+(z) \quad \text{for a.a. } z \in \Omega. \end{aligned}$$

Then from (5.187) and using the Lebesgue dominated convergence theorem (see Theorem I.3.94), we obtain

$$\int_\Omega f_+(z) \widehat{u}_1\, dz = 0,$$

which contradicts both cases in hypothesis **(iii)**. Therefore the sequence $\{u_n\}_{n\geqslant 1} \subseteq H_0^1(\Omega)$ is bounded and so we may assume that

$$u_n \xrightarrow{w} u \quad \text{in } H_0^1(\Omega) \quad \text{and} \quad u_n \longrightarrow u \quad \text{in } L^2(\Omega). \tag{5.188}$$

In (5.176), we choose $h = u_n - u \in H_0^1(\Omega)$, pass to the limit as $n \to +\infty$ and use (5.188). Then

$$\|Du_n\|_2 \longrightarrow \|Du\|_2,$$

so $u_n \longrightarrow u$ in $H_0^1(\Omega)$ (see (5.188) and use the Kadec–Klee property of Hilbert spaces; see Corollary 1.26), thus φ satisfies the PS-condition (see Definition 5.45(a)).

Solution of Problem 5.78

Since u_0 is a weak solution of the Dirichlet boundary value problem, we have

$$\int\limits_\Omega (Du_0, Dh)_{\mathbb{R}^N}\, dz - \widehat{\lambda}_1 \int\limits_\Omega u_0 h\, dz = \int\limits_\Omega f(z, u_0) h\, dz \qquad \forall\, h \in H_0^1(\Omega).$$

Let $h = \widehat{u}_1 \in H_0^1(\Omega)$. Then we have

$$0 = \int\limits_\Omega f(z, u_0)\widehat{u}_1\, dz.$$

Since $\widehat{u}_1(z) > 0$ for all $z \in \Omega$, from the assumption that $f_-(z) \leqslant f(z,x) \leqslant f_+(z)$ for almost all $z \in \Omega$, all $x \in \mathbb{R}$ and because the inequalities are strict on sets of positive Lebesgue measure, we have

$$\int\limits_\Omega f_-(z)\widehat{u}_1\, dz < 0 < \int\limits_\Omega f_+(z)\widehat{u}_1\, dz.$$

Solution of Problem 5.79

We choose $\varepsilon_1, \varepsilon_2 \in (0, \varepsilon)$ and $\eta, \lambda > 0$ such that

$$\varphi(u_0) - \inf_X \varphi < \varepsilon_2 < \varepsilon_1, \tag{5.189}$$

$$0 < \eta < 1 - \tfrac{\varepsilon_1}{\varepsilon}, \tag{5.190}$$

$$0 < \tfrac{\lambda}{\eta} < \tfrac{\varepsilon_1}{(\sqrt{\varepsilon_1} + \sqrt{\eta\varepsilon_2})^2} < 1. \tag{5.191}$$

We set $\delta = (1-\eta)\varepsilon$. Let $\varphi_0 = \varphi$ and then define

$$\varphi_1(u) = \varphi_0(u) + \delta\|u - u_0\|_X^2 \qquad \forall\, u \in X.$$

Inductively we set

$$\varphi_{n+1}(u) = \varphi_n(u) + \delta\eta^n \|u - u_n\|_X^2 \qquad \forall\, n \geqslant 1, \tag{5.192}$$

with $u_n \in X$ chosen so that

$$\varphi_n(u_n) \leqslant \lambda\varphi_{n-1}(u_{n-1}) + (1-\lambda)\inf_X \varphi_{n-1}. \tag{5.193}$$

Let $\beta_n = \varphi_n(u_n)$ and $m_{n+1} = \inf_X \varphi_n$ for $n \geqslant 0$. From (5.193), we have

$$m_n \leqslant m_{n+1} \leqslant \beta_{n+1} \leqslant \lambda\beta_n + (1-\lambda)m_{n+1} \leqslant \beta_n,$$

so

$$\beta_{n+1} - m_{n+1} \leqslant \lambda(\beta_n - m_{n+1}) \leqslant \lambda(\beta_n - m_n) \leqslant \lambda^{n+1}(\beta_0 - m_0). \tag{5.194}$$

In (5.192), we choose $u = u_{n+1}$. Then

$$\begin{aligned} \beta_{n+1} &= \varphi_{n+1}(u_{n+1}) = \varphi_n(u_{n+1}) + \beta\eta^n \|u_{n+1} - u_n\|_X^2 \\ &\geqslant m_{n+1} + \beta\eta^n \|u_{n+1} - u_n\|_X^2, \end{aligned}$$

so

$$\delta\eta^n \|u_{n+1} - u_n\|_X^2 \leqslant \lambda^{n+1}(\beta_0 - m_0) \leqslant \lambda^{n+1}\varepsilon_2 \tag{5.195}$$

(see (5.194) and (5.189)). Then for $m > n$, we have

$$\begin{aligned} \|u_m - u_n\|_X &\leqslant \sum_{k=1}^{m-n} \|u_{n+k} - u_n\|_X \\ &\leqslant \Big(\frac{\lambda}{\eta}\Big)^{\frac{n}{2}}\Big(\frac{\lambda\varepsilon_2}{\delta}\Big)^{\frac12} + \ldots + \Big(\frac{\lambda}{\eta}\Big)^{\frac{m-1}{2}}\Big(\frac{\lambda\varepsilon_2}{\delta}\Big)^{\frac12} \\ &= \Big(\frac{\lambda\varepsilon_2}{\delta}\Big)^{\frac12}\Big(\frac{\lambda}{\eta}\Big)^{\frac{\eta}{2}} \frac{1 - (\frac{\lambda}{\eta})^{\frac{m-n}{2}}}{1 - (\frac{\lambda}{\eta})^{\frac12}} < \Big(\frac{\lambda\varepsilon_2}{\delta}\Big)^{\frac12} \frac{1}{1 - (\frac{\lambda}{\eta})^{\frac12}} < \Big(\frac{\varepsilon_1}{\delta}\Big)^{\frac12} \end{aligned} \tag{5.196}$$

(see (5.195) and (5.191)), so $\{u_n\}_{n\geqslant 1} \subseteq X$ is a Cauchy sequence in X, hence convergent in X to $u_\varepsilon \in X$. Also note that $\eta < \frac{\varepsilon - \varepsilon_1}{\varepsilon}$ (see (5.194)), so

$$\delta = (1-\eta)\varepsilon = \varepsilon - \eta\varepsilon > \varepsilon_1,$$

thus

$$\|u_m - u_n\|_X < 1 \qquad \forall\, m > n$$

(see (5.196)). We set

$$g(u) = \frac12 \sum_{n\geqslant 1} \eta^n (1-\eta) \|u - u_n\|_X^2.$$

Then $g \in \mathcal{Y}$. Passing to the limit as $m \to +\infty$, we obtain

$$\|u_\varepsilon - u_n\|_X \leqslant 1 \qquad \forall\, n \geqslant 0,$$

so

$$\|u_\varepsilon - u_0\|_X \leqslant 1.$$

Also, from (5.194) and the lower semicontinuity of φ_n, we have

$$\begin{aligned}\varphi(u) + 2\varepsilon g(u) &= \sup_{n\geqslant 0} \varphi_n(u) \geqslant \lim_{n\to+\infty} m_n \\ = \lim_{n\to+\infty} \varphi_n(u_n) &\geqslant \sup_{k\geqslant 1} \liminf_{n\to+\infty} \varphi_k(u_n) \\ \geqslant \sup_{k\geqslant 1} \varphi_k(u_\varepsilon) &= \varphi(u_\varepsilon) + 2\varepsilon g(u_\varepsilon)\end{aligned}$$

and the inequality is strict if $u \neq u_\varepsilon$ (see (5.194)).

Remark. The result of this problem is known as the ***Borwein–Preiss smooth variational principle***. In general it is not possible to assume that the functions $g \in \mathcal{Y}$ consist of a single summand. This is true if X is a reflexive Banach space. In fact these are the only Banach spaces for which this can be done (see Problem 5.80).

Solution of Problem 5.80

Let $u^* \in X^*$ with $\|u^*\|_* = 1$. We define

$$\varphi(u) = \begin{cases} \langle u^*, u\rangle & \text{if } \|u\|_X \leqslant 1, \\ +\infty & \text{if } \|u\|_X > 1, \end{cases}$$

that is, $\varphi = u^* + i_{\overline{B}_1}$, with $\overline{B}_1 = \{u \in X : \|u\|_X \leqslant 1\}$ and $i_{\overline{B}_1}$ is the indicator function of the set $\overline{B}_1$ (see Definition 3.55). Then $\varphi \in \Gamma_0(X)$ (see Remark 3.56) and $\inf_X f = -1$. We use Problem 5.79 with $u_0 = 0$ and $\varepsilon = 1$. So, we can find $u_\varepsilon \in \overline{B}_1$ and $g \in \mathcal{Y}$ such that

$$\varphi(u_\varepsilon) + 2g(u_\varepsilon) \leqslant \varphi(u) + 2g(u) \qquad \forall\, u \in X.$$

We have $u_\varepsilon \in \operatorname{dom}\varphi$ and

$$0 \in \partial(\varphi + 2g)(u_\varepsilon) = u^* + 2\partial g(u_\varepsilon) = u^* + 2\mathcal{F}(u_\varepsilon - h) \qquad \forall\, h \neq u_\varepsilon$$

(see Proposition 3.97 and Problem 3.54), so

$$-u^* \in 2\mathcal{F}(u_\varepsilon - h),$$

thus

$$\|u_\varepsilon - h\|_X = \tfrac{1}{2} \quad \text{and} \quad \langle u^*, h - u_\varepsilon\rangle = \|u_\varepsilon - h\|_X.$$

Invoking the James theorem (see Theorem I.5.74), we conclude that X is reflexive.

Solution of Problem 5.81

From Problem 5.79 (see also the Remark following the solution of that problem), given $\varepsilon > 0$, we can find $u_\varepsilon \in H$ such that

$$\varphi(u_\varepsilon) \leqslant \inf_H \varphi + \varepsilon, \tag{5.197}$$

$$\|u_\varepsilon - u_0\|_H \leqslant 1, \tag{5.198}$$

$$\varphi(u_\varepsilon) + \varepsilon\big(\|u_\varepsilon - u_0\|_H^2 - \|u - u_0\|_H^2\big) \leqslant \varphi(u) \qquad \forall\, u \in H. \tag{5.199}$$

Choosing $u = u_\varepsilon + th$ with $t > 0$, $h \in X$ in (5.199) we have

$$\varepsilon\big(-2t\,(h, u_\varepsilon - u_0)_H - t^2\|h\|_H\big) \leqslant \varphi(u_\varepsilon + th) - \varphi(u_\varepsilon), \tag{5.200}$$

with $(\cdot,\cdot)_H$ being the inner product of H. Dividing (5.200) with $t > 0$ and letting $t \searrow 0$, we obtain

$$-2\varepsilon\,(h, u_\varepsilon - u_0)_H \leqslant \langle \varphi'(u_\varepsilon), h\rangle \qquad \forall\, h \in H,$$

so

$$\|\varphi'(u_\varepsilon)\|_* \leqslant 2\varepsilon \tag{5.201}$$

(see (5.198)). Moreover, since by hypothesis $\varphi \in C^2(H;\mathbb{R})$, we have

$$\tfrac{1}{t^2}\big(\varphi(u+th) + \varphi(u-th) - 2\varphi(u)\big) \longrightarrow \langle \varphi''(u)h, h\rangle \quad \text{as } t \to 0. \tag{5.202}$$

We use (5.199) first with $u = u_\varepsilon + th$ and then with $u = u_\varepsilon - th$. We add the resulting inequalities and obtain

$$-2\varepsilon t^2\|u\|_H^2 \leqslant \varphi(u_\varepsilon + th) + \varphi(u_\varepsilon - th) - 2\varphi(u_\varepsilon).$$

Multiplying with $\frac{1}{t^2}$, letting $t \to 0$ and using (5.202), we obtain

$$-2\varepsilon\|h\|_H \leqslant \langle \varphi''(u_\varepsilon)h, h\rangle \qquad \forall\, h \in H. \tag{5.203}$$

Finally, let $\varepsilon = \frac{1}{n}$ and let $u_n = u_{\varepsilon_n}$. From (5.197), (5.201), and (5.203), we see that $\{u_n\}_{n\geqslant 1}$ is the desired sequence.

Solution of Problem 5.82

Note that $\frac{2\lambda}{\lambda+\mu} < 1$. So, if $u \in X \setminus K_\varphi$, then from the definition of the dual norm, we can find $h_u \in X$ with $\|h_u\|_X = 1$ such that

$$\left\langle \varphi'(u), h_u \right\rangle \ > \ \tfrac{2\lambda}{\lambda+\mu}\|\varphi'(u)\|_*. \tag{5.204}$$

Let $y_u = \frac{\lambda+\mu}{2}\|\varphi'(u)\|_* h_u$. We have

$$\|y_u\|_X \ < \ \mu\|\varphi'(u)\|_* \tag{5.205}$$

and

$$\lambda\|\varphi'(u)\|_*^2 \ < \ \left\langle \varphi'(u), y_u \right\rangle \tag{5.206}$$

(see (5.204)). The continuity of the map $u \longmapsto \varphi'(u)$ implies that we can find an open neighborhood U_u of u such that

$$\|y_u\|_X \ \leqslant \ \mu\|\varphi'(v)\|_* \quad \text{and} \quad \lambda\|\varphi'(u)\|_*^2 \ \leqslant \ \left\langle \varphi'(v), y_u \right\rangle \qquad \forall\, v \in U_u \tag{5.207}$$

(see (5.205) and (5.206)). We consider the family $\{U_u\}_{u \in X\setminus K_\varphi}$ of open sets. This is an open cover of $X \setminus K_\varphi$. So, it admits a locally finite refinement $\{V_u\}_{u\in X\setminus K_\varphi}$. There is a locally Lipschitz partition of unity $\{\xi_u\}_{u\in X\setminus K_\varphi}$ subordinate to $\{V_u\}_{u\in X\setminus K_\varphi}$. We set

$$\widehat{g}(h) \ = \ \sum_{u\in X\setminus K_\varphi} \xi_u(h) y_u \qquad \forall\, h \in X \setminus K_\varphi.$$

Evidently $\widehat{g}\colon X \setminus K_\varphi \longrightarrow X$ is a locally Lipschitz map which satisfies

$$\|\widehat{g}(h)\|_X \ \leqslant \ \sum_{u\in X\setminus K_\varphi} \xi_u(h)\|y_u\|_X \ \leqslant \ \mu\|\varphi'(h)\|_*$$

and

$$\left\langle \varphi'(h), \widehat{g}(h) \right\rangle \ = \ \sum_{u\in X\setminus K_\varphi} \xi_u(h) \left\langle \varphi'(h), y_u \right\rangle \ \geqslant \ \lambda\|\varphi'(h)\|_*^2 \qquad \forall\, h \in X \setminus K_\varphi$$

(see (5.207)).

Solution of Problem 5.83

As in the solution of Problem 5.82, since $\frac{2\lambda}{\lambda+\mu} < 1$, from the definition of the dual norm, we see that for every $u \in X \setminus K_\varphi$, we can find $h_u \in X$ with $\|h_u\|_X = 1$ such that

$$\tfrac{2\lambda}{\lambda+\mu}\|\varphi'(u)\|_{X^*} \ < \ \left\langle \varphi'(u), h_u \right\rangle.$$

Let $\widehat{y}_u = \frac{\lambda+\mu}{2} h_u$ and $\widetilde{y}_u = \frac{\lambda+\mu}{2\|\varphi'(u)\|_*}$. We have

$$\lambda\|\varphi'(u)\|_* \ < \ \left\langle \varphi'(u), \widehat{y}_u \right\rangle \quad \text{and} \quad \|\widehat{y}_u\|_X \ < \ \mu \tag{5.208}$$

and

$$\lambda \ < \ \left\langle \varphi'(u), \widetilde{y}_u \right\rangle \ \leqslant \ \|\varphi'(u)\|_*\|\widetilde{y}_u\|_X \ < \ \mu. \tag{5.209}$$

The continuity of the map $u \longmapsto \varphi'(u)$, implies that we can find an open neighborhood U_u of u such that

$$\lambda\|\varphi'(h)\|_* \leqslant \langle \varphi'(h), \widehat{y}_u \rangle$$

and

$$\lambda \leqslant \langle \varphi'(h), \widetilde{y}_u \rangle \leqslant \|\varphi'(h)\|_* \|\widetilde{y}_u\|_X \leqslant \mu \qquad \forall\, h \in U_u$$

(see (5.208) and (5.209)). Then we consider a locally finite refinement $\{V_u\}_{u \in X \setminus K_\varphi}$ of $\{U_u\}_{u \in X \setminus K_\varphi}$ and a locally Lipschitz partition of unity $\{\xi_u\}_{u \in X \setminus K_\varphi}$ subordinate to the cover $\{V_u\}_{u \in X \setminus K_\varphi}$. As in the solution of Problem 5.82, we produce the desired locally Lipschitz maps $\widehat{g}, \widetilde{g} \colon X \setminus K_\varphi \longrightarrow X$.

Solution of Problem 5.84

We claim that there exist $\varepsilon_0 > 0$ and $\eta > 0$ such that

$$\text{if } |\varphi(u) - c| \leqslant \varepsilon_0, \text{ then } \|\varphi'(u)\|_* \geqslant \eta \vartheta(\|u\|_X). \tag{5.210}$$

Indeed, if (5.210) does not hold, then we can find a sequence $\{u_n\}_{n \geqslant 1} \subseteq X$ such that

$$|\varphi(u_n) - c| \leqslant \tfrac{1}{n} \quad \text{and} \quad \|\varphi'(u_n)\|_* \leqslant \tfrac{1}{n}\vartheta(\|u_n\|_X) \qquad \forall\, n \geqslant 1. \tag{5.211}$$

Since by hypothesis φ satisfies the PS_c^θ-condition, by passing to a suitable subsequence if necessary, we may assume that

$$u_n \longrightarrow u \quad \text{in } X,$$

so $\varphi(u) = c$ and $\varphi'(u) = 0$ (see (5.211)), thus $u \in K_\varphi^c$, a contradiction to the hypothesis that $K_\varphi^c = \emptyset$.

So, (5.210) holds. Let $\varepsilon < \min\{\varepsilon_0, \eta\}$. From (5.210), we have

$$\text{if } |\varphi(u) - c| \leqslant \varepsilon, \text{ then } \|\varphi'(u)\|_* \geqslant \varepsilon \vartheta(\|u\|_X). \tag{5.212}$$

Because of the hypothesis on ϑ, we can find $\beta > 0$ such that

$$4 < \int\limits_r^{r+\beta} \vartheta(s)\, ds. \tag{5.213}$$

We introduce the sets

$$\begin{aligned} A &= \{u \in X : |\varphi(u) - c| \geqslant 2\varepsilon\} \cup K_\varphi, \\ C &= \{u \in X : |\varphi(u) - c| \leqslant \varepsilon\}. \end{aligned}$$

These are closed sets and $A \cap C = \emptyset$ (see (5.212)). So, by the Urysohn lemma (see Theorem I.2.136), we can find a continuous map $\xi \colon X \longrightarrow [0,1]$ such that $\xi|_A \equiv 0$ and $\xi|_C \equiv 1$ (for example, we can take $\xi(u) = \frac{\operatorname{dist}(u,A)}{\operatorname{dist}(u,A)+\operatorname{dist}(u,C)}$ for all $u \in X$). Using Problem 5.83, we can find a locally Lipschitz map $\widehat{g} \colon X \setminus K_\varphi \longrightarrow X$ such that

$$\|\varphi'(u)\|_* \leqslant \langle \varphi'(u), \widehat{g}(u) \rangle \quad \text{and} \quad \|\widehat{g}(u)\|_X \leqslant 2 \qquad \forall\, u \in X \setminus K_\varphi. \tag{5.214}$$

We set

$$v(u) = \xi(u)\widehat{g}(u) \qquad \forall\, u \in X \setminus K_\varphi$$

and consider the following abstract Cauchy problem:

$$\begin{cases} \sigma'(t) = v(\sigma(t)), \ t \geqslant 0, \\ \sigma(0) = u \in X \setminus K_\varphi. \end{cases} \tag{5.215}$$

The local Lipschitzness of v implies that problem (5.215) has a unique local flow. In fact since $\|v(u)\|_X \leqslant 2$ for all $u \in X \setminus K_\varphi$ (see (5.214)), it follows that the flow is in fact global (that is, it is valid for all $t \geqslant 0$). We have

$$\|\sigma(t) - u\|_X \leqslant \int_0^t \|\sigma'(s)\|_X \, ds \leqslant \int_0^t \|\widehat{g}(\sigma(s))\|_X \, ds \leqslant 2t$$

(see (5.214)) and

$$\begin{aligned} \tfrac{d}{dt}\varphi(\sigma(t)) &= \langle \varphi'(\sigma(t)), \sigma'(t) \rangle \\ &= \xi(\sigma(t)) \langle \varphi'(\sigma(t)), \widehat{g}(\sigma(t)) \rangle \geqslant 0 \end{aligned} \tag{5.216}$$

(by the chain rule; see Theorem 3.19). Since $\xi|_A = 0$, we see that

$$\text{if } |\varphi(u) - c| \geqslant 2\varepsilon, \text{ then } \sigma(t) = u \qquad \forall\, t \geqslant 0. \tag{5.217}$$

Let $h \in \varphi_{c-\varepsilon} \cap \overline{B}_r$ be such that $\varphi(\sigma(t,h)) < c+\varepsilon$ for all $t \in [0,\beta]$, where $\sigma(\cdot,h)$ is the flow of (5.215) starting at h (that is, $\sigma(0,h) = h$). Then $\sigma(t,h) \in C$ for all $t \in [0,\beta]$

and we have

$$\begin{aligned}
\varphi(\sigma(\beta,h))-\varphi(h) &= \int_0^\beta \tfrac{d}{dt}\varphi(\sigma(s,h))\,ds \\
= \int_0^\beta \langle\varphi'(\sigma(s,h)),\sigma'(s,h)\rangle\,ds &= \int_0^\beta \langle\varphi'(\sigma(s,h)),\widehat{g}(\sigma(s,h))\rangle\,ds \\
\geqslant \int_0^\beta \|\varphi'(\sigma(s,h))\|_*\,ds &\geqslant \varepsilon\int_0^\beta \vartheta(\|\sigma(s,h)\|_X)\,ds \\
\geqslant \varepsilon\int_0^\beta \vartheta(\|h\|_X+2\varepsilon)\,ds &\geqslant \varepsilon\int_0^\beta \vartheta(r+2s)\,ds \\
= \tfrac{\varepsilon}{2}\int_r^{r+2\beta}\vartheta(s)\,ds &> 2\varepsilon
\end{aligned}$$

(using the chain rule (see Theorem 3.19), the fact that $\sigma(s,h)\in C$ for all $s\in[0,\beta]$, exploiting (5.214), (5.212), the fact that ϑ is nonincreasing and since $h\in\overline{B}_r$), so

$$c+\varepsilon > \varphi(\sigma(\beta,h)) > \varphi(h)+2\varepsilon > c-\varepsilon+2\varepsilon = c+\varepsilon,$$

a contradiction. So, we have proved that $h\in\varphi_{c-\varepsilon}\cap\overline{B}_r$, then there exists $t^*\in[0,\beta]$ such that

$$\varphi(\sigma(t^*,h)) \geqslant c+\varepsilon.$$

Therefore, if we set

$$h(t,u) = \sigma(\beta t,u) \qquad \forall\, t\in[0,1],\ u\in X\setminus K_\varphi,$$

then this continuous map h satisfies statements **(a)**, **(b)** and **(c)** of the problem (see (5.216)).

Solution of Problem 5.85

If the claim of the problem is not true, then we can find $\varepsilon_n \searrow 0$ and $u_n\in\varphi_{c-\varepsilon}^{c+\varepsilon}\setminus U$ such that

$$(1+\|u\|_X)\|\varphi'(u)\|_* < \eta\varepsilon_n \qquad \forall\, n\geqslant 1,$$

so

$$(1+\|u_n\|_X)\varphi'(u_n) \longrightarrow 0 \quad \text{in } X^*.$$

Also, we have $\varphi(u_n)\longrightarrow c$. Since by hypothesis φ satisfies the C_c-condition, we may assume that

$$u_n \longrightarrow u \quad \text{in } X,$$

with $u \in X \setminus U$, so $\varphi(u) = c$ and $\varphi'(u) = 0$, a contradiction.

Solution of Problem 5.86

From the hypotheses on φ it is clear that $c \geqslant m > 0$. Arguing by contradiction, suppose that $K_\varphi^c = \emptyset$. Let $\varepsilon \in (0, \frac{m}{2})$ and choose $U_1 \in \mathcal{D}$ such that

$$c - \varepsilon \leqslant \inf_{u \in \partial U_1} \varphi(u) \tag{5.218}$$

(see the definition of c). Since by definition $U_0 \in \mathcal{D}$ is bounded, we can find $r > 0$ such that $U_0 \subseteq \overline{B}_r$. From Problem 5.84, we know that there exists a continuous map $h\colon [0,1] \times X \longrightarrow X$ such that:

(a) $h(0,u) = u$ for all $u \in X$, for every $t \in [0,1]$, the map $h(t,\cdot)\colon X \longrightarrow X$ is homeomorphism and

$$\varphi(h(t,u)) \leqslant \varphi(h(s,u)) \qquad \forall\, t,s \in [0,1],\ t \leqslant s,\ u \in X;$$

(b) if $|\varphi(u) - c| \geqslant 2\varepsilon$, then $h(t,u) = u$ for all $t \in [0,1]$;

(c) $h(1, \varphi_{c-\varepsilon} \cap U_0) \subseteq \varphi_{c+\varepsilon}$.

Let $U_2 = h(1, U_1)$. From **(a)** above, we have that U_2 is open and $\partial U_2 = h(1, \partial U_1)$. Because of the choice of $\varepsilon > 0$, we have

$$\varphi(u_0), \varphi(0) \leqslant 0 < c - 2\varepsilon,$$

so

$$u_0, 0 \in \{u \in X : |\varphi(u) - c| \geqslant 2\varepsilon\},$$

thus

$$0 = h(1,0) \in U_2 \quad \text{and} \quad u_0 = h(1,u_0) \notin \overline{U}_2$$

(see **(b)** above), hence $U_2 \in \mathcal{D}$. Note that

$$\text{if } \varphi(u) \geqslant c - \varepsilon, \text{ then } \varphi(h(1,u)) \geqslant c + \varepsilon \tag{5.219}$$

(see **(c)**), so

$$\partial U_1 \subseteq \varphi_{c-\varepsilon} \tag{5.220}$$

(see (5.218)). So, we have

$$\inf_{u \in \partial U_2} \varphi(u) = \inf_{u \in \partial U_1} \varphi(h(1,u)) \geqslant m + \varepsilon$$

(see (5.219) and (5.220)). This contradicts the definition of c, since $U_2 \in \mathcal{D}$.

Solution of Problem 5.87
Since by hypothesis $u = 0$ is a local minimizer of φ, we can find $\varepsilon \in (0, \|u_0\|_X)$ such that

$$\varphi(0) = 0 \leqslant \varphi(u) \qquad \forall\, u \in \overline{B}_\varepsilon,$$

where $\overline{B}_\varepsilon = \{v \in X : \|v\|_X \leqslant \varepsilon\}$. One of the following conditions is true:

(a) there exists $\widehat{r} \in (0, \varepsilon)$ such that

$$\varphi(0) = 0 < \inf\{\varphi(u) : \|u\|_X = \widehat{r}\} = \widehat{m};$$

or

(b) for every $r \in (0, \varepsilon)$, we have

$$\inf\{\varphi(u) : \|u\|_X = r\} = 0.$$

Suppose that **(a)** holds. Then $B_{\widehat{r}} \in \mathcal{D}$ and we have

$$c = \sup_{U \in \mathcal{D}} \inf_{u \in \partial U} \varphi(u) \geqslant \widehat{m} > 0$$

(see **(a)**). Then we can use Problem 5.86 and conclude that c is a critical value of φ. Hence we can find $\widehat{u} \in K_\varphi$ with $\varphi(\widehat{u}) = c > 0 = \varphi(0) \geqslant \varphi(u_0)$, thus $\widehat{u} \notin \{0, u_0\}$.

Now suppose that **(b)** holds. We argue as in the solution of Problem 5.86. So, let $0 < r < \eta < \varrho < \varepsilon$ and let $\{u_n\}_{n \geqslant 1}$ be a minimizing sequence of φ on ∂B_η. We set

$$\psi(u) = \begin{cases} \varphi(\varrho \frac{u}{\|u\|_X}) & \text{if } \|u\|_X \geqslant \varrho, \\ \varphi(u) & \text{if } \|u\|_X < \varrho. \end{cases} \tag{5.221}$$

Using the Ekeland variational principle (see Corollary 5.38) on ψ, we can find a sequence $\{v_n\}_{n \geqslant 1} \subseteq X$ such that

$$\psi(v_n) \leqslant \psi(u_n), \quad \|v_n - u_n\|_X \leqslant \tfrac{1}{n} \qquad \forall\, n \geqslant 1 \tag{5.222}$$

and

$$\psi(v_n) \leqslant \psi(y) + \tfrac{1}{n}\|y - v_n\| \qquad \forall\, y \in X,\ n \geqslant 1. \tag{5.223}$$

Then $v_n \in \operatorname{int} R_{r,\varrho}$ for all $n \geqslant n_0$, where $R_{r,\varrho} = \{u \in X : r \leqslant \|u\|_X \leqslant \varrho\}$. Hence

$$\psi(v_n) = \varphi(u_n) \qquad \forall\, n \geqslant n_0$$

(see (5.221)). In (5.222)–(5.223) we choose $y = v_n + th$ with $\|h\|_X = 1$. Then letting $t \to 0$, we obtain

$$\|\varphi'(v_n)\|_* \leqslant \tfrac{1}{n} \qquad \forall\, n \geqslant n_0,$$

so

$$\frac{\|\varphi'(v_n)\|_*}{\vartheta(\|v_n\|_X)} \leqslant \frac{\|\varphi'(v_n)\|_*}{\vartheta(\varrho)} \leqslant \frac{1}{\vartheta(\varrho)} \frac{1}{n} \qquad \forall\, n \geqslant n_0,$$

thus

$$\frac{\varphi'(v_n)}{\vartheta(\|v_n\|_X)} \longrightarrow 0 \quad \text{in } X^*.$$

Since φ satisfies the PS_0^θ-condition (recall that $0 = \inf\{\psi(u) = \varphi(u) : \|u\|_X = \eta\}$), by passing to a suitable subsequence if necessary, we may assume that

$$v_n \longrightarrow \widehat{u} \quad \text{in } X,$$

so

$$\varphi(\widehat{u}) = 0 \quad \text{and} \quad \varphi'(\widehat{u}) = 0.$$

So, for every $\eta \in (r, \varrho)$ there exists a critical point of φ on ∂B_η.

Solution of Problem 5.88

Let $\Gamma = \{\gamma \in C([0,1]; X) : \gamma(0) = 0, \ \gamma(1) = u_0\}$. We furnish Γ with the supremum distance d_∞ defined by

$$d_\infty(\gamma_1, \gamma_2) = \max_{t \in [0,1]} \|\gamma_1(t) - \gamma_2(t)\|_X \qquad \forall \gamma_1, \gamma_2 \in \Gamma.$$

Evidently (Γ, d_∞) is a complete metric space.

Let $\vartheta \colon \Gamma \longrightarrow \mathbb{R}$ be the functional defined by

$$\vartheta(\gamma) = \max_{t \in [0,1]} \varphi(\gamma(t)) \qquad \forall \gamma \in \Gamma. \tag{5.224}$$

From the Claim in the solution of Problem 5.51, we know that ϑ is continuous. Also, we have

$$m_r \leqslant \vartheta(\gamma) \qquad \forall \gamma \in \Gamma. \tag{5.225}$$

Indeed note that for $\gamma \in \Gamma$, we can find $t_r \in [0,1]$ such that $\|\gamma(t_r)\|_X = r$ (see hypothesis **(ii)**). Therefore

$$m_r \leqslant \varphi(\gamma(t_r)) \leqslant \vartheta(\gamma) \qquad \forall \gamma \in \Gamma.$$

Given $\varepsilon > 0$, by the Ekeland variational principle (see Corollary 5.38), we can find $\gamma_\varepsilon \in \Gamma$ such that

$$\vartheta(\gamma_\varepsilon) \leqslant \inf_\Gamma \vartheta + \varepsilon \tag{5.226}$$

and

$$\vartheta(\gamma_\varepsilon) \leqslant \vartheta(\gamma) + \varepsilon d_\infty(\gamma, \gamma_\varepsilon) \qquad \forall \gamma \in \Gamma. \tag{5.227}$$

Let $\xi \in C_0([0,1]; X) = \{\mu \in C([0,1]; X) : \mu(0) = 0, \ \mu(1) = 0\}$. We use (5.227) with $\gamma = \gamma_\varepsilon + \lambda\xi$, $\lambda > 0$. Then

$$-\varepsilon d_\infty(\gamma_\varepsilon + \lambda\xi, \gamma_\varepsilon) \leqslant \vartheta(\gamma_\varepsilon + t\xi) - \vartheta(\gamma_\varepsilon),$$

so

$$-\varepsilon\|\xi\|_\infty \leqslant \tfrac{1}{\lambda}\big(\vartheta(\gamma_\varepsilon + \lambda\xi) - \vartheta(\gamma_\varepsilon)\big), \tag{5.228}$$

where $\|\xi\|_\infty = \max_{t\in[0,1]} \|\xi(t)\|_X$. We have

$$\begin{aligned}\vartheta(\gamma_\varepsilon + \lambda\xi) - \vartheta(\gamma_\varepsilon) &= \max_{t\in[0,1]} \varphi((\gamma_\varepsilon + \lambda\xi)(t)) - \max_{t\in[0,1]} \varphi(\gamma_\varepsilon(t)) \\ = \max_{t\in[0,1]} \big(\varphi(\gamma_\varepsilon(t)) + \lambda \langle \varphi'(\gamma_\varepsilon(t)), \xi(t)\rangle \big) &+ \beta(\lambda) - \max_{t\in[0,1]} \varphi(\gamma_\varepsilon(t)),\end{aligned} \tag{5.229}$$

with $\frac{\beta(\lambda)}{\lambda} \longrightarrow 0$ as $\lambda \to 0$. Consider the functional $\psi\colon C([0,1]) \longrightarrow \mathbb{R}$ defined by

$$\psi(\sigma) = \max_{t\in[0,1]} \sigma(t) \qquad \forall\, \sigma \in C([0,1]).$$

We see that ψ is convex and

$$\partial\psi(\sigma) = \{\nu \in M_b[0,1] : \nu \geqslant 0,\ \nu([0,1]) = 1,\ \operatorname{supp}(\nu) \subseteq M(\sigma)\},$$

where $M(\sigma) = \{t \in [0,1] : \psi(\sigma) = \sigma(t)\}$. Let

$$\sigma_1(t) = \varphi(\gamma_\varepsilon(t)) \quad \text{and} \quad \sigma_2(t) = \langle \varphi'(\gamma_\varepsilon(t)), \xi(t)\rangle \qquad \forall\, t \in [0,1].$$

Evidently $\sigma_1, \sigma_2 \in C([0,1])$. Then from (5.228) and (5.229), we have

$$\begin{aligned} -\varepsilon\|\xi\|_\infty &\leqslant \lim_{\lambda\to 0} \tfrac{1}{\lambda}\big(\vartheta(\gamma_\varepsilon + \lambda\xi) - \vartheta(\gamma_\varepsilon)\big) \\ &\leqslant \lim_{\lambda\to 0} \tfrac{1}{\lambda}\big(\psi(\sigma_1 + \lambda\sigma_2) - \psi(\sigma_1)\big) \\ &= \max_{\nu\in\partial\psi(\sigma_1)} \langle \nu, \sigma_2\rangle_b = \max_{\nu\in\partial\psi(\sigma)} \int_0^1 \langle \varphi'(\gamma_\varepsilon(t)), \xi(t)\rangle \, d\nu(t),\end{aligned}$$

where by $\langle\cdot,\cdot\rangle_b$ we denote the duality brackets for the pair $(M_b[0,1], C([0,1]))$ (see Theorem 1.88). Consider $\xi \in C_0([0,1];X)$ with $\|\xi\|_\infty \leqslant 1$. Then

$$\begin{aligned} -\varepsilon &\leqslant \inf_{\substack{\xi \in C_0([0,1];X)\\ \|\xi\|_\infty \leqslant 1}} \max_{\nu\in\partial\psi(\sigma)} \int_0^1 \langle \varphi'(\gamma_\varepsilon(t)), \xi(t)\rangle \, d\nu(t) \\ &= \max_{\nu\in\partial\psi(\sigma)} \inf_{\substack{\xi \in C_0([0,1];X)\\ \|\xi\|_\infty \leqslant 1}} \int_0^1 \langle \varphi'(\gamma_\varepsilon(t)), \xi(t)\rangle \, d\nu(t) \\ &= \max_{\nu\in\partial\psi(\sigma)} \Big(- \int_0^1 \|\varphi'(\gamma_\varepsilon(t))\|_* \, d\nu(t)\Big) \\ &= - \min_{t\in M(\varphi\circ\gamma_\varepsilon)} \|\varphi'(\gamma_\varepsilon(t))\|_* \end{aligned}$$

(by the von Neumann minimax theorem; see Theorem 5.146). So, there exists $t_\varepsilon \in [0,1]$ such that

$$\varphi(\gamma_\varepsilon(t_\varepsilon)) = \max_{t\in[0,1]} \varphi(\gamma_\varepsilon(t)) \quad \text{and} \quad \|\varphi'(\gamma_\varepsilon(t_\varepsilon))\|_* \leqslant \varepsilon. \tag{5.230}$$

Let $\varepsilon_n = \frac{1}{n}$, $\gamma_n = \gamma_{\varepsilon_n}$, $t_n = t_{\varepsilon_n}$ and $u_n = \gamma_n(t_n)$. From (5.225), (5.227), and (5.230), we have

$$m_r \leqslant \varphi(u_n) \leqslant \inf_X \varphi + \tfrac{1}{n} \quad \text{and} \quad \varphi'(u_n) \longrightarrow 0 \quad \text{in } X^*.$$

Since φ satisfies the WC-condition (see Problem 5.41), we can find $\widehat{u} \in K_\varphi$ with $m_r \leqslant \liminf_{n\to+\infty} \varphi(u_n) \leqslant \varphi(\widehat{u})$.

Solution of Problem 5.89

Let B_1, B_2 be open balls centered at the origin such that

$$\overline{B}_1 \subseteq B_2, \quad K_\varphi \cap B_1 = \{0\}, \quad \varphi'|_{B_2} \text{ is Lipschitz continuous.}$$

Also, let $\overline{B} \subseteq B_1$ be a closed ball centered at the origin of radius less then r, where $r > 0$ is as in the definition of local linking (see Definition 5.77). Note that $\overline{B}$ and B_1^c are disjoint closed sets. So, there is a locally Lipschitz function $\xi\colon H \longrightarrow [0,1]$ such that

$$\xi|_{\overline{B}} \equiv 1 \quad \text{and} \quad \xi|_{B_1^c} \equiv 0. \tag{5.231}$$

Let $\nabla\varphi$ be the gradient of φ, that is,

$$\langle \varphi'(u), u \rangle = (\nabla\varphi(u), u)_H \qquad \forall\, u, h \in H,$$

where $(\cdot,\cdot)_H$ denotes the inner product of H. We introduce the vector filed

$$g(u) = \xi(u)\|\mathrm{proj}_V(u)\|_X \nabla\varphi(u) \qquad \forall\, u \in H, \tag{5.232}$$

where $\mathrm{proj}_V \in \mathcal{L}(H;V)$ is the orthogonal projection onto V. Evidently g is locally Lipschitz and bounded on H. We consider the following abstract Cauchy problem

$$\begin{cases} \sigma'(t) = g(\sigma(t)) & t \geqslant 0, \\ \sigma(0) = u. \end{cases}$$

This problem has a global flow $\sigma(t,u)$. Let $h(u) = \sigma(1,u)$. Then since $h|_{B_1^c} = I_{B_1^c}$ (see (5.231)) and h is an injection, we have that $h(\overline{B}) \subseteq B_1$ and so

$$K_\varphi \cap h(\overline{B}) = \{0\}.$$

Also, clearly we have $h|_{\overline{B}\cap Y} = I_{\overline{B}\cap Y}$ (see (5.232)). Finally, if $u \in \overline{B} \cap V$, $u \neq 0$, then

$$\begin{aligned} \varphi(h(u)) &= \varphi(u) + \int_0^1 \tfrac{d}{dt}\varphi(\sigma(t))\,dt \\ &= \varphi(u) + \int_0^1 \langle \varphi'(\sigma(t)), \sigma'(t) \rangle\,dt = \varphi(u) + \int_0^1 \big(\nabla\varphi(\sigma(t)), \sigma'(t)\big)_H \\ &= \varphi(u) + \int_0^1 \xi(\sigma(t))\|\mathrm{proj}_V(\sigma(t))\|_X \|\nabla\varphi(\sigma(t))\|_X^2\,dt > 0 \end{aligned}$$

(by the chain rule (see Theorem 3.19) and use Definition 5.77 and (5.231)).

Solution of Problem 5.90

Clearly $\varphi(0,0)=0$ and $(0,0)$ is a local minimizer of φ and is the only critical point of φ. Also, $\varphi(3,-4)<0$. So, the mountain pass geometry is satisfied (see Theorem 5.63). Let

$$c = \inf_{\gamma\in\Gamma}\max_{t\in[0,1]}\varphi(\gamma(t)),$$

where $\Gamma = \{\gamma\in C([0,1];\mathbb{R}^2):\ \gamma(0)=(0,0),\ \gamma(1)=(3,-4)\}$. Then $c>0=\varphi(0,0)>\varphi(3,-4)$ and so we can use Problem 5.53 and find a sequence $\{u_n\}_{n\geqslant 1}\subseteq\mathbb{R}^2$ such that

$$\varphi(u_n)\longrightarrow c \quad\text{and}\quad \varphi'(u_n)\longrightarrow 0 \quad\text{in } \mathbb{R}^2.$$

Since $K_\varphi=\{(0,0)\}$, we must have $|u_n|\longrightarrow+\infty$.

Solution of Problem 5.91

"$\mathrm{NPS}'_c \Longrightarrow \mathrm{NPS}_c$":

Let $u_n^*\in\partial\varphi(u_n)$ be such that $m(u_n)=\|u_n^*\|_*\longrightarrow 0$ (see Proposition 3.128). Thus, given $\varepsilon>0$, we can find $n_0=n_0(\varepsilon)\geqslant 1$ such that

$$m(u_n)\leqslant\varepsilon \qquad \forall\, n\geqslant n_0. \tag{5.233}$$

We have

$$\langle u_n^*,h\rangle \leqslant \varphi^0(u_n;h) \qquad \forall\, n\geqslant 1,\ h\in X$$

(see Definition 3.126), so

$$-\varepsilon\|h\|_X \leqslant \varphi^0(u_n;h) \qquad \forall\, n\geqslant n_0,\ h\in X$$

(see (5.233)), thus

$$0\leqslant \varphi^0(u_n;h)+\varepsilon\|h\|_X \qquad \forall\, n\geqslant n_0. \tag{5.234}$$

Also, $\varphi(u_n)\longrightarrow c$. This fact, (5.234) and the hypothesis that φ satisfies the NPS'_c-condition imply that the sequence $\{u_n\}_{n\geqslant1}\subseteq X$ admits a strongly convergent subsequence and so the NPS_c-condition holds.

"$\mathrm{NPS}_c \Longrightarrow \mathrm{NPS}'_c$":

Suppose that the sequence $\{u_n\}_{n\geqslant1}\subseteq X$ satisfies

$$\varphi(u_n)\longrightarrow c \tag{5.235}$$

and for every $\varepsilon > 0$, there exists $n_0 = n_0(\varepsilon) \geqslant 1$ such that

$$0 \leqslant \varphi^0(u_n; h) + \varepsilon\|h\|_X \qquad \forall\, n \geqslant n_0,\ h \in X. \tag{5.236}$$

From (5.236) and Proposition 3.128, we have

$$\begin{aligned} 0 &\leqslant \sigma_{\partial\varphi(u_n)}(h) + \varepsilon\sigma_{\overline{B}_1^*}(h) = \sigma_{\partial\varphi(u_n)}(h) + \sigma_{\varepsilon\overline{B}_1^*}(h) \\ &= \sigma_{\partial\varphi(u_n)+\varepsilon\overline{B}_1^*}(h) \qquad \forall\, n \geqslant n_0,\ h \in X \end{aligned}$$

(see Definition 2.46), where $\overline{B}_1^* = \{u^* \in X^* :\ \|u^*\|_* \leqslant 1\}$, so

$$0 \in \partial\varphi(u_n) + \varepsilon\overline{B}_1^* \qquad \forall\, n \geqslant n_0.$$

Therefore we can find $u_n^* \in \partial\varphi(u_n)$ and $e_n^* \in \overline{B}_1^*$ such that

$$u_n^* + \varepsilon e_n^* = 0 \qquad \forall\, n \geqslant n_0,$$

so

$$m(u_n) \leqslant \|u_n^*\|_* \leqslant \|\varepsilon e_n^*\|_* \leqslant \varepsilon \qquad \forall\, n \geqslant n_0,$$

thus

$$m(u_n) \longrightarrow 0. \tag{5.237}$$

From (5.235), (5.236) and since by hypothesis φ satisfies the NPS$_c$-condition, we infer that the sequence $\{u_n\}_{n\geqslant 1} \subseteq X$ admits a strongly convergent subsequence and so the NPS$_c'$-condition holds.

Solution of Problem 5.92

Using the Ekeland variational principle (see Theorem 5.35), we can find $\widetilde{u} \in X$ such that

$$\varphi(\widetilde{u}) + \varepsilon\|\widetilde{u} - \widehat{u}\|_X \leqslant \varphi(\widehat{u}),$$

and

$$\varphi(\widetilde{u}) \leqslant \varphi(u) + \varepsilon\|u - \widetilde{u}\| \qquad \forall\, u \in X. \tag{5.238}$$

In (5.238) we choose $u = \widetilde{u} + th$ with $t > 0$, $h \in X$. Then

$$-\varepsilon\|h\|_X \leqslant \tfrac{1}{t}\big(\varphi(\widetilde{u} + th) - \varphi(\widetilde{u})\big),$$

so

$$-\varepsilon\|h\|_X \leqslant \varphi^0(\widetilde{u}; h)$$

(see Definition 3.122), thus

$$0 \leqslant \varphi^0(\widetilde{u}; h) + \varepsilon\|h\|_X \qquad \forall\, h \in X.$$

Solution of Problem 5.93

We argue by contradiction. So, suppose that the claim of the problem is not true and $c = \liminf\limits_{\|u\|_X \to +\infty} \varphi(u)$ is finite. Then for every $n \geqslant 1$ we can find $u_n \in X$ such that

$$\varphi(u_n) \leqslant c + \tfrac{1}{n} \quad \text{and} \quad \|u_n\|_X \geqslant 2n. \tag{5.239}$$

We have

$$\varphi(u_n) \leqslant \inf_X \varphi + \big(c + \tfrac{1}{n} - \inf_X \varphi\big). \tag{5.240}$$

Given $\varepsilon > 0$, we can find $n_0 = n_0(\varepsilon) \geqslant 1$ such that $c + \frac{1}{n} - \inf_X \varphi \leqslant \varepsilon$. We use the Ekeland variational principle (see Theorem 5.36) with $\varepsilon > 0$ and $\lambda = n \geqslant n_0$. We can find $v_n \in X$ such that

$$\varphi(v_n) \leqslant \varphi(u_n), \quad \|v_n - u_n\|_X \leqslant n \tag{5.241}$$

and

$$\varphi(v_n) \leqslant \varphi(u_n) + \tfrac{\varepsilon}{n}\|u - u_n\| \qquad \forall\, u \in X. \tag{5.242}$$

In (5.242) we choose $u = v_n + th$ with $t > 0$, $h \in X$. Then

$$0 \leqslant \tfrac{1}{t}\big(\varphi(v_n + th) - \varphi(v_n)\big) + \tfrac{\varepsilon}{n}\|h\|_X \qquad \forall\, n \geqslant n_0,\ t > 0,$$

so

$$0 \leqslant \varphi^0(v_n; h) + \tfrac{\varepsilon}{h}\|h\|_X \leqslant \varphi^0(v_n; h) + \varepsilon\|h\|_X \qquad \forall\, n \geqslant n_0,\ h \in X. \tag{5.243}$$

We have

$$\|v_n\|_X = \|v_n - u_n + u_n\| \geqslant \|u_n\|_X - \|v_n - u_n\|_X \geqslant 2n - n = n$$

(see (5.239) and (5.241)), so

$$\|v_n\|_X \longrightarrow +\infty. \tag{5.244}$$

On the other hand, from (5.240) and (5.241), we have that the sequence $\{\varphi(u_n)\}_{n \geqslant 1} \subseteq \mathbb{R}$ is bounded. This fact, together with (5.243) and Problem 5.91 (recall that by hypothesis φ satisfies the NPS_c-condition for every $c \in \mathbb{R}$), imply that the sequence $\{v_n\}_{n\geqslant 1}$ admits a strongly convergent subsequence, which contradicts (5.244). We conclude that the claim of the problem is true.

Solution of Problem 5.94

By replacing $\varphi(u)$ be $\widehat{\varphi}(u) = \varphi(u_0 + u) - \varphi(u_0)$ for all $u \in X$, if necessary, we may assume without any loss of generality that $u_0 = 0$ and $\varphi(0) = 0$. Also, we assume that $\varphi(u_0) \leqslant 0 = \varphi(0)$ (the reasoning is similar if the opposite inequality holds). Let $\varepsilon \in (0, \|u_1\|_X)$ be such that

$$0 = \varphi(0) \leqslant \varphi(u) \qquad \forall\, u \in \overline{B}_\varepsilon = \{y \in X : \|y\|_X \leqslant \varepsilon\}. \tag{5.245}$$

One of the following conditions holds

(a) There exists $r \in (0, \varepsilon)$ such that $\inf\big\{\varphi(u) : \|u\|_X = r\big\} = m_r > 0 = \varphi(0)$; or
(b) For every $r \in (0, \varepsilon)$, we have $\inf\{\varphi(u) : \|u\|_X = r\} = 0$.

If condition **(a)** holds, then the claim of the problem follows from the nonsmooth mountain pass theorem (see Theorem 5.75 and recall that the NPS$_c$-condition implies the nonsmooth C$_c$-condition). So, suppose that condition **(b)** holds. Let $r, \varrho \in (0, \varepsilon)$ with $r < \varrho$ and let $\{u_n\}_{n\geqslant 1} \subseteq \partial B_r$ be a minimizing sequence of $\varphi|_{\partial B_r}$ (here $\partial B_r = \{y \in X : \|y\|_X = r\}$). Consider the function $\psi \colon X \longrightarrow \mathbb{R}$ defined by

$$\psi(u) = \begin{cases} \varphi(u) & \text{if } \|u\|_X \leqslant \varrho, \\ \varphi(\varrho\frac{u}{\|u\|_X}) & \text{if } \|u\|_X > \varrho. \end{cases} \tag{5.246}$$

Clearly ψ is locally Lipschitz and bounded below. Using the Ekeland variational principle (see Theorem 5.36) as in the solution of Problem 5.93, we can find a sequence $\{v_n\}_{n\geqslant 1} \subseteq X$ such that

$$\psi(v_n) \leqslant \psi(u_n), \quad \|v_n - u_n\|_X \leqslant \tfrac{1}{n} \qquad \forall\, n \geqslant 1 \tag{5.247}$$

and

$$0 \leqslant \psi^0(v_n; h) + \tfrac{1}{n}\|h\|_X \qquad \forall\, n \geqslant 1,\ h \in X. \tag{5.248}$$

We have that

$$\psi(u_n) = \varphi(u_n) \longrightarrow 0$$

(see (5.247) and (5.248)), so the sequence $\{\psi(v_n)\}_{n\geqslant 1} \subseteq \mathbb{R}$ is bounded.

Note that for $n \geqslant 1$ big, we have that

$$v_n \in \overline{B}_\varrho \ \text{ and } \ \varphi|_{\overline{B}_\varrho} = \psi|_{\overline{B}_\varrho},$$

where $\overline{B}_\varrho = \{y \in X : \|y\|_X \leqslant \varrho\}$. By hypothesis φ satisfies the NPS$_c$-condition for every $c \in \mathbb{R}$. This fact together with Problem 5.91 imply that by passing to a suitable subsequence if necessary, we may assume that $v_n \longrightarrow \widehat{u}$ in X and $\|\widehat{u}\|_X = r$

(see (5.247) and (5.248)). Also, exploiting the upper semicontinuity of $\varphi^0(\cdot,\cdot)$ (see Proposition 3.123(c)), we have

$$0 \leqslant \varphi^0(\widehat{u};h) \qquad \forall\, h \in X,$$

so $0 \in \partial\varphi(\widehat{u})$. In fact we can say that for every $r \in (0,\varepsilon)$, $K_\varphi \cap \partial B_r \neq \emptyset$.

Solution of Problem 5.95

First let $c \in \mathbb{R} \setminus D$ and consider a sequence $\{u_n\}_{n\geqslant 1} \subseteq H_0^1(\Omega)$ such that

$$\varphi(u_n) \longrightarrow c \quad \text{and} \quad \varphi'(u_n) \longrightarrow 0 \quad \text{in } H^{-1}(\Omega) = H_0^1(\Omega)^*. \tag{5.249}$$

We claim that the sequence $\{u_n\}_{n\geqslant 1} \subseteq H_0^1(\Omega)$ is bounded. Arguing indirectly, suppose that at least for a subsequence, we have

$$\|u_n\| \longrightarrow +\infty. \tag{5.250}$$

We know that we have the following orthogonal direct sum decomposition

$$H_0^1(\Omega) = \overline{H}_{k-1} \oplus E(\widehat{\lambda}_k) \oplus \widehat{H}_{k+1},$$

where

$$\overline{H}_{k-1} = \bigoplus_{i=1}^{k-1} E(\widehat{\lambda}_i) \quad \text{and} \quad \widehat{H}_{k+1} = \overline{\bigoplus_{i\geqslant k+1} E(\widehat{\lambda}_i)}.$$

So, for all $n \geqslant 1$, we have

$$u_n = \overline{u}_n + u_n^0 + \widehat{u}_n, \quad \text{with } \overline{u}_n \in \overline{H}_{k-1},\ u_n^0 \in E(\widehat{\lambda}_k),\ \widehat{u}_n \in \widehat{H}_{k+1}.$$

From (5.249), we have

$$\Big|\langle A(u_n),h\rangle - \widehat{\lambda}_k \int_\Omega u_n h\,dz - \int_\Omega f(z,u_n)h\,dz\Big| \leqslant \varepsilon_n\|h\| \qquad \forall\, h \in H_0^1(\Omega), \tag{5.251}$$

with $\varepsilon_n \searrow 0$, where $A \in \mathcal{L}(H_0^1(\Omega), H^{-1}(\Omega))$ is defined by

$$\langle A(u),h\rangle = \int_\Omega (Du,Dh)_{\mathbb{R}^N}\,dz \qquad \forall\, u,h \in H_0^1(\Omega).$$

In (5.251) we choose $h = \widehat{u}_n \in H_0^1(\Omega)$. Exploiting the orthogonality of the subspaces, we obtain

$$\Big|\|D\widehat{u}_n\|_2^2 - \widehat{\lambda}_k\|\widehat{u}_n\|_2^2 - \int_\Omega f(z,u_n)\widehat{u}_n\,dz\Big| \leqslant \varepsilon_n\|\widehat{u}_n\| \qquad \forall\, n \geqslant 1. \tag{5.252}$$

Hypotheses **(i)** and **(ii)** imply that

$$|f(z,x)| \leqslant \widehat{a}(z) \quad \text{for a.a. } z \in \Omega, \text{ all } x \in \mathbb{R}, \tag{5.253}$$

with $\widehat{a} \in L^q(\Omega)$. Note that $\frac{2N}{N+2}$ is the conjugate exponent of $2^* = \frac{2N}{N-2}$ if $N \geqslant 3$ (that is, $\frac{N+2}{2N} + \frac{N-2}{2N} = 1$). Using the Rellich–Kondrachov embedding theorem (see Theorem 1.135) the Hölder inequality (see Theorem 1.3) and (5.253), we have

$$\Big|\int_\Omega f(z,u_n)\widehat{u}_n\,dz\Big| \leqslant \|\widehat{a}\|_{\frac{2N}{N-2}}\|\widehat{u}\|_{2^*} \leqslant c_1\|\widehat{u}_n\| \qquad \forall\, n \geqslant 1, \tag{5.254}$$

for some $c_1 > 0$. Also, from Remark 5.130, we have

$$\|D\widehat{u}_n\|_2^2 - \widehat{\lambda}_k\|\widehat{u}_n\|_2^2 \geqslant c_2\|\widehat{u}_n\|^2 \qquad \forall\, n \geqslant 1, \tag{5.255}$$

for some $c_2 > 0$. Returning to (5.251) and using (5.254) and (5.255), we infer that

$$c_2\|\widehat{u}_n\|^2 \leqslant c_3\|\widehat{u}_n\| \qquad \forall\, n \geqslant 1,$$

for some $c_3 > 0$, so the sequence $\{\widehat{u}_n\}_{n\geqslant 1} \subseteq H_0^1(\Omega)$ is bounded.

In a similar fashion, using $h = \overline{u}_n \in H_0^1(\Omega)$ in (5.251) and reasoning as above, we show that the sequence $\{\overline{u}_n\}_{n\geqslant 1} \subseteq H_0^1(\Omega)$ is bounded.

From (5.250) and the boundedness of the sequences $\{\widehat{u}_n\}_{n\geqslant 1}, \{\overline{u}_n\}_{n\geqslant 1} \subseteq H_0^1(\Omega)$, it follows that

$$\|u_n^0\| \longrightarrow +\infty. \tag{5.256}$$

Let $y_n = \frac{u_n}{\|u_n^0\|}$ for $n \geqslant 1$. Then because of the boundedness of the sequences $\{\widehat{u}_n\}_{n\geqslant 1}, \{\overline{u}_n\}_{n\geqslant 1} \subseteq H_0^1(\Omega)$, we get that the sequence $\{y_n\}_{n\geqslant 1} \subseteq H_0^1(\Omega)$ is bounded and passing to a subsequence if necessary, we may assume that

$$y_n \overset{w}{\longrightarrow} y \quad \text{in } H_0^1(\Omega) \quad \text{and} \quad y_n \longrightarrow y \quad \text{in } L^2(\Omega) \tag{5.257}$$

with $y \in E(\widehat{\lambda}_k)$, $\|y\| = 1$. Because of (5.257) we may assume that

$$y_n(z) \longrightarrow y(z) \quad \text{for a.a. } z \in \Omega,$$

so

$$\begin{cases} u_n(z) \longrightarrow +\infty & \text{for a.a. } z \in \{y > 0\}, \\ u_n(z) \longrightarrow -\infty & \text{for a.a. } z \in \{y < 0\}. \end{cases} \tag{5.258}$$

Then hypothesis **(iii)** and the Lebesgue dominated convergence theorem (see Theorem I.3.94) imply that

$$\int_\Omega F(z,u_n)\,dz \longrightarrow \int_{\{y>0\}} F_+\,dz + \int_{\{y<0\}} F_-\,dz. \tag{5.259}$$

Note that

$$
\begin{aligned}
& \Big| \int_\Omega f(z,u_n)\widehat{u}\,dz \Big| \\
& \leqslant \|f(\cdot,u_n(\cdot))\|_{\frac{2N}{N+2}} \|\widehat{u}_n\|_{2^*} \leqslant c_4 \|f(\cdot,u_n(\cdot))\|_{\frac{2N}{N+2}} \qquad \forall\, n \geqslant 1,
\end{aligned}
$$

for some $c_4 > 0$ (by the Hölder inequality (see Theorem 1.3), the Rellich–Kondrachov embedding theorem (see Theorem 1.135) and using the boundedness of the sequence $\{\widehat{u}_n\}_{n\geqslant 1} \subseteq H_0^1(\Omega)$).

But hypothesis **(i)** and (5.258) imply

$$
\|f(\cdot,u_n(\cdot))\|_{\frac{2N}{N+2}} \longrightarrow 0,
$$

so

$$
\int_\Omega f(z,u_n)\widehat{u}_n\,dz \longrightarrow 0. \tag{5.260}
$$

Then from (5.251) with $h = \widehat{u}_n \in H_0^1(\Omega)$, we have

$$
\|D\widehat{u}_n\|_2^2 - \widehat{\lambda}_k \|\widehat{u}\|_2^2 \longrightarrow 0
$$

(see (5.260)), so

$$
\|\widehat{u}_n\| \longrightarrow 0 \tag{5.261}
$$

(recall that $c_5\|\widehat{u}_n\|^2 \leqslant \|D\widehat{u}_n\|_2^2 - \widehat{\lambda}_k\|\widehat{u}_n\|_2^2$ for all $n \geqslant 1$, for some $c_5 > 0$).

In a similar fashion, we show that

$$
\|\overline{u}_n\| \longrightarrow 0. \tag{5.262}
$$

Then from (5.249), (5.259), (5.261), and (5.262), we have

$$
c = -\int_{\{y>0\}} F_+\,dz - \int_{\{y<0\}} F_-\,dz,
$$

so $c \in D$, a contradiction. Therefore the sequence $\{u_n\}_{n\geqslant 1} \subseteq H_0^1(\Omega)$ is bounded and so we may assume that

$$
u_n \xrightarrow{w} u \quad \text{in } H_0^1(\Omega) \quad \text{and} \quad u_n \longrightarrow u \quad \text{in } L^{q'}(\Omega) \tag{5.263}
$$

(with $\frac{1}{q} + \frac{1}{q'} = 1$). So, if in (5.251) we choose $h = u_n - u$, pass to the limit as $n \to +\infty$ and use (5.263), we obtain

$$
\|Du_n\|_2 \longrightarrow \|Du\|_2,
$$

so by the Kadec–Klee property in Hilbert spaces (see Corollary 1.26)

$$
u_n \longrightarrow u \quad \text{in } H_0^1(\Omega)
$$

and thus φ satisfies the PS_c-condition for all $c \in \mathbb{R} \setminus D$.

Next suppose that $c \in D$. Let $y \in E(\widehat{\lambda}_k)$ with $\|y\| = 1$. We have

$$\varphi(ty) = -\int_\Omega F(z, ty)\, dz \tag{5.264}$$

and

$$\langle \varphi'(ty), h \rangle = -\int_\Omega f(z, ty) h\, dz \qquad \forall\, h \in H_0^1(\Omega). \tag{5.265}$$

Letting $t \to +\infty$, we have

$$\varphi(ty) \longrightarrow -\int_{\{y>0\}} F_+\, dz - \int_{\{y<0\}} F_-\, dz$$

(see (5.264) and hypothesis **(iii)**), so $\varphi'(ty) \longrightarrow 0$ (see (5.265) and the hypothesis).

Let $t_n = n$ and $u_n = ny$. Then the sequence $\{u_n\}_{n \geqslant 1}$ has no strongly convergent subsequence and so we conclude that φ does not satisfy the PS_c-condition when $c \in D$.

Solution of Problem 5.96

Let $\{y_n\}_{n \geqslant 1} \subseteq X$ be a sequence such that

$$\xi(y_n) \leqslant c + \tfrac{1}{n} \quad \text{and} \quad \|y_n\|_X \geqslant 2n \qquad \forall\, n \geqslant 1. \tag{5.266}$$

Since ξ is lower semicontinuous and bounded below, we can use the Ekeland variational principle (see Theorem 5.36) with $\varepsilon = c + \frac{1}{n} - \inf\limits_X \xi$ and $\lambda = n$. Then we have a sequence $\{u_n\}_{n \geqslant 1} \subseteq X$ such that

$$\xi(u_n) \leqslant \xi(y_n) \leqslant c + \tfrac{1}{n} \qquad \forall\, n \geqslant 1, \tag{5.267}$$

$$\xi(u_n) \leqslant \xi(u) + \tfrac{1}{n}\varepsilon \|u - u_n\|_X \qquad \forall\, u \in X,\ n \geqslant 1, \tag{5.268}$$

$$\|u_n - y_n\|_X \leqslant n \qquad \forall\, n \geqslant 1. \tag{5.269}$$

In (5.268) we choose $u = u_n + th$ with $t > 0$, $h \in X$. We have

$$\varphi(u_n) + \psi(u_n) \leqslant \varphi(u_n + th) + \psi(u_n + th) + \tfrac{\varepsilon t}{n} \|h\|_X,$$

so

$$\begin{aligned}
-\tfrac{\varepsilon t}{n} \|h\|_X &\leqslant \varphi(u_n + th) - \varphi(u_n) + \psi(u_n + th) - \psi(u_n) \\
&= \varphi(u_n + th) - \varphi(u_n) + \psi((1-t)u_n + t(u_n + h)) - \psi(u_n) \\
&\leqslant \varphi(u_n + th) - \varphi(u_n) + t(\psi(u_n + h) - \psi(u_n))
\end{aligned}$$

(since $\psi \in \Gamma_0(X)$), thus

$$-\tfrac{\varepsilon}{n}\|h\|_X \leqslant \tfrac{1}{t}\big(\varphi(u_n+th)-\varphi(u_n)\big)+\psi(u_n+th)-\psi(u_n)$$

and hence

$$-\tfrac{\varepsilon}{n}\|h\|_X \leqslant \varphi^0(u_n;h)+\psi(u_n+th)-\psi(u_n) \qquad \forall\, h\in X,\ n\geqslant 1. \tag{5.270}$$

We have

$$\begin{aligned}\|u_n\|_X &= \|u_n-y_n+y_n\|_X\\ &\geqslant \|y_n\|_X-\|u_n-y_n\|_X \geqslant 2n-n = n \qquad \forall\, n\geqslant 1\end{aligned}$$

(see (5.266) and (5.269)), so

$$\|u_n\|_X \longrightarrow +\infty. \tag{5.271}$$

From (5.267) and (5.271), we see that

$$\xi(u_n) \longrightarrow c. \tag{5.272}$$

Then from (5.270), (5.271), and (5.272), we see that $\{u_n\}_{n\geqslant 1}\subseteq X$ satisfies the desired condition.

Solution of Problem 5.97

Arguing by contradiction, suppose that $c = \liminf\limits_{\|u\|_X\to+\infty}\xi(u)\in\mathbb{R}$. Then according to Problem 5.96, we can find a sequence $\{u_n\}_{n\geqslant 1}\subseteq X$ such that

$$\|u_n\|_X \longrightarrow +\infty,\quad \xi(u_n)\longrightarrow c$$

and

$$-\varepsilon_n\|u-u_n\|_X \leqslant \varphi^0(u_n;u-u_n)+\psi(u)+\psi(u_n) \qquad \forall\, u\in X,$$

with $\varepsilon_n \searrow 0$. But this contradicts the hypothesis that ξ satisfies the SPS-condition.

Solution of Problem 5.98

Let $\{u_n\}_{n\geqslant 1}\subseteq X$ be a sequence such that

$$\varphi(u_n)\longrightarrow c \quad\text{and}\quad m(u_n)\longrightarrow 0. \tag{5.273}$$

Because φ satisfies the NPS$_w$-condition, passing to the subsequence if necessary, we may assume that $u_n \xrightarrow{w} u$ in X. Due to the compactness of the embedding $X\hookrightarrow Y$, we have

$$u_n \longrightarrow u \quad \text{in } Y. \tag{5.274}$$

Let $u_n^* \in \partial\varphi(u_n)$ be such that $m(u_n) = \|u_n^*\|_*$ for all $n \geqslant 1$. We have

$$u_n^* = \|u_n\|_X^{p-2}\mathcal{F}(u_n) - v_n^*,$$

with $\mathcal{F}\colon X \longrightarrow X^*$ being the duality map of X (see Definition 2.112) and $v_n^* \in \partial\psi(u_n)$ for all $n \geqslant 1$ (see Problem 3.40). From the homogeneity of $\mathcal{F}$, we have

$$\mathcal{F}(\|u_n\|_X^{p-2}u_n) = u_n^* + v_n^* \qquad \forall\, n \geqslant 1,$$

so

$$\|u_n\|_X^{p-2}u_n = \mathcal{F}^{-1}(u_n^* + v_n^*) \qquad \forall\, n \geqslant 1. \tag{5.275}$$

We know that $\mathcal{F}^{-1}\colon X^* \longrightarrow X$ is the duality map and it is continuous since both X and X^* are uniformly convex Banach spaces (see Proposition 2.114). From (5.273), we have

$$u_n^* \longrightarrow 0 \quad \text{in } X^*. \tag{5.276}$$

From Problem 3.124, we know that $\partial\psi(u_n) = \partial\widehat{\psi}|_X(u_n)$. Then Proposition 3.128(b) and (5.274) imply that the sequence $\{v_n^*\}_{n\geqslant 1} \subseteq Y^*$ is relatively w-compact and so by the Eberlein–Smulian theorem (see Theorem I.5.78), we may assume that

$$v_n^* \xrightarrow{w} v^* \quad \text{in } Y^*.$$

But from Problem 1.103(a), we know that Y^* is embedded into X^* compactly and densely. Hence

$$v_n^* \longrightarrow v^* \quad \text{in } X^*. \tag{5.277}$$

Then from (5.276), (5.277) and the continuity of $\mathcal{F}^{-1}$, we have

$$\mathcal{F}^{-1}(u_n^*, v_n^*) \longrightarrow 0 \quad \text{in } X,$$

so

$$\|u_n\|_X^{p-2}u_n \longrightarrow 0 \quad \text{in } X$$

(see (5.275)). Since the map $\xi\colon X \longrightarrow X$ defined by $\xi(u) = \|u\|_X^{p-2}u$ is a homeomorphism, we conclude that $u_n \longrightarrow u$ in X and this proves that φ satisfies the NPS_c-condition for every $c \in \mathbb{R}$.

Solution of Problem 5.99

Because of hypothesis **(iii)**, we can apply the Ekeland variational principle (see Theorem 5.35) on the function $y \longmapsto \varphi(\widehat{u}, y)$. So, we can find $\widetilde{u} \in C$ such that

$$\varphi(\widehat{u}, \widetilde{u}) + \varepsilon d_X(\widetilde{u}, \widehat{u}) \leqslant \varphi(\widehat{u}, \widehat{u}) = 0$$

(see hypothesis **(i)**) and

$$\varphi(\widehat{u}, \widetilde{u}) < \varphi(\widehat{u}, u) + \varepsilon d_X(u, \widetilde{u}) \qquad \forall\, u \in C,\ u \neq \widetilde{u}. \tag{5.278}$$

From hypothesis **(ii)**, we have

$$\varphi(\widehat{u}, u) \leqslant \varphi(\widehat{u}, \widetilde{u}) + \varphi(\widetilde{u}, u). \tag{5.279}$$

Using (5.279) in (5.278), we obtain

$$\varphi(\widehat{u}, \widetilde{u}) < \varphi(\widehat{u}, \widetilde{u}) + \varphi(\widetilde{u}, u) + \varepsilon d_X(u, \widetilde{u}) \qquad \forall\, u \in C,\ u \neq \widetilde{u},$$

so

$$0 < \varphi(\widetilde{u}, u) + \varepsilon d_X(u, \widetilde{u}) \qquad \forall\, u \in C,\ u \neq \widetilde{u}.$$

Solution of Problem 5.100

Using Problem 5.99, for every $n \geqslant 1$, we can find $u_n \in C$ such that

$$-\tfrac{1}{n}\|u_n - y\|_X \leqslant \varphi(u_n, y) \qquad \forall\, y \in C. \tag{5.280}$$

Since the set C is w-compact and $\{u_n\}_{n\geqslant 1} \subseteq C$, by the Eberlein–Smulian theorem (see Theorem I.5.78), passing to a subsequence if necessary, we may assume that

$$u_n \xrightarrow{w} \widehat{u} \quad \text{in } X. \tag{5.281}$$

Then, from (5.280), we have

$$0 = \limsup_{n\to+\infty} \big(-\tfrac{1}{n}\|u_n - y\|_X\big) \leqslant \limsup_{n\to+\infty} \varphi(u_n, y) \leqslant \varphi(\widehat{u}, y) \qquad \forall\, y \in C$$

(see hypothesis **(iv)** and (5.281)).

Solution of Problem 5.101

For every $u \in C$, let

$$S(u) = \{y \in C : \|y\|_X \leqslant \|u\|_X,\ \varphi(u, y) \leqslant 0\}.$$

Note that $S(u) \neq \emptyset$ (since $u \in S(u)$; see hypothesis **(i)**). Evidently $S(u)$ is bounded and so by the reflexivity of X, the set $S(u)$ is relatively w-compact. Moreover, if $y \in S(u)$, then $S(y) \subseteq S(u)$ (see hypothesis **(ii)**).

On the function $\varphi|_{K\times K}$ we apply Problem 5.100 and we can find $\widehat{u} \in K$ such that

$$0 \leqslant \varphi(\widehat{u}, y) \qquad \forall\, y \in K. \tag{5.282}$$

Suppose that $\widehat{u}$ is not the desired element. Then we can find $y \in C$ such that

$$\varphi(\widehat{u}, y) < 0. \tag{5.283}$$

Exploiting the weak lower semicontinuity of the norm functional (see Proposition I.5.56(c)), we can find $v_y \in \overline{S(y)}^w$ such that $\|v_y\|_X = \min\limits_{v \in \overline{S(y)}^w} \|v\|_X$.

First we assume that $v_y \in K$. Let $\varepsilon > 0$ be such that

$$\varphi(\widehat{u}, y) \leqslant -\varepsilon \tag{5.284}$$

(see (5.283)). Let $\{v_n\}_{n \geqslant 1} \subseteq S(y)$ be a sequence such that $v_n \xrightarrow{w} v_y$ (recall that the space X is reflexive and the set $S(y)$ is bounded). We have $\{v_n\}_{n \geqslant 1} \subseteq K$ (see hypothesis **(iii)**) and

$$\varphi(y, v_n) \leqslant 0 \qquad \forall\, n \geqslant 1,$$

so

$$\varphi(\widehat{u}, v_n) \leqslant \varphi(\widehat{u}, y) + \varphi(y, v_n) \leqslant -\varepsilon \qquad \forall\, n \geqslant 1$$

(see hypothesis **(ii)** and (5.284)), thus

$$\varphi(\widehat{u}, v_y) \leqslant -\varepsilon$$

(see hypothesis **(iv)**). Since $v_y \in K$, we contradict (5.282). So, we must have $v_y \in C \setminus K$. Then according to hypothesis **(iii)**, we can find $\overline{v} \in C$ such that $\|\overline{v}\|_X \leqslant \|v_y\|_X$ and $\varphi(v_y, \overline{v}) \leqslant 0$. Hence

$$\overline{v} \in S(v_y) \subseteq S(y) \quad \text{and} \quad \|\overline{v}\|_X < \|v_y\|_X = \min_{v \in \overline{S(y)}^w} \|v\|_X,$$

a contradiction. We conclude that $\widehat{u} \in C$ is the desired solution.

Solution of Problem 5.102

First we show the existence of a positive solution. To this end, let $f_+(z, x) = f(z, x^+)$, where $x^+ = \max\{x, 0\}$ and let $F_+(z, x) = \int\limits_0^x f_+(x, s)\, ds$. We consider the C^1-functional $\varphi_+ \colon H_0^1(\Omega) \longrightarrow \mathbb{R}$ define by

$$\varphi_+(u) = \tfrac{1}{2}\|Du\|_2^2 - \int\limits_\Omega F_+(z, u)\, dz \qquad \forall\, z \in H_0^1(\Omega).$$

Hypotheses **(i)** and **(ii)** imply that given $\varepsilon > 0$, we can find $c_\varepsilon > 0$ such that

$$F_+(z, x) \leqslant \tfrac{1}{2}(\vartheta(z) + \varepsilon)x^2 + c_\varepsilon \quad \text{for a.a. } z \in \Omega, \text{ all } x \in \mathbb{R}. \tag{5.285}$$

Then for all $u \in H_0^1(\Omega)$, we have

$$\begin{aligned} \varphi_+(u) &\geqslant \tfrac{1}{2}\big(\|Du\|_2^2 - \int\limits_\Omega \vartheta(z)u^2\, dz\big) - \tfrac{\varepsilon}{2}\|u\|^2 - c_\varepsilon|\Omega|_N \\ &\geqslant \tfrac{1}{2}(c_0 - \varepsilon)\|u\|^2 - c_\varepsilon|\Omega|_N \end{aligned}$$

for some $c_0 > 0$ (see (5.285) and Problem 5.67). Choosing $\varepsilon \in (0, c_0)$, we see that

$$\varphi_+(u) \longrightarrow +\infty \quad \text{as } \|u\| \to +\infty.$$

Also, using the Rellich–Kondrachov embedding theorem (see Theorem 1.135), we see that φ_+ is sequentially weakly lower semicontinuous. So, Theorem 5.6 implies the existence of $u_0 \in H_0^1(\Omega)$ such that

$$\varphi_+(u_0) = \inf_{u \in H_0^1(\Omega)} \varphi_+(u). \tag{5.286}$$

Hypothesis **(iii)** implies that given $\varepsilon > 0$, we can find $\delta = \delta(\varepsilon) > 0$ such that

$$F_+(z,x) \geqslant \tfrac{1}{2}(\eta(z) - \varepsilon)x^2 \quad \text{for a.a. } z \in \Omega, \text{ all } x \in [0,\delta]. \tag{5.287}$$

Let $\widehat{u}_1 \in \operatorname{int} C_+$ be the L^2-normalized (that is, $\|\widehat{u}_1\|_2 = 1$) eigenfunction corresponding to $\widehat{\lambda}_1 > 0$ (recall that $\widehat{\lambda}_1$ is simple). Let $t \in (0,1)$ be sufficiently small such that $t\widehat{u}_1(z) \in [0,\delta]$ for all $z \in \overline{\Omega}$. Then

$$\begin{aligned} \varphi_+(t\widehat{u}_1) &\leqslant \tfrac{t^2}{2}\Big(\|D\widehat{u}_1\|_2^2 - \int_\Omega \eta(z)\widehat{u}_1^2\,dz\Big) + \tfrac{t^2}{2}\varepsilon \\ &= \tfrac{t^2}{2}\Big(\int_\Omega (\widehat{\lambda}_1 - \eta(z))\widehat{u}_1^2\,dz + \varepsilon\Big) \end{aligned} \tag{5.288}$$

(see (5.287) and recall that $\|\widehat{u}_1\|_2 = 1$). Note that

$$\xi^* = \int_\Omega (\eta(z) - \widehat{\lambda}_1)\widehat{u}_1^2\,dz > 0$$

(see hypothesis **(iii)** and recall that $\widehat{u}_1(z) > 0$ for all $z \in \Omega$). So, from (5.288), we have

$$\varphi_+(t,\widehat{u}_1) \leqslant \tfrac{t^2}{2}(-\xi^* + \varepsilon).$$

Choosing $\varepsilon \in (0,\xi^*)$ we see that

$$\varphi_+(t\widehat{u}_1) < 0 = \varphi_+(0),$$

so

$$\varphi_+(u_0) < 0 = \varphi_+(0)$$

(see (5.286)), thus $u_0 \neq 0$. From (5.288) we have $\varphi_+'(u_0) = 0$, so

$$A(u_0) = N_{f_+}(u_0), \tag{5.289}$$

where $A \in \mathcal{L}(H_0^1(\Omega), H^{-1}(\Omega))$ is defined by

$$\langle A(u), h\rangle \;=\; \int_\Omega (Du, Dh)_{\mathbb{R}^N}\, dz \qquad \forall\, u, h \in H_0^1(\Omega)$$

and $N_{f_+}(u_0)(\cdot) = f_+(\cdot, u_0(\cdot))$. On (5.289) we act with $-u_0^- \in H_0^1(\Omega)$ and obtain $\|Du_0^-\|_2^2 = 0$, so $u_0 \geqslant 0$, $u_0 \neq 0$. Thus, u_0 is a positive solution of the Dirichlet problem. Standard regularity theory implies that $u_0 \in C_+ \setminus \{0\}$. Let $\varrho = \|u_0\|_\infty$ and let $\xi_\varrho > 0$ be as postulated by hypothesis **(v)**. Then

$$-\Delta u_0(z) + \xi_\varrho u_0(z) \;=\; f(z, u_0(z)) + \xi_\varrho u_0(z) \;\geqslant\; 0 \quad \text{for a.a. } z \in \Omega,$$

so

$$\Delta u_0(z) \;\leqslant\; \xi_\varrho u_0(z) \quad \text{for a.a. } z \in \Omega$$

and thus $u_0 \in \operatorname{int} C_+$ (by the nonlinear maximum principle (see Theorem 5.139)). We show that this is the unique positive solution of the Dirichlet problem. So, let $\widehat{u} \in H_0^1(\Omega)$ be another positive solution of the Dirichlet problem. As above we show that $\widehat{u} \in \operatorname{int} C_+$. Let $t_0 > 0$ be the biggest positive real number such that $t_0\widehat{u} \leqslant u_0$ (see Problem 4.182). Suppose that $t_0 < 1$. Let $\varrho = \|u_0\|_\infty$ and let $\xi_\varrho > 0$ be as in hypothesis **(v)**. Then

$$\begin{aligned} -\Delta(t_0\widehat{u}) + \xi_\varrho(t_0\widehat{u}) \;&=\; t_0\big(f(z,\widehat{u}) + \xi_\varrho\widehat{u}\big) \;\leqslant\; f(z, t_0\widehat{u}) + \xi_\varrho(t_0\widehat{u}) \\ \leqslant\; f(z, u_0) + \xi_\varrho u_0 \;&=\; -\Delta u_0 + \xi_\varrho u_0 \end{aligned}$$

(use hypothesis **(iv)** and **(v)** and the facts that $t_0 < 1$ and $t_0\widehat{u} \leqslant u_0$), so

$$\Delta(u_0 - t_0\widehat{u}) \;\leqslant\; \xi_\varrho(u_0 - t_0\widehat{u}),$$

so by the nonlinear maximum principle (see Theorem 5.139), we have $u_0 - t_0\widehat{u} \in \operatorname{int} C_+$. This contradicts the maximality of $t_0 > 0$. Therefore $t_0 \geqslant 1$ and so $\widehat{u} \leqslant u_0$. Interchanging the roles of u_0 and $\widehat{u}$ in the above argument, we have that $u_0 \leqslant \widehat{u}$ and thus $u_0 = \widehat{u}$. So $u_0 \in \operatorname{int} C_+$ is the unique positive solution of the Dirichlet problem.

Solution of Problem 5.103

Suppose that u has constant sign and to fix things we assume that $u \geqslant 0$. Then $u \in C_+ \setminus \{0\}$ (see Theorem 5.129(e) and Remarks 5.133, 5.135). Using the nonlinear maximum principle (see Theorem 5.139) we have that $u \in \operatorname{int} C_+$. From the Picone identity (see Theorem 5.142) and integrating by parts, we have

$$\begin{aligned} 0 \leqslant R(\widehat{u}, u) &= \|D\widehat{u}_1\|_p^p - \int\limits_\Omega |Du|^{p-2}\big(Du, D(\tfrac{\widehat{u}_1^p}{u^{p-1}})\big)_{\mathbb{R}^N}\,dz \\ &= \|D\widehat{u}_1\|_p^p - \int\limits_\Omega (-\Delta_p u)\tfrac{\widehat{u}_1^p}{u^{p-1}}\,dz \\ &= \widehat{\lambda}_1 - \int\limits_\Omega \lambda u^{p-1}\tfrac{\widehat{u}_1^p}{u^{p-1}}\,dz = \widehat{\lambda}_1 - \lambda < 0, \end{aligned}$$

a contradiction.

Solution of Problem 5.104

Let $\{\lambda_n\}_{n\geqslant 1} \subseteq \sigma_0(p)$ be a sequence such that $\lambda_n \longrightarrow \lambda$. Consider $u_n \in E(\lambda_n)$ for all $n \geqslant 1$. Then

$$\|u_n\| = 1 \quad \text{and} \quad A(u_n) = \lambda_n |u_n|^{p-2}u_n \qquad \forall\, n \geqslant 1, \tag{5.290}$$

where $A\colon W_0^{1,p}(\Omega) \longrightarrow W^{-1,p'}(\Omega) = W_0^{1,p}(\Omega)^*$ is the nonlinear map defined by

$$\langle A(u), h\rangle = \int\limits_\Omega |Du|^{p-2}\,(Du, Dh)_{\mathbb{R}^N}\,dz \qquad \forall\, u, h \in W_0^{1,p}(\Omega).$$

We may assume that

$$u_n \xrightarrow{w} u \quad \text{in } W_0^{1,p}(\Omega) \quad \text{and} \quad u_n \longrightarrow u \quad \text{in } L^p(\Omega). \tag{5.291}$$

On (5.290) we act with $u_n - u \in W_0^{1,p}(\Omega)$, pass to the limit as $n \to +\infty$ and use (5.291). Then

$$\lim_{n\to+\infty} \langle A(u_n), u_n - u\rangle = 0,$$

so

$$u_n \longrightarrow u \quad \text{in } W_0^{1,p}(\Omega), \quad \text{with } \|u\| = 1 \tag{5.292}$$

(see Problem 2.192). So, if in (5.290) we pass to the limit as $n \to +\infty$ and use (5.292), then

$$A(u) = \lambda|u|^{p-2}u,$$

so $\lambda \in \sigma_0(p)$ and thus the set $\sigma_0(p)$ is closed. From the above argument it is clear that $E(\lambda) \subseteq W_0^{1,p}(\Omega)$ is compact.

Solution of Problem 5.105

Problem 5.103 implies that $u^+, u^- \neq 0$. We have

$$\int_\Omega |Du|^{p-2} (Du, Dh)_{\mathbb{R}^N}\, dz \;=\; \lambda \int_\Omega |u|^{p-2} uh\, dz \qquad \forall\, h \in W_0^{1,p}(\Omega). \tag{5.293}$$

In (5.293) we choose $h = u^+ \in W_0^{1,p}(\Omega)$ (see Problem 1.152). We obtain

$$\|Du^+\|_p^p \;=\; \lambda \|u^+\|_p^p \tag{5.294}$$

(see Problem 1.152). Using the Rellich–Kondrachov embedding theorem (see Theorem 1.135) and the Hölder inequality (see Theorem 1.3), we obtain

$$\lambda \|u^+\|_p^p \;\leqslant\; \lambda \|u^+\|_{p^*}^p |\Omega_+|_N^{1-\frac{p}{p^*}}, \tag{5.295}$$

where $p^* = \frac{Np}{N-p}$ if $p < N$ and $p^* = 2p$ if $N \leqslant p$. Also recall that

$$\|u^+\|_{p^*} \;\leqslant\; c\|Du^+\|_p \tag{5.296}$$

for some $c > 0$ depending on p and N. So, finally from (5.294), (5.295), and (5.296), we have

$$1 \;\leqslant\; \lambda c^p |\Omega_+|_N^{1-\frac{p}{p^*}}. \tag{5.297}$$

Similarly, we obtain

$$1 \;\leqslant\; \lambda c^p |\Omega_-|_N^{1-\frac{p}{p^*}}. \tag{5.298}$$

From (5.297) and (5.298) we infer that

$$\min\left\{|\Omega_+|_N,\ |\Omega_-|_N\right\} \;\geqslant\; (\lambda c^p)^r,$$

with $c > 0$ depending only on p, N and $r = \frac{N}{p}$ if $p < N$, $r = 2$ if $N \leqslant p$.

Solution of Problem 5.106

From Proposition 5.136, we have that $u \in L^\infty(\Omega)$ and then we can apply Proposition 5.138 and infer that $u \in C_+ \setminus \{0\}$ (see Theorem 5.129(e) and Remark 5.133). We have

$$\Delta_p u(z) \;\leqslant\; \|\xi\|_\infty u(z)^{p-1} \quad \text{for a.a. } z \in \Omega,$$

thus $u \in \operatorname{int} C_+$ (see the nonlinear maximum principle; Theorem 5.139).

Solution of Problem 5.107

Let $\eta \geqslant \|\xi\|_\infty$. Then replacing $\xi(z)$ with $\xi(z)+\eta$ and λ with $\lambda+\eta$, we see that without any loss of generality, we may assume that $\xi(z) \geqslant 0$ for almost all $z \in \Omega$. We consider the C^1-functional $\varphi\colon W_0^{1,p}(\Omega) \longrightarrow \mathbb{R}$ defined by

$$\varphi(u) = \|Du\|_p^p + \int_\Omega \xi(z)|u|^p\,dz \qquad \forall\, u \in W_0^{1,p}(\Omega).$$

Since $\xi \geqslant 0$ and using the Poincaré inequality (see Theorem 1.131), we see that $\varphi(u) \longrightarrow +\infty$ as $\|u\| \to +\infty$. Also, using the Rellich–Kondrachov embedding theorem (see Theorem 1.135), we see that φ is sequentially weakly lower semicontinuous. Let

$$M = \big\{u \in W_0^{1,p}(\Omega) : \|u\|_p = 1\big\}.$$

Again the Rellich–Kondrachov embedding theorem implies that the set M is w-closed. So, using Theorem 5.6, we can find $\widehat{u}_1 \in M$ such that

$$\varphi(\widehat{u}_1) = \inf_{u \in M} \varphi(u) = \widehat{\lambda}_1(\xi),$$

so

$$\varphi'(\widehat{u}_1) = \widehat{\lambda}_1(\xi)|\widehat{u}_1|^{p-2}\widehat{u}_1$$

(by the Lagrange multiplier rule; see Proposition 5.31), thus

$$A(\hat{u}_1) + \xi(z)|\hat{u}_1|^{p-2}\hat{u}_1 = \widehat{\lambda}_1(\xi)|\hat{u}_1|^{p-2}\hat{u}_1, \tag{5.299}$$

where $A\colon W_0^{1,p}(\Omega) \longrightarrow W^{-1,p'}(\Omega) = W_0^{1,p}(\Omega)^*$ (with $\frac{1}{p}+\frac{1}{p'}=1$) is defined by

$$\langle A(u), h\rangle = \int_\Omega |Du|^{p-2}\,(Du, Dh)_{\mathbb{R}^N}\,dz \qquad \forall\, u,h \in W_0^{1,p}(\Omega).$$

So, $\widehat{\lambda}_1(\xi)$ is the smallest eigenvalue of $(-\Delta_p + \xi(z)I, W_0^{1,p}(\Omega))$ and is expressed variationally by

$$\widehat{\lambda}_1(\xi) = \inf_{u \in W_0^{1,p}(\Omega)\setminus\{0\}} \frac{\|Du\|_p^p + \int_\Omega \xi(x)|u|^p\,dz}{\|u\|_p^p}. \tag{5.300}$$

From the nonlinear regularity theory (see Propositions 5.136 and 5.138) we see that every eigenfunction corresponding to $\widehat{\lambda}_1(\xi)$ belongs to $C_0^1(\overline{\Omega})$.

First we show that such an eigenfunction has constant sign (that is, it is either positive or negative). So, let $u \in W_0^{1,p}(\Omega)$ be an eigenfunction corresponding to $\widehat{\lambda}_1(\xi)$.

We have $u^+, u^- \in W_0^{1,p}(\Omega)$ (see Problem 1.152). Acting on (5.299) first with u^+ and then with u^- we obtain

$$\varphi(u^+) = \widehat{\lambda}_1(\xi)\|u^+\|_p^p \tag{5.301}$$

$$\varphi(u^-) = \widehat{\lambda}_1(\xi)\|u^-\|_p^p. \tag{5.302}$$

Suppose that u is nodal (that is, it is sign changing). Then $u^+ \neq 0$ and $u^- \neq 0$. From (5.300), (5.301), and (5.302), it follows that u^+ and u^- are eigenfunctions corresponding to $\widehat{\lambda}_1(\xi)$. Hence by the nonlinear regularity and the nonlinear maximum principle, we have that $u^+, u^- \in \operatorname{int} C_+$ (see Theorem 5.129(e) and Remark 5.133), a contradiction. So, u must be of constant sign.

Next we show that $\widehat{\lambda}_1(\xi)$ is simple. So, suppose that u, v are two eigenfunctions corresponding to $\widehat{\lambda}_1(\xi)$. Then from the previous part of the solution we have that $u, v \in \operatorname{int} C_+$. Using the Picone identity (see Theorem 5.142), we have

$$\begin{aligned} 0 &\leqslant \|Du\|_p^p - \int_\Omega |Dv|^{p-2}\left(Dv, D\left(\tfrac{u^p}{v^{p-1}}\right)\right)_{\mathbb{R}^N} dz \\ &= \|Du\|_p^p - \int_\Omega (-\Delta_p v)\tfrac{u^p}{v^{p-1}}\, dz \\ &= \|Du\|_p^p + \int_\Omega \xi(z)u^p\, dz - \widehat{\lambda}_1(\xi)\|u\|_p^p = 0 \end{aligned}$$

(by integration by parts), so

$$L(u,v)(z) = 0 \qquad \forall\, z \in \overline{\Omega}$$

(see the Picone identity; Theorem 5.142), and thus $u = \mu v$ for some $\mu > 0$ and this proves the simplicity of $\widehat{\lambda}_1(\xi)$.

Finally we show the isolation of $\widehat{\lambda}_1(\xi)$. Arguing by contradiction, suppose that we can find a sequence of eigenvalues $\{\widehat{\lambda}_n(\xi)\}_{n\geqslant 1}$ such that $\widehat{\lambda}_n(\xi) \searrow \widehat{\lambda}_1(\xi)$. Let $\{u_n\}_{n\geqslant 1} \subseteq W_0^{1,p}(\Omega)$ be a sequence of corresponding normalized eigenfunctions. Then Propositions 5.137 and 5.138 imply that we can find $\alpha \in (0,1)$ and $M > 0$ such that

$$u_n \in C_0^{1,\alpha}(\overline{\Omega}) \quad \text{and} \quad \|u_n\|_{C_0^{1,\alpha}(\overline{\Omega})} \leqslant M \qquad \forall\, n \geqslant 1.$$

Exploiting the compactness of the embedding $C_0^{1,\alpha}(\overline{\Omega}) \hookrightarrow C_0^1(\overline{\Omega})$ (by the Arzela–Ascoli theorem; see Theorem I.2.181), we may assume that

$$u_n \longrightarrow u \quad \text{in } C_0^1(\overline{\Omega}), \quad \text{with } \|u\| = 1.$$

Evidently u is an eigenvalue of $\widehat{\lambda}_1(\xi)$ and so we may assume that $u \in \operatorname{int} C_+$. Hence $u_n \in \operatorname{int} C_+$ for all $n \geqslant n_0$. But as in Problem 5.103, using the Picone identity (see

Theorem 5.142), we can show that each u_n must be nodal, a contradiction.

Solution of Problem 5.108

Arguing by contradiction, suppose that $\widehat{\lambda}_1(\xi) \leqslant 0$. Let $\widehat{u}_1 \in \operatorname{int} C_+$ (see Theorem 5.129(e)) be an eigenfunction corresponding to $\widehat{\lambda}_1(\xi)$ (see Problem 5.107). Let $\widehat{v}_1 = -\widehat{u}_1$. We have

$$-\Delta_p \widehat{v}_1(z) + \xi(z)|\widehat{v}_1(z)|^{p-2}\widehat{v}_1(z) \geqslant 0 \quad \text{for a.a. } z \in \Omega,$$

so by hypothesis, we get

$$\widehat{v}_1(z) > 0 \qquad \forall\, z \in \Omega$$

and so

$$\widehat{u}_1(z) < 0 \qquad \forall\, z \in \Omega,$$

a contradiction.

Solution of Problem 5.109

First note that the nonlinear regularity theory and the nonlinear maximum principle (see Propositions 5.136, 5.138 and Theorem 5.139) imply that $v \in \operatorname{int} C_+$. Let $\widehat{u}_1 \in \operatorname{int} C_+$ be an eigenfunction corresponding to $\widehat{\lambda}_1(\xi)$. Arguing by contradiction, suppose that $\widehat{\lambda}_1(\xi) \leqslant 0$. Using Theorem 5.142, we have

$$\begin{aligned} 0 \;&\leqslant\; \int\limits_\Omega R(\widehat{u}_1, v)\,dz \;=\; \|D\widehat{u}_1\|_p^p - \int\limits_\Omega (-\Delta_p v)\frac{\widehat{u}_1^p}{v^{p-1}}\,dz \\ &<\; \|D\widehat{u}_1\|_p^p + \int\limits_\Omega \xi(z)\widehat{u}_1^p\,dz \;=\; \widehat{\lambda}_1(\xi)\|\widehat{u}_1\|_p^p \;\leqslant\; 0 \end{aligned}$$

(by integration by parts and using the facts that $g \geqslant 0$, $g \neq 0$ and $\frac{\widehat{u}_1^p}{v^{p-1}}(z) > 0$ for all $z \in \Omega$), a contradiction. So, we must have that $\widehat{\lambda}_1(\xi) > 0$.

Solution of Problem 5.110

From the nonlinear regularity theory (see Propositions 5.137 and 5.138), we know that $u \in C_0^1(\overline{\Omega})$. From the equation in the Dirichlet problem, for all $h \in W_0^{1,p}(\Omega)$, we have

$$\int\limits_\Omega |Du|^{p-2}\,(Du, Dh)_{\mathbb{R}^N}\,dz + \int\limits_\Omega \xi(z)|u|^{p-2}uh\,dz \;=\; \int\limits_\Omega g(z)h\,dz.$$

Choosing $h = -u^- \in W_0^{1,p}(\Omega)$, we obtain

$$\|Du\|_p^p + \int_\Omega \xi(z)|u^-|^p\,dz \leqslant 0$$

(recall that $g \geqslant 0$), so $\widehat{\lambda}_1(\xi) \leqslant 0$ (see Problem 5.107), a contradiction, unless $u^- = 0$. Therefore $u \geqslant 0$, hence $u \in C_+$.

Solution of Problem 5.111

Consider the functional $\varphi\colon W_0^{1,p}(\Omega) \longrightarrow \mathbb{R}$ defined by

$$\varphi(u) = \tfrac{1}{p}\|Du\|_p^p + \tfrac{1}{p}\int_\Omega \xi(z)|u|^p\,dz - \int_\Omega g(z)u\,dz \qquad \forall\, u \in W_0^{1,p}(\Omega).$$

We claim that $\varphi(u) \longrightarrow +\infty$ when $\|u\| \longrightarrow +\infty$. Arguing by contradiction suppose that we can find a sequence $\{u_n\}_{n\geqslant 1} \subseteq W_0^{1,p}(\Omega)$ and $M > 0$ such that

$$\|u_n\| \longrightarrow +\infty \quad \text{and} \quad \varphi(u_n) \leqslant M \qquad \forall\, n \geqslant 1. \tag{5.303}$$

Let $y_n = \frac{u_n}{\|u_n\|}$ for all $n \geqslant 1$. Then $\|y_n\| = 1$ for all $n \geqslant 1$. So, we may assume that

$$y_n \xrightarrow{w} y \quad \text{in } W_0^{1,p}(\Omega) \quad \text{and} \quad y_n \longrightarrow y \quad \text{in } L^p(\Omega) \tag{5.304}$$

(see the Rellich–Kondrachov embedding theorem; Theorem 1.135). From the inequality in (5.303), we have

$$\begin{aligned} &\tfrac{1}{p}\|Dy_n\|_p^p + \tfrac{1}{p}\int_\Omega \xi(z)|y_n|^p\,dz \\ &\qquad - \frac{1}{\|u_n\|^{p-1}}\int_\Omega g(z)y_n\,dz \leqslant \frac{M}{\|u_n\|^p} \qquad \forall\, n \geqslant 1. \end{aligned} \tag{5.305}$$

Passing to the limit in (5.305) and using (5.304), (5.303) and the fact that $p > 1$, we obtain

$$\|Dy\|_p^p + \int_\Omega \xi(z)|y|^p\,dz \leqslant 0,$$

so

$$\widehat{\lambda}_1(\xi)\|y\|_p^p \leqslant 0$$

(see Problem 5.107), thus $y = 0$ (recall that by hypothesis $\widehat{\lambda}_1(\xi) > 0$). Then from (5.304) and (5.305) we infer that

$$\|Dy_n\|_p \longrightarrow 0,$$

so

$$y_n \longrightarrow 0 \quad \text{in } W_0^{1,p}(\Omega),$$

a contradiction to the fact that $\|y_n\| = 1$ for all $n \geqslant 1$. So, we have

$$\varphi(u) \longrightarrow +\infty \quad \text{as } \|u\| \to +\infty. \tag{5.306}$$

Also using the Rellich–Kondrachov embedding theorem (see Theorem 1.135), we see that φ is sequentially weakly lower semicontinuous. This fact and (5.306) permit the use of Theorem 5.6. So, we can find $u_0 \in W_0^{1,p}(\Omega)$ such that

$$\varphi(u_0) = \inf_{u \in W_0^{1,p}(\Omega)} \varphi(u),$$

hence $\varphi'(u_0) = 0$ and thus

$$A(u_0) + \xi(z)|u_0|^{p-2}u_0 = g \quad \text{in } W^{-1,p'}(\Omega) = W_0^{1,p}(\Omega)^*$$

(with $\frac{1}{p} + \frac{1}{p'} = 1$), where $A\colon W_0^{1,p}(\Omega) \longrightarrow W^{-1,p'}(\Omega) = W_0^{1,p}(\Omega)^*$ is the nonlinear map defined by

$$\langle A(u), h \rangle = \int_\Omega |Du|^{p-2} (Du, Dh)_{\mathbb{R}^N}\, dz \qquad \forall\, u, h \in W_0^{1,p}(\Omega).$$

Hence u_0 is a solution of the Dirichlet problem.

From Problem 5.110 we have that $u_0 \in C_+$.

Next we show the uniqueness of u_0. First suppose that $g \neq 0$. Then any nonnegative solution of the Dirichlet problem is nontrivial and by the nonlinear maximum principle (see Theorem 5.139) must be in $\operatorname{int} C_+$. So, if u_0, v are nonnegative solutions of the Dirichlet problem, then $u_0, v \in \operatorname{int} C_+$ and using the Picone identity (see Theorem 5.142), we have

$$0 \leqslant \|Du_0\|_p^p + \int_\Omega \xi(z)|u_0|^p\, dz - \int_\Omega g(z)\frac{u_0^p}{v^{p-1}}\, dz,$$

$$0 \leqslant \|Dv\|_p^p + \int_\Omega \xi(z)|v|^p\, dz - \int_\Omega g(z)\frac{v^p}{u_0^{p-1}}\, dz,$$

so

$$0 \leqslant \int_\Omega g(z)u_0\frac{v^{p-1}-u_0^{p-1}}{v^{p-1}}\, dz,$$

$$0 \leqslant \int_\Omega g(z)v\frac{u_0^{p-1}-v^{p-1}}{u_0^{p-1}}\, dz.$$

Adding the last two inequalities, we obtain

$$0 \leqslant \int_\Omega g(z)(v^{p-1} - u_0^{p-1})\frac{u_0^{p-1}-v^{p-1}}{u_0^{p-1}v^{p-1}},$$

so $u_0 = v$.

If $g \equiv 0$, then from variational characterization of $\widehat{\lambda}_1$ (see Problem 5.107) and since by hypothesis $\widehat{\lambda}_1(z) > 0$, $u = 0$ is the only solution.

Solution of Problem 5.112

We know that $\min\{u_0, \overline{u}\} \in W^{1,p}(\Omega)$ (see Problem 1.153). Also we have

$$u_0 - \min\{u_0, \overline{u}\} \in W_0^{1,p}(\Omega) \quad \text{and} \quad u_0 - \min\{u_0, \overline{u}\} \geqslant 0.$$

So, we have

$$\begin{aligned} 0 &\leqslant \int_\Omega \left(|D\overline{u}|^{p-2} D\overline{u} - |Du_0|^{p-2} Du_0, Du_0 - D\min\{u_0, \overline{u}\} \right)_{\mathbb{R}^N} dz \\ &= \int_\Omega \left(|D\min\{u_0, \overline{u}\}|^{p-2} D\min\{u_0, \overline{u}\} - |Du_0|^{p-2} Du_0, Du_0 \right. \\ &\qquad \left. - D\min\{u_0, \overline{u}\} \right)_{\mathbb{R}^N} dz \\ &\leqslant 0, \end{aligned}$$

so $u_0 = \min\{u_0, \overline{u}\}$, hence $u_0(z) \leqslant \overline{u}(z)$ for almost all $z \in \Omega$.

Solution of Problem 5.113

Consider the C^1-functional $\varphi \colon W_0^{1,p}(\Omega) \longrightarrow \mathbb{R}$ defined by

$$\varphi(u) = \tfrac{1}{p}\|Du\|_p^p + \int_\Omega F(z, u(z))\, dz - \langle g, u \rangle \qquad \forall\, u \in W_0^{1,p}(\Omega),$$

with $F(z,x) = \int\limits_0^x f(z,s)\, ds$. Note that

$$|F(z,x)| \leqslant c(1 + |x|^p) \quad \text{for a.a. } z \in \Omega, \text{ all } x \in \mathbb{R}, \tag{5.307}$$

with $c > 0$. Then using the Rellich–Kondrachov embedding theorem (see Theorem 1.135) and the Lebesgue dominated convergence theorem (see Theorem I.3.94), we see that the map $u \longmapsto \int\limits_\Omega F(z,u)\, dz$ is sequentially continuous on $W_0^{1,p}(\Omega)$ with the weak topology. So, φ is sequentially weakly lower semicontinuous. We will show that

$$\varphi(u) \longrightarrow +\infty \quad \text{as } \|u\| \to +\infty. \tag{5.308}$$

Arguing by contradiction, suppose that (5.308) is not true. Then we can find a sequence $\{u_n\}_{n\geqslant 1} \subseteq W_0^{1,p}(\Omega)$ and $M>0$ such that

$$\|u_n\| \longrightarrow +\infty \quad \text{and} \quad \varphi(u_n) \leqslant M \qquad \forall\, n \geqslant 1. \tag{5.309}$$

We set $y_n = \frac{u_n}{\|u_n\|}$ for all $n \geqslant 1$ and so we may assume that

$$y_n \xrightarrow{w} y \quad \text{in } W_0^{1,p}(\Omega) \quad \text{and} \quad y_n \longrightarrow y \quad \text{in } L^p(\Omega) \tag{5.310}$$

(see Theorem 1.135). From the inequality in (5.309), we have

$$\tfrac{1}{p}\|Du\|_p^p + \int_\Omega \tfrac{F(z,u_n)}{\|u_n\|^p}\,dz - \tfrac{1}{\|u_n\|^{p-1}}\langle g, y_n\rangle \leqslant \tfrac{M}{\|u_n\|^p} \qquad \forall\, n \geqslant 1.$$

Passing to the limit as $n \to +\infty$ and using (5.310), we have

$$\tfrac{1}{p}\|Dy\|_p^p + \liminf_{n\to+\infty} \int_\Omega \tfrac{F(z,u_n)}{\|u_n\|^p}\,dz \leqslant 0. \tag{5.311}$$

Note that from the hypothesis on the function $f(z,x)$, we have

$$\xi_\pm(z) \leqslant \liminf_{x\to\pm\infty} \tfrac{pF(z,x)}{|x|^p} \tag{5.312}$$

uniformly for almost all $z\in\Omega$. Using (5.312) and the Fatou lemma (see Theorem I.3.95) as in the solution of Problem 5.68, we have

$$\liminf_{n\to+\infty} \int_\Omega \tfrac{F(z,u_n)}{\|u_n\|^p}\,dz \geqslant \tfrac{1}{p}\int_\Omega \xi_+(z)(y^+)^p\,dz + \tfrac{1}{p}\int_\Omega \xi_-(z)(y^-)^p\,dz.$$

Using this and the variational characterization of $\widehat{\lambda}_1$ (see Problem 5.107) in (5.311) we obtain

$$\widehat{\lambda}_1(\xi_+)\|y^+\|_p^p + \widehat{\lambda}_1(\xi_-)\|y^-\|_p^p \leqslant 0,$$

so $y=0$ (recall that by hypothesis $\widehat{\lambda}_1(\xi_\pm)>0$), thus

$$y_n \longrightarrow 0 \quad \text{in } W_0^{1,p}(\Omega)$$

(see (5.311)), a contradiction to the fact that $\|y_n\|=1$ for all $n\geqslant 1$. This shows that (5.308) holds. Therefore, we can use Theorem 5.6 and find $u_0 \in W_0^{1,p}(\Omega)$ such that

$$\varphi(u_0) = \inf_{u\in W_0^{1,p}(\Omega)} \varphi(u),$$

so $\varphi'(u_0)=0$ and thus

$$A(u_0) + N_f(u_0) = g \quad \text{in } W^{-1,p'}(\Omega) = W^{1,p}(\Omega)^* \tag{5.313}$$

(with $\frac{1}{p}+\frac{1}{p'}=1$). Recall that $A\colon W_0^{1,p}(\Omega) \longrightarrow W^{1,p'}(\Omega)$ is defined by

$$\langle A(u),h\rangle \;=\; \int\limits_\Omega |Du|^{p-2}\,(Du,Dh)_{\mathbb{R}^N}\,dz \qquad \forall\, u,h\in W_0^{1,p}(\Omega)$$

and $N_f\colon L^p(\Omega)\longrightarrow L^{p'}(\Omega)$ is defined by

$$N_f(u)(\cdot) \;=\; f(\cdot,u(\cdot)) \qquad \forall\, u\in L^p(\Omega).$$

From (5.313) we conclude that u_0 is a weak solution of the problem.

Solution of Problem 5.114

"$\Longrightarrow$": Assume that the Dirichlet problem has a solution u. Let $\widehat{u}_1$ be the positive eigenfunction corresponding to $\widehat{\lambda}_1>0$. We know that $\widehat{u}_1\in C^\infty(\Omega)$ and $\widehat{u}_1(z)>0$ for all $z\in\Omega$. From the equality in the Dirichlet problem, we have

$$\int\limits_\Omega (-\Delta u)\widehat{u}_1\,dz \;=\; \lambda\int\limits_\Omega u\widehat{u}_1\,dz+\int\limits_\Omega u^{p-1}\widehat{u}_1\,dz,$$

so

$$\int\limits_\Omega u(-\Delta\widehat{u}_1)\,dz \;>\; \lambda\int\limits_\Omega u\widehat{u}_1\,dz,$$

thus

$$\widehat{\lambda}_1\int\limits_\Omega u\widehat{u}_1\,dz \;>\; \lambda\int\limits_\Omega u\widehat{u}_1\,dz$$

and hence $\lambda<\widehat{\lambda}_1$ (since $\int\limits_\Omega u\widehat{u}_1\,dz>0$).

"$\Longleftarrow$": Suppose that $\lambda<\widehat{\lambda}_1$. Consider the C^1-functional $\varphi_\lambda\colon H_0^1(\Omega)\longrightarrow\mathbb{R}$ defined by

$$\varphi_\lambda(u) \;=\; \tfrac{1}{2}\|Du\|_2^2-\tfrac{\lambda}{2}\|u\|_2^2-\tfrac{1}{p}\|u^+\|_2^2 \qquad \forall\, u\in H_0^1(\Omega).$$

We show that φ_λ satisfies the PS-condition (see Definition 5.45(a)). So, we consider a sequence $\{u_n\}_{n\geqslant1}\subseteq H_0^1(\Omega)$ such that

$$|\varphi_\lambda(u_n)| \;\leqslant\; M \qquad \forall\, n\geqslant 1, \tag{5.314}$$

for some $M>0$ and

$$\varphi_\lambda'(u_n) \;\longrightarrow\; 0 \quad \text{in } H^{-1}(\Omega)=H_0^1(\Omega)^*. \tag{5.315}$$

From (5.314) and (5.315) for $n \geqslant 1$ big, we have

$$\varphi_\lambda(u_n) - \tfrac{1}{p}\langle \varphi_\lambda'(u_n), u_n\rangle \leqslant M + \|u_n\|,$$

so

$$\left(\tfrac{1}{2} - \tfrac{1}{p}\right)\left(\|Du_n\|_2^2 - \lambda\|u_n\|_2^2\right) \leqslant M + \|u_n\|,$$

thus

$$\left(\tfrac{1}{2} - \tfrac{1}{p}\right)\left(1 - \tfrac{\lambda}{\widehat{\lambda}_1}\right)\|Du_n\|_2^2 \leqslant M + \|u_n\|. \tag{5.316}$$

Since $p > 2$ and $\lambda < \widehat{\lambda}_1$, from (5.316) and the Poincaré inequality (see Theorem 1.131), we infer that the sequence $\{u_n\}_{n\geqslant 1} \subseteq H_0^1(\Omega)$ is bounded. Then Problem 5.32 implies that φ satisfies the PS-condition.

We have

$$\varphi_\lambda(u) \geqslant \tfrac{1}{2}\left(1 - \tfrac{\lambda}{\widehat{\lambda}_1}\right)\|Du\|_2^2 - c\|Du\|_2^p$$

for some $c > 0$ (see Theorems 1.131 and 1.135). Since $p > 2$ and recalling that $\|Du\|_2$ is an equivalent norm on $H_0^1(\Omega)$ (see Theorem 1.131), we can find $\varrho \in (0,1)$ small such that

$$\varphi_\lambda(0) = 0 < \inf\{\varphi_\lambda(u) : \|u\| = \varrho\} = m_\varphi. \tag{5.317}$$

Since $p > 2$, we see that

$$\varphi_\lambda(t\widehat{u}_1) \longrightarrow -\infty, \quad \text{as } t \to +\infty. \tag{5.318}$$

Because of (5.317) and (5.318) and recalling that φ_λ satisfies the PS-condition, we see that we can apply the mountain pass theorem (see Theorem 5.63) and obtain $u_0 \in H_0^1(\Omega)$ such that

$$\varphi_\lambda(0) = 0 < m_\varrho \leqslant \varphi_\lambda(u_0) \tag{5.319}$$

and

$$\varphi_\lambda'(u_0) = 0. \tag{5.320}$$

From (5.319), we see that $u_0 \neq 0$. From (5.320), we have

$$A(u_0) = \lambda u_0 + (u_0^+)^{p-1}, \tag{5.321}$$

with $A \in \mathcal{L}(H_0^1(\Omega); H^{-1}(\Omega))$ defined by

$$\langle A(u), h\rangle = \int_\Omega (Du, Dh)_{\mathbb{R}^N}\, dz \qquad \forall\, u, h \in H_0^1(\Omega).$$

On (5.321) we act with $-u_0^- \in H_0^1(\Omega)$. Then

$$\|Du_0^-\|_2^2 = \lambda\|u_0^-\|_2^2.$$

If $u_0^- \neq 0$, then

$$\widehat{\lambda}_1 \leqslant \lambda < \widehat{\lambda}_1,$$

a contradiction. So, $u_0 \geqslant 0$, $u_0 \neq 0$ and from (5.321), we see that u_0 is a weak solution of the considered Dirichlet problem.

Solution of Problem 5.115

We have

$$-\Delta(u_1 - u_2)(z) = f(z, u_1(z)) - f(z, u_2(z)) \quad \text{in } \Omega, \quad (u_1 - u_2)|_{\partial\Omega} = 0. \quad (5.322)$$

We define

$$m(z) = \begin{cases} \frac{f(z,u_1(z))-f(z,u_2(z))}{u_1(z)-u_2(z)} & \text{if} \quad u_1(z) \neq u_2(z), \\ f'_x(z, u_1(z)) & \text{if} \quad u_1(z) = u_2(z). \end{cases}$$

Since the solutions are in $L^\infty(\Omega)$, from the mean value theorem and hypothesis **(i)**, we see that $m \in L^\infty(\Omega)$. Let $v(z) = u_1(z) - u_2(z)$. Then from (5.322), we have

$$-\Delta v(z) = m(z)v(z) \quad \text{in } \Omega, \quad v|_{\partial\Omega} = 0. \quad (5.323)$$

If $m(z) \leqslant 0$ for almost all $z \in \Omega$, then

$$\|Dv\|_2^2 = \int_\Omega m(z)v^2\,dz \leqslant 0,$$

so $v = 0$ and thus $u_1 = u_2$.

So, suppose that $m^+ \neq 0$. By hypothesis **(ii)**, we have $m(z) \leqslant \widehat{\lambda}_2$ for almost all $z \in \Omega$ and the inequality is strict on a set of positive measure. If $v \neq 0$, then from (5.323) and Proposition 5.131, we have

$$1 = \widehat{\lambda}_k(m) > \widehat{\lambda}_k(\widehat{\lambda}_2) = \frac{\widehat{\lambda}_k}{\widehat{\lambda}_2},$$

for some $k \geqslant 1$, thus $k = 1$. So, v is the eigenfunction corresponding to the first eigenvalue $\widehat{\lambda}_1(m)$. Hence v has constant sign (see Theorem 5.129) and thus

$$v = u_1 - u_2 \geqslant 0 \quad \text{or} \quad v = u_1 - u_2 \leqslant 0.$$

Solution of Problem 5.116

Let $\{u_n\}_{n\geqslant 1} \subseteq H_0^1(\Omega)$ be a sequence such that the sequence $\{\varphi(u_n)\}_{n\geqslant 1} \subseteq \mathbb{R}$ is bounded and

$$\varphi'(u_n) \longrightarrow 0 \quad \text{in } H^{-1}(\Omega) = H_0^1(\Omega)^*.$$

Suppose that $\|u_n\| \longrightarrow +\infty$ and let $y_n = \frac{u_n}{\|u_n\|}$ for all $n \geqslant 1$. Then $\|y_n\| = 1$ for all $n \geqslant 1$ and so, passing to a subsequence if necessary, we may assume that

$$y_n \xrightarrow{w} y \quad \text{in } H_0^1(\Omega) \quad \text{and} \quad y_n \longrightarrow y \quad \text{in } L^2(\Omega). \tag{5.324}$$

We have

$$\left|\langle\varphi'(u_n), h\rangle\right| \leqslant \varepsilon_n\|h\| \qquad \forall\, h \in H_0^1(\Omega),$$

with $\varepsilon_n \searrow 0$, so

$$\left|\, \langle A(u_n), h\rangle - \int_\Omega f(z, u_n)h\, dz\right| \leqslant \varepsilon_n\|h\| \qquad \forall\, h \in H_0^1(\Omega),\ n \geqslant 1,$$

with $A \in \mathcal{L}(H_0^1(\Omega); H^{-1}(\Omega))$ defined by

$$\langle A(u), h\rangle = \int_\Omega (Du, Dh)_{\mathbb{R}^N}\, dz \qquad \forall\, u, h \in H_0^1(\Omega).$$

Multiplying with $\frac{1}{\|u_n\|}$, we obtain

$$\left|\, \langle A(y_n), h\rangle - \int_\Omega \frac{f(z,u_n)}{\|u_n\|} h\, dz\right| \leqslant \frac{\varepsilon_n}{\|u_n\|}\|h\| \qquad \forall\, h \in H_0^1(\Omega),\ n \geqslant 1. \tag{5.325}$$

From hypothesis **(i)** and **(ii)**, we have that

$$|f(z, x)| \leqslant c(1 + |x|) \quad \text{for a.a. } z \in \Omega, \text{ all } x \in \mathbb{R},$$

with $c > 0$. So, reasoning as in the solution of Problem 5.68 and using hypothesis **(ii)**, we have

$$\frac{N_f(u_n)}{\|u_n\|} \xrightarrow{w} \xi_+(z)y^+ - \xi_-(z)y^- \quad \text{in } L^2(\Omega) \tag{5.326}$$

(where $N_f(u_n)(\cdot) = f(\cdot, u_n(\cdot))$). So, if in (5.325) we pass to the limit as $n \to +\infty$ and use (5.324) and (5.326), then

$$\langle A(y), h\rangle = \int_\Omega (\xi_+(z)y^+ - \xi_-(z)y^-)h\, dz \qquad \forall\, h \in H_0^1(\Omega),$$

so

$$A(y) = \xi_+(z)y^+ - \xi_-(z)y^-,$$

thus

$$\begin{cases} -\Delta y(z) = \xi_+(z)y(z)^+ - \xi_-(z)y(z)^- & \text{in } \Omega, \\ y|_{\partial\Omega} = 0, \end{cases}$$

hence $y = 0$ (by hypothesis). Then from (5.325) with $h = y_n \in H_0^1(\Omega)$, we have

$$\|Dy_n\|_2^2 \longrightarrow 0,$$

so $y_n \longrightarrow 0$ in $H_0^1(\Omega)$, which contradicts the fact that $\|y_n\| = 1$ for all $n \geqslant 1$. This shows that the sequence $\{u_n\}_{n\geqslant 1} \subseteq H_0^1(\Omega)$ is bounded. Thus φ satisfies the

PS-condition (see Problem 5.32).

Solution of Problem 5.117

(a) Let

$$\psi_1(u) = \|Du\|_2^2 - \int_\Omega \eta(z)u^2\,dz \qquad \forall\, u \in H_0^1(\Omega).$$

From Theorem 5.129, we know that $\psi_1|_{\widehat{H}_m} \geqslant 0$. Arguing by contradiction, suppose that the claim of the problem is not true. Exploiting the 2-homogeneity of ψ_1, we can find a sequence $\{u_n\}_{n\geqslant 1} \subseteq \widehat{H}_m$ such that

$$\|u_n\| = 1 \quad \text{and} \quad \psi_1(u_n) < \tfrac{1}{n} \qquad \forall\, n \geqslant 1. \tag{5.327}$$

Passing to a subsequence if necessary, we may assume that

$$u_n \xrightarrow{w} u \in \widehat{H}_m \quad \text{in } H_0^1(\Omega) \quad \text{and} \quad u_n \longrightarrow u \quad \text{in } L^2(\Omega). \tag{5.328}$$

So, if in the inequality in (5.327) we pass to the limit as $n \to +\infty$ and use (5.328), then

$$\|Du\|_2^2 \leqslant \int_\Omega \eta(z)u^2\,dz \leqslant \widehat{\lambda}_m\|u\|_2^2, \tag{5.329}$$

so

$$\|Du\|_2^2 = \widehat{\lambda}_m\|u\|_2^2,$$

(since $u \in \widehat{H}_m$), thus $u \in E(\widehat{\lambda}_m)$.
If $u = 0$, then $\|Du\|_2 \longrightarrow 0$, so $u_n \longrightarrow 0$ in $H_0^1(\Omega)$, a contradiction to the fact that $\|u_n\| = 1$ for all $n \geqslant 1$. So, $u \neq 0$. From Theorem 5.129(f), we know that $E(\widehat{\lambda}_m)$ has the unique continuation property, hence $u(z) \neq 0$ for almost all $z \in \Omega$. Then from the first inequality in (5.329) and the hypothesis on the function η, we have

$$\|Du\|_2^2 \leqslant \int_\Omega \eta(z)u^2\,dz < \widehat{\lambda}_m\|u\|_2^2,$$

which contradicts the fact that $u \in E(\widehat{\lambda}_m)$. This proves part **(a)**.

(b) In this case we consider the functional

$$\psi_2(u) = \|Du\|_2^2 - \int_\Omega \vartheta(z)u^2\,dz \qquad \forall\, u \in H_0^1(\Omega).$$

We have $\psi_2|_{\overline{H}_m} \leqslant 0$ (see Theorem 5.129). Again we argue by contradiction. So, because of the 2-homogeneity of ψ_2, we can find a sequence $\{u_n\}_{n\geqslant 1} \subseteq \overline{H}_m$ such that

$$\|u_n\| = 1 \quad \text{and} \quad \psi_2(u_n) > -\tfrac{1}{n} \qquad \forall\, n \geqslant 1.$$

Due to the finite dimensionality of $\overline{H}_m$, we may assume that

$$u_n \longrightarrow u \in \overline{H}_m \quad \text{in } H_0^1(\Omega), \quad \|u\| = 1, \tag{5.330}$$

so $\psi_2(u) \geqslant 0$, thus

$$\|Du\|_2^2 \geqslant \int_\Omega \vartheta(z)u^2\,dz \geqslant \widehat{\lambda}_m\|u\|_2^2 \tag{5.331}$$

and hence $u \in E(\widehat{\lambda}_m)$ (see Theorem 5.129). As before since $E(\widehat{\lambda}_m)$ has the unique continuation property and using (5.331) and the hypothesis on the function ϑ, we obtain

$$\|Du\|_2^2 > \widehat{\lambda}_m\|u\|_2^2$$

(see (5.330)), a contradiction. This proves part **(b)**.

Solution of Problem 5.118

Let $\varphi\colon H_0^1(\Omega) \longrightarrow \mathbb{R}$ be the C^1-functional defined by

$$\varphi(u) = \tfrac{1}{2}\|Du\|_2^2 - \int_\Omega F(z,u)\,dz \qquad \forall\, u \in H_0^1(\Omega).$$

Evidently the critical points of φ are the weak solutions of the considered Dirichlet problem. From the solution of Problem 5.68, we know that hypotheses **(i)**, **(ii)** and **(iii)** imply that φ satisfies

$$\text{if } \|u\| \longrightarrow +\infty, \text{ then } \varphi(u) \longrightarrow +\infty.$$

So, φ is bounded below and satisfies the PS-condition (see Definition 5.45(a)).

Let

$$\overline{H}_m = \bigoplus_{k=1}^m E(\widehat{\lambda}_k) \quad \text{and} \quad \widehat{H}_{m+1} = \overline{H}_m^\perp = \overline{\bigoplus_{k\geqslant m+1} E(\widehat{\lambda}_k)}.$$

We have $H_0^1(\Omega) = \overline{H}_m \oplus \widehat{H}_{m+1}$. Hypotheses **(i)** and **(iv)** imply that given $\varepsilon > 0$ and $r > 2$, we can find $c_1 = c_1(\varepsilon, r) > 0$ such that

$$F(z,x) \leqslant \tfrac{1}{2}(\vartheta(z)+\varepsilon)x^2 + c_1|x|^r \quad \text{for a.a. } z \in \Omega, \text{ all } v \in \mathbb{R}. \tag{5.332}$$

Then for all $u \in \widehat{H}_{m+1}$, we have

$$\begin{aligned}
\varphi(u) &\geqslant \tfrac{1}{2}\|Du\|_2^2 - \tfrac{1}{2}\int_\Omega \vartheta(z)|u|^2\,dz - \tfrac{\varepsilon}{2}\|u\|^2 - c_2\|u\|^r \\
&\geqslant \tfrac{1}{2}(c_3 - \varepsilon)\|u\|^2 - c_2\|u\|^r
\end{aligned}$$

for some $c_2, c_3 > 0$ (see Problem 5.117). Choosing $\varepsilon \in (0, c_3)$, we have

$$\varphi(u) \geqslant c_4\|u\|^2 - c_2\|u\|^r \qquad \forall\, u \in \widehat{H}_{m+1},$$

for some $c_4 > 0$. Since $r > 2$, we can find $\delta_1 \in (0,1)$ small such that

$$\varphi(u) \geqslant 0 \qquad \forall\, u \in \widehat{H}_{m+1},\ \|u\| \leqslant \delta_1. \tag{5.333}$$

The subspace $\overline{H}_m$ is finite dimensional and so all norms are equivalent. Since $\overline{H}_m \subseteq L^\infty(\Omega)$, we can find $\delta_2 > 0$ such that

$$|u(z)| \leqslant \delta_0 \quad \text{for a.a. } z \in \Omega, \text{ all } u \in \overline{H}_m,\ \|u\| \leqslant \delta_2.$$

Then using hypothesis **(iv)**, we have that

$$\varphi(u) \leqslant \tfrac{1}{2}\|Du\|_2^2 - \tfrac{\lambda_m}{2}\|u\|_2^2 \leqslant 0 \qquad \forall\, u \in \overline{H}_m,\ \|u\| \leqslant \delta_2 \tag{5.334}$$

(see Theorem 5.129). From (5.333) and (5.334), it follows that φ has a local linking at 0. If $\inf\limits_{H_0^1(\Omega)} \varphi = 0$, then all the elements $u \in \overline{H}_m$, $\|u\| \leqslant \delta_2$ are critical points of φ, hence weak solutions of the Dirichlet problem. If $\inf\limits_{H_0^1(\Omega)} \varphi < 0$, then we can apply Theorem 5.79 and infer that φ has at least two nontrivial critical points $u_0, \widehat{u} \in H_0^1(\Omega)$. These are nontrivial weak solutions of the considered Dirichlet problem.

Solution of Problem 5.119

From Proposition 5.132, we see that $\widehat{\lambda}_1 \leqslant \widehat{\lambda}_V$. Suppose that $\widehat{\lambda}_1 = \widehat{\lambda}_V$. Then we can find a sequence $\{u_n\}_{n\geqslant 1} \subseteq V$ such that

$$\|u_n\|_p = 1 \qquad \forall\, n \geqslant 1 \quad \text{and} \quad \|Du_n\|_p^p \longrightarrow \widehat{\lambda}_1 = \widehat{\lambda}_V. \tag{5.335}$$

Evidently the sequence $\{u_n\}_{n\geqslant 1} \subseteq W_0^{1,p}(\Omega)$ is bounded and so we may assume that

$$u_n \xrightarrow{w} u \in V \quad \text{in } W_0^{1,p}(\Omega) \quad \text{and} \quad u_n \longrightarrow u \quad \text{in } L^p(\Omega). \tag{5.336}$$

We have

$$\widehat{\lambda}_1 \leqslant \|Du\|_p^p \leqslant \liminf_{n\to+\infty} \|Du_n\|_p^p = \widehat{\lambda}_V = \widehat{\lambda}_1$$

(see (5.336) and (5.335)), so

$$\|Du\|_p^p = \widehat{\lambda}_1,$$

thus $u = \pm\widehat{u}_1$ (see Remark 5.133 and recall that $\|u\|_p = 1$; see (5.335) and (5.336)). But $u \in V$, a contradiction. This proves that $\widehat{\lambda}_1 < \widehat{\lambda}_V$.

Next we show that $\widehat{\lambda}_V \leqslant \widehat{\lambda}_2$. Arguing by contradiction, suppose that $\widehat{\lambda}_2 < \widehat{\lambda}_V$. According to Proposition 5.136, we can find $\widehat{\gamma} \in \widehat{\Gamma}$ such that

$$\|D\widehat{\gamma}(t)\|_p^p < \widehat{\lambda}_V \qquad \forall\, t \in [0,1]. \tag{5.337}$$

Since $\widehat{\gamma} \in \widehat{\Gamma}$, we have $\widehat{\gamma}(-1) = -\widehat{u}_1$ and $\widehat{\gamma}(1) = \widehat{u}_1$. Consider the function $\xi\colon [-1,1] \longrightarrow \mathbb{R}$ defined by

$$\xi(t) = \int_\Omega \widehat{\gamma}(t)(z)\widehat{u}_1(z)^{p-1}\,dz \qquad \forall\, t \in [-1,1].$$

Clearly ξ is continuous and $\xi(-1) < 0 < \xi(1)$. So by the Bolzano theorem, we can find $t_0 \in (-1,1)$ such that $\xi(t_0) = 0$. We have

$$\int_\Omega \widehat{\gamma}(t_0)\widehat{u}_1^{p-1}\,dz = 0,$$

so $\widehat{\gamma}(t_0) \in V$ and thus $\|D\widehat{\gamma}_0(t_0)\| \geqslant \widehat{\lambda}_V$, a contradiction (see (5.337)).

Solution of Problem 5.120

Let $C = \{u \in C^1([0,1]) : u(0) = 1,\ u(1) = 0\}$ and

$$\varphi(u) = \int_0^1 tu'(t)^2\,dt \qquad \forall\, u \in W^{1,2}(0,1).$$

Then $m = \inf\limits_C \varphi$. Evidently $m \geqslant 0$. Suppose for the moment that $m = 0$ and that the infimum is realized by $u \in C$. Then $\varphi(u) = 0$, so

$$u'(t) = 0 \qquad \forall\, t \in [0,1]$$

and thus $u \equiv 0$ (since $u(1) = 0$), a contradiction, since $u \in C$. So, if $m = 0$, then the infimum cannot be realized.

Therefore, we need to show that $m = 0$. Define

$$C_* = \{u \in W^{1,2}(0,1) : u(1) = 0,\ u(0) = 1\}, \quad m_* = \inf_{u \in C_*} \varphi(u).$$

For every $n \geqslant 1$, let

$$u_n(t) = \begin{cases} 1 & \text{if } t \in [0, \frac{1}{n}], \\ \frac{-\ln t}{\ln n} & \text{if } t \in (\frac{1}{n}, 1]. \end{cases}$$

Then $u_n \in C_*$ for all $n \geqslant 1$ and we have

$$\varphi(u_n) = \tfrac{1}{\ln n} \longrightarrow 0,$$

so $m_* = 0$. Given $\varepsilon > 0$ and $v \in C_*$, we can find $u \in C$ such that

$$\|u - v\|_{W^{1,2}(0,1)} \leqslant \varepsilon. \tag{5.338}$$

We have

$$\begin{aligned}|\varphi(u)-\varphi(v)| &\leqslant \int_0^1 t|u'(t)^2-v'(t)^2|\,dt \\ &\leqslant \int_0^1 t\big(|u'(t)|+|v'(t)|\big)|u'(t)-v'(t)|\,dt \\ &\leqslant \big(\|u'\|_2+\|v'\|_2\big)\|u'-v'\|_2 \leqslant c\varepsilon\end{aligned}$$

for some $c \in \mathbb{R}$ (see (5.338)), so

$$\varphi(u) \leqslant \varphi(v) + c\varepsilon.$$

Taking infimum over $u \in C$ and $v \in C_*$ and recalling that $m_* = 0$, we obtain that

$$0 \leqslant m \leqslant c\varepsilon.$$

Since $\varepsilon > 0$ is arbitrary, we let $\varepsilon \searrow 0$ to conclude that $m = 0$.

Remark. This is a variant of the so-called ***Weierstrass counterexample*** with which Weierstrass criticized the universality of the Dirichlet principle back in 1872.

Solution of Problem 5.121

(a) Let $\lambda > 1$ and recall that for almost all $z \in \Omega$, $f(z,0) = 0$ and $f(z,\cdot)$ is convex. So, we have

$$f(z,\tfrac{1}{\lambda}x) = f(z,\tfrac{1}{\lambda}x+(1-\tfrac{1}{\lambda})0) \leqslant \tfrac{1}{\lambda}f(z,x) \quad \text{for a.a. } z \in \Omega, \text{ all } x \in \mathbb{R}. \tag{5.339}$$

Let $\overline{u} = \lambda u_0$. For every $h \in H_0^1(\Omega)$ with $h \geqslant 0$, we have

$$\begin{aligned}\int_\Omega (-\Delta\overline{u})h\,dz &= \int_\Omega (D\overline{u},Dh)_{\mathbb{R}^N}\,dz = \int_\Omega \lambda\,(Du_0,Dh)_{\mathbb{R}^N}\,dz \\ &= \int_\Omega \lambda f(z,u_0)h\,dz \leqslant \int_\Omega f(z,\overline{u})h\,dz\end{aligned}$$

(since u_0 is a solution of the considered Dirichlet problem, using (5.339) with $x = \overline{u}(z) = \lambda u_0(z)$).

(b) Let $0 < \lambda < 1$. As before, we have

$$f(z, \lambda x) = f(z, \lambda x + (1-\lambda)0) \leqslant \lambda f(z, x) \quad \text{for a.a. } z \in \Omega, \text{ all } x \in \mathbb{R}. \tag{5.340}$$

Let $\underline{u} = \lambda u_0$. For every $h \in H_0^1(\Omega)$ with $h \geqslant 0$, we have

$$\begin{aligned} \int_\Omega (-\Delta \underline{u}) h \, dz &= \int_\Omega (D\underline{u}, Dh)_{\mathbb{R}^N} \, dz = \int_\Omega \lambda (Du_0, Dh)_{\mathbb{R}^N} \, dz \\ &= \int_\Omega \lambda f(z, u_0) h \, dz \geqslant \int_\Omega f(z, \underline{u}) h \, dz \end{aligned}$$

(since u_0 is a solution of the considered Dirichlet problem, using (5.340) and recalling that $\underline{u} = \lambda u_0$).

Solution of Problem 5.122

Let $u \in X$ be a critical point of φ. From the convexity of φ, we have

$$\langle \varphi'(u), v - u \rangle \leqslant \varphi(v) - \varphi(u) \qquad \forall v \in X,$$

so

$$0 \leqslant \varphi(v) - \varphi(u) \qquad \forall v \in X$$

(since $u \in K_\varphi$), thus

$$\varphi(u) \leqslant \varphi(v) \qquad \forall v \in X.$$

So, u is a global minimizer of φ.

Solution of Problem 5.123

Note that hypotheses **(i)** and **(ii)** imply that

$$|f(z, x)| \leqslant c_1(1 + |x|) \quad \text{for a.a. } z \in \Omega, \text{ all } x \in \mathbb{R}, \tag{5.341}$$

with $c_1 > 0$. Hence, if $F(z, x) = \int\limits_0^x f(z, s) \, ds$, then

$$F(z, x) \leqslant c_2(1 + |x|^2) \quad \text{for a.a. } z \in \Omega, \text{ all } x \in \mathbb{R},$$

with $c_2 > 0$ and we can have the C^1-functional $\varphi\colon H_0^1(\Omega) \longrightarrow \mathbb{R}$ defined by

$$\varphi(u) = \frac{1}{2}\|Du\|_2^2 - \int_\Omega F(z,u)\,dz - \int_\Omega g(z)u\,dz \qquad \forall\, u \in H_0^1(\Omega).$$

We show that φ satisfies the PS-condition (see Definition 5.45(a)). So, suppose that $\{u_n\}_{n\geqslant 1} \subseteq H_0^1(\Omega)$ is a sequence such that the sequence $\{\varphi(u_n)\}_{n\geqslant 1} \subseteq \mathbb{R}$ is bounded and

$$\varphi'(u_n) \longrightarrow 0 \quad \text{in } H^{-1}(\Omega) = H_0^1(\Omega)^*. \tag{5.342}$$

Suppose that the sequence $\{u_n\}_{n\geqslant 1} \subseteq H_0^1(\Omega)$ is unbounded. Then, passing to a suitable subsequence, we may assume that $\|u_n\| \longrightarrow +\infty$. We set $y_n = \frac{u_n}{\|u_n\|}$ for $n \geqslant 1$. Since $\|y_n\| = 1$ for all $n \geqslant 1$, by passing to a suitable subsequence if necessary, we can say that

$$y_n \xrightarrow{w} u \quad \text{in } H_0^1(\Omega) \quad \text{and} \quad y_n \longrightarrow y \quad \text{in } L^2(\Omega). \tag{5.343}$$

From (5.342), we have

$$\Big|\langle A(y_n), h\rangle - \int_\Omega \frac{f(z,u_n)}{\|u_n\|} h\,dz - \int_\Omega \frac{g(z)}{\|u_n\|} h\,dz\Big| \leqslant \frac{\varepsilon_n\|h\|}{\|u_n\|}. \tag{5.344}$$

Here $A \in \mathcal{L}(H_0^1(\Omega), H^{-1}(\Omega))$ is defined by

$$\langle A(u), h\rangle = \int_\Omega (Du, Dh)_{\mathbb{R}^N}\,dz \qquad \forall\, u,h \in H_0^1(\Omega)$$

From (5.341), we see that the sequence $\left\{\frac{N_f(u_n)}{\|u_n\|}\right\}_{n\geqslant 1} \subseteq L^2(\Omega)$ is bounded (recall that $N_f(u)(\cdot) = f(\cdot, u(\cdot))$ for all $u \in H_0^1(\Omega)$). As in the solution of Problem 5.68, using hypotheses **(ii)**$_1$ or **(ii)**$_2$, we can say that

$$\frac{N_f(u_n)}{\|u_n\|} \xrightarrow{w} \xi y \quad \text{in } L^2(\Omega), \tag{5.345}$$

with $\xi \in L^\infty(\Omega)$ such that

$$\eta(z) \leqslant \xi(z) \leqslant \widehat{\eta}(z) \quad \text{for a.a. } z \in \Omega \tag{5.346}$$

if hypothesis **(ii)**$_1$ holds and

$$\eta_0(z) \leqslant \xi(z) \leqslant \widehat{\eta}_0(z) \quad \text{for a.a. } z \in \Omega \tag{5.347}$$

if hypothesis **(ii)**$_2$ holds. Then, passing to the limit in (5.344) and using (5.343) and (5.345), we obtain

$$\langle A(y), h\rangle = \int_\Omega \xi(z)h\,dz \qquad \forall\, h \in H_0^1(\Omega). \tag{5.348}$$

so

$$A(y) = \xi(y) \quad \text{in } H^{-1}(\Omega) = H_0^1(\Omega)^*,$$

thus

$$\begin{cases} -\Delta y(z) = \xi(z)y(z) \text{ for a.a.} z \in \Omega, \\ u|_{\partial\Omega} = 0. \end{cases} \tag{5.349}$$

If hypothesis **(ii)**$_1$ holds, then ξ satisfies (5.346). Using Proposition 5.131, we have

$$\widehat{\lambda}_m(\widehat{\lambda}_m) = 1 > \widehat{\lambda}_m(\xi) \quad \text{and} \quad \widehat{\lambda}_{m+1}(\widehat{\lambda}_{m+1}) = 1 < \widehat{\lambda}_{m+1}(\xi). \tag{5.350}$$

From (5.349) and (5.350), it follows that $y = 0$. But then from (5.344)) with $h = y_n \in H_0^1(\Omega)$, we have

$$\|Dy_n\|_2 \longrightarrow 0,$$

so

$$y_n \longrightarrow 0 \quad \text{in } H_0^1(\Omega),$$

a contradiction, since $\|y_n\| = 1$ for all $n \geqslant 1$.

If hypothesis **(ii)**$_2$ holds, then ξ satisfies (5.347). We have

$$\|Dy\|_2^2 = \int_\Omega \xi(z)y^2\,dz$$

(use (5.348) with $h = y \in H_0^1(\Omega)$), so

$$c_0\|y\|^2 \leqslant 0$$

for some $c_0 > 0$ (see Problem 5.67), thus $y = 0$ and hence

$$y_n \longrightarrow 0 \quad \text{in } H_0^1(\Omega)$$

(as before), a contradiction.

So, we infer that the sequence $\{u_n\}_{n\geqslant 1} \subseteq H_0^1(\Omega)$ is bounded. Problem 5.32 implies that φ satisfies the PS-condition.

Suppose that hypothesis **(ii)**$_1$ holds and let

$$\overline{H}_m = \bigoplus_{k=1}^m E(\widehat{\lambda}_k), \quad \widehat{H}_{m+1} = \overline{H}_m^\perp = \overline{\bigoplus_{k\geqslant m+1} E(\widehat{\lambda}_k)}.$$

We have $H_0^1(\Omega) = \overline{H}_m \oplus \widehat{H}_{m+1}$. Hypotheses **(i)** and **(ii)**$_1$ imply that given $\varepsilon > 0$, we can find $c_3 = c_3(\varepsilon)$, $c_4 = c_4(\varepsilon) > 0$ such that, for almost all $z \in \Omega$ and all $x \in \mathbb{R}$, we have

$$\tfrac{1}{2}(\eta(z) - \varepsilon)x^2 - c_3 \leqslant F(z,x) \leqslant \tfrac{1}{2}(\widehat{\eta}(z) + \varepsilon)x^2 + c_4. \tag{5.351}$$

Then for all $u \in \overline{H}_m$, we have

$$\begin{aligned} \varphi(u) &\leqslant \tfrac{1}{2}\|Du\|_2^2 - \tfrac{1}{2}\int_\Omega \eta(z)u^2\,dz + \tfrac{\varepsilon}{2}\|u\|_2^2 + c_3|\Omega|_N + \|g\|_2\|u\|_2 \\ &\leqslant \tfrac{1}{2}(\varepsilon - c_5)\|u\|^2 + c_3|\Omega|_N + \|g\|_2\|u\|_2, \end{aligned} \tag{5.352}$$

for some $c_5 > 0$ (see (5.351) and Problem 5.117). Choosing $\varepsilon \in (0, c_5)$, from (5.352) we see that

$$\varphi(u) \longrightarrow -\infty \quad \text{as } \|u\| \to +\infty,\ u \in \overline{H}_m. \tag{5.353}$$

On the other hand, if $u \in \widehat{H}_{m+1}$, then

$$\begin{aligned} \varphi(u) &\geqslant \tfrac{1}{2}\|Du\|_2^2 - \tfrac{1}{2}\int_\Omega \widehat{\eta}(z)u^2\,dz - \tfrac{\varepsilon}{2}\|u\|_2^2 - c_4|\Omega|_N - \|g\|_2\|u\|_2 \\ &\geqslant \tfrac{1}{2}(c_6 - \varepsilon)\|u\|^2 + c_4|\Omega|_N - \|g\|_2\|u\|_2, \end{aligned} \tag{5.354}$$

for some $c_6 > 0$ (see Problem 5.117). Choosing $\varepsilon \in (0, c_6)$, from (5.354) we see that

$$\varphi(u) \longrightarrow +\infty \quad \text{as } \|u\| \to +\infty,\ u \in \widehat{H}_{m+1}. \tag{5.355}$$

From (5.353) and (5.355), we see that we can find $r > 0$ such that

$$\max_{u \in \partial B_r \cap \overline{H}_m} \varphi(u) < \inf_{u \in \widehat{H}_{m+1}} \varphi(u) = \mu_m. \tag{5.356}$$

Then (5.356) permits the use of the saddle point theorem (see Theorem 5.64). So, we can find $u_0 \in K_\varphi$ such that $\mu_m \leqslant \varphi(u_0)$. Evidently u_0 is a weak solution of the considered Dirichlet problem.

Next suppose that hypothesis **(ii)**$_2$ holds. Hypotheses **(i)** and **(ii)**$_2$ imply that given $\varepsilon > 0$, we can find $c_7 = c_7(\varepsilon) > 0$ such that

$$F(z, x) \leqslant \tfrac{1}{2}(\widehat{\eta}_0(z) + \varepsilon)x^2 + c_7 \quad \text{for a.a. } z \in \Omega,\ \text{all } x \in \mathbb{R}. \tag{5.357}$$

Then for all $u \in H_0^1(\Omega)$, we have

$$\begin{aligned} \varphi(u) &\geqslant \tfrac{1}{2}\|Du\|_2^2 - \tfrac{1}{2}\int_\Omega \widehat{\eta}_0(z)u^2\,dz - \tfrac{\varepsilon}{2}\|u\|_2^2 - c_7|\Omega|_N - \|g\|_2\|u\|_2 \\ &\geqslant \tfrac{1}{2}(c_8 - \varepsilon)\|u\|^2 - c_7|\Omega|_N - c_9\|u\|, \end{aligned} \tag{5.358}$$

for some $c_8, c_9 > 0$ (see Problem 5.67). Choosing $\varepsilon \in (0, c_8)$, from (5.358), we see that

$$\varphi(u) \longrightarrow +\infty \quad \text{as } \|u\| \to +\infty. \tag{5.359}$$

Also, φ is sequentially weakly lower semicontinuous. This fact and (5.359) permit the use of Theorem 5.6. So, we can find $u_0 \in H_0^1(\Omega)$ such that

$$\varphi(u_0) = \inf_{u \in H_0^1(\Omega)} \varphi(u),$$

thus $u_0 \in K_\varphi$ and so u_0 is a weak solution of the considered Dirichlet problem.

Solution of Problem 5.124

From Problem 5.123, we already know that for every $g \in L^2(\Omega)$, the considered Dirichlet problem has a solution $u_0 \in H_0^1(\Omega)$. As in the solution of Problem 5.123, let $\varphi\colon H_0^1(\Omega) \longrightarrow \mathbb{R}$ be defined by

$$\varphi(u) = \tfrac{1}{2}\|Du\|_2^2 - \int_\Omega F(z,u)\,dz - \int_\Omega gu\,dz \qquad \forall\, u \in H_0^1(\Omega).$$

In this case $\varphi \in C^2(H_0^1(\Omega))$ (see Hypothesis **(i)**) and for all $u \in H_0^1(\Omega)$, we have

$$\langle \varphi''(u)v, h\rangle = \int_\Omega (Dv, Dh)_{\mathbb{R}^N}\,dz - \int_\Omega f'_x(z,u)vh\,dz \qquad \forall\, v,h \in H_0^1(\Omega).$$

Then for all $u \in H_0^1(\Omega)$, we have

$$\langle \varphi''(u)v, v\rangle = \|Dv\|_2^2 - \int_\Omega f'_x(z,u(z))v^2\,dz \geqslant c\|v\|^2,$$

for some $c > 0$ (see Problem 5.67), so φ is strictly convex (see Problem 3.62) and thus φ has a unique critical point which is a global minimizer of φ (see Problem 5.122).

Solution of Problem 5.125

Since Ω is star-shaped with respect to $0 \in \mathbb{R}^N$ and the boundary $\partial\Omega$ is Lipschitz (thus the outward unit normal n exists σ-almost everywhere on $\partial\Omega$), we have

$$(z, n(z))_{\mathbb{R}^N} > 0 \quad \text{for } \sigma\text{-a.e. on } \partial\Omega$$

(σ being the surface measure ($(N-1)$-dimensional Hausdorff measure) on $\partial\Omega$). Using the Pohozaev identity (see Theorem 5.143), we have

$$\left(\tfrac{N}{p} - \tfrac{N-2}{p}\right)\|u\|_p^p = \int_{\partial\Omega} \left(\tfrac{\partial u}{\partial n}\right)^2 (z,n)_{\mathbb{R}^N}\,d\sigma > 0,$$

so $\frac{N}{p} > \frac{N-2}{2}$ and thus $2^* = \frac{2N}{N-2} > p$.

Solution of Problem 5.126

Let $u_0 \in H_0^1(\Omega)$ be the solution of the considered Dirichlet problem. Then from

Remark 5.140, we know that $u_0 \in C^2(\Omega) \cap C^1(\overline{\Omega})$. Setting $F(x) = \frac{\lambda}{2}x^2 + \frac{1}{2^*}|x|^{2^*}$ for all $x \in \mathbb{R}$, from the Pohozaev identity (see Theorem 5.143), we have

$$\|Du_0\|_2^2 - 2^* \int_\Omega F(u_0)\, dz + \frac{1}{2} \int_{\partial\Omega} \left(\frac{\partial u_0}{\partial n}\right)^2 (z, n)_{\mathbb{R}^N}\, d\sigma = 0,$$

so

$$\|Du_0\|_2^2 - \|u_0\|_{2^*}^{2^*} - \frac{\lambda N}{N-2}\|u\|_2^2 + \frac{1}{2} \int_{\partial\Omega} \left(\frac{\partial u_0}{\partial n}\right)^2 (z, n)_{\mathbb{R}^N}\, d\sigma = 0,$$

thus

$$\lambda\left(1 - \frac{N}{N-2}\right)\|u\|_2^2 + \frac{1}{2} \int_{\partial\Omega} \left(\frac{\partial u_0}{\partial n}\right)^2 (z, n)_{\mathbb{R}^N}\, d\sigma = 0$$

(see the first equality in the Dirichlet problem), hence

$$|\lambda| \frac{2}{N-2}\|u_0\|_2^2 + \frac{1}{2} \int_{\partial\Omega} \left(\frac{\partial u_0}{\partial n}\right)^2 (z, n)_{\mathbb{R}^N}\, d\sigma = 0$$

and finally $u_0 = 0$ (since Ω is star-shaped).

Solution of Problem 5.127

We know that $\widehat{u} \in C^2(\Omega) \cap C^1(\overline{\Omega})$ (see Remark 5.140). Using the Pohozaev identity (see Theorem 5.143), we have

$$\frac{\widehat{\lambda} N}{2}\|\widehat{u}\|_2^2 - \frac{N-2}{N}\|D\widehat{u}\|_2^2 = \frac{1}{2} \int_{\partial\Omega} \left(\frac{\partial \widehat{u}}{\partial n}\right)^2 (z, n)_{\mathbb{R}^N}\, d\sigma,$$

so

$$\frac{\widehat{\lambda} N}{2}\|\widehat{u}\|_2^2 - \frac{\widehat{\lambda}(N-2)}{2}\|\widehat{u}\|_2^2 = \frac{1}{2} \int_{\partial\Omega} \left(\frac{\partial \widehat{u}}{\partial n}\right)^2 (z, n)_{\mathbb{R}^N}\, d\sigma$$

(since $-\Delta\widehat{u} = \widehat{\lambda}\widehat{u}$), thus

$$\widehat{\lambda}\|\widehat{u}\|_2^2 = \frac{1}{2} \int_{\partial\Omega} \left(\frac{\partial \widehat{u}}{\partial n}\right)^2 (z, n)_{\mathbb{R}^N}\, d\sigma.$$

Solution of Problem 5.128

"$\Longrightarrow$": Arguing by contradiction, suppose that we can find

$$u^* \in \mathrm{Fix}_G(X^*) \cap \mathrm{Fix}_G(X)^\perp, \quad u^* \neq 0.$$

Let $\varphi(u) = \langle u^*, u\rangle$ for all $u \in X$. We have $\varphi \in C^1(X)$ and $\varphi' = u^*$. Also for all $g \in G$, we have

$$\varphi(gu) = \langle u^*, gu\rangle = \langle g^{-1}u^*, u\rangle = \langle u^*, u\rangle = \varphi(u)$$

(since $u^* \in \mathrm{Fix}_G(X^*)$), so φ is G-invariant. Since $\varphi' = u^* \neq 0$, we see that $K_\varphi = \emptyset$. Because $u^* \in \mathrm{Fix}_G(X)^\perp$, we have $u^*|_{\mathrm{Fix}_G(X)} = 0$, so $\varphi'(u) = 0$ for all $u \in \mathrm{Fix}_G(X)$, contradicting the hypothesis.
"$\Longleftarrow$": Let $\varphi \in C^1(X)$ be G-invariant and let $u_0 \in K_{\varphi|_{\mathrm{Fix}_G(X)}}$. We have

$$\varphi(u_0) = \varphi|_{\mathrm{Fix}_G(X)}(u_0)$$

and

$$\varphi(u_0 + h) = \varphi|_{\mathrm{Fix}_G(X)}(u_0 + h) \qquad \forall\, h \in \mathrm{Fix}_G(X).$$

It follows that

$$\langle \varphi'(u_0), h\rangle = \Big\langle (\varphi|_{\mathrm{Fix}_G(X)})'(u_0), h\Big\rangle_0 \qquad \forall\, h \in \mathrm{Fix}_G(X), \tag{5.360}$$

where $\langle\cdot,\cdot\rangle_0$ denotes the duality brackets for the pair $(\mathrm{Fix}_G(X^*), \mathrm{Fix}_G(X))$. From (5.360) and since $u_0 \in K_{\varphi|_{\mathrm{Fix}_G(X)}}$, we have that

$$\varphi'(u_0) \in \mathrm{Fix}_G(X)^\perp. \tag{5.361}$$

Exploiting the G-invariance of φ, for all $g \in G$ and all $u, h \in X$, we have

$$\begin{aligned} \langle \varphi'(gu), h\rangle &= \lim_{\lambda\to 0} \tfrac{\varphi(gu+\lambda h)-\varphi(gu)}{\lambda} \\ = \lim_{\lambda\to 0} \tfrac{\varphi(u+\lambda g^{-1}h)-\varphi(u)}{\lambda} &= \langle \varphi'(u), g^{-1}h\rangle = \langle g(\varphi'(u), h)\rangle, \end{aligned}$$

so φ' is equivariant (see Definition 5.84). Since $u_0 \in \mathrm{Fix}_G(X)$, we have

$$g\varphi'(u_0) = \varphi'(gu_0) = \varphi'(u_0) \qquad \forall\, g \in G,$$

so

$$\varphi'(u_0) \in \mathrm{Fix}_G(X^*),$$

thus

$$\varphi'(u_0) \in \mathrm{Fix}_G(X^*) \cap \mathrm{Fix}_G(X)^\perp$$

(see (5.361)) and hence $\varphi'(u_0) = 0$ (by the hypothesis). So, we conclude that $K_{\varphi|_{\mathrm{Fix}_G(X)}} \subseteq K_\varphi$.

Solution of Problem 5.129
For every $g \in G$ and every $u^* \in X^*$, we have

$$\begin{aligned} \|gu^*\|_* &= \sup\{\langle gu^*, h\rangle : \|h\|_X \leqslant 1\} \\ &= \sup\{\langle u^*, g^{-1}h\rangle : \|h\|_X \leqslant 1\} \\ &\leqslant \sup\{\|u^*\|_*\|g^{-1}h\|_X : \|h\|_X \leqslant 1\} \\ &= \sup\{\|u^*\|_*\|h\|_X : \|h\|_X \leqslant 1\} = \|u^*\|_* \end{aligned}$$

(since G acts on X isometrically). Using this fact, for every $g \in G$ and $u^* \in X^*$, we have

$$\|u^*\|_* = \|g^{-1}(gu^*)\|_* \leqslant \|gu^*\|_*,$$

so

$$\|u^*\|_* = \|gu^*\|_*.$$

Thus, the action of G on X^* is isometric.

Solution of Problem 5.130
$\mathcal{F}^{-1}$ is the duality map for X^* and since X is strictly convex, $\mathcal{F}^{-1}$ is single-valued (see Proposition 2.114). Given $u^* \in \mathrm{Fix}_G(X^*)$, we have

$$\|g\mathcal{F}^{-1}(u^*)\|_X = \|\mathcal{F}^{-1}(u^*)\|_X = \|u^*\|_* \tag{5.362}$$

(since the action G on X is isometric and using Definition 2.112) and

$$\begin{aligned} \langle u^*, g\mathcal{F}^{-1}(u^*)\rangle &= \langle g^{-1}u^*, \mathcal{F}^{-1}(u^*)\rangle \\ = \langle u^*, \mathcal{F}^{-1}(u^*)\rangle &= \|u^*\|_*^2 = \|\mathcal{F}^{-1}(u^*)\|_X^2 \end{aligned}$$

(see Problem 5.129 and (5.362)). Therefore $g\mathcal{F}^{-1}(u^*) = \mathcal{F}^{-1}(u^*)$ for all $g \in G$, hence $\mathcal{F}^{-1}(u^*) \in \mathrm{Fix}_G(X)$.

Solution of Problem 5.131
Let μ be the Haar probability measure on G. An easy application of the closed graph theorem implies that for every $u \in X$, there exists a unique element $A(u) \in X$ such that

$$\langle u^*, A(u)\rangle = \int_G \langle u^*, gu\rangle \, d\mu(g) \qquad \forall\, u^* \in X^*.$$

Then $A \in \mathcal{L}(X)$ and is a projection on $\mathrm{Fix}_G(X)$ and if $C \subseteq X$ is a closed, convex, and G-invariant set, then $A(C) \subseteq C$. According to Problem 5.128, in order to establish the claim of the problem, it suffices to show that

$$\mathrm{Fix}_G(X^*) \cap \mathrm{Fix}_G(X)^{\perp} = \{0\}.$$

Arguing by contradiction suppose that we can find

$$u^* \in \mathrm{Fix}_G(X^*) \cap \mathrm{Fix}_G(X)^{\perp} \setminus \{0\}.$$

Let $C = \{u \in X : \langle u^*, u\rangle = 1\}$. Then the set C is closed and convex (in fact a hyperplane) and G-invariant. For any $u \in C$, we have $A(u) \in C \cap \mathrm{Fix}_G(X)$ and so $\langle u^*, A(u)\rangle = 0$. But since $A(u) \in C$, we have $\langle u^*, A(u)\rangle = 1$, a contradiction.

Solution of Problem 5.132
Let $\{u_n\}_{n\geqslant 1} \subseteq W_0^{1,p}(\Omega)$ be a sequence such that

$$|\varphi(u_n)| \leqslant M_0 \qquad \forall\, n \geqslant 1, \tag{5.363}$$

for some $M_0 > 0$ and

$$\varphi'(u_n) \longrightarrow 0 \quad \text{in } W^{-1,p'}(\Omega) = W_0^{1,p}(\Omega)^*. \tag{5.364}$$

From (5.363), we have

$$\tfrac{q}{p}\|Du_n\|_p^p - \int_\Omega qF(z,u_n)\,dz \leqslant qM_0 \qquad \forall\, n \geqslant 1. \tag{5.365}$$

From (5.364), we have

$$\Big|\langle A(u_n), h\rangle - \int_\Omega f(z,u_n)h\,dz\Big| \leqslant \varepsilon_n\|h\| \qquad \forall\, h \in W_0^{1,p}(\Omega), \tag{5.366}$$

with $\varepsilon_n \searrow 0$. Here $A\colon W_0^{1,p}(\Omega) \longrightarrow W^{-1,p'}(\Omega)$ is the nonlinear map defined by

$$\langle A(u), h\rangle = \int_\Omega |Du|^{p-2}\,(Du, Dh)_{\mathbb{R}^N} \qquad \forall\, u,h \in W_0^{1,p}(\Omega).$$

In (5.366) we choose $h = u_n \in W_0^{1,p}(\Omega)$ and obtain

$$-\|Du_n\|_p^p + \int_\Omega f(z,u_n)u_n\,dz \leqslant \varepsilon\|u_n\| \qquad \forall\, n \geqslant 1. \tag{5.367}$$

Adding (5.366) and (5.367), we obtain

$$\left(\tfrac{q}{p}-1\right)\|Du_n\|_p^p + \int\limits_\Omega \big(f(z,u_n)u_n - qF(z,u_n)\big)\,dz \leqslant qM_0 + \varepsilon_n\|u_n\| \qquad \forall\, n \geqslant 1.$$

Using hypotheses **(i)** and **(ii)**, we see that

$$\left(\tfrac{q}{p}-1\right)\|Du_n\|_p^p \leqslant c(1+\|u_n\|) \qquad \forall\, n \geqslant 1, \tag{5.368}$$

for some $c > 0$. Since $1 < p < q$ and using the Poincaré inequality (see Theorem 1.131), from (5.368), we infer that the sequence $\{u_n\}_{n\geqslant 1} \subseteq W_0^{1,p}(\Omega)$ is bounded. So, passing to a subsequence if necessary, we may assume that

$$u_n \xrightarrow{w} u \quad \text{in } W_0^{1,p}(\Omega) \quad \text{and} \quad u_n \longrightarrow u \quad \text{in } L^p(\Omega). \tag{5.369}$$

In (5.366), we choose $h = u_n - u \in W_0^{1,p}(\Omega)$, pass to the limit as $n \to +\infty$ and use hypothesis **(i)** and (5.369). Then

$$\lim_{n\to+\infty} \langle A(u_n), u_n - u\rangle = 0,$$

so $u_n \longrightarrow u$ in $W_0^{1,p}(\Omega)$ (see Problem 2.192). Therefore φ satisfies the PS-condition.

Solution of Problem 5.133
From hypothesis **(i)** we have

$$\tfrac{q}{x} \leqslant \tfrac{f(t,x)}{F(t,x)} \quad \text{for } \mu\text{-a.a. } t \in T, \text{ all } x \geqslant M,$$

so

$$\int\limits_M^x \tfrac{q}{s}\,ds \leqslant \int\limits_M^x \tfrac{F'_x(t,s)}{F(t,s)}\,ds \qquad \forall\, x \geqslant M,$$

thus

$$q(\ln x - \ln M) \leqslant \ln F(t,x) - \ln F(t,M) \qquad \forall\, x \geqslant M,$$

hence

$$\ln\left(\tfrac{x}{M}\right)^q \leqslant \ln\left(\tfrac{F(t,x)}{F(t,M)}\right) \qquad \forall\, x \geqslant M$$

and finally

$$F(t,M)\tfrac{x^q}{M^q} \leqslant F(t,x) \quad \text{for } \mu\text{-a.a. } t \in T, \text{ all } x \geqslant M.$$

In a similar fashion, we show that

$$F(t,-M)\tfrac{|x|^q}{M^q} \leqslant F(t,x) \quad \text{for } \mu\text{-a.a. } t \in T, \text{ all } x \leqslant -M.$$

So, finally we have

$$c|x|^q \leqslant F(t,x) \quad \text{for } \mu\text{-a.a. } t \in T, \text{ all } |x| \geqslant M,$$

with $c = \frac{1}{M^q} \min\{\operatorname*{ess\,inf}_T F(\cdot, M), \operatorname*{ess\,inf}_T F(\cdot, -M)\} > 0$ (see hypothesis **(ii)**). Therefore

$$c|x|^{q-p} \leqslant \frac{F(t,x)}{|x|^p} \quad \text{for } \mu\text{-a.a. } t \in T, \text{ all } |x| \geqslant M,$$

so

$$\lim_{x\to\pm\infty} \frac{F(t,x)}{|x|^p} = +\infty \quad \text{uniformly for } \mu\text{-a.a. } t \in T,$$

thus

$$\lim_{x\to\pm\infty} \frac{f(t,x)}{|x|^{p-2}x} = +\infty \quad \text{uniformly for } \mu\text{-a.a. } t \in T$$

(see hypothesis **(i)**).

The converse is not true. To see this, consider the function

$$f(x) = |x|^{p-2}x\big(\ln|x| + \tfrac{1}{p}\big)$$

(for the sake of simplicity we drop the t-dependence). We have

$$\lim_{x\to\pm\infty} \frac{f(x)}{|x|^{p-2}x} = +\infty.$$

However, $F(x) = \frac{1}{p}|x|^p \ln|x|$ and so we cannot find $q > p$ and $M > 0$ satisfying hypothesis **(i)**.

Solution of Problem 5.134

We introduce the C^1-functional $\varphi\colon H_0^1(\Omega) \longrightarrow \mathbb{R}$ defined by

$$\varphi(u) = \tfrac{1}{2}\|Du\|_2^2 - \int_\Omega F(z,u(z))\,dz \qquad \forall\, u \in H_0^1(\Omega).$$

Because of hypothesis **(i)**, φ is even. From Problem 5.132, we know that φ satisfies the PS-condition (see Definition 5.45(a)). For $m \geqslant 1$, let

$$\overline{H}_m = \bigoplus_{k=1}^{m} E(\widehat{\lambda}_k) \quad \text{and} \quad \widehat{H}_{m+1} = \overline{H}_m^{\perp} = \overline{\bigoplus_{m+1} E(\widehat{\lambda}_k)}.$$

We have

$$H_0^1(\Omega) = \overline{H}_m \oplus \widehat{H}_{m+1}.$$

For every $u \in \widehat{H}_{m+1}$ with $\|u\| = \varrho$, we have

$$\varphi(u) \geqslant \tfrac{1}{2}\|Du\|_2^2 - c\|u\|_r^r - c|\Omega|_N \tag{5.370}$$

(see hypothesis **(ii)**). If $N \geqslant 3$ and since $2 < r < 2^*$, we can find $t \in (0,1)$ such that

$$\tfrac{1}{r} = \tfrac{t}{2} + \tfrac{1-t}{2^*}$$

and so from the interpolation inequality (see Problem 1.30), we have

$$\|u\|_r \leqslant \|u\|_2^t \|u\|_{2^*}^{1-t},$$

so

$$\|u\|_r^r \leqslant \|u\|_2^{tr}\|u\|_{2^*}^{(1-t)r} \leqslant \frac{c_1}{\widehat{\lambda}_{m+1}^{\frac{tr}{2}}}\|Du\|_2^{tr}\|Du\|_2^{(1-t)r}, \tag{5.371}$$

for some $c_1 > 0$. Here we have used the variational characterization of $\widehat{\lambda}_{m+1}$ (see Theorem 5.129) and the Rellich–Kondrachov embedding theorem (see Theoremt.1.133), together with the Poincaré inequality (see Theorem 1.131). Using (5.371) in (5.370), we obtain

$$\begin{aligned} \varphi(u) &\geqslant \Big(\tfrac{1}{2} - c_1\frac{1}{\widehat{\lambda}_{m+1}^{\frac{tr}{2}}}\|Du\|_2^{r-2}\Big)\|Du\|_2^2 - c|\Omega|_N \\ &= \Big(\tfrac{1}{2} - c_1\frac{1}{\widehat{\lambda}_{m+1}^{\frac{tr}{2}}}\varrho^{r-2}\Big)\|Du\|_2^2 - c|\Omega|_N. \end{aligned} \tag{5.372}$$

Choosing $m \geqslant 1$ such that $c_1\frac{1}{\widehat{\lambda}_{m+1}^{\frac{tr}{2}}}\varrho^{r-2} < \frac{1}{2}$, from (5.372), we see that

$$\varphi(u) \geqslant \beta \qquad \forall\, u \in H_{m+1},\ \|u\| = \varrho,$$

for some $\beta \in \mathbb{R}$. From the solution of Problem 5.133, we have

$$c_2|x|^q \leqslant F(z,x) \quad \text{for a.a. } z \in \Omega, \text{ all } |x| \geqslant M,$$

with $c_2 > 0$. Combining this with hypothesis **(ii)**, we infer that

$$c_2|x|^q - c_3 \leqslant F(z,x) \quad \text{for a.a. } z \in \Omega, \text{ all } x \in \mathbb{R}, \tag{5.373}$$

with $c_3 > 0$. Since $\overline{H}_m$ is finite dimensional, all norms are equivalent and so for all $u \in \overline{H}_m$ with $\|u\| = \varrho$, we have

$$\varphi(u) \leqslant \tfrac{1}{2}\varrho^2 - c_4\varrho^q + c_5,$$

for some $c_4, c_5 > 0$, thus

$$\varphi(u) \longrightarrow -\infty \quad \text{as } \|u\| \to +\infty,\ u \in \overline{H}_m.$$

So, we can use Theorem 5.94 and infer that φ has an unbounded sequence of pairs

$\{\pm u_n\}_{n\geqslant 1}$ of critical points which clearly are weak solutions of the considered Dirichlet problem.

Solution of Problem 5.135

First note that the regularity theory (see Remark 5.140) implies that every weak solution of the considered parametric Dirichlet problem belongs in $C_0^1(\overline{\Omega})$. Arguing by contradiction, suppose that the claim of the problem is not true. Then we can find a sequence $\{\lambda_n\}_{n\geqslant 1}$ such that $\lambda_n \searrow \widehat{\lambda}_1$ and a sequence of corresponding solutions $\{u_n\}_{n\geqslant 1} \subseteq C_0^1(\overline{\Omega})$ of the considered parametric Dirichlet problem such that $u_n \notin -\operatorname{int} C_+$ (that is, $u_n(z_n) > 0$ for some $z_n \in \Omega$). We consider two distinct cases.

Case 1. The sequence $\{u_n\}_{n\geqslant 1} \subseteq L^2(\Omega)$ is bounded.

From the equality in the Dirichlet problem, we see that the sequence $\{u_n\}_{n\geqslant 1} \subseteq H_0^1(\Omega)$ is bounded and so, passing to a subsequence if necessary, we my assume

$$u_n \xrightarrow{w} u \quad \text{in } H_0^1(\Omega) \quad \text{and} \quad u_n \longrightarrow u \quad \text{in } L^2(\Omega). \tag{5.374}$$

We have

$$\begin{cases} -\Delta u_n(z) &= \lambda_n u_n(z) + g(z) \quad \text{for a.a. } z \in \Omega, \\ u_n|_{\partial\Omega} &= 0, \ n \geqslant 1. \end{cases} \tag{5.375}$$

From (5.374) and (5.375), in the limit we have

$$\begin{cases} -\Delta u(z) &= \widehat{\lambda}_1 u(z) + g(z) \quad \text{for a.a. } z \in \Omega, \\ u|_{\partial\Omega} &= 0. \end{cases} \tag{5.376}$$

We multiply (5.376) with $\widehat{u}_1(z)$ ($\widehat{u}_1 \in \operatorname{int} C_+$) and integrate over Ω. Then

$$\int_\Omega (-\Delta u)\widehat{u}_1\,dz = \widehat{\lambda}_1 \int_\Omega u\widehat{u}_1\,dz + \int_\Omega g\widehat{u}_1\,dz,$$

so

$$\int_\Omega u(-\Delta \widehat{u}_1)\,dz = \widehat{\lambda}_1 \int_\Omega u\widehat{u}_1\,dz + \int_\Omega g\widehat{u}_1\,dz$$

(see Theorem 1.130), thus

$$\widehat{\lambda}_1 \int_\Omega u\widehat{u}_1\,dz = \widehat{\lambda}_1 \int_\Omega u\widehat{u}_1\,dz + \int_\Omega g\widehat{u}_1\,dz$$

and hence

$$\int_\Omega g\widehat{u}_1\,dz = 0,$$

a contradiction, since $g \neq 0$, $g \geqslant 0$, $\widehat{u}_1 \in \operatorname{int} C_+$.

Case 2. The sequence $\{u_n\}_{n\geqslant 1} \subseteq L^2(\Omega)$ is unbounded.

Passing to a subsequence if necessary, we may assume that $\|u_n\|_2 \longrightarrow +\infty$. Let $y_n = \frac{u_n}{\|u_n\|_2}$ for all $n \geqslant 1$. Then

$$\begin{cases} -\Delta y_n(z) = \lambda_n y_n(z) + \frac{g(z)}{\|u_n\|_2} & \text{for a.a. } z \in \Omega, \\ y_n|_{\partial\Omega} = 0, \ n \geqslant 1. \end{cases}$$

From Proposition 5.138, we know that we can find $\alpha \in (0,1)$ and $M > 0$ such that

$$y_n \in C_0^{1,\alpha}(\overline{\Omega}) \quad \text{and} \quad \|y_n\|_{C_0^{1,\alpha}(\overline{\Omega})} \leqslant M \qquad \forall\, n \geqslant 1. \tag{5.377}$$

Exploiting the compactness of the embedding $C_0^{1,\alpha}(\overline{\Omega}) \hookrightarrow C^1(\overline{\Omega})$, from (5.377), we have

$$y_n \longrightarrow y \quad \text{in } C_0^1(\overline{\Omega}), \quad \|y\|_2 = 1 \tag{5.378}$$

and

$$-\Delta y(z) = \widehat{\lambda}_1 y(z) \quad \text{for a.a. } z \in \Omega.$$

So, $y = \pm\widehat{u}_1$ (see (5.378)). If $y = \widehat{u}_1$, then from (5.378), we see that $y_n \in \operatorname{int} C_+$ for all $n \geqslant n_0$. We fix $n \geqslant n_0$ and write $y_n = e^{-v}$ with $v \in C^1(\overline{\Omega})$. We take $h \in C^1(\overline{\Omega})$ and use $e^v h^2$ as a test function in the equation

$$\begin{cases} -\Delta y_n(z) = \lambda_n y_n(z) + g_n(z) & \text{for a.a. } z \in \Omega, \\ y_n|_{\partial\Omega} = 0, \end{cases}$$

with $g_n = \frac{g}{\|u_n\|}$. We obtain

$$\int_\Omega \big(Dy_n, D(e^v h^2)\big)_{\mathbb{R}^N}\, dz + \int_\Omega 2e^v h\, (Dy_n, Dv)_{\mathbb{R}^N}\, dz$$
$$= \lambda_n \int_\Omega h^2\, dz + \int_\Omega g_n e^v h^2\, dz,$$

so

$$-\int_\Omega h^2 |Dv|\, dz - 2\int_\Omega h\, (Dv, Dh)_{\mathbb{R}^N}\, dz = \lambda_n \int_\Omega h^2\, dz + \int_\Omega g_n e^v h^2\, dz,$$

thus

$$\|Dh\|_2^2 - \|Dh + Dv\|_2^2 = \lambda_n \int_\Omega h^2\, dz + \int_\Omega g_n e^v h^2\, dz$$

and hence

$$\lambda_n \int_\Omega h^2\, dz \leqslant \|Dh\|_2^2 \tag{5.379}$$

(recall that $g \geqslant 0$). Since the embedding $C_0^1(\overline{\Omega}) \hookrightarrow H_0^1(\Omega)$ is dense and $h \in C_0^1(\overline{\Omega})$ is arbitrary, from (5.379) and the variational characterization of $\widehat{\lambda}_1$ (see Theorem 5.129 and Problem 5.107), we have that $\lambda_n \leqslant \widehat{\lambda}_1$, a contradiction. If $y = -\widehat{u}_1$, then $y_n \in -\operatorname{int} C_+$ for all $n \geqslant n_1$ and so $u_n \in -\operatorname{int} C_+$ for all $n \geqslant n_1$, again a contradiction.

Solution of Problem 5.136

If in the definition of the Krasnoselskii genus (see Definition 5.91) we use $h = I_{\mathbb{R}^N}$, then we see that $\gamma(\partial\Omega) \leqslant N$. Arguing by contradiction, suppose that $\gamma(\partial\Omega) = m < N$. Then we can find $h \in C(\partial\Omega; \mathbb{R}^m \setminus \{0\})$ odd. In fact the Dugundji extension theorem (see Theorem 2.8) allows us to assume that h is defined on all $\mathbb{R}^N$, that is, $h \in C(\mathbb{R}^N; \mathbb{R}^m)$ with $h(u) \neq 0$ for all $u \in \partial\Omega$. Moreover, we can think of h as having values in $\mathbb{R}^N$ by taking the last $N - m$ components to be equal to zero. Then we can define the Brouwer degree $d(h, \Omega, 0)$ and by Theorem 4.18, we have

$$d(h, \Omega, 0) \;\neq\; 0.$$

From the continuity of the Brouwer degree on the reference point (see Theorem 4.11(f)), we can find $\varepsilon > 0$ such that

$$d(h, \Omega, y) \;\neq\; 0 \qquad \forall\, y \in B_\varepsilon,$$

where $B_\varepsilon = \{y \in \mathbb{R}^N : |y| < \varepsilon\}$. By the existence property of the Brouver degree (see Theorem 4.11(d)) for all $y \in B_\varepsilon$ the equation $h(u) = y$ has a solution $u \in \Omega$. Hence $B_\varepsilon \subseteq h(\Omega) \subseteq \mathbb{R}^m$ with $m < N$, a contradiction. So, we have $\gamma(\partial\Omega) = N$.

Solution of Problem 5.137

Let $\{u_n\}_{n\geqslant 1} \subseteq X$ be a dense sequence (it exists since by hypothesis X is separable). Let $X_k = \operatorname{span}\{u_1, \ldots, u_k\}$. Then $X = \overline{\bigcup_{k\geqslant 1} X_k}$. From Problem 5.136, we have

$$\gamma(\partial B_1 \cap X_k) \;=\; k \qquad \forall\, k \geqslant 1.$$

Since $\gamma(\partial B_1 \cap X_k) \leqslant \gamma(\partial B_1)$ (see Definition 5.88(b)), we conclude that $\gamma(\partial B_r) = +\infty$.

Solution of Problem 5.138

Consider the continuous odd function $h\colon C \cup (-C) \longrightarrow \mathbb{R} \setminus \{0\}$

$$h(u) \;=\; \begin{cases} 1 & \text{if } \; u \in C, \\ -1 & \text{if } \; u \in -C. \end{cases}$$

Then from Definition 5.91, we have $\gamma(C \cup (-C)) = 1$

Solution of Problem 5.139

For $\delta > 0$, we define

$$E_\delta \;=\; \{u \in \partial\Omega : \; |h(u)| \geqslant \delta\}.$$

Since $h \in C(E_\delta; \mathbb{R}^m \setminus \{0\})$ is odd, from Definition 5.91, we have

$$\gamma(E_\delta) \;\leqslant\; m. \tag{5.380}$$

We have

$$\partial\Omega \;=\; E_\delta \cup (\partial\Omega \setminus E_\delta),$$

so

$$\overline{\partial\Omega} \;=\; \overline{E_\delta \cup (\partial\Omega \setminus E_\delta)} \;=\; \overline{E}_\delta \cup \overline{\partial\Omega \setminus E_\delta},$$

thus

$$\partial\Omega \;=\; E_\delta \cup \overline{\partial\Omega \setminus E_\delta},$$

hence

$$\gamma(\partial\Omega) \;\leqslant\; \gamma(E_\delta) + \gamma(\,\overline{\partial\Omega \setminus E_\delta}\,)$$

(from the subadditivity property of γ; see Definition 5.88(c)) and finally

$$N - m \;\leqslant\; \gamma(\,\overline{\partial\Omega \setminus E_\delta}\,) \tag{5.381}$$

(see Problem 5.139 and (5.380)). Note that

$$\overline{\partial\Omega \setminus E_\delta} \;=\; F_\delta \;=\; \big\{u \in \partial\Omega : \; |h(u)| \leqslant \delta\big\}.$$

Using the continuity property of γ (see Definition 5.88(d)), we find $\varepsilon > 0$ such that, if $D_\varepsilon = \big\{u \in \partial\Omega : \; d_{\mathbb{R}^N}(u, D) \leqslant \varepsilon\big\}$, then

$$\gamma(D_\varepsilon) \;=\; \gamma(D).$$

We claim that for $\delta > 0$ small, we have $F_\delta \subseteq D_\varepsilon$. Indeed, if this is not possible, then we can find a sequence $\{u_n\}_{n \geqslant 1} \subseteq \partial\Omega$ such that

$$|h(u_n)| \;\leqslant\; \tfrac{1}{n} \quad \text{and} \quad d_{\mathbb{R}^N}(u_n, D) \;>\; \varepsilon \qquad \forall\, n \geqslant 1. \tag{5.382}$$

Since $\partial\Omega$ is compact, by passing to a suitable subsequence if necessary, we may assume that $u_n \longrightarrow \widetilde{u} \in \partial\Omega$. Passing to the limit as $n \to +\infty$ in (5.382), we obtain

$$h(\widetilde{u}) \;=\; 0 \quad \text{and} \quad d_{\mathbb{R}^N}(\widetilde{u}, D) \;\geqslant\; 0,$$

a contradiction. So, for $\delta > 0$ small we have

$$D \subseteq F_\delta \subseteq D_\varepsilon,$$

thus

$$\gamma(D) = \gamma(F_\delta \geqslant N-m)$$

(see (5.381) and Definition 5.88(b)).

Solution of Problem 5.140

Let $\mathrm{proj}_V \in \mathcal{L}(X;V)$ be a projection onto V and let μ be the Haar probability measure on G. We set

$$E = \int_G L(g^{-1})\mathrm{proj}_V(L(g))\,d\mu,$$

with L being the group representation (see Definition 5.84). Then $E \in \mathcal{L}(X;V)$ and if $v \in V$, we have

$$\begin{aligned} E(v) &= \int_G L(g^{-1})\mathrm{proj}_V L(g)v\,d\mu = \int_G L(g^{-1})L(g)\mathrm{proj}_V(v)\,d\mu \\ &= \int_G L(g^{-1}g)v\,d\mu = \int_G v\,d\mu \end{aligned}$$

(see Definition 5.84 and recall that $v \in V$ and $\mu(G) - 1$), so E is a projection onto V.

Also clearly we have $L(g) \circ E = E \circ L(g)$ for all $g \in G$ (see Problem 1.66). So, E is equivariant. Therefore $(I_X - E)(X)$ is a G-invariant topological complement of V.

Solution of Problem 5.141

As before let μ be the Haar probability measure on G and consider the quadratic form

$$q(u) = \int_G |L(g)(u)|^2\,d\mu \qquad \forall\, u \in \mathbb{R}^N.$$

Evidently q is positive definite and so there exists an invertible matrix $A\colon \mathbb{R}^N \longrightarrow \mathbb{R}^N$ such that $q(u) = |A(u)|^2$. Evidently $\{AL(g)A^{-1}\}_{g\in G}$ is a representation of G over

$\mathbb{R}^N$. For fixed $g \in G$ and $u \in \mathbb{R}^N$, we have

$$\begin{aligned}
|AL(g)A^{-1}(u)|^2 &= q(L(g)A^{-1}(u)) = \int_G |L(g')L(g)A^{-1}(u)|^2 \, d\mu(g') \\
&= \int_G |L(g'g)A^{-1}(u)|^2 \, d\mu(g') = \int_G |L(\widehat{g})A^{-1}(u)|^2 \, d\mu(\widehat{g}) \\
&= q(A^{-1}(u)) = |AA^{-1}(u)|^2 = |u|^2
\end{aligned}$$

(see Definition 5.88(b) and recall that μ is right-invariant), so

$$|AL(g)A^{-1}(u)| = |u|,$$

thus $\{AL(g)A^{-1}\}_{g \in G}$ is an isometric representation.

Solution of Problem 5.142

(a) We have

$$\begin{aligned}
\langle \varphi'(gu), v \rangle &= \lim_{\lambda \to 0} \tfrac{1}{\lambda} \big(\varphi(g(u + \lambda g^{-1} v)) - \varphi(gu) \big) \\
&= \lim_{\lambda \to 0} \tfrac{1}{\lambda} \big(\varphi(u + \lambda g^{-1} v) - \varphi(u) \big) = \langle \varphi'(u), g^{-1} v \rangle \qquad \forall\, g \in G,\ u, v \in X
\end{aligned}$$

(since φ is G-invariant).

(b) From part **(a)**, we have

$$\langle \varphi'(gu), v \rangle = \langle \varphi'(u), g^{-1} v \rangle \qquad \forall\, g \in G,\ u, v \in X. \tag{5.383}$$

Since the action of G on X is isometric, we have $\|v\|_X = \|g^{-1}v\|_X$ for all $g \in G$, $v \in X$. Taking the supremum over $\overline{B}_1 = \{v \in X : \|v\|_X \leqslant 1\}$ of both sides in (5.383), we obtain

$$\|\varphi'(gu)\|_* = \|\varphi'(u)\|_* \qquad \forall\, u \in X.$$

Solution of Problem 5.143

From Proposition 5.52, we already have a pseudogradient vector field $v \colon X \setminus K_\varphi \longrightarrow X$. Let μ be the unique Haar probability measure on G. We set

$$\widehat{v}(u) = \int_G g v(g^{-1}u) \, d\mu \qquad \forall\, u \in X \setminus K_\varphi.v \tag{5.384}$$

For every $g' \in G$, we have

$$\widehat{v}(g'u) = \int\limits_G gv(g^{-1}g'u)\, d\mu(g)$$

$$= \int\limits_\Omega g'(g')^{-1}gv(((g')^{-1}g)^{-1}u)\, d\mu(g) = g'\widehat{v}(u),$$

so $\widehat{v}$ is equivariant. We verify that $\widehat{v}$ is a pseudogradient vector field for φ. To this end, note that

$$\|\widehat{v}(u)\|_X \leqslant \int\limits_G \|v(g^{-1}u)\|_X\, d\mu \leqslant \int\limits_\Omega 2\|\varphi'(g^{-1}u)\|_*\, d\mu = 2\|\varphi'(u)\|_*$$

(since the action of G is isometric, using Definition 5.50, Problem 5.142 and the fact that $\mu(G) = 1$). Using Problem 5.142, we have

$$\langle \varphi'(u), \widehat{v}(u)\rangle = \int\limits_G \langle \varphi'(u), gv(g^{-1}u)\rangle\, d\mu$$

$$= \int\limits_G \langle \varphi'(g^{-1}u), v(g^{-1}u)\rangle\, d\mu$$

$$\geqslant \int\limits_G \|\varphi'(g^{-1}u)\|_*\, d\mu = \int\limits_G \|\varphi'(u)\|_*^2\, d\mu = \|\varphi'(u)\|_*$$

(since φ' is equivariant (see the solution of Problem 5.128), the action is isometric and using Definition 5.50 and the fact that $\mu(G) = 1$).

It remains to show that $\widehat{v}$ is locally Lipschitz. For $u \in X \setminus K_\varphi$, let $O(u) = \{gv :\ g \in G\}$. The compactness of G implies that $O(u) \subseteq X$ is compact (see Definition 5.84(c)). So, we can find $\delta > 0$ such that if

$$O(u)_\delta = \{y \in X \setminus K_\varphi :\ \operatorname{dist}(y, O(u)) \leqslant \delta\},$$

then $v|_{O(u)_\delta}$ is Lipschitz continuous (recall that v is locally Lipschitz; see Definition 5.50). Let $k > 0$ be the Lipschitz constant of v on $O(u)_\delta$. Since $O(u)_\delta$ is clearly G-invariant, for all $y, h \in B_\delta(u)$ (where $B_\delta(u) = \{w \in X :\ \|w - u\|_X < \delta\}$), we have

$$\|\widehat{v}(y) - \widehat{v}(h)\|_X \leqslant \int\limits_\Omega \|gv(g^{-1}y) - gv(g^{-1}h)\|_X\, d\mu$$

$$= \int\limits_\Omega \|v(g^{-1}y) - v(g^{-1}h)\|_X\, d\mu$$

$$\leqslant k\int\limits_\Omega \|g^{-1}y - g^{-1}h\|_X\, d\mu = k\|y - h\|_X$$

(since the action is isometric and $\mu(G) = 1$). So, $\widehat{v}$ is locally Lipschitz and so it is an equivalent pseudogradient vector field.

Solution of Problem 5.144

Let $u^* \in \partial\varphi(gu)$. Then from Definition 3.70, we have

$$\begin{aligned}\varphi(h) - \varphi(u) = \varphi(gh) - \varphi(gu) &\geqslant \langle u^*, gh - gu\rangle \\ &= \langle g^{-1}u^*, h - u\rangle \qquad \forall\, h \in X\end{aligned}$$

(since φ is G-invariant and using Problem 5.129), so $g^{-1}u^* \in \partial\varphi(u)$, thus

$$\partial\varphi(gu) \subseteq g\partial\varphi(u) \qquad \forall\, g \in G,\ u \in X.$$

In this inclusion, we replace u be gu and g by g^{-1}. Then

$$\partial\varphi(u) \subseteq g^{-1}\partial\varphi(gu),$$

so

$$g\partial\varphi(u) \subseteq \partial\varphi(gu),$$

thus

$$g\partial\varphi(u) = \partial\varphi(gu) \qquad \forall\, g \in G,\ u \in X.$$

Solution of Problem 5.145

Let $\{u_n\}_{n\geqslant 1} \subseteq H_0^1(\Omega)$ be a sequence such that

$$\varphi_\lambda(u_n) \longrightarrow c < \frac{S^{\frac{N}{2}}}{N} \tag{5.385}$$

and

$$\varphi_\lambda'(u_n) \longrightarrow 0 \quad \text{in } H^{-1}(\Omega) = H_0^1(\Omega)^*. \tag{5.386}$$

From (5.386), we have

$$\left|\left\langle \varphi_\lambda'(u_n), h\right\rangle\right| \leqslant \varepsilon_n\|h\| \qquad \forall\, h \in H_0^1(\Omega),$$

with $\varepsilon_n \searrow 0$, so

$$\Big|\int_\Omega (Du_n, Dh)\, dz - \lambda\int_\Omega u_n h\, dz - \int_\Omega |u_n|^{2^*-2}u_n h\, dz\Big| \leqslant \varepsilon_n\|h\| \tag{5.387}$$

for all $n \geqslant 1$. Choosing $h = u_n \in H_0^1(\Omega)$ in (5.387), we obtain

$$-\|Du_n\|_2^2 + \lambda\|u_n\|_2^2 + \|u_n\|_{2^*}^{2^*} \leqslant c_1\|u_n\| \qquad \forall\, n \geqslant 1, \tag{5.388}$$

for some $c_1 > 0$. From (5.385), we have

$$\frac{2^*}{2}\|Du_n\|_2^2 - \frac{2^*\lambda}{2}\|u_n\|_2^2 - \|u_n\|_{2^*}^{2^*} \leqslant M \qquad \forall\, n \geqslant 1, \tag{5.389}$$

for some $M > 0$. Adding (5.388) and (5.389), we have

$$\left(\frac{2^*}{2} - 1\right)\left(\|Du_n\|_2^2 - \lambda\|u_n\|_2^2\right) \leqslant c_2(1 + \|u_n\|) \qquad \forall\, n \geqslant 1,$$

for some $c_2 > 0$, so

$$\|u_n\|^2 \leqslant c_3(1 + \|u_n\|) \qquad \forall\, n \geqslant 1,$$

for some $c_3 > 0$ (recall that $\lambda \in (0, \widehat{\lambda}_1)$), thus the sequence $\{u_n\}_{n\geqslant 1} \subseteq H_0^1(\Omega)$ is bounded. So, passing to a subsequence if necessary, we may assume that

$$u_n \xrightarrow{w} u \quad \text{in } H_0^1(\Omega) \text{ and in } L^{2^*}(\Omega) \quad \text{and} \quad u_n \longrightarrow u \quad \text{in } L^2(\Omega) \tag{5.390}$$

(see the Rellich–Kondrachov embedding theorem; Theorem 1.135). Note that

$$|u_n|^{2^*-2}u_n \xrightarrow{w} |u|^{2^*-2}u \quad \text{in } L^{\frac{2N}{N+2}}(\Omega)$$

(we have $\frac{1}{2^*} + \frac{N+2}{2N} = 1$) and so, if in (5.387) we pass to the limit as $n \to +\infty$ and use also (5.390), we obtain

$$\int_\Omega (Du, Dh)_{\mathbb{R}^N}\, dz = \lambda \int_\Omega uh\, dz + \int_\Omega |u|^{2^*-2}uh\, dz \qquad \forall\, h \in H_0^1(\Omega).$$

Taking $h = u \in H_0^1(\Omega)$, we have

$$\|Du\|_2^2 = \lambda\|u\|_2^2 + \|u\|_{2^*}^{2^*}. \tag{5.391}$$

Using (5.391), we see that

$$\varphi_\lambda(u) = \left(\frac{1}{2} - \frac{1}{2^*}\right)\|u\|_{2^*}^{2^*} = \frac{1}{N}\|u\|_{2^*}^{2^*}. \tag{5.392}$$

From the Brézis–Lieb lemma (see Theorem I.3.98), we have

$$\|u_n\|_{2^*}^{2^*} = \|u_n - u\|_{2^*}^{2^*} + \|u\|_{2^*}^{2^*} + d_n \qquad \forall\, n \geqslant 1, \tag{5.393}$$

with $d_n \searrow 0$. Also, using (5.390), we have

$$\|u_n\|_2^2 = \|u\|_2^2 + d_n' \qquad \forall\, n \geqslant 1, \tag{5.394}$$

with $d_n' \searrow 0$. Moreover, note that

$$\|Du_n - Du\|_2^2 = \|Du_n\|_2^2 - 2\int_\Omega (Du_n, Du)_{\mathbb{R}^N}\, dz + \|Du\|_2^2,$$

so

$$\|Du_n\|_2^2 \;=\; \|Du_n - Du\|_2^2 + \|Du\|_2^2 + d_n'' \qquad \forall\, n \geqslant 1, \tag{5.395}$$

with $d_n'' \searrow 0$ (see (5.390)). It follows that

$$\begin{aligned} \varphi_\lambda(u_n) \;&=\; \tfrac{1}{2}\|Du_n\|_2^2 - \tfrac{\lambda}{2}\|u_n\|_2^2 - \tfrac{1}{2^*}\|u_n\|_{2^*}^{2^*} \\ &=\; \varphi_\lambda(u) + \tfrac{1}{2}\|Du_n - Du\|_2^2 - \tfrac{1}{2^*}\|u_n\|_{2^*}^{2^*} + d_n''' \qquad \forall\, n \geqslant 1, \end{aligned} \tag{5.396}$$

with $d_n''' \searrow 0$ (see (5.393), (5.394), and (5.395)). Note that

$$\langle \varphi_\lambda'(u_n), u_n \rangle \;\longrightarrow\; 0$$

(see (5.386)), so

$$\|Du_n\|_2^2 - \lambda\|u_n\|_2^2 - \|u_n\|_{2^*}^{2^*} \;\longrightarrow\; 0. \tag{5.397}$$

Using once more (5.393), (5.394), and (5.395), from (5.397), we infer that

$$\|Du_n - Du\|_2^2 + \|Du\|_2^2 - \lambda\|u\|_2^2 - \big(\|u_n - u\|_{2^*}^{2^*} + \|u\|_{2^*}^{2^*}\big) \;=\; e_n \qquad \forall\, n \geqslant 1,$$

with $e_n \searrow 0$, so

$$\|Du_n - Du\|_2^2 - \|u_n - u\|_{2^*}^{2^*} \;=\; e_n \qquad \forall\, n \geqslant 1$$

(see (5.391)), thus

$$\lim_{n\to+\infty} \|Du_n - Du\|_2^2 \;=\; \lim_{n\to+\infty} \|u_n - u\|_{2^*}^{2^*} \;=\; \xi. \tag{5.398}$$

From the definition of the embedding constant S, we have

$$S\|u_n - u\|_{2^*}^2 \;\leqslant\; \|Du_n - Du\|_2^2 \qquad \forall\, n \geqslant 1,$$

so $S\xi^{\frac{2}{2^*}} \leqslant \xi$ (see (5.398)).

If $\xi = 0$, then from (5.398) we see that $u_n \longrightarrow u$ in $H_0^1(\Omega)$ and so we are done.

If $\xi > 0$, then $S^{\frac{N}{2}} \leqslant \xi$. From (5.385), (5.396), and (5.398), we have

$$c \;=\; \varphi_\lambda(u) + \big(\tfrac{1}{2} - \tfrac{1}{2^*}\big)\xi + e_n' \;=\; \varphi_\lambda(u) + \tfrac{1}{N}\xi + e_n', \tag{5.399}$$

with $e_n' \searrow 0$. But from (5.392) we have that $\varphi_\lambda(u) \geqslant 0$. So, from (5.399) in the limit as $n \to +\infty$, we obtain

$$c \;\geqslant\; \tfrac{1}{N}\xi \;\geqslant\; \tfrac{1}{N}S^{\frac{N}{2}},$$

which contradicts (5.385). So, $\xi = 0$ and we conclude that φ_λ satisfies the PS_c-condition (see Definition 5.45(a)).

Solution of Problem 5.146

We consider the triple (X_{m+1}, X_m, X_1) and define the long exact sequence of singular homology groups corresponding to this triple

$$\begin{aligned} \ldots \longrightarrow H_k(X_m, X_1) &\xrightarrow{i_*} H_k(X_{m+1}, X_1) \\ &\xrightarrow{j_*} H_k(X_{m+1}, X_m) \longrightarrow \ldots \end{aligned} \tag{5.400}$$

with i_* and j_* being the homomorphisms induced by the inclusions

$$i\colon (X_m, X_1) \longrightarrow (X_{m+1}, X_1) \quad \text{and} \quad j\colon (X_{m+1}, X_1) \longrightarrow (X_{m+1}, X_m)$$

(see Theorem I.2.277). From the rank theorem, we have

$$\begin{aligned} \operatorname{rank} H_k(X_{m+1}, X_1) &= \operatorname{rank}\ker j_* + \operatorname{rank}\operatorname{im} j_* \\ = \operatorname{rank}\operatorname{im} i^* + \operatorname{rank}\operatorname{im} j_* &\leqslant \operatorname{rank} H_k(X_m, X_1) + \operatorname{rank} H_k(X_{m+1}, X_m) \end{aligned}$$

(due to the exactness of (5.400)). So, the claim of the problem is true for $m = 1$ and then by induction it is true for all $m \geqslant 1$.

Solution of Problem 5.147

As always, when $b = +\infty$, we have $\varphi^b = X$. Since $K_\varphi \cap [a, b] = \{c\}$, from the second deformation lemma (see Theorem 5.56), we have that φ^c is a strong deformation retract of φ^b and φ^a is a strong deformation retract of $\varphi^c \setminus K_\varphi^c$. Let $\delta > 0$ be small so that the open balls $B_\delta(u) = \{v \in X : \|v - u\|_X < \delta\}$ for $u \in K_\varphi^c$ are mutually disjoint. Then from the excision property of singular homology, we have

$$H_k(\varphi^c, \varphi^c \setminus K_\varphi^c) = \bigoplus_{u \in K_\varphi^c} H_k(\varphi^c \cap B_\delta(u), \varphi^c \cap B_\delta \setminus \{u\}) = \bigoplus_{u \in K_\varphi^c} C_k(\varphi, u).$$

Solution of Problem 5.148

Let $\{c_1, \ldots, c_m\} \subseteq (a, b)$ be all the critical values of φ in $[a, b]$. Let

$$a = a_1 < c_1 < a_2 < c_2 < \ldots < c_{m-1} < a_m < c_m < a_{m+1} = b.$$

Then using Problem 5.146 with $X_1 = \varphi^{a_i}$ for $i \in \{1, \ldots, m\}$, we have

$$\begin{aligned} \operatorname{rank} H_k(\varphi^b, \varphi^a) &\leqslant \sum_{i=1}^m \operatorname{rank} H_k(\varphi^{a_{i+1}}, \varphi^{a_i}) \\ = \sum_{i=1}^m \sum_{u \in K_\varphi^{c_i}} \operatorname{rank} C_k(\varphi, u) &= \sum_{u \in K_\varphi[a,b]} C_k(\varphi, u) \end{aligned}$$

(see Problem 5.147).

Solution of Problem 5.149

Let $a < \inf \varphi(K_\varphi)$ and $b = +\infty$. From Problem 5.148, we have

$$\operatorname{rank} H_k(X, \varphi^a) \leqslant \sum_{u \in K_\varphi[a,+\infty)} \operatorname{rank} C_k(\varphi, u)$$

(recall that $\varphi^{+\infty} = X$), so

$$\operatorname{rank} C_k(\varphi, \infty) \leqslant \sum_{u \in K_\varphi} \operatorname{rank} C_k(\varphi, u)$$

(see Definition 5.113). Since $C_k(\varphi, \infty) \neq 0$, it follows that there exists $u \in K_\varphi$ such that $C_k(\varphi, u) \neq 0$.

Solution of Problem 5.150

Since u is an isolated critical point of φ and a local minimizer of φ, we can find an open neighborhood U of u such that

$$K_\varphi \cap U = \{u\} \quad \text{and} \quad c = \varphi(u) < \varphi(v) \qquad \forall\, v \in U \setminus \{u\}. \tag{5.401}$$

Then, from Definition 5.102, we have

$$C_k(\varphi, u) = H_k(\{u\}, \emptyset) = H_k(\{u\}) = \delta_{k,0}\mathbb{Z}$$

(see (5.401)).

Solution of Problem 5.151

Since u is a local maximizer and an isolated critical point of φ, we can find $\varrho > 0$ such that

$$K_\varphi \cap \overline{B}_\varrho(u) = \{u\} \quad \text{and} \quad c = \varphi(u) > \varphi(v) \qquad \forall\, v \in \overline{B}_\varrho(u) \setminus \{u\}. \tag{5.402}$$

We have

$$C_k(\varphi, u) = H_k(\overline{B}_\varrho(u), \overline{B}_\varrho(u) \setminus \{u\}) \qquad \forall\, k \geqslant 0 \tag{5.403}$$

(see (5.402) and Definition 5.102). We know that $\overline{B}_\varrho(u) \setminus \{u\}$ is homotopy equivalent to $\partial B_\varrho(u)$ (see Problem I.2.146), hence it is homotopy equivalent to $S^{d-1} = \{u \in \mathbb{R}^d : |u| = 1\}$ with $d = \dim X$. Also, $\overline{B}_\varrho(u)$ is clearly homotopy equivalent to $\overline{B}_1^d = \{u \in \mathbb{R}^d : |u| \leqslant 1\}$. Hence from (5.403), we have

$$C_k(\varphi, u) = H_k(\overline{B}_1^d, S^{d-1}) = H_k(S^{d-1}, \star) = \delta_{k,d}\mathbb{Z} \qquad \forall\, k \geqslant 0,$$

with $\star \in S^{d-1}$ (see Proposition I.2.276).

When X is infinite dimensional, $\overline{B}_\varrho(u)$ and $\overline{B}_\varrho(u) \setminus \{u\}$ are both contractible and so from Proposition 5.144(f),(g), we have

$$H_k(\overline{B}_\varrho(u), \overline{B}_\varrho(u) \setminus \{u\}) = 0 \qquad \forall\, k \geqslant 0,$$

thus $C_k(\varphi, u) = 0$ for all $k \geqslant 0$.

Solution of Problem 5.152

There are three distinct possibilities:

(a) u is a local minimizer of φ. Then Problem 5.150 implies that $C_k(\varphi, u) = \delta_{k,0}\mathbb{Z}$ for all $k \geqslant 0$.

(b) u is a local maximizer of φ. Then Problem 5.151 implies that $C_k(\varphi, u) = \delta_{k,1}\mathbb{Z}$ for all $k \geqslant 0$.

(c) u is neither a local minimizer nor a local maximizer. We can find $\delta > 0$ such that $[u - \delta, u + \delta] \cap K_\varphi = \{u\}$. Since u is not a local extremum of φ, then φ is either increasing or decreasing on $[u - \delta, u + \delta]$. To fix things, we assume that φ is increasing on $[u - \delta, u + \delta]$ (the other case is treated similarly). Then

$$[u - \delta, u + \delta] \cap \varphi^{\varphi(u)} = [u - \delta, u].$$

Hence, from Definition 5.102, we have

$$C_k(\varphi, u) = H_k([u - \delta, u], [u - \delta, u]) = 0 \qquad \forall\, k \geqslant 0$$

(see Proposition 5.144(f),(g)).

Solution of Problem 5.153

The existence of a global minimizer $u_0 \in X$ follows from Problem 5.46. Also, from Problem 5.150, we have

$$C_k(\varphi, u_0) = \delta_{k,0}\mathbb{Z} \qquad \forall\, k \geqslant 0.$$

Let $a < \inf\limits_X \varphi$ (recall that φ is bounded below). From Definition 5.113, we have

$$C_k(\varphi, \infty) = H_k(X, \varphi^a) = H_k(X) = \delta_{k,0}\mathbb{Z} \qquad \forall\, k \geqslant 0$$

(since $\varphi^a = \emptyset$ and X is contractible).

Solution of Problem 5.154

Let $a < \varphi(0) < b = +\infty$. Using Problem 5.147, we have

$$H_k(X, \varphi^a) = C_k(\varphi, 0) \qquad \forall\, k \geqslant 0$$

(recall that $\varphi^\infty = X$), so $C_k(\varphi, 0) = C_k(\varphi, \infty)$ for all $k \geqslant 0$ (see Definition 5.113).

Solution of Problem 5.155

Arguing by contradiction, suppose that $K_\varphi = \{0\}$. Then according to Problem 5.154, we have

$$C_k(\varphi, \infty) = C_k(\varphi, 0) \qquad \forall\, k \geqslant 0,$$

a contradiction, since for $k = k_0$, by hypothesis $C_{k_0}(\varphi, 0) \neq C_{k_0}(\varphi, \infty)$.

Solution of Problem 5.156

(a) From Problem 5.149, we know that there exists $u \in K_\varphi$ such that $C_k(\varphi, u) \neq 0$. Since by hypothesis $C_k(\varphi, 0) = 0$, we infer that $u \neq 0$.

(b) Choose $\varepsilon > 0$ small such that $[-\varepsilon, \varepsilon] \cap \varphi(K_\varphi) = \{0\}$ and let $a < \inf \varphi(K_\varphi)$, $a < -\varepsilon$. From Problem 5.147, we have

$$\operatorname{rank} C_k(\varphi, 0) \leqslant \operatorname{rank} H_k(\varphi^\varepsilon, \varphi^{-\varepsilon}). \tag{5.404}$$

Also, from Definition 5.113, we have

$$C_k(\varphi, \infty) = H_k(X, \varphi^a). \tag{5.405}$$

We apply Proposition 5.145 to the sets $\varphi^a \subseteq \varphi^{-\varepsilon} \subseteq \varphi^\varepsilon \subseteq X$. We have

$$\begin{aligned}\operatorname{rank} H_k(\varphi^{-\varepsilon}, \varphi^\varepsilon) \leqslant &\operatorname{rank} H_{k-1}(\varphi^{-\varepsilon}, \varphi^a) + \operatorname{rank} H_{k+1}(X, \varphi^\varepsilon)\\ &+ \operatorname{rank} H_k(X, \varphi^a),\end{aligned}$$

so

$$\begin{aligned}\operatorname{rank} C_k(\varphi,0) \leqslant \operatorname{rank} H_{k-1}(\varphi^{-\varepsilon},\varphi^a) + \operatorname{rank} H_{k+1}(X,\varphi^{\varepsilon}) \\ + \operatorname{rank} C_k(\varphi,\infty)\end{aligned} \tag{5.406}$$

(see (5.404) and (5.405)). By hypothesis $\operatorname{rank} C_k(\varphi,0) \neq 0$ and $\operatorname{rank} C_k(\varphi,\infty) = 0$. So, from (5.406), we see that

$$H_{k-1}(\varphi^{-\varepsilon},\varphi^a) \neq 0 \quad \text{or} \quad H_{k+1}(X,\varphi^{\varepsilon}) \neq 0.$$

Then Problem 5.148 implies that we can find $u \in K_\varphi$, $u \neq 0$ such that

$$\varphi(u) \leqslant -\varepsilon < 0 \quad \text{and} \quad C_{k-1}(\varphi,u) \neq 0$$

or

$$\varphi(u) \geqslant -\varepsilon > 0 \quad \text{and} \quad C_{k+1}(\varphi,u) \neq 0.$$

Solution of Problem 5.157
From Problem 5.156, we know that we can find $u \in K_\varphi$ such that $\varphi(u) > 0$ and $C_1(\varphi,u) \neq 0$. Then Proposition 5.124 implies that $u \in K_\varphi$ is of mountain pass type.

Solution of Problem 5.158
From Problem 5.153 we know that $C_k(\varphi,\infty) = 0$ and φ has a global minimizer $u_0 \in X$ such that $C_k(\varphi,u_0) = \delta_{k,0}\mathbb{Z}$ for all $k \geqslant 0$. Since by hypothesis $C_k(\varphi,0) \neq 0$ with $k \geqslant 1$, we infer that $u_0 \neq 0$. If $k \geqslant 2$, then Problem 5.156 implies that we can find $\widehat{u} \in K_\varphi$, $\widehat{u} \neq 0$ such that

$$\big[\varphi(\widehat{u}) < 0 \text{ and } C_{k-1}(\varphi,\widehat{u}) \neq 0\big] \quad \text{or} \quad \big[\varphi(\widehat{u}) > 0 \text{ and } C_{k+1}(\varphi,\widehat{u}) \neq 0\big]. \tag{5.407}$$

Since $k \geqslant 2$, from (5.407) we see that $\widehat{u} \neq u_0$.

Solution of Problem 5.159

From Remark 5.140, we know that $u_0 \in C_0^1(\overline{\Omega})$. For every $u \in H_0^1(\Omega)$, we set

$$\begin{aligned}\psi(u) &= \varphi(u_0+u)-\varphi(u_0)\\ &= \tfrac{1}{2}\|D(u_0+u)\|_2^2-\tfrac{1}{2}\|Du_0\|_2^2-\int_\Omega \big(F(z,u_0+u)\\ &\quad -F(z,u_0)\big)\,dz\\ &= \tfrac{1}{2}\|Du\|_2^2+\int_\Omega (Du_0,Du)_{\mathbb{R}^N}\,dz-\int_\Omega \big(F(z,u_0+u)\\ &\quad -F(z,u_0)\big)\,dz\\ &= \tfrac{1}{2}\|Du\|_2^2-\int_\Omega \big(F(z,u_0+u)-F(z,u_0)-f(z,u_0)u\big)\,dz\end{aligned} \tag{5.408}$$

(since $u_0 \in K_\varphi$). Let $h(z,x)=f(z,u_0(z)+x)-f(z,u_0(z))$. Then h is a Carathéodory function,

$$|h(z,x)| \leqslant \widehat{c}(1+|x|^{r-1}) \quad \text{for a.a. } z\in\Omega, \text{ all } x\in\mathbb{R},$$

with $\widehat{c}>0$ and if $H(z,x)=\int\limits_0^x h(z,s)\,ds$, then

$$H(z,x) = F(z,u_0(z)+x)-F(z,u_0(z))-f(z,u_0(z))x.$$

So, from (5.408), we have

$$\psi(u) = \tfrac{1}{2}\|Du\|_2^2-\int_\Omega H(z,u)\,dz \qquad \forall\, u\in H_0^1(\Omega).$$

Note that

$$\langle\psi'(u),v\rangle = \langle\varphi'(u_0+u),v\rangle \qquad \forall\, v\in H_0^1(\Omega),$$

so $u=0$ is an isolated critical point of ψ. Therefore from the definition of h, we see that without any loss of generality, we may assume that $u_0=0$. Let $\delta>0$ be as in the hypothesis on f and g and let $\xi\colon\mathbb{R}\longrightarrow[-\delta,\delta]$ be a C^1-function, which is nondecreasing, $\xi(s)=-\delta$ if $s\leqslant-\delta$, $\xi(s)=s$ if $s\in[-\frac{\delta}{2},\frac{\delta}{2}]$ and $\xi(s)=\delta$ if $s\geqslant\delta$. We set

$$\widehat{\psi}(u) = \tfrac{1}{2}\|Du\|_2^2-\int_\Omega F(z,\xi(u))\,dz.$$

To establish the claim of the problem, it suffices to show that $u=0$ is an isolated critical point of $\widehat{\psi}$ and $C_k(\psi,0)=C_k(\widehat{\psi},0)$. We consider the one-parameter family of functions

$$\psi_t(t) = \tfrac{1}{2}\|Du\|_2^2-\int_\Omega F(z,(1-t)u+t\xi(u))\,dz \qquad \forall\, u\in H_0^1(\Omega),\ t\in[0,1].$$

According to Problem 5.32 each ψ_t satisfies the PS-condition on $\overline{B}_\varrho = \{u \in H_0^1(\Omega) : \|u\| \leqslant \varrho\}$ (see Definition 5.45(a)). Also, $t \longmapsto \psi_t(u)$ is clearly continuous. So, in order to apply Theorem 5.125, it suffices to show that for $\varrho > 0$ small, $\overline{B}_\varrho \cap K_{\psi_t} = \{0\}$ for all $t \in [0,1]$. Arguing by contradiction, suppose that we could find two sequences $\{t_n\}_{n\geqslant 1} \subseteq [0,1]$ and $\{u_n\}_{n\geqslant 1} \subseteq H_0^1(\Omega) \setminus \{0\}$ such that

$$u_n \longrightarrow 0 \quad \text{in } H_0^1(\Omega) \quad \text{and} \quad \psi'_{t_n}(u_n) = 0 \qquad \forall\, n \geqslant 1. \tag{5.409}$$

From (5.409), we have

$$-\Delta u_n(z) = (1 - t_n + t_n\xi'(u_n))f(z, (1-t_n)u_n + t_n\xi(u_n)) \quad \text{in } \Omega.$$

From Remark 5.140, we know that there exist $\alpha \in (0,1)$ and $M > 0$ such that

$$u \in C_0^{1,\alpha}(\overline{\Omega}) \quad \text{and} \quad \|u_n\|_{C_0^{1,\alpha}(\overline{\Omega})} \leqslant M \qquad \forall\, n \geqslant 1.$$

Exploiting the compactness of the embedding $C_0^{1,\alpha}(\overline{\Omega}) \hookrightarrow C_0^1(\overline{\Omega})$, we have

$$u_n \longrightarrow 0 \quad \text{in } C_0^1(\overline{\Omega})$$

(see (5.409)), so

$$|u_n(z)| \leqslant \tfrac{\delta}{2} \qquad \forall\, z \in \overline{\Omega},\ n \geqslant n_0,$$

thus $\{u_n\}_{n\geqslant 1}$ are critical points of ψ, a contradiction to the fact that $u = 0$ is an isolated critical point of φ. So, we can apply Theorem 5.125 and have

$$C_k(\psi, 0) = C_k(\widehat{\psi}, 0) \qquad \forall\, k \geqslant 0,$$

thus

$$C_k(\varphi, u_0) = C_k(\widehat{\varphi}, u_0) \qquad \forall\, k \geqslant 0.$$

Remark. This problem reveals the local character of the critical groups.

Solution of Problem 5.160
(a) We have

$$\varphi_\lambda(u) \geqslant \tfrac{1}{2}\|Du\|_2^2 - \tfrac{1}{2}\tfrac{\lambda}{\widehat{\lambda}_1}\|Du\|_2^2 = \tfrac{1}{2}(1 - \tfrac{\lambda}{\widehat{\lambda}_1})\|Du\|_2^2 \qquad \forall\, u \in H_0^1(\Omega),$$

so $u = 0$ is the unique global minimizer of φ_λ (recall that $\lambda < \widehat{\lambda}_1$), thus

$$C_k(\varphi_\lambda, 0) = \delta_{k,0}\mathbb{Z} \qquad \forall\, k \geqslant 0$$

(see Problem 5.150).

(b) Note that $\varphi_\lambda \in C^2(H_0^1(\Omega))$ and

$$\langle \varphi_\lambda''(u)v, h\rangle = \int_\Omega (Dv, Dh)_{\mathbb{R}^N}\, dz - \lambda \int_\Omega vh\, dz \qquad \forall\, u, v, h \in H_0^1(\Omega).$$

We consider the orthogonal direct sum decomposition

$$H_0^1(\Omega) = \overline{H}_m \oplus \widehat{H}_{m+1},$$

where $\overline{H}_m = \bigoplus\limits_{i=1}^{m} E(\widehat{\lambda}_i)$ and $\widehat{H}_{m+1} = \overline{H}_m^{\perp} = \overline{\bigoplus\limits_{i\geqslant m+1} E(\widehat{\lambda}_i)}$. Since $\lambda \in (\widehat{\lambda}_m, \widehat{\lambda}_{m+1})$, we have

$$\begin{aligned} \langle \varphi_\lambda''(u)v, v\rangle &< 0 \qquad \forall\, v \in \overline{H}_m \setminus \{0\},\ u \in H_0^1(\Omega), \\ \langle \varphi_\lambda''(u)v, v\rangle &> 0 \qquad \forall\, v \in \widehat{H}_{m+1} \setminus \{0\},\ u \in H_0^1(\Omega). \end{aligned}$$

Using Theorem 5.106, we conclude that

$$C_k(\varphi_\lambda, 0) = \delta_{k,d_m}\mathbb{Z} \qquad \forall\, k \geqslant 0,$$

with $d_m = \dim \overline{H}_m$.

Solution of Problem 5.161

(a) Let $g\colon \Omega \times \mathbb{R} \longrightarrow \mathbb{R}$ be the Carathéodory function defined by

$$g(z,x) = \begin{cases} f(z,x) & \text{if } |x| \leqslant \delta, \\ f(z,\delta) & \text{if } |x| > \delta. \end{cases} \tag{5.410}$$

Set $G(z,x) = \int\limits_0^x g(z,s)\, ds$ and consider the C^1-functional $\widehat{\varphi}\colon H_0^1(\Omega) \longrightarrow \mathbb{R}$ defined by

$$\widehat{\varphi}(u) = \tfrac{1}{2}\|Du\|_2^2 - \int_\Omega F(z,u)\, dz \qquad \forall\, u \in H_0^1(\Omega).$$

Because of (5.410) and Problem 5.159, $u = 0$ is an isolated critical point of $\widehat{\varphi}$ and

$$C_k(\varphi, 0) = C_k(\widehat{\varphi}, 0) \qquad \forall\, k \geqslant 0. \tag{5.411}$$

Note that

$$\widehat{\varphi}(u) \geqslant \tfrac{1}{2}\|Du\|_2^2 - \tfrac{\widehat{\lambda}_1}{2}\|u\|_2^2 \geqslant 0 \qquad \forall\, u \in H_0^1(\Omega),$$

so $u = 0$ is a global minimizer of φ, thus

$$C_k(\varphi, 0) = \delta_{k,0}\mathbb{Z} \qquad \forall\, k \geqslant 0$$

(see Problem 5.150).

(b) Let $\widehat{\varphi} \in C^1(H_0^1(\Omega))$ be as in part **(a)** of the problem. We consider the orthogonal direct sum decomposition

$$H_0^1(\Omega) = \overline{H}_m \oplus \widehat{H}_{m+1},$$

with $\overline{H}_m = \bigoplus_{i=1}^{m} E(\widehat{\lambda}_i)$ and $\widehat{H}_{m+1} = \overline{H}_m^{\perp} = \overline{\bigoplus_{i \geqslant m+1} E(\widehat{\lambda}_i)}$. Then every $u \in H_0^1(\Omega)$ can be written in a unique fashion as

$$u = \overline{u} + \widehat{u}, \quad \text{with } \overline{u} \in \overline{H}_m,\ \widehat{u} \in \widehat{H}_{m+1}.$$

We consider the C^2-functional $\widehat{\psi}\colon H_0^1(\Omega) \longrightarrow \mathbb{R}$ defined by

$$\widehat{\psi}(u) = \tfrac{1}{2}\|D\widehat{u}\|_2^2 - \tfrac{1}{2}\|D\overline{u}\|_2^2 \qquad \forall\, u \in H_0^1(\Omega).$$

We consider the map $h(t,u) = (1-t)\widehat{\varphi}(u) + t\widehat{\psi}(u)$. As in the solution of Problem 5.161 in order to use Theorem 5.125, we need to show that we cannot find two sequences $\{t_n\}_{n\geqslant 1} \subseteq [0,1]$ and $\{u_n\}_{n\geqslant 1} \subseteq H_0^1(\Omega)$ such that

$$t_n \longrightarrow t, \quad u_n \longrightarrow 0 \quad \text{and} \quad h_u'(t_n,u_n) = 0. \tag{5.412}$$

Arguing by contradiction, suppose that (5.412) holds. Then we have

$$(1-t_n)A(u_n) + t_nA(\widehat{u}_n) - tA(\overline{u}_n) = (1-t_n)N_f(u_n) \qquad \forall\, n \geqslant 1. \tag{5.413}$$

Recall that

$$\langle A(u), v\rangle = \int_\Omega (Du, Dv)_{\mathbb{R}^N}\, dz \qquad \forall\, u,v \in H_0^1(\Omega)$$

and

$$N_f(u)(\cdot) = f(\cdot, u(\cdot)) \qquad \forall\, u \in H_0^1(\Omega).$$

From (5.413), we have

$$\begin{aligned} &-(1-t_n)\Delta u_n(z) - t_n\Delta\widehat{u}_n(z) + t_n\Delta\overline{u}_n(z) \\ &= (1-t_n)f(z,u_n(z)) \quad \text{for a.a. } z \in \Omega. \end{aligned}$$

So, we can find $\alpha \in (0,1)$ and $M > 0$ such that

$$u_n \in C_0^{1,\alpha}(\overline{\Omega}) \quad \text{and} \quad \|u_n\|_{C_0^{1,\alpha}(\overline{\Omega})} \leqslant M \qquad \forall\, n \geqslant 1 \tag{5.414}$$

(see Remark 5.140). Due to the compactness of the embedding $C_0^{1,\alpha}(\overline{\Omega}) \hookrightarrow C_0^1(\overline{\Omega})$, from (5.412) and (5.414), we have

$$u_n \longrightarrow 0 \quad \text{in } C_0^1(\overline{\Omega}).$$

So, we can find $n_0 \geqslant 1$ such that

$$|u_n(z)| \leqslant \delta \qquad \forall\, z \in \overline{\Omega},\ n \geqslant n_0. \tag{5.415}$$

On (5.415) we act with $v = \widehat{u}_n - \overline{u}_n \in H_0^1(\Omega)$. Due to the orthogonality of the component spaces, we have

$$\begin{aligned} &(1-t_n)\|D\widehat{u}_n\|_2^2 - (1-t_n)\|D\overline{u}_n\|_2^2 + t_n\|D\widehat{u}_n\|_2^2 - t_n\|D\overline{u}_n\|_2^2 \\ = \; &(1-t_n)\int\limits_\Omega \tfrac{f(z,u_n)}{u_n} u_n(\widehat{u}_n - \overline{u}_n)\,dz, \end{aligned}$$

so

$$\|D\widehat{u}_n\|_2^2 - \|D\overline{u}_n\|_2^2 \leqslant (1-t_n)\big(\widehat{\lambda}_{m+1}\|\widehat{u}_n\|_2^2 - \widehat{\lambda}_m\|\overline{u}_n\|_2^2\big)$$

(see (5.415) and the hypothesis), thus for all $n \geqslant n_0$, we have

$$\|D\widehat{u}_n\|_2^2 - (1-t_n)\widehat{\lambda}_{m+1}\|\widehat{u}_n\|_2^2 - \big(\|D\overline{u}_n\|_2^2 - (1-t_n)\widehat{\lambda}_m\|\overline{u}_n\|_2^2\big) \leqslant 0. \quad (5.416)$$

We know that $u = 0$ is an isolated critical point of $\widehat{\varphi}$. Hence in (5.412), we have that $t_n \in [0,1)$ for all $n \geqslant n_0$. This fact and (5.416) leads to a contradiction (see Theorem 5.129). Therefore we can use Theorem 5.125 and we have

$$C_k(\widehat{\varphi},0) = C_k(\widehat{\psi},0) \qquad \forall\, k \geqslant 0. \quad (5.417)$$

But from Proposition 5.124, we have

$$C_k(\widehat{\psi},0) = \delta_{k,d_m}\mathbb{Z} \qquad \forall\, k \geqslant 0,$$

with $d_m = \dim \overline{H}_m$, so

$$C_k(\widehat{\varphi},0) = \delta_{k,d_m}\mathbb{Z} \qquad \forall\, k \geqslant 0$$

(see (5.417)), thus

$$C_k(\varphi,0) = \delta_{k,d_m}\mathbb{Z} \qquad \forall\, k \geqslant 0$$

(see (5.411)).

Solution of Problem 5.162

Recall that $\widehat{\lambda}_1 > 0$ (see Theorem 5.129). From the sign condition on f, we have

$$\tfrac{f(z,x)}{x} \leqslant 0 < \widehat{\lambda}_1 \quad \text{for a.a. } z \in \Omega, \text{ all } |x| \leqslant \delta.$$

So, using Problem 5.161(a), we conclude that $C_k(\varphi,0) = \delta_{k,0}\mathbb{Z}$ for all $k \geqslant 0$.

Solution of Problem 5.163

Let $\eta = \inf\limits_{u \in \overline{U} \setminus K} |\varphi'(u)| > 0$. Consider a function $\vartheta \in C^\infty(\mathbb{R}^N)$ such that

$$\vartheta(u) = \begin{cases} 1 & \text{if} \quad u \in K, \\ 0 & \text{if} \quad u \notin U. \end{cases}$$

Also, we pick $R, \beta > 0$ such that

$$U \subseteq B_R(0) = \{u \in \mathbb{R}^N : |u| < R\},$$

$$\beta R \|\vartheta\|_\infty \leqslant \tfrac{\varepsilon}{2} \quad \text{and} \quad \beta\|\vartheta\|_\infty + \beta R \|\vartheta'\|_\infty \leqslant \tfrac{1}{2}\min\{\eta, \varepsilon\}.$$

Using the Sard theorem (see Theorem 4.2), we can find $h_0 \in \mathbb{R}^N$ with $|h_0| \leqslant \beta$ such that $-h_0$ is not a critical value of φ' (that is, $\varphi''(u)$ is invertible if $\varphi'(u) = -h_0$). We define $\widehat{\varphi} \in C^2(\mathbb{R}^N)$ by setting

$$\widehat{\varphi}(u) = \varphi(u) + \vartheta(u)\,(u, h_0)_{\mathbb{R}^N} \qquad \forall\, u \in \mathbb{R}^N.$$

We have

$$\widehat{\varphi}'(u) = \varphi'(u) + \vartheta(u) h_0 + \vartheta'(u)\,(u, h_0)_{\mathbb{R}^N} \qquad \forall\, u \in \mathbb{R}^N. \tag{5.418}$$

The choices of ϑ and h_0 imply that $\widehat{\varphi}$ has properties (a) and **(b)**. Also

$$\inf_{u \in \overline{U} \setminus K} |\widehat{\varphi}'(u)| = \tfrac{\eta}{2}. \tag{5.419}$$

Let $u \in K_{\widehat{\varphi}} \cap \overline{U}$. Then from (5.419) it follows that $u \in \operatorname{int} K$. Then from (5.418), we have

$$0 = \widehat{\varphi}'(u) = \varphi'(u) + h_0.$$

Since $-h_0$ is not a critical value of φ', it follows that

$$\widehat{\varphi}''(u) = \varphi''(u) \text{ is invertible,}$$

so $K_{\widehat{\varphi}} \cap \overline{U}$ has nondegenerate elements, thus isolated.

Recall that $K_{\widehat{\varphi}} \cap \overline{U} \subseteq K$. So, it follows that the set $K_{\widehat{\varphi}} \cap \overline{U}$ is finite. This proves **(c)** and finishes the solution of the problem.

Solution of Problem 5.164

First suppose that **(ii)(I)** holds. Hypothesis **(i)** and **(ii)** imply that given $\varepsilon > 0$, we can find $c_\varepsilon > 0$ such that

$$F(z, x) \leqslant \tfrac{1}{2}(\widehat{\eta}(z) + \varepsilon)x^2 + c_\varepsilon |x|^r \quad \text{for a.a. } z \in \Omega, \text{ all } x \in \mathbb{R}. \tag{5.420}$$

For all $u \in H_0^1(\Omega)$, we have

$$\begin{aligned} \varphi(u) &\geqslant \tfrac{1}{2}\Big(\|Du\|_2^2 - \int_\Omega \eta(z)u^2\,dz\Big) - \tfrac{\varepsilon}{2}\|u\|_2^2 - c_\varepsilon\|u\|_r^r \\ &\geqslant \tfrac{1}{2}\big(c_0 - \tfrac{\varepsilon}{\widehat{\lambda}_1}\big)\|Du\|_2^2 - c_1\|u\|^r, \end{aligned}$$

for some $c_0, c_1 > 0$ (see (5.420) and Problem 5.67). Choosing $\varepsilon \in (0, \widehat{\lambda}_1 c_0)$, we obtain

$$\varphi(u) \geqslant c_2\|u\|^2 - c_1\|u\|^r.$$

Since $r > 2$, we can find $\varrho \in (0,1)$ small such that

$$\varphi(u) \geqslant 0 = \varphi(0) \qquad \forall\, u \in H_0^1(\Omega),\ \|u\| \leqslant \varrho,$$

so $u = 0$ is a local minimizer of φ, thus $u \in K_\varphi$.

We show that $0 \in K_\varphi$ is isolated. Suppose that this is not the case. Then we can find a sequence $\{u_n\}_{n\geqslant 1} \subseteq H_0^1(\Omega)$ such that

$$u_n \longrightarrow 0 \quad \text{in } H_0^1(\Omega) \quad \text{and} \quad \varphi'(u_n) = 0 \qquad \forall\, n \geqslant 1.$$

We have

$$A(u_n) = N_f(u_n) \qquad \forall\, n \geqslant 1, \tag{5.421}$$

where $A \in \mathcal{L}(H_0^1(\Omega), H^{-1}(\Omega))$ is defined by

$$\langle A(u), h\rangle = \int_\Omega (Du, Dh)_{\mathbb{R}^N}\,dz \qquad \forall\, u, h \in H_0^1(\Omega)$$

and N_f is given by

$$N_f(u)(\cdot) = f(\cdot, u(\cdot)) \qquad \forall\, u \in H_0^1(\Omega).$$

Let $y_n = \frac{u_n}{\|u_n\|}$ for all $n \geqslant 1$. Then $\|y_n\| = 1$ for all $n \geqslant 1$ and so we may assume that

$$y_n \xrightarrow{w} y \quad \text{in } H_0^1(\Omega) \quad \text{and} \quad y_n \longrightarrow y \quad \text{in } L^r(\Omega). \tag{5.422}$$

From (5.421), we have

$$A(y_n) = \frac{N_f(u_n)}{\|u_n\|} \qquad \forall\, n \geqslant 1. \tag{5.423}$$

Hypothesis **(ii)(I)** implies that

$$\frac{N_f(u_n)}{\|u_n\|} \xrightarrow{w} \vartheta y \quad \text{in } L^{r'}(\Omega) \tag{5.424}$$

(where $\frac{1}{r} + \frac{1}{r'} = 1$), with $\vartheta \in L^\infty(\Omega)$ such that

$$\eta(z) \leqslant \vartheta(z) \leqslant \widehat{\eta}(z) \quad \text{for a.a. } z \in \Omega \tag{5.425}$$

(see the solution of Problem 5.68). In (5.423) we pass to the limit as $n \to +\infty$ and use (5.422) and (5.424). Then

$$A(y) = \vartheta y,$$

so

$$\begin{cases} -\Delta y(z) = \vartheta(z)y(z) & \text{for a.a. } z \in \Omega, \\ y|_{\partial\Omega} = 0. \end{cases} \tag{5.426}$$

From Proposition 5.131 and (5.425), we have

$$\widehat{\lambda}_1(\vartheta) > \widehat{\lambda}_1(\widehat{\lambda}_1) = 1,$$

so $y = 0$ (see (5.426)). On the other hand, if on (5.423) we act with $y_n - y \in H_0^1(\Omega)$ and pass to the limit as $n \to +\infty$, then

$$\lim_{n\to+\infty} \int_\Omega (Dy_n, Dy_n - Dy)_{\mathbb{R}^N}\, dz = 0,$$

so

$$\|Dy_n\|_2 \longrightarrow \|Dy\|_2,$$

thus

$$y_n \longrightarrow y \quad \text{in } H_0^1(\Omega)$$

(by the Kadec–Klee property; see Corollary 1.26), hence $\|y\| = 1$, a contradiction. So, $0 \in K_\varphi$ is isolated and then Problem 5.150 implies that

$$C_k(\varphi, 0) = \delta_{k,0}\mathbb{Z} \qquad \forall\, k \geqslant 0.$$

Now suppose that hypothesis **(ii)(II)** holds. Fix $\lambda \in (\widehat{\lambda}_m, \widehat{\lambda}_{m+1})$ and let $\psi\colon H_0^1(\Omega) \longrightarrow \mathbb{R}$ be the C^2-functional defined by

$$\psi(u) = \tfrac{1}{2}\|Du\|_2^2 - \tfrac{\lambda}{2}\|u\|_2^2 \qquad \forall\, u \in H_0^1(\Omega).$$

We consider the function $h(t,u) = (1-t)\varphi(u) + t\psi(u)$ for all $(t,u) \in [0,1] \times H_0^1(\Omega)$. Suppose that we could find two sequences $\{t_n\}_{n\geqslant 1} \subseteq [0,1]$ and $\{u_n\}_{n\geqslant 1} \subseteq H_0^1(\Omega)$ such that

$$t_n \longrightarrow t, \quad u_n \longrightarrow 0 \ \text{ in } H_0^1(\Omega) \ \text{ and } \ h'_u(t_n, u_n) = 0 \qquad \forall\, n \geqslant 1.$$

Setting as before $y_n = \frac{u_n}{\|u_n\|}$ for $n \geqslant 1$, we have $\|y_n\| = 1$ for all $n \geqslant 1$ and (5.422), (5.423) hold. In the limit as $n \to +\infty$, we obtain

$$\begin{cases} -\Delta y(z) = \vartheta(z)y(z) & \text{for a.a. } z \in \Omega, \\ y|_{\partial\Omega} = 0. \end{cases} \tag{5.427}$$

We have

$$\widehat{\lambda}_m(\widehat{\lambda}_m) = 1 > \widehat{\lambda}_m(\vartheta) \ \text{ and } \ \widehat{\lambda}_{m+1}(\widehat{\lambda}_{m+1}) = 1 < \widehat{\lambda}_{m+1}(\vartheta),$$

so $y = 0$ (see (5.427)). But as before the Kadec–Klee property gives that

$$y_n \longrightarrow y \quad \text{in } H_0^1(\Omega),$$

so $\|y\| = 1$, a contradiction. So, we have

$$C_k(\varphi, 0) = C_k(\psi, 0) \qquad \forall\, k \geqslant 0$$

and thus

$$C_k(\varphi, 0) = \delta_{k,d_m}\mathbb{Z} \qquad \forall\, k \geqslant 0,$$

with $d_m = \dim \bigoplus\limits_{i=1}^{m} E(\widehat{\lambda}_i)$ (see Problem 5.160).

Solution of Problem 5.165

Since by hypothesis $\lambda \in \mathbb{R} \setminus \sigma(2)$, we have $K_\varphi = \{0\}$ and so Problem 5.155 implies that

$$C_k(\varphi_\lambda, \infty) = C_k(\varphi_\lambda, 0) \qquad \forall\, k \geqslant 0. \tag{5.428}$$

From Problem 5.160, we have:

- If $\lambda < \widehat{\lambda}_1$, then $C_k(\varphi_\lambda, 0) = \delta_{k,0}\mathbb{Z}$ for all $k \geqslant 0$. Hence

  $$C_k(\varphi_\lambda, \infty) = \delta_{k,0}\mathbb{Z} \qquad \forall\, k \geqslant 0$$

  (see (5.428)).

- If $\lambda \in (\widehat{\lambda}_m, \widehat{\lambda}_{m+1})$ for some $m \geqslant 1$, then $C_k(\varphi_\lambda, 0) = \delta_{k,d_m}$ for all $k \geqslant 0$, with $d_m = \dim \bigoplus\limits_{i=1}^{m} E(\widehat{\lambda}_i)$. Hence

  $$C_k(\varphi_\lambda, \infty) = \delta_{k,d_m}\mathbb{Z} \qquad \forall\, k \geqslant 0$$

  (see (5.428)).

Solution of Problem 5.166

First we show that

$$\text{if } \|u\| \longrightarrow +\infty, \text{ then } \varphi(u) \longrightarrow +\infty. \tag{5.429}$$

Arguing by contradiction, suppose that (5.429) is not true. Then we can find a sequence $\{u_n\}_{n\geqslant 1} \subseteq W_0^{1,p}(\Omega)$ and $M > 0$ such that

$$\|u_n\| \longrightarrow +\infty \quad \text{and} \quad \varphi(u_n) \leqslant M \qquad \forall\, n \geqslant 1. \tag{5.430}$$

From (5.430), we have

$$\frac{1}{p}\|Du_n\|_p^p - \int_\Omega F(z,u_n)\,dz \leqslant M \qquad \forall\, n \geqslant 1. \tag{5.431}$$

Let $y_n = \frac{u_n}{\|u_n\|}$ for all $n \geqslant 1$. Then $\|y_n\| = 1$ for all $n \geqslant 1$ and so we may assume that

$$y_n \xrightarrow{w} y \quad \text{in } W_0^{1,p}(\Omega) \quad \text{and} \quad y_n \longrightarrow y \quad \text{in } L^p(\Omega) \tag{5.432}$$

(see the Rellich–Kondrachov embedding theorem; Theorem 1.135). From (5.431), we have

$$\frac{1}{p}\|Dy_n\|_p^p - \int_\Omega \frac{F(z,u_n)}{\|u_n\|^p}\,dz \leqslant \frac{M}{\|u_n\|^p} \qquad \forall\, n \geqslant 1. \tag{5.433}$$

Hypothesis **(i)** implies that

$$|F(z,x)| \leqslant \widehat{c}(1+|x|^p) \quad \text{for a.a. } z \in \Omega, \text{ all } x \in \mathbb{R},$$

with $\widehat{c} > 0$. From this growth estimate and (5.432), we see that the sequence $\left\{\frac{F(\cdot,u_n(\cdot))}{\|u_n\|^p}\right\}_{n\geqslant 1} \subseteq L^1(\Omega)$ is uniformly integrable. So, using the Dunford–Pettis theorem (see Theorem 1.28), we may assume that

$$\frac{F(\cdot,u_n(\cdot))}{\|u_n\|^p} \xrightarrow{w} \frac{1}{p}\vartheta|y|^p \quad \text{in } L^1(\Omega), \tag{5.434}$$

with $\vartheta \in L^\infty(\Omega)$, $\vartheta(z) \leqslant \widehat{\lambda}_1$ for almost all $z \in \Omega$ (see hypothesis **(ii)** and the solution of Problem 5.68). Hence, if in (5.433) we pass to the limit as $n \to +\infty$ and use (5.432) and (5.434), then

$$\|Dy\|_p^p - \int_\Omega \vartheta(z)|y|^p\,dz \leqslant 0. \tag{5.435}$$

First suppose that the inequality $\vartheta(z) \leqslant \widehat{\lambda}_1$ is strict on a set of positive measure. Then from (5.435) and Problem 5.67, we have

$$c_0\|Dy\|_p^p \leqslant 0,$$

so $y = 0$. Thus, from (5.433) and (5.434), we have

$$y_n \longrightarrow y \quad \text{in } W_0^{1,p}(\Omega),$$

a contradiction to the fact that $\|y_n\| = 1$ for all $n \geqslant 1$.

Next we assume that $\vartheta(z) = \widehat{\lambda}_1$ for almost all $z \in \Omega$. From (5.435) and Proposition 5.132, we have

$$y = \xi\widehat{u}_1,$$

with $\xi \in \mathbb{R}$. If $\xi = 0$, then as above we have $y_n \longrightarrow 0$ in $W_0^{1,p}(\Omega)$, a contradiction to the fact that $\|y_n\| = 1$ for all $n \geqslant 1$. So, $\xi \neq 0$. To fix things, we assume $\xi > 0$

(the reasoning is similar if $\xi < 0$). We have $y \in \operatorname{int} C_+$ (see Theorem 5.129(e)) and so $y(z) > 0$ for all $z \in \Omega$. This means that

$$u_n(z) \longrightarrow +\infty \quad \text{for a.a. } z \in \Omega. \tag{5.436}$$

For almost all $z \in \Omega$ and all $x > 0$, we have

$$\tfrac{d}{dx}\left(\tfrac{F(z,x)}{x^p}\right) = \tfrac{f(z,x)x^p - px^{p-1}F(z,x)}{x^{2p}} = \tfrac{f(z,x)x - pF(z,x)}{x^{p+1}}. \tag{5.437}$$

Hypothesis **(iii)** implies that given any $\eta > 0$, we can find $\widehat{M} = \widehat{M}(\eta) > 0$ such that

$$f(z,x)x - pF(z,x) \geqslant \eta \quad \text{for a.a. } z \in \Omega, \text{ all } x \geqslant \widehat{M}. \tag{5.438}$$

Returning to (5.437) and using (5.438), we have

$$\tfrac{d}{dx}\left(\tfrac{F(z,x)}{x^p}\right) \geqslant \tfrac{\eta}{x^{p+1}} \quad \text{for a.a. } z \in \Omega, \text{ all } x \geqslant \widehat{M},$$

so

$$\tfrac{F(z,x)}{x^p} - \tfrac{F(z,v)}{v^p} \geqslant -\tfrac{\eta}{p}\left(\tfrac{1}{x^p} - \tfrac{1}{v^p}\right) \quad \text{for a.a. } z \in \Omega, \text{ all } x \geqslant v \geqslant \widehat{M}. \tag{5.439}$$

Let $x \to +\infty$. Then from (5.439) and using hypothesis **(ii)**, we have

$$\tfrac{\widehat{\lambda}_1}{p} - \tfrac{F(z,v0}{v^p} \geqslant \tfrac{\eta}{p}\tfrac{1}{v^p} \quad \text{for a.a. } z \in \Omega, \text{ all } v \geqslant M,$$

so

$$\widehat{\lambda}_1 v^p - pF(z,v) \geqslant \eta \quad \text{uniformly for a.a. } z \in \Omega, \text{ all } v \geqslant M.$$

Since $\eta > 0$ is arbitrary, we infer that

$$\widehat{\lambda}_1 v^p - pF(z,v) \longrightarrow +\infty \quad \text{uniformly for a.a. } z \in \Omega, \text{ as } v \to +\infty.$$

In a similar fashion, we show that

$$\widehat{\lambda}_1 |v|^p - pF(z,v) \longrightarrow +\infty \quad \text{uniformly for a.a. } z \in \Omega, \text{ as } v \to -\infty.$$

Therefore

$$\widehat{\lambda}_1 |v|^p - pF(z,v) \longrightarrow +\infty \quad \text{uniformly for a.a. } z \in \Omega, \text{ as } |v| \to +\infty. \tag{5.440}$$

From (5.436), (5.440), and the Fatou lemma, we have

$$\int_\Omega \left(\widehat{\lambda}_1 |u_n|^p - pF(z,u_n)\right) dz \longrightarrow +\infty. \tag{5.441}$$

On the other hand, from (5.440), we have

$$\|Du_n\|_p^p - \int_\Omega pF(z,u_n)\,dz \leqslant pM \qquad \forall\, n \geqslant 1,$$

so

$$\widehat{\lambda}_1 \|u_n\|_p^p - \int_\Omega pF(z, u_n)\, dz \leqslant pM \qquad \forall\, n \geqslant 1$$

(see Proposition 5.132), thus

$$\int_\Omega \big(\widehat{\lambda}_1 |u_n|^p - pF(z, u_n)\big)\, dz \leqslant pM \qquad \forall\, n \geqslant 1. \tag{5.442}$$

Comparing (5.441) and (5.442) we reach a contradiction. This means that (5.430) is true and so φ is bounded below. Also, every Cerami sequence is bounded and so using Problem 2.192, we infer that φ satisfies the C-condition. Therefore, we can use Problem 5.153 and conclude that $C_k(\varphi, \infty) = \delta_{k,0}\mathbb{Z}$ for all $k \geqslant 0$.

Solution of Problem 5.167

Let $i\colon (E, E_0) \longrightarrow (X, X \setminus D)$ and $j\colon (h(E), h(E_0)) \longrightarrow (X, X \setminus D)$ be the inclusion maps. We consider the following commutative diagram

$$\begin{array}{ccc} H_m(E, E_0) & \xrightarrow{\ i_*\ } & H_m(X, X \setminus D) \\ \Big\downarrow{\scriptstyle h_*} & & \Big\downarrow{\scriptstyle h_*} \\ H_m(h(E), h(E_0)) & \xrightarrow[\ j_*\ ]{} & H_m(X, X \setminus h(D)) \end{array} .$$

Then from this diagram and since i_* is nontrivial (see Definition 5.118), it follows that j_* is nontrivial and this by Definition 5.118 means that $\{h(E_0), h(E)\}$ and $h(D)$ homologically link in dimension m.

Solution of Problem 5.168

Let $\star \in E_0$. For every $m \geqslant 0$, we have the following commutative diagram of homomorphisms

$$\begin{array}{ccc} H_m(E, E_0) & \xrightarrow{\ \partial_*^1\ } & H_{m-1}(E_0, \star) \\ \Big\downarrow{\scriptstyle i_*} & & \Big\downarrow{\scriptstyle j_*} \\ H_m(X, X \setminus D) & \xrightarrow[\ \partial_*^2\ ]{} & H_{m-1}(X \setminus D, \star) \end{array} , \tag{5.443}$$

where ∂_*^1 and ∂_*^2 are the boundary homomorphisms and j_* is the homomorphism induced by the inclusion map $j\colon (E_0, \star) \longrightarrow (X \setminus D, \star)$. When E is contractible, using the long exact sequence from Theorem I.2.277, we see that ∂_*^1 and ∂_*^2 are isomorphisms. So, from (5.443), it follows that

$$i_* \text{ is nontrivial} \quad \Longleftrightarrow \quad j_* \text{ is nontrivial},$$

so "$\{E, E_0\}$ and D homologically link in dimension m" if and only if "$\{E_0, \star\}$ and D homologically link in dimension $m-1$."

Solution of Problem 5.169

Let $j\colon (E_0, \{u_1\}) \longrightarrow (X \setminus D_1, \{u_1\})$ be the inclusion map and let

$$\xi\colon (X \setminus D, \{u_1\}) \longrightarrow (E_0, \{u_1\})$$

be the map defined by

$$\xi(u) = \begin{cases} u_0 & \text{if } u \in U, \\ u_1 & \text{if } u \in X \setminus \overline{U}. \end{cases}$$

We have $\xi \circ j = I_{(E_0,\{u_1\})}$ and so $j_*\colon H_0(E_0, \{u_1\}) \longrightarrow H_0(X \setminus D, \{u_1\})$ is injective. But $H_0(E_0, \{u_1\}) = \mathbb{Z}$ and so j_* is nontrivial. Since E is contractible, Problem 5.168 implies that $\{E_0, E\}$ and D homologically link in dimension 1 (see the solution of Problem 5.168).

Solution of Problem 5.170

Let $a < \min\{\inf\limits_V \varphi, \inf \varphi(K_\varphi)\}$ (see hypothesis **(i)**). Also, because of hypothesis **(ii)**, we can find $M > 0$ big such that $\varphi(y) \leqslant a$ for all $y \in Y$ with $\|y\| \geqslant M$. Let

$$C_M^Y = \{y \in Y : \|y\| = M\}.$$

We have $C_M^Y \subseteq \varphi^a \subseteq X \setminus V \subseteq X$. Let $h\colon [0,1] \times (X \setminus V) \longrightarrow X \setminus V$ be defined by

$$h(t, y+v) = (1-t)(y+v) + Mt\tfrac{y}{\|y\|} \qquad \forall\, x = y+v \in X \setminus V$$

(here $y \in Y$, $v \in V$ since $X = Y \oplus V$). Note that for all $t \in [0,1]$, $h(t,\cdot)|_{C_M^Y} = I_{C_M^Y}$ and $h(1, X \setminus V) \subseteq C_M^Y$. Hence C_M^Y is a strong deformation retract of $X \setminus V$ (see Definition 4.112(b)). It follows that

$$H_k(X \setminus V, C_M^Y) = 0 \qquad \forall\, k \geqslant 0 \tag{5.444}$$

(see Proposition 5.144(b)). We have the following commutative diagram of homomorphisms

$$\begin{array}{ccccc} H_k(\varphi^a, C_M^Y) & \xrightarrow{\;i_*\;} & H_k(X, C_M^Y) & \xrightarrow{\;j_*\;} & H_k(X, \varphi^a) = C_k(\varphi, \infty) \\ & \searrow^{\eta_*} & \uparrow{\scriptstyle \vartheta_*} & & \\ & & H_k(X \setminus V, C_M^Y) = 0 & & \end{array} \tag{5.445}$$

(see (5.444)), with i_*, j_*, η_*, ϑ_* being the homomorphisms induced by the corresponding inclusion maps. In (5.445) the top row is exact. We have $i_* = 0$ (see (5.445)) and so j_* is an injection for all $k \geqslant 0$. From Theorem I.2.277 (the long exact sequence for the triple $\star \in C_M^Y \subseteq X$), we have

$$H_d(X, C_M^Y) \;=\; H_{d-1}(C_M^Y, \star) \;=\; \mathbb{Z} \tag{5.446}$$

(see Problem I.2.183). From (5.445) and (5.446) we conclude that $C_d(\varphi, \infty) \neq 0$.

Solution of Problem 5.171
According to Definition 5.113, we have

$$C_k(\varphi, \infty) \;=\; H_k(X, \varphi^a) \qquad \forall\, k \geqslant 0. \tag{5.447}$$

From Theorem 5.48, we know that φ^b is a strong deformation retract of X. So, Proposition 5.144(b) implies that

$$H_k(X, \varphi^b) \;=\; 0 \qquad \forall\, k \geqslant 0. \tag{5.448}$$

Using the long exact sequence of singular homology groups for the triple $\varphi^a \subseteq \varphi^b \subseteq X$ (see Theorem I.2.277) together with (5.447), (5.448), we conclude that

$$C_k(\varphi, \infty) \;=\; H_k(\varphi^b, \varphi^a) \qquad \forall\, k \geqslant 0.$$

Solution of Problem 5.172
We have $a = \inf \varphi(K_\varphi) = +\infty$, so $\varphi^a = X$. Hence from Definition 5.113, we have

$$C_k(\varphi, \infty) \;=\; H_k(X, \varphi^a) \;=\; H_k(X, X) \;=\; 0 \qquad \forall\, k \geqslant 0$$

(see Proposition 5.144(b)).

Solution of Problem 5.173

First we show that hypothesis $C_0(\varphi, \infty) \neq 0$ implies that φ is bounded below. To see this, we argue indirectly. So, suppose that for any $a < \inf \varphi(K_\varphi)$, we have that $\varphi^a \neq \emptyset$. From Definition 5.113, we have

$$C_0(\varphi, \infty) \;=\; H_0(X, \varphi^a) \;=\; 0$$

(since X is path-connected), a contradiction. So, φ is bounded below and then for $a < 0$ with $|a| > 0$ big, we will have $\varphi^a = \emptyset$ and thus

$$C_k(\varphi, \infty) \;=\; H_k(X, \varphi^a) \;=\; H_k(X) \;=\; \delta_{k,0}\mathbb{Z} \qquad \forall\, k \geqslant 0.$$

Solution of Problem 5.174

Hypotheses **(i)** and **(ii)** imply that given $\varepsilon > 0$, we can find $c_\varepsilon > 0$ such that

$$F(z, x) \;\leqslant\; \tfrac{1}{p}(\vartheta(z) + \varepsilon)|x|^p + c_\varepsilon \quad \text{for a.a. } z \in \Omega, \text{ all } x \in \mathbb{R}. \tag{5.449}$$

Using (5.449), we have

$$\begin{aligned} \varphi(u) \;&\geqslant\; \tfrac{1}{p}\Big(\|Du\|_p^p - \int_\Omega \vartheta(z)|u|^p\, dz\Big) - \tfrac{\varepsilon}{p\widehat{\lambda}_1}\|Du\|_p^p - c_\varepsilon|\Omega|_N \\ &\geqslant\; \tfrac{1}{p}\big(c_0 - \tfrac{\varepsilon}{\widehat{\lambda}1}\big)\|Du\|_p^p - c_\varepsilon|\Omega|_N \end{aligned} \tag{5.450}$$

(see Proposition 5.132 and Problem 5.67). Choosing $\varepsilon \in (0, \widehat{\lambda}_1 c_0)$, from (5.450), we see that

$$\varphi(u) \;\longrightarrow\; +\infty \quad \text{as } \|u\| \to +\infty.$$

Since φ is sequentially weakly lower semicontinuous (see the Rellich–Kondrachov embedding theorem; Theorem 1.135), by Theorem 5.6, we can find $u_0 \in W_0^{1,p}(\Omega)$ such that

$$\varphi(u_0) \;=\; \inf_{u \in W_0^{1,p}(\Omega)} \varphi(u),$$

so $\varphi'(u_0) = 0$ and u_0 is isolated (since K_φ is discrete). Hence u_0 is a weak solution of the considered Dirichlet problem and Problem 5.150 implies that

$$C_k(\varphi, u_0) \;=\; \delta_{k,0}\mathbb{Z} \qquad \forall\, k \geqslant 0.$$

Solution of Problem 5.175

Since φ is not bounded below, we have $\varphi^a \neq \emptyset$. From Definition 5.113, we have

$$C_k(\varphi,\infty) \;=\; H_k(X,\varphi^a) \qquad \forall\, k \geqslant 0. \tag{5.451}$$

Fix $\star \in \varphi^a$ and recall that since X is contractible, we have

$$H_k(X,\star) \;=\; 0 \qquad \forall\, k \geqslant 0 \tag{5.452}$$

(see Proposition 5.144(f)). From the long exact sequence of singular homology groups for the triple $x \in \varphi^a \subseteq X$, we have

$$0 \longrightarrow H_k(X,\varphi^a) = C_k(\varphi,\infty) \xrightarrow{\partial_*} H_{k-1}(\varphi^a,\star) \longrightarrow 0$$

(see (5.451) and (5.452)), so ∂_* is a group isomorphism, thus

$$C_k(\varphi,\infty) \;=\; H_{k-1}(\varphi^a,\star) \qquad \forall\, k \geqslant 0.$$

Solution of Problem 5.176

First we produce the positive solution. So, let $\varphi_\lambda^+ \colon H_0^1(\Omega) \longrightarrow \mathbb{R}$ be the C^1-functional defined by

$$\varphi_\lambda^+(u) \;=\; \tfrac{1}{2}\|Du\|_2^2 + \int\limits_\Omega F(z,u^+)\,dz - \tfrac{\lambda}{2}\|u^+\|_2^2 \qquad \forall\, u \in H_0^1(\Omega),$$

with $F(z,x) = \int\limits_0^x f(z,s)\,ds$. Hypotheses **(i)** and **(ii)** imply that given any $\eta > 0$, we can find $c_1 = c_1(\eta) > 0$ such that

$$F(z,x) \;\geqslant\; \tfrac{\eta}{2}x^2 - c_1 \quad \text{for a.a. } z \in \Omega, \text{ all } x \in \mathbb{R}.$$

We have

$$\varphi_\lambda^+(u) \;\geqslant\; \tfrac{1}{2}\|Du\|_2^2 + \tfrac{1}{2}(\eta-\lambda)\|u^+\|_2^2 - c_1|\Omega|_N. \tag{5.453}$$

Since $\eta > 0$ is arbitrary, we choose $\eta > \lambda$ and so from (5.453) we infer that

$$\varphi_\lambda^+(u) \;\longrightarrow\; +\infty \quad \text{as } \|u\| \to +\infty.$$

Evidently φ_λ^+ is se sequentially weakly lower semicontinuous. So, Theorem 5.6 implies the existence of $u_0 \in H_0^1(\Omega)$ such that

$$\varphi_\lambda^+(u_0) \;=\; \inf_{u \in H_0^1(\Omega)} \varphi_\lambda^+(u). \tag{5.454}$$

Hypothesis **(iii)** implies that given $\varepsilon > 0$, we can find $\delta = \delta(\varepsilon) > 0$ such that

$$F(z,x) \leqslant \tfrac{\varepsilon}{2}x^2 \quad \text{for a.a. } z \in \Omega, \text{ all } |x| \leqslant \delta. \tag{5.455}$$

Recall that $\widehat{u}_1 \in \operatorname{int} C_+$. So, we can find $t \in (0,1)$ small such that

$$0 \leqslant t\widehat{u}_1(z) \leqslant \delta \qquad \forall\, z \in \overline{\Omega}. \tag{5.456}$$

Then we have

$$\begin{aligned} \varphi_\lambda^+(t\widehat{u}_1) &\leqslant \tfrac{t^2}{2}\|D\widehat{u}_1\|_2^2 + \tfrac{\varepsilon t^2}{2}\|\widehat{u}_1\|_2^2 - \tfrac{\lambda t^2}{2}\|\widehat{u}_1\|_2^2 \\ &\leqslant \tfrac{t^2}{2}\big(\widehat{\lambda}_1 + \varepsilon - \lambda\big) \end{aligned}$$

(see (5.455), (5.456) and use the facts that $\|D\widehat{u}_1\|_2^2 = \widehat{\lambda}_1\|\widehat{u}_1\|_2^2$ and $\|\widehat{u}_1\|_2 = 1$). Because $\lambda > \widehat{\lambda}_1$, we choose $\varepsilon > 0$ small such that $\widehat{\lambda}_1 + \varepsilon < \lambda$. Then

$$\varphi_\lambda^+(t\widehat{u}_1) < 0 = \varphi_\lambda^+(0) = 0,$$

so

$$\varphi_\lambda^+(u_0) < 0 = \varphi_\lambda^+(0)$$

and so $u_0 \neq 0$. From (5.454), we have

$$(\varphi_\lambda^+)'(u_0) = 0,$$

so

$$A(u_0) = N_f(u_0^+) \quad \text{in } H^{-1}(\Omega) = H_0^1(\Omega)^*, \tag{5.457}$$

where $A \in \mathcal{L}(H_0^1(\Omega); H^{-1}(\Omega))$ is defined by

$$\langle A(u), h\rangle = \int_\Omega (Du, Dh)_{\mathbb{R}^N}\, dz \qquad \forall\, u, h \in H_0^1(\Omega)$$

and $N_f(u)(\cdot) = f(\cdot, u(\cdot))$ for all $u \in H_0^1(\Omega)$. On (5.457) we act with $-u_0^- \in H_0^1(\Omega)$ and obtain

$$\|Du_0^-\|_2^2 = 0,$$

so $u_0 \geqslant 0$, $u_0 \neq 0$. We have

$$\begin{cases} -\Delta u_0(z) = f(z, u_0(z)) & \text{for a.a. } z \in \Omega, \\ u_0|_{\partial\Omega} = 0 \end{cases} \tag{5.458}$$

(see (5.457)), so $u_0 \in C_+ \setminus \{0\}$ (see Remark 5.140). Hypotheses **(i)** and **(iii)** imply that

$$|f(z,x)| \leqslant c_2|x| \quad \text{for a.a. } z \in \Omega, \text{ all } x \in \mathbb{R},$$

for some $c_2 > 0$. Using this growth estimate in (5.458), via the nonlinear maximum principle (see Theorem 5.139), we conclude that $u_0 \in \operatorname{int} C_+$.

If we consider the C^1-functional $\varphi_\lambda^- \colon H_0^1(\Omega) \longrightarrow \mathbb{R}$ defined by

$$\varphi_\lambda^-(u) = \tfrac{1}{2}\|Du\|_2^2 + \int_\Omega F(z, -u^-)\, dz + \tfrac{\lambda}{2}\|u^-\|_2^2 \qquad \forall\, u \in H_0^1(\Omega),$$

then reasoning as above, via the direct method of the calculus of variations, we obtain a solution $v_0 \in -\operatorname{int} C_+$ of the considered Dirichlet problem which is also a minimizer of φ_λ^-.

Let $\varphi_\lambda \colon H_0^1(\Omega) \longrightarrow \mathbb{R}$ be the energy functional corresponding to the considered Dirichlet problem, namely

$$\varphi_\lambda(u) = \tfrac{1}{2}\|Du\|_2^2 + \int_\Omega F(z,u)\,dz - \tfrac{\lambda}{2}\|u\|_2^2 \qquad \forall\, u \in H_0^1(\Omega).$$

We have that $\varphi_\lambda \in C^1(H_0^1(\Omega))$ and the critical points of φ_λ are the weak solutions of the Dirichlet problem. If

$$H_0^+ = \{u \in H_0^1(\Omega) : \ u(z) \geqslant 0 \quad \text{for a.a. } z \in \Omega\},$$

then clearly

$$\varphi_\lambda\big|_{H_0^+} = \varphi_\lambda^+\big|_{H_0^+} \quad \text{and} \quad \varphi_\lambda\big|_{-H_0^+} = \varphi_\lambda^-\big|_{-H_0^+}.$$

Since $u_0 \in \operatorname{int} C_+$ and $v_0 \in -\operatorname{int} C_+$, it follows that u_0, v_0 are local $C_0^1(\overline{\Omega})$-minimizers of φ_λ, thus also

$$u_0, v_0 \text{ are local } H_0^1(\Omega)\text{-minimizers of } \varphi_\lambda \tag{5.459}$$

(see Proposition 5.141). Without any loss of generality, we assume that $\varphi_\lambda(v_0) \leqslant \varphi_\lambda(u_0)$ (the reasoning is similar if the opposite inequality holds). Suppose that $K_{\varphi_\lambda} = \{0, u_0, v_0\}$ (otherwise we already have a third nontrivial solution, which by Remark 5.140 belongs in $C_0^1(\overline{\Omega})$ and we are done). Then (5.459) and Problem 5.27 imply that there exists $\varrho \in (0,1)$ small such that

$$\varphi_\lambda(v_0) \leqslant \varphi_\lambda(u_0) < \inf\{\varphi_\lambda(u) : \ \|u - u_0\| = \varrho\} = m_\varrho^\lambda, \quad \|v_0 - u_0\| > \varrho. \tag{5.460}$$

As we did for φ_λ^+ in the beginning of the solution, we can show that

$$\varphi_\lambda(u) \longrightarrow +\infty \quad \text{as } \|u\| \to +\infty,$$

so φ_λ satisfies the PS-condition (see Definition 5.45(a) and Problem 5.32). From this and (5.460), we see that we can apply Theorem 5.63 and obtain $y_0 \in H_0^1(\Omega)$ such that

$$y_0 \in K_{\varphi_\lambda} \quad \text{and} \quad m_\varrho^\lambda \leqslant \varphi_\lambda(y_0). \tag{5.461}$$

From (5.460) and (5.461), we see that y_0 is a weak solution of the considered Dirichlet problem and $y_0 \notin \{u_0, v_0\}$, thus $y_0 \in C_0^1(\overline{\Omega})$ (regularity theory; see Remark 5.140). We need to show that y_0 is nontrivial. By Theorem 5.63, we have

$$\varphi_\lambda(y_0) = \inf_{\gamma \in \Gamma} \max_{0 \leqslant t \leqslant 1} \varphi_\lambda(\gamma(t)), \tag{5.462}$$

where $\Gamma = \{\gamma \in C([0,1]; H_0^1(\Omega)) : \ \gamma(0) = v_0, \ \gamma(1) = u_0\}$. Then according to (5.462) in order to have the nontriviality of y_0, it suffices to find $\gamma_* \in \Gamma$ such that

$\varphi_\lambda|_{\gamma_*} < 0 = \varphi_\lambda(0)$. Let

$$\begin{aligned} \partial B_1^{L^2} &= \{u \in L^2(\Omega) : \|u\|_2 = 1\}, \\ M &= H_0^1(\Omega) \cap \partial B_1^{L^2}, \quad M_c = M \cap C_0^1(\overline{\Omega}), \\ \widehat{\Gamma} &= \{\widehat{\gamma} \in C([-1,1]; M) : \gamma(-1) = -\widehat{u}_1, \ \gamma(1) = \widehat{u}_1\}, \\ \widehat{\Gamma}_c &= \{\widehat{\gamma} \in C([-1,1]; M_c) : \gamma(-1) = -\widehat{u}_1, \ \gamma(1) = \widehat{u}_1\}. \end{aligned}$$

Using Theorem 2.58, we show that $\widehat{\Gamma}_c$ is dense in $\widehat{\Gamma}$. Then using Proposition 5.136, we see that given $\varepsilon > 0$, we can find $\widehat{\gamma}_0 \in \widehat{\Gamma}_c$ such that

$$\max_{-1 \leqslant t \leqslant 1} \|D\widehat{\gamma}_0(t)\|_2^2 \leqslant \widehat{\lambda}_2 + \varepsilon. \tag{5.463}$$

Hypothesis **(iii)** implies that we can find $\delta = \delta(\varepsilon) > 0$ such that

$$F(z,x) \leqslant \tfrac{\varepsilon}{2} x^2 \quad \text{for a.a. } z \in \Omega, \text{ all } |x| \leqslant \delta. \tag{5.464}$$

Since $\widehat{\gamma}_0 \in \widehat{\Gamma}_c$, we can find $\vartheta \in (0,1)$ small such that

$$\vartheta |\widehat{\gamma}_0(t)(z)| \leqslant \delta \qquad \forall\, t \in [-1,1], \ z \in \overline{\Omega}. \tag{5.465}$$

We have

$$\varphi_\lambda(\vartheta \widehat{\gamma}_0(t)) \leqslant \tfrac{\vartheta^2}{2}(\widehat{\lambda}_2 + \varepsilon) + \tfrac{\vartheta^2 \varepsilon}{2} - \tfrac{\lambda \vartheta^2}{2} \tag{5.466}$$

(see (5.463), (5.464), (5.465) and recall that $\|\widehat{\gamma}_0(t)\| = 1$ for all $t \in [-1,1]$). Since $\lambda > \widehat{\lambda}_2$, choosing $\varepsilon > 0$ small such that $\widehat{\lambda}_2 + 2\varepsilon < \lambda$, from (5.466), we see that

$$\varphi_\lambda|_{\gamma_0} < 0, \tag{5.467}$$

where $\gamma_0 = \vartheta \widehat{\gamma}_0$. Let $a = \varphi_\lambda^+(u_0) < 0 = \varphi_\lambda^+(0)$. Since $K_{\varphi_\lambda} = \{0, u_0, v_0\}$, we see that $K_{\varphi_\lambda^+} = \{0, u_0\}$. Hence $(\varphi_\lambda^+)^a = \{u_0\}$. Using the second deformation theorem (see Theorem 5.56), we can find a deformation $h\colon [0,1] \times ((\varphi_\lambda^+)^0 \setminus K^0_{\varphi_\lambda^+}) \longrightarrow (\varphi_\lambda^+)^0$ such that

$$h(1, (\varphi_\lambda^+)^0 \setminus K^0_{\varphi_\lambda^+}) = \{u_0\}, \tag{5.468}$$

$$\varphi_\lambda^+(h(t,u)) \leqslant \varphi_\lambda^+(h(s,u)) \qquad \forall\, t,s \in [0,1], \ t \leqslant s, \ u \in (\varphi_\lambda^+)^0 \setminus K^0_{\varphi_\lambda^+}. \tag{5.469}$$

Let $\gamma_+(t) = h(t, \vartheta \widehat{u}_1)^+$ for all $t \in [0,1]$. Then this is a continuous path in $H_0^1(\Omega)$ and

$$\begin{aligned} \gamma_+(0) &= h(0, \vartheta \widehat{u}_1) = \vartheta \widehat{u}_1^+ = \vartheta \widehat{u}_1, \\ \gamma_+(1) &= h(1, \vartheta \widehat{u}_1) = u_0, \\ \varphi_\lambda(\gamma_+(t)) &= \varphi_\lambda^+(\gamma_+(t)) \leqslant \varphi_\lambda^+(\vartheta \widehat{u}_1) = \varphi_\lambda(\vartheta \widehat{u}_1) < 0 \qquad \forall\, t \in [0,1] \end{aligned}$$

(see (5.469) and (5.467)). Therefore γ_+ is a path in $H_0^1(\Omega)$ connecting $\vartheta \widehat{u}_1$ and u_0 and such that

$$\varphi_\lambda|_{\gamma_+} < 0. \tag{5.470}$$

Similarly we produce a path γ_- in $H_0^1(\Omega)$ connecting $-\vartheta\widehat{u}_1$ and v_0 and such that

$$\varphi_\lambda\big|_{\gamma_-} < 0. \tag{5.471}$$

Concatenating γ_-, γ_0 and γ_+ we produce $\gamma_* \in \Gamma$ such that $\varphi_\lambda\big|_{\gamma_*} < 0$, so $y_0 \neq 0$.

Remark. A careful reading of the above solution reveals that in order to produce the two constant sign solutions $u_0 \in \operatorname{int} C_+$ and $v_0 \in -\operatorname{int} C_+$, it is enough to assume that $\lambda > \widehat{\lambda}_1$.

Bibliography

[1] Ambrosetti, A., Arcoya, D.: An Introduction to Nonlinear Functional Analysis and Elliptic Problems. Birkhäuser, Boston (2011)

[2] Ambrosetti, A., Malchiodi, A.: Nonlinear Analysis and Semilinear Elliptic Problems. Cambridge University Press, Cambridge (2007)

[3] Chang, K.-C.: Infinite-Dimensional Morse Theory and Multiple Solution Problems. Birkhäuser, Boston (1993)

[4] Dacorogna, B.: Direct Methods in the Calculus of Variations. Springer, New York (2008)

[5] Ekeland, I.: Convexity Methods in Hamiltonian Mechanics. Springer, Berlin (1990)

[6] Gasiński, L., Papageorgiou, N.S.: Nonsmooth Critical Point Theory and Nonlinear Boundary Value Problems. Chapman & Hall/CRC, Boca Raton (2005)

[7] Gasiński, L., Papageorgiou, N.S.: Nonlinear Analysis. Chapman & Hall/CRC, Boca Raton (2006)

[8] Ghoussoub, N.: Duality and Perturbation Methods in Critical Point Theory. Cambridge University Press, Cambridge (1993)

[9] Gilbarg, D., Trudinger, N.S.: Elliptic Partial Differential Equations of Second Order. Springer, Berlin (2001)

[10] Mawhin, J., Willem M.: Critical Point Theory and Hamiltonian Systems. Springer, New York (1989)

[11] Motreanu, D., Motreanu, V.V., Papageorgiou, N.S.: Topological and Variational Methods with Applications to Nonlinear Boundary Value Problems. Springer, New York (2014)

[12] Palais, R.S., Smale, S.: A generalized Morse theory. Bull. Am. Math. Soc. **70**, 165–172 (1964)

[13] Perera, K., Agarwal, R.P., O'Regan, D.: Morse Theoretic Aspects of p-Laplacian Type Operators. American Mathematical Society, Providence (2010)

[14] Rabinowitz, P.H.: Minimax Methods in Critical Point Theory with Applications to Differential Equations. CBMS Regional Conference Series in Mathematics, vol. 65. American Mathematical Society, Providence (1986)

[15] Struwe, M.: Variational Methods. Applications to Nonlinear Partial Differential Equations and Hamiltonian Systems. Springer, Berlin (2008)

[16] Willem, M.: Minimax Theorems. Birkhäuser, Boston (1996)

[17] Zeidler, E.: Nonlinear Functional Analysis and Its Applications, vol. III. Springer, New York (1990)

List of Symbols

Symbol	Page	Meaning
$\mathcal{L}^p(\Omega)$	p. 1	space of measurable functions with finite L^p-norm
$f \sim g$	p. 1	$f(\omega) = g(\omega)$ for μ-a.a. $\omega \in \Omega$
$L^p(\Omega)$	p. 1	$L^p(\Omega) = \mathcal{L}^p(\Omega)/_\sim$
$\|f\|_p$	p. 1	$\|f\|_p = \big(\int_\Omega \lvert f(\omega)\rvert^p \, d\mu\big)^{\frac{1}{p}}$
$\mathcal{L}^\infty(\Omega)$	p. 1	set of measurable functions $f\colon \Omega \longrightarrow \mathbb{R}$ for which there exists $M > 0$ such that $\mu\big(\{\omega \in \Omega : \lvert f(\omega)\rvert > M\}\big) = 0$
$L^\infty(\Omega)$	p. 1	$L^\infty(\Omega) = \mathcal{L}^\infty(\Omega)/_\sim$
$\|f\|_\infty$	p. 1	$\|f\|_\infty = \inf\big\{M > 0 : \mu\big(\{\omega \in \Omega : \lvert f(\omega)\rvert > M\} > 0\big)\big\}$
$C_b(X)$	p. 2	space of all bounded continuous functions $f\colon X \longrightarrow \mathbb{R}$
$L^p_{\text{loc}}(\Omega)$	p. 2	space of functions $f\colon \Omega \longrightarrow \mathbb{R} \cup \{\pm\infty\}$ such that $f \in L^p(K)$ for every compact set $K \subseteq \Omega$
$u_n \xrightarrow{\mu} u$	p. 3	convergence in μ-measure
$u_n \xrightarrow{au} u$	p. 3	convergence almost uniform
$L^0(\Omega)$	p. 4	space of all equivalence classes of measurable functions
$M(u)(x)$	p. 7	$M(u)(x) = \sup\limits_{r>0} \frac{1}{\lambda^N(B_r(x))} \int_{B_r(x)} \lvert u(z)\rvert \, dz$
$M_r(u)(x)$	p. 7	$M_r(u)(x) = \sup\limits_{0<r<R} \frac{1}{\lambda^N(B_r(x))} \int_{B_r(x)} \lvert u(z)\rvert \, dz$
$L \log L(A)$	p. 8	space of measurable functions $u\colon A \longrightarrow \mathbb{R}$ such that $\int_A \lvert u(z)\rvert \log(2 + \lvert u(z)\rvert) \, dz < +\infty$
$\int_A u \, d\mu$	p. 9	Bochner integral of u over A

L. Gasiński, N.S. Papageorgiou, *Exercises in Analysis: Part 2*, Problem Books in Mathematics, DOI 10.1007/978-3-319-27817-9

Symbol	Page	Description
i_C	p. 420	indicator function of C
$\Gamma_0(X)$	p. 421	cone of proper, lower semicontinuous, and convex functions
$\mathrm{cl}\,\varphi$	p. 421	closure of φ
$\overline{\mathrm{conv}}\,\varphi$	p. 421	convex closure of φ
$\mathrm{core}\,C$	p. 422	core of C
$\varphi'(u;h)$	p. 422	directional derivative of φ at u in direction h
$\partial\varphi(u_0)$	p. 423	subdifferential of φ at u_0
φ^*	p. 425	conjugate of φ
φ^{**}	p. 425	second conjugate of φ
$\varphi \boxdot \psi$	p. 426	infimal convolution
$\partial_\varepsilon\varphi$	p. 428	ε-subdifferential of φ
$T_C(u)$	p. 431	tangent cone to C at u
$N_C(u)$	p. 431	normal cone to C at u
$\varphi^0(u;h)$	p. 433	generalized directional derivative of φ at u in direction h
$\partial\varphi(u)$	p. 434	generalized (Clarke) subdifferential
$\mathrm{dist}(\cdot, C)$	p. 435	distance function from C
$T'_C(u)$	p. 436	Clarke tangent cone to C at u
$N'_C(u)$	p. 436	Clarke normal cone to C at u
$T^b_C(u)$	p. 436	contingent (or Bouligand) tangent cone to C at u
$\varphi_n \xrightarrow{\Gamma} \varphi$	p. 437	Γ-convergence
$\big(\Gamma\text{-}\liminf_{n\to+\infty} \varphi_n\big)(u)$	p. 437	Γ-limit inferior
$\big(\Gamma\text{-}\limsup_{n\to+\infty} \varphi_n\big)(u)$	p. 437	Γ-limit superior
$C_n \xrightarrow{M} C$	p. 440	convergence of sets in Mosco sense
$\varphi_n \xrightarrow{M} \varphi$	p. 440	convergence of functions in Mosco sense
$M_b(T)_+$	p. 442	Radon measures on T
$M^1_+(T)$	p. 442	probability measures on T
$SM^1_+(T)$	p. 442	subprobability measures on T
$\widehat{R}(\Omega;T)$	p. 442	space of transition probabilities

$\widehat{SR}(\Omega;T)$	p. 442	space of transition subprobabilities
w_n	p. 443	narrow topology
$\lambda\circ\vartheta^{-1}$	p. 443	image measure of λ under ϑ
$\mathcal{Y}(\Omega,T;\mu)$	p. 443	space of Young measures with respect to μ
$S\mathcal{Y}(\Omega,T;\mu)$	p. 443	space of Young submeasures with respect to μ
$R(\Omega;T)$	p. 444	$R(\Omega;T)=\widehat{R}(\Omega;T)/_\sim$
$SR(\Omega;T)$	p. 444	$R(\Omega;T)=\widehat{SR}(\Omega;ST)/_\sim$
$\xi_n\xrightarrow{n}\xi$	p. 445	narrow convergence
$M_n(\mathbb{R})$	p. 451	space of all $n\times n$-real matrices
$\mathrm{End}_n(\mathbb{R})$	p. 451	set of all invertible elements of $M_n(\mathbb{R})$
K^0	p. 457	polar of K
$CU_b(T)$	p. 477	space of uniformly continuous and bounded functions on T
$J_\varphi(u_0)$	p. 617	Jacobian of φ at u_0
S_φ	p. 617	critical set of φ
$\varphi(S_\varphi)$	p. 617	crease of φ
$d(\varphi,\Omega,h)$	p. 618	degree of φ at h with respect to Ω
sgn	p. 618	sign function
$d(\varphi,\Omega,h)$	p. 619	degree of φ at h with respect to Ω
$d(\varphi,\Omega,h)$	p. 619	degree of φ at h with respect to Ω
$i_\infty(\varphi,h)$	p. 620	index of φ at infinity
$d_1(\psi)$	p. 621	Brouwer degree of ψ
$d_{LS}(\varphi,\Omega,h)$	p. 624	Leray–Schauder degree
$d_{SC}(I_X-f,\Omega,0)$	p. 627	degree of $SC_k(\overline{\Omega};X)$-function
$d_C(I_X-f,\Omega,0)$	p. 628	degree of $S(\overline{\Omega};X)$-function
$\mathcal{Y}_{(S_+)}$	p. 630	family of demicontinuous, $(S)_+$-maps
$d_{(S)_+}$	p. 630	degree of $\mathcal{Y}_{(S_+)}$-maps
$\mathcal{Y}_M$	p. 632	family of admissible triples to define degree d_M
d_M	p. 632	degree on $\mathcal{Y}_M$
$\mathcal{Y}_{SV}$	p. 634	family of admissible triples to define degree d_{SV}
d_{SV}	p. 634	degree of $\mathcal{Y}_{SV}$-maps

K_φ^c	p. 824	$\{u \in K_\varphi : \varphi(u) = c\}$
K_Φ	p. 824	$\{u \in X : -\varphi'(u) \in \partial\psi(u)\}$ (critical set of $\Phi = \varphi + \psi$)
$\{L(g)\}_{g\in G}$	p. 827	representation of G over X
$\mathrm{Fix}_G(X)$	p. 827	set of invariant (or fixed) points
$\mathcal{Y}$	p. 827	$\{D \subseteq X : D$ is closed and G-invariant$\}$
$\mathcal{Y}_k$	p. 828	$\{D \subseteq X : D$ is compact, invariant and $i(D) \geqslant k\}$
$\gamma(D)$	p. 828	Krasnoselskii genus of D
$cat_Y(D)$	p. 829	Ljusternik–Schnirelmann category of D in Y
$\sigma_0(p)$	p. 839	set of eigenvalues of $(-\Delta_p, W_0^{1,p}(\Omega))$
$K_\varphi[a,b]$	p. 878	$\{u \in K_\varphi : a \leqslant \varphi(u) \leqslant b\}$
$\sigma(2)$	p. 880	set of eigenvalues of $(-\Delta, H_0^1(\Omega))$

Index

L. Gasiński, N.S. Papageorgiou, *Exercises in Analysis: Part 2*, Problem Books in Mathematics, DOI 10.1007/978-3-319-27817-9

U